Graduate Texts in Contemporary Physics

Series Editors:

R. Stephen Berry
Joseph L. Birman
Jeffrey W. Lynn
Mark P. Silverman
H. Eugene Stanley
Mikhail Voloshin

Graduate Texts in Contemporary Physics

S.T. Ali, J.P. Antoine, and J.P. Gazeau: **Coherent States, Wavelets and Their Generalizations**

A. Auerbach: **Interacting Electrons and Quantum Magnetism**

B. Felsager: **Geometry, Particles, and Fields**

P. Di Francesco, P. Mathieu, and D. Sénéchal: **Conformal Field Theories**

A. Gonis and W.H. Butler: **Multiple Scattering in Solids**

K.T. Hecht: **Quantum Mechanics**

J.H. Hinken: **Superconductor Electronics: Fundamentals and Microwave Applications**

J. Hladik: **Spinors in Physics**

Yu.M. Ivanchenko and A.A. Lisyansky: **Physics of Critical Fluctuations**

M. Kaku: **Introduction to Superstrings and M-Theory, 2nd Edition**

M. Kaku: **Strings, Conformal Fields, and M-Theory, 2nd Edition**

H.V. Klapdor (ed.): **Neutrinos**

J.W. Lynn (ed.): **High-Temperature Superconductivity**

H.J. Metcalf and P. van der Straten: **Laser Cooling and Trapping**

R.N. Mohapatra: **Unification and Supersymmetry: The Frontiers of Quark-Lepton Physics, 2nd Edition**

H. Oberhummer: **Nuclei in the Cosmos**

G.D.J. Phillies: **Elementary Lectures in Statistical Mechanics**

R.E. Prange and S.M. Girvin (eds.): **The Quantum Hall Effect**

B.M. Smirnov: **Clusters and Small Particles: In Gases and Plasmas**

M. Stone: **The Physics of Quantum Fields**

F.T. Vasko and A.V. Kuznetsov: **Electronic States and Optical Transitions in Semiconductor Heterostructures**

A.M. Zagoskin: **Quantum Theory of Many-Body Systems: Techniques and Applications**

K.T. Hecht

Quantum Mechanics

With 101 Illustrations

Springer

K. T. Hecht
Department of Physics
University of Michigan
2409 Randall Laboratory
Ann Arbor, MI 48109
USA

Series Editors

R. Stephen Berry
Department of Chemistry
University of Chicago
Chicago, IL 60637
USA

Joseph L. Birman
Department of Physics
City College of CUNY
New York, NY 10031
USA

Jeffery W. Lynn
Department of Physics
University of Maryland
College Park, MD 20742
USA

Mark P. Silverman
Department of Physics
Trinity College
Hartford, CT 06106
USA

H. Eugene Stanley
Center for Polymer Studies
Physics Department
Boston University
Boston, MA 02215
USA

Mikhail Voloshin
Theoretical Physics Institute
Tate Laboratory of Physics
The University of Minnesota
Minneapolis, MN 55455
USA

Library of Congress Cataloging-in-Publication Data
Hecht, K.T. (Karl Theodor), 1926-
Quantum mehanics / Karl T. Hecht
p. cm.
Includes bibliographical references and index.
ISBN 0-387-98919-6
1. Quantum theory. I. Title. II. Series
QC174.12H433 2000
530.14'3—dc21 99-42830

ISBN 0-387-21386-4

All rights reserved. This work may not be translated or copied in whole or in part without the written permission of the publisher (Springer-Verlag New York, Inc., 175 Fifth Avenue, New York, NY 10010, USA), except for brief excerpts in connection with reviews or scholarly analysis. Use in connection with any form of information storage and retrieval, electronic adaptation, computer software, or by similar or dissimilar methodology now known or hereafter developed is forbidden.
The use of general descriptive names, trade names, trademarks, etc., in this publication even if the former are identified, is not to be taken as a sign that such names, as understood by the Trade Marks and Merchandise Marks Act, may accordingly be used freely by anyone.

Hecht: Quantum Mechanics

© 2000 Springer-Verlag New York, Inc.

All rights reserved. No part of this publication may be reproduced, stored in any electronic or mechanical form, including photocopy, recording or otherwise, without the prior written permission of the publisher.

First Indian Reprint 2009

ISBN 978-81-8489-182-9

This edition is manufactured in India for sale in India, Pakistan, Bangladesh, Nepal and Sri Lanka and any other country as authorized by the publisher.

This edition is published by Springer (India) Private Limited,
A part of Springer Science+Business Media, Registered Office: 906-907, Akash Deep Building, Barakhamba Road, New Delhi – 110 001, India.

Printed in India by Rajkamal Electric Press, Delhi.

Preface

This book is an outgrowth of lectures given at the University of Michigan at various times from 1966–1996 in a first-year graduate course on quantum mechanics. It is meant to be at a fairly high level. On the one hand, it should provide future research workers with the tools required to solve real problems in the field. On the other hand, the beginning graduate courses at the University of Michigan should be self-contained. Although most of the students will have had an undergraduate course in quantum mechanics, the lectures are intended to be such that a student with no previous background in quantum mechanics (perhaps an undergraduate mathematics or engineering major) can follow the course from beginning to end.

Part I of the course, Introduction to Quantum Mechanics, thus begins with a brief background chapter on the duality of nature, which hopefully will stimulate students to take a closer look at the two references given there. These references are recommended for every serious student of quantum mechanics. Chapter 1 is followed by a review of Fourier analysis before we meet the Schrödinger equation and its interpretation. The dual purpose of the course can be seen in Chapters 4 and 5, where an introduction to simple square well problems and a first solution of the one-dimensional harmonic oscillator by Fuchsian differential equation techniques are followed by an introduction to the Bargmann transform, which gives us an elegant tool to show the completeness of the harmonic oscillator eigenfunctions and enables us to solve some challenging harmonic oscillator problems, (e.g., the case of general n for problem 11). Early chapters (7 through 12) on the eigenvalue problem are based on the coordinate representation and include detailed solutions of the spherical harmonics and radial functions of the hydrogen atom, as well as many of the soluble, one-dimensional potential problems. These chapters are based on the factorization method. It is hoped the ladder step-up and step-down operator approach of this method will help to lead the student naturally from the Schrödinger equation approach to the more modern algebraic techniques of quantum mechanics, which are introduced in Chapters 13 to 19. The full Dirac bra, ket notation is introduced in Chapter 13. These chapters also give the full algebraic approach to the general angular momentum problem, SO(3) or SU(2), the harmonic oscillator algebra, and the SO(2,1) algebra. The solution for the latter is given in problem 23, which is used in considerable detail in later chapters. The problems often amplify the material of the course.

Part II of the course, Chapters 20 to 26, on time-independent perturbation theory, is based on Fermi's view that most of the important problems of quantum mechanics can be solved by perturbative techniques. This part of the course shows how various types of degeneracies can be handled in perturbation theory, particularly the case in which a degeneracy is not removed in lowest order of perturbation theory so that the lowest order perturbations do not lead naturally to the symmetry-adapted basis; a case ignored in many books on quantum mechanics and perhaps particularly important in the case of accidental near-degeneracies. Chapters 25 and 26 deal with magnetic-field perturbations, including a short section on the Aharanov–Bohm effect, and a treatment on fine structure and Zeeman perturbations in one-electron atoms.

Part III of the course, Chapters 27 to 35, then gives a detailed treatment of angular momentum and angular momentum coupling theory, including a derivation of the matrix elements of the general rotation operator, Chapter 29; spherical tensor operators, Chapter 31; the Wigner–Eckart theorem, Chapter 32; angular momentum recoupling coefficients and their use in matrix elements of coupled tensor operators in an angular-momentum-coupled basis, Chapter 34; as well as the use of an SO(2,1) algebra and the stretched Coulombic basis and its power in hydrogenic perturbation theory without the use of the infinite sum and continuum integral contributions of the conventional hydrogenic basis, Chapter 35.

Since the full set of chapters is perhaps too much for a one-year course, some chapters or sections, and, in particular, some mathematical appendices, are marked in the table of contents with an asterisk (*). This symbol designates that the chapter can be skipped in a first reading without loss of continuity for the reader. Chapters 34 and 35 are such chapters with asterisks. Because of their importance, however, an alternative is to skip Chapters 36 and 37 on the WKB approximation. These chapters are therefore placed at this point in the book, although they might well have been placed in Part II on perturbation theory.

Part IV of the lectures, Chapters 38 to 40, gives a first introduction to systems of identical particles, with the emphasis on the two-electron atom and a chapter on variational techniques.

Parts I through IV of the course deal with bound-state problems. Part V on scattering theory, which might constitute the beginning of a second semester, begins the treatment of continuum problems with Chapters 41 through 56 on scattering theory, including a treatment of inelastic scattering processes and rearrangement collisions, and the spin dependence of scattering cross sections. The polarization of particle beams and the scattering of particles with spin are used to introduce density matrices and statistical distributions of states.

Part VI of the course gives a conventional introduction to time-dependent perturbation theory, including a chapter on magnetic resonance and an application of the sudden and adiabatic approximations in the reversal of magnetic fields.

Part VII on atom–photon interactions includes an expansion of the quantized radiation field in terms of the full set of vector spherical harmonics, leading to a detailed derivation of the general electric and magnetic multipole-transition matrix elements needed in applications to nuclear transitions, in particular.

Parts V through IX may again be too much material for the second semester of a one-year course. At the University of Michigan, curriculum committees have at various times insisted that the first-year graduate course include *either* an introduction to Dirac theory of relativistic spin $\frac{1}{2}$-particles *or* an introduction to many-body theory. Part VIII of the course on relativistic quantum mechanics, Chapters 69 through 77, and Part IX, an introduction to many-body theory, Chapters 78 and 79, are therefore written so that a lecturer could choose *either* Part VIII *or* Part IX to complete the course.

The problems are meant to be an integral part of the course. They are often meant to build on the material of the lectures and to be real problems (rather than small exercises, perhaps to derive specific equations). They are, therefore, meant to take considerable time and often to be somewhat of a challenge. In the actual course, they are meant to be discussed in detail in problem sessions. For this reason, detailed solutions for a few key problems, particularly in the first part of the course, are given in the text as part of the course (e.g., the results of problem 23 are very much used in later chapters, and problem 34, actually a very simple problem in perturbation theory is used to illustrate how various types of degeneracies can be handled properly in perturbation theory in a case in which the underlying symmetry leading to the degeneracy *might not* be easy to recognize). In the case of problem 34, the underlying symmetry *is* easy to recognize. The solution therefore also shows how this symmetry should be exploited.

The problems are not assigned to specific chapters, but numbered 1 through 55 for Parts I through IV of the course, and, again, 1 through 51 for Parts V through IX, the second semester of the course. They are placed at the point in the course where the student should be ready for a particular set of problems.

The applications and assigned problems of these lectures are taken largely from the fields of atomic and molecular physics and from nuclear physics, with a few examples from other fields. This selection, of course, shows my own research interests, but I believe, is also because these fields are fertile for the applications of nonrelativistic quantum mechanics.

I first of all want to acknowledge my own teachers. I consider myself extremely fortunate to have learned the subject from David M. Dennison and George E. Uhlenbeck. Among the older textbooks used in the development of these lectures, I acknowledge the books by Leonard I. Schiff, *Quantum Mechanics*, McGraw-Hill, 1949; Albert Messiah, *Quantum Mechanics, Vol. I and II*, John Wiley and Sons, 1965; Eugene Merzbacher, *Quantum Mechanics*, John Wiley and Sons, 1961; Kurt

Gottfried, *Quantum Mechanics*, W. A. Benjamin, Inc., 1966; and L. D. Landau and E. M. Lifshitz, *Quantum Mechanics. Nonrelativistic Theory. Vol.* 3. *Course of Theoretical Physics*, Pergamon Press 1958. Hopefully, the good features of these books have found their way into my lectures.

References to specific books, chapters of books, or research articles are given throughout the lectures wherever they seemed to be particularly useful or relevant. Certainly, no attempt is made to give a complete referencing. Each lecturer in a course on quantum mechanics must give the student his own list of the many textbooks a student should consult. The serious student of the subject, however, must become familiar with the two classics: P. A. M. Dirac, *The Principles of Quantum Mechanics*, Oxford University Press, first ed. 1930; and Wolfgang Pauli, *General Principles of Quantum Mechanics*, Springer-Verlag, 1980, (an English translation of the 1933 Handbuch der Physik article in Vol. 24 of the Handbuch).

Finally, I want to thank the many students at the University of Michigan who have contributed to these lectures with their questions. In fact, it was the encouragement of former students of this course that has led to the idea these lectures should be converted into book form. I also thank Prof. Yasuyuki Suzuki for his many suggestions after a careful reading of an early version of the manuscript. Particular thanks are due to Dr. Sudha Swaminathan and Dr. Frank Lamelas for their great efforts in making all of the figures.

Contents

Part I

Introduction to Quantum Mechanics

1

Background: The Duality of Nature

(Good references for this chapter on the historical background are the article by Niels Bohr, entitled "Discussions with Einstein. Epistemological Problems in Atomic Physics." In *Albert Einstein. Philosopher-Scientist.* Vol. VII of Library of Living Philosophers, Paul A. Schilpp, ed., Evanston, Illinois, 1949; and the little book by Werner Heisenberg, *The Physical Principles of the Quantum Theory*, Dover Publications, 1949.)

The results of the experimental developments of the late nineteenth and early twentieth century led us to a picture of nature that showed the *duality* of nature on the atomic scale. *Both* material particles and electromagnetic radiation show *both* particle-like and wave-like aspects. However, particles can be localized in space-time. In classical physics, x,y,z,t for a particle can be specified exactly. Particles are also indivisible. Half an electron, or a fractional part of an electron, does not exist. On the other hand, waves *cannot* be localized. They must be somewhat extended in space-time to give a meaning to wavelength, λ, and frequency, ν. Waves are always divisible. Partial reflection and transmission of a wave at an interface between two media can exist.

This duality poses a real dilemma: The particle picture seems incompatible with that of waves, in particular, the interference effects. Yet, it is precisely the interference effects that determine λ and ν, which via the deBroglie relation, $p = h/\lambda$, and the Bohr relation, $E = h\nu$, determine the dynamical attributes of the particle.

A The Young Double Slit Experiment

To illustrate the paradoxical situation, consider the classical interference experiment of the Young double slit. We could think either of light waves, electromagnetic radiation, or of matter waves, electron deBroglie waves, going through the double slit arrangement.

The incident beam can be made so weak that, on average, only one photon (or electron) at a time will pass through the apparatus and be incident on the photographic plate. Because only one photon at a time goes through the apparatus, the possibility of interference between different photons is eliminated. An interference pattern will still be on the photographic plate, however. Clearly, a photon that has reached the photographic plate must have passed through *either* slit 1 *or* slit 2. Imagine it was slit 1; then, if slit 2 had been closed, no interference pattern would have occurred. Hence, the seemingly terrible paradox that the behavior of the photon is influenced by the presence of a slit, through which it cannot have passed.

The resolution of the paradox rests on the fact that the classical causal space-time description of nature which rests on the "clear-cut separability between the phenomena and the means of observing these phenomena," does not apply. On the atomic scale, an "uncontrollable interaction between the object and the measuring instrument" exists. (The words in quotation marks are those of Niels Bohr.) As a result, the above experiment can be set up in either of two "complementary" ways; as above, and as shown in Fig. 1.1, to exhibit the interference fringes but in a setup that makes it impossible to answer experimentally the question: "Through which slit did the photon pass?". Alternatively, we could alter the experimental setup to answer experimentally the question: "Through which slit did the photon pass?" In setting up the experiment in this second way, however, we have lost the possibility of a precise wavelength measurement through the interference pattern. The interference pattern will have been wiped out.

Because our knowledge of phenomena on the atomic scale is restricted, a wave-field must be associated with the particle motion. For any wavefield, an uncertainty relation exists connecting position and wavelength; $\Delta k \Delta x \approx 2\pi$, where $k = 2\pi/\lambda$. This relation follows from straightforward Fourier analysis of a wave packet of finite extent in space. Now, with the deBroglie relation, $p = h/\lambda$, this leads to $\Delta p \Delta x \approx h$, the Heisenberg uncertainty relation.

B More Detailed Analysis of the Double Slit Experiment

In region I, to the left of the single slit (see Fig. 1.1), we assume we have a plane wave effectively of infinite extent in the y direction and proceeding in the x direction. Then, in region I, $p_y = 0$; i. e., p_y is known precisely, so $\Delta p_y = 0$;

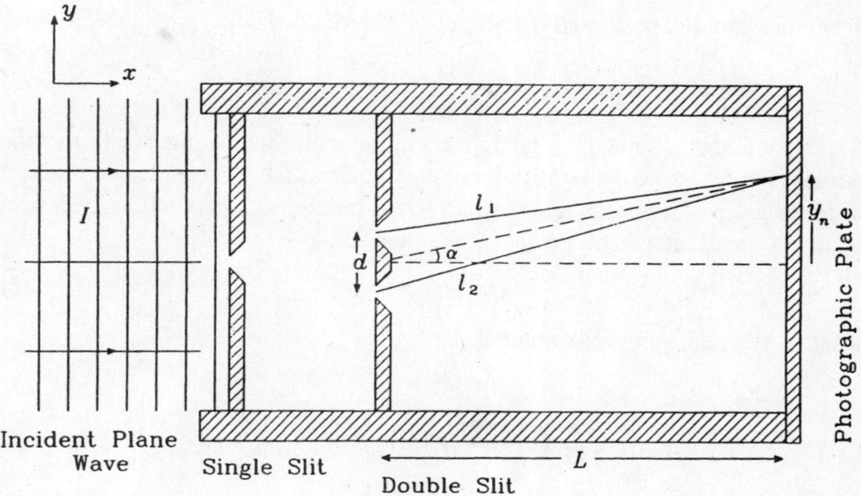

FIGURE 1.1. Conventional double slit experimental setup. Rigidly fixed slits.

now, we have no knowledge of where the photon is located in space, $\Delta y = \infty$. The y position of the particle is completely uncertain.

At the position of the double slit (with massive, rigidly fixed slits bolted to massive apparatus; see the figures in the article by Bohr), no experimental means of determining through which slit the photon is passing exist. It must go through either of the two slits. Hence, at this x-position, $\Delta y \approx d$. In its passage through one of the slits, the photon will interact with the slit jaw, which is massive, bolted firmly to a huge apparatus, and it can absorb recoil momentum without moving. Most of the photons end up within the first few bright fringes near the central maximum. We cannot predict which fringe, however. Let us assume the photon ends up in the nth bright fringe, where n is a small integer. Then, the change in p_y at the double slit position is

$$\Delta p_y \approx p\alpha = \frac{h}{\lambda}\alpha \approx \frac{h}{\lambda}\frac{y_n}{L}. \tag{1}$$

Note

$$\begin{aligned} l_1^2 &= (\tfrac{1}{2}d - y_n)^2 + L^2 \\ l_2^2 &= (\tfrac{1}{2}d + y_n)^2 + L^2, \end{aligned} \tag{2}$$

so

$$l_2^2 - l_1^2 = 2y_n d \approx (l_2 - l_1)2L, \tag{3}$$

and

$$\frac{y_n}{L} = \frac{(l_2 - l_1)}{d} = \frac{n\lambda}{d}. \tag{4}$$

Therefore, at the x-position of the double slit, with $\Delta y \approx d$,

$$\Delta y \Delta p_y \approx d \frac{h}{\lambda} \frac{y_n}{L} = d \frac{h}{\lambda} \frac{n\lambda}{d} = nh, \tag{5}$$

where the uncertainty in the y-component of the momentum at the position of the double slit must be of the same order of magnitude as the change of y-component of momentum of a photon that ends up at the interference maximum given by a relatively small integer, n. Thus, at the double slit,

$$\Delta p_y \Delta y \approx nh, \tag{6}$$

where n is a relatively small number.

C Complementary Experimental Setup

The question now arises: Could we have modified the experimental setup at the x-position of the double slit to narrow the uncertainty in y at this x-position? In particular, could we have modified the experimental setup to answer experimentally the question: Through which slit did the photon pass? We could do this by making the slit jaws movable, so the momentum exchange between photon and slit jaw could be detected. Bohr imagines the very light slit jaws being suspended from springs, so the photon will jiggle the slit as it goes through the slit opening. (See Fig. 1.2, drawn in the style of the Bohr article.) Imagine the photon ends up at the position where the central interference maximum would have occured in the conventional double slit experiment, (with the apparatus of Fig. 1.1, that is, at the most likely final position of the photon for the experimental setup of Fig. 1.1). Then,

$$p_{y_{\text{final}}} - p_{y_{\text{initial}}} \approx p \times \theta = \frac{h}{\lambda} \times \frac{d}{2L} = \frac{h}{\lambda} \times \frac{\lambda}{2(y_1 - y_0)} = \frac{h}{2(y_1 - y_0)}, \tag{7}$$

where $(y_1 - y_0)$ is the distance between the first and zeroth (central) bright fringe in the interference pattern of the conventional experimental setup. This actual change in the photon's momentum at the position of the slits would now lead to a recoil in the slit jaws, which can be detected. An uncertainty will still exist in the y-position of the photon as it passes through the slit, because of this jiggling of the slit; even though our jiggle detectors can tell us through which slit the photon has passed (so $\Delta y \ll d$). Now, let us use the uncertainty relation to determine the best possible Δy caused by to the jiggling of the slit,

$$\Delta y \approx \frac{h}{\Delta p_y} \approx \frac{h \times 2(y_1 - y_0)}{h} = 2(y_1 - y_0). \tag{8}$$

That is, Δy is of the order of the distance between bright fringes in the conventional double slit setup. This Δy, however, is now due to the jiggling of the slit. If the slit jiggles on average by an amount equal to the distance between interference fringes, the interference pattern on the photographic plate will surely be washed out

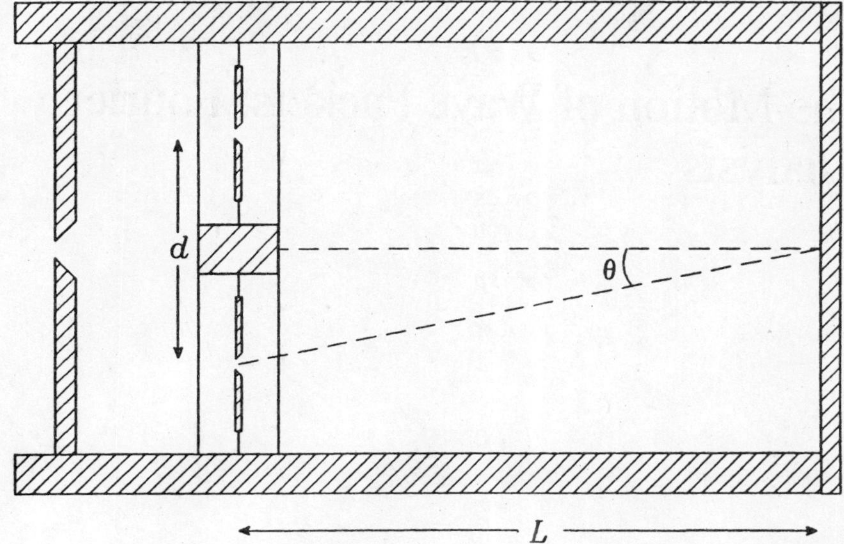

FIGURE 1.2. Complementary double slit experimental setup. Movable slits.

completely. In answering experimentally the question through which slit did the photon pass, we have by altering the experimental setup destroyed those features of the setup that previously made the precise wavelength measurement possible. This illustrates Bohr's complementarity principle. We can set up the double slit experiment to get very precise wavelength (hence, momentum) information about the photon. In this case, we cannot make a position measurement of the photon precise enough to tell us through which slit the photon passed. Alternatively, if we use the complementary experimental setup, which can answer this question about the position of the photon experimentally, we cannot determine the wavelength (hence, momentum) of the photon with sufficient accuracy.

Because we can have only partial position and partial momentum information about a photon or a material particle on the atomic scale, it becomes natural to associate a wave packet with the motion of the particle (either photon or material particle). A wave packet can give us partial position and wavelength (or wave-number k) information through the wave packet relation, $\Delta k_y \Delta y \approx 2\pi$, which follows from the Fourier analysis of the wave packet, the subject of the next chapter.

2

The Motion of Wave Packets: Fourier Analysis

Because we will need to work with wave packets of finite extent, it will be very useful to first give a brief review of Fourier analysis.

A Fourier Series

We shall start by studying periodic functions of infinite extent in space. First consider periodic functions $f(x)$ with a periodicity interval 2π, such that $f(x+2\pi) = f(x)$. For real functions $f(x)$, we usually use Fourier expansions in cosine and sine functions. For the complex functions of quantum theory, it will be advantageous to use a Fourier expansion in exponential functions.

1. Fourier Expansion:

$$f(x) = \sum_{n=-\infty}^{\infty} a_n e^{inx}, \tag{1}$$

where we will exploit the orthogonality of the exponential functions.

2. Orthogonality:

$$\int_{-\pi}^{\pi} dx' e^{i(n-m)x'} = 2\pi\, \delta_{nm}, \tag{2}$$

which is expressed in terms of the usual Kronecker delta. With this orthogonality relation, the expansion coefficients, a_n, can be determined via the Fourier inversion theorem. If we multiply $f(x)$ by the complex conjugate of a specific exponential, say, e^{-imx}, with some specific, fixed m, and integrate both sides of the resultant equation over the periodicity interval, say, from $-\pi$ to $+\pi$, the orthogonality property will pick out one specific a_m, with value given by the Fourier coefficients.

3. Fourier coefficients:

$$a_m = \frac{1}{2\pi}\int_{-\pi}^{\pi} dx' f(x') e^{-imx'}. \tag{3}$$

Substituting this coefficient back into the Fourier expansion, we get the

4. Fourier expression for $f(x)$:

$$f(x) = \sum_{n=-\infty}^{\infty} \frac{1}{2\pi}\int_{-\pi}^{\pi} dx' f(x') e^{in(x-x')}. \tag{4}$$

It will be convenient to introduce orthonormal functions, $\phi_n(x)$,

$$\phi_n(x) = \frac{1}{\sqrt{2\pi}} e^{inx}. \tag{5}$$

The four basic Fourier equations can then be rewritten as

$$f(x) = \sum_{n=-\infty}^{\infty} b_n \phi_n(x), \tag{6}$$

$$\int_{-\pi}^{\pi} dx' \phi_n^*(x') \phi_m(x') = \delta_{nm}, \tag{7}$$

$$b_n = \int_{-\pi}^{\pi} dx' f(x') \phi_n^*(x'), \tag{8}$$

$$f(x) = \sum_{n=-\infty}^{\infty} \int_{-\pi}^{\pi} dx' f(x') \phi_n(x) \phi_n^*(x'). \tag{9}$$

Finally, it will be convenient to use a periodicity interval of length $(2l)$, where l has the dimension of a length, where now $f(x+2l) = f(x)$ and the orthonormal functions can be expressed as

$$\frac{1}{\sqrt{2l}} e^{\frac{in\pi x}{l}}.$$

The four basic Fourier equations can then be rewritten as

$$f(x) = \sum_{n=-\infty}^{\infty} c_n \frac{1}{\sqrt{2l}} e^{\frac{in\pi x}{l}}, \tag{10}$$

$$\frac{1}{2l}\int_{-l}^{+l} dx' e^{i(n-m)\frac{\pi x'}{l}} = \delta_{nm}, \tag{11}$$

$$c_n = \frac{1}{\sqrt{2l}}\int_{-l}^{+l} dx' f(x') e^{-i\frac{n\pi x'}{l}}, \tag{12}$$

$$f(x) = \frac{1}{2l} \sum_{n=-\infty}^{\infty} \int_{-l}^{+l} dx' f(x') e^{i\frac{n\pi}{l}(x-x')}. \tag{13}$$

It will now be useful to introduce the wavenumber, k_n

$$k_n = \frac{n\pi}{l} = \frac{2\pi}{\lambda_n}; \qquad \text{with} \quad \lambda_n = \frac{2l}{n}, \tag{14}$$

so $\phi_n(x) = \frac{1}{\sqrt{2l}} e^{ik_n x}$. This relation will be particularly useful in making the transition from the Fourier series to the Fourier integral for a wave packet of finite extent.

B Fourier Integrals

Now suppose the repeating function, with periodicity interval $(2l)$, has the form of a wave packet of extent $\sim a$, with $a < l$, which repeats from $-\infty$ to $+\infty$, as shown in Fig. 2.1. Now, suppose we let $l \to \infty$, keeping the wave packet unchanged, with a fixed. Then, by taking the limit $l \to \infty$, provided $f(x) \to 0$ sufficiently rapidly as $x \to \pm\infty$, we can make the transition from a periodic function to a nonperiodic one, i.e., a transition from an infinite wave train to a wave packet of finite extent in space. As $l \to \infty$, the spectrum of possible k_n goes from a discrete spectrum to a continuous one, because

$$k_{n+1} - k_n = \frac{\pi}{l} \to 0 \qquad \text{as} \quad l \to \infty. \tag{15}$$

Because the number of spectral terms in a k-space interval dk is (see Fig. 2.2)

$$\frac{dk}{(\text{interval between successive } \mathrm{k_n})} = \frac{dk}{\pi/l},$$

the discrete sum over n in the Fourier series goes over to a continuous integral

$$\sum_{n=-\infty}^{\infty} \to \int_{-\infty}^{\infty} \frac{dk}{\pi/l}.$$

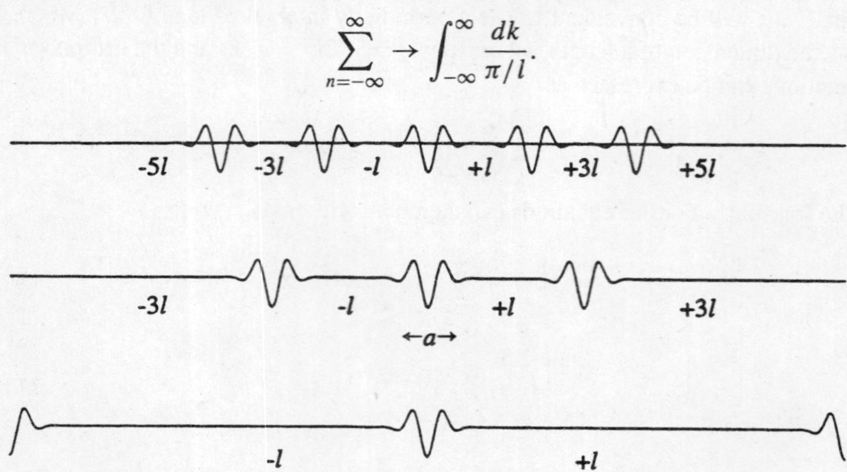

FIGURE 2.1. Periodic wave form, $l \to \infty$, a fixed.

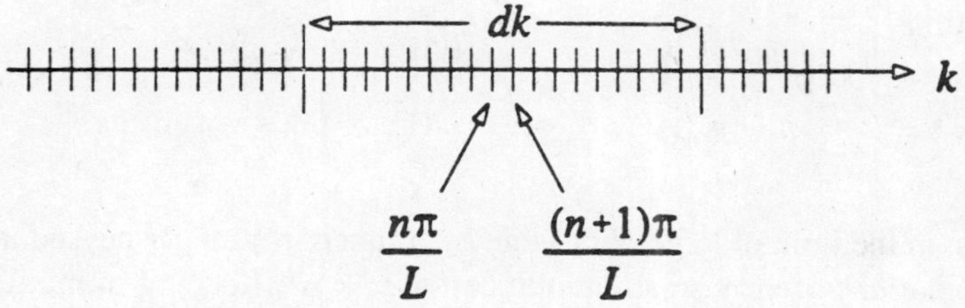

FIGURE 2.2. The spectrum of k values, $k_n = n\pi/L$. The number of spectral terms in the dk interval $= \left[dk/\frac{\pi}{L}\right]$.

Thus, the Fourier expression for $f(x)$ becomes

$$f(x) = \frac{1}{2l}\frac{l}{\pi}\int_{-\infty}^{\infty} dk \int_{-\infty}^{\infty} dx' f(x') e^{ik(x-x')}. \tag{16}$$

We can then think of the Fourier development in terms of a Fourier amplitude function, $g(k)$, as

$$f(x) = \frac{1}{\sqrt{2\pi}}\int_{-\infty}^{\infty} dk g(k) e^{ikx}, \tag{17}$$

with amplitude function $g(k)$, the so-called Fourier transform of $f(x)$, given by

$$g(k) = \frac{1}{\sqrt{2\pi}}\int_{-\infty}^{\infty} dx' f(x') e^{-ikx'}. \tag{18}$$

Note, however, the orthonormality integral becomes divergent when $k = k'$,

$$\frac{1}{2\pi}\int_{-\infty}^{\infty} dx' e^{ix'(k-k')} = \delta(k - k'). \tag{19}$$

The Kronecker delta becomes a Dirac delta function.

C The Dirac Delta Function

If we rewrite the Fourier series in terms of a limit of a sum over a finite number of terms,

$$f(x) = \lim_{N\to\infty}\int_{-\pi}^{\pi} dx' f(x') \sum_{n=-N}^{+N} \phi_n(x)\phi_n^*(x'); \tag{20}$$

or, similarly, if we rewrite the Fourier integral as

$$f(x) = \lim_{k_0\to\infty}\int_{-\infty}^{\infty} dx' f(x') \frac{1}{2\pi}\int_{-k_0}^{k_0} dk e^{ik(x-x')}, \tag{21}$$

the function

$$K(x,x') = \sum_{n=-N}^{+N} \phi_n(x)\phi_n^*(x') \qquad \text{or} \qquad K(x,x') = \frac{1}{2\pi}\int_{-k_0}^{+k_0} dk e^{ik(x-x')} \tag{22}$$

becomes, in the limit of large N or large k_0, a function strongly peaked at $x = x'$ with oscillations of very small amplitude for $x \neq x'$. Keeping in mind that the real limiting processes should be those expressed by eqs. (20) and (21), physicists blithely interchange the infinite sum or the infinte k-integral with the x'-integral, through the definition of the Dirac delta "function"

$$\sum_{n=-\infty}^{\infty} \phi_n(x)\phi_n^*(x') = \delta(x-x')$$
$$\frac{1}{2\pi}\int_{-\infty}^{\infty} dk e^{ik(x-x')} = \delta(x-x'), \tag{23}$$

where the Dirac delta "function" is not at all a function in the mathematician's sense. It is what mathematicians call a "distribution" (see, e.g., an appendix in Vol. I of the books by Messiah). The Dirac delta function "picks out" the value $x' = x$ for the function being integrated. It has meaning only through the integrals. By itself, it diverges at the value $x' = x$. The Dirac delta function is defined through the following properties:

$$\delta(x-x') = 0 \qquad \text{for} \qquad x' \neq x. \tag{24}$$

For $x' = x$, the Dirac delta function becomes ∞ in such a way that

$$\int_a^b dx'\delta(x-x') = 1, \qquad \text{if} \quad x' = x \text{ is in the interval } (a,b), \tag{25}$$

and

$$\int_{-\infty}^{\infty} dx' f(x')\delta(x-x') = f(x). \tag{26}$$

Our limiting process, given through eq. (21), e.g., would give

$$\delta(x-x') = \lim_{k_0\to\infty} \frac{1}{2\pi}\int_{-k_0}^{k_0} dk e^{ik(x-x')} = \lim_{k_0\to\infty} \frac{\sin k_0(x-x')}{\pi(x-x')}. \tag{27}$$

See Fig. 2.3 for a plot of this diffraction-like peaked function for finite k_0. This representation of the Dirac delta function is not, however, unique. Another example (of the infinite number of possibilities) would be

$$\delta(x-x') = \frac{1}{\pi}\lim_{\epsilon\to 0}\frac{\epsilon}{[(x-x')^2+\epsilon^2]}. \tag{28}$$

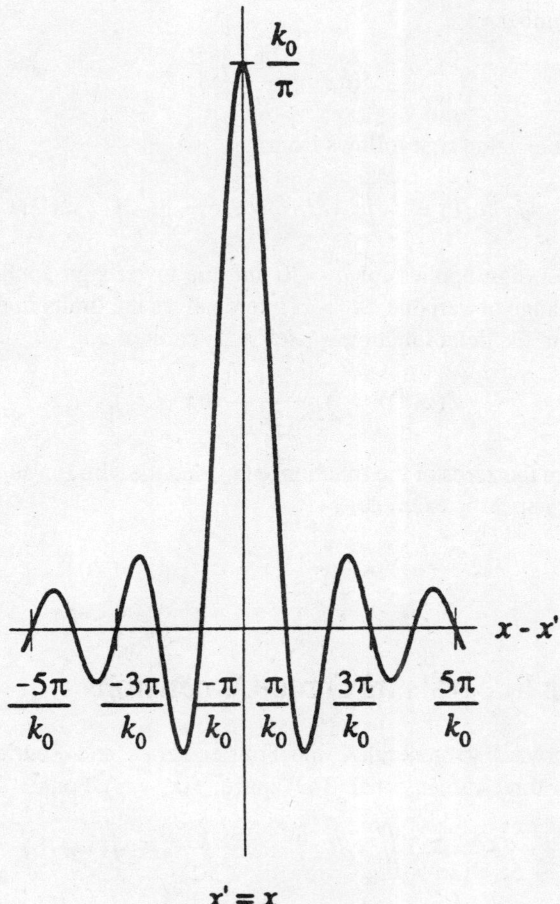

FIGURE 2.3. The function $\frac{\sin[k_0(x-x')]}{\pi(x-x')}$.

D Properties of the Dirac Delta Function

The Dirac delta function is an even function of its argument

$$\delta(-x) = \delta(x). \tag{29}$$

Other properties, such as

$$x\frac{d}{dx}\delta(x) = -\delta(x), \tag{30}$$

follow by integration by parts, because delta function relations have meaning only through their applications within integrals

$$\int_a^b dx x\delta'(x) = \Big[x\delta(x)\Big]_a^b - \int_a^b dx\delta(x) = -\int_a^b dx\delta(x). \tag{31}$$

If a is a real number,

$$\delta(ax) = \frac{1}{|a|}\delta(x). \tag{32}$$

Note the absolute value sign follows from

$$\int_{-\infty}^{\infty} dx\delta(ax) = \frac{1}{a}\int_{-\infty}^{\infty} d(ax)\delta(ax) = \pm\frac{1}{a}\int_{-\infty}^{\infty} dx'\delta(x'), \tag{33}$$

where the upper sign applies for $a > 0$ and the lower sign applies for $a < 0$, because the change of variable $ax = x'$ interchanges the limits in this latter case. If the variable in the delta function is itself a function of x,

$$\delta(\phi(x)) = \sum_n \frac{1}{|(\frac{d\phi}{dx})_{x_n}|}\delta(x - x_n), \tag{34}$$

where the x_n are the zeros of the function, $\phi(x)$, and the sum is a sum over all such zeros. As a very specific example,

$$\delta(x^2 - a^2) = \frac{1}{2|a|}\big[\delta(x-a) + \delta(x+a)\big]. \tag{35}$$

E Fourier Integrals in Three Dimensions

It is straightforward to generalize the Fourier series and Fourier integrals to functions in our three-dimensional (3-D) space, $f(x, y, z)$. For a 3-D wave packet,

$$f(x, y, z) = \frac{1}{(2\pi)^3}\int_{-\infty}^{\infty} dk_x \int_{-\infty}^{\infty} dk_y \int_{-\infty}^{\infty} dk_z \int_{-\infty}^{\infty} dx' \int_{-\infty}^{\infty} dy' \times \int_{-\infty}^{\infty} dz' f(x', y', z') e^{i[k_x(x-x')+k_y(y-y')+k_z(z-z')]}. \tag{36}$$

It will be useful to introduce the following shorthand notation for this Fourier integral expression

$$f(\vec{r}) = \frac{1}{(2\pi)^3}\int d\vec{k}\int d\vec{r}\,' f(\vec{r}\,') e^{i\vec{k}\cdot(\vec{r}-\vec{r}\,')}, \tag{37}$$

where

$$f(\vec{r}) = \frac{1}{(2\pi)^{\frac{3}{2}}}\int d\vec{k} g(\vec{k}) e^{i(\vec{k}\cdot\vec{r})}, \tag{38}$$

$$g(\vec{k}) = \frac{1}{(2\pi)^{\frac{3}{2}}}\int d\vec{r}\,' f(\vec{r}\,') e^{-i(\vec{k}\cdot\vec{r}\,')}. \tag{39}$$

(Note, in particular, the symbol, $d\vec{r}$, when it follows an integral sign, is merely a shorthand notation for $d\vec{r} \equiv dxdydz$ and the single integral sign preceding $d\vec{r}$ is shorthand for a triple integral over all of 3-D space.)

F The Operation $\frac{1}{i}\frac{\partial}{\partial x}$

We note

$$\frac{1}{i}\frac{\partial}{\partial x}f(\vec{r}) = \frac{1}{(2\pi)^{\frac{3}{2}}}\int d\vec{k}g(\vec{k})k_x e^{i(\vec{k}\cdot\vec{r})}. \tag{40}$$

Thus, we see, if $g(\vec{k})$ is the Fourier transform of $f(\vec{r})$, $\vec{k}g(\vec{k})$ is the Fourier transform of $\frac{1}{i}\vec{\nabla} f(\vec{r})$, similarly, $-(\vec{k}\cdot\vec{k})g(\vec{k})$ is the Fourier transform of $\nabla^2 f(x, y, z)$, and so on.

G Wave Packets

A plane scalar wave propagating in the direction of the $\vec{k}$ vector can be given by the scalar function

$$\psi(\vec{r}, t) = Ae^{i(\vec{k}\cdot\vec{r}-\omega t)}, \tag{41}$$

with constant amplitude, A, where the circular frequency, ω, is in general related to $\vec{k}$ through the dispersion law

$$\omega = f(\vec{k}), \qquad \text{or} \quad \omega = f(k), \quad \text{with} \quad k = |\vec{k}|, \tag{42}$$

where the latter is valid for an isotropic medium. Moreover, in a nondispersive medium, in vacuum, e.g., $\omega = ck$.

To go from this infinite wave train to a wave packet of finite extent in space, we need to form the wave packet from a superposition of amplitudes with different $\vec{k}$-values. For a 3-D wave packet,

$$\psi(\vec{r}, t) = \frac{1}{(2\pi)^{\frac{3}{2}}}\int d\vec{k}A(\vec{k})e^{i(\vec{k}\cdot\vec{r}-\omega t)}. \tag{43}$$

To simplify the discussion, assume the wave packet proceeds in one dimension only, say, the x-direction. Then,

$$\psi(x, t) = \frac{1}{\sqrt{2\pi}}\int_{-\infty}^{\infty} dkA(k)e^{i(kx-\omega t)}. \tag{44}$$

To use a very simple example, assume $A(k)$ is different from zero only in an interval, $k_0 - \frac{1}{2}\Delta k \le k \le k_0 + \frac{1}{2}\Delta k$, and moreover, assume $A(k)$ has the constant value, A, in this k-space interval. If the interval Δk is not too large, we can expand $\omega(k)$ about k_0, and retain only the dominant terms,

$$\omega(k) = \omega(k_0) + (k - k_0)\Big(\frac{d\omega}{dk}\Big)_0 + \cdots, \tag{45}$$

and the wave function can be written as

$$\psi(x, t) = \frac{A}{\sqrt{2\pi}}e^{i[k_0x-\omega(k_0)t]}\int_{k_0-\frac{1}{2}\Delta k}^{k_0+\frac{1}{2}\Delta k} dke^{i(k-k_0)[x-(\frac{d\omega}{dk})_0t]}$$

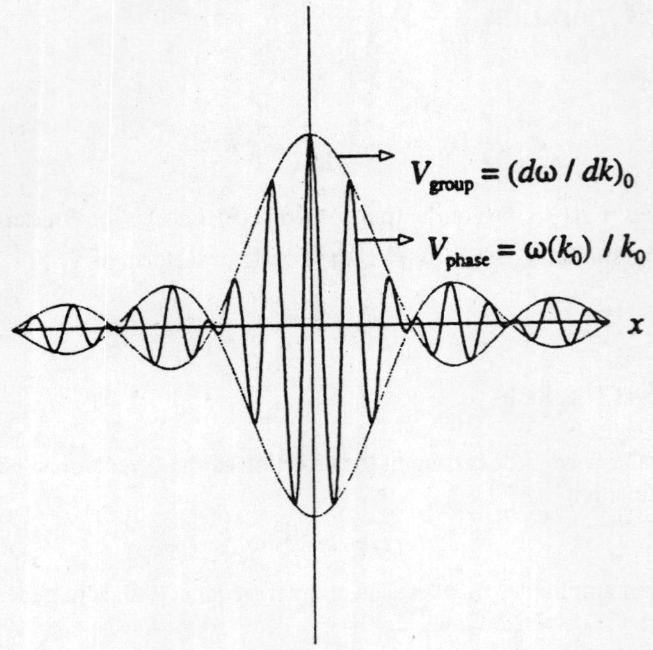

FIGURE 2.4. The wave packet of eq. (46).

$$= \sqrt{\frac{2}{\pi}} A e^{i[k_0 x - \omega(k_0)t]} \frac{\sin\left(\frac{\Delta k}{2}[x - (\frac{d\omega}{dk})_0 t]\right)}{[x - (\frac{d\omega}{dk})_0 t]}. \tag{46}$$

This wave packet is shown in Fig. 2.4. We note, in particular, the individual wavelets travel with the phase velocity

$$v_{\text{phase}} = \frac{\omega(k_0)}{k_0}. \tag{47}$$

The wave train itself, the envelope of the packet, however, travels with the group velocity

$$v_{\text{group}} = \left(\frac{d\omega}{dk}\right)_0. \tag{48}$$

If we assume most of the energy of the wave train lies in the large central peak of the wave envelope, we can take the extent of the wave packet to be $\Delta x \approx 2\frac{2\pi}{\Delta k}$. Even for more sophisticated functions, $A(k)$, we will find the Fourier integral analysis always gives

$$\Delta x \Delta k \approx 2\pi, \tag{49}$$

neglecting factors of order 2 in this approximation. This is the uncertainty relation for a wave packet. Note, in particular, it follows for all wave packets, merely from the Fourier analysis.

H Propagation of Wave Packets: The Wave Equation

The wave equation, the propagation law for the wave, is intimately related to the dispersion law

$$\omega = f(k). \tag{50}$$

In one dimension, with

$$\psi(x,t) = \frac{1}{\sqrt{2\pi}} \int dk A(k) e^{i(kx-\omega t)}, \tag{51}$$

$$-\frac{1}{i}\frac{\partial}{\partial t}\psi = \frac{1}{\sqrt{2\pi}} \int dk \omega A(k) e^{i(kx-\omega t)}, \tag{52}$$

$$\left(\frac{1}{i}\right)^n \frac{\partial^n}{\partial x^n}\psi = \frac{1}{\sqrt{2\pi}} \int dk k^n A(k) e^{i(kx-\omega t)}. \tag{53}$$

For functions $f(k)$ that can be given by Taylor expansions,

$$f(k) = \sum_{n=0} \alpha_n k^n,$$

we then have

$$f\left(\frac{1}{i}\frac{\partial}{\partial x}\right)\psi = \frac{1}{\sqrt{2\pi}} \int dk f(k) A(k) e^{i(kx-\omega t)}. \tag{54}$$

Eqs. (52) and (54) then lead to

$$\left[-\frac{1}{i}\frac{\partial}{\partial t} - f\left(\frac{1}{i}\frac{\partial}{\partial x}\right)\right]\psi = \frac{1}{\sqrt{2\pi}} \int dk[\omega - f(k)] A(k) e^{i(kx-\omega t)} = 0, \tag{55}$$

so the dispersion law, $\omega = f(k)$, leads to the wave equation

$$\left[-\frac{1}{i}\frac{\partial}{\partial t} - f\left(\frac{1}{i}\frac{\partial}{\partial x}\right)\right]\psi = 0. \tag{56}$$

For the special case of a nondispersive medium, with $\omega = ck$, we would have

$$-\frac{1}{i}\frac{\partial \psi}{\partial t} - \frac{c}{i}\frac{\partial \psi}{\partial x} = \int dk[\omega - ck] A(k) e^{i(kx-\omega t)} = 0. \tag{57}$$

So that, seemingly, the wave equation in this simple case of a nondispersive medium becomes

$$\frac{1}{c}\frac{\partial \psi}{\partial t} + \frac{\partial \psi}{\partial x} = 0. \tag{58}$$

This equation looks like a strange wave equation, however. Its solutions would be $\psi(x,t) = F(x - ct)$, where F is any arbitrary function. That is, this wave equation would permit wave propagation only in the positive x-direction and, hence, would correspond to a nonisotropic medium. The difficulty here is not with our method of arriving at the wave equation, but that we have written the dispersion law in a way that builds in this anisotropy. For a nondispersive, isotropic medium, we have to express the dispersion law in the form

$$\omega^2 - c^2k^2 = 0, \tag{59}$$

or in three dimensions

$$\omega^2 - c^2(k_x^2 + k_y^2 + k_z^2) = 0. \tag{60}$$

The technique we have used to arrive at the wave equation would then give us

$$\frac{1}{c^2}\frac{\partial^2\psi}{\partial t^2} - \frac{\partial^2\psi}{\partial x^2} = 0 \tag{61}$$

in one dimension, and

$$\frac{1}{c^2}\frac{\partial^2\psi}{\partial t^2} - \nabla^2\psi = 0 \tag{62}$$

in three dimensions.

Note, finally, our method of arriving at the wave equation from the dispersion law is not a derivation of the wave equation. Our method may also not give a unique expression for the wave equation.

3

The Schrödinger Wave Equation and Probability Interpretation

A The Wave Equation

With the Bohr relation for the energy, $E = \hbar\omega$, and the deBroglie relation for the momentum vector, $\vec{p} = \hbar\vec{k}$, we see the dispersion relation for waves, $\omega = f(\vec{k})$, goes over to a relation between energy and momentum. For a conservative system, this relation can be expressed through $E = H(\vec{p}, \vec{r})$, where H is the Hamiltonian function. In particular, for a free, nonrelativistic particle, of mass m, this "dispersion relation" becomes

$$E = \frac{(\vec{p}\cdot\vec{p})}{2m}. \tag{1}$$

Now, convert our wave packet expansion from an expansion in $\vec{k}$ to one in $\vec{p}$

$$\Psi(\vec{r}, t) = \frac{1}{(2\pi\hbar)^{\frac{3}{2}}}\int d\vec{p}A(\vec{p})e^{\frac{i}{\hbar}(\vec{p}\cdot\vec{r}-Et)}, \tag{2}$$

so

$$-\frac{\hbar}{i}\frac{\partial\Psi}{\partial t} = \frac{1}{(2\pi\hbar)^{\frac{3}{2}}}\int d\vec{p}EA(\vec{p})e^{\frac{i}{\hbar}(\vec{p}\cdot\vec{r}-Et)}, \tag{3}$$

$$-\frac{\hbar^2}{2m}\nabla^2\Psi = \frac{1}{(2\pi\hbar)^{\frac{3}{2}}}\int d\vec{p}\frac{(\vec{p}\cdot\vec{p})}{2m}A(\vec{p})e^{\frac{i}{\hbar}(\vec{p}\cdot\vec{r}-Et)}. \tag{4}$$

As a result, we get

$$-\frac{\hbar}{i}\frac{\partial\Psi}{\partial t} + \frac{\hbar^2}{2m}\nabla^2\Psi = \frac{1}{(2\pi\hbar)^{\frac{3}{2}}}\int d\vec{p}\Big[E - \frac{(\vec{p}\cdot\vec{p})}{2m}\Big]e^{\frac{i}{\hbar}(\vec{p}\cdot\vec{r}-Et)} = 0, \tag{5}$$

and the relation between E and $\vec{p}$ leads us to the wave equation for a free particle, the Schrödinger equation for a free particle,

$$-\frac{\hbar}{i}\frac{\partial\Psi}{\partial t} = -\frac{\hbar^2}{2m}\nabla^2\Psi(\vec{r}, t). \tag{6}$$

The group velocity of a wave packet now becomes

$$\vec{v}_{\text{group}} = \frac{d\omega}{d\vec{k}} = \frac{dE}{d\vec{p}} = \frac{\vec{p}}{m} = \vec{v}_{\text{particle}}. \tag{7}$$

The uncertainty relations for waves go over to the Heisenberg uncertainty relations

$$\begin{array}{lcl} \Delta k_x \Delta x \approx 2\pi & \rightarrow & \Delta p_x \Delta x \approx h \\ \Delta k_y \Delta y \approx 2\pi & \rightarrow & \Delta p_y \Delta y \approx h \\ \Delta k_z \Delta z \approx 2\pi & \rightarrow & \Delta p_z \Delta z \approx h. \end{array} \tag{8}$$

For a particle subject to a conservative force derivable from a potential $V(x, y, z)$, with

$$E = H(\vec{p}, \vec{r}) = \frac{(\vec{p}\cdot\vec{p})}{2m} + V(x, y, z), \tag{9}$$

this relation between E and $\vec{p}$ gives the wave equation

$$-\frac{\hbar}{i}\frac{\partial\Psi}{\partial t} = -\frac{\hbar^2}{2m}\nabla^2\Psi + V(x, y, z)\Psi. \tag{10}$$

Finally, we end with a remark about relativistic wave equations. The relation between energy and momentum for a relativistic particle, of rest mass m_0,

$$\frac{E^2}{c^2} - p^2 = m_0^2c^2, \tag{11}$$

would lead us to a wave equation both second order in time and space derivatives, and again involving a single wave function $\Psi(\vec{r}, t)$. (See Problem 2). This equation leads to the so-called Klein–Gordon equation. An alternative solution for the relativistic wave equation was given by Dirac, whose wave equation is first order in both time and space derivatives. Essentially, it comes from the square root of the above dispersion relation and, therefore, leads to both positive and negative energy solutions. It is based not on a single Ψ, but on a number of ψ_α, actually, with $\alpha = 1, \ldots, 4$. We shall come back to the Dirac equation near the end of the book.

B The Probability Axioms

To use the Schrödinger wave equation, we need to understand the physical meaning of the wave function, Ψ. We begin with a few remarks:

1. The particle and the wavefield are equally real. (The wavefield is not a ghost field guiding the particle.) Both are pictures in the human mind to account for physical reality; both, however, have their limitations.

2. In practice, the wavefield is used in the following way: The result of a certain experiment lets us represent the particle motion by a wave packet at a certain time. The wave equation is then used to predict how the experiment evolves in time.

3. Because of the uncontrollable interaction of object and measuring instrument, we are led to a probability description.

The probability of finding a particle within a volume element $dxdydz$ about a point x, y, z at a time t, will be given by

$$W(x, y, z; t)dxdydz,$$

where $W(x, y, z; t)$ is a probability density. This probability density must satisfy certain sensibility restrictions; i.e., $W(x, y, z; t)$ must be a sensible probability density:

$$W \geq 0. \tag{12}$$

A negative probability density makes no sense. To make W patently positive, it makes sense to let the probability density be given by

$$W(\vec{r}; t) = \Psi^* \Psi. \tag{13}$$

Note, in optics or electromagnetism, physically measurable quantities, such as energy or intensity, are proportional to the square of the amplitude of the wave, or if the amplitude can be complex to the absolute value squared. Also,

$$\frac{d}{dt} \int_{\text{all space}} d\vec{r} \;\; W(\vec{r}, t) = 0. \tag{14}$$

The probability of finding the particle somewhere is independent of time. In non-relativistic quantum mechanics, we are building a theory of indestructible particles. Also,

$$\int_{\text{all space}} d\vec{r} \;\; W \quad \text{is Galilean invariant.} \tag{15}$$

All observers agree one particle exists somewhere.

The conservation of probability leads to a continuity equation. Because Ψ can be complex, the Schrödinger equation is really two equations

$$\begin{aligned} -\frac{\hbar}{i}\frac{\partial \Psi}{\partial t} &= -\frac{\hbar^2}{2m}\nabla^2\Psi + V\Psi \\ +\frac{\hbar}{i}\frac{\partial \Psi^*}{\partial t} &= -\frac{\hbar^2}{2m}\nabla^2\Psi^* + V\Psi^*. \end{aligned} \tag{16}$$

Multiplying the first of these equations by $-\Psi^*$, the second equation by Ψ, and adding the two resultant equations, assuming also the potential function is real, we get

$$\frac{\hbar}{i}\frac{\partial}{\partial t}(\Psi^*\Psi) = \frac{\hbar^2}{2m}\text{div}(\Psi^* \vec{\nabla} \Psi - \Psi \vec{\nabla} \Psi^*). \tag{17}$$

This relation has the form of a continuity equation

$$\frac{\partial W}{\partial t} + \mathrm{div}\vec{S} = 0, \tag{18}$$

if we choose

$$\vec{S} = \frac{\hbar}{2mi}(\Psi^* \vec{\nabla} \Psi - \Psi \vec{\nabla} \Psi^*), \tag{19}$$

where this equation must be interpreted as a probability density current; i.e., as the probability per second per unit area normal to the direction of $\vec{S}$ that the particle be streaming in the direction of $\vec{S}$. The integral form of the continuity equation could be written as

$$\frac{d}{dt}\int_{\text{Vol.}} d\vec{r}\Psi^*\Psi + \int_{\text{Surf.}} dA(\vec{S}\cdot\vec{n}) = 0, \tag{20}$$

where the volume integral is over a finite volume and the surface integral is over the surface surrounding this finite volume, dA being an element of surface area and $\vec{n}$ being the outward normal to the surface. In integral form, this equation says: The time rate of change of probability of finding the particle within the finite volume must be the negative of the probability of the net outflow of the particle. If we let the volume grow to include all of our 3-D space and if we assume Ψ and $\Psi^* \to 0$ sufficiently fast as a function of r as the surface recedes to infinity, the surface integral will go to zero and our second probability restriction is satisfied.

Finally, examine the Galilean invariance. Suppose observers in a primed reference frame are moving with velocity, v, parallel to the x-direction. Then,

$$x = x' + vt', \qquad y = y', \qquad z = z', \qquad t = t', \tag{21}$$

with

$$\frac{\partial}{\partial t'} = 1\frac{\partial}{\partial t} + v\frac{\partial}{\partial x}, \quad \frac{\partial}{\partial x'} = 1\frac{\partial}{\partial x} + 0\frac{\partial}{\partial t}, \quad \frac{\partial}{\partial y'} = \frac{\partial}{\partial y}, \quad \frac{\partial}{\partial z'} = \frac{\partial}{\partial z}. \tag{22}$$

In addition, the wave function must also change under the Galilean transformation, according to

$$\Psi'(x', y', z'; t') = \Psi(x, y, z; t)e^{-\frac{i}{\hbar}(mvx - \frac{1}{2}mv^2 t)}. \tag{23}$$

Comparing the Schrödinger equations in the primed and unprimed frames, we are led to

$$W' = W; \qquad \vec{S}' = \vec{S} - \vec{v}\Psi^*\Psi. \tag{24}$$

If observers in the unprimed frame see no streaming of probability to right or left, $(\vec{S}) = 0$, observers in the primed frame will see a streaming to the left, as they should because a particle in a region with $W \neq 0$ will appear to be moving in the direction of $-\vec{v}$ to an observer in the primed frame.

A final remark: If Ψ is real everywhere, $\vec{S}$ is zero everywhere; then, no transport of probability exists. A probability density with a real Ψ corresponds to a situation in which particles will stream in the $+x$ and $-x$ directions with equal probability.

Note, we need complex Ψ's to describe beams of particles streaming toward or away from a target.

C The Calculation of Average Values of Dynamical Quantities

On the atomic scale, we cannot give an exact orbit description, e.g., $x(t)$, for the motion of a particle. We can, however, give the probability theory average value of dynamical variables, such as x, as functions of the time. Define this average or expectation value of x through

$$\langle x \rangle = \int d\vec{r}\, x\, \Psi^*(\vec{r},t)\Psi(\vec{r},t); \tag{25}$$

or, similarly,

$$\langle x^n \rangle = \int d\vec{r}\, x^n\, \Psi^*(\vec{r},t)\Psi(\vec{r},t). \tag{26}$$

The question then arises, how do we define the corresponding expectation value of a momentum component, $\langle p_x \rangle$? It will be convenient to define a momentum space probability density, so the probability of finding a particle in the momentum range $dp_x dp_y dp_z$ about some value p_x, p_y, p_z is given by

$$\phi^*(\vec{p},t)\phi(\vec{p},t)dp_x dp_y dp_z,$$

so

$$\langle p_x \rangle = \int d\vec{p}\; p_x \phi^*(\vec{p},t)\phi(\vec{p},t), \tag{27}$$

where $\phi(\vec{p},t)$ is the Fourier transform of $\Psi(\vec{r},t)$. Comparing with eq. (2), and letting $\phi(\vec{p},t) = A(\vec{p})e^{-\frac{i}{\hbar}Et}$,

$$\phi(\vec{p},t) = \frac{1}{(2\pi\hbar)^{\frac{3}{2}}}\int d\vec{r}\,'\Psi(\vec{r}\,',t)e^{-\frac{i}{\hbar}\vec{p}\cdot\vec{r}\,'}, \tag{28}$$

$$\Psi(\vec{r},t) = \frac{1}{(2\pi\hbar)^{\frac{3}{2}}}\int d\vec{p}\phi(\vec{p},t)e^{\frac{i}{\hbar}\vec{p}\cdot\vec{r}}, \tag{29}$$

$$\Psi(\vec{r},t) = \frac{1}{(2\pi\hbar)^3}\int d\vec{p}\int d\vec{r}\,'\Psi(\vec{r}\,',t)e^{\frac{i}{\hbar}\vec{p}\cdot(\vec{r}-\vec{r}\,')}. \tag{30}$$

Note

$$\begin{aligned}\int d\vec{p}\phi^*(\vec{p},t)\phi(\vec{p},t) &= \underline{\frac{1}{(2\pi\hbar)^3}\int d\vec{p}}\int d\vec{r}\Psi^*(\vec{r},t)\underline{\int d\vec{r}\,'\Psi(\vec{r}\,',t)e^{\frac{i}{\hbar}\vec{p}\cdot(\vec{r}-\vec{r}\,')}}\\ &= \int d\vec{r}\Psi^*(\vec{r},t)\underline{\Psi(\vec{r},t)},\end{aligned} \tag{31}$$

where the underlined quantities in the first line of this equation give $\Psi(\vec{r}, t)$ via the use of eq. (30). This relation between $\Psi(\vec{r}, t)$ and its Fourier transform $\phi(\vec{p}, t)$ is known as Parseval's theorem. Note, the probability of finding the particle *somewhere*, with *some* momentum, is equal to 1 for a theory of one particle.

The same type of Fourier transformation can also be used to express the expectation value $\langle p_x \rangle$ in terms of a space rather than a momentum integral

$$\begin{aligned} \langle p_x \rangle &= \int d\vec{p}\ \ p_x \phi^*(\vec{p}, t)\phi(\vec{p}, t) \\ &= \frac{1}{(2\pi\hbar)^3}\int d\vec{p} \int d\vec{r}\Psi^*(\vec{r}, t)\ p_x \int d\vec{r}\,'\Psi(\vec{r}\,', t)e^{\frac{i}{\hbar}\vec{p}\cdot(\vec{r}-\vec{r}\,')} \\ &= \underline{\frac{1}{(2\pi\hbar)^3}\int d\vec{p}} \int d\vec{r}\Psi^*(\vec{r}, t)\left(\frac{\hbar}{i}\frac{\partial}{\partial x}\underline{\int d\vec{r}\,'\Psi(\vec{r}\,', t)e^{\frac{i}{\hbar}\vec{p}\cdot(\vec{r}-\vec{r}\,')}}\right) \\ &= \int d\vec{r}\Psi^*(\vec{r}, t)\frac{\hbar}{i}\frac{\partial}{\partial x}\underline{\Psi(\vec{r}, t)}, \end{aligned} \tag{32}$$

so $\langle p_x \rangle$ can also be evaluated through

$$\langle p_x \rangle = \int d\vec{r}\Psi^*(\vec{r}, t)\frac{\hbar}{i}\left(\frac{\partial}{\partial x}\Psi(\vec{r}, t)\right). \tag{33}$$

Similarly,

$$< p_x^2 >= \int d\vec{r}\Psi^*(\vec{r}, t)\left(-\hbar^2\frac{\partial^2}{\partial x^2}\Psi(\vec{r}, t)\right). \tag{34}$$

Finally, by the same technique, we could express $\langle x \rangle$ in terms of momentum rather than space integrals

$$\langle x \rangle = \int d\vec{p}\phi^*(\vec{p}, t)i\hbar\left(\frac{\partial}{\partial p_x}\phi(\vec{p}, t)\right). \tag{35}$$

D Precise Statement of the Uncertainty Principle

Now that we have defined $\langle x \rangle$, $< x^2 >$, and so on precisely in terms of the probability densities, we can formulate the Heisenberg uncertainty principle more precisely. Taking the usual statistical definition of the uncertainty, Δx,

$$\begin{aligned} (\Delta x)^2 &= < (x - \langle x \rangle)^2 >=< x^2 > -2\langle x \rangle\langle x \rangle + \langle x \rangle^2 \\ &= < x^2 > -\langle x \rangle^2, \end{aligned} \tag{36}$$

and, similarly, for Δp_x. The precise statement of the Heisenberg uncertainty principle is then

$$\begin{aligned} \Delta p_x \Delta x &\geq \tfrac{1}{2}\hbar \\ \Delta p_y \Delta y &\geq \tfrac{1}{2}\hbar \\ \Delta p_z \Delta z &\geq \tfrac{1}{2}\hbar. \end{aligned} \tag{37}$$

For simplicity, give a derivation only for one-dimensional (1-D) motion and consider the motion to be in the x-direction. To prove the uncertainty relation, consider the following integral, a function of a real parameter λ,

$$I(\lambda) \equiv \int_{-\infty}^{\infty} dx \left| (x - \langle x \rangle)\Psi + i\lambda\Big(\frac{\hbar}{i}\frac{\partial \Psi}{\partial x} - \langle p_x \rangle \Psi\Big) \right|^2. \tag{38}$$

Note, through its definition, $I(\lambda) \geq 0$. Writing out all of the terms

$$\begin{aligned} I(\lambda) = &\int_{-\infty}^{\infty} dx \Psi^*(x - \langle x \rangle)^2 \Psi \\ &+ \lambda\hbar \int_{-\infty}^{\infty} dx \big(\Psi^* \frac{\partial \Psi}{\partial x} + \Psi \frac{\partial \Psi^*}{\partial x}\big)(x - \langle x \rangle) \\ &+ \lambda^2 \hbar^2 \int_{-\infty}^{\infty} dx \frac{\partial \Psi^*}{\partial x}\frac{\partial \Psi}{\partial x} \\ &+ \lambda^2 \langle p_x \rangle \frac{\hbar}{i} \int_{-\infty}^{\infty} dx \Big[\Psi \frac{\partial \Psi^*}{\partial x} - \Psi^* \frac{\partial \Psi}{\partial x} \Big] \\ &+ \lambda^2 \langle p_x \rangle^2 \int_{-\infty}^{\infty} dx \Psi^* \Psi. \end{aligned} \tag{39}$$

Now, note term (2) (in the second line) can be rewritten

$$\int_{-\infty}^{\infty} dx \frac{\partial(\Psi^*\Psi)}{\partial x}(x - \langle x \rangle) = \Big[(x - \langle x \rangle)\Psi^*\Psi \Big]_{-\infty}^{\infty} - \int_{-\infty}^{\infty} dx \Psi^* \Psi = -1, \tag{40}$$

where we have assumed $\Psi \to 0$ sufficiently fast as $x \to \pm\infty$, so the integrated term is zero. Similarly, term (3) can be rewritten as

$$\hbar^2 \int_{-\infty}^{\infty} dx \frac{\partial \Psi^*}{\partial x}\frac{\partial \Psi}{\partial x} = \hbar^2 \Big[\Psi^* \frac{\partial \Psi}{\partial x} \Big]_{-\infty}^{\infty} + \int_{-\infty}^{\infty} dx \Psi^* \Big(-\hbar^2 \frac{\partial^2 \Psi}{\partial x^2} \Big) = < p_x^2 > . \tag{41}$$

Finally, in term (4), rewrite

$$\frac{\hbar}{i} \int_{-\infty}^{\infty} dx \frac{\partial \Psi^*}{\partial x} \Psi = \frac{\hbar}{i} \Big[\Psi^* \Psi \Big]_{-\infty}^{\infty} - \int dx \Psi^* \frac{\hbar}{i}\frac{\partial \Psi}{\partial x} = -\langle p_x \rangle, \tag{42}$$

so the full expression of term (4) can be rewritten as $-2\langle p_x \rangle^2$. Putting together all of the terms, we then get

$$I(\lambda) = (\Delta x)^2 - \hbar\lambda + (\Delta p_x)^2 \lambda^2 \geq 0. \tag{43}$$

With $I(\lambda) = a\lambda^2 + b\lambda + c$, the requirement, $I(\lambda) \geq 0$, is met if $b^2 - 4ac \leq 0$. Thus, $\hbar^2 - 4(\Delta x)^2(\Delta p_x)^2 \leq 0$, and, therefore,

$$\Delta p_x \Delta x \geq \tfrac{1}{2}\hbar. \tag{44}$$

Thus, in the most favorable wave packet, the so-called minimum wave packet, we can have the minimum uncertainty, $\frac{1}{2}\hbar$. We shall see later this is true for a Gaussian wave packet.

E Ehrenfest's Theorem: Equations of Motion

Classically, the Hamiltonian for a single particle of mass m,

$$H = \frac{\vec{p}^2}{2m} + V(x, y, z), \tag{45}$$

leads to the equations of motion

$$\frac{dx}{dt} = \frac{\partial H}{\partial p_x}; \qquad \frac{dp_x}{dt} = -\frac{\partial H}{\partial x} = -\frac{\partial V}{\partial x}. \tag{46}$$

Quantum mechanically the velocity of the particle no longer has a precise meaning, but we can ask: How does the expectation value of x change with time? Calculate $\frac{d\langle x\rangle}{dt}$:

$$\frac{d\langle x\rangle}{dt} = \int d\vec{r}\, x \frac{\partial}{\partial t}\Big(\Psi^*(\vec{r}, t)\Psi(\vec{r}, t)\Big). \tag{47}$$

(Note, in particular, the quantity x in the integrand is not a function of the time. It is merely the dummy integration variable, which weights the time-dependent function $\Psi^*\Psi$. It says we must weight all values of x from $-\infty$ to $+\infty$ with x and the probability density function to obtain $\langle x\rangle$.) Using the continuity equation, we can rewrite this as

$$\begin{aligned}\frac{d\langle x\rangle}{dt} &= -\int d\vec{r}\;\; x\operatorname{div}\vec{S} = -\int d\vec{r}\;\; x\left(\frac{\partial S_x}{\partial x} + \frac{\partial S_y}{\partial y} + \frac{\partial S_z}{\partial z}\right)\\ &= -\int_{-\infty}^{\infty} dy \int_{-\infty}^{\infty} dz \Big[xS_x\Big]_{x=-\infty}^{\infty} - \int_{-\infty}^{\infty} dx \int_{-\infty}^{\infty} dz \Big[xS_y\Big]_{y=-\infty}^{\infty}\\ &\quad - \int_{-\infty}^{\infty} dx \int_{-\infty}^{\infty} dy \Big[xS_z\Big]_{z=-\infty}^{\infty} + \int d\vec{r}\;\; S_x\\ &= \frac{\hbar}{2mi}\int d\vec{r}\left(\Psi^*\frac{\partial\Psi}{\partial x} - \Psi\frac{\partial\Psi^*}{\partial x}\right)\\ &= \frac{1}{m}\int d\vec{r}\;\; \Psi^*\left(\frac{\hbar}{i}\frac{\partial\Psi}{\partial x}\right) = \frac{1}{m}\langle p_x\rangle,\end{aligned} \tag{48}$$

where all integrated terms disappear, and we have done one more integration by parts on the $\Psi\frac{\partial\Psi^*}{\partial x}$ term in the last step. Therefore, we see

$$\frac{d\langle x\rangle}{dt} = \frac{1}{m}\langle p_x\rangle, \tag{49}$$

which is the first equation of motion, provided we replace x and p_x by their expectation values $\langle x\rangle$ and $\langle p_x\rangle$. In exactly the same fashion,

$$\begin{aligned}\frac{d\langle p_x\rangle}{dt} &= \frac{d}{dt}\int d\vec{r}\Psi^*\frac{\hbar}{i}\frac{\partial\Psi}{\partial x}\\ &= \int d\vec{r}\left(\frac{\hbar}{i}\frac{\partial\Psi^*}{\partial t}\right)\left(\frac{\partial\Psi}{\partial x}\right) + \int d\vec{r}\Psi^*\frac{\hbar}{i}\left(\frac{\partial^2\Psi}{\partial t\partial x}\right)\\ &= \int d\vec{r}\left(\frac{\hbar}{i}\frac{\partial\Psi^*}{\partial t}\right)\left(\frac{\partial\Psi}{\partial x}\right) + \int_{-\infty}^{\infty} dy \int_{-\infty}^{\infty} dz \frac{\hbar}{i}\Psi^*\frac{\partial\Psi}{\partial t}\Bigg]_{x=-\infty}^{\infty}\end{aligned}$$

$$
\begin{aligned}
&+\int d\vec{r}\frac{\partial\Psi^*}{\partial x}\Big(-\frac{\hbar}{i}\frac{\partial\Psi}{\partial t}\Big)\\
&=-\frac{\hbar^2}{2m}\int d\vec{r}\Big[\nabla^2\Psi^*\frac{\partial\Psi}{\partial x}+\frac{\partial\Psi^*}{\partial x}\nabla^2\Psi\Big]+\int d\vec{r}\Big[\Psi^*V\frac{\partial\Psi}{\partial x}+\frac{\partial\Psi^*}{\partial x}V\Psi\Big]\\
&=-\frac{\hbar^2}{2m}[\ 0\ ]+\int d\vec{r}\ V\frac{\partial(\Psi^*\Psi)}{\partial x}\\
&=\int_{-\infty}^{\infty}dy\int_{-\infty}^{\infty}dz\Psi^*\Psi V\Big]_{x=-\infty}^{\infty}-\int d\vec{r}\ \Psi^*\Psi\Big(\frac{\partial V}{\partial x}\Big),
\end{aligned}
\tag{50}
$$

where the term with ∇^2 operators disappears via integrations by parts, similarly to the integrated terms shown explicitly. Thus,

$$
\frac{d\langle p_x\rangle}{dt}=-<\Big(\frac{\partial V}{\partial x}\Big)>,
\tag{51}
$$

which is the quantum-mechanical analogue of the second classical equation of motion, where again the classical quantities, p_x, and, $\frac{\partial V}{\partial x}$, have been replaced with their quantum-mechanical expectation values.

F Operational Calculus, The Linear Operators of Quantum Mechanics, Hilbert Space

In the last few sections, we have met many operators acting on the Schrödinger Ψ. In this section, we want to make a more systematic study of the linear operators of quantum theory. The operator, O, is a command. Acting on a function, it produces another function, $(O\Psi(x))$. Examples include $O = f(x)$. Acting on the function $\Psi(x)$, it produces the new function, $f(x)\Psi(x)$. The command is: Multiply the old function by $f(x)$. The second common example is the operator $\frac{d}{dx}$. When acting on the function Ψ, it produces the new function

$$
\frac{d\Psi(x)}{dx}.
$$

The operators of quantum theory are linear operators. When acting on functions $\Psi(x)$, which are linear combinations of functions, $\Psi(x) = \lambda_1\Psi_1(x) + \lambda_2\Psi_2(x)$, where λ_1 and λ_2 are arbitrary complex numbers, the linear operator O yields

$$
(O\Psi(x)) = \lambda_1(O\Psi_1(x)) + \lambda_2(O\Psi_2(x)).
\tag{52}
$$

Another type of operator is the parity or space-inversion operator, P. It is the command: Change x to $-x$, y to $-y$, and z to $-z$ in the function on which it acts

$$
P\Psi(x, y, z) = \Psi(-x, -y, -z).
$$

Products of operators are operators acting in succession on our functions:

$$
O_2O_1\Psi = O_2(O_1\Psi).
$$

Because we deal with wave functions and operators acting on wave functions, we need to define the function space on which the operators act. The wave functions of quantum mechanics are square-integrable in coordinate space.

$$\int d\vec{r}\Psi^*\Psi = \text{finite}. \tag{53}$$

The space of all square-integrable functions is known as a Hilbert space. It will be useful to define a Scalar Product of the functions Φ and Ψ by

$$\int d\vec{r}\Phi^*\Psi = <\Phi, \Psi>. \tag{54}$$

The functions Φ and Ψ are said to be orthogonal to each other if $<\Phi, \Psi> = 0$. Note, the scalar product has the property

$$<\Phi, \Psi> = <\Psi, \Phi>^*. \tag{55}$$

Note also, the scalar product is linear in Ψ,

$$<\Phi, \lambda_1\Psi_1 + \lambda_2\Psi_2> = \lambda_1 <\Phi, \Psi_1> + \lambda_2 <\Phi, \Psi_2>, \tag{56}$$

but is antilinear in Φ,

$$<\lambda_1\Phi_1 + \lambda_2\Phi_2, \Psi> = \lambda_1^* <\Phi_1, \Psi> + \lambda_2^* <\Phi_2, \Psi>. \tag{57}$$

Eq. (56) follows from the linear character of the unit operator.

An important concept is the Adjoint of an Operator, written as $O^\dagger$.

$$\int d\vec{r}\Phi^*(O\Psi) = \int d\vec{r}(O^\dagger\Phi)^*\Psi \tag{58}$$

for any arbitrary pair of functions Φ and Ψ of our function space.

A Hermitian Operator, O, is a self-adjoint operator. If

$$O = O^\dagger, \quad \text{then } O \text{ is hermitian}. \tag{59}$$

Note, the operator

$$\frac{d}{dx} \quad \text{is not hermitian},$$

but the operator

$$\frac{\hbar}{i}\frac{d}{dx} \quad \text{is hermitian},$$

because

$$\begin{aligned}\int d\vec{r}\Psi^*\left(\frac{\hbar}{i}\frac{d\Phi}{dx}\right) &= \frac{\hbar}{i}\int\int dydz\Psi^*\Phi\Big]_{x=-\infty}^{\infty} - \frac{\hbar}{i}\int d\vec{r}\frac{d\Psi^*}{dx}\Phi \\ &= \int d\vec{r}\left(\frac{\hbar}{i}\frac{d\Psi}{dx}\right)^*\Phi,\end{aligned} \tag{60}$$

where the integrated term must be zero for square-integrable functions, Ψ and Φ. Note also, the operator $\frac{\hbar}{i}\frac{\partial}{\partial x}$ is the momentum operator, p_x, when acting on a function of the coordinates.

$$(p_x)_{\text{op.}}\Psi(\vec{r},t) = \frac{\hbar}{i}\frac{\partial}{\partial x}\Psi(\vec{r},t). \tag{61}$$

The product of two hermitian operators is in general not hermitian,

$$\int d\vec{r}\,\Psi^* O_2 O_1 \Phi = \int d\vec{r}(O_2^\dagger \Psi)^* O_1 \Phi = \int d\vec{r}(O_1^\dagger O_2^\dagger \Psi)^* \Phi, \tag{62}$$

so

$$(O_2 O_1)^\dagger = O_1^\dagger O_2^\dagger. \tag{63}$$

The product of the two hermitian operators is hermitian only if the operators commute. In the general case (for noncommuting operators), if O_1 and O_2 are both hermitian,

$$(O_1 O_2 + O_2 O_1) \qquad \text{and} \qquad i(O_1 O_2 - O_2 O_1)$$

are hermitian.

The commutator of two operators is very important in quantum mechanics. It is defined as

$$[O_1, O_2] = (O_1 O_2 - O_2 O_1). \tag{64}$$

G The Heisenberg Commutation Relations

The commutator

$$[p_x, x] = \frac{\hbar}{i}. \tag{65}$$

This commutator relation follows from

$$(p_x x - x p_x)\Psi = \frac{\hbar}{i}\frac{\partial}{\partial x}(x\Psi) - x\frac{\hbar}{i}\frac{\partial}{\partial x}(\Psi) = \frac{\hbar}{i}\Psi \tag{66}$$

for *all* Ψ of the Hilbert space. Eq. (65) is known as the Heisenberg commutation relation. Although we have demonstrated it here with the use of the wave function, it was introduced into quantum theory by Heisenberg without the concept of a wave function (see chapter 6D). Similarly, again using the technique of eq. (66),

$$[p_x, F(x,y,z)] = \frac{\hbar}{i}\frac{\partial F}{\partial x}. \tag{67}$$

On the other hand,

$$[p_x, G(p_x, p_y, p_z)] = 0, \tag{68}$$

but

$$[G(p_x, p_y, p_z), x] = \frac{\hbar}{i}\frac{\partial G}{\partial p_x}. \tag{69}$$

In particular, if H is a Hamiltonian, a function of the operators, p_x, p_y, p_z, x, y, z,

$$[p_x, H] = \frac{\hbar}{i}\frac{\partial H}{\partial x}; \qquad [x, H] = -\frac{\hbar}{i}\frac{\partial H}{\partial p_x}. \tag{70}$$

Expectation values of Hermitian operators are real. If $O = O^\dagger$,

$$< \Psi, O\Psi >=< O^\dagger\Psi, \Psi >=< O\Psi, \Psi >=< \Psi, O\Psi >^*, \tag{71}$$

so $< O >=< O >^*$.

If $O = O(\vec{p}, \vec{r}, t)$,

$$\frac{\hbar}{i}\frac{d}{dt} < O >=< [H, O] > +\frac{\hbar}{i} < \frac{\partial O}{\partial t} >, \tag{72}$$

where H is the Hamiltonian of the system, a hermitian operator, $H = H^\dagger$.

$$\begin{aligned}
\frac{d}{dt} < \Psi, O\Psi > &= \left(< \frac{\partial \Psi}{\partial t}, O\Psi > + < \Psi, O\frac{\partial \Psi}{\partial t} > + < \Psi, \frac{\partial O}{\partial t}\Psi > \right) \\
&= \frac{i}{\hbar}[< H\Psi, O\Psi > + < \Psi, O(-H\Psi) >] + < \frac{\partial O}{\partial t} > \\
&= \frac{i}{\hbar} < \Psi, (H^\dagger O - OH)\Psi > + < \frac{\partial O}{\partial t} > \\
&= \frac{i}{\hbar} < \Psi, (HO - OH)\Psi > + < \frac{\partial O}{\partial t} > .
\end{aligned} \tag{73}$$

H Generalized Ehrenfest Theorem

In the above relation, let $O = q_s$, where q_s is a generalized coordinate, and the Hamiltonian of the system is expressed in terms of generalized coordinates, q_s, and their canonically conjugate momenta, p_s. Then,

$$\begin{aligned}
\frac{\hbar}{i}\frac{d}{dt} < q_s > &= < [H, q_s] >= \frac{\hbar}{i} < \frac{\partial H}{\partial p_s} > \\
\frac{\hbar}{i}\frac{d}{dt} < p_s > &= < [H, p_s] >= -\frac{\hbar}{i} < \frac{\partial H}{\partial q_s} > .
\end{aligned} \tag{74}$$

These relations are the quantum analogues of the Hamiltonian form of the equations of motion.

$$\begin{aligned}
\frac{d}{dt} < q_s > &= < \frac{\partial H}{\partial p_s} > \\
\frac{d}{dt} < p_s > &= - < \frac{\partial H}{\partial q_s} > .
\end{aligned} \tag{75}$$

I Conservation Theorems: Angular Momentum, Runge–Lenz Vector, Parity

In the last section, we showed

$$\frac{d}{dt} < O >= \frac{i}{\hbar} < [H, O] > + < \frac{\partial O}{\partial t} > . \tag{76}$$

Thus, if an operator, O, commutes with the Hamiltonian, H, and is not an explicit function of the time, the time derivative of its expectation value in any state, Ψ, is equal to zero. This operator is the quantum-mechanical analogue of a classical "integral" of the motion. The operator, O, is conserved.

The simplest example, as in classical physics, is the Hamiltonian operator itself, provided it is not an explicit function of the time. For such a Hamiltonian, the conserved value of H is the energy E, as in classical physics. As a second example, consider single-particle motion in a central force field, with

$$H = \frac{1}{2m}(\vec{p} \cdot \vec{p}) + V(r), \tag{77}$$

where the potential function is a function of the scalar r only, where $r^2 = x^2 + y^2 + z^2$. (Note, we could also have chosen the two-body system with a central interaction, provided we replace m by the reduced mass and $\vec{r}$ stands for the relative vector, $\vec{r}_1 - \vec{r}_2$; see the next section). In this case, the three components of the orbital angular momentum vector,

$$\vec{L} = [\vec{r} \times \vec{p}], \tag{78}$$

are conserved quantities. For example,

$$\begin{aligned}
[H, L_x] &= [H, (yp_z - zp_y)] = y[H, p_z] + [H, y]p_z - z[H, p_y] - [H, z]p_y \\
&= y[V, p_z] + \tfrac{1}{2m}[p_y^2, y]p_z - z[V, p_y] - \tfrac{1}{2m}[p_z^2, z]p_y \\
&= -\frac{\hbar}{i} y \frac{\partial V}{\partial z} + \frac{\hbar}{i}\frac{p_y}{m} p_z - z\Big(-\frac{\hbar}{i}\frac{\partial V}{\partial y}\Big) - \frac{\hbar}{i}\frac{p_z}{m} p_y \\
&= \frac{\hbar}{i}\Big(-y\frac{dV}{dr}\frac{z}{r} + z\frac{dV}{dr}\frac{y}{r}\Big) = 0.
\end{aligned} \tag{79}$$

Similarly,

$$[H, L_y] = [H, L_z] = 0. \tag{80}$$

In the above calculation, we have made use of the trivial but useful commutator identities

$$[A, BC] = B[A, C] + [A, B]C; \quad \text{also} \quad [AB, C] = A[B, C] + [A, C]B. \tag{81}$$

From the commutators, $[H, L_k] = 0$, we also have that H commutes with the operator $\vec{L}^2$

$$[H, \vec{L}^2] = \sum_{\alpha=x,y,z} L_\alpha[H, L_\alpha] + [H, L_\alpha]L_\alpha = 0. \tag{82}$$

As a third example, consider the hydrogen atom Hamiltonian

$$H = \frac{1}{2\mu}(\vec{p} \cdot \vec{p}) - \frac{e^2}{r}. \tag{83}$$

For the Hamiltonian with this $1/r$ potential, the three components of the Runge–Lenz vector, $\vec{\mathcal{R}}$, are conserved, where classically $\vec{\mathcal{R}} = \frac{1}{\mu}[\vec{p} \times \vec{L}] - \frac{e^2}{r}\vec{r}$. Quantum mechanically, we must convert this into a hermitian operator by using the symmetrized form of the first term. Remembering the interchange of the order of the two vectors in a vector product introduces a minus sign, the symmetrized form for $\vec{\mathcal{R}}$ is

$$\vec{\mathcal{R}} = \frac{1}{2\mu}\Big([\vec{p} \times \vec{L}] - [\vec{L} \times \vec{p}]\Big) - \frac{e^2}{r}\vec{r}. \tag{84}$$

Note,

$$\begin{aligned} \tfrac{1}{2}\big([\vec{p} \times \vec{L}]_x - [\vec{L} \times \vec{p}]_x\big) &= \tfrac{1}{2}\Big((p_y L_z - p_z L_y) - (L_y p_z - L_z p_y)\Big) \\ &= \tfrac{1}{2}\Big((p_y L_z + L_z p_y) - (p_z L_y + L_y p_z)\Big). \end{aligned} \tag{85}$$

Note, $p_y L_z$ is not a hermitian operator, because p_y does not commute with $L_z = (xp_y - yp_x)$. The symmetrized form of this operator, $\frac{1}{2}(p_y L_z + L_z p_y)$, however, is hermitian. In making the transition from classical physics quantities to quantum-mechanical operators, the hermitian, symmetrized form of the classical quantities will often give the needed quantum-mechanical operators. The proof that $[H, \mathcal{R}_k] = 0$, where H is the hydrogen atom Hamiltonian, will be left as an exercise (part of problem 13).

As a final example of a conserved operator, consider the space inversion or parity operator, P, where

$$P\Psi(x, y, z, t) = \Psi(-x, -y, -z, t). \tag{86}$$

For a Hamiltonian,

$$H = -\frac{\hbar^2}{2m}\nabla^2 + V(x, y, z), \qquad \text{with} \quad V(-x, -y, -z) = V(x, y, z), \tag{87}$$

that is, with a potential that is space-inversion invariant, we have

$$\begin{aligned} HP\Psi(\vec{r}, t) &= H\Big(\frac{\hbar}{i}\vec{\nabla}, \vec{r}\Big)\Psi(-\vec{r}, t) \\ PH\Psi(\vec{r}, t) &= H\Big(-\frac{\hbar}{i}\vec{\nabla}, -\vec{r}\Big)\Psi(-\vec{r}, t) = H\Big(\frac{\hbar}{i}\vec{\nabla}, \vec{r}\Big)\Psi(-\vec{r}, t), \end{aligned} \tag{88}$$

so

$$(HP - PH)\Psi(\vec{r}, t) = 0 \tag{89}$$

for all Ψ of our Hilbert space. Hence, $[H, P] = 0$, and P is a conserved quantity. Finally, note also, P is a hermitian operator, because

$$\int d\vec{r}\,\Psi^*(\vec{r}, t)P\Psi(\vec{r}, t) = \int d\vec{r}\,\Psi^*(\vec{r}, t)\Psi(-\vec{r}, t) = \int d\vec{r}\,\Psi^*(-\vec{r}, t)\Psi(\vec{r}, t)$$

$$= \int d\vec{r} \Big(P^{\dagger} \Psi(\vec{r}, t) \Big)^{*} \Psi(\vec{r}, t),$$

where we have made the change of variables, $x \to -x$, $y \to -y$, $z \to -z$, and have changed the order of the integration limits in the three integrals implied by our shorthand notation in the last step of the first line. We see $P^{\dagger} = P$. Therefore, the expectation value of the operator P must also be real. Finally, because

$$P^2 \Psi(\vec{r}, t) = \Psi(\vec{r}),$$

the operator P^2 has an expectation value of 1. The real expectation value of the operator P can thus be only either $+1$ or -1. The solutions of the Schrödinger equation for the space-inversion invariant Hamiltonian of eq. (87) must thus either be unchanged or change sign under the space-inversion operation. The wave function must have even or odd parity.

J Quantum-Mechanical Hamiltonians for More General Systems

As an example of a slightly more general system, consider a particle of mass, m, and charge, e, moving in an electromagnetic field derivable from a vector potential, $\vec{A}(\vec{r})$, and a scalar potential, $\Phi(\vec{r})$. The classical Hamiltonian is given by

$$H = \frac{(\vec{p} - \frac{e}{c}\vec{A})^2}{2m} + e\Phi. \tag{90}$$

This relation can be written in the form

$$H = \frac{(\vec{p} \cdot \vec{p})}{2m} - \frac{e}{2mc}(\vec{p} \cdot \vec{A} + \vec{A} \cdot \vec{p}) + \frac{e^2}{2mc^2}\vec{A} \cdot \vec{A} + e\Phi. \tag{91}$$

Note, because in general $[p_x, A_x] \neq 0$, we have written the scalar product of $\vec{A}$ with $\vec{p}$ in symmetrized, hermitian form, so the Hamiltonian in the second form is a candidate for the quantum-mechanical Hamiltonian of this system. This relation is indeed the correct quantum-mechanical Hamiltonian. The predictions based on this form of the Hamiltonian are in agreement with the experiment! (We shall study this system in more detail later, where we will discuss the role of the gauge of the potentials, the gauge transformation, the Aharanov–Bohm effect, etc.) For more complicated systems, however, the simple process of symmetrization of a classical Hamiltonian may not be sufficient, particularly if the physically relevant coordinates are a set of complicated curvilinear coordinates, say, in the case of a many-body system in which some of the degrees of freedom are "frozen" or do not come into play at a region of low-energy excitations.

K The Schrödinger Equation for an n-particle System

For the general n-particle system, with the Hamiltonian

$$H = \sum_{k=1}^{n} \frac{(\vec{p}_k \cdot \vec{p}_k)}{2m_k} + V(\vec{r}_1, \vec{r}_2, \ldots, \vec{r}_n), \tag{92}$$

we are led (as in the case of the single-particle system) to the Schrödinger equation

$$-\frac{\hbar}{i}\frac{\partial \Psi}{\partial t} = -\sum_k \frac{\hbar^2}{2m_k} \nabla_k^2 \Psi + V(\vec{r}_1, \vec{r}_2, \ldots, \vec{r}_n)\Psi. \tag{93}$$

For the two-particle system with no external forces, in particular, it will be useful to make a transformation to relative and center of mass coordinates

$$\vec{r} = \vec{r}_1 - \vec{r}_2, \qquad \vec{R} = \frac{(m_1\vec{r}_1 + m_2\vec{r}_2)}{(m_1 + m_2)}, \tag{94}$$

and

$$-\frac{\hbar}{i}\frac{\partial \Psi}{\partial t} = -\frac{\hbar^2}{2(m_1 + m_2)} \nabla^2_{\text{C.M.}} \Psi - \frac{\hbar}{2\mu} \nabla^2_{\text{rel.}} \Psi + V(\vec{r})\Psi, \tag{95}$$

where μ is the reduced mass, $\mu = m_1 m_2/(m_1 + m_2)$ and the potential is a function of the relative $\vec{r}$ only (no external fields). In this case, the center of mass motion separates. With $\Psi(\vec{R}, \vec{r}, t) = \Psi_{\text{C.M.}}(\vec{R}, t)\Psi_{\text{rel.}}(\vec{r}, t)$, the center of mass term leads to a plane wave solution

$$\Psi_{\text{C.M.}}(\vec{R}, t) = Ae^{\frac{i}{\hbar}(\vec{P}\cdot\vec{R} - E_{\text{transl.}} t)}, \tag{96}$$

where $\vec{P}$ is the linear momentum associated with the center of mass motion of the system of mass $(m_1 + m_2)$ and $E_{\text{transl.}}$ is the translational energy associated with the center of mass motion, $E_{\text{transl.}} = P^2/2(m_1 + m_2)$. The Schrödinger equation is then effectively an equation equivalent to a single-particle equation, provided the mass is replaced by the reduced mass.

$$-\frac{\hbar}{i}\frac{\partial \Psi_{\text{rel.}}}{\partial t} = -\frac{\hbar^2}{2\mu} \nabla^2_{\text{rel.}} \Psi_{\text{rel.}}(\vec{r}, t) + V(\vec{r})\Psi_{\text{rel.}}(\vec{r}, t). \tag{97}$$

In n-particle systems, it will be useful to transform from the coordinates $x_1, \ldots, z_n$ to a set of generalized coordinates q_s. Often, some of these will not come into play. In a polyatomic molecule, e.g., the ammonia molecule, NH_3, we have a system with 4 atomic nuclei and 10 electrons, a 14-particle system with 42 degrees of freedom (assuming we can neglect the electron and nuclear spins). These coordinates could be chosen as the 3 center of mass coordinates, which merely describe the free-particle translation of the whole system in space: 3 angular coordinates, say, three Euler angles, ϕ, θ, χ, which describe the orientation of the molecule in space and describe the rotational motion of the molecule; 6 relative coordinates, which describe the relative motions of the atomic nuclei, that is, the vibrational motions of the molecule; and finally 3×10 electronic coordinates (again ignoring for the moment the electron spins), which describe the electron motions of the molecule. At very low excitation energies, only the rotational degrees of freedom, ϕ, θ, χ, may need to be considered. Therefore, if we can transform the general $3n = 42$-dimensional Laplacian operator of this 14-particle system from the 3×14 Cartesian coordinates to the physically relevant 42 generalized coordinates, q_s, including ϕ, θ, χ, we arrive at the desired Schrödinger equation, if the variations with all q_s are neglected, except for the variations with the needed ϕ, θ, χ. Even for the 1-particle system, it will generally be useful to express the Schrödinger equation not in terms of the Cartesian coordinates x, y, z, but in terms of some set of curvilinear coordinates, e.g., spherical coordinates r, θ, ϕ.

L The Schrödinger Equation in Curvilinear Coordinates

If we transform from the 3n Cartesian coordinates $x_1, \dots, z_n$ to a new set of 3n generalized coordinates q_s, with $s = 1, \dots, 3n$, through

$$x_1 = f_1(q_1, q_2, \dots, q_{3n}), \quad \cdots, \quad \cdots, \quad z_n = f_{3n}(q_1, q_2, \dots, q_{3n}) \tag{98}$$

the classical kinetic energy expression can be written as a homogeneous quadratic function of the $\dot{q}_i$,

$$T = \tfrac{1}{2}\sum_{i,j} g_{ij}\dot{q}_i\dot{q}_j, \tag{99}$$

where the g_{ij} are in general functions of the q_s. The classical Hamiltonian expressed in the generalized momenta, p_s, canonically conjugate to these q_s, can then be written as

$$H_{\text{class.}} = \tfrac{1}{2}\sum_{ij} g^{ij} p_i p_j + V(q_1, q_2, \dots, q_{3n}), \tag{100}$$

where the g^{ij} matrix, that is, the superscripted g-matrix, is the inverse of the g_{ij} matrix, that is, the subscripted g-matrix

$$\sum_\alpha g_{i\alpha} g^{\alpha j} = \delta_i^j, \qquad \text{and} \qquad \sum_\alpha g^{i\alpha} g_{\alpha j} = \delta_j^i. \tag{101}$$

Note, the g^{ij} are in general complicated functions of the q_s and do not commute with the p_s. A large number of ways would exist of making the kinetic energy term hermitian, so the hermiticity requirement alone does not lead to the correct quantum-mechanical Hamiltonian. We know, however, how to transform the 3n-dimensional Laplacian operator from its Cartesian form to the form involving partial derivatives with respect to the new generalized q_s. Therefore, we can write the proper Schrödinger equation. We need, in addition to the g^{ij}, the function g, given by the determinant of the subscripted g-matrix,

$$g = |g_{ij}| = \det(g_{ij}). \tag{102}$$

Writing the 3n-dimensional Laplacian ∇^2 in the curvilinear coordinates, we arrive at the Schrödinger equation

$$-\frac{\hbar}{i}\frac{\partial \Psi}{\partial t} = -\frac{\hbar^2}{2}\sum_{ij}\frac{1}{\sqrt{g}}\frac{\partial}{\partial q_i}\left(g^{ij}\sqrt{g}\frac{\partial \Psi}{\partial q_j}\right) + V\Psi. \tag{103}$$

If Ψ is assumed to be a function of only a few of the q_i, the equation will simplify.

Problems

1. A free particle moving in the x-direction, (1-D motion) has a momentum distribution given by

$$\phi(p,t) = \sqrt{\frac{1}{\alpha\sqrt{\pi}}}e^{-\frac{1}{2}\frac{(p-p_0)^2}{\alpha^2}}e^{-\frac{i}{\hbar}\frac{p^2}{2m}t}$$

$$\phi(p,t) = \sqrt{\frac{4}{3\alpha^5\sqrt{\pi}}}(p-p_0)^2 e^{-\frac{1}{2}\frac{(p-p_0)^2}{\alpha^2}}e^{-\frac{i}{\hbar}\frac{p^2}{2m}t}.$$

For both cases, calculate the spatial probability density amplitude function, $\Psi(x,t)$, for this particle. Calculate $< p >$, $< p^2 >$, $< \Delta p >$, $\langle x \rangle$, $< x^2 >$, and $< \Delta x >$, and verify the uncertainty principle. Give an interpretation of Δx, in the limit $\hbar \to 0$, in classical terms.

2. From the "dispersion law,"

$$\frac{E^2}{c^2} - (\vec{p}\cdot\vec{p}) = m_0^2 c^2$$

for a relativistic free particle, derive a wave equation. (This equation is known as the Klein–Gordon equation.) If the probability density current is to have the form

$$\vec{S} = \frac{\hbar}{2mi}\left(\Psi^*\vec{\nabla}\Psi - \Psi\vec{\nabla}\Psi^*\right)$$

with

$$div\vec{S} + \frac{\partial W}{\partial t} = 0$$

to preserve conservation of probability, show how the probability density, W, must be related to Ψ, $\frac{\partial \Psi}{\partial t}$, Ψ^*, $\frac{\partial \Psi^*}{\partial t}$. Is this an acceptable W? Is $W \geq 0$ everywhere, for all t?

3. A particle of charge, e, and mass, m, in an electromagnetic field, derivable from vector and scalar potentials, $\vec{A}$, Φ, has a Hamiltonian

$$H = \frac{\vec{p}^2}{2m} - \frac{e}{2mc}(\vec{p}\cdot\vec{A} + \vec{A}\cdot\vec{p}) + \frac{e^2}{2mc^2}\vec{A}^2 + e\Phi.$$

(Note the symmetrized form of the second term.) Write the Schrödinger equation for this case. Find an expression for the probability density current, $\vec{S}$, with $W = \Psi^*\Psi$. Calculate

$$\frac{d\langle x\rangle}{dt}, \quad \text{and} \quad \frac{d\langle p_x\rangle}{dt},$$

and show how these are related to the classical equations of motion.

Also, show the wave equation is gauge invariant, and under the transformation

$$\vec{A} \to \vec{A}' = \vec{A} + \vec{\nabla}\chi, \qquad \Phi \to \Phi' - \frac{1}{c}\frac{\partial \chi}{\partial t},$$

where $\chi = \chi(x, y, z; t)$, the wave equation remains unchanged, provided

$$\Psi \to \Psi' = \Psi e^{\frac{ie}{\hbar c}\chi(x,y,z;t)}.$$

4. In describing scattering processes of complex projectiles from nuclei, it is sometimes useful to use a fictitious complex potential

$$V = V_1 + iV_2,$$

where V_1 and V_2 are both real. Assume V_2 = constant = W, inside a sphere of radius, $r_0 = 10^{-12}$ cm, and $V_2 = 0$ for $r > r_0$. Determine, W, magnitude in eV and sign, so the probability is 0.1 per 10^{-21} seconds for the *loss* of flux of incoming projectile particles. (Incoming α particles, e.g., can be "lost" by conversion to ^{3}He and neutrons, etc. Note, 10^{-21} seconds is a typical traversal time for a fast but nonrelativistic nuclear particle through a heavy nucleus.)

5. The classical kinetic energy for a rigid rotator, e.g., a polyatomic molecule such as H_2O to very good approximation, is given in terms of the three Euler angles, ϕ, θ, χ, and the three principal moments of inertia, A, B, C, by

$$2T = A(\dot{\theta}\cos\chi + \dot{\phi}\sin\theta\sin\chi)^2 + B(-\dot{\theta}\sin\chi + \dot{\phi}\sin\theta\cos\chi)^2 + C(\dot{\chi} + \dot{\phi}\cos\theta)^2.$$

Assuming other degrees of freedom, such as vibrational, translational, and electronic, in the case of the polyatomic molecule, can be neglected, then $V = 0$. Write the Schrödinger equation for the rigid rotator (asymmetric case, $A \neq B \neq C$). For the symmetric rotator, with $A = B$, show the time-independent wave function separates via the Ansatz

$$\psi(\phi, \theta, \chi) = \frac{1}{2\pi} e^{iM\phi} e^{iK\chi}\Theta(\theta),$$

and write the differential equation for $\Theta(\theta)$. (For the asymmetric case, the differential equation approach may not be the best way to solve this problem.)

6. Transpose the Schrödinger equation for the hydrogenic atom, with

$$H = \frac{\vec{p}\cdot\vec{p}}{2\mu} - \frac{Ze^2}{r},$$

where the above $\vec{p}$ and $\vec{r}$, which are $\vec{p}_{\text{physical}}$, and $\vec{r}_{\text{physical}}$ are transcribed into dimensionless $\vec{p}$ and $\vec{r}$ via

$$\vec{r}_{\text{phys.}} = a_0\vec{r}, \quad \vec{p}_{\text{phys.}} = \frac{\hbar}{a_0}\vec{p}, \qquad \text{with } a_0 = \frac{\hbar^2}{\mu e^2 Z},$$

$$H_{\text{phys.}} = \frac{\mu Z^2 e^4}{\hbar^2}H, \qquad E_{\text{phys.}} = \frac{\mu Z^2 e^4}{\hbar^2}\epsilon.$$

Transpose the Schrödinger equation for the hydrogenic atom, further, into an equation written in terms of "stretched parabolic coordinates," μ, ν, ϕ, defined in terms of the dimensionless $\vec{r} = (x, y, z)$ by

$$\mu = \sqrt{(r+z)}[-2\epsilon]^{\frac{1}{4}}, \qquad \nu = \sqrt{(r-z)}[-2\epsilon]^{\frac{1}{4}}, \qquad \phi = \tan^{-1}\left(\frac{y}{x}\right),$$

where $r = \sqrt{x^2 + y^2 + z^2}$, and $\epsilon = E/(\mu Z^2 e^4/\hbar^2)$, or

$$x = \frac{\mu\nu}{[-2\epsilon]^{\frac{1}{2}}}\cos\phi, \qquad y = \frac{\mu\nu}{[-2\epsilon]^{\frac{1}{2}}}\sin\phi, \qquad z = \frac{(\mu^2 - \nu^2)}{2[-2\epsilon]^{\frac{1}{2}}},$$

where this transformation is useful for bound states, with $\epsilon < 0$. Transform the Laplacian into these stretched parabolic coordinates.

Another set of useful coordinates for the hydrogenic atom are the conventional parabolic coordinates, defined in terms of the dimensionless $\vec{r} = (x, y, z)$, by

$$\xi = r - z, \qquad \eta = r + z, \qquad \phi = \tan^{-1}\left(\frac{y}{x}\right),$$

$$x = \sqrt{\xi\eta}\cos\phi, \qquad y = \sqrt{\xi\eta}\sin\phi, \qquad z = \tfrac{1}{2}(\eta - \xi).$$

Transform the Laplacian into these curvilinear coordinates, and write the Schrödinger equation for the hydrogenic atom in these parabolic coordinates.

4

Schrödinger Theory: The Existence of Discrete Energy Levels

A The Time-Independent Schrödinger Equation

The Schrödinger equation for a single particle moving under the influence of a time-independent conservative force

$$-\frac{\hbar}{i}\frac{\partial \Psi}{\partial t} = -\frac{\hbar^2}{2m}\nabla^2\Psi + V\Psi = H\Psi \tag{1}$$

can be converted to the time-independent equation for the function $\psi(x, y, z)$ by assuming

$$\Psi = f(t)\psi(x, y, z), \tag{2}$$

with

$$-\frac{1}{f}\left\{\frac{\hbar}{i}\frac{df}{dt}\right\} = \frac{1}{\psi}\left\{-\frac{\hbar^2}{2m}\nabla^2\psi + V\psi\right\} = \frac{1}{\psi}(H\psi) = \text{const}, \tag{3}$$

where we have converted a function of the time only on the left-hand side of the equation into a function of x, y, z only on the right-hand side. Because this must hold for all values of t and all x, y, z, the left-hand side and the right-hand side must be equal to a constant. We have separated the equation. The physical significance of the constant can be seen to be the energy, E.

$$f(t) = \exp(-\frac{i}{\hbar}Et) \tag{4}$$

and

$$-\frac{\hbar^2}{2m}\nabla^2\psi + V\psi = E\psi. \tag{5}$$

For a 1-D problem, in particular,

$$\frac{d^2\psi}{dx^2} + \frac{2m}{\hbar^2}(E - V(x))\psi = 0. \tag{6}$$

In this chapter, we shall show the Schrödinger equation, eq. (5), for potentials V, for which the classical motion would be restricted to a bound region of space, will lead to allowed (square-integrable) solutions, the so-called characteristic functions, or "eigenfunctions," of the equation only for certain discrete allowed energies, the so-called "eigenvalues" or characteristic values of the energies.

B The Simple, Attractive Square Well

The 1-D Schrödinger equation, eq. (6), has particularly simple solutions in regions $a < x < b$, where $V(x)$ can be replaced by a constant, with simple sinusoidal solutions for regions with $E > V$ and simple exponentials for regions with $E < V$. The simplest 1-D problem is that of a single, attractive square well of width $2a$, with

$$V(x) = 0 \quad \text{for} \quad -a \le x \le +a, \qquad V(x) = +V_0 \quad \text{for } |x| > a, \tag{7}$$

(see Fig. 4.1). The Schrödinger equation becomes

$$\begin{aligned} &\frac{d^2\psi}{dx^2} + k^2\psi(x) = 0, \quad \text{with } k^2 = \frac{2mE}{\hbar^2} \quad \text{for } -a \le x \le +a, \\ &\frac{d^2\psi}{dx^2} - \kappa^2\psi(x) = 0, \quad \text{with } \kappa^2 = \frac{2m(V_0 - E)}{\hbar^2} \quad \text{for } |x| \ge a. \end{aligned} \tag{8}$$

In order to have square-integrable solutions, the $\psi(x)$ must be restricted to exponentially decaying solutions outside the potential well; i.e.,

$$\psi(x) = Ce^{-\kappa x} \quad \text{for } x > +a, \qquad \psi(x) = De^{+\kappa x} \quad \text{for } x < -a. \tag{9}$$

In the interior, for $-a \le x \le +a$, the most general solution is

$$\psi(x) = A\cos kx + B\sin kx.$$

(For the moment, we have not made use of the symmetry of the potential.) In order to have solutions with sensible probability densities, both the probability density and the probability density currents must be continuous functions of x. This solution can be ensured by requiring the continuity of both $\psi(x)$ and its first derivative at the discontinuities of the potential at $x = \pm a$. The continuity of $\psi(x)$ and its first derivative at $x = +a$ leads to the boundary conditions

$$A\cos ka + B\sin ka = Ce^{-\kappa a}, \tag{10}$$

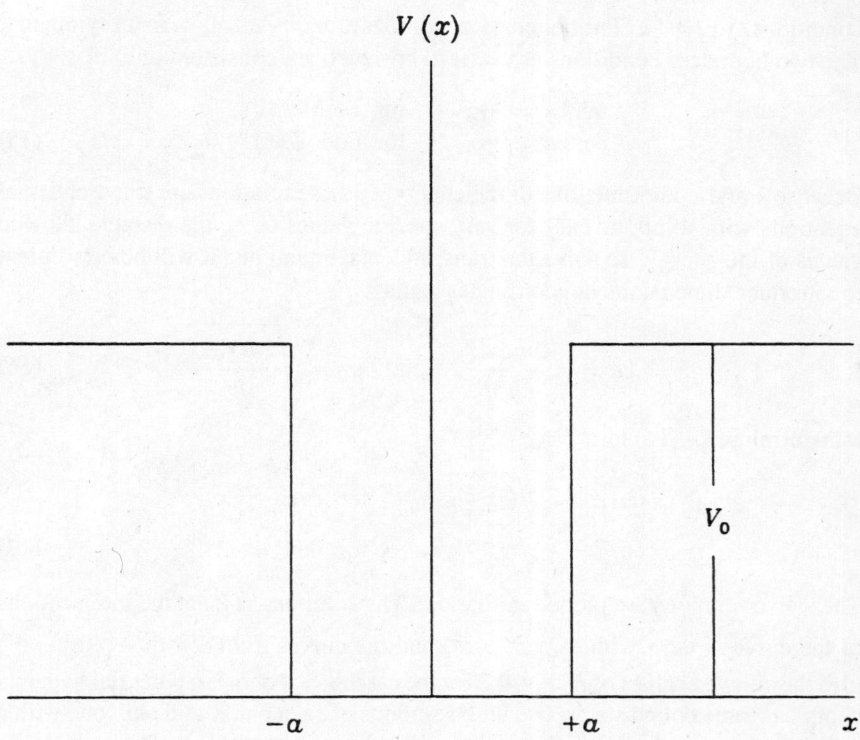

FIGURE 4.1. The attractive square well potential.

$$-Ak\sin ka + Bk\cos ka = -\kappa Ce^{-\kappa a}, \tag{11}$$

and at $x = -a$, we are led to the boundary conditions

$$A\cos ka - B\sin ka = De^{-\kappa a}, \tag{12}$$

$$Ak\sin ka + Bk\cos ka = \kappa De^{-\kappa a}. \tag{13}$$

Eliminating the constant C from eqs. (10) and (11) and the constant D from eqs. (12) and (13), we are led to the further restriction

$$AB = 0,$$

with the two possible solutions:

$$B = 0, \quad D = C, \qquad \text{or} \qquad A = 0, \quad D = -C. \tag{14}$$

We see the solutions are either even or odd functions of x. This could have been seen at once from the space-reflection symmetry of our potential, $V(-x) = V(x)$. Because the Schrödinger equation is invariant under the 1-D space-inversion operation, $x \rightarrow -x$, our solutions must have good parity; see Section I of Chapter 3. The $\psi(x)$ must be either even or odd functions of x; either $\cos kx$ or $\sin kx$ functions in the region $|x| < a$. It would have been sufficient to apply the boundary conditions at $x = +a$. The boundary conditions at $x = -a$ follow from symmetry. The two boundary conditions at $x = +a$, however, are consistent only if

$$\begin{aligned} k \tan ka &= +\kappa, \qquad \text{for even } \psi(x), \\ k \cot ka &= -\kappa, \qquad \text{for odd } \psi(x). \end{aligned} \tag{15}$$

Because k and κ are functions of the energy E, these relations are transcendental equations with solutions only for very specific values of E, the discrete allowed values of the energy. To solve the transcendental equations, it will be convenient to introduce dimensionless coordinates z and z_0,

$$z = ka = \sqrt{\frac{2mEa^2}{\hbar^2}}, \quad \text{and} \quad z_0 = \sqrt{\frac{2mV_0a^2}{\hbar^2}}, \tag{16}$$

transforming eq. (15) into

$$\begin{aligned} z \tan z &= +\sqrt{(z_0^2 - z^2)} \qquad \text{for even } \psi(x), \\ z \cot z &= -\sqrt{(z_0^2 - z^2)} \qquad \text{for odd } \psi(x). \end{aligned} \tag{17}$$

These two relations are plotted in Fig. 4.2. The solutions $z(E)$ at the intersections of the curves $z \tan z$ with $+\sqrt{(z_0^2 - z^2)}$ and the curves $z \cot z$ with $-\sqrt{(z_0^2 - z^2)}$ give the allowed values of E. Fig. 4.2 for the case $z_0 = 4$ shows a potential with this depth has three bound states, two with solutions of even parity and only one with a solution of odd parity. Note also, only one even bound state exists if $z_0 < \pi/2$, but at least this one bound state always exists, even in the limit of a shallow potential well, with $V_0 \to 0$. Note, also, in the limit of an infinitely deep well, as $V_0 \to \infty$, the solutions are

$$\begin{aligned} &z \to (2N+1)\frac{\pi}{2}, \quad N = 0, 1, 2, \ldots, \qquad \text{for even } \psi \\ &z \to 2N\frac{\pi}{2}, \quad N = 1, 2, \ldots, \qquad \text{for odd } \psi \end{aligned} \tag{18}$$

or

$$z_n = n\frac{\pi}{2}, \quad \text{thus,} \qquad E_n = \frac{n^2\pi^2\hbar^2}{2m(2a)^2}, \quad \text{with} \quad n = 1, 2, \ldots. \tag{19}$$

Note, in the case, $V_0 \to \infty$, the wave functions are exactly 0 in the region $|x| > a$, and the interior solutions obey $\psi(\pm a) = 0$. In this case, the derivatives of the wave function are discontinuous at $x = \pm a$. Both the probability density and the probability density current, however, have the value zero at the boundaries and are therefore still continuous at the boundaries.

Note, also, so far we have considered only bound states with $E < V_0$. For $E > V_0$, the solutions of the Schrödinger equation are oscillatory for all values of x, from $-\infty \to +\infty$. Merely, a change of wavelength ("index of refraction") occurs as the wave traverses the region of the potential well. Because they reach from $x = -\infty$ to $x = +\infty$, the wave functions are no longer square integrable. They still have, however, a sensible probability interpretation. The amplitudes of

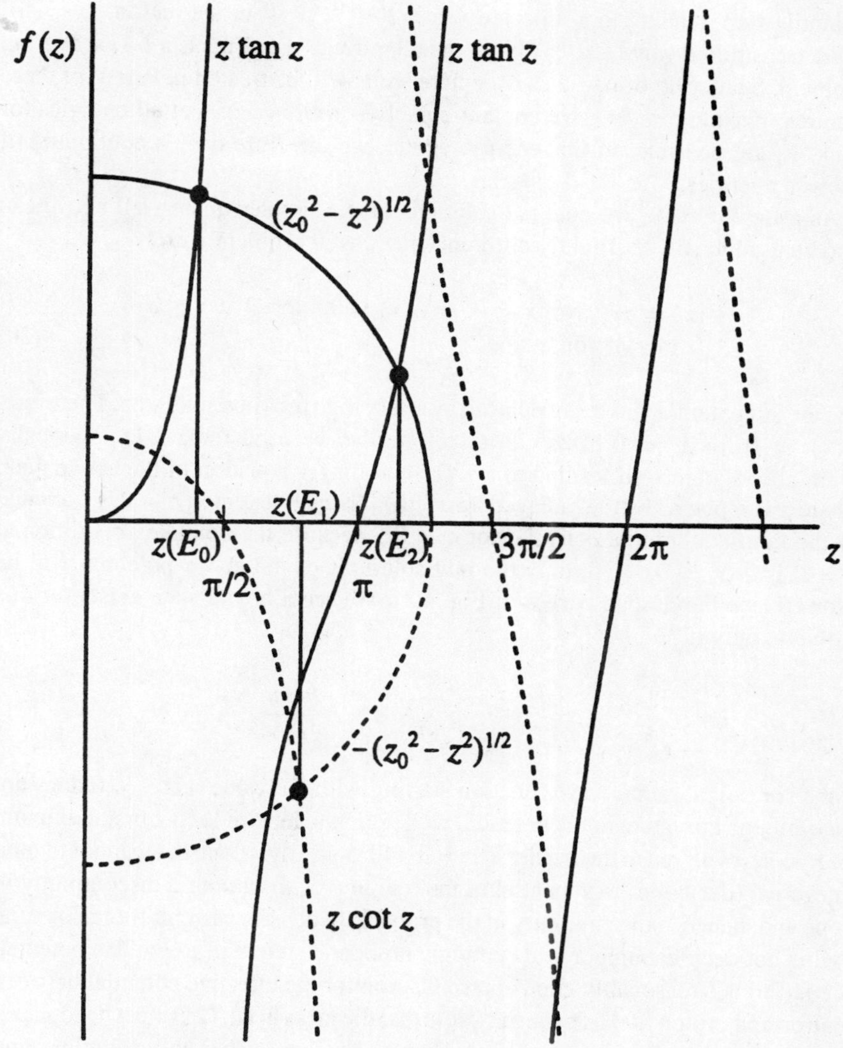

FIGURE 4.2. The transcendental eqs. (17) for the case $z_0 = 4$. Solid lines for even $\psi(x)$. Dashed lines for odd $\psi(x)$.

the sinusoidal waves give the strengths of the probability density current and can therefore be determined from the experimental flux of particles. Note, however, $\vec{S}$ is zero for real $\psi(x)$. The physics of the problem dictates we use complex solutions of the type

$$e^{i(\pm kx - Et)}$$

for particles moving in the $\pm x$ direction. The amplitude of the wave $Ae^{i(k_0 x - Et)}$, for $x < -a$, with $k_0 = [2m(E - V_0)/\hbar^2]^{1/2}$, is determined by the flux of particles from a source at $x = -\infty$. These particles can be reflected or transmitted by the potential step, leading to a reflected wave, $Be^{(-ik_0x-iEt)}$ in the region, $x < -a$, and a transmitted wave $Ce^{i(k_0x-Et)}$ in the region $x > +a$. This is a 1-D scattering problem. Scattering of particles by square wells will be treated in Part V of these lectures. For the moment, we content ourselves with noting that all energies for $E > V_0$ are possible. In this energy regime, we therefore have a continuum of allowed energies.

As a final remark, we note the solutions of the simple square well above can also be used to solve a slightly different square well problem with

$$\begin{aligned}&V(x) = \infty, \quad \text{for } x < 0, \quad V(x) = 0, \quad \text{for } 0 < x \le a,\\ &V(x) = V_0 \quad \text{for } x > a;\end{aligned} \tag{20}$$

i.e., the left potential has been replaced by a very high (∞) potential step. Therefore, $\psi(x) = 0$ for $x < 0$. This can therefore also be used for a 3-D spherically symmetric square well leading to a 1-D Schrödinger equation of the above type, where x is replaced by the radial coordinate, r. (Note, the region $r < 0$ is excluded by the fictitious infinite potential for $r < 0$.) Because the boundary condition at $r = 0$ is $\psi(r = 0) = 0$, only the odd solutions of the above potential will be allowed (see the dashed curves of Fig. 4.2). We see a bound state exists for this problem, only if

$$z_0 > \frac{\pi}{2}, \qquad \text{or} \quad V_0 a^2 > \frac{\hbar^2}{2m}\frac{\pi^2}{4}. \tag{21}$$

If the sinusoidal radial wave function starting with the value 0 at $r = 0$ does not have enough curvature in the potential well region to have at least a first maximum for $r < a$, it will reach the barrier at $r = a$ with a positive slope that cannot fit onto a negative (decaying) exponential in the region $r > a$ without a discontinuity in slope and, hence, a discontinuity in the probability density current. If the potential well is not deep enough or wide enough, no bound state will exist. The potential of eq. (20) is a reasonably good approximation for the effective potential between neutron and proton in the deuteron. [Note that the mass in eq. (21) must be replaced by the reduced mass of this 2-body problem.] The deuteron has only a single bound state in its 2-particle spin triplet ($S = 1$) state. Moreover, the binding energy of this state, of 2.22 MeV (with $E = V_0 - 2.22$ MeV) is small compared with the expected value of V_0. The deuteron is therefore a barely bound system with

$$V_0 a^2 \approx \frac{\hbar^2}{2\mu}\frac{\pi^2}{4}.$$

The deuteron has no bound states with 2-particle spin $S = 0$. The potential must therefore be spin dependent. The $S = 0$ potential just misses having a bound state. This property makes itself felt in a large scattering cross section for $E - V_0 \approx 0$, a low-energy resonance. For a detailed discussion of proton–neutron scattering and the bound or nearly bound states of the deuteron, see Chapter 44.

Problems 7–8: Square Well Problems

More complicated square well problems can often be used to gain qualitative solutions for more sophisticated problems. The following two problems can be used to illustrate some interesting physics.

7. The double-minimum potential problem. The square well double-minimum potential, shown in (c) of Fig. P7, can be used as a rough approximation for the potential governing the motion of the N atom relative to the H_3 plane, one of the vibrational degrees of freedom of the ammonia molecule, NH_3 (the degree of freedom responsible for the transition used in the NH_3 MASER, the historical forerunner of all LASERS and MASERS).

$$V = V_0 \quad \text{for } |x| < a \qquad \text{Region II},$$

$$V = 0 \quad \text{for } a < |x| < b \qquad \text{Regions I, III},$$

$$V = \infty \quad \text{for } |x| > b \qquad \text{Regions IV}.$$

The mass, μ, is the reduced mass for the N–H_3 pair:

$$\mu = \frac{3m_H m_N}{(3m_H + m_N)}.$$

Exploit $V(x)$ is an even function of x, so the solutions, $\psi(x)$, must be either even or odd functions of x. It is therefore sufficient to find acceptable solutions for $x \geq 0$ and continue these appropriately into the region, $x < 0$. Find the transcendental equations from which the eigenvalues of E, corresponding to both the even and odd eigenfunctions, can be found for the states with $E < V_0$. Show graphically how the solutions can be found. Show, in particular, that for $E \ll V_0$, the solutions follow from

$$k_n(b - a) = n\pi - \Delta\phi, \quad \text{with } \Delta\phi \ll 1, \quad k_n^2 = \frac{2\mu}{\hbar^2} E_n,$$

with slightly different $\Delta\phi$ for the eigenvalues associated with the even and odd solutions, so the eigenvalues of E occur in nearly degenerate pairs, when $E \ll V_0$.

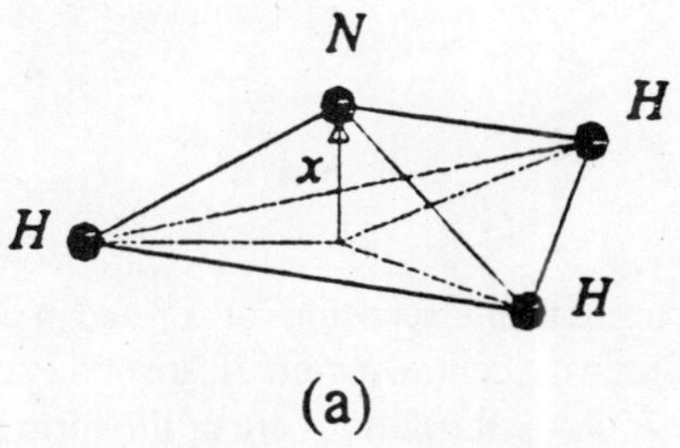

(a)

FIGURE P7. (a) The NH_3 inversion coordinate, x.

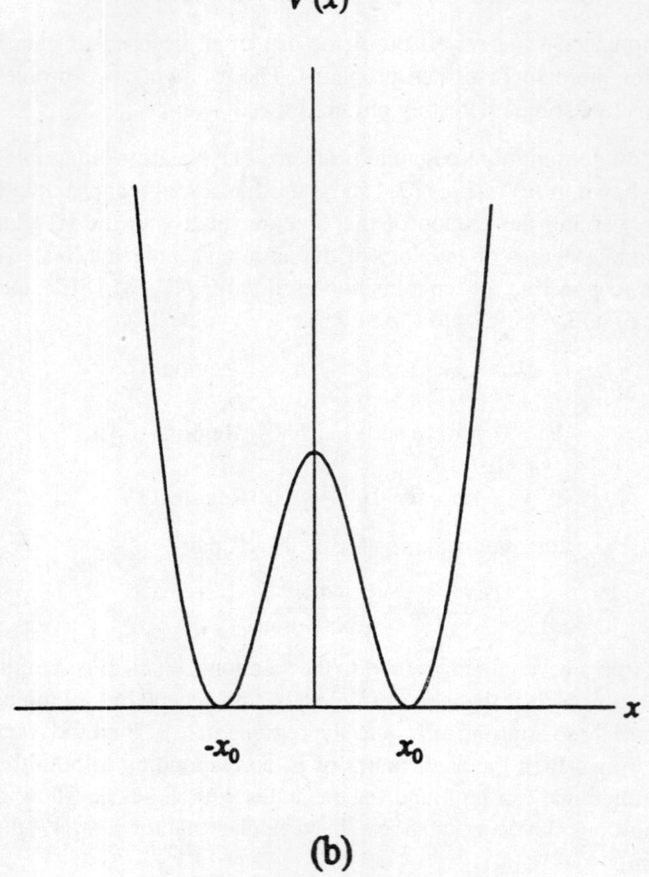

FIGURE P7. (b) Realistic $V(x)$.

Show, in this case, the splitting, ΔE_n, of the nearly degenerate pair is given by

$$\Delta E_n = (E_n^{\text{odd}} - E_n^{\text{even}}) = \frac{\hbar^3 n^2 \pi^2 \sqrt{8}}{(b-a)^3 \sqrt{\mu^3 (V_0 - E_n)}} e^{-\frac{2a}{\hbar}\sqrt{2\mu(V_0 - E_n)}},$$

where

$$E_n \approx \frac{n^2 \pi^2 \hbar^2}{2\mu(b-a)^2}.$$

Retain only dominant terms in all expansions of $\Delta\phi$ and in powers of E_n / V_0. Hints: The even (odd) solutions in the central region, II, are of the form $\cosh \kappa x$, ($\sinh \kappa x$), where $\kappa^2 = 2\mu(V_0 - E)/\hbar^2$. All solutions are of the form $\sin[k(x - b)]$ in region III.

8. Virtually bound states. Assume the potential, $V(r)$, shown in (a) of the Fig. P8, which is an effective potential for the motion of an α-particle relative to a heavy nucleus, can be approximated by the simpler square well potential of (b). Find solutions, $\psi(r)$, for this square well problem for energies, $E > 0$. The boundary condition at $r = 0$ is $\psi(r = 0) = 0$. Note, all energies, $E > 0$, lead to acceptable oscillatory solutions in region III.

Show, in general, for arbitrary positive energies, E,

$$\left|\frac{\psi_{\text{III}}}{\psi_{\text{I}}}\right|^2 \text{ is of order } e^G, \qquad \text{with } G = \frac{2(b-a)}{\hbar}\sqrt{2\mu(V_1 - E)}.$$

For $|\psi_{\text{I}}|$ and $|\psi_{\text{III}}|$, take the amplitudes of the oscillatory functions in regions I and III.

Show, however, the ratio

$$\left|\frac{\psi_{\text{III}}}{\psi_{\text{I}}}\right|^2 \text{ can be of order } e^{-G}$$

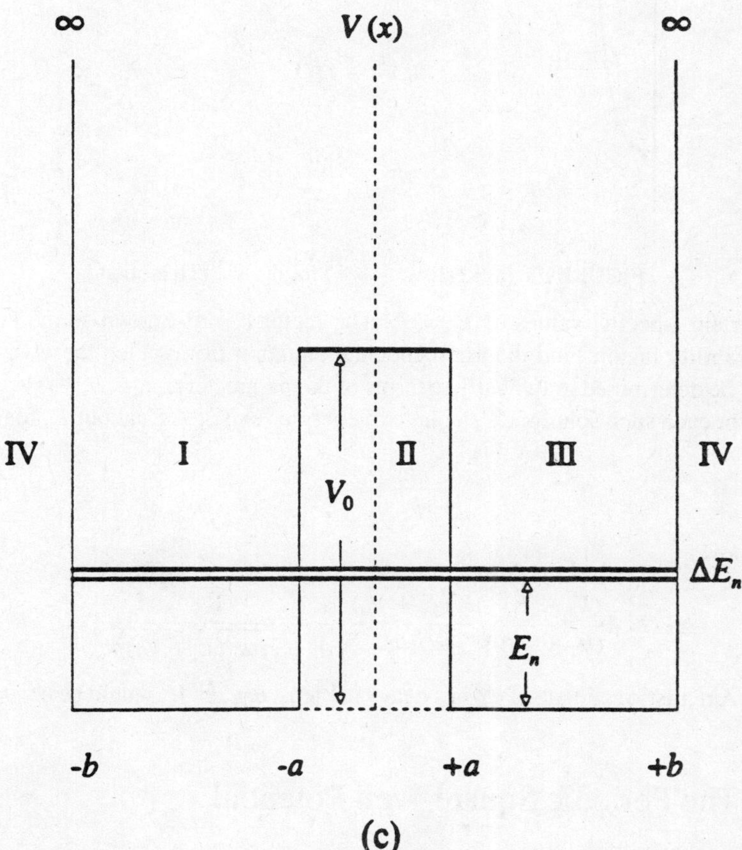

FIGURE P7. (c) Square well analogue.

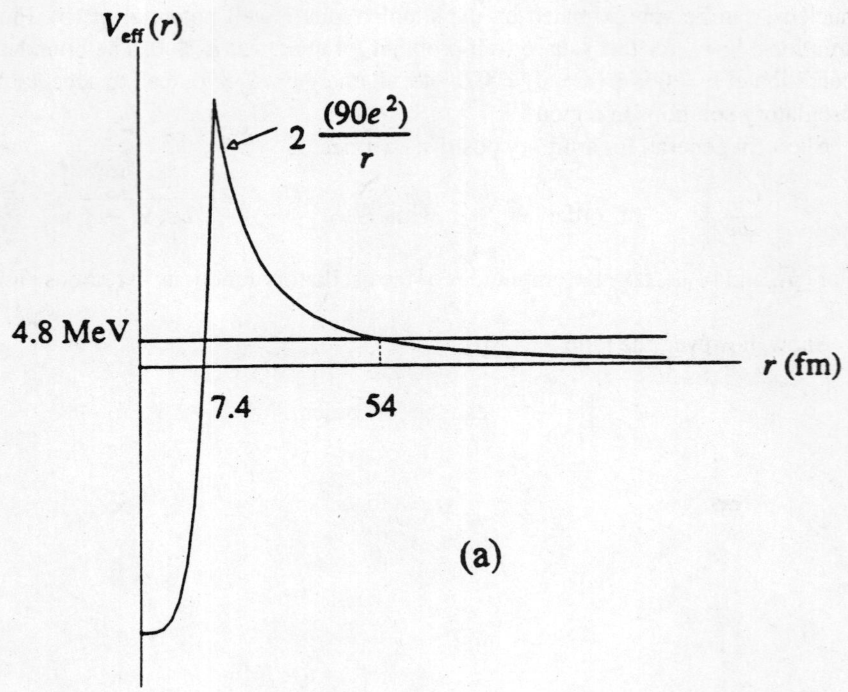

FIGURE P8. (a) Realistic $V_{eff.}(r)$ for the α-^{234}Th motion.

for certain, specific values of $E = \bar{E}$. The factor e^{-G} is known as the Gamow penetrability factor. Find the transcendental equation from which these values of $\bar{E}$ can be determined graphically in terms of the parameters, μ, a, b, V_0, V_1. Show also, for each such solution, $\bar{E}$, a range of energies exists, ΔE, about $\bar{E}$, for which

$$\left|\frac{\psi_{\text{III}}}{\psi_{\text{I}}}\right|^2 \approx e^{-G},$$

and show

$$\Delta E \approx 4\frac{(V_1 - \bar{E})}{(V_1 + V_0)}\sqrt{\frac{\hbar^2(\bar{E} + V_0)}{2\mu a^2}}\,\frac{1}{\cos\sqrt{2\mu a^2(\bar{E} + V_0)/\hbar^2}}\,e^{-G}.$$

Note: A realistic estimate of e^G in a heavy nucleus, e.g., ^{238}U, would be $e^G \approx 10^{38}$.

C The Periodic Square Well Potential

Another interesting case in which a square well approximation may shed considerable light on an important physical problem is that of an N-fold periodic potential. For very large N, this leads to a basic problem in condensed matter physics, the motion of an electron in a crystalline lattice with N lattice points. For very small

N, such as $N = 2$ or $N = 3$, examples of motions in an N-fold periodic potential may be found in the hindered internal rotation of one atomic unit relative to another in a molecule. A symmetrical X_2Y_4 molecule, such as ethylene, C_2H_4, e.g., has one degree of freedom, ϕ, which describes the highly hindered rotational motion of one essentially rigid CH_2 unit relative to the other on a circle in a plane perpendicular to the C – C symmetry axis, as shown in Fig. 4.3. The wave equation separates approximately, so the hindered internal rotation can be described by the one degree of freedom Schrödinger equation

$$-\frac{\hbar^2}{2I}\frac{d^2\psi}{d\phi^2} + V(\phi)\psi(\phi) = E\psi(\phi), \tag{22}$$

with $I = I_1I_2/(I_1 + I_2)$, and $I_1 = I_2 = 2m_Y r_Y^2$. The potential, $V(\phi)$, could be approximated by a purely sinusoidal potential,

$$V(\phi) = \tfrac{1}{2}V_0(1 - \cos 2\phi),$$

or, on the other hand, by a square well potential with potential valleys of $V = 0$ and widths $2a$ centered at $\phi = 0$ and at $\phi = \pi$, and potential barriers of constant heights of V_0 and widths w centered at $\phi = \frac{1}{2}\pi$ and at $\phi = \frac{3}{2}\pi$, where $4a + 2w = 2\pi r_Y$.

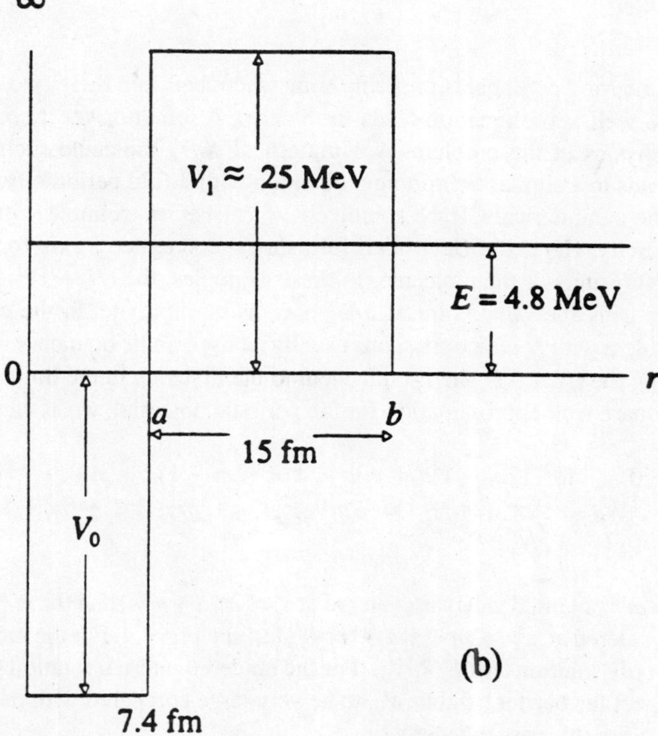

FIGURE P8. (b) Square well analogue.

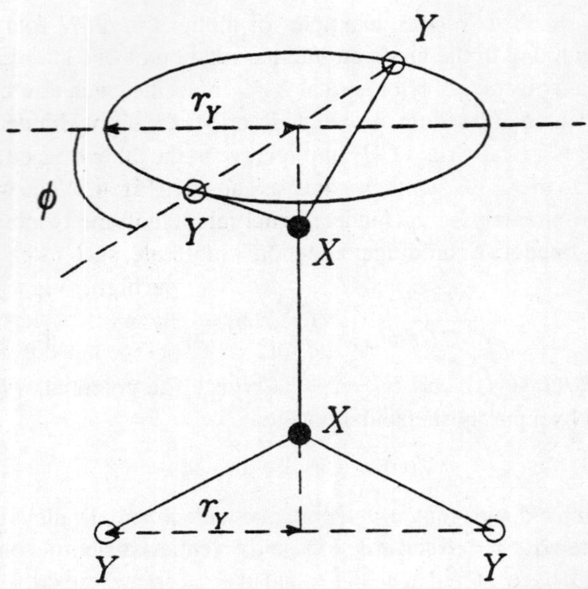

FIGURE 4.3. The X_2Y_4 molecule and its internal rotational coordinate, ϕ.

The true hindering potential is probably somewhere between these two extremes. The square well approximation leads to the easiest solution; yet it contains the essential physics of the problem. A symmetrical X_2Y_6 molecule such as C_2H_6 (ethane) leads to a similar Schrödinger equation with 3-fold periodicity, i.e., with $N = 3$. The symmetrical CH_3NO_2 molecule furnishes an example with $N = 6$-fold periodicity. Here, the C–N bond furnishes the symmetry axis for both the CH_3 and NO_2 units of the molecule. In these examples, the $(N+1)^{st}$ site of the potential is truly the same point in 3-D space as the first site. In the condensed matter problem with N lattice sites, one usually takes periodic boundary conditions by assuming the $(N+1)^{st}$ site is equivalent to the first site, in the limit $N \to \infty$.

In the square well approximation for the periodic potential, we assume

$$\begin{aligned} &V = 0 \qquad \text{for } (2m-1)a + mw < x < (2m+1)a + mw; \\ &V = +V_0 \qquad \text{for } (2m+1)a + mw < x < (2m+1)a + (m+1)w; \\ &m = 0, 1, \ldots, N. \end{aligned} \tag{23}$$

We see the mth potential valley is centered at $x = 2ma + mw$, and the mth potential barrier is centered at $x = (2m+1)a + (m + \frac{1}{2}w)$; see Fig. 4.4. For the moment, we shall seek only solutions for $E < V_0$. (For the hindered internal rotation problems, we can expect the barrier heights, V_0, to be very large compared with the energies of interest.) For this case, we define

$$k^2 = \frac{2\mu E}{\hbar^2}, \qquad \kappa^2 = \frac{2\mu(V_0 - E)}{\hbar^2},$$

where μ is an effective mass for the problem. We expect the following solutions. In the mth valley centered at $x = 2ma + mw$:

$$\psi(x) = C_m \cos k[x - m(2a + w)] + D_m \sin k[x - m(2a + w)];$$

under the mth potential hill, centered at $x = (2m + 1)a + (m + \frac{1}{2}w)$:

$$\psi(x) = A_m \cosh \kappa \Big(x - [(2m + 1)a + (m + \tfrac{1}{2}w)]\Big)$$

$$+B_m \sinh \kappa \Big(x - [(2m + 1)a + (m + \tfrac{1}{2})w]\Big).$$

The potential is invariant under reflections in the planes centered at $x = 2ma + mw$ and at $x = (2m+1)a+(m+\frac{1}{2}w)$. We might thus be tempted to assume our solutions are either even or odd under these reflection operations and that either $A_m = 0$ or $B_m = 0$, and, similarly, either $C_m = 0$ or $D_m = 0$. These assumptions would be good if all allowed energies were nondegenerate. We shall find, however, most of the allowed energy values are doubly degenerate, with two allowed solutions. We therefore retain the above linear combinations of even and odd functions. To ensure the continuity of the probability density and the probability density currents at the discontinuities of the potential, we shall again require the continuity of the wave functions and their first derivatives at the boundaries between the potential hills and valleys. With the solution under the $(m - 1)^{\text{st}}$ potential hill given by

$$\psi(x) = A_{m-1} \cosh \kappa \Big(x - [(2m - 1)a + (m - \tfrac{1}{2})w]\Big)$$

$$+B_{m-1} \sinh \kappa \Big(x - [(2m - 1)a + (m - \tfrac{1}{2})w]\Big),$$

the continuity of ψ and its first derivative at the left boundary of the mth valley, i.e., at $x = (2m - 1)a + mw$, leads to

$$A_{m-1} \cosh \kappa \tfrac{w}{2} + B_{m-1} \sinh \kappa \tfrac{w}{2} = C_m \cos ka - D_m \sin ka; \tag{24}$$

$$\kappa\big(A_{m-1} \sinh \kappa \tfrac{w}{2} + B_{m-1} \cosh \kappa \tfrac{w}{2}\big) = k\big(C_m \sin ka + D_m \cos ka\big). \tag{25}$$

The continuity of ψ and its first derivative at the right boundary of the mth valley, at $x = (2m + 1)a + mw$, leads to

$$C_m \cos ka + D_m \sin ka = A_m \cosh \kappa \tfrac{w}{2} - B_m \sinh \kappa \tfrac{w}{2}, \tag{26}$$

$$k\big(-C_m \sin ka + D_m \cos ka\big) = \kappa\big(-A_m \sinh \kappa \tfrac{w}{2} + B_m \cosh \kappa \tfrac{w}{2}\big). \tag{27}$$

Solving eqs. (24) and (25) for C_m and D_m and substituting into eqs. (26) and (27) leads to the relation

$$\begin{pmatrix} A_m \\ B_m \end{pmatrix} = \mathbf{M} \begin{pmatrix} A_{m-1} \\ B_{m-1} \end{pmatrix}, \tag{28}$$

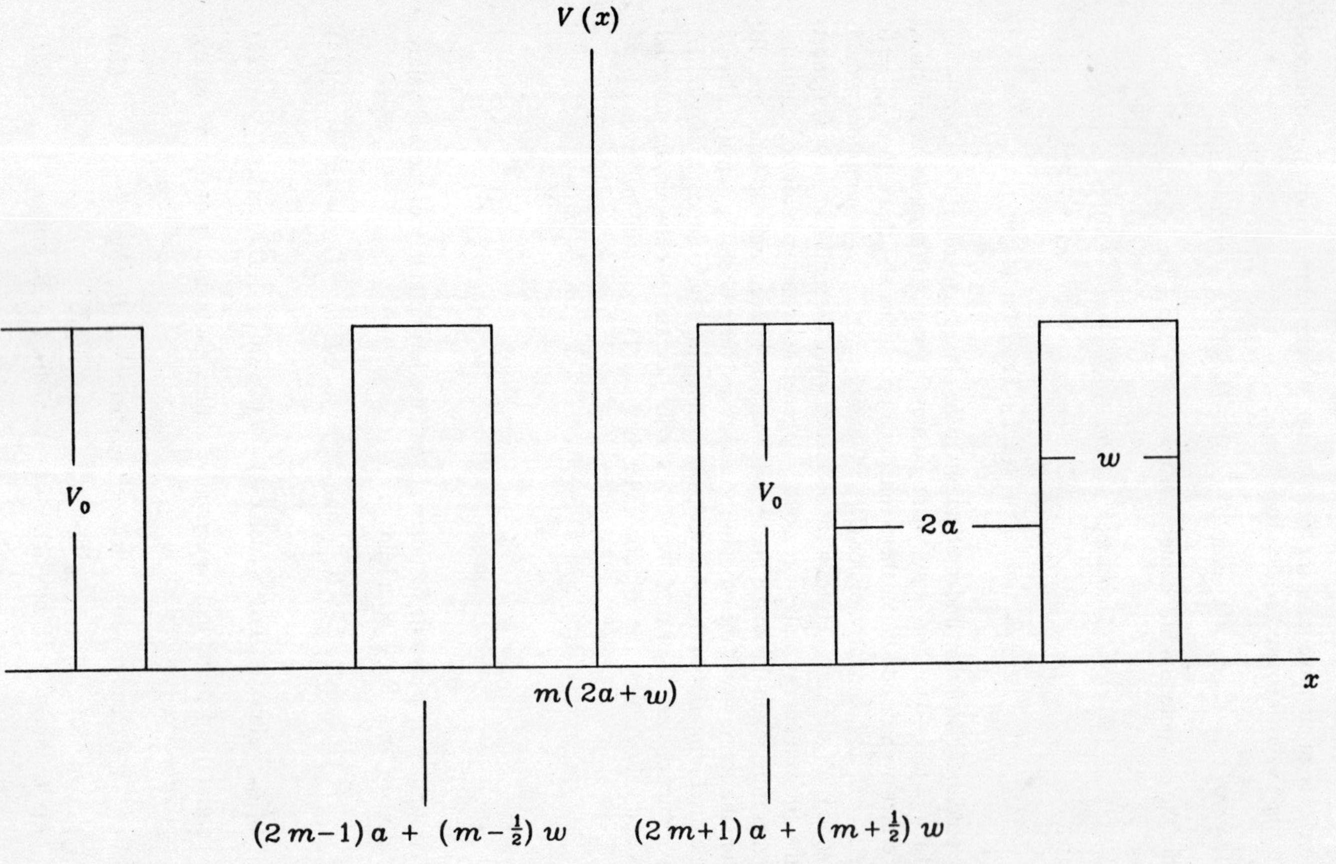

FIGURE 4.4. The periodic square well potential with barrier height, V_0, width, w, and potential valleys of width, $2a$, with $V = 0$.

where the 2×2 matrix, $\mathbf{M}$, is given by

$$\mathbf{M} = \cos 2ka \begin{pmatrix} P & Q+\gamma \\ Q-\gamma & P \end{pmatrix}, \tag{29}$$

with

$$\begin{aligned} P &= \cosh \kappa w + \tfrac{1}{2}\Big(\frac{\kappa}{k} - \frac{k}{\kappa}\Big) \tan 2ka \sinh \kappa w \\ &= \tfrac{1}{2} e^{\kappa w}\Big[1 + \tfrac{1}{2}\Big(\frac{\kappa}{k} - \frac{k}{\kappa}\Big) \tan 2ka\Big] + \tfrac{1}{2} e^{-\kappa w}\Big[1 - \tfrac{1}{2}\Big(\frac{\kappa}{k} - \frac{k}{\kappa}\Big) \tan 2ka\Big], \\ Q &= \sinh \kappa w + \tfrac{1}{2}\Big(\frac{\kappa}{k} - \frac{k}{\kappa}\Big) \tan 2ka \cosh \kappa w \\ &= \tfrac{1}{2} e^{\kappa w}\Big[1 + \tfrac{1}{2}\Big(\frac{\kappa}{k} - \frac{k}{\kappa}\Big) \tan 2ka\Big] - \tfrac{1}{2} e^{-\kappa w}\Big[1 - \tfrac{1}{2}\Big(\frac{\kappa}{k} - \frac{k}{\kappa}\Big) \tan 2ka\Big], \\ \gamma &= \tfrac{1}{2}\Big(\frac{\kappa}{k} + \frac{k}{\kappa}\Big) \tan 2ka. \end{aligned} \tag{30}$$

The continuity of the probability density and the probability density current require

$$\begin{pmatrix} A_N \\ B_N \end{pmatrix} = \mathbf{M}^N \begin{pmatrix} A_0 \\ B_0 \end{pmatrix} = \pm \begin{pmatrix} A_0 \\ B_0 \end{pmatrix}. \tag{31}$$

In particular, the *wave function* is not single valued for the case of the minus sign in the $\pm$ above. The probability density and the probability density current, however, are single valued. Also, the wave function would diverge as $e^{pN\kappa w}$ as $x \to pN(2a+w)$ in the above, or as $\phi \to pN(2\pi)$ in the wave function of eq. (22), as $p \to \infty$; unless the coefficients of the $e^{+\kappa w}$ terms of P and Q above are precisely equal to zero, or at most of order $e^{-\kappa w}$. We are thus led to the requirement

$$\tfrac{1}{2}\Big(\frac{\kappa}{k} - \frac{k}{\kappa}\Big) \tan 2ka = -1 + 2\beta e^{-\kappa w}, \tag{32}$$

where the new parameter, β, may, like k and κ, in general, also be a function of the energy E. Eq. (32) will thus lead to a transcendental equation for the determination of the allowed values of the energy, E, where the parameter, β, must also be fixed to satisfy eq. (31). It will be instructive to examine first the case of high potential barriers, $V_0 \gg E$. This case will actually be of interest for the problems of internal hindered rotations in most molecules. In the limit $V_0 \to \infty$, we have a problem with N-potential wells with infinitely high walls. In that case we saw [eq. (19)] $ka = \frac{1}{2} n\pi$. For the N-fold periodic square well with large V_0, we shall therefore try

$$2ka = n\pi + 2(\Delta k)a, \tag{33}$$

where $(\Delta k)a$ are small quantities, dependent on the integer n. Terms of second order in these small quantities will be negligible. Thus,

$$\cos 2ka \approx (-1)^n, \qquad \tan 2ka \approx 2(\Delta k)a = -2\frac{1}{\big(\frac{\kappa}{k} - \frac{k}{\kappa}\big)}(1 - 2\beta e^{-\kappa w}). \tag{34}$$

In the high barrier approximation, we have

$$e^{-\kappa w} \ll \frac{k}{\kappa} \ll 1.$$

With $\beta \approx$ order(1), we might thus expect the $\beta e^{-\kappa w}$ term to be negligible and obtain an energy shift, given by $(\Delta k)a \approx (-k/\kappa)$, of the $2N$-fold degenerate zeroth-order energy of $E_n^{(0)} = (\hbar^2 n^2 \pi^2 / 8\mu a^2)$. [The factor 2 in the degeneracy factor, $2N$, comes from the $\pm$ sign in the boundary condition of eq. (31).] Even though the splitting of the $2N$-fold degenerate levels with $E \ll V_0$ will be smaller than the above shifts by a factor of $e^{-\kappa w}$, this splitting is of primary interest. We will therefore retain this factor in eq. (34). In the high barrier limit, $e^{-\kappa w} \ll (k/\kappa) \ll 1$, the 2×2 matrix $\mathbf{M}$ reduces to

$$\mathbf{M} = (-1)^n \begin{pmatrix} \beta & \beta - 1 \\ \beta + 1 & \beta \end{pmatrix}. \tag{35}$$

For $N = 2$, the matrix needed for eq. (31) is

$$\mathbf{M}^2 = \begin{pmatrix} (2\beta^2 - 1) & 2\beta(\beta - 1) \\ 2\beta(\beta + 1) & (2\beta^2 - 1) \end{pmatrix}. \tag{36}$$

Eq. (31) then has allowed solutions for

$$\begin{aligned}
&\beta = +1, \quad \text{with} \begin{pmatrix} A_0 \\ B_0 \end{pmatrix} = \begin{pmatrix} 0 \\ 1 \end{pmatrix}; \quad \begin{pmatrix} A_2 \\ B_2 \end{pmatrix} = \begin{pmatrix} 1 & 0 \\ 4 & 1 \end{pmatrix} \begin{pmatrix} 0 \\ 1 \end{pmatrix} = +1 \begin{pmatrix} 0 \\ 1 \end{pmatrix}, \\
&\beta = -1, \quad \text{with} \begin{pmatrix} A_0 \\ B_0 \end{pmatrix} = \begin{pmatrix} 1 \\ 0 \end{pmatrix}; \quad \begin{pmatrix} A_2 \\ B_2 \end{pmatrix} = \begin{pmatrix} 1 & 4 \\ 0 & 1 \end{pmatrix} \begin{pmatrix} 1 \\ 0 \end{pmatrix} = +1 \begin{pmatrix} 1 \\ 0 \end{pmatrix}, \\
&\beta = 0, \quad \text{with} \begin{pmatrix} A_2 \\ B_2 \end{pmatrix} = \begin{pmatrix} -1 & 0 \\ 0 & -1 \end{pmatrix} \begin{pmatrix} A_0 \\ B_0 \end{pmatrix} = -1 \begin{pmatrix} A_0 \\ B_0 \end{pmatrix}, \\
&\qquad \text{where now} \begin{pmatrix} A_0 \\ B_0 \end{pmatrix} = \begin{pmatrix} 1 \\ 0 \end{pmatrix} \text{ or } \begin{pmatrix} 0 \\ 1 \end{pmatrix}.
\end{aligned} \tag{37}$$

For $N = 2$, three solutions exist for the allowed energies: two of them corresponding to $\beta = +1$ and $\beta = -1$ with but a single eigenfunction, corresponding to nondegenerate energy eigenvalues; and one with $\beta = 0$ with two independent solutions (which could be any linear combination of the above solutions), corresponding to a double degeneracy of this energy level. Expanding eq. (34) in powers of (k/κ), but retaining the dominant energy splitting term, we obtain for $N = 2$

$$E_n = \frac{\hbar^2}{2\mu a^2} \left(\frac{n^2\pi^2}{4} - n\pi \frac{k_n}{\kappa_n} + \begin{matrix} +2n\pi (k_n/\kappa_n) e^{-\kappa_n w} \\ 0 \\ -2n\pi (k_n/\kappa_n) e^{-\kappa_n w} \end{matrix} \right), \tag{38}$$

where we can approximate (k_n/κ_n) by $\sqrt{(E_n^{(0)}/V_0)}$, but will retain the $E_n^{(0)}$ term in the exponential factor,

$$e^{-\kappa_n w} \approx e^{-[2\mu(V_0 - E_n^{(0)}) w^2/\hbar^2]^{\frac{1}{2}}},$$

because of the sensitivity of the exponential factor on its exponent.

Next, for the three-fold periodic potential with $N = 3$, we have

$$\mathbf{M}^3 = (-1)^n \begin{pmatrix} \beta(4\beta^2 - 3) & (4\beta^2 - 1)(\beta - 1) \\ (4\beta^2 - 1)(\beta + 1) & \beta(4\beta^2 - 3) \end{pmatrix}. \tag{39}$$

The boundary condition of eq. (31) is satisfied for $\beta = +1$ and $\beta = -1$, with nondegenerate solutions

$$\begin{pmatrix} A_0 \\ B_0 \end{pmatrix} = \begin{pmatrix} 0 \\ 1 \end{pmatrix} \quad \text{and} \quad \begin{pmatrix} 1 \\ 0 \end{pmatrix}, \quad \text{respectively,}$$

and for $\beta = +\frac{1}{2}$ and $\beta = -\frac{1}{2}$, where both of these lead to doubly degenerate levels with a linear combination of the above two solutions. The energy $E_n^{(0)}$ is thus split into four levels, a highest and a lowest nondegenerate level and two intermediate doubly degenerate levels.

At this stage, it should be mentioned that the splittings of the ground-state $n = 0$ internal rotation energies in the molecules, C_2H_4 and C_2H_6, are too small to be observable. The factors, $e^{-\kappa_0 w}$, are too small to be observable in these molecules. In the methyl alcohol molecule, CH_3OH, however, the splittings of the $n = 0$ and higher levels are observable and have been studied extensively by microwave spectroscopy. In this molecule, the internal rotation degree of freedom is strongly coupled with the rotational degrees of freedom of the whole molecule. Since this molecule is an asymmetric rotator, (see Chapter 15), the combined rotation–internal rotation spectrum is very complicated.

For the case of general N, it will be convenient to introduce the new parameter α, via

$$\beta = \cos\alpha.$$

In terms of this new parameter, we have

$$\mathbf{M}^N = (-1)^{nN} \begin{pmatrix} \cos N\alpha & \frac{\sin N\alpha}{\sin\alpha}(\cos\alpha - 1) \\ \frac{\sin N\alpha}{\sin\alpha}(\cos\alpha + 1) & \cos N\alpha \end{pmatrix}. \tag{40}$$

Eqs. (36) and (39) show this is satisfied for $N = 2$ and $N = 3$. Also,

$$\begin{pmatrix} \cos(N-1)\alpha & \frac{\sin(N-1)\alpha}{\sin\alpha}(\cos\alpha - 1) \\ \frac{\sin(N-1)\alpha}{\sin\alpha}(\cos\alpha + 1) & \cos(N-1)\alpha \end{pmatrix} \begin{pmatrix} \cos\alpha & (\cos\alpha - 1) \\ (\cos\alpha + 1) & \cos\alpha \end{pmatrix}$$
$$= \begin{pmatrix} \cos N\alpha & \frac{\sin N\alpha}{\sin\alpha}(\cos\alpha - 1) \\ \frac{\sin N\alpha}{\sin\alpha}(\cos\alpha + 1) & \cos N\alpha \end{pmatrix}, \tag{41}$$

so that the relation (40) is proved by induction. In this general case, two nondegenerate levels again exist, with $\alpha = 0$, and $\alpha = \pi$, and now $(N - 1)$ doubly degenerate levels with

$$\alpha = \frac{\ell\pi}{N}; \qquad \ell = 1, 2, \ldots, (N-1).$$

The energies for the case $E_n^{(0)} \ll V_0$ are given by

$$E_n = \frac{\hbar^2}{2\mu a^2}\left(\frac{n^2\pi^2}{4} - n\pi\sqrt{\frac{E_n^{(0)}}{V_0}}\left[1 - 2\cos\frac{\ell\pi}{N}e^{-[2\mu(V_0 - E_n^{(0)})w^2/\hbar^2]^{\frac{1}{2}}}\right]\right)$$

$$\ell = 0, 1, \ldots, N. \tag{42}$$

For very large N in a crystalline lattice, therefore, we have a set of $(N + 1)$ finely spaced, discrete, allowed energy values, centered about a slightly downward-shifted $E_n^{(0)}$. In the limit, $N \to \infty$, this becomes a continuous narrow band of allowed energies of bandwidth

$$\Delta E = \frac{\hbar^2}{ma^2} n^2 \pi^2 \sqrt{\frac{\hbar^2}{2ma^2 V_0}} e^{-[2m(V_0 - E_n^{(0)}) w^2/\hbar^2]^{\frac{1}{2}}}, \tag{43}$$

where we have set $\mu = m$, the electron mass. These continuous bands of allowed energies are separated by energy gaps of order $(E_{n+1}^{(0)} - E_n^{(0)})$. As $E_n^{(0)}$ approaches V_0, the bandwidths become larger and the gaps smaller. Of course, as $E_n^{(0)} \to V_0$, our high V_0 approximations are no longer valid. The bandwidth and gap structure, however, survives even into the region $E > V_0$. (For details, see, e.g., C. Kittel, *Introduction to Solid State Physics*, New York: John Wiley, 1956.) In a real solid, we must of course also deal with a 3-D structure. It is therefore perhaps interesting to note that in a cooler ring of some modern generation heavy ion accelerators we may approximate a truly 1-D crystal of cold (hence, nearly monoenergetic) heavy ions. In the limit of temperature, $T \to 0$, these form a 1-D crystal of equally spaced monoenergetic heavy ions. Here, indeed, the $(N + 1)^{\text{st}}$ ion *is* the 1st ion, and the periodic boundary condition of eq. (31) is no longer an approximation. Although the square well solution has all of the qualitative features found with a more realistic potential, an approximate solution for a more realistic $V(x)$ can be found through the WKB approximation to be treated in Chapters 36 and 37 (see, in particular, problem 55).

D The Existence of Discrete Energy Levels: General $V(x)$

For a $V(x)$ that is such that $V \to \infty$ for both large positive and large negative values of x, the existence of a discrete set of allowed energy levels follows in a general way from the requirement that the solutions be square-integrable, i.e.,

$$\int_{-\infty}^{+\infty} \psi^* \psi \, dx = \text{finite}, \tag{44}$$

and that ψ and $\frac{d\psi}{dx}$ be continuous. For the type of potential function shown in Fig. 4.5, with an arbitrary E, but $E > V_{min.}$, we have in region I, with $E > V(x)$, between the left and right classical turning points,

$$\frac{d^2\psi}{dx^2} + k^2(x)\psi = 0. \tag{45}$$

In region I, therefore, the solutions are oscillatory, but with a variable (x-dependent) wavelength because $k(x) = \frac{2\pi}{\lambda}$; i.e., the curvature is always toward the x-axis. In

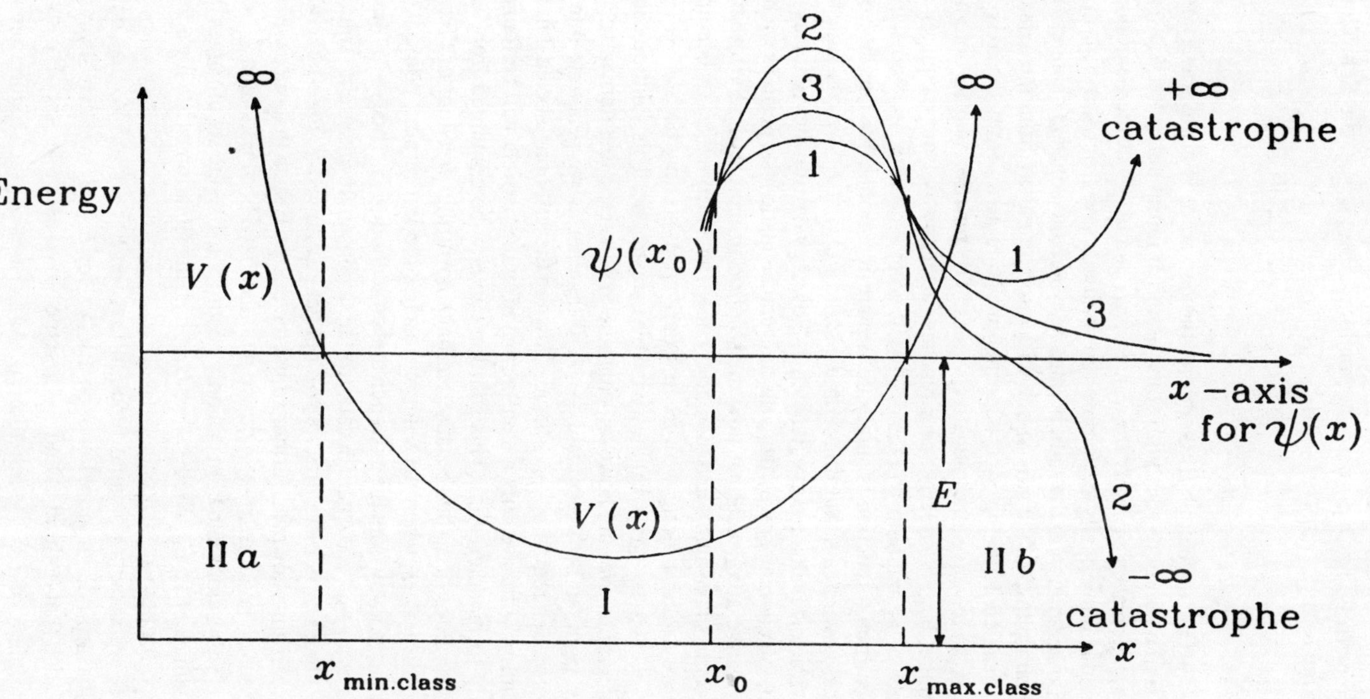

FIGURE 4.5. Solutions, $\psi(x)$, for three different initial conditions at x_0.

regions II, conversely, for $x > x_{\text{max.class.}}$, or for $x < x_{\text{min.class.}}$, the Schrödinger equation has the form

$$\frac{d^2\psi}{dx^2} - \kappa^2(x)\psi = 0, \tag{46}$$

because $E < V(x)$. In regions II, therefore, the curvature is always away from the axis. To find a solution, start with some assumed initial value for $\psi(x_0)$ and $\frac{d\psi}{dx}|_{x_0}$. The equation then gives us the value of $\psi(x)$ and $\frac{d\psi}{dx}$ at the neighboring points. We could numerically determine the solution, say, from some x_0 in the classically allowed region to the right boundary, where the solution changes from one with curvature toward the axis to one with curvature away from the axis. For the solution, labeled 1, in Fig. 4.5, e.g., the curvature away from the axis will be such that $\psi(x)$ never reaches negative values. The function and its derivative will thus both get larger and larger as x reaches further away from the classically allowed values of x; and both $\psi(x)$ and $\frac{d\psi}{dx}$ will go to $+\infty$ as $x \to +\infty$. This is a catastrophe. Such a function is surely not square-integrable. We can, however, start the process over again. Starting with the same $\psi(x_0)$ at x_0, we can adjust the first derivative at x_0, as in the curve, labeled 2. Now, as we reach the right classical turning point, the curvature away from the axis can be made less; perhaps we have chosen a derivative at x_0 such that now the solution in the classically forbidden region, IIa, reaches the value 0 and thereafter curves away from the axis becoming more and more negative along with its first derivative, so now both $\psi \to -\infty$ and $\frac{d\psi}{dx} \to -\infty$ as $x \to +\infty$. Again, we have a catastrophe. This solution cannot be square-integrable. We can, however, continue to adjust the first derivative at x_0 until it is just right, so both $\psi(x)$ and $\frac{d\psi}{dx} \to 0$ together as we penetrate into the classically forbidden region, $x \to +\infty$, as shown in the solution, labeled 3 in Fig. 4.5. This solution will have only a small probability the particle will be found in the classically forbidden region. This solution can now be continued from x_0 to more negative values of x, but because we have no further freedom of "fixing" the first derivative at x_0, when the solution reaches the left turning point, it will undoubtedly curve away from the axis such that either both $\psi(x)$ and $\frac{d\psi}{dx} \to +\infty$ or both $\to -\infty$ as $x \to -\infty$. Again, a catastrophe: a nonsquare-integrable solution. For arbitrary values of E, therefore, we will not get an allowed (square-integrable) solution. We can now, however, further adjust the energy E such that once we have fixed the proper behavior as $x \to +\infty$ we will also have both $\psi(x)$ and $\frac{d\psi}{dx} \to 0$ as $x \to -\infty$. This unique situation can only occur for a discrete set of values of E, the allowed values of E: $E_0, E_1, E_2, \ldots, E_n, \ldots$. This situation exists for a $V(x)$, which $\to +\infty$ for both $x \to \pm\infty$.

In Fig. 4.6, we show a potential function that for $E > \bar{V} = V_\infty$ has only a left classically forbidden region. For such a $V(x)$, for $E > \bar{V}$, we can always fix the solution such that both $\psi(x)$ and $\frac{d\psi}{dx}$ together $\to 0$ as $x \to -\infty$. For such a potential, all values of $E > \bar{V}$ are allowed. As $x \to +\infty$, the solution remains oscillatory. The $\psi(x)$ is not square integrable, but the solution has a sensible probability interpretation. It now corresponds to a particle with finite kinetic energy coming in from $+\infty$ being reflected near $x = 0$ and going back out to $+\infty$. This

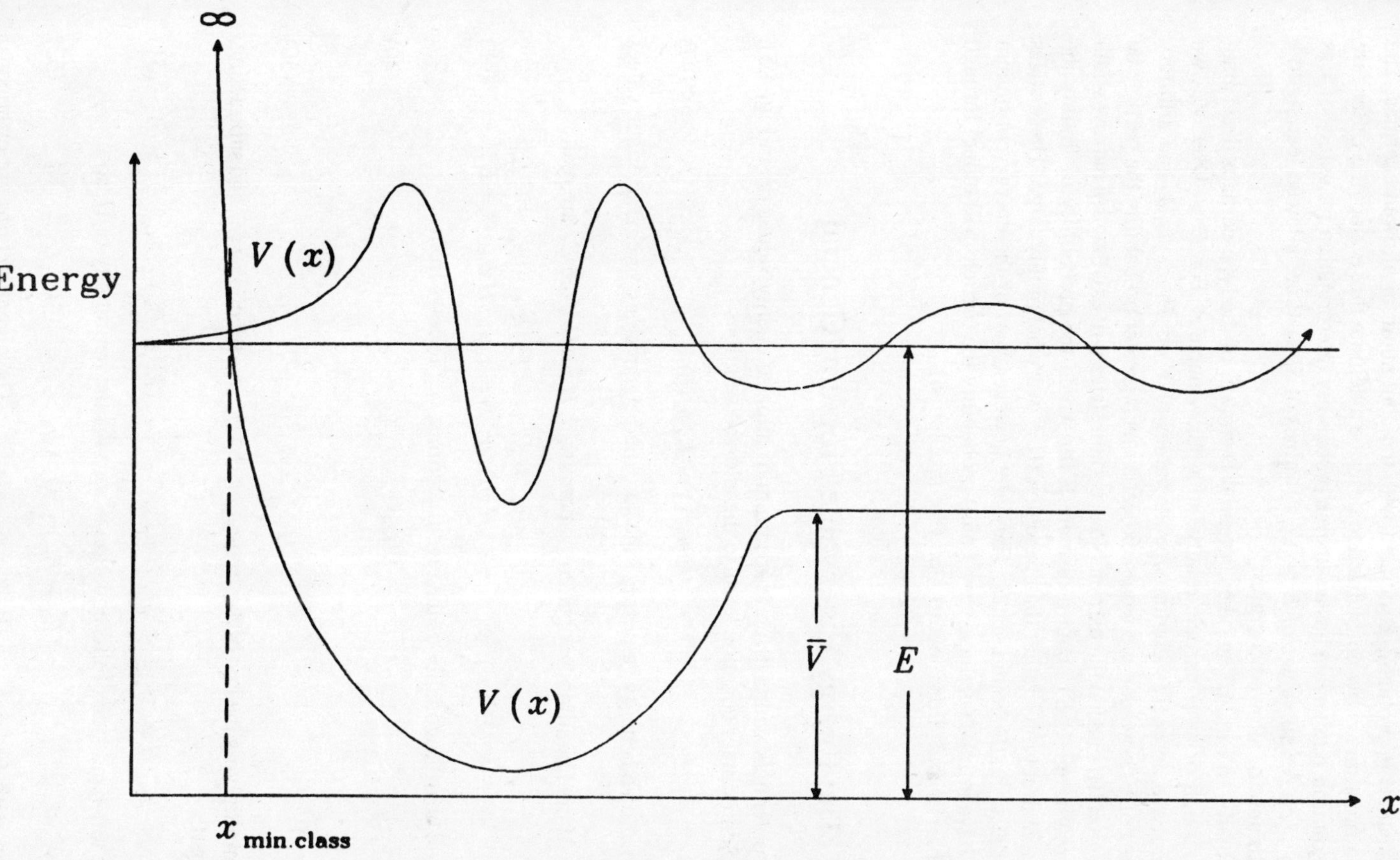

FIGURE 4.6. $V(x)$ with a continuous spectrum for $E > \bar{V} = V_\infty$.

is a scattering problem, associated with the continuum of allowed energies. The wave function is now normalized to describe a definite flux (value of $\vec{S}$). Because the solution $\psi(x)$ again has a sensible probability interpretation for values of x as $x \to +\infty$ for all values of $E > \bar{V}$, all values of $E > \bar{V}$ are allowed leading to a continuum of allowed energies.

Finally, in Fig. 4.7, another potential is shown, of the type perhaps describing the motion of an α-particle relative to a heavy nucleus. For $E > \bar{V} = V_\infty$, we again have an energy continuum; for values of $\bar{V} < E < V_{\text{barrier}}$ and arbitrary values of E, however, we would expect a much greater probability the particle be in region III, outside the barrier. Now, certain states will exist, with a narrow width (narrow range ΔE) about a discrete E for which the probability of finding the particle in region I rather than in region III is overwhelmingly large. These states are the virtually bound states. They are, however, part of the energy continuum and have a finite (perhaps very small) probability the particle will tunnel through the barrier and stream out to $+\infty$ (see problem 8).

E The Energy Eigenvalue Problem: General

For potentials with a discrete spectrum of allowed energy values (as in Fig. 4.5), the Schrödinger equation leads to the allowed solutions

$$H\psi_n(x) = E_n\psi_n(x), \tag{47}$$

with allowed energy values, E_n, the so-called eigenvalues, or characteristic values of E.

1. If the Hamiltonian operator is hermitian, $H = H^\dagger$, the E_n are real.

$$\begin{aligned} &< \psi_n, H\psi_n >= E_n < \psi_n, \psi_n >= E_n \\ =&< (H^\dagger\psi_n), \psi_n >=< (H\psi_n), \psi_n >=< \psi_n, H\psi_n >^*= E_n^*. \end{aligned} \tag{48}$$

2. The orthogonality of the eigenfunctions, ψ_n, follows from

$$H\psi_n = E_n\psi_n, \tag{49}$$

and

$$H\psi_m^* = E_m\psi_m^*. \tag{50}$$

By multiplying the first of these equations by ψ_m^*, the second by ψ_n, and subtracting, we get

$$< \psi_m, H\psi_n > - < \psi_n, H\psi_m >^* \;=\; (E_n - E_m) < \psi_m, \psi_n > . \tag{51}$$

The left-hand side of this equation is zero via the hermiticity of H, so

$$(E_n - E_m) < \psi_m, \psi_n > \;=\; 0. \tag{52}$$

Thus, with $E_n \neq E_m$, $\quad < \psi_m, \psi_n >= 0.\quad$ The eigenfunctions are orthogonal to each other.

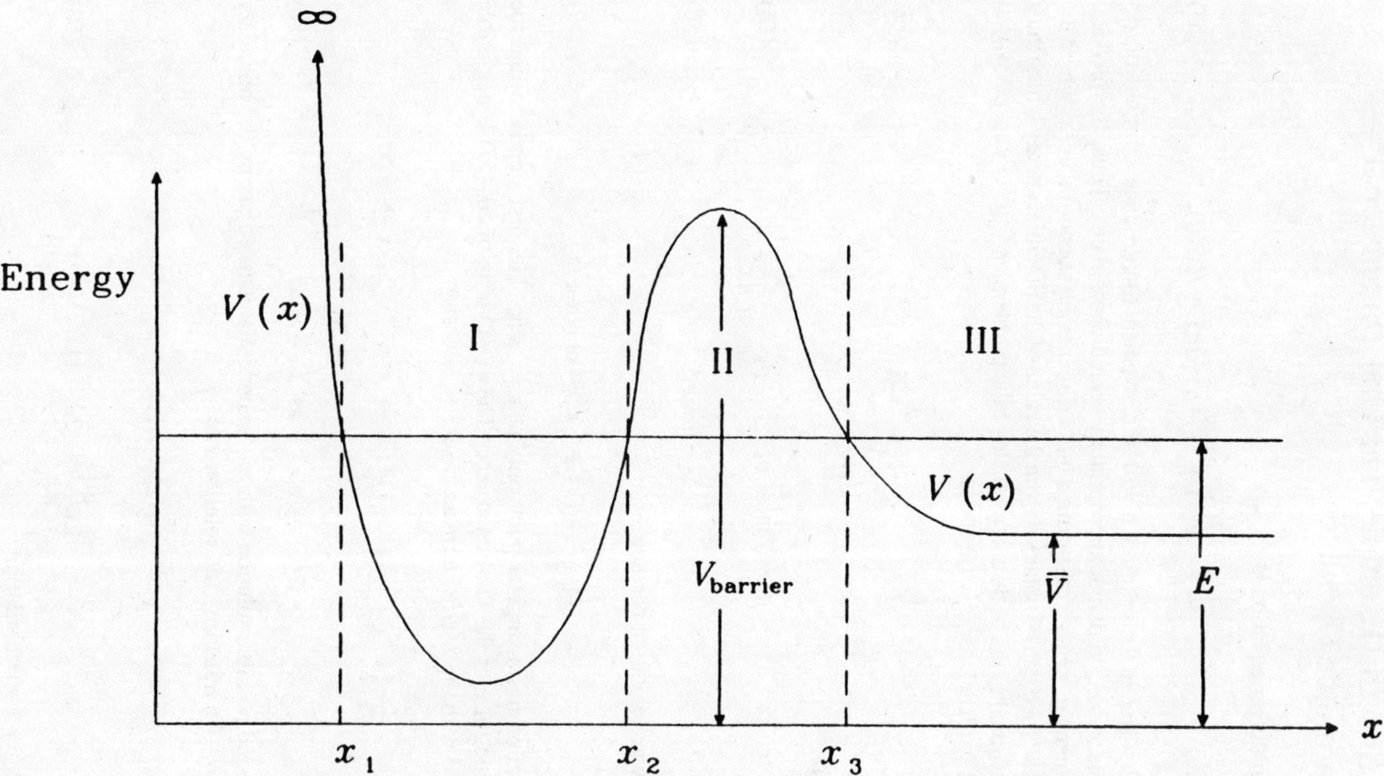

FIGURE 4.7. $V(x)$ with possible virtually bound states for $\bar{V} < E < V_{\text{barrier}}$.

F A Specific Example: The One-Dimensional Harmonic Oscillator

The Schrödinger equation is

$$-\frac{\hbar^2}{2m}\frac{d^2}{dq^2}\psi(q)+\frac{m\omega_0^2}{2}q^2\psi(q)=E\psi(q). \tag{53}$$

As a first step in the solution, it will be convenient to introduce a dimensionless coordinate x; i.e., to define appropriately scaled coordinates. Thus, the physical displacement, q, will be transformed into a dimensionless coordinate, x, where the "yardstick" for q can be obtained from the value of the potential energy which must be proportional to the basic energy scale of the problem, $m\omega_0^2q^2=\text{const.}(\hbar\omega_0)$, so a natural yardstick for q is $\sqrt{\hbar/m\omega_0}$:

$$q=\sqrt{\frac{\hbar}{m\omega_0}}x. \tag{54}$$

Similarly,

$$p=\sqrt{\hbar m\omega_0}\,p_x,\qquad p_x=\frac{1}{i}\frac{d}{dx}, \tag{55}$$

$$E=\hbar\omega_0\epsilon, \tag{56}$$

and the wave equation becomes

$$-\frac{1}{2}\frac{d^2}{dx^2}\psi(x)+\frac{1}{2}x^2\psi(x)=\epsilon\psi(x). \tag{57}$$

The equation has a singular point only at $x=\pm\infty$. The first step is to find the asymptotic form of the solution at $\pm\infty$. The $\epsilon\psi(x)$ term of eq. (57) is negligible compared with the $x^2\psi(x)$ term as $x\to\pm\infty$. Because

$$\frac{d^2}{dx^2}\left(e^{-\frac{x^2}{2}}\right)=(x^2-1)e^{-\frac{x^2}{2}}\to x^2e^{-\frac{x^2}{2}}\quad \text{as } x\to\pm\infty,$$

$$\psi(x)\to e^{-\frac{x^2}{2}}\quad \text{as } x\to\pm\infty. \tag{58}$$

(The second possible solution with a + exponential is ruled out by the boundary condition.) We transform the solution into

$$\psi(x)=u(x)e^{-\frac{x^2}{2}}, \tag{59}$$

$$\frac{d^2u}{dx^2}-2x\frac{du}{dx}+(2\epsilon-1)u(x)=0. \tag{60}$$

For $u(x)$, try a series solution

$$u(x)=\sum_{k=0}a_kx^k, \tag{61}$$

where

$$\sum_{k=2} a_k k(k-1)x^{k-2} = \sum_{k=0} (2k+1-2\epsilon)a_k x^k. \tag{62}$$

Changing the dummy summation index on the left-hand side from $k \to (k+2)$ and equating coefficients of the kth term leads to the two-term recursion relation

$$\frac{a_{k+2}}{a_k} = \frac{2k+1-2\epsilon}{(k+2)(k+1)}. \tag{63}$$

To examine the behavior of this infinite series at large values of x, look at the asymptotic form as $k \to \infty$. This form is

$$\frac{a_{k+2}}{a_k} \to \frac{2}{k} \tag{64}$$

or

$$a_{2m} \to \frac{1}{m!}, \tag{65}$$

so we would have

$$u(x) \to e^{+x^2}. \tag{66}$$

Thus, for general values of ϵ, $\psi(x) \to \infty$, we do not have a square-integrable solution. For the special value

$$2\epsilon = (2n+1), \tag{67}$$

the infinite series of eq. (61) can terminate at the nth term. If n is an even integer and $a_0 \neq 0$, the recursion formula of eq. (63) yields $a_{n+2} = 0$, and, therefore, $a_m = 0$ with $m = n+2k$. If n is an even integer, and if we had $a_1 \neq 0$, however, all a_m with odd integers m would survive up to $m \to \infty$, and the infinite series would again diverge as e^{+x^2}. If n is an even integer, we must therefore have $a_1 = 0$. Similarly, if n is an odd integer, we must have $a_0 = 0$. The series therefore terminates

$$\begin{aligned} &\text{with} \quad \text{n} = \text{even}, \quad a_0 \neq 0, \quad a_1 = 0, \quad a_{n+2} = 0, \\ &\text{with} \quad \text{n} = \text{odd}, \quad a_1 \neq 0, \quad a_0 = 0, \quad a_{n+2} = 0. \end{aligned} \tag{68}$$

For these cases, the wave functions of eqs. (59) and (61) are square-integrable and lead to the discrete set of allowed energy eigenvalues

$$2\epsilon = (2n+1); \qquad E = \hbar\omega_0(n+\frac{1}{2}). \tag{69}$$

To find the coefficients of the polynomial of degree n, invert the recursion relation:

$$\frac{a_{k-2}}{a_k} = -\frac{k(k-1)}{2(n-k+2)}, \tag{70}$$

leading to

$$\frac{a_{n-2j}}{a_n} = (-1)^j \frac{n(n-1)\cdots(n-2j+2)(n-2j+1)}{2^j 2\cdot 4\cdots(2j-2)2j}$$

$$= (-1)^j \frac{n!}{(n-2j)!2^{2j}j!}. \tag{71}$$

With $a_n = 2^n$, this solution is the standard Hermite polynomial, $H_n(x)$,

$$H_n(x) = \sum_{j=0}^{[\frac{n}{2}]} (-1)^j \frac{2^{n-2j}n!}{j!(n-2j)!} x^{n-2j}. \tag{72}$$

The Hermite polynomial can be defined in three ways:
1. Through the regular solutions of the differential equation:

$$H_n''(x) - 2xH_n'(x) + 2nH_n(x) = 0. \tag{73}$$

2. Through a generating function, where the parameter, s, may be an arbitrary complex number:

$$e^{-s^2+2sx} = \sum_{n=0}^{\infty} \frac{H_n(x)}{n!} s^n. \tag{74}$$

3. Through a differential relation, or a Rodrigues-type formula:

$$H_n(x) = (-1)^n e^{x^2} \frac{d^n}{dx^n}(e^{-x^2}). \tag{75}$$

Thus,

$$E_n = \hbar\omega_0(n + \tfrac{1}{2}), \qquad \psi_n(x) = N_n H_n(x) e^{-\frac{x^2}{2}}. \tag{76}$$

The normalization constant, N_n, can be evaluated most simply through the Rodrigues-type formula

$$\begin{aligned} |N_n|^2 \int_{-\infty}^{\infty} dx H_n^2(x) e^{-x^2} &= |N_n|^2 \int_{-\infty}^{\infty} dx H_n(x)(-1)^n \frac{d^n}{dx^n}(e^{-x^2}) \\ = |N_n|^2 \int_{-\infty}^{\infty} dx (e^{-x^2})(\frac{d^n}{dx^n} H_n(x)) &= |N_n|^2 \int_{-\infty}^{\infty} dx e^{-x^2} n! a_n \\ &= |N_n|^2 \sqrt{\pi} n! 2^n = 1. \end{aligned} \tag{77}$$

Here, we have integrated by parts n times and have used the fact that the integrated parts, to be evaluated at $\pm\infty$, are all dominated by the factor e^{-x^2}. Choosing N_n to be real

$$N_n = \sqrt{\frac{1}{n!2^n\sqrt{\pi}}}. \tag{78}$$

Final note: Sometimes it is necessary to normalize the wave function in real, physical space, i.e., with

$$\int_{-\infty}^{\infty} dq\, \psi^* \psi = 1. \tag{79}$$

Then,

$$N_n = \sqrt{\frac{1}{n!2^n} \sqrt{\frac{m\omega_0}{\hbar\pi}}}. \tag{80}$$

5

Harmonic Oscillator Calculations

A The Bargmann Transform

For many calculations involving 1-D harmonic oscillator wave functions, it is useful to introduce the Bargmann transform through the kernel function

$$A(k, x) = \frac{1}{\pi^{\frac{1}{4}}} exp(-\tfrac{1}{2}k^2 + \sqrt{2}kx - \tfrac{1}{2}x^2), \tag{1}$$

where k is a complex number. Given a square-integrable function, $\psi(x)$, its Bargmann transform, $F(k)$, is given by

$$F(k) = \int_{-\infty}^{\infty} dx \psi(x) A(k, x), \tag{2}$$

where

$$\psi(x) = \frac{1}{\pi} \int d^2k e^{-kk^*} A(k^*, x) F(k), \tag{3}$$

and the integral is over the 2-D complex k-plane; i.e., with $k = a + ib$,

$$\int d^2k \equiv \int_{-\infty}^{\infty} da \int_{-\infty}^{\infty} db. \tag{4}$$

Now, from the definition of $A(k, x)$ and the generating function definition for the Hermite polynomials, with $s = k/\sqrt{2}$,

$$A(k, x) = \sum_{n=0} \left(\frac{H_n(x)}{\sqrt{n! 2^n \sqrt{\pi}}} e^{-\frac{1}{2}x^2} \right) \frac{k^n}{\sqrt{n!}} = \sum_{n=0} \psi_n(x) \frac{k^n}{\sqrt{n!}}. \tag{5}$$

We therefore see

$$\psi_n(x) \qquad \text{has} \quad \text{Bargmann} \quad \text{transform} \qquad \frac{k^n}{\sqrt{n!}}. \tag{6}$$

We can transform a scalar product from x-space into k-space, or from k-space into x-space:

$$\begin{aligned}
&\frac{1}{\pi}\int d^2k e^{-kk^*} F_1^*(k)F_2(k) \\
&= \frac{1}{\pi}\int d^2k e^{-kk^*}\int dx\psi^{(1)}(x)^* A(k^*,x)\int dx'\psi^{(2)}(x')A(k,x') \\
&= \int dx\psi^{(1)}(x)^*\psi^{(2)}(x),
\end{aligned} \tag{7}$$

where we have used, again with $k = a + ib$,

$$\begin{aligned}
&\frac{1}{\pi}\int d^2k e^{-kk^*} A(k^*,x)A(k,x') \\
&= \frac{1}{\sqrt{\pi}}\int_{-\infty}^{\infty} da e^{-[\sqrt{2}a - \frac{(x+x')}{2}]^2} e^{-[\frac{(x-x')}{2}]^2}\frac{1}{\pi}\int_{-\infty}^{\infty} db e^{i\sqrt{2}b(x'-x)} \\
&= e^{-[\frac{(x-x')}{2}]^2}\frac{1}{2\pi}\int_{-\infty}^{\infty} db' e^{ib'(x'-x)} = \delta(x'-x).
\end{aligned} \tag{8}$$

In the last step, we have used

$$f(x)\delta(x) = f(0)\delta(x). \tag{9}$$

Eq. (7) thus permits us to evaluate a scalar product either in x-space or in k-space. At times, the latter may lead to the easier integral.

B Completeness Relation

The delta-function property of the integral of eq. (8) is also useful to prove the completeness of the set of harmonic oscillator eigenfunctions $\psi_n(x)$. An arbitrary 1-D square-integrable function, $\Psi(x)$, can be expanded in a generalized Fourier series in oscillator eigenfunctions, $\psi_n(x)$,

$$\Psi(x) = \sum_{n=0}^{\infty} c_n\psi_n(x), \tag{10}$$

with coefficient, c_n, evaluated by the Fourier inversion theorem

$$c_n = \int_{-\infty}^{\infty} dx'\Psi(x'), \psi_n^*(x'); \tag{11}$$

so

$$\Psi(x) = \sum_{n=0}^{\infty}\int_{-\infty}^{\infty} dx'\Psi(x')\psi_n(x)\psi_n^*(x'). \tag{12}$$

This requires

$$\sum_{n=0}^{\infty}\psi_n(x)\psi_n^*(x') = \delta(x-x'). \tag{13}$$

This is the completeness relation we want to prove. Substituting eq. (5) into the left-hand side of eq. (8), we get

$$\begin{aligned}&\frac{1}{\pi}\int d^2k e^{-kk^*}\sum_{n=0}^{\infty}\psi_n^*(x)\frac{k^{*n}}{\sqrt{n!}}\sum_{m=0}^{\infty}\psi_m(x')\frac{k^m}{\sqrt{m!}} = \\ &\frac{1}{\pi}\sum_{n=0}^{\infty}\sum_{m=0}^{\infty}\frac{\psi_n^*(x)\psi_m(x')}{\sqrt{n!m!}}\int d^2k e^{-kk^*}k^{*n}k^m. \end{aligned} \tag{14}$$

Doing the integral in polar coordinates, with $k = \rho e^{i\phi}$,

$$\begin{aligned}\int d^2k e^{-k^*k}k^{*n}k^m &= \int_0^{\infty} d\rho\rho e^{-\rho^2}\rho^{n+m}\int_0^{2\pi} d\phi e^{i(m-n)\phi} \\ &= \int_0^{\infty} d\rho\rho e^{-\rho^2}\rho^{n+m}2\pi\delta_{nm} = \pi n!\delta_{nm}. \end{aligned} \tag{15}$$

Combining eqs. (8) and (14) leads to the needed completeness relation given by eq. (13).

C A Second Useful Application: The matrix $(x)_{nm}$

As a second example, we will use the Bargmann kernel to calculate the following useful integral

$$\int_{-\infty}^{\infty} dx\psi_n^*(x)x\psi_m(x) = <\psi_n^*, x\psi_m> \equiv (x)_{nm}. \tag{16}$$

With $m = n$ this integral would be needed to calculate the expectation value of x in the nth eigenstate. Because in that case the integrand is an odd function of x, this expectation value is zero. The particle is equally likely to be in the right half or the left half of the x domain. For general n, m, the two-index quantity defined as $(x)_{nm}$ in eq. (16) will be shown to be a matrix in Chapter 6. For general n, m, we can evaluate the needed integral by considering the integral

$$\int_{-\infty}^{\infty} dx A(l^*, x)xA(k, x) \tag{17}$$

as a function of the arbitrary complex parameters k and l^* in two ways

$$\begin{aligned}&\int_{-\infty}^{\infty} dx A(l^*, x)xA(k, x) = \sum_{n,m}\frac{l^{*n}}{\sqrt{n!}}\frac{k^m}{\sqrt{m!}}\int_{-\infty}^{\infty} dx\psi_n^*(x)x\psi_m(x) \\ &= \frac{1}{\sqrt{\pi}}e^{l^*k}\int_{-\infty}^{\infty} dx x e^{-[x-\frac{1}{\sqrt{2}}(l^*+k)]^2} = \frac{1}{\sqrt{\pi}}e^{l^*k}\int_{-\infty}^{\infty} dx'\big(x' + \frac{1}{\sqrt{2}}(l^*+k)\big)e^{-x'^2}\end{aligned}$$

$$= \frac{1}{\sqrt{2}}(l^* + k)e^{l^*k} = \frac{1}{\sqrt{2}}\sum_{n=0} \frac{(l^{*n+1}k^n + l^{*n}k^{n+1})}{n!}$$
$$= \frac{1}{\sqrt{2}}\left(\sum_{n=1} \frac{l^{*n}k^{n-1}}{(n-1)!} + \sum_{n=0} \frac{l^{*n}k^{n+1}}{n!}\right). \tag{18}$$

The only surviving terms are those in which the powers of k, viz., m, differ from n by ± 1. Therefore

$$\int_{-\infty}^{\infty} dx\psi_n^*(x)x\psi_m(x) = 0 \qquad \text{for} \quad m \neq (n \pm 1)$$
$$= \sqrt{\frac{(n+1)}{2}} \quad \text{for} \quad m = (n+1)$$
$$= \sqrt{\frac{n}{2}} \quad \text{for} \quad m = (n-1). \tag{19}$$

Problems

9. For the 1-D harmonic oscillator, calculate all nonzero matrix elements of q^2, p, and p^2 (for the nonzero matrix elements of q, see eq. (19) above). For a general state

$$\Psi(q, t) = \sum_n c_n \psi_n(q) e^{-\frac{i}{\hbar}E_n t},$$

calculate $< q >, < q^2 >, < p >, < p^2 >, \Delta q$, and Δp as functions of the c_n. Try to determine values of c_n for which the product $(\Delta p)(\Delta q)$ is a minimum. (Hint: Try $c_0 = 1$, all other $c_n = 0$.) For the special case

$$c_0 = \frac{1}{\sqrt{2}}, \qquad c_1 = \frac{i}{\sqrt{2}},$$

calculate $< q >, < q^2 >$, and Δq as functions of the time, t. For this special case, also calculate $\vec{S}$; that is, calculate as a function of q and t the probability per unit time and unit area normal to the displacement, q, that the particle is streaming in the direction of q.

10. A particle of mass, m, in a 1-D harmonic oscillator potential has a probability density amplitude at $t = 0$, specified by the initial value

$$\Psi(q, t = 0) = \left[\frac{m\omega_0}{\hbar\pi}\right]^{\frac{1}{4}} e^{-\frac{1}{2}\frac{m\omega_0}{\hbar}(q-q_0)^2},$$

that is, by the $n = 0$ eigenfunction displaced by a distance q_0. Calculate $P(E_n)$, the probability the particle is in an energy eigenstate with energy E_n at $t = 0$ as a function of $x_0 = q_0/\sqrt{\hbar/m\omega_0}$ and n. [Check that $P(E_n) \to \delta_{n0}$, as $q_0 \to 0$.] Calculate $\Psi\Psi^*$ at a later time, t, and discuss the motion of the particle. Note: You may be able to sum an infinite series by using the generating function definition of the Hermite polynomials, $H_n(x)$, or the Bargmann kernel expansion

$$A(k, x) = \sum_{n=0}^{\infty} \psi_n(x) \frac{k^n}{\sqrt{n!}}.$$

11. Repeat problem 10 for the case when

$$\Psi(q, t = 0) = \left[\frac{m\omega_0}{\hbar\pi}\right]^{\frac{1}{4}} \sqrt{\frac{2m\omega_0}{\hbar}}(q - q_0)e^{-\frac{1}{2}\frac{m\omega_0}{\hbar}(q-q_0)^2},$$

that is, by the $n = 1$ eigenfunction displaced by a distance q_0, or for the case when

$$\Psi(q, t = 0) = \psi_n(q - q_0)$$

for arbitrary n.

Solution for Problem 11

a. The case $n = 1$: Let us write Ψ at $t = 0$ in terms of the dimensionless x and x_0

$$\Psi(x, 0) = \frac{2(x - x_0)}{\sqrt{2}\pi^{\frac{1}{4}}} e^{-\frac{1}{2}(x-x_0)^2} = \sqrt{2}(x - x_0)e^{-\frac{1}{4}x_0^2} \frac{e^{-\frac{1}{2}\left(\frac{x_0}{\sqrt{2}}\right)^2 + \sqrt{2}x\frac{x_0}{\sqrt{2}} - \frac{1}{2}x^2}}{\pi^{\frac{1}{4}}}$$
$$= \sqrt{2}(x - x_0)e^{-\frac{1}{4}x_0^2} A(\frac{x_0}{\sqrt{2}}, x) = \sqrt{2}(x - x_0)e^{-\frac{1}{4}x_0^2} \sum_{n=0}^{\infty} \frac{\psi_n(x)}{\sqrt{n!}} \left(\frac{x_0}{\sqrt{2}}\right)^n, \quad (1)$$

where we have used the expansion of the Bargmann kernel function, $A(k, x)$, with $k = x_0/\sqrt{2}$, in terms of the normalized $\psi_n(x)$. If we further use

$$x\psi_n(x) = \sqrt{\frac{(n+1)}{2}} \psi_{n+1}(x) + \sqrt{\frac{n}{2}} \psi_{n-1}(x), \quad (2)$$

the above yields

$$\Psi(x, 0) = e^{-\frac{1}{4}x_0^2} \Bigg(-\sqrt{2}x_0 \sum_{n=0}^{\infty} \frac{\psi_n(x)}{\sqrt{n!}} \left(\frac{x_0}{\sqrt{2}}\right)^n$$
$$+ \sum_{n=0}^{\infty} \sqrt{(n+1)} \frac{\psi_{n+1}(x)}{\sqrt{n!}} \left(\frac{x_0}{\sqrt{2}}\right)^n + \sum_{n=1}^{\infty} \sqrt{n} \frac{\psi_{n-1}(x)}{\sqrt{n!}} \left(\frac{x_0}{\sqrt{2}}\right)^n \Bigg)$$
$$= e^{-\frac{1}{4}x_0^2} \sum_{n=0}^{\infty} \Bigg(-\sqrt{2}x_0 \frac{\psi_n(x)}{\sqrt{n!}} \left(\frac{x_0}{\sqrt{2}}\right)^n$$
$$+ n \frac{\psi_n(x)}{\sqrt{n!}} \left(\frac{x_0}{\sqrt{2}}\right)^{n-1} + \frac{x_0}{\sqrt{2}} \frac{\psi_n(x)}{\sqrt{n!}} \left(\frac{x_0}{\sqrt{2}}\right)^n \Bigg), \quad (3)$$

where we have shifted indices $n \to (n - 1)$ and $n \to (n + 1)$ in the last two sums above. (The second sum is proportional to a factor n and thus begins at $n = 1$.) We have therefore expanded our $\Psi(x, 0)$ in terms of the $\psi_n(x)$

$$\Psi(x, 0) = \sum_n c_n \psi_n(x), \quad \text{with } c_n = \frac{e^{-\frac{1}{4}x_0^2}}{\sqrt{n!}} \left(\frac{x_0}{\sqrt{2}}\right)^{n-1} \left(n - \tfrac{1}{2}x_0^2\right). \quad (4)$$

At a later time, we have

$$\Psi(x,t) = \sum_n c_n \psi_n(x) e^{-i\omega_0(n+\frac{1}{2})t}$$
$$= e^{-\frac{i}{2}\omega_0 t} \sum_n e^{-\frac{1}{4}x_0^2} \frac{\sqrt{2}}{x_0}\left(n - \frac{x_0^2}{2}\right)\left(\frac{x_0 e^{-i\omega_0 t}}{\sqrt{2}}\right)^n \frac{\psi_n(x)}{\sqrt{n!}}. \quad (5)$$

We can now sum these infinite series, using the expansion of the Bargmann kernel function, through

$$\sum_{n=0}^{\infty} \frac{\psi_n(x)}{\sqrt{n!}} k^n = A(k,x) = \frac{1}{\pi^{\frac{1}{4}}} e^{-\frac{1}{2}k^2 + \sqrt{2}kx - \frac{1}{2}x^2}$$
$$\sum_{n=1}^{\infty} n \frac{\psi_n(x)}{\sqrt{n!}} k^{n-1} = \frac{dA(k,x)}{dk} = (-k + \sqrt{2}x)A(k,x), \quad (6)$$

now with $k = (x_0 e^{-i\omega_0 t}/\sqrt{2})$. This yields

$$\Psi(x,t) = e^{-\frac{i}{2}\omega_0 t}\left(\left[-\left(\frac{x_0}{\sqrt{2}} e^{-i\omega_0 t}\right) + \sqrt{2}x\right] e^{-i\omega_0 t} - \frac{x_0}{\sqrt{2}}\right)$$
$$\times e^{-\frac{1}{4}x_0^2}\left[\frac{e^{[-\frac{1}{4}x_0^2 e^{-2i\omega_0 t} + x x_0 e^{-i\omega_0 t} - \frac{1}{2}x^2]}}{\pi^{\frac{1}{4}}}\right]$$
$$= e^{-\frac{i}{2}\omega_0 t}\left(e^{-i\omega_0 t}\left[\frac{2(x - x_0 \cos\omega_0 t)}{\sqrt{2}}\right]\right)$$
$$\times \frac{e^{-\frac{1}{2}(x - x_0\cos\omega_0 t)^2} e^{-i x_0 \sin\omega_0 t (x - \frac{1}{2}x_0 \cos\omega_0 t)}}{\pi^{\frac{1}{4}}}$$
$$= e^{-i\frac{3}{2}\omega_0 t} \frac{2(x - x_0\cos\omega_0 t)}{\sqrt{2}\pi^{\frac{1}{4}}} e^{-\frac{1}{2}(x - x_0\cos\omega_0 t)^2} e^{-i x_0 \sin\omega_0 t (x - \frac{1}{2}x_0\cos\omega_0 t)}. \quad (7)$$

Therefore,

$$|\Psi(x,t)|^2 = 2\frac{(x - x_0\cos\omega_0 t)^2}{\sqrt{\pi}} e^{-(x - x_0\cos\omega_0 t)^2} = |\psi_1(x - x_0\cos\omega_0 t)|^2; \quad (8)$$

that is, the probability density is that of the $n = 1$ state, but it oscillates about the origin with the oscillator frequency, (ω_0), with amplitude x_0, and without change of shape.

Our derivation so far has made use of some simple properties of harmonic oscillator eigenfunctions, [see eq. (2)], and the expansion of the Bargmann kernel function in terms of the $\psi_n(x)$, or, what would be equivalent, the generating function definition of the Hermite polynomials.

b. The case of arbitrary n: To generalize our result to a function $\Psi(x,0) = \psi_n(x - x_0)$ with arbitrary n, it may prove more convenient to work with the Bargmann transform of $\psi_n(x - x_0)$:

$$F_n(k) = \int_{-\infty}^{\infty} dx\, \psi_n(x - x_0) A(k,x) = \int_{-\infty}^{\infty} dx'\, \psi_n(x') A(k, x' + x_0)$$

$$= e^{\sqrt{2}kx_0 - \frac{1}{2}x_0^2}\int_{-\infty}^{\infty} dx'\psi_n(x')e^{-\frac{1}{2}k^2-\frac{1}{2}x'^2+\sqrt{2}(k-\frac{x_0}{\sqrt{2}})x'}$$
$$= e^{-\frac{1}{4}x_0^2+\frac{1}{\sqrt{2}}kx_0}\int_{-\infty}^{\infty} dx'\psi_n(x')A\big((k-\frac{x_0}{\sqrt{2}}), x'\big)$$
$$= e^{-\frac{1}{4}x_0^2+\frac{1}{\sqrt{2}}kx_0}\int_{-\infty}^{\infty} dx'\psi_n(x')\sum_{N=0}^{\infty}\frac{\psi_N(x')}{\sqrt{N!}}\Big(k-\frac{x_0}{\sqrt{2}}\Big)^N$$
$$= e^{-\frac{1}{4}x_0^2+\frac{1}{\sqrt{2}}kx_0}\frac{\Big(k-\frac{x_0}{\sqrt{2}}\Big)^n}{\sqrt{n!}}, \tag{9}$$

where we have used the reality and the orthonormality of the harmonic oscillator eigenfunctions. To obtain the expansion coefficients, $c_m^{(n)}$,

$$c_m^{(n)} = \int_{-\infty}^{\infty} dx\psi_m^*(x)\psi_n(x-x_0) = \frac{1}{\pi}\int d^2k e^{-kk^*}F_n(k)\frac{k^{*m}}{\sqrt{m!}}, \tag{10}$$

it is sufficient to expand $F_n(k)$ in powers of k and use the k-space orthonormality relation

$$\frac{1}{\pi}\int d^2k e^{-kk^*}\frac{k^m}{\sqrt{m!}}\frac{k^{*l}}{\sqrt{l!}} = \delta_{ml}. \tag{11}$$

For this purpose, therefore, we expand

$$F_n(k) = e^{-\frac{1}{4}x_0^2+\frac{x_0}{\sqrt{2}}k}\frac{\Big(k-\frac{x_0}{\sqrt{2}}\Big)^n}{\sqrt{n!}}$$
$$= e^{-\frac{1}{4}x_0^2}\frac{1}{\sqrt{n!}}\sum_{b=0}^{\infty}\sum_{a=0}^{n}\frac{1}{b!}\Big(\frac{x_0}{\sqrt{2}}\Big)^b\frac{n!}{a!(n-a)!}\Big(-\frac{x_0}{\sqrt{2}}\Big)^{n-a}k^{a+b}$$
$$= \sum_{m=0}^{\infty}\Big[\frac{1}{\sqrt{n!}}e^{-\frac{1}{4}x_0^2}\sum_{a=0}^{n}\frac{n!}{a!(n-a)!}\frac{(-1)^{n-a}}{(m-a)!}\Big(\frac{x_0}{\sqrt{2}}\Big)^{n+m-2a}\Big]k^m, \tag{12}$$

so

$$c_m^{(n)} = \frac{1}{\sqrt{n!}}\frac{1}{\sqrt{m!}}e^{-\frac{1}{4}x_0^2}\sum_{a=0}^{n}(-1)^{n-a}\frac{n!}{a!(n-a)!}\frac{m!}{(m-a)!}\Big(\frac{x_0}{\sqrt{2}}\Big)^{n+m-2a}. \tag{13}$$

For the time-dependent function, we then get

$$\Psi(x,t) = \sum_m c_m^{(n)}\psi_m(x)e^{-i\omega_0(m+\frac{1}{2})t}, \tag{14}$$

and we could perform the m sums via

$$\sum_m\frac{m!}{(m-a)!}\psi_m(x)\frac{k^m}{\sqrt{m!}} = \frac{d^a}{dk^a}A(k,x), \qquad \text{now with } k = \frac{x_0e^{-i\omega_0 t}}{\sqrt{2}}. \tag{15}$$

For small values of n, where the a sum contributes only $n+1$ terms, this method works well, as you could again verify for the special case, $n = 1$. For arbitrary values of n, particularly for large values of n, we could use the summed form of the Bargmann transform, $F_n(k)$. From the above expansion in powers of k,

$$\text{If}\ \ \Psi(x,0)\ \ \text{has Bargmann transform}\qquad F_n(k),$$

$$\text{then}\ \ \Psi(x,t)\ \ \text{has Bargmann transform}\quad F_n(k,t) = e^{-\frac{1}{2}\omega_0 t} F_n(ke^{-i\omega_0 t}).$$

Therefore, for us,

$$\begin{aligned} F_n(k,t) &= e^{-\frac{1}{2}\omega_0 t} e^{-\frac{1}{4}x_0^2} e^{[\frac{x_0}{\sqrt{2}} ke^{-i\omega_0 t}]} \frac{\left(ke^{-i\omega_0 t} - \frac{x_0}{\sqrt{2}}\right)^n}{\sqrt{n!}} \\ &= e^{-i(n+\frac{1}{2})\omega_0 t} e^{-\frac{1}{4}x_0^2} e^{[\frac{x_0}{\sqrt{2}} ke^{-i\omega_0 t}]} \frac{\left(k - \frac{x_0 e^{+i\omega_0 t}}{\sqrt{2}}\right)^n}{\sqrt{n!}}. \end{aligned} \tag{16}$$

The function, $\Psi(x,t)$, then follows at once from the inverse Bargmann transform

$$\Psi(x,t) = \frac{1}{\pi}\int d^2k e^{-kk^*} A(k^*,x) F_n(k,t). \tag{17}$$

To do this integral, it will now be convenient to make the substitution

$$k' = \left(k - \frac{x_0 e^{i\omega_0 t}}{\sqrt{2}}\right),$$

so

$$\begin{aligned} \Psi(x,t) &= \frac{e^{-i(n+\frac{1}{2})\omega_0 t}}{\pi} \int d^2k' e^{-(k' + \frac{x_0 e^{i\omega_0 t}}{\sqrt{2}})(k'^* + \frac{x_0 e^{-i\omega_0 t}}{\sqrt{2}})} \\ &\times \frac{e^{-\frac{1}{2}(k'^* + \frac{x_0 e^{-i\omega_0 t}}{\sqrt{2}})^2} e^{\sqrt{2}k'^* x + x x_0 e^{-i\omega_0 t}} e^{-\frac{1}{2}x^2}}{\pi^{\frac{1}{4}}} e^{[\frac{x_0}{\sqrt{2}} k' e^{-i\omega_0 t}]} e^{+\frac{x_0^2}{4}} \left[\frac{k'^n}{\sqrt{n!}}\right] \\ &= \frac{e^{-i(n+\frac{1}{2})\omega_0 t}}{\pi} \int d^2k' e^{-k'^* k'} \frac{e^{[-\frac{1}{2}k'^{*2} + \sqrt{2}k'^*(x - x_0\cos\omega_0 t) - \frac{1}{2}(x - x_0 \cos\omega_0 t)^2]}}{\pi^{\frac{1}{4}}} \\ &\times \left[\frac{k'^n}{\sqrt{n!}}\right] e^{-ix_0 \sin\omega_0 t(x - \frac{x_0}{2}\cos\omega_0 t)} \\ &= e^{-i(n+\frac{1}{2})\omega_0 t} \left[\frac{1}{\pi}\int d^2k' e^{-k'^* k'} A\big(k'^*, (x - x_0\cos\omega_0 t)\big)\left[\frac{k'^n}{\sqrt{n!}}\right]\right] \\ &\times e^{-ix_0 \sin\omega_0 t(x - \frac{x_0}{2}\cos\omega_0 t)} \\ &= \psi_n(x - x_0\cos\omega_0 t) e^{-i(n+\frac{1}{2})\omega_0 t} e^{-ix_0\sin\omega_0 t(x - \frac{x_0}{2}\cos\omega_0 t))}. \end{aligned} \tag{18}$$

For arbitrary n,

$$|\Psi(x,t)|^2 = |\psi_n(x - x_0\cos\omega_0 t)|^2, \qquad \text{if}\quad \Psi(x,0) = \psi_n(x - x_0). \tag{19}$$

For arbitrary n, therefore, the probability density oscillates without change of shape about the origin with the oscillator frequency, (ω_0), and with amplitude x_0, if the initial state is the nth oscillator state displaced in the x-direction through a distance x_0. This extremely simple property is unique for the harmonic oscillator and does not follow for the energy eigenstates of more complicated Hamiltonians. Also, the use of the Bargmann transform greatly facilitated the proof for general, n.

12. The 2-D isotropic harmonic oscillator with Hamiltonian

$$H + \frac{1}{2m}(p_x^2 + p_y^2) + \frac{m\omega_0^2}{2}(x^2 + y^2)$$

has eigenfunctions

$$\psi_{n_1 n_2}(x, y) = \psi_{n_1}(x)\psi_{n_2}(y),$$

with eigenvalues, $E_{n_1 n_2} = \hbar\omega_0(n_1 + n_2 + 1)$. Show that H is invariant to rotations

$$x' = x\cos\theta + y\sin\theta,$$

$$y' = -x\sin\theta + y\cos\theta,$$

where θ is a constant. Show by means of the Bargmann kernel that an eigenfunction in which only the x' degree of freedom is excited can be expanded in terms of the above $\psi_{n_1 n_2}$; i.e., find the expansion coefficients, $c^{(N)}_{n_1 n_2}$:

$$\psi_N(x')\psi_0(y') = \sum_{n_1 n_2} c^{(N)}_{n_1 n_2}\psi_{n_1}(x)\psi_{n_2}(y).$$

13. For the conservation laws for the hydrogen atom, the three components of the Runge–Lenz vector are

$$\vec{\mathcal{R}} = \frac{1}{2\mu}\left([\vec{p} \times \vec{L}] - \vec{L} \times \vec{p}]\right) - \frac{Ze^2}{r}\vec{r}.$$

Show that they are hermitian when written in the above form. Also, show that they commute with the hydrogen atom Hamiltonian

$$H = \frac{(\vec{p} \cdot \vec{p})}{2\mu} - \frac{Ze^2}{r}.$$

In the above, $\vec{\mathcal{R}}$ and H are expressed in terms of the physical quantities, $\vec{r}_{\text{phys.}}$, $\vec{p}_{\text{phys.}}$, $\vec{L}_{\text{phys.}}$, and $H_{\text{phys.}}$. If these are expressed in terms of dimensionless quantities, $\vec{r}$, $\vec{p}$, $\vec{L}$, H, through

$$\vec{r}_{\text{phys.}} = a_0\vec{r}, \qquad \vec{p}_{\text{phys.}} = \frac{\hbar}{a_0}\vec{p}, \qquad \vec{L}_{\text{phys.}} = \hbar\vec{L},$$

$$H_{\text{phys.}} = \frac{\mu Z^2 e^4}{\hbar^2}H, \qquad \text{with } a_0 = \frac{\hbar^2}{\mu Z e^2},$$

the Runge vector in physical units, as given above, can be expressed in terms of a dimensionless $\vec{\mathcal{R}}$ by

$$\vec{\mathcal{R}}_{\text{phys.}} = Ze^2\vec{\mathcal{R}}.$$

Show that this dimensionless $\vec{\mathcal{R}}$ can also be expressed as

$$\vec{\mathcal{R}} = [\vec{p} \times \vec{L}] - i\vec{p} - \frac{\vec{r}}{r} = \vec{r}(\vec{p} \cdot \vec{p}) - \vec{p}(\vec{r} \cdot \vec{p}) - \frac{\vec{r}}{r}.$$

Also,

$$(\vec{\mathcal{R}} \cdot \vec{\mathcal{R}}) = \left(\vec{p} \cdot \vec{p} - \frac{2}{r}\right)(\vec{L} \cdot \vec{L} + 1) + 1 = 2H(\vec{L} \cdot \vec{L} + 1) + 1,$$

and

$$\vec{\mathcal{R}} \cdot \vec{L} = \vec{L} \cdot \vec{\mathcal{R}} = 0.$$

6

Further Interpretation of the Wave Function

Consider a quantum-mechanical system with a Hamiltonian that has a discrete spectrum only, with allowed energies, E_n, and eigenfunctions, ψ_n. In general, the state of this quantum system can be specified by a wave function

$$\Psi(\vec{r}, t) = \sum_n c_n \psi_n(\vec{r}), e^{-\frac{i}{\hbar}E_n t} \tag{1}$$

describing a system for which the energy is not uniquely specified. If it is a single particle,

$$< \Psi, \Psi >= 1 = \sum_n c_n^* c_n < \psi_n, \psi_n > + \sum_{n \neq m} c_n^* c_m < \psi_n, \psi_m > e^{\frac{i}{\hbar}(E_n - E_m)t}. \tag{2}$$

Because $< \psi_n, \psi_m >= 0$ for $n \neq m$,

$$< \Psi, \Psi >= \sum_n |c_n|^2. \tag{3}$$

Similarly,

$$< E >=< \Psi, H\Psi >= \sum_n |c_n|^2 E_n, \tag{4}$$

$$< E^k >=< \Psi, H^k \Psi >= \sum_n |c_n|^2 E_n^k. \tag{5}$$

It is natural to interpret $|c_n|^2$ as $P(E_n)$, the probability the particle be found in the state with energy E_n. Without a coupling of our system to an outside field, that is, without an outside perturbation, these $P(E_n)$ are independent of the time.

A Application 1: Tunneling through a Barrier

As a simplest application, consider the NH_3 molecule system, where the motion of the N-atom relative to the H_3 plane is governed by the double-minimum potential of problem 7. The lowest-energy eigenfunction was an even function of x; the first excited state, at an excitation energy, ΔE, above the ground state (see Fig. 6.1), has an eigenfunction that is an odd function of x, but otherwise almost identical with the lowest-energy eigenfunction (see Fig. 6.2). If we actually make a measurement of the position of the N-atom when the x-vibrational motion is not excited, we will find the N-atom either above the H_3 plane, $x > 0$, or below it, $x < 0$. Suppose at $t = 0$ we make a measurement telling us the N-atom is above the H_3 plane. Then,

$$\Psi(x, t = 0) = \frac{1}{\sqrt{2}}\big(\psi_{0,\text{even}} + \psi_{0,\text{odd}}\big) = \psi_{\text{Right}}. \tag{6}$$

Note $\sum_n |c_n|^2 = 1$. At any later time,

$$\begin{aligned}\Psi(x, t) &= \frac{1}{\sqrt{2}}\big(\psi_{0,\text{even}} e^{-\frac{i}{\hbar}E_{0,\text{even}}t} + \psi_{0,\text{odd}} e^{-\frac{i}{\hbar}E_{0,\text{odd}}t}\big)\\ &= \frac{1}{\sqrt{2}} e^{-\frac{i}{\hbar}E_{0,\text{even}}t}\big(\psi_{0,\text{even}} + \psi_{0,\text{odd}} e^{-\frac{i}{\hbar}\Delta E t}\big).\end{aligned} \tag{7}$$

In particular, when

$$t = \frac{\hbar\pi}{\Delta E}, \tag{8}$$

$$\Psi(x, t) = \frac{1}{\sqrt{2}} e^{-\frac{i}{\hbar}E_{0,\text{even}}t}(\psi_{0,\text{even}} - \psi_{0,\text{odd}}), \qquad |\Psi(x, t)| = |\psi_{\text{Left}}|. \tag{9}$$

In this time, therefore, the N-atom has tunneled from the right potential minimum through the barrier to the left potential minimum. In twice this time, we will again find the N-atom in the right minimum. The N-atom tunnels back and forth through the potential barrier with a frequency given by

$$\nu_{\text{tunneling}} = \frac{\Delta E}{2\pi\hbar}. \tag{10}$$

From the solution of problem 7, ΔE is proportional to the Gamow factor, e^{-G}, with $G = 2a\sqrt{[2\mu(V_0 - E)/\hbar^2]}$ for the square well barrier of width $2a$. For a more general $V(x)$, this would be replaced by

$$G = \int_{-a}^{+a} dx \sqrt{[2\mu(V(x) - E)/\hbar^2]},$$

as will be shown in Chapter 37, where the Gamow factor, e^{-G}, is the most crucial part for the probability of tunneling through the barrier.

We end with a parenthetic remark: In the above discussion, we have used another result of problem 7. The lowest-state eigenfunction for our symmetric potential, with $V(-x) = V(x)$, is an even function of x. This result seems to be universally

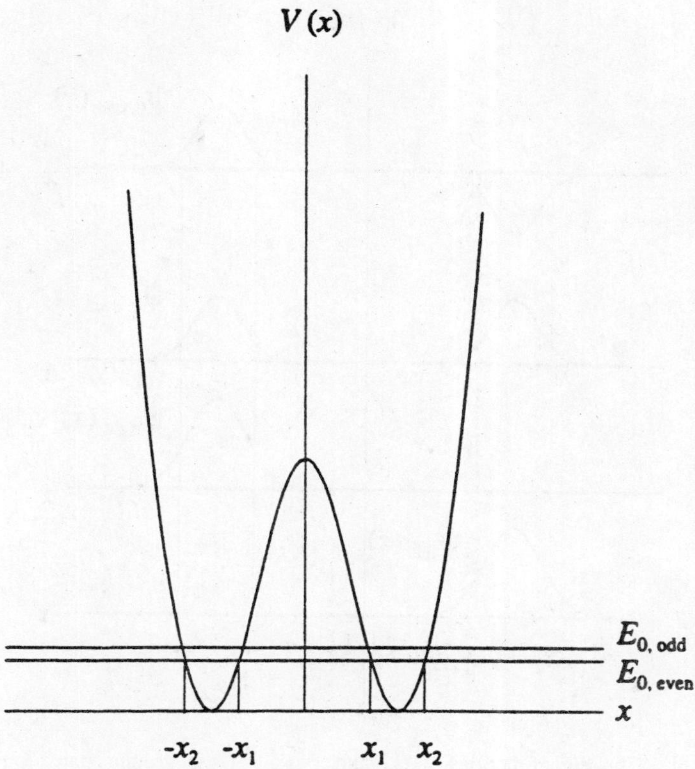

FIGURE 6.1. The ground-state doublet of the NH_3 double minimum potential.

true for symmetric potentials, which can be understood in terms of the curvature of the eigenfunctions. For two similar eigenfunctions of opposite parity, the function of odd parity must have a node in the center at $x = 0$ and must therefore have a somewhat greater curvature to "fit" into the potential, leading to a greater positive value of the expectation value of the kinetic energy. It is, however, not completely clear this property could be negated by the expectation value of the potential energy for a "pathological" potential perhaps having a large contribution to $< V >$ from the region near $x = 0$. In fact, the double minimum potential of this section, in which the central potential barrier is sufficiently infinite, is such a "pathological" case for which the ground state wave function is a degenerate doublet of an even and odd function. The ground-state wave function is no longer a pure even function of x.

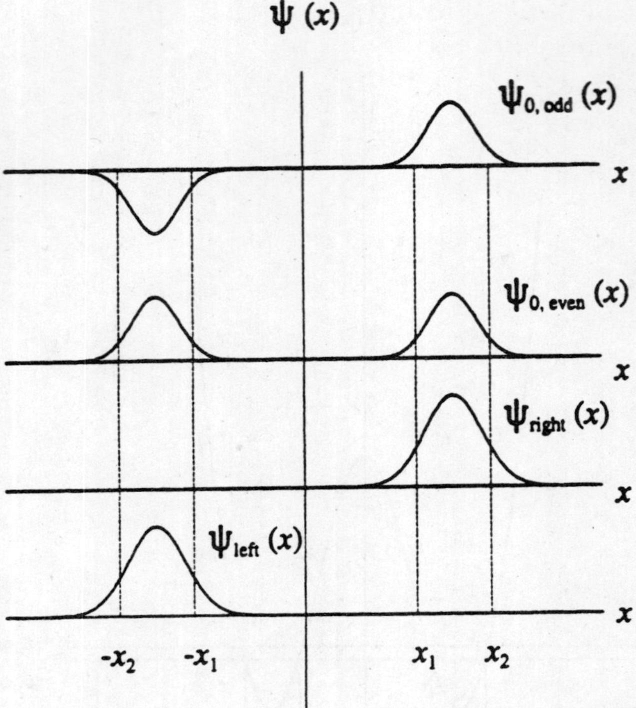

FIGURE 6.2. The eigenfunctions $\psi_{0,\text{odd}}$ and $\psi_{0,\text{even}}$ of the ground state doublet and $\psi_{\text{right/left}} = \sqrt{\frac{1}{2}}(\psi_{0,\text{even}} \pm \psi_{0,\text{odd}})$.

B Application 2: Time-dependence of a general oscillator $< q >$

The probability amplitudes c_n can tell us the quantum mechanical expectation value of any operator, O, through

$$< O >= \sum_{n,m} c_n^* c_m < \psi_n, O\psi_m > e^{\frac{i}{\hbar}(E_n - E_m)t}. \tag{11}$$

As a particular example, let $O = q$, the physical displacement coordinate of the 1-D harmonic oscillator. The two-index quantities $(q)_{nm} =< \psi_n, q\psi_m >$ for the harmonic oscillator were nonzero only for $m = n \pm 1$ (see Chapter 5). Thus,

$$\begin{aligned} < q > &= \sqrt{\frac{\hbar}{m\omega_0}}\langle x \rangle = \sqrt{\frac{\hbar}{2m\omega_0}} \sum_n \left[c_n^* c_{n-1}\sqrt{n} e^{i\omega_0 t} + c_n^* c_{n+1}\sqrt{n+1} e^{-i\omega_0 t}\right] \\ &= \sum_n \sqrt{\frac{\hbar n}{2m\omega_0}} \left[c_n^* c_{n-1} e^{i\omega_0 t} + c_{n-1}^* c_n e^{-i\omega_0 t}\right] \end{aligned}$$

$$= \sum_n \sqrt{\frac{\hbar n}{2m\omega_0}} \left[2\mathcal{R}eal(c_n^* c_{n-1}) \cos \omega_0 t - 2\mathcal{I}m(c_n^* c_{n-1}) \sin \omega_0 t\right]$$

$$= \sum_n \sqrt{\frac{\hbar n}{2m\omega_0}} A_n \cos(\omega_0 t + \phi_n), \tag{12}$$

where we have defined $2c_n^* c_{n-1} \equiv A_n e^{i\phi_n}$. If we use $< q >$ to describe the quantum-mechanical motion of the simple 1-D harmonic oscillator, the result is very similar to the classical motion.

C Matrix Representations

For the expectation value, $< O >$, of eq. (11), it is tempting to interpret the two-index quantity, $O_{nm} \equiv < \psi_n, O\psi_m >$, as the nm^{th} matrix element of an infinite-dimensional matrix. (The set of numbers, O_{nm}, contain all experimentally observable information about the dynamical quantity represented by the operator, O.) To prove O_{nm} is a matrix, all laws of matrix algebra must be satisfied:

1) multiplication by a scalar (a complex number or "c-number"), λO:

$$< \psi_n, \lambda O \psi_m > = \lambda O_{nm}. \tag{13}$$

2) addition of two matrices, $O_1 + O_2$:

$$< \psi_n, (O_1 + O_2)\psi_m > = (O_1)_{nm} + (O_2)_{nm}. \tag{14}$$

3) matrix multiplication, $O_2 O_1$:

$$< \psi_n, O_2 O_1 \psi_m > = ? \tag{15}$$

To prove the law of matrix-multiplication, the new function obtained by acting with O_1 on ψ_m can be expanded in terms of the ψ_k in a generalized Fourier series

$$O_1 \psi_m = \sum_k c_k \psi_k, \qquad \text{with} \quad c_k = < \psi_k, O_1 \psi_m > = (O_1)_{km}, \tag{16}$$

so

$$< \psi_n, O_2 O_1 \psi_m > = \sum_k < \psi_n, O_2 \psi_k >< \psi_k, O_1 \psi_m >, \tag{17}$$

or

$$\left(O_2 O_1\right)_{nm} = \sum_k \left(O_2\right)_{nk} \left(O_1\right)_{km}. \tag{18}$$

This relation is the familiar law of matrix multiplication. The matrix elements,

$$O_{nm} = < \psi_n, O\psi_m > = \int dx \psi_n^*(x) O \psi_m(x),$$

were introduced here with the concept of the Schrödinger wave equation and the use of the energy eigenfunctions of this wave equation. Heisenberg first introduced such matrix elements of dynamical quantities entirely without the concept of a wave equation or a wave function.

D Heisenberg Matrix Mechanics

We are therefore now at a stage where we can make a short historical remark about the Heisenberg derivation of the laws of quantum mechanics. Heisenberg did not think in terms of a wave equation or in terms of a wave function. He started by thinking about the laws of classical dynamics for a periodic (or more generally a multiple-periodic or quasiperiodic system) in terms of a Fourier analysis of the classical generalized coordinates, q. For a simple periodic system,

$$q(t) = \sum_n q_n e^{in\omega t}, \quad \text{with} \quad \omega = 2\pi\nu, \tag{19}$$

where q_n is the Fourier amplitude for the nth overtone of the classical fundamental ω. (For a multiple-periodic system, $n\omega$ would be replaced by $n_1\omega_1 + n_2\omega_2 + \cdots + n_f\omega_f$, and the sum would be over f overtone indices, and the Fourier coefficients would depend on f integers, $q_{\vec{n}} = q_{n_1 n_2 \cdots n_f}$.)

Now, Heisenberg reasoned: Because the nth overtone has to be replaced by a two-index quantity, via the Bohr frequency relation,

$$n\omega \quad \rightarrow \quad \frac{E_n - E_m}{\hbar} = \omega_{nm} \qquad \text{Bohr}, \tag{20}$$

the Fourier coefficient q_n should also be replaced by a two-index quantity

$$q_n \quad \rightarrow \quad q_{nm} \qquad \text{Heisenberg} \quad \text{matrix}. \tag{21}$$

Moreover, these q_{nm} are the only observable (physically meaningful) quantities. In addition, because matrices do not commute, the quantum mechanically meaningful p and q matrices do not commute. In particular, Heisenberg introduced the Planck constant into his matrix algebra with the simple *assumption*

$$\sum_k \left(p_{nk}q_{km} - q_{nk}p_{km}\right) = \frac{\hbar}{i}\delta_{nm}. \tag{22}$$

In the limit, $\hbar \rightarrow 0$, p and q do commute as they should in the classical limit, when $\hbar$ becomes too small to matter. The genius of the Heisenberg approach is contained in this *Heisenberg relation*, which we have already met in Section 3G in the framework of the Schrödinger approach.

Using the p, q matrix commutation relation, the relation between $H(p, q)$ and E, and the commutators $[p, H(p, q)]$ and $[q, H(p, q)]$, which follow from eq. (22), Heisenberg found the allowed energy values and the matrix elements of p and q via matrix algebra for the 1-D harmonic oscillator and other simple dynamical systems, *without* the use of a wave equation. The equivalence between the Heisenberg q_{nm} and the Schrödinger $< \psi_n, q\psi_m >$ was demonstrated by Schrödinger in 1926. (Wolfgang Pauli in an unpublished letter to P. Jordan is reputed to have shown this equivalence even earlier). In these lectures, we shall give the Heisenberg derivation for the energy eigenvalues, E_n, and the matrix elements q_{nm} and p_{nm} of the simple harmonic oscillator in Chapter 19 after we have gained some facility in the calculation of matrix elements of dynamical quantities by both Schrödinger and algebraic (Heisenberg) techniques.

7
The Eigenvalue Problem

One of the basic problems needing to be solved in quantum theory is the general eigenvalue problem, for some hermitian operator, say, A, with $A^\dagger = A$,

$$A\psi_a(x) = \lambda_a \psi_a(x). \tag{1}$$

We shall learn how to solve such problems by purely algebraic techniques,without introducing wave functions and differential equations. For the moment, however, let us go back to the coordinate representation, and, in particular, let us choose $A = H$, where H is the Hamiltonian for a single particle in three dimensions, or for the two-particle problem after transformation to center of mass and relative coordinates. Keeping the center of mass fixed, the eigenvalue problem for the relative motion of the two-particle system is given by the Schrödinger equation

$$-\frac{\hbar^2}{2\mu}\nabla^2\psi + V(x, y, z)\psi = E\psi. \tag{2}$$

If the potential is a function of the scalar distance r only, spherical coordinates will be natural and

$$-\left(\frac{\partial^2}{\partial r^2} + \frac{2}{r}\frac{\partial}{\partial r} + \frac{1}{r^2}\left[\frac{1}{\sin\theta}\frac{\partial}{\partial\theta}\sin\theta\frac{\partial}{\partial\theta} + \frac{1}{\sin^2\theta}\frac{\partial^2}{\partial\phi^2}\right]\right)\psi + \frac{2\mu}{\hbar^2}(V(r) - E)\psi = 0. \tag{3}$$

Now, let

$$\psi(r, \theta, \phi) = R(r)Y(\theta, \phi) = R(r)\Theta(\theta)\Phi(\phi). \tag{4}$$

Substituting into the equation, and subsequently dividing by $\psi = R\Theta\Phi$, and then multiplying from the left with r^2, leads to a separation of the wave equation

$$\frac{r^2}{R}\left[\frac{d^2R}{dr^2}+\frac{2}{r}\frac{dR}{dr}\right]+\frac{2\mu r^2}{\hbar^2}(E-V(r)) = -\frac{1}{\Theta}\left[\frac{1}{\sin\theta}\frac{\partial}{\partial\theta}\sin\theta\frac{\partial\Theta}{\partial\theta}\right]-\frac{1}{\Phi\sin^2\theta}\frac{\partial^2\Phi}{\partial\phi^2}. \tag{5}$$

Now, we have a function of r only, on the left-hand side of the equation, equaling a function of θ and ϕ only on the right. Hence, each function must be equal to the same constant, to be named, λ_0. By multiplying the right-hand side by $\sin^2\theta$, we can further separate the θ and ϕ-dependent pieces. Letting the new separation constant be named m^2, we get the three separated equations

$$-\frac{\hbar^2}{2\mu}\left[\frac{d^2R}{dr^2}+\frac{2}{r}\frac{dR}{dr}\right]+\left[\frac{\hbar^2}{2\mu r^2}\lambda_0+V(r)\right]R(r) = ER(r), \tag{6}$$

$$-\frac{d^2\Theta}{d\theta^2}-\cot\theta\frac{d\Theta}{d\theta}+\frac{m^2}{\sin^2\theta}\Theta(\theta) = \lambda_0\Theta(\theta), \tag{7}$$

$$-\frac{d^2\Phi}{d\phi^2} = m^2\Phi(\phi). \tag{8}$$

The solution to the last equation is trivial

$$\Phi(\phi) = e^{\pm im\phi}. \tag{9}$$

We shall prove later the separation constant, m, must be an integer. We shall defer the proof to later, but we note that it does *not* follow from the requirement that the wave function be single valued. It is $\psi\psi^*$ and the probability density current, $\vec{S}$, that must be single valued, i.e., have the same value at ϕ and $(\phi+2\pi)$. The r and θ equations can be simplified by eliminating the first derivative term to make them have the form of a 1-D Schrödinger equation. Because the volume element in spherical coordinates has the weighting factor $r^2\sin\theta$, and we require the normalization

$$\int_0^\infty dr r^2|R(r)|^2\int_0^\pi d\theta\sin\theta|\Theta|^2\int_0^{2\pi} d\phi|\Phi|^2 = 1, \tag{10}$$

(we will find it convenient to make each integral separately equal to unity), it will be useful to "one-dimensionalize" by transforming to new 1-D functions, u,

$$rR(r) = u(r), \qquad \sqrt{\sin\theta}\,\Theta(\theta) = u(\theta), \qquad \Phi(\phi) = u(\phi). \tag{11}$$

The 1-D equations are then

$$\left(-\frac{d^2}{dr^2}+\frac{2\mu}{\hbar^2}\left[V(r)+\frac{\hbar^2\lambda_0}{2\mu r^2}\right]\right)u(r) = \frac{2\mu}{\hbar^2}Eu(r) = \lambda u(r), \tag{12}$$

$$\left(-\frac{d^2}{d\theta^2}+\frac{(m^2-\frac{1}{4})}{\sin^2\theta}\right)u(\theta) = (\lambda_0+\frac{1}{4})u(\theta) = \lambda u(\theta), \tag{13}$$

$$-\frac{d^2u}{d\phi^2} = m^2u(\phi) = \lambda u(\phi). \tag{14}$$

The generic eigenvalue problem we want to solve has the form

$$(-\frac{d^2}{dx^2} + r(x,m))u_{\lambda m}(x) = \lambda u_{\lambda m}(x). \tag{15}$$

The effective potential term often contains a parameter, named m in the generic equation, such as the parameter, m, in the θ equation, or the parameter λ_0 in the r equation.

One of the methods used by Schrödinger to solve this type of problem is the so-called factorization method, which naturally leads to a constructive process via ladder operators. The introduction of such ladder operators will ease the transition to the algebraic techniques, which we will use later to solve such eigenvalue problems, beginning with Chapter 14, where we reexamine many of these problems in a new light.

A The Factorization Method: Ladder Operators

[A good reference for this method is: L. Infeld and T. E. Hull, Reviews of Modern Physics **23** (1951) 21. The table of factorizations at the end of the article gives a listing of 31 wave equations for which solutions are known in analytic form.]

In the factorization method, an attempt is made to solve the eigenvalue problem of eq. (15) by factoring the Schrödinger operator containing a second derivative operator into a product of two factors, each containing only a first derivative operator. Defining

$$\begin{aligned} O_+(m) &= -\frac{d}{dx} + k(x,m), \\ O_-(m) &= +\frac{d}{dx} + k(x,m), \end{aligned} \tag{16}$$

which through the basic second-order equation, eq. (15), satisfy the two equations

$$\begin{aligned} \text{I}: &\quad O_+(m)O_-(m)u_{\lambda m}(x) = [\lambda - \mathcal{L}(m)]u_{\lambda m}(x), \\ \text{II}: &\quad O_-(m+1)O_+(m+1)u_{\lambda m}(x) = [\lambda - \mathcal{L}(m+1)]u_{\lambda m}(x). \end{aligned} \tag{17}$$

For the specific case of the θ equation, our eq. (13), the function

$$k(\theta, m) = (m - \frac{1}{2})\cot\theta \tag{18}$$

will do the trick. Our equation (I) becomes

$$\begin{aligned} &\left(-\frac{d}{d\theta} + (m - \frac{1}{2})\cot\theta\right)\left(+\frac{d}{d\theta} + (m - \frac{1}{2})\cot\theta\right)u_{\lambda m}(\theta) \\ &= \left(-\frac{d^2}{d\theta^2} + \frac{(m^2 - \frac{1}{4})}{\sin^2\theta} - (m - \frac{1}{2})^2\right)u_{\lambda m}(\theta) \end{aligned}$$

$$= [\lambda - (m - \frac{1}{2})^2]u_{\lambda m}(\theta). \tag{19}$$

Equation (II) becomes

$$\left(+\frac{d}{d\theta} + (m + \frac{1}{2})\cot\theta\right)\left(-\frac{d}{d\theta} + (m + \frac{1}{2})\cot\theta\right)u_{\lambda m}(\theta)$$
$$= \left(-\frac{d^2}{d\theta^2} + \frac{(m^2 - \frac{1}{4})}{\sin^2\theta} - (m + \frac{1}{2})^2\right)u_{\lambda m}(\theta)$$
$$= [\lambda - (m + \frac{1}{2})^2]u_{\lambda m}(\theta). \tag{20}$$

The proposed factorization works for the θ equation and leads in this case to

$$\mathcal{L}(m) = (m - \tfrac{1}{2})^2. \tag{21}$$

We will postpone the question, treated in detail by Infeld and Hull, for which "potentials" does the factorization work? Let us first prove a number of theorems.

Theorem I:

If $u_{\lambda m}(x)$ is an eigenfunction of the generic equation with parameter, m, and eigenvalue λ, then $\left[O_-(m)u_{\lambda m}(x)\right]$ is an eigenfunction of the equation with parameter, $m-1$, and the same eigenvalue λ, and $\left[O_+(m+1)u_{\lambda m}(x)\right]$ is an eigenfunction of the equation with parameter, $m+1$, and the same eigenvalue λ.

That is,

$$O_-(m)u_{\lambda m}(x) = \text{const.}u_{\lambda(m-1)}(x),$$
$$O_+(m+1)u_{\lambda m}(x) = \text{const.}u_{\lambda(m+1)}(x). \tag{22}$$

To see the first, act on equation (I) from the left with $O_-(m)$ to give

$$O_-(m)O_+(m)\left[O_-(m)u_{\lambda m}\right] = [\lambda - \mathcal{L}(m)]\left[O_-(m)u_{\lambda m}\right]; \tag{23}$$

that is, $O_-(m)u_{\lambda m}$ is a solution of equation (II), with m replaced by $(m - 1)$. Similarly, acting on equation (II) from the left with $O_+(m + 1)$ gives

$$O_+(m+1)O_-(m+1)\left[O_+(m+1)u_{\lambda m}\right] = [\lambda - \mathcal{L}(m+1)]\left[O_+(m+1)u_{\lambda m}\right]; \tag{24}$$

that is, $O_+(m + 1)u_{\lambda m}$ is a solution of equation (I), now with m replaced by $m + 1$. Thus, $O_-(m)$ and $O_+(m + 1)$ are m step-down, or step-up, operators that can ladder from a known solution to other solutions. Still to be answered: Are the new functions square-integrable if the original $u_{\lambda m}$ were square-integrable? Do the m-ladders continue indefinitely to smaller or larger values? These questions still need to be answered. To see these, we need additional theorems.

Theorem II:

$$O_-(m) = O_+(m)^\dagger, \qquad O_+(m) = O_-(m)^\dagger. \tag{25}$$

These relations follow from the adjoint properties of the two parts of the operators

$$[-\frac{d}{dx}] = [+\frac{d}{dx}]^\dagger; \qquad k(x, m) = k(x, m)^\dagger. \tag{26}$$

We can use this theorem to investigate the square-integrability of $u_{\lambda(m\pm 1)}$. Assuming $u_{\lambda m}$ is square-integrable, over an interval from a to b, consider

$$\begin{aligned}
&\int_a^b dx u^*_{\lambda m+1}(x) u_{\lambda m+1}(x) \\
&= |\text{const.}|^2 \int_a^b dx \big[O_+(m+1)u_{\lambda m}(x)\big]^* \big[O_+(m+1)u_{\lambda m}(x)\big] \\
&= |\text{const.}|^2 \int_a^b dx u^*_{\lambda m}(x) O_-(m+1) O_+(m+1) u_{\lambda m}(x) \\
&= |\text{const.}|^2 [\lambda - \mathcal{L}(m+1)] \int_a^b dx u^*_{\lambda m}(x) u_{\lambda m}(x).
\end{aligned} \tag{27}$$

If the number $[\lambda - \mathcal{L}(m+1)]$ is a positive number, the final result is a patently positive quantity, and $u_{\lambda m+1}$ is square-integrable and can be normalized to one by an appropriate choice of the constant. If $\mathcal{L}(m)$ is an increasing function of m (see Fig. 7.1), however, an m-value will come such that $\mathcal{L}(m+1)$ will be greater than λ. Eq. (27) then would say that a patently positive quantity on the left-hand side of the equation would have to be a patently negative quantity on the right-hand side. This cannot be. Hence, the assumption that the solution $u_{\lambda m}$ was square-integrable must have been wrong. The only way out of the soup comes if the m step-up process quits; i.e., if a maximum possible value of m exists, $m_{\max}$, such that

$$O_+(m_{\max}+1) u_{\lambda m_{\max}}(x) = 0, \tag{28}$$

which would require

$$\lambda = \mathcal{L}(m_{\max}+1). \tag{29}$$

Eq. (28) is a first-order equation, which can in principle always be integrated

$$-\frac{du_{\lambda m_{\max}}}{dx} + k(x, m_{\max}+1) u_{\lambda m_{\max}} = 0. \tag{30}$$

For example, in the case of our θ equation,

$$\frac{du_{\lambda m_{\max}}}{u_{\lambda m_{\max}}} = (m_{\max} + \frac{1}{2}) \cot\theta d\theta, \tag{31}$$

leading to

$$ln u_{\lambda_{max}} = \big[ln(\sin\theta)\big]^{m_{\max}+\frac{1}{2}}. \tag{32}$$

If we name

$$m_{\max} = l, \tag{33}$$

we can write this solution

$$u(\theta) = N_l \sin^{(l+\frac{1}{2})}\theta, \qquad \Theta(\theta) = N_l \sin^l\theta; \tag{34}$$

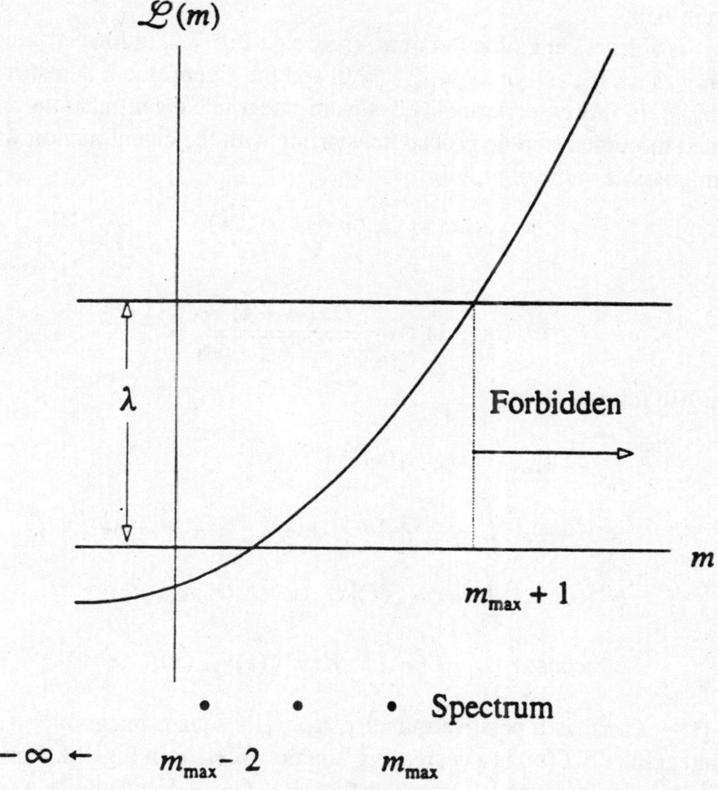

FIGURE 7.1. Case 1. A monotonically increasing $\mathcal{L}(m)$.

recalling that $\sqrt{\sin\theta}\Theta(\theta) = u(\theta)$. The normalization constant, N_l, can be evaluated to be

$$|N_l| = \sqrt{\frac{(2l+1)!!}{2(2l)!!}} = \sqrt{\frac{1 \cdot 3 \cdot 5 \cdot \cdots \cdot (2l+1)}{2[2 \cdot 4 \cdot 6 \cdots 2l]}}. \tag{35}$$

These considerations lead us to theorem IIIa.

Theorem IIIa:

If $\mathcal{L}(m)$ is an increasing function of m, a highest value of m exists, $m_{\max}$, such that $O_+(m_{\max}+1)u_{\lambda m_{\max}} = 0$, and the eigenvalue, λ, is restricted by $\lambda = \mathcal{L}(m_{\max}+1)$. In this case, normalized square-integrable eigenfunctions $u_{\lambda m}$ can be obtained from

$$u_{\lambda m-1}(x) = \mathcal{O}_-(m)u_{\lambda m}(x), \tag{36}$$

where

$$\mathcal{O}_-(m) \equiv \frac{O_-(m)}{\sqrt{[\lambda - \mathcal{L}(m)]}}. \tag{37}$$

That is, we can use a laddering process to ladder down from the eigenfunction with maximum possible m to arbitrary m, by repeated application of this operation.

Theorem IIIb:

If $\mathcal{L}(m)$ is a decreasing function of m, (see, e.g., Fig. 7.2), a lowest value of m exists, $m_{\min}$, such that $O_-(m_{\min})u_{\lambda m_{\min}} = 0$, and the eigenvalue λ is restricted by $\lambda = \mathcal{L}(m_{\min})$. In this case, normalized, square-integrable eigenfunctions $u_{\lambda m}$ can be obtained through a step-up procedure, starting with the eigenfunction with the minimum possible value of m, via

$$u_{\lambda m+1}(x) = \mathcal{O}_+(m+1)u_{\lambda m}(x), \tag{38}$$

where

$$\mathcal{O}_+(m+1) \equiv \frac{O_+(m+1)}{\sqrt{[\lambda - \mathcal{L}(m+1)]}}. \tag{39}$$

Theorem IIIb follows from

$$\begin{aligned}
&\int_a^b dx u^*_{\lambda m-1}(x)u_{\lambda m-1}(x) \\
&= |\text{const.}|^2 \int_a^b dx\big[O_-(m)u_{\lambda m}(x)\big]^* O_-(m)u_{\lambda m}(x) \\
&= |\text{const.}|^2 \int_a^b dx u^*_{\lambda m}(x)\big[O_+(m)O_-(m)u_{\lambda m}(x)\big] \\
&= |\text{const.}|^2[\lambda - \mathcal{L}(m)] \int_a^b dx u^*_{\lambda m}(x)u_{\lambda m}(x).
\end{aligned} \tag{40}$$

Now if $[\lambda - \mathcal{L}(m)]$ is a positive quantity, $u_{\lambda m-1}$ is square-integrable, if $u_{\lambda m}$ is square-integrable. If $\mathcal{L}(m)$ is a decreasing function of m, as in Fig. 7.2, however, a value of m would (in general) come such that $[\lambda - \mathcal{L}(m-1)]$ would be a negative quantity, and again we would have a patently positive quantity on the left-hand side of the equation equal to a patently negative quantity on the right. The initial assumption that $u_{\lambda m}$ be square-integrable must have been wrong. In the special case when $\lambda = \mathcal{L}(m_{\min})$, however, the laddering process quits at the value $m_{\min}$, and now no inconsistency exist.

In this case,

$$O_-(m_{\min})u_{\lambda m_{\min}} = 0, \tag{41}$$

$$+\frac{du_{\lambda m_{\min}}}{dx} + k(x, m_{\min})u_{\lambda m_{\min}}(x) = 0. \tag{42}$$

In this case, if $\mathcal{L}(m)$ is a *monotonic*, decreasing function of m, (see Fig. 7.2), the spectrum of allowed m values runs from $m_{\min}, m_{\min}+1, m_{\min}+2, \ldots,$ on to $+\infty$; the functions with higher m values being generated by repeated action with $O_+(m+1)$.

So far, we have considered cases with $\mathcal{L}(m)$ being monotonic increasing or decreasing functions of m. Our special example of the θ equation, however, with $\mathcal{L}(m) = (m - \frac{1}{2})^2$, see Fig. 7.3, is an increasing function of m for positive m values and a decreasing function of m for negative m values. In this case, the laddering process will lead to square-integrable functions only if both a minimum value of

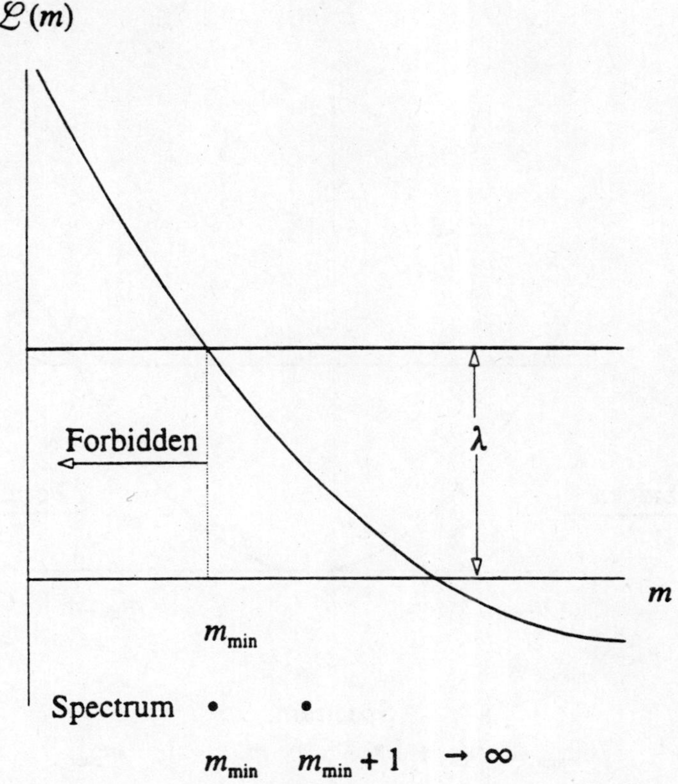

FIGURE 7.2. Case 2. A monotonically decreasing $\mathcal{L}(m)$.

m and a maximum value of m exist. The spectrum of allowed m values is restricted to a finite number = $(m_{max} - m_{min} + 1)$.

In the special case of the θ-equation, we have both

$$\lambda = \mathcal{L}(m_{min}) = \mathcal{L}(m_{max} + 1) = (m_{min} - \tfrac{1}{2})^2 = (m_{max} + \tfrac{1}{2})^2, \tag{43}$$

$$\text{and thus} \qquad m_{min}^2 - m_{min} = m_{max}^2 + m_{max}. \tag{44}$$

This quadratic equation for m_{min} has the two roots, $m_{min} = -m_{max}$ and $m_{min} = +(m_{max} + 1)$. Clearly, the last equation violates the meaning of m_{min}. Thus, with $m_{max} \equiv l$, the allowed m values range from $+l$ in steps of one down to $-l$. Because $(m_{max} - m_{min}) = 2l$ must be an integer, we have the result, $2l$ must be an integer. Thus, seemingly l can be either an integer or a $\frac{1}{2}$-integer. Later, we shall prove only the integer values are allowed for the orbital or θ equation.

Finally, the function $\mathcal{L}(m)$ could be a decreasing function of m for large positive values of m and an increasing function of m for negative values of m (see Fig. 7.4). In this case for a $\lambda < \mathcal{L}_{max}$, now two ranges of m values exist, one beginning at an m_{min} and going in integer steps on to $+\infty$, and a second beginning at an m_{max} and going in integer steps onto $-\infty$. If $\lambda > \mathcal{L}_{max}$, then all m values would

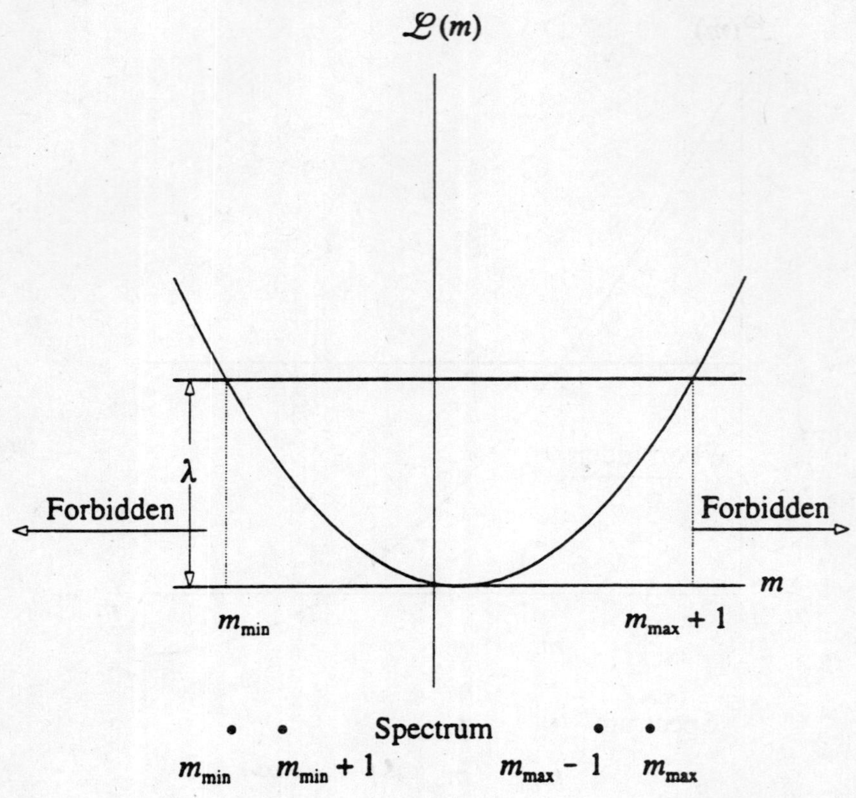

FIGURE 7.3. Case 3. An $\mathcal{L}(m)$ with an allowed spectrum such that $m_{\min.} \leq m \leq m_{\max.}$.

be allowed. In this last case, therefore, λ also has a continuous spectrum for all values of $\lambda > \mathcal{L}_{\max}$. In this case, the normalization integral should have the delta function form

$$\int_{-\infty}^{\infty} dx u^*_{\lambda' m}(x) u_{\lambda m}(x) = \delta(\lambda' - \lambda). \tag{45}$$

With $\lambda > \mathcal{L}(m)$ for all possible m, the normalized ladder operators, $\mathcal{O}_+(m+1)$ and $\mathcal{O}_-(m)$ exist. Moreover, they will preserve this normalization. If the $u_{\lambda m}(x)$ are normalized according to eq. (45), then

$$\int_{-\infty}^{\infty} dx u^*_{\lambda'(m-1)}(x) u_{\lambda(m-1)}(x) = \frac{\int_{-\infty}^{\infty} dx u^*_{\lambda' m} \Big[O_+(m) O_-(m) u_{\lambda m} \Big]}{\sqrt{[\lambda' - \mathcal{L}(m)][\lambda - \mathcal{L}(m)]}}$$
$$= \sqrt{\frac{[\lambda - \mathcal{L}(m)]}{[\lambda' - \mathcal{L}(m)]}} \int_{-\infty}^{\infty} dx u^*_{\lambda' m}(x) u_{\lambda m}(x) = \delta(\lambda' - \lambda). \tag{46}$$

In this case, however, it may be difficult to find a solution for a starting value, $u_{\lambda m_0}(x)$.

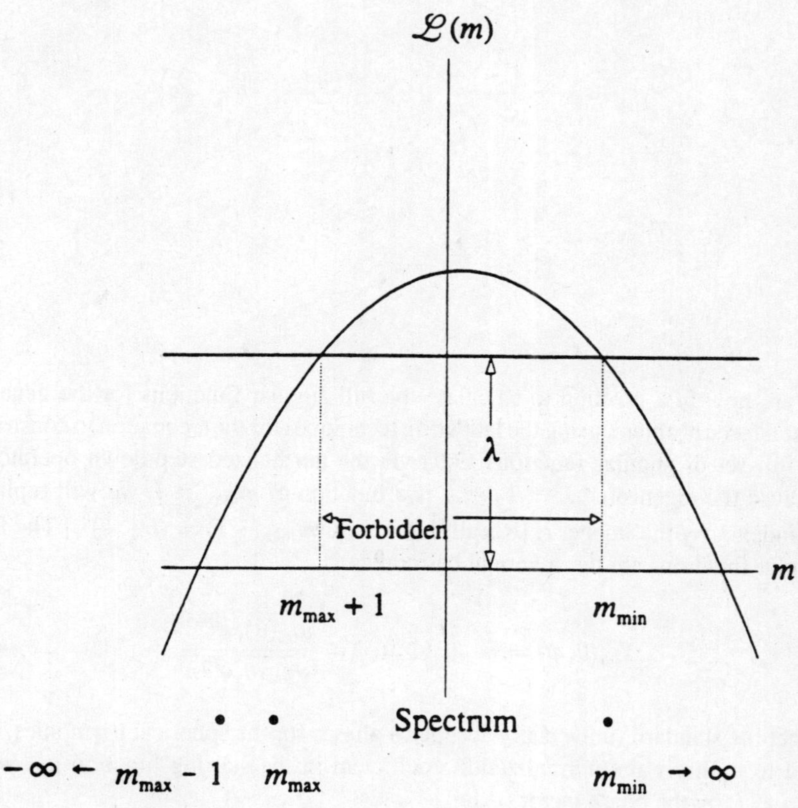

FIGURE 7.4. Case 4. An $\mathcal{L}(m)$ with two allowed branches: $m = m_{\max} \rightarrow -\infty$, and $m = m_{\min} \rightarrow +\infty$.

8

Spherical Harmonics, Orbital Angular Momentum

We are now in a position to calculate the full angular functions for the general central force problem, using the laddering techniques for the θ equation to construct the full set of angular functions $\Theta(\theta)$ via the normalized step-down operators. Because the eigenvalue $\lambda = \lambda_0 + \frac{1}{4}$ is a function of $m_{\max} \equiv l$, we will replace the index λ by the integer l. [Recall that $\lambda = \mathcal{L}(m_{\max} + 1) = (l + \frac{1}{2})^2$.] The full angular functions are the spherical harmonics

$$Y_{lm}(\theta, \phi) = \Theta_{lm}(\theta)\Phi_m(\phi) = \frac{u_{lm}(\theta)}{\sqrt{\sin\theta}} \frac{e^{im\phi}}{\sqrt{2\pi}}. \tag{1}$$

To get the standard (universally accepted) phases for the spherical harmonics, we need to multiply the normalization coefficient in the starting function u_{ll}, with $m_{\max} = l$, by the phase factor $(-1)^l$

$$u_{ll}(\theta) = (-1)^l \sqrt{\frac{(2l+1)!!}{2[2l!!]}} \sin^{l+\frac{1}{2}}(\theta). \tag{2}$$

In addition, we need to multiply the normalized step-operators $\mathcal{O}_-(m)$ and $\mathcal{O}_+(m+1)$ of eqs. (37) and (39) of Chapter 7 by a phase factor (-1). Thus,

$$Y_{l(m-1)} = -\frac{e^{-i\phi}}{\sqrt{\sin\theta}} \frac{\left[\left(\frac{d}{d\theta} + (m - \frac{1}{2})\cot\theta\right)u_{lm}(\theta)\right]}{\sqrt{(l+\frac{1}{2})^2 - (m-\frac{1}{2})^2}} \frac{e^{im\phi}}{\sqrt{2\pi}}. \tag{3}$$

Setting $u_{lm}(\theta) = \sqrt{\sin\theta}\,\Theta_{lm}(\theta)$ in this equation, this becomes

$$Y_{l(m-1)} = \frac{e^{-i\phi}}{\sqrt{(l+m)(l-m+1)}}\left[\left(-\frac{d}{d\theta} - m\cot\theta\right)\Theta_{lm}(\theta)\right]\frac{e^{im\phi}}{\sqrt{2\pi}}. \quad (4)$$

Finally, putting

$$me^{im\phi} = -i\frac{\partial}{\partial\phi}e^{im\phi}, \quad (5)$$

we obtain

$$Y_{l(m-1)}(\theta,\phi) = \frac{e^{-i\phi}}{\sqrt{(l+m)(l-m+1)}}\left[-\frac{\partial}{\partial\theta} + i\cot\theta\frac{\partial}{\partial\phi}\right]Y_{lm}(\theta,\phi). \quad (6)$$

Similarly, using the normalized, standard-phase step-up operator $-\mathcal{O}_+(m+1)$,

$$Y_{l(m+1)}(\theta,\phi) = \frac{e^{+i\phi}}{\sqrt{(l-m)(l+m+1)}}\left[+\frac{\partial}{\partial\theta} + i\cot\theta\frac{\partial}{\partial\phi}\right]Y_{lm}(\theta,\phi). \quad (7)$$

A Angular Momentum Operators

It will now be useful to express the operators converting the Y_{lm} into $Y_{l(m\pm1)}$ in terms of dimensionless angular momentum operators, such as

$$\frac{L_z}{\hbar} = \frac{1}{i}\left(x\frac{\partial}{\partial y} - y\frac{\partial}{\partial x}\right). \quad (8)$$

Transforming to spherical coordinates

$$x = r\sin\theta\cos\phi, \qquad y = r\sin\theta\sin\phi, \qquad z = r\cos\theta, \quad (9)$$

and using

$$\begin{aligned}
&\frac{\partial r}{\partial x} = \sin\theta\cos\phi, && \frac{\partial r}{\partial y} = \sin\theta\sin\phi, && \frac{\partial r}{\partial z} = \cos\theta,\\
&\frac{\partial\theta}{\partial x} = \frac{\cos\theta\cos\phi}{r}, && \frac{\partial\theta}{\partial y} = \frac{\cos\theta\sin\phi}{r}, && \frac{\partial\theta}{\partial z} = -\frac{\sin\theta}{r},\\
&\frac{\partial\phi}{\partial x} = -\frac{\sin\phi}{r\sin\theta}, && \frac{\partial\phi}{\partial y} = \frac{\cos\phi}{r\sin\theta}, && \frac{\partial\phi}{\partial z} = 0,
\end{aligned} \quad (10)$$

we get

$$\frac{L_z}{\hbar} \equiv L_0 = \frac{1}{i}\frac{\partial}{\partial\phi}, \quad (11)$$

$$\frac{(L_x \pm iL_y)}{\hbar} \equiv L_\pm = e^{\pm i\phi}\left(\pm\frac{\partial}{\partial\theta} + i\cot\theta\frac{\partial}{\partial\phi}\right), \quad (12)$$

and

$$\frac{(\vec{L}\cdot\vec{L})}{\hbar^2} = L_0^2 + \frac{1}{2}(L_+L_- + L_-L_+) = -\left(\frac{1}{\sin\theta}\frac{\partial}{\partial\theta}\sin\theta\frac{\partial}{\partial\theta} + \frac{1}{\sin^2\theta}\frac{\partial^2}{\partial\phi^2}\right). \quad (13)$$

Hence, the spherical harmonics are simultaneous eigenfunctions of the operators, $(\vec{L}\cdot\vec{L})$, and L_z, with

$$\begin{aligned} L_z Y_{lm}(\theta,\phi) &= \hbar m Y_{lm}(\theta,\phi), \\ (\vec{L}\cdot\vec{L}) Y_{lm}(\theta,\phi) &= \hbar^2 \lambda_0 Y_{lm}(\theta,\phi) = \hbar^2 l(l+1) Y_{lm}(\theta,\phi). \end{aligned} \tag{14}$$

In addition, eqs. (6) and (7) can be put into the form

$$L_- Y_{lm} = \sqrt{(l+m)(l-m+1)}\, Y_{l(m-1)}, \tag{15}$$

$$L_+ Y_{lm} = \sqrt{(l-m)(l+m+1)}\, Y_{l(m+1)}. \tag{16}$$

The $Y_{lm}(\theta,\phi)$ form an orthonormal complete set over the surface of the unit sphere. Thus, the matrix elements of the operators $L_\pm$ are

$$\langle Y_{l'm'}, L_- Y_{lm}\rangle = \delta_{l'l}\delta_{m'(m-1)}\sqrt{(l+m)(l-m+1)}, \tag{17}$$

$$\langle Y_{l'm'}, L_+ Y_{lm}\rangle = \delta_{l'l}\delta_{m'(m+1)}\sqrt{(l-m)(l+m+1)}, \tag{18}$$

and

$$\langle Y_{l'm'}, L_0 Y_{lm}\rangle = \delta_{l'l}\delta_{m'm} m. \tag{19}$$

These matrix elements can also be used to obtain the matrix elements of L_x and L_y.

$$\begin{aligned} \langle Y_{l'm'}, L_x Y_{lm}\rangle &= \frac{\hbar}{2}\langle Y_{l'm'}, (L_+ + L_-) Y_{lm}\rangle \\ &= \delta_{l'l}\frac{\hbar}{2}\Big(\delta_{m'(m+1)}\sqrt{(l-m)(l+m+1)} \\ &\qquad + \delta_{m'(m-1)}\sqrt{(l+m)(l-m+1)}\Big). \end{aligned} \tag{20}$$

Similarly,

$$\begin{aligned} \langle Y_{l'm'}, L_y Y_{lm}\rangle &= \frac{\hbar}{2}\langle Y_{l'm'}, (-iL_+ + iL_-) Y_{lm}\rangle \\ &= \delta_{l'l}\frac{\hbar}{2}\Big(-i\delta_{m'(m+1)}\sqrt{(l-m)(l+m+1)} \\ &\qquad + i\delta_{m'(m-1)}\sqrt{(l+m)(l-m+1)}\Big). \end{aligned} \tag{21}$$

The infinite-dimensional matrices for L_x, L_y, and L_z thus factor into $(2l+1)$ by $(2l+1)$ submatrices. As a simple, specific example, the submatrices for $l = 1$ are (in units of $\hbar$),

$$\mathbf{L}_x = \begin{pmatrix} 0 & \frac{1}{\sqrt{2}} & 0 \\ \frac{1}{\sqrt{2}} & 0 & \frac{1}{\sqrt{2}} \\ 0 & \frac{1}{\sqrt{2}} & 0 \end{pmatrix},$$

$$\mathbf{L}_y = \begin{pmatrix} 0 & \frac{-i}{\sqrt{2}} & 0 \\ \frac{+i}{\sqrt{2}} & 0 & \frac{-i}{\sqrt{2}} \\ 0 & \frac{+i}{\sqrt{2}} & 0 \end{pmatrix},$$

$$\mathbf{L}_z = \begin{pmatrix} +1 & 0 & 0 \\ 0 & 0 & 0 \\ 0 & 0 & -1 \end{pmatrix},$$

where rows and columns are labeled in the conventional order, $m = +1, 0, -1$.

Because the spherical harmonics form a complete orthonormal set, we can translate the operators $L_\pm$ into the following functional forms. For example,

$$L_+ = \sum_{l=0}^{\infty} \sum_{m=-l}^{m=+l} Y_{l(m+1)}(\theta, \phi) Y_{lm}^*(\theta, \phi) \left(\sqrt{(l-m)(l+m+1)} \right). \tag{22}$$

In our method of constructing the $(2l + 1)$ spherical harmonics for a particular l, we have started with the eigenfunction with $m = m_{\max} = l$, and we have then used the normalized step-down operators, $\mathcal{O}_-(m)$, to calculate the remaining $2l$ eigenfunctions. Alternatively, we could have started with $m = m_{\min}$ and laddered with $\mathcal{O}_+(m+1)$. A third possibility would be to start with the spherical harmonics with $m = 0$ and use successive application of $L_\pm$ to calculate the spherical harmonics with $\pm m$.

$$\begin{aligned} Y_{lm} &= \frac{(L_+)^m Y_{l0}}{\sqrt{l(l-1)\cdots(l-m+1)(l+1)(l+2)\cdots(l+m)}} \\ &= \sqrt{\frac{(l-m)!}{(l+m)!}} (L_+)^m Y_{l0}, \end{aligned} \tag{23}$$

and

$$\begin{aligned} Y_{l-m} &= \frac{(L_-)^m Y_{l0}}{\sqrt{l(l-1)\cdots(l-m+1)(l+1)(l+2)\cdots(l+m)}} \\ &= \sqrt{\frac{(l-m)!}{(l+m)!}} (L_-)^m Y_{l0}. \end{aligned} \tag{24}$$

Now, because

$$(L_+)^* = -(L_-), \tag{25}$$

we see

$$Y_{l-m}(\theta, \phi) = (-1)^m Y_{lm}^*(\theta, \phi). \tag{26}$$

Thus, it is sufficient to calculate the spherical harmonics with $m \geq 0$. As a final remark, the three operators L_x, L_y, L_z are all hermitian, and hence,

$$L_+^\dagger = L_-; \qquad L_-^\dagger = L_+. \tag{27}$$

9

ℓ-Step operators for the θ Equation

In the last section, we calculated the matrix elements of the operators, L_x, L_y, L_z. These operators are functions only of θ, ϕ, and $\frac{\partial}{\partial\theta}$, $\frac{\partial}{\partial\phi}$. It will also be extremely useful to have the matrix elements of the angular parts of the position vector of the particle, viz. $\frac{x}{r} = \sin\theta\cos\phi$, $\frac{y}{r} = \sin\theta\sin\phi$, and $\frac{z}{r} = \cos\theta$. To get these, it would be useful to interchange the role of the quantum numbers l, and m, in the factorization method and derive expressions for ladder operators changing l to $l \pm 1$, keeping m fixed.

For this purpose, rewrite the 1-D θ equation in the form

$$-\sin^2\theta\frac{d^2\Theta}{d\theta^2} - \sin\theta\cos\theta\frac{d\Theta}{d\theta} - l(l+1)\sin^2\theta\Theta = -m^2\Theta. \tag{1}$$

If we can find a change of variable transforming the derivative operators into 1-D form, we will have succeeded, because $-m^2$ can then play the role of the fixed λ, whereas the parameter l is in a position to be stepped. The transformation achieving the desired result is

$$z = ln\left(\tan\frac{\theta}{2}\right), \qquad \theta = 0 \to z = -\infty, \quad \theta = \pi \to z = +\infty. \tag{2}$$

Note,

$$\frac{d}{d\theta} = \frac{dz}{d\theta}\frac{d}{dz} = \frac{1}{\sin\theta}\frac{d}{dz}, \qquad \frac{d^2}{d\theta^2} = -\frac{\cos\theta}{\sin^2\theta}\frac{d}{dz} + \frac{1}{\sin^2\theta}\frac{d^2}{dz^2}, \tag{3}$$

so

$$-\sin^2\theta\frac{d^2}{d\theta^2} - \sin\theta\cos\theta\frac{d}{d\theta} = -\frac{d^2}{dz^2}. \tag{4}$$

Also,

$$\cosh z = \frac{1}{\sin\theta}, \qquad \tanh z = -\cos\theta. \tag{5}$$

Now, with

$$\Theta(\theta(z)) = v(z), \tag{6}$$

eq. (1) is transformed into

$$\left[-\frac{d^2}{dz^2} - \frac{l(l+1)}{\cosh^2(z)}\right] v_{\lambda l}(z) = -m^2 v_{\lambda l}(z) = \lambda v_{\lambda l}(z). \tag{7}$$

Now, with $\lambda = -m^2$, the role of m and l have been interchanged, l being a parameter in the "potential function." This equation is now factorizable with the factors

$$O_\pm(l) = \big(\mp\frac{d}{dz} + l\tanh z\big), \tag{8}$$

with

$$\begin{aligned} O_+(l)O_-(l)v_{\lambda l} &= \left[-\frac{d}{dz} + l\tanh z\right]\left[+\frac{d}{dz} + l\tanh z\right] v_{\lambda l} \\ &= \left(\left[-\frac{d^2}{dz^2} - \frac{l(l+1)}{\cosh^2 z}\right] + l^2\right) v_{\lambda l} \\ &= [\lambda - \mathcal{L}(l)]v_{\lambda l}, \end{aligned} \tag{9}$$

and

$$\begin{aligned} O_-(l+1)O_+(l+1)v_{\lambda l} &= \left[+\frac{d}{dz} + (l+1)\tanh z\right]\left[-\frac{d}{dz} + (l+1)\tanh z\right] v_{\lambda l} \\ &= \left(\left[-\frac{d^2}{dz^2} - \frac{l(l+1)}{\cosh^2 z}\right] + (l+1)^2\right) v_{\lambda l} \\ &= [\lambda - \mathcal{L}(l+1)]v_{\lambda l}, \end{aligned} \tag{10}$$

so

$$\mathcal{L}(l) = -l^2. \tag{11}$$

$\mathcal{L}(l)$ is a decreasing function of l and, with the negative $\lambda = -m^2$, must be such that a minimum value of l exists, with

$$\lambda = -m^2 = \mathcal{L}(l_{min}) = -l_{min}^2. \tag{12}$$

The starting function, with $l = l_{min} = m$, is obtained from

$$O_-(l_{min})v_{\lambda l_{min}=m} = \big(\frac{d}{dz} + m\tanh z\big)v_{\lambda m}, \tag{13}$$

so

$$\frac{dv_{\lambda m}}{v_{\lambda m}} = -m\tanh z dz, \qquad \text{with} \quad ln v_{\lambda m} = ln(\cosh z)^{-m}. \tag{14}$$

This leads to

$$v_{\lambda m} = \frac{N_m}{(\cosh z)^m} = N_m(\sin\theta)^m, \tag{15}$$

which agrees with our earlier solution for $\Theta(\theta)$, with $l = m$. The remaining solutions with $l > m$ can be obtained through the normalized step-up operators, via

$$\begin{aligned} v_{\lambda(l+1)} &= \frac{1}{\sqrt{[\lambda - \mathcal{L}(l+1)]}}\left(-\frac{d}{dz} + (l+1)\tanh z\right)v_{\lambda l} \\ &= \frac{1}{\sqrt{[-m^2 + (l+1)^2]}}\left(-\frac{d}{dz} + (l+1)\tanh z\right)v_{\lambda l}. \end{aligned} \tag{16}$$

Similarly,

$$v_{\lambda(l-1)} = \frac{1}{\sqrt{[-m^2 + l^2]}}\left(+\frac{d}{dz} + l\tanh z\right)v_{\lambda l}. \tag{17}$$

Note, however, the normalization preserved by these $v_{\lambda l}(z)$ is

$$\int_{-\infty}^{\infty} dz|v_{\lambda l}(z)|^2 = 1, \tag{18}$$

whereas, with $\Theta_{lm}(\theta(z)) = v_{\lambda l}(z)$, $d\theta = \sin\theta dz$, and $\cosh z = \frac{1}{\sin\theta}$, we should have normalized with a weighting factor in z-space

$$\int_0^\pi d\theta\sin\theta|\Theta(\theta)|^2 = \int_{-\infty}^{\infty} dz(\cosh^{-2}(z))|v_{\lambda l}(z)|^2 = 1. \tag{19}$$

The lack of the weighting factor $\cosh^{-2}(z)$ means our functions $\Theta_{lm}(\theta(z))$ can be identified with the $v_{\lambda m}(z)$ only with the inclusion of an additional normalization factor, c_{lm}, via

$$\Theta_{lm}(\theta) = c_{lm}v_{\lambda l}(z). \tag{20}$$

With $\frac{d}{dz} = \sin\theta\frac{d}{d\theta}$, and $\tanh z = -\cos\theta$, eqs. (16) and (17) translate into

$$\Theta_{(l+1)m}(\theta) = -\frac{(c_{(l+1)m}/c_{lm})}{\sqrt{(l+1+m)(l+1-m)}}\left[-\sin\theta\frac{d}{d\theta} - (l+1)\cos\theta\right]\Theta_{lm}(\theta), \tag{21}$$

$$\Theta_{(l-1)m}(\theta) = -\frac{(c_{(l-1)m}/c_{lm})}{\sqrt{(l+m)(l-m)}}\left[\sin\theta\frac{d}{d\theta} - l\cos\theta\right]\Theta_{lm}(\theta). \tag{22}$$

We have again introduced an extra minus sign to agree with the standard phase conventions for spherical harmonics. This minus sign is the analog of that introduced in eqs. (4) and (7) of Chapter 8.

To calculate the ratios of c_{lm} coefficients, we shall calculate $\Theta_{(l+1)(m+1)}(\theta)$ in two ways, by stepping from (l, m) to $(l+1, m+1)$ along two different paths in the l, m parameter space.

For path (1), step first from (l, m) to $(l, m+1)$ via the m step-up operator of eq. (7) of Chapter 8:

$$\frac{1}{\sqrt{(l-m)(l+m+1)}}\left[\frac{d}{d\theta} - m\cot\theta\right], \tag{23}$$

and follow this relation with a step from $(l, m+1)$ to $(l+1, m+1)$ with the l step-up operator.

For path (2), step first from (l, m) to $(l+1, m)$ and follow this with a step from $(l+1, m)$ to $(l+1, m+1)$.

Path (1) leads to:

$$\begin{aligned}\Theta_{(l+1)(m+1)} &= \frac{c_{(l+1)(m+1)}}{c_{l(m+1)}} \frac{\left[\sin\theta\frac{d}{d\theta} + (l+1)\cos\theta\right]}{\sqrt{(l+m+2)(l-m)}} \frac{\left[\frac{d}{d\theta} - m\cot\theta\right]}{\sqrt{(l-m)(l+m+1)}}\Theta_{lm}\\ &= \frac{(c_{(l+1)(m+1)}/c_{l(m+1)})}{(l-m)\sqrt{(l+m+1)(l+m+2)}}\left(\sin\theta\left[\frac{d^2}{d\theta^2} + \cot\theta\frac{d}{d\theta}\right]\right.\\ &\quad \left. + (l-m)\cos\theta\frac{d}{d\theta} - m(l+1)\frac{\cos^2\theta}{\sin\theta} + \frac{m}{\sin\theta}\right)\Theta_{lm}.\end{aligned} \tag{24}$$

Now use

$$\left[\frac{d^2}{d\theta} + \cot\theta\frac{d}{d\theta}\right] = \left[\frac{m^2}{\sin^2\theta} - l(l+1)\right] \tag{25}$$

via the θ equation, and simplify the above by factoring out the factor $(l-m)$ to yield

$$\Theta_{(l+1)(m+1)} = \frac{(c_{(l+1)(m+1)}/c_{l(m+1)})}{\sqrt{(l+m+1)(l+m+2)}}\left[\cos\theta\frac{d}{d\theta} - \frac{m}{\sin\theta} - (l+1)\sin\theta\right]\Theta_{lm}. \tag{26}$$

Similarly, using path (2), we get

$$\Theta_{(l+1)(m+1)} = \frac{(c_{(l+1)m}/c_{lm})}{\sqrt{(l+m+1)(l+m+2)}}\left[\cos\theta\frac{d}{d\theta} - \frac{m}{\sin\theta} - (l+1)\sin\theta\right]\Theta_{lm}. \tag{27}$$

We see

$$\frac{c_{(l+1)(m+1)}}{c_{l(m+1)}} = \frac{c_{(l+1)m}}{c_{lm}}. \tag{28}$$

That is, the ratio is independent of m, and we can calculate it by setting $m = l$

$$\frac{c_{(l+1)m}}{c_{lm}} = \frac{c_{(l+1)l}}{c_{ll}}. \tag{29}$$

Setting $m = l$ in eq. (26), and using eqs. (28) and (29), leads to

$$\begin{aligned}\Theta_{(l+1)(l+1)} &= N_{l+1}\sin^{l+1}\theta\\ &= \frac{(c_{(l+1)l}/c_{ll})}{\sqrt{(2l+1)(2l+2)}}\left[\cos\theta\frac{d}{d\theta} - \frac{l}{\sin\theta} - (l+1)\sin\theta\right]N_l\sin^l\theta\end{aligned}$$

$$= -\frac{c_{(l+1)l}}{c_{ll}}\sqrt{\frac{(2l+1)}{(2l+2)}}N_l \sin^{l+1}\theta. \tag{30}$$

Therefore, with

$$\frac{N_{l+1}}{N_l} = -\sqrt{\frac{(2l+3)}{(2l+2)}}, \tag{31}$$

[see eq. (2) of Chapter 8],

$$\frac{c_{(l+1)l}}{c_{ll}} = \sqrt{\frac{(2l+3)}{(2l+1)}} = \frac{c_{(l+1)m}}{c_{lm}}. \tag{32}$$

Using this result, we can now rewrite the l-step equations [eqs. (21) and (22)], as

$$\Theta_{(l+1)m} = \sqrt{\frac{(2l+3)/(2l+1)}{(l+1+m)(l+1-m)}}\left[\sin\theta\frac{d}{d\theta} + (l+1)\cos\theta\right]\Theta_{lm}, \tag{33}$$

$$\Theta_{(l-1)m} = \sqrt{\frac{(2l-1)/(2l+1)}{(l+m)(l-m)}}\left[-\sin\theta\frac{d}{d\theta} + l\cos\theta\right]\Theta_{lm}. \tag{34}$$

Adding these two equations, after multiplication of each by the inverse of the square root factor, leads to

$$\cos\theta\,\Theta_{lm} = \sqrt{\frac{(l+1+m)(l+1-m)}{(2l+1)(2l+3)}}\Theta_{(l+1)m} + \sqrt{\frac{(l+m)(l-m)}{(2l+1)(2l-1)}}\Theta_{(l-1)m}. \tag{35}$$

Also, using the c-ratio of eq. (32), eq. (26) can now be rewritten explicitly as

$$\Theta_{(l+1)(m+1)} = \sqrt{\frac{(2l+3)/(2l+1)}{(l+m+1)(l+m+2)}}\left[\cos\theta\frac{d}{d\theta} - \frac{m}{\sin\theta} - (l+1)\sin\theta\right]\Theta_{lm}. \tag{36}$$

Similarly, using step-operations from (l, m) to $(l, m+1)$ and then from $(l, m+1)$ to $(l-1, m+1)$, eliminating the second derivative term via the differential equation, as for eq. (26), we get the companion equation

$$\Theta_{(l-1)(m+1)} = \sqrt{\frac{(2l-1)/(2l+1)}{(l-m)(l-m-1)}}\left[\cos\theta\frac{d}{d\theta} - \frac{m}{\sin\theta} + l\sin\theta\right]\Theta_{lm}. \tag{37}$$

Eqs. (36) and (37) can now be combined to give

$$\sin\theta\,\Theta_{lm} = -\sqrt{\frac{(l+m+2)(l+m+1)}{(2l+3)(2l+1)}}\Theta_{(l+1)(m+1)} + \sqrt{\frac{(l-m)(l-m-1)}{(2l+1)(2l-1)}}\Theta_{(l-1)(m+1)}. \tag{38}$$

Finally, using the products of m step-down, and l step-up or down operators, we get

$$\Theta_{(l+1)(m-1)} = \sqrt{\frac{(2l+3)/(2l+1)}{(l+2-m)(l+1-m)}}\left[-\cos\theta\frac{d}{d\theta} - \frac{m}{\sin\theta} + (l+1)\sin\theta\right]\Theta_{lm}, \tag{39}$$

$$\Theta_{(l-1)(m-1)} = \sqrt{\frac{(2l-1)/(2l+1)}{(l+m)(l+m-1)}}\left[-\cos\theta\frac{d}{d\theta} - \frac{m}{\sin\theta} - l\sin\theta\right]\Theta_{lm}. \tag{40}$$

Eqs. (39) and (40) can now be combined to give

$$\sin\theta\,\Theta_{lm} =$$

$$\sqrt{\frac{(l+2-m)(l+1-m)}{(2l+3)(2l+1)}}\Theta_{(l+1)(m-1)} - \sqrt{\frac{(l+m)(l+m-1)}{(2l+1)(2l-1)}}\Theta_{(l-1)(m-1)}. \tag{41}$$

Finally, by multiplying eq. (35) by $e^{im\phi}/\sqrt{2\pi}$, eq. (38) by $e^{i(m+1)\phi}/\sqrt{2\pi}$, and eq. (41) by $e^{i(m-1)\phi}/\sqrt{2\pi}$, these operators can be converted to equations involving the full spherical harmonics, $Y_{lm}(\theta,\phi)$:

$$\cos\theta\, Y_{lm} = \sqrt{\frac{(l+1+m)(l+1-m)}{(2l+1)(2l+3)}}Y_{(l+1)m} + \sqrt{\frac{(l+m)(l-m)}{(2l+1)(2l-1)}}Y_{(l-1)m}, \tag{42}$$

$$e^{i\phi}\sin\theta\, Y_{lm} =$$

$$-\sqrt{\frac{(l+m+2)(l+1+m)}{(2l+3)(2l+1)}}Y_{(l+1)(m+1)} + \sqrt{\frac{(l-m)(l-m-1)}{(2l+1)(2l-1)}}Y_{(l-1)(m+1)}, \tag{43}$$

$$e^{-i\phi}\sin\theta\, Y_{lm} =$$

$$\sqrt{\frac{(l+2-m)(l+1-m)}{(2l+3)(2l+1)}}Y_{(l+1)(m-1)} - \sqrt{\frac{(l+m)(l+m-1)}{(2l+1)(2l-1)}}Y_{(l-1)(m-1)}. \tag{44}$$

Now, because the $Y_{lm}(\theta,\phi)$ form a complete orthonormal set over the surface of a sphere, we can use eqs. (42)–(44) to find the matrix elements of the operators $\cos\theta$, $e^{\pm i\phi}\sin\theta$. These operators give us the matrix elements of the angular parts of x, y, z, because

$$\frac{(x\pm iy)}{r} = e^{\pm i\phi}\sin\theta, \qquad \frac{z}{r} = \cos\theta. \tag{45}$$

Matrix elements of higher powers of x, y, and z (angular parts) can then be obtained from these operators by matrix multiplication.

Final remark: The angular functions $\Theta_{lm}(\theta)$ are not by themselves orthogonal for different values of m. Thus, in general,

$$\int_0^\pi d\theta\,\sin\theta\,\Theta^*_{lm'}(\theta)\Theta_{lm}(\theta) \neq \delta_{m'm}. \tag{46}$$

l m	$Y_{lm}(\theta,\phi)$	$\mathcal{Y}_{lm}=r^lY_{lm}$
0 0	$\frac{1}{\sqrt{4\pi}}$	$\frac{1}{\sqrt{4\pi}}$
1 0	$\sqrt{\frac{3}{4\pi}}\cos\theta$	$\sqrt{\frac{3}{4\pi}}z$
1 ± 1	$\mp\sqrt{\frac{3}{8\pi}}\sin\theta e^{\pm i\phi}$	$\mp\sqrt{\frac{3}{8\pi}}(x\pm iy)$
2 0	$\sqrt{\frac{5}{16\pi}}(3\cos^2\theta-1)$	$\sqrt{\frac{5}{16\pi}}(2z^2-x^2-y^2)$
2 ± 1	$\mp\sqrt{\frac{15}{8\pi}}\cos\theta\sin\theta e^{\pm i\phi}$	$\mp\sqrt{\frac{15}{8\pi}}z(x\pm iy)$
2 ± 2	$\sqrt{\frac{15}{32\pi}}\sin^2\theta e^{\pm 2i\phi}$	$\sqrt{\frac{15}{32\pi}}(x\pm iy)^2$
3 0	$\sqrt{\frac{7}{16\pi}}(5\cos^3\theta-3\cos\theta)$	$\sqrt{\frac{7}{16\pi}}z[2z^2-3(x^2+y^2)]$
3 ± 1	$\mp\sqrt{\frac{21}{64\pi}}(5\cos^2\theta-1)\sin\theta e^{\pm i\phi}$	$\mp\sqrt{\frac{21}{64\pi}}(4z^2-x^2-y^2)(x\pm iy)$
3 ± 2	$\sqrt{\frac{105}{32\pi}}\cos\theta\sin^2\theta e^{\pm 2i\phi}$	$\sqrt{\frac{105}{32\pi}}z(x\pm iy)^2$
3 ± 3	$\mp\sqrt{\frac{35}{64\pi}}\sin^3\theta e^{\pm 3i\phi}$	$\mp\sqrt{\frac{35}{64\pi}}(x\pm iy)^3$

The orthogonality in m comes via the functions $\Phi(\phi)$. For this reason, eqs. (38) and (41) were set up so that both terms on the right-hand sides have the *same* values of m.

A table of some of the simplest spherical harmonics is included here. These harmonics can be calculated very trivially through eqs. (1) and (2) of Chapter 8 with a few applications of the laddering operations of eq. (6) of Chapter 8 and the use of the symmetry property, eq. (26) of Chapter 8. They are given here so they can be combined with a tabulation of solid harmonics, defined by

$$\mathcal{Y}_{lm}=r^lY_{lm}(\theta,\phi). \tag{47}$$

These can be expressed as homogeneous polynomials of degree l in x, y, z by acting on

$$\mathcal{Y}_{ll}=N_l(x+iy)^l, \qquad \text{with} \quad N_l=(-1)^l\sqrt{\frac{(2l+1)!!}{2(2l)!!2\pi}}, \tag{48}$$

$(l-m)$ times in succession with the m step-down operator, $\mathcal{O}_-(m)$, where

$$\begin{aligned}\mathcal{O}_-(m)&=-\frac{e^{-i\phi}}{\sqrt{(l+m)(l-m+1)}}\left[\frac{d}{d\theta}+m\cot\theta\right]\\ &=-\frac{(x+iy)^{-1}}{\sqrt{(l+m)(l-m+1)}}\left[z(x\frac{d}{dx}+y\frac{d}{dy}+m)-(x^2+y^2)\frac{d}{dz}\right],\end{aligned} \tag{49}$$

and

$$\mathcal{O}_-(m)\mathcal{Y}_{lm} = \mathcal{Y}_{l,m-1}. \tag{50}$$

This relation leads to

$$\mathcal{Y}_{lm} = (-1)^{l-m} N_{ll} \sqrt{\frac{(l+m)!}{(2l)!(l-m)!}} \sum_{\alpha=0}^{[\frac{l-m}{2}]} c_\alpha^{(m)} (x+iy)^m z^{l-m-2\alpha}(x^2+y^2)^\alpha \tag{51}$$

for the states with $m \geq 0$. The coefficients are related by

$$c_{\alpha=0}^{(m-1)} = 2mc_{\alpha=0}^{(m)}, \tag{52}$$

and, for $\alpha = 1, \ldots, [\frac{l-m}{2}]$,

$$c_\alpha^{(m-1)} = 2(m+\alpha)c_\alpha^{(m)} - (l-m-2\alpha+2)c_{\alpha-1}^{(m)}. \tag{53}$$

The solution gives

$$\mathcal{Y}_{lm} = (-1)^m \sqrt{\frac{(l+m)!(2l+1)!!}{(l-m)!(2l)!2(2l)!!2\pi}} \times \sum_{\alpha=0}^{[\frac{l-m}{2}]} (-1)^\alpha \frac{2^{l-m-2\alpha} l!(l-m)!}{\alpha!(m+\alpha)!(l-m-2\alpha)!} (x+iy)^m z^{l-m-2\alpha}(x^2+y^2)^\alpha. \tag{54}$$

As an additional footnote, this result can also be used to find a solution to the following useful problem: Express the solid spherical harmonics in the relative motion vector, $\vec{r}_1 - \vec{r}_2$, as functions of the solid harmonics in $\vec{r}_1$ and $\vec{r}_2$; i.e., express the $\mathcal{Y}_{lm}(\vec{r}_1 - \vec{r}_2)$ as functions of $\mathcal{Y}_{lm}(\vec{r}_1)$ and $\mathcal{Y}_{lm}(\vec{r}_2)$. It will be useful first to define $\mathcal{Z}_{lm}$ via

$$\mathcal{Y}_{lm} \equiv -(1)^m \sqrt{\frac{(l+m)!(2l+1)!!}{(l-m)!(2l)!2^{l+1}l!2\pi}} \mathcal{Z}_{lm}, \tag{55}$$

so

$$\mathcal{Z}_{lm} = \sum_{\alpha=0}^{[\frac{(l-m)}{2}]} (-1)^\alpha \frac{2^{l-m-2\alpha} l!(l-m)!}{\alpha!(m+\alpha)!(l-m-2\alpha)!} (x+iy)^{m+\alpha}(x-iy)^\alpha z^{l-m-2\alpha}, \tag{56}$$

or

$$\begin{aligned} \mathcal{Z}_{lm}(\vec{r}_1 - \vec{r}_2) &= \sum_{\alpha=0}^{[\frac{l-m}{2}]} (-1)^\alpha \frac{2^{l-m-2\alpha} l!(l-m)!}{\alpha!(m+\alpha)!(l-m-2\alpha)!} \\ &\times [(x_1+iy_1)-(x_2+iy_2)]^{m+\alpha}[(x_1-iy_1)-(x_2-iy_2)]^\alpha (z_1-z_2)^{l-m-2\alpha} \\ &= \sum_{\alpha=0}^{[\frac{l-m}{2}]} \sum_{\beta=0}^{m+\alpha} \sum_{\gamma=0}^{\alpha} \sum_{\delta=0}^{l-m-2\alpha} (-1)^{\alpha+\beta+\gamma+\delta} \\ &\times \frac{2^{l-m-2\alpha} l!(l-m)!}{\beta!(m+\alpha-\beta)!\gamma!(\alpha-\gamma)!\delta!(l-m-2\alpha-\delta)!} \\ &\times (x_1+iy_1)^{m+\alpha-\beta}(x_1-iy_1)^{\alpha-\gamma} z_1^{l-m-2\alpha-\delta}(x_2+iy_2)^\beta (x_2-iy_2)^\gamma z_2^\delta, \end{aligned} \tag{57}$$

where, renaming $\gamma = \alpha_2$, we have $\gamma = \alpha_2$, $\beta = m_2 + \alpha_2$; and $\delta = l_2 - m_2 - 2\alpha_2$. In addition, with $m = m_1 + m_2$, and $l = l_1 + l_2$, and, defining α_1 through $\alpha = \alpha_1 + \alpha_2$, we have

$$m + \alpha - \beta = m_1 + \alpha_1, \qquad \alpha - \gamma = \alpha_1, \qquad l - m - 2\alpha - \delta = l_1 - m_1 - 2\alpha_1.$$

The above expression for $\mathcal{Z}_{lm}(\vec{r}_1 - \vec{r}_2)$ can then be rewritten as

$$\begin{aligned}
\mathcal{Z}_{lm}(\vec{r}_1 - \vec{r}_2) &= \sum_{l_2(l_1)}\sum_{m_2(m_1)} \sum_{\alpha_1=0}^{[\frac{l_1-m_1}{2}]} \sum_{\alpha_2=0}^{[\frac{l_2-m_2}{2}]} (-1)^{l_2+\alpha_1+\alpha_2} \\
&\times \frac{l!(l-m)!2^{l_1-m_1-2\alpha_1+l_2-m_2-2\alpha_2}}{\alpha_1!\alpha_2!(m_1+\alpha_1)!(m_2+\alpha_2)!(l_1-m_1-2\alpha_1)!(l_2-m_2-2\alpha_2)!} \\
&\times (x_1+iy_1)^{m_1+\alpha_1}(x_1-iy_1)^{\alpha_1} z_1^{l_1-m_1-2\alpha_1} \\
&\times (x_2+iy_2)^{m_2+\alpha_2}(x_2-iy_2)^{\alpha_2} z_2^{l_2-m_2-2\alpha_2} \\
&= \sum_{l_2(l_1)}\sum_{m_2(m_1)} (-1)^{l_2} \frac{l!(l-m)!}{l_1!(l_1-m_1)!l_2!(l_2-m_2)!} \mathcal{Z}_{l_1m_1}(\vec{r}_1)\mathcal{Z}_{l_2m_2}(\vec{r}_2). \qquad (58)
\end{aligned}$$

Finally, using the definition of $\mathcal{Z}_{lm}$, we get

$$\mathcal{Y}_{lm}(\vec{r}_1 - \vec{r}_2) = \sum_{l_2(l_1)}\sum_{m_2(m_1)}(-1)^{l_2}\mathcal{Y}_{l_1m_1}(\vec{r}_1)\mathcal{Y}_{l_2m_2}(\vec{r}_2)\times$$

$$\left[\frac{(l+m)!(l-m)!l!(2l+1)!!(2l_1)!(2l_2)!4\pi}{(2l)!(l_1+m_1)!(l_1-m_1)!l_1!(2l_1+1)!!(l_2+m_2)!(l_2-m_2)!l_2!(2l_2+1)!!}\right]^{\frac{1}{2}}. \qquad (59)$$

10

The Radial Functions for the Hydrogenic Atom

Because we have solved the angular part of the one-body problem for a spherically symmetric V(r) (or, equivalently, the angular part for the relative motion of a two-body problem), it would be good to provide a detailed example for a particular potential, V(r). Because the Coulomb problem is soluble via the factorization method, let us solve the radial problem for the general hydrogenic atom, i.e., the one electron atom (with $Z = 1, 2, 3, \ldots$) for hydrogen, once-ionized Helium, twice-ionized Lithium, and so on,where

$$V(r) = -\frac{Ze^2}{r}. \tag{1}$$

The one-dimensionalized radial equation is

$$\left(-\frac{\hbar^2}{2\mu}\frac{d^2}{dr^2} + \left[-\frac{Ze^2}{r} + \frac{\hbar^2 l(l+1)}{2\mu r^2}\right]\right)u(r) = Eu(r), \tag{2}$$

where the coordinate r in this equation is the "physical" r, measured in centimeters or Angstrom units and E is the energy measured in eV, for example. Let us first switch to dimensionless quantities, and let the $r_{phys.}$ and E of the above equation be replaced by dimensionless quantities r, and ϵ

$$r_{phys.} = \frac{a_0}{Z}r, \qquad \text{with} \quad a_0 = \frac{\hbar^2}{\mu e^2}, \qquad E = \frac{\mu Z^2 e^4}{\hbar^2}\epsilon, \tag{3}$$

leading to the radial equation in dimensionless quantities

$$\left(-\frac{d^2}{dr^2} - \frac{2}{r} + \frac{l(l+1)}{r^2}\right)u_{\lambda l}(r) = 2\epsilon u_{\lambda l}(r) = \lambda u_{\lambda l}(r). \tag{4}$$

This equation is factorizable via the factors

$$O_\pm(l) = \left(\mp\frac{d}{dr} + \frac{l}{r} - \frac{1}{l}\right), \tag{5}$$

with

$$O_+(l)O_-(l)u_{\lambda l} = \left(-\frac{d^2}{dr^2} - \frac{2}{r} + \frac{l(l+1)}{r^2} + \frac{1}{l^2}\right)u_{\lambda l} = \left[\lambda + \frac{1}{l^2}\right]u_{\lambda l}, \tag{6}$$

$$\begin{aligned} O_-(l+1)O_+(l+1)u_{\lambda l} &= \left(-\frac{d^2}{dr^2} - \frac{2}{r} + \frac{l(l+1)}{r^2} + \frac{1}{(l+1)^2}\right)u_{\lambda l} \\ &= \left[\lambda + \frac{1}{(l+1)^2}\right]u_{\lambda l}, \end{aligned} \tag{7}$$

so the factorization works, and

$$\mathcal{L}(l) = -\frac{1}{l^2}. \tag{8}$$

Because $\mathcal{L}$ is an increasing function of l for positive l, and because ϵ and, hence, λ must be a negative quantity for bound states, $[\lambda - \mathcal{L}(l+1)]$ will be a positive quantity only up through a maximum l-value. Thus,

$$\lambda = \mathcal{L}(l_{max} + 1) = -\frac{1}{(l_{max}+1)^2} = 2\epsilon. \tag{9}$$

Renaming the integer l_{max}: $l_{max} + 1 = n$, or $l_{max} = (n-1)$, we obtain the hydrogen result

$$\epsilon = -\frac{1}{2n^2}, \qquad E = -\frac{\mu Z^2 e^4}{\hbar^2}\frac{1}{2n^2}. \tag{10}$$

The starting function is obtained from

$$O_+(l_{max}+1)u_{\lambda l_{max}} = O_+(n)u_{n,l=(n-1)} = \left(-\frac{d}{dr} + \frac{n}{r} - \frac{1}{n}\right)u_{n,n-1} = 0, \tag{11}$$

leading to the normalized solution

$$u_{n,l=n-1} = N_n r^n e^{-\frac{r}{n}}, \qquad \text{with} \qquad N_n = \sqrt{\left(\frac{2}{n}\right)^{2n+1}\frac{1}{(2n)!}}. \tag{12}$$

The radial functions for the lower l values for a definite n can be obtained from these by action with the normalization-preserving step-down operators, $\mathcal{O}_-(l)$,

$$\mathcal{O}_-(l) = \frac{1}{\sqrt{[\lambda - \mathcal{L}(l)]}}\left(\frac{d}{dr} + \frac{l}{r} - \frac{1}{l}\right) = \frac{nl}{\sqrt{(n-l)(n+l)}}\left(\frac{d}{dr} + \frac{l}{r} - \frac{1}{l}\right). \tag{13}$$

For example, for $n = 2$, the starting function with $l = 1$ is given by

$$u_{n=2,l=1}(r) = \frac{1}{2\sqrt{6}}r^2 e^{-\frac{r}{2}}. \tag{14}$$

The eigenfunction with $l = 0$ is obtained via

$$u_{n=2,l=0}(r) = \frac{2}{\sqrt{3}}\left(\frac{d}{dr} + \frac{1}{r} - 1\right)\frac{1}{2\sqrt{6}}r^2 e^{-\frac{r}{2}}$$
$$= \frac{1}{2\sqrt{2}}r(2-r)e^{-\frac{r}{2}}. \tag{15}$$

We tabulate a few of the radial eigenfunctions obtained in this way for the lower n values. With $rR(r) = u(r)$, the $R(r)$ are given by

$$\begin{aligned}
&\text{For} \quad n = 1, l = 0 : R_{10}(r) = 2e^{-r},\\
&\text{For} \quad n = 2, l = 1 : R_{21}(r) = \frac{1}{2\sqrt{6}}re^{-\frac{r}{2}},\\
&\text{For} \quad n = 2, l = 0 : R_{20}(r) = \frac{1}{2\sqrt{2}}(2-r)e^{-\frac{r}{2}},\\
&\text{For} \quad n = 3, l = 2 : R_{32}(r) = \frac{2\sqrt{2}}{3^4\sqrt{15}}r^2 e^{-\frac{r}{3}},\\
&\text{For} \quad n = 3, l = 1 : R_{31}(r) = \frac{2\sqrt{2}}{3^4\sqrt{3}}r(6-r)e^{-\frac{r}{3}},\\
&\text{For} \quad n = 3, l = 0 : R_{30}(r) = \frac{2}{3^4\sqrt{3}}(27 - 18r + 2r^2)e^{-\frac{r}{3}}.
\end{aligned} \tag{16}$$

Here, the dimensionless r is $r = (Zr_{\text{phys.}}/a_0)$. To convert to a normalization in physical space $\int_0^\infty dr_{\text{phys.}} r_{\text{phys.}}^2 |R(r_{\text{phys.}})|^2 = 1$, the above results must be multiplied with the additional normalization factor $(Z/a_0)^{\frac{3}{2}}$.

11

Shape-Invariant Potentials: Soluble One-Dimensional Potential Problems

Having seen and used a number of examples, let us now look at the factorization method in a more general way. For the factorization method to work, we must have

$$[O_+(m)O_-(m) + \mathcal{L}(m)]u_{\lambda m} = \lambda u_{\lambda m}, \tag{1}$$

$$\left(\left[-\frac{d}{dx} + k(x,m)\right]\left[\frac{d}{dx} + k(x,m)\right] + \mathcal{L}(m)\right)u_{\lambda m} = \lambda u_{\lambda m}, \tag{2}$$

$$\begin{aligned}&\left(-\frac{d^2}{dx^2} + [k^2(x,m) - k'(x,m) + \mathcal{L}(m)]\right)u_{\lambda m} = \\ &\left(-\frac{d^2}{dx^2} + V(x,m)\right)u_{\lambda m} = \lambda u_{\lambda m},\end{aligned} \tag{3}$$

where the potential function $V(x, m)$ is expressed in terms of $k(x, m)$ and its first derivative is expressed by a prime. We must also have

$$[O_-(m+1)O_+(m+1) + \mathcal{L}(m+1)]u_{\lambda m} = \lambda u_{\lambda m}. \tag{4}$$

Eqs. (1) and (4) are the two conditions, I and II, of eq. (17) of Chapter 7, which must be satisfied for the factorization method to work. Now, shifting the index m to $(m - 1)$ in eq. (4)

$$\begin{aligned}&[O_-(m)O_+(m) + \mathcal{L}(m)]u_{\lambda(m-1)} = \lambda u_{\lambda(m-1)} \\ &= \left(\left[\frac{d}{dx} + k(x,m)\right]\left[-\frac{d}{dx} + k(x,m)\right] + \mathcal{L}(m)\right)u_{\lambda(m-1)} = \lambda u_{\lambda(m-1)}\end{aligned} \tag{5}$$

$$\left(-\frac{d^2}{dx^2} + [k^2(x,m) + k'(x,m) + \mathcal{L}(m)]\right)u_{\lambda(m-1)}$$
$$= (-\frac{d^2}{dx^2} + V(x, m-1))u_{\lambda(m-1)} = \lambda u_{\lambda(m-1)}, \qquad (6)$$

where

$$V(x,m) = k^2(x,m) - k'(x,m) + \mathcal{L}(m), \qquad (7)$$

$$V(x,m-1) = k^2(x,m) + k'(x,m) + \mathcal{L}(m), \qquad (8)$$

or

$$V(x,m) - V(x,m-1) = -2k'(x,m). \qquad (9)$$

In general, of course, this equation will not be satisfied for arbitrary $k(x,m)$. The factorization method works only if this condition is satisfied. Iterating this equation for $m-1, m-2, \ldots$, down to $m=1$, we are lead to the relation

$$V(x,m) - V(x,0) = [k^2(x,m) - k'(x,m) + \mathcal{L}(m) - k^2(x,0)$$
$$+ k'(x,0) - \mathcal{L}(0)] = -2\sum_{n=1}^{n=m} k'(x,n). \qquad (10)$$

Infeld and Hull studied the question: What kind of $k(x,m)$ can satisfy this equation? Trying first a Taylor series in m

$$k(x,m) = k_0(x) + k_1(x)m + \cdots, \qquad (11)$$

they found the potential collapses to a constant independent of x (hence, a trivial unimportant case), if terms quadratic in m or higher powers of m are included. Nevertheless, the possible functions $k_0(x)$ and $k_1(x)$ lead to a number of interesting equations. Similarly, trying Laurent series in m

$$k(x,m) = \cdots + \frac{k_{-1}(x)}{m} + k_0(x) + k_1(x)m + \cdots, \qquad (12)$$

they again found inverse quadratic and higher inverse powers of m lead to potentials independent of x and, hence, trivial. With the inverse first power in m, however, they found a number of new interesting cases.

It would of course be much nicer if we could immediately answer the question: Given a potential, $V(x,m)$, can we find solutions for the Schrödinger equation by the factorization method, or, what is equivalent: Can we find expressions for its eigenfunctions and eigenvalues in simple analytic form? Because this question has no simple general answer, we shall be content to follow the backward approach of Infeld and Hull, and starting with a set of possible $k(x,m)$ discover quite a number of soluble problems. Recall again that the factorization method involves nothing more mathematically challenging than the integration of a first-order differential equation and the taking of first derivatives in the laddering process.

A Shape-Invariant Potentials

If the potentials $V(x, m)$ and $V(x, m-1)$ are related as of eq. (9), the following is true.

1. The factorization method works.

2. The spectrum of allowed eigenvalues, λ, for the potential $V(x, m-1)$ is the same as that for $V(x, m)$, except the eigenvalue, $\lambda = \mathcal{L}(m_{\text{min}})$, does not exist in the spectrum for $V(x, m-1)$, because $u_{\lambda m_{\text{min}}-1}$ does not exist, assuming for now we are dealing with a case for which $\mathcal{L}(m)$ is a decreasing function of m. This follows because the eigenvalue λ does not change when we shift m to $m-1$ in equation (II).

3. The potentials $V(x, m)$ and $V(x, m-1)$ are said to have the same shape, because the dependence on x is the same, and only the value of m is replaced by $m-1$ ("Shape invariance" of the potential).

Now, had we written equation (I) of the factorized form with m replaced by $m-1$, and then shifted $m-1$ to $m-2$ in equation (II), we see the equation for $V(x, m-2)$ has the same spectrum as that for $V(x, m-1)$, except the eigenvalue $\lambda = \mathcal{L}(m_{\text{min}} - 1)$ is now missing. Thus, we can have a whole set of potentials with the same shape, all with the same spectrum, except the lowest eigenvalue of $V(x, m)$ is missing in $V(x, m-1)$, the lowest eigenvalue of $V(x, m-1)$, and hence the two lowest eigenvalues of $V(x, m)$ are missing for $V(x, m-2)$, and so on. Thus, the spectrum for $V(x, m-n)$ is the same as that for $V(x, m)$, except the lowest n eigenvalues of $V(x, m)$ are missing in the spectrum for $V(x, m-n)$, provided the factorization is such that the eigenvalues are given by $\lambda = \mathcal{L}(m_{\text{min}})$, that is, cases for which $\mathcal{L}(m)$ is a decreasing function of m. Similar arguments can be made for the other case, i.e., if $\mathcal{L}(m)$ is an increasing function of m. In that case, setting $m \to m+1$ in eqs. (3) and (6), we see $V(x, m+1)$ has the same spectrum of λ values, now with $\lambda = \mathcal{L}(m_{\text{max.}}+1)$, except $\lambda = \mathcal{L}(m_{\text{max.}}+1)$, which exists in the spectrum for $V(x, m)$ with eigenfunction $u_{\lambda, m_{\text{max.}}}$, does not exist in the spectrum for $V(x, m+1)$, because $u_{\lambda, m_{\text{max}}+1}$ does not exist. Similarly, in the spectrum for $V(x, m+n)$, the eigenvalues $\lambda = \mathcal{L}(m_{\text{max.}}+1), \mathcal{L}(m_{\text{max.}}+2), \ldots, \mathcal{L}(m_{\text{max.}}+n)$ do not exist. The lowest eigenvalue for $V(x, m+n)$ is $\lambda = \mathcal{L}(m_{\text{max.}}+n+1)$, which is also the nth eigenvalue for $V(x, m)$.

B A Specific Example

As a very specific example, consider a 1-D Schrödinger equation for a particle moving in the domain $0 \leq x \leq a$ under the potential

$$\begin{aligned} V(x) &= \frac{\mathcal{V}_0}{\sin^2(\frac{\pi x}{a})}, \qquad 0 \leq x \leq a, \\ &= \infty, \qquad x \leq 0, \qquad x \geq a, \end{aligned} \tag{13}$$

$$-\frac{\hbar}{2m}\frac{d^2}{dx^2}u(x)+\frac{\mathcal{V}_0}{\sin^2(\frac{\pi x}{a})}u(x)=Eu(x), \tag{14}$$

or, with

$$\frac{\pi x}{a}=\theta, \qquad E=\epsilon\left(\frac{\hbar^2\pi^2}{2ma^2}\right), \qquad \mathcal{V}_0=V_0\left(\frac{\hbar^2\pi^2}{2ma^2}\right), \tag{15}$$

$$-\frac{d^2u}{d\theta^2}+\frac{V_0}{\sin^2\theta}u=\epsilon u(\theta), \tag{16}$$

to be compared with our factorizable equation

$$-\frac{d^2u}{d\theta^2}+\frac{[m_0^2-\frac{1}{4}]}{\sin^2\theta}u=\lambda u=\epsilon u. \tag{17}$$

Now, we let

$$V_0(\theta,m_0)=\frac{[m_0^2-\frac{1}{4}]}{\sin^2\theta}, \qquad \text{with} \quad V_0+\frac{1}{4}=m_0^2. \tag{18}$$

In order to work in the m region near $m_{\min}$, we shall choose the negative root for m_0

$$m_0=-\sqrt{V_0+\frac{1}{4}}=-|m_0|. \tag{19}$$

(We will subsequently investigate the region of m values near the positive root to show these give the same result.) For the above θ equation, we found $\mathcal{L}(m)=(m-\frac{1}{2})^2$, and with $m_{\min}=m_0=-|m_0|$, we get the lowest eigenvalue, $\lambda_0=\epsilon_0$,

$$\lambda_0=\mathcal{L}(m_0)=(m_0-\frac{1}{2})^2=(|m_0|+\frac{1}{2})^2=(\sqrt{V_0+\frac{1}{4}}+\frac{1}{2})^2. \tag{20}$$

The eigenfunction for the ground state of V_0 is obtained from

$$O_-(m_0)u_{\lambda_0 m_0}=\frac{du}{d\theta}+(m_0-\frac{1}{2})\cot\theta u(\theta)=0, \tag{21}$$

with the solution

$$u_{\lambda_0 m_0}=N\big(\sin(\theta)\big)^{\frac{1}{2}-m_0}. \tag{22}$$

This is a square-integrable function, with $m_0=-|m_0|$. The companion potential $V(\theta,m_0-1)=V(\theta,(-|m_0|-1))$ has ground-state eigenvalue $\lambda=(-|m_0|-1-\frac{1}{2})^2=(|m_0|+1+\frac{1}{2})^2$. Let us name this potential V_1, its ground state eigenvalue λ_1, and note that this is the first excited state, λ_1, for the potential V_0. Similarly, the nth-companion potential $V(\theta,m_0-n)=V(\theta,(-|m_0|-n))$ has lowest eigenvalue λ_n, where we name this potential V_n:

$$\lambda_n=(m_0-n-\frac{1}{2})^2=(|m_0|+n+\frac{1}{2})^2. \tag{23}$$

This is the ground state for the potential V_n and the nth excited state for V_0. The ground state wave function for the potential V_n is

$$u_{\lambda_n(m_0-n)} = \sqrt{\frac{\pi}{a}\frac{\Gamma(|m_0|+n+\frac{3}{2})}{\Gamma(\frac{1}{2})\Gamma(|m_0|+n+1)}}\sin^{(|m_0|+n+\frac{1}{2})}(\theta) = N_n \sin^{(|m_0|+n+\frac{1}{2})}(\theta), \tag{24}$$

where we have now included the normalization factor explicitly. Recall

$$\int_0^\pi d\theta \sin^{2\alpha}\theta = B(\tfrac{1}{2}, \alpha+\tfrac{1}{2}) = \frac{\Gamma(\frac{1}{2})\Gamma(\alpha+\frac{1}{2})}{\Gamma(\alpha+1)},$$

where B is the Beta function expressed in terms of Γ functions. To get the eigenfunction for the first excited state of the potential V_{n-1}, with this energy λ_n, we need to act with the normalized step-up operator $\mathcal{O}_+(m_0-n+1)$,

$$u_{\lambda_n(m_0-n+1)} = \frac{O_+(-|m_0|-n+1)}{\sqrt{[\lambda_n - \mathcal{L}(m_0-n+1)]}} u_{\lambda_n(m_0-n)}, \tag{25}$$

with

$$\lambda_n = (|m_0|+n+\frac{1}{2})^2, \qquad \mathcal{L}(m_0-n+1) = (|m_0|+n-\tfrac{1}{2})^2. \tag{26}$$

Finally, to get the eigenfunction for the nth excited state with this energy λ_n in the potential V_0, we need to act n-times with such step-up operators (laddering along the horizontal λ_n-line in Fig. 11.1):

$$\begin{aligned} u_{\lambda_n m_0}(\theta) &= \mathcal{O}_+(-|m_0|)\cdots\mathcal{O}_+((-|m_0|-n+2)\mathcal{O}_+(-|m_0|-n+1) \\ &\quad\times N_n \sin^{(|m_0|+n+\frac{1}{2})}(\theta) \\ &= \frac{\left(-\frac{d}{d\theta}-(|m_0|+\frac{1}{2})\cot\theta\right)}{\sqrt{[\lambda_n-(|m_0|+\frac{1}{2})^2]}}\cdots\frac{\left(-\frac{d}{d\theta}-(|m_0|+n-\frac{3}{2})\cot\theta\right)}{\sqrt{[\lambda_n-(|m_0|+n-\frac{3}{2})^2]}} \\ &\quad\times\frac{\left(-\frac{d}{d\theta}-(|m_0|+n-\frac{1}{2})\cot\theta\right)}{\sqrt{[\lambda_n-(|m_0|+n-\frac{1}{2})^2]}} N_n \sin^{(|m_0|+n+\frac{1}{2})}(\theta). \end{aligned} \tag{27}$$

In Fig. 11.1, a family of shape-invariant potentials of this $(1/\sin^2\theta)$ shape are shown, where we have chosen $m_0 = -1.1$, so the strength of V_0 is $(-1.1)^2 - \frac{1}{4} = 0.96$, leading to potentials V_1, V_2, V_3, V_4 with strengths of 4.16, 9.36, 16.56, 25.76, respectively. The energies given by eq. (23) are shown in the figure.

In particular, if we had tried to continue the laddering process of eq. (27) one more time from $m_0 = -1.1$ to an $m_0 = -0.1$, we would be led to a potential of strength $(-0.1)^2 - \frac{1}{4} = -0.24$, of the opposite sign from the potentials shown, i.e., a repulsive potential, with no bound states. Therefore, the process has to stop at V_0. No connection can exist from the problem with negative m values to the branch with positive m values, as for the θ equation for the spherical harmonics. The negative and positive m values are connected only in two special cases: if m_0

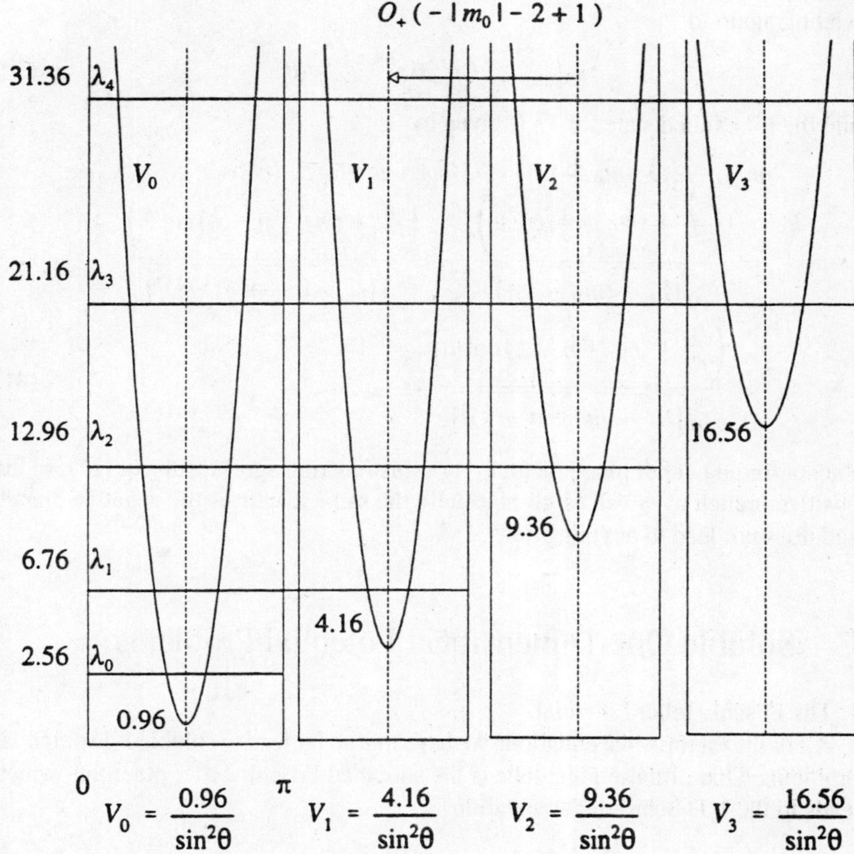

FIGURE 11.1. The family of shape-invariant $1/\sin^2\theta$ potentials, with $m_0 = -1.1$; $m = m_{\text{min.}\to+\infty}$.

is integer or $\frac{1}{2}$-integer. For an arbitrary value of V_0, it remains to be shown that the positive root, $m_0 = +\sqrt{V_0 + \frac{1}{4}}$, gives the same spectrum of eigenvalues and eigenfunctions. In the region of positive m values, $\mathcal{L}(m) = (m-\frac{1}{2})^2$ is an increasing function of m and $\lambda = \mathcal{L}(m_{\text{max.}} + 1)$. The shape-invariant partner potentials are $V_0(\theta, m_0)$, $V_1(\theta, m_0 + 1), \dots, V_n(\theta, m_0 + n)$, with $m_{\text{max.}} = m_0 = +\sqrt{V_0 + \frac{1}{4}}$ for V_0 and $m_{\text{max.}} = m_0 + n$ for V_n, so λ_n, which is the ground-state eigenvalue for V_n and the nth excited state for V_0, is given by

$$\lambda_n = (m_0 + n + \tfrac{1}{2})^2 \qquad \text{with} \quad m_0 > 0, \tag{28}$$

in agreement with eq. (23). Now the ground-state eigenfunction for V_n is given by

$$O_+(m_0 + n + 1)u_{\lambda_n, m_0+n} = \left(-\frac{d}{d\theta} + (m_0 + n + \tfrac{1}{2})\cot\theta\right)u_{\lambda_n, m_0+n} = 0, \tag{29}$$

leading again to

$$u_{\lambda_n, m_0+n} = N_n \sin^{(m_0+n+\frac{1}{2})}(\theta), \tag{30}$$

and the nth excited state for V_0 is given by

$$\begin{aligned} u_{\lambda_n m_0} &= \mathcal{O}_-(m_0+1) \cdots \mathcal{O}_-(m_0+n-1)\mathcal{O}_-(m_0+n) u_{\lambda_n(m_0+n)} \\ &= \frac{\left(\frac{d}{d\theta} + (m_0+\frac{1}{2})\cot\theta\right)}{\sqrt{[\lambda_n - (m_0+\frac{1}{2})^2]}} \cdots \frac{\left(\frac{d}{d\theta} + (m_0+n-\frac{3}{2})\cot\theta\right)}{\sqrt{[\lambda_n - (m_0+n-\frac{3}{2})^2]}} \\ &\times \frac{\left(\frac{d}{d\theta} + (m_0+n-\frac{1}{2})\cot\theta\right)}{\sqrt{[\lambda_n - (m_0+n-\frac{1}{2})^2]}} u_{\lambda_n(m_0+n)}. \end{aligned} \tag{31}$$

Except for an overall phase factor $(-1)^n$, this function agrees with eq. (27), so the positive branch of m values gives exactly the same results as the negative branch and does not lead to anything new.

C Soluble One-Dimensional Potential Problems

1. The Pöschl–Teller Potential.

All of the factorizable equations we have met so far lead to soluble 1-D potential problems. One of these potentials is the so-called Pöschl–Teller potential, which leads to the 1-D Schrödinger equation

$$-\frac{\hbar^2}{2m}\frac{d^2u(x)}{dx^2} - \frac{\mathcal{V}}{\cosh^2(x/a)} u(x) = Eu(x), \tag{32}$$

or introducing dimensionless quantities

$$z = \frac{x}{a}, \qquad V = \mathcal{V}\frac{2ma^2}{\hbar^2}, \qquad \epsilon = E\frac{2ma^2}{\hbar^2},$$

$$-\frac{d^2u(z)}{dz^2} - \frac{V}{\cosh^2 z} u(z) = \epsilon u(z) = \lambda u(z), \tag{33}$$

where the case $V > 0$ leads to an attractive potential. With

$$V = l(l+1), \quad \text{or} \quad l = -\tfrac{1}{2} \pm \sqrt{V + \tfrac{1}{4}},$$

Chapter 9 tells us

$$O_\pm = \mp \frac{d}{dz} + l \tanh z, \qquad \text{with} \quad \mathcal{L}(l) = -l^2. \tag{34}$$

This equation corresponds to an $\mathcal{L}(l)$ of case 4 of Chapter 7 with allowed negative values of $\epsilon = \lambda$ only for positive values of $l = l_{\text{min.}}, (l_{\text{min.}}+1), \ldots, (l_{\text{min.}}+n), \ldots,$

and for negative values of $l = l_{\text{max.}}, (l_{\text{max.}} - 1), \ldots, (l_{\text{max.}} - n), \ldots$. If we choose the positive branch, with $l = -\frac{1}{2} + \sqrt{V + \frac{1}{4}}$,

$$l = l_{\text{min.}} + n = -\tfrac{1}{2} + \sqrt{V + \tfrac{1}{4}}, \quad \text{so with} \quad \epsilon = \mathcal{L}(l_{\text{min.}}),$$

we have

$$\epsilon_n = \lambda_n = \mathcal{L}(l_{\text{min.}}) = -(l - n)^2 = -\left(\sqrt{V + \tfrac{1}{4}} - (n + \tfrac{1}{2})\right)^2, \qquad (35)$$

with shape-invariant potential partners $V_0(z, l_0), V_1(z, l_0 - 1), \ldots, V_n(z, l_0 - n)$, where now $l_0 = -\frac{1}{2} + \sqrt{V + \frac{1}{4}}$. Now a maximum possible value of $n = n_{\text{max.}}$ exists, however, for which $(l_0 - n_{\text{max.}})$ is such that

$$(l_0 - n_{\text{max.}})(l_0 - n_{\text{max.}} + 1) > 0; \quad \text{but } 0 < (l_0 - n_{\text{max.}}) < 1.$$

In that case,

$$\left(l_0 - (n_{\text{max.}} + 1)\right)\left(l_0 - (n_{\text{max.}} + 1) + 1\right) < 0,$$

and this implies the potential $V(z, l_0 - (n_{\text{max.}} + 1))$ is repulsive and therefore has no bound states. The condition $0 \leq (l_0 - n_{\text{max.}}) \leq 1$ determines $n_{\text{max.}}$ through

$$n_{\text{max.}} + \tfrac{1}{2} \leq \sqrt{V + \tfrac{1}{4}} \leq n_{\text{max.}} + \tfrac{3}{2}, \qquad \text{or}$$

$$n_{\text{max.}}(n_{\text{max.}} + 1) \leq V \leq (n_{\text{max.}} + 1)(n_{\text{max.}} + 2).$$

For $0 \leq V \leq 2$, $n_{\text{max.}} = 0$, and therefore only a single bound state with $\epsilon_0 = -(-\frac{1}{2} + \sqrt{V + \frac{1}{4}})^2$ exists, but always at least this one bound state exists, even as $V \to 0$. Note the similarity in this regard between the Pöschl–Teller potential and the square well potential with $V = -V_0$ for $|z| \leq a$, and $V = 0$ for $|z| > a$ (see section B of Chapter 4).

Finally, $u_{\lambda_n, l_{\text{min.}} = (l_0 - n)}$ is determined from

$$\frac{d}{dz} u_{\lambda_n (l_0 - n)} = -[(l_0 - n) \tanh z] u_{\lambda_n (l_0 - n)} = 0, \qquad \text{so}$$

$$u_{\lambda_n (l_0 - n)} = \frac{N_n}{(\cosh z)^{l_0 - n}} = \sqrt{\frac{\Gamma(l_0 - n + 1)}{a\Gamma(\frac{1}{2})\Gamma(l_0 - n)}} \frac{1}{(\cosh z)^{l_0 - n}}, \qquad (36)$$

where this is the ground-state eigenfunction for the potential, $V_n(z, l_0 - n)$, with $\epsilon_n = -(\sqrt{V + \frac{1}{4}} - (n + \frac{1}{2}))^2$, which is also the energy of the nth excited state for the potential, $V_0 = -V/(\cosh^2 z)$. The normalized eigenfunction for this n^{th} excited state of V_0 is again given by

$$u_{\lambda_n l_0} = \frac{O_+(l_0)}{\sqrt{[-(l_0 - n)^2 + l_0^2]}} \cdots \frac{O_+(l_0 - n + 2)}{\sqrt{[(-(l_0 - n)^2 + (l_0 - n + 2)^2]}}$$
$$\times \frac{O_+(l_0 - n + 1)}{\sqrt{[-(l_0 - n)^2 + (l_0 - n + 1)^2]}} u_{\lambda_n (l_0 - n)}. \qquad (37)$$

Finally, the negative branch of allowed l values, with $l = -\frac{1}{2} - \sqrt{V + \frac{1}{4}}$, i.e., with $l < 0$, gives no additional eigenvalues or eigenvectors. In this case, $l = l_{\text{max.}} - n$, and $\lambda = \mathcal{L}(l_{\text{max.}} + 1) = -(l + n + 1)^2 = -(-\sqrt{V + \frac{1}{4}} + n + \frac{1}{2})^2$, in agreement with the result of eq. (35) for the positive branch of allowed l values. Except for a possible overall phase factor, the eigenfunctions again agree with those of the other branch. We could again show this explicitly as in the previous example, but also we note: Because we are dealing with a 1-D eigenvalue problem, we do not expect degeneracies for the general ϵ_n.

The inverse $\sin^2\theta$ potential, $V/(\sin^2\theta)$, and the Pöschl–Teller potential, $-V/(\cosh^2 z)$, are special cases of the general factorizable case, for which the 1-D Schrödinger equation can be written as

$$\left(-\frac{d^2}{dz^2} + V(z, m)\right)u(z) = \lambda u(z), \qquad \text{with}$$

$$V(z, m) = \frac{b^2(m + c)(m + c + 1) + d^2 + 2bd(m + c + \frac{1}{2})\cos b(z + p)}{\sin^2 b(z + p)} \tag{38}$$

and with

$$O_\pm(m) = \mp\frac{d}{dz} + (m + c)b\cot b(z + p) + \frac{d}{\sin b(z + p)},$$
$$\mathcal{L}(m) = b^2(m + c)^2, \tag{39}$$

where b, c, d, and p are arbitrary constants. For example, the Pöschl–Teller potential is obtained by setting $b = -i, c = 0, d = 0, p = (i\pi)/2$. Other specializations of this general case are listed by Infeld and Hull; see also problem 16, which treats the θ equation for the symmetric rigid rotator.

2. One-Dimensionalized Hydrogenic Potential.

The factorization of the radial equation for the hydrogen atom leads to a factorizable Schrödinger equation for a 1-D hydrogen-like potential

$$-\frac{\hbar^2}{2m}\frac{d^2u}{dx^2} + V(x)u(x) = Eu(x), \quad \text{with} \quad V(x) = -\frac{A}{x} + \frac{B}{x^2} \quad \text{for } |x| \geq 0, \tag{40}$$

where we set $V = \infty$ for $x < 0$ and $V(x)$ has a minimum for positive values of $x = 2B/A$ if both $A > 0$, $B > 0$. With dimensionless quantities

$$z = \frac{x}{(\hbar^2/mA)}, \quad \epsilon = E\left(\frac{\hbar^2}{mA^2}\right), \quad l(l + 1) = \frac{2m}{\hbar^2}B,$$

the Schrödinger equation becomes

$$\left(-\frac{d^2}{dz^2} - \frac{2}{z} + \frac{l(l + 1)}{z^2}\right)u(z) = 2\epsilon u(z) = \lambda u(z), \qquad \text{with} \tag{41}$$

$$l = -\frac{1}{2} \pm \sqrt{\frac{2mB}{\hbar^2} + \frac{1}{4}}. \tag{42}$$

The results of Chapter 10 tell us

$$O_{\pm}(l) = \mp\frac{d}{dz} + \left(\frac{l}{z} - \frac{1}{l}\right), \qquad \text{with } \mathcal{L}(l) = -\frac{1}{l^2}. \tag{43}$$

From Chapter 10, we also know this equation will have bound states with $\lambda = 2\epsilon < 0$. For the branch of $\mathcal{L}(l)$, which is an increasing function of l, i.e., the positive branch with $l_0 = -\frac{1}{2} + \sqrt{(2mB)/\hbar^2 + \frac{1}{4}}$, the allowed l values range from $l_{\text{max.}}$, $(l_{\text{max.}} - 1), \ldots, (l_{\text{max.}} - n), \ldots$, and $\lambda = \mathcal{L}(l_{\text{max.}} + 1)$, so, with $\lambda_n = 2\epsilon_n$,

$$2\epsilon_n = -\frac{1}{(l_{\text{max.}} + 1)^2} = -\frac{1}{(l + n + 1)^2} = -\frac{1}{\left[\sqrt{\frac{2mB}{\hbar^2} + \frac{1}{4}} + (n + \frac{1}{2})\right]^2}. \tag{44}$$

The shape-invariant partner potentials are $V_0(z, l_0)$, $V_1(l_0 + 1), \ldots, V_n(z, l_0 + n)$. The ground-state eigenfunction of $V_n(z, l_0 + n)$, with eigenvalue ϵ_n, is given by

$$O_+(l_0 + n + 1)u_{\lambda_n(l_0+n)} = \left(-\frac{d}{dz} + \frac{(l_0 + n + 1)}{z} - \frac{1}{(l_0 + n + 1)}\right)u_{\lambda_n(l_0+n)} = 0 \tag{45}$$

leading to a normalized

$$u_{\lambda_n(l_0+n)} = \sqrt{\left[\frac{2}{(l_0 + n + 1)}\right]^{(2l_0+n+3)} \frac{1}{\Gamma(2l_0 + 2n + 3)}} z^{l_0+n+1} e^{-\frac{z}{(l_0+n+1)}}. \tag{46}$$

The ground-state eigenfunction of $V_0(z, l_0)$ is obtained from this equation by setting $n = 0$. The eigenfunction of the nth excited state of V_0, with ϵ_n, is again given by

$$u_{\lambda_n l_0} = \frac{O_-(l_0 + 1)}{\sqrt{[\lambda_n - \mathcal{L}(l_0 + 1)]}} \cdots \frac{O_-(l_0 + n - 1)}{\sqrt{[\lambda_n - \mathcal{L}(l_0 + n - 1)]}} \cdot \times \frac{O_-(l_0 + n)}{\sqrt{[\lambda_n - \mathcal{L}((l_0 + n)]}} u_{\lambda_n(l_0+n)}. \tag{47}$$

For an arbitrary value of l_0, not equal to an integer or $\frac{1}{2}$-integer, and a fixed $l_{\text{max.}}$, an n value will exist such that $l(l + 1) = (l_{\text{max.}} - n)(l_{\text{max.}} - n + 1)$ becomes a negative quantity. Because the generalized hydrogenic potential remains attractive even for this case, the value of the integer, n, can go to arbitrarily high values, and an infinite number of bound states exist. The values for $(l_0 + n)$ are positive for all positive integers n in the shape-invariant partner potentials, $V_n(z, l_0 + n)$. The action of the n stepdown operators, O_-, on $u_{\lambda_n(l_0+n)}$ produce an eigenfunction of the form, $z^{l_0+1}\mathcal{P}_n(z)e^{-\frac{z}{(l_0+n+1)}}$, where $\mathcal{P}_n(z)$ is a polynomial of degree n. This function is square-integrable over the interval, $0 \leq z \leq \infty$, for all positive integers, n. Because a second branch of allowed $\mathcal{L}(l)$ values for negative values of l exists, with $l = -\frac{1}{2} - \sqrt{2mB/\hbar^2 + \frac{1}{4}} = -(l_0 + 1)$ and $l = l_{\text{min.}}$, $(l_{\text{min.}} + 1), \ldots, (l_{\text{min.}} + n), \ldots$, we again need to examine the possibility this branch would lead to new eigenvalues. For $l < 0$, $\mathcal{L}(l)$ is a decreasing function of l. Therefore, now, with $\lambda_n = 2\epsilon_n$,

$$2\epsilon_n = \mathcal{L}(l_{\text{min.}}) = -\frac{1}{l_{\text{min.}}^2} = -\frac{1}{(l - n)^2} = -\frac{1}{\left[-\frac{1}{2} - \sqrt{2mB/\hbar^2 + \frac{1}{4}} - n\right]^2}, \tag{48}$$

exactly the same result as that already obtained for the positive l branch. Both the eigenvalues and eigenfunctions obtained from this negative l branch, thus, do not give anything new.

3. The Morse Potential.

Another 1-D potential leading to a factorizable Schrödinger equation is the Morse potential (see Fig. 11.2),

$$V(x) = D(e^{-2(x/a)} - 2e^{-(x/a)}), \tag{49}$$

$$\left[-\frac{\hbar^2}{2\mu}\frac{d^2}{dx^2} + D(e^{-2(x/a)} - 2e^{-(x/a)})\right]u(x) = Eu(x). \tag{50}$$

D gives the classical ionization or dissociation energy. The potential has a minimum value, $V_{\text{min.}} = -D$, at $x = 0$. For $E \geq 0$, a continuous spectrum exists. The particle can proceed to $x \to +\infty$. With the introduction of dimensionless quantities,

$$z = \frac{x}{a}, \quad \epsilon = E\frac{2\mu a^2}{\hbar^2}, \quad \delta^2 = D\frac{2\mu a^2}{\hbar^2}, \quad \text{this function leads to}$$

$$-\frac{d^2u}{dz^2} + (\delta^2 e^{-2z} + 2\delta(m + \frac{1}{2})e^{-z})u(z) = \epsilon u(z) = \lambda u(z), \tag{51}$$

where the parameter, m, with

$$\frac{(m+\frac{1}{2})}{\delta} = -1, \tag{52}$$

has been introduced to put the equation into factorizable form, with

$$O_\pm(m) = \mp\frac{d}{dz} + (\delta e^{-z} + m), \qquad \text{and} \quad \mathcal{L}(m) = -m^2. \tag{53}$$

Note, $-\delta - \frac{1}{2} = -\sqrt{(2\mu a^2 D/\hbar^2)} - \frac{1}{2}$, and hence, m, is a patently negative quantity. For $m < 0$, the above $\mathcal{L}(m)$ is an increasing function of m. For bound states, with $\lambda < 0$, a maximum possible value of $m = m_{\text{max.}}$ exists. The allowed m values are $m = m_{\text{max.}}, (m_{\text{max.}} - 1), \ldots, (m_{\text{max.}} - n), \ldots$, with

$$\begin{aligned}\lambda_n = \epsilon_n = \mathcal{L}(m_{\text{max.}} + 1) &= -(m+n+1)^2 = -(-\delta + n + \tfrac{1}{2})^2 \\ &= -\delta^2 + 2\delta(n+\tfrac{1}{2}) - (n+\tfrac{1}{2})^2,\end{aligned} \tag{54}$$

so

$$E_n = -D + 2\sqrt{\frac{\hbar^2 D}{2\mu a^2}}(n+\frac{1}{2}) - \frac{\hbar^2}{2\mu a^2}(n+\frac{1}{2})^2. \tag{55}$$

For the case $\delta \gg 1$, the last term, quadratic in $(n + \frac{1}{2})$, will be much smaller than the linear term, and the excitation energy is that of a slightly anharmonic oscillator, with

$$E_n + D \approx \hbar\omega(n+\tfrac{1}{2}), \qquad \text{with } \hbar\omega = 2\sqrt{\frac{\hbar^2 D}{2\mu a^2}}. \tag{56}$$

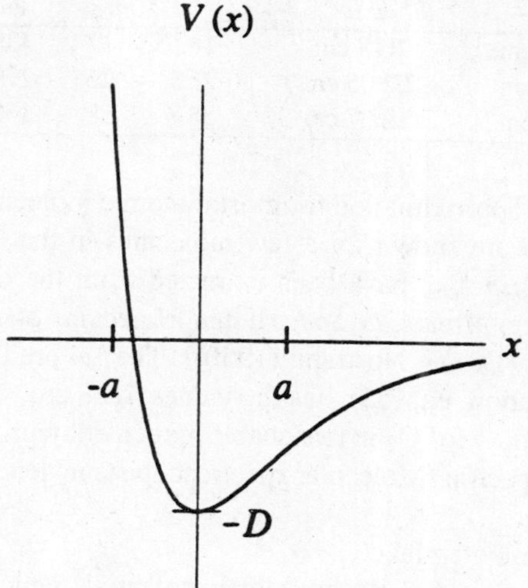

FIGURE 11.2. The Morse potential.

The shape-invariant partner potentials are $V_0(z, m_0)$, $V_1(z, m_0+1), \ldots, V_n(z, m_0+n)$, with $m_0 = -(\delta + \frac{1}{2})$, so

$$V_n(z, m_0 + n) = ((\delta^2 e^{-2z} + 2\delta(-\delta + n)e^{-z}). \tag{57}$$

This potential has the form shown in Fig. 11.2, with an attractive minimum, only for $n < \delta$; and the number of vibrational states is therefore limited. A maximum possible n value, $n_{\text{max.}}$ exists.

The ground-state eigenfunction of V_n is given by

$$O_+(m_0 + n + 1)u_{\lambda_n(m_0+n)} = \left[-\frac{d}{dz} + \left(\delta e^{-z} + (n - \delta + \tfrac{1}{2})\right)\right]u_{\lambda_n(m_0+n)} = 0, \tag{58}$$

$$u_{\lambda_n(m_0+n)} = N_n e^{-[(\delta-n-\frac{1}{2})z+\delta e^{-z}]}. \tag{59}$$

Successive action with the normalized step-down operators, $\mathcal{O}_-(m)$, with $m = (m_0 + n), (m_0 + n - 1), \ldots, (m_0 + 1)$ yields the needed n^{th} excited-state eigenfunction of V_0. These $u_{\lambda_n,m}$ are normalized in the interval $-\infty \le z \le +\infty$. Also, the $u_{\lambda_n,m}$ are normalizable only for integers n such that $n < (\delta - \frac{1}{2})$, which determines, $n_{\text{max.}}$.

In the actual applications, the Morse potential is used for the relative motion of the two atoms in a diatomic molecule, i.e., for the radial function of this two-body problem. In that case, therefore, μ is the reduced mass of the diatomic molecule, and $x = (r - r_e)$, where, r_e is the equilibrium value of the interatomic distance, r. Thus, the eigenfunctions *should* apply to the interval, $-(r_e/a) \le z \le +\infty$. For realistic parameters for most diatomic molecules, however, the Morse potential has such a large positive value at $z = -(r_e/a)$ that the Morse eigenfunctions are effectively zero for $z < -(r_e/a)$. The 1-D solutions found above for the full z-space

Molecule	$D_{\text{obs.}}\frac{1}{hc}$	$(\hbar^2/2\mu a^2)_{\text{obs.}}\frac{1}{hc}$	δ	$n_{\text{max.}}$	$(\hbar\omega)_{\text{obs.}}\frac{1}{hc}$
H_2	38,276 cm^{-1}	118 cm^{-1}	18.0	17	4395 cm^{-1}
HCl	37,257 cm^{-1}	52.05 cm^{-1}	26.7	26	2990 cm^{-1}
O_2	41,758 cm^{-1}	12.07 cm^{-1}	58.8	58	1580 cm^{-1}

are therefore a good approximation for most diatomic molecules. The parameters $D, (\hbar^2/2\mu a^2)$, and δ are shown for a few molecules in the table provided here. The $D_{\text{obs.}}$ and $(\hbar^2/2\mu a^2)_{\text{obs.}}$ have been extracted from the observed vibrational spectra, [G. Herzberg; *Molecular Spectra and Molecular Structure*. I. *Spectra of Diatomic Molecules*, D. van Nostrand (1950)]. The $\hbar\omega$ predicted by the Morse-potential energy relation, eq. (55), has the values 4128 cm^{-1} for H_2, 2728 cm^{-1} for HCl, and 1407 cm^{-1} for O_2 in reasonable agreement with the values extracted from the observed spectra. (Molecular spectroscopists in general give (energy/hc) in wavenumbers, cm^{-1}.)

4. The Rosen–Morse Potential.

A similar potential is the Rosen–Morse potential, which leads to the 1-D Schrödinger equation

$$-\frac{\hbar^2}{2\mu}\frac{d^2u}{dx^2}+\Big(-\frac{V_1}{\cosh^2(x/a)}+2V_2\tanh\Big(\frac{x}{a}\Big)\Big)u(x)=Eu(x). \tag{60}$$

With dimensionless quantities,

$$z=\frac{x}{a},\qquad \epsilon=E\frac{2\mu a^2}{\hbar^2},\qquad m(m+1)=V_1\frac{2\mu a^2}{\hbar^2},\qquad q=V_2\frac{2\mu a^2}{\hbar^2},$$

this equation leads to

$$-\frac{d^2u}{dz^2}+\Big(-\frac{m(m+1)}{\cosh^2 z}+2q\tanh z\Big)u(z)=\epsilon u(z). \tag{61}$$

This potential has an attractive well with a minimum at z_0, given by

$$\tanh z_0=-\frac{V_2}{V_1}=-\frac{q}{m(m+1)}, \tag{62}$$

which has a solution only for

$$|q/m(m+1)|<1. \tag{63}$$

The equation is factorizable, with

$$O_\pm(m)=\Big(\mp\frac{d}{dz}+m\tanh z+\frac{q}{m}\Big),\qquad \text{and}\quad \mathcal{L}(m)=-m^2-\frac{q^2}{m^2}. \tag{64}$$

Choosing the branch of $\mathcal{L}(m)$ with positive values of m, this $\mathcal{L}(m)$ belongs to case 4. The maximum of the function $\mathcal{L}(m)$ occurs at $m=\sqrt{|q|}$, where $\mathcal{L}(m)$ has the value $-2|q|$, which is also the ionization or dissociation value of $V(z)$. Thus, bound states will exist if $\epsilon<-2|q|$, and the requirement $[\lambda-\mathcal{L}(m)]\geq 0$, together with the requirement of eq. (63) leads to an allowed branch of m values with $m>\sqrt{|q|}$, where $\mathcal{L}(m)$ is a decreasing function of m, so $m=m_{\text{min.}}, (m_{\text{min.}}+1), \dots, (m_{\text{min.}}+$

$n)$, ..., with

$$\epsilon_n = \lambda_n = \mathcal{L}(m_{\text{min.}}) = -(m-n)^2 - \frac{q^2}{(m-n)^2}$$
$$= -\Big(\sqrt{\frac{2\mu a^2 V_1}{\hbar^2} + \tfrac{1}{4}} - (n+\tfrac{1}{2})\Big)^2 - \frac{q^2}{\Big(\sqrt{\frac{2\mu a^2 V_1}{\hbar^2} + \tfrac{1}{4}} - (n+\tfrac{1}{2})\Big)^2}. \quad (65)$$

The shape-invariant partner potentials are $V_0(z, m_0)$, $V_1(z, m_0-1)$, ..., $V_n(z, m_0-n)$, with $m_0 = -\frac{1}{2} + \sqrt{(2\mu a^2 V_1/\hbar^2) + \frac{1}{4}}$. The ground-state eigenfunction of $V(z, m_0 - n)$ is determined by

$$\Big(\frac{d}{dz} + (m_0 - n)\tanh z + \frac{q}{(m_0 - n)}\Big)u_{\lambda_n(m_0-n)} = 0,$$
$$u_{\lambda_n(m_0-n)} = N_n \frac{1}{(\cosh z)^{(m_0-n)}} e^{-\frac{q}{(m_0-n)}z}, \quad (66)$$

and the eigenfunction of the nth excited state of V_0 is obtained from this function by the action of n operators $\mathcal{O}_+(m)$ with m running from $(m_0 - n + 1)$ to m_0. Again, a maximum n value exists beyond which the potential $V(z, m_0 - n)$ ceases to have an attractive minimum with bound states and square-integrable bound-state eigenfunctions, so V_0 again has only a finite number of bound states.

5. The one-dimensional harmonic oscillator revisited.

The 1-D harmonic oscillator Schrödinger equation

$$\Big(-\frac{d^2}{dx^2} + x^2\Big)u(x) = 2\epsilon u(x), \quad (67)$$

with dimensionless, x and ϵ, is factorizable, with

$$O_\pm = \Big(\mp\frac{d}{dx} + x\Big). \quad (68)$$

The only parameter, however, is the energy, ϵ, itself, and the factors, $O_\pm$, are not functions of ϵ. Now,

$$\begin{aligned} &\text{I} \quad O_+O_-u(x) = (2\epsilon - 1)u(x) = [-1 + 2\epsilon]u(x), \\ &\text{II} \quad O_-O_+u(x) = (2\epsilon + 1)u(x) = [-1 + 2(\epsilon + 1)]u(x), \end{aligned} \quad (69)$$

are to be compared with

$$\begin{aligned} &\text{I} \quad O_+O_-u_{\lambda m} = [\lambda - \mathcal{L}(m)]u_{\lambda m}, \\ &\text{II} \quad O_-O_+u_{\lambda m} = [\lambda - \mathcal{L}(m+1)]u_{\lambda m}. \end{aligned} \quad (70)$$

Therefore, λ has the single-fixed eigenvalue, $\lambda = -1$, and the parameter, m, is replaced by ϵ, with $\mathcal{L}(\epsilon) = -2\epsilon$. Because $\mathcal{L}(\epsilon)$ is a decreasing function of ϵ, an $\epsilon_{\text{min.}}$ exists, with

$$\lambda = -1 = \mathcal{L}(\epsilon_{\text{min.}}) = -2\epsilon_{\text{min.}}, \qquad \text{so } \epsilon_{\text{min.}} = \tfrac{1}{2}. \quad (71)$$

The allowed values of ϵ are $\epsilon = \epsilon_{\text{min.}}, (\epsilon_{\text{min.}} + 1), \ldots, (\epsilon_{\text{min.}} + n) = (\frac{1}{2} + n), \ldots$. Therefore, $\epsilon_n = (n + \frac{1}{2})$. The starting eigenfunction is obtained from

$$\begin{aligned} O_- u_{-1,\epsilon_{\text{min}}} &= \left(\frac{d}{dx} + x\right) u_{-1,\epsilon_{\text{min.}}} = 0, \\ u_{-1,\epsilon_{\text{min}}} &= \frac{1}{\pi^{\frac{1}{4}}} e^{-\frac{1}{2}x^2}. \end{aligned} \tag{72}$$

The excited-state eigenfunctions are obtained with the normalized step-up operators

$$\mathcal{O}_+(n+1) = \frac{(-\frac{d}{dx} + x)}{\sqrt{[-1 + 2\epsilon_{(n+1)}]}} = \frac{(-\frac{d}{dx} + x)}{\sqrt{2(n+1)}},$$

so

$$u_{-1,\epsilon_n}(x) = \frac{1}{\sqrt{2^n n! \sqrt{\pi}}} \left(-\frac{d}{dx} + x\right)^n e^{-\frac{1}{2}x^2}. \tag{73}$$

Using the identities

$$\left(-\frac{d}{dx} + x\right) = e^{\frac{1}{2}x^2}\left[e^{-\frac{1}{2}x^2}\left(-\frac{d}{dx} + x\right)e^{\frac{1}{2}x^2}\right]e^{-\frac{1}{2}x^2} = e^{\frac{1}{2}x^2}\left[-\frac{d}{dx}\right]e^{-\frac{1}{2}x^2}, \tag{74}$$

the normalized n^{th} eigenfunction becomes

$$\begin{aligned} u_{-1,\epsilon_n}(x) &= \frac{1}{\sqrt{2^n n! \sqrt{\pi}}} e^{\frac{1}{2}x^2}\left(-\frac{d}{dx}\right)^n e^{-\frac{1}{2}x^2} e^{-\frac{1}{2}x^2} \\ &= \frac{e^{-\frac{1}{2}x^2}}{\sqrt{2^n n! \sqrt{\pi}}}\left[e^{x^2}\left(-\frac{d}{dx}\right)^n e^{-x^2}\right] = \frac{e^{-\frac{1}{2}x^2} H_n(x)}{\sqrt{2^n n! \sqrt{\pi}}}, \end{aligned} \tag{75}$$

where we have used the Rodriguez-type definition of the Hermite polynomial, $H_n(x)$, [see eq. (75) of Chapter 4].

Finally, the radial equation for the 3-D harmonic oscillator can also be solved by the factorization method. For details, see problem 15.

Altogether, Infeld and Hull list 31 generalizations or specializations of the Pöschl–Teller, hydrogenic, Morse, Rosen–Morse, 1-D harmonic oscillator, or 3-D harmonic oschillator Schrödinger equations, which lead to eigenvalues and eigenfunctions in analytic form, where the eigenfunctions correspond to many of the well-known functions of classical analysis. The question now arises: Do additional potentials exist for which the 1-D Schrödinger problem can be solved exactly? This question will be partially answered in the next chapter.

Problems

14. (a) Find the eigenvalues, ϵ_n, and the normalized eigenfunctions for all of the bound states of a Pöschl–Teller potential with dimensionless

$$V(z) = -\frac{V_0}{\cosh^2 z}, \qquad \text{with } V_0 = 7.2.$$

(b) A particle of mass μ moves in one dimension subject to the Schrödinger equation

$$-\frac{\hbar^2}{2\mu a^2}\Big(-\frac{d^2}{dz^2}+\frac{m(m+1)}{\sinh^2 z}-2\nu\coth z\Big)u(z)=Eu(z),$$

where z is a dimensionless variable, restricted to $z \geq 0$, and m and ν are dimensionless potential constants. Find the conditions that must be satisfied by these constants, so the potential has an attractive minimum for $z > 0$, and find an expression for the eigenvalues, E_n, as a function of n. Does a maximum possible value of n exist?

15. The 3-D harmonic oscillator. With $u(r) = rR(r)$, and $r = \sqrt{\hbar/m\omega_0}\rho$, $E = \hbar\omega_0\epsilon$, the radial wave equation for the 3-D harmonic oscillator (with $l = 0, 1, 2, \ldots$) takes the form

$$-\frac{d^2u}{d\rho^2}+\Big[\frac{l(l+1)}{\rho^2}+\rho^2\Big]u(\rho)=2\epsilon u(\rho).$$

Show that this equation is factorizable via

$$O_\pm(l)=-\Big(\mp\frac{d}{d\rho}+(\frac{l}{\rho}-\rho)\Big),$$

but the standard λ must be interpreted as $\lambda = 2\epsilon + 2l$; so $O_\pm(l)$ steps both l and ϵ. Show that this equation is also factorizable via

$$\bar{O}_\pm(l)=\Big(\mp\frac{d}{d\rho}+(\frac{l}{\rho}+\rho)\Big),$$

but now with $\lambda = 2\epsilon - 2l$.

Use these results to show that

$$E=\hbar\omega_0(N+\tfrac{3}{2}),\quad \text{with } N=0,1,2,\ldots$$

and that the allowed l values for a particular, N, are

$$l=N, N-2, N-4, \ldots, 0(\text{or } 1),\quad \text{for } N=\text{even (odd)}.$$

Find the normalized eigenfunctions for the special states with $l = N$. Construct the four normalized step operators, which convert normalized u_{Nl} into normalized $u_{N+1,l-1}$, $u_{N-1,l+1}$, $u_{N+1,l+1}$, and $u_{N-1,l-1}$. Construct all normalized radial eigenfunctions for $N \leq 3$

Find relations giving ρu_{Nl} as a linear combination of (i) $u_{N+1,l+1}$ and $u_{N-1,l+1}$, and as a linear combination of (ii) $u_{N+1,l-1}$ and $u_{N-1,l-1}$. Use these relations to find matrix elements of the operators, $\rho\cos\theta$ and $\rho\sin\theta e^{\pm i\phi}$ in the complete 3-D oscillator basis, ψ_{Nlm}. (The u_{Nl} and $u_{N'l'}$, with $l' \neq l$, are not orthogonal to each other in ρ-space, but the full energy eigenfunctions, $u_{Nl}(\rho)Y_{lm}(\theta,\phi)$ form a complete orthogonal set.)

Find all nonzero matrix elements of the operator, ρ^2.

Note: The above matrix elements of $\rho\cos\theta$ and $\rho\sin\theta e^{\pm i\phi}$ give the matrix elements of the dimensionless z and $(x \pm iy)$. The corresponding matrix elements of the dimensionless p_z and $(p_x \pm ip_y)$ can be obtained by utilizing the commutator relations

$$p_z = i[H, z], \qquad (p_x \pm ip_y) = i[H, (x \pm iy)],$$

together with the known matrix elements of the dimensionless H and z, $(x \pm iy)$. Use this technique to find the expressions for the nonzero matrix elements of p_z.

Solution for Problem 15

With

$$O_\pm(l) = -\Big(\mp\frac{d}{d\rho} + (\frac{l}{\rho} - \rho)\Big),$$

(where the extra overall minus sign in this definition is added merely for convenience to gain phases for the final matrix elements in best agreement with the "standard" phases for the 3-D oscillator), we have the two basic equations

$$\begin{aligned} O_+(l)O_-(l) &= -\frac{d^2}{d\rho^2} + \frac{l(l+1)}{\rho^2} + \rho^2 - (2l-1), \\ O_-(l+1)O_+(l+1) &= -\frac{d^2}{d\rho^2} + \frac{l(l+1)}{\rho^2} + \rho^2 - (2l+3), \end{aligned} \tag{1}$$

and

$$\begin{aligned} O_+(l)O_-(l)u_{\lambda l} &= [2\epsilon - (2l-1)]u_{\lambda l} = [\lambda - \mathcal{L}(l)]u_{\lambda l}, \\ O_-(l+1)O_+(l+1)u_{\lambda l} &= [2\epsilon - (2l+3)]u_{\lambda l} = [\lambda - \mathcal{L}(l+1)]u_{\lambda l}. \end{aligned} \tag{2}$$

These two equations are satisfied only if $\mathcal{L}(l+1) - \mathcal{L}(l) = 4$, and have a proper solution only if

$$\mathcal{L}(l) = 4l + c, \qquad \lambda = 2\epsilon + 2l + c + 1, \qquad c = \text{a constant}. \tag{3}$$

We will find it convenient to choose, $c = -1$ (this choice is quite arbitrary and will not affect final results). With this choice,

$$\mathcal{L}(l) = (4l - 1), \qquad \lambda = 2\epsilon + 2l. \tag{4}$$

Because our $l \geq 0$, this $\mathcal{L}(l)$ is an increasing function of l. Thus, an $l_{\text{max.}}$ exists (to be named N), $l_{\text{max.}} = N$, with $\lambda = \mathcal{L}(l_{\text{max.}} + 1) = (4l_{\text{max.}} + 3) = (4N + 3)$, and therefore

$$2\epsilon = (2N + 3), \qquad E_N = \hbar\omega_0(N + \tfrac{3}{2}). \tag{5}$$

The starting functions of $u_{\lambda l_{\text{max.}}}$ are given by

$$O_+(l_{\text{max.}} + 1)u_{\lambda l_{\text{max.}}}(\rho) = 0, \qquad \Big(\frac{d}{d\rho} - \frac{N+1}{\rho} + \rho\Big)u_{\lambda N}(\rho) = 0, \tag{6}$$

where this first-order differential equation has the solution

$$u_{\lambda, l=N}(\rho) = \mathcal{N}\rho^{N+1}e^{-\frac{1}{2}\rho^2}, \tag{7}$$

$$\text{with } |\mathcal{N}|^2 \int_0^\infty d\rho \rho^{2N+2} e^{-\rho^2} = \tfrac{1}{2}|\mathcal{N}|^2 \int_0^\infty d\eta \eta^{N+\frac{1}{2}} e^{-\eta} = \tfrac{1}{2}|\mathcal{N}|^2 \Gamma(N+\tfrac{3}{2}) = 1.$$

$$\Big[\text{Note, } \tfrac{1}{2}\Gamma(N+\tfrac{3}{2}) = \tfrac{1}{2}(N+\tfrac{1}{2})(N-\tfrac{1}{2})\cdots\tfrac{3}{2}\tfrac{1}{2}\sqrt{\pi} = \frac{(2N+1)!!}{2^{N+2}}\sqrt{\pi}\Big].$$

$O_-(l)$ changes $l \to (l-1)$, but because it keeps λ invariant, and $\lambda = 2\epsilon + 2l$, it must simultaneously raise ϵ by one unit, and hence, shifts $N \to (N+1)$. Similarly, $O_+(l+1)$ simultaneously changes $l \to (l+1)$, $N \to (N-1)$. To obtain the possible l values for a fixed N, we first examine the action of the operators

$$\bar{O}_\pm(l) = \Big(\mp\frac{d}{d\rho} + (\frac{l}{\rho} + \rho)\Big),$$

with

$$\begin{aligned}
\bar{O}_+(l)\bar{O}_-(l)u_{\bar{\lambda}l} &= \Big(-\frac{d^2}{d\rho^2} + \frac{l(l+1)}{\rho^2} + \rho^2 + (2l-1)\Big)u_{\bar{\lambda}l} \\
&= (2\epsilon + 2l - 1)u_{\bar{\lambda}l} = [\bar{\lambda} - \bar{\mathcal{L}}(l)]u_{\bar{\lambda}l} \\
\bar{O}_-(l+1)\bar{O}_+(l+1)u_{\bar{\lambda}l} &= \Big(-\frac{d^2}{d\rho^2} + \frac{l(l+1)}{\rho^2} + \rho^2 + (2l+3)\Big)u_{\bar{\lambda}l} \\
&= (2\epsilon + 2l + 3)u_{\bar{\lambda}l} = [\bar{\lambda} - \bar{\mathcal{L}}(l+1)]u_{\bar{\lambda}l},
\end{aligned} \tag{8}$$

which can be satisfied by

$$\bar{\mathcal{L}}(l) = -(4l-1), \qquad \bar{\lambda} = (2\epsilon - 2l). \tag{9}$$

$\bar{\lambda} - \bar{\mathcal{L}}(l)$ remains positive for all possible values of l, even as l is increased indefinitely. No new limits are set on l by the operators $\bar{O}_\pm$. Also,

$\bar{O}_-(l)$ changes $l \to (l-1)$ and simultaneously $N \to (N-1)$, and

$\bar{O}_+(l+1)$ changes $l \to (l+1)$ and simultaneously $N \to (N+1)$.

Starting with the maximum l value for a particular N, $l_{\text{max.}} = N$, successive action with $O_-(l)$ followed by $\bar{O}_-(l-1)$, or equally well $\bar{O}_-(l)$ followed by $O_-(l-1)$, will change a state with quantum numbers, N, l to a state with quantum numbers $N, (l-2)$, skipping states with $N, (l-1)$. The four operators, $O_\pm$, $\bar{O}_\pm$, do not change the parity of $(N+l)$. For a fixed energy (fixed N), the possible l values are

$$l = N, (N-2), (N-4), \cdots, 0 \text{ (or } 1), \qquad \text{for } N = \text{even (or odd)}.$$

It will now be convenient to define the four step operators preserving the normalization of the u_{Nl}, which will be denoted by $\mathcal{O}$. [Also, we will characterize the radial eigenfunctions by the quantum numbers, N, l; that is, we will replace the λ (or $\bar{\lambda}$) with the quantum number N, which gives the energy eigenvalue, ϵ.]

$$\mathcal{O}_-(l) = \frac{O_-(l)}{\sqrt{[\lambda - \mathcal{L}(l)]}}, \quad \mathcal{O}_+(l+1) = \frac{O_+(l+1)}{\sqrt{[\lambda - \mathcal{L}(l+1)]}},$$

$$\text{with } [\lambda - \mathcal{L}(l)] = (2\epsilon + 2l) - (4l - 1) = (2N + 3 - 2l + 1).$$

$$\bar{\mathcal{O}}_-(l) = \frac{\bar{O}_-(l)}{\sqrt{[\bar{\lambda} - \bar{\mathcal{L}}(l)]}}, \quad \bar{\mathcal{O}}_+(l+1) = \frac{\bar{O}_+(l+1)}{\sqrt{[\bar{\lambda} - \bar{\mathcal{L}}(l+1)]}},$$

with $= [\bar{\lambda} - \bar{\mathcal{L}}(l)] = (2\epsilon - 2l) + (4l - 1) = (2N + 3 + 2l - 1)$.

Thus,

$$u_{(N+1)(l-1)} = \mathcal{O}_-(l)u_{Nl} = \frac{1}{\sqrt{2(N+2-l)}}\left(-\frac{d}{d\rho} - \frac{l}{\rho} + \rho\right)u_{Nl},$$

$$u_{(N-1)(l+1)} = \mathcal{O}_+(l+1)u_{Nl} = \frac{1}{\sqrt{2(N-l)}}\left(\frac{d}{d\rho} - \frac{(l+1)}{\rho} + \rho\right)u_{Nl},$$

$$u_{(N-1)(l-1)} = \bar{\mathcal{O}}_-(l)u_{Nl} = \frac{1}{\sqrt{2(N+1+l)}}\left(\frac{d}{d\rho} + \frac{l}{\rho} + \rho\right)u_{Nl},$$

$$u_{(N+1)(l+1)} = \bar{\mathcal{O}}_+(l+1)u_{Nl} = \frac{1}{\sqrt{2(N+3+l)}}\left(-\frac{d}{d\rho} + \frac{(l+1)}{\rho} + \rho\right)u_{Nl}.$$

Combining the first and third of these relations, we get

$$\rho u_{Nl} = \sqrt{\frac{(N+2-l)}{2}}u_{(N+1)(l-1)} + \sqrt{\frac{(N+1+l)}{2}}u_{(N-1)(l-1)}. \tag{10}$$

Similarly, combining the second and fourth relation, we get

$$\rho u_{Nl} = \sqrt{\frac{(N-l)}{2}}u_{(N-1)(l+1)} + \sqrt{\frac{(N+3+l)}{2}}u_{(N+1)(l+1)}. \tag{11}$$

If we left-multiply the first of these equations with $u^*_{(N+1)(l-1)}$ and integrate over ρ, and use the orthonormality of the u_{Nl} with the *same* l value, we get

$$\int_0^\infty d\rho u^*_{(N+1)(l-1)}\rho u_{Nl} = \int_0^\infty d\rho\rho^2 R^*_{(N+1)(l-1)}\rho R_{Nl} = \sqrt{\frac{(N+2-l)}{2}},$$

where we have used

$$\int_0^\infty d\rho u^*_{(N+1)(l-1)}u_{(N-1)(l-1)} = 0.$$

Both functions have the same l value, viz., $(l-1)$, and where we recall that the 1-D $u_{NL}(\rho)$ is related to the radial function, $R_{Nl}(\rho)$, via $u_{Nl}(\rho) = \rho R_{Nl}(\rho)$, where we also recall ρ is the dimensionless radial coordinate $\rho = r_{\text{phys.}}/\sqrt{\hbar/m\omega_0}$. Finally, if we combine the dimensionless ρ with the angular functions, we get the components of the (dimensionless) vector $\vec{r}$: $z = \rho\cos\theta$; $(x \pm iy) = \rho\sin\theta e^{\pm i\phi}$. With the matrix elements of the angular functions given through eqs. (42)–(44) of Chapter 9, we have, e.g.,

$$\langle\psi_{(N+1)(l-1)m}, \rho\cos\theta\psi_{Nlm}\rangle = \sqrt{\frac{(N+2-l)}{2}}\sqrt{\frac{(l^2-m^2)}{(2l+1)(2l-1)}},$$

$$\langle\psi_{(N-1)(l-1)m}, \rho\cos\theta\psi_{Nlm}\rangle = \sqrt{\frac{(N+1+l)}{2}}\sqrt{\frac{(l^2-m^2)}{(2l+1)(2l-1)}},$$

$$\langle\psi_{(N+1)(l+1)m}, \rho\cos\theta\psi_{Nlm}\rangle = \sqrt{\frac{(N+3+l)}{2}}\sqrt{\frac{[(l+1)^2-m^2]}{(2l+1)(2l+3)}},$$

$$\langle\psi_{(N-1)(l+1)m}, \rho\cos\theta\psi_{Nlm}\rangle = \sqrt{\frac{(N-l)}{2}}\sqrt{\frac{[(l+1)^2-m^2]}{(2l+1)(2l+3)}}, \tag{12}$$

where the similar matrix elements of $\rho\sin\theta e^{\pm i\phi}$ differ only in the l, m dependent square root factors coming from the angular parts [which now also change m to $(m \pm 1)$].

To get the matrix elements of ρ^2, we can combine eqs. (10) and (11)

$$\rho^2 u_{Nl} = \sqrt{\frac{(N+2-l)}{2}}\left(\sqrt{\frac{(N+2-l)}{2}}u_{Nl} + \sqrt{\frac{(N+3+l)}{2}}u_{(N+2)l}\right) + \sqrt{\frac{(N+1+l)}{2}}\left(\sqrt{\frac{(N-l)}{2}}u_{(N-2)l.} + \sqrt{\frac{(N+1+l)}{2}}u_{Nl}\right). \tag{13}$$

This equation leads to the matrix elements

$$\begin{aligned}\langle\psi_{Nlm}, \rho^2\psi_{Nlm}\rangle &= (N+\tfrac{3}{2}),\\ \langle\psi_{(N+2)lm}, \rho^2\psi_{Nlm}\rangle &= \tfrac{1}{2}\sqrt{(N+2-l)(N+l+3)},\\ \langle\psi_{(N-2)lm}, \rho^2\psi_{Nlm}\rangle &= \tfrac{1}{2}\sqrt{(N-l)(N+l+1)}.\end{aligned} \tag{14}$$

Finally, to obtain matrix elements of p_z and $(p_x \pm ip_y)$, we can use the commutator relations

$$p_z = i[H, z], \qquad (p_x \pm ip_y) = i[H, (x \pm iy)],$$

so, e.g.,

$$\langle\psi_{N'l'm}, p_z\psi_{Nlm}\rangle = i[(N'+\tfrac{3}{2}) - (N+\tfrac{3}{2})]\langle\psi_{N'l'm}, z\psi_{Nlm}\rangle, \tag{15}$$

giving

$$\langle\psi_{(N+1)(l-1)m}, p_z\psi_{Nlm}\rangle = i\sqrt{\frac{(N+2-l)}{2}}\sqrt{\frac{(l^2-m^2)}{(2l+1)(2l-1)}},$$

$$\langle\psi_{(N-1)(l-1)m}, p_z\psi_{Nlm}\rangle = -i\sqrt{\frac{(N+1+l)}{2}}\sqrt{\frac{(l^2-m^2)}{(2l+1)(2l-1)}},$$

$$\langle\psi_{(N+1)(l+1)m}, p_z\psi_{Nlm}\rangle = i\sqrt{\frac{(N+3+l)}{2}}\sqrt{\frac{[(l+1)^2-m^2]}{(2l+1)(2l+3)}},$$

$$\langle\psi_{(N-1)(l+1)m}, p_z\psi_{Nlm}\rangle = -i\sqrt{\frac{(N-l)}{2}}\sqrt{\frac{[(l+1)^2-m^2]}{(2l+1)(2l+3)}}. \tag{16}$$

As our last result, we shall obtain explicit expressions for the normalized radial eigenfunctions for $N \leq 3$. The functions with $l = l_{\text{max.}} = N$ are given through eq. (7). Functions with lower l values can be obtained with actions of $\mathcal{O}_-$ or $\bar{\mathcal{O}}_-$:

$$u_{N=3,l=3} = \sqrt{\frac{2}{\Gamma(\frac{9}{2})}}\rho^4 e^{-\frac{1}{2}\rho^2} = \sqrt{\frac{2^5}{105\sqrt{\pi}}}\rho^4 e^{-\frac{1}{2}\rho^2},$$

$$u_{N=2,l=2} = \sqrt{\frac{2}{\Gamma(\frac{7}{2})}}\rho^3 e^{-\frac{1}{2}\rho^2} = \sqrt{\frac{2^4}{15\sqrt{\pi}}}\rho^3 e^{-\frac{1}{2}\rho^2},$$

$$u_{N=1,l=1} = \sqrt{\frac{2}{\Gamma(\frac{5}{2})}}\rho^2 e^{-\frac{1}{2}\rho^2} = \sqrt{\frac{2^3}{3\sqrt{\pi}}}\rho^2 e^{-\frac{1}{2}\rho^2},$$

$$u_{N=0,l=0} = \sqrt{\frac{2}{\Gamma(\frac{3}{2})}}\rho e^{-\frac{1}{2}\rho^2} = \sqrt{\frac{2^2}{\sqrt{\pi}}}\rho e^{-\frac{1}{2}\rho^2},$$

$$u_{N=3,l=1} = \mathcal{O}_-(2)u_{N=2,l=2} = \frac{1}{2}\Big(-\frac{d}{d\rho} - \frac{2}{\rho} + \rho\Big)u_{N=2,l=2},$$

$$= \sqrt{\frac{2^2}{15\sqrt{\pi}}}(2\rho^4 - 5\rho^2)e^{-\frac{1}{2}\rho^2},$$

$$u_{N=2,l=0} = \mathcal{O}_-(1)u_{N=1,l=1} = \frac{1}{2}\Big(-\frac{d}{d\rho} - \frac{1}{\rho} + \rho\Big)u_{N=1,l=1},$$

$$= \sqrt{\frac{2}{3\sqrt{\pi}}}(2\rho^3 - 3\rho)e^{-\frac{1}{2}\rho^2}.$$

16. The symmetric top rigid rotator. In problem 5, the Schrödinger equation for the symmetric top rigid rotator, with $A = B \neq C$, led to the θ equation via the assumed form of the solution

$$\psi_{JMK}(\phi, \theta, \chi) = \frac{e^{iM\phi}}{\sqrt{2\pi}}\frac{e^{iK\chi}}{\sqrt{2\pi}}\Theta_{JMK}(\theta).$$

This θ equation is one-dimensionalized via

$$u_{JMK}(\theta) = \sqrt{\sin\theta}\,\Theta_{JMK}(\theta)$$

to give

$$\left(-\frac{d^2}{d\theta^2} + \frac{M^2 + K^2 - 2MK\cos\theta}{\sin^2\theta}\right)u_{\lambda MK}(\theta) = \lambda u_{\lambda MK}(\theta),$$

where

$$E = \frac{\hbar^2}{2A}(\lambda - \tfrac{1}{4} - K^2) + \frac{\hbar^2}{2C}K^2.$$

Show that this equation can be factorized in two ways, via

$$O_{\pm}(M) = \left(\mp\frac{d}{d\theta} + (M - \tfrac{1}{2})\cot\theta - \frac{K}{\sin\theta}\right)$$

or

$$O_{\pm}(K) = \left(\mp\frac{d}{d\theta} + (K - \tfrac{1}{2})\cot\theta - \frac{M}{\sin\theta}\right)$$

where $\lambda = (J + \frac{1}{2})^2$, with $J = M_{\text{max.}} = K_{\text{max.}}$. Assume M and K can only be integers, so J is an integer.

Convert the above to normalized M step- and K step-operators, which preserve the normalization

$$\int_0^{\pi} d\theta \sin\theta |\Theta_{JMK}(\theta)|^2 = 1.$$

Find the normalized $\Theta_{JJK}(\theta)$ with $M = J$, but arbitrary allowed K, and $\Theta_{JMJ}(\theta)$ with $K = J$ but arbitrary allowed M.

Find the normalized J step-operators that step $J \to (J \pm 1)$, but keep M and K fixed. These operators will require new normalization factor ratios, c_{J+1MK}/c_{JMK}, as for the corresponding spherical harmonic problem. Prove these ratios are independent of K and, hence, can be taken over from the known case with $K = 0$.

Find all nonzero matrix elements of $\cos\theta$ and $\sin\theta e^{\pm i\phi}$, $\sin\theta e^{\pm i\chi}$:

$$\langle \psi_{J'M'K'}, \cos\theta\, \psi_{JMK}\rangle,$$

$$\langle \psi_{J'M'K'}, \sin\theta e^{\pm i\phi}\, \psi_{JMK}\rangle,$$

$$\langle \psi_{J'M'K'}, \sin\theta e^{\pm i\chi}\, \psi_{JMK}\rangle.$$

12

The Darboux Method: Supersymmetric Partner Potentials

Even if the two partner potentials, $V(x, m)$ and $V(x, m-1)$, of the form $[k^2(x, m)\mp k'(x, m)+\mathcal{L}(m)]$ of eqs. (7) and (8) of the last chapter do *not* have the same shape, it may still be possible to say something about the eigenvalue spectrum of the partner potential if the eigenvalues of one of the potentials are known. This will be true whether or not the potentials are functions of a parameter, m. This has been known since 1882 through the work of G. Darboux; (Comptes Rendus Acad. de Sci. (Paris) **94**(1882)1456). This 19th century work has only recently been rediscovered by quantum theorists in connection with work in particle physics on supersymmetry. Hence, the partner potentials are known as supersymmetric partner potentials.

Suppose we have an eigenvalue problem with a potential $V_1(x)$

$$\left(-\frac{d^2}{dx^2} + V_1(x)\right)u_\lambda(x) = \left(A^\dagger A + \text{const.}\right)u_\lambda(x) = \lambda u_\lambda(x), \tag{1}$$

which is a solved problem and can be put in the form

$$\left(\left[-\frac{d}{dx} + k(x)\right]\left[\frac{d}{dx} + k(x)\right] + \text{const.}\right)u_\lambda(x) = \lambda u_\lambda(x). \tag{2}$$

Because we have no parameter, m, we have named the two operators, A, and $A^\dagger$,

$$A = \left[\frac{d}{dx} + k(x)\right]; \qquad A^\dagger = \left[-\frac{d}{dx} + k(x)\right]. \tag{3}$$

In addition, let $u_{\bar{\lambda}}$ be any solution of eq. (1), perhaps *not* a square-integrable solution,

$$\left(-\frac{d^2}{dx^2} + V_1(x)\right)u_{\bar{\lambda}}(x) = \bar{\lambda}u_{\bar{\lambda}}(x). \tag{4}$$

This equation is satisfied if we choose

$$k(x) = -\left(\frac{u'_{\bar{\lambda}}}{u_{\bar{\lambda}}}\right), \qquad \text{and} \qquad \text{const.} = \bar{\lambda}, \tag{5}$$

because

$$A^{\dagger}A = -\frac{d^2}{dx^2} + \left(\frac{u'_{\bar{\lambda}}}{u_{\bar{\lambda}}}\right)^2 + \frac{d}{dx}\left(\frac{u'_{\bar{\lambda}}}{u_{\bar{\lambda}}}\right) = -\frac{d^2}{dx^2} + \left(\frac{u''_{\bar{\lambda}}}{u_{\bar{\lambda}}}\right), \tag{6}$$

so

$$A^{\dagger}Au_{\bar{\lambda}} = -\frac{d^2u_{\bar{\lambda}}}{dx^2} + u''_{\bar{\lambda}} = 0 = (\bar{\lambda} - \text{const.})u_{\bar{\lambda}}(x). \tag{7}$$

Therefore, the constant in the original equation must be $\bar{\lambda}$ and the potential $V_1(x)$ is given by

$$V_1(x) = \left(\frac{u''_{\bar{\lambda}}}{u_{\bar{\lambda}}}\right) + \bar{\lambda}. \tag{8}$$

Now, let us look at a different eigenvalue problem, with a different potential, and different eigenfunctions, but with the same eigenvalues λ

$$AA^{\dagger}w_{\lambda}(x) = (\lambda - \bar{\lambda})w_{\lambda}(x). \tag{9}$$

The order of the operators, A, and $A^{\dagger}$, is reversed from that in the original equation, which was

$$A^{\dagger}Au_{\lambda}(x) = (\lambda - \bar{\lambda})u_{\lambda}(x). \tag{10}$$

Now, because

$$\begin{aligned}\left(AA^{\dagger} + \bar{\lambda}\right)w_{\lambda}(x) &= \lambda w_{\lambda}(x)\\ &= \left[-\frac{d^2}{dx^2} + 2\left(\frac{u'_{\bar{\lambda}}}{u_{\bar{\lambda}}}\right)^2 - \left(\frac{u''_{\bar{\lambda}}}{u_{\bar{\lambda}}}\right) + \bar{\lambda}\right]w_{\lambda}(x)\\ &= \left(-\frac{d^2}{dx^2} + V_2(x)\right)w_{\lambda},\end{aligned} \tag{11}$$

we have

$$\begin{aligned}V_2(x) &= 2\left(\frac{u'_{\bar{\lambda}}}{u_{\bar{\lambda}}}\right)^2 - \left(\frac{u''_{\bar{\lambda}}}{u_{\bar{\lambda}}}\right) + \bar{\lambda}\\ &= 2\left(\frac{u'_{\bar{\lambda}}}{u_{\bar{\lambda}}}\right)^2 + 2\bar{\lambda} - V_1(x),\end{aligned} \tag{12}$$

where problem 1, with potential $V_1(x)$, is given by

$$A^\dagger A u_\lambda = (\lambda - \bar{\lambda}) u_\lambda, \tag{13}$$

whereas problem 2, with potential $V_2(x)$, is given by

$$A A^\dagger w_\lambda = (\lambda - \bar{\lambda}) w_\lambda. \tag{14}$$

Now, acting on eq. (13) from the left with A yields

$$A A^\dagger (A u_\lambda) = (\lambda - \bar{\lambda})(A u_\lambda). \tag{15}$$

Thus, we see: If u_λ is an eigenfunction of problem 1, with eigenvalue λ, (Au_λ) is an eigenfunction of problem 2, with the same eigenvalue, λ. The question remains: Is w_λ square-integrable, if u_λ is square-integrable ? To answer this question, calculate

$$\begin{aligned}\int_{-\infty}^{+\infty} dx w_\lambda^* w_\lambda &= \int_{-\infty}^{+\infty} dx (A u_\lambda)^* A u_\lambda = \int_{-\infty}^{+\infty} dx u_\lambda^* (A^\dagger A u_\lambda) \\ &= (\lambda - \bar{\lambda}) \int_{-\infty}^{+\infty} dx u_\lambda^* u_\lambda.\end{aligned} \tag{16}$$

Thus, if u_λ is square-integrable over the domain from $-\infty$ to $+\infty$ (as assumed here, or over some domain from a to b), and *if* the value $\bar{\lambda}$ lies below the lowest allowed eigenvalue λ of the original problem, the right-hand side is positive, and

$$w_\lambda(x) = \left[\frac{d}{dx} - \left(\frac{u'_{\bar{\lambda}}}{u_{\bar{\lambda}}}\right)\right] u_\lambda(x) \tag{17}$$

will also be square-integrable, even if $u_{\bar{\lambda}}$ is not. A word of caution is needed here. The above derivation required the property $A^\dagger = (A)^\dagger$, which required an integration by parts over the domain from $-\infty$ to $+\infty$. For the needed integrals to exist, the logarithmic derivative, $(u'_{\bar{\lambda}}/u_{\bar{\lambda}})$, which arises through the function $k(x)$, must not have any infinities; i.e., the function $u_{\bar{\lambda}}$ must not have any zeros. This will be true in general if $V(x)$ has both a left and a right classical turning point, and if $\bar{\lambda}$ lies below the lowest eigenvalue λ. For the lowest possible eigenvalue, the eigenfunction u_λ will have just enough curvature away from the x-axis in the classically forbidden regions so both u_λ and its first derivative will go to zero together as $x \to \pm\infty$, as required for a square-integrable function. Moreover, the lowest allowed eigenfunction will have no zeros. For a $\bar{\lambda}$ below the lowest allowed λ, the curvature away from the x-axis in the classically forbidden regions will be too great and the function $u_{\bar{\lambda}}(x)$ will approach ∞ for both $x \to \pm\infty$ before $u_{\bar{\lambda}}(x)$ can reach the value zero (see Fig. 12.1). Thus, for every eigenvalue λ of the potential $V_1(x)$, a square-integrable (calculable) eigenfunction of the potential $V_2(x)$ exists. The potential $V_2(x)$, however, has an additional eigenvalue, $\bar{\lambda}$, below the lowest λ. Two candidates exist for square-integrable eigenfunctions associated with this additional eigenvalue

$$w_{\bar{\lambda}}^{(1)} = \frac{1}{u_{\bar{\lambda}}}, \qquad w_{\bar{\lambda}}^{(2)} = \frac{1}{u_{\bar{\lambda}}} \int_0^x d\xi [u_{\bar{\lambda}}(\xi)]^2. \tag{18}$$

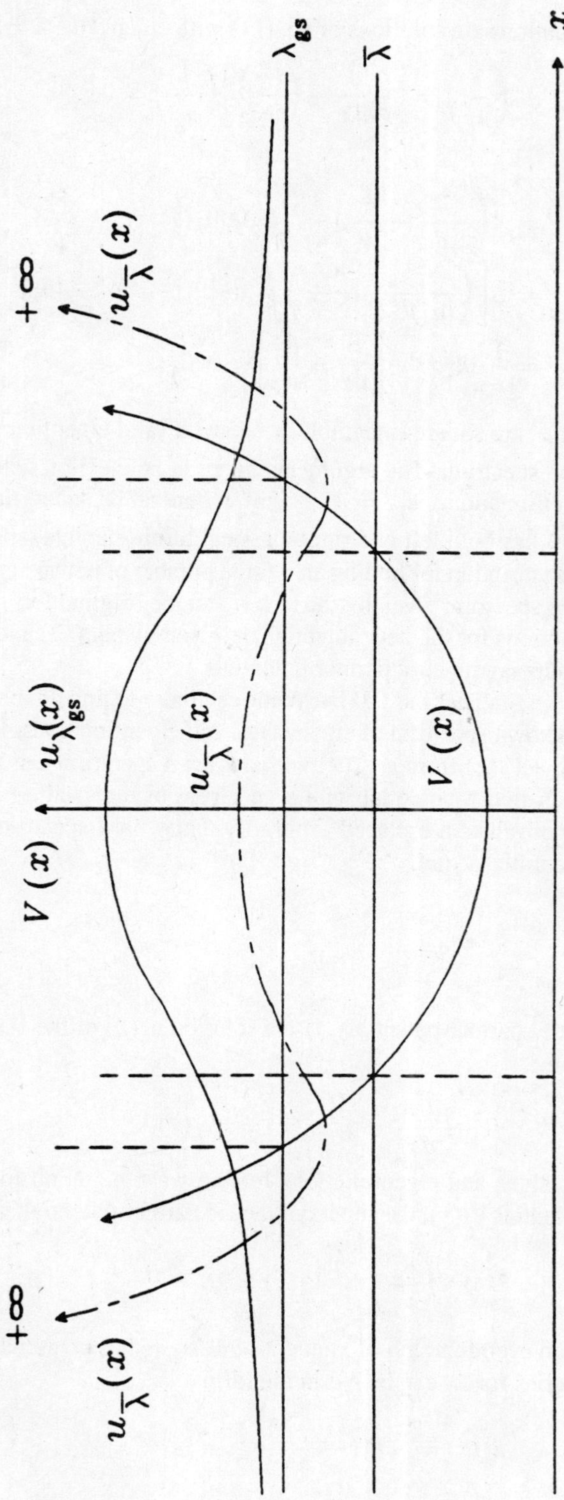

FIGURE 12.1. A $V(x)$ with a single deep minimum showing the allowed ground-state eigenfunction with $\lambda = \lambda_{gs}$ and a solution for $\bar{\lambda} < \lambda_{gs}$.

To show these equations are solutions of eq. (11) with eigenvalue $\lambda = \bar{\lambda}$, note

$$A^{\dagger}\left(\frac{1}{u_{\bar{\lambda}}}\right) = \left[-\frac{d}{dx} - \left(\frac{u'_{\bar{\lambda}}}{u_{\bar{\lambda}}}\right)\right]\left(\frac{1}{u_{\bar{\lambda}}}\right) = 0. \tag{19}$$

Also,

$$\begin{aligned} AA^{\dagger}w^{(2)}_{\bar{\lambda}} &= A\left(-\frac{d}{dx} - \frac{u'_{\bar{\lambda}}}{u_{\bar{\lambda}}}\right)\frac{1}{u_{\bar{\lambda}}}\int_0^x d\xi\,[u_{\bar{\lambda}}(\xi)]^2 \\ &= A\left[\left(\frac{u'_{\bar{\lambda}}}{(u_{\bar{\lambda}})^2} - \frac{u'_{\bar{\lambda}}}{(u_{\bar{\lambda}})^2}\right)\int_0^x d\xi\,[u_{\bar{\lambda}}(\xi)]^2 - \frac{1}{u_{\bar{\lambda}}}[u_{\bar{\lambda}}]^2\right] \\ &= -Au_{\bar{\lambda}} = -\left[\frac{d}{dx} - \frac{u'_{\bar{\lambda}}}{u_{\bar{\lambda}}}\right]u_{\bar{\lambda}} = 0. \end{aligned} \tag{20}$$

If either $w^{(1)}_{\bar{\lambda}}$ or $w^{(2)}_{\bar{\lambda}}$ are square-integrable, we have a valid eigenfunction for the additional $\bar{\lambda}$ of the spectrum. The arguments given in connection with Fig. 12.1 show the new eigenfunction , $w^{(1)}_{\bar{\lambda}} = 1/u_{\bar{\lambda}}$ will in general be square-integrable if the potential $V_1(x)$ has both left and right classical turning points, and if $\bar{\lambda} < \lambda$. Thus, we have a prescription for finding an infinite number of new potentials $V_2(x)$ with an eigenvalue spectrum given by the new $\bar{\lambda}$ and the original full spectrum of λ's. The eigenfunctions for the new potential are given by eq. (17) and (18). This is the method of supersymmetric partner potentials.

In problem 17, we shall use the 1-D harmonic oscillator to find double minimum potentials with a known spectrum of eigenvalues and eigenfunctions. Because the process of finding a $V_2(x)$ from a $V_1(x)$ with a known spectrum can in principle be iterated, we can find a potential with a spectrum of eigenvalues such that a few low-lying eigenvalues are placed arbitrarily, but with a spectrum of higher eigenvalues of the initial $V_1(x)$.

Problems

17. Supersymmetric partner potentials. Use a solution $u_{\bar{\lambda}}(x)$ of the 1-D harmonic oscillator equation

$$-\frac{d^2u_{\bar{\lambda}}}{dx^2} + x^2u_{\bar{\lambda}}(x) = \bar{\lambda}u_{\bar{\lambda}}(x)$$

to find the eigenvalues and eigenfunctions for the wave equation for a particle moving in the potential $V_2(x)$, the supersymmetric partner potential, where

$$V_2(x) = 2\bar{\lambda} + 2\left(\frac{u'_{\bar{\lambda}}}{u_{\bar{\lambda}}}\right)^2 - V_1(x) = 2\bar{\lambda} + 2\left(\frac{u'_{\bar{\lambda}}}{u_{\bar{\lambda}}}\right)^2 - x^2$$

in the case $u_{\bar{\lambda}}$ is an even function of x, and $\bar{\lambda} < 1$, so $u_{\bar{\lambda}}$ has no zeros. Show first that the infinite series for $u_{\bar{\lambda}}$ can be put in the form

$$u_{\bar{\lambda}}(x) = {}_1F_1\left(\frac{(1-\bar{\lambda})}{4}; \frac{1}{2}; x^2\right)e^{-\frac{1}{2}x^2},$$

where

$$ {}_1F_1(a;b;x^2) = \sum_{n=0}^{\infty} \frac{(a)_n}{(b)_n}\frac{x^{2n}}{n!}, $$

$$ \text{and } (a)_n = a(a+1)(a+2)\cdots(a+n-1), \quad (a)_0 = 1, $$

and show that

$$ u'_{\bar{\lambda}} = x\left((1-\bar{\lambda})\,{}_1F_1(\frac{(5-\bar{\lambda})}{4};\frac{3}{2};x^2) - {}_1F_1(\frac{(1-\bar{\lambda})}{4};\frac{1}{2};x^2)\right)e^{-\frac{1}{2}x^2}. $$

The case with two nearly degenerate levels near $\lambda = 1$ is of particular interest. Plot the potential $V_2(x)$ together with the eigenvalue spectrum for the two cases:

$$ \bar{\lambda} = 1 - \tfrac{1}{3}, \qquad \bar{\lambda} = 1 - \tfrac{1}{256}. $$

Also, plot the eigenfunctions for the two lowest energy eigenvalues.

18. Find the hydrogenic expectation values of $(1/r, 1/r^2, 1/r^3)$:

(a) Use

$$ \frac{dO}{dt} = \frac{i}{\hbar}[H,O] + \frac{\partial O}{\partial t} $$

to derive the quantum-mechanical form of the virial theorem, for an N-particle system, including $N = 1$:

$$ \frac{1}{4}\frac{d}{dt}\langle\psi, \sum_{i=1}^{N}(\vec{r}_i\cdot\vec{p}_i + \vec{p}_i\cdot\vec{r}_i)\psi\rangle = \langle\psi, \sum_{i=1}^{N}\frac{\vec{p}_i^{\,2}}{2m}\psi\rangle - \frac{1}{2}\langle\psi, \sum_{i=1}^{N}(\vec{r}_i\cdot\vec{\nabla}_i V)\psi\rangle. $$

Use this theorem to find the expectation value of $1/r$ for a state ψ_{nlm} of the hydrogen atom.

(b) Derive the Hellmann–Feynman theorem which applies to a system whose Hamiltonian is a function of a parameter, ν, and states

$$ \frac{\partial E_n}{\partial \nu} = \langle\psi_n, \frac{\partial H}{\partial \nu}\psi_n\rangle. $$

Use this theorem to calculate the expectation value of $1/r^2$ for a state ψ_{nlm} of the hydrogen atom. Use l as the parameter, $(l \equiv \nu)$. The quantum number, n, depends on this parameter through $n = (n_r + l + 1)$; $(n_r = 0, 1, 2, \ldots)$.

(c) The radial functions, $u_{nl}(r), u_{n'l'}(r)$, with $l' \neq l$, are by themselves not orthogonal to each other

$$ \int_0^\infty dr u^*_{nl}(r)u_{n'l'}(r) = \int_0^\infty dr r^2 R^*_{nl}(r)R_{n'l'}(r) \neq 0, $$

but show that, with $n' = n$,

$$ \int_0^\infty dr \frac{1}{r^2}u^*_{nl}(r)u_{nl'}(r) = \int_0^\infty dr R^*_{nl}(r)R_{nl'}(r) = 0, \qquad \text{with } l' \neq l. $$

Hint: Use the radial equation to evaluate

$$\frac{\hbar^2}{2\mu}[l'(l'+1) - l(l+1)] \int_0^\infty dr \frac{1}{r^2} u_{nl}^* u_{nl'}.$$

Now eliminate $\frac{d}{dr}$ from the hydrogen step-up and down operators, $O_+(l+1)$, $O_-(l)$, and combine the resulting expression for $\frac{1}{r}u_{nl}(r)$ with the above to evaluate the expectation value of $1/r^3$ in the state ψ_{nlm}.

19. Commutator algebra for the hydrogen atom.

(a) Use the dimensionless angular momentum and Runge–Lenz vectors, $\vec{L}$ and $\vec{\mathcal{R}}$, as well as the dimensionless $\vec{p}$, $\vec{r}$, H, and ϵ of problem 13 to show these dimensionless operators satisfy the commutation relations

$$[L_j, L_k] = i\epsilon_{jk\alpha} L_\alpha, \qquad [L_j, \mathcal{R}_k] = i\epsilon_{jk\alpha}\mathcal{R}_\alpha,$$

$$[\mathcal{R}_j, \mathcal{R}_k] = (-\vec{p}^{\,2} + \frac{2}{r}) i\epsilon_{jk\alpha} L_\alpha = (-2H) i\epsilon_{jk\alpha} L_\alpha.$$

(b) Define

$$\vec{V} = \frac{\vec{\mathcal{R}}}{\sqrt{(-2\epsilon)}}$$

to show in the subspace of a fixed, n,

$$[L_j, L_k] = i\epsilon_{jk\alpha} L_\alpha, \qquad [L_j, V_k] = i\epsilon_{jk\alpha} V_\alpha, \qquad [V_j, V_k] = i\epsilon_{jk\alpha} L_\alpha.$$

If we define

$$L_{ij} = \frac{1}{i}\left(x_i \frac{\partial}{\partial x_j} - x_j \frac{\partial}{\partial x_i} \right)$$

so $L_1 = L_{23}$; $L_2 = L_{31}$; $L_3 = L_{12}$, V_j can be defined through

$$V_j = \frac{1}{i}\left(x_j \frac{\partial}{\partial x_4} - x_4 \frac{\partial}{\partial x_j} \right).$$

Show that these operators satisfy the above commutation relations. That is, show the six operators, $\vec{L}$ and $\vec{V}$ can be related to the six operators, L_{ij}, with $i, j = 1, \ldots, 4$, which are the angular momentum operators in an abstract 4-D space generating rotations in this abstract 4-D space, x_1, x_2, x_3, x_4, where x_1, x_2, x_3 are our 3-D real space.

(c) Show that $\vec{M}$ and $\vec{N}$, defined by

$$\vec{M} = \tfrac{1}{2}(\vec{L} + \vec{V}), \qquad \vec{N} = \tfrac{1}{2}(\vec{L} - \vec{V}),$$

satisfy the commutation relations of two commuting angular momentum operators

$$[M_j, M_k] = i\epsilon_{jk\alpha} M_\alpha, \qquad [N_j, N_k] = i\epsilon_{jk\alpha} N_\alpha, \qquad [M_j, N_k] = 0.$$

(d) Show that

$$\vec{M}^2 - \vec{N}^2 = \tfrac{1}{2}\Big((\vec{L}\cdot\vec{V}) + (\vec{V}\cdot\vec{L}) \Big) = 0, \qquad \text{see Problem 13,}$$

$$2(\vec{M}^2 + \vec{N}^2) + 1 = (\vec{L}^2 + \vec{V}^2 + 1) = -\frac{1}{2\epsilon} = n^2.$$

(e) Show that the double angular momentum eigenfunctions, $\psi_{j_1 m_1 j_2 m_2}$, with

$$\vec{M}^2 \psi_{j_1 m_1 j_2 m_2} = j_1(j_1 + 1)\psi_{j_1 m_1 j_2 m_2}, \qquad M_3 \psi_{j_1 m_1 j_2 m_2} = m_1 \psi_{j_1 m_1 j_2 m_2},$$

$$\vec{N}^2 \psi_{j_1 m_1 j_2 m_2} = j_2(j_2 + 1)\psi_{j_1 m_1 j_2 m_2}, \qquad N_3 \psi_{j_1 m_1 j_2 m_2} = m_2 \psi_{j_1 m_1 j_2 m_2},$$

are also eigenvectors of the hydrogen atom H, provided

$$j_1 = j_2 = \frac{(n-1)}{2}.$$

13

The Vector Space Interpretation of Quantum-Mechanical Systems

A Different "Representations" of the State of a Quantum-Mechanical System

So far, we have specified the state of a quantum-mechanical system by the wave function, $\Psi(\vec{r}, t)$, i.e., by specifying the value of the scalar function, Ψ, for all values of x, y, z, at a particular time, t. Ψ could also be specified, however, at a particular time by the infinite set of numbers, $c_n(t)$, in the expansion of $\Psi(\vec{r}, t)$ in the compleete set of energy eigenfunctions of the system,

$$c_n e^{-\frac{i}{\hbar}E_n t} = <\psi_n, \Psi(\vec{r}, t)> \equiv c_n(t), \tag{1}$$

with

$$c_n = <\psi_n, \Psi(\vec{r}, t = 0)> = \int d\vec{r} \psi_n^*(\vec{r}) \Psi(\vec{r}, t = 0). \tag{2}$$

Alternatively, Ψ could just as well be specified by another set of numbers, a set of Fourier coefficients of some other (complete) set of generalized Fourier functions. For example, we might use the eigenfunctions of some other Hamiltonian, $\overline{H}$, not the Hamiltonian of *our* system, or possibly $\overline{H}$ could be some other Hermitian operator, not a Hamiltonian. Let us assume, in particular, $\overline{H}$ has both a discrete and a continuous spectrum, where the discrete spectrum is numbered by an index, $i = 1, 2, \ldots, n$, perhaps a finite number or perhaps an infinite number, whereas the continuous eigenvalue spectrum is parameterized by a continuous variable, α, such that

$$\overline{H} u_i(\vec{r}) = E_i u_i(\vec{r}) \tag{3}$$

$$\overline{H} w_\alpha(\vec{r}) = E(\alpha) w_\alpha(\vec{r}), \tag{4}$$

where the eigenfunctions form an orthonormal set, with

$$< u_i, u_j >= \delta_{ij}, \qquad < u_i, w_\alpha >= 0, \qquad < w_\alpha, w_{\alpha'} >= \delta(\alpha - \alpha'). \tag{5}$$

The eigenfunctions must of course also form a complete set, where the completeness relation is now

$$\sum_i u_i^*(\vec{r}\,') u_i(\vec{r}) + \int d\alpha w_\alpha^*(\vec{r}\,') w_\alpha(\vec{r}) = \delta(\vec{r}\,' - \vec{r}). \tag{6}$$

Now, Ψ can be expanded in terms of this complete set via

$$\Psi(\vec{r}, t) = \sum_i c_i u_i(\vec{r}) + \int d\alpha c(\alpha) w_\alpha(\vec{r}), \tag{7}$$

where the c_i and $c(\alpha)$ are given by

$$c_i =< u_i, \Psi >= \int d\vec{r}\,' u_i^*(\vec{r}\,') \Psi(\vec{r}\,', t), \tag{8}$$

$$c(\alpha) =< w_\alpha, \Psi >= \int d\vec{r}\,' w_\alpha^*(\vec{r}\,') \Psi(\vec{r}\,', t). \tag{9}$$

The c_i and $c(\alpha)$ are now implicitly time dependent. These c_i and $c(\alpha)$ now give us still another alternative for the description of our quantum-mechanical system.

So far, we have used Fourier expansions in a set of orthonormal functions that themselves were square-integrable functions; i.e., they were themselves part of our Hilbert space. The basis functions in the Fourier expansions, however, need not themselves be square-integrable. In fact, our original expansion in terms of ordinary Fourier plane-wave functions was of this type. The plane-wave functions (in the general notation of this chapter) are

$$u_{\vec{p}}(\vec{r}) = \frac{1}{(2\pi\hbar)^{3/2}} e^{\frac{i}{\hbar}(\vec{p}\cdot\vec{r})}. \tag{10}$$

These functions are eigenfunctions simultaneously of the three operators $(p_x)_{\text{op.}}$, $(p_y)_{\text{op.}}$, $(p_z)_{\text{op.}}$, with eigenvalues p_x, p_y, p_z, with, e.g.,

$$\frac{\hbar}{i} \frac{\partial}{\partial x} u_{\vec{p}}(\vec{r}) = p_x u_{\vec{p}}(\vec{r}). \tag{11}$$

Because the spectrum of possible $\vec{p}$ is continuous, the orthogonality is expressed in terms of a Dirac delta function in place of a Kronecker delta

$$< u_{\vec{p}\,'}, u_{\vec{p}} >= \delta(\vec{p}\,' - \vec{p}), \tag{12}$$

and the completeness relation is given by

$$\int d\vec{p}\, u_{\vec{p}}^*(\vec{r}\,') u_{\vec{p}}(\vec{r}) = \delta(\vec{r}\,' - \vec{r}). \tag{13}$$

The Fourier expansion of Ψ is the standard Fourier one, with expansion coefficients, $\phi(\vec{p}, t)$, now the standard Fourier transform

$$\Psi(\vec{r}, t) = \int d\vec{p}\, \phi(\vec{p}, t) u_{\vec{p}}(\vec{r}), \tag{14}$$

with

$$\phi(\vec{p}, t) = < u_{\vec{p}}, \Psi > . \tag{15}$$

The $\phi(\vec{p}, t)$ now give us another alternative way of specifying the state of our quantum-mechanical system.

Finally, we could even use a basis of Dirac delta functions for the generalized Fourier expansion of our Ψ, where these can be written, in analogy with the $u_{\vec{p}}$, or the ψ_n, as

$$u_{\vec{r}_0}(\vec{r}) = \delta(\vec{r} - \vec{r}_0), \tag{16}$$

where these u's are simultaneously the eigenfunctions of the three operators x, y, z; with eigenvalues x_0, y_0, z_0. For example,

$$x u_{\vec{r}_0}(\vec{r}) = x\delta(\vec{r} - \vec{r}_0) = x_0\delta(\vec{r} - \vec{r}_0), \tag{17}$$

where we have used the delta function relation

$$f(x)\delta(x) = f(0)\delta(x). \tag{18}$$

The orthogonality of the u's is now given by the delta function relation

$$\int d\vec{r} u^*_{\vec{r}_0}(\vec{r}) u_{\vec{r}_0'}(\vec{r}) = \int d\vec{r}\, \delta(\vec{r} - \vec{r}_0)\delta(\vec{r} - \vec{r}_0') = \delta(\vec{r}_0 - \vec{r}_0'). \tag{19}$$

The completeness relation is now given by the integral

$$\int d\vec{r}_0 u^*_{\vec{r}_0}(\vec{r}\,') u_{\vec{r}_0}(\vec{r}) = \int d\vec{r}_0 \delta(\vec{r}_0 - \vec{r}\,')\delta(\vec{r}_0 - \vec{r}) = \delta(\vec{r}\,' - \vec{r}). \tag{20}$$

The Fourier expansion of a ψ (let us make it time independent for simplicity), in terms of Fourier amplitudes $c_{\vec{r}_0}$, now leads to

$$\psi(\vec{r}) = \int d\vec{r}_0 c_{\vec{r}_0} u_{\vec{r}_0}(\vec{r}) = \int d\vec{r}_0 \psi(\vec{r}_0)\delta(\vec{r} - \vec{r}_0). \tag{21}$$

Now the Fourier coefficients $c_{\vec{r}_0}$, the analogs of the c_n, are just the wave functions $\psi(\vec{r}_0)$. Hence, this description of the state of our system is just the wave function description, where we specify ψ at every point x_0, y_0, z_0 in space time.

To summarize, we have given a number of alternative ways to give a complete description of the state of our system:

1) through specification of Ψ at every point in space time;

2) through the coefficients c_n through an expansion of Ψ in terms of the eigenfunctions of the Hamiltonian of our system;

3) through the coefficients, c_i and $c(\alpha)$, through an expansion of Ψ in terms of the eigenfunctions of an hermitian operator H with both a discrete and a continuous eigenvalue spectrum;

4) through the $\phi(\vec{p}, t)$ in a standard Fourier expansion of plane waves.

There are thus many "representations" of our quantum-mechanical system. This will lead us to an introduction to the Dirac notation in the next section.

B The Dirac Notation

Many different "representations" exist that can specify the state of an atomic quantum-mechanical system, described by the wave function, $\Psi(\vec{r}, t)$. We can specify Ψ at every point in space time; we can specify it through the c_n, or through another set c_i and $c(\alpha)$, or through the $\phi(p, t)$, and so on.

Because our $\Psi(\vec{r}, t)$ belong to the space of square-integrable functions, a quantum system specified by a Ψ can be thought of as a vector in infinite-dimensional vector space. The coefficients, c_n, can be thought of as the components of this vector along a particular set of coordinate axes, similarly, for the c_i and $c(\alpha)$ for a different set of coordinate axes. In the same way, the $\phi(\vec{p}, t)$ can be thought of as the components of the state vector along the set of axes specified by the momentum values, and the $\Psi(\vec{r}, t)$ as the components of the state vector along a set of axes specified by the values of the coordinates. The vector space of our Ψ with a well-defined complex scalar product is also called a Hilbert space. Just as in ordinary finite-dimensional vector analysis, it will be convenient to specify a vector by a generic symbol, not always by its coordinates along particular axes. Dirac proposed to do this through his "ket" symbol. Thus, a state vector is specified by $|\Psi\rangle$. Because scalar products $< \Phi, \Psi >$ are complex numbers, linear in Ψ but antilinear in Φ, with

$$
\begin{aligned}
&< \Phi, (\lambda_1 \Psi_1 + \lambda_2 \Psi_2) >= \lambda_1 < \Phi, \Psi_1 > +\lambda_2 < \Phi, \Psi_2 >, \qquad \text{but} \\
&< (\lambda_1 \Phi_1 + \lambda_2 \Phi_2), \Psi >= \lambda_1^* < \Phi_1, \Psi > +\lambda_2^* < \Phi_2, \Psi >,
\end{aligned} \tag{22}
$$

where λ_1 and λ_2 are complex numbers, Dirac defines for every "ket" vector, $|\Psi\rangle$, a dual vector, the so-called "bra," denoted by $\langle\Psi|$, where

$$
\begin{aligned}
|\lambda_1 \Psi_1 + \lambda_2 \Psi_2\rangle &= \lambda_1 |\Psi_1\rangle + \lambda_2 |\Psi_2\rangle, \\
\langle\lambda_1 \Psi_1 + \lambda_2 \Psi_2| &= \lambda_1^* \langle\Psi_1| + \lambda_2^* \langle\Psi_2|.
\end{aligned} \tag{23}
$$

The "bra" vector then permits us to write the scalar product of two vectors in terms of a "bracket"

$$
\langle\Phi, \Psi\rangle \equiv \langle\Phi|\Psi\rangle. \tag{24}
$$

In the new language, the Hilbert space is an infinite-dimensional vector space of all vectors $|\Psi\rangle$ with a finite norm,

$$
\langle\Psi|\Psi\rangle = \text{finite real positive number}.
$$

Because we are dealing with an infinite-dimensional vector space, we will also insist on the absolute convergence of expansions, such as

$$\Psi_N = \sum_{n=0}^{N} c_n \psi_n \qquad \text{as} \quad N \to \infty.$$

A linear operator, O, acting on a "ket," $|\Psi\rangle$, converts this into a new "ket," $|\Psi'\rangle = O|\Psi\rangle$, with a dual "bra," given by $\langle\Psi'| = \langle\Psi|O^\dagger$, so

$$\langle\Phi|\Psi'\rangle = \langle\Phi|O|\Psi\rangle = \langle\Psi|O^\dagger|\Phi\rangle^*. \tag{25}$$

$\langle\Phi|\Psi\rangle$ is a number, usually a complex number.

In this new language, an operator can be written as, e.g.,

$$|\Phi\rangle\langle\Lambda| \tag{26}$$

because

$$|\Phi\rangle\langle\Lambda|\Psi\rangle = (\text{a complex number})|\Phi\rangle; \tag{27}$$

that is, the operator $|\Phi\rangle\langle\Lambda|$, acting on the vector $|\Psi\rangle$, converts it into a new vector $|\Phi\rangle$, multiplied by the complex number $\langle\Lambda|\Psi\rangle$.

A very important type of operator is the projection operator, which projects an arbitrary state vector, $|\Psi\rangle$, onto a basis vector, such as $|\psi_n\rangle$. We will assume that $|\psi_n\rangle$ is normalized such that

$$\langle\psi_n|\psi_n\rangle = 1. \tag{28}$$

Then,

$$P_n = |\psi_n\rangle\langle\psi_n|. \tag{29}$$

Note,

$$P_n^2 = |\psi_n\rangle\langle\psi_n|\psi_n\rangle\langle\psi_n| = |\psi_n\rangle\langle\psi_n| = P_n, \qquad \text{via} \quad \langle\psi_n|\psi_n\rangle = 1. \tag{30}$$

Once we have projected the arbitrary vector onto the nth basis vector, projecting once more will not alter this result, so $P_n^2 = P_n$.

The completeness relation for the functions ψ_n can be translated into the "closure" relation in Dirac notation

$$\sum_n |\psi_n\rangle\langle\psi_n| = 1. \tag{31}$$

If the nth energy eigenvalue is degenerate, i.e., if g_n independent eigenfunctions $\psi_n^{(i)}$ with $i = 1, 2, \ldots, g_n$ exist (assuming we have orthonormalized the states with the same n, viz., $\langle\psi_n^{(i)}|\psi_n^{(j)}\rangle = \delta_{ij}$), it may be useful to define a projection operator

$$P_n = \sum_{i=1}^{i=g_n} |\psi_n^{(i)}\rangle\langle\psi_n^{(i)}| \tag{32}$$

that projects onto the subspace of states with definite energy E_n. Now the closure relation (completeness condition) becomes

$$\sum_n \sum_{i=1}^{i=g_n} |\psi_n^{(i)}\rangle\langle\psi_n^{(i)}| = 1. \tag{33}$$

For simplicity, let us for the moment assume no degeneracies exist. Then, P_n acting on an arbitrary state vector $|\Psi\rangle$ yields

$$P_n|\Psi\rangle = |\psi_n\rangle\langle\psi_n|\Psi\rangle = |\psi_n\rangle c_n, \tag{34}$$

or, using the closure relation, an arbitrary state vector, $|\Psi\rangle$, can be expanded in terms of base vectors, $|\psi_n\rangle$,

$$|\Psi\rangle = 1|\Psi\rangle = \sum_n |\psi_n\rangle\langle\psi_n|\Psi\rangle = \sum_n |\psi_n\rangle c_n. \tag{35}$$

This equation is the analog in the infinite-dimensional Hilbert space of the expansion of a vector, $\vec{V}$, in ordinary vector analysis in terms of base vectors, $\vec{e}_i$ (unit vectors along the ith direction),

$$\vec{V} = \sum_i \vec{e}_i V_i. \tag{36}$$

If we use some of the other representations introduced in the previous section, we can construct similar projection operators and similar projections. In a basis of eigenvectors of the operator, $\overline{H}$, with

$$\overline{H}|u_i\rangle = E_i|u_i\rangle, \qquad \overline{H}|w_\alpha\rangle = E(\alpha)|w_\alpha\rangle. \tag{37}$$

The closure relation is given by

$$\sum_i |u_i\rangle\langle u_i| + \int d\alpha |w_\alpha\rangle\langle w_\alpha| = 1, \tag{38}$$

and a state vector $|\Psi\rangle$ can be expanded in this basis by

$$|\Psi\rangle = 1|\Psi\rangle = \sum_i |u_i\rangle\langle u_i|\Psi\rangle + \int d\alpha |w_\alpha\rangle\langle w_\alpha|\Psi\rangle = \sum_i |u_i\rangle c_i + \int d\alpha |w_\alpha\rangle c(\alpha). \tag{39}$$

Similarly, in the basis of eigenvectors of the momentum operator, $\vec{p}$,

$$|\Psi\rangle = \int d\vec{p}|u_{\vec{p}}\rangle\langle u_{\vec{p}}|\Psi\rangle = \int d\vec{p}|u_{\vec{p}}\rangle\phi(\vec{p}, t). \tag{40}$$

Finally, in the basis of eigenvectors of the position operator, $\vec{r}$,

$$|\psi\rangle = \int d\vec{r}_0|u_{\vec{r}_0}\rangle\langle u_{\vec{r}_0}|\psi\rangle. \tag{41}$$

The projection onto the base vector $|u_{\vec{r}_0}\rangle$ is now just the Schrödinger wave function, $\langle u_{\vec{r}_0}|\psi\rangle = \psi(\vec{r}_0)$, where we have used the orthonormality of the $u_{\vec{r}_0}$, eq. (19).

In all of the above examples, we have expanded ket vectors in terms of ket base vectors. We could of course have done the same with the bra versions of the vectors. For example, multiplying with the unit operator from the right,

$$\langle\Psi| = \sum_n \langle\Psi|\psi_n\rangle\langle\psi_n| = \sum_n c_n^*\langle\psi_n|. \tag{42}$$

Finally, operators can be represented through their matrix elements via

$$O = 1O1 = \sum_{n,m} |\psi_n\rangle\langle\psi_n|O|\psi_m\rangle\langle\psi_m| = \sum_{n,m} |\psi_n\rangle\langle\psi_m|(O)_{nm}, \tag{43}$$

$$O|\Psi\rangle = \sum_{n,m} |\psi_n\rangle\langle\psi_n|O|\psi_m\rangle\langle\psi_m|\Psi\rangle = \sum_{n,m} |\psi_n\rangle O_{nm}c_m = |\Psi'\rangle = \sum_n |\psi_n\rangle c_n', \tag{44}$$

so

$$c_n' = \sum_m O_{nm}c_m. \tag{45}$$

C Notational Abbreviations

Often, we shall abbreviate the ket $|\psi_n\rangle$ simply by $|n\rangle$. If we have a one-degree of freedom problem, a single quantum number n, associated with the eigenvalue E_n, is sufficient to specify the ket. For a particle moving in three dimensions, three quantum numbers would be needed. For example, for a single particle (without spin) moving in a 3-D harmonic oscillator well, the ket would be specified by $|n_1n_2n_3\rangle$. For a single particle with spin moving in a central potential $V(r)$, the specification $|nlm_lm_s\rangle$ would be natural. Thus, we abbreviate

$$|\psi_n\rangle \quad \rightarrow \quad |nlm_lm_s\rangle. \tag{46}$$

Similarly, for the momentum representation

$$|u_{\vec{p}}\rangle \quad \rightarrow \quad |\vec{p}\rangle = |p_xp_yp_z\rangle, \tag{47}$$

and, in the coordinate representation

$$|u_{\vec{r}_0}\rangle \quad \rightarrow \quad |\vec{r}_0\rangle = |x_0y_0z_0\rangle. \tag{48}$$

14

The Angular Momentum Eigenvalue Problem (Revisited)

A Simultaneous Eigenvectors of Commuting Hermitian Operators

So far, we have solved the angular momentum eigenvalue problem very specifically in the coordinate representation for the case of the orbital angular momentum eigenfunctions, the well-known spherical harmonics. Let us look at this problem once more from a much more general point of view, which can be taken over for *any* angular momentum problem, or even more generally for any problem involving three hermitian operators with the same commutation relations as the L_x, L_y, and L_z. We want to solve the problem of finding the simultaneous eigenvalues and eigenvectors of the two (commuting) operators $\vec{L} \cdot \vec{L}$, and L_z. We will now write the eigenequations in the new language

$$\begin{aligned}(\vec{L} \cdot \vec{L})|\lambda m\rangle &= \lambda|\lambda m\rangle \\ L_z|\lambda m\rangle &= m|\lambda m\rangle,\end{aligned} \tag{1}$$

where we have purposely used the Dirac notation for the eigenvectors, so no implication is made as to a choice of representation.

The problem of finding the simultaneous eigenvectors of a pair (or more generally a number) of commuting hermitian operators is a very general one, because the complete specification of a base vector for an n-degree of freedom problem will in general involve n quantum numbers, associated with the eigenvalues of n commuting, hermitian operators. Let us first look at the case with $n = 2$. Let us first prove a theorem as follows.

Theorem: Two hermitian operators A and B have the same set of eigenvectors if and only if they commute.

First, assume the set of vectors $|\alpha\rangle$ are eigenvectors of both A and B:

$$\begin{aligned} A|\alpha\rangle &= a_\alpha|\alpha\rangle, \\ B|\alpha\rangle &= b_\alpha|\alpha\rangle. \end{aligned} \tag{2}$$

Now, let $[A, B] = (AB - BA)$ act on an arbitrary state vector $|\psi\rangle$ of our vector space. We shall assume the states $|\alpha\rangle$ form a complete set, and also note that a particular $|\alpha\rangle$ may require more than one label for a complete specification, so α may be a shorthand for two labels. Then, with

$$\sum_\alpha |\alpha\rangle\langle\alpha| = 1 \tag{3}$$

in the subspace in which A, and B act, we can project an arbitrary state vector $|\psi\rangle$ onto the $|\alpha\rangle$ basis to get

$$(AB - BA)|\psi\rangle = \sum_\alpha (AB - BA)|\alpha\rangle\langle\alpha|\psi\rangle = \sum_\alpha (a_\alpha b_\alpha - b_\alpha a_\alpha)|\alpha\rangle\langle\alpha|\psi\rangle = 0, \tag{4}$$

because the a_α and b_α are ordinary real numbers.

Conversely, if the $|\alpha\rangle$ are eigenvectors of the operator A, and if $[A, B] = 0$,

$$\begin{aligned} (AB - BA)|\psi\rangle &= 0 \\ &= (AB - BA)\sum_\alpha |\alpha\rangle\langle\alpha|\psi\rangle \\ &= \sum_\alpha (A - a_\alpha)B|\alpha\rangle\langle\alpha|\psi\rangle \\ &= \sum_{\alpha,\alpha'} (A - a_\alpha)|\alpha'\rangle\langle\alpha'|B|\alpha\rangle\langle\alpha|\psi\rangle \\ &= \sum_{\alpha,\alpha'} (a_{\alpha'} - a_\alpha)\langle\alpha'|B|\alpha\rangle\langle\alpha|\psi\rangle. \end{aligned} \tag{5}$$

Because $|\psi\rangle$ is an arbitrary vector and the $|\alpha\rangle$ are assumed to form a basis for the subspace of our vector space, we must have, for each pair of basis states $|\alpha\rangle$, $|\alpha'\rangle$,

$$(a_{\alpha'} - a_\alpha)\langle\alpha'|B|\alpha\rangle = 0. \tag{6}$$

Hence,

$$\langle\alpha'|B|\alpha\rangle = 0, \qquad \text{if} \qquad a_{\alpha'} \neq a_\alpha. \tag{7}$$

If the eigenvalues a_α have no degeneracies, i.e., if but a single eigenvector associated with each a_α exists, the matrix of B is diagonal in the $|\alpha\rangle$ basis, and $\langle\alpha'|B|\alpha\rangle = \delta_{\alpha,\alpha'}b_\alpha$. The more common situation, however, is one in which degeneracies associated with the eigenvalues of A exist. For example, if $A = \vec{L}^2$, $(2l + 1)$ eigenvectors associated with each eigenvalue of A exist. Then, with

$$A|\alpha^{(i)}\rangle = a_\alpha|\alpha^{(i)}\rangle, \qquad \text{with} \quad i = 1, 2, \ldots g_{a_\alpha}, \tag{8}$$

$$\langle\alpha'^{(j)}|B|\alpha^{(i)}\rangle = \delta_{\alpha,\alpha'}\langle\alpha^{(j)}|B|\alpha^{(i)}\rangle. \tag{9}$$

In this case, it is still possible to take a linear combination of the $|\alpha^{(i)}\rangle$, with the same eigenvalue, a_α of the operator A, to make the matrix of the operator B diagonal in this g_{a_α}- dimensional subspace. To specify the basis completely, we then need the simultaneous eigenvalues of both operators A and B.

B The Angular Momentum Algebra

With this introduction, let us look at the eigenvalue problem of eq. (1) for the operators L_z, and $\vec{L}^2$, built from the three (dimensionless) operators, L_x, L_y, L_z, where these satisfy the commutation relations

$$[L_x, L_y] = iL_z, \quad \text{and cyclically,} \quad \text{or} \quad [L_j, L_k] = i\epsilon_{jk\nu}L_\nu. \tag{10}$$

Our results will depend only on these commutation relations. Hence, other operators with these same commutation relations will also be interesting. The other angular momentum example involves the three components of the spin operator, S_x, S_y, S_z. In the case of the orbital angular momentum operator, we were able to write the operators in terms of functions of θ, ϕ, $\frac{\partial}{\partial\theta}$, and $\frac{\partial}{\partial\phi}$, i.e., in terms of explicit functions of the orbital coordinates. In the case of the electron, and other fundamental particles, we know nothing about the intrinsic or internal coordinates of such particles. In 1924, G. E. Uhlenbeck and S. Goudsmit had a picture in their mind of an electron that was like a little rotating sphere, but to the best of our ability to measure anything today to many significant figures, the electron is still a point particle (with no observable internal structure). Hence, we cannot relate the spin operators to internal "angles." In the case of a rotating molecule, we can write the rotational, internal angular momentum in terms of three Euler angles and their partial derivatives (as we shall see). Even though we know nothing about the internal structure of an electron, however, it will be reasonable to assume the three components of $\vec{S}$ obey the same commutation relations as the three components of $\vec{L}$:

$$[S_j, S_k] = i\epsilon_{jk\nu}S_\nu. \tag{11}$$

In addition, because the spin-degree of freedom will involve only internal or intrinsic degrees of freedom of the electron, it will be natural to assume all components of $\vec{S}$ commute with all components of $\vec{L}$. With this additional relation, the three components of $\vec{J}$, the total angular momentum, with

$$\vec{J} = \vec{L} + \vec{S}, \tag{12}$$

will also have the basic commutation relations of angular momentum

$$[J_j, J_k] = i\epsilon_{jk\nu}J_\nu. \tag{13}$$

Still other operators exist, which may not be at all the three components of an angular momentum, but have the same commutation relations. For example, the three operators

$$M_1 = \frac{1}{4}[(p_x^2 + x^2) - (p_y^2 + y^2)],$$
$$M_2 = \frac{1}{2}(p_x p_y + xy),$$
$$M_3 = \frac{1}{2}(x p_y - y p_x), \tag{14}$$

where x and p_x are dimensionless coordinate and momenta, with x and y measured in appropriate length units, such that $[p_j, x_k] = -i\delta_{jk}$. From these commutation relations, it follows that

$$[M_j, M_k] = i\epsilon_{jk\nu}M_\nu. \tag{15}$$

Because we shall prove the eigenvalues of M_3 (like those of S_z, or J_z) can only be either integers or $\frac{1}{2}$-integers, and because $M_3 = \frac{1}{2}L_z$, we now have the proof that the orbital angular momentum quantum number m must be an integer. (It cannot be a $\frac{1}{2}$-integer, as the corresponding m quantum number of an arbitrary "spin.") The operators M_j of eq. (14), which are single-particle operators, can be generalized to operators for a many-body system by summing over N particle indices, e.g.,

$$M_3 = \frac{1}{2}\sum_{n=1}^{n=N}\left(x_n p_{y_n} - y_n p_{x_n}\right). \tag{16}$$

Thus, the result can be generalized to the m quantum number of an N-body system.

C General Angular Momenta

Let us consider the generic operators J_x, J_y, J_z. Again, we define

$$J_\pm = (J_x \pm iJ_y), \qquad J_0 = J_z. \tag{17}$$

The commutation relations translate to

$$[J_0, J_+] = +J_+, \qquad [J_0, J_-] = -J_-, \qquad [J_+, J_-] = 2J_0. \tag{18}$$

It will be useful to rewrite $\vec{J}^2$ in various ways

$$\vec{J}^2 = \frac{1}{2}(J_x + iJ_y)(J_x - iJ_y) + \frac{1}{2}(J_x - iJ_y)(J_x + iJ_y) + J_z^2$$
$$= \frac{1}{2}(J_+J_- + J_-J_+) + J_0^2$$
$$= J_-J_+ + J_0^2 + J_0$$
$$= J_+J_- + J_0^2 - J_0, \tag{19}$$

where we have used $[J_+, J_-] = 2J_0$ to write the operator in two basic forms. Now, let us assume $|\lambda m\rangle$ is simultaneously an eigenvector of $\vec{J}^2$ and J_0, with eigenvalues λ and m, respectively, where these are (so far) arbitrary real numbers:

$$\vec{J}^2|\lambda m\rangle = \lambda|\lambda m\rangle,$$

$$J_0|\lambda m\rangle = m|\lambda m\rangle. \tag{20}$$

Acting on these two equations from the left with the operator J_+ and using the commutation relations, $[J_+, \vec{J}^2] = 0$ and $[J_+, J_0] = -J_+$, we get

$$\vec{J}^2\Big(J_+|\lambda m\rangle\Big) = \lambda\Big(J_+|\lambda m\rangle\Big),$$
$$J_0\Big(J_+|\lambda m\rangle\Big) = (m+1)\Big(J_+|\lambda m\rangle\Big). \tag{21}$$

Similarly, acting on both equations from the left with J_-, we get

$$\vec{J}^2\Big(J_-|\lambda m\rangle\Big) = \lambda\Big(J_-|\lambda m\rangle\Big),$$
$$J_0\Big(J_-|\lambda m\rangle\Big) = (m-1)\Big(J_-|\lambda m\rangle\Big). \tag{22}$$

Thus, if $|\lambda m\rangle$ is an eigenvector of $\vec{J}^2$ and J_0 with eigenvalues λ and m, two possibilities exist. Either $(J_+|\lambda m\rangle)$ is an eigenvector of $\vec{J}^2$ and J_0, with eigenvalues λ and $(m+1)$, or $J_+|\lambda m\rangle = 0$. Similarly, either $(J_-|\lambda m\rangle)$ is an eigenvector of $\vec{J}^2$ and J_0, with eigenvalues λ and $(m-1)$, or $J_-|\lambda m\rangle = 0$.

Let us assume the first possibility for the operator J_+. Using eq. (19), let us evaluate the diagonal matrix element of J_-J_+ for the eigenstate $|\lambda m\rangle$

$$\begin{aligned}
\langle\lambda m|J_-J_+|\lambda m\rangle &= \langle\lambda m|(\vec{J}^2 - J_0^2 - J_0)|\lambda m\rangle = [\lambda - m(m+1)]\langle\lambda m|\lambda m\rangle \\
&= \sum_{\lambda',m'}\langle\lambda m|J_-|\lambda' m'\rangle\langle\lambda' m'|J_+|\lambda m\rangle \\
&= \sum_{\lambda',m'}\langle\lambda' m'|J_-^\dagger|\lambda m\rangle^*\langle\lambda' m'|J_+|\lambda m\rangle = \sum_{\lambda',m'}|\langle\lambda' m'|J_+|\lambda m\rangle|^2 \\
&= [\lambda - m(m+1)].
\end{aligned} \tag{23}$$

The state $J_+|\lambda m\rangle$ can exist if $\lambda > m(m+1)$. In that case, we could act with J_+ to make the new state with $(m+1)$ and calculate the diagonal matrix element in the state $|\lambda(m+1)\rangle$. Moreover, we could repeat this process to ladder our way up to a state with $(m+n)$, but the same λ, so

$$\sum_{\lambda',m'}|\langle\lambda' m'|J_+|\lambda(m+n)\rangle|^2 = [\lambda - (m+n)(m+n+1)]. \tag{24}$$

An integer n will then be large enough that the negative quantity $-(m+n)(m+n+1)$ will overwhelm the positive λ, and we will have a patently positive quantity on the left-hand side of the equation equal to a negative quantity on the right. Therefore, the assumption that $\langle\lambda m|\lambda m\rangle = 1$ for the starting value of m must have been wrong. If the starting value of m is such that the laddering process has an upper bound, however, we must come to a maximum value of m, such that $(m+n) = m_{\max}$, with

$$J_+|\lambda m_{\max}\rangle = 0, \tag{25}$$

and, hence,

$$\lambda = m_{\max}(m_{\max}+1). \tag{26}$$

Similarly, using the last form of eq. (19), $J_+J_- = (\vec{J}^2 - J_0^2 + J_0)$, we arrive in the same way at the result

$$\langle\lambda m|J_+J_-|\lambda m\rangle = \sum_{\lambda''m''}|\langle\lambda''m''|J_-|\lambda m\rangle|^2 = [\lambda - m(m-1)]. \tag{27}$$

Now, we can repeat the step-down process and step-down m until it would become such a large negative number, so the negative quantity $-m(m-1) = -|m|(|m|+1)$ would overwhelm the positive λ. Thus, again the step-down ladder must quit at a value $m = m_{\rm min}$ for which

$$J_-|\lambda m_{\rm min}\rangle = 0, \tag{28}$$

and

$$\lambda = m_{\rm min}(m_{\rm min} - 1). \tag{29}$$

Hence,

$$\lambda = m_{\rm max}(m_{\rm max} + 1) = m_{\rm min}(m_{\rm min} - 1), \tag{30}$$

leading to $m_{\rm min.} = +\frac{1}{2} \pm \sqrt{(m_{\rm max.} + \frac{1}{2})^2)}$, so $m_{\rm min.} = -m_{\rm max}$. The other root, $m_{\rm min.} = (m_{\rm max.} + 1)$, is of course meaningless. Let us now name $m_{\rm max.} = j$. Because now $(m_{\rm max.} - m_{\rm min.}) = 2m_{\rm max.} = 2j =$ integer, the quantum number j can only be an integer or a $\frac{1}{2}$-integer; with

$$\lambda = j(j+1), \qquad \text{where} \qquad m = +j, (j-1), \ldots, -j. \tag{31}$$

In addition, because the operators $J_\pm$ do not change the eigenvalue λ, the sums over λ' or λ'' in eqs. (24) and (27) collapse to the single value λ. Also, if the eigenvectors are such that the state $|\lambda m_{\rm max.}\rangle$ is nondegenerate, the state $(J_-|\lambda m_{\rm max.}\rangle)$ will also be nondegenerate; similarly for states with even lower m values. Then, the sum over m' in eq. (23) and m'' in eq. (27) collapses to a single value. Thus, replacing the label λ with the quantum number j,

$$|\langle j(m+1)|J_+|jm\rangle|^2 = [j(j+1) - m(m+1)], \tag{32}$$

$$|\langle j(m-1)|J_-|jm\rangle|^2 = [j(j+1) - m(m-1)]. \tag{33}$$

Choosing the matrix elements themselves to be real, we get

$$\langle j(m+1)|J_+|jm\rangle = \sqrt{(j-m)(j+m+1)}, \tag{34}$$

$$\langle j(m-1)|J_-|jm\rangle = \sqrt{(j+m)(j-m+1)}, \tag{35}$$

and, with $J_x = \frac{1}{2}(J_+ + J_-)$, $J_y = \frac{i}{2}(J_- - J_+)$,

$$\begin{aligned}
&\langle jm'|J_x|jm\rangle = \\
&\tfrac{1}{2}\Big(\delta_{m'(m+1)}\sqrt{(j-m)(j+m+1)} + \delta_{m'(m-1)}\sqrt{(j+m)(j-m+1)}\Big),\\
&\langle jm'|J_y|jm\rangle = \\
&\tfrac{1}{2}\Big(-i\delta_{m'(m+1)}\sqrt{(j-m)(j+m+1)} + i\delta_{m'(m-1)}\sqrt{(j+m)(j-m+1)}\Big),\\
&\langle jm|J_z|jm\rangle = m.
\end{aligned} \tag{36}$$

For $j = \frac{1}{2}$, these matrices are very simple 2×2 matrices with rows and columns labeled by $m = +\frac{1}{2}, m = -\frac{1}{2}$, in that order. It is convenient to factor out the factor $\frac{1}{2}$, and to define the operator $\vec{\sigma}$, via

$$\vec{J} = \frac{1}{2}\vec{\sigma}, \tag{37}$$

where the matrices have the simple form

$$\sigma_x = \begin{pmatrix} 0 & 1 \\ 1 & 0 \end{pmatrix}, \qquad \sigma_y = \begin{pmatrix} 0 & -i \\ i & 0 \end{pmatrix}, \qquad \sigma_z = \begin{pmatrix} 1 & 0 \\ 0 & -1 \end{pmatrix}. \tag{38}$$

These equations are the famous Pauli spin-matrices. They satisfy

$$\sigma_j \sigma_k = i\epsilon_{jk\alpha}\sigma_\alpha + \delta_{jk}. \tag{39}$$

As a final small exercise, let us calculate ΔJ_x for an angular momentum system is in a definite angular momentum eigenstate $|jm\rangle$. We note

$$\langle jm|J_x|jm\rangle = 0, \tag{40}$$

so the expectation value of this perpendicular component of $\vec{J}$ is zero in the eigenstate with definite value of J_z. Also,

$$\begin{aligned} \langle jm|J_x^2|jm\rangle &= \sum_{m'=(m\pm 1)} \langle jm|J_x|jm'\rangle\langle jm'|J_x|jm\rangle \\ &= \sum_{m'=(m\pm 1)} |\langle jm'|J_x|jm\rangle|^2 \\ &= \frac{1}{4}\big([j(j+1) - m(m+1)] + [j(j+1) - m(m-1)]\big) \\ &= \frac{1}{2}[j(j+1) - m^2]. \end{aligned} \tag{41}$$

Furthermore, the diagonal matrix elements of J_y and J_y^2 have the same values as those for the x component. Thus, converting now to physical components with angular momentum in units of $\hbar$,

$$\Delta J_x = \Delta J_y = \hbar\sqrt{\frac{j(j+1) - m^2}{2}}, \tag{42}$$

in the state $|jm\rangle$ for which J_z has the precise value $\hbar m$, so $\Delta J_z = 0$. This is illustrated with the semiclassical vector model in which the vector $\vec{J}$, now of length $\hbar\sqrt{j(j+1)}$, is pictured to precess about its z-component with precise value $\hbar m$, which is less than $\hbar\sqrt{j(j+1)}$ even in the state with $m = j$. Also, for $j = \frac{1}{2}, m = \pm\frac{1}{2}$, we have $\Delta J_x = \Delta J_y = \frac{1}{2}\hbar$, the minimum quantum-mechanical uncertainty.

Final Remark: With our choice of phase for eqs. (34) and (35), we have made the matrix elements of J_y pure imaginary, and the matrix elements for J_x real. This is the standard angular momentum phase convention. All three components, however, are of course equivalent. We could, e.g., have used a basis in which J_x is diagonal, J_y real and off-diagonal, and J_z pure imaginary and off-diagonal.

15

Rigid Rotators: Molecular Rotational Spectra

A The Diatomic Molecule Rigid Rotator

For a rotating molecule the angular momentum, associated with the rotation of this "nearly rigid" body can be expressed in terms of Euler angles and their partial derivatives. Hence, this may be a good first example. Consider the simplest case: a diatomic molecule, e.g., the HCl molecule with one hydrogen and one Chlorine nucleus and 1 + 17 electrons. The full 20-body problem is extremely complicated, but at very low energies no excitations associated with the electron degrees of freedom will come into play. The electron cloud binds the two atomic nuclei into a nearly rigid structure. The position of the diatomic molecule in 3-D space can be described by a radial coordinate, r, giving the distance between the H and Cl nuclei, and two angles, θ, and ϕ, giving the orientation in space of the molecule axis, or H–Cl line. The wave function can be written as $\psi(r,\theta,\phi) = R(r)Y_{lm}(\theta,\phi)$. The electron cloud gives rise to a potential, $V(r)$, with a deep (nearly parabolic) well with a minimum ar $r = r_e$, where this is the equilibrium distance between the two atomic nuclei. The radial problem is associated with the vibrational motion of the molecule, a nearly harmonic oscillator motion to good first approximation. The energy associated with this vibration, $\hbar\omega_0$, is approximately 30 times that associated with the lower rotational excitations. Thus, at sufficiently low energies, we can replace the radial coordinate with its constant equilibrium value, r_e, and the Hamiltonian collapses to

$$H = -\frac{\hbar^2}{2\mathcal{I}_e}\left(\frac{\partial^2}{\partial\theta^2} + \cot\theta\frac{\partial}{\partial\theta} + \frac{1}{\sin^2\theta}\frac{\partial^2}{\partial\phi^2}\right) = \frac{\hbar^2}{2\mu r_e^2}\vec{L}^2, \tag{1}$$

with corresponding Schrödinger equation

$$\frac{\hbar^2}{2\mu r_e^2}\vec{L}^2 Y_{lm}(\theta,\phi) = EY_{lm}(\theta,\phi) = \frac{\hbar^2}{2\mu r_e^2} l(l+1)Y_{lm}(\theta,\phi). \tag{2}$$

Here, μ is the H-Cl reduced mass, and $\mathcal{I}_e$ is the moment of inertia about an axis perpendicular to the molecular axis, through the center of mass of the system. The energies are

$$E_l = \frac{\hbar^2}{2\mu r_e^2} l(l+1). \tag{3}$$

The eigenfunctions are the standard spherical harmonics. Each level is $(2l+1)$-fold degenerate, because the energy does not depend on m.

B The Polyatomic Molecule Rigid Rotator

For a polyatomic molecule, such as H_2O, with an isosceles triangle equilibrium structure, the rotational Hamiltonian is more complicated. We now need three Euler angles to specify the orientation in space of the nearly rigid molecule: two angles, θ and ϕ, to give the direction of the triangle's symmetry axis, and a third angle, χ, to describe the "spinning" of the two H atoms about this symmetry axis. Again, assuming the energies to be considered are so low vibrational excitations can be neglected, we can replace the coordinates of the atomic nuclei by their (constant) equilibrium values and are led to the rigid rotator Hamiltonian

$$H = \frac{1}{2A}P_{x'}^2 + \frac{1}{2B}P_{y'}^2 + \frac{1}{2C}P_{z'}^2, \tag{4}$$

where $P_{z'}$ is the component of the rotational angular momentum vector along the z', body-fixed principal axis, the H_2O symmetry axis; similarly, the x' and y' axes can be taken as the remaining principal axes, one perpendicular to the plane of the triangle, the other lying in the triangle plane, all going through the center of mass of the molecule. The constants A, B, C are the three principal moments of inertia in the equilibrium configuration: $I_{xx} = A$, $I_{yy} = B$, $I_{zz} = C$. The principal or primed axes components of the rotational angular momentum vector, $\vec{P}$, must be translated to operator form to write the above Hamiltonian in quantum-mechanical form. Using the techniques of problem 5, these components are [converting from the physical angular momentum components of eq. (4) to dimensionless ones, e.g., $(P_{z'})_{\text{phys.}} = \hbar P_{z'}$],

$$\begin{aligned}
P_{x'} &= \frac{1}{i}\left(\frac{\sin\chi}{\sin\theta}\frac{\partial}{\partial\phi} + \cos\chi\frac{\partial}{\partial\theta} - \sin\chi\cot\theta\frac{\partial}{\partial\chi}\right),\\
P_{y'} &= \frac{1}{i}\left(\frac{\cos\chi}{\sin\theta}\frac{\partial}{\partial\phi} - \sin\chi\frac{\partial}{\partial\theta} - \cos\chi\cot\theta\frac{\partial}{\partial\chi}\right),\\
P_{z'} &= \frac{1}{i}\frac{\partial}{\partial\chi},
\end{aligned} \tag{5}$$

with

$$\vec{P}^2 = -\left(\frac{\partial^2}{\partial\theta^2} + \cot\theta\frac{\partial}{\partial\theta} + \frac{1}{\sin^2\theta}\left[\frac{\partial^2}{\partial\phi^2} - 2\cos\theta\frac{\partial^2}{\partial\phi\partial\chi} + \cos^2\theta\frac{\partial^2}{\partial\chi^2}\right] + \frac{\partial^2}{\partial\chi^2}\right). \tag{6}$$

We can of course also write the space-fixed, x, y, and z components of the rotational angular momentum in this operator form. The component of greatest interest to us is the space-fixed z-component

$$P_z = \frac{1}{i}\frac{\partial}{\partial\phi}. \tag{7}$$

The three operators P_z, $P_{z'}$, and $\vec{P}^2$, form a set of three commuting operators. In addition, straightforward calculation gives the commutator algebra of the three body-fixed or principal axis components of the rotational angular momentum operator

$$[P_{x'}, P_{y'}] = -iP_{z'}, \qquad [P_{y'}, P_{z'}] = -iP_{x'}, \qquad [P_{z'}, P_{x'}] = -iP_{y'}. \tag{8}$$

Note the minus signs! These signs are the complex conjugates of the standard angular momentum commutators. If we had taken the space-fixed components P_x, P_y, and P_z, we would have been led to the standard angular momentum commutator algebra. In translating the standard results to their complex conjugates (needed for the primed components), we must merely interchange P'_+ and P'_-, where now

$$P'_+ = (P_{x'} + iP_{y'}), \qquad P'_- = (P_{x'} - iP_{y'}). \tag{9}$$

Now

$$[P_{z'}, P'_+] = -P'_+, \qquad [P_{z'}, P'_-] = +P'_-. \tag{10}$$

(Note the difference in sign compared with the standard angular momentum algebra!) Also, now, the simultaneous eigenvectors of the three commuting operators, P_z, $P_{z'}$, and $\vec{P}^2$, will yield a complete basis of the subspace of our Hilbert space, corresponding to the three rotational degrees of freedom. (We need three quantum numbers, and three commuting, hermitian operators.) The needed eigenvector equations are

$$\begin{aligned} \vec{P}^2|JMK\rangle &= \lambda|JMK\rangle = J(J+1)|JMK\rangle, \\ P_z|JMK\rangle &= M|JMK\rangle, \\ P_{z'}|JMK\rangle &= K|JMK\rangle, \end{aligned} \tag{11}$$

where the commutator algebra of the unprimed angular momentum components leads to $M_{\max} = -M_{\min} = J$, and the commutator algebra of the primed angular momentum components leads to $K_{\max} = -K_{\min} = J$, where $\lambda = J(J+1)$. Because we are dealing with orbital degrees of freedom of a many-body system, the quantum numbers, J, M, K must all be integers, with $M = J, (J-1), \ldots, -J$, and, similarly, $K = J, (J-1), \ldots, -J$. (We use capital letters for the J, M, K quantum numbers according to the usual convention by which capital letters are used for many-body systems.)

We will now write the rigid rotator Hamiltonian for the asymmetric top with $A \neq B \neq C$ (valid for the H_2O molecule), by first introducing the energy constants

$$a = \frac{\hbar^2}{2A}, \qquad b = \frac{\hbar^2}{2B}, \qquad c = \frac{\hbar^2}{2C}. \tag{12}$$

$$\begin{aligned} H &= aP_{x'}^2 + bP_{y'}^2 + cP_{z'}^2 = \frac{1}{2}(a+b)(P_{x'}^2 + P_{y'}^2) + \frac{1}{2}(a-b)(P_{x'}^2 - P_{y'}^2) + cP_{z'}^2 \\ &= \frac{1}{2}(a+b)(\vec{P}^2 - P_{z'}^2) + \frac{1}{4}(a-b)(P'_+P'_+ + P'_-P'_-) + cP_{z'}^2. \end{aligned} \tag{13}$$

In the $|JMK\rangle$ basis, the nonzero matrix elements of the primed components of $\vec{P}$ are

$$\begin{aligned} \langle JM(K+1)|P'_-|JMK\rangle &= \sqrt{(J-K)(J+K+1)}, \\ \langle JM(K-1)|P'_+|JMK\rangle &= \sqrt{(J+K)(J-K+1)}, \\ \langle JMK|P_{z'}|JMK\rangle &= K. \end{aligned} \tag{14}$$

(P'_- is now a K step-up operator, and P'_+ is a K step-down operator. This results because the commutation relations of the primed components of the rotational angular-momentum operators are the complex conjugates of the standard ones.) The Hamiltonian is not diagonal in the $|JMK\rangle$ basis. In this basis, the diagonal matrix elements of the Hamiltonian are

$$\langle JMK|H|JMK\rangle = \frac{1}{2}(a+b)(J(J+1) - K^2) + cK^2. \tag{15}$$

With $a \neq b$, off-diagonal terms exist. For the special case of a symmetric rotator, with $a = b$, however, these terms vanish, and the $|JMK\rangle$ are eigenstates of this symmetric rotator. The rotational energies are

$$E_{JK} = \frac{1}{2}(a+b)(J(J+1) - K^2) + cK^2. \tag{16}$$

For the asymmetric rotator, with $a \neq b$, we need the matrix elements

$$\begin{aligned} &\langle JMK'|P'_-P'_-|JMK\rangle \\ &= \delta_{K'(K+2)}\sqrt{(J-K)(J-K-1)(J+K+1)(J+K+2)}, \\ &\langle JMK'|P'_+P'_+|JMK\rangle \\ &= \delta_{K'(K-2)}\sqrt{(J+K)(J+K-1)(J-K+1)(J-K+2)}. \end{aligned} \tag{17}$$

When multiplied by (a-b)/4, these give the nonzero off-diagonal matrix elements. To get the energy eigenvectors, we now need to make a transformation from the $|JMK\rangle$ basis to a basis of the type $|JME_\alpha\rangle$, where these base vectors are simultaneously eigenvectors of the three commuting operators, $\vec{P}^2$, P_z, and H, with

$$\begin{aligned} \vec{P}^2|JME_\alpha\rangle &= J(J+1)|JME_\alpha\rangle, \\ P_z|JME_\alpha\rangle &= M|JME_\alpha\rangle, \\ H|JME_\alpha\rangle &= E_\alpha|JME_\alpha\rangle, \end{aligned} \tag{18}$$

where α is a label that simply orders the energy eigenvalues for a particular $J\ M$. To find the energy eigenvalues and eigenvectors, we need to make a transformation from the $|JMK\rangle$ basis to the $|JME_\alpha\rangle$ basis:

$$|JME_\alpha\rangle = \sum_{K=-J}^{K=+J} |JMK\rangle\langle JMK|JME_\alpha\rangle = \sum_{K=-J}^{K=+J} c_K(E_\alpha)|JMK\rangle. \qquad (19)$$

If we substitute this linear combination of the $|JMK\rangle$ into the energy eigenequation

$$H|JME_\alpha\rangle = E_\alpha|JME_\alpha\rangle, \qquad (20)$$

we get

$$\sum_K H|JMK\rangle\langle JMK|JME_\alpha\rangle = E_\alpha|JME_\alpha\rangle. \qquad (21)$$

Taking the scalar product of this with a particular $\langle JMK'|$, i.e., with left-multiplication by the bra for a particular value of K', we get

$$\sum_K \langle JMK'|H|JMK\rangle\langle JMK|JME_\alpha\rangle = E_\alpha\langle JMK'|JME_\alpha\rangle. \qquad (22)$$

If we use the shorthand notation

$$\langle JMK'|H|JMK\rangle = H_{K'K}, \qquad c_K = \langle JMK|JME_\alpha\rangle, \qquad (23)$$

the above equation can be written as

$$\sum_K H_{K'K}c_K = E_\alpha c_{K'}, \qquad \text{or} \qquad \sum_K \big(H_{K'K} - E_\alpha\delta_{K'K}\big)c_K = 0. \qquad (24)$$

The H submatrix for a particular J, and some fixed M, has been abbreviated by its matrix elements $H_{K'K}$, where the common quantum numbers, J and M, have been suppressed. [From eqs. (15) and (17), these matrix elements are functions only of J and K and are completely independent of M.] Eq. (24) is a set of $(2J+1)$ linear equations in the unknown coefficients, c_K, with $K' = +J, +(J-1), \ldots, -J$. These linear equations have solutions for the c_K if/only if the determinant of the coefficients is zero:

$$\det|H_{K'K} - E_\alpha\delta_{K'K}| = 0. \qquad (25)$$

This determinantal relation leads to a polynomial in the unknown E_α of degree $(2J+1)$, which must be set equal to zero, leading to (2J+1) roots E_α, with $\alpha = 1, 2, \ldots, (2J+1)$.

For example, for $J = 1$, the linear equations are

$$\begin{aligned} (H_{+1+1} - E)c_{+1} + H_{+10}c_0 + H_{+1-1}c_{-1} &= 0,\\ H_{0+1}c_{+1} + (H_{00} - E)c_0 + H_{0-1}c_{-1} &= 0,\\ H_{-1+1}c_{+1} + H_{-10}c_0 + (H_{-1-1} - E)c_{-1} &= 0, \end{aligned} \qquad (26)$$

with the determinantal relation

$$\begin{vmatrix} (H_{+1+1} - E) & H_{+10} & H_{+1-1}, \\ H_{0+1} & (H_{00} - E) & H_{0-1}, \\ H_{-1+1} & H_{-10} & (H_{-1-1} - E) \end{vmatrix} = 0.$$

The Hamiltonian matrix for $J = 1$ follows from eqs. (15) and (17). The $J = 1$ matrix is

$$\langle JMK'|(H-E)|JMK\rangle =$$

$$\begin{array}{c} \\ K=+1 \\ K=0 \\ K=-1 \end{array} \begin{array}{c} \begin{array}{ccc} K=+1 & K=0 & K=-1 \end{array} \\ \begin{pmatrix} (\frac{a+b}{2}+c-E) & 0 & \frac{a-b}{2} \\ 0 & (a+b-E) & 0 \\ \frac{a-b}{2} & 0 & (\frac{a+b}{2}+c-E), \end{pmatrix} \end{array}$$

where it would have been advantageous to rearrange the columns and rows (taking first all even K values, followed by all odd K values), because the matrix elements of H are nonzero only for $\Delta K = \pm 2$. The determinant of the $(H - E)$ matrix will then always factor into two subdeterminants. For $J = 1$, the determinantal relation leads to the requirement

$$\Big[(a+b-E)\Big]\Big[(\frac{a+b}{2}+c-E)^2 - (\frac{a-b}{2})^2\Big] = 0, \tag{27}$$

with the three roots

$$\begin{aligned} E_1 &= (a+b), \\ E_2 &= (\frac{a+b}{2}+c) + (\frac{a-b}{2}) = (a+c), \\ E_3 &= (\frac{a+b}{2}+c) - (\frac{a-b}{2}) = (b+c). \end{aligned} \tag{28}$$

For $E = E_1 = (a + b)$, the allowed c's are given by $c_{+1} = c_{-1} = 0$, $c_0 = 1$. For $E = E_2$ or $E = E_3$, we must have $c_0 = 0$, and the remaining c's follow from the equations

$$\begin{aligned} (\frac{a+b}{2}+c-E)c_{-1} + (\frac{a-b}{2})c_{+1} &= 0 \\ (\frac{a-b}{2})c_{-1} + (\frac{a+b}{2}+c-E)c_{+1} &= 0 \end{aligned} \tag{29}$$

For $E = E_2 = (a + c)$, these equations have the solution $c_{-1} = c_{+1} = \sqrt{\frac{1}{2}}$. Conversely, for $E = E_3 = (b + c)$, these equations lead to $-c_{-1} = c_{+1} = \sqrt{\frac{1}{2}}$. We have normalized the solutions such that $\sum_K |c_K|^2 = 1$.

Thus, the energy eigenvalues and eigenvectors of the asymmetric rotator, with $J = 1$ are

$$E_1 = (a+b), \qquad |J=1\,M\,E_1\rangle = |J=1\,M\,K=0\rangle,$$

and for

$$E_2 = (a+c), \qquad \text{and} \quad E_3 = (b+c),$$

$$\begin{aligned} |J=1\,M\,E_2\rangle &= \frac{1}{\sqrt{2}}\big(|J=1\,M\,K=+1\rangle + |J=1\,M\,K=-1\rangle\big), \\ |J=1ME_3\rangle &= \frac{1}{\sqrt{2}}\big(|J=1MK=+1\rangle - |J=1MK=-1\rangle\big). \end{aligned} \tag{30}$$

For arbitrary J, we are led to a $(2J+1)\times(2J+1)$ determinantal problem. The $(2J+1)$ roots of eq. (24) will give us the eigenvalues E_α and the eigenvectors for arbitrary J. The energies are independent of M. All states are therefore still $(2J+1)$-fold degenerate. This degeneracy can be lifted only by an external field. For the H_2O molecule, which has a permanent electric dipole moment directed along its symmetry axis, the degeneracy could be removed if the molecule is placed in an external electric field (Stark effect).

In condensed-matter physics, effective Hamiltonians of the type of eq. (13) are often useful. These may be functions, e.g., of the spin operators of an impurity ion and have the general form

$$H = aS_x^2 + bS_y^2 + cS_z^2 + d(S_xS_y + S_yS_x) + e(S_xS_z + S_zS_x) + f(S_yS_z + S_zS_y). \quad (31)$$

The combinations, such as $(S_xS_y + S_yS_x)$, are hermitian. Problems of this type can be solved by the techniques illustrated in this section by the asymmetric rotator.

Problems

20. Find the allowed energies for the $J = 2$ states of the asymmetric rigid rotator with Hamiltonian

$$H = aP_{x'}^2 + bP_{y'}^2 + cP_{z'}^2 = \tfrac{1}{2}(a+b)(\vec{P}^2 - P_{z'}^2) + \tfrac{1}{4}(a-b)(P'_+P'_+ + P'_-P'_-) + cP_{z'}^2$$

as functions of a, b, c. Find the eigenvectors of these states as linear combinations of the $|JMK\rangle$; i.e., find the coefficients, c_K, for the allowed $J = 2$ states in the expansions

$$|JME_\alpha\rangle = \sum_K c_K^{(\alpha)} |JMK\rangle.$$

21. For the asymmetric rigid rotator of problem 20, show from the symmetry of the Hamiltonian, H, the eigenvectors split into four classes of the form

$$|JME_\alpha\rangle_{\pm,e(o)} = \sum_K \frac{1}{\sqrt{2}}(|JMK\rangle \pm |JM-K\rangle),$$

where the $e(o)$ states involve a sum over even or odd K values only. Using this $e+, e-, o+, o-$ basis, show the 7×7 matrix of the Hamiltonian matrix for $J = 3$ factors into three 2×2 submatrices and one 1×1 submatrix, and find the allowed energies for $J = 3$ as functions of a, b, c.

22. An impurity ion with a spin, $S = \frac{3}{2}$, is imbedded in a magnetic crystal and is subject to the local effective Hamiltonian

$$H = a(S_xS_y + S_yS_x) + bS_z^2,$$

where S_x, S_y, S_z are the three components of the spin operator and a and b are constants. Find the Hamiltonian matrix in the basis, $|SM_S\rangle$, where M_S is the eigenvalue of S_z. Find the energy eigenvalues, E_α, and the energy eigenvectors as linear combinations of the $|SM_S\rangle$.

16

Transformation Theory

A General

In our example of the diagonalization of the asymmetric rotator Hamiltonian in the last chapter, we encountered a very special case of a very general problem in quantum theory, the transformation from one basis in Hilbert space to another. In our example, it was a transformation from the $|JMK\rangle$ basis to the $|JME_\alpha\rangle$ basis, involving two different complete sets of commuting operators to specify the two different bases. In our specific example of the $J = 1$ energy eigenstates, the transformation was a very simple one, in a 3-D subspace of the asymmetric rotator subspace of the full Hilbert space of our problem, and thus it involved a 3×3 transformation.

For a general state vector, $|\psi\rangle$, in the asymmetric rotator subspace of the Hilbert space of the polyatomic molecule system, we could use either the representation

$$|\psi\rangle = \sum_{JMK} |JMK\rangle\langle JMK|\psi\rangle \tag{1}$$

or the representation

$$|\psi\rangle = \sum_{JM\alpha} |JM\alpha\rangle\langle JM\alpha|\psi\rangle, \tag{2}$$

where we have used the abbreviation α for E_α, and

$$\langle JM\alpha|\psi\rangle = \sum_{K} \langle JM\alpha|JMK\rangle\langle JMK|\psi\rangle, \tag{3}$$

and

$$\langle JMK|\psi\rangle = \sum_{\alpha}\langle JMK|JM\alpha\rangle\langle JM\alpha|\psi\rangle. \tag{4}$$

Here, eqs. (1) and (2) are the analogs in Hilbert space of the relations in ordinary n-dimensional vector space that give the specification of a vector $\vec{V}$ in terms of the components along a set of axes defined by unit vectors $\vec{e}_i$, or in terms of components along a set of $\vec{e}^{\,\prime}_{\alpha}$, which are unit vectors along a set of rotated coordinate axes.

$$\vec{V} = \sum_{k}\vec{e}_k V_k, \tag{5}$$

$$\vec{V} = \sum_{\alpha}\vec{e}^{\,\prime}_{\alpha} V'_{\alpha}, \tag{6}$$

with

$$V'_{\alpha} = \sum_{k} O_{\alpha k} V_k, \qquad \text{where} \qquad O_{\alpha k} = \vec{e}^{\,\prime}_{\alpha}\cdot\vec{e}_k. \tag{7}$$

The inverse gives

$$V_k = \sum_{\alpha}\left(O^{-1}\right)_{k\alpha} V'_{\alpha}, \qquad \text{with} \quad \left(O^{-1}\right)_{k\alpha} = O_{\alpha k}, \tag{8}$$

showing the orthogonal character of the transformation matrix, i.e., the $O_{\alpha k}$ satisfy the orthogonality relations

$$\sum_{k} O_{\alpha k} O_{\beta k} = \delta_{\alpha\beta}, \qquad \sum_{\alpha} O_{\alpha k} O_{\alpha j} = \delta_{kj}. \tag{9}$$

Now, eqs. (3) and (4) are the analogs in Hilbert space of the ordinary vector relations, eqs. (7) and (8). If we name $\langle JM\alpha|\psi\rangle \equiv c'_{\alpha}$ and $\langle JMK|\psi\rangle \equiv c_K$,

$$\begin{aligned} c'_{\alpha} &= \sum_{K}\langle JM\alpha|JMK\rangle c_K = \sum_{K} U_{\alpha K} c_K, \\ c_K &= \sum_{\alpha}\langle JMK|JM\alpha\rangle c'_{\alpha} = \sum_{\alpha}\left(U^{-1}\right)_{K\alpha} c'_{\alpha}, \end{aligned} \tag{10}$$

where now

$$\left(U^{-1}\right)_{K\alpha} = \langle JMK|JM\alpha\rangle = \langle JM\alpha|JMK\rangle^* = U^*_{\alpha K}. \tag{11}$$

That is, the transformation is now unitary, rather than just orthogonal. The U matrix elements satisfy the unitary conditions

$$\sum_{K} U_{\alpha K} U^*_{\beta K} = \delta_{\alpha\beta}, \qquad \sum_{\alpha} U_{\alpha K} U^*_{\alpha K'} = \delta_{KK'}. \tag{12}$$

Now, the inverse matrix is the complex conjugate of the transposed matrix. We can also think of the U, not as a matrix, but as an operator, where

$$U = \sum_{\alpha,K} |JM\alpha\rangle\langle JM\alpha|JMK\rangle\langle JMK|. \tag{13}$$

That is, U is the operator converting a $|JMK\rangle$ into a $|JM\alpha\rangle$ and multiplying it by the complex number $U_{\alpha K}$. Similarly,

$$U^{-1} = \sum_{K,\alpha} |JMK\rangle\langle JMK|JM\alpha\rangle\langle JM\alpha|, \tag{14}$$

where now

$$U^{-1} = U^\dagger \tag{15}$$

For the very specific case of the $J = 1$ states of the asymmetric rotator, using the ordering of energies of eqs. (28)–(30) of Chapter 15 for the index α, we have

$$U = \langle\alpha|K\rangle = \begin{matrix} \\ \alpha = 1 \\ \alpha = 2 \\ \alpha = 3 \end{matrix} \begin{matrix} K=+1 & K=0 & K=-1 \\ \left(\begin{matrix} 0 \\ \frac{1}{\sqrt{2}} \\ \frac{1}{\sqrt{2}} \end{matrix}\right. & \begin{matrix} 1 \\ 0 \\ 0 \end{matrix} & \left.\begin{matrix} 0 \\ \frac{1}{\sqrt{2}} \\ -\frac{1}{\sqrt{2}} \end{matrix}\right) \end{matrix}$$

and

$$U^\dagger = \langle K|\alpha\rangle = \begin{matrix} \\ K = +1 \\ K = 0 \\ K = -1 \end{matrix} \begin{matrix} \alpha=1 & \alpha=2 & \alpha=3 \\ \left(\begin{matrix} 0 \\ 1 \\ 0 \end{matrix}\right. & \begin{matrix} \frac{1}{\sqrt{2}} \\ 0 \\ \frac{1}{\sqrt{2}} \end{matrix} & \left.\begin{matrix} \frac{1}{\sqrt{2}} \\ 0 \\ -\frac{1}{\sqrt{2}} \end{matrix}\right) \end{matrix}.$$

Combining this with the matrix, $H_{KK'}$ of Chapter 15, we can see by straightforward matrix multiplication that

$$\sum_{K,K'} \langle\alpha'|K'\rangle\langle K'|H|K\rangle\langle K|\alpha\rangle = \langle\alpha'|H|\alpha\rangle, \tag{16}$$

or, in matrix form,

$$\sum_{K,K'} U_{\alpha'K'} H_{K'K} (U^\dagger)_{K\alpha} = H_{\alpha'\alpha} = E_\alpha \delta_{\alpha'\alpha}. \tag{17}$$

B Note on Generators of Unitary Operators and the Transformation $UHU^\dagger = H'$

A unitary operator, U, can be generated by a hermitian operator, $G = G^\dagger$, by exponentiation

$$U = e^{i\epsilon G}, \tag{18}$$

where ϵ is a real finite number and the operator G is called the generator of the unitary transformation. To prove the unitary character of U, consider first an infinitesimal transformation, with $\epsilon = \epsilon_0 \ll 1$, so

$$U = 1 + i\epsilon_0 G, \qquad U^\dagger = 1 - i\epsilon_0 G, \tag{19}$$

so

$$UU^\dagger = (1 + i\epsilon_0 G)(1 - i\epsilon_0 G) = 1 + \text{Order}(\epsilon_0^2) = 1. \tag{20}$$

To convert this to a transformation with a finite ϵ, write the exponential in the limiting form

$$\begin{aligned} e^{i\epsilon G} &= \lim_{N\to\infty}\left(1 + i\frac{\epsilon}{N}G\right)^N \\ &= \lim_{N\to\infty}\sum_k \frac{(i\epsilon)^k}{N^k}\frac{G^k}{k!}\frac{N!}{(N-k)!} = \sum_k \frac{(i\epsilon)^k}{k!}G^k. \end{aligned} \tag{21}$$

To prove the product of N factors $(1 + i\frac{\epsilon}{N}G)$ is unitary, we still need to show the product of two unitary operators is unitary

$$\begin{aligned} (U_1U_2)^\dagger &= U_2^\dagger U_1^\dagger \\ = (U_1U_2)^{-1} &= U_2^{-1}U_1^{-1} = U_2^\dagger U_1^\dagger \end{aligned} \tag{22}$$

if $U_1^{-1} = U_1^\dagger$ and $U_2^{-1} = U_2^\dagger$. Finally, it will be very useful to have a series expansion in powers of ϵ of the transformed Hamiltonian, $H' = UHU^\dagger$,

$$\begin{aligned} H' = e^{i\epsilon G}He^{-i\epsilon G} &= \left(1 + i\epsilon G + \frac{(i\epsilon)^2}{2!}G^2 + \cdots\right)H\left(1 - i\epsilon G + \frac{(-i\epsilon)^2}{2!}G^2 + \cdots\right) \\ &= H + i\epsilon[G,H] + \frac{(i\epsilon)^2}{2!}[G,[G,H]] \\ &\quad + \cdots + \frac{(i\epsilon)^n}{n!}[G,[G,[G,\ldots[G,H]]]]_n + \cdots \end{aligned} \tag{23}$$

where we have used $(G^2H - 2GHG + HG^2) = [G,[G,H]]$ for the second term. The nth term, involving n commutators, can be seen to follow from the $(n-1)^{th}$ term from the Taylor expansion in ϵ,

$$f(\epsilon) = e^{i\epsilon G}He^{-i\epsilon G} = \sum_n \frac{\epsilon^n}{n!}\left(\frac{d^n f(\epsilon)}{d\epsilon^n}\right)_{\epsilon=0}, \tag{24}$$

where

$$\frac{df(\epsilon)}{d\epsilon} = e^{i\epsilon G}i[G,H]e^{-i\epsilon G},$$

and with

$$\begin{aligned} \frac{d^{(n-1)}f(\epsilon)}{d\epsilon^{(n-1)}} &= e^{i\epsilon G}i^{n-1}[G,[G,\cdots[G,H]]]_{n-1}e^{-i\epsilon G}, \\ \frac{d^n f(\epsilon)}{d\epsilon^n} &= iG\frac{d^{(n-1)}f(\epsilon)}{d\epsilon^{(n-1)}} - i\frac{d^{(n-1)}f(\epsilon)}{d\epsilon^{(n-1)}}G = i[G, \frac{d^{(n-1)}f(\epsilon)}{d\epsilon^{(n-1)}}]. \end{aligned} \tag{25}$$

This expansion in multiple commutators of G with H is particularly useful, if (1) the nth commutator is zero for a relatively small n; (2) if H and H' differ only by a small term (perturbation theory), so the infinite series can, in good approximation, be terminated after a few terms; or (3) if the nth commutator is so simple the series can be summed.

17

Another Example: Successive Polarization Filters for Beams of Spin $s = \frac{1}{2}$ Particles

So far, our first example of a unitary transformation from one basis to another involved a finite-dimensional unitary submatrix. Let us consider one more example of this type, an even simpler example involving spin $s = \frac{1}{2}$ particles, hence, a 2×2-dimensional transformation. Suppose we have a beam of spin $s = \frac{1}{2}$ particles. They can be prepared, so all are in a state of definite spin orientation, say, with $m_s = +\frac{1}{2}$, or with $m_s = -\frac{1}{2}$, along some specific z-direction in 3-D space by passing the beam through a polarization filter. The historically first such filter is that employed by Stern and Gerlach involving a set of three magnets, with nonuniform magnetic fields, placed in succession along the beam line, so a set of baffles can eliminate the particles with one of the two spin orientations. Other types of sophisticated polarization filters exist. (For a reference to modern polarization filters, see, e.g., *Polarized Beams and Polarized Gas Targets*, Hans Paetz gen. Schieck and Lutz Sydow, eds. World Scientific, 1996). We will assume the filter is perfect and prepares particles in a pure state of very definite m_s along a specific z-direction. Suppose the first such filter is followed with a second filter, identical to the first, but now with its new z' axis oriented along some new direction, given by polar and azimuth angles, θ and ϕ, relative to the original x, y, z axes, and set for some definite m'_s along the new direction. What fraction of the $s = \frac{1}{2}$-particles will pass through the second filter?

The first filter prepares particles in eigenstates $|m = \pm\frac{1}{2}\rangle$, which are eigenstates of $\vec{S}^2$ and S_z:

$$\vec{S}^2|\tfrac{1}{2}m\rangle = \tfrac{3}{4}|\tfrac{1}{2}m\rangle, \qquad S_z|\tfrac{1}{2}m\rangle = m|\tfrac{1}{2}m\rangle. \tag{1}$$

The second filter passes particles in the eigenstates $|\alpha = \pm\frac{1}{2}\rangle$, which are eigenstates of $\vec{S}^2$ and $S_{z'}$:

$$\vec{S}^2|\tfrac{1}{2}\alpha\rangle = \tfrac{3}{4}|\tfrac{1}{2}\alpha\rangle, \qquad S_{z'}|\tfrac{1}{2}\alpha\rangle = \alpha|\tfrac{1}{2}\alpha\rangle. \tag{2}$$

The answer to our problem is as follows: The probability a particle in the beam with definite $|\frac{1}{2}m\rangle$ will pass through the second filter set to pass a specific $|\frac{1}{2}\alpha\rangle$ is

$$P(m, \alpha) = |\langle\tfrac{1}{2}\alpha|\tfrac{1}{2}m\rangle|^2, \tag{3}$$

so we need to calculate the transformation coefficient $\langle\frac{1}{2}\alpha|\frac{1}{2}m\rangle = U_{\alpha m}$. It will be advantageous to switch to the operator $\vec{\sigma}$, via $\vec{S} = \frac{1}{2}\vec{\sigma}$, and to omit the common s quantum number of $\frac{1}{2}$ in all equations. Here, $\vec{\sigma}$ is the Pauli spin operator, whose components $\sigma_x, \sigma_y, \sigma_z$ lead to the three Pauli spin matrices, we have already met through eq. (38) of Chapter 14. Thus,

$$\sigma_z|m\rangle = \lambda_m|m\rangle, \qquad \sigma_{z'}|\alpha\rangle = \lambda_\alpha|\alpha\rangle, \tag{4}$$

with $\lambda_m = \pm 1$ for states with $m = \pm\frac{1}{2}$, and $\lambda_\alpha = \pm 1$ for states with $\alpha = \pm\frac{1}{2}$. Now, we shall rewrite the relation $\sigma_{z'}|\alpha\rangle = \lambda_\alpha|\alpha\rangle$ as

$$\sum_m \sigma_{z'}|m\rangle\langle m|\alpha\rangle = \lambda_\alpha|\alpha\rangle. \tag{5}$$

Left-multiplying by a specific $\langle m'|$ leads to

$$\sum_m \langle m'|\sigma_{z'}|m\rangle\langle m|\alpha\rangle = \lambda_\alpha\langle m'|\alpha\rangle. \tag{6}$$

Now we can express $\sigma_{z'}$ in terms of the original x, y, z components of $\vec{\sigma}$

$$\sigma_{z'} = \sin\theta\cos\phi\sigma_x + \sin\theta\sin\phi\sigma_y + \cos\theta\sigma_z, \tag{7}$$

and use the 2×2 matrices

$$\sigma_x = \begin{pmatrix} 0 & 1 \\ 1 & 0 \end{pmatrix}, \qquad \sigma_y = \begin{pmatrix} 0 & -i \\ +i & 0 \end{pmatrix}, \qquad \sigma_z = \begin{pmatrix} +1 & 0 \\ 0 & -1 \end{pmatrix},$$

to evaluate

$$\langle m'|\sigma_{z'}|m\rangle = \begin{pmatrix} \cos\theta & \sin\theta e^{-i\phi} \\ \sin\theta e^{i\phi} & -\cos\theta \end{pmatrix}.$$

Then, eq. (6) can be rewritten in matrix form, where we shall also use the shorthand notation $\langle m|\alpha\rangle = c_\pm$ for c_m with $m = \pm\frac{1}{2}$,

$$\begin{pmatrix} \cos\theta & \sin\theta e^{-i\phi} \\ \sin\theta e^{i\phi} & -\cos\theta \end{pmatrix}\begin{pmatrix} c_+ \\ c_- \end{pmatrix} = \lambda_\alpha\begin{pmatrix} c_+ \\ c_- \end{pmatrix},$$

leading to the two linear equations

$$\begin{aligned} (\cos\theta - \lambda_\alpha)c_+ + \sin\theta e^{-i\phi}c_- &= 0, \\ \sin\theta e^{i\phi}c_+ - (\cos\theta + \lambda_\alpha)c_- &= 0. \end{aligned} \tag{8}$$

For $\lambda_\alpha = +1$, these equations lead to

$$\frac{c_+}{c_-} = \frac{\sin\theta e^{-i\phi}}{(1-\cos\theta)} = \frac{(1+\cos\theta)}{\sin\theta e^{i\phi}} = \frac{\cos(\frac{\theta}{2})e^{-i\frac{\phi}{2}}}{\sin(\frac{\theta}{2})e^{+i\frac{\phi}{2}}}, \tag{9}$$

so, for $\lambda = +1$:

$$c_+ = \cos(\frac{\theta}{2})e^{-i\frac{\phi}{2}}, \qquad c_- = \sin(\frac{\theta}{2})e^{+i\frac{\phi}{2}}, \tag{10}$$

where the undetermined normalization factor has been chosen, so $\sum_m |c_m|^2 = 1$. These two numbers give us the first column of the unitary 2×2 matrix $\langle m|\alpha\rangle$, with $\alpha = +\frac{1}{2}$. In the same way, putting $\lambda_\alpha = -1$, we get the second column of the $\langle m|\alpha\rangle$ matrix with $\alpha = -\frac{1}{2}$ to give

$$\langle m|\alpha\rangle = \begin{pmatrix} \cos(\frac{\theta}{2})e^{-i\frac{\phi}{2}} & -\sin(\frac{\theta}{2})e^{-i\frac{\phi}{2}} \\ \sin(\frac{\theta}{2})e^{i\frac{\phi}{2}} & \cos(\frac{\theta}{2})e^{i\frac{\phi}{2}} \end{pmatrix}.$$

In our notation, this is the matrix for $U^\dagger$, viz., $U^\dagger_{m\alpha}$. To obtain the matrix for U, $\langle\alpha|m\rangle = U_{\alpha m}$, we need to transpose and complex conjugate the above matrix to get

$$\langle\alpha|m\rangle = \begin{pmatrix} \cos(\frac{\theta}{2})e^{+i\frac{\phi}{2}} & \sin(\frac{\theta}{2})e^{-i\frac{\phi}{2}} \\ -\sin(\frac{\theta}{2})e^{i\frac{\phi}{2}} & \cos(\frac{\theta}{2})e^{-i\frac{\phi}{2}} \end{pmatrix}.$$

Finally, this U matrix can be written as

$$U = \begin{pmatrix} \cos\frac{\theta}{2} & \sin\frac{\theta}{2} \\ -\sin\frac{\theta}{2} & \cos\frac{\theta}{2} \end{pmatrix} \begin{pmatrix} e^{i\frac{\phi}{2}} & 0 \\ 0 & e^{-i\frac{\phi}{2}} \end{pmatrix} = e^{i\frac{\theta}{2}\sigma_y} e^{i\frac{\phi}{2}\sigma_z}.$$

The last operator form of this unitary transformation follows, for the z component, from

$$\mathbf{1} = \sigma_z^2 = \sigma_z^4 = \cdots = \sigma_z^{2n},$$

$$\sigma_z = \sigma_z^3 = \sigma_z^5 = \cdots = \sigma_z^{2n+1}, \tag{11}$$

and, for the y component, from the similar relation

$$\mathbf{1} = \sigma_y^2 = \sigma_y^4 = \cdots \sigma_y^{2n},$$

$$\sigma_y = \sigma_y^3 = \sigma_y^5 = \cdots \sigma_y^{2n+1}, \tag{12}$$

so

$$e^{i\frac{\theta}{2}\sigma_y} = \begin{pmatrix} 1 & 0 \\ 0 & 1 \end{pmatrix} \left(1 - \left(\frac{\theta}{2}\right)^2 \frac{1}{2!} + \left(\frac{\theta}{2}\right)^4 \frac{1}{4!} + \cdots\right)$$

$$+ i \begin{pmatrix} 0 & -i \\ i & 0 \end{pmatrix} \left(\left(\frac{\theta}{2}\right) - \left(\frac{\theta}{2}\right)^3 \frac{1}{3!} + \left(\frac{\theta}{2}\right)^5 \frac{1}{5!} + \cdots\right).$$

Finally, $U^\dagger$ can be written in similar operator form as

$$U^\dagger = e^{-i\frac{\phi}{2}\sigma_z} e^{-i\frac{\theta}{2}\sigma_y}. \tag{13}$$

Also, note the appearance of the half-angles, associated with the $s = \frac{1}{2}$ character of the particles. Thus, e.g., if both polarization filters are set for the spin-projection $m = +\frac{1}{2}$, the fraction of the incoming particles that will pass through the second filter is

$$P(m = +\tfrac{1}{2}, \alpha = +\tfrac{1}{2}) = \cos^2(\frac{\theta}{2}). \tag{14}$$

18

Transformation Theory for Systems with Continuous Spectra

So far, we have studied transformations from one basis to another only for basis vectors that are the eigenvectors of a set of commuting hermitian operators whose spectra are discrete. Moreover, we have studied finite-dimensional subspaces of these vector spaces, so our unitary transformation matrices were finite-dimensional. Commuting operators, however, such as the operators x, y, z or the operators p_x, p_y, p_z exist with continuous spectra. Still other operators have both discrete and continuous spectra. We need to study the transformation theory for the base vectors of this type. In particular, the coordinate and momentum representations are of great importance.

In the coordinate representation, we project state vectors $|\psi\rangle$ onto the base vectors $|\vec{r}_0\rangle$ that are simultaneous eigenvectors of the three operators x, y, z,

$$x|\vec{r}_0\rangle = x_0|\vec{r}_0\rangle, \qquad y|\vec{r}_0\rangle = y_0|\vec{r}_0\rangle, \qquad z|\vec{r}_0\rangle = z_0|\vec{r}_0\rangle. \tag{1}$$

The spectrum is continuous, so the orthogonality relation is given by a Dirac delta function

$$\langle\vec{r}_0'|\vec{r}_0\rangle = \delta(\vec{r}_0' - \vec{r}_0), \tag{2}$$

and the completeness condition is given by the closure relation

$$\int d\vec{r}_0|\vec{r}_0\rangle\langle\vec{r}_0| = 1, \tag{3}$$

so

$$|\psi\rangle = \int d\vec{r}_0|\vec{r}_0\rangle\langle\vec{r}_0|\psi\rangle, \tag{4}$$

and

$$O|\psi\rangle = \int\int d\vec{r}_0{}' d\vec{r}_0 |\vec{r}_0{}'\rangle\langle\vec{r}_0{}'|O|\vec{r}_0\rangle\langle\vec{r}_0|\psi\rangle. \tag{5}$$

Now, if the operator, O, is a function of the coordinate operators x, y, z, only,

$$\text{with} \qquad O = F(x, y, z), \qquad \langle\vec{r}_0{}'|O|\vec{r}_0\rangle = F(x_0, y_0, z_0)\delta(\vec{r}_0{}' - \vec{r}_0), \tag{6}$$

so

$$\langle\psi_1|O|\psi_2\rangle = \int d\vec{r}_0 \psi_1^*(\vec{r}_0)F(x_0, y_0, z_0)\psi_2(\vec{r}_0). \tag{7}$$

If, on the other hand, an operator, O, is a function of the momentum operators p_x, p_y, p_z, it will be useful to transform to the momentum representation, with base vectors that are the eigenvectors of the momentum operators

$$p_x|\vec{p}_0\rangle = p_{x_0}|\vec{p}_0\rangle, \qquad p_y|\vec{p}_0\rangle = p_{y_0}|\vec{p}_0\rangle, \qquad p_z|\vec{p}_0\rangle = p_{z_0}|\vec{p}_0\rangle. \tag{8}$$

We shall be particularly interested in the unitary transformation matrix $\langle\vec{p}_0|\vec{r}_0\rangle$. We could, of course, use Fourier integral analysis, from which we know the result. Let us, however, rederive this transformation matrix to learn how to deal with base vectors with continuous spectra. It will, in particular, be useful to introduce a unitary operator, the translation operator, which will serve our purpose.

A The Translation Operator

Consider the operator

$$U = e^{-\frac{i}{\hbar}\vec{c}\cdot\vec{p}} = U(\vec{c}), \tag{9}$$

where $\vec{c} = (c_1, c_2, c_3)$ and the c_j are ordinary numbers, so-called "c-numbers," the components of $\vec{p}$ are operators. We then have the commutator relations

$$[x, U] = -\frac{\hbar}{i}\frac{\partial U}{\partial p_x} = c_1 U, \tag{10}$$

with similar equations for the commutators $[y, U]$, $[z, U]$. These relations lead to

$$xU|\vec{r}_0\rangle = (Ux + c_1 U)|\vec{r}_0\rangle = (x_0 + c_1)U|\vec{r}_0\rangle. \tag{11}$$

Thus, the new eigenvector obtained by acting with U on $|\vec{r}_0\rangle$ is an eigenvector of the operator, x, with eigenvalue $(x_0 + c_1)$, similarly, for the y and z components. Therefore,

$$U(\vec{c})|\vec{r}_0\rangle = |\vec{r}_0 + \vec{c}\rangle, \tag{12}$$

or

$$U(\vec{r}_0)|0\rangle = |\vec{r}_0\rangle. \tag{13}$$

We see the translation character of the operator U. The translation operator, $U = e^{-\frac{i}{\hbar}\vec{c}\cdot\vec{p}}$, can be regarded from two points of view as follows.

(1) Passive point of view. Action on base vectors. From this point of view, of eq. (12), the operator $U(\vec{c})$ acts on a base vector $|\vec{r}_0\rangle$ and converts it to a new base vector $|\vec{r}_0 + \vec{c}\rangle$, but the physical system is not translated. This process is illustrated by Fig. 18.1(a), where the physical system is illustrated by the maximum in its probability density at $x = a$. By shifting the origin of the coordinate system to the left a distance c_1, the maximum of the probability density lies at $a + c_1$ in the translated coordinate system.

(2) Active point of view. Action on the physical system. In this point of view, the coordinate axes remain fixed, but the physical system is translated by the action of the operator $U(\vec{c})$ from its original position given by a vector $\vec{a}$ to a new position described by the vector $\vec{a} + \vec{c}$. Now, we let U act on the original state vector $|\psi\rangle$ to make a new, translated, state vector $|\psi'\rangle$

$$U(\vec{c})|\psi\rangle = |\psi'\rangle, \tag{14}$$

so the original probability amplitude function $\psi(\vec{r})$ is shifted to a new probability amplitude function $\psi'(\vec{r})$, with

$$\begin{aligned}\psi'(\vec{r}) &= \langle\vec{r}|\psi'\rangle = \langle\vec{r}|U(\vec{c})|\psi\rangle = \langle\psi|U^{\dagger}|\vec{r}\rangle^* \\ &= \langle\psi|e^{+\frac{i}{\hbar}\vec{c}\cdot\vec{p}}|\vec{r}\rangle^* = \langle\psi|\vec{r} - \vec{c}\rangle^* = \langle\vec{r} - \vec{c}|\psi\rangle = \psi(\vec{r} - \vec{c}).\end{aligned} \tag{15}$$

Now, if the original $\psi(\vec{r})$ had a maximum at $\vec{r} = \vec{a}$, the translated $\psi'(\vec{r}) = \psi(\vec{r} - \vec{c})$ will have a maximum at $(\vec{r} - \vec{c}) = \vec{a}$, or at $\vec{r} = (\vec{a} + \vec{c})$; i.e., the physical system has been translated along the positive direction of the vector $\vec{c}$. This process is illustrated in Fig. 18.1(b).

B Coordinate Representation Matrix Elements of p_x

Having discovered the properties of the translation operator, $U(\vec{c})$, we shall now use it to calculate the matrix elements of p_x in the coordinate representation, $\langle\vec{r}_0'|p_x|\vec{r}_0\rangle$.

In the relation

$$\langle\vec{r}_0'|U(\vec{c})|\vec{r}_0\rangle = \langle\vec{r}_0'|\vec{r}_0 + \vec{c}\rangle, \tag{16}$$

consider a vector $\vec{c} = \vec{\epsilon} = (\epsilon, 0, 0)$, where ϵ is an infinitesimal, $\epsilon \ll 1$, so eq. (16) becomes, retaining only first-order quantities in ϵ,

$$\begin{aligned}&\langle\vec{r}_0'|(1 - \frac{i\epsilon}{\hbar}p_x)|\vec{r}_0\rangle = \delta(\vec{r}_0' - \vec{r}_0) - \frac{i\epsilon}{\hbar}\langle\vec{r}_0'|p_x|\vec{r}_0\rangle \\ &= \langle\vec{r}_0'|\vec{r}_0 + \vec{\epsilon}\rangle = \delta(\vec{r}_0' - \vec{r}_0 - \vec{\epsilon}) = \delta(\vec{r}_0' - \vec{r}_0) - \epsilon\left[\frac{\partial}{\partial x}\delta(\vec{r})\right]_{\vec{r}=\vec{r}_0'-\vec{r}_0},\end{aligned} \tag{17}$$

so

$$\langle\vec{r}_0'|p_x|\vec{r}_0\rangle = \frac{\hbar}{i}\left[\frac{\partial}{\partial x}\delta(\vec{r})\right]_{\vec{r}=\vec{r}_0'-\vec{r}_0}$$

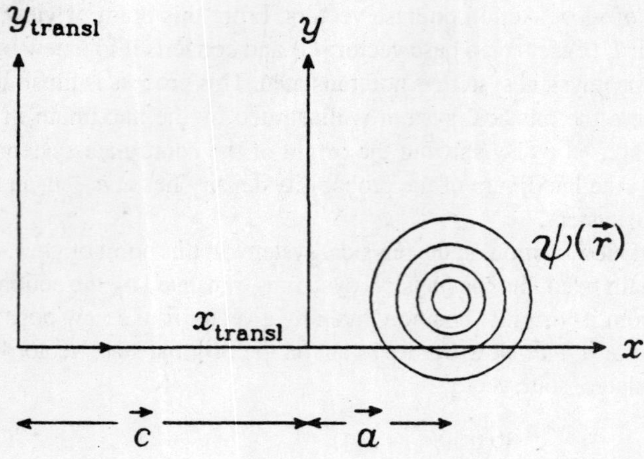

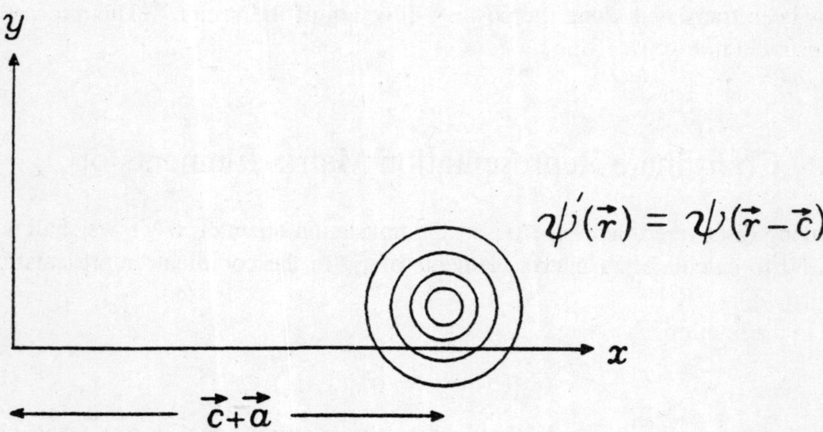

FIGURE 18.1. The translation operator. (a) Passive point of view. (b) Active point of view.

$$= \frac{\hbar}{i}\frac{\partial}{\partial x_0'}\delta(\vec{r}_0' - \vec{r}_0) = -\frac{\hbar}{i}\frac{\partial}{\partial x_0}\delta(\vec{r}_0' - \vec{r}_0). \tag{18}$$

Note, in particular, the minus sign in the last entry. The Dirac delta function is an even function of its argument, but its derivative is an odd function. We are now in a position to calculate the quantity $\langle\psi_1|p_x|\psi_2\rangle$ by the use of the coordinate representation. Projecting the states $|\psi_2\rangle$, and similarly $\langle\psi_1|$, onto their coordinate-space base vectors

$$
\begin{aligned}
\langle \psi_1 | p_x | \psi_2 \rangle &= \int d\vec{r}_0' \int d\vec{r}_0 \langle \psi_1 | \vec{r}_0' \rangle \langle \vec{r}_0' | p_x | \vec{r}_0 \rangle \langle \vec{r}_0 | \psi_2 \rangle \\
&= \int d\vec{r}_0' \psi_1^*(\vec{r}_0') \int d\vec{r}_0 \left(-\frac{\hbar}{i} \frac{\partial}{\partial x_0} \delta(\vec{r}_0' - \vec{r}_0) \right) \psi_2(\vec{r}_0) \\
&= -\frac{\hbar}{i} \int d\vec{r}_0' \psi_1^*(\vec{r}_0') \left[\int dy_0 \int dz_0 \delta(\vec{r}_0' - \vec{r}_0) \psi_2(\vec{r}_0) \right]_{x_0=-\infty}^{x_0=+\infty} \\
&+ \frac{\hbar}{i} \int d\vec{r}_0' \psi_1^*(\vec{r}_0') \int d\vec{r}_0 \delta(\vec{r}_0' - \vec{r}_0) \frac{\partial}{\partial x_0} \psi_2(\vec{r}_0) \\
&= \int d\vec{r}_0' \psi_1^*(\vec{r}_0') \frac{\hbar}{i} \frac{\partial}{\partial x_0'} \psi_2(\vec{r}_0'). \qquad (19)
\end{aligned}
$$

This relation is of course just our previous result. (As always, the integrated term in the integration by parts has the value zero at the limits $x_0 = \pm\infty$.)

C Calculation of the Transformation Matrix $\langle \vec{r}_0 | \vec{p}_0 \rangle$

From the known matrix elements of p_x in the coordinate representation, we can now also calculate the unitary matrix $\langle \vec{p}_0 | \vec{r}_0 \rangle$ or its inverse $\langle \vec{r}_0 | \vec{p}_0 \rangle$. We shall proceed as in Chapters 16–17. From the eigenvector equations

$$
p_x | \vec{p}_0 \rangle = p_{0_x} | \vec{p}_0 \rangle, \qquad p_y | \vec{p}_0 \rangle = p_{0_y} | \vec{p}_0 \rangle, \qquad p_z | \vec{p}_0 \rangle = p_{0_z} | \vec{p}_0 \rangle, \qquad (20)
$$

we get, e.g.,

$$
\int d\vec{r}_0 p_x | \vec{r}_0 \rangle \langle \vec{r}_0 | \vec{p}_0 \rangle = p_{0_x} | \vec{p}_0 \rangle, \qquad (21)
$$

which is the exact analog of eq. (5) of Chapter 17, except the unit operator in the form $\sum_m |m\rangle\langle m|$ has been replaced by the unit operator in the form $\int d\vec{r}_0 |\vec{r}_0\rangle\langle \vec{r}_0|$. As in that case, left-multiplying with a specific $\langle \vec{r}_0'|$ now leads to

$$
\int d\vec{r}_0 \langle \vec{r}_0' | p_x | \vec{r}_0 \rangle \langle \vec{r}_0 | \vec{p}_0 \rangle = p_{0_x} \langle \vec{r}_0' | \vec{p}_0 \rangle. \qquad (22)
$$

Using the result of eq. (18), this relation leads, with an integration by parts, to

$$
\begin{aligned}
& -\frac{\hbar}{i} \int d\vec{r}_0 \left(\frac{\partial}{\partial x_0} \delta(\vec{r}_0' - \vec{r}_0) \right) \langle \vec{r}_0 | \vec{p}_0 \rangle \\
&= +\frac{\hbar}{i} \int d\vec{r}_0 \delta(\vec{r}_0' - \vec{r}_0) \left(\frac{\partial}{\partial x_0} \langle \vec{r}_0 | \vec{p}_0 \rangle \right) \\
&= +\frac{\hbar}{i} \frac{\partial}{\partial x_0'} \left(\langle \vec{r}_0' | \vec{p}_0 \rangle \right) \\
&= p_{0_x} \langle \vec{r}_0' | \vec{p}_0 \rangle. \qquad (23)
\end{aligned}
$$

Assuming we can factor the scalar product

$$
\langle \vec{r}_0' | \vec{p}_0 \rangle = \langle x_0' | p_{0_x} \rangle \langle y_0' | p_{0_y} \rangle \langle z_0' | p_{0_z} \rangle, \qquad (24)
$$

the equation for $\langle x_0'|p_{0_x}\rangle$ becomes an ordinary differential equation

$$\frac{\hbar}{i}\frac{d}{dx_0'}\langle x_0'|p_{0_x}\rangle = p_{0_x}\langle x_0'|p_{0_x}\rangle. \tag{25}$$

This equation can be integrated at once to yield

$$\langle x_0'|p_{0_x}\rangle = \text{const.}e^{+\frac{i}{\hbar}x_0'p_{0_x}}, \tag{26}$$

with similar results for $\langle y_0'|p_{0_y}\rangle$ and $\langle z_0'|p_{0_z}\rangle$, to give

$$\langle \vec{r}_0'|\vec{p}_0\rangle = \frac{1}{(2\pi\hbar)^{\frac{3}{2}}}e^{\frac{i}{\hbar}\vec{r}_0'\cdot\vec{p}_0}, \tag{27}$$

where the integration constant has been chosen to satisfy the Dirac delta function orthonormality

$$\langle \vec{p}_0'|\vec{p}_0\rangle = \int d\vec{r}_0\langle p_0'|\vec{r}_0\rangle\langle\vec{r}_0|\vec{p}_0\rangle = \frac{1}{(2\pi\hbar)^3}\int d\vec{r}_0 e^{\frac{i}{\hbar}\vec{r}_0\cdot(\vec{p}_0-\vec{p}_0')} = \delta(\vec{p}_0'-\vec{p}_0). \tag{28}$$

The results for $\langle\vec{r}_0|\vec{p}_0\rangle$ and $\langle\vec{p}_0|\vec{r}_0\rangle$ are of course well known from Fourier analysis. We have rederived them here to show how they can be derived from transformation theory.

19

Time-Dependence of State Vectors, Algebraic Techniques, Coherent States

Before examining the time-dependence of state vectors it will be useful to recapitulate and summarize the postulates of quantum theory.

A Recapitulation: The Postulates of Quantum Theory

I. The state of a physical system is specified by a vector, $|\psi\rangle$, of the infinite-dimensional Hilbert space. (We assume $\langle\psi|\psi\rangle = 1$.)

II. Every physically observable quantity is described by a hermitian operator, A.

III. The only possible result of the actual measurement of this physically observable quantity is one of the eigenvalues of the corresponding operator, A. (a) If a_n is a nondegenerate eigenvalue of A (part of the discrete spectrum of A), with

$$A|\alpha_n\rangle = a_n|\alpha_n\rangle, \qquad \text{with} \quad \langle\alpha_n|\alpha_n\rangle = 1, \tag{1}$$

the probability a measurement of the physical observable, A, of the system specified by the state $|\psi\rangle$ will yield the value a_n is given by

$$P(a_n) = |\langle\alpha_n|\psi\rangle|^2. \tag{2}$$

(b) If a_m is a degenerate eigenvalue of A (again part of the discrete spectrum of A), with

$$A|\alpha_m^{(i)}\rangle = a_m|\alpha_m^{(i)}\rangle, \qquad \text{with} \quad i = 1, 2, \dots, g_{a_m}, \tag{3}$$

the corresponding probability a measurement of the observable A will yield the value a_m is

$$P(a_m) = \sum_{i=1}^{i=g_{a_m}} |\langle \alpha_m^{(i)} | \psi \rangle|^2. \tag{4}$$

(c) If A has a continuous spectrum (or if part of the spectrum of A is continuous), with

$$A|\alpha\rangle = \alpha|\alpha\rangle, \tag{5}$$

the probability a measurement of the physical observable, A, will yield a value between α and $\alpha + d\alpha$ is given by

$$P(\alpha)d\alpha = d\alpha |\langle \alpha | \psi \rangle|^2 \tag{6}$$

for the system specified by the state $|\psi\rangle$. Note: In the basis in which A is diagonal

$$A = \sum_{n,i} |\alpha_n^{(i)}\rangle a_n \langle \alpha_n^{(i)}| + \int d\alpha |\alpha\rangle \alpha \langle \alpha| \tag{7}$$

and

$$\begin{aligned} \langle \psi | A | \psi \rangle &= \sum_{n,i} a_n |\langle \alpha_n^{(i)} | \psi \rangle|^2 + \int d\alpha \alpha |\langle \alpha | \psi \rangle|^2 \\ &= \sum_n a_n P(a_n) + \int d\alpha \alpha P(\alpha). \end{aligned} \tag{8}$$

The operators

$$\sum_i |\alpha_m^{(i)}\rangle\langle \alpha_m^{(i)}|, \qquad \text{or} \qquad d\alpha|\alpha\rangle\langle\alpha|,$$

are projection operators onto a_m or α, respectively.

IV. Immediately after a measurement of the physical observable, A, the state of the system is specified by a new state vector. If the system was originally specified by the state vector $|\psi\rangle$, and if the measurement of A performed on the system specified by $|\psi\rangle$ yielded the specific value a_m, immediately after this measurement, the state of the system is specified by the new state vector

$$\frac{\sum_i |\alpha_m^{(i)}\rangle\langle\alpha_m^{(i)}|\psi\rangle}{\sqrt{\sum_j |\langle\alpha_m^{(j)}|\psi\rangle|^2}}.$$

For a nondegenerate a_n, on the other hand, the new state vector is simply $|\alpha_n\rangle$. The measurement of A disturbs the system! In the case of the two successive polarization filters of Chapter 17, after passing through the first filter, the state of the system was specified by $|m\rangle$. If a measurement of the polarization along the new direction z' then yielded the value λ_α, the particle comes out of the second apparatus in the state $|\alpha\rangle$.

B Time Evolution of a state $|\psi\rangle$

If at time t_0 the state of the system is specified by $|\psi(t_0)\rangle$, what is the state of the system at a later time, t; i.e., what is $|\psi(t)\rangle$? If a measurement is made on the system of some observable A (or B, etc.), the time-evolution of the quantum system is noncausal. The measurement of A (between times t_0 and t) disturbs the system. The measurement process itself must be taken into account. If we enlarge the system to include the whole measurement apparatus, this larger system would have to be studied, and it may not be practical to consider this larger system. For an isolated system (undisturbed by an observer and his apparatus in the time interval from t_0 to t), however, the evolution in time from $|\psi(t_0)\rangle$ to $|\psi(t)\rangle$ is completely causal (though we are still tied to the probability description). The time evolution is given by

$$-\frac{\hbar}{i}\frac{d}{dt}|\psi(t)\rangle = H(t)|\psi(t)\rangle. \tag{9}$$

The Hamiltonian will often not be an explicit function of the time, but we have here allowed for the possibility of an explicit time-dependence. We can also think of the $|\psi(t)\rangle$ being produced by the action of a unitary operator acting on $|\psi(t_0)\rangle$

$$|\psi(t)\rangle = U(t, t_0)|\psi(t_0)\rangle, \qquad \text{with} \quad U(t_0, t_0) = 1, \tag{10}$$

so

$$-\frac{\hbar}{i}\frac{d}{dt}U(t, t_0)|\psi(t_0)\rangle = H(t)U(t, t_0)|\psi(t_0)\rangle, \tag{11}$$

and because this is valid for any arbitrary $|\psi(t_0)\rangle$,

$$-\frac{\hbar}{i}\frac{d}{dt}U(t, t_0) = H(t)U(t, t_0). \tag{12}$$

This equation has the solution

$$U(t, t_0) = 1 - \frac{i}{\hbar}\int_{t_0}^{t} dt' H(t')U(t', t_0). \tag{13}$$

If H is an explicit function of the time, it may be difficult to do the integral. If H is not an explicit function of the time, the equation for U can be integrated and yields

$$U(t, t_0) = e^{-\frac{i}{\hbar}(t-t_0)H}. \tag{14}$$

U is unitary if H is hermitian, $H^\dagger = H$. The operator H is then the generator of this unitary operator, which now gives a shift in time or a "time-translation." Recall the operator p_x was the generator, G, for a space translation in the x-direction. Now the time shift, $(t - t_0)$, has taken the place of the space shift, c_1, of Chapter 18. For an H, which is not an explicit function of time,

$$|\psi(t)\rangle = e^{-\frac{i}{\hbar}(t-t_0)H}|\psi(t_0)\rangle. \tag{15}$$

The description given here is the so-called *Schrödinger picture* of quantum theory, in which (a) the state vectors $|\psi\rangle$ change with time, but (b) all physical observables A, such as p_x, or x, or functions of $\vec{r}$, $\vec{p}$, such as $H(\vec{p},\vec{r})$, are assumed to have no explicit time dependence.

An alternative point of view was taken by Heisenberg and is known by the name, the *Heisenberg picture*. Now, (a) state vectors are assumed constant in time, but (b) the physical observables are taken to vary with time. To make the transition from the Schrödinger picture to the Heisenberg picture, consider the expectation value of an operator, $O(\vec{p},\vec{r})$, where we use the subscript, S, to designate this operator, *not* explicitly a function of t, i.e., $O = O_S$, similarly for the time-dependent or Schrödinger state vector $|\psi(t)\rangle = |\psi_S\rangle$. Then,

$$\begin{aligned} &\langle\psi(t)|O|\psi(t)\rangle \equiv \langle\psi_S|O_S|\psi_S\rangle \\ &= \langle\psi(t_0)|U^\dagger(t,t_0)OU(t,t_0)|\psi(t_0)\rangle \equiv \langle\psi_H|O_H|\psi_H\rangle, \end{aligned} \tag{16}$$

where the subscript H now stands for Heisenberg. The time-independent state vector is $|\psi(t_0)\rangle = |\psi_H\rangle$, and the time-dependent Heisenberg operator, O_H, is given by

$$O_H = U^\dagger(t,t_0)O_S U(t,t_0). \tag{17}$$

Note,

$$\begin{aligned} -\frac{\hbar}{i}\frac{dO_H}{dt} &= \left(-\frac{\hbar}{i}\frac{dU^\dagger}{dt}\right)O_S U + U^\dagger O_S\left(-\frac{\hbar}{i}\frac{dU}{dt}\right) \\ &= -U^\dagger H O_S U + U^\dagger O_S H U \\ &= -(U^\dagger H U)(U^\dagger O_S U) + (U^\dagger O_S U)(U^\dagger H U) \\ &= -HO_H + O_H H, \end{aligned} \tag{18}$$

where we have used $O_H = U^\dagger O_S U$ and that the Hamiltonian operator, H, commutes with U, which is a function only of pure numbers and the operator H itself. Thus, $U^\dagger H U = HU^\dagger U = H$. The time dependence of O_H is then given by the equation

$$-\frac{\hbar}{i}\frac{dO_H}{dt} = [O_H, H]. \tag{19}$$

This relation is known as the Heisenberg equation. Even in the Schrödinger picture, we sometimes introduce artificially an explicit time dependence, so $O_S = O_S(t)$. In that case, the corresponding equation would be

$$-\frac{\hbar}{i}\frac{dO_H}{dt} = [O_H, H] + \left(-\frac{\hbar}{i}\frac{\partial O}{\partial t}\right)_H, \quad \text{with} \quad \left(-\frac{\hbar}{i}\frac{\partial O}{\partial t}\right)_H = U^\dagger\left(-\frac{\hbar}{i}\frac{\partial O_S}{\partial t}\right)U. \tag{20}$$

In the Heisenberg picture, the kets and bras are time independent. We shall take matrix elements of eq. (19) (assuming that $\frac{\partial O}{\partial t} = 0$) in the energy representation, i.e., between eigenstates of the Hamiltonian, $|n\rangle$, assuming for the moment H has only a discrete spectrum. Then, eq. (19) leads to

$$-\frac{\hbar}{i}\frac{d}{dt}\langle n|O_H(t)|m\rangle = \langle n|O_H(t)|m\rangle(E_m - E_n), \tag{21}$$

$$\frac{d(\langle n|O_H(t)|m\rangle)}{\langle n|O_H(t)|m\rangle} = +\frac{i}{\hbar}(E_n - E_m)dt, \tag{22}$$

$$\langle n|O_H(t)|m\rangle = \langle n|O|m\rangle e^{\frac{i}{\hbar}(E_n - E_m)t}, \tag{23}$$

where we have named the integration constant $\langle n|O|m\rangle$, which is time independent. The Heisenberg matrix elements, $O_{nm} \equiv \langle n|O|m\rangle$, are then just the $\langle n|O_S|m\rangle$. Eq. (23) was essentially the starting point in Heisenberg's thinking. He started with the Fourier time analysis of the classical quantity, O, and replaced the nth overtone of the classical ω with the two-index Bohr quantity $\omega_{nm} = (E_n - E_m)/\hbar$. Similarly, he replaced the nth Fourier coefficient, O_n, in the Fourier time expansion with a two-index quantity he interpreted as the nm^{th} matrix element of O, $\langle n|O|m\rangle$.

C The Heisenberg Treatment of the One-Dimensional Harmonic Oscillator: Oscillator Annihilation and Creation Operators

Let us now briefly follow Heisenberg's analysis of one of his simplest examples, the 1-D harmonic oscillator. Again, we introduce scale factors to transform the physical coordinate, momentum, energy, and so on, to dimensionless x, p_x, ..., $x_{phys.} = x\sqrt{\hbar/m\omega_0}$, $p_{phys.} = p_x\sqrt{\hbar m\omega_0}$, $H_{phys.} = H\hbar\omega_0$, so

$$H = \frac{1}{2}(p_x^2 + x^2). \tag{24}$$

Introduce the two operators

$$a = \frac{1}{\sqrt{2}}(x + ip_x), \qquad a^\dagger = \frac{1}{\sqrt{2}}(x - ip_x). \tag{25}$$

From the commutation relation, $[p_x, x] = -i$, we get the commutator

$$[a, a^\dagger] = 1, \tag{26}$$

and the Hamiltonian can be expressed as

$$H = \frac{1}{2}\left(a^\dagger a + aa^\dagger\right) = \left(a^\dagger a + \frac{1}{2}\right). \tag{27}$$

It will be useful to name the operator $a^\dagger a = N$. Then, we have the family of three commutation relations

$$[N, a^\dagger] = +a^\dagger, \qquad [N, a] = -a, \qquad [a, a^\dagger] = 1. \tag{28}$$

These relations should be compared and contrasted with the standard angular momentum commutation relations

$$[L_0, L_+] = +L_+, \qquad [L_0, L_-] = -L_-, \qquad [L_-, L_+] = -2L_0. \tag{29}$$

The commutator algebra of a, $a^\dagger$, N, is known as the Heisenberg algebra. Note, in particular, the difference from the angular momentum [or SO(3)] algebra in the last entry. If the eigenvectors of the operator, N, are named $|n\rangle$, and its eigenvalues N_n, these $|n\rangle$ are also the eigenvectors of H, because $H = N + \frac{1}{2}$,

$$N_{\text{op.}}|n\rangle = N_n|n\rangle, \qquad \text{and} \qquad E_n = \hbar\omega_0(N_n + \frac{1}{2}), \tag{30}$$

or

$$(1) \quad a^\dagger a|n\rangle = N_n|n\rangle, \qquad (2) \quad aa^\dagger|n\rangle = (N_n + 1)|n\rangle. \tag{31}$$

Acting with a on (1), and using $aa^\dagger = N_{\text{op.}} + 1$, we get

$$aa^\dagger\big(a|n\rangle\big) = N_n\big(a|n\rangle\big), \qquad \text{or} \quad N_{\text{op.}}\big(a|n\rangle\big) = (N_n - 1)\big(a|n\rangle\big). \tag{32}$$

Similarly, acting with $a^\dagger$ on (2), we get

$$a^\dagger a\big(a^\dagger|n\rangle\big) = (N_n + 1)\big(a^\dagger|n\rangle\big), \qquad \text{or} \quad N_{\text{op.}}\big(a^\dagger|n\rangle\big) = (N_n + 1)\big(a^\dagger|n\rangle\big). \tag{33}$$

Eq. (32) tells us

$$\text{either} \qquad (a|n\rangle) = \text{const.}|(n-1)\rangle, \qquad \text{or} \qquad (a|n\rangle) = 0. \tag{34}$$

Similarly, eq. (33) tells us

$$\text{either} \qquad (a^\dagger|n\rangle) = \text{const.}'|(n+1)\rangle, \qquad \text{or} \qquad (a^\dagger|n\rangle) = 0, \tag{35}$$

where $|(n \pm 1)\rangle$ are shorthand notation for the eigenvectors of $N_{\text{op.}}$ with eigenvalues $(N_n \pm 1)$. Now we take the diagonal matrix element of the operator, $N_{\text{op.}}$ (note the similarity of the procedure for the angular momentum algebra!),

$$\begin{aligned} N_n &= \langle n|a^\dagger a|n\rangle \\ &= \sum_k \langle n|a^\dagger|k\rangle\langle k|a|n\rangle \\ &= \sum_k |\langle k|a|n\rangle|^2 \\ &\geq 0. \end{aligned} \tag{36}$$

Thus, N_n is positive definite. Now, if $(a|n\rangle)$ exists, repeat this process by taking the diagonal matrix element of $N_{\text{op.}}$ between states $|(n-1)\rangle$. We conclude $(N_n - 1) \geq 0$. We can repeat this process j times to conclude $(N_n - j) \geq 0$, if $(a|n-j+1\rangle) \neq 0$. If N_n is positive, however, an integer j will eventually come, which is big enough such that $(N_n - j)$ would be negative unless we hit a state $|n_{\text{min.}}\rangle$, such that $(a|n_{\text{min.}}\rangle) = 0$. For this state, eq. (31) tells us $N_{n_{\text{min.}}} = 0$. Now, if we act on this state, with eigenvalue $N_{n_{\text{min}}} = 0$, n times in succession with $a^\dagger$, we get a state with eigenvalue $(0+n)$, i.e., with $N_n = n$. Also, continued operation with $a^\dagger$ leaves the eigenvalue of $N_{\text{op.}}$ positive, no upper bound to this discrete set of eigenvalues exists. Thus,

$$E_n = \hbar\omega_0(n + \frac{1}{2}). \tag{37}$$

All that remains is the job of calculating the matrix elements of x and p_x. From eq. (36)

$$N_n = n = \sum_k |\langle k|a|n\rangle|^2 \\ = |\langle (n-1)|a|n\rangle|^2, \tag{38}$$

where we have assumed the states for this one-degree-of-freedom problem are nondegenerate, so there is just one state, with $k = (n-1)$. Except, for an arbitrary phase, we have determined the matrix element of the operator, a. Choosing the simplest (positive, real) value for this matrix element, we have

$$\langle (n-1)|a|n\rangle = \sqrt{n}. \tag{39}$$

Hermitian conjugation gives us

$$\langle n|a^\dagger|(n-1)\rangle = \sqrt{n}, \qquad \text{or} \qquad \langle (n+1)|a^\dagger|n\rangle = \sqrt{n+1}. \tag{40}$$

$a^\dagger$ is an oscillator quantum creation operator, and a is an oscillator quantum annihilation operator. Now, using

$$x = \frac{1}{\sqrt{2}}(a + a^\dagger), \qquad \text{and} \qquad p_x = \frac{i}{\sqrt{2}}(-a + a^\dagger), \tag{41}$$

we also get

$$\langle m|x|n\rangle = \frac{1}{\sqrt{2}}\left(\delta_{m(n-1)}\sqrt{n} + \delta_{m(n+1)}\sqrt{n+1}\right), \tag{42}$$

$$\langle m|p_x|n\rangle = \frac{1}{\sqrt{2}}\left(-i\delta_{m(n-1)}\sqrt{n} + i\delta_{m(n+1)}\sqrt{n+1}\right). \tag{43}$$

These relations are of course results we have obtained before; but the Heisenberg method of calculation required no knowledge of wave functions or differential equations. As before, we can build more complicated operators from powers of x and p_x, and then express any operator in Heisenberg (time-dependent) form

$$O_H(t) = \sum_{n,m} |n\rangle\langle n|O|m\rangle\langle m| e^{\frac{i}{\hbar}(E_n - E_m)t}. \tag{44}$$

For x and p_x, we could write the Heisenberg (time-dependent) form of these operators in terms of the Schrödinger (time-independent) oscillator quantum annihilation and creation operators, a and $a^\dagger$, as

$$x_H(t) = \frac{1}{\sqrt{2}}\left(a e^{-i\omega_0 t} + a^\dagger e^{+i\omega_0 t}\right), \\ (p_x)_H(t) = \frac{1}{\sqrt{2}}\left(-ia e^{-i\omega_0 t} + ia^\dagger e^{+i\omega_0 t}\right), \tag{45}$$

where we have used eq. (44) to get the time dependence, but have subsequently left off the unit operators, $\sum_n |n\rangle\langle n| = 1$ and $\sum_m |m\rangle\langle m| = 1$.

D Oscillator Coherent States

Remembering the physical displacement and momentum coordinates of the harmonic oscillator are related to the dimensionless x and p_x of the last section by $q = \sqrt{\hbar/m\omega_0}x$ and $p = \sqrt{\hbar m\omega_0}p_x$, we can express the time-dependent displacement and momentum operators through eqs. (41) and (45) by

$$q(t) = \sqrt{\frac{\hbar}{2m\omega_0}}\Big(ae^{-i\omega_0 t} + a^\dagger e^{+i\omega_0 t}\Big),$$
$$p(t) = m\omega_0\sqrt{\frac{\hbar}{2m\omega_0}}\Big(-iae^{-i\omega_0 t} + ia^\dagger e^{+i\omega_0 t}\Big). \tag{46}$$

If we compare this with the classical solution for the harmonic oscillator, $q(t) = q_0\cos(\omega_0 t + \phi)$, with

$$q(t) = \Big(\frac{q_0 e^{-i\phi}}{2}e^{-i\omega_0 t} + \frac{q_0 e^{i\phi}}{2}e^{+i\omega_0 t}\Big),$$
$$p(t) = m\omega_0\Big(-i\frac{q_0 e^{-i\phi}}{2}e^{-i\omega_0 t} + i\frac{q_0 e^{i\phi}}{2}e^{-i\omega_0 t}\Big), \tag{47}$$

we are led to the idea that it might be very useful to replace the eigenvectors, $|n\rangle$, of the harmonic oscillator hamiltonian (or the oscillator quantum number operator), with eigenvectors of the operator, a (or alternatively $a^\dagger$), if we want to study the transition from the quantum oscillator to the classical oscillator. Moreover, we might expect the eigenvalue of the operator, a, to be given by a complex number, where the square of the absolute value of this number is related to the energy or the number of oscillator quanta, and the argument of this complex number is related to the classical phase. The new oscillator representation in terms of the eigenvectors of the oscillator annihilation operator, a, might be particularly useful if the physics of interest involves a statistical distribution of states with different numbers of oscillator quanta, particularly, if the average oscillator excitation number is large, as in the classical limit. Later, when we quantize the electromagnetic field (see Chapter 60), we shall write the hamiltonian of the electromagnetic field

$$H = \sum_{\vec{k},\mu}\hbar\omega(a^\dagger_{\vec{k}\mu}a_{\vec{k}\mu} + \tfrac{1}{2})$$

in terms of an infinite number of oscillators with annihilation and creation operators, $a_{\vec{k}\mu}$ and $a^\dagger_{\vec{k}\mu}$. These operators are interpreted as photon annihilation and creation operators, where the vector $\vec{k}$ is the wave vector, corresponding to the circular frequency $\omega = kc = |\vec{k}|c$, and μ is a polarization index for the photon. The generalization of the single-mode coherent state to be studied in this section to the multimode coherent state of the electromagnetic field will be useful in the study of optical beams whose proper quantum-mechanical description is given by a statistical distribution of quantum states with different numbers of oscillator quanta. (The seminal papers by R. J. Glauber on quantum optics and optical coherent states are all reprinted in an introduction to coherent states by John R.

Klauder and Bo-Sture Skagerstam, *Coherent States. Applications in Physics and Mathematical Physics*, Singapore: World Scientific, 1985.)

For the single-mode harmonic oscillator, two slightly different definitions of the coherent states can be found in the literature, to be denoted by $|z\rangle$, where z is a complex number giving the eigenvalue of the oscillator annihilation operator. In the first definition of the coherent state

$$|z\rangle_{\mathrm{I}} = e^{z^* a^\dagger - za}|0\rangle = U(z)|0\rangle, \tag{48}$$

where $|0\rangle$, the "vacuum state," is the oscillator ground state and $U(z)$ is a unitary operator, because

$$U(z)^\dagger = e^{-(z^* a^\dagger - za)} = U(-z) = U(z)^{-1}. \tag{49}$$

The noncommuting operators, $A \equiv z^* a^\dagger$ and $B = -za$, have a very simple commutator, $[A, B] = zz^*$, a c number commuting with both A and B. In this very special case, when

$$[A, [A, B]] = 0 \qquad \text{and} \qquad [B, [A, B]] = 0, \tag{50}$$

we can write the operator relation

$$e^{(A+B)} = e^{-\frac{1}{2}[A,B]} e^A e^B, \tag{51}$$

as can be shown by direct verification. Note the so-called normal order of the operators on the right-hand side, with creation operators, $a^\dagger$, sitting to the left of annihilation operators, a. The above result leads to

$$|z\rangle_{\mathrm{I}} = e^{-\frac{1}{2}z^* z} e^{z^* a^\dagger} e^{-za}|0\rangle = e^{-\frac{1}{2}z^* z} e^{z^* a^\dagger}|0\rangle, \tag{52}$$

because $a|0\rangle = 0$. For some purposes, a somewhat simpler second definition of the coherent state may therefore be useful

$$|z\rangle_{\mathrm{II}} = e^{z^* a^\dagger}|0\rangle, \tag{53}$$

which differs from the definition $|z\rangle_{\mathrm{I}}$ by the simple c number function, $e^{-zz^*/2}$. This second definition may be a particularly useful definition for generalized coherent states, such as the angular momentum coherent states to be introduced in the next section for which the commutator algebra no longer satisfies the simple relations of eq. (50). To avoid confusion, we will denote the complex number z in the two different definitions by two different symbols

$$\begin{aligned} |z\rangle_{\mathrm{I}} &\equiv |\alpha\rangle = e^{-\frac{1}{2}\alpha^*\alpha} e^{\alpha^* a^\dagger}|0\rangle = e^{-\frac{1}{2}\alpha^*\alpha} \sum_{n=0} \frac{(\alpha^*)^n}{\sqrt{n!}}|n\rangle, \\ |z\rangle_{\mathrm{II}} &\equiv |z\rangle = e^{z^* a^\dagger}|0\rangle = \sum_{n=0} \frac{(z^*)^n}{\sqrt{n!}}|n\rangle. \end{aligned} \tag{54}$$

Both are eigenvectors of the oscillator annihilation operator, a, with eigenvalue given by the complex number α^* or z^*.

$$\begin{aligned} a|\alpha\rangle &= \alpha^*|\alpha\rangle, \\ a|z\rangle &= z^*|z\rangle. \end{aligned} \tag{55}$$

This follows, for the type II coherent state, from

$$a|z\rangle = ae^{z^*a^\dagger}|0\rangle = a\sum_{n=0}\frac{(z^*a^\dagger)^n}{n!}|0\rangle = z^*\sum_{n=1}\frac{(z^*a^\dagger)^{n-1}}{(n-1)!}|0\rangle = z^*e^{z^*a^\dagger}|0\rangle. \quad (56)$$

For the type I coherent state, the c number, $e^{-\alpha\alpha^*/2}$, is merely carried along in the analagous derivation.

The coherent states $|\alpha\rangle$ (or $|z\rangle$) then give us another continuous representation of an arbitrary state vector $|\psi\rangle$ of a physical system. Besides the discrete oscillator quanta representation, $\langle n|\psi\rangle$, we already have two continuous representations, the coordinate representation $\langle x|\psi\rangle$ and the momentum representation $\langle p_x|\psi\rangle$, where the operators x and p_x have a continuous range of eigenvalues from $-\infty \to +\infty$. We can now add the coherent state representation $\langle\alpha|\psi\rangle$, where α is a complex number,

$$\alpha = \xi + i\eta = \rho e^{i\phi},$$

and ξ and η range from $-\infty \to +\infty$, and ρ ranges from $0 \to \infty$ and ϕ from $0 \to 2\pi$. Unlike $\langle x|\psi\rangle$ and $\langle p_x|\psi\rangle$, however, which are orthonormal, continuous representations of $|\psi\rangle$, with

$$\langle x'|x\rangle = \delta(x'-x), \qquad \langle p'_x|p_x\rangle = \delta(p'_x - p_x),$$

the scalar product $\langle\alpha'|\alpha\rangle$ does not lead to δ-functions. Instead,

$$\begin{aligned}\langle\alpha'|\alpha\rangle &= e^{-\frac{1}{2}|\alpha'|^2}e^{-\frac{1}{2}|\alpha|^2}\langle 0|e^{\alpha' a}e^{\alpha^* a^\dagger}|0\rangle = e^{-\frac{1}{2}|\alpha'|^2}e^{-\frac{1}{2}|\alpha|^2}\sum_{n,m}\frac{\alpha'^m}{\sqrt{m!}}\frac{\alpha^{*n}}{\sqrt{n!}}\langle m|n\rangle \\ &= e^{-\frac{1}{2}|\alpha'|^2}e^{-\frac{1}{2}|\alpha|^2}e^{\alpha'\alpha^*}, \end{aligned} \quad (57)$$

so $\langle\alpha'|\alpha\rangle$ is a complicated function of α and α', even though $\langle\alpha|\alpha\rangle = 1$. Also,

$$\langle z'|z\rangle = e^{z'z^*}, \qquad \text{with} \quad \langle z|z\rangle = e^{|z|^2} \neq 1. \quad (58)$$

The coherent states, however, are complete. In fact, with relations (57) or (58), in place of the Dirac δ functions, the coherent states are overcomplete. The completeness can be seen from the existence of the unit operators. For the type I coherent state, the unit operator is

$$\mathbf{1} = \frac{1}{\pi}\int \mathbf{d}^2\alpha|\alpha\rangle\langle\alpha| = \frac{1}{\pi}\int_{-\infty}^{\infty}\mathbf{d}\xi\int_{-\infty}^{\infty}\mathbf{d}\eta|\alpha\rangle\langle\alpha| = \frac{1}{\pi}\int_0^{\infty}\rho\mathbf{d}\rho\int_0^{2\pi}\mathbf{d}\phi|\alpha\rangle\langle\alpha|. \quad (59)$$

With the use of eq. (54), this unit operator transforms into

$$\begin{aligned}\mathbf{1} &= \frac{1}{\pi}\int d^2\alpha e^{-\alpha\alpha^*}\sum_{n=0}^{\infty}\sum_{m=0}^{\infty}\frac{(\alpha^*)^n}{\sqrt{n!}}\frac{\alpha^m}{\sqrt{m!}}|n\rangle\langle m| \\ &= \sum_{n,m}\frac{1}{\pi}\int_0^{\infty} d\rho\frac{e^{-\rho^2}\rho^{n+m+1}}{\sqrt{n!m!}}\int_0^{2\pi} d\phi e^{i(m-n)\phi}|n\rangle\langle m| \\ &= \sum_{n,m}2\int_0^{\infty} d\rho\frac{e^{-\rho^2}\rho^{n+m+1}}{\sqrt{n!m!}}\delta_{nm}|n\rangle\langle m| = \sum_{n=0}^{\infty}2\int_0^{\infty} d\rho\frac{e^{-\rho^2}\rho^{2n+1}}{n!}|n\rangle\langle n|\end{aligned}$$

$$= \sum_{n=0}^{\infty} |n\rangle\langle n|, \tag{60}$$

where the completeness relation, $\sum_n |n\rangle\langle n| = 1$, was in fact proved in detail in Chapter 5, eq. (13), via

$$\sum_{n=0}^{\infty} \langle x'|n\rangle\langle n|x\rangle = \sum_{n=0}^{\infty} \psi_n(x')\psi_n^*(x) = \delta(x'-x) = \langle x'|x\rangle. \tag{61}$$

For the type II coherent state, conversely, the unit operator requires the Bargmann weighting factor, e^{-zz^*}, so

$$\begin{aligned} \mathbf{1} &= \frac{1}{\pi}\int d^2z e^{-zz^*}|z\rangle\langle z| = \frac{1}{\pi}\int d^2z e^{-zz^*}\sum_{n,m}\frac{z^{*n}z^m}{\sqrt{n!m!}}|n\rangle\langle m| \\ &= \sum_{n=0}^{\infty}|n\rangle\langle n|, \end{aligned} \tag{62}$$

as above.

The type II coherent state realization $\langle z|\psi\rangle$ of an arbitrary state vector is simply the Bargmann transform of $\psi(x)$: $\langle z|\psi\rangle = F(z)$.

$$\langle z|\psi\rangle = \sum_{n=0}^{\infty}\langle n|\frac{z^n}{\sqrt{n!}}|\psi\rangle = \sum_{n=0}^{\infty}\frac{z^n}{\sqrt{n!}}\int_{-\infty}^{\infty} dx\psi_n^*(x)\psi(x) = \sum_{n=0}^{\infty}\frac{z^n}{\sqrt{n!}}c_n, \tag{63}$$

where $\psi(x) = \langle x|\psi\rangle$ is the coordinate representation of $|\psi\rangle$, so

$$\langle z|\psi\rangle = \int dx A(x,z)\psi(x) = F(z),$$

where $A(x,z)$ is the Bargmann kernel function with the simple expansion in terms of the *real* $\psi_n(x)$ [see eq. (5) of chapter 5],

$$A(x,z) = \sum_{n=0}^{\infty}\psi_n(x)\frac{z^n}{\sqrt{n!}}.$$

We therefore again have two possible forms of the scalar product of two state vectors

$$\langle\psi_a|\psi_b\rangle = \int_{-\infty}^{\infty} dx\psi_a^*(x)\psi_b(x) = \frac{1}{\pi}\int d^2z e^{-zz^*}F_a^*(z)F_b(z). \tag{64}$$

The complex k of Chapter 5 has here been renamed z, because k for the multimode electromagnetic oscillator is reserved for the wavenumber of the mode. Note, further, the natural appearance of the Bargmann measure, e^{-zz^*}/π, in the scalar product. Finally, the coherent state realizations $F(z) = \langle z|\psi\rangle$ are analytic functions of z: We have mapped the coordinate functions, $\sum_n \psi_n(x)c_n$, into analytic functions, $\sum_n (z^n/\sqrt{n!})c_n$, in a 2-D complex domain.

Let us examine some further properties of the coherent state $|\alpha\rangle$. From eq. (54), $\langle n|\alpha\rangle = e^{-\frac{1}{2}\alpha^*\alpha}(\alpha^*)^n/\sqrt{n!}$. The probability of finding the oscillator in the nth level in the coherent state $|\alpha\rangle$ is, therefore,

$$P_n(\alpha) = \frac{e^{-|\alpha|^2}|\alpha|^{2n}}{n!}. \tag{65}$$

The expectation value of the oscillator quantum number operator in the state $|\alpha\rangle$ is

$$< N >= \langle\alpha|a^\dagger a|\alpha\rangle = \alpha\alpha^* = |\alpha|^2, \tag{66}$$

so the probability

$$P_n(\alpha) = \frac{e^{-<N>}(< N >)^n}{n!} \tag{67}$$

is given by the familiar Poisson distribution. We also have the expectation values

$$\begin{aligned}
\langle\alpha|x|\alpha\rangle &= \sqrt{\tfrac{1}{2}}(\langle\alpha|a + a^\dagger|\alpha\rangle = \sqrt{\tfrac{1}{2}}(\alpha^* + \alpha) = \sqrt{2}\xi,\\
\langle\alpha|x^2|\alpha\rangle &= \tfrac{1}{2}\langle\alpha|(aa + a^\dagger a^\dagger + 2a^\dagger a + 1)|\alpha\rangle = \tfrac{1}{2}[(\alpha^* + \alpha)^2 + 1],\\
\langle\alpha|p_x|\alpha\rangle &= -i\sqrt{\tfrac{1}{2}}\langle\alpha|a - a^\dagger|\alpha\rangle = -i\sqrt{\tfrac{1}{2}}(\alpha^* - \alpha) = -\sqrt{2}\eta,\\
\langle\alpha|p_x^2|\alpha\rangle &= \tfrac{1}{2}\langle\alpha|(-aa - a^\dagger a^\dagger + 2a^\dagger a + 1)|\alpha\rangle = \tfrac{1}{2}[1 - (\alpha^* - \alpha)^2],
\end{aligned} \tag{68}$$

so

$$(\Delta x)^2_\alpha = \tfrac{1}{2}, \qquad (\Delta p_x)^2_\alpha = \tfrac{1}{2}. \tag{69}$$

In the coherent state, $|\alpha\rangle$, we therefore have

$$\Delta p_x \Delta x = \tfrac{1}{2} \qquad \text{and, thus,} \quad \Delta p \Delta q = \frac{\hbar}{2}, \tag{70}$$

where dimensionless x and p_x have been converted to physical q and p. The coherent state is therefore a state with the minimum possible uncertainty, despite the seemingly complicated probability distribution, $P_n(\alpha)$, spread over a range of states $|n\rangle$ about the most likely $n =< N >= |\alpha|^2$. To understand this, we note that the unitary operator, $U(\alpha) = e^{(\alpha^* a^\dagger - \alpha a)}$, shifts the operators, a and $a^\dagger$, hence, x and p_x, according to

$$U(\alpha)aU^{-1}(\alpha) = a - \alpha^*, \qquad U(\alpha)a^\dagger U^{-1}(\alpha) = a^\dagger - \alpha, \tag{71}$$

where we have used

$$\sum_{n=0}^{\infty}[a, \frac{(\alpha a - \alpha^* a^\dagger)^n}{n!}] = -\alpha^* \sum_{n=1}^{\infty} \frac{(\alpha a - \alpha^* a^\dagger)^{n-1}}{(n-1)!}.$$

Eq. (71) thus leads to

$$\begin{aligned}
U(\alpha)xU^{-1}(\alpha) &= x - \sqrt{\tfrac{1}{2}}(\alpha + \alpha^*) = (x - \sqrt{2}\xi),\\
U(\alpha)p_x U^{-1}(\alpha) &= p_x + i\sqrt{\tfrac{1}{2}}(\alpha^* - \alpha) = (p_x + \sqrt{2}\eta).
\end{aligned} \tag{72}$$

$U(\alpha)$ is a displacement operator shifting coordinate, x, and momentum, p_x,

$$x \to x - \sqrt{2}\Re(\alpha), \qquad p_x \to p_x + \sqrt{2}\Im(\alpha).$$

Also, the eigenvalue equation, $a|\alpha\rangle = \alpha^*|\alpha\rangle$, leads to the following simple differential equations in coordinate and momentum space:

$$\frac{1}{\sqrt{2}}\Big(x + \frac{d}{dx}\Big)\langle x|\alpha\rangle = \alpha^*\langle x|\alpha\rangle,$$
$$\frac{i}{\sqrt{2}}\Big(\frac{d}{dp_x} + p_x\Big)\langle p_x|\alpha\rangle = \alpha^*\langle p_x|\alpha\rangle, \tag{73}$$

with solutions

$$\langle x|\alpha\rangle = \mathcal{N}e^{-\frac{1}{2}(x-\sqrt{2}\alpha^*)^2}, \qquad \text{with } \mathcal{N} = e^{-\eta^2}/\pi^{\frac{1}{4}},$$
$$\langle p_x|\alpha\rangle = \mathcal{N}'e^{-\frac{1}{2}(p_x+i\sqrt{2}\alpha^*)^2}, \qquad \text{with } \mathcal{N}' = e^{+\xi^2}/\pi^{\frac{1}{4}}, \tag{74}$$

so

$$|\langle x|\alpha\rangle|^2 = \frac{1}{\sqrt{\pi}}e^{-(x-\sqrt{2}\xi)^2}, \qquad |\langle p_x|\alpha\rangle|^2 = \frac{1}{\sqrt{\pi}}e^{-(p_x+\sqrt{2}\eta)^2}. \tag{75}$$

The minimum uncertainties for the coherent state can now be understood because these are the probabilities for the space and momentum distributions of a displaced oscillator in its $n = 0$ lowest state, with a displacement $+\sqrt{2}\Re(\alpha)$ in coordinate space and $-\sqrt{2}\Im(\alpha)$ in momentum space. Moreover, such an oscillator wave packet will move with time *without* change of shape (see problems 10 and 11). This is the motion of a coherent wave packet.

Just as state vectors can be given in the coherent state representation, $\langle\alpha|\psi\rangle$, physical quantities represented by operators, O, can also be given in the coherent state representation. In the oscillator quantum number representation, we multiplied an operator, O, by unit operators on the left and on the right to yield, $O = \sum_{n,m}|n\rangle\langle n|O|m\rangle\langle m|$. Similarly, we could represent O by

$$O = \frac{1}{\pi^2}\int d^2\alpha \int d^2\beta |\alpha\rangle\langle\alpha|O|\beta\rangle\langle\beta|, \tag{76}$$

where $|\alpha\rangle$ and $|\beta\rangle$ are coherent states. Using eq. (54), we have

$$O = \frac{1}{\pi^2}\int d^2\alpha \int d^2\beta e^{-\frac{1}{2}(|\alpha|^2+|\beta|^2)}|\alpha\rangle \sum_{n,m}\Big(\frac{\alpha^n}{\sqrt{n!}}\langle n|O|m\rangle\frac{\beta^{*m}}{\sqrt{m!}}\Big)\langle\beta|. \tag{77}$$

The operator, O, in this form is two-sided, made up of operators, $|\alpha\rangle\langle\beta|$, involving two different coherent states. In our earlier continuous representations, such as the coordinate representation, e.g., operators were expressible in terms of functions of a single x and its derivative. In coordinate representation, e.g., an operator, $O(x, p_x)$, which is a function of the basic operators, x and p_x, was expressible as $O(x, \frac{1}{i}\frac{\partial}{\partial x})$, where we were able to use

$$\langle x'|O(x, p_x)|x\rangle = \delta(x' - x)O(x, \frac{1}{i}\frac{\partial}{\partial x}),$$

so O becomes expressible as a function of a single x and its derivative through the δ function relation, $\langle x'|x\rangle = \delta(x' - x)$. Because this relation does not exist in the coherent state representation, the analagous operator relations must be handled with some care. For the harmonic oscillator, an operator, O, can be expressed as a function of the basic operators, $a^\dagger$ and a, $O(a^\dagger, a)$. We want to express such an operator as a function of z and $\partial/\partial z$. We have purposely chosen z in place of α because a type II coherent state will be somewhat simpler for this purpose. For type II coherent states, z-space realizations of state vectors, $\langle z|\psi\rangle$, are given by the Bargmann transforms, $F(z) = \sum_n c_n z^n/\sqrt{n!}$. The scalar product in the complex z-space involves the Bargmann measure, e^{-zz^*}/π. In this scalar product, the operator $\partial/\partial z$ is the adjoint of the operator z. Given two Bargmann-space functions,

$$F_a(z) = \langle z|\psi_a\rangle = \sum_n a_n \frac{z^n}{\sqrt{n!}} \quad \text{and} \quad F_b(z) = \langle z|\psi_b\rangle = \sum_n b_n \frac{z^n}{\sqrt{n!}},$$

we have (using the orthonormality of the $z^n/\sqrt{n!}$)

$$\frac{1}{\pi}\int d^2z e^{-zz^*} F_a^*(z)\Big(\frac{\partial}{\partial z}F_b(z)\Big) = \sum_{n=0} a_n^* b_{n+1}\sqrt{(n+1)}$$
$$= \frac{1}{\pi}\int d^2z e^{-zz^*}\Big(zF_a(z)\Big)^* F_b(z). \tag{78}$$

Also, from

$$z\Big(\frac{z^n}{\sqrt{n!}}\Big) = \sqrt{(n+1)}\Big(\frac{z^{n+1}}{\sqrt{(n+1)!}}\Big) \quad \text{and} \quad \frac{\partial}{\partial z}\Big(\frac{z^n}{\sqrt{n!}}\Big) = \sqrt{n}\Big(\frac{z^{n-1}}{\sqrt{(n-1)!}}\Big), \tag{79}$$

we have

$$\langle n'|z|n\rangle = \sqrt{(n+1)}\delta_{n'(n+1)} = \langle n'|a^\dagger|n\rangle,$$
$$\langle n'|\frac{\partial}{\partial z}|n\rangle = \sqrt{n}\delta_{n'(n-1)} = \langle n'|a|n\rangle. \tag{80}$$

The z-space realizations of the operators, $a^\dagger$ and a, are therefore

$$\gamma(a^\dagger) = z \quad \text{and} \quad \gamma(a) = \frac{\partial}{\partial z}. \tag{81}$$

A more complicated operator, such as $x^2 = \frac{1}{2}(aa + a^\dagger a^\dagger + 2a^\dagger a + 1)$, e.g., has the z-space realization

$$\gamma(x^2) = \frac{1}{2}\Big(\frac{\partial^2}{\partial z^2} + z^2 + 2z\frac{\partial}{\partial z} + 1\Big).$$

Finally, with

$$\langle z|\psi\rangle = \langle 0|e^{za}|\psi\rangle = F(z)$$

$\langle z|\psi'\rangle$, with $|\psi'\rangle = O|\psi\rangle$, is given by $\langle 0|e^{za}O|\psi\rangle$, which we want to write in the form $\gamma(O)F(z)$, where $\gamma(O)$ is to be determined as a function of operators, z, and $\partial/\partial z$. For this purpose, we rewrite $\langle z|O|\psi\rangle$ with the use of the unit operator, $e^{-za}e^{za} = 1$, as

$$\langle 0|e^{za}O|\psi\rangle = \langle 0|(e^{za}Oe^{-za})e^{za}|\psi\rangle$$

$$= \langle 0| \Big(O + z[a, O] + \frac{z^2}{2!}[a, [a, O]] + \cdots \Big) e^{za} |\psi\rangle, \tag{82}$$

where we have used eq. (23) of Chapter 16 (renaming $i\epsilon \equiv z$) for the expansion of $e^{za} O e^{-za}$. In particular, with $O = a$, we have

$$\begin{aligned} \langle z|a|\psi\rangle &= \langle 0|a e^{za} \psi\rangle = \langle 0|\frac{\partial}{\partial z} e^{za} |\psi\rangle = \frac{\partial}{\partial z} \langle 0| e^{za} |\psi\rangle \\ &= \frac{\partial}{\partial z} \langle z|\psi\rangle, \end{aligned} \tag{83}$$

leading to

$$\gamma(a) = \frac{\partial}{\partial z}. \tag{84}$$

Similarly, with $O = a^\dagger$, we have

$$\langle z|a^\dagger|\psi\rangle = \langle 0|(a^\dagger + z) e^{za} |\psi\rangle = \langle 0| z e^{za} |\psi\rangle = z\langle z|\psi\rangle, \tag{85}$$

where we have used $\langle 0|a^\dagger = (a|0\rangle)^\dagger = 0$, leading again to

$$\gamma(a^\dagger) = z. \tag{86}$$

E Angular Momentum Coherent States

Because we can make an analogy between the operators

$a, a^\dagger, 1 = [a, a^\dagger]$ of the oscillator algebra and

$J_-, J_+, J_0 = -\frac{1}{2}[J_-, J_+]$ of the angular momentum algebra,

it is possible to define angular momentum coherent states in analogy with the oscillator coherent states. Again, we will distinguish between two slightly different definitions. For the type I coherent state, using a complex variable α, we define

$$|\alpha\rangle_{\rm I} \equiv |\alpha\rangle = e^{\alpha^* J_+ - \alpha J_-} |J, M = -J\rangle. \tag{87}$$

For the type II coherent state, using the complex variable z, we define

$$|z\rangle_{\rm II} \equiv |z\rangle = e^{z^* J_+} |J, M = -J\rangle. \tag{88}$$

A generic $|J, M\rangle$ has been used for the angular momentum eigenvectors, and the oscillator ground state, $|0\rangle$, has been replaced with the angular momentum eigenvector with the lowest possible eigenvalue of J_0, $M = -J$, so $J_-|J, M = -J\rangle = 0$ in analogy with $a|0\rangle = 0$. For the type I coherent state, it will be useful to relate the complex number α to the real angle variables θ and ϕ, via

$$-\alpha = \frac{\theta}{2} e^{i\phi},$$

where θ, ϕ are polar and azimuth angles giving the standard orientation of a unit vector $\vec{r}/r$ in our 3-D world. With this choice of parameterization of the complex variable, α, the type I coherent state becomes

$$\begin{aligned}|\alpha\rangle &= e^{-i\theta(-\sin\phi J_x + \cos\phi J_y)}|M = -J\rangle \\ &= e^{-i\theta(\vec{n}\cdot\vec{J})}|M = -J\rangle, \end{aligned} \tag{89}$$

where $\vec{n}$ is a unit vector in the x, y-plane making an angle ϕ with the y-axis and $(\frac{\pi}{2} - \phi)$ with the negative x-axis; i.e., $\vec{n}$ is a unit vector in the direction of the y' axis after a rotation about the z-axis through an angle ϕ. We shall return to the type I coherent state in chapter 29 after studying rotation operators in our 3-D world in greater generality.

For the type II coherent state, we can expand $|z\rangle$ in terms of angular momentum eigenstates $|JM\rangle$, via

$$\begin{aligned}|z\rangle = e^{z^* J_+}|M = -J\rangle &= \sum_{n=0}^{2J} \frac{z^{*n}}{n!}(J_+)^n|M = -J\rangle \\ &= \sum_{n=0}^{2J} \frac{z^{*n}}{\sqrt{n!}}\sqrt{\frac{(2J)!}{(2J-n)!}}|M = -J + n\rangle. \end{aligned} \tag{90}$$

An arbitrary state vector $|\psi\rangle$ in the subspace of Hilbert space appropriate to our angular momentum operator, $\vec{J}$, can now be specified through its z-space realization, $\langle z|\psi\rangle$,

$$\begin{aligned}\langle z|\psi\rangle = \langle M = -J|e^{zJ_-}|\psi\rangle &= \sum_{n=0}^{2J} \frac{z^n}{\sqrt{n!}}\sqrt{\frac{(2J)!}{(2J-n)!}}\langle M = -J + n|\psi\rangle \\ &= \sum_{n=0}^{2J} \frac{z^n}{\sqrt{n!}} K_n \langle M = -J + n|\psi\rangle. \end{aligned} \tag{91}$$

Except for a new numerical factor, K_n, we have expanded the coherent state in terms of the orthonormal z-space oscillator basis. The orthonormality of the $z^n/\sqrt{n!}$ requires the Bargmann weighting function e^{-zz^*}/π in the complex z-plane. We will therefore find it convenient to use the unit operator in the Bargmann form

$$1 = \frac{1}{\pi}\int d^2z e^{-zz^*}|z\rangle\langle z|$$

for z-space scalar products. We have thus mapped the angular momentum states onto oscillator states in the complex z-space realization. The oscillator excitation, however, is now limited to $n \leq 2J$. Also, the angular momentum coherent state $|z\rangle$ is *not* an eigenvector of the operator J_- because of the additional n-dependent numerical factors, K.

To get the z-space realizations of operators, we use

$$\begin{aligned}\langle z|O|\psi\rangle &= \langle -J|e^{zJ_-}O|\psi\rangle = \langle -J|(e^{zJ_-}Oe^{-zJ_-})e^{zJ_-}|\psi\rangle \\ &= \langle -J|\Big(O + z[J_-, O] + \frac{z^2}{2!}[J_-, [J_-, O]] + \cdots\Big)e^{zJ_-}|\psi\rangle, \end{aligned} \tag{92}$$

where we have used the abbreviation, $|J, M = -J\rangle \equiv |-J\rangle$, for the state of lowest possible M. Let us choose $O = J_-, J_0, J_+$ in turn to get the z-space realizations of the angular momentum operators themselves.

$$\begin{aligned}
\langle z|J_-|\psi\rangle &= \langle -J|J_- e^{zJ_-}\psi\rangle = \langle -J|\frac{\partial}{\partial z}e^{zJ_-}|\psi\rangle = \frac{\partial}{\partial z}\langle -J|e^{zJ_-}|\psi\rangle \\
&= \frac{\partial}{\partial z}\langle z|\psi\rangle, \\
\langle z|J_0|\psi\rangle &= \langle -J|(J_0 + zJ_-)e^{zJ_-}|\psi\rangle = \langle -J|(-J + z\frac{\partial}{\partial z})e^{zJ_-}|\psi\rangle, \\
&= (-J + z\frac{\partial}{\partial z})\langle z|\psi\rangle; \\
\langle z|J_+|\psi\rangle &= \langle -J|(J_+ - 2J_0 z - z^2 J_-)e^{zJ_-}|\psi\rangle = \langle -J|\Big(2Jz - z^2\frac{\partial}{\partial z}\Big)e^{zJ_-}|\psi\rangle \\
&= \Big(2Jz - z^2\frac{\partial}{\partial z}\Big)\langle z|\psi\rangle,
\end{aligned} \tag{93}$$

where we have used $\langle -J|J_0 = (J_0| - J\rangle)^\dagger = -J\langle -J|$, and $\langle -J|J_+ = 0$, via the hermitian conjugate of $J_-| - J\rangle = 0$. We have thus found z-space realizations of the operators, J_-, J_0, J_+,

$$\begin{aligned}
\Gamma(J_-) &= \frac{\partial}{\partial z}, \\
\Gamma(J_0) &= (-J + z\frac{\partial}{\partial z}), \\
\Gamma(J_+) &= z(2J - z\frac{\partial}{\partial z}).
\end{aligned} \tag{94}$$

It is easy to verify these $\Gamma(J_i)$ satisfy the angular momentum commutation rules, which of course is just a check of our arithmetic. In addition,

$$\Gamma(\vec{J}^2) = \tfrac{1}{2}\big[\Gamma(J_+)\Gamma(J_-) + \Gamma(J_-)\Gamma(J_+)\big] + \Gamma(J_0)^2 = J(J+1). \tag{95}$$

$\Gamma(J_+)$, however, is not the adjoint of $\Gamma(J_-)$ with respect to the Bargmann measure, where, as we have seen, $\partial/\partial z$ is the adjoint of z. This is related to the fact that our z-space realization of the angular momentum operators is a nonunitary one. It is the reason why we have used $\Gamma(J_i)$ for the above z-space realization of the J_i, reserving $\gamma(J_i)$ for the unitary one. To calculate matrix elements of an operator, O, through its z-space realization, built from operators z and $\partial/\partial z$, we see from the expansion of $\langle z|\psi\rangle$ of eq. (91) that such a $\Gamma(O)$, acting on the nth term of the expansion will in general create Bargmann space orthonormal $(z^{n'}/\sqrt{n'!})$ not multiplied by the proper $K_{n'}$. To attain the proper $(K_{n'}z^{n'}/\sqrt{n'!})$, we can multiply the resultant obtained from the action of $\Gamma(O)$ by $(K_{n'}^{-1} \times K_{n'}) = 1$ and thereby transform the nonunitary form of the operator, $\Gamma(O)$, into a unitary form, to be denoted by $\gamma(O)$, where

$$\gamma(O) = K^{-1}\Gamma(O)K = \Big(\gamma(O^\dagger)\Big)^\dagger, \tag{96}$$

thus making $\gamma(O)$ unitary. The operators, K, are merely the command: Multiply an orthonormal Bargmann space function $z^n/\sqrt{n!}$ by the appropriate factor, K_n.

For the most general, O, eq. (95) can be put in the form

$$\gamma(O) = K^{-1}\Gamma(O)K = \big(\gamma(O^\dagger)\big)^\dagger = \big(K^{-1}\Gamma(O^\dagger)K\big)^\dagger = K^\dagger\big(\Gamma(O^\dagger)\big)^\dagger (K^{-1})^\dagger, \tag{97}$$

or, via left-multiplication by K and right-multiplication by $K^\dagger$,

$$\Gamma(O)KK^\dagger = KK^\dagger\big(\Gamma(O^\dagger)\big)^\dagger. \tag{98}$$

For the specific operator, $O = J_+$, of the angular momentum algebra, eq. (96) becomes

$$\gamma(J_+) = K^{-1}\Gamma(J_+)K = \big(\gamma(J_-)\big)^\dagger = K^\dagger\big(\Gamma(J_-)\big)^\dagger (K^{-1})^\dagger, \tag{99}$$

and eq. (98) becomes

$$z\big(2J - z\frac{\partial}{\partial z}\big)KK^\dagger = KK^\dagger\Big(\frac{\partial}{\partial z}\Big)^\dagger = KK^\dagger z. \tag{100}$$

In eq. (91), the factor K_n was evaluated from the known matrix elements of J_+ acting n times in succession on the state $|J, -J\rangle$. If we had not had prior knowledge of these matrix elements, we could now evaluate these by using eq. (100) to first evaluate $(KK^\dagger)_n$. In particular, both the operator z and the operator $\Gamma(J_+) = z(2J - z\partial/\partial z)$ convert a z-space function, z^n, into a z-space function, z^{n+1}, so eq. (100) becomes

$$z(2J - n)(KK^\dagger)_n = (KK^\dagger)_{n+1} z.$$

On the right-hand side of the equation, the action of $KK^\dagger$ follows the action of the operator z and $z\partial/\partial z(z^n) = n(z^n)$. We therefore have

$$\frac{(KK^\dagger)_{n+1}}{(KK^\dagger)_n} = (2J - n).$$

Because the Bargmann state with $n = 0$ has the same normalization as the angular momentum eigenstate, $|J, M = -J\rangle$, we have $(KK^\dagger)_0 = 1$. Iterating the above recursion relation for $KK^\dagger$, starting with $n = 0$, we obtain

$$(KK^\dagger)_n = 2J(2J-1)\cdots(2J+1-n) = \frac{(2J)!}{(2J-n)!}.$$

The hermitian operator, $KK^\dagger$, must have real eigenvalues. In our special case, K is the simple command: Multiply $(z^n/\sqrt{n!})$ by an n-dependent factor. We can make this renormalization factor real without loss of generality, so K_n becomes the real number

$$K_n = (K^\dagger)_n = \sqrt{\frac{(2J)!}{(2J-n)!}}. \tag{101}$$

We can therefore rederive the matrix elements of J_-, J_0, and J_+. Eqs. (94) and (96) lead to

$$\langle n-1|J_-|n\rangle = (K^{-1})_{n-1}\big(n-1|\frac{\partial}{\partial z}|n\big)K_n = \sqrt{(2J+1-n)n}$$

$$
\begin{aligned}
&= \sqrt{(J+1-M)(J+M)}, \\
\langle n|J_0|n\rangle n &= (K^{-1})_n \big(n|(-J + z\frac{\partial}{\partial z}|n\big) K_n = (-J+n) \\
&= M, \\
\langle n+1|J_+|n\rangle &= (K^{-1})_{n+1}\big(n+1|z(2J - z\frac{\partial}{\partial z})|n\big) K_n \\
&= \sqrt{\frac{(2J-n-1)!}{(2J-n)!}}\sqrt{n+1}(2J-n) = \sqrt{(2J-n)(n+1)} \\
&= \sqrt{(J-M)(J+M+1)}, \qquad (102)
\end{aligned}
$$

where the Bargmann space matrix elements

$$
(n'|\Gamma(O)|n) = \frac{1}{\pi}\int d^2 z e^{-zz^*} \frac{z^{*n'}}{\sqrt{n'!}} \Gamma(O) \frac{z^n}{\sqrt{n!}}
$$

have been denoted by round parentheses and we have used $M = -J + n$ to express all matrix elements in their standard form.

Final Notes:

(1). The technique used here to derive the matrix elements of the angular momentum operators can be used to derive the matrix elements of more complicated families of operators with more complicated commutator algebras. For coherent state techniques of such generalized coherent states, see, e.g., A. Perelomov, *Generalized Coherent States and Their Applications*. Springer-Verlag, 1986, or K. T. Hecht, *The Vector Coherent State Method and its Application to Problems of Higher Symmetries*. Lecture Notes in Physics **290**. Springer-Verlag, 1987.

(2). The technique used here, which involved a mapping of angular momentum eigenstates onto orthonormal harmonic oscillator z-space Bargmann eigenstates, $z^n/\sqrt{n!}$, is useful if we have no a priori knowledge of the numerical values of the K operator or the $KK^\dagger$ eigenvalues. Alternatively, we could have used a different z-space measure to make the $(z^n/\sqrt{n!})K_n$ into an orthonormal set in the complex z-space domain. This would have involved a change of measure

$$
\frac{1}{\pi}e^{-zz^*} \rightarrow \frac{(2J+1)}{\pi}\frac{1}{(1+zz^*)^{2J+2}},
$$

as can be seen from the orthonormality integral

$$
\begin{aligned}
&\frac{(2J+1)}{\pi}\int d^2 z \frac{1}{(1+zz^*)^{2J+2}} \frac{z^{*m}K_m^*}{\sqrt{m!}}\frac{z^n K_n}{\sqrt{n!}} \\
&= \frac{(2J+1)}{\pi}\int_0^\infty \frac{d\rho\rho^{n+m+1}}{(1+\rho^2)^{2J+2}}\int_0^{2\pi} d\phi e^{i(n-m)\phi}\frac{K_m^* K_n}{\sqrt{m!n!}} \\
&= \delta_{nm}\frac{(2J+1)(2J)!}{n!(2J-n)!}2\int_0^\infty \frac{d\rho\rho^{2n+1}}{(1+\rho^2)^{2J+2}} = \delta_{nm}, \qquad (103)
\end{aligned}
$$

via the integral

$$
2\int_0^\infty \frac{d\rho\rho^{2n+1}}{(1+\rho^2)^{2J+2}} = \int_0^\infty \frac{d\tau\tau^n}{(1+\tau)^{2J+2}} = B(n+1, 2J+1-n)
$$

$$= \frac{\Gamma(n+1)\Gamma(2J+1-n)}{\Gamma(2J+2)} = \frac{n!(2J-n)!}{(2J+1)!}, \tag{104}$$

where $B(p,q)$ is the Beta function expressed by Γ functions and in terms of factorials because $2J$ must be an integer.

Problems

23. Given three hermitian operators, T_1, T_2, T_3, with commutation relations

$$[T_2, T_3] = iT_1, \qquad [T_3, T_1] = iT_2, \qquad [T_1, T_2] = -iT_3,$$

which differ from the angular momentum commutator algebra because of the minus sign in the last commutation relation! Show that the three T_j all commute with the operator

$$\mathcal{T}^2 = T_3^2 - T_1^2 - T_2^2.$$

Again, note the minus signs and the difference from the angular momentum case. Convert the operators, T_j, to the new set

$$T_\pm = (T_1 \pm iT_2), \qquad T_3 = T_0,$$

and show these equations satisfy the commutation relations

$$[T_0, T_\pm] = \pm T_\pm, \qquad [T_+, T_-] = -2T_0.$$

Again, note the minus sign in the last commutation relation. Solve the simultaneous eigenvalue problem

$$\begin{aligned} \mathcal{T}^2|\lambda m\rangle &= \lambda|\lambda m\rangle = j(j+1)|\lambda m\rangle = j(j+1)|jm\rangle, \\ T_0|\lambda m\rangle &= m|\lambda m\rangle = m|jm\rangle, \end{aligned} \tag{1}$$

where we have named $\lambda = j(j+1)$. (No implication exists that j be an integer or half-integer.) Show, in particular, that now:

(1) If T_3 has positive eigenvalues, a minimum possible, $m_{\text{min.}}$ exists, such that

$$T_-|\lambda m_{\text{min.}}\rangle = 0, \qquad m = m_{\text{min.}} + n, \quad \text{with } n = 0, 1, 2, \ldots, \to \infty,$$

where $m_{\text{min.}} = (j+1)$.

(2) If T_3 has negative eigenvalues, a maximum possible $m_{\text{max.}} = -|m_{\text{max.}}|$ exists, such that

$$T_+|\lambda m_{\text{max.}}\rangle = 0, \qquad m = m_{\text{max.}} - n = -(|m_{\text{max.}}| + n), \quad \text{with } n = 0, 1, \ldots \to \infty,$$

where now $m_{\text{max.}} = -(j+1)$ and we assume j is positive.

Find the nonzero matrix elements of T_+ and T_-,

$$\langle jm'|T_+|jm\rangle = \langle j(m+1)|T_+|jm\rangle,$$

$$\langle jm'|T_-|jm\rangle = \langle j(m-1)|T_-|jm\rangle,$$

and note the differences and similarities with the corresponding angular momentum case.

Note: The angular momentum operators, J_k, generate the group SO(3) (the special orthogonal transformations in three dimensions, with determinant = +1), in the case when j are integers, or SU(2) (the special unitary transformations in two dimensions) and in the case when j are half-integers. The three operators, T_k, on the other hand, generate the group SO(2,1). Note the two minus signs and the one plus sign in the operator $\mathcal{T}^2$.

Solution for Problem 23: The SO(2,1) Algebra

The various commutator relations follow from the given commutation relations by simple commutator algebra. For example,

$$\begin{aligned}[T_3, \mathcal{T}^2] &= -T_1[T_3, T_1] - [T_3, T_1]T_1 - T_2[T_3, T_2] - [T_3, T_2]T_2 \\ &= -iT_1T_2 - iT_2T_1 + iT_2T_1 + iT_1T_2 = 0. \end{aligned} \tag{2}$$

We are interested in the simultaneous eigenvectors of the two commuting hermitian operators, T_3 and $\mathcal{T}^2$,

$$\begin{aligned}\mathcal{T}^2|\lambda m\rangle &= \lambda|\lambda m\rangle, \\ T_3|\lambda m\rangle &= m|\lambda m\rangle. \end{aligned} \tag{3}$$

Let us consider the new vectors, $T_\pm|\lambda m\rangle$. Acting on either of these with both $\mathcal{T}^2$ and T_3, we get (with the use of the commutation relations),

$$\begin{aligned}\mathcal{T}^2\Big(T_+|\lambda m\rangle\Big) &= T_+\mathcal{T}^2|\lambda m\rangle = \lambda\Big(T_+|\lambda m\rangle\Big), \\ T_3\Big(T_+|\lambda m\rangle\Big) &= T_+T_3|\lambda m\rangle + T_+|\lambda m\rangle = (m+1)\Big(T_+|\lambda m\rangle\Big). \end{aligned} \tag{4}$$

Thus, if $|\lambda m\rangle$ is simultaneously an eigenvector of $\mathcal{T}^2$ and T_3, with eigenvalues λ and m, either $\Big(T_+|\lambda m\rangle\Big)$ is simultaneously an eigenvector of $\mathcal{T}^2$ and T_3 with eigenvalues λ and $(m+1)$ or $\Big(T_+|\lambda m\rangle\Big) = 0$. Similarly,

$$\begin{aligned}\mathcal{T}^2\Big(T_-|\lambda m\rangle\Big) &= T_-\mathcal{T}^2|\lambda m\rangle = \lambda\Big(T_-|\lambda m\rangle\Big), \\ T_3\Big(T_-|\lambda m\rangle\Big) &= T_-T_3|\lambda m\rangle - T_-|\lambda m\rangle = (m-1)\Big(T_-|\lambda m\rangle\Big). \end{aligned} \tag{5}$$

Thus, if $|\lambda m\rangle$ is simultaneously an eigenvector of $\mathcal{T}^2$ and T_3, with eigenvalues λ and m, either $\Big(T_-|\lambda m\rangle\Big)$ is simultaneously an eigenvector of $\mathcal{T}^2$ and T_3 with eigenvalues λ and $(m-1)$; or $\Big(T_-|\lambda m\rangle\Big) = 0$. To investigate these two possibilities, let us rewrite $\mathcal{T}^2$

$$\begin{aligned}\mathcal{T}^2 = T_3^2 - T_1^2 - T_2^2 = T_0^2 - \tfrac{1}{2}(T_+T_- + T_-T_+) &= T_0^2 - T_+T_- - T_0 \\ &= T_0^2 - T_-T_+ + T_0, \end{aligned} \tag{6}$$

so $$T_+T_- = T_0^2 - T_0 - \mathcal{T}^2$$

$$\text{and} \qquad T_-T_+ = T_0^2 + T_0 - \mathcal{T}^2. \tag{7}$$

Now, let us take the diagonal matrix element of these two relations between states with the same λ, m, assuming a state $|\lambda m\rangle$ exists, viz., it leads to a square-integrable eigenfunction. First,

$$\begin{aligned}
\langle \lambda m|T_+T_-|\lambda m\rangle &= \Big((m^2 - m) - \lambda\Big) \\
&= \sum_{m'} \langle \lambda m|T_+|\lambda m'\rangle\langle \lambda m'|T_-|\lambda m\rangle \\
&= \sum_{m'} |\langle \lambda m'|T_-|\lambda m\rangle|^2 \geq 0,
\end{aligned} \tag{8}$$

where we have used $\langle \lambda m|T_+|\lambda m'\rangle = \langle \lambda m'|T_-|\lambda m\rangle^*$ and have summed over a complete set of intermediate states. We must have $m' = (m-1)$. By including a sum over states with $m' = (m-1)$, we have allowed for the possibility more than one independent state with that restriction exists. Similarly, we have

$$\begin{aligned}
\langle \lambda m|T_-T_+|\lambda m\rangle &= \Big((m^2 + m) - \lambda\Big) \\
&= \sum_{m'} |\langle \lambda m'|T_+|\lambda m\rangle|^2 \geq 0.
\end{aligned} \tag{9}$$

We have the two patently positive quantities of eqs. (7) and (8) only if the two functions, $f(m) = \big((m^2-m)-\lambda\big)$ or $\big((m^2+m)-\lambda\big)$ are equal to or greater than zero. The two functions of m, $(m^2 \mp m)$, have minima at $m = \pm\frac{1}{2}$, both with a minimum value of $-\frac{1}{4}$. Unlike the corresponding operator of the angular momentum algebra, with its slightly different commutation relations, the operator, $\mathcal{T}^2$, is no longer a sum of positive hermitian operators. Thus, the eigenvalue, λ, could be either positive or negative. In particular, if $\lambda < -\frac{1}{4}$, the quantities $[(m^2 \mp m) - \lambda]$ are positive for all values of m, positive or negative. Thus, all values of $\pm m$ are possible, and for any λ, such that $\lambda < -\frac{1}{4}$, we have a continuous spectrum of allowed values for both λ and m. Conversely, if $\lambda > 0$, and if an eigenvalue, $m_0 > 0$ exists, such that $((m_0^2 - m_0) - \lambda) > 0$, the step-down action of n operations with T_- could eventually lead to an $(m_0 - n)$ such that $\big((m_0 - n)(m_0 - n - 1) - \lambda\big) < 0$, and eq. (7) would lead to an inconsistency. A patently positive quantity on one side of eq. (7) would be equal to a negative quantity on the other side. Hence, our assumption of the existence of a square-integrable $|\lambda m_0\rangle$ must have been incorrect. If m_0, however, is such that an integer n exists such that $(m_0 - n) \equiv m_{\text{min.}}$, so

$$\Big(T_-|\lambda m_{\text{min.}}\rangle\Big) = 0 \qquad \text{and} \quad [m_{\text{min.}}(m_{\text{min.}} - 1) - \lambda] = 0, \tag{10}$$

an inconsistency never exists. If we name

$$\lambda = j(j+1), \tag{11}$$

with $j \geq 0$, to be as close as possible to the language of the angular momentum algebra, the solution to $[m_{\text{min.}}(m_{\text{min.}} - 1) - \lambda] = 0$ gives us $m_{\text{min.}} = (j+1)$. Only the positive root has meaning in this case. Also, the quantum number j is only a language to give us the eigenvalue λ. In this case, j may not be an integer or half-integer. The actual values of $m_{\text{min.}} = (j+1)$ and $\lambda = j(j+1)$ will depend on the detailed properties of the operators $\mathcal{T}^2$ and T_3, i.e., on the specific nature of the physics of the problem. If the nature of the problem is such that the eigenvalues of $\mathcal{T}^2$ and T_3 must all be positive definite, we are done. The spectrum is given by

$$m = m_{\text{min.}}, (m_{\text{min.}} + 1), \ldots, (m_{\text{min.}}, +n), \ldots, \to \infty,$$

$$m = (j+1), (j+2), \ldots, (j+1+n), \ldots, \to \infty, \qquad \text{with } \lambda = j(j+1).$$

[The possible m-values can go to ∞. This follows because for $m_{\text{min.}} > 0$, the quantity $(m^2_{\text{min.}} + m_{\text{min.}} - \lambda) = 2m_{\text{min.}} > 0$ and thus $[(m_{\text{min.}}+n)(m_{\text{min.}}+n+1) - \lambda] > 0$ for any positive integer n, so the state $(T_+|\lambda(m_{\text{min.}} + n)\rangle)$ exists; i.e., the state $|\lambda(m_{\text{min.}} + n + 1)\rangle$ also leads to a square-integrable eigenfunction.]

Let us next examine the possibility the operators $\mathcal{T}^2$ and T_3 are such that $\lambda > 0$, but $m < 0$. Now let us suppose some $m_0 = -|m_0|$ exists; i.e., the state $|\lambda, -|m_0|\rangle$ leads to square-integrable eigenfunctions. Now, if eq. (8) is satisfied for $m = -|m_0|$, i.e., $|m_0|(|m_0| + 1) - \lambda > 0$, the state with $m = (-|m_0| - n)$ leads to an $m(m-1) - \lambda = (|m_0| + n)(|m_0| + n + 1) - \lambda$ also > 0, so n actions with T_- would lead to another allowed state. However, n actions with the step-up operator, T_+, would lead to a state with $m = -|m_0| + n$, for which the function $[(m^2 + m) - \lambda]$ of eq. (9) would lead to the value $[(|m_0| - n)(|m_0| - n - 1) - \lambda]$, which for large enough n could now be negative. Thus, eq. (9) would say that a patently positive quantity is equal to a function that can become a negative quantity for a large enough n. Now, a state $|\lambda, -|m_0|\rangle$ can be an allowed state only if an integer n exists, such that $m = (-|m_0| + n) \equiv m_{\text{max.}}$ and

$$\Big(T_+|\lambda m_{\text{max.}}\rangle\Big) = 0, \qquad \text{so that} \quad (m_{\text{max.}}(m_{\text{max.}} + 1) - \lambda) = 0, \qquad (12)$$

where now $m_{\text{max.}}$ must be the negative root of the equation: $(m_{\text{max.}}(m_{\text{max.}} + 1) - j(j+1)) = 0$; that is, $m_{\text{max.}} = -\frac{1}{2} - \sqrt{\lambda + \frac{1}{4}} = -(j+1)$. In this case, therefore, the spectrum of possible m-values is

$$-\infty, \ldots, -(j+1+n), \ldots, -(j+2), -(j+1) = m_{\text{max.}}, \quad \text{again with } \lambda = j(j+1).$$

Now, however, for general $\lambda = j(j+1) > 0$, the two branches of allowed m values are unconnected, so the commutator algebra of the T_i does not lead to additional restrictions on λ and, hence, j. In particular, j need not be an integer or a half-integer. The physics of the operators T_3, $\mathcal{T}^2$, dictate the nature of the eigenvalues m and λ. Thus, for some physical applications for which the eigenvalues of T_3 can be positive only, only the positive branch of allowed m values can exist.

Let us now finally use eqs. (7) and (8) to find the matrix elements of the operators $T_\pm$. We shall look at the simple case, in which the states

$$|\lambda m_{\text{min.}}\rangle \qquad \text{or} \qquad |\lambda m_{\text{max.}}\rangle$$

are nondegenerate; i.e., the two relations

$$T_-|\lambda m_{\text{min.}}\rangle = 0 \qquad \text{and} \qquad T_+|\lambda, -|m_{\text{max.}}|\rangle = 0$$

are assumed to have only one allowed solution. This would of course be automatic if these two relations lead to first-order differential equations. In this case, action with T_+ or T_-, respectively, would lead to a single (nondegenerate) state with the appropriate m value. Thus, all states in the ladder of either positive or negative m values might be expected to be nondegenerate, and the sums over m' in eqs. (8) and (9) would collapse to a single term with $m' = (m-1)$ or $m' = (m+1)$, respectively. For branches of allowed m values, eq. (9) tells us

$$|\langle j(m+1)|T_+|jm\rangle|^2 = m(m+1) - j(j+1) = (m-j)(m+j+1). \tag{13}$$

This relation leads to no upper limit for positive m values with $m \geq (j+1)$, but leads to a zero matrix element for $m = -(j+1)$ within the branch of negative m values. Similarly, eq. (8) leads to

$$|\langle j(m-1)|T_-|jm\rangle|^2 = m(m-1) - j(j+1) = (m+j)(m-j-1). \tag{14}$$

Now, no zero matrix elements exist for the negative branch with $m \leq -(j+1)$, but this matrix element is automatically zero if $m = (j+1)$. As for the corresponding angular momentum problem, eqs. (13) and (14) do not fix the phases of these matrix elements. If we choose these phases such that the matrix elements of T_+, T_- are real, we have

$$\begin{aligned}\langle j(m+1)|T_+|jm\rangle &= \sqrt{(m-j)(m+j+1)},\\ \langle j(m-1)|T_-|jm\rangle &= \sqrt{(m+j)(m-j-1)}.\end{aligned} \tag{15}$$

(In particular, these equations satisfy $\langle jm'|T_-|jm\rangle = \langle jm|T_+|jm'\rangle^*$.)

Final remark: The operators $J_x = J_1$, $J_y = J_2$, $J_z = J_3$ of the angular-momentum algebra, which commute with the operator $\vec{J}^2 = J_1^2 + J_2^2 + J_3^2$ are the generators of infinitesimal rotations in three-space about the x, y, and z axes and thus connected with the group SO(3), the "special orthogonal group in three dimensions" (where the "special" means the 3×3 orthogonal rotation matrices have determinant $+1$, leading to pure rotations and not including rotation reflections). The operators T_1, T_2, T_3, which commute with the operator, $\mathcal{T}^2 = T_3^2 - T_1^2 - T_2^2$, conversely, are related in a similar way with a 3-D space with two space-like and one time-like dimension and thus connected with the group SO(2,1). As for the SO(3) group, most of the properties of the group SO(2,1) follow from the matrix elements of the T_i, which are now known to us.

Finally, the angular momentum operators, $J_\pm$ are related to the ladder operators of case 3 $\mathcal{L}(m)$ of the factorization method, whereas the $T_\pm$ of the SO(2,1) algebra are related to the ladder operators of case 4 $\mathcal{L}(m)$ of the factorization method.

Problems 24–26 give some actual physical examples of the SO(2,1) algebra. The signs of λ and m are usually dictated by the nature of the problem.

24. The 3-D harmonic oscillator via the SO(2,1) algebra. For the 3-D harmonic oscillator, define dimensionless quantities, $\vec{r}$, H, etc., via

$$\vec{r}_{\text{phys.}} = \sqrt{\hbar/m\omega_0}\,\vec{r} \qquad \text{and} \qquad H_{\text{phys.}} = \hbar\omega_0 H; \ldots .$$

Show that the three operators

$$T_1 = \tfrac{1}{4}(r^2 - \vec{p}^{\,2}) = \tfrac{1}{4}(r^2 + \nabla^2),$$

$$T_2 = -\tfrac{1}{4}(\vec{r}\cdot\vec{p} + \vec{p}\cdot\vec{r}) = \frac{i}{2}\Big(r\frac{\partial}{\partial r} + \tfrac{3}{2}\Big),$$

$$T_3 = \tfrac{1}{4}(r^2 + \vec{p}^{\,2}) = \tfrac{1}{4}(r^2 - \nabla^2),$$

where

$$\nabla^2 = \Big(\frac{\partial^2}{\partial r^2} + \frac{2}{r}\frac{\partial}{\partial r}\Big) - \frac{\vec{L}^2}{r^2} = -p_r^2 - \frac{\vec{L}^2}{r^2}$$

[with $\vec{L}^2$ given in terms of θ, ϕ-dependent operators through eq. (13) of Chapter 8] are hermitian with respect to the conventional volume element and satisfy the SO(2,1) commutation relations of problem 23.

Note that, with $\psi = R(r)Y_{lm}(\theta,\phi)$, the operator ∇^2 is hermitian with respect to the usual measure, $r^2\sin\theta$,

$$\int_0^{2\pi} d\phi \int_0^{\pi} d\theta \sin\theta \int_0^{\infty} dr r^2 \psi_2^*(\nabla^2\psi_1) = \int_0^{2\pi} d\phi \int_0^{\pi} d\theta \sin\theta \int_0^{\infty} dr r^2 \psi_1^*(\nabla^2\psi_2).$$

Also,

$$H = \frac{H_{\text{phys}}}{\hbar\omega_0} = \tfrac{1}{2}(r^2 - \nabla^2) = 2T_3 = 2T_0$$

with known (positive) eigenvalues, $(N + \frac{3}{2})$. Show that the functions

$$r^l e^{-\frac{1}{2}r^2} Y_{lm}(\theta,\phi)$$

satisfy the equations

$$T_-\Big(r^l e^{-\frac{1}{2}r^2} Y_{lm}(\theta,\phi)\Big) = 0 \text{ and } T_0\Big(r^l e^{-\frac{1}{2}r^2} Y_{lm}(\theta,\phi)\Big) = \tfrac{1}{2}(l+\tfrac{3}{2})\Big(r^l e^{-\frac{1}{2}r^2} Y_{lm}(\theta,\phi)\Big),$$

so

$$m_{\text{min.}} = \tfrac{1}{2}(l + \tfrac{3}{2}) \equiv (j+1),$$

where m and j refer to the quantum numbers as defined in problem 23. (m and j are $\frac{1}{4}$-integers here!)

Use these results to show

$$E = \hbar\omega_0(l + 2n + \tfrac{3}{2}),$$

and the dimensionless operator r^2 is given by

$$r^2 = 2T_0 + T_+ + T_-.$$

Find the nonzero matrix elements of r^2 as functions of n and l or N and l. (Consult the results of problem 15.)

25. (a) For the SO(2,1) algebra with operators, T_-, T_+, T_0, satisfying the commutation relations

$$[T_0, T_\pm] = \pm T_\pm \qquad \text{and} \qquad [T_+.T_-] = -2T_0,$$

with eigenvalues $(T_0)_{\text{eigen}} = m = m_{\text{min.}} + n = (j+1+n)$, where $n = 0, 1, \ldots, \to \infty$, show that coherent state z-space realizations of these operators can be given by

$$\Gamma(T_-) = \frac{\partial}{\partial z}, \qquad \Gamma(T_0) = j + 1 + z\frac{\partial}{\partial z},$$

$$\Gamma(T_+) = 2(j+1)z + z^2\frac{\partial}{\partial z}.$$

Find the eigenvalues, K_n, of the operator, K, which converts the above nonunitary $\Gamma(T_i)$ into unitary $\gamma(T_i)$ for this algebra, and rederive the general expressions found in problem 23 for the matrix elements of T_-, T_+, T_0, in a $|jm\rangle$ basis.

(b) For the 1-D harmonic oscillator, the oscillator annihilation and creation operators, $a_x, a_x^\dagger$, expressed in terms of the dimensionless x and p_x, are

$$a_x = \tfrac{1}{\sqrt{2}}(x + ip_x) \qquad \text{and} \qquad a_x^\dagger = \tfrac{1}{\sqrt{2}}(x - ip_x).$$

Show that the three operators

$$T_+ = \tfrac{1}{4}a_x^\dagger a_x^\dagger, \qquad T_- = \tfrac{1}{4}a_x a_x, \qquad T_0 = \tfrac{1}{2}(a_x^\dagger a_x + \tfrac{1}{2}),$$

satisfy the SO(2,1) commutation relations

$$[T_0, T_\pm] = \pm T_\pm \qquad \text{and} \qquad [T_+, T_-] = -2T_0.$$

Show that the two oscillator states, $|0\rangle$, and $|1\rangle = a_x^\dagger|0\rangle$, satisfy

$$(1): T_-|0\rangle = 0, \qquad T_0|0\rangle = \tfrac{1}{4}|0\rangle,$$

$$(2): T_-|1\rangle = 0, \qquad T_0|1\rangle = \tfrac{3}{4}|1\rangle,$$

so $m_{\text{min.}} = \frac{1}{4}$ for case (1) and $m_{\text{min.}} = \frac{3}{4}$ for case (2). Use these results together with the results of problem 23, to calculate the matrix elements

$$\langle n'|a_x^\dagger a_x^\dagger|n\rangle,$$

$$\langle n'|a_x a_x|n\rangle.$$

26. For the hydrogen atom in stretched parabolic coordinates, μ, ν, ϕ (see problem 6), the following operators are useful

$$T_1 = \frac{1}{4}\left(\frac{\partial^2}{\partial\mu^2} + \frac{1}{\mu}\frac{\partial}{\partial\mu}\right) - \frac{m^2}{4\mu^2} + \frac{\mu^2}{4},$$

$$T_2 = \frac{i}{2}\left(\mu\frac{\partial}{\partial\mu} + 1\right),$$

$$T_3 = -\frac{1}{4}\left(\frac{\partial^2}{\partial\mu^2} + \frac{1}{\mu}\frac{\partial}{\partial\mu}\right) + \frac{m^2}{4\mu^2} + \frac{\mu^2}{4},$$

$$T_1' = \frac{1}{4}\left(\frac{\partial^2}{\partial\nu^2} + \frac{1}{\nu}\frac{\partial}{\partial\nu}\right) - \frac{m^2}{4\nu^2} + \frac{\nu^2}{4},$$

$$T_2' = \frac{i}{2}\left(\nu\frac{\partial}{\partial\nu} + 1\right),$$

$$T_3' = -\frac{1}{4}\left(\frac{\partial^2}{\partial\nu^2} + \frac{1}{\nu}\frac{\partial}{\partial\nu}\right) + \frac{m^2}{4\nu^2} + \frac{\nu^2}{4}.$$

Show that both the T_j and the T_j' satisfy the SO(2,1) commutation relations of problem 23. Show that the operators are hermitian with respect to the scalar product

$$\int_0^\infty d\mu\mu U_1^*(\mu)U_2(\mu) \qquad \text{for the } T_j,$$

and with respect to the scalar product

$$\int_0^\infty d\nu\nu V_1^*(\nu)V_2(\nu) \qquad \text{for the } T_j'.$$

Note: With $\psi(\mu, \nu, \phi) = U(\mu)V(\nu)\Phi(\phi)$, with $\Phi_m(\phi) = e^{im\phi}/\sqrt{2\pi}$, the standard scalar product would have been

$$\int d\vec{r}\psi_1^*\psi_2 = \int_0^{2\pi} d\phi \int_0^\infty d\mu \int_0^\infty d\nu \frac{(\mu^2+\nu^2)\mu\nu}{[-2\epsilon]^{\frac{3}{2}}}$$

$$U_1^*(\mu)V_1^*(\nu)\Phi_1^*(\phi)U_2(\mu)V_2(\nu)\Phi_2(\phi).$$

Show how the Schrödinger equation for the hydrogenic atom can be rewritten in terms of the operators, T_j, T_j'. For this purpose, rewrite the Schrödinger equation

$$(H - \epsilon)\psi = 0,$$

or

$$-\frac{1}{2}\frac{(-2\epsilon)}{(\mu^2+\nu^2)}\left(\frac{\partial^2}{\partial\mu^2} + \frac{1}{\mu}\frac{\partial}{\partial\mu} + \frac{\partial^2}{\partial\nu^2} + \frac{1}{\nu}\frac{\partial}{\partial\nu} + \left(\frac{1}{\mu^2} + \frac{1}{\nu^2}\right)\frac{\partial^2}{\partial\phi^2}\right)\psi$$

$$-\frac{2\sqrt{(-2\epsilon)}}{(\mu^2+\nu^2)}\psi - \epsilon\psi = 0,$$

by left-multiplying with $\frac{1}{2}n^2(\mu^2+\nu^2)$ to gain

$$-\frac{1}{4}\left(\frac{\partial^2}{\partial\mu^2} + \frac{1}{\mu}\frac{\partial}{\partial\mu} + \frac{\partial^2}{\partial\nu^2} + \frac{1}{\nu}\frac{\partial}{\partial\nu} + \left(\frac{1}{\mu^2} + \frac{1}{\nu^2}\right)\frac{\partial^2}{\partial\phi^2}\right)\psi$$

$$+\frac{1}{4}(\mu^2+\nu^2)\psi - \frac{1}{\sqrt{(-2\epsilon)}}\psi = 0.$$

Show

$$\mathcal{T}^2 = T_3^2 - T_1^2 - T_2^2 = (T_3 - T_1)(T_3 + T_1) + [T_1, T_3] - T_2^2,$$

or

$$\mathcal{T}^2 = (T_3 - T_1)(T_3 + T_1) - iT_2 - T_2^2 = \frac{(m^2-1)}{4}.$$

Similarly,

$$\mathcal{T}'^2 = \frac{(m^2-1)}{4},$$

where m is the eigenvalue of the operator

$$\frac{1}{i}\frac{\partial}{\partial\phi}.$$

Show that for positive values of m:

$$(T_3)_{\text{eigen}} = \tfrac{1}{2} + \tfrac{1}{2}m + n_1, \qquad n_1 = 0, 1, 2, \ldots, \to \infty,$$

$$(T_3')_{\text{eigen}} = \tfrac{1}{2} + \tfrac{1}{2}m + n_2, \qquad n_2 = 0, 1, 2, \ldots, \to \infty.$$

Find the corresponding ranges for $(T_3)_{\text{eigen}}$ and $(T_3')_{\text{eigen}}$, valid for negative values of m.

Find the energy, ϵ, for the hydrogenic atom as a function of the quantum numbers, m, n_1, n_2.

In an $|mn_1n_2\rangle$ basis, find expressions for the nonzero matrix elements of the dimensionless variables

$$(r+z) = \frac{\mu^2}{[-2\epsilon]^{\frac{1}{2}}} \qquad \text{and} \qquad (r-z) = \frac{\nu^2}{[-2\epsilon]^{\frac{1}{2}}}.$$

Part II

Time-Independent Perturbation Theory

20
Perturbation Theory

A Introductory Remarks

If we have a Hamiltonian for which we cannot find exact eigenvalues and eigenvectors, we can in principle use the technique employed for the asymmetric rotator for more challenging problems. If for the moment $|n\rangle$ is shorthand for the eigenvectors for a complete set of commuting operators, *including* the Hamiltonian in question, and $|\alpha\rangle$ is shorthand for the eigenvectors of another complete set of commuting operators, spanning the same subspace of Hilbert space, but now *including* a simpler Hamiltonian, H_0, for which we *do* know the eigenvalues and eigenvectors, we can expand the unknown eigenvectors $|n\rangle$ in terms of the known $|\alpha\rangle$. From a knowledge of the matrix elements, $\langle\alpha'|H|\alpha\rangle$, we can in principle diagonalize the matrix in this basis for H_0 to find the eigenvalues and eigenvectors for the needed H. The difficulty, of course, is that in general this matrix will be infinite-dimensional. With modern computers, however, it may be possible to diagonalize this matrix in an $N \times N$ limit, where N is taken to be a large number, and then test the possible convergence as N grows even larger. This method will be particularly successful if H differs from a known H_0 by terms that can be classified by a parameter of smallness, λ, with $\lambda \ll 1$. Then, we can study the "corrections" to the eigenvalues and eigenvectors in a very systematic way as a power series in λ. This will be the first detailed study of the next chapter, which will include:

(1) stationary-state or time-independent perturbation theory,

(a) Rayleigh–Schrödinger expansion,

(b) Wigner–Brillouin expansion.

This study will be in contrast to time-dependent perturbation theory, which will be covered in a later chapter; where we will make a similar expansion of the time-dependent integral of eq. (13) of Chapter 19. Other approximation techniques, also to be covered in later chapters, are the WKB or semiclassical approximation and variational methods.

The WKB approximation will be treated only after a long excursion on angular momentum theory, Part III of the course. Variational methods useful for the n electron atom will be discussed in Part IV of the course on systems of identical particles.

B Transition Probabilities

Before proceeding with stationary-state perturbation theory, it will be advantageous to give a very brief first discussion of transition probabilities to answer the question: What is the probability an atomic system in an excited eigenstate, E_n, will make a transition to a lower state, E_m, through the spontaneous emission of a photon? To answer this question in a rigorous way, we will have to study both the atomic system and the electromagnetic field quantum mechanically; i.e., we would have to quantize the electromagnetic field and then study the interaction of the quantized electromagnetic field (photon) with the atomic system. Because we will save the study of time-dependent perturbation theory for a later chapter, we will do this in a rigorous fashion then. To have a formula for the transition probability for the spontaneous emission of a photon, however, let us give a very brief "plausibility" argument for the transition probability formula now. This will actually be the historically first (the so-called Bohr correspondence principle) argument for this formula. (Keep in mind, however, the rigorous derivation will come later. Historically, it also came later with a famous paper by Dirac on the quantization of the electromagnetic field.)

The Bohr argument goes as follows: Classically, a charged particle in motion will emit electromagnetic radiation only if the particle is accelerated. Quantitatively, the *classical* result is given by the Larmor formula, which calculates the energy *loss* per unit time of the charged particle (or a system of N charged particles) to the emission of electromagnetic radiation (in c.g.s. units),

$$-\frac{dE}{dt} = \frac{2e^2}{3c^3}\vec{a}^2 \quad \text{or} \quad -\frac{dE}{dt} = \frac{2}{3c^3}\sum_{i=1}^{N}(e_i\vec{a}_i)^2, \tag{1}$$

where $\vec{a}_i$ is the acceleration vector of the ith particle with charge e_i. The classical recipe for calculating this energy loss for a system of N moving particles involves the Fourier time analysis for the three components of the electric dipole moment

$$\sum_{i=1}^{N} e_i x_i = \sum_{n}^{\infty} \mu_n^{(x)}(e^{in\omega t} + e^{-in\omega t}),$$

$$\sum_{i=1}^{N} e_i y_i = \sum_{n}^{\infty} \mu_n^{(y)} \left(e^{in\omega t} + e^{-in\omega t}\right),$$
$$\sum_{i=1}^{N} e_i z_i = \sum_{n}^{\infty} \mu_n^{(z)} \left(e^{in\omega t} + e^{-in\omega t}\right), \qquad (2)$$

assuming for the moment the time-dependent functions are real. Then, according to classical theory, the frequencies radiated are the classical mechanical frequency and its overtones, $\nu = \omega/2\pi$, and $n\nu$ (or in the case of multiple-periodic systems, the classical frequencies and their overtones or combination tones, $n_1\nu_1 + n_2\nu_2$, etc.). Taking the second time derivatives of the x_i, y_i, z_i of eq. (2), the classically predicted energy loss to the nth *overtone* (time-averaged over one cycle) would have been

$$-\overline{\left[\frac{dE}{dt}\right]_{n\nu}} = \frac{2}{3c^3} \times 2(n\omega)^4 \left(\left(\mu_n^{(x)}\right)^2 + \left(\mu_n^{(y)}\right)^2 + \left(\mu_n^{(z)}\right)^2\right). \qquad (3)$$

Of course, this classical result is incorrect. This is what troubled Niels Bohr from 1913 to 1925. First, this result does not predict the observed frequencies. The hydrogen spectrum is *not* a fundamental frequency and its n overtones. Even worse, this classical result predicts the frequencies should change with time. As the system loses energy and the mechanical energy becomes more negative, the Kepler frequencies (or the Bohr "1913 frequencies") would increase with time; the electron would spiral in to the proton and suffer a catastrophe in a time of the order of 10^{-8} seconds. Bohr argued, however, the classically predicted frequency, $n\omega$, should be replaced by the Bohr frequency, ω_{nm}, and the classically predicted Fourier coefficients, $\mu_n^{(j)}$, should be replaced by a two-index quantity, $\mu_{nm}^{(j)}$, which he identified with the Heisenberg matrix element. Thus,

$$\begin{array}{lll} \text{classical} & n\nu \to (E_n - E_m)/h & \text{(Bohr)}, \\ \text{classical} & \mu_n^{(x)} \to \langle n|\mu_x|m\rangle & \text{(Heisenberg)}, \\ \text{classical} & \mu_n^{(y)} \to \langle n|\mu_y|m\rangle & \text{(Heisenberg)}, \\ \text{classical} & \mu_n^{(z)} \to \langle n|\mu_z|m\rangle & \text{(Heisenberg)}. \end{array} \qquad (4)$$

This argument is the Bohr correspondence principle, which yields the result

$$-\left[\frac{dE}{dt}\right]_{\nu_{nm}} = \frac{64\pi^4}{3c^3}\nu_{nm}^4 \left[|\langle n|\mu_x|m\rangle|^2 + |\langle n|\mu_y|m\rangle|^2 + |\langle n|\mu_z|m\rangle|^2\right], \qquad (5)$$

where the operators are $\mu_x = \sum_i^N e_i x_i$, and so on. To convert this principle to a transition probability, introduce the "Einstein A,"

$$-\left[\frac{dE}{dt}\right]_{\nu_{nm}} = h\nu_{nm} A_{n\to m} N_n, \qquad (6)$$

where $h\nu_{nm}$ is the energy of the emitted photon, $A_{n\to m}$, gives the probability per unit time for the spontaneous emission of a photon with this frequency, and N_n is the number of atoms in the initial state, n. The transition probability per second the atom makes a transition from an excited state, n, to a lower state, m, is therefore given in terms of the matrix elements of the three components of the electric dipole moment operator by

$$A_{n\to m} = \frac{64\pi^4}{3hc^3}\nu_{nm}^3\Big[|\langle n|\mu_x|m\rangle|^2 + |\langle n|\mu_y|m\rangle|^2 + |\langle n|\mu_z|m\rangle|^2\Big]. \tag{7}$$

If the matrix elements of the three electric dipole moment components are all zero, the transition is "forbidden." The correspondence principle argument then gives the correct result. A rigorous derivation will be given in a later chapter, when we shall quantize the electromagnetic field and introduce the interaction between the quantized electromagnetic field (photons) and the isolated atomic system. The above electric dipole result, however, is only the dominant term in an expansion involving a series in powers of a/λ, where a gives the size of the atomic system and λ is the wavelength of the emitted photon. Higher order terms will involve matrix elements of magnetic moment operators, electric quadrupole moment operators, and even higher magnetic and electric multipole moments. In nuclei, these higher order terms are often important.

Finally, we make a remark about induced absorption and emission processes. If the atomic system is in a beam or a bath of photons, the probability for induced absorption and emission processes is given by the "Einstein B"s and by $\rho(\nu_{nm})$, the energy density of the electromagnetic radiation (photon beam). Through his study of the black-body radiation spectrum, Einstein found the relation

$$B_{m\to n} = B_{n\to m} = \frac{c^3}{8\pi h\nu_{nm}^3}A_{n\to m}, \tag{8}$$

where the probability per unit time of an induced absorption process is

$$\rho(\nu_{nm})B_{m\to n}N_m, \tag{9}$$

and the probability per unit time of an induced emission process is

$$\rho(\nu_{nm})B_{n\to m}N_n, \tag{10}$$

where $\rho(\nu_{nm})$ is the energy per unit volume of the photon beam at the transition frequency, and N_n and N_m are the number of atoms in state n and m, respectively.

Problems

27. Find the lifetime τ in seconds of the 2p state of the hydrogen atom

$$\tau = \frac{1}{A_{2p\to 1s}},$$

where $A_{2p\to 1s}$ is the Einstein A coefficient. Assume the three substates with $m = 0, \pm 1$ are populated with equal probability initially.

28. For the diatomic molecule rigid rotator, the space-fixed components of the electric dipole moment operator are

$$\mu_x^{(\text{el.})} = \mu_e \sin\theta\cos\phi, \qquad \mu_y^{(\text{el.})} = \mu_e \sin\theta\sin\phi, \qquad \mu_z^{(\text{el.})} = \mu_e \cos\theta,$$

where μ_e is the permanent electric dipole moment of the molecule, oriented along the molecular symmetry axis. (Homonuclear diatomic molecules, such as H_2 or N_2, have no permanent electric dipole moment. Their pure rotational transitions can therefore only be seen in Raman spectroscopy.)

Recall the rotational energies and wave functions are given by

$$E_J = \frac{\hbar^2}{2I_e} J(J+1), \quad \text{with } I_e = \mu r_e^2, \qquad \psi_{JM}(\theta, \phi) = Y_{JM}(\theta, \phi).$$

Give a general formula for the energy loss per second for an emission line for a transition $J \to (J-1)$, assuming the molecule is in a gaseous sample in thermal equilibrium at temperature T, where the number of molecules in the state with energy E_J is

$$N_J = \frac{(2J+1)e^{-(E_J/kT)}}{\sum_J (2J+1)e^{-(E_J/kT)}} N_{\text{total}} \approx \frac{\hbar^2}{2I_e kT}(2J+1)e^{-(E_J/kT)} N_{\text{total}}.$$

For example, for the HBr molecule with $r_e = 1.414 \times 10^{-8}$ cm, and $\mu_e = e(0.17 \times 10^{-8}$ cm), make an estimate for the $J = 3 \to 2$ transition in terms of the number of photons emitted per second. Your numerical answer should explain why such spectra are observed as absorption spectra rather than emission spectra. That is, the spontaneous emission process is very unlikely, so diatomic molecule rotational transitions are observed by utilizing the stimulated absorption and emission process for incident radiation of the appropriate frequency in the far infrared or microwave region. To come to the same conclusion, also calculate the lifetime in seconds of the first excited rotational state, with $J = 1$, and compare this result for the HBr molecule with the result of problem 27 for the first excited state of the hydrogen atom.

21

Stationary-State Perturbation Theory

A Rayleigh–Schrödinger Expansion

If a Hamiltonian, H, differs very little from a Hamiltonian, H_0, for which an exact solution is known, an expansion procedure may give a good approximation for the eigenvalues and eigenvectors of the full H. Many physically interesting problems, e.g., the Stark or Zeeman effects in an atom or a molecule, may involve a small perturbation of an exactly soluble problem. Let us assume the full Hamiltonian can be expanded about an H_0, with a known set of eigenvalues and eigenvectors, through a parameter of smallness, λ, with $\lambda \ll 1$

$$H = H^{(0)} + \lambda H^{(1)} + \lambda^2 H^{(2)} + \lambda^3 H^{(3)} + \cdots \quad . \tag{1}$$

Sometimes, only a first-order term may exist, say, a perturbing potential, V. Then,

$$H = H^{(0)} + \lambda V, \qquad \text{with} \quad H^{(2)} = H^{(3)} = \cdots = 0. \tag{2}$$

We want a systematic solution to the problem

$$H|n\rangle = E_n|n\rangle, \tag{3}$$

assuming we know the solution

$$H^{(0)}|n^{(0)}\rangle = E_n^{(0)}|n^{(0)}\rangle. \tag{4}$$

B Case 1: Nondegenerate State

Let us first treat the simplest case in which the state $|n^{(0)}\rangle$ is nondegenerate. We will expand both E_n and $|n\rangle$ in a power series in λ.

$$E_n = E_n^{(0)} + \lambda E_n^{(1)} + \lambda^2 E_n^{(2)} + \cdots, \tag{5}$$

$$|n\rangle = |n^{(0)}\rangle + \lambda|n^{(1)}\rangle + \lambda^2|n^{(2)}\rangle + \cdots. \tag{6}$$

The equation, $(H - E_n)|n\rangle = 0$, or

$$\begin{aligned}\Big((H^{(0)} - E_n^{(0)}) + \lambda(H^{(1)} - E_n^{(1)}) + \lambda^2(H^{(2)} - E_n^{(2)}) + \cdots\Big)& \\ \times \Big(|n^{(0)}\rangle + \lambda|n^{(1)}\rangle + \lambda^2|n^{(2)}\rangle + \cdots\Big) &= 0,\end{aligned} \tag{7}$$

can now be solved order by order, peeling off one term at a time. Thus,

$$\begin{aligned}\lambda^1 :\quad & (H^{(0)} - E_n^{(0)})|n^{(1)}\rangle + (H^{(1)} - E_n^{(1)})|n^{(0)}\rangle = 0, \\ \lambda^2 :\quad & (H^{(0)} - E_n^{(0)})|n^{(2)}\rangle + (H^{(1)} - E_n^{(1)})|n^{(1)}\rangle + (H^{(2)} - E_n^{(2)})|n^{(0)}\rangle = 0, \\ & \qquad \cdots \qquad \cdots \qquad \cdots \\ \lambda^j :\quad & (H^{(0)} - E_n^{(0)})|n^{(j)}\rangle + (H^{(1)} - E_n^{(1)})|n^{(j-1)}\rangle + \cdots \\ & \qquad + (H^{(j)} - E_n^{(j)})|n^{(0)}\rangle = 0.\end{aligned} \tag{8}$$

To solve the first-order equation, first left-multiply this equation with $\langle n^{(0)}|$, to convert it to matrix element form:

$$\langle n^{(0)}|(H^{(0)} - E_n^{(0)})|n^{(1)}\rangle + \langle n^{(0)}|(H^{(1)} - E_n^{(1)})|n^{(0)}\rangle = 0. \tag{9}$$

The first term can be seen to be zero, because

$$\langle n^{(0)}|H^{(0)} - E_n^{(0)}|n^{(1)}\rangle = \langle n^{(1)}|H^{(0)\dagger} - E_n^{(0)}|n^{0}\rangle^* = 0, \tag{10}$$

via $(H^{(0)} - E_n^{(0)})|n^{(0)}\rangle = 0$. Eq. (9) then gives the first-order correction to the energy:

$$E_n^{(1)} = \langle n^{(0)}|H^{(1)}|n^{(0)}\rangle. \tag{11}$$

That is, $E_n^{(1)}$ is given by the simple diagonal matrix element of $H^{(1)}$. For the first-order corrections to the state vector, consider the first-order equation

$$(H^{(0)} - E_n^{(0)})|n^{(1)}\rangle = -(H^{(1)} - E_n^{(1)})|n^{(0)}\rangle, \tag{12}$$

which is of the form, $\mathcal{O}|n^{(1)}\rangle = |v\rangle$, where the right-hand side is a known vector that can be calculated. In coordinate representation, this equation could be written in the form of an inhomogeneous differential equation for the unknown function, $\psi_n^{(1)}(\vec{r})$, with a known (calculable) function $\phi_v(\vec{r}) = \langle\vec{r}|v\rangle$ on the right-hand side;

$$(H^{(0)} - E_n^{(0)})\psi_n^{(1)}(\vec{r}) = \phi_v(\vec{r}).$$

The solution is of the form: a particular solution of the inhomogeneous equation to which a solution of the homogeneous equation can be added. The latter is just a solution of the zeroth order Schrödinger equation. In vector (ket) form, this equation is

$$|n^{(1)}\rangle = c_n^{(1)}|n^{(0)}\rangle + Q_n^{(0)}|n^{(1)}\rangle, \tag{13}$$

where the projection operator,

$$Q_n^{(0)} = 1 - P_n^{(0)} = 1 - |n^{(0)}\rangle\langle n^{(0)}| = \sum_{k\neq n}|k^{(0)}\rangle\langle k^{(0)}|, \tag{14}$$

projects onto the subspace of the Hilbert space excluding the state vector, $|n^{(0)}\rangle$. [The sum over k may have to include an integral, if the spectrum of $H^{(0)}$ includes a continuum.] Now left-multiply eq. (11) with a specific $\langle k^{(0)}| \neq \langle n^{(0)}|$ to yield

$$\langle k^{(0)}|(H^{(0)} - E_n^{(0)})|n^{(1)}\rangle = -\langle k^{(0)}|H^{(1)}|n^{(0)}\rangle, \tag{15}$$

leading to

$$\langle k^{(0)}|n^{(1)}\rangle = \frac{\langle k^{(0)}|H^{(1)}|n^{(0)}\rangle}{(E_n^{(0)} - E_k^{(0)})}, \tag{16}$$

so

$$|n^{(1)}\rangle = c_n^{(1)}|n^{(0)}\rangle + \sum_{k\neq n}|k^{(0)}\rangle\frac{\langle k^{(0)}|H^{(1)}|n^{(0)}\rangle}{E_n^{(0)} - E_k^{(0)}}. \tag{17}$$

To determine $c_n^{(1)}$, normalize the state vector $|n\rangle$ through first-order terms. With $|n\rangle = |n^{(0)}\rangle + \lambda|n^{(1)}\rangle$, and $|m\rangle = |m^{(0)}\rangle + \lambda|m^{(1)}\rangle$,

$$\begin{aligned}
\langle m|n\rangle &= \delta_{mn} \\
&= \langle m^{(0)}|n^{(0)}\rangle + \lambda\Big(\langle m^{(0)}|n^{(1)}\rangle + \langle n^{(0)}|m^{(1)}\rangle^*\Big) \\
&= \delta_{mn}\Big(1 + \lambda[c_n^{(1)} + c_m^{(1)*}]\Big) \\
&+ \lambda\left(\frac{\langle m^{(0)}|H^{(1)}|n^{(0)}\rangle}{E_n^{(0)} - E_m^{(0)}} + \frac{\langle n^{(0)}|H^{(1)}|m^{(0)}\rangle^*}{E_m^{(0)} - E_n^{(0)}}\right)\langle m^{(0)}|Q_n^{(0)}|n^{(0)}\rangle \\
&= \delta_{mn}\Big(1 + \lambda[c_n^{(1)} + c_m^{(1)*}]\Big),
\end{aligned} \tag{18}$$

where the δ_{mn} term contributes only when $n = m$, whereas the second term, which could in principle have contributed when $m \neq n$, is automatically zero. Hence, we have orthonormality through first order in λ, provided $c_n^{(1)} + c_n^{(1)*} = 0$. The simplest choice is to make $c_n^{(1)}$ real and, hence, $c_n^{(1)} = 0$. Therefore,

$$|n^{(1)}\rangle = \sum_{k\neq n}|k^{(0)}\rangle\frac{\langle k^{(0)}|H^{(1)}|n^{(0)}\rangle}{E_n^{(0)} - E_k^{(0)}}. \tag{19}$$

We could also have written this in operator form, in terms of the projection operator, $Q_n^{(0)}$,

$$|n^{(1)}\rangle = Q_n^{(0)}\frac{1}{(E_n^{(0)} - H^{(0)})}H^{(1)}|n^{(0)}\rangle. \tag{20}$$

Alternatively, using the properties of the projection operator

$$(Q_n^{(0)})^2 = Q_n^{(0)}, \qquad [Q_n^{(0)}, H^{(0)}] = 0, \qquad Q_n^{(0)}|n^{(0)}\rangle = 0, \tag{21}$$

we could also have written this as

$$|n^{(1)}\rangle = Q_n^{(0)} \frac{1}{E_n^{(0)} - H^{(0)}} Q_n^{(0)} H^{(1)} |n^{(0)}\rangle, \tag{22}$$

where the extra (redundant) $Q_n^{(0)}$ has been added as a safety factor to make sure the inverse operator

$$\frac{1}{E_n^{(0)} - H^{(0)}}$$

has a meaning. Because this operator can never act on $|n^{(0)}\rangle$, either to the left or to the right, we will never be plagued by zero denominators. This operator is a function of $H^{(0)}$ and can be thought of as being expanded in a Taylor series in $H^{(0)}$, where $H^{(0)}$ acting on a $|k^{(0)}\rangle$ will simply yield $E_k^{(0)}|k^{(0)}\rangle$.

C Second-Order Corrections

To get the second-order corrections to the energy, let us (always the first step!) left-multiply the second-order equation by $\langle n^{(0)}|$ to get

$$-\langle n^{(0)}|(H^{(0)} - E_n^{(0)})|n^{(2)}\rangle = \langle n^{(0)}|H^{(1)} - E_n^{(1)}|n^{(1)}\rangle + \langle n^{(0)}|H^{(2)}|n^{(0)}\rangle - E_n^{(2)}. \tag{23}$$

The left-hand side is again zero [think of "left action" of $(H^{(0)} - E_n^{(0)})$ on $\langle n^{(0)}|$], so, substituting for $|n^{(1)}\rangle$ in the right-hand side,

$$E_n^{(2)} = \langle n^{(0)}|H^{(2)}|n^{(0)}\rangle + \sum_{k \neq n} \frac{\langle n^{(0)}|H^{(1)}|k^{(0)}\rangle\langle k^{(0)}|H^{(1)}|n^{(0)}\rangle}{E_n^{(0)} - E_k^{(0)}}, \tag{24}$$

or

$$E_n^{(2)} = \langle n^{(0)}|H^{(2)}|n^{(0)}\rangle + \sum_{k \neq n} \frac{|\langle k^{(0)}|H^{(1)}|n^{(0)}\rangle|^2}{E_n^{(0)} - E_k^{(0)}}. \tag{25}$$

This formula, together with the result for the first-order correction, $E_n^{(1)}$, eq. (11), was named by Fermi as "Golden Rule I." (We shall meet an analagous "Golden Rule II" in time-dependent perturbation theory!), Fermi said most of the interesting results of quantum theory can be calculated with these Golden Rules. (Note also: If $|n\rangle$ is the ground state of the system, the perturbations caused by the higher states will "push down" and lower the ground-state energy. The second term then has patently positive numerators and negative denominators.)

To obtain the second-order correction to the state vector, $|n^{(2)}\rangle$, we again decompose this into a part proportional to $|n^{(0)}\rangle$ and a part orthogonal to $|n^{(0)}\rangle$,

$$|n^{(2)}\rangle = c_n^{(2)}|n^{(0)}\rangle + Q_n^{(0)}|n^{(2)}\rangle. \tag{26}$$

The $c_n^{(2)}$ will again be determined in the end to preserve orthonormality through second order. The piece $Q_n^{(0)}|n^{(2)}\rangle$ can be determined from the second-order equation

$$(E_n^{(0)} - H^{(0)})|n^{(2)}\rangle = (H^{(1)} - E_n^{(1)})|n^{(1)}\rangle + (H^{(2)} - E_n^{(2)})|n^{(0)}\rangle \tag{27}$$

by inverting the operator, *after* action with the projection operator $Q_n^{(0)}$,

$$\begin{aligned} Q_n^{(0)}|n^{(2)}\rangle \\ = Q_n^{(0)}\frac{1}{E_n^{(0)} - H^{(0)}}Q_n^{(0)}\Big((H^{(1)} - E_n^{(1)})|n^{(1)}\rangle + (H^{(2)} - E_n^{(2)})|n^{(0)}\rangle\Big) \\ = Q_n^{(0)}\frac{1}{E_n^{(0)} - H^{(0)}}Q_n^{(0)}(H^{(1)} - E_n^{(1)})Q_n^{(0)}\frac{1}{E_n^{(0)} - H^{(0)}}Q_n^{(0)}H^{(1)}|n^{(0)}\rangle \\ + Q_n^{(0)}\frac{1}{E_n^{(0)} - H^{(0)}}Q_n^{(0)}H^{(2)}|n^{(0)}\rangle. \end{aligned} \tag{28}$$

Now, substituting for $Q_n^{(0)}$, in the form

$$Q_n^{(0)} = \sum_{k\neq n}|k^{(0)}\rangle\langle k^{(0)}| = \sum_{l\neq n}|l^{(0)}\rangle\langle l^{(0)}|, \tag{29}$$

and using the result, $E_n^{(1)} = \langle n^{(0)}|H^{(1)}|n^{(0)}\rangle$, we get

$$Q_n^{(0)}|n^{(2)}\rangle = \sum_{k\neq n}|k^{(0)}\rangle\langle k^{(0)}|n^{(2)}\rangle, \tag{30}$$

with

$$\begin{aligned} \langle k^{(0)}|n^{(2)}\rangle = \sum_{l\neq n}\frac{\langle k^{(0)}|H^{(1)}|l^{(0)}\rangle\langle l^{(0)}|H^{(1)}|n^{(0)}\rangle}{(E_n^{(0)} - E_k^{(0)})(E_n^{(0)} - E_l^{(0)})} \\ - \langle n^{(0)}|H^{(1)}|n^{(0)}\rangle\frac{\langle k^{(0)}|H^{(1)}|n^{(0)}\rangle}{(E_n^{(0)} - E_k^{(0)})^2} \\ + \frac{\langle k^{(0)}|H^{(2)}|n^{(0)}\rangle}{(E_n^{(0)} - E_k^{(0)})}. \end{aligned} \tag{31}$$

Now, we still need to evaluate $c_n^{(2)} = \langle n^{(0)}|n^{(2)}\rangle$, which is again chosen to normalize $|n\rangle$ through second order. In calculating $\langle n|n\rangle$ through second order, we get a contribution $1 + c_n^{(2)} + c_n^{(2)*}$, no contributions from $\langle n^{(0)}|Q_n^{(0)}|n^{(2)}\rangle$, but now a contribution from $\langle n^{(1)}|n^{(1)}\rangle$. Again, letting $c_n^{(2)}$ be real, we get

$$c_n^{(2)} = -\frac{1}{2}\sum_{k\neq n}\frac{|\langle k^{(0)}|H^{(1)}|n^{(0)}\rangle|^2}{(E_n^{(0)} - E_k^{(0)})^2}. \tag{32}$$

An alternative way to normalize the ket $|n\rangle$ through some order, which may be particularly convenient if we want to go through some relatively high order in λ, would be to set all $c_n^{(j)} = 0$ and then normalize the final result for $|n^{(0)}\rangle + \sum_{j=1}\lambda^j Q_n^{(0)}|n^{(j)}\rangle$. Through second order, this procedure would give the overall normalization constant as

$$N_n = \frac{1}{\sqrt{1 + \lambda^2 \sum_{k \neq n} \frac{|\langle k^{(0)}|H^{(1)}|n^{(0)}\rangle|^2}{(E_n^{(0)} - E_k^{(0)})^2}}}. \tag{33}$$

When the square root is expanded in powers of λ^2, this result agrees with eq. (32).

By "turning the handle of the crank," we can generalize the results of eq. (25) and (31) for second order to arbitrarily high order. Clearly, the results become more and more complicated.

D The Wigner–Brillouin Expansion

This slightly different expansion becomes particularly useful in the special (but very common) case when $H^{(1)} = V$ and $H^{(2)} = H^{(3)} = \cdots = 0$, and particularly if we want to go to very high order. In the Wigner–Brillouin expansion, the needed E_n is at first not expanded. The equation to be solved by successive approximation is then

$$(E_n - H^{(0)})|n\rangle = \lambda V|n\rangle \tag{34}$$

and is to be solved by

$$|n\rangle = |n^{(0)}\rangle + Q_n^{(0)}|n\rangle, \tag{35}$$

with the state vector to be normalized in the very end. By inverting the operator $(E_n - H^{(0)})$, after multiplication of eq. (34) by $Q_n^{(0)}$, we get

$$Q_n^{(0)}|n\rangle = Q_n^{(0)} \frac{1}{E_n - H^{(0)}} Q_n^{(0)} \lambda V|n\rangle. \tag{36}$$

Now, we substitute for $|n\rangle$ through eq. (35) and iterate this process over and over to get the expansion

$$\begin{aligned} Q_n^{(0)}|n\rangle = {} & Q_n^{(0)} \frac{1}{E_n - H^{(0)}} Q_n^{(0)} \lambda V|n^{(0)}\rangle \\ & + Q_n^{(0)} \frac{1}{E_n - H^{(0)}} Q_n^{(0)} \lambda V Q_n^{(0)} \frac{1}{E_n - H^{(0)}} Q_n^{(0)} \lambda V|n^{(0)}\rangle \\ & + Q_n^{(0)} \frac{1}{E_n - H^{(0)}} Q_n^{(0)} \lambda V Q_n^{(0)} \frac{1}{E_n - H^{(0)}} Q_n^{(0)} \lambda V \\ & \qquad \times Q_n^{(0)} \frac{1}{E_n - H^{(0)}} Q_n^{(0)} \lambda V|n^{(0)}\rangle + \cdots . \end{aligned} \tag{37}$$

To get an expression for the correction to the energy, $(E_n - E_n^{(0)})$, left-multiply eq. (34) by $\langle n^{(0)}|$ to get

$$\langle n^{(0)}|(E_n - H^{(0)})|n\rangle = \langle n^{(0)}|\lambda V|n\rangle, \tag{38}$$

leading to

$$E_n - E_n^{(0)} = \langle n^{(0)}|\lambda V|n^{(0)}\rangle$$

$$+\sum_{k\neq n}\frac{\langle n^{(0)}|\lambda V|k^{(0)}\rangle\langle k^{(0)}|\lambda V|n^{(0)}\rangle}{(E_n-E_k^{(0)})}$$
$$+\sum_{k\neq n}\sum_{l\neq n}\frac{\langle n^{(0)}|\lambda V|k^{(0)}\rangle\langle k^{(0)}|\lambda V|l^{(0)}\rangle\langle l^{(0)}|\lambda V|n^{(0)}\rangle}{(E_n-E_k^{(0)})(E_n-E_l^{(0)})}$$
$$+\cdots. \tag{39}$$

Clearly, this can be generalized very easily to an arbitrarily high number of terms in the expansion in powers of λ. The unknown, E_n, for which we are solving, however, now appears in all energy denominators. To find this as a power series in λ, we must substitute

$$E_n = E_n^{(0)} + \lambda E_n^{(1)} + \lambda^2 E_n^{(2)} + \cdots \tag{40}$$

in all energy denominators. Then, expand these in powers of λ peeling off, first the first-order, then the second-order, and higher order terms. The final result will, of course, be the same as that given by the Rayleigh–Schrödinger expansion, but the simplicity of the first step may make the Wigner–Brillouin expansion useful in the case when very high orders are needed.

22

Example 1: The Slightly Anharmonic Oscillator

In Chapter 15, we discussed the diatomic molecule, a complicated many-body system. At low energies, however, we can neglect specific treatment of the electron degrees of freedom. The electron cloud can in first approximation be taken as the source of a potential binding the two atomic nuclei into a nearly rigid, vibrating structure. The position of the atomic nuclei of the diatomic molecule in 3-D space can be described by the three coordinates: r, the radial distance between the two atomic nuclei, and θ, and ϕ, the two angles describing the orientation of the molecule axis in our 3-D space. The electron cloud gives rise to a potential (see Fig. 22.1) with a deep minimum at $r = r_e$, where r_e is the equilibrium distance between the two nuclei. For very small values of r, the potential becomes strongly repulsive and rises to ∞. For very large values of r, the potential approaches a constant value of $V_{\text{diss.}}$. If we can raise the energy above this value, i.e., if $E > V_{\text{diss.}}$, the molecule will dissociate into two atomic fragments. For $E \ll V_{\text{diss.}}$, however, the potential will be nearly parabolic and can be expanded about the value $r = r_e$,

$$V(r) = V(r_e) + \frac{1}{2}\left(\frac{d^2V}{dr^2}\right)_{r_e}(r - r_e)^2 + \frac{1}{3!}\left(\frac{d^3V}{dr^3}\right)_{r_e}(r - r_e)^3 + \cdots . \tag{1}$$

The Schrödinger equation for the wave function $\psi(r, \theta, \phi) = u(r)Y_{lm}(\theta, \phi)$ separates approximately (for the vibration–rotation perturbations, see problem 30) into a radial equation describing the vibration of the molecule and an angular equation describing the rotation of the molecule (see Chapter 15). The 1-D radial equation can be described by the Hamiltonian

$$H = -\frac{\hbar^2}{2\mu}\frac{\partial^2}{\partial r^2} + V(r). \tag{2}$$

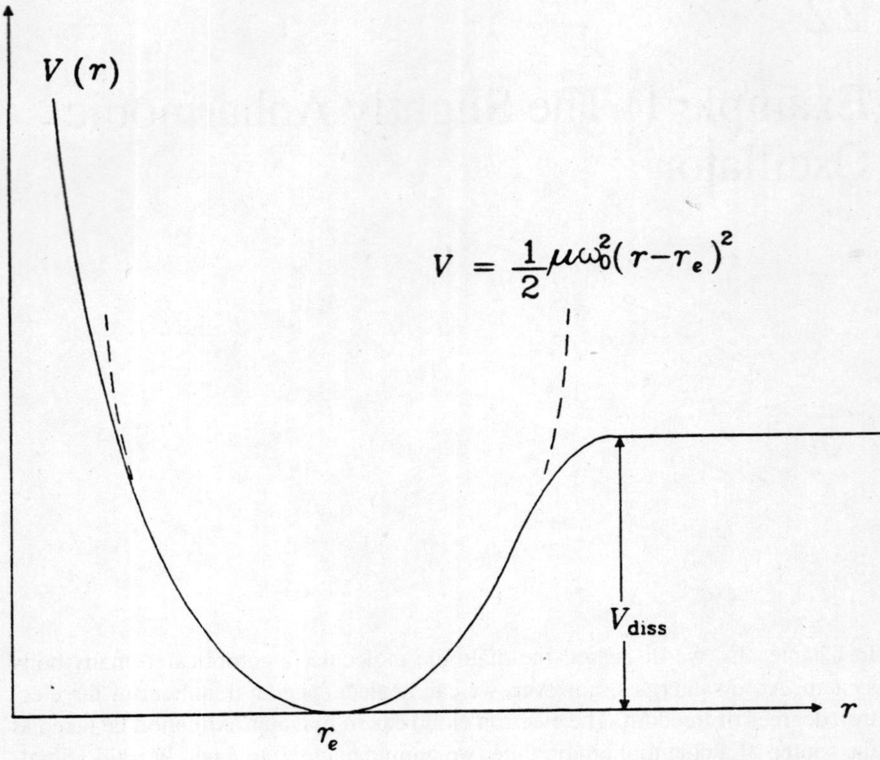

FIGURE 22.1. Diatomic molecule potential function.

For low-energy vibrational excitations, $V(r)$ can be approximated by the first few terms of the above Taylor expansion. Then, if we replace the vibrational coordinate, $(r - r_e)$, by a dimensionless coordinate, x,

$$(r - r_e) = \sqrt{\frac{\hbar}{\mu\omega_0}}x, \tag{3}$$

the vibrational Hamiltonian can be rewritten as

$$H = \hbar\omega_0\Big(\frac{1}{2}[-\frac{d^2}{dx^2} + x^2] + \lambda_3 x^3 + \lambda_4 x^4\Big), \tag{4}$$

with

$$\lambda_3 = \frac{1}{3!\hbar\omega_0}\left(\frac{\hbar}{\mu\omega_0}\right)^{\frac{3}{2}}\left(\frac{d^3V}{dr^3}\right)_{r_e}, \qquad \lambda_4 = \frac{1}{4!\hbar\omega_0}\left(\frac{\hbar}{\mu\omega_0}\right)^{2}\left(\frac{d^4V}{dr^4}\right)_{r_e}. \tag{5}$$

In most molecules $\lambda_3 \ll 1$, and $\lambda_4 \simeq$ order $(\lambda_3)^2$. Thus, we can write the Hamiltonian of eq. (4)

$$\begin{aligned} H &= H^{(0)} + \lambda H^{(1)} + \lambda^2 H^{(2)} \\ &= H^{(0)} + \hbar\omega_0\lambda_3 x^3 + \hbar\omega_0\lambda_4 x^4. \end{aligned} \tag{6}$$

The zeroth-order Hamiltonian is the simple 1-D harmonic oscillator Hamiltonian. Its eigenvectors are all nondegenerate, so we can use the machinery of stationary-state perturbation theory for an arbitrary nondegenerate state $|n\rangle$. We merely need to calculate the matrix elements of the two operators x^3 and x^4. These operators follow by matrix multiplication from the simple (known) matrix elements of x. Calculating first the matrix elements of x^2, we can use these together with the matrix elements of x to evaluate all needed matrix elements. We list some of the needed results, as follows:

$$\begin{aligned}
\langle n-2|x^2|n\rangle &= \tfrac{1}{2}\sqrt{n(n-1)} \\
\langle n+2|x^2|n\rangle &= \tfrac{1}{2}\sqrt{(n+1)(n+2)} \\
\langle n|x^2|n\rangle &= (n+\tfrac{1}{2}) \\
\langle m|x^2|n\rangle &= 0; \qquad \text{for} \quad m \neq n, (n \pm 2);
\end{aligned} \tag{7}$$

$$\begin{aligned}
\langle n-3|x^3|n\rangle &= \tfrac{1}{2}\sqrt{[n(n-1)(n-2)/2]} \\
\langle n+3|x^3|n\rangle &= \tfrac{1}{2}\sqrt{[(n+1)(n+2)(n+3)/2]} \\
\langle n-1|x^3|n\rangle &= \tfrac{3}{2}n\sqrt{[n/2]} \\
\langle n+1|x^3|n\rangle &= \tfrac{3}{2}(n+1)\sqrt{[(n+1)/2]} \\
\langle m|x^3|n\rangle &= 0; \qquad \text{for} \quad m \neq (n \pm 3), (n \pm 1).
\end{aligned} \tag{8}$$

Through second order, we shall only need the diagonal matrix element of x^4. This matrix element has the value

$$\langle n|x^4|n\rangle = \tfrac{3}{2}(n^2 + n + \tfrac{1}{2}). \tag{9}$$

The terms cubic in x have no diagonal matrix elements. Therefore, $H^{(1)}$ has no diagonal matrix element, and $E_n^{(1)} = 0$. The first correction to the energy is given by the second-order term

$$\begin{aligned}
\lambda^2 E_n^{(2)} &= \langle n^{(0)}|H^{(2)}|n^{(0)}\rangle + \sum_{k\neq n} \frac{|\langle k^{(0)}|H^{(1)}|n^{(0)}\rangle|^2}{(E_n^{(0)} - E_k^{(0)})} \\
&= \lambda_4 \langle n^{(0)}|\hbar\omega_0 x^4|n^{(0)}\rangle + \sum_{k\neq n} \frac{|\langle k^{(0)}|\lambda_3 \hbar\omega_0 x^3|n^{(0)}\rangle|^2}{\hbar\omega_0(n-k)} \\
&= \lambda_4 \hbar\omega_0 \tfrac{3}{2}(n^2 + n + \tfrac{1}{2}) + \frac{\lambda_3^2}{8}\hbar\omega_0 \left(\frac{9n^3}{+1} + \frac{9(n+1)^3}{-1} \right. \\
&\qquad \left. + \frac{n(n-1)(n-2)}{+3} + \frac{(n+1)(n+2)(n+3)}{-3} \right).
\end{aligned} \tag{10}$$

The final result gives

$$\lambda^2 E_n^{(2)} = \hbar\omega_0 \left[\tfrac{3}{2}\lambda_4(n^2 + n + \tfrac{1}{2}) - \lambda_3^2(30n^2 + 30n + 11) \right]. \tag{11}$$

We will also need the corrected state vectors. Often, it is sufficient to know these vectors to first order. For the slightly anharmonic oscillator, we have (to first order),

$$|n\rangle = |n^{(0)}\rangle + \lambda_3 \left(\sum_{k\neq n} |k^{(0)}\rangle \frac{\langle k^{(0)}|x^3|n^{(0)}\rangle}{(n-k)} \right)$$

$$
\begin{aligned}
&= |n^{(0)}\rangle + \lambda_3 \Bigg(|(n-3)^{(0)}\rangle \frac{\sqrt{n(n-1)(n-2)}}{6\sqrt{2}} \\
&\quad - |(n+3)^{(0)}\rangle \frac{\sqrt{(n+3)(n+2)(n+1)}}{6\sqrt{2}} \\
&\quad + |(n-1)^{(0)}\rangle \frac{3n\sqrt{n}}{2\sqrt{2}} - |(n+1)^{(0)}\rangle \frac{3(n+1)\sqrt{(n+1)}}{2\sqrt{2}} \Bigg). \qquad (12)
\end{aligned}
$$

These corrected state vectors will be needed to calculate the corrections to the transition probabilities. For a diatomic molecule, the electric dipole moment along the direction of the molecular symmetry axis can be given by

$$
\mu^{(el.)} = \mu_e + \left(\frac{d\mu}{dr}\right)_e (r - r_e) + \cdots = \mu_e + e_{\text{eff.}} \sqrt{\frac{\hbar}{\mu\omega_0}} x + \cdots \, , \qquad (13)
$$

where the dipole moment derivative is expressed in terms of an effective charge, $e_{\text{eff.}}$. Also, both the permanent electric dipole moment, μ_e, and the effective charge (which gives the strength of the dipole moment change during the harmonic oscillation) are zero for a homonuclear diatomic molecule, such as H_2, O_2, or N_2. Off-diagonal matrix elements of the electric dipole moment operator are given by

$$
\langle m^{(0)} | \mu^{(el.)} | n^{(0)} \rangle = \delta_{m(n\pm1)} e_{\text{eff.}} \sqrt{\frac{\hbar}{\mu\omega_0}} \langle m^{(0)} | x | n^{(0)} \rangle . \qquad (14)
$$

This equation leads to the zeroth-order vibrational selection rule, $\Delta n = \pm 1$ and a zeroth-order transition probability given by the Einstein A

$$
A_{n\to(n-1)} = \frac{8\pi\omega_0^3}{3hc^3} e_{\text{eff.}}^2 \frac{\hbar}{\mu\omega_0} \frac{n}{2}. \qquad (15)
$$

If first-order anharmonic corrections are included in the state vector, the vibrational selection rule $\Delta n = \pm 1$ is partially relaxed. For example, now transitions $n \to (n-2)$ may become possible. If we write the analog of eq. (12) for the bra $\langle (n-2)|$ through first order

$$
\begin{aligned}
\langle (n-2)| &= \langle (n-2)^{(0)}| + \lambda_3 \Bigg(\langle (n-3)^{(0)}| \frac{3(n-2)\sqrt{(n-2)}}{2\sqrt{2}} \\
&\quad - \langle (n-1)^{(0)}| \frac{3(n-1)\sqrt{(n-1)}}{2\sqrt{2}} + \langle (n-5)^{(0)}| \frac{\sqrt{(n-2)(n-3)(n-4)}}{6\sqrt{2}} \\
&\quad - \langle (n+1)^{(0)}| \frac{\sqrt{(n+1)n(n-1)}}{6\sqrt{2}} \Bigg), \qquad (16)
\end{aligned}
$$

to first order in λ_3, the matrix element of $\langle (n-2)|x|n\rangle$ gets contributions from the zeroth-order component of $|n\rangle$ with the first-order components $\langle (n-1)^{(0)}|$ and $\langle (n+1)^{(0)}|$ of $\langle (n-2)|$, and from the zeroth-order component of $\langle (n-2)|$ with the first-order components $|(n-3)^{(0)}\rangle$ and $|(n-1)^{(0)}\rangle$ of $|n\rangle$, leading to

$$
\langle (n-2)| \mu^{(el.)} |n\rangle = e_{\text{eff.}} \sqrt{\frac{\hbar}{\mu\omega_0}} \lambda_3
$$

$$\times\left(-\frac{3(n-1)\sqrt{(n-1)}}{2\sqrt{2}}\sqrt{\frac{n}{2}}-\frac{\sqrt{(n+1)n(n-1)}}{6\sqrt{2}}\sqrt{\frac{(n+1)}{2}}\right.$$
$$\left.+\frac{\sqrt{n(n-1)(n-2)}}{6\sqrt{2}}\sqrt{\frac{(n-2)}{2}}+\frac{3n\sqrt{n}}{2\sqrt{2}}\sqrt{\frac{(n-1)}{2}}\right)$$
$$=e_{\text{eff.}}\sqrt{\frac{\hbar}{\mu\omega_0}}\lambda_3\frac{1}{2}\sqrt{n(n-1)}. \tag{17}$$

This equation leads to the transition probability given by the Einstein A

$$A_{n\to(n-2)}=\frac{8\pi(2\omega_0)^3}{3hc^3}e^2_{\text{eff.}}\frac{\hbar}{\mu\omega_0}\lambda_3^2\frac{n(n-1)}{4}. \tag{18}$$

This transition probability is weaker by a factor of λ_3^2 compared with the zeroth-order allowed transition $n\to(n-1)$.

A number of remarks are in order: (1) The formula given here for the transition probabilities is for the true 1-D anharmonic oscillator. It therefore assumes the diatomic molecule remains oriented in a specific direction in space, say, the x-direction in a crystalline environment. In a free diatomic molecule, say, in a gaseous sample in a microwave wave guide, the molecule is of course free to both rotate and vibrate. To get the transition probabilities, we would need the matrix elements of

$$\mu_x^{(el.)}=\mu_r\sin\theta\cos\phi,\qquad \mu_y^{(el.)}=\mu_r\sin\theta\cos\phi,\qquad \mu_z^{(el.)}=\mu_r\cos\theta, \tag{19}$$

with radial part, μ_r, given by eq. (13). The matrix elements given by eqs. (14) and (17) are just the radial (vibrational) part of the full electric dipole moment matrix element. This must be augmented by the angular (rotational) matrix elements of $\sin\theta\cos\phi$, $\sin\theta\sin\phi$, and $\cos\theta$. These matrix elements were actually evaluated in Chapter 9. These matrix elements lead to the rotational selection rule, $\Delta l=\pm1$. Therefore, the actual transition in a free diatomic molecule involves both a change in vibrational quantum number, $\Delta n=\pm1$, and a change in the rotational quantum number, $\Delta l=\pm1$, leading to a vibration–rotation rather than to a pure vibrational transition.

(2): The actual numerical values of a vibrational transition probability, such as that given by eq. (15), is very small, corresponding to inverse times of the order of seconds or minutes, compared with an atomic electronic transition probability corresponding to lifetimes of the order of 10^{-8} seconds. Molecular vibrational or vibrational–rotational transitions are thus usually too weak to be seen in spontaneous emission. They are easily observed, however, in induced absorption processes, by placing the gaseous molecular sample in an electromagnetic beam of the appropriate infrared or microwave frequency. The transition probabilities will then be given by the Einstein B coefficients and the energy density of the incident beam. Because the Einstein B coefficients are proportional to the Einstein A coefficients, the results of this section will still be useful. Also, the induced emission probability for a transition $n\to(n-1)$ will be less than the induced absorption probability for the transition$(n-1)\to n$ if the number of molecules in the lower state, $N_{(n-1)}$, is greater than the number of molecules in the upper state, N_n, the usual situation for a gaseous sample in thermal equilibrium.

23

Perturbation Theory for Degenerate Levels

A Diagonalization of $H^{(1)}$: Transformation to Proper Zeroth-Order Basis

Assume the eigenvector for the energy state with eigenvalue E_n is degenerate; i.e., assume g_n independent eigenvectors exist such that

$$H^{(0)}|nr^{(0)}\rangle = E_n^{(0)}|nr^{(0)}\rangle, \qquad \text{with} \quad r = 1, 2, \ldots, g_n, \tag{1}$$

where the label r may just be an ordinal label identifying the different eigenvectors, or it may be a shorthand notation for additional quantum numbers. Clearly, our previous method might lead to difficulties, because now zeros could be in the energy denominators $(E_n^{(0)} - E_k^{(0)})$. A state $|k^{(0)}\rangle$ different from $|nr^{(0)}\rangle$ could now include a state with the same zeroth-order energy. Note, however, if $H^{(1)}$ is made diagonal within the g_n-dimensional subspace, $|nr^{(0)}\rangle$, with $r = 1, 2, \ldots, g_n$, this difficulty will never arise. Therefore, the first step of the perturbation expansion will involve a transformation from the subbasis $|nr^{(0)}\rangle$ to the new subbasis $|nr'^{(0)}\rangle$, with

$$|nr'^{(0)}\rangle = \sum_{r=1}^{g_n} |nr^{(0)}\rangle\langle nr^{(0)}|nr'^{(0)}\rangle = \sum_{r=1}^{g_n} |nr^{(0)}\rangle c_r, \tag{2}$$

such that

$$\langle ns'^{(0)}|H^{(1)}|nr'^{(0)}\rangle = \delta_{rs} E_{nr'}^{(1)}. \tag{3}$$

Rewriting the first-order equation of the perturbation expansion for the transformed state $|nr'^{(0)}\rangle$ yields

$$(E_n^{(0)} - H^{(0)})|nr'^{(1)}\rangle = (H^{(1)} - E_n^{(1)})|nr'^{(0)}\rangle. \tag{4}$$

Left-multiplying by $\langle nr^{(0)}|$ leads to a zero for the left-hand side of this equation, so

$$\langle nr^{(0)}|H^{(1)}|nr'^{(0)}\rangle - E_n^{(1)}\langle nr^{(0)}|nr'^{(0)}\rangle = 0. \tag{5}$$

Introducing the unit operator $\sum_s |ns^{(0)}\rangle\langle ns^{(0)}|$ for the subspace in question, we can transform this equation into

$$\sum_{s=1}^{g_n}\langle nr^{(0)}|H^{(1)}|ns^{(0)}\rangle\langle ns^{(0)}|nr'^{(0)}\rangle = E_n^{(1)}\sum_{s=1}^{g_n}\delta_{rs}\langle ns^{(0)}|nr'^{(0)}\rangle, \tag{6}$$

or, in shorthand form, with $\langle ns^{(0)}|nr'^{(0)}\rangle \equiv c_s$,

$$\sum_{s}^{g_n}(H_{rs}^{(1)} - E_n^{(1)}\delta_{rs})c_s = 0, \qquad \text{with} \quad r = 1, \ldots, g_n. \tag{7}$$

This system of g_n linear equations will have a solution for the c_s if and only if the determinant of the coefficients is equal to zero

$$|H_{rs}^{(1)} - E_n^{(1)}\delta_{rs}| = 0. \tag{8}$$

This equation leads to an equation of degree g_n in the unknown $E_n^{(1)}$ with g_n solutions $E_{nr'}^{(1)}$, such that

$$H^{(1)}|nr'^{(0)}\rangle = E_{nr'}^{(1)}|nr'^{(0)}\rangle, \qquad \text{with} \quad |nr'^{(0)}\rangle = \sum_s |ns^{(0)}\rangle c_s. \tag{9}$$

The first task in the case of degenerate-level perturbation theory then is to find those linear combinations of the zeroth-order eigenvectors $|ns^{(0)}\rangle$ that diagonalize $H^{(1)}$. These so-called *proper* or *stabilized* zeroth-order eigenvectors will be the basis for the subsequent steps in the perturbation expansion. Three possibilities need to be considered.

B Three Cases of Degenerate Levels

Case (1): The initial basis $|nr^{(0)}\rangle$ may already be such that $H^{(1)}$ is diagonal in this basis. (If higher order terms exist, such as $H^{(2)}$, we assume they are also diagonal in this basis.) This may not be such a fortuitous accident. Often, we choose the initial basis to be adapted to the symmetry of the problem. Both $H^{(0)}$ and $H^{(1)}$ may have symmetries that naturally lead to a *proper* choice of basis. The choice of this *proper* or *symmetry-adapted* basis may obviate the first step in degenerate-level perturbation theory, the diagonalization of $H^{(1)}$ in the initial zeroth-order basis. Effectively, therefore, such a case can be treated by nondegenerate perturbation theory.

Other cases now exist in which the nondegenerate perturbation theory formulae of Chapter 21 are sufficient to some order in the parameter of smallness, λ. For example, suppose two (or more) degenerate states $|nr^{(0)}\rangle$ and $|ns^{(0)}\rangle$ are not connected with each other through second order. Assume they are connected with each other only in third order, either through the action of $H^{(1)}$ three times, or through terms such as

$$\langle ns^{(0)}|H^{(1)}|n''u^{(0)}\rangle\langle n''u^{(0)}|H^{(1)}|n't^{(0)}\rangle\langle n't^{(0)}|H^{(1)}|nr^{0)}\rangle$$

in the perturbation expansion (or similar terms combining a single action of both $H^{(1)}$ and $H^{(2)}$). In that case, if we are interested only in second-order corrections to the energy, the degenerate states $|nr^{(0)}\rangle$ and $|ns^{(0)}\rangle$ are effectively unconnected in second order and the nondegenerate perturbation theory formulae of Chapter 21 apply.

Case (2): The diagonalization of $H^{(1)}$ in the initial zeroth-order basis may lead to a set of g_n-distinct eigenvalues $E^{(1)}_{nr'}$, with $E^{(1)}_{nr'} \neq E^{(1)}_{ns'}$, for $r \neq s$. In this case, the subsequent steps in the perturbation expansion closely parallel those for nondegenerate-level perturbation theory.

Case (3): The diagonalization of $H^{(1)}$ may not remove the zeroth-order degeneracy completely. In this case, a special treatment is necessary. This treatment is similar to that discussed in a next chapter for two (or several) nearly (or precisely) degenerate levels, the toughest of all cases. Again, this case is not as uncommon as might have been thought. For example, the symmetries of $H^{(0)}$ and $H^{(1)}$ could be such that all matrix elements of $H^{(1)}$ are zero in the $|nr^{(0)}\rangle$ sub-basis. Perhaps the $|nr^{(0)}\rangle$ all have the same parity and $H^{(1)}$ has the opposite parity. In this case, we cannot diagonalize $H^{(1)}$ to find the *proper* zeroth-order basis.

C Higher Order Corrections with Proper Zeroth-Order Basis

Let us consider case (2). In this case, the first-order equation

$$(E^{(0)}_n - H^{(0)})|nr'^{(1)}\rangle = (H^{(1)} - E^{(1)}_{nr'})|nr'^{(0)}\rangle \tag{10}$$

leads to

$$\begin{aligned} Q^{(0)}_n|nr'^{(1)}\rangle &= Q^{(0)}_n \frac{1}{E^{(0)}_n - H^{(0)}} Q^{(0)}_n H^{(1)}|nr'^{(0)}\rangle \\ &= \sum_{k\neq n}\sum_{s} |ks^{(0)}\rangle \frac{\langle ks^{(0)}|H^{(1)}|nr'^{(0)}\rangle}{(E^{(0)}_n - E^{(0)}_k)}. \end{aligned} \tag{11}$$

Upon left-multiplication with $\langle nr'^{(0)}|$, the second-order equation

$$(E^{(0)}_n - H^{(0)})|nr'^{(0)}\rangle = (H^{(1)} - E^{(1)}_{nr'})|nr'^{(1)}\rangle + (H^{(2)} - E^{(2)}_{nr'})|nr'^{(0)}\rangle \tag{12}$$

leads to

$$0 = \sum_{k\neq n}\sum_{s} \frac{\langle nr'^{(0)}|H^{(1)}|ks^{(0)}\rangle\langle ks^{(0)}|H^{(1)}|nr'^{(0)}\rangle}{(E_n^{(0)} - E_k^{(0)})} + \langle nr'^{(0)}|H^{(2)}|nr'\rangle - E_{nr'}^{(2)}, \tag{13}$$

so

$$E_{nr'}^{(2)} = \langle nr'^{(0)}|H^{(2)}|nr'^{(0)}\rangle + \sum_{k\neq n}\sum_{s} \frac{|\langle ks^{(0)}|H^{(1)}|nr'^{(0)}\rangle|^2}{(E_n^{(0)} - E_k^{(0)})}. \tag{14}$$

All steps in this and subsequent steps in the perturbation formalism parallel the earlier nondegenerate-state perturbation theory, except state vectors $|n^{(0)}\rangle$ or their conjugate bras must be replaced by $|nr'^{(0)}\rangle$ and the projection operators $Q_n^{(0)}$ will have to include besides the sum over $k \neq n$ a sum over the degeneracy label s. Note, in particular, that the states $|ks^{(0)}\rangle$ need *not* be transformed to primed form.

D Application 1: Stark Effect in the Diatomic Molecule Rigid Rotator

Let us consider a nonhomonuclear diatomic molecule, with a permanent electric dipole moment, which is perturbed by an external electric field, $\vec{\mathcal{E}}$. Let us consider only the lowest energies of this system, so the molecule can be considered as a rigid rotator, with zeroth-order energy eigenvalues and eigenfunctions (or eigenvectors)

$$E_l^{(0)} = \frac{\hbar^2}{2I_e} l(l+1), \qquad Y_{lm}(\theta,\phi) = \langle\theta,\phi|lm\rangle, \tag{15}$$

with $I_e = \mu r_e^2$ and

$$H^{(0)}|lm^{(0)}\rangle = E_l^{(0)}|lm^{(0)}\rangle, \qquad \text{with} \quad m = +l, \dots, -l. \tag{16}$$

The lth level, thus, has a $(2l+1)$-fold degeneracy, the general degeneracy associated with rotationally invariant sytems. In the presence of an external electric field, $\vec{\mathcal{E}}$, we must add a term

$$H^{(1)} = -\vec{\mu}^{(el.)} \cdot \vec{\mathcal{E}} = -\mu_e \cos\theta \mathcal{E}, \tag{17}$$

where we have assumed the electric field $\vec{\mathcal{E}}$ is in the space-fixed z direction and the symmetry axis of the molecule makes a polar angle θ with this z direction. This Stark perturbation (Stark effect) is often used to identify the l values of initial states in purely rotational transitions. To show $H^{(1)}$ is a weak perturbation, assume the electric field is 1,000 Volts/cm, a strong field. The permanent electric dipole moments, μ_e, of diatomic molecules are of order $e \times 10^{-8}$ cm. Thus, even in this strong field, $H^{(1)}$ can be expected to be of order 10^{-5}eV. Take the HCl molecule as a specific example, so $\mu \approx m_{\text{proton}}$, with $\mu c^2 \approx 10^9 eV$. Take $r_e \approx 10^{-8}$cm (the precise value is known from the rotational spectrum to be 1.2746×10^{-8}cm). Then,

$$\frac{\hbar^2}{2\mu r_e^2} = \frac{(\hbar c)^2}{2(\mu c^2)r_e^2} \approx \frac{(1.973 \times 10^{-5} eV cm)^2}{2 \times 10^9 eV \times 10^{-16} cm^2} \approx 2 \times 10^{-3} eV, \tag{18}$$

showing the Stark term, $H^{(1)}$, can indeed be treated as a perturbation. To calculate the matrix elements of $H^{(1)}$, recall (Chapter 9)

$$\cos\theta\, Y_{lm} = \sqrt{\frac{[(l+1)^2 - m^2]}{(2l+1)(2l+3)}}\, Y_{(l+1)m} + \sqrt{\frac{[l^2 - m^2]}{(2l+1)(2l-1)}}\, Y_{(l-1)m}, \tag{19}$$

so

$$\langle (l+1)m|\cos\theta|lm\rangle = \sqrt{\frac{[(l+1)^2 - m^2]}{(2l+1)(2l+3)}},$$

$$\langle (l-1)m|\cos\theta|lm\rangle = \sqrt{\frac{[l^2 - m^2]}{(2l+1)(2l-1)}}. \tag{20}$$

Our $H^{(1)}$ does not have matrix elements diagonal in the quantum number, l. No first-order contribution to the Stark energy shift occurs. At first, it appears this belongs to case (3) of section B of this chapter and might require further treatment. All matrix elements of the perturbing Hamiltonian, however, are diagonal in m. States of a particular m are therefore completely unconnected from states of different m. We can therefore treat states of a particular m by themselves, as if they were unconnected from the rest, hence, by nondegenerate perturbation theory. This is of course connected to the symmetry of our Hamiltonian, even, including the full perturbation, our Hamiltonian has axial symmetry. Our zeroth-order state vectors, of good eigenvalue m, are automatically the *proper symmetry-adapted* zeroth-order state vectors. (If we had chosen to call the direction of the $\vec{\mathcal{E}}$ field the x rather than the z direction, our $H^{(1)}$ would have been $-\mu_e \sin\theta \cos\phi \mathcal{E}$. If we had diagonalized this $H^{(1)}$, we would essentially have effected a rotation from our original x direction to a new z direction. By choosing our z rather than our x direction along the direction of the outside $\vec{\mathcal{E}}$, our zeroth-order state vectors have automatically become the *proper* ones for the perturbation calculation.

The second-order contributions to the energy are then simply

$$\begin{aligned}
\lambda^2 E_{lm}^{(2)} &= \sum_{l'=l\pm 1} \frac{|\langle l'm^{(0)}|H^{(1)}|lm^{(0)}\rangle|^2}{E_l^{(0)} - E_{l'}^{(0)}} = \frac{\mu_e^2 \mathcal{E}^2}{\hbar^2/2I_e} \\
&\times \left(\frac{[l^2 - m^2]/(2l+1)(2l-1)}{[l(l+1) - (l-1)l]} + \frac{[(l+1)^2 - m^2]/(2l+3)(2l+1)}{[l(l+1) - (l+1)(l+2)]} \right) \\
&= \frac{2I_e \mu_e^2 \mathcal{E}^2}{\hbar^2} \left(\frac{[l^2 - m^2]}{2l(2l+1)(2l-1)} - \frac{[(l+1)^2 - m^2]}{2(l+1)(2l+1)(2l+3)} \right) \\
&= \frac{2I_e \mu_e^2 \mathcal{E}^2}{\hbar^2} \frac{[l(l+1) - 3m^2]}{2l(l+1)(2l-1)(2l+3)}.
\end{aligned} \tag{21}$$

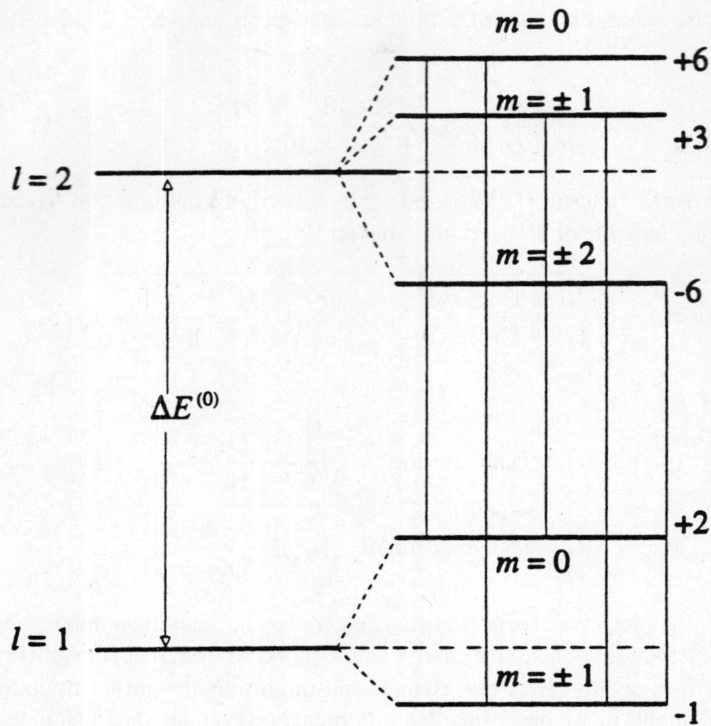

FIGURE 23.1. Second order Stark splitting of the $l = 2$ and $l = 1$ states of the diatomic molecule rigid rotator. The energy shifts are in units of $[I_e \mu_e^2 \mathcal{E}^2 / \hbar^2 l(l+1)(2l-1)(2l+3)]$. $\Delta E^{(0)} = 2\hbar^2 / I_e$.

The Stark splitting of the $l = 2$ and $l = 1$ levels is shown in Fig. 23.1., where energy shifts are shown in units of $(I_e \mu_e^2 \mathcal{E}^2 / \hbar^2 l(l+1)(2l-1)(2l+3)$. Note that the lth rotational level is split into only $(l+1)$ levels since the second order energy shift depends only on m^2. The transition probabilities for the $l = 2 \rightarrow l = 1$ transitions are now given by the matrix elements of

$$\mu_x^{(el.)} = \mu_e \sin\theta \cos\phi; \qquad \mu_y^{(el.)} = \mu_e \sin\theta \sin\phi; \qquad \mu_z^{(el.)} = \mu_e \cos\theta. \quad (22)$$

These lead to the selection rules, $\Delta m = \pm 1$ for the x and y components, and $\Delta m = 0$ for the z component. Thus the $l = 2 \rightarrow l = 1$ transition is split into five components, corresponding to the transitions $m = 2 \rightarrow 1, 1 \rightarrow 0, 1 \rightarrow 1, 0 \rightarrow 1, 0 \rightarrow 0$. (Actually, the transition would be observed through an induced absorption process). The line pattern, including relative intensities, is shown in Fig. 23.2, where the shifts in frequency from the unperturbed frequency, $(h/2\pi^2 I_e)$, are given in units of $\Delta = (I_e \mu_e^2 \mathcal{E}^2 / 2\pi\hbar^3 210)$.

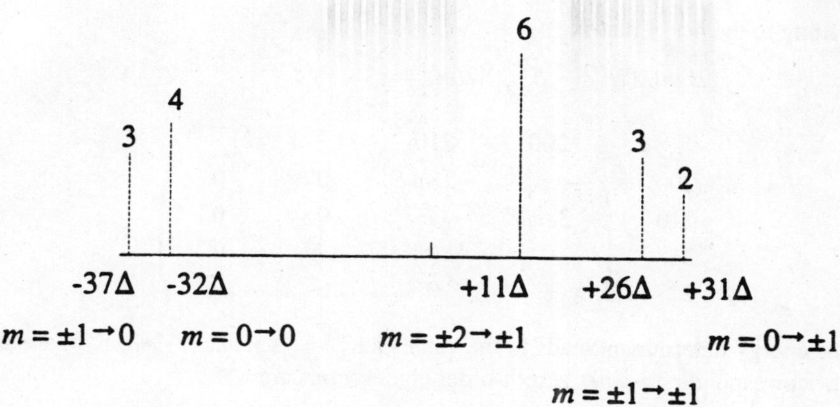

FIGURE 23.2. The Stark splitting of the $l = 2 \rightarrow l = 1$ transition. The numbers above the lines give the relative intensities. Frequency shifts from the zeroth-order frequency are given in units of $\Delta = (I_e \mu_e^2 \mathcal{E}^2 / 2\pi\hbar^3 210)$.

E Application 2: Stark Effect in the Hydrogen Atom

In an external electric field, $\vec{\mathcal{E}}$, the perturbing Hamiltonian is now

$$H^{(1)} = -(-e\vec{r}_1 + e\vec{r}_2) \cdot \vec{\mathcal{E}} = e(\vec{r}_1 - \vec{r}_2) \cdot \vec{\mathcal{E}} = ea_0 r \cos\theta \mathcal{E}, \tag{23}$$

where a_0 is the Bohr radius and we have again introduced a dimensionless r via $|\vec{r}_{\text{relative}}| = a_0 r$. For an electric field $\mathcal{E}$ of 1,000 Volts/cm, $H^{(1)}$ is again of order 10^{-5}eV, but now the zeroth-order energy is the Bohr energy $\mu e^4/\hbar^2 = 27$eV, so $|H^{(1)}/H^{(0)}| \ll 1$, the ratio now being of order 10^{-6}. Although the m quantum number is again a good quantum number to all orders, because the perturbation can not change m, several l values now exist for a given m [with the exception of the states with $m = \pm(n-1)$]. Let us take the four-fold degenerate state with $n = 2$ as a special example. Now, a nonzero $H^{(1)}$ matrix element connecting the $m = 0$ states with $l = 0$ and $l = 1$ exists,

$$\begin{aligned} \langle nlm = 210|\lambda H^{(1)}|nlm = 200\rangle &= ea_0\mathcal{E}\langle 210|r\cos\theta|200\rangle \\ &= ea_0\mathcal{E} I^{\text{rad.}}_{21,20} \frac{1}{\sqrt{3}}, \end{aligned} \tag{24}$$

where the angular part of the matrix element, with value $1/\sqrt{3}$, follows from eq. (19) and the radial part is given by

$$\begin{aligned} I^{\text{rad.}}_{21,20} &= \int_0^\infty drr^2 R_{n=2,l=1}(r) r R_{n=2,l=0}(r) \\ &= \frac{1}{4\sqrt{3}} \int_0^\infty drr^4 (1 - \frac{r}{2}) e^{-r} \\ &= \frac{1}{4\sqrt{3}} (4! - \frac{5!}{2}) = -3\sqrt{3}, \end{aligned} \tag{25}$$

leading to the 4 × 4 matrix

$$\langle 2l'm|\lambda(H^{(1)} - E^{(1)}_{n=2})|2lm\rangle =$$

$$\begin{array}{c} \\ 200 \\ 210 \\ 21+1 \\ 21-1 \end{array} \begin{array}{c} \begin{array}{cccc} 200 & 210 & 21+1 & 21-1 \end{array} \\ \begin{pmatrix} -\lambda E_2^{(1)} & -3ea_0\mathcal{E} & 0 & 0 \\ -3ea_0\mathcal{E} & -\lambda E_2^{(1)} & 0 & 0 \\ 0 & 0 & -\lambda E_2^{(1)} & 0 \\ 0 & 0 & 0 & -\lambda E_2^{(1)} \end{pmatrix} \end{array}.$$

The energy determinant leads to the values $\lambda E_2^{(1)} = +3ea_0\mathcal{E}, -3ea_0\mathcal{E}, 0$, and 0. The corresponding *proper* zeroth-order eigenvectors are

$$\begin{aligned} &\text{For} \quad +3ea_0\mathcal{E}: && \frac{1}{\sqrt{2}}(|200\rangle - |210\rangle) \\ &\text{For} \quad -3ea_0\mathcal{E}: && \frac{1}{\sqrt{2}}(|200\rangle + |210\rangle) \\ &\text{For} \quad 0: && |21+1\rangle \\ &\text{For} \quad 0: && |21-1\rangle. \end{aligned} \tag{26}$$

Because the first-order Stark effect in atomic hydrogen is very small, these first-order results may be sufficient. Because we have determined the *proper* zeroth-order eigenvectors, we could now use eq. (14) to calculate the Stark energy corrections to second order. The sum over states with $k \neq n$, however, is now a sum over an infinite number of discrete states and in fact includes a continuum sum (i.e., an integral) over the hydrogenic continuum states, because the operator $r\cos\theta$ has nonzero matrix elements connecting a state nlm to states $n'(l \pm 1)m$, with *all* possible values of $n' \neq n$. (We shall find the stretched parabolic coordinates of problems 6 and 26 will give us an elegant way out of this computational difficulty; see problem 37.)

24

The Case of Nearly Degenerate Levels

[Alternatively, the results of this section can also be used for precisely degenerate levels in case (3), when $H^{(1)}$ does not remove the degeneracy and hence does not give the *proper* or *stabilized* zeroth-order state vectors.]

If for some specific pair of levels, n and m, $(E_n^{(0)} - E_m^{(0)})$ is accidentally very small (the case of an accidental near degeneracy), particularly if $\langle m^{(0)}|H^{(1)}|n^{(0)}\rangle$ is of the same order of magnitude as $(E_n^{(0)} - E_m^{(0)})$, our perturbation theory formulae would give a very poor approximation for this pair of levels. A technique that can deal with this situation is the following: We shall make a unitary transformation on the original perturbed Hamiltonian, H, to transform it to a new Hamiltonian, H', to eliminate the off-diagonal matrix elements that connect the nearly degenerate levels to all other levels, or at least make these off-diagonal matrix elements small enough in orders of powers of λ, so they will not contribute to the energies of states n and m to some particular order in λ. An elegant way to achieve this follows in the next section.

A Perturbation Theory by Similarity Transformation

We shall try to find a unitary operator, U, generated by a hermitian operator, G, such that H is transformed into H'

$$H' = UHU^{\dagger} = U(H^{(0)} + \lambda H^{(1)} + \lambda^2 H^{(2)} + \cdots)U^{\dagger}, \qquad \text{with} \quad U = e^{i\lambda G}, \quad (1)$$

where the parameter, λ, in U is the parameter of smallness in the perturbation expansion. (The eigenvalues of a hermitian operator are invariant to similarity transformations.) In particular, if we succeed in choosing a G such that the first-order matrix elements of H' connecting states n and m to states $k \neq n, m$ are all equal to zero, the 2×2 matrix for H' in the n, m subspace will give us the energies correct to order λ^2. The surviving off-diagonal matrix elements (of order λ^2) connecting states n and m to states $k \neq n, m$ would contribute to the energies E_n and E_m only through their squares, divided by zeroth-order energy differences. The strategy then will be to find a G with matrix elements such that $\langle k^{(0)}|H'^{(1)}|n^{(0)}\rangle = 0$, and $\langle k^{(0)}|H'^{(1)}|m^{(0)}\rangle = 0$ (with similar zeros for the transposed matrix elements of $H'^{(1)}$) for all $k \neq n, m$.

$$\begin{aligned} H' &= (1 + i\lambda G - \frac{\lambda^2}{2}G^2 + \cdots)(H^{(0)} + \lambda H^{(1)} + \lambda^2 H^{(2)} + \cdots) \\ &\quad \times (1 - i\lambda G - \frac{\lambda^2}{2}G^2 + \cdots) \\ &= H^{(0)} + \lambda(H^{(1)} + i[G, H^{(0)}]) + \lambda^2(H^{(2)} + i[G, H^{(1)}] \\ &\qquad - \tfrac{1}{2}[G, [G, H^{(0)}]]) + \cdots \\ &= H^{(0)} + \lambda H'^{(1)} + \lambda^2 H'^{(2)} + \cdots \; . \end{aligned} \tag{2}$$

Now, we shall choose G such that, with $k \neq n, m$,

$$\begin{aligned} \langle k^{(0)}|H'^{(1)}|n^{(0)}\rangle &= \langle k^{(0)}|H^{(1)}|n^{(0)}\rangle + i\langle k^{(0)}|[G, H^{(0)}]|n^{(0)}\rangle = 0 \\ &= \langle k^{(0)}|H^{(1)}|n^{(0)}\rangle + i(E_n^{(0)} - E_k^{(0)})\langle k^{(0)}|G|n^{(0)}\rangle. \end{aligned} \tag{3}$$

With a similar relation for the km^{th} matrix element, this equation leads to

$$\begin{aligned} \langle k^{(0)}|G|n^{(0)}\rangle &= i\frac{\langle k^{(0)}|H^{(1)}|n^{(0)}\rangle}{(E_n^{(0)} - E_k^{(0)})}, \quad \langle k^{(0)}|G|m^{(0)}\rangle = i\frac{\langle k^{(0)}|H^{(1)}|m^{(0)}\rangle}{(E_m^{(0)} - E_k^{(0)})}, \\ \langle n^{(0)}|G|k^{(0)}\rangle &= i\frac{\langle n^{(0)}|H^{(1)}|k^{(0)}\rangle}{(E_k^{(0)} - E_n^{(0)})}, \quad \langle m^{(0)}|G|k^{(0)}\rangle = i\frac{\langle m^{(0)}|H^{(1)}|k^{(0)}\rangle}{(E_k^{(0)} - E_m^{(0)})}. \end{aligned} \tag{4}$$

All remaining matrix elements of G will be set equal to zero. In particular,

$$\langle n^{(0)}|G|n^{(0)}\rangle = \langle m^{(0)}|G|m^{(0)}\rangle = \langle n^{(0)}|G|m^{(0)}\rangle = 0. \tag{5}$$

Now,

$$\begin{aligned} &\langle n^{(0)}|H'^{(1)}|n^{(0)}\rangle = \langle n^{(0)}|H^{(1)}|n^{(0)}\rangle \\ &+ i\sum_{k\neq n,m}\Big[\langle n^{(0)}|G|k^{(0)}\rangle\langle k^{(0)}|H^{(0)}|n^{(0)}\rangle - \langle n^{(0)}|H^{(0)}|k^{(0)}\rangle\langle k^{(0)}|G|n^{(0)}\rangle\Big] \\ &= \langle n^{(0)}|H^{(1)}|n^{(0)}\rangle. \end{aligned} \tag{6}$$

Similarly,

$$\begin{aligned} \langle m^{(0)}|H'^{(1)}|m^{(0)}\rangle &= \langle m^{(0)}|H^{(1)}|m^{(0)}\rangle, \\ \langle n^{(0)}|H'^{(1)}|m^{(0)}\rangle &= \langle n^{(0)}|H^{(1)}|m^{(0)}\rangle. \end{aligned} \tag{7}$$

For

$$H'^{(2)} = H^{(2)} + i(GH^{(1)} - H^{(1)}G) - \frac{1}{2}\Big(G^2H^{(0)} - 2GH^{(0)}G + H^{(0)}G^2\Big), \tag{8}$$

let us first calculate the nm^{th} matrix element

$$\langle n^{(0)}|H'^{(2)}|m^{(0)}\rangle = \langle n^{(0)}|H^{(2)}|m^{(0)}\rangle$$
$$+\, i\sum_{k\neq n,m}\left[\langle n^{(0)}|G|k^{(0)}\rangle\langle k^{(0)}|H^{(1)}|m^{(0)}\rangle - \langle n^{(0)}|H^{(1)}|k^{(0)}\rangle\langle k^{(0)}|G|m^{(0)}\rangle\right]$$
$$-\tfrac{1}{2}\sum_{k\neq n,m}\Big[\langle n^{(0)}|G|k^{(0)}\rangle\langle k^{(0)}|G|m^{(0)}\rangle\langle m^{(0)}|H^{(0)}|m^{(0)}\rangle$$
$$-\,2\langle n^{(0)}|G|k^{(0)}\rangle\langle k^{(0)}|H^{(0)}|k^{(0)}\rangle\langle k^{(0)}|G|m^{(0)}\rangle$$
$$+\,\langle n^{(0)}|H^{(0)}|n^{(0)}\rangle\langle n^{(0)}|G|k^{(0)}\rangle\langle k^{(0)}|G|m^{(0)}\rangle\Big]$$
$$= \langle n^{(0)}|H^{(2)}|m^{(0)}\rangle + \sum_{k\neq n,m}\langle n^{(0)}|H^{(1)}|k^{(0)}\rangle\langle k^{(0)}|H^{(1)}|m^{(0)}\rangle\times$$
$$\left[\left(\frac{1}{(E_n^{(0)}-E_k^{(0)})}+\frac{1}{(E_m^{(0)}-E_k^{(0)})}\right)-\frac{1}{2}\left(\frac{E_m^{(0)}-2E_k^{(0)}+E_n^{(0)}}{(E_n^{(0)}-E_k^{(0)})(E_m^{(0)}-E_k^{(0)})}\right)\right]. \tag{9}$$

Now, using the trivial identity

$$\frac{1}{(E_n^{(0)}-E_k^{(0)})}+\frac{1}{(E_m^{(0)}-E_k^{(0)})}=\frac{(E_n^{(0)}+E_m^{(0)}-2E_k^{(0)})}{(E_n^{(0)}-E_k^{(0)})(E_m^{(0)}-E_k^{(0)})}, \tag{10}$$

and defining the average energy for the pair of levels $\bar{E}_{n,m}^{(0)} = \frac{1}{2}(E_n^{(0)} + E_m^{(0)})$, we obtain

$$\langle n^{(0)}|H'^{(2)}|m^{(0)}\rangle = \langle n^{(0)}|H^{(2)}|m^{(0)}\rangle$$
$$+\sum_{k\neq n,m}\langle n^{(0)}|H^{(1)}|k^{(0)}\rangle\langle k^{(0)}|H^{(1)}|m^{(0)}\rangle\frac{(\bar{E}_{n,m}^{(0)}-E_k^{(0)})}{(E_n^{(0)}-E_k^{(0)})(E_m^{(0)}-E_k^{(0)})}. \tag{11}$$

By setting $m = n$ in this expression, we can also immediately get the matrix element $\langle n^{(0)}|H'^{(2)}|n^{(0)}\rangle$, and similarly by setting $n = m$, we get $\langle m^{(0)}|H'^{(2)}|m^{(0)}\rangle$. With these results, the 2×2 submatrix of H' connecting the two states $|n\rangle$ and $|m\rangle$ is

$$\begin{pmatrix} H'_{nn} & H'_{nm} \\ H'_{mn} & H'_{mm} \end{pmatrix},$$

where, with an obvious shorthand matrix notation for the matrix elements, we have (through second order)

$$H'_{nn} = E_n^{(0)} + \lambda H_{nn}^{(1)} + \lambda^2\left[H_{nn}^{(2)} + \sum_{k\neq n,m}\frac{|H_{kn}^{(1)}|^2}{(E_n^{(0)}-E_k^{(0)})}\right]$$
$$H'_{nm} = \lambda H_{nm}^{(1)} + \lambda^2\left[H_{nm}^{(2)} + \sum_{k\neq n,m}\frac{H_{nk}^{(1)}H_{km}^{(1)}(\bar{E}_{n,m}^{(0)}-E_k^{(0)})}{(E_n^{(0)}-E_k^{(0)})(E_m^{(0)}-E_k^{(0)})}\right]$$
$$H'_{mn} = \lambda H_{mn}^{(1)} + \lambda^2\left[H_{mn}^{(2)} + \sum_{k\neq n,m}\frac{H_{mk}^{(1)}H_{kn}^{(1)}(\bar{E}_{n,m}^{(0)}-E_k^{(0)})}{(E_n^{(0)}-E_k^{(0)})(E_m^{(0)}-E_k^{(0)})}\right]$$
$$H'_{mm} = E_m^{(0)} + \lambda H_{mm}^{(1)} + \lambda^2\left[H_{mm}^{(2)} + \sum_{k\neq n,m}\frac{|H_{km}^{(1)}|^2}{(E_m^{(0)}-E_k^{(0)})}\right]. \tag{12}$$

To get the energies to second order, it is now only necessary to diagonalize this 2×2 matrix, leading to

$$E = E_{\pm} = \frac{1}{2}\left((H'_{nn} + H'_{mm}) \pm \sqrt{(H'_{nn} - H'_{mm})^2 + 4|H'_{nm}|^2}\right). \tag{13}$$

This result is quite general. Note: If the two levels n and m are not nearly degenerate, the result applies for a *single* level, say, the nth one, and in that case, we have simply regained the result of nondegenerate-level perturbation theory. The result also applies to a pair of exactly degenerate levels. In that case, with $\bar{E}^{(0)}_{nm} = E^{(0)}_n = E^{(0)}_m$, the off-diagonal matrix element has a term

$$\sum_{k \neq n,m} \frac{H^{(1)}_{nk} H^{(1)}_{km}}{(E^{(0)}_n - E^{(0)}_k)}.$$

This equation will be important in case (3), in which $H^{(1)}_{nm}$ is zero and does not remove the degeneracy in first order, and in the case in which $H^{(1)}$ does not lead to the *proper* zeroth-order state vectors. Finally, the diagonal and off-diagonal matrix elements given by eq. (12) can be used in the case in which the degeneracy or near degeneracy is greater than two-fold.

B An Example: Two Coupled Harmonic Oscillators with $\omega_1 \approx 2\omega_2$

Let us consider the Hamiltonian for two coupled nearly harmonic oscillators with cubic and quartic coupling terms

$$\begin{aligned} H &= \tfrac{1}{2}\hbar\omega_1(p_x^2 + x^2) + \tfrac{1}{2}\hbar\omega_2(p_y^2 + y^2) \\ &\quad + \lambda\hbar\omega_c x y^2 + \lambda^2(\hbar\omega_d x^4 + \hbar\omega_e y^4 + \hbar\omega_f x^2 y^2), \end{aligned} \tag{14}$$

where x, p_x, y, p_y, are dimensionless variables as for the 1-D oscillator. It is assumed $\omega_1 \approx 2\omega_2$. States with $|n_1 n_2\rangle$ are then nearly degenerate with states $|(n_1 - 1)(n_2 + 2)\rangle$. Using matrix elements of x, x^2, x^4 from earlier chapters, and combining these to yield, e.g.,

$$\langle (n_1 - 1)(n_2 + 2)|xy^2|n_1 n_2\rangle = \sqrt{\frac{n_1(n_2 + 1)(n_2 + 2)}{8}}, \tag{15}$$

we get, for the nearly degenerate levels $n_1 n_2 = 10$ and $n_1 n_2 = 02$, the 2×2 matrix for H':

$$\begin{aligned} H'_{10,10} &= \tfrac{3}{2}\hbar\omega_1 + \tfrac{1}{2}\hbar\omega_2 - \lambda^2 \frac{(\hbar\omega_c)^2}{\hbar\omega_1}\left(\frac{1}{8} + \frac{\omega_1}{2(\omega_1 + 2\omega_2)}\right), \\ &\quad + \lambda^2 \tfrac{1}{4}(15\hbar\omega_d + 3\hbar\omega_e + 3\hbar\omega_f), \\ H'_{10,02} &= H'_{02,10} = \tfrac{1}{2}\lambda\hbar\omega_c, \\ H'_{02,02} &= \tfrac{1}{2}\hbar\omega_1 + \tfrac{5}{2}\hbar\omega_2 - \lambda^2 \frac{(\hbar\omega_c)^2}{\hbar\omega_1}\left(\frac{25}{8} + \frac{3\omega_1}{2(\omega_1 + 2\omega_2)}\right), \end{aligned}$$

$$+\lambda^2 \tfrac{1}{4}(3\hbar\omega_d + 39\hbar\omega_e + 5\hbar\omega_f). \tag{16}$$

In particular, in this special example, $H^{(1)}$ does not contribute a second-order term to H'_{nm}, but it does contribute to the two diagonal terms, via

$$\sum_{k_1k_2\neq 10,02} \frac{|\langle k_1k_2|H^{(1)}|10\rangle|^2}{(E^{(0)}_{10} - E^{(0)}_{k_1k_2})} =$$
$$\frac{|\langle 20|H^{(1)}|10\rangle|^2}{-\hbar\omega_1} + \frac{|\langle 22|H^{(1)}|10\rangle|^2}{(-\hbar\omega_1 - 2\hbar\omega_2)} + \frac{|\langle 00|H^{(1)}|10\rangle|^2}{\hbar\omega_1}$$
$$= \frac{(\hbar\omega_c)^2}{\hbar\omega_1}\left(\frac{1}{4}\frac{1}{(-1)} + \frac{1}{8}\frac{1}{(+1)} + \frac{1}{2}\frac{\omega_1}{(-(\omega_1 + 2\omega_2))}\right) \tag{17}$$

and

$$\sum_{k_1k_2\neq 10,02} \frac{|\langle k_1k_2|H^{(1)}|02\rangle|^2}{(E^{(0)}_{02} - E^{(0)}_{k_1k_2})} = \frac{|\langle 12|H^{(1)}|02\rangle|^2}{-\hbar\omega_1} + \frac{|\langle 14|H^{(1)}|02\rangle|^2}{(-\hbar\omega_1 - 2\hbar\omega_2)}$$
$$= \frac{(\hbar\omega_c)^2}{\hbar\omega_1}\left(\frac{25}{8}\frac{1}{(-1)} + \frac{3}{2}\frac{\omega_1}{(-(\omega_1 + 2\omega_2))}\right). \tag{18}$$

This example has been chosen as a simplified model for a real near degeneracy. The linear symmetrical CO_2 molecule, with an O-C-O configuration, has three vibrational frequencies, an in-phase and an out-of-phase stretching of the two CO bonds with frequencies, named ω_1 and ω_3, and a two-fold degenerate oscillation in which the C atom moves in a direction perpendicular to the equilibrium line relative to the O–O group, where this two-fold degenerate frequency has been named ω_2. For CO_2, the three observed frequencies are

$$\frac{\hbar\omega_1}{hc} = 1351.2cm^{-1}, \qquad \frac{\hbar\omega_2}{hc} = 672.2cm^{-1}, \qquad \frac{\hbar\omega_3}{hc} = 2396.4cm^{-1}.$$

(In molecular spectroscopy, "frequencies" are usually given in "wavenumbers," i.e., in cm^{-1}, in waves per centimeter.) Note that $\hbar\omega_1 - 2\hbar\omega_2 = 6.8cm^{-1}$. This difference is much less than the experimentally deduced coupling term $\frac{1}{2}\hbar\omega_c = 50cm^{-1}$. The problem of this near degeneracy was first solved by Fermi. The near degeneracy in CO_2 is known as the Fermi resonance. (Finally, we have made our simplified Hamiltonian such that $V(y) = +V(-y)$, so it mimicks the real potential of CO_2.)

Finally, we need to have a more explicit expression for the perturbed state vectors $|n\rangle$ and $|m\rangle$. We have converted $H|n\rangle$ and $H|m\rangle$ into $UH|n\rangle = UHU^\dagger(U|n\rangle)$ and $UH|m\rangle = UHU^\dagger(U|m\rangle)$, where we now have $H' = UHU^\dagger$ acting on $|n^{(0)}\rangle = U|n\rangle$ (similar for $U|m\rangle$). Thus, we have

$$|n\rangle = U^{-1}|n^{(0)}\rangle = (1 - i\lambda G - \frac{\lambda^2}{2}G^2)|n^{(0)}\rangle, \tag{19}$$

leading to

$$|n\rangle = |n^{(0)}\rangle - i\lambda \sum_{k\neq n,m} |k^{(0)}\rangle\langle k^{(0)}|G|n^{(0)}\rangle$$

$$
\begin{aligned}
&-\frac{\lambda^2}{2}\sum_{k\neq n,m}\Big(|n^{(0)}\rangle\langle n^{(0)}|G|k^{(0)}\rangle\langle k^{(0)}|G|n^{(0)}\rangle \\
&\qquad + |m^{(0)}\rangle\langle m^{(0)}|G|k^{(0)}\rangle\langle k^{(0)}|G|n^{(0)}\rangle\Big) \\
&= |n^{(0)}\rangle\left(1-\frac{\lambda^2}{2}\sum_{k\neq n,m}\frac{|\langle k^{(0)}|H^{(1)}|n^{(0)}\rangle|^2}{(E_n^{(0)}-E_k^{(0)})^2}\right) \\
&+ |m^{(0)}\rangle\left(-\frac{\lambda^2}{2}\sum_{k\neq n,m}\frac{\langle m^{(0)}|H^{(1)}|k^{(0)}\rangle\langle k^{(0)}|H^{(1)}|n^{(0)}\rangle}{(E_n^{(0)}-E_k^{(0)})(E_m^{(0)}-E_k^{(0)})}\right) \\
&+ \lambda\sum_{k\neq n,m}|k^{(0)}\rangle\frac{\langle k^{(0)}|H^{(1)}|n^{(0)}\rangle}{(E_n^{(0)}-E_k^{(0)})},
\end{aligned}
\tag{20}
$$

with a similar expression for $|m\rangle$. The final expression for the eigenvectors associated with the energy eigenstates $|E_\pm\rangle$ of eq. (13) will be

$$
\begin{aligned}
|E_+\rangle &= c|n\rangle + s|m\rangle, \\
|E_-\rangle &= -s|n\rangle + c|m\rangle,
\end{aligned}
\tag{21}
$$

$$
\text{with}\quad \frac{c}{s} = \frac{H'_{nm}}{(E_+ - H'_{nn})}, \qquad \text{with}\quad c^2+s^2=1. \tag{22}
$$

25

Magnetic Field Perturbations

So far, we have looked at a number of perturbation problems involving an external electric field, $\vec{\mathcal{E}}$. We would like to look at similar problems involving external magnetic fields.

A The Quantum Mechanics of a Free, Charged Particle in a Magnetic Field

Classically, the Hamiltonian of a charged particle, of charge e and mass m, in a magnetic field, $\vec{B}$, derivable from a vector potential, $\vec{A}$, via $\vec{B} = \text{curl}\vec{A}$, is given by

$$H = \frac{1}{2m}(\vec{p} - \frac{e}{c}\vec{A}) \cdot (\vec{p} - \frac{e}{c}\vec{A}). \tag{1}$$

In a uniform field, $\vec{B}_0$, e.g., with $\vec{A} = \frac{1}{2}[\vec{B}_0 \times \vec{r}]$, the classical equations of motion in Hamiltonian formalism lead to

$$m\frac{d^2\vec{r}}{dt^2} = \frac{e}{c}[\vec{v} \times \vec{B}_0]. \tag{2}$$

The Schrödinger equation follows from eq. (1) via $\vec{p} \to \frac{\hbar}{i}\vec{\nabla}$. At first glance, this equation seems to be dependent on the choice of gauge of the vector potential. [With the choice of the so-called symmetric gauge for an electron in a uniform magnetic field in the z-direction, see eq. (15) below.] If $\vec{A}$ gives rise to a magnetic induction $\vec{B}$, a vector potential $\vec{A}'$ with $\vec{A}' = \vec{A} + \vec{\nabla} f$, where f is any function $f(x, y, z)$, will give rise to the *same* field $\vec{B}$. To keep the Schrödinger equation form invariant to this gauge transformation, we must gauge not only $\vec{A}$, but the Schrödinger wave function as well. The quantum-mechanical form of eq. (1) is form invariant under the gauge transformation (see problem 3),

$$\begin{aligned} \vec{A} \to \vec{A}' &= \vec{A} + \vec{\nabla} f(x, y, z), \\ \psi \to \psi' &= \psi e^{\frac{ie}{\hbar c} f(x,y,z)}. \end{aligned} \tag{3}$$

B Aharanov–Bohm Effect

The above gauge transformation was exploited by Bohm and Aharanov in a famous paper (Phys. Rev. **115** (1959) 485) to show that the quantum-mechanical wave function describing the motion of electrons can be influenced by the presence of magnetic fields, even if the electron trajectories are such that the electrons do not experience the Lorentz force $\frac{e}{c}[\vec{v} \times \vec{B}]$, if the trajectories are limited to regions in which $\vec{B} = 0$. They proposed the following experiment: An electron beam from an electron source is split into two identical beams, subsequently reflected from identical reflectors to end in a detector, as shown in Fig. 25.1. If the electrons in beam 1 are described by the wave function ψ_1 and the electrons in beam 2 are described by the wave function ψ_2, the number of electrons arriving at the common detector will be proportional to $|\psi_1 + \psi_2|^2$. Bohm and Aharanov proposed to place a tightly wound, infinitely long solenoid of small radius, a, behind the screen. A magnetic field $\vec{B}_0$ exists *inside* the solenoid, parallel to the solenoid axis, but the field outside the solenoid is precisely zero. The electrons therefore traverse only regions of space where the $\vec{B}$ field is precisely zero. (Of course, this infinitely long, tightly wound solenoid is a "theorist's" solenoid, but it can be approximated very well in the actual experiment.) Because the electrons are always in regions of zero field, choose a gauge in which the vector potential $\vec{A}'$ is also zero outside the solenoid; i.e., choose a gauge for which

$$\begin{aligned} \vec{A}' = 0 &= \vec{A} + \vec{\nabla} f(x, y, z) \\ \psi' &= \psi e^{\frac{ie}{\hbar c} f(x,y,z)}. \end{aligned} \tag{4}$$

Therefore, taking a line integral of $\vec{A}'$ around the exterior contour shown in Fig. 25.1,

$$\begin{aligned} 0 &= \oint \vec{A} \cdot d\vec{l} + \oint \vec{\nabla} f \cdot d\vec{l} \\ &= \int_0^a r dr \int_0^{2\pi} d\phi [\vec{\nabla} \times \vec{A}] \cdot \vec{n} + \int_0^{2\pi} \frac{\partial f(r, \theta, \phi)}{r \partial \phi} r d\phi \\ &= B_0 \pi a^2 + \Big(f(r, \theta, \phi = 2\pi) - f(r, \theta, \phi = 0) \Big), \end{aligned} \tag{5}$$

where we have converted the first line integral to a surface integral via Stokes's theorem. The unit vector $\vec{n}$ is parallel to the solenoid axis. The above therefore leads to

$$B_0 \pi a^2 = -\Big(f(2\pi) - f(0) \Big). \tag{6}$$

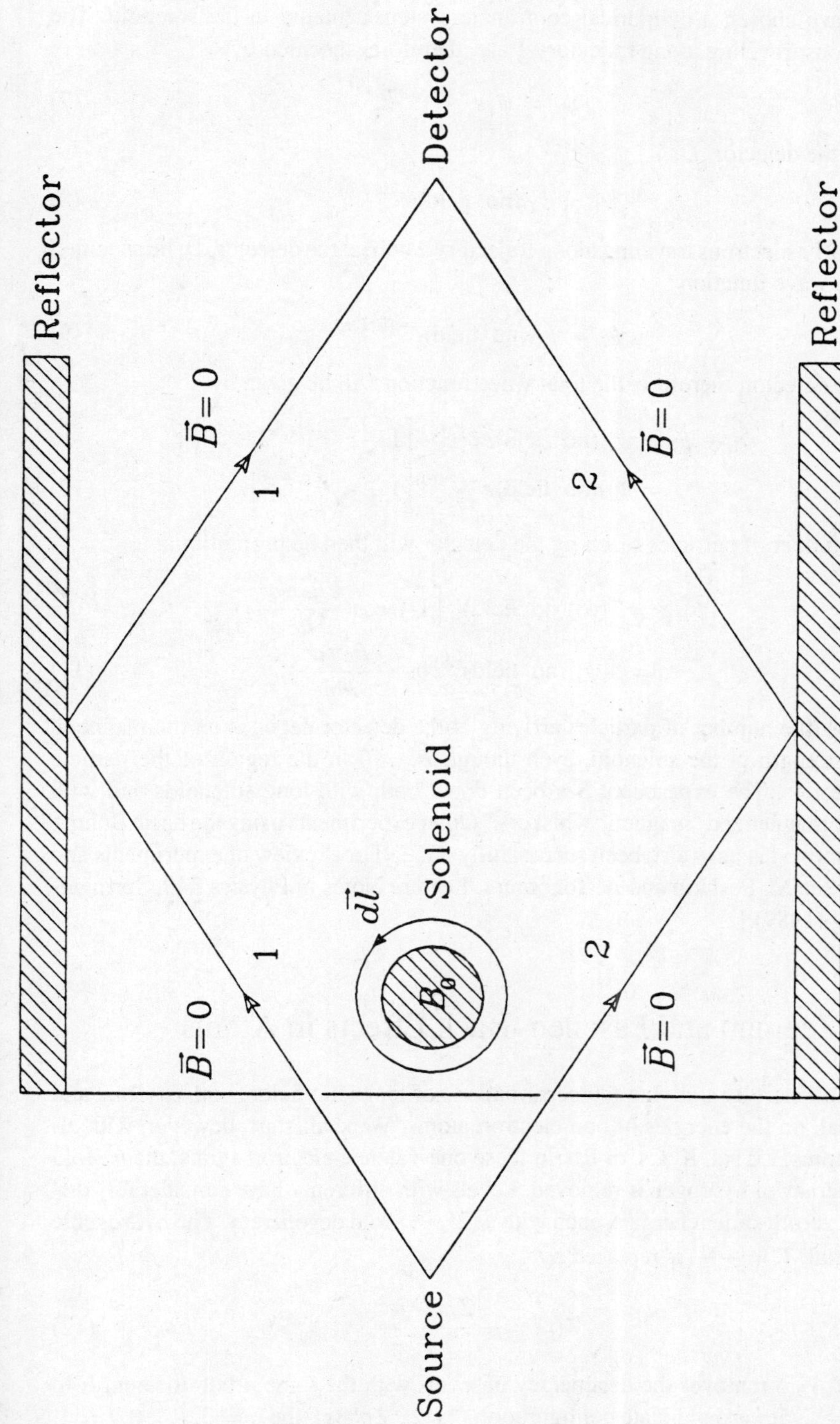

FIGURE 25.1. The Aharanov–Bohm effect

We have chosen a cylindrical coordinate system centered in the solenoid. The electrons traveling along trajectory 1 are, therefore, specified by

$$\psi_1 = \psi_1' e^{-\frac{ie}{\hbar c} f(\phi=0)}, \tag{7}$$

so, at the detector, D,

$$\psi_1 = \psi_D(\text{no field}) e^{-\frac{ie}{\hbar c} f(0)}. \tag{8}$$

Similarly, electrons traveling along trajectory 2 will, at the detector, D, be specified by the wave function

$$\psi_2 = \psi_D(\text{no field}) e^{-\frac{ie}{\hbar c} f(2\pi)}. \tag{9}$$

At the detector, therefore, the total wave function will be given by

$$\begin{aligned} \psi_1 + \psi_2 &= \psi_D(\text{no field}) e^{-\frac{ie}{\hbar c} f(0)} \Big[1 + e^{-\frac{ie}{\hbar c}(f(2\pi) - f(0))} \Big] \\ &= \psi_D(\text{no field}) e^{-\frac{ie}{\hbar c} f(0)} \Big[1 + e^{+\frac{ie}{\hbar c} B_0 \pi a^2} \Big]. \end{aligned} \tag{10}$$

The number of particles reaching the detector will then be proportional to

$$\begin{aligned} |\psi_D|^2 &= 2|\psi_D(\text{no field})|^2 \Big[1 + \cos(\frac{eB_0 \pi a^2}{\hbar c}) \Big] \\ &= 4|\psi_D(\text{no field})|^2 \cos^2(\frac{eB_0 \pi a^2}{2\hbar c}). \end{aligned} \tag{11}$$

That is, the number of particles arriving at the detector depends on the magnetic field strength in the solenoid, even though $\vec{B} = 0$ in the region of the particle trajectories. The experiment has been done, both with long solenoids and with highly magnetized "magnetic whiskers." Other experiments using the basic Bohm–Aharanov idea have also been successfully done. [For a review of experiments and ideas, see M. Peshkin and A. Tonomura, Lecture Notes in Physics **340**, Springer-Verlag (1989).]

C Zeeman and Paschen–Back Effects in Atoms

We shall start by studying the perturbations of magnetic fields, both external and internal, on the energies of one-electron atoms. We shall start, however, with alkali atoms, Li, Na, K, Cs, or Rb. In these one-valence-electron atoms, the n^2-fold degeneracy of hydrogen is removed. Levels with different l have considerably different zeroth-order energies, each with a $(2l+1)$-fold degeneracy. The hydrogenic potential, $V = -\frac{e^2}{r}$, is replaced by

$$V(r) = -\frac{Z_{\text{eff.}}(r) e^2}{r}. \tag{12}$$

The $Z_{\text{eff.}}(r)$ removes the degeneracy of levels with the same n but different l. In Na, e.g., with ground-state configuration $(1s^2 2s^2 2p^6 3s)$, the $n = 3, l = 0, l = 1,$

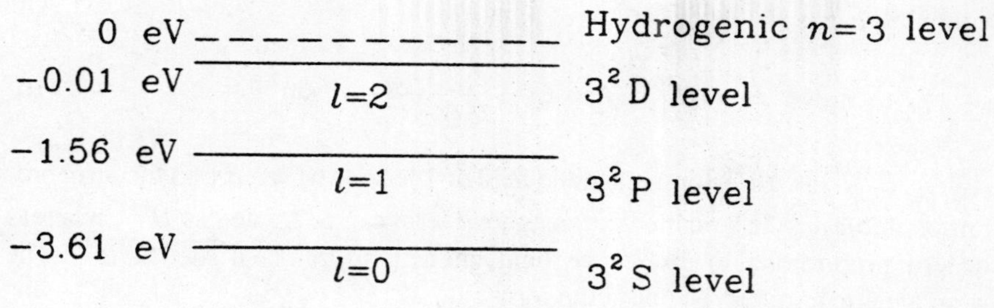

0 eV — Hydrogenic $n=3$ level
−0.01 eV — $l=2$ — 3^2D level
−1.56 eV — $l=1$ — 3^2P level
−3.61 eV — $l=0$ — 3^2S level

FIGURE 25.2. The Na-valence electron spectrum.

and $l = 2$ valence electron levels $(3^2S, 3^2P, 3^2D)$ are split in zeroth order, as shown in Fig. 25.2. The s ($l = 0$) and p ($l = 1$) orbits are penetrating orbits. They penetrate the spherically symmetric innershell electron cloud and see effectively a $Z > 1$; hence, they lie at lower energy, with the s orbit having a considerably larger $Z_{\text{eff.}}$ than the p orbit. The d ($l = 2$) electron spends most of its time outside the innershell electron cloud and thus sees an effective charge very nearly equal to $(11 - 10) = 1$. This level has been shifted to lower energy by only $-.01$ eV, relative to a purely hydrogenic value with $Z = 1$.

In a uniform, external magnetic field, $\vec{B}_0$, with $\vec{A} = \frac{1}{2}[\vec{B}_0 \times \vec{r}]$ (where we have chosen a specific gauge, the so-called symmetric gauge), the Hamiltonian (ignoring for the moment the spin of the electron) has the form

$$H = \frac{1}{2m}(\vec{p} - \frac{e}{c}\vec{A}) \cdot (\vec{p} - \frac{e}{c}\vec{A}) + V(r). \tag{13}$$

Choosing $\vec{B}_0$ along the z-direction, so

$$A_x = -\frac{1}{2}B_0 y, \qquad A_y = +\frac{1}{2}B_0 x. \tag{14}$$

$$\begin{aligned} H &= \frac{p_x^2 + p_y^2 + p_z^2}{2m} - \frac{eB_0}{2mc}(xp_y - yp_x) + \frac{m}{2}\left(\frac{eB_0}{2mc}\right)^2 (x^2 + y^2) + V(r) \\ &= \frac{p_x^2 + p_y^2}{2m} + \frac{m}{2}\omega_L^2(x^2 + y^2) + \hbar\omega_L L_z + \frac{p_z^2}{2m} + V(x, y, z), \end{aligned} \tag{15}$$

where we have used the Larmor frequency

$$\omega_L = \frac{|e|B_0}{2mc}, \qquad \text{with } \hbar\omega_L = 5.8 \times 10^{-5}\frac{B_0}{\text{tesla}} eV,$$

and we have converted to a dimensionless, orbital angular momentum operator, L_z. (Remember the electron charge is negative.) For a free particle, with $V(x, y, z) = 0$, L_z and p_z commute with the Hamiltonian of eq. (15) and can be replaced by their eigenvalues, m_l, and$\hbar k_z$. The problem of a free particle in a uniform magnetic field therefore reduces to a 2-D harmonic oscillator problem (see problem 42). For an atomic problem with a central $V(r)$, it will of course be useful to convert to spherical coordinates and dimensionless atomic units, with a physical $\vec{r} = (x, y, z)$, with $|\vec{r}| = a_0 r$, so

$$H = H^{(0)} + \hbar\omega_L L_z + \frac{1}{2}\frac{(\hbar\omega_L)^2}{(me^4/\hbar^2)} r^2 \sin^2\theta. \tag{16}$$

$(\hbar\omega_L)/(\frac{me^4}{\hbar^2}) \approx 10^{-6}$ for a magnetic field of 1 tesla = 10^4 gauss, so the term proportional to $\hbar\omega_L$, the so-called paramagnetic term can be treated as $H^{(1)}$, whereas the term proportional to $(\hbar\omega_L)^2$, the diamagnetic term, can be treated as $H^{(2)}$. The paramagnetic term can be put in the form

$$H^{(1)} = -(\vec{\mu}^{\,\text{(magn.)}}_{\text{orbital}} \cdot \vec{B}) , \qquad \text{with} \quad \vec{\mu}^{\,\text{(magn.)}}_{\text{orbital}} = \frac{e\hbar}{2mc}\vec{L} . \tag{17}$$

In 1924, when Uhlenbeck and Goudsmit postulated the existence of electron spin, with an associated spin magnetic moment, they added empirically a spin magnetic moment interaction to this paramagnetic term, but they found (empirically by fitting the predicted Zeeman spectra to the experimentally observed data) the gyromagnetic ratio for the spin magnetic moment requires an additional factor $g_s = 2$ relative to that predicted for the orbital magnetic moment, so

$$\vec{\mu}^{\,\text{(magn.)}}_{\text{spin}} = \frac{e\hbar}{2mc} g_s \vec{S}. \tag{18}$$

Although introduced empirically in 1924, the factor $g_s = 2$ comes out automatically from the Dirac relativistic quantum theory of the electron. Thus,

$$H^{(1)} = -(\vec{\mu}^{\,\text{(magn.)}}_{\text{orbital}} \cdot \vec{B}_0) - (\vec{\mu}^{\,\text{(magn.)}}_{\text{spin}} \cdot \vec{B}_0) = \hbar\omega_L(L_z + 2S_z). \tag{19}$$

This perturbation has extremely simple matrix elements in the $|nlm_l m_s\rangle$ basis, where we have added the eigenvalue m_s of the operator S_z to complete the basis, leading to the first-order magnetic field correction to the energy

$$E^{(1)}_{nlm_l m_s} = \hbar\omega_L(m_l + 2m_s). \tag{20}$$

This formula would be correct in the limit in which the external field $\vec{B}_0$ is large compared with the internal atomic magnetic fields and their effects on the spin magnetic moment. We shall look at these effects next.

D Spin-Orbit Coupling and Thomas Precession

Because of the motion of the valence electron, the electron sees an effective internal magnetic field that can interact with the spin magnetic moment of the electron. To first order in $\frac{v}{c}$, the magnetic field at the electron is

$$\vec{B} = \frac{1}{c}[\vec{\mathcal{E}} \times \vec{v}] = -\frac{1}{c}[\vec{\nabla}\Phi^{(\text{el.})} \times \frac{\vec{p}}{m}] = +\frac{1}{|e|mc}\frac{dV}{dr}\frac{1}{r}[\vec{r} \times \vec{p}], \tag{21}$$

where we have converted the electric scalar potential to the potential function $V(r)$ and have used the fact that the electron charge is negative. We could of course also think of an observer "sitting on the electron" seeing the nuclear and innerelectron charge moving with a velocity $-\vec{v}$ relative to the electron, therefore setting up a current giving rise to the magnetic field at the site of the electron (see Fig. 25.3). When the electron's spin magnetic moment interacts with this internal magnetic field, we get a new contribution to the perturbed Hamiltonian of the one-electron atom

$$H^{(1)}_{\text{spin-orbit}} = -\vec{\mu}_s \cdot \vec{B} = +\frac{|e|\hbar g_s}{2mc}\frac{1}{r}\frac{dV}{dr}\frac{\hbar}{|e|mc}(\vec{S}\cdot\vec{L}) = \frac{\hbar^2}{m^2c^2}\Big(\frac{1}{r}\frac{dV}{dr}\Big)(\vec{S}\cdot\vec{L}), \quad (22)$$

where $\vec{L}$ and $\vec{S}$ are dimensionless. Besides this magnetic spin-orbit term, a second purely relativistic correction term exists, the Thomas precession term, which has exactly the same form, but has an additional numerical factor of $-\frac{1}{2}$,

$$H^{(1)}_{\text{Thomas}} = -\frac{1}{2}H^{(1)}_{\text{spin-orbit}}. \quad (23)$$

This purely relativistic term follows because two successive Lorentz transformations along different successive directions in the orbit are equivalent to a single Lorentz transformation plus a rotation in 3-D space. This rotation causes a precession of the intrinsic spin vector of the electron, the so-called Thomas precession,

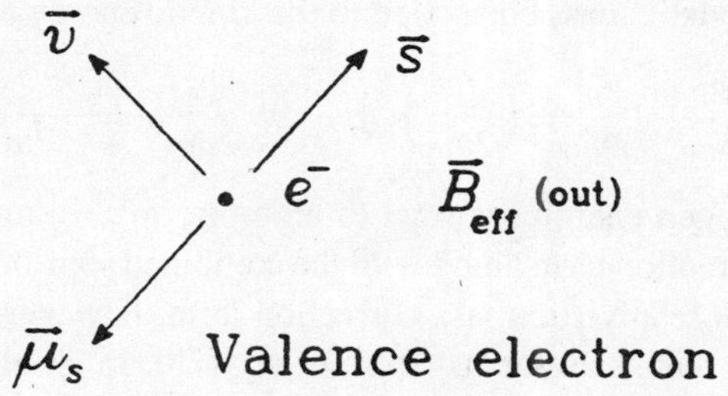

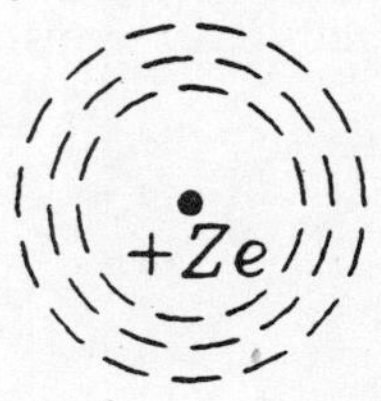

Inner shell of electron cloud

FIGURE 25.3. Model of an alkali atom.

which cancels half of the magnetic spin orbit term. Thus,

$$H^{(1)}_{\text{spin-orbit}} + H^{(1)}_{\text{Thomas}} = \frac{\hbar^2}{2m^2c^2}\left(\frac{1}{r}\frac{dV}{dr}\right)(\vec{S}\cdot\vec{L}) = \frac{\hbar^2}{2m^2c^2}\left(\frac{\bar{Z}_{\text{eff.}}e^2}{r^3}\right)(\vec{S}\cdot\vec{L}), \quad (24)$$

where $\bar{Z}_{\text{eff.}}(r) = Z_{\text{eff.}}(r) - r\frac{dZ_{\text{eff.}}}{dr}$ in an alkali atom; but $\bar{Z}_{\text{eff.}}$ can be replaced by 1 in hydrogen. Converting the physical r in this equation to a dimensionless r via $r_{\text{phys.}} = a_0 r$, we have

$$H^{(1)} = \frac{1}{2}\frac{\left(\frac{me^4}{\hbar^2}\right)^2}{mc^2}\left(\frac{\bar{Z}_{\text{eff.}}}{r^3}\right)(\vec{S}\cdot\vec{L}) = \frac{1}{2}mc^2\alpha^4\left(\frac{\bar{Z}_{\text{eff.}}}{r^3}\right)(\vec{S}\cdot\vec{L}), \quad (25)$$

where r, $\vec{S}$, and $\vec{L}$ are now all dimensionless and α is the fine structure constant. This term is of order

$$\frac{1}{mc^2}\left(\frac{me^4}{\hbar^2}\right)^2 = mc^2\left(\frac{e^2}{\hbar c}\right)^4 = mc^2\alpha^4 \approx 0.5MeV(\frac{1}{137})^4 \approx 10^{-3}eV. \quad (26)$$

To this order of magnitude, we must also consider the first-order relativistic mass correction. From

$$W = \sqrt{m_0^2c^4 + p^2c^2} = m_0c^2 + \frac{p^2}{2m_0} - \frac{p^4}{8m_0^3c^2} + \cdots \ , \quad (27)$$

we get the relativistic mass correction to the kinetic energy term

$$W - m_0c^2 = \frac{p^2}{2m_0} - \frac{1}{2m_0c^2}\left(\frac{p^2}{2m_0}\right)^2 + \cdots = \frac{p^2}{2m_0} - \frac{(E^{(0)} - V(r))^2}{2m_0c^2} + \cdots \ . \quad (28)$$

To get the hydrogen energies correct to order $mc^2\alpha^4$, we must include this relativistic mass correction term along with the combined spin-orbit and Thomas term of eq. (25). This relativistic mass correction term, however, has been converted to a function of r only in the last form of eq. (28). In an alkali atom, therefore, it can be simply absorbed into the $Z_{\text{eff.}}(r)/r$ term. This term merely establishes the zeroth-order energies of the separated $l = 0, l = 1, l = 2, \ldots$ terms. These terms are essentially taken from experiment and not calculated very precisely. We shall calculate accurately the splitting of such a $2(2l+1)$-fold degenerate term into fine structure and Zeeman components, but take the zeroth-order positions of the different l levels from experiment. For the alkali atoms, therefore, this perturbation term can be absorbed into the zeroth-order terms.

26

Fine Structure and Zeeman Perturbations in Alkali Atoms

We shall now look in detail at the fine structure (magnetic spin orbit + Thomas) and Zeeman perturbation terms; i.e., we shall diagonalize the first-order Hamiltonian,

$$H^{(1)} = \frac{mc^2\alpha^4}{2}\left(\frac{\bar{Z}_{\text{eff.}}(r)}{r^3}\right)(\vec{S}\cdot\vec{L}) + \hbar\omega_L(L_z + 2S_z), \tag{1}$$

in the $2(2l+1)$ degenerate subspace of a particular l sublevel of the valence n in an alkali atom. The fine structure term is of order 10^{-3}eV, and the Zeeman term would be of this order of magnitude only for very strong fields, of the order of $\sim$ 20 tesla, but both terms are small compared with the zeroth-order separation of different l substates. (In Na, these terms were of the order of $1-2$eV.) Matrix elements are easy to calculate in the $|nlm_lm_s\rangle$ basis. Nonzero matrix elements are

$$\langle nlm_lm_s|L_z + 2S_z|nlm_lm_s\rangle = m_l + 2m_s. \tag{2}$$

Writing $\vec{L}\cdot\vec{S} = \frac{1}{2}(L_+S_- + L_-S_+) + L_0S_0$,

$$\begin{aligned}
&\langle nlm_lm_s|\vec{L}\cdot\vec{S}|nlm_lm_s\rangle = m_lm_s,\\
&\langle nl(m_l+1)(m_s-1)|\vec{L}\cdot\vec{S}|nlm_lm_s\rangle\\
&= \tfrac{1}{2}\sqrt{(l-m_l)(l+m_l+1)(s+m_s)(s-m_s+1)},\\
&\langle nl(m_l-1)(m_s+1)|\vec{L}\cdot\vec{S}|nlm_lm_s\rangle\\
&= \tfrac{1}{2}\sqrt{(l+m_l)(l-m_l+1)(s-m_s)(s+m_s+1)},
\end{aligned} \tag{3}$$

with $s = \frac{1}{2}$. Now, if we introduce

$$\vec{J} = \vec{L} + \vec{S}, \qquad \text{with} \qquad J_z = L_0 + S_0, \tag{4}$$

where J_z has eigenvalue $m_j = m_l + m_s$, all terms of our $H^{(1)}$ do not change the quantum number m_j. (In passing, lowercase letters are usually used for *single-*

particle angular momentum quantum numbers.) Because the values $m_j = l + \frac{1}{2}$ and $m_j = -(l + \frac{1}{2})$ can each be made in only one way, whereas all other possible m_j values can be made in two ways, the full $2(2l + 1)$ matrix of $H^{(1)}$ will split into 2 (1×1) and $2l$ (2×2) submatrices. It is best to convert to m_j and $m_s = \pm\frac{1}{2}$ in the matrix element expressions, so, e.g.,

$$\langle nl, m_l' = (m_j \mp \tfrac{1}{2}), m_s = \pm\tfrac{1}{2}|\vec{L} \cdot \vec{S}|nl, m_l = (m_j \pm \tfrac{1}{2}), m_s = \mp\tfrac{1}{2}\rangle =$$
$$\tfrac{1}{2}\sqrt{(l \pm m_j + \tfrac{1}{2})(l \mp m_j + \tfrac{1}{2})1 \cdot 1} = \tfrac{1}{2}\sqrt{(l + \tfrac{1}{2})^2 - m_j^2}. \tag{5}$$

In addition, we introduce the radial matrix element integral

$$\frac{mc^2\alpha^4}{2}\int_0^\infty drr^2|R_{nl}(r)|^2 \frac{\bar{Z}_{\text{eff.}}(r)}{r^3} \equiv \beta_{nl}. \tag{6}$$

This number is common for all matrix elements of the full $2(2l + 1) \times 2(2l + 1)$ matrix of $H^{(1)}$. The general 2×2 submatrix for a given m_j is

$$\begin{pmatrix} H_{++} & H_{+-} \\ H_{-+} & H_{--} \end{pmatrix},$$

where we have used the subscript + for the state with $m_s = +\frac{1}{2}$, and $m_l = m_j - \frac{1}{2}$, and the subscript - for the state with $m_s = -\frac{1}{2}$, and $m_l = m_j + \frac{1}{2}$. In the matrix,

$$H_{++} = \tfrac{1}{2}\beta_{nl}(m_j - \tfrac{1}{2}) + \hbar\omega_L(m_j + \tfrac{1}{2}),$$
$$H_{+-} = H_{-+} = \frac{\beta_{nl}}{2}\sqrt{\left[(l + \tfrac{1}{2})^2 - m_j^2\right]},$$
$$H_{--} = -\tfrac{1}{2}\beta_{nl}(m_j + \tfrac{1}{2}) + \hbar\omega_L(m_j - \tfrac{1}{2}). \tag{7}$$

The 2×2 energy determinant leads to

$$E = \tfrac{1}{2}\left[(H_{++} + H_{--}) \pm \sqrt{(H_{++} - H_{--})^2 + 4|H_{+-}|^2}\right], \tag{8}$$

or

$$E_{\pm}^{(1)} = (-\frac{\beta_{nl}}{4} + \hbar\omega_L m_j) \pm \frac{1}{2}\sqrt{\beta_{nl}^2(l + \tfrac{1}{2})^2 + (\hbar\omega_L)^2 + 2m_j\beta_{nl}(\hbar\omega_L)}. \tag{9}$$

For the 1×1 submatrices with $m_j = \pm(l + \frac{1}{2})$, the energies are given by the diagonal matrix elements

$$E^{(1)} = \beta_{nl}\frac{l}{2} \pm \hbar\omega_L(l + 1). \tag{10}$$

With no external magnetic field, i.e., $\hbar\omega_L = 0$, we have

$$E_{nl}^{(1)} = +\beta_{nl}\frac{l}{2} \quad \text{with} \quad (2l + 2)\text{–fold degeneracy}$$
$$= -\beta_{nl}\frac{(l + 1)}{2} \quad \text{with} \quad 2l\text{–fold degeneracy}. \tag{11}$$

These two levels correspond to the j values $j = (l \pm \frac{1}{2})$, respectively. With

$$\vec{J} = \vec{L} + \vec{S}, \quad \text{so} \quad \vec{L} \cdot \vec{S} = \tfrac{1}{2}(\vec{J} \cdot \vec{J} - \vec{L} \cdot \vec{L} - \vec{S} \cdot \vec{S}), \tag{12}$$

$\vec{L} \cdot \vec{S}$ has eigenvalues

$$\tfrac{1}{2}[j(j+1) - l(l+1) - \tfrac{3}{4}],$$

leading to the eigenvalue $+\frac{1}{2}l$ for $j = (l + \frac{1}{2})$, and the eigenvalue $-\frac{1}{2}(l+1)$ for $j = (l - \frac{1}{2})$.

For the weak-field case, $\hbar\omega_L \ll \beta_{nl}$, the energies are (expanding the square roots to first order)

$$E_{nl}^{(1)} = \beta_{nl}\frac{l}{2} + \hbar\omega_L m_j\left(1 + \frac{1}{(2l+1)}\right), \qquad \text{for} \quad j = (l + \tfrac{1}{2}),$$
$$E_{nl}^{(1)} = -\beta_{nl}\frac{(l+1)}{2} + \hbar\omega_L m_j\left(1 - \frac{1}{(2l+1)}\right), \qquad \text{for} \quad j = (l - \tfrac{1}{2}). \quad (13)$$

For the huge field case, conversely, with $\hbar\omega_L \gg \beta_{nl}$,

$$E_{nl}^{(1)} = \hbar\omega_L(m_j \pm \tfrac{1}{2}) \pm \frac{\beta_{nl}}{2}m_j - \frac{\beta_{nl}}{4}. \quad (14)$$

These energies are shown as a function of the external field strength, B_0, (or $\hbar\omega_L$), for an $l = 2$ state in Fig. 26.1.

Finally, we need to find the eigenvectors as linear combinations of the two states with $m_s = +\frac{1}{2}$, $m_l = (m_j - \frac{1}{2})$, to be denoted by +, and $m_s = -\frac{1}{2}$, $m_l = (m_j + \frac{1}{2})$, to be denoted by -. In particular, for the special case with $B_0 = 0$ (hence, $\hbar\omega_L = 0$), we get

$$[\frac{\beta_{nl}}{2}(m_j - \tfrac{1}{2}) - E^{(1)}]c_+ + \frac{\beta_{nl}}{2}\sqrt{(l + \tfrac{1}{2} + m_j)(l + \tfrac{1}{2} - m_j)}c_- = 0. \quad (15)$$

With $E_{nl}^{(1)} = +\beta_{nl}l/2$, for the state with $j = (l + \frac{1}{2})$, we get

$$\frac{c_+}{c_-} = \frac{\sqrt{(l + \frac{1}{2} + m_j)}}{\sqrt{(l + \frac{1}{2} - m_j)}}. \quad (16)$$

This equation leads to the normalized coefficients

$$c_+ = \sqrt{\frac{(l + \frac{1}{2} + m_j)}{(2l+1)}}, \qquad c_- = \sqrt{\frac{(l + \frac{1}{2} - m_j)}{(2l+1)}}. \quad (17)$$

With $E_{nl} = -\beta_{nl}(l+1)/2$, i.e., for the state with $j = (l - \frac{1}{2})$, we get in the same way

$$c_+ = -\sqrt{\frac{(l + \frac{1}{2} - m_j)}{(2l+1)}}, \qquad c_- = \sqrt{\frac{(l + \frac{1}{2} + m_j)}{(2l+1)}}. \quad (18)$$

In this special case, we have calculated the transformation coefficients from a basis $|nlm_l sm_s\rangle$ that are eigenvectors of the four commuting operators

$$\vec{L} \cdot \vec{L}, \qquad L_z, \qquad \vec{S} \cdot \vec{S}, \qquad S_z,$$

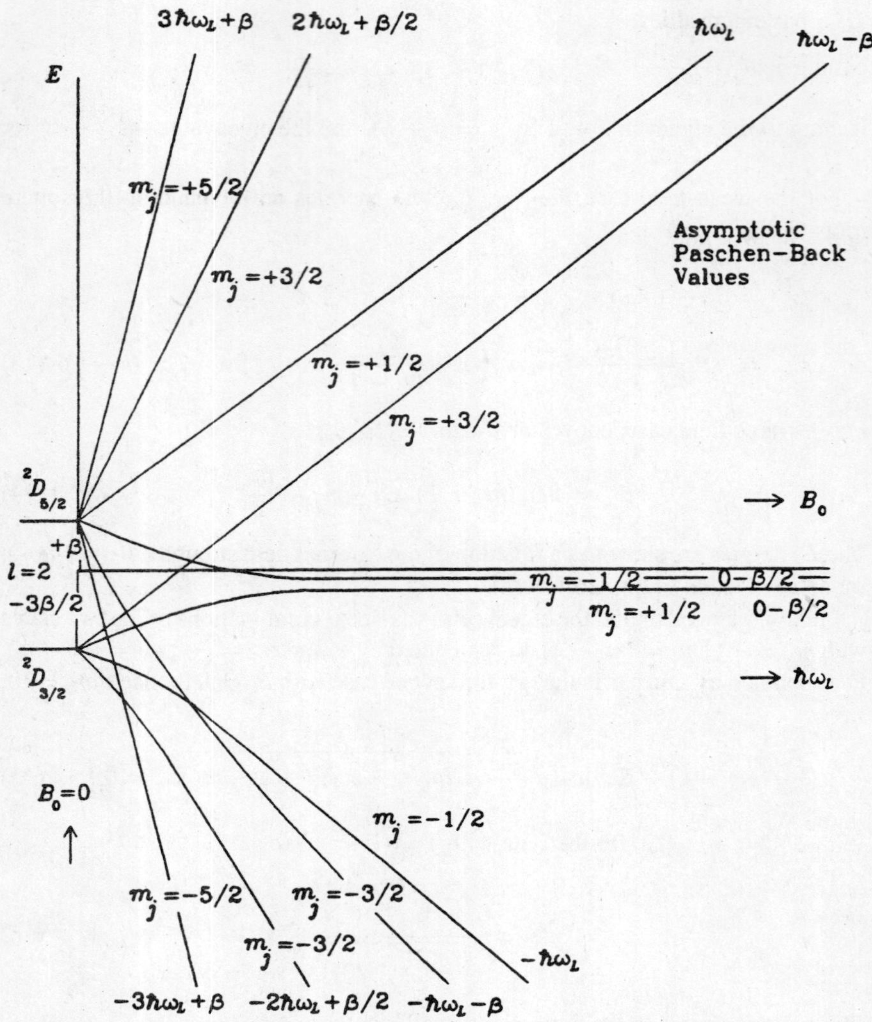

FIGURE 26.1. Magnetic field splitting of an alkali atom ^2D level.

to a basis $|nlsjm_j\rangle$ that are eigenvectors of the four commuting operators

$$\vec{L}\cdot\vec{L}, \quad \vec{S}\cdot\vec{S}, \quad \vec{J}\cdot\vec{J}, \quad J_z;$$

that is, we have calculated the transformation coefficients

$$\langle lm_l sm_s | lsjm_j\rangle.$$

These are known as angular momentum coupling coefficients or Clebsch–Gordan coefficients. For the specific case with $s = \frac{1}{2}$, these coefficients are given by

$$\begin{array}{c} \\ m_s = +\frac{1}{2} \\ m_s = -\frac{1}{2} \end{array} \begin{array}{c} \quad j = (l+\frac{1}{2}) \qquad j = (l-\frac{1}{2}) \\ \begin{pmatrix} \sqrt{\frac{(l+\frac{1}{2}+m_j)}{(2l+1)}} & -\sqrt{\frac{(l+\frac{1}{2}-m_j)}{(2l+1)}} \\ \sqrt{\frac{(l+\frac{1}{2}-m_j)}{(2l+1)}} & \sqrt{\frac{(l+\frac{1}{2}+m_j)}{(2l+1)}} \end{pmatrix} \end{array}$$

We shall study this type of transformation coefficient in a much more general way in the next chapter.

Problems

29. For a Hamiltonian (with parameter $\lambda \ll 1$),

$$H = H^{(0)} + \lambda H^{(1)} + \lambda^2 H^{(2)} + \lambda^3 H^{(3)},$$

(a) derive expressions for

$$E_n^{(3)} \qquad \text{and} \qquad Q_n^{(0)}|n^{(3)}\rangle$$

for a nondegenerate state, $|n\rangle$.

(b) Specialize the result of (a) to the Hamiltonian

$$H = H^{(0)} + \lambda V.$$

Use the Wigner–Brillouin expansion for this case, and verify the result of (a) as applied to this simpler case.

(c) For the Hamiltonian of (b), prove the second-order shift of the ground-state energy is bounded by

$$|E_{n=0}^{(2)}| \leq \frac{|\ \langle 0^{(0)}|V^2|0^{(0)}\rangle - (\langle 0^{(0)}|V|0^{(0)}\rangle)^2\ |}{(E_{n=1}^{(0)} - E_{n=0}^{(0)})}.$$

Note, $E_{n=0}^{(2)} \leq 0$.

30. The vibrating–rotating diatomic molecule. The Hamiltonian for a vibrating–rotating diatomic molecule is given by

$$-\frac{\hbar^2}{2\mu}\frac{\partial^2}{\partial r^2} - \frac{\hbar^2}{2\mu r^2}\left(\frac{1}{\sin\theta}\frac{\partial}{\partial\theta}\sin\theta\frac{\partial}{\partial\theta} + \frac{1}{\sin^2\theta}\frac{\partial^2}{\partial\phi^2}\right) + V(r)$$

$$= -\frac{\hbar^2}{2\mu}\frac{\partial^2}{\partial r^2} + V(r) + \frac{\hbar^2}{2\mu r^2}\left(\frac{(\vec{J}\cdot\vec{J})}{\hbar^2}\right).$$

Assume $V(r)$ can be approximated by the quadratic term

$$V(r) = \frac{\mu\omega^2}{2}(r - r_e)^2, \qquad \text{with } (r - r_e) = \sqrt{\frac{\hbar}{\mu\omega}}x.$$

If the $1/r^2$ term of the angular part is expanded as

$$\frac{1}{r^2} = \frac{1}{r_e^2}\left(1 - 2\frac{(r - r_e)}{r_e} + 3\frac{(r - r_e)^2}{r_e^2} + \cdots,\right.$$

the above Hamiltonian can be written as

$$H = H^{(0)}_{\text{vibrator}} + H^{(0)}_{\text{rotator}} + \lambda H^{(1)} + \lambda^2 H^{(2)} + \cdots,$$

where

$$H^{(0)}_{\text{vib.}} = \frac{\hbar\omega}{2}\left(-\frac{\partial^2}{\partial x^2} + x^2\right), \qquad H^{(0)}_{\text{rot.}} = \frac{\hbar^2}{2\mu r_e^2}\left(\frac{(\vec{J}\cdot\vec{J})}{\hbar^2}\right),$$

with

$$H^{(0)}|nJM\rangle = \left(\hbar\omega(n + \tfrac{1}{2}) + \frac{\hbar^2 J(J+1)}{2\mu r_e^2}\right)|nJM\rangle.$$

(Note: We have used capital letters, $\vec{J}$, and J, M, in place of the l, m, used earlier for the diatomic molecule rigid rotator. This is in agreement with the convention that lowercase letters are reserved for the single-particle angular momentum quantum numbers, whereas capital letters are used for the angular momentum quantum numbers of many-particle systems.) Take the parameter of smallness, λ, as

$$\lambda = \sqrt{\frac{\hbar^2}{2\mu r_e^2}\frac{1}{\hbar\omega}},$$

so the vibration–rotation interaction terms are perturbations, with

$$H^{(1)} = -\frac{\hbar^2}{2\mu r_e^2}2\sqrt{2}x\left(\frac{(\vec{J}\cdot\vec{J})}{\hbar^2}\right), \qquad H^{(2)} = \frac{\hbar^2}{2\mu r_e^2}6x^2\left(\frac{(\vec{J}\cdot\vec{J})}{\hbar^2}\right).$$

[To investigate the smallness of the parameter, λ, take the HCl molecule as a typical example. For HCl, (with Cl isotope 35), $\hbar\omega = hc(2989.74\text{cm}^{-1})$; $\hbar^2/(2\mu r_e^2) = hc(10.5909\text{cm}^{-1})$, so $\lambda = .06$.]

Show that the $|nJM\rangle$, with $\langle\vec{r}|nJM\rangle = \psi_n(x)Y_{JM}(\theta,\phi)$, are "proper" zeroth-order eigenvectors, and find corrections to the zeroth-order energies, including terms of order $\lambda^2(\hbar^2/2\mu r_e^2)$.

Warning: We have taken as our zeroth-order Hamiltonian both the soluble vibrator and rotator Hamiltonians. Because their coefficients differ by the factor, λ^2, second-order perturbation theory will here give both terms of order $\lambda^2(\hbar\omega)$ and terms of order $\lambda^2(\hbar^2/2\mu r_e^2)$.

31. A diatomic molecule rigid rotator with a permanent electric dipole moment along the molecule axis is placed in a *non*uniform electric field, $\vec{\mathcal{E}}$, so the zeroth-order rigid rotator solutions are perturbed by

$$\lambda H^{(1)} = -\vec{\mu}^{(\text{el.})}\cdot\vec{\mathcal{E}} = 2k\sin^2\theta\sin 2\phi, \qquad \text{with } k = \lambda\left(\frac{\hbar^2}{2I_e}\right); \quad \lambda \ll 1.$$

For the states with $J = 0$, and $J = 1$, find the perturbed energies correct to order $\lambda^2(\hbar^2/2I_e)$, and show how the $J = 0 \to J = 1$ absorption transition is split by this perturbation. Give the relative intensities of the Stark fine structure components of this transition. (Calculate relative intensities only in dominant, zeroth-order approximation.)

32. A diatomic molecule rigid rotator with a permanent electric dipole moment is placed in a uniform electric field in the x-direction, so

$$\lambda H^{(1)} = -\vec{\mu}^{(\text{el.})} \cdot \vec{\mathcal{E}} = -\mu_e^{(\text{el.})} |\mathcal{E}| \sin\theta \cos\phi.$$

Because the Stark energy is independent of the direction of the electric field, you know the result for the second-order energy correction (see Chapter 23). To test yourself on your knowledge of perturbation theory, find the second-order corrections for the energies of the $J = 1$ states by using the above $H^{(1)}$ and a zeroth-order basis in which $\vec{J}^2$ and J_z, perpendicular to the direction of $\vec{\mathcal{E}}$, are diagonal. This method is admittedly the hard way to do an easy problem. The $|JM^{(0)}\rangle$ are now not proper zeroth-order eigenvectors, because the above $H^{(1)}$ now connects states with different M. Also, $H^{(1)}$ now does not remove the zeroth-order degeneracy in first order. We are therefore dealing with case (3), as enumerated in Chapter 23. Hint: The best way to solve such a problem is with the use of our formulae for three nearly degenerate (or exactly degenerate) states and diagonalize the 3×3 H'-matrix for $J = 1$ which to order λ^2 is effectively disconnected from states with $J' \neq J$.

33. For a slightly asymmetric rotator, with

$$H = H^{(0)} + \lambda H^{(1)} = \tfrac{1}{2}(a+b)\vec{P}^2 + \tfrac{1}{2}(2c-a-b)P_{z'}^2 + \tfrac{1}{4}(a-b)(P'_+P'_+ + P'_-P'_-),$$

use the parameter of smallness, $\lambda \ll 1$,

$$\lambda = \frac{(a-b)}{(2c-a-b)}, \qquad \text{with } \lambda H^{(1)} = \tfrac{1}{4}(2c-a-b)\lambda(P'_+P'_+ + P'_-P'_-).$$

Note,

$$E_{JK}^{(0)} = \tfrac{1}{2}\Big[(a+b)J(J+1) + (2c-a-b)K^2\Big].$$

To order λ^2, zeroth-order states with $K = 0$ and $|K| \geq 3$ can be treated by nondegenerate perturbation theory. These states effectively belong to case (1), as enumerated in Chapter 23. For these states, find E_{JK} correct to second order as general functions of J and K. For states with $|K| = 1$ and $|K| = 2$, use degenerate-level perturbation theory. Show, in particular, states with $|K| = 1$ belong to case (2), as enumerated in Chapter 23, but states with $|K| = 2$ belong to case (3). For the latter, therefore, use the treatment for two nearly (or exactly) degenerate levels to find the energies correct to order λ^2. Use the results of problems 20 and 21 to expand the exact expressions for these energies in powers of λ for states with $J = 2$ and $J = 3$ to compare with the perturbation theory results. Also, verify your perturbation theory results give the correct values for $J = 1$.

34. Two identical diatomic units on opposite ends of a long-chain molecule are constrained to move on identical parallel circles of equal radius, but are almost free to rotate on these circles of equal radius, so they are subject to a Hamiltonian

$$H = H^{(0)} + \lambda H^{(1)} = -\frac{\hbar^2}{2I_e}\left(\frac{\partial^2}{\partial\phi_1^2} + \frac{\partial^2}{\partial\phi_2^2}\right) + V_0\cos(\phi_1 - \phi_2),$$

where I_e is a constant and $V_0 = \lambda(\hbar^2/2I_e)$, with $\lambda \ll 1$, so the V_0 term can be treated as a perturbation. The zeroth-order solutions are

$$E^{(0)}_{m_1m_2} = \frac{\hbar^2}{2I_e}(m_1^2 + m_2^2), \qquad \text{with } \psi^{(0)}_{m_1m_2}(\phi_1, \phi_2) = \frac{1}{2\pi}e^{im_1\phi_1}e^{im_2\phi_2},$$

with

$$m_1 = 0, \pm1, \pm2, \ldots, \qquad m_2 = 0, \pm1, \pm2, \ldots.$$

Enumerate all states with zeroth-order energies, $E^{(0)} \leq 5(\hbar^2/2I_e)$, and find their degeneracies. Indicate which states belong to cases (1), (2), or (3) of degenerate-state perturbation theory, and find the perturbed energies for the above states correct to order $\lambda^2(\hbar^2/2I_e)$. [An alternative method of solution: Make use of the symmetry of the Hamiltonian to find the *proper* zeroth-order wave functions as linear combinations of the above $\psi^{(0)}_{m_1m_2}$, and show that these *proper* zeroth-order wave functions reduce the calculation of the degenerate states to case (1) automatically.]

Solution for Problem 34

The Schrödinger equation in the dimensionless quantities, $\epsilon = E/(\hbar^2/2I_e)$ and $\lambda = V_0/(\hbar^2/2I_e)$, has the simple form

$$-\left(\frac{\partial^2}{\partial\phi_1^2} + \frac{\partial^2}{\partial\phi_2^2}\right)\psi(\phi_1, \phi_2) + \lambda\cos(\phi_1 - \phi_2)\psi(\phi_1, \phi_2) = \epsilon\psi(\phi_1, \phi_2), \tag{1}$$

with zeroth-order solutions

$$\psi^{(0)}(\phi_1, \phi_2) = \frac{1}{2\pi}e^{im_1\phi_1}e^{im_2\phi_2}, \qquad \text{with} \quad \epsilon^{(0)} = (m_1^2 + m_2^2). \tag{2}$$

The needed matrix elements of $H^{(1)} = \cos(\phi_1 - \phi_2)$ are extremely simple in this basis. The only nonzero matrix elements are

$$\begin{aligned}\langle(m_1+1)(m_2-1)|H^{(1)}|m_1m_2\rangle &= \tfrac{1}{2},\\ \langle(m_1-1)(m_2+1)|H^{(1)}|m_1m_2\rangle &= \tfrac{1}{2}.\end{aligned} \tag{3}$$

All other matrix elements are zero. Note: $H^{(1)}$ is diagonal in the quantum number $M = (m_1 + m_2)$, so this quantum number is conserved to all orders of the perturbation. We list in the table below the possible quantum numbers m_1, m_2, M, as well as the perturbation type and the total degeneracy number, g_ϵ, for all states with $\epsilon^{(0)} \leq 5$. Degeneracies of 4 and 8 are common to most of the $\epsilon^{(0)}$ for this problem. Because $H^{(1)}$, however, does not connect states of different total M, most of the states of this system are effectively doubly degenerate or nondegenerate.

Among the states with a single m_1, m_2-combination for a fixed M are the states with $\epsilon^{(0)} = 2$ and $M = +2$, or $M = -2$, as well as the ground state. Also, a double action with $H^{(1)}$ on a state $|m_1 m_2\rangle$ can convert it only to states with

$$m_1, m_2 \rightarrow m_1', m_2' = (m_1 + 2), (m_2 - 2) \quad \text{or} \quad m_1, m_2 \quad \text{or} \quad (m_1 - 2), (m_2 + 2).$$

Therefore, states such as the pair of states with $\epsilon^{(0)} = 5$, $M = +1$, with $m_1, m_2 = +2, -1$, or $-1, +2$, with $\Delta m_1, \Delta m_2 = -3, +3$, are unconnected through second order and can thus effectively be treated as if they were nondegenerate (if only corrections through second order are significant). These states are therefore listed as belonging to type (1), using the characterization of degenerate states given in Chapter 23. For type (1) states, the m_1, m_2 basis is effectively the *proper* basis, and the states can be treated as if they were nondegenerate.

$\epsilon^{(0)}$	g_ϵ	m_1	m_2	M	type
0	1	0	0	0	(1)
1	4	+1	0	+1	(2)
		0	+1	+1	(2)
		−1	0	−1	(2)
		0	−1	−1	(2)
2	4	+1	+1	+2	(1)
		+1	−1	0	(3)
		−1	+1	0	(3)
		−1	−1	−2	(1)

$\epsilon^{(0)}$	g_ϵ	m_1	m_2	M	type
4	4	+2	0	+2	(3)
		0	+2	+2	(3)
		−2	0	−2	(3)
		0	−2	−2	(3)
5	8	+2	+1	+3	(2)
		+1	+2	+3	(2)
		−2	−1	−3	(2)
		−1	−2	−3	(2)
		+2	−1	+1	(1)
		−1	+2	+1	(1)
		−2	+1	−1	(1)
		+1	−2	−1	(1)

Doubly degenerate states with a $\Delta m_1, \Delta m_2 = \pm 1, \mp 1$, conversely, have their degeneracy removed in first order. For such states, diagonalization of $H^{(1)}$ to find the $\epsilon^{(1)}$ will also yield the *proper* linear combinations of the zeroth-order eigenvectors to carry forward the higher order perturbations. Such states are characterized as type (2) according to the catalog of Chapter 23.

Finally, doubly degenerate states with a $\Delta m_1, \Delta m_2$ of $\pm 2, \mp 2$ have their degeneracy removed only in second order. Such states are best treated by transforming the 2×2 Hamiltonian matrix H into a new $H' = UHU^\dagger$, as in Chapter 24 [see, in particular, eq. (12) of Chapter 24]. These are states characterized as type (3) in Chapter 23.

For states of type (1), we get the second-order corrections, $\epsilon^{(2)}$, via

$$\epsilon^{(2)} = \frac{|\langle (m_1+1)(m_2-1)|H^{(1)}|m_1 m_2\rangle|^2}{(\epsilon^{(0)}_{m_1 m_2} - \epsilon^{(0)}_{(m_1+1)(m_2-1)})} + \frac{|\langle (m_1-1)(m_2+1)|H^{(1)}|m_1 m_2\rangle|^2}{(\epsilon^{(0)}_{m_1 m_2} - \epsilon^{(0)}_{(m_1-1)(m_2+1)})}.$$

Thus, for the state with $\epsilon^{(0)} = 0$,

$$\epsilon^{(2)} = \tfrac{1}{4}\left(\frac{1}{(0-2)} + \frac{1}{(0-2)}\right) = -\tfrac{1}{4}.$$

For the state with $\epsilon^{(0)} = 2$, $M = +2$; and similarly for $M = -2$,

$$\epsilon^{(2)} = \tfrac{1}{4}\left(\frac{1}{(2-4)} + \frac{1}{(2-4)}\right) = -\tfrac{1}{4}.$$

For the state with $\epsilon^{(0)} = 5, m_1, m_2 = +2, -1, M = +1$ (similarly for the other state with $M = +1$, and the two states with $M = -1$),

$$\epsilon^{(2)} = \tfrac{1}{4}\left(\frac{1}{(5-13)} + \frac{1}{(5-1)}\right) = +\tfrac{1}{32}.$$

For states of type (2), we first diagonalize $H^{(1)}$. For example, for the states with $\epsilon^{(0)} = 1, M = +1$, the 2×2 matrix for $(H^{(1)} - \epsilon^{(1)})$ is

$$\begin{array}{c c} & \begin{array}{cc} +1\,0 & 0+1 \end{array} \\ \begin{array}{c} +1\,0 \\ 0+1 \end{array} & \begin{pmatrix} -\epsilon^{(1)} & \frac{1}{2} \\ \frac{1}{2} & -\epsilon^{(1)} \end{pmatrix}, \end{array}$$

with eigenvalues and eigenvectors, given by

$$\epsilon^{(1)} = +\tfrac{1}{2}; \quad |M = +1, (+)\rangle = \tfrac{1}{\sqrt{2}}\big(|+10\rangle + |0+1\rangle\big),$$

$$\epsilon^{(1)} = -\tfrac{1}{2}; \quad |M = +1, (-)\rangle = \tfrac{1}{\sqrt{2}}\big(|+10\rangle - |0+1\rangle\big),$$

where, now,

$$\epsilon^{(2)} = \sum_{m_1' m_2'}{}' \frac{|\langle m_1' m_2' | H^{(1)} | M = +1, (\pm)\rangle|^2}{(1 - \epsilon^{(0)}_{m_1' m_2'})}, \tag{4}$$

and the primed sum excludes the states with $\epsilon^{(0)} = 1$. For the $M = +1$ state, therefore, $m_1' m_2'$ can only take the values $m_1' m_2' = +2, -1$ [with only the $+10$ piece of $|M = +1(\pm)\rangle$ contributing to the matrix element], and $m_1' m_2' = -1, +2$, [with only the $0 + 1$ piece of $|M = +1(\pm)\rangle$ contributing to the matrix element]. Thus,

$$\epsilon^{(2)}_{(\pm)} = |\tfrac{1}{2\sqrt{2}}|^2 \frac{1}{(1-5)} + |\pm \tfrac{1}{2\sqrt{2}}|^2 \frac{1}{(1-5)} = -\frac{1}{16},$$

with the same result for the two states with $M = -1$. The 2×2 matrix $(H^{(1)} - \epsilon^{(1)})$, has exactly the same form for any pair of states of type (2), so $\epsilon^{(1)} = \pm\frac{1}{2}$ for all type (2) states. For states with $\epsilon^{(0)} = 5, M = +3$, we have

$$\epsilon^{(2)}_{(\pm)} = |\frac{1}{2\sqrt{2}}|^2 \frac{1}{(5-9)} + |\pm \frac{1}{2\sqrt{2}}|^2 \frac{1}{(5-9)} = -\frac{1}{16}.$$

Finally, for states of type (3), the matrix for $H' = UHU^\dagger$ with zeroth-order basis states $|m_1 m_2\rangle$ and $|(m_1 - 2)(m_2 + 2)\rangle$ now has the matrix elements

$$\langle m_1 m_2 | H' | m_1 m_2\rangle = \sum_{m_1' m_2'}{}' \frac{|\langle m_1' m_2' | H^{(1)} | m_1 m_2\rangle|^2}{(\epsilon^{(0)}_{m_1 m_2} - \epsilon^{(0)}_{m_1' m_2'})},$$

$$\langle m_1 m_2 | H' | (m_1 - 2)(m_2 + 2)\rangle = \langle (m_1 - 2)(m_2 + 2) | H' | m_1 m_2\rangle =$$

$$\frac{\langle m_1m_2|H^{(1)}|(m_1-1)(m_2+1)\rangle\langle(m_1-1)(m_2+1)|H^{(1)}|(m_1-2)(m_2+2)\rangle}{(\epsilon^{(0)}_{m_1m_2}-\epsilon^{(0)}_{(m_1-1)(m_2+1)})},$$

$$\langle(m_1-2)(m_2+2)|H'|(m_1-2)(m_2+2)\rangle=\sum_{m_1'm_2'}{}'\frac{|\langle m_1'm_2'|H^{(1)}|(m_1-2)(m_2+2)\rangle|^2}{(\epsilon^{(0)}_{m_1m_2}-\epsilon^{(0)}_{m_1'm_2'})},$$

where the primed sums again exclude the states with $\epsilon^{(0)}_{m_1'm_2'}=\epsilon^{(0)}_{m_1m_2}$.

For states with $\epsilon^{(0)}=2$, $M=0$, therefore, the matrix $(H'^{(2)}-\epsilon^{(2)})$ is

$$\begin{array}{c} \\ +1-1 \\ -1+1 \end{array}\begin{array}{c} \begin{array}{cc} +1-1 & \qquad\qquad -1+1 \end{array} \\ \begin{pmatrix} \frac{1}{4}\left(\frac{1}{(2-8)}+\frac{1}{(2-0)}\right)-\epsilon^{(2)} & \frac{1}{4}\frac{1}{(2-0)} \\ \frac{1}{4}\frac{1}{(2-0)} & \frac{1}{4}\left(\frac{1}{(2-8)}+\frac{1}{(2-0)}\right)-\epsilon^{(2)} \end{pmatrix}\end{array}=$$

$$\begin{array}{c} \\ +1-1 \\ -1+1 \end{array}\begin{array}{c} \begin{array}{cc} +1-1 & -1+1 \end{array} \\ \begin{pmatrix} \frac{1}{12}-\epsilon^{(2)} & \frac{1}{8} \\ \frac{1}{8} & \frac{1}{12}-\epsilon^{(2)} \end{pmatrix}\end{array},$$

with eigenvalues and eigenvectors

$$\epsilon^{(2)}=+\frac{5}{24},\qquad |M=0(+)\rangle=\tfrac{1}{\sqrt{2}}(|+1,-1\rangle+|-1,+1\rangle),$$

$$\epsilon^{(2)}=-\frac{1}{24},\qquad |M=0(-)\rangle=\tfrac{1}{\sqrt{2}}(|+1,-1\rangle-|-1,+1\rangle).$$

Finally, for type (3) states with $\epsilon^{(0)}=4$ and $M=+2$,(or $M=-2$), we also have $\epsilon^{(2)}=+5/24,-1/24$, with similar $(\pm)$ eigenvectors.

Through second order, therefore, the five lowest zeroth-order energy states of our problem are split into 11 energies, with

$$\begin{aligned}
\epsilon&=0-\tfrac{1}{4}\lambda^2, && g_\epsilon=1,\\
\epsilon&=1+\tfrac{1}{2}\lambda-\tfrac{1}{16}\lambda^2, && g_\epsilon=2\quad\text{with } M=\pm1,\\
\epsilon&=1-\tfrac{1}{2}\lambda-\tfrac{1}{16}\lambda^2, && g_\epsilon=2\quad\text{with } M=\pm1,\\
\epsilon&=2-\tfrac{1}{4}\lambda^2, && g_\epsilon=2\quad\text{with } M=\pm2,\\
\epsilon&=2+\tfrac{5}{24}\lambda^2, && g_\epsilon=1\quad\text{with } M=0\ (+),\\
\epsilon&=2-\tfrac{1}{24}\lambda^2, && g_\epsilon=1\quad\text{with } M=0\ (-),\\
\epsilon&=4+\tfrac{5}{24}\lambda^2, && g_\epsilon=2\quad\text{with } M=\pm2,\\
\epsilon&=4-\tfrac{1}{24}\lambda^2, && g_\epsilon=2\quad\text{with } M=\pm2,\\
\epsilon&=5+\tfrac{1}{2}\lambda-\tfrac{1}{16}\lambda^2, && g_\epsilon=2\quad\text{with } M=\pm3,\\
\epsilon&=5-\tfrac{1}{2}\lambda-\tfrac{1}{16}\lambda^2, && g_\epsilon=2\quad\text{with } M=\pm3,\\
\epsilon&=5+\tfrac{1}{32}\lambda^2, && g_\epsilon=4\quad\text{with } (M=\pm1)^2.
\end{aligned}\tag{5}$$

Although the matrix elements of our $H^{(1)}$ were extremely simple in the $|m_1m_2\rangle$ basis, the method seems to be somewhat complicated because we had to pay attention to the perturbation type. Note, We could have written, however, a general matrix for the transformed Hamiltonian, $H' = UHU^\dagger$, for which the 2×2 matrix for the general case would split into two 1×1 matrices for states of type (1) (see e.g., the states with $\epsilon^{(0)} = 5$, $M = +1$). Also, by including the off-diagonal matrix elements of $H^{(1)}$, the full 2×2 matrix for H' at once gives both the eigenvalues through second order and the eigenvectors as the correct linear combination of the zeroth-order state vectors. For example, for the most general case of type (2), the 2×2 matrix for $(H' - \epsilon^{(0)}_{m_1(m_1+1)})$ is

$$(H' - \epsilon^{(0)}\mathbf{1}) = \begin{array}{c} \\ (m_1+1)m_1 \\ m_1(m_1+1) \end{array} \begin{array}{cc} (m_1+1)m_1 & m_1(m_1+1) \\ \left(\begin{array}{c} -\frac{1}{16}\lambda^2 \\ \frac{1}{2}\lambda \end{array}\right. & \left.\begin{array}{c} \frac{1}{2}\lambda \\ -\frac{1}{16}\lambda^2 \end{array}\right) \end{array},$$

where we have used the fact that $\epsilon^{(0)}_{m_1(m_1+1)} - \epsilon^{(0)}_{(m_1-1)(m_1+2)} = -4$ is independent of m_1. The above matrix leads to the eigenvalues $\pm\frac{1}{2}\lambda - \frac{1}{16}\lambda^2$, with eigenvectors

$$\frac{1}{\sqrt{2}}(|(m_1+1)m_1\rangle \pm |m_1(m_1+1)\rangle),$$

as seen in the special cases above.

Similarly, for the most general case of type (3), the 2×2 matrix for $(H' - \epsilon^{(0)}_{m_1(m_1+2)})$ is

$$(H' - \epsilon^{(0)}\mathbf{1}) = \begin{array}{c} \\ (m_1+2)m_1 \\ m_1(m_1+2) \end{array} \begin{array}{cc} (m_1+2)m_1 & m_1(m_1+2) \\ \left(\begin{array}{c} +\frac{1}{12}\lambda^2 \\ +\frac{1}{8}\lambda^2 \end{array}\right. & \left.\begin{array}{c} \frac{1}{8}\lambda^2 \\ +\frac{1}{12}\lambda^2 \end{array}\right) \end{array},$$

where we have used the fact that $\epsilon^{(0)}_{m_1(m_1+2)} - \epsilon^{(0)}_{(m_1-1)(m_1+3)} = -6$ and $\epsilon^{(0)}_{m_1(m_1+2)} - \epsilon^{(0)}_{(m_1+1)(m_1+1)} = +2$, are both independent of m_1. Diagonalization of this matrix leads to eigenvalues, $(\frac{1}{12} \pm \frac{1}{8})\lambda^2$, with eigenvectors

$$\frac{1}{\sqrt{2}}(|m_1(m_1+2)\rangle \pm |(m_1+2)m_1\rangle),$$

as seen in the special type (3) cases above.

The general 2×2 matrix for $(H' - \epsilon^{(0)}_{m_1(m_1+n)})$, with $n \geq 3$, will factor into two 1×1 matrices

$$(H' - \epsilon^{(0)}\mathbf{1}) = \begin{array}{c} \\ (m_1+n)m_1 \\ m_1(m_1+n) \end{array} \begin{array}{cc} (m_1+n)m_1 & m_1(m_1+n) \\ \left(\begin{array}{c} +\frac{1}{4}\lambda^2\frac{1}{(n^2-1)} \\ 0 \end{array}\right. & \left.\begin{array}{c} 0 \\ +\frac{1}{4}\lambda^2\frac{1}{(n^2-1)} \end{array}\right) \end{array},$$

again in agreement with our special case above, with $\epsilon^{(0)}_{-1,2} = 5$, with $n = 3$. In this case, the degeneracy is not removed through second order. The *proper* linear combination of zeroth-order state vectors would be discovered only in higher order of perturbation theory

Alternative Method: Symmetry-Adapted Eigenfunctions

Although the diagonalization of the H' matrices is extremely simple, the question arises: Is there a simpler way of discovering the *proper* linear combination of zeroth-order state vectors that would automatically reduce the problem to nondegenerate perturbation theory. In general, degeneracies (unless "accidental") arise because of some underlying symmetry. Sometimes, of course, this symmetry may be very sophisticated and not so easy to discover. Therefore, if we have used the wrong coordinates for our Schrödinger equation (which are not "symmetry adapted"), we may have missed some of the simplicity of the problem. Conversely, if we start with symmetry-adapted or *proper* zeroth-order wave functions or state vectors, the perturbation problem may be reduced to nondegenerate perturbation theory, in spite of the degeneracies of the zeroth-order problem. In our simple example, we have not at all made use of the fact that the perturbing potential is an *even* function of the relative coordinate, $(\phi_1 - \phi_2)$. It will therefore be useful to transform from the "single-particle" coordinates, ϕ_1, ϕ_2, to the relative coordinate $(\phi_1 - \phi_2)$, and a "center of mass" coordinate, Φ, via the transformation

$$\phi = (\phi_1 - \phi_2), \qquad \Phi = \tfrac{1}{2}(\phi_1 + \phi_2). \tag{6}$$

We have made the Jacobian of this transformation equal to one. In the new variables, the Schrödinger equation is

$$\left[-\left(2\frac{\partial^2}{\partial\phi^2} + \frac{1}{2}\frac{\partial^2}{\partial\Phi^2}\right) + \lambda\cos\phi\right]\psi(\phi, \Phi) = \epsilon\psi(\phi, \Phi), \tag{7}$$

with zeroth-order solutions

$$\psi^{(0)}(\phi, \Phi) = \frac{1}{2\pi}e^{im\phi}e^{iM\Phi} \quad \text{and} \quad \epsilon^{(0)} = 2m^2 + \tfrac{1}{2}M^2. \tag{8}$$

From the inverse of the above transformation, we have

$$(m_1\phi_1 + m_2\phi_2) = \tfrac{1}{2}(m_1 - m_2)\phi + (m_1 + m_2)\Phi = m\phi + M\Phi. \tag{9}$$

Thus, $\quad m = \frac{1}{2}(m_1 - m_2) \quad$ and $\quad M = (m_1 + m_2)$,

So now $\quad m = 0, \pm\frac{1}{2}, \pm 1, \pm\frac{3}{2}, \pm 2, \ldots \quad$ and $\quad M = 0, \pm 1, \pm 2, \ldots$.

The above exponentials, $e^{im\phi}$, however, are not yet symmetry adapted. We need to replace them with functions that are even or odd functions of ϕ, viz., with

$$\psi^{(0)}_{\text{even}}(\phi) = \frac{1}{\sqrt{2\pi}} \text{ for } m = 0, \qquad \psi^{(0)}_{\text{even}}(\phi) = \frac{\cos m\phi}{\sqrt{\pi}} \quad \text{for } m = \tfrac{1}{2}, 1, \tfrac{3}{2}, 2, \ldots,$$

$$\psi^{(0)}_{\text{odd}}(\phi) = \frac{\sin m\phi}{\sqrt{\pi}} \quad \text{for } m = \tfrac{1}{2}, 1, \tfrac{3}{2}, 2, \ldots$$

(We have now restricted the quantum number m such that $m \geq 0$.) Because our perturbing Hamiltonian is independent of Φ, it again conserves the quantum number, M. Because it is an even function of ϕ, it cannot connect even functions to odd functions. The only non-zero matrix elements of $H^{(1)}$ are now, first for $m > 0$, but excluding cases for which $(m-1) \leq 0$:

$$\langle (m+1), \text{even}|H^{(1)}|m, \text{even}\rangle = \langle (m+1), \text{odd}|H^{(1)}|m, \text{odd}\rangle = \tfrac{1}{2},$$
$$\langle (m-1), \text{even}|H^{(1)}|m, \text{even}\rangle = \langle (m-1), \text{odd}|H^{(1)}|m, \text{odd}\rangle = \tfrac{1}{2}. \qquad (10)$$

The additional special cases are

$$\langle 1, \text{even}|H^{(1)}|0, \text{even}\rangle = \langle 0, \text{even}|H^{(1)}|1, \text{even}\rangle = \tfrac{1}{\sqrt{2}},$$
$$\langle \tfrac{1}{2}, \text{even}|H^{(1)}|\tfrac{1}{2}, \text{even}\rangle = +\tfrac{1}{2},$$
$$\langle \tfrac{1}{2}, \text{odd}|H^{(1)}|\tfrac{1}{2}, \text{odd}\rangle = -\tfrac{1}{2}. \qquad (11)$$

All full matrix elements must be diagonal in M. Nondegenerate perturbation theory now gives

$$\begin{aligned}
\epsilon_{m=0,\text{even},M} &= \epsilon^{(0)} - \tfrac{1}{4}\lambda^2 + \cdots, \\
\epsilon_{m=\frac{1}{2},\text{even},M} &= \epsilon^{(0)} + \tfrac{1}{2}\lambda - \tfrac{1}{16}\lambda^2 + \cdots, \\
\epsilon_{m=\frac{1}{2},\text{odd},M} &= \epsilon^{(0)} - \tfrac{1}{2}\lambda - \tfrac{1}{16}\lambda^2 + \cdots, \\
\epsilon_{m=1,\text{even},M} &= \epsilon^{(0)} + \tfrac{5}{24}\lambda^2 + \cdots, \\
\epsilon_{m=1,\text{odd},M} &= \epsilon^{(0)} - \tfrac{1}{24}\lambda^2 + \cdots, \\
\epsilon_{m=\frac{n}{2},\text{even},M} &= \epsilon^{(0)} + \tfrac{1}{4}\lambda^2 \tfrac{1}{(n^2-1)} + \cdots \quad \text{for } n \geq 3, \\
\epsilon_{m=\frac{n}{2},\text{odd}.M} &= \epsilon^{(0)} + \tfrac{1}{4}\lambda^2 \tfrac{1}{(n^2-1)} + \cdots \quad \text{for } n \geq 3.
\end{aligned} \qquad (12)$$

In particular, the eigenvalue for $m = 1$ for the even case gets contributions from off-diagonal elements with $m = 0$ missing for the odd case. It is also easy to show that λ^3 terms contribute only to the energies of states with $m = \frac{1}{2}$ and $m = \frac{3}{2}$, with contributions $\mp\frac{1}{128}\lambda^3$ for the even (odd) states with $m = \frac{1}{2}$, but $\pm\frac{1}{128}\lambda^3$ for the even (odd) states with $m = \frac{3}{2}$.

35. An atom of mass M in a long complicated molecule is constrained to move on a circle of radius, r_e, but is essentially free to move on this circle, with Hamiltonian

$$H^{(0)} = -\frac{\hbar^2}{2I_e}\frac{\partial^2}{\partial\phi^2}, \qquad \text{with } I_e = Mr_e^2,$$

with zeroth-order energies and eigenfunctions

$$E_m^{(0)} = \frac{\hbar^2}{2I_e}m^2, \qquad \text{and} \qquad \psi_m^{(0)}(\phi) = \frac{1}{\sqrt{2\pi}}e^{im\phi},$$

with $m = 0, \pm 1, \pm 2, \ldots$. If this free rotational motion is perturbed by a potential of the form

$$V(\phi) = V_0\cos(2\phi), \qquad \text{with } V_0 = \lambda\frac{\hbar^2}{2I_e}, \quad \text{and } \lambda \ll 1,$$

find corrections to the energy good through order $\lambda^2(\hbar^2/2I_e)$. Pay particular attention to the $|m|$ values that may require special treatment. [Footnote: In this case, some $|m|$ values belonging to cases (2) or (3) will exist. We could again have reduced all calculations to case (1) by making use of the symmetry of the Hamiltonian to find *proper* symmetry-adapted zeroth-order wave functions.]

36. (a) For the hydrogen atom, $Z = 1$, find the perturbation corrections to the energy to order, $mc^2\alpha^4$, caused by the magnetic spin orbit, Thomas $(\vec{l} \cdot \vec{s})$ term, and the relativistic mass correction to the kinetic energy, but no external magnetic fields, as a function of n, l, j, and show

$$E_{nj} = \frac{1}{2}\mu c^2\alpha^4\left(-\frac{1}{n^2} + \frac{\alpha^2}{n^3}\left[-\frac{1}{(j+\frac{1}{2})} + \frac{3}{4n}\right] + \cdots\right);$$

that is, show the states with the same j but different l are still degenerate. Here, α is the fine structure constant. Use the results of problem 18 to get some of the needed matrix elements. Show $|nljm_j\rangle$ are the *proper* zeroth-order states and use these for the calculation. [Footnote: For the states with $l = 0$, $j = (l + \frac{1}{2}) = \frac{1}{2}$, your method of calculation may not be rigorously correct, because it will involve a factor $(l/l) = (0/0)$. The rigorous derivation of this case will have to wait for Dirac theory, Chapter 74, and the so-called Darwin term.]

(b) Repeat for the once-ionized helium atom (one-electron atom with $Z = 2$). The helium nucleus, unlike the proton, has no nuclear spin, hence, no nuclear magnetic moment and no so-called hyperfine perturbation terms. In this case, find the additional first-order energy perturbation caused by a uniform external magnetic field, B_0, assuming$\hbar\omega_L \ll mc^2\alpha^4$. For the special case, $n = 2$, calculate, in addition, corrections of order $(\hbar\omega_L)^2/mc^2\alpha^4$.

37. The perturbed hydrogen atom in stretched parabolic coordinates: Stark effect.

In stretched parabolic coordinates, we showed in problem 6 $(H^{(0)} - \epsilon)\psi = 0$ can be rewritten as

$$\left((-2\epsilon)\frac{2}{(\mu^2+\nu^2)}\left[-\frac{1}{4}\left(\frac{\partial^2}{\partial\mu^2} + \frac{1}{\mu}\frac{\partial}{\partial\mu} - \frac{m^2}{\mu^2}\right) - \frac{1}{4}\left(\frac{\partial^2}{\partial\nu^2} + \frac{1}{\nu}\frac{\partial}{\partial\nu} - \frac{m^2}{\nu^2}\right)\right] - \frac{2\sqrt{(-2\epsilon)}}{(\mu^2+\nu^2)} - \epsilon\right)\psi = 0. \tag{1}$$

For the hydrogen atom perturbed by a uniform external electric field, $\vec{\mathcal{E}}$, we have

$$H = H^{(0)} + \lambda z = H^{(0)} + \lambda\frac{(\mu^2 - \nu^2)}{2\sqrt{(-2\epsilon)}}, \qquad \text{with } \lambda = \frac{ea_0\mathcal{E}}{(me^4/\hbar^2)},$$

all in dimensionless units, so

$$(H^{(0)} + \lambda H^{(1)} - \epsilon)\psi = 0$$

becomes

$$\left((-2\epsilon)\frac{2}{(\mu^2+\nu^2)}\left[-\frac{1}{4}\left(\frac{\partial^2}{\partial\mu^2} + \frac{1}{\mu}\frac{\partial}{\partial\mu}\right) + \frac{m^2}{4\mu^2} - \frac{1}{4}\left(\frac{\partial^2}{\partial\nu^2} + \frac{1}{\nu}\frac{\partial}{\partial\nu}\right) + \frac{m^2}{4\nu^2}\right] - \frac{2\sqrt{(-2\epsilon)}}{(\mu^2+\nu^2)}\right.$$

$$+ \lambda H^{(1)} - \epsilon \Bigg) \psi = 0. \tag{2}$$

Show, from the results of problem 26, we can rewrite this as

$$\left(T_3 + T_3' - \frac{1}{\sqrt{(-2\epsilon)}} + \lambda H^{(1)} \frac{(\mu^2 + \nu^2)}{2(-2\epsilon)} \right) \psi = 0, \tag{3}$$

$$\text{with} \qquad \lambda H^{(1)} \frac{(\mu^2 + \nu^2)}{2(-2\epsilon)} = \frac{\lambda}{4} \frac{(\mu^4 - \nu^4)}{(-2\epsilon)^{\frac{3}{2}}}. \tag{4}$$

To carry through the perturbation formalism, expand

$$\epsilon = \epsilon^{(0)} + \lambda \epsilon^{(1)} + \lambda^2 \epsilon^{(2)} + \cdots = -\frac{1}{2n^2} + \lambda \epsilon^{(1)} + \lambda^2 \epsilon^{(2)} + \cdots, \tag{5}$$

$$\begin{aligned} \frac{1}{\sqrt{(-2\epsilon)}} &= \frac{n}{\sqrt{1 - 2\lambda n^2 \epsilon^{(1)} - 2\lambda^2 n^2 \epsilon^{(2)} + \cdots}} \\ &= n + \lambda n^3 \epsilon^{(1)} + \lambda^2 [n^3 \epsilon^{(2)} + \tfrac{3}{2} n^5 (\epsilon^{(1)})^2] + \cdots, \end{aligned} \tag{6}$$

$$\frac{\lambda}{(-2\epsilon)^{\frac{3}{2}}} = \lambda n^3 + \lambda^2 3 n^5 \epsilon^{(1)} + \cdots, \tag{7}$$

and show eq. (3) can be rewritten as

$$\begin{aligned} \Bigg((T_3 + T_3' - n) &+ \lambda \left[\tfrac{1}{4} n^3 (\mu^4 - \nu^4) - n^3 \epsilon^{(1)} \right] + \lambda^2 \Big[\tfrac{1}{4} (\mu^4 - \nu^4) 3 n^5 \epsilon^{(1)} \\ &- n^3 \epsilon^{(2)} - \tfrac{3}{2} n^5 \left(\epsilon^{(1)}\right)^2 \Big] + \cdots \Bigg) \Big(|n^{(0)}\rangle + \lambda |n^{(1)}\rangle + \lambda^2 |n^{(2)}\rangle + \cdots \Big) = 0 \end{aligned} \tag{8}$$

in analogy with the standard perturbation expansion

$$\begin{aligned} &\Big((H^{(0)} - E_n^{(0)}) + \lambda (H^{(1)} - E_n^{(1)}) + \lambda^2 (H^{(2)} - E_n^{(2)}) + \cdots \Big) \\ &\qquad \times \Big(|n^{(0)}\rangle + \lambda |n^{(1)}\rangle + \lambda^2 |n^{(2)}\rangle + \cdots \Big) = 0. \end{aligned} \tag{9}$$

In eq. (8), the notation $|n^{(0)}\rangle$ is shorthand for $|m n_1 n_2\rangle$ (see problem 26) with $n = (|m| + n_1 + n_2 + 1)$.

Express $\frac{1}{4}(\mu^4 - \nu^4)$ in terms of the operators, T_3, T_+, T_-, and T_3', T_+', T_-'.

(a) Derive the expression for the first-order Stark energy

$$\epsilon^{(1)} = \frac{3}{2} n (n_1 - n_2).$$

(b) For the nondegenerate ground state, with $n = 1$, $(m = n_1 = n_2 = 0)$, find the second-order Stark correction; i.e., calculate $\epsilon^{(2)}_{n=1}$.

(c) States with $|m| = (n - 1)$ (arbitrary n) can also be treated by nondegenerate perturbation theory, and find $\epsilon^{(2)}$ for such states as a function of n.

[Footnote: The evaluation of $E_n^{(2)}$ by standard perturbation theory, using eq. (1) in the conventional $|n^{(0)}\rangle = |nlm\rangle$ basis, would have required an infinite sum over discrete states with $n' \neq n$ and an integral over the continuum states of the hydrogen atom, even to obtain the simple ground state, $n = 1$, result of (b).]

38. The Stark effect in the hydrogen atom for degenerate levels (arbitrary n, m) in stretched parabolic coordinates.

(a) Using the parallel between

$$H^{(0)} \rightarrow T_3 + T_3', \qquad E_n^{(0)} \rightarrow n,$$

$$H^{(1)} \rightarrow \tfrac{1}{4}(\mu^4 - \nu^4)n^3, \qquad E_n^{(1)} \rightarrow n^3\epsilon^{(1)}, \quad \text{etc.},$$

found in problem 37, make the parallel of the unitary transformation $UHU^\dagger = H'$, such that the $(n - |m|)$-fold degenerate states for fixed n and m (taking $m \geq 0$ without loss of generality) are "unhooked" from states with $n' \neq n$ to within second order in λ (cf., Chapter 24).

(b) Show that sums such as

$$\sum_{n_1'}\sum_{n_2', n' \neq n} \frac{\langle m(n_1 + k)(n_2 - k)|H^{(1)}_{\text{eff.}}|mn_1'n_2'\rangle\langle mn_1'n_2'|H^{(1)}_{\text{eff.}}|mn_1n_2\rangle}{(n - n')},$$

with $k = \pm 1, \pm 2$ are zero, so, effectively, to second order in λ, we can get $\epsilon_n^{(2)}$ from the diagonal matrix elements of $H'_{\text{eff.}}$ in the $|mn_1n_2\rangle$ basis. Note: In the above, $H^{(1)}_{\text{eff.}} = \frac{1}{4}(\mu^4 - \nu^4)n^3$.

(c) Calculate the Stark energy corrections, $\epsilon_n^{(1)}$, $\epsilon_n^{(2)}$, as functions of m, n_1, n_2, for an arbitrary excited state of hydrogen. In particular, show

$$\epsilon_n^{(1)} = \lambda\tfrac{3}{2}n(n_1 - n_2),$$

and

$$\epsilon_n^{(2)} = -\lambda^2\frac{n^4}{8}\Big(34(n_1^2 + n_2^2 - n_1n_2) + 17(n_1 + n_2) + 18$$
$$+17m(n_1 + n_2 + 1) + 4m^2 - 27(n_1 - n_2)^2\Big),$$

or

$$\epsilon_n^{(2)} = -\lambda^2\frac{n^4}{16}\Big(17n^2 - 3(n_1 - n_2)^2 - 9m^2 + 19\Big).$$

Part III

Angular Momentum Theory

27

Angular Momentum Coupling Theory

In the last chapter, we calculated (for the special case $s = \frac{1}{2}$) the transformation coefficient $\langle nlm_l sm_s | nlsjm_j \rangle = \langle nlsjm_j | nlm_l sm_s \rangle^*$ from a basis $|nlm_l sm_s\rangle$ in which L_z and S_z are diagonal, which was a good basis for the case in which the external B_0 field was the dominant perturbation, to the basis $|nlsjm_j\rangle$, the proper basis in the limit $B_0 = 0$ in which the $(\vec{L} \cdot \vec{S})$ term is the dominant perturbation and in which the operators $(\vec{J} \cdot \vec{J})$ and J_z are diagonal. This was a special case of an important and common problem, met in many applications of quantum theory.

Given two commuting angular momentum operators, $\vec{J}_1$ and $\vec{J}_2$, each with standard angular momentum commutation relations, i.e., with

$$[\vec{J}_1, \vec{J}_2] = 0, \qquad \text{and}$$
$$[J_0, J_\pm] = \pm J_\pm, \qquad [J_+, J_-] = 2J_0, \qquad \text{for both} \quad \vec{J}_1, \ \vec{J}_2. \tag{1}$$

We construct the coupled angular momentum vector

$$\vec{J} = \vec{J}_1 + \vec{J}_2, \tag{2}$$

which also satisfies the standard angular momentum commutation relations of eq. (1). We will often need to make the transformation from the $|j_1 m_1 j_2 m_2\rangle$ basis, where these are simultaneously eigenvectors of the four commuting operators

$$(\vec{J}_1 \cdot \vec{J}_1), \quad (J_1)_z, \quad (\vec{J}_2 \cdot \vec{J}_2), \quad (J_2)_z,$$

to the $|j_1 j_2 jm\rangle$ basis, where these are simultaneously eigenvectors of the four commuting operators

$$(\vec{J}_1 \cdot \vec{J}_1), \quad (\vec{J}_2 \cdot \vec{J}_2), \quad (\vec{J} \cdot \vec{J}), \quad J_z.$$

The possible m values are

$$\begin{aligned} m_1 &= j_1,\ (j_1 - 1),\ (j_1 - 2),\ \ldots,\ -j_1, \\ m_2 &= j_2,\ (j_2 - 1),\ (j_2 - 2),\ \ldots,\ -j_2, \\ m &= j,\ (j - 1),\ (j - 2),\ \ldots,\ -j, \end{aligned} \tag{3}$$

where m is an additive quantum number

$$m = m_1 + m_2. \tag{4}$$

First, we need to find the possible j values for a given j_1, j_2. We shall name the j_1, j_2, such that $j_1 \geq j_2$. To find the possible values of j, we simply count the number of occurences for each possible value of m.

The maximum possible m value is $m = j_1 + j_2$, which can be made in only one way, with $m_1 = j_1$ and $m_2 = j_2$. Hence, one j value must exist with $j = j_1 + j_2$.

There are two ways of making $m = (j_1 + j_2 - 1)$; either with $m_1 = j_1, m_2 = (j_2 - 1)$, or with $m_1 = (j_1 - 1), m_2 = j_2$. One linear combination of these two states will be the state with $j = j_1 + j_2$ and $m = (j_1 + j_2 - 1)$. The other linear combination of these two states must be a state with $m = j = (j_1 + j_2 - 1)$. Thus, one j value must exist with $j = (j_1 + j_2 - 1)$.

There are three ways of making states with $m = (j_1 + j_2 - 2)$, viz., with $m_1, m_2 = j_1, (j_2 - 2), (j_1 - 1), (j_2 - 1)$, or $(j_1 - 2), j_2$. One linear combination of these three states is needed to make the state with $j = j_1 + j_2$ and $m = (j_1 + j_2 - 2)$. A second linear combination of these three states is needed to make the state with $j = (j_1 + j_2 - 1)$ and $m = (j_1 + j_2 - 2)$. This leaves one linear combination to make a single state with $m = j = (j_1 + j_2 - 2)$, so one j value exists with $j = (j_1 + j_2 - 2)$.

This process can be continued for $k \leq 2j_2$ (recall we chose $j_1 \geq j_2$), so that there are $(k + 1)$ ways of making states with $m = (j_1 + j_2 - k)$. Of these, k linear combinations are needed to make the states with this m value, but with one of the allowed j values with $j > (j_1 + j_2 - k)$, leaving but a single linear combination of these states with $m = j = (j_1 + j_2 - k)$, so a single j value exists with $(j_1 + j_2 - k)$.

This process quits with $k = 2j_2$. For $k \geq 2j_2$ and $m \geq 0$, there are only $(2j_2 + 1)$ ways of making the m value $m = (j_1 + j_2 - k)$ with $k \geq 2j_2$, and all $(2j_2 + 1)$ independent linear combinations of these states are needed to make the states with $j \geq (j_1 - j_2)$, so no new j values arise, with $j < (j_1 - j_2)$. Finally, symmetry exists between the positive and negative m values.

Thus, the possible j values are

$$(j_1 + j_2),\ (j_1 + j_2 - 1),\ (j_1 + j_2 - 2),\ \ldots,\ |(j_1 - j_2)|,$$

with each j value occuring once only. The total number of states is

$$\begin{aligned} \sum_{|(j_1-j_2)|}^{(j_1+j_2)} (2j+1) &= \sum_{k=0}^{k=2j_2} [2(j_1 + j_2 - k) + 1] \\ &= (2j_2 + 1)(2j_1 + 2j_2 + 1) - 2\frac{(2j_2 + 1)2j_2}{2} \\ &= (2j_1 + 1)(2j_2 + 1), \end{aligned} \tag{5}$$

as it should be.

A General Properties of Vector Coupling Coefficients

We shall need the transformation coefficients of the unitary transformation

$$|j_1 j_2 jm\rangle = \sum_{m_1,(m_2)} |j_1 m_1 j_2 m_2\rangle\langle j_1 m_1 j_2 m_2 | j_1 j_2 jm\rangle, \tag{6}$$

where $\langle j_1 m_1 j_2 m_2 | j_1 j_2 jm\rangle$ is the unitary transformation matrix, which could be written as,

$$U_{m_1 m_2, jm},$$

where the row label is given by $m_1 m_2$, and the column label is specified by jm. We sum over both m_1 and m_2 in eq. (6), but m is fixed and because $m = m_1 + m_2$, m_2 is determined by m_1 and m, so it gets "dragged along" in the sum. This is why we have put m_2 in parentheses in the summation symbol. Because j_1 and j_2 is common to both bases, the unitary transformation coefficient is often abbreviated by

$$\langle j_1 m_1 j_2 m_2 | jm\rangle.$$

It is known as a "Clebsch–Gordan coefficient," or as a "Wigner coefficient," or as a "vector coupling coefficient." Slightly different notations are used by different people. Other commonly used notations are $\langle j_1 j_2 m_1 m_2 | jm\rangle$ (note the different order of the labels in the left-hand side), or $C^{j_1 j_2 j}_{m_1 m_2 m}$, or several others.

The inverse of the above transformation, eq. (6), is in Dirac notation

$$|j_1 m_1 j_2 m_2\rangle = \sum_j |j_1 j_2 jm\rangle\langle j_1 j_2 jm | j_1 m_1 j_2 m_2\rangle, \tag{7}$$

where the summation is one over j only, because m is fixed by the fixed values of m_1 and m_2 and

$$\begin{aligned}\langle j_1 j_2 jm | j_1 m_1 j_2 m_2\rangle &= (U^{-1})_{jm,m_1 m_2} = (U_{m_1 m_2, jm})^* \\ &= \langle j_1 m_1 j_2 m_2 | j_1 j_2 jm\rangle^* = \langle j_1 m_1 j_2 m_2 | jm\rangle^*.\end{aligned} \tag{8}$$

The Clebsch–Gordan coefficients can all be made real. (This is the "world" standard to which everyone adheres!) Therefore, the complex conjugate sign is not needed for the inverse transformation, and we can write the inverse transformation, in terms of the Clebsch–Gordan coefficient notation, as

$$|j_1 m_1 j_2 m_2\rangle = \sum_j |j_1 j_2 jm\rangle\langle j_1 m_1 j_2 m_2 | jm\rangle. \tag{9}$$

(From the point of view of the Dirac notation, the transformation coefficient appears to have bra and ket inverted. This inversion is because we have made use of the unitary property of this *real* transformation coefficient. We shall always write the Clebsch–Gordan coefficient in the Dirac-like notation, but with the m_1, m_2

labels always on the left!) Using the unitary property of this real transformation coefficient, we get the: Orthogonality relations of the Clebsch–Gordan coefficients:

$$\sum_{m_1,(m_2)} \langle j_1 m_1 j_2 m_2 | jm\rangle \langle j_1 m_1 j_2 m_2 | j' m'\rangle = \delta_{jj'}\delta_{mm'},$$
$$\sum_j \langle j_1 m_1 j_2 m_2 | jm\rangle \langle j_1 m_1' j_2 m_2' | jm\rangle = \delta_{m_1 m_1'}\delta_{m_2 m_2'}. \tag{10}$$

B Methods of Calculation

For $j_2 = \frac{1}{2}$, we have already found one method: diagonalize the operator $(\vec{J}_1 \cdot \vec{J}_2)$ in the $|j_1 m_1 j_2 m_2\rangle$ basis. For $j_2 \geq 1$, however, this method would lead to a diagonalization of 3×3, $4 \times 4, \ldots,$ matrices. Hence, we shall look for better methods. One of these methods involves recursion formulae for the Clebsch–Gordan coefficients. We shall derive a recursion formula for the Clebsch–Gordan coefficients by acting on the state vector $|j_1 j_2 jm\rangle$ with the operator,

$$J_+ = (J_1)_+ + (J_2)_+,$$

$$J_+ |j_1 j_2 jm\rangle = \sum_{m_1',(m_2')} \Big((J_1)_+ + (J_2)_+\Big) |j_1 m_1' j_2 m_2'\rangle \langle j_1 m_1' j_2 m_2' | jm\rangle, \tag{11}$$

or

$$\begin{aligned}
&\sqrt{(j-m)(j+m+1)}|j_1 j_2 j (m+1)\rangle \\
&= \sum_{m_1',(m_2')} \Big(\sqrt{(j_1 - m_1')(j_1 + m_1' + 1)} |j_1 (m_1'+1) j_2 m_2'\rangle \\
&+ \sqrt{(j_2 - m_2')(j_2 + m_2' + 1)} |j_1 m_1' j_2 (m_2'+1)\rangle \Big) \langle j_1 m_1' j_2 m_2' | jm\rangle.
\end{aligned} \tag{12}$$

Now, expanding $|j_1 j_2 j (m+1)\rangle$ on the left-hand side of this relation, and renaming the dummy summation indices $m_1', m_2' = (m_1 - 1), m_2$ in the first term of the right-hand side, and making the change $m_1', m_2' = m_1, (m_2 - 1)$ in the second term of the right-hand side, we get

$$\begin{aligned}
&\sqrt{(j-m)(j+m+1)} \sum_{m_1,(m_2)} |j_1 m_1 j_2 m_2\rangle \langle j_1 m_1 j_2 m_2 | j(m+1)\rangle \\
&= \sum_{m_1,(m_2)} |j_1 m_1 j_2 m_2\rangle \Big(\sqrt{(j_1 - m_1 + 1)(j_1 + m_1)} \langle j_1 (m_1 - 1) j_2 m_2 | jm\rangle \\
&+ \sqrt{(j_2 - m_2 + 1)(j_2 + m_2)} \langle j_1 m_1 j_2 (m_2 - 1) | jm\rangle \Big).
\end{aligned} \tag{13}$$

Now, with left-multiplication by a $\langle j_1 m_1 j_2 m_2|$ with a specific, fixed m_1 and m_2, this equation is converted to the recursion relation, as follows.

Recursion Formula I:

$$\sqrt{(j-m)(j+m+1)} \langle j_1 m_1 j_2 m_2 | j(m+1)\rangle$$

$$= \sqrt{(j_1+m_1)(j_1-m_1+1)}\langle j_1(m_1-1)j_2m_2|jm\rangle + \sqrt{(j_2+m_2)(j_2-m_2+1)}\langle j_1m_1j_2(m_2-1)|jm\rangle. \quad (14)$$

For states with $m = j$, this three-term recursion formula is reduced to a two-term recursion formula, which leads to

$$\frac{\langle j_1(m_1-1)j_2m_2|jj\rangle}{\langle j_1m_1j_2(m_2-1)|jj\rangle} = -\sqrt{\frac{(j_2+m_2)(j_2-m_2+1)}{(j_1+m_1)(j_1-m_1+1)}}. \quad (15)$$

We can use this successively, starting with $m_1 = j_1$, and, hence, $m_2 = j - j_1 + 1$, to relate

$$\langle j_1m_1j_2(m_2 = j - m_1)|jj\rangle \qquad \text{to} \qquad \langle j_1j_1j_2(j-j_1)|jj\rangle,$$

and, thus, get

$$\frac{\langle j_1m_1j_2(j-m_1)|jj\rangle}{\langle j_1j_1j_2(j-j_1)|jj\rangle} = (-1)^{j_1-m_1}\sqrt{\frac{(j_2+j-j_1+1)(j_2+j-j_1+2)\cdots(j_2+j-m_1)}{2j_1(2j_1-1)\cdots(j_1+m_1+1)}} \times \sqrt{\frac{(j_2-j+j_1)(j_2-j+j_1-1)\cdots(j_2-j+m_1+1)}{1\cdot 2\cdots(j_1-m_1)}}$$
$$= (-1)^{j_1-m_1}\sqrt{\frac{(j_2+j-m_1)!}{(j_2+j-j_1)!}\frac{(j_2-j+j_1)!}{(j_2-j+m_1)!}\frac{(j_1+m_1)!}{2j_1!(j_1-m_1)!}}. \quad (16)$$

Now we can calculate $|\langle j_1j_1j_2(j-j_1)|jj\rangle|$ by using the orthonormality

$$\sum_{m_1}|\langle j_1m_1j_2(j-m_1)|jj\rangle|^2 = 1. \quad (17)$$

To do the sum, we will need an addition theorem for binomial coefficients

$$\sum_{m_1}\frac{(a+m_1)!(b-m_1)!}{(c+m_1)!(d-m_1)!} = \frac{(a+b+1)!(a-c)!(b-d)!}{(c+d)!(a+b-c-d+1)!}. \quad (18)$$

Because this relation will only give us the absolute value of the starting coefficient $\langle j_1j_1j_2(j-j_1)|jj\rangle$, its phase must be chosen. The choice universally accepted, the so-called Condon and Shortley phase convention, is the following: This starting coefficient is chosen to be real and positive. Then,

$$\langle j_1j_1j_2(j-j_1)|jj\rangle = \sqrt{\frac{2j_1!(2j+1)!}{(j_1+j_2+j+1)!(j_1-j_2+j)!}}, \quad (19)$$

and, finally,

$$\langle j_1m_1j_2(j-m_1)|jj\rangle = (-1)^{j_1-m_1}\times$$
$$\sqrt{\frac{(j_1+m_1)!(j_2+j-m_1)!(j_1+j_2-j)!(2j+1)!}{(j_1-m_1)!(j_2-j+m_1)!(j_2-j_1+j)!(j_1-j_2+j)!(j_1+j_2+j+1)!}}. \quad (20)$$

Now we need to calculate coefficients with $m < j$. We can accomplish this by deriving a recursion formula that steps down in m. By repeating the steps of eqs. (11)–(14) by acting on the coupled state $|j_1 j_2 jm\rangle$ with the step-down operator $J_- = (J_1)_- + (J_2)_-$, we arrive at the analogue of eq. (14) as follows.

Recursion Formula II:

$$\begin{aligned} &\sqrt{(j+m)(j-m+1)}\langle j_1 m_1 j_2 m_2 | j(m-1)\rangle \\ &= \sqrt{(j_1-m_1)(j_1+m_1+1)}\langle j_1(m_1+1) j_2 m_2 | jm\rangle \\ &+ \sqrt{(j_2-m_2)(j_2+m_2+1)}\langle j_1 m_1 j_2 (m_2+1) | jm\rangle. \end{aligned} \tag{21}$$

Repeated application of this recursion formula II will give us the coefficients with arbitrary m, starting with the known coefficient with $m = j$. In practice, the most widely used tables are those in which one of the angular momenta is reasonably small, say, $j_2 = \frac{1}{2}, 1, \frac{3}{2}, 2$. (See problem 39.) To calculate some of these it will be useful to first study the symmetries of the Clebsch–Gordan coefficients.

28

Symmetry Properties of Clebsch–Gordan Coefficients

Clebsch–Gordan coefficients in which the three angular momenta, j_1, j_2, and $j \equiv j_3$, are reordered may be simply related to each other. The most trivial case involves the exchange of the order of the quantum numbers, $j_1 m_1$ and $j_2 m_2$. The state vector $|j_1 m_1 j_2 m_2\rangle$ is a direct product of two vectors involving separate subspaces of the full Hilbert space, or in terms of the coordinate representation, the wave function $\psi_{j_1 m_1} \psi_{j_2 m_2}$ is a product of functions involving different variables. For example, $\psi_{j_1 m_1}$ might be a function of orbital variables and $\psi_{j_2 m_2}$ might be a function of spin variables. Thus, the product of these two functions should not depend on the order in which we write the two functions. Therefore, when we expand this product function in terms of the total angular momentum eigenfunctions $\Psi_{j_1 j_2 j m}$, the result must be independent of the order in which we write the original product function, $\psi_{j_1 m_1} \psi_{j_2 m_2}$, or $\psi_{j_2 m_2} \psi_{j_1 m_1}$, with the possible exception of an overall phase factor. This phase factor comes in because our phase convention fixing the overall sign of the Clebsch–Gordan coefficients gives preference to the angular momenta sitting in the number 1 and number 3 positions of the Clebsch–Gordan coefficient. Thus, $\langle j_1 j_1 j_2 m_2 | j_3 j_3\rangle$ must be positive by our phase convention. Similarly, $\langle j_2 j_2 j_1 m_1 | j_3 j_3\rangle$ must also be positive. On the contrary, the Clebsch–Gordan coefficient $\langle j_1 m_1 j_2 j_2 | j_3 j_3\rangle$ has the sign $(-1)^{j_1 - m_1}$ with $m_1 = j_3 - j_2$. Hence, its sign is $(-1)^{j_1 + j_2 - j_3}$. Thus, the coefficients in which the order of j_1 and j_2 is exchanged will differ by this phase factor for all possible m's. Thus, we have our first symmetry property:

$$\langle j_1 m_1 j_2 m_2 | j_3 m_3\rangle = (-1)^{j_1 + j_2 - j_3} \langle j_2 m_2 j_1 m_1 | j_3 m_3\rangle. \tag{1}$$

Next, if we rearrange the vector addition equation,

$$\vec{J}_1 + \vec{J}_2 = \vec{J}_3, \tag{2}$$

to read

$$\vec{J}_3 - \vec{J}_2 = \vec{J}_1, \tag{3}$$

we can see that the Clebsch–Gordan coefficient $\langle j_1 m_1 j_2 m_2 | j_3 m_3 \rangle$ must be related to the coefficient $\langle j_3 m_3 j_2 - m_2 | j_1 m_1 \rangle$. In particular, if we make the substitution $j_1 m_1 \leftrightarrow j_3 m_3$ and $m_2 \to -m_2$ in recursion formula I (or II), we obtain recursion formula II (or I), provided the transformed coefficients are related to the original ones via a phase factor proportional to $(-1)^{m_2}$ and an m-independent factor; i.e., we expect

$$\langle j_1 m_1 j_2 m_2 | j_3 m_3 \rangle = (-1)^{m_2} K(j_1, j_2, j_3) \langle j_3 m_3 j_2 - m_2 | j_1 m_1 \rangle, \tag{4}$$

where $K(j_1, j_2, j_3)$ is the m-independent overall factor. This factor can be determined via the orthonormality of the Clebsch–Gordan coefficients

$$\begin{aligned} \sum_{m_3} \sum_{m_1,(m_2)} |\langle j_1 m_1 j_2 m_2 | j_3 m_3 \rangle|^2 &= \sum_{m_3} 1 = (2j_3 + 1) \\ &= \sum_{m_1} \sum_{m_3,(m_2)} |K(j_1, j_2, j_3)|^2 |\langle j_3 m_3 j_2 - m_2 | j_1 m_1 \rangle|^2 = \sum_{m_1} |K(j_1, j_2, j_3)|^2 \\ &= (2j_1 + 1)|K(j_1, j_2, j_3)|^2. \end{aligned} \tag{5}$$

Thus,

$$K(j_1, j_2, j_3) = (-1)^{\phi(j_1, j_2, j_3)} \sqrt{\frac{(2j_3 + 1)}{(2j_1 + 1)}}, \tag{6}$$

where the j_i-dependent phase ϕ can be determined because the coefficients with both $m_1 = j_1$ and $m_3 = j_3$ and hence $m_2 = j_3 - j_1$ must both be positive, and hence $\phi = j_1 - j_3$. Thus, we get a second symmetry property

$$\langle j_1 m_1 j_2 m_2 | j_3 m_3 \rangle = (-1)^{j_1 - j_3 + m_2} \sqrt{\frac{(2j_3 + 1)}{(2j_1 + 1)}} \langle j_3 m_3 j_2 - m_2 | j_1 m_1 \rangle. \tag{7}$$

By combining this symmetry property with the first one, $(1 \leftrightarrow 2)$ exchange, we get

$$\langle j_1 m_1 j_2 m_2 | j_3 m_3 \rangle = (-1)^{j_2 + m_2} \sqrt{\frac{(2j_3 + 1)}{(2j_1 + 1)}} \langle j_2 - m_2 j_3 m_3 | j_1 m_1 \rangle. \tag{8}$$

This process is a cyclic exchange of the type $123 \to (-2)31$. If we follow this by the cyclic exchange $(-2)31 \to (-3)1(-2)$ and subsequently by the cyclic exchange $(-3)1(-2) \to (-1)(-2)(-3)$, we obtain

$$\langle j_1 m_1 j_2 m_2 | j_3 m_3 \rangle = (-1)^{j_1 + j_2 - j_3} \langle j_1 - m_1 j_2 - m_2 | j_3 - m_3 \rangle, \tag{9}$$

where we have used the identity $j_1+m_1+j_2+m_2+j_3+m_3 = j_1+j_2+j_3+2m_3$, and $j_1+j_2+j_3+2m_3 = j_1+j_2-j_3+$ even integer, because $2j_3+2m_3$ is always an even integer. (Either j_3 and m_3 are both integers or are both $\frac{1}{2}$-integers.)

We have now derived several symmetry properties, involving interchanges such as 123 → 213, or cyclic interchanges such as 123 → (−2)31, or changes of sign in all m's, 123 → (−1)(−2)(−3). Twelve such symmetry properties exist altogether. These properties are much easier to remember by introducing the 3-j symbol, defined by

$$\begin{pmatrix} j_1 & j_2 & j_3 \\ m_1 & m_2 & -m_3 \end{pmatrix} = \frac{(-1)^{j_1-j_2+m_3}}{\sqrt{(2j_3+1)}} \langle j_1 m_1 j_2 m_2 | j_3 m_3 \rangle. \tag{10}$$

This 3-j symbol has the following symmetry properties: The 3-j symbol is invariant under any even permutation of columns. The 3-j symbol changes sign by the factor $(-1)^{j_1+j_2+j_3}$ under either an odd permutation of columns or under the transformation $m_i \to -m_i$ for all $i = 1, 2, 3$.

Although the symmetry properties are easier to remember in terms of the 3-j symbol, this symbol does not have simple orthonormality properties. The orthonormality relations for the Clebsch–Gordan coefficients are so useful most authors prefer to use the Clebsch–Gordan coefficients.

References: several little books on angular momentum coupling in quantum mechanics exist: (1) D. M. Brink and G. R. Satchler. *Angular Momentum*. Oxford: Clarendon Press, 1968; (2) M. E. Rose. *Elementary Theory of Angular Momentum*. New York: John Wiley, 1957; (3) A. R. Edmonds. *Angular Momentum in Quantum Mechanics*. Princeton University Press, 1974.

Tables: *The 3-j and 6-j Symbols*. M. Rotenberg, R. Bivins, N. Metropolis, and J. K. Wooten. Cambridge, Mass.: MIT Press, 1959. *Tables of Clebsch-Gordan Coefficients*, Peking: Science Press, 1965.

TABLE 28.1. $\langle j_1 m_1 \frac{1}{2} m_2 | jm \rangle$

$j =$	$m_2 = +\frac{1}{2}$	$m_2 = -\frac{1}{2}$
$j_1 + \frac{1}{2}$	$\sqrt{\frac{(j_1+m+\frac{1}{2})}{(2j_1+1)}}$	$\sqrt{\frac{(j_1-m+\frac{1}{2})}{(2j_1+1)}}$
$j_1 - \frac{1}{2}$	$-\sqrt{\frac{(j_1-m+\frac{1}{2})}{(2j_1+1)}}$	$\sqrt{\frac{(j_1+m+\frac{1}{2})}{(2j_1+1)}}$

TABLE 28.2. $\langle j_1 m_1 1 m_2 | jm \rangle$

$j =$	$m_2 = +1$	$m_2 = 0$	$m_2 = -1$
$j_1 + 1$	$\sqrt{\frac{(j_1+m)(j_1+m+1)}{(2j_1+1)(2j_1+2)}}$	$\sqrt{\frac{(j_1-m+1)(j_1+m+1)}{(2j_1+1)(j_1+1)}}$	$\sqrt{\frac{(j_1-m)(j_1-m+1)}{(2j_1+1)(2j_2+2)}}$
j_1	$-\sqrt{\frac{(j_1+m)(j_1-m+1)}{2j_1(j_1+1)}}$	$\frac{m}{\sqrt{j_1(j_1+1)}}$	$\sqrt{\frac{(j_1-m)(j_1+m+1)}{2j_1(j_1+1)}}$
$j_1 - 1$	$\sqrt{\frac{(j_1-m)(j_1-m+1)}{2j_1(2j_1+1)}}$	$-\sqrt{\frac{(j_1-m)(j_1+m)}{j_1(2j_1+1)}}$	$\sqrt{\frac{(j_1+m+1)(j_1+m)}{2j_1(2j_1+1)}}$

The most useful Clebsch–Gordan coefficients are those in which one of the angular momenta, say, j_2, are small. Such coefficients, with $j_2 \leq 4$, can be found in general algebraic form in the last reference (Peking: Science Press, 1965). Coefficients with $j_2 = \frac{1}{2}$ and $j_2 = 1$ are appended.

29

Invariance of Physical Systems Under Rotations

Before going further with our study of angular momentum, it will be advantageous to study the general behavior of physical systems under rotations in our 3-D space. If a state vector which describes the state of a physical system is specified by $|\psi\rangle$, the state vector for the rotated system will be specified by $|\psi_{\text{rot.}}\rangle = R|\psi\rangle$. (We use the subscript, rot., in place of a prime, which is often used for the rotated state, because primes are also often used on quantum labels.) The operator, R, is the operator that rotates the system. Recall from the theory of translation operators, two possible points of view exist for such operators: (1) The active point of view, in which R is used to rotate the system. (2) The passive point of view, in which the system is left unchanged and R is used to rotate the coordinate system (in the opposite sense) to view the system from a rotated reference frame. We shall use the active point of view in this chapter.

The operator R is a linear, unitary operator:

$$R^{-1} = R^{\dagger}, \tag{1}$$

$$R(\lambda_1|\psi_1\rangle + \lambda_2|\psi_2\rangle) = \lambda_1(R|\psi_1\rangle) + \lambda_2(R|\psi_2\rangle). \tag{2}$$

Also, note the following properties.

1. If $|\psi\rangle \rightarrow |\psi_{\text{rot.}}\rangle = R|\psi\rangle$, then $\langle\psi| \rightarrow \langle\psi_{\text{rot.}}| = \langle\psi|R^{\dagger}$. (3)

2. If $|\chi\rangle = O|\psi\rangle$, then $|\chi_{\text{rot.}}\rangle = R|\chi\rangle = ROR^{\dagger}(R|\psi\rangle)$,
   so $O_{\text{rot.}} = ROR^{\dagger}$. (4)

If $[R, O] = 0$, then $O_{\text{rot.}} = O$, and if O is hermitian, $O_{\text{rot.}}$ is hermitian. Also, if $\langle\chi|\psi\rangle$ are observable amplitudes, then

$$3. \qquad \langle\chi_{\text{rot.}}|\psi_{\text{rot.}}\rangle = \langle\chi|R^\dagger R|\psi\rangle = \langle\chi|\psi\rangle. \tag{5}$$

Matrix elements of operators are also invariant:

$$4. \qquad \langle\chi_{\text{rot.}}|O_{\text{rot.}}|\psi_{\text{rot.}}\rangle = \langle\chi|O|\psi\rangle. \tag{6}$$

Relations among operators are preserved under rotations:

$$5. \quad \text{If} \quad [A, B] = iC, \qquad \text{then} \quad [A_{\text{rot.}}, B_{\text{rot.}}] = iC_{\text{rot.}}. \tag{7}$$

A Rotation Operators

We shall begin by studying a single-particle system and assume for the moment that the particle has no spin. We shall construct the rotation operator for a rotation through an angle, α, about a specific axis. We shall also take the z axis of our coordinate system along the direction of the rotation axis. Then, in analogy with the translation operator, $T = e^{-\frac{i}{\hbar}c_1 p_x}$, we shall try

$$R_z(\alpha) = e^{-\frac{i}{\hbar}\alpha L_{z\text{phys.}}} = e^{-i\alpha L_z}, \tag{8}$$

where we have converted the physical angular momentum operator (z component) into the dimensionless L_z in the last step. To study the action of $R_z(\alpha)$ on a general $|\psi\rangle$, expand $|\psi\rangle$ in terms of angular momentum eigenfunctions.

$$|\psi\rangle = \sum_{nlm}|nlm\rangle\langle nlm|\psi\rangle \qquad \text{or}$$

$$\langle\vec{r}|\psi\rangle = \psi(r,\theta,\phi) = \sum_{nlm}\langle\vec{r}|nlm\rangle\langle nlm|\psi\rangle$$

$$= \sum_{nlm} R_{nl}(r)\Theta_{lm}(\theta)\frac{e^{im\phi}}{\sqrt{2\pi}}c_{nlm}. \tag{9}$$

Then,

$$|\psi_{\text{rot.}}\rangle = R|\psi\rangle = \sum_{nlm} e^{-i\alpha L_z}|nlm\rangle\langle nlm|\psi\rangle = \sum_{nlm} e^{-i\alpha m}|nlm\rangle\langle nlm|\psi\rangle$$

$$\text{or} \quad \langle\vec{r}|\psi_{\text{rot.}}\rangle = \psi_{\text{rot.}}(r,\theta,\phi) = \sum_{nlm} R_{nl}(r)\Theta_{lm}(\theta)e^{im(\phi-\alpha)}\frac{c_{nlm}}{\sqrt{2\pi}}. \tag{10}$$

Thus,

$$\psi_{\text{rot.}}(r,\theta,\phi) = \psi(r,\theta,\phi_{\text{rot.}}) = \psi(r,\theta,\phi-\alpha). \tag{11}$$

We see (Fig. 29.1), if the original $\psi(r,\theta,\phi)$ has a maximum at some angle $\phi = \phi_0$, the rotated wave function, $\psi_{\text{rot.}}$, has a maximum where $(\phi-\alpha) = \phi_0$, that is, where $\phi = \phi_0 + \alpha$. In other words, the physical system has been rotated in the positive sense through an angle α. Note: The prime is often used to designate $\phi_{\text{rot.}}$, i.e., $\phi_{\text{rot.}} \equiv \phi' = (\phi - \alpha)$, and note the last minus sign.

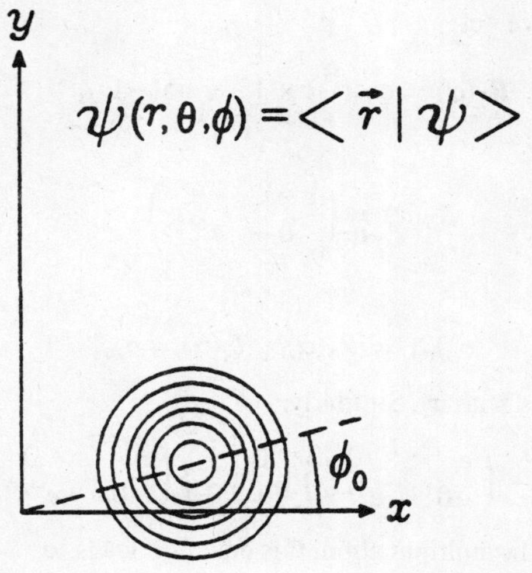

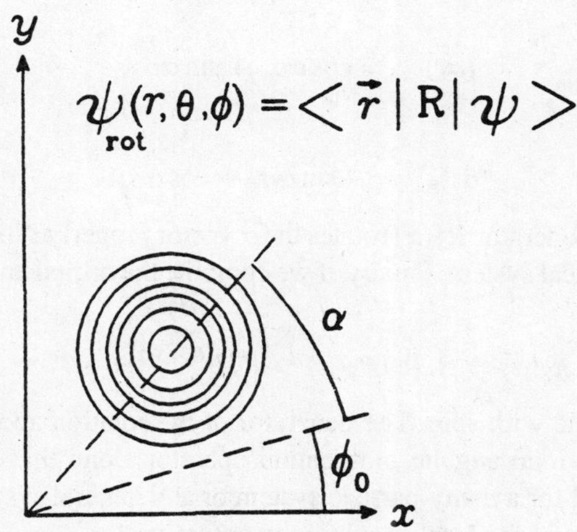

FIGURE 29.1. The rotation operation, $e^{-i\alpha L_2}$.

Next, we shall look at a single-particle system, but now we assume the particle is a spin $\frac{1}{2}$-particle, like the electron. Because spin and orbital operators commute, we shall try

$$R = R_L R_S, \tag{12}$$

with

$$R_S(\alpha) = e^{-i\alpha S_z} = e^{-i\frac{\alpha}{2}\sigma_z}. \tag{13}$$

Now, using $\sigma_z^2 = 1$, we get

$$R_S(\alpha) = \cos(\frac{\alpha}{2}) \times 1 - i \sin(\frac{\alpha}{2})\sigma_z, \tag{14}$$

or

$$R_S(\alpha) = \begin{pmatrix} e^{-i\frac{\alpha}{2}} & 0 \\ 0 & e^{+i\frac{\alpha}{2}} \end{pmatrix}, \tag{15}$$

leading to

$$(\sigma_z)_{\text{rot.}} = R_S(\alpha)\sigma_z R_S^\dagger(\alpha) = \sigma_z, \tag{16}$$

because R_S commutes with σ_z. Similarly,

$$(\sigma_x)_{\text{rot.}} = \begin{pmatrix} e^{-i\frac{\alpha}{2}} & 0 \\ 0 & e^{+i\frac{\alpha}{2}} \end{pmatrix} \begin{pmatrix} 0 & 1 \\ 1 & 0 \end{pmatrix} \begin{pmatrix} e^{+i\frac{\alpha}{2}} & 0 \\ 0 & e^{-i\frac{\alpha}{2}} \end{pmatrix}. \tag{17}$$

Carrying out the matrix multiplication, this equation leads to

$$(\sigma_x)_{\text{rot.}} = \cos\alpha \begin{pmatrix} 0 & 1 \\ 1 & 0 \end{pmatrix} + \sin\alpha \begin{pmatrix} 0 & -i \\ +i & 0 \end{pmatrix}, \tag{18}$$

or

$$(\sigma_x)_{\text{rot.}} = \cos\alpha\sigma_x + \sin\alpha\sigma_y, \tag{19}$$

and, similarly,

$$(\sigma_y)_{\text{rot.}} = -\sin\alpha\sigma_x + \cos\alpha\sigma_y. \tag{20}$$

Thus, the rotation operator $R_S(\alpha)$ rotates the $\vec{\sigma}$ vector properly. Note: This $\vec{\sigma}$ vector is part of the physical system. Finally, if we combine the orbital and spin operators, we get

$$R_z(\alpha) = R_L(\alpha)R_S(\alpha) = e^{-i\alpha(L_z+S_z)} = e^{-i\alpha J_z} \tag{21}$$

for a single particle with spin. The generator of the rotation about an axis is the component of the total angular momentum operator along that axis. This result holds equally well for a many-particle system or any general system, provided J_z is the z component of the *total* angular momentum vector.

B General Rotations, $R(\alpha, \beta, \gamma)$

The most general rotation will be parameterized by the three Euler angles, α, β, and γ, and will be built from the three successive rotations as follows.

The first rotation through the angle α about the original space-fixed z axis will take the (x, y, z) coordinate system to a rotated (x_1, y_1, z_1) system, with $z_1 = z$.

The second rotation through an angle β about the new y_1 axis will take the (x_1, y_1, z_1) coordinate system to a new $(x_2, y_2 z_2)$ system, with $y_2 = y_1$.

The third rotation through an angle γ about the z_2 axis will take the (x_2, y_2, z_2) coordinate system to the final rotated (x', y', z') system, with $z' = z_2$.

We shall think of the coordinate systems as being attached to our physical system. Thus, the general rotation can be expressed through

$$R(\alpha, \beta, \gamma) = e^{-i\gamma J_{z'}} e^{-i\beta J_{y_1}} e^{-i\alpha J_z}, \tag{22}$$

where this is not a very handy form because the three generators of the unitary transformations, $R(\alpha)$, $R(\beta)$, and $R(\gamma)$, are expressed in terms of angular momentum components along three different coordinate systems. Using $O_{\text{rot.}} = ROR^{\dagger}$, however, and noting the operator J_{y_1} is reached from J_y via the rotation $R(\alpha)$, we have

$$e^{-i\beta J_{y_1}} = R(\alpha) e^{-i\beta J_y} R(\alpha)^{\dagger} = e^{-i\alpha J_z} e^{-i\beta J_y} e^{+i\alpha J_z}. \tag{23}$$

Similarly, noting the operator $J_{z'} = J_{z_2}$ is reached from J_{z_1} via the rotation $R(\beta)$, we have

$$e^{-i\gamma J_{z'}} = R(\beta) e^{-i\gamma J_{z_1}} R(\beta)^{\dagger}. \tag{24}$$

Thus, we can write

$$R(\alpha, \beta, \gamma) = R(\beta) e^{-i\gamma J_{z_1}} R(\beta)^{-1} R(\beta) e^{-i\alpha J_z} = R(\beta) e^{-i\gamma J_{z_1}} e^{-i\alpha J_z}. \tag{25}$$

Now, noting $J_{z_1} = J_z$, the two rotation operators on the extreme right commute with each other, and we can write

$$R(\alpha, \beta, \gamma) = e^{-i\beta J_{y_1}} R(\alpha) e^{-i\gamma J_z} = R(\alpha) e^{-i\beta J_y} R(\alpha)^{-1} R(\alpha) e^{-i\gamma J_z}. \tag{26}$$

This process leads to the final result,

$$R(\alpha, \beta, \gamma) = e^{-i\alpha J_z} e^{-i\beta J_y} e^{-i\gamma J_z}. \tag{27}$$

Now, all generators are expressed with respect to components along the original axes, but seemingly the order of the rotations is "backwards," we start on the right with the γ rotation, followed by β, and last the α-term.

C Transformation of Angular Momentum Eigenvectors or Eigenfunctions

Having derived a useful expression for the most general rotation operator, we can now give an expression for a rotated state vector that is an eigenvector of both $\vec{J}^2$ and J_z (and other operators commuting with these two), in terms of the original eigenvectors of this type

$$|(JM)_{\text{rot.}}\rangle = R(\alpha, \beta, \gamma)|JM\rangle = \sum_{\mu} |J\mu\rangle\langle J\mu|R(\alpha, \beta, \gamma)|JM\rangle, \tag{28}$$

where, for simplicity of notation, we have omitted all quantum numbers other than J and M (associated with the remaining operators). Also, the matrix elements of operators, J_z and J_y are diagonal in the quantum number J. Thus, the J sum disappears from the unit operator, $\sum_{J,\mu} |J\mu\rangle\langle J\mu|$. The rotation matrix is usually denoted by the symbol D (for the German word "Darstellung," or representation), because this rotation matrix is an irreducible representation matrix of the rotation group $SO(3)$ for integral angular momenta or the unitary group $SU(2)$ for $\frac{1}{2}$-integral spins.

$$\langle J\mu|e^{-i\alpha J_z}e^{-i\beta J_y}e^{-i\gamma J_z}|JM\rangle = D^J_{\mu M}(\alpha,\beta,\gamma)^*. \tag{29}$$

The * is added so the simple exponential factors have + signs; (but this notation is not universal!)

$$D^J_{\mu M}(\alpha,\beta,\gamma) = e^{+i\mu\alpha}d^J_{\mu M}(\beta)e^{+iM\gamma}, \tag{30}$$

where

$$d^J_{\mu M}(\beta) = \langle J\mu|e^{-i\beta J_y}|JM\rangle \tag{31}$$

is a *real* function of β, because the matrix elements of iJ_y are always real in the standard angular momentum conventions. (Therefore, no * was needed on the d function.)

We can also convert eq. (28) into an equation for angular-momentum eigenfunctions

$$\begin{aligned}\langle\vec{r}|(JM)_{\text{rot.}}\rangle &= (\psi_{\text{rot.}})_{JM}(r,\theta,\phi,\vec{\sigma}) = \psi_{JM}(r,\theta',\phi',\vec{\sigma}')\\ &= \sum_\mu \psi_{J\mu}(r,\theta,\phi,\vec{\sigma})D^J_{\mu M}(\alpha,\beta,\gamma)^*,\end{aligned} \tag{32}$$

where we have now used primed angles for the angles in the rotated angular momentum eigenfunction. Recall $\theta',\phi' = \theta,(\phi-\alpha)$ for the simple z rotation with $\beta=0,\gamma=0$. In general, θ',ϕ' are complicated functions of $\theta,\phi,\alpha,\beta,\gamma$. We can write the inverse of this transformation

$$\psi_{JM}(r,\theta,\phi,\vec{\sigma}) = \sum_\mu \psi_{J\mu}(r,\theta',\phi',\vec{\sigma}')((D^{-1})^J_{\mu M})^*. \tag{33}$$

Now, making use of the unitary property of the D matrix,

$$\psi_{JM}(r,\theta,\phi,\vec{\sigma}) = \sum_\mu D^J_{M\mu}(\alpha,\beta,\gamma)\psi_{J\mu}(r,\theta',\phi',\vec{\sigma}'). \tag{34}$$

The job of calculating the $d^J_{\mu M}(\beta)$ of eq. (31) remains. For the smallest J values, this process is quite straightforward. For example, for $J=\frac{1}{2}$, with

$$\mathbf{1} = \sigma_y^2 = \sigma_y^4 = \cdots, \qquad \sigma_y = \sigma_y^3 = \sigma_y^5\cdots, \tag{35}$$

so

$$e^{-i\frac{\beta}{2}\sigma_y} = \sum_n\left(\frac{-i\beta}{2}\right)^n\frac{(\sigma_y)^n}{n!} = \mathbf{1}\cos(\frac{\beta}{2}) - i\sigma_y\sin(\frac{\beta}{2}). \tag{36}$$

Substituting the matrices

$$\mathbf{1} = \begin{pmatrix}1 & 0\\ 0 & 1\end{pmatrix}, \quad \text{and} \quad \sigma_y = \begin{pmatrix}0 & -i\\ +i & 0\end{pmatrix}, \tag{37}$$

we get the 2×2 d matrix

$$d^{\frac{1}{2}}_{m'm}(\beta) = \begin{pmatrix} \cos\frac{\beta}{2} & -\sin\frac{\beta}{2} \\ +\sin\frac{\beta}{2} & \cos\frac{\beta}{2} \end{pmatrix}. \tag{38}$$

Similarly, for $J = 1$, using the 3×3 matrix relations in this case,

$$J_y = \begin{pmatrix} 0 & \frac{-i}{\sqrt{2}} & 0 \\ \frac{+i}{\sqrt{2}} & 0 & \frac{-i}{\sqrt{2}} \\ 0 & \frac{+i}{\sqrt{2}} & 0 \end{pmatrix} = J_y^3 = J_y^5 = \cdots, \tag{39}$$

and

$$J_y^2 = \begin{pmatrix} \frac{1}{2} & 0 & -\frac{1}{2} \\ 0 & 1 & 0 \\ -\frac{1}{2} & 0 & \frac{1}{2} \end{pmatrix} = J_y^4 = J_y^6 = \cdots, \tag{40}$$

we get

$$e^{-i\beta J_y} = \mathbf{1} - iJ_y \sin\beta + J_y^2(\cos\beta - 1), \tag{41}$$

leading to

$$d^{J=1}_{M'M} = \begin{pmatrix} \frac{1+\cos\beta}{2} & -\frac{\sin\beta}{\sqrt{2}} & \frac{1-\cos\beta}{2} \\ \frac{\sin\beta}{\sqrt{2}} & \cos\beta & -\frac{\sin\beta}{\sqrt{2}} \\ \frac{1-\cos\beta}{2} & \frac{\sin\beta}{\sqrt{2}} & \frac{1+\cos\beta}{2} \end{pmatrix}. \tag{42}$$

For higher values of J, this direct method of calculating the d matrices will of course become more and more difficult, and we shall have to find a better method.

D General Expression for the Rotation Matrices

Although we know the matrix elements of $J_y = -\frac{i}{2}(J_+ - J_-)$, the calculation of $(J_y)^n$ (in the expansion of the exponential $e^{-i\beta J_y}$) is complicated because of the noncommutability of the operators J_+ and J_-. The calculation of the d matrix would be straightforward if we could restructure the rotation operator in the form,

$$e^{\beta_+ J_+} e^{\beta_0 J_0} e^{\beta_- J_-}, \qquad \text{or} \qquad e^{\gamma_- J_-} e^{\gamma_0 J_0} e^{\gamma_+ J_+}.$$

In principle, the transformation of an operator product of the form $e^A e^B$ into the form e^C, where A and B are noncommuting operators (or their matrix realizations) can be achieved by the so-called Baker–Campbell–Hausdorff relation

$$e^A e^B = e^C, \qquad \text{with}$$
$$C = A + B + \tfrac{1}{2}[A, B] + \tfrac{1}{12}\Big([A, [A, B]] + [B, [B, A]]\Big) + \cdots, \tag{43}$$

where the $\cdots$ involves triple and quadruple and ever higher commutators of the operators A and B. The Baker–Campbell–Hausdorff expansion is very useful in cases in which the multiple commutators are all zero, say, after the second or third term. For the angular momentum algebra, unfortunately, the series is an infinite one, with ever more complicated coefficients. The desired final result, however, depends only on the angular momentum commutator algebra. The coefficients, $\beta_\pm, \beta_0$, or $\gamma_\pm, \gamma_0$, depend only on the commutator algebra of J_+, J_-, and J_0, not on the quantum number, J. It will therefore be sufficient to use the simplest nontrivial representation of the rotation operator, viz., the representation for $J = \frac{1}{2}$, where we deal with extremely simple 2×2 matrices. It will be useful to solve a slightly more general problem, and "disentangle" the more general operator, $(a_+ J_+ + a_0 J_0 + a_- J_-)$ through

$$e^{(a_+ J_+ + a_0 J_0 + a_- J_-)} = e^{b_+ J_+} e^{(lnb_0) J_0} e^{b_- J_-} = e^{c_- J_-} e^{(lnc_0) J_0} e^{c_+ J_+}. \tag{44}$$

(We have renamed $\beta_0 = lnb_0$, $\beta_\pm = b_\pm$, for convenience, similarly for the $\gamma_\pm, \gamma_0$. Now,

$$\text{for } J = \tfrac{1}{2},$$

$$J_+ = \begin{pmatrix} 0 & 1 \\ 0 & 0 \end{pmatrix}, \qquad J_- = \begin{pmatrix} 0 & 0 \\ 1 & 0 \end{pmatrix}, \qquad J_0 = \begin{pmatrix} \frac{1}{2} & 0 \\ 0 & -\frac{1}{2} \end{pmatrix}, \tag{45}$$

so $(J_+)^n$ and $(J_-)^n$, with $n \geq 2$, are all null matrices. With

$$(a_+ J_+ + a_0 J_0 + a_- J_-) = \begin{pmatrix} \frac{1}{2}a_0 & a_+ \\ a_- & -\frac{1}{2}a_0 \end{pmatrix} \equiv \mathbf{a}, \qquad \text{we have} \tag{46}$$

$$\mathbf{a}^2 = \begin{pmatrix} (\frac{1}{4}a_0^2 + a_+ a_-) & 0 \\ 0 & (\frac{1}{4}a_0^2 + a_+ a_-) \end{pmatrix} = a^2 \begin{pmatrix} 1 & 0 \\ 0 & 1 \end{pmatrix} = a^2 \mathbf{1}, \tag{47}$$

and with $\mathbf{a}^{2n} = a^{2n}\mathbf{1}$ and $\mathbf{a}^{2n+1} = a^{2n}\mathbf{a}$, we have

$$\begin{aligned} e^{(a_+ J_+ + a_0 J_0 + a_- J_-)} &= \begin{pmatrix} \cosh a + \frac{1}{2}\frac{a_0}{a}\sinh a & \frac{a_+}{a}\sinh a \\ \frac{a_-}{a}\sinh a & \cosh a - \frac{1}{2}\frac{a_0}{a}\sinh a \end{pmatrix} \\ &\equiv \cosh a \mathbf{1} + \frac{\sinh a}{a}\mathbf{a}. \end{aligned} \tag{48}$$

[The scalar a is defined through $a^2 = (\frac{1}{4}a_0^2 + a_+ a_-)$.] Similarly,

$$e^{b_+ J_+} e^{(lnb_0) J_0} e^{b_- J_-} = \frac{1}{\sqrt{b_0}} \begin{pmatrix} (b_0 + b_+ b_-) & b_+ \\ b_- & 1 \end{pmatrix}, \tag{49}$$

and

$$e^{c_- J_-} e^{(lnc_0) J_0} e^{c_+ J_+} = \sqrt{c_0} \begin{pmatrix} 1 & c_+ \\ c_- & (\frac{1}{c_0} + c_+ c_-) \end{pmatrix}. \tag{50}$$

Now, the coefficients $b_\pm$, b_0 or $c_\pm$, c_0 can be evaluated in terms of the $a_\pm$, a_0 by comparing eqs. (49) or (50) with eq. (48). We are interested in the rotation operator, $e^{-i\beta J_y} = e^{-\frac{1}{2}\beta(J_+ - J_-)}$. In our special case, therefore, $a_\pm = \mp\frac{1}{2}\beta$, $a_0 = 0$. For this case, we have

$$b_\pm = \mp \tan\tfrac{\beta}{2}, \quad b_0 = \frac{1}{\cos^2(\frac{\beta}{2})}, \qquad c_\pm = \mp \tan\tfrac{\beta}{2}, \quad c_0 = \cos^2(\tfrac{\beta}{2}), \tag{51}$$

so

$$
\begin{aligned}
e^{-i\beta J_y} &= e^{+\tan\frac{\beta}{2}J_-}\left(\cos^2(\tfrac{\beta}{2})\right)^{J_0}e^{-\tan\frac{\beta}{2}J_+} \\
&= e^{-\tan\frac{\beta}{2}J_+}\frac{1}{\left(\cos^2(\frac{\beta}{2})\right)^{J_0}}e^{+\tan\frac{\beta}{2}J_-}.
\end{aligned}
\tag{52}
$$

With the first form of this result, we can write

$$
\begin{aligned}
d^J_{M'M}(\beta) &= \langle JM'|e^{+\tan\frac{\beta}{2}J_-}\left(\cos^2(\tfrac{\beta}{2})\right)^{J_0}e^{-\tan\frac{\beta}{2}J_+}|JM\rangle \\
&= \sum_n \frac{\left(\tan\frac{\beta}{2}\right)^{M+n-M'}\left(\cos\frac{\beta}{2}\right)^{2(M+n)}\left(-\tan\frac{\beta}{2}\right)^n}{n!(M+n-M')!} \\
&\quad\times \langle JM'|(J_-)^{M+n-M'}(J_+)^n|JM\rangle,
\end{aligned}
\tag{53}
$$

where both n and $M+n-M'$ must be positive integers (including zero) and restrictions on magnetic quantum numbers limit the sum over n to a sum from $n \geq (M'-M)$ to $n \leq (J-M)$ for the case $M'-M \geq 0$, and to a sum from $n=0$ to $n \leq (J-M)$ for the case $M'-M<0$. Using the simple known matrix elements of $J_\pm$, with

$$
(J_+)^n|JM\rangle = \sqrt{\frac{(J-M)!(J+M+n)!}{(J-M-n)!(J+M)!}}|J(M+n)\rangle \qquad \text{and} \tag{54}
$$

$$
(J_-)^{M+n-M'}|J(M+n)\rangle = \sqrt{\frac{(J+M+n)!(J-M')!}{(J+M')!(J-M-n)!}}|JM'\rangle. \tag{55}
$$

We therefore get

$$
\begin{aligned}
d^J_{M'M}(\beta) &= \left[\frac{(J-M)!(J-M')!}{(J+M)!(J+M')!}\right]^{\frac{1}{2}}\sum_n(-1)^n \\
&\times \frac{(J+M+n)!}{(J-M-n)!(M+n-M')!n!}(\cos\tfrac{\beta}{2})^{M+M'}(\sin\tfrac{\beta}{2})^{M-M'+2n},
\end{aligned}
\tag{56}
$$

where the sum over n ranges from $n = \max.[0, M'-M]$ to $n=(J-M)$. In the very special case $M=J$, the integer n is restricted to $n=0$, and

$$
d^J_{M'J}(\beta) = \left[\frac{(2J)!}{(J+M')!(J-M')!}\right]^{\frac{1}{2}}(\cos\tfrac{\beta}{2})^{J+M'}(\sin\tfrac{\beta}{2})^{J-M'}. \tag{57}
$$

In the further special case $M=-J$, it will be useful to rename the summation index $n = n' + J + M'$, so eq. (56) yields

$$
\begin{aligned}
d^J_{M',-J}(\beta) &= (-1)^{J+M'}\left[\frac{2J)!}{(J+M')!(J-M')!}\right]^{\frac{1}{2}}(\cos\tfrac{\beta}{2})^{-J+M'}(\sin\tfrac{\beta}{2})^{J+M'} \\
&\times \sum_{n'=0}^{J-M'}\frac{(J-M')!(-1)^{n'}}{(J-M'-n')!n'!}(\sin^2\tfrac{\beta}{2})^{n'}
\end{aligned}
$$

$$= (-1)^{J+M'} \left[\frac{(2J)!}{(J+M')!(J-M')!} \right]^{\frac{1}{2}} (\cos\tfrac{\beta}{2})^{J-M'} (\sin\tfrac{\beta}{2})^{J+M'}, \tag{58}$$

where we have used the binomial expansion of $(1 - \sin^2(\beta/2))^{J-M'}$.

E Rotation Operators and Angular Momentum Coherent States

In Chapter 19, we defined two slightly different types of angular momentum coherent states $|\alpha\rangle$ and $|z\rangle$, giving us two slightly different continuous representations of state vectors $|\psi\rangle$ of our 3-D world in terms of functions of the complex variables, α and z, where these complex numbers are related to the orientation of the physical system in our 3-D world. We are now in a position to see the relationship between these type I and type II coherent states. Such angular coherent states will again be very useful for physical systems best described through a statistical distribution of states with different orientations in our laboratory.

The type I coherent state was defined through

$$|\alpha\rangle = e^{\alpha^* J_+ - \alpha J_-}|J, M = -J\rangle = e^{-i\theta(\vec{J}\cdot\vec{n})}|J, M = -J\rangle, \tag{59}$$

where the complex variable, α, is related to the angles θ, ϕ through $\alpha = -(\theta/2)e^{i\phi}$, so the unit vector, $\vec{n}$, lies in the x, y-plane and is rotated forward from the laboratory y axis through an angle ϕ. The coherent state $|\alpha\rangle$ can thus be expressed through a rotation operator, with Euler angles, ϕ, θ, and $\gamma = 0$:

$$|\alpha\rangle \equiv |\alpha(\theta, \phi)\rangle = R(\phi, \theta, 0)|J, M = -J\rangle. \tag{60}$$

We can therefore expand the type I angular coherent state through

$$\begin{aligned}
|\alpha(\theta, \phi)\rangle &= \sum_{M=-J}^{+J} |JM\rangle D^J_{M,-J}(\phi, \theta, 0)^* \\
&= \sum_{M=-J}^{+J} |JM\rangle c_{J,M} (-1)^{J+M} \left(\sin\tfrac{\theta}{2}\right)^{J+M} \left(\cos\tfrac{\theta}{2}\right)^{J-M} e^{-iM\phi} \\
&= \sum_{M=-J}^{+J} |JM\rangle c_{J,M} \frac{(-1)^{J+M}\left(\tan\tfrac{\theta}{2}\right)^{J+M}}{\left(1+\tan^2\tfrac{\theta}{2}\right)^J} e^{-iM\phi}, \\
\text{with} \quad c_{J,M} &= \sqrt{\frac{(2J)!}{(J+M)!(J-M)!}},
\end{aligned} \tag{61}$$

where we have used eq. (58) of the last section and the trivial identity,

$$\cos^2\tfrac{\theta}{2} = (1 + \tan^2\tfrac{\theta}{2})^{-1}.$$

The above expansion suggests the unit operator appropriate for this type of coherent state is

$$\begin{aligned}1 &= \frac{(2J+1)}{4\pi}\int\!\!\int d\Omega|\alpha(\theta,\phi)\rangle\langle\alpha(\theta,\phi)| \\ &= \sum_{M,M'}\frac{(2J+1)}{4\pi}\int\!\!\int d\Omega D^J_{M,-J}(\phi,\theta,0)^* D^J_{M',-J}(\phi,\theta,0)|JM\rangle\langle JM'| \\ &= \sum_M |JM\rangle\langle JM|, \end{aligned} \tag{62}$$

where $d\Omega = \sin\theta d\theta d\phi$. The angular ranges in the integrals have their usual values, and we have made use of the orthonormality integral

$$\int\!\!\int d\Omega D^J_{M,\mu}(\phi,\theta,0)^* D^{J'}_{M',\mu'}(\phi,\theta,0) = \delta_{JJ'}\delta_{MM'}\delta_{\mu\mu'}\frac{4\pi}{(2J+1)}. \tag{63}$$

[The derivation of this integral is given in detail through eq. (30) of the next chapter; note the D functions with Euler angle $\gamma = 0$ are of course independent of this third angle.]

The type II angular momentum coherent state, conversely, was defined by

$$\begin{aligned}|z\rangle = e^{z^* J_+}|J,-J\rangle &= \sum_{n=0}^{2J}\frac{z^{*n}}{\sqrt{n!}}\sqrt{\frac{(2J)!}{(2J-n)!}}|J, M=-J+n\rangle \\ &= \sum_{M=-J}^{+J}(z^*)^{J+M}\sqrt{\frac{(2J)!}{(J+M)!(J-M)!}}|JM\rangle. \end{aligned} \tag{64}$$

In Chapter 19, it was shown that the z space functions

$$\frac{z^n}{\sqrt{n!}}\sqrt{\frac{(2J)!}{(2J-n)!}}$$

formed an orthonormal set with respect to the measure

$$\frac{(2J+1)}{\pi}\frac{d^2 z}{(1+zz^*)^{2J+2}}.$$

We can now complete our discussion of these angular momentum coherent states by showing explicitly the relationship between the two types of coherent states. Eq. (61) could have been obtained directly by putting the operator $e^{\alpha^* J_+ - \alpha J_-}$ into normal ordered form through the comparison of eqs. (49) and (48) of the last section, yielding

$$\begin{aligned}|\alpha\rangle &= e^{b_+ J_+}\frac{1}{\left(\cos^2\frac{\theta}{2}\right)^{J_0}}e^{b_- J_-}|-J\rangle \\ &= e^{b_+ J_+}\left(\cos^2\tfrac{\theta}{2}\right)^{J}|-J\rangle, \\ \text{with}\quad & b_\pm = \mp\tan\tfrac{\theta}{2}e^{\mp i\phi}. \end{aligned} \tag{65}$$

Expansion of the exponential $e^{b_+ J_+}$ again leads to eq. (61), but the present form suggests a change from the complex variable, α, to the new complex variable z, where

$$z = \rho e^{i\phi} = -\tan\tfrac{\theta}{2} e^{i\phi},$$

so

$$|\alpha\rangle = e^{z^* J_+} \frac{1}{(1 + zz^*)^J} |-J\rangle. \tag{66}$$

Also, with

$$\rho^2 = zz^* = \tan^2\tfrac{\theta}{2},$$

we have

$$\sin\theta d\theta d\phi = 4\rho d\rho d\phi \frac{1}{(1+\rho^2)^2}.$$

With this relation and eq. (66), the unit operator can be transformed into

$$\begin{aligned}
&\frac{(2J+1)}{4\pi} \int\!\!\int d\Omega\, |\alpha\rangle\langle\alpha| \\
&= \frac{(2J+1)}{\pi} \int_0^{2\pi} d\phi \int_0^\infty \frac{d\rho\rho}{(1+\rho^2)^2} \frac{e^{z^* J_+}}{(1+\rho^2)^J} |-J\rangle\langle -J| \frac{e^{z J_-}}{(1+\rho^2)^J} \\
&= \frac{(2J+1)}{\pi} \int \frac{d^2 z}{(1+zz^*)^{2J+2}} |z\rangle\langle z|.
\end{aligned} \tag{67}$$

This relation is precisely the unit operator needed for the type II coherent states, $|z\rangle$, with a measure making the z space functions

$$\frac{z^n}{\sqrt{n!}} \sqrt{\frac{(2J)!}{(2J-n)!}}$$

into an orthonormal set, as shown in Chapter 19.

30

The Clebsch–Gordan Series

For a system built from state vectors, $|j_1 m_1\rangle$ and $|j_2 m_2\rangle$ express the angular momentum–coupled state vector $|j_1 j_2 j m\rangle$ for the rotated system in terms of the uncoupled state vectors, also in the rotated system

$$
\begin{aligned}
|(j_1 j_2 j m)_{\text{rot.}}\rangle &= \sum_{m_1,(m_2)} |(j_1 m_1)_{\text{rot}}\rangle |(j_2 m_2)_{\text{rot.}}\rangle \langle j_1 m_1 j_2 m_2 | j m\rangle \\
&= \sum_{\mu} |j_1 j_2 j \mu\rangle D^{j*}_{\mu m} \\
&= \sum_{m_1,(m_2)} \sum_{\mu_1,(\mu_2)} |j_1 \mu_1\rangle |j_2 \mu_2\rangle D^{j_1 *}_{\mu_1 m_1} D^{j_2 *}_{\mu_2 m_2} \langle j_1 m_1 j_2 m_2 | j m\rangle. \qquad (1)
\end{aligned}
$$

Now, expanding

$$
|j_1 \mu_1\rangle |j_2 \mu_2\rangle = \sum_{j'} |j_1 j_2 j' \mu\rangle \langle j_1 \mu_1 j_2 \mu_2 | j' \mu\rangle, \qquad (2)
$$

left-multiplying by $\langle j_1 j_2 j \mu|$, and using the orthonormality of the coupled vectors $\langle j_1 j_2 j \mu|$ and $|j_1 j_2 j' \mu\rangle$, we get (after complex conjugation of this equation, using the reality of the Clebsch–Gordan coefficients)

$$
D^{j}_{\mu m} = \sum_{m_1,(m_2)} \sum_{\mu_1,(\mu_2)} D^{j_1}_{\mu_1 m_1} D^{j_2}_{\mu_2 m_2} \langle j_1 m_1 j_2 m_2 | j m\rangle \langle j_1 \mu_1 j_2 \mu_2 | j \mu\rangle. \qquad (3)
$$

This relation is the so-called Clebsch–Gordan series. This relation could be used in a build-up process to calculate the D functions for $j = \frac{3}{2}$ from the known D functions for $j = 1$ and $j = \frac{1}{2}$, and so on for D functions of higher j values.

Finally, from the inverse of the process used here, we get the second relation

$$D^{j_1}_{\mu_1 m_1} D^{j_2}_{\mu_2 m_2} = \sum_j D^j_{\mu m} \langle j_1 m_1 j_2 m_2 | jm \rangle \langle j_1 \mu_1 j_2 \mu_2 | j\mu \rangle. \tag{4}$$

A Addition Theorem for Spherical Harmonics

Let us replace the angular momentum eigenfunctions $\psi_{jm}(r, \theta, \phi, \vec{\sigma})$ by the spherical harmonics, $Y_{lm}(\theta\ \phi)$ and use the general rotation relation

$$Y_{lm}(\theta', \phi') = \sum_\mu Y_{l\mu}(\theta, \phi) D^l_{\mu m}(\alpha, \beta, \gamma)^*. \tag{5}$$

Suppose we have two particles, with position vectors $\vec{r}_1 = (r_1, \theta_1, \phi_1)$ and $\vec{r}_2 = (r_2, \theta_2, \phi_2)$, (see Fig. 30.1), we first note

$$\sum_m Y^*_{lm}(\theta_1, \phi_1) Y_{lm}(\theta_2 \phi_2) = \mathcal{I}, \tag{6}$$

where $\mathcal{I}$ is a rotationally invariant quantity, depending only on the relative position of the two particles.

$$\sum_m Y^*_{lm}(\theta'_1, \phi'_1) Y_{lm}(\theta'_2, \phi'_2) =$$
$$\sum_m \sum_{\mu\nu} Y^*_{l\mu}(\theta_1, \phi_1) D^l_{\mu m}(\alpha, \beta, \gamma) Y_{l\nu}(\theta_2, \phi_2) D^l_{\nu m}(\alpha, \beta, \gamma)^*. \tag{7}$$

Now, using the unitarity of the D functions,

$$\sum_m D^l_{\mu m} D^{l*}_{\nu m} = \delta_{\mu\nu}, \tag{8}$$

we have

$$\sum_m Y^*_{lm}(\theta'_1, \phi'_1) Y_{lm}(\theta'_2, \phi'_2) = \sum_\mu Y^*_{l\mu}(\theta_1, \phi_1) Y_{l\mu}(\theta_2, \phi_2) = \mathcal{I}. \tag{9}$$

To evaluate the invariant, choose the x', y', z' coordinate system, such that $\theta'_1 = 0$; i.e., choose the z' axis along $\vec{r}_1$. Then,

$$Y_{lm}(0, \phi'_1) = \delta_{m0} \sqrt{\frac{(2l+1)}{4\pi}} \Big(P_l(\cos 0) \Big) = \delta_{m0} \sqrt{\frac{(2l+1)}{4\pi}}, \tag{10}$$

so

$$\mathcal{I} = \sqrt{\frac{(2l+1)}{4\pi}} Y_{l0}(\theta'_2 = \theta_{12}, 0) = \frac{(2l+1)}{4\pi} P_l(\cos\theta_{12}), \tag{11}$$

and

$$\sum_m Y^*_{lm}(\theta_1, \phi_1) Y_{lm}(\theta_2, \phi_2) = \frac{(2l+1)}{4\pi} P_l(\cos\theta_{12}). \tag{12}$$

That is, this invariant is expressed in terms of the Legendre polynomial, expressed in terms of the angle θ_{12} between the two vectors $\vec{r}_1$ and $\vec{r}_2$.

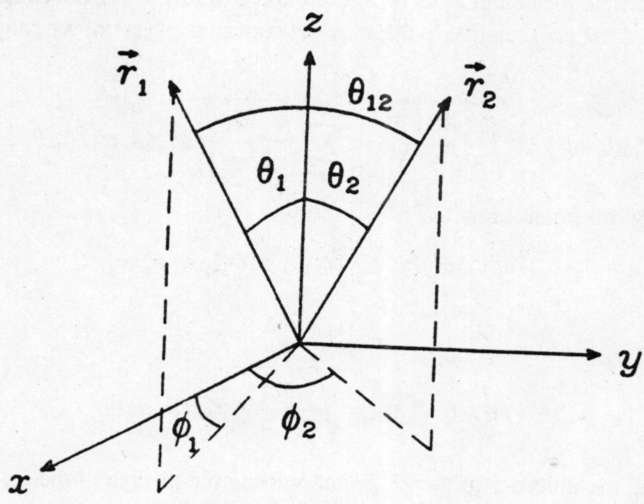

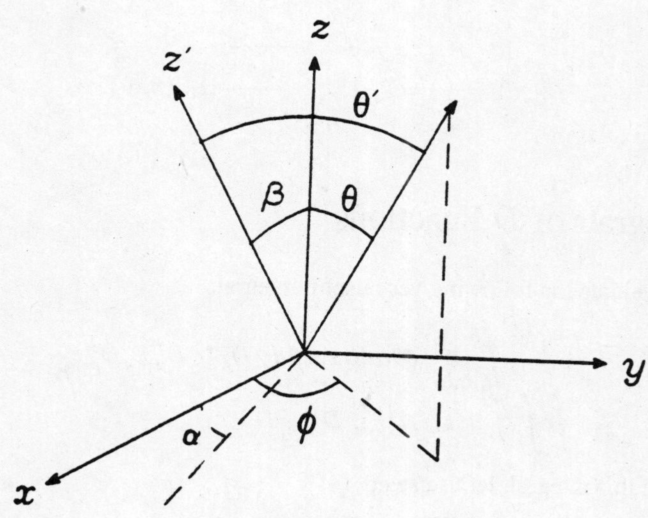

FIGURE 30.1.

By renaming $\theta_2, \phi_2 = \theta, \phi$, and $\theta_1, \phi_1 = \beta, \alpha$, and $\theta_{12} = \theta'$, i. e., by thinking of a single particle located at $\vec{r}_2 \equiv (r, \theta, \phi)$ relative to the x, y, z coordinate system, while $\vec{r}_2 \equiv (r, \theta', \phi')$ relative to the x', y', z' coordinate system, we can rewrite eq. (11) as

$$Y_{l0}(\theta', \phi') = \sqrt{\frac{4\pi}{(2l+1)}}\mathcal{I} = \sum_m \sqrt{\frac{4\pi}{(2l+1)}} Y^*_{lm}(\beta, \alpha) Y_{lm}(\theta, \phi). \tag{13}$$

Comparing this equation with

$$Y_{l0}(\theta', \phi') = \sum_m Y_{lm}(\theta, \phi) D^{l*}_{m0}(\alpha, \beta, 0), \tag{14}$$

we get

$$D^l_{m0}(\alpha, \beta, 0) = \sqrt{\frac{4\pi}{(2l+1)}} Y_{lm}(\beta, \alpha). \tag{15}$$

Also, using the unitarity of the $D^l_{\mu m}$, and writing the inverse to the Euler rotation transformation (α, β, γ) as $(-\gamma, -\beta, -\alpha)$,

$$\begin{aligned} D^l_{0m}(0, \beta, \gamma) &= D^{l*}_{m0}(-\gamma, -\beta, 0) = \sqrt{\frac{4\pi}{(2l+1)}} Y^*_{lm}(-\beta, -\gamma) \\ &= (-1)^m \sqrt{\frac{4\pi}{(2l+1)}} Y_{lm}(\beta, \gamma), \end{aligned} \tag{16}$$

so

$$D^l_{0m}(0, \beta, \gamma) = (-1)^m \sqrt{\frac{4\pi}{(2l+1)}} Y_{lm}(\beta, \gamma). \tag{17}$$

B Integrals of D Functions

We shall evaluate the following very useful integral:

$$\begin{aligned} I &= \int_0^{2\pi} \int_0^{\pi} \int_0^{2\pi} d\alpha d\beta \sin\beta d\gamma \, D^{j_3 *}_{\mu_3 m_3} D^{j_1}_{\mu_1 m_1} D^{j_2}_{\mu_2 m_2} \\ &= \int\!\!\int\!\!\int d\Omega_{\alpha\beta\gamma} \, D^{j_3 *}_{\mu_3 m_3} D^{j_1}_{\mu_1 m_1} D^{j_2}_{\mu_2 m_2}. \end{aligned} \tag{18}$$

To evaluate this integral, let us use eq. (4)

$$D^{j_1}_{\mu_1 m_1} D^{j_2}_{\mu_2 m_2} = \sum_j \langle j_1 m_1 j_2 m_2 | j m \rangle \langle j_1 \mu_1 j_2 \mu_2 | j \mu \rangle D^j_{\mu m} \tag{19}$$

and the relation

$$D^{j_3 *}_{\mu_3 m_3} = \left(e^{i\mu_3 \alpha} d^{j_3}_{\mu_3 m_3}(\beta) e^{i m_3 \gamma} \right)^* = e^{-i\mu_3 \alpha} d^{j_3}_{\mu_3 m_3}(\beta) e^{-i m_3 \gamma}, \tag{20}$$

where we have used the reality of the d function. In addition the d functions have the property

$$d^{j_3}_{\mu_3 m_3}(\beta) = (-1)^{\mu_3 - m_3} d^{j_3}_{-\mu_3,-m_3}(\beta). \tag{21}$$

For the values of $j = \frac{1}{2}$ and $j = 1$, this equation follows by inspection, (see Chapter 29). For values of $j \geq \frac{3}{2}$, this equation follows from the build-up relation, eq. (3), and the fact that sign change under the transformation $m_i \to -m_i$ in the m_i-dependent Clebsch–Gordan coefficient is balanced by the same sign change under the transformation $\mu_i \to -\mu_i$ in the μ_i-dependent Clebsch–Gordan coefficient. Our integral can then be transformed into

$$I = \int\int\int d\Omega_{\alpha\beta\gamma} D^{j_3 *}_{\mu_3 m_3} D^{j_1}_{\mu_1 m_1} D^{j_2}_{\mu_2 m_2} = \sum_j \langle j_1 m_1 j_2 m_2 | jm\rangle \langle j_1 \mu_1 j_2 \mu_2 | j\mu\rangle$$
$$\times (-1)^{\mu_3 - m_3} \int\int\int d\Omega_{\alpha\beta\gamma} D^{j_3}_{-\mu_3,-m_3} D^{j}_{\mu m}. \tag{22}$$

Using the Clebsch–Gordan series once more on the product of two D functions in our integral, we obtain

$$I = \sum_j \sum_{j'} \langle j_1 m_1 j_2 m_2 | jm\rangle \langle j_1 \mu_1 j_2 \mu_2 | j\mu\rangle (-1)^{\mu_3 - m_3} \langle j\mu j_3 - \mu_3 | j'\mu - \mu_3\rangle$$
$$\times \langle jm j_3 - m_3 | j'm - m_3\rangle \int\int\int d\Omega_{\alpha\beta\gamma} D^{j'}_{\mu-\mu_3, m-m_3}. \tag{23}$$

Now, we can use explicitly

$$\int\int\int d\Omega_{\alpha\beta\gamma} D^{L}_{\mathcal{M}M} = \int_0^{2\pi} d\alpha e^{i\mathcal{M}\alpha} \int_0^{2\pi} d\gamma e^{iM\gamma} \int_0^{\pi} d\beta \sin\beta d^L_{\mathcal{M}M}$$
$$= 2\pi \delta_{\mathcal{M}0} 2\pi \delta_{M0} \int_0^{\pi} d\beta \sin\beta d^L_{00}(\beta). \tag{24}$$

We also use

$$d^L_{00}(\beta) = \sqrt{\frac{4\pi}{(2L+1)}} Y_{L0}(\beta, -) \qquad \text{and} \qquad Y_{00} = \frac{1}{\sqrt{4\pi}} \tag{25}$$

to get

$$\int\int\int d\Omega_{\alpha\beta\gamma} D^{L}_{\mathcal{M}M} =$$
$$2\pi \delta_{\mathcal{M}0} \delta_{M0} \int_0^{2\pi} d\gamma \int_0^{\pi} d\beta \sin\beta \Big(\sqrt{\frac{4\pi}{(2L+1)}} Y_{L0}\Big)\Big(\sqrt{4\pi} Y^*_{00}\Big)$$
$$= 8\pi^2 \delta_{\mathcal{M}0} \delta_{M0} \delta_{L0}, \tag{26}$$

where we have made use of the orthonormality of the spherical harmonics. Thus, in eq. (23), we must have $j' = 0$, $\mu - \mu_3 = 0$, and $m - m_3 = 0$. This result simplifies two of the Clebsch–Gordan coefficients. From the symmetry property

(123) → (3 − 21) of the Clebsch–Gordan coefficients, we get

$$\langle jmj_3 - m_3|00\rangle = \frac{(-1)^{j-m_3}\langle 00j_3m_3|jm\rangle}{\sqrt{(2j+1)}} = \frac{(-1)^{j_3-m_3}}{\sqrt{(2j_3+1)}}\delta_{jj_3}\delta_{mm_3}. \tag{27}$$

Similarly,

$$\langle j\mu j_3 - \mu_3|00\rangle = \frac{(-1)^{\mu_3-j_3}}{\sqrt{(2j_3+1)}}\delta_{jj_3}\delta_{\mu\mu_3}. \tag{28}$$

With all of these relations, we get our final very simple result

$$\iiint d\Omega_{\alpha\beta\gamma} D^{j_3*}_{\mu_3m_3} D^{j_1}_{\mu_1m_1} D^{j_2}_{\mu_2m_2} = 8\pi^2 \frac{\langle j_1\mu_1j_2\mu_2|j_3\mu_3\rangle\langle j_1m_1j_2m_2|j_3m_3\rangle}{(2j_3+1)}. \tag{29}$$

A special case follows from the above by setting $j_2 = \mu_2 = m_2 = 0$. This leads to the orthonormality integral for the D functions

$$\iiint d\Omega_{\alpha\beta\gamma} D^{j*}_{\mu m} D^{j'}_{\mu'm'} = \frac{8\pi^2}{(2j+1)}\delta_{jj'}\delta_{\mu\mu'}\delta_{mm'}. \tag{30}$$

Another important special case follows by setting $\mu_1 = 0$, $\mu_2 = 0$, $\mu_3 = 0$, and now setting $j_i = l_i$, where l_i denotes the angular momentum quantum number is an integer. Using our result,

$$D^l_{m0}(\alpha,\beta,\gamma) = \sqrt{\frac{4\pi}{(2l+1)}}Y_{lm}(\beta,\alpha),$$

we get as a special case of the above

$$\int_0^{2\pi} d\alpha \int_0^{\pi} d\beta \sin\beta Y^*_{l_3m_3} Y_{l_1m_1} Y_{l_2m_2}$$
$$= \sqrt{\frac{(2l_1+1)(2l_2+1)}{(2l_3+1)4\pi}}\langle l_1m_1l_2m_2|l_3m_3\rangle\langle l_10l_20|l_30\rangle. \tag{31}$$

We could also write this formula as

$$\langle nl'm'|Y_{l_0m_0}|nlm\rangle = \sqrt{\frac{(2l+1)(2l_0+1)}{(2l'+1)4\pi}}\langle lml_0m_0|l'm'\rangle\langle l0l_00|l'0\rangle. \tag{32}$$

In earlier chapters, we worked hard to calculate these matrix elements for the special case with $l_0 = 1$. Now, everything follows from a knowledge of Clebsch–Gordan coefficients. Eqs. (31) and (32) tell us in particular these matrix elements are zero, unless l, l_0, l' satisfy the triangle condition of angular momentum addition, i.e., unless $l' = |l - l_0|, |l - l_0| + 1, \cdots, (l + l_0)$. Also, these matrix elements are zero unless $l + l_0 - l' =$ even integer. This parity selection rule follows because

$$\langle l_10l_20|l_30\rangle = (-1)^{l_1+l_2-l_3}\langle l_10l_20|l_30\rangle \tag{33}$$

a special case of the symmetry property

$$\langle j_1 - m_1j_2 - m_2|j_3 - m_3\rangle = (-1)^{j_1+j_2-j_3}\langle j_1m_1j_2m_2|j_3m_3\rangle. \tag{34}$$

Because the parity of the spherical harmonics, Y_{lm}, is given by $(-1)^l$, operators of odd parity can connect levels of even parity only to levels of odd parity and vice versa, whereas operators of even parity must preserve the parity of the states. In terms of the polar and azimuth angles, θ, ϕ, the space inversion operation can be achieved by the transformation $\theta, \phi \rightarrow (\pi - \theta), (\phi + \pi)$. The parity of the spherical harmonics follows from

$$Y_{lm}(\pi - \theta, \phi + \pi) = (-1)^l Y_{lm}(\theta, \phi). \tag{35}$$

The result of eq. (31) or (32) can be generalized to a much wider class of operators, which transform under rotations in our three-space like the spherical harmonics. These operators are named spherical tensor operators and will be the subject of the next chapter. Their matrix elements between the appropriate angular momentum eigenstates can be shown to have a form similar to that of eq. (31).

Problems

39. Calculate the Clebsch–Gordan coefficients, $\langle j_1 m_1 1 m_2 | jm \rangle$, for the three values, $m_2 = +1, 0, -1$, and $j = (j_1 + 1), j_1, (j_1 - 1)$, as functions of j_1 and m. Calculate first the coefficients $\langle j_1 m_1 j - m | 1 - m_2 \rangle$ with $m_2 = 0, +1$ from the known coefficients with $m_2 = -1$. [See eq. (20) of Chapter 27 for the latter.] Then use symmetry properties of the Clebsch–Gordan coefficients to relate these coefficients to $\langle j_1 m_1 1 m_2 | jm \rangle$.

40. Calculate $D^{\frac{3}{2}}_{\mu m}(\alpha, \beta, \gamma)$ from a knowledge of $d^{\frac{1}{2}}_{\mu_1 m_1}(\beta)$ and $d^1_{\mu_2 m_2}(\beta)$ and the Clebsch–Gordan series.

A beam of particles with spin $s = \frac{3}{2}$ is polarized by a filter so the particles are in a state with $m_s = +\frac{3}{2}$, with respect to the beam axis. If there are N (particles/cm^2 sec.) in the beam, how many particles will pass through a second filter set for $m_s = +\frac{1}{2}$, but with its axis rotated about a direction perpendicular to the beam through an angle of 30°?

41. The 2-D isotropic harmonic oscillator and the D functions.

The eigenvectors $|n_1 n_2\rangle$ of the 2-D isotropic harmonic oscillator Hamiltonian

$$H = \hbar\omega_0(\tfrac{1}{2}p_x^2 + \tfrac{1}{2}x^2 + \tfrac{1}{2}p_y^2 + \tfrac{1}{2}y^2), \quad \text{with} \quad H|n_1 n_2\rangle = \hbar\omega_0(n_1 + n_2 + 1)|n_1 n_2\rangle,$$

can be expressed in terms of oscillator creation operators

$$a_x^\dagger = \sqrt{\tfrac{1}{2}}(x - ip_x); \qquad a_y^\dagger = \sqrt{\tfrac{1}{2}}(y - ip_y),$$

by

$$|n_1 n_2\rangle = \frac{(a_x^\dagger)^{n_1}}{\sqrt{n_1!}} \frac{(a_y^\dagger)^{n_2}}{\sqrt{n_2!}} |00\rangle.$$

Show that the operators, Λ_i, defined by

$$\Lambda_i = \tfrac{1}{2}\,(a_x^\dagger \quad a_y^\dagger)\,(\sigma_i) \begin{pmatrix} a_x \\ a_y \end{pmatrix},$$

where σ_i are 2×2 Pauli spin matrices, satisfy standard angular momentum commutation relations, such that, if we define

$$\Lambda_\pm = (\Lambda_1 \pm i\Lambda_2), \qquad \Lambda_0 = \Lambda_3,$$

$$[\Lambda_0, \Lambda_\pm] = \pm\Lambda_\pm, \qquad [\Lambda_+, \Lambda_-] = 2\Lambda_0,$$

with

$$\Lambda_+ = a_x^\dagger a_y, \qquad \Lambda_- = a_y^\dagger a_x, \qquad \Lambda_0 = \tfrac{1}{2}(a_x^\dagger a_x - a_y^\dagger a_y).$$

Show that the $|n_1 n_2\rangle$ are eigenvectors of the operators $\vec{\Lambda}^2$ and Λ_0, $|\Lambda M_\Lambda\rangle$, with

$$\Lambda = \tfrac{1}{2}(n_1 + n_2), \qquad M_\Lambda = \tfrac{1}{2}(n_1 - n_2).$$

We therefore have

$$a_x^\dagger|0\rangle = |\Lambda = \tfrac{1}{2}, M_\Lambda = +\tfrac{1}{2}\rangle, \qquad a_y^\dagger|0\rangle = |\Lambda = \tfrac{1}{2}, M_\Lambda = -\tfrac{1}{2}\rangle,$$

and

$$e^{-i\beta\Lambda_2}a_x^\dagger|0\rangle = \sum_{\bar{M}_\Lambda}|\tfrac{1}{2}\bar{M}_\Lambda\rangle d^{\frac{1}{2}}_{\bar{M}_\Lambda,+\frac{1}{2}}(\beta) = \left(a_x^\dagger \cos(\frac{\beta}{2}) + a_y^\dagger \sin(\frac{\beta}{2})\right)|0\rangle,$$

$$e^{-i\beta\Lambda_2}a_y^\dagger|0\rangle = \sum_{\bar{M}_\Lambda}|\tfrac{1}{2}\bar{M}_\Lambda\rangle d^{\frac{1}{2}}_{\bar{M}_\Lambda,-\frac{1}{2}}(\beta) = \left(-a_x^\dagger \sin(\frac{\beta}{2}) + a_y^\dagger \cos(\frac{\beta}{2})\right)|0\rangle.$$

If we now rename

$$\Lambda \to j, \qquad M_\Lambda \to m, \qquad \text{so} \qquad n_1 = j + m; \;\; n_2 = j - m,$$

we have

$$e^{-i\beta\Lambda_2}|n_1 n_2\rangle = \sum_{\bar{n}_1,\bar{n}_2}|\bar{n}_1\bar{n}_2\rangle d^{\frac{1}{2}(n_1+n_2)}_{\frac{1}{2}(\bar{n}_1-\bar{n}_2),\frac{1}{2}(n_1-n_2)}(\beta),$$

or

$$e^{-i\beta\Lambda_2}|jm\rangle = \sum_{m'}|jm'\rangle d^j_{m',m}(\beta)$$

$$= \frac{\left(a_x^\dagger \cos(\frac{\beta}{2}) + a_y^\dagger \sin(\frac{\beta}{2})\right)^{j+m}}{\sqrt{(j+m)!}} \frac{\left(-a_x^\dagger \sin(\frac{\beta}{2}) + a_y^\dagger \cos(\frac{\beta}{2})\right)^{j-m}}{\sqrt{(j-m)!}}|0\rangle.$$

Use this expression to derive a general expression for the d function:

$$d^j_{m'm}(\beta) = \sqrt{\frac{(j+m')!(j-m')!}{(j+m)!(j-m)!}}\sum_\alpha \frac{(-1)^{j-m-\alpha}(j+m)!(j-m)!}{(m+m'+\alpha)!(j-m'-\alpha)!\alpha!(j-m-\alpha)!}$$

$$\times\left(\sin(\frac{\beta}{2})\right)^{2j-2\alpha-m-m'}\left(\cos(\frac{\beta}{2})\right)^{2\alpha+m+m'}$$

42. Show that the Hamiltonian for an electron in a uniform external magnetic field, $\vec{B}_0$, in the z direction, with $A_x = -\frac{1}{2} y_{\text{phys.}} B_0$, $A_y = \frac{1}{2} x_{\text{phys.}} B_0$, can be written in terms of dimensionless x and y, defined by

$$x_{\text{phys.}} = x\sqrt{\frac{\hbar}{m\omega_L}}, \qquad y_{\text{phys.}} = y\sqrt{\frac{\hbar}{m\omega_L}}, \qquad \text{with} \quad \omega_L = \frac{|e|B_0}{2mc},$$

as

$$H = \hbar\omega_L\Big((a_x^\dagger a_x + a_y^\dagger a_y + 1) + i(a_y^\dagger a_x - a_x^\dagger a_y)\Big) + \frac{p_z^2}{2m},$$

where

$$a_x = \sqrt{\tfrac{1}{2}}(x + ip_x), \qquad a_x^\dagger = \sqrt{\tfrac{1}{2}}(x - ip_x),$$

$$a_y = \sqrt{\tfrac{1}{2}}(y + ip_y), \qquad a_y^\dagger = \sqrt{\tfrac{1}{2}}(y - ip_y),$$

are 1-D harmonic oscillator annihilation and creation operators. Use the commutator algebra of the operators, $\Lambda_+ = a_x^\dagger a_y$, $\Lambda_- = a_y^\dagger a_x$, and $\Lambda_0 = \frac{1}{2}(a_x^\dagger a_x - a_y^\dagger a_y)$ (see problem 41) and standard angular momentum theory results to find the energy eigenvalues for $H(x, y)$. In particular, find the degeneracy, g_n, for the nth level, with energy E_n.

For this purpose, use a basis in which the operators $\vec{\Lambda}^2$ and Λ_2, rather than the conventional $\vec{\Lambda}^2$ and Λ_3, are diagonal. Also, the full energy includes a contribution from the free particle motion in the z direction,

$$E_{\text{total}} = E_n + \frac{\hbar^2 k^2}{2m}, \qquad \text{with } k = 0 \to \infty,$$

where $\hbar^2 k^2$ gives the eigenvalue of p_z^2. Also, try to give an explanation for the value of the degeneracy, g_n, in terms of the possible classical orbits of the electron.

31

Spherical Tensor Operators

From the previous discussion, it is clear it would be advantageous to give vectors, such as $\vec{r}$, not in Cartesian component, but in spherical component form. Recalling

$$\begin{pmatrix} rY_{1+1} \\ rY_{10} \\ rY_{1-1} \end{pmatrix} = \sqrt{\frac{3}{4\pi}} \times \begin{pmatrix} -\frac{1}{\sqrt{2}} r \sin\theta e^{+i\phi} \\ r\cos\theta \\ +\frac{1}{\sqrt{2}} r \sin\theta e^{-i\phi} \end{pmatrix} = \sqrt{\frac{3}{4\pi}} \times \begin{pmatrix} -\frac{1}{\sqrt{2}}(x+iy) \\ z \\ +\frac{1}{\sqrt{2}}(x-iy) \end{pmatrix},$$

it will be useful to write the vector $\vec{r}$ in terms of the spherical components (r_{+1}, r_0, r_{-1}), with

$$r_{+1} = -\tfrac{1}{\sqrt{2}}(x+iy), \qquad r_0 = z, \qquad r_{-1} = +\tfrac{1}{\sqrt{2}}(x-iy). \tag{1}$$

Note, in particular, the differences between r_{+1} and $r_+ = (x+iy)$, and r_{-1} and $r_- = (x-iy)$. Before generalizing this vector result to higher rank tensor components in spherical form, let us look at second rank tensors, T_{ij}, in Cartesian component form. Write the general second rank tensor in terms of a symmetric, traceless part, S_{ij}, an antisymmetric part, A_{ij}, and the trace $\sum_\alpha T_{\alpha\alpha}$.

$$\begin{aligned} T_{ij} &= \left(\tfrac{1}{2}(T_{ij}+T_{ji}) - \tfrac{1}{3}\delta_{ij}\sum_\alpha T_{\alpha\alpha}\right) + \tfrac{1}{2}(T_{ij}-T_{ji}) + \tfrac{1}{3}\delta_{ij}\sum_\alpha T_{\alpha\alpha} \\ &= S_{ij} + A_{ij} + \tfrac{1}{3}\delta_{ij}\sum_\alpha T_{\alpha\alpha}, \end{aligned} \tag{2}$$

where $S_{ij} = S_{ji}$ and $\sum_\alpha S_{\alpha\alpha} = 0$, and $A_{ij} = -A_{ji}$. Under this decomposition of the nine components of the tensor, the 9×9 rotation matrix $O_{ij,\alpha\beta}$ that gives the rotated tensor components T'_{ij} in terms of the original $T_{\alpha\beta}$ via

$$T'_{ij} = \sum_{\alpha,\beta} O_{ij,\alpha\beta} T_{\alpha\beta} \tag{3}$$

will split into three submatrices. The five independent components of the traceless, symmetric tensor, S_{ij}, transform only among themselves. The three components of the antisymmmetric tensor, A_{ij}, will transform only among themselves, and the trace of the tensor is a rotationally invariant quantity. The five independent components of S_{ij} transform like the five components of a spherical harmonic, Y_{2m}. The three components of A_{ij} transform like the three components of a vector. We can see this at once if we build the tensor T_{ij} from two vectors (x, y, z) and (X, Y, Z), with $A_{12} = (xY - yX)$, $A_{31} = (zX - xZ)$, and $A_{23} = (yZ - zY)$, where these are the z, y, and x components of the vector product $\vec{r} \times \vec{R}$. The trace of the tensor transforms like the spherical harmonic Y_{00}. We have thus succeeded in finding second rank tensor components that transform like spherical harmonics, with $l = 2$, 1, and 0. As we go to higher rank tensors, this type of decomposition will become more difficult. For example, the 27 components of a third rank tensor will have 10 totally symmetric components. These components could be split further into seven components that transform like spherical harmonics, with $l = 3$, and three components that transform like the three components of a spherical harmonic, with $l = 1$. A single, totally antisymmetric tensor component exists, which is rotationally invariant; i.e., it transforms like a Y_{00}. The 16 remaining components of mixed symmetry could be split into two sets of five components that transform like spherical harmonics with $l = 2$ and two sets of three components that transform like spherical harmonics, with $l = 1$.

A Definition: Spherical Tensors

The set of $(2k + 1)$ components of a spherical tensor T^k_q with $q = +k, (k - 1), \ldots, -k$ and $k =$ integer or $\frac{1}{2}$-integer are a set of $(2k + 1)$ operators that under rotations transform like the components of an angular momentum eigenfunction, ψ_{kq}. Recalling $O_{\text{rot.}} = ROR^{-1}$,

$$(T^k_q)_{\text{rot.}} = R(\alpha, \beta, \gamma)T^k_q R^{-1}(\alpha, \beta, \gamma) = \sum_{\nu=-k}^{k} T^k_\nu D^{k*}_{\nu q}(\alpha, \beta, \gamma). \tag{4}$$

B Alternative Definition

The components of a spherical tensor can also be defined through their commutator relations with the components of the total angular momentum vector of the system on which the tensor components act. In particular,

$$\begin{aligned} [J_0, T^k_q] &= qT^k_q, \\ [J_\pm, T^k_q] &= \sqrt{(k \mp q)(k \pm q + 1)}T^k_{(q\pm1)}, \end{aligned} \tag{5}$$

where this definition essentially just involves infinitesimal rotation operators in place of the finite rotation operators of eq. (4). Let R correspond to an infinitesimal rotation about the z, x, y axes. For example, with

$$R = e^{-i\alpha J_z}, \qquad \text{and with} \qquad \alpha \ll 1, \tag{6}$$

we get

$$\begin{aligned} RT_q^k R^{-1} &= (1 - i\alpha J_z + \cdots)T_q^k(1 + i\alpha J_z + \cdots) = T_q^k - i\alpha[J_z, T_q^k] + \cdots \\ &= \sum_\nu D_{\nu q}^{k*} T_q^k = \delta_{\nu q}(1 - i\alpha q + \cdots)T_q^k, \end{aligned} \tag{7}$$

where we have used

$$D_{\nu q}^{k*} = \langle k\nu | e^{-i\alpha J_z} | kq \rangle = \delta_{\nu q}(1 - i\alpha q + \cdots), \tag{8}$$

thus leading to the first relation, $[J_z, T_q^k] = qT_q^k$. Similarly, combining infinitesimal rotations about the x and y axes, we are lead to the remaining two relations of eq. (5). We can use these relations to show the r_q of eq.(1) are spherical tensor components of rank $k = 1$. We can build higher rank spherical tensors from spherical vectors like these by a build-up process.

C Build-up Process

If $V_{q_1}^{k_1}$ and $U_{q_2}^{k_2}$ are spherical tensors, T_q^k, defined by

$$T_q^k = \sum_{q_1,(q_2)} V_{q_1}^{k_1} U_{q_2}^{k_2} \langle k_1 q_1 k_2 q_2 | kq \rangle, \tag{9}$$

are spherical tensors of rank k. This relation follows from

$$\begin{aligned} RT_q^k R^{-1} &= \sum_{q_1,q_2} RV_{q_1}^{k_1} R^{-1} RU_{q_2}^{k_2} R^{-1} \langle k_1 q_1 k_2 q_2 | kq \rangle \\ &= \sum_{q_1,q_2} \sum_{q_1',q_2'} V_{q_1'}^{k_1} U_{q_2'}^{k_2} D_{q_1' q_1}^{k_1 *} D_{q_2' q_2}^{k_2 *} \langle k_1 q_1 k_2 q_2 | kq \rangle \\ &= \sum_j \sum_{q_1' q_2'} V_{q_1'}^{k_1} U_{q_2'}^{k_2} \left(\sum_{q_1 q_2} \langle k_1 q_1 k_2 q_2 | kq \rangle \langle k_1 q_1 k_2 q_2 | jq \rangle \right) \langle k_1 q_1' k_2 q_2' | jq' \rangle D_{q'q}^{j*} \\ &= \sum_j \sum_{q_1' q_2'} V_{q_1'}^{k_1} U_{q_2'}^{k_2} \left(\delta_{jk} \right) \langle k_1 q_1' k_2 q_2' | jq' \rangle D_{q'q}^{j*} \\ &= \sum_{q_1' q_2'} V_{q_1'}^{k_1} U_{q_2'}^{k_2} \langle k_1 q_1' k_2 q_2' | kq' \rangle D_{q'q}^{k*} = \sum_{q'} T_{q'}^k D_{q'q}^{k*}. \end{aligned} \tag{10}$$

We shall often also use the shorthand notation

$$T_q^k = \sum_{q_1,(q_2)} V_{q_1}^{k_1} U_{q_2}^{k_2} \langle k_1 q_1 k_2 q_2 | kq \rangle \equiv [V^{k_1} \times U^{k_2}]_q^k. \tag{11}$$

Let us now use this build-up process to construct tensors from two vectors each of spherical rank, $l = 1$. Let us choose the coordinate vector of eq. (1), with $r_q = (r_{+1}, r_0, r_{-1})$, and as our second vector, the momentum vector with spherical components $p_q = (p_{+1}, p_0, p_{-1})$, and let us construct the spherical tensors

$$T_q^k = \sum_{q_1 q_2} r_{q_1} p_{q_2} \langle 1q_1 1q_2 | kq \rangle, \tag{12}$$

with $k = 0, 1, 2$. With $k = 0$, using

$$\langle 1q_1 1 - q_1 | 00 \rangle = \frac{1}{\sqrt{3}}(-1)^{1-q_1}, \tag{13}$$

we get

$$\begin{aligned} T_0^0 &= -\frac{1}{\sqrt{3}} \sum_q (-1)^q r_{+q} p_{-q} = -\frac{1}{\sqrt{3}} (\vec{r} \cdot \vec{p}) \\ &= -\frac{1}{\sqrt{3}} \Big(\tfrac{1}{2}(x + iy)(p_x - ip_y) + \tfrac{1}{2}(x - iy)(p_x + ip_y) + zp_z \Big). \end{aligned} \tag{14}$$

Note,

$$(\vec{r} \cdot \vec{p}) = \sum_q (-1)^q r_q^1 p_{-q}^1. \tag{15}$$

We can now generalize this scalar product to the more general scalar product of two tensors of rank k,

$$(T^k \cdot T^k) = \sum_m (-1)^m T_m^k T_{-m}^k. \tag{16}$$

We shall continue by constructing next the coupled spherical tensor operator, constructed from the vectors $\vec{r}$ and $\vec{p}$ to make the spherical tensor of rank $l = 1$,

$$T_q^1 = [r^1 \times p^1]_q^1. \tag{17}$$

$$\begin{aligned} T_{+1}^1 &= \langle 1110 | 11 \rangle r_{+1} p_0 + \langle 1011 | 11 \rangle r_0 p_{+1} \\ &= \tfrac{1}{\sqrt{2}} r_{+1} p_0 + (-\tfrac{1}{\sqrt{2}}) r_0 p_{+1} \\ &= -\tfrac{1}{2}(x + iy)p_z + \tfrac{1}{2} z(p_x + ip_y) = \tfrac{i}{\sqrt{2}} \Big(-\tfrac{1}{\sqrt{2}} (L_x + iL_y) \Big). \end{aligned} \tag{18}$$

In general,

$$[r^1 \times p^1]_m^1 = \tfrac{i}{\sqrt{2}} L_m, \qquad \text{with} \quad m = +1, 0, -1. \tag{19}$$

Also, in general,

$$\begin{aligned} [V^1 \times U^1]_{+1}^1 &= \tfrac{1}{2}(V_0 U_+ - V_+ U_0), \\ [V^1 \times U^1]_0^1 &= -\tfrac{1}{2\sqrt{2}}(V_+ U_- - V_- U_+), \\ [V^1 \times U^1]_{-1}^1 &= \tfrac{1}{2}(V_0 U_- - V_- U_0), \end{aligned} \tag{20}$$

where $V_+ = (V_x + iV_y)$, $V_- = (V_x - iV_y)$. Note again the difference from the spherical components $V_{+1} = -\frac{1}{\sqrt{2}} V_+$ and $V_{-1} = +\frac{1}{\sqrt{2}} V_-$. Finally, we build a spherical tensor of rank $l = 2$ from two vectors. The results for the five components are

$$[V^1 \times U^1]_{+2}^2 = \tfrac{1}{2} V_+ U_+,$$

$$
\begin{aligned}
[V^1 \times U^1]^2_{+1} &= -\tfrac{1}{2}(V_0U_+ + V_+U_0), \\
[V^1 \times U^1]^2_0 &= -\frac{1}{\sqrt{6}}\Big[\tfrac{1}{2}(V_+U_- + V_-U_+) - 2V_0U_0\Big], \\
[V^1 \times U^1]^2_{-1} &= \tfrac{1}{2}(V_0U_- + V_-U_0), \\
[V^1 \times U^1]^2_{-2} &= \tfrac{1}{2}V_-U_-.
\end{aligned} \tag{21}
$$

32
The Wigner–Eckart Theorem

The Wigner–Eckart theorem gives the M dependence of matrix elements of spherical tensor operators in a basis of good angular momentum. If J and M are the angular momentum quantum numbers for the total angular momentum operator appropriate for the space of a particular T^k_q,

$$\langle \alpha' J' M' | T^k_q | \alpha J M \rangle = F(\alpha, J, k, \alpha', J') \langle J M k q | J' M' \rangle, \tag{1}$$

where α is shorthand for all quantum numbers other than J, M, the total angular momentum quantum numbers for the system. The factor $F(\alpha, J, k, \alpha', J')$ is independent of the quantum numbers M, q, M'. The dependence on M, q, M' sits solely in the Clebsch–Gordan coefficient. The factor $F(\alpha, J, k, \alpha', J')$ is usually written in terms of the so-called "double-barred" or "reduced" matrix element,

$$F(\alpha, J, k, \alpha', J') = \frac{\langle \alpha' J' \| T^k \| \alpha J \rangle}{\sqrt{(2J'+1)}}, \tag{2}$$

so

$$\langle \alpha' J' M' | T^k_q | \alpha J M \rangle = \frac{\langle \alpha' J' \| T^k \| \alpha J \rangle}{\sqrt{(2J'+1)}} \langle J M k q | J' M' \rangle. \tag{3}$$

The factor $1/\sqrt{(2J'+1)}$ is factored out of the M, q, M'-independent factor for convenience. (It makes the absolute value of the reduced or double-barred matrix element invariant to bra, ket interchange). Also, the Wigner–Eckart theorem greatly reduces the labor of calculating matrix elements. We need to choose only one (convenient!) set M, q, M' to calculate the full matrix element. The rest then follow through the Clebsch–Gordan coefficient. A very important special case is the one for rotationally invariant or scalar, $k = 0$, operators. Because $\langle J M 0 0 | J' M' \rangle =$

$1\delta_{JJ'}\delta_{MM'}$, the matrix element of a scalar ($k = 0$) operator is *independent* of M. The important case is that of a Hamiltonian operator of an isolated system. Because the properties of the system cannot depend on the orientation of the frame of reference from which it is viewed, the Hamiltonian must be a spherical tensor of rank $l = 0$. The energies will be independent of the quantum number M. The $(2J + 1)$-fold degeneracy of the energy levels can be removed only through the action of external perturbations. More important for calculational purposes, we can choose any convenient M, such as $M = J$, to calculate the energies of the isolated (unperturbed) system.

We now briefly give a few double-barred matrix elements of some simple operators. The "extra" $\sqrt{(2J'+1)}$ factor, part of the definition of the double barred matrix element, leads to the fact that the reduced matrix element of the simple unit operator is not unity.

$$\langle J' \| 1 \| J\rangle = \sqrt{(2J+1)}\delta_{JJ'}. \tag{4}$$

For brevity of notation, we omit the additional quantum numbers α. Next, from the matrix element

$$\langle JM|J_0|JM\rangle = M, \tag{5}$$

and the value of the Clebsch–Gordan coefficient,

$$\langle JM10|JM\rangle = \frac{M}{\sqrt{J(J+1)}}, \tag{6}$$

we get the double-barred matrix element for $\vec{J} = J^{k=1}$:

$$\langle J' \| J^1 \| J\rangle = \sqrt{(2J+1)J(J+1)}\delta_{JJ'}. \tag{7}$$

As a third example, from our result for the matrix element of $Y_{l_0 m_0}$ in an orbital angular momentum basis, we can read off at once (from eq. (32) of Chapter 30)

$$\langle l' \| Y^{l_0} \| l\rangle = \sqrt{\frac{(2l+1)(2l_0+1)}{4\pi}}\langle l0l_00|l'0\rangle. \tag{8}$$

Before proving the Wigner–Eckart theorem, let us give a useful, special case as follows.

A Diagonal Matrix Elements of Vector Operators

The matrix element of any vector operator, $\vec{V} = V^1_q$, *diagonal* in J, is given by

$$\langle \alpha' J(M+q)|V_q|\alpha JM\rangle = \frac{1}{J(J+1)}\langle \alpha' J(M+q)|(\vec{V}\cdot\vec{J})J_q|\alpha JM\rangle. \tag{9}$$

(The theorem does *not* apply, however, to matrix elements off-diagonal in J, which may also be nonzero!). To prove the theorem, we make repeated use of the Wigner–Eckart theorem

$$\langle \alpha' J(M+q)|(\vec{V}\cdot\vec{J})J_q|\alpha JM\rangle =$$

$$
\begin{aligned}
&\sum_m (-1)^m \langle\alpha' J(M+q)|V_{-m}|\alpha J(M+m+q)\rangle \\
&\times \langle\alpha J(M+m+q)|J_m|\alpha J(M+q)\rangle\langle\alpha J(M+q)|J_q|\alpha JM\rangle \\
&= \sum_m (-1)^m \frac{\langle\alpha' J\|V^1\|\alpha J\rangle}{\sqrt{(2J+1)}} \langle J(M+m+q)1-m|J(M+q)\rangle \\
&\times \langle J(M+q)1m|J(M+m+q)\rangle\sqrt{J(J+1)}\langle JM1q|J(M+q)\rangle\sqrt{J(J+1)} \\
&= J(J+1)\sum_m \left(\langle J(M+m+q)1-m|J(M+q)\rangle\right)^2 \\
&\times \langle JM1q|J(M+q)\rangle\frac{\langle\alpha' J\|V^1\|\alpha J\rangle}{\sqrt{(2J+1)}} \\
&= J(J+1)\langle JM1q|J(M+q)\rangle\frac{\langle\alpha' J\|V^1\|\alpha J\rangle}{\sqrt{(2J+1)}} \\
&= J(J+1)\langle\alpha' J(M+q)|V_q^1|\alpha JM\rangle, \qquad (10)
\end{aligned}
$$

where we have used $(-1)^{2m} = +1$ and have done the m sum using the orthonormality of the Clebsch–Gordan coefficients.

B Proof of the Wigner–Eckart Theorem

To derive the Wigner–Eckart theorem, consider the ket

$$\sum_{M,(q)} T_q^k|\alpha JM\rangle\langle JMkq|J'M'\rangle \equiv |\gamma\rangle, \qquad (11)$$

where the ket $|\gamma\rangle$ is a function of the quantum numbers J', M', as well as α, J, and k. First, this ket is an eigenvector of the angular momentum operators with eigenvalues, J', M'. To show this, act on $|\gamma\rangle$ first with J_+ (subsequently with J_- and J_0).

$$
\begin{aligned}
J_+|\gamma\rangle &= \sum_{M,(q)} J_+T_q^k|\alpha JM\rangle\langle JMkq|J'M'\rangle \\
&= \sum_{M,(q)} \left(T_q^k J_+ + [J_+, T_q^k]\right)|\alpha JM\rangle\langle JMkq|J'M'\rangle. \qquad (12)
\end{aligned}
$$

From the spherical tensor character of T_q^k, we have

$$[J_+, T_q^k] = \sqrt{(k-q)(k+q+1)}T_{q+1}^k, \qquad (13)$$

so

$$
\begin{aligned}
J_+|\gamma\rangle = \sum_{M,(q)} \Big(&\sqrt{(J-M)(J+M+1)}\langle JMkq|J'M'\rangle T_q^k|\alpha J(M+1)\rangle \\
&+ \sqrt{(k-q)(k+q+1)}\langle JMkq|J'M'\rangle T_{q+1}^k|\alpha JM\rangle\Big). \qquad (14)
\end{aligned}
$$

Now, rename the dummy summation indices as follows: In the first term, change $M \to (M-1)$; in the second term, change $q \to (q-1)$, noting, however, M' is a fixed number, so

$$J_+|\gamma\rangle = \sum_{M,(q)} \Big(\sqrt{(J-M+1)(J+M)}\langle J(M-1)kq|J'M'\rangle + \sqrt{(k-q+1)(k+q)}\langle JMk(q-1)|J'M'\rangle \Big) T_q^k|\alpha JM\rangle. \quad (15)$$

The terms in large parentheses can be replaced, using recursion formula I for Clebsch–Gordan coefficients (eq. (14) of Chapter 27), to yield

$$J_+|\gamma\rangle = \sqrt{(J'-M')(J'+M'+1)} \sum_{M,(q)} T_q^k|\alpha JM\rangle\langle JMkq|J'(M'+1)\rangle; \quad (16)$$

i.e., the vector $|\gamma\rangle$ with eigenvalues J', M' has been converted to a new vector $|\gamma'\rangle$ in which the eigenvalue M' has been replaced by $(M'+1)$, all other labels, such as α, J, K, J', being the same. In addition, the numerical multiplicative factor is the standard matrix element of the operator J_+ acting on an eigenvector with quantum numbers J', M'. Similarly,

$$J_-|\gamma\rangle = \sqrt{(J'+M')(J'-M'+1)} \sum_{M,(q)} T_q^k|\alpha JM\rangle\langle JMkq|J'(M'-1)\rangle, \quad (17)$$

and

$$J_0|\gamma\rangle = M' \sum_{M,(q)} T_q^k|\alpha JM\rangle\langle JMkq|J'M'\rangle. \quad (18)$$

Thus, the ket $|\gamma\rangle$ is an angular momentum eigenvector with eigenvalues J', M', which is also dependent on the labels (α, J, k),

$$|\gamma\rangle = |(\alpha, J, k)J'M'\rangle. \quad (19)$$

Therefore,

$$\langle\alpha'' J'' M''|\gamma\rangle = \delta_{J'J''}\delta_{M'M''} F(\alpha, J, k; \alpha'' J''), \quad (20)$$

or (using barred labels for the summed quantum numbers),

$$\sum_{\bar{M},\bar{q}} \langle\alpha'' J'' M''|T_{\bar{q}}^k|\alpha J\bar{M}\rangle\langle J\bar{M}k\bar{q}|J'M'\rangle = \delta_{J'J''}\delta_{M'M''} F(\alpha, J, k; \alpha'', J''). \quad (21)$$

Multiplying both sides of this equation by $\langle JMkq|J'M'\rangle$, summing over J', and using the orthogonality for Clebsch–Gordan coefficients, yields

$$\begin{aligned} &\sum_{\bar{M},\bar{q}} \Big(\sum_{J'} \langle J\bar{M}k\bar{q}|J'M'\rangle\langle JMkq|J'M'\rangle \Big) \langle\alpha'' J'' M''|T_{\bar{q}}^k|\alpha J\bar{M}\rangle \\ &= \sum_{\bar{M},\bar{q}} \delta_{M\bar{M}}\delta_{q\bar{q}} \langle\alpha'' J'' M''|T_{\bar{q}}^k|\alpha J\bar{M}\rangle \\ &= \sum_{J'} \delta_{J'J''}\delta_{M'M''} F(\alpha, J, k; \alpha'', J'')\langle JMkq|J'M'\rangle, \end{aligned} \quad (22)$$

leading to the desired result

$$\langle\alpha'' J'' M''|T_q^k|\alpha JM\rangle = F(\alpha, J, k; \alpha'', J'')\langle JMkq|J''M''\rangle. \quad (23)$$

33

Nuclear Hyperfine Structure in One-Electron Atoms

We shall use the magnetic nuclear hyperfine structure calculation in one-electron atoms to illustrate how the Wigner–Eckart theorem can be exploited in actual calculations. So far, we have calculated the perturbed one-electron spectrum (in hydrogen or in alkali atoms) only under the assumption the nucleus (the proton in hydrogen or the alkali atomic nucleus) have a charge but no magnetic properties. Because the proton has a spin (with spin quantum number $\frac{1}{2}$), it also has a magnetic moment, similarly for alkali atoms. The spin of the stable isotope of Na, e.g., is $\frac{3}{2}$, and hence, it also has a nuclear magnetic moment. These nuclear magnetic moments will have a magnetic interaction with the spin magnetic moment of the electron. In addition, the motion of the electron relative to the nucleus will set up an effective magnetic field at the nucleus (proportional to the orbital angular momentum of the electron). The magnetic moment of the nucleus will also lead to a $-(\vec{\mu}^{\text{magn.}} \cdot \vec{B})$ magnetic interaction caused by this field. These magnetic interactions of nuclear origin will give rise to the so-called hyperfine structure splitting of the $nlsj$ states of the one-electron atoms. If we call the electron magnetic moment $\vec{\mu}_s$ and the nuclear magnetic moment $\vec{\mu}_I$,

$$\vec{\mu}_s = \frac{e\hbar}{2mc} 2\vec{s} = -\frac{|e|\hbar}{2mc} 2\vec{s}, \qquad \vec{\mu}_I = \frac{|e|\hbar}{2Mc} g_I \vec{I}, \tag{1}$$

where we use the vector, $\vec{I}$, for the nuclear spin operator, with quantum number, I, e.g., $I = \frac{1}{2}$ for the proton, $I = \frac{3}{2}$ for ^{23}Na. Also, m is the electron mass, whereas M is the proton mass. (Even in the Na nucleus, the magnetic moment is caused by the last odd valence proton in the nucleus, so it is the proton mass in the so-called nuclear magneton, $\frac{|e|\hbar}{2Mc}$, which gives a natural measure for nuclear magnetic moments.) The g factor for the nuclear magnetic moment is not simply equal to

2.0, as for the point electron (as follows from Dirac theory), but is "anomalous," e.g., $g_I = 2 \times 2.793$ for the proton. This "anomalous" factor is now rather well understood in terms of the quark structure of the nucleon.

The magnet–magnet interaction between the magnetic moments of the electron and the nucleus is given by [see, e.g., J. D. Jackson, 2nd ed., eq. (5.73)]

$$\begin{aligned} H_{\text{magn.-magn.}} &= \left(\left[(\vec{\mu}_s \cdot \vec{\mu}_I) - 3(\frac{\vec{r}}{r} \cdot \vec{\mu}_s)(\frac{\vec{r}}{r} \cdot \vec{\mu}_I) \right] \frac{1}{r^3} - \frac{8\pi}{3} (\vec{\mu}_s \cdot \vec{\mu}_I) \delta(\vec{r}) \right) \\ &= \frac{e^2\hbar^2}{2mMc^2} g_I \left(\left[-(\vec{s} \cdot \vec{I}) + 3(\frac{\vec{r}}{r} \cdot \vec{s})(\frac{\vec{r}}{r} \cdot \vec{I}) \right] \frac{1}{r^3} + \frac{8\pi}{3} (\vec{s} \cdot \vec{I}) \delta(\vec{r}) \right), \end{aligned} \tag{2}$$

including the delta-function term that comes into play only if the two magnets can be on top of each other. This so-called contact term, (part of *classical* electrodynamics according to J. D. Jackson, *Classical Electrodynamics*, New York: John Wiley, 1967), comes into play only in atomic s-states, $l = 0$ states, where the electron has a finite probability of sitting "on" the nucleus. The additional $-(\vec{\mu}_I \cdot \vec{B}_{\text{eff.}})$ term, with

$$\vec{B}_{\text{eff.}} = \frac{e}{cr^3} [\vec{r} \times \frac{\vec{p}}{m}] = -\frac{|e|\hbar}{mcr^3} \vec{l}, \tag{3}$$

caused by the effective magnetic field of the moving electron at the nucleus, leads to an additional interaction term

$$H_{\text{orb.-nucl. magn.}} = \frac{e^2\hbar^2}{2mMc^2} g_I (\vec{l} \cdot \vec{I}) \frac{1}{r^3}. \tag{4}$$

In these relations, the quantity r is $r_{\text{phys.}} = a_0 r$, where the last r is our usual dimensionless r. In terms of this dimensionless r, the combined magnetic interactions give

$$H_{\text{int.}} = \frac{mc^2\alpha^4}{2} \frac{m}{M} g_I \left(\left[-(\vec{s} \cdot \vec{I}) + 3(\frac{\vec{r}}{r} \cdot \vec{s})(\frac{\vec{r}}{r} \cdot \vec{I}) \right] \frac{1}{r^3} + \frac{8\pi}{3} (\vec{s} \cdot \vec{I}) \delta(\vec{r}) + (\vec{l} \cdot \vec{I}) \frac{1}{r^3} \right), \tag{5}$$

where we have also used

$$\delta(a_0 \vec{r}) = \frac{1}{a_0^3} \delta(\vec{r}), \tag{6}$$

and $\alpha = (1/137)$ is the fine structure constant. This hyperfine structure term is $\frac{m}{M} g_I$ times the fine structure term. Now, express the magnetic moment–magnetic moment interaction in terms of spherical tensor operators. In terms of the coupled spherical tensors, defined by

$$[V^{k_1} \times U^{k_2}]^k_q \equiv \sum_{q_1, q_2} V^{k_1}_{q_1} U^{k_2}_{q_2} \langle k_1 q_1 k_2 q_2 | kq \rangle, \tag{7}$$

we need the spherical tensors $[r^1 \times r^1]^2_q \equiv r^2_q$, and $[s^1 \times I^1]^2_q$. These have spherical components

$$r^2_{\pm 2} = \tfrac{1}{2}(x \pm iy)^2, \qquad r^2_{\pm 1} = \mp z(x \pm iy), \qquad r^2_0 = \frac{1}{\sqrt{6}}(2z^2 - x^2 - y^2), \tag{8}$$

and

$$[s^1 \times I^1]^2_{\pm 2} = \tfrac{1}{2} s_\pm I_\pm; \qquad [s^1 \times I^1]^2_{\pm 1} = \mp \tfrac{1}{2}(s_\pm I_0 + s_0 I_\pm),$$
$$[s^1 \times I^1]^2_0 = -\frac{1}{\sqrt{6}}\Big(\tfrac{1}{2}(s_+ I_- + s_- I_+) - 2 s_0 I_0\Big). \tag{9}$$

Also, r^2_q can be expressed in terms of standard spherical harmonics by

$$r^2_q = r^2 \sqrt{\frac{8\pi}{15}}\, Y_{2,q}(\theta, \phi). \tag{10}$$

In terms of these spherical tensors of rank, $k = 2$, the magnetic dipole–magnetic dipole interaction term can be expressed through

$$\Big(-(\vec{s} \cdot \vec{I}) + 3(\vec{s} \cdot \frac{\vec{r}}{r})(\vec{I} \cdot \frac{\vec{r}}{r})\Big) = 3\sqrt{\frac{8\pi}{15}} \sum_q (-1)^q Y_{2,q} [s^1 \times I^1]^2_{-q}$$

$$= 3\sqrt{\frac{8\pi}{15}} (Y^2 \cdot [s^1 \times I^1]^2), \tag{11}$$

where the dot in the last form stands for the generalized scalar product of two spherical tensors each of spherical rank 2. This type of interaction is therefore sometimes called a "tensor" interaction, but the term is somewhat of a misnomer. The full interaction is a scalar, a rotationally invariant $k = 0$ spherical tensor, but the term "tensor" is used because it involves the scalar product of two spherical tensors each of spherical rank 2. Because we want to calculate matrix elements of this interaction in an $|n[[l \times s]j \times I]FM_F\rangle$ basis, corresponding to the vector coupling

$$(\vec{l} + \vec{s}) + \vec{I} = \vec{j} + \vec{I} = \vec{F} \tag{12}$$

basis, rather than a

$$\vec{l} + (\vec{s} + \vec{I}) = \vec{l} + \vec{S} = \vec{F} \tag{13}$$

basis, we need to tailor the magnet–magnet interaction to the coupling scheme of eq. (12), where $\vec{l}$ and $\vec{s}$ are coupled to resultant $\vec{j}$, which is then coupled to the nuclear spin $\vec{I}$ to resultant total angular momentum $\vec{F}$. The form of eq. (11) is tailored more to the coupling scheme of eq. (13), where the electron spin, $\vec{s}$, and the nuclear spin, $\vec{I}$, are coupled to total spin $\vec{S}$, which is then coupled with $\vec{l}$ to resultant total angular momentum $\vec{F}$. By rearranging the order of the couplings, we can also express the magnet–magnet interaction through

$$\Big(-(\vec{s} \cdot \vec{I}) + 3(\vec{s} \cdot \frac{\vec{r}}{r})(\vec{I} \cdot \frac{\vec{r}}{r})\Big)$$

$$= -\sqrt{8\pi}([Y^2 \times s^1]^1 \cdot I^1) = -\sqrt{8\pi} \sum_q (-1)^q [Y^2 \times s^1]^1_q I^1_{-q}. \tag{14}$$

As a specific example, we shall now calculate the hyperfine splitting of the $2p_{\frac{3}{2}}$, i.e., the $n = 2, l = 1, j = \frac{3}{2}$, fine structure level of hydrogen. Because the proton has a spin $I = \frac{1}{2}$, this level will be split into two hyperfine components, with quantum numbers $F = 2$ and $F = 1$. Because the contact term, with the delta function, can contribute only in s states, with $l = 0$, for which the electron has a nonzero probability of being on the proton, this term cannot contribute to the p-state hyperfine splitting. Thus, both the nonzero terms have a $\frac{1}{r^3}$ radial dependence and lead to the radial integral

$$\beta_{\text{h.f.s.}} = \frac{mc^2}{2}\alpha^4 \frac{m}{M} g_I \int_0^\infty dr r^2 |R_{nl}(r)|^2 \frac{1}{r^3}. \tag{15}$$

This hyperfine structure integral, $\beta_{\text{h.f.s.}} = \frac{m}{M} g_I \beta_{nl}$, where $\beta_{nl} = \beta_{\text{f.s.}}$ gives the stength of the fine structure splitting.

We shall now exploit the Wigner–Eckart theorem in two ways: (1) Because $H_{\text{int.}}$ is a $k = 0$ scalar operator, it will be sufficient to calculate the diagonal matrix element of this interaction for states with $M_F = F$. (2) In addition, the calculation can be reduced to a calculation of the reduced or double-barred matrix elements of the two operators, $[Y^2 \times s^1]^1_q$ and l^1_q, in the $|n[l \times s]jm_j\rangle$ basis. The needed matrix elements of I^1_q are very simple and can then be combined with these after application of the Wigner–Eckart theorem. To calculate the double-barred matrix elements of $[Y^2 \times s^1]^1$ and l^1, let us use the state with $m_j = +j = +\frac{3}{2}$,

$$|[l = 1 \times s = \tfrac{1}{2}]j = \tfrac{3}{2} m_j = +\tfrac{3}{2}\rangle = |l = 1 \;\; m_l = +1 \;\; s = \tfrac{1}{2} \;\; m_s = +\tfrac{1}{2}\rangle, \tag{16}$$

for the calculation, so

$$\begin{aligned}
\langle j = m_j = \tfrac{3}{2}|[Y^2 \times s^1]^1_0|j = m_j = \tfrac{3}{2}\rangle &= \langle \tfrac{3}{2}\tfrac{3}{2}10|\tfrac{3}{2}\tfrac{3}{2}\rangle \frac{\langle \frac{3}{2}\|[Y^2 \times s^1]^1\|\frac{3}{2}\rangle}{\sqrt{4}} \\
&= \langle l = m_l = 1, s = m_s = \tfrac{1}{2}|[Y^2 \times s^1]^1_0|l = 1 m_l = 1, s = m_s = \tfrac{1}{2}\rangle \\
&= \langle 1 + 1\tfrac{1}{2} + \tfrac{1}{2}|Y_{20}\ s_0|1 + 1\tfrac{1}{2} + \tfrac{1}{2}\rangle\langle 2010|10\rangle \\
&= \langle 1 \;\; m_l = +1|Y_{20}|1 \;\; m_l = +1\rangle(+\tfrac{1}{2})\langle 2010|10\rangle \\
&= \sqrt{\frac{3 \cdot 5}{3 \cdot 4\pi}}\langle 1120|11\rangle\langle 1020|10\rangle\langle 2010|10\rangle(+\tfrac{1}{2}) \\
&= \frac{1}{5}\sqrt{\frac{1}{8\pi}},
\end{aligned} \tag{17}$$

where only the $Y_{20}s_0$ component of $[Y^2 \times s^1]^1_0$ contributes to our matrix element because it is diagonal in *both* m_l and m_s. This simplification is related to our "clever" choice of $m_j = +\frac{3}{2}$ in the full matrix element. In the above, we have used eq. (32) of Chapter 30 for the matrix element of Y_{20} and the values of the Clebsch–Gordan coefficients

$$\begin{aligned}
\langle 2010|10\rangle = \langle 1020|10\rangle &= -\sqrt{\frac{2}{5}}, \\
\langle 1120|11\rangle &= +\sqrt{\frac{1}{10}}.
\end{aligned} \tag{18}$$

Finally, using

$$\langle \tfrac{3}{2}\tfrac{3}{2}10|\tfrac{3}{2}\tfrac{3}{2}\rangle = \sqrt{\frac{3}{5}} \tag{19}$$

for the first step in eq. (17), we get the desired reduced matrix element

$$\langle \tfrac{3}{2}\|[Y^2 \times s^1]^1\|\tfrac{3}{2}\rangle = \frac{1}{\sqrt{30\pi}}. \tag{20}$$

Similarly, we can get the reduced matrix element of l_q^1 from

$$\begin{aligned}
\langle j = \tfrac{3}{2}\ m_j = +\tfrac{3}{2}|l_0^1|j = \tfrac{3}{2}\ m_j = +\tfrac{3}{2}\rangle &= \langle \tfrac{3}{2}\tfrac{3}{2}10|\tfrac{3}{2}\tfrac{3}{2}\rangle \frac{\langle \tfrac{3}{2}\|l^1\|\tfrac{3}{2}\rangle}{\sqrt{4}} \\
= \langle l = 1\ m_l = +1\ s = m_s = \tfrac{1}{2}|l_0|l = 1\ m_l = +1\ s = m_s = \tfrac{1}{2}\rangle \\
= +1,
\end{aligned} \tag{21}$$

so

$$\langle \tfrac{3}{2}\|l^1\|\tfrac{3}{2}\rangle = 2\sqrt{\frac{5}{3}}. \tag{22}$$

Although the calculation of this reduced matrix element in the state with $j = (l+\frac{1}{2})$ was extremely trivial, we note an alternative method that might be useful for states with $j = (l - \frac{1}{2})$ or for states with $s > \frac{1}{2}$ and several possible j values. From eq. (9) of Chapter 32, we could also have written

$$\begin{aligned}
\langle [l \times s]j\|\vec{l}\|[l \times s]j\rangle &= \frac{\langle [l \times s]j\|(\vec{l}\cdot\vec{j})\vec{j}\|[l \times s]j\rangle}{j(j+1)} \\
&= \frac{[j(j+1) + l(l+1) - s(s+1)]}{2j(j+1)}\langle j\|\vec{j}\|j\rangle \\
&= \tfrac{1}{2}[j(j+1) + l(l+1) - s(s+1)]\sqrt{\frac{(2j+1)}{j(j+1)}},
\end{aligned} \tag{23}$$

where we have used eq. (7) of Chapter 32 for the reduced matrix element of $\vec{j} \equiv j^1$, and have also used

$$(\vec{l}\cdot\vec{j}) = (\vec{l}\cdot(\vec{l}+\vec{s})) = (\vec{l}\cdot\vec{l}) + (\vec{l}\cdot\vec{s}), \tag{24}$$

with

$$(\vec{l}\cdot\vec{s}) = \tfrac{1}{2}[(\vec{j}\cdot\vec{j}) - (\vec{l}\cdot\vec{l}) - (\vec{s}\cdot\vec{s})]. \tag{25}$$

With $l = 1$, $s = \frac{1}{2}$, and $j = \frac{3}{2}$, this relation immediately gives the result of eq. (22). Now that we have the reduced matrix elements of $[Y^2 \times s^1]^1$ and $l^1 \equiv \vec{l}$, we can calculate the matrix elements of $H_{\text{int.}}$ in the needed states with $M_F = F$. First, for $F = 2$, with

$$|[j \times I]F\ M_F\rangle = |[\tfrac{3}{2} \times \tfrac{1}{2}]F = 2\ M_F = +2\rangle = |j = m_j = \tfrac{3}{2}\quad I = M_I = \tfrac{1}{2}\rangle, \tag{26}$$

we get

$$
\begin{aligned}
&\langle F = M_F = 2|H_{\text{int.}}|F = M_F = 2\rangle = \beta_{\text{h.f.s.}}(-\sqrt{8\pi})\\
&\times \langle F = M_F = 2|([Y_2 \times s^1]^1 \cdot I^1) + (\vec{l}\cdot\vec{I})|F = M_F = 2\rangle\\
&= \beta_{\text{h.f.s.}}\Big(-\sqrt{8\pi}\langle j = m_j = \tfrac{3}{2}|[Y^2 \times s^1]^1_0|j = m_j = \tfrac{3}{2}\rangle\\
&+\ \langle j = m_j = \tfrac{3}{2}|l_0|j = m_j = \tfrac{3}{2}\rangle\Big)\langle I = M_I = \tfrac{1}{2}|I_0|I = M_I = \tfrac{1}{2}\rangle\\
&= \beta_{\text{h.f.s.}}\Big(-\sqrt{8\pi}\frac{\langle\frac{3}{2}\|[Y^2 \times s^1]^1\|\frac{3}{2}\rangle}{\sqrt{4}} + \frac{\langle\frac{3}{2}\|l^1\|\frac{3}{2}\rangle}{\sqrt{4}}\Big)\langle\tfrac{3}{2}\tfrac{3}{2}10|\tfrac{3}{2}\tfrac{3}{2}\rangle(+\tfrac{1}{2})\\
&= \beta_{\text{h.f.s.}}\Big(-\sqrt{8\pi}\frac{1}{2\sqrt{30\pi}} + 2\sqrt{\frac{5}{3}}\cdot\tfrac{1}{2}\Big)\sqrt{\frac{3}{5}}(+\tfrac{1}{2})\\
&= \beta_{\text{h.f.s.}}\Big(-\frac{1}{5} + 1\Big)(+\tfrac{1}{2}) = \frac{2}{5}\beta_{\text{h.f.s.}}.
\end{aligned}
\tag{27}
$$

For the state with $F = 1$, the calculation is somewhat more complicated. We shall again choose the state with $M_F = F$ to do the calculation, where now

$$
\begin{aligned}
&|[j = \tfrac{3}{2} \times I = \tfrac{1}{2}]F = M_F = 1\rangle = \sum_{m_j M_I}|\tfrac{3}{2}m_j\tfrac{1}{2}M_I\rangle\langle\tfrac{3}{2}m_j\tfrac{1}{2}M_I|11\rangle\\
&= |\tfrac{3}{2} + \tfrac{3}{2}\tfrac{1}{2} - \tfrac{1}{2}\rangle\langle\tfrac{3}{2} + \tfrac{3}{2}\tfrac{1}{2} - \tfrac{1}{2}|1 + 1\rangle + |\tfrac{3}{2} + \tfrac{1}{2}\tfrac{1}{2} + \tfrac{1}{2}\rangle\langle\tfrac{3}{2} + \tfrac{1}{2}\tfrac{1}{2} + \tfrac{1}{2}|1 + 1\rangle\\
&= |\tfrac{3}{2} + \tfrac{3}{2}\tfrac{1}{2} - \tfrac{1}{2}\rangle\frac{\sqrt{3}}{2} + |\tfrac{3}{2} + \tfrac{1}{2}\tfrac{1}{2} + \tfrac{1}{2}\rangle(-\tfrac{1}{2}).
\end{aligned}
\tag{28}
$$

Calculating first the matrix element for the $([Y^2 \times s^1]^1 \cdot I^1)$ term, we get

$$
\begin{aligned}
&\langle[\tfrac{3}{2} \times \tfrac{1}{2}]F = 1\ \ M_F = 1|([Y^2 \times s^1]^1 \cdot I^1)|[\tfrac{3}{2} \times \tfrac{1}{2}]F = 1\ \ M_F = 1\rangle\\
&= \frac{3}{4}\langle\tfrac{3}{2} + \tfrac{3}{2}\tfrac{1}{2} - \tfrac{1}{2}|[Y^2 \times s^1]^1_0\ \ I_0|\tfrac{3}{2} + \tfrac{3}{2}\tfrac{1}{2} - \tfrac{1}{2}\rangle\\
&+ \frac{1}{4}\langle\tfrac{3}{2} + \tfrac{1}{2}\tfrac{1}{2} + \tfrac{1}{2}|[Y^2 \times s^1]^1_0\ \ I_0|\tfrac{3}{2} + \tfrac{1}{2}\tfrac{1}{2} + \tfrac{1}{2}\rangle\\
&- \frac{\sqrt{3}}{4}\langle\tfrac{3}{2} + \tfrac{3}{2}\tfrac{1}{2} - \tfrac{1}{2}|\Big(-[Y^2 \times s^1]^1_{+1}\ \ I_{-1}\Big)|\tfrac{3}{2} + \tfrac{1}{2}\tfrac{1}{2} + \tfrac{1}{2}\rangle\\
&- \frac{\sqrt{3}}{4}\langle\tfrac{3}{2} + \tfrac{1}{2}\tfrac{1}{2} + \tfrac{1}{2}|\Big(-[Y^2 \times s^1]^1_{-1}\ \ I_{+1}\Big)|\tfrac{3}{2} + \tfrac{3}{2}\tfrac{1}{2} - \tfrac{1}{2}\rangle,
\end{aligned}
\tag{29}
$$

where the state vectors in the four terms are given in the $|j\ m_j\ I\ M_I\rangle$ basis. Noting the matrix elements of the spherical components of $\vec{I}$ are given by

$$
\langle\tfrac{1}{2}M_I|I_0|\tfrac{1}{2}M_I\rangle = M_I \qquad \text{and} \qquad \langle\tfrac{1}{2} \pm \tfrac{1}{2}|I_{\pm 1}|\tfrac{1}{2} \mp \tfrac{1}{2}\rangle = \mp\frac{1}{\sqrt{2}},
\tag{30}
$$

and using the Wigner–Eckart theorem for the matrix elements of the operators $[Y^2 \times s^1]^1_q$ in the $|j\ m_j\rangle$ basis, we get

$$
\begin{aligned}
&-\sqrt{8\pi}\langle[\tfrac{3}{2} \times \tfrac{1}{2}]F = M_F = 1|([Y^2 \times s^1]^1 \cdot I^1)|[\tfrac{3}{2} \times \tfrac{1}{2}]F = M_F = 1\rangle\\
&= -\sqrt{8\pi}\Big[\frac{3}{4}\langle\tfrac{3}{2}\tfrac{3}{2}10|\tfrac{3}{2}\tfrac{3}{2}\rangle(-\tfrac{1}{2}) + \frac{1}{4}\langle\tfrac{3}{2}\tfrac{1}{2}10|\tfrac{3}{2}\tfrac{1}{2}\rangle(+\tfrac{1}{2})\\
&+ \frac{\sqrt{3}}{4}\langle\tfrac{3}{2}\tfrac{1}{2}1 + 1|\tfrac{3}{2}\tfrac{3}{2}\rangle(\frac{1}{\sqrt{2}}) + \frac{\sqrt{3}}{4}\langle\tfrac{3}{2}\tfrac{3}{2}1 - 1|\tfrac{3}{2}\tfrac{1}{2}\rangle(-\frac{1}{\sqrt{2}})\Big]\frac{1}{\sqrt{4}}\langle\tfrac{3}{2}\|[Y^2 \times s^1]^1\|\tfrac{3}{2}\rangle
\end{aligned}
$$

$$= -\sqrt{8\pi}\left[\frac{3}{4}\sqrt{\frac{3}{5}}(-\tfrac{1}{2}) + \frac{1}{4}\frac{1}{\sqrt{15}}(+\tfrac{1}{2}) + \frac{\sqrt{3}}{4}(-\sqrt{\frac{2}{5}})(\frac{1}{\sqrt{2}}) + \frac{\sqrt{3}}{4}(+\sqrt{\frac{2}{5}})(-\frac{1}{\sqrt{2}})\right]\frac{1}{\sqrt{4}}\frac{1}{\sqrt{30\pi}}$$
$$= +\frac{1}{6}. \tag{31}$$

The matrix element of the $(\vec{l}\cdot\vec{I})$ term differs from this matrix element only by the ratio of the reduced matrix elements, which is a factor of -5. Thus,

$$\langle F = M_F = 1|H_{\text{int.}}|F = M_F = 1\rangle = \beta_{\text{h.f.s.}}(\frac{1}{6} - \frac{5}{6}) = -\frac{2}{3}\beta_{\text{h.f.s.}}. \tag{32}$$

Thus, the hyperfine levels of the hydrogen $2p_{\frac{3}{2}}$ fine structure level, with $F = 2$ and $F = 1$ are shifted by $+\frac{2}{5}\beta_{\text{h.f.s.}}$ and $-\frac{2}{3}\beta_{\text{h.f.s.}}$, leading to a splitting of $\frac{16}{15}\beta_{\text{h.f.s.}}$. We leave as an exercise the splitting for the $2p_{\frac{1}{2}}$ fine structure level with a shift of $+\frac{2}{3}\beta_{\text{h.f.s.}}$ for the $F = 1$ hyperfine level and a shift of $-2\beta_{\text{h.f.s.}}$ for the $F = 0$ hyperfine level.

Problems

43. In a one-electron atom, such as sodium, find the relative intensities for the transitions, $n[l\frac{1}{2}]j \to n'[(l-1)\frac{1}{2}]j'$ for the four possibilities: $j = (l \pm \frac{1}{2})$; $j' = (l-1) \pm \frac{1}{2}$. Express your answers in terms of general functions of l. Show first that

$$\sum_{\alpha=x,y,z} |\langle n'l'\tfrac{1}{2}j'm'|\mu_\alpha^{(\text{el.})}|nl\tfrac{1}{2}jm\rangle|^2 = \sum_{q=\pm1,0} |\langle n'l'\tfrac{1}{2}j'm'|\mu_q^{(\text{el.})}|nl\tfrac{1}{2}jm\rangle|^2,$$

where $\mu_q^{(\text{el.})}$ are spherical components of the electric dipole moment operator. Also, show that

$$\sum_{m,q,m'} |\langle n'l'\tfrac{1}{2}j'm'|\mu_q^{(\text{el.})}|nl\tfrac{1}{2}jm\rangle|^2 = |\langle n'l'\tfrac{1}{2}j'\|\vec{\mu}^{(\text{el.})}\|nl\tfrac{1}{2}j\rangle|^2,$$

so only the reduced (or double-barred) matrix elements of $\vec{\mu}^{(\text{el.})}$, a spherical tensor of rank 1, need to be calculated. Also, use the fact that the m sublevels of the initial state are all populated with an equal probability of $1/(2j+1)$.

44. In an atom with an atomic nucleus, with a nuclear spin, $I \geq 1$, a hyperfine interaction exists, caused by the electrostatic interaction between the nuclear electric quadrupole moment and the atomic electrons. In a one-electron atom, such as Na, this interaction gives rise to a hyperfine perturbation

$$H_{\text{int.}} = -e^2\sum_m(-1)^m Q_{2,m}\sqrt{\frac{4\pi}{5}}Y_{2,-m}(\theta,\phi)\frac{1}{r^3},$$

where the $Q_{2,m}$ are the (spherical) laboratory components of the nuclear quadrupole operator, a spherical tensor of rank 2, and r, θ, ϕ give the position of the valence electron. The ${}^{23}_{11}\mathrm{Na}_{12}$ nucleus has a spin $I = \frac{3}{2}$ and a nuclear quadrupole moment, $Q = 0.14$ barns (1 barn $\equiv 10^{-24}\mathrm{cm}^2$). "The" nuclear quadrupole moment, Q, is defined as

$$Q = 2\langle IM_I = I|Q_{2,0}|IM_I = I\rangle = 2\langle II20|II\rangle\frac{\langle I\|Q_2\|I\rangle}{\sqrt{(2I+1)}}$$

$$= 2\sqrt{\frac{I(2I-1)}{(2I+1)(I+1)(2I+3)}}\langle I\|Q_2\|I\rangle.$$

We can evaluate the reduced matrix element of Q_2 in terms of the experimental Q through this relation. Also, Q is related to the nuclear charge density, $\rho_{\text{nucl.}}$, through

$$Q = \frac{1}{|e|}\int d\vec{r}_{\text{nucl.}} r^2_{\text{nucl}}(3\cos^2\theta_{\text{nucl.}} - 1)\rho_{\text{nucl.}},$$

so Q has the dimension cm^2, because the charge $|e|$ is factored out in the definition.

Calculate the hyperfine splitting caused by this perturbation for the $3p_{\frac{3}{2}}$ and $3p_{\frac{1}{2}}$ levels of the Na atom. Show, in particular, the F=1, 2 hyperfine levels for $p_{\frac{1}{2}}$ are not affected by this perturbation, whereas the $p_{\frac{3}{2}}$ state is split into hyperfine sublevels with F = 0, 1, 2, 3. Show, however, the F=1 and F=3 sublevels remain (accidentally) degenerate. Express the hyperfine splitting in terms of Q and the quantity

$$\beta_{nl} = \frac{me^4}{\hbar^2}\alpha^2\int_0^\infty drr^2|R_{nl}(r)|^2\frac{1}{r^3},$$

where r is the dimensionless radial coordinate measured in atomic units.

45. The hyperfine splitting of the $1s_{\frac{1}{2}}$ ground state of the hydrogen atom arises solely through the delta function contact term of

$$H_{\text{h.f.int.}} = \frac{mc^2\alpha^4}{2}\frac{m}{M}g_I\left(\left[-(\vec{s}\cdot\vec{I})+3(\frac{\vec{r}}{r}\cdot\vec{s})(\frac{\vec{r}}{r}\cdot\vec{I})\right]\frac{1}{r^3}+\frac{8\pi}{3}(\vec{s}\cdot\vec{I})\delta(\vec{r})+(\vec{l}\cdot\vec{I})\frac{1}{r^3}\right).$$

Calculate the hyperfine splitting of the hydrogen atom ground state. Find the numerical value of the energy difference between the $F = 1$ and $F = 0$ hyperfine sublevels. (Recall $g_I = 2 \times 2.793$ for the proton.)

46. In a diatomic molecule with one atomic nucleus with a spin, $I \geq 1$, a hyperfine interaction exists through the electrostatic interaction between the nuclear electric quadrupole moment and the electric field at the nucleus caused by the molecular electrons and the second nuclear charge. This interaction gives rise to a quadrupole hyperfine perturbation

$$H_{\text{h.f.int.}} = eq\sum_m(-1)^m Q_{2,m}Y_{2,-m}(\theta,\phi)\sqrt{\frac{4\pi}{5}},$$

where $Q_{2,m}$ are the (spherical) laboratory components of the nuclear quadrupole operator (see problem 44) and the angles θ, ϕ give the orientation in the laboratory of the diatomic molecule symmetry axis. The number, eq, gives the θ, ϕ-independent part of the molecular matrix element of the inhomogeneous electric field at the nucleus, using the molecular electronic wave function. Here, e is the electronic charge, and q is a commonly used notation. Assume the nuclear spin is $I = 1$. Find the hyperfine splitting of the first excited rotational state with $J = 1$ and zeroth-order rotational wave function, $\psi^{\text{rot.}}_{JM} = Y_{J=1,M}(\theta, \phi)$, as a function of eq and Q, where Q is "the" nuclear quadrupole moment, as defined in problem 44. That is, find the splitting of the rotational level with $J = 1$ into hyperfine levels with $F = 0, 1, 2$, where $\vec{F} = \vec{J} + \vec{I}$.

47. A diatomic molecule rigid rotator with zeroth-order rotational energies, $\hbar^2 J(J+1)/(2\mu r_e^2)$, and zeroth-order rotational eigenfunctions, $Y_{JM}(\theta, \phi)$, has atomic nuclei with nuclear spins, $I_1 = \frac{3}{2}$ and $I_2 = \frac{1}{2}$. Assume the nuclei have no electric quadrupole moments, so the hyperfine interaction is caused by a magnet–magnet type interaction leading to an interaction Hamiltonian of the form

$$H_{\text{int.}} = a(\vec{I}_1 \cdot \vec{I}_2) + b\Big[3(\vec{I}_1 \cdot \frac{\vec{r}}{r})(\vec{I}_2 \cdot \frac{\vec{r}}{r}) - (\vec{I}_1 \cdot \vec{I}_2)\Big],$$

where $\vec{r}$ is the vector pointing from nucleus 2 to nucleus 1, with angular coordinates, θ, ϕ, and a and b are constants. Assume, however, $a \gg b$ [but both $a, b \ll \hbar^2/(2\mu r_e^2)$]. For these values of the constants, the $|[J[I_1 I_2]I]FM_F\rangle$ basis is a good basis for the hyperfine structure calculation. Here, $\vec{I}_1 + \vec{I}_2 = \vec{I}$, where $\vec{I}$ is the total nuclear spin vector, and $\vec{J} + \vec{I} = \vec{F}$, where $\vec{F}$ is the total angular momentum vector. Find the hyperfine energies as functions of a and b for the rotational state with $J = 1$ and hyperfine multiplet with $I = 1$, $F = 0, 1, 2$, and $I = 2$, $F = 1, 2, 3$.

34

Angular Momentum Recoupling: Matrix Elements of Coupled Tensor Operators in an Angular Momentum Coupled Basis

In Chapter 33, we evaluated the matrix elements of a vector-coupled spherical tensor of type $[U^{k_1} \times V^{k_2}]^k_q$ in terms of the reduced matrix elements of both U^{k_1} and V^{k_2} in the appropriate angular momentum–coupled basis,

$$\langle [j_1' \times j_2']J'M'|[U^{k_1} \times V^{k_2}]^k_q|[j_1 \times j_2]JM\rangle,$$

where we managed to simplify the calculation by judicious use of the Wigner–Eckart theorem and convenient choices of magnetic quantum numbers. In some cases, however, the calculation still involved fairly tedious m sums of expressions involving products of Clebsch–Gordan coefficients. It is the purpose of this chapter to show matrix elements of the above type can be expressed in terms of so-called angular momentum recoupling coefficients. Because values of these recoupling coefficients are available through tabulations or computer codes, and in many cases through algebraic expressions, it will be valuable to study these recoupling coefficients.

A The Recoupling of Three Angular Momenta: Racah Coefficients or 6-j Symbols

So far, we have studied transformations of states involving two commuting angular momentum operators from a basis

$$|j_1 m_1 j_2 m_2\rangle \quad \text{simultaneous eigenvectors of} \quad \vec{j}_1^{\,2}, j_{1_z}, \vec{j}_2^{\,2}, j_{2_z},$$

to a basis with good total angular momentum, $\vec{J} = \vec{j}_1 + \vec{j}_2$,

$$|j_1 j_2 J M\rangle \quad \text{simultaneous eigenvectors of} \quad \vec{j}_1^{\,2}, \vec{j}_2^{\,2}, \vec{J}^{\,2}, J_z.$$

For states involving three commuting angular momentum operators, we require six angular momentum quantum numbers, hence, six commuting operators, e.g.,

$$\vec{j}_1^{\,2}, j_{1_z}, \vec{j}_2^{\,2}, j_{2_z}, \vec{j}_3^{\,2}, j_{3_z},$$

for a complete specification of the basis:

$$|j_1 m_1 j_2 m_2 j_3 m_3\rangle.$$

As for the case of two angular momenta, it will often be advantageous to use a basis with good total angular momentum in which $\vec{J}^{\,2}$ is diagonal. Now, however, the five commuting operators $\vec{j}_1^{\,2}, \vec{j}_2^{\,2}, \vec{j}_3^{\,2}, \vec{J}^{\,2}$, and $J_z = j_{1_z} + j_{2_z} + j_{3_z}$ will be insufficient. A sixth operator is needed. To find this sixth operator, we could couple to total $\vec{J}$ in two ways

$$\begin{aligned} \vec{J} &= (\vec{j}_1 + \vec{j}_2) + \vec{j}_3 = \vec{J}_{12} + \vec{j}_3 \\ &= \vec{j}_1 + (\vec{j}_2 + \vec{j}_3) = \vec{j}_1 + \vec{J}_{23}, \end{aligned} \tag{1}$$

and use the eigenvectors of the six commuting operators

$$\vec{j}_1^{\,2}, \vec{j}_2^{\,2}, \vec{j}_3^{\,2}, \vec{J}_{12}^{\,2}; \vec{J}^{2}, J_z :$$

$$|[[j_1 j_2] J_{12} j_3] J M\rangle,$$

where we first couple the two angular momenta $\vec{j}_1, \vec{j}_2$ to a state with good $\vec{J}_{12}^{\,2}$, or alternatively, we could use the eigenvectors of the six commuting operators

$$\vec{j}_1^{\,2}, \vec{j}_2^{\,2}, \vec{j}_3^{\,2}, \vec{J}_{23}^{\,2}; \vec{J}^{2}, J_z :$$

$$|[j_1 [j_2 j_3] J_{23}] J M\rangle,$$

where we couple the angular momentum $\vec{j}_1$ to a state in which $\vec{j}_2$ and $\vec{j}_3$ have been coupled to a state of good $\vec{J}_{23}^{\,2}$. The transformation from the one basis to the other must be unitary

$$\begin{aligned} &|[[j_1 j_2] J_{12} j_3] J M\rangle \\ &\quad = \sum_{J_{23}} |[j_1 [j_2 j_3] J_{23}] J M\rangle \langle [j_1 [j_2 j_3] J_{23}] J M | [[j_1 j_2] J_{12} j_3] J M\rangle \\ &\quad = \sum_{J_{23}} |[j_1 [j_2 j_3] J_{23}] J M\rangle U(j_1 j_2 J j_3; J_{12} J_{23}), \end{aligned} \tag{2}$$

where we have renamed the unitary transformation coefficient a U coefficient,

$$\langle [j_1 [j_2 j_3] J_{23}] J M | [[j_1 j_2] J_{12} j_3] J M\rangle = U(j_1 j_2 J j_3; J_{12} J_{23}) \equiv U_{J_{12} J_{23}}(j_1 j_2 J j_3), \tag{3}$$

where we have anticipated this unitary transformation is independent of the magnetic quantum number M. It is a matrix element of the unit operator, a spherical tensor of rank 0, in the total angular momentum basis, hence, M-independent by the Wigner–Eckart theorem. The rather strange order of the angular momentum quantum numbers is that first introduced by Racah through his W coefficient, related to the above U coefficient via

$$U(j_1 j_2 J j_3; J_{12} J_{23}) = \sqrt{(2J_{12}+1)(2J_{23}+1)}\, W(j_1 j_2 J j_3; J_{12} J_{23}). \tag{4}$$

Finally, we have indicated this unitary transformation could be expressed through the elements of a matrix with row index J_{12} and column index J_{23}, where the matrix elements are functions of the quantum numbers j_1, j_2, J, j_3, common to both bases. The unitary character of the transformation gives us

$$U^{-1}_{J_{23}J_{12}} = U^*_{J_{12}J_{23}} = U_{J_{12}J_{23}}. \tag{5}$$

Because the U coefficients can be expressed in terms of Clebsch–Gordan coefficients, which are real, the U coefficients are real, so the * has been omitted in the last step, and the inverse matrix is given by the transposed matrix, $U^{-1} = \tilde{U}$. Therefore,

$$\begin{aligned} |[j_1[j_2 j_3]J_{23}]JM\rangle &= \sum_{J_{12}} U^{-1}_{J_{23}J_{12}} |[[j_1 j_2]J_{12} j_3]JM\rangle \\ &= \sum_{J_{12}} U_{J_{12}J_{23}} |[[j_1 j_2]J_{12} j_3]JM\rangle \\ &= \sum_{J_{12}} U(j_1 j_2 J j_3; J_{12} J_{23}) |[[j_1 j_2]J_{12} j_3]JM\rangle. \end{aligned} \tag{6}$$

Because the signs of angular momentum–coupled functions depend on the order of the couplings, it is important to keep the order of the various couplings, as indicated by the order of the angular momentum couplings inside the [ ] brackets. For this purpose a pictorial representation of eqs. (2) and (6) is also useful. See Figs. 34.2(a) and (b), where the arrow also helps to indicate the order of the couplings. See also Figs. 34.1(a) and (b) for a pictorial representation of the coupling of two angular momenta.

B Relations between U Coefficients and Clebsch–Gordan Coefficients

By expanding the angular momentum coupled states of eqs. (2) and (6) in terms of uncoupled states via the appropriate Clebsch–Gordan coefficients, we can express the U coefficients in terms of sums of products of Clebsch–Gordan coefficients. Thus,

$$\begin{aligned} &|[[j_1 j_2]J_{12} j_3]JM\rangle \\ &= \sum_{m_1 m_2 m_3} |j_1 m_1\rangle |j_2 m_2\rangle |j_3 m_3\rangle \langle j_1 m_1 j_2 m_2 | J_{12} M_{12}\rangle \langle J_{12} M_{12} j_3 m_3 | JM\rangle \\ &= \sum_{m_1 m_2 m_3} |j_1 m_1\rangle \sum_{J'_{23}} |[j_2 j_3] J'_{23} M_{23}\rangle \langle j_2 m_2 j_3 m_3 | J'_{23} M_{23}\rangle \\ &\qquad \times \langle j_1 m_1 j_2 m_2 | J_{12} M_{12}\rangle \langle J_{12} M_{12} j_3 m_3 | JM\rangle \end{aligned}$$

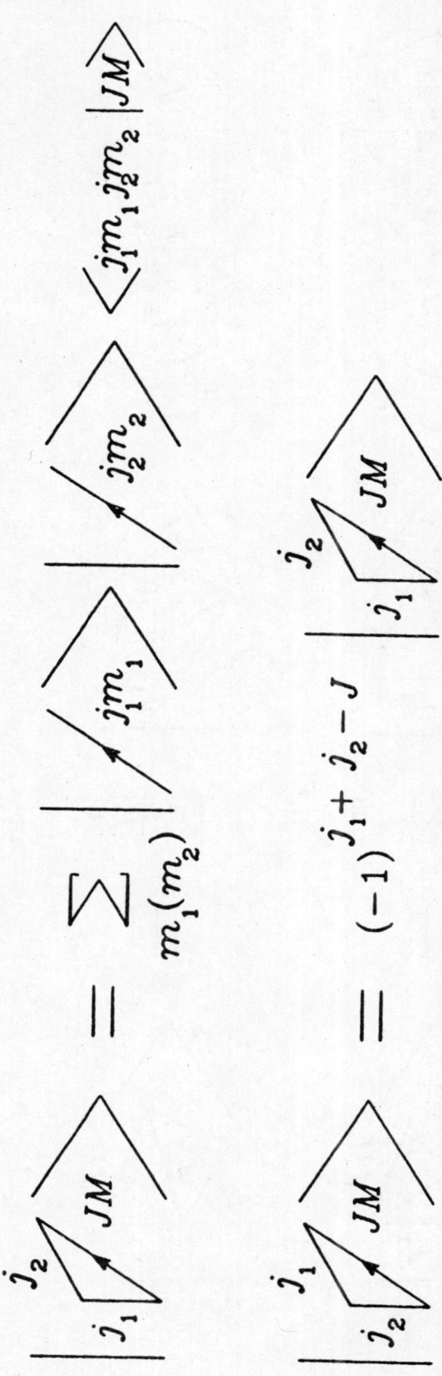

FIGURE 34.1. Coupling of two angular momenta.

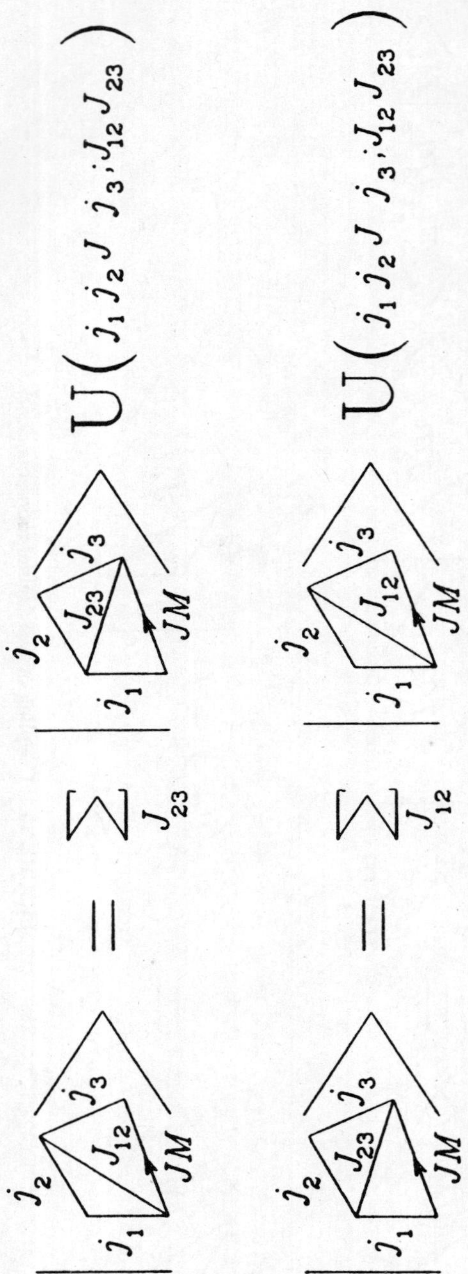

FIGURE 34.2. Recoupling of three angular momenta.

$$= \sum_{m_1 m_2 m_3} \sum_{J'_{23}} \sum_{J'} |[j_1[j_2 j_3]J'_{23}]J'M\rangle \langle j_1 m_1 J'_{23} M_{23}|J'M\rangle$$
$$\times \langle j_2 m_2 j_3 m_3 | J'_{23} M_{23}\rangle \langle j_1 m_1 j_2 m_2 | J_{12} M_{12}\rangle \langle J_{12} M_{12} j_3 m_3 | JM\rangle. \quad (7)$$

Now, left-multiplying this relation by $\langle [j_1[j_2 j_3]J_{23}]JM|$ and using,

$$\langle [j_1[j_2 j_3]J_{23}]JM | [j_1[j_2 j_3]J'_{23}]J'M\rangle = \delta_{J'_{23} J_{23}} \delta_{J'J}, \quad (8)$$

in the last step of the above relation, we are led to

$$U(j_1 j_2 J j_3; J_{12} J_{23}) = \sum_{m_1 m_2 m_3} \langle j_1 m_1 j_2 m_2 | J_{12} M_{12}\rangle$$
$$\times \langle J_{12} M_{12} j_3 m_3 | JM\rangle \langle j_1 m_1 J_{23} M_{23} | JM\rangle \langle j_2 m_2 j_3 m_3 | J_{23} M_{23}\rangle \quad (9)$$

where it is important to remember M is fixed at some specific value, $M = m_1 + m_2 + m_3$, when taking the m_i sums, so that there are essentially only two m_i sums to perform. A second relation can be obtained from

$$|j_1 m_1\rangle |[j_2 j_3]J_{23} M_{23}\rangle = \sum_J |[j_1[j_2 j_3]J_{23}]JM\rangle \langle j_1 m_1 J_{23} M_{23} | JM\rangle$$
$$= \sum_J \sum_{J_{12}} |[[j_1 j_2]J_{12} j_3]JM\rangle U(j_1 j_2 J j_3; J_{12} J_{23}) \langle j_1 m_1 J_{23} M_{23} | JM\rangle$$
$$= \sum_{m_2} |j_1 m_1\rangle |j_2 m_2\rangle |j_3 m_3\rangle \langle j_2 m_2 j_3 m_3 | J_{23} M_{23}\rangle$$
$$= \sum_{m_2} \sum_{J'_{12}} \sum_{J'} |[[j_1 j_2]J'_{12} j_3]J'M\rangle \langle j_1 m_1 j_2 m_2 | J'_{12} M_{12}\rangle$$
$$\times \langle J'_{12} M_{12} j_3 m_3 | J'M\rangle \langle j_2 m_2 j_3 m_3 | J_{23} M_{23}\rangle. \quad (10)$$

Now, using the orthonormality of the states

$$\langle [[j_1 j_2]J_{12} j_3]JM | [[j_1 j_2]J'_{12} j_3]J'M\rangle = \delta_{J_{12} J'_{12}} \delta_{JJ'}, \quad (11)$$

we get

$$U(j_1 j_2 J j_3; J_{12} J_{23}) \langle j_1 m_1 J_{23} M_{23} | JM\rangle =$$
$$\sum_{m_2} \langle j_1 m_1 j_2 m_2 | J_{12} M_{12}\rangle \langle J_{12} M_{12} j_3 m_3 | JM\rangle \langle j_2 m_2 j_3 m_3 | J_{23} M_{23}\rangle. \quad (12)$$

In this relation, m_1, M_{23}, and M, are all fixed. Because the right-hand side, therefore, involves only a single m sum, this relation is particularly useful for the evaluation of U coefficients.

Finally, we get a third relation, by starting with

$$|j_1 m_1 j_2 m_2 j_3 m_3\rangle =$$
$$\sum_{J_{12}} \sum_J |[[j_1 j_2]J_{12} j_3]JM\rangle \langle j_1 m_1 j_2 m_2 | J_{12} M_{12}\rangle \langle J_{12} M_{12} j_3 m_3 | JM\rangle$$
$$= \sum_{J_{23}} \sum_{J'} |[j_1[j_2 j_3]J_{23}]J'M\rangle \langle j_2 m_2 j_3 m_3 | J_{23} M_{23}\rangle \langle j_1 m_1 J_{23} M_{23} | J'M\rangle$$
$$= \sum_{J'_{12} J'} \sum_{J_{23}} |[[j_1 j_2]J'_{12} j_3]J'M\rangle U(j_1 j_2 J' j_3; J'_{12} J_{23})$$
$$\times \langle j_2 m_2 j_3 m_3 | J_{23} M_{23}\rangle \langle j_1 m_1 J_{23} M_{23} | J'M\rangle, \quad (13)$$

leading to

$$\langle j_1 m_1 j_2 m_2 | J_{12} M_{12} \rangle \langle J_{12} M_{12} j_3 m_3 | JM \rangle = \sum_{J_{23}} U(j_1 j_2 J j_3; J_{12} J_{23}) \langle j_2 m_2 j_3 m_3 | J_{23} M_{23} \rangle \langle j_1 m_1 J_{23} M_{23} | JM \rangle. \quad (14)$$

In this relation, all the m_i and, hence, the M_{ij} and M are fixed.

All three relations, eqs. (9), (12), and (14) may be useful in the evaluation of U coefficients.

As a very simple example, let us evaluate the U coefficients for the case $j_1 = l, j_2 = \frac{1}{2}, j_3 = \frac{1}{2}$. This recoupling coefficient might be needed in two-electron configurations in which one electron has arbitrary orbital angular momentum, l, whereas the second electron is an s electron with $l = 0$. Here, a U coefficient is needed in the transformation from LS to jj coupling and has the form $U(l\frac{1}{2}J\frac{1}{2}; jS)$, where S is the total two-electron spin and J is the total angular momentum quantum number. Here, J can have the values $l + 1, l$, and $l - 1$. With $J = l + 1$, however, S is fixed uniquely at $S = 1$, and j is fixed uniquely at $j = (l + \frac{1}{2})$. The U transformation matrix is a 1×1 matrix. For any 1×1 transformation, the U coefficient has the value $+1$. Thus,

$$U(l\tfrac{1}{2}(l+1)\tfrac{1}{2}; (l+\tfrac{1}{2})1) = +1. \quad (15)$$

Similarly,

$$U(l\tfrac{1}{2}(l-1)\tfrac{1}{2}; (l-\tfrac{1}{2})1) = +1. \quad (16)$$

When $J = l$, however, S has the two possible values, $S = 0, 1$, and j has the two possible values $j = (l + \frac{1}{2}), (l - \frac{1}{2})$. In this case, the U transformation matrix is a 2×2 matrix. With the simple table of Clebsch–Gordan coefficients of Chapter 28, eqs. (9) or (12) yield

$$U(l\tfrac{1}{2}l\tfrac{1}{2}; jS) = \begin{matrix} \\ j=(l+\frac{1}{2}) \\ j=(l-\frac{1}{2}) \end{matrix} \begin{matrix} S=0 \qquad\qquad S=1 \\ \begin{pmatrix} \sqrt{\frac{l+1}{2l+1}} & \sqrt{\frac{l}{2l+1}} \\ -\sqrt{\frac{l}{2l+1}} & \sqrt{\frac{l+1}{2l+1}} \end{pmatrix} \end{matrix}. \quad (17)$$

C Alternate Forms for the Recoupling Coefficients for Three Angular Momenta

The recoupling coefficients for three commuting angular momentum operators were first introduced by Racah through his W coefficient. The relation between the Racah W coefficient and the unitary U coefficient has been given in eq. (4). Because the Clebsch–Gordan coefficients are subject to $2 \times 3!$ symmetry relations most easily expressed via the 3-j symbol (see Chapter 28) there will of course be similar symmetry relations for the U coefficients. To see the symmetry relations most easily, without factors of $\sqrt{(2J+1)}$ or complicated phase factors, it is useful to introduce the so-called 6-j coefficient or 6-j symbol, conventionally written between curly brackets in two rows,

$$U(j_1 j_2 J j_3; J_{12} J_{23}) = (-1)^{j_1+j_2+j_3+J} \sqrt{(2J_{12}+1)(2J_{23}+1)} \begin{Bmatrix} j_1 & j_2 & J_{12} \\ j_3 & J & J_{23} \end{Bmatrix}. \tag{18}$$

The 6-j symbol must satisfy four angular momentum addition triangle relations: The three angular momenta in the first row must satisfy an angular momentum addition triangle relation. Any angular momentum quantum number from the first row satisfies such a triangle relation with two partners from the second row, which must lie in columns different from the column of the symbol in the first row. The 6-j symbol is invariant under the following symmetry transformations:

(1) The 6-j symbol is invariant under any permutation of columns (i.e., six symmetry operations).

(2) The 6-j symbol is invariant under an exchange of the j's in rows 1 and 2 for any two columns:

$$\begin{Bmatrix} a & b & c \\ d & e & f \end{Bmatrix} = \begin{Bmatrix} d & e & c \\ a & b & f \end{Bmatrix} = \begin{Bmatrix} a & e & f \\ d & b & c \end{Bmatrix} = \begin{Bmatrix} d & b & f \\ a & e & c \end{Bmatrix}. \tag{19}$$

In actual calculations, the unitary form of the recoupling coefficients is often the most useful. In order to make use of its symmetry properties, however, it is clearly advantageous to convert it to 6-j form first. As a simple application of such symmetry properties, let us evaluate the U coefficient in which the quantum number, $J_{23} = 0$, and therefore $J = j_1$, and $j_3 = j_2$. If one of the labels, j_1, j_2, j_3, or J, has the value zero, the U matrix is a 1×1 matrix and the U coefficient has the value $+1$. Therefore,

$$\begin{aligned} U(j_1 j_2 j_1 j_2; J_{12} 0) &= \sqrt{(2J_{12}+1)}(-1)^{2j_1+2j_2} \begin{Bmatrix} j_1 & j_2 & J_{12} \\ j_2 & j_1 & 0 \end{Bmatrix} \\ &= \sqrt{(2J_{12}+1)}(-1)^{2j_1+2j_2} \begin{Bmatrix} J_{12} & j_1 & j_2 \\ 0 & j_2 & j_1 \end{Bmatrix} \\ &= \sqrt{\frac{(2J_{12}+1)}{(2j_1+1)(2j_2+1)}}(-1)^{j_1+j_2-J_{12}} \Big(U(J_{12} j_1 j_2 0; j_2 j_1) = +1 \Big). \end{aligned} \tag{20}$$

Therefore,

$$U(j_1 j_2 j_1 j_2; J_{12} 0) = (-1)^{j_1+j_2-J_{12}} \sqrt{\frac{(2J_{12}+1)}{(2j_1+1)(2j_2+1)}}. \tag{21}$$

Similarly,

$$U(j_1 j_1 j_2 j_2; 0 J_{23}) = (-1)^{j_1+j_2-J_{23}} \sqrt{\frac{(2J_{23}+1)}{(2j_1+1)(2j_2+1)}}. \tag{22}$$

D Matrix Element of $(U^k(1)\cdot V^k(2))$ in a Vector-Coupled Basis

To illustrate the usefulness of the recoupling coefficients of Racah type, let us calculate the matrix element of a scalar operator of type

$$(U^k(1)\cdot V^k(2)) = \sum_q (-1)^q U^k_q(1) V^k_{-q}(2) \tag{23}$$

in a $|[j_1 j_2]JM\rangle$ basis, where $U^k_q(1)$ are spherical tensors of rank k built from operators acting in the space of variables of type $(1) \equiv (\vec{r}_1, \vec{\sigma}_1, \ldots)$, similarly for $V^k_{-q}(2)$ and space (2) and where the vectors $|j_1 m_1\rangle$ are angular momentum eigenvectors for the subspace (1), similarly for $|j_2 m_2\rangle$ and space (2). That is, we want to calculate matrix elements of type

$$\langle [j_1' j_2']JM|(U^k(1)\cdot V^k(2))|[j_1 j_2]JM\rangle.$$

These are matrix elements of the type met in Chapter 33. For simplicity, we have left off additional quantum numbers that may be needed for a full specification of the states in question. Expanding the angular momentum coupled states in ket and bra and using the Wigner–Eckart theorem to express the matrix elements of $U^k_q(1)$ and $V^k_{-q}(2)$ in terms of their reduced matrix elements, we have

$$\begin{aligned}
&\langle [j_1' j_2']JM|(U^k(1)\cdot V^k(2))|[j_1 j_2]JM\rangle = \\
&\sum_q \sum_{m_1(m_2)} \sum_{m_1'(m_2')} \langle j_1 m_1 j_2 m_2|JM\rangle \langle j_1' m_1' j_2' m_2'|JM\rangle \langle j_1 m_1 kq|j_1' m_1'\rangle \\
&\times \frac{\langle j_1'\|U^k(1)\|j_1\rangle}{\sqrt{(2j_1'+1)}} \langle j_2 m_2 k - q|j_2' m_2'\rangle (-1)^q \frac{\langle j_2'\|V^k(2)\|j_2\rangle}{\sqrt{(2j_2'+1)}}.
\end{aligned} \tag{24}$$

Now, let us use a symmetry property of the Clebsch–Gordan coefficients to reexpress

$$\langle j_2 m_2 k - q|j_2' m_2'\rangle = (-1)^{k-q} \sqrt{\frac{(2j_2'+1)}{(2j_2+1)}} \langle kq j_2' m_2'|j_2 m_2\rangle. \tag{25}$$

The above matrix element then can be rewritten as

$$\begin{aligned}
&\langle [j_1' j_2']JM|(U^k(1)\cdot V^k(2))|[j_1 j_2]JM\rangle = (-1)^k \frac{\langle j_1'\|U^k(1)\|j_1\rangle \langle j_2'\|V^k(2)\|j_2\rangle}{\sqrt{(2j_1'+1)(2j_2+1)}} \\
&\times \sum_q \sum_{m_1(m_2)} \sum_{m_1'(m_2')} \langle j_1 m_1 kq|j_1' m_1'\rangle \langle j_1' m_1' j_2' m_2'|JM\rangle \\
&\quad \times \langle j_1 m_1 j_2 m_2|JM\rangle \langle kq j_2' m_2'|j_2 m_2\rangle.
\end{aligned} \tag{26}$$

Comparing the sum over q, m_1, m_1' of the product of the four Clebsch–Gordan coefficients in the last lines with eq. (9), the identification of the six angular momenta in these Clebsch–Gordan coefficients with those of eq. (9) yields

$$\sum_{q,m_1,m_1'} \langle j_1 m_1 kq|j_1' m_1'\rangle \langle j_1' m_1' j_2' m_2'|JM\rangle \langle j_1 m_1 j_2 m_2|JM\rangle \langle kq j_2' m_2'|j_2 m_2\rangle$$

$$= U(j_1 k J j_2'; j_1' j_2), \tag{27}$$

so

$$\begin{aligned}
&\langle [j_1' j_2'] JM | (U^k(1) \cdot V^k(2)) | [j_1 j_2] JM \rangle \\
&= (-1)^k \frac{\langle j_1' \| U^k(1) \| j_1 \rangle \langle j_2' \| V^k(2) \| j_2 \rangle}{\sqrt{(2j_1'+1)(2j_2+1)}} U(j_1 k J j_2'; j_1' j_2) \\
&= (-1)^{j_1+j_2'+J} \langle j_1' \| U^k(1) \| j_1 \rangle \langle j_2' \| V^k(2) \| j_2 \rangle \begin{Bmatrix} j_1 & k & j_1' \\ j_2' & J & j_2 \end{Bmatrix}.
\end{aligned} \tag{28}$$

E Recoupling of Four Angular Momenta: 9-j Symbols

In a two-electron configuration of an atom, we may be interested in a transformation from an LS-coupled basis to a jj-coupled basis,

$$|[[l_1 l_2]L[s_1 s_2]S]JM\rangle \longrightarrow |[[l_1 s_1]j_1[l_2 s_2]j_2]JM\rangle.$$

This is a special case (with $s_1 = s_2 = \frac{1}{2}$) of a recoupling of four angular momenta; j_1, j_2, j_3, j_4, if we name $j_1 \equiv l_1,\ \ j_2 \equiv l_2$, and $j_3 \equiv s_1,\ \ j_4 \equiv s_2$. We need the transformation from a basis in which J_{12} and J_{34} are good quantum numbers to a basis in which J_{13} and J_{24} are good quantum numbers

$$\begin{aligned}
|[[j_1 j_2]J_{12}[j_3 j_4]J_{34}]JM\rangle &= \sum_{J_{13}J_{24}} |[[j_1 j_3]J_{13}[j_2 j_4]J_{24}]JM\rangle \\
&\quad \langle [[j_1 j_3]J_{13}[j_2 j_4]J_{24}]JM | [[j_1 j_2]J_{12}[j_3 j_4]J_{34}]JM\rangle \\
&= \sum_{J_{13}J_{24}} |[[j_1 j_3]J_{13}[j_2 j_4]J_{24}]JM\rangle \ U \begin{pmatrix} j_1 & j_2 & J_{12} \\ j_3 & j_4 & J_{34} \\ J_{13} & J_{24} & J \end{pmatrix}.
\end{aligned} \tag{29}$$

See Fig. 34.3 for a pictorial representation of this relation. The $U(\cdots)$ symbol involving the nine j's is again a unitary transformation matrix, again independent of M; now with row and column indices specified by two quantum numbers each,

$$U \begin{pmatrix} j_1 & j_2 & J_{12} \\ j_3 & j_4 & J_{34} \\ J_{13} & J_{24} & J \end{pmatrix} = U_{J_{12}J_{34}, J_{13}J_{24}}. \tag{30}$$

For example, with $j_1 \equiv l_1 = 1$, $j_2 \equiv l_2 = 2$, and $j_3 \equiv s_1 = \frac{1}{2}$, $j_4 \equiv s_2 = \frac{1}{2}$, and resultant total $J = 1$, this would be a 3×3 transformation matrix, where the row labels $J_{12}J_{34} \equiv LS$ have the three possible values 10, 11, 21, and the column labels $J_{13}J_{24} \equiv jj'$ have the three possible values $\frac{1}{2}\frac{3}{2}$, $\frac{3}{2}\frac{3}{2}$, $\frac{3}{2}\frac{5}{2}$. Because the above matrix is again both unitary and real, we have

$$\left(U^{-1}\right)_{J_{13}J_{24}, J_{12}J_{34}} = U_{J_{12}J_{34}, J_{13}J_{24}}. \tag{31}$$

Therefore, the inverse transformation is given by

$$|[[j_1 j_3]J_{13}[j_2 j_4]J_{24}]JM\rangle$$

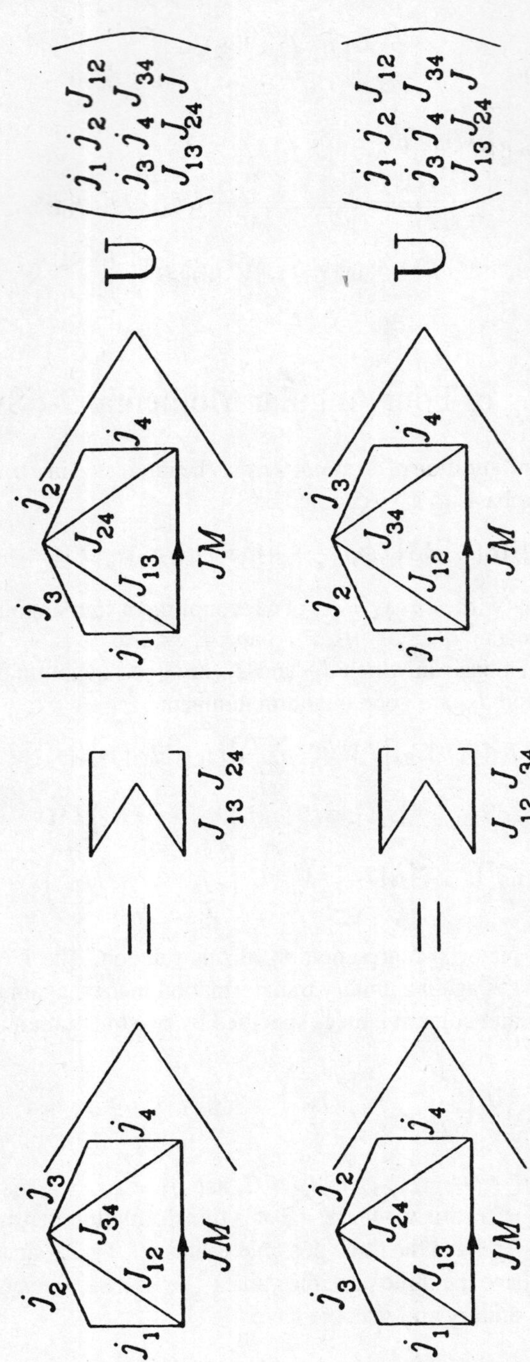

FIGURE 34.3. Recoupling of four angular momenta

$$= \sum_{J_{12}J_{34}} |[[j_1 j_2]J_{12}[j_3 j_4]J_{34}]JM\rangle \; U\begin{pmatrix} j_1 & j_2 & J_{12} \\ j_3 & j_4 & J_{34} \\ J_{13} & J_{24} & J \end{pmatrix}. \tag{32}$$

See Fig. 34.3(b). In the arrangement of the positions of the nine j's in the $U(\cdots)$ symbol, each row and each column corresponds to a coupling of two angular momenta to a resultant. The 9-j U coefficient can thus be expressed in terms of sums over products of six Clebsch–Gordan coefficients

$$U\begin{pmatrix} j_1 & j_2 & J_{12} \\ j_3 & j_4 & J_{34} \\ J_{13} & J_{24} & J \end{pmatrix} =$$

$$\sum_{m_i} \langle j_1 m_1 j_2 m_2 | J_{12} M_{12}\rangle \langle j_3 m_3 j_4 m_4 | J_{34} M_{34}\rangle \langle J_{12} M_{12} J_{34} M_{34} | JM\rangle$$
$$\times \langle j_1 m_1 j_3 m_3 | J_{13} M_{13}\rangle \langle j_2 m_2 j_4 m_4 | J_{24} M_{24}\rangle \langle J_{13} M_{13} J_{24} M_{24} | JM\rangle, \tag{33}$$

where the sum is over all m_i, but with $M = m_1 + m_2 + m_3 + m_4$ fixed at a specific value. Clearly, the symmetry properties of the Clebsch–Gordan coefficients will again lead to many symmetry properties of the unitary 9-j transformation coefficients. These will again have their simplest form not in terms of the unitary 9-j transformation coefficients, but in terms of the so-called 9-j symbol, always written in curly brackets, which is defined by

$$U\begin{pmatrix} j_1 & j_2 & J_{12} \\ j_3 & j_4 & J_{34} \\ J_{13} & J_{24} & J \end{pmatrix}$$
$$= \sqrt{(2J_{12}+1)(2J_{34}+1)(2J_{13}+1)(2J_{24}+1)} \begin{Bmatrix} j_1 & j_2 & J_{12} \\ j_3 & j_4 & J_{34} \\ J_{13} & J_{24} & J \end{Bmatrix}. \tag{34}$$

This $\{\cdots\}$ 9-j symbol has the following symmetry properties:

(1) The 9-j symbol is invariant under any even permutation of rows or columns.

(2) The 9-j symbol is invariant under reflection in either diagonal.

(3) The 9-j symbol changes sign by the factor, $(-1)^{j_1+j_2+j_3+j_4+J_{12}+J_{34}+J_{13}+J_{24}+J}$, involving all nine j's, under any odd permutation of rows or columns.

The 9-j transformation coefficients can be expressed in terms of products of 6-j U coefficients. For example,

$$|[[j_1 j_2]J_{12}[j_3 j_4]J_{34}]JM\rangle$$
$$= \sum_{J_{234}} |[j_1[j_2[j_3 j_4]J_{34}]J_{234}]JM\rangle U(j_1 j_2 J J_{34}; J_{12} J_{234})$$
$$= \sum_{J_{234}} (-1)^{j_3+j_4-J_{34}} |[j_1[j_2[j_4 j_3]J_{34}]J_{234}]JM\rangle U(j_1 j_2 J J_{34}; J_{12} J_{234})$$
$$= \sum_{J_{234}} \sum_{J_{24}} (-1)^{j_3+j_4-J_{34}} |[j_1[[j_2 j_4]J_{24} j_3]J_{234}]JM\rangle$$
$$\times U(j_2 j_4 J_{234} j_3; J_{24} J_{34}) U(j_1 j_2 J J_{34}; J_{12} J_{234})$$
$$= \sum_{J_{234}} \sum_{J_{24}} (-1)^{j_3+j_4-J_{34}} (-1)^{j_3+J_{24}-J_{234}} |[j_1[j_3[j_2 j_4]J_{24}]J_{234}]JM\rangle$$
$$\times U(j_2 j_4 J_{234} j_3 : J_{24} J_{34}) U(j_1 j_2 J J_{34}; J_{12} J_{234})$$

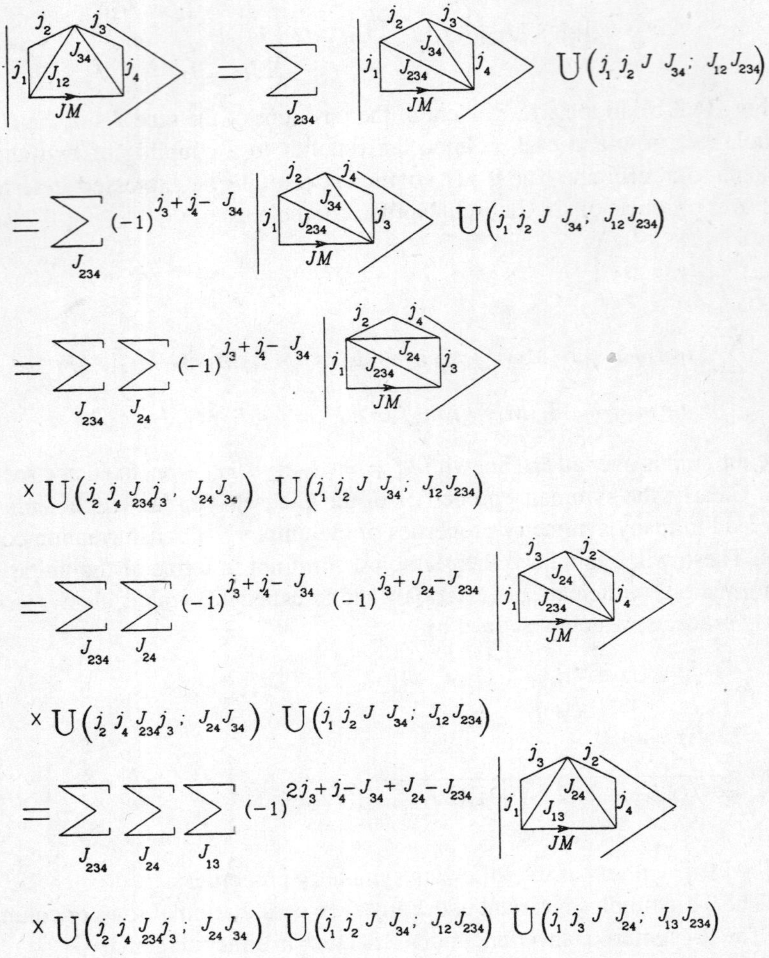

$$= \sum_{J_{234}} U(j_1 j_2 J J_{34}; J_{12} J_{234})$$

$$= \sum_{J_{234}} (-1)^{j_3+j_4-J_{34}} U(j_1 j_2 J J_{34}; J_{12} J_{234})$$

$$= \sum_{J_{234}} \sum_{J_{24}} (-1)^{j_3+j_4-J_{34}}$$

$$\times U(j_2 j_4 J_{234} j_3; J_{24} J_{34})\, U(j_1 j_2 J J_{34}; J_{12} J_{234})$$

$$= \sum_{J_{234}} \sum_{J_{24}} (-1)^{j_3+j_4-J_{34}} (-1)^{j_3+J_{24}-J_{234}}$$

$$\times U(j_2 j_4 J_{234} j_3; J_{24} J_{34})\, U(j_1 j_2 J J_{34}; J_{12} J_{234})$$

$$= \sum_{J_{234}} \sum_{J_{24}} \sum_{J_{13}} (-1)^{2j_3+j_4-J_{34}+J_{24}-J_{234}}$$

$$\times U(j_2 j_4 J_{234} j_3; J_{24} J_{34})\, U(j_1 j_2 J J_{34}; J_{12} J_{234})\, U(j_1 j_3 J J_{24}; J_{13} J_{234})$$

FIGURE 34.4. Pictorial version of eq. (35).

$$= \sum_{J_{234}} \sum_{J_{24}} \sum_{J_{13}} (-1)^{2j_3+j_4-J_{34}-J_{234}+J_{24}} |[[j_1 j_3]J_{13}[j_2 j_4]J_{24}]JM\rangle$$
$$\times U(j_2 j_4 J_{234} j_3; J_{24} J_{34}) U(j_1 j_2 J J_{34}; J_{12} J_{234}) U(j_1 j_3 J J_{24}; J_{13} J_{234})$$
$$= \sum_{J_{13}, J_{24}} |[[j_1 j_3]J_{13}[j_2 j_4]J_{24}]JM\rangle \; U\begin{pmatrix} j_1 & j_2 & J_{12} \\ j_3 & j_4 & J_{34} \\ J_{13} & J_{24} & J \end{pmatrix}. \tag{35}$$

The various steps in this relation are easier to follow in a pictorial representation. See Fig. 34.4. Comparing the last two lines of this relation, we get

$$U\begin{pmatrix} j_1 & j_2 & J_{12} \\ j_3 & j_4 & J_{34} \\ J_{13} & J_{24} & J \end{pmatrix} =$$

$$\sum_{J_{234}}(-1)^{2j_3+j_4-J_{34}-J_{234}+J_{24}}U(j_2j_4J_{234}j_3 : J_{24}J_{34})$$
$$\times U(j_1j_2JJ_{34}; J_{12}J_{234})U(j_1j_3JJ_{24}; J_{13}J_{234})$$
$$= \sum_{J_{234}}(-1)^{2J_{234}}(2J_{234}+1)\sqrt{(2J_{12}+1)(2J_{34}+1)(2J_{13}+1)(2J_{24}+1)}$$
$$\times \begin{Bmatrix} j_2 & j_4 & J_{24} \\ j_3 & J_{234} & J_{34} \end{Bmatrix} \begin{Bmatrix} j_1 & j_2 & J_{12} \\ J_{34} & J & J_{234} \end{Bmatrix} \begin{Bmatrix} j_1 & j_3 & J_{13} \\ J_{24} & J & J_{234} \end{Bmatrix}. \tag{36}$$

This relation is particularly useful if one of the angular momenta has the value zero. For example, if $j_2 = 0$. Then, with $J_{24} = j_4$ and $J_{234} = J_{34}$, we have $U(0j_4J_{34}j_3; j_4J_{34}) = +1$ and with $J_{12} = j_1$ and again $J_{234} = J_{34}$, we have $U(j_10JJ_{34}; j_1J_{34}) = +1$. The above relation collapses to

$$U\begin{pmatrix} j_1 & 0 & j_1 \\ j_3 & j_4 & J_{34} \\ J_{13} & j_4 & J \end{pmatrix} = U(j_1j_3Jj_4; J_{13}J_{34}). \tag{37}$$

This relation can also be seen directly from the pictorial representation of Fig. 34.5, from which we see the triangle, coupling j_1 with 0 to resultant j_1, rides along on the back of a single Racah type of recoupling transformation involving the recoupling of j_1, j_3, and j_4. Similarly, we have

$$U\begin{pmatrix} j_1 & j_2 & J_{12} \\ j_3 & 0 & j_3 \\ J_{13} & j_2 & J \end{pmatrix} = (-1)^{j_1+J-J_{12}-J_{13}}U(j_2j_1Jj_3; J_{12}J_{13}). \tag{38}$$

See also Fig. 34.6.

F Matrix Element of a Coupled Tensor Operator, $[U^{k_1}(1) \times V^{k_2}(2)]^k_q$ in a Vector-Coupled Basis

To appreciate how the 9-j transformation coefficient can facilitate calculations in an angular momentum coupled basis, let us calculate the matrix element of a vector-coupled tensor operator,

$$[U^{k_1}(1) \times V^{k_2}(2)]^k_q = \sum_{q_1(q_2)} \langle k_1q_1k_2q_2|kq\rangle U^{k_1}_{q_1}(1)V^{k_2}_{q_2}(2), \tag{39}$$

in a $|[j_1j_2]JM\rangle$ basis, where again, the $U^{k_1}_{q_1}(1)$ are spherical tensors of rank k_1 acting on variables of type (1) and where the $|j_1m_1\rangle$ are angular momentum eigenvectors of the space (1), similarly for $V^{k_2}_{q_2}(2)$ and the $|j_2m_2\rangle$ and space (2). The needed matrix element can be expressed in terms of a reduced matrix element via the Wigner–Eckart theorem and can be expanded in terms of Clebsch–Gordan coefficients via

$$\langle[j_1'j_2']J'M'|[U^{k_1}(1) \times V^{k_2}(2)]^k_q|[j_1j_2]JM\rangle$$
$$= \langle JMkq|J'M'\rangle \frac{\langle[j_1'j_2']J'\|[U^{k_1}(1) \times V^{k_2}(2)]^k\|[j_1j_2]J\rangle}{\sqrt{(2J'+1)}}$$

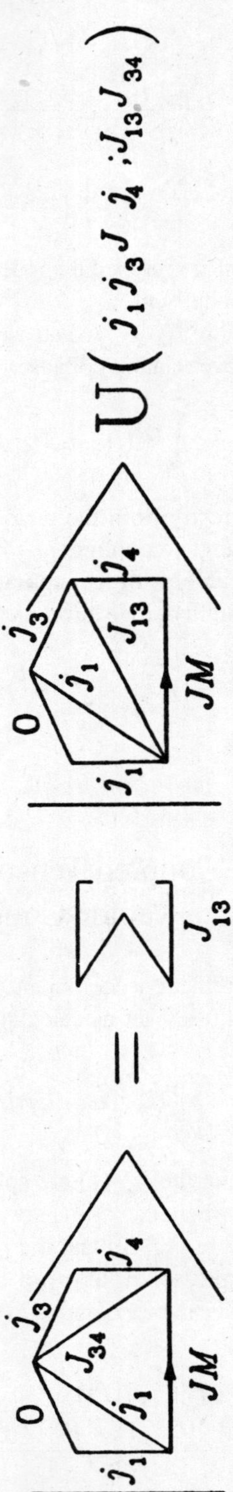

FIGURE 34.5. Pictorial version of Eq. (37).

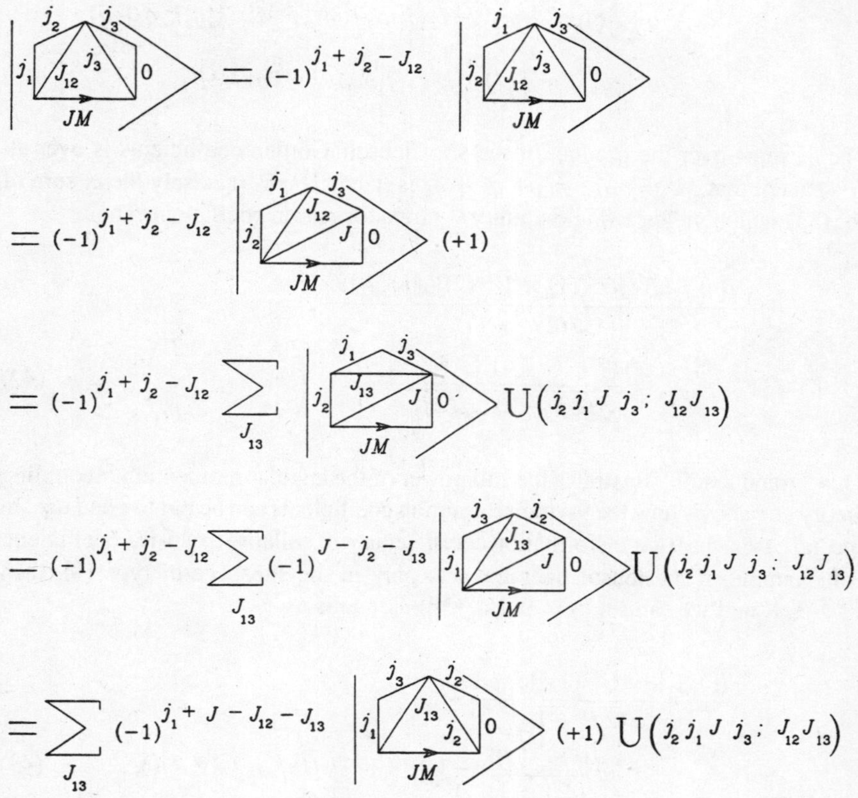

FIGURE 34.6. Pictorial version of eq. (38).

$$= \sum_{m_1 m_1' q_1} \langle j_1 m_1 j_2 m_2 | JM \rangle \langle j_1' m_1' j_2' m_2' | J'M' \rangle \langle k_1 q_1 k_2 q_2 | kq \rangle$$

$$\times \langle j_1 m_1 k_1 q_1 | j_1' m_1' \rangle \frac{\langle j_1' \| U^{k_1} \| j_1 \rangle}{\sqrt{(2j_1'+1)}} \langle j_2 m_2 k_2 q_2 | j_2' m_2' \rangle \frac{\langle j_2' \| V^{k_2} \| j_2 \rangle}{\sqrt{(2j_2'+1)}}, \quad (40)$$

where the magnetic quantum numbers, M, q, and M' are fixed at specific values. If we multiply this equation by $\langle JMkq|J'M'\rangle$, and, keeping M' fixed, sum over all possible values of M and $q = (M' - M)$, the orthogonality relation,

$$\sum_{M,(q)} \langle JMkq | J'M' \rangle^2 = 1,$$

will pick out the reduced matrix element for our coupled tensor operator

$$\frac{\langle [j_1' j_2'] J' \| [U^{k_1}(1) \times V^{k_2}(2)]^k \| [j_1 j_2] J \rangle}{\sqrt{(2J'+1)}}$$

$$= \frac{\langle j_1' \| U^{k_1} \| j_1 \rangle}{\sqrt{(2j_1'+1)}} \frac{\langle j_2' \| V^{k_2} \| j_2 \rangle}{\sqrt{(2j_2'+1)}}$$

$$\times \sum_{m_1,m'_1} \sum_{q_1,q_2} \langle j_1 m_1 j_2 m_2 | J M \rangle \langle j'_1 m'_1 j'_2 m'_2 | J' M' \rangle \langle k_1 q_1 k_2 q_2 | k q \rangle$$
$$\times \langle j_1 m_1 k_1 q_1 | j'_1 m'_1 \rangle \langle j_2 m_2 k_2 q_2 | j'_2 m'_2 \rangle \langle J M k q | J' M' \rangle. \tag{41}$$

The m sums over the product of the six Clebsch–Gordan coefficients is over all m's except that $M' = m_1 + m_2 + q_1 + q_2$ is fixed. This is precisely the m sum of eq. (33) which yields a single unitary 9-j transformation coefficient. Thus,

$$\frac{\langle [j'_1 j'_2] J' \| [U^{k_1}(1) \times V^{k_2}(2)]^k \| [j_1 j_2] J \rangle}{\sqrt{(2J'+1)}} = \frac{\langle j'_1 \| U^{k_1} \| j_1 \rangle}{\sqrt{(2j'_1+1)}} \frac{\langle j'_2 \| V^{k_2} \| j_2 \rangle}{\sqrt{(2j'_2+1)}} \, U \begin{pmatrix} j_1 & j_2 & J \\ k_1 & k_2 & k \\ j'_1 & j'_2 & J' \end{pmatrix}. \tag{42}$$

This "grand result" illustrates the full power of the angular momentum recoupling theory and shows how the 9-j transformation coefficients can be put to good use. In special cases, the 9-j transformation coefficients will collapse to a 6-j U coefficient. For example, if the tensor operator acts only in the subspace of type (1), then $V^{k_2}_{q_2} \to 1$, and we can set $k_2 = 0$. Eq. (38) then tells us

$$\frac{\langle [j'_1 j'_2] J' \| U^{k_1}(1) \| [j_1 j_2] J \rangle}{\sqrt{(2J'+1)}} = \frac{\langle j'_1 \| U^{k_1} \| j_1 \rangle}{\sqrt{(2j'_1+1)}} (-1)^{j_1 + J' - J - j'_1} U(j_2 j_1 J' k_1; J j'_1), \tag{43}$$

where we have also used

$$\langle j'_2 \| 1 \| j_2 \rangle = \delta_{j_2 j'_2} \sqrt{(2j'_2+1)}.$$

An example of this would be the matrix element of an electric dipole moment operator, which is independent of spin variables, in an $|[ls]jm_j\rangle$ basis

$$\frac{\langle [l's]j' \| \vec{\mu}^{(\text{el.})} \| [ls]j \rangle}{\sqrt{(2j'+1)}} = \frac{\langle l' \| \vec{\mu}^{(\text{el.})} \| l \rangle}{\sqrt{(2l'+1)}} (-1)^{l - l' + j' - j} U(\tfrac{1}{2} l j' 1; j l'). \tag{44}$$

Another special case would be the case of a scalar operator, with $k = 0$, hence $k_2 = k_1$, and $J' = J$. We leave it as an exercise in the symmetry properties of the 9-j symbol to show our general result of eq. (42) then collapses to the result already derived in eq. (28). Recall $(U^k \cdot V^k) = (-1)^k [(2k+1)]^{\frac{1}{2}} [U^k \times V^k]^0_0$.

Although tabulations of 9-j symbols are not readily available, computer codes are easy to construct. For tabulations of 6-j symbols, see the references to tabulations at the end of Chapter 28. For algebraic expressions for 6-j symbols with at least one $j \leq 2$, see the references at the end of Chapter 28.

G An Application: The Nuclear Hyperfine Interaction in a One-Electron Atom Revisited

In Chapter 33, we calculated the nuclear hyperfine interaction in a one-electron atom with the use of a few Wigner coefficients by making judicious use of the Wigner–Eckart theorem. In particular, we calculated the nuclear hyperfine splitting of a one-electron $p_{3/2}$ state, where our nucleus had a nuclear spin $I = \frac{1}{2}$ as in hydrogen, and we also quoted the result for the partner $p_{1/2}$ state. We can now repeat this calculation in a much more general way by making full use of the angular momentum recoupling machinery of this addendum.

The nuclear hyperfine interaction for a hydrogenic s-state, with $l = 0$, was given in Chapter 33 as

$$H_{\text{h.f.int}} = \frac{mc^2\alpha^4}{2}\frac{m}{M}g_I\frac{8\pi}{3}(\vec{s}\cdot\vec{I})\delta(\vec{r}). \tag{45}$$

For a hydrogenic state with $l \neq 0$, conversely, it was shown to be

$$H_{\text{h.f.int}} = \frac{mc^2\alpha^4}{2}\frac{m}{M}g_I\left[\frac{1}{r^3}\left((\vec{l}\cdot\vec{I}) - \sqrt{8\pi}([Y^2\times s^1]^1\cdot I^1)\right)\right], \tag{46}$$

[see eqs. (5) and (14) of Chapter 33]. If we define the needed hydrogenic integrals via

$$\begin{aligned}\beta'_{\text{h.f.s.}}(l=0) &= \frac{mc^2\alpha^4}{2}\frac{m}{M}g_I\int d\vec{r}|\psi(\vec{r})_{n00}|^2\frac{8\pi}{3}\delta(\vec{r}) = \frac{mc^2\alpha^4}{2}\frac{m}{M}g_I\frac{2}{3}|R_{n0}(0)|^2,\\ \beta_{\text{h.f.s.}}(l\neq 0) &= \frac{mc^2\alpha^4}{2}\frac{m}{M}g_I\int_0^\infty drr^2|R_{nl}(r)|^2\frac{1}{r^3},\end{aligned} \tag{47}$$

then we have an effective hyperfine interaction Hamiltonian that can be written

$$\begin{aligned}H_{\text{h.f.int}} &= \beta_{\text{h.f.s.}}\left[(\vec{l}\cdot\vec{I}) - \sqrt{8\pi}([Y^2\times s^1]^1\cdot I^1)\right], &&\text{for } l\neq 0;\\ H_{\text{h.f.int}} &= \beta'_{\text{h.f.s.}}(\vec{s}\cdot\vec{I}), &&\text{for } l = 0.\end{aligned} \tag{48}$$

All terms in these electron-nuclear spin interactions are of the form $(U^{k=1}(1)\cdot V^{k=1}(2))$, where the space (1) is that of the electron and includes both electron orbital and electron spin variables, whereas the space (2) is that of the nuclear intrinsic variables characterized by the nuclear spin vector, $\vec{I}$. We will of course need the coupling scheme $\vec{l}+\vec{s} = \vec{j}$ for the electron variables, and we will assume the fine structure splitting is much greater than the hyperfine structure splitting, so j is a good quantum number. Finally, we will couple electron $\vec{j}$ with the nuclear spin, $\vec{I}$, to resultant total angular momentum, $\vec{F}$: $\vec{j}+\vec{I} = \vec{F}$. All of the needed matrix elements are then given in terms of electron reduced matrix elements, the nuclear spin reduced matrix element, and a single 6-j symbol by formula (28) of this chapter. For example,

$$\langle[[ls]jI]FM_F|(\vec{l}\cdot\vec{I})|[[ls]jI]FM_F\rangle$$

$$= (-1)^{j+I+F} \langle [ls]j \| \vec{l} \| [ls]j \rangle \langle I \| \vec{I} \| I \rangle \begin{Bmatrix} j & 1 & j \\ I & F & I \end{Bmatrix}, \tag{49}$$

with similar expressions for matrix elements of the operators $(\vec{s} \cdot \vec{I})$ and $([Y^2 \times s^1]^1 \cdot \vec{I})$. The reduced matrix elements of the electronic operators, $\vec{l}, \vec{s}$, and $[Y^2 \times s^1]^1_q$ in the $|[ls]jm_j\rangle$ basis of the $\vec{l} + \vec{s} = \vec{j}$-coupled scheme are all given through formula (42) of this chapter. For example,

$$\frac{\langle [ls]j \| \vec{s} \| [ls]j \rangle}{\sqrt{(2j+1)}} = \frac{\langle l \| 1 \| l \rangle}{\sqrt{(2l+1)}} \frac{\langle s \| \vec{s} \| s \rangle}{\sqrt{(2s+1)}} U \begin{pmatrix} l & \frac{1}{2} & j \\ 0 & 1 & 1 \\ l & \frac{1}{2} & j \end{pmatrix}. \tag{50}$$

Because we are interested in the reduced matrix element of $\vec{s}$ mainly for states with orbital angular momentum, $l = 0$, the needed unitary 9-j coefficient, with all zeros in the first column, corresponds to a 1×1 unitary transformation and thus has the value, $+1$. The reduced matrix elements for the unit operator and the angular momentum vector operator are given for any angular momentum basis by eqs. (4) and (7) of Chapter 32. Thus,

$$\langle l \| 1 \| l \rangle = \sqrt{(2l+1)}, \qquad \langle s \| \vec{s} \| s \rangle = \sqrt{(2s+1)s(s+1)} = \sqrt{\tfrac{3}{2}};$$

so, with $[ls]j = [0\frac{1}{2}]\frac{1}{2}$

$$\langle [0\tfrac{1}{2}]\tfrac{1}{2} \| \vec{s} \| [0\tfrac{1}{2}]\tfrac{1}{2} \rangle = \sqrt{\tfrac{3}{2}}.$$

Similarly,

$$\frac{\langle [ls]j \| \vec{l} \| [ls]j \rangle}{\sqrt{(2j+1)}} = \frac{\langle l \| \vec{l} \| l \rangle}{\sqrt{(2l+1)}} \frac{\langle s \| 1 \| s \rangle}{\sqrt{(2s+1)}} U \begin{pmatrix} l & \frac{1}{2} & j \\ 1 & 0 & 1 \\ l & \frac{1}{2} & j \end{pmatrix}$$
$$= \sqrt{l(l+1)} U(\tfrac{1}{2} l j 1; j l), \tag{51}$$

where we have used eq. (38) to convert the unitary 9-j coefficient with one zero to a unitary Racah coefficient. This can be read from tables of 6-j symbols. For $l = 1$, this U coefficient has the values: $\sqrt{\frac{5}{6}}$ for $j = \frac{3}{2}$, and $\sqrt{\frac{2}{3}}$ for $j = \frac{1}{2}$. Therefore,

$$\langle [1\tfrac{1}{2}]\tfrac{3}{2} \| \vec{l} \| [1\tfrac{1}{2}]\tfrac{3}{2} \rangle = 2\sqrt{\tfrac{5}{3}}; \qquad \langle [1\tfrac{1}{2}]\tfrac{1}{2} \| \vec{l} \| [1\tfrac{1}{2}]\tfrac{1}{2} \rangle = 2\sqrt{\tfrac{2}{3}}.$$

Finally,

$$\langle [ls]j \| [Y^2 \times s^1]^1 \| [ls]j \rangle = \sqrt{(2j+1)} \frac{\langle l \| Y^2 \| l \rangle}{\sqrt{(2l+1)}} \frac{\langle \frac{1}{2} \| \vec{s} \| \frac{1}{2} \rangle}{\sqrt{2}} U \begin{pmatrix} l & \frac{1}{2} & j \\ 2 & 1 & 1 \\ l & \frac{1}{2} & j \end{pmatrix}, \tag{52}$$

where the reduced matrix element of Y^2 was given in Chapter 32. It has the value

$$\frac{\langle l \| Y^2 \| l \rangle}{\sqrt{(2l+1)}} = \sqrt{\frac{5}{4\pi}} \langle l020 | l0 \rangle = -\frac{1}{\sqrt{2\pi}} \quad \text{for } l = 1.$$

The unitary 9-j coefficients above can be given in terms of 6-j symbols through eq. (36). For $l = 1$, and both $j = \frac{3}{2}$ and $j = \frac{1}{2}$, the sum over J_{234} in this relation collapses to a single term: J_{234} has the unique value $\frac{3}{2}$. The above 9-j U coefficients have the values $-1/(3\sqrt{5})$ and $+(2/3)$ for $j=\frac{3}{2}$ and $j=\frac{1}{2}$, respectively. Thus,

$$\langle[1\tfrac{1}{2}]\tfrac{3}{2}\|[Y^2\times s^1]^1\|[1\tfrac{1}{2}]\tfrac{3}{2}\rangle=\frac{1}{\sqrt{30\pi}},\qquad \langle[1\tfrac{1}{2}]\tfrac{1}{2}\|[Y^2\times s^1]^1\|[1\tfrac{1}{2}]\tfrac{1}{2}\rangle=-\frac{1}{\sqrt{3\pi}}.$$

With these electron reduced matrix elements, together with the nuclear spin reduced matrix element,

$$\langle I\|\vec{I}\|I\rangle=\sqrt{(2I+1)I(I+1)}, \tag{53}$$

the nuclear hyperfine interaction in a one-electron atom is then given by, first for the case $l\neq 0$,

$$\begin{aligned}&\langle[[ls]jI]FM_F|H_{\text{h.f.int.}}|[[ls]jI]FM_F\rangle=\beta_{\text{h.f.s.}}\sqrt{(2I+1)I(I+1)}\\&\times\Big[\langle[ls]j\|\vec{l}\|[ls]j\rangle-\sqrt{8\pi}\langle[ls]j\|[Y^2\times s^1]^1\|[ls]j\rangle\Big]\\&\times(-1)^{j+I+F}\begin{Bmatrix} j & 1 & j\\ I & F & I\end{Bmatrix},\end{aligned} \tag{54}$$

and for the case, $l=0$, with $j=\frac{1}{2}$,

$$\begin{aligned}&\langle[[ls]jI]FM_F|H_{\text{h.f.int.}}|[[ls]jI]FM_F\rangle=\beta'_{\text{h.f.s.}}\sqrt{(2I+1)I(I+1)}\\&\times\langle[ls]j\|\vec{s}\|[ls]j\rangle(-1)^{j+I+F}\begin{Bmatrix} j & 1 & j\\ I & F & I\end{Bmatrix}.\end{aligned} \tag{55}$$

For the cases, $l=1$ and $l=0$, with $I=\frac{1}{2}$, we will need the 6-j symbols

$$\begin{Bmatrix} \frac{3}{2} & 1 & \frac{3}{2}\\ \frac{1}{2} & 2 & \frac{1}{2}\end{Bmatrix}=\frac{1}{2\sqrt{10}},\qquad \begin{Bmatrix} \frac{3}{2} & 1 & \frac{3}{2}\\ \frac{1}{2} & 1 & \frac{1}{2}\end{Bmatrix}=\frac{\sqrt{5}}{6\sqrt{2}},$$

$$\begin{Bmatrix} \frac{1}{2} & 1 & \frac{1}{2}\\ \frac{1}{2} & 1 & \frac{1}{2}\end{Bmatrix}=\frac{1}{6},\qquad \begin{Bmatrix} \frac{1}{2} & 1 & \frac{1}{2}\\ \frac{1}{2} & 0 & \frac{1}{2}\end{Bmatrix}=\frac{1}{2},$$

so, with $l=1$, and both $I=\frac{1}{2}$ and $s=\frac{1}{2}$,

$$\begin{aligned}\langle[[1s]jI]FM_F|H_{\text{h.f.int.}}|[[1s]jI]FM_F\rangle&=+\tfrac{2}{5}\beta_{\text{h.f.s.}}\ \text{ for }\ j=\tfrac{3}{2},\ F=2,\\ \langle[[1s]jI]FM_F|H_{\text{h.f.int.}}|[[1s]jI]FM_F\rangle&=-\tfrac{2}{3}\beta_{\text{h.f.s.}}\ \text{ for }\ j=\tfrac{3}{2},\ F=1,\\ \langle[[1s]jI]FM_F|H_{\text{h.f.int.}}|[[1s]jI]FM_F\rangle&=+\tfrac{2}{3}\beta_{\text{h.f.s.}}\ \text{ for }\ j=\tfrac{1}{2},\ F=1,\\ \langle[[1s]jI]FM_F|H_{\text{h.f.int.}}|[[1s]jI]FM_F\rangle&=-2\beta_{\text{h.f.s.}}\ \text{ for }\ j=\tfrac{1}{2},\ F=0,\end{aligned} \tag{56}$$

and for $l=0$ states,

$$\begin{aligned}\langle[[0s]sI]FM_F|H_{\text{h.f.int.}}|[[0s]sI]FM_F\rangle&=+\tfrac{1}{4}\beta'_{\text{h.f.s.}},\qquad \text{for }\ F=1,\\ \langle[[0s]sI]FM_F|H_{\text{h.f.int.}}|[[0s]sI]FM_F\rangle&=-\tfrac{3}{4}\beta'_{\text{h.f.s.}},\qquad \text{for }\ F=0.\end{aligned} \tag{57}$$

35

Perturbed Coulomb Problems via SO(2,1) Algebra

A Perturbed Coulomb Problems: The Conventional Approach with its Infinite Sums and Continuum Integrals: An Example: The Second-Order Stark Effect of the Hydrogen Atom Ground State

So far, we have not solved many perturbed Coulomb problems by conventional perturbation theory, that is, by the conventional radial and angular functions. The difficulty here is that perturbation terms which are functions of the radial coordinate, r, lead to an infinite number of nonzero matrix elements, connecting a bound state of definite, n, to all other bound states, as well as to the full spectrum of continuum states. The complete set of states includes both the bound states and the continuum states. The unit operator in this conventional basis is given by

$$1 = \sum_{n=1}^{\infty}\sum_{l,m}|nlm\rangle\langle nlm| + \int d\Omega_k \int_0^{\infty} dk k^2 |\vec{k}\rangle\langle\vec{k}|, \tag{1}$$

where the continuum states are the continuum solutions of the Coulomb problem, with energy, $\hbar^2 k^2/(2\mu)$, which in the limit $k \to \infty$ go over to plane wave states $\to \langle\vec{r}|\vec{k}\rangle = e^{i\vec{k}\cdot\vec{r}}/(2\pi)^{\frac{3}{2}}$. (These continuum Coulomb states will be discussed in detail in Chapter 42 in connection with our study of scattering theory.)

To illustrate the difficulties with the conventional radial Coulomb functions, let us look at a very simple perturbation problem: the second-order Stark effect of the hydrogen atom ground state. The Hamiltonian is

$$H = H^{0)} + H^{(1)} = H_{\text{Coulomb}} - e\mathcal{E}r\cos\theta, \tag{2}$$

where $\vec{\mathcal{E}}$ is the external electric field in the z direction. Conventional second-order perturbation theory gives

$$E^{(2)}_{n=1,l=0,m=0} = \sum_{n=2}^{\infty} \frac{|\langle n10|H^{(1)}|100\rangle|^2}{(E_1^{(0)} - E_n^{(0)})} + \int d\Omega_k \int_0^{\infty} dkk^2 \frac{|\langle \vec{k}|H^{(1)}|100\rangle|^2}{[E_1^{(0)} - \frac{\hbar^2k^2}{2\mu}]}. \tag{3}$$

The needed matrix element between the hydrogen ground state and an arbitrary discrete excited state will be calculated later in this chapter. It is

$$\langle nlm|H^{(1)}|100\rangle = -\delta_{l1}\delta_{m0}e\mathcal{E}a_0 2^4 n^3 \frac{(n-1)^{n-3}}{(n+1)^{n+3}} \left[\frac{(n+1)n(n-1)}{3}\right]^{\frac{1}{2}}. \tag{4}$$

The needed matrix element between the hydrogenic ground state and a continuum state will be calculated in the mathematical appendix to Chapter 42. It is

$$\langle \vec{k}|H^{(1)}|100\rangle = -\delta_{l1}\delta_{m0}e\mathcal{E}\frac{a_0^{\frac{5}{2}}}{(2\pi)^{\frac{3}{2}}} \frac{4\pi}{\sqrt{3}} Y_{lm}(\theta_k, \phi_k)$$
$$\times \left[\frac{(\gamma^2+1)2\pi\gamma}{(1-e^{-2\gamma\pi})}\right]^{\frac{1}{2}} \frac{8i\gamma^5}{(\gamma^2+1)^3} e^{-2\gamma\tan^{-1}(1/\gamma)}, \tag{5}$$

where γ is the Coulomb parameter

$$\gamma = \frac{1}{ka_0}, \qquad a_0 = \frac{\hbar^2}{Z\mu e^2}, \tag{6}$$

and $\vec{k}$ is given by its magnitude, k, and direction specified by polar and azimuth angles, θ_k, ϕ_k. With these results, the second-order correction to the energy is

$$E^{(2)} = -\Big(\frac{e^2a_0^2\mathcal{E}^2}{\mu Z^2e^4/\hbar^2}\Big)\Big(\sum_{n=2}^{\infty}\frac{2^9n^9}{3}\frac{(n-1)^{2n-6}}{(n+1)^{2n+6}}$$
$$+ \int d\Omega_k Y^{\star}_{10}(\theta_k,\phi_k)Y_{10}(\theta_k,\phi_k)\frac{512}{3}\int_0^{\infty} d\gamma \frac{\gamma^9}{(\gamma^2+1)^6}\frac{e^{-4\gamma\tan^{-1}(1/\gamma)}}{(1-e^{-2\gamma\pi})}\Big)$$
$$= -\Big(\frac{e^2a_0^2\mathcal{E}^2}{\mu Z^2e^4/\hbar^2}\Big)\Big(\sum_{n=2}^{\infty}\frac{2^9n^9}{3}\frac{(n-1)^{2n-6}}{(n+1)^{2n+6}} + .418371\Big)$$
$$= -\Big(\frac{e^2a_0^2\mathcal{E}^2}{\mu Z^2e^4/\hbar^2}\Big)2.25, \tag{7}$$

where the γ integral is best done numerically, and, as we shall see, the final result is given, in the appropriate units, by -(9/4), exactly. The infinite discrete sum converges quite rapidly. For example, the first five terms, through $n = 6$, give 1.792758, or a total of 2.211129 with the continuum contribution, which is within 1.7% of the exact result. We shall see in this chapter, however, we can arrive at the exact final result very simply by a perturbation expansion requiring a discrete sum of just two terms.

B The Runge–Lenz Vector as an ℓ Step Operator and the SO(4) Algebra of the Coulomb Problem

Before introducing the operators of an SO(2,1) algebra, which will be used to simplify the perturbation expansions for a perturbed Coulomb problem, let us look first at the commutator algebra of the quantum-mechanical operators arising from the known classical integrals for the Coulomb potential, the orbital angular momentum vector, $\vec{L}$, and the Runge–Lenz vector, $\vec{\mathcal{R}}$. It will be convenient to express all operators in dimensionless quantities through atomic units. Thus,

$$\vec{r}_{\text{phys.}} = a_0\vec{r}, \qquad \vec{p}_{\text{phys.}} = \frac{\hbar}{a_0}\vec{p}, \qquad \vec{L}_{\text{phys.}} = \hbar\vec{L}, \qquad \text{with } a_0 = \frac{\hbar^2}{Z\mu e^2},$$
$$H_{\text{phys.}} = \frac{Z^2\mu e^4}{\hbar^2}H, \qquad E_{\text{phys.}} = \frac{Z^2\mu e^4}{\hbar^2}\epsilon. \tag{8}$$

In these units, the Runge–Lenz vector (see problems 13 and 19) is

$$\vec{\mathcal{R}} = \frac{1}{2}\Big([\vec{p}\times\vec{L}] - [\vec{L}\times\vec{p}]\Big) - \frac{\vec{r}}{r} = \vec{r}(\vec{p}\cdot\vec{p}) - \vec{p}(\vec{r}\cdot\vec{p}) - \frac{\vec{r}}{r}, \tag{9}$$

with the properties

$$(\vec{\mathcal{R}}\cdot\vec{\mathcal{R}}) = 2H(\vec{L}\cdot\vec{L}+1)+1, \tag{10}$$

and

$$\vec{\mathcal{R}}\cdot\vec{L} = \vec{L}\cdot\vec{\mathcal{R}} = 0. \tag{11}$$

It was useful to define

$$\vec{V} = \frac{\vec{\mathcal{R}}}{\sqrt{(-2\epsilon)}} = n\vec{\mathcal{R}},$$

valid for negative-energy bound states. With this definition, eq. (10) becomes

$$(\vec{V}\cdot\vec{V}) + (\vec{L}\cdot\vec{L}) + 1 = \frac{1}{(-2\epsilon)} = n^2, \tag{12}$$

and the commutator algebra of the operators, $\vec{V}$, $\vec{L}$, is given by

$$[L_j, L_k] = i\epsilon_{jka}L_a, \qquad [L_j, V_k] = i\epsilon_{jka}V_a, \qquad [V_j, V_k] = i\epsilon_{jka}L_a. \tag{13}$$

If we define $L_{jk} = -L_{kj}$, with $j, k = 1, \ldots, 4$ through

$$L_{jk} = \frac{1}{i}\left(x_j\frac{\partial}{\partial x_k} - x_k\frac{\partial}{\partial x_j}\right), \quad \text{with } L_1 = L_{23},\ L_2 = L_{31},\ L_3 = L_{12},$$
$$\text{and } V_j = L_{j4}, \tag{14}$$

the six operators, $\vec{L}$, $\vec{V}$, satisfy the same commutation relations as the six L_{jk}; i.e., the six operators, $\vec{L}$, $\vec{V}$, constitute infinitesimal rotation generators in an abstract 4-D space; i.e., they generate an SO(4) group, with a subgroup, SO(3), generated by the three components of $\vec{L}$.

It is useful to define the vector operators, $\vec{M}$ and $\vec{N}$, through

$$\vec{M} = \tfrac{1}{2}(\vec{L} + \vec{V}), \qquad \vec{N} = \tfrac{1}{2}(\vec{L} - \vec{V}), \tag{15}$$

where these operators satisfy the commutation relations of two *commuting* angular momentum operators

$$[M_j, M_k] = i\epsilon_{jka} M_a, \qquad [N_j, N_k] = i\epsilon_{jka} N_a, \qquad [M_j, N_k] = 0. \tag{16}$$

Eqs. (11) and (12) then lead to

$$\begin{aligned}(\vec{M}^2 - \vec{N}^2) &= \tfrac{1}{2}\Big(\vec{L}\cdot\vec{V} + \vec{V}\cdot\vec{L}\Big) = 0,\\ (\vec{M}^2 + \vec{N}^2) &= \tfrac{1}{2}(n^2 - 1).\end{aligned} \tag{17}$$

If the eigenvalues of the operators $\vec{M}^2$ and $\vec{N}^2$ are denoted by $j_1(j_1+1)$ and $j_2(j_2+1)$, respectively, these relations require

$$j_1 = j_2 = \tfrac{1}{2}(n-1). \tag{18}$$

These angular momentum quantum numbers can thus be integral or half-integral. Two commuting angular momentum operators of this type generate a group which is a direct product of two SU(2) groups: $SU(2) \times SU(2)$. Because $\vec{L} = (\vec{M} + \vec{N})$, and we want to construct states of good orbital angular momentum, we want to use the vector-coupled basis $|[j_1 \times j_2]lm\rangle = |[\tfrac{1}{2}(n-1) \times \tfrac{1}{2}(n-1)]lm\rangle$, which is an eigenvector simultaneously of $\vec{L}^2$, L_0, and H, the latter with eigenvalue, $\epsilon = -1/2n^2$. The quantum number, n, also gives the eigenvalues of $\vec{M}^2$ and $\vec{N}^2$. Angular momentum vector coupling rules tell us the possible quantum numbers l range from 0, 1, ... to (n-1).

To understand the significance of the vector $\vec{V} = (\vec{M} - \vec{N})$, let us rewrite

$$\vec{V} = n\Big(\vec{r}(\vec{p}\cdot\vec{p} - \frac{1}{r}) - \vec{p}(\vec{r}\cdot\vec{p})\Big)$$

in terms of spherical coordinates, r, θ, ϕ. It will be sufficient to choose one component, e.g., the z component, which then has the form

$$\begin{aligned}\frac{V_0}{n} &= r\cos\theta\left(-\frac{\partial^2}{\partial r^2} - \frac{2}{r}\frac{\partial}{\partial r} + \frac{l(l+1)}{r^2} - \frac{1}{r}\right)\\ &\quad + \left(\cos\theta\Big(\frac{\partial}{\partial r}\Big) + \frac{\sin\theta e^{i\phi}}{2r}L_- - \frac{\sin\theta e^{-i\phi}}{2r}L_+\right)\Big(r\frac{\partial}{\partial r}\Big).\end{aligned} \tag{19}$$

Using the known matrix elements of $L_\pm$ and of $\cos\theta$ and $\sin\theta e^{\pm i\phi}$, see e.g., eqs. (42)–(44) of Chapter 9, we see that action of this operator on a Coulomb eigenfunction yields

$$\begin{aligned}&V_0 R_{nl}(r) Y_{lm}(\theta,\phi) =\\ &n(l+1)\left[\left(-\frac{\partial}{\partial r} + \frac{l}{r} - \frac{1}{(l+1)}\right)R_{nl}(r)\right]\sqrt{\frac{[(l+1)^2 - m^2]}{(2l+1)(2l+3)}}\,Y_{(l+1)m}\\ &+ nl\left[\left(+\frac{\partial}{\partial r} + \frac{(l+1)}{r} - \frac{1}{l}\right)R_{nl}(r)\right]\sqrt{\frac{[l^2 - m^2]}{(2l+1)(2l-1)}}\,Y_{(l-1)m}.\end{aligned} \tag{20}$$

Rewriting the radial function in terms of the one-dimensionalized radial function, $R_{nl}(r) = u_{nl}(r)/r$, and introducing the Clebsch–Gordan coefficients, $\langle lm10|(l \pm 1)m\rangle$ (see the table at the end of Chapter 28), this can be rewritten as

$$
\begin{aligned}
&V_0 R_{nl}(r) Y_{lm}(\theta, \phi) = \\
&n(l+1)\frac{1}{r}\left[\left(-\frac{\partial}{\partial r} + \frac{(l+1)}{r} - \frac{1}{(l+1)}\right)u_{nl}(r)\right] \\
&\times \sqrt{\frac{(l+1)}{(2l+3)}}\langle lm10|(l+1)m\rangle Y_{(l+1)m} \\
&- nl\frac{1}{r}\left[\left(+\frac{\partial}{\partial r} + \frac{l}{r} - \frac{1}{l}\right)u_{nl}(r)\right]\sqrt{\frac{l}{(2l-1)}}\langle lm10|(l-1)m\rangle Y_{(l-1)m} \\
&= n(l+1)\frac{1}{r}\Big(O_+(l+1)u_{nl}(r)\Big)\sqrt{\frac{(l+1)}{(2l+3)}}\langle lm10|(l+1)m\rangle Y_{(l+1)m} \\
&- nl\frac{1}{r}\Big(O_-(l)u_{nl}(r)\Big)\sqrt{\frac{l}{(2l-1)}}\langle lm10|(l-1)m\rangle Y_{(l-1)m},
\end{aligned} \tag{21}
$$

where the operators $O_+(l+1)$ and $O_-(l)$ are the l step operators introduced in Chapter 10 to construct the radial Coulomb eigenfunctions. Finally, using the operators, $\mathcal{O}_+(l+1)$ and $\mathcal{O}_-(l)$, which construct normalized radial functions [see eq. (13) of Chapter 10], we have

$$
\begin{aligned}
&V_0\left[\frac{1}{r}u_{nl}(r)\right]Y_{lm}(\theta, \phi) = \\
&\sqrt{(l+1)[n^2-(l+1)^2]}\frac{1}{r}\mathcal{O}_+(l+1)u_{nl}(r)\frac{\langle lm10|(l+1)m\rangle}{\sqrt{(2l+3)}}Y_{(l+1)m} \\
&- \sqrt{l[n^2-l^2]}\frac{1}{r}\mathcal{O}_-(l)u_{nl}(r)\frac{\langle lm10|(l-1)m\rangle}{\sqrt{(2l-1)}}Y_{(l-1)m},
\end{aligned} \tag{22}
$$

where $(1/r)[\mathcal{O}_+(l+1)u_{nl}(r)]$ gives the normalized $R_{n(l+1)}(r)$, and similarly for $(1/r)[\mathcal{O}_-(l)u_{nl}(r)]$. Recalling the definition of the reduced matrix elements

$$
\langle nl'm|V_0|nlm\rangle = \frac{\langle lm10|l'm\rangle}{\sqrt{(2l'+1)}}\langle nl'\|\vec{V}\|nl\rangle,
$$

we see

$$
\begin{aligned}
\langle n(l+1)\|\vec{V}\|nl\rangle &= \sqrt{(l+1)[n^2-(l+1)^2]}, \\
\langle n(l-1)\|\vec{V}\|nl\rangle &= -\sqrt{l[n^2-l^2]}.
\end{aligned} \tag{23}
$$

These equations give the only nonzero matrix elements of $\vec{V}$. We note that the properly normalized Runge–Lenz vector operator gives the l step operators in the subspace of a definite principal quantum number, n. Because the components of $\vec{L}$ furnish the m step operators for a fixed l; the operators $\vec{V}$ and $\vec{L}$ can serve to construct all n^2 states of a definite n, starting with the maximum possible $l = m = (n-1)$.

Although we have achieved our aim of finding the meaning of $\vec{V}$, we could have arrived at the above reduced matrix elements of $\vec{V}$ by using the angular momentum–coupled basis $|[\frac{1}{2}(n-1) \times \frac{1}{2}(n-1)]lm\rangle$, in which the angular momentum eigenvectors of $\vec{M}$ are coupled to the angular momentum eigenvectors of $\vec{N}$ to resultant good orbital angular momentum of definite l, m. With $\vec{L} = \vec{M} + \vec{N}$, and $\vec{V} = \vec{M} - \vec{N}$, the reduced matrix elements of $\vec{L}$ and $\vec{V}$ follow from the reduced matrix elements of $\vec{M}$ and $\vec{N}$ via a simple unitary 9-j coefficient [see eq. (42) of Chapter 34]. Recall the vector $\vec{M}$ is a spherical tensor of spherical rank 1 in the $\vec{M}$ space, but because it does not act on the $\vec{N}$ space, it can be thought of as multiplied by the number 1, a spherical tensor of rank 0 in the $\vec{N}$ space. Thus,

$$\begin{aligned}
&\langle [j_1 j_2]l' \| \vec{M} \pm \vec{N} \| [j_1 j_2]l\rangle \\
&\equiv \langle [\tfrac{1}{2}(n-1)\ \tfrac{1}{2}(n-1)]l' \| \vec{M} \pm \vec{N} \| [\tfrac{1}{2}(n-1)\ \tfrac{1}{2}(n-1)]l\rangle \\
&= \tfrac{1}{2}\sqrt{(2l'+1)(n-1)(n+1)}\left[U\begin{pmatrix} \frac{1}{2}(n-1) & \frac{1}{2}(n-1) & l \\ 1 & 0 & 1 \\ \frac{1}{2}(n-1) & \frac{1}{2}(n-1) & l' \end{pmatrix} \right. \\
&\left. \pm U\begin{pmatrix} \frac{1}{2}(n-1) & \frac{1}{2}(n-1) & l \\ 0 & 1 & 1 \\ \frac{1}{2}(n-1) & \frac{1}{2}(n-1) & l' \end{pmatrix} \right] \\
&= \tfrac{1}{2}\sqrt{(2l'+1)(n-1)(n+1)}\left(1 \pm (-1)^{l'+l}\right) \\
&\quad \times \left[U\begin{pmatrix} \frac{1}{2}(n-1) & \frac{1}{2}(n-1) & l \\ 1 & 0 & 1 \\ \frac{1}{2}(n-1) & \frac{1}{2}(n-1) & l' \end{pmatrix} \right],
\end{aligned} \tag{24}$$

where we have used the reduced matrix elements,

$$\langle j' \| 1 \| j\rangle = \delta_{jj'}\sqrt{(2j+1)}, \qquad \langle j' \| \vec{J} \| j\rangle = \delta_{jj'}\sqrt{(2j+1)j(j+1)},$$

valid for any angular momentum operator [see eqs. (4) and (7) of Chapter 32], where now $\vec{J}$ is either $\vec{M}$ or $\vec{N}$ and $j = \frac{1}{2}(n-1)$ for both. In the last step of the above equation, we have also used a symmetry property of the 9-j coefficent, involving interchange of columns 1 and 2. Finally, the unitary 9-j coefficient with one angular momentum, j_4, of zero can be replaced by a unitary Racah coefficient [see eq. (38) of Chapter 34], to yield

$$\begin{aligned}
&\langle [\tfrac{1}{2}(n-1)\ \tfrac{1}{2}(n-1)]l' \| \vec{M} \pm \vec{N} \| [\tfrac{1}{2}(n-1)\ \tfrac{1}{2}(n-1)]l\rangle = \\
&\tfrac{1}{2}\sqrt{(2l'+1)(n-1)(n+1)}(-1)^{l'-l}U(\tfrac{1}{2}(n-1)\tfrac{1}{2}(n-1)l'1; l\tfrac{1}{2}(n-1)) \\
&\quad \times \left(1 \pm (-1)^{l+l'}\right).
\end{aligned} \tag{25}$$

We see at once $\vec{L} = \vec{M} + \vec{N}$ has only diagonal matrix elements with $l' = l$, whereas $\vec{V} = \vec{M} - \vec{N}$ has zero diagonal matrix element with $l' = l$. (This can, of course, also be seen from the negative parity of the operator $\vec{V}$.) The Racah coefficients, with one $j = 1$, are tabulated as algebraic functions of the possible angular momentum quantum numbers (see the angular momentum references at the end of Chapter 28). Putting in these values, we obtain the above results for $\langle nl' \| \vec{V} \| nl\rangle$, and the expected

$$\langle nl' \| \vec{L} \| nl\rangle = \delta_{l'l}\sqrt{(2l+1)l(l+1)}. \tag{26}$$

C The SO(2,1) Algebra

So far, the six operators, $\vec{V}$ and $\vec{L}$, can be used to generate the states and matrix elements in the n^2-dimensional subspace of a definite bound state of the hydrogen atom. To obtain general expressions for the matrix elements of radial functions, we still need to construct operators that can raise or lower the principal quantum number, n. For this purpose, we introduce the three operators, useful also for general central force potentials,

$$\begin{aligned} T_1 &= \tfrac{1}{2}\Big(rp_r^2 + \frac{l(l+1)}{r} - r\Big) \\ T_2 &= rp_r \\ T_3 &= \tfrac{1}{2}\Big(rp_r^2 + \frac{l(l+1)}{r} + r\Big), \end{aligned} \tag{27}$$

with

$$p_r = \frac{1}{i}\Big(\frac{\partial}{\partial r} + \frac{1}{r}\Big), \tag{28}$$

where r is again the dimensionless $r = r_{\text{phys.}}/a_0$. These T_i satisfy the SO(2,1) commutation relations

$$[T_1, T_2] = -iT_3, \qquad [T_2, T_3] = iT_1, \qquad [T_3, T_1] = iT_2, \tag{29}$$

or in terms of $T_\pm = (T_1 \pm iT_2)$

$$[T_3, T_\pm] = \pm T_\pm, \qquad \text{but} \quad [T_+, T_-] = -2T_3. \tag{30}$$

Except for one minus sign, these operators would be standard angular momentum operators. (The eigenvectors and operator matrix elements of these T_i were studied in problem 23.) These T_i, however, are hermitian operators not with respect to the standard volume element measure, $d\Omega dr r^2$, but instead with respect to the measure, $d\Omega dr r$. For example,

$$\begin{aligned} &\int\!\!\int d\Omega \int_0^\infty dr r \psi_1^* \big(T_2\psi_2\big) = \int\!\!\int d\Omega \int_0^\infty dr r \psi_1^* \frac{1}{i}\Big(r\frac{\partial}{\partial r} + 1\Big)\psi_2 \\ &= \int\!\!\int d\Omega \Big(\frac{1}{i} r^2 \psi_1^*\psi_2|_0^\infty + \Big(-\int_0^\infty dr \psi_2 \frac{1}{i}\frac{\partial}{\partial r}\big(r^2\psi_1^*\big) + \int_0^\infty dr \psi_2 \frac{r}{i}\psi_1^*\Big) \\ &= -\int\!\!\int d\Omega \Big(\int_0^\infty dr \psi_2 \frac{r}{i}\psi_1^* + \int_0^\infty dr r^2 \psi_2 \frac{1}{i}\frac{\partial \psi_1^*}{\partial r}\Big) \\ &= \int\!\!\int d\Omega \Big(\int_0^\infty dr r \psi_2 \big(T_2\psi_1\big)\Big)^*. \end{aligned} \tag{31}$$

The three T_i commute with the generalized $\vec{T}^2$ or SO(2,1) Casimir operator, $\mathcal{C}$, defined by

$$\begin{aligned} \mathcal{C} &= T_3^2 - T_1^2 - T_2^2 = (T_3 - T_1)(T_3 + T_1) - [T_3, T_1] - T_2^2 \\ &= r^2 p_r^2 + l(l+1) - irp_r - (rp_r)(rp_r) = l(l+1). \end{aligned} \tag{32}$$

In a basis in which both $\mathcal{C}$ and T_3 are diagonal,

$$\mathcal{C}|ql\rangle = l(l+1)|ql\rangle,$$

$$T_3 = q|ql), \tag{33}$$

the spectrum of allowed q values is given by a ladder of values starting with a minimum q value (see problem 23)

$$q = q_{\text{min.}} + n_r = (l+1) + n_r, \qquad \text{with} \quad n_r = 0, 1, 2, \dots, \infty, \tag{34}$$

so the eigenvalue, q, of the operator T_3 can be identified with the principal quantum number $n = l + 1 + n_r$, and the nonzero matrix elements of $\vec{T}$ are given by

$$\begin{aligned} (ql|T_3|ql) &= q = n, \\ ((q-1)l|T_-|ql) &= \sqrt{(q+l)(q-l-1)}, \\ ((q+1)l|T_+|ql) &= \sqrt{(q-l)(q+l+1)}. \end{aligned} \tag{35}$$

(See problem 23.) Note: T_- annihilates the state with $q = (l+1)$, and T_+ can act as step-up operator to create states with $q = n > (l+1)$.

D The Dilation Property of the Operator, T_2

For the ordinary angular momentum algebra, the operator $e^{i\phi L_2}$ acted as a rotation operator. For the SO(2,1) algebra, the corresponding operator, e^{iaT_2}, acts as a dilation operator. Here, the parameter, a, like ϕ, is a real number. This dilation or stretching property follows because e^{iaT_2} acting on an arbitrary function of r yields

$$e^{iaT_2} F(r) = e^{a(r\frac{d}{dr}+1)} F(r) = e^a F(e^a r). \tag{36}$$

This relation follows if $F(r)$ can be expanded in an infinite series,

$$F(r) = c_0 + \sum_{n=1}^{\infty} (c_{-n}/r^n + c_n r^n),$$

because

$$e^{ar\frac{d}{dr}} r^{\pm n} = \sum_k \frac{a^k}{k!}\left(r\frac{d}{dr}\right)^k r^{\pm n} = \Big(\sum_k \frac{(\pm an)^k}{k!}\Big) r^{\pm n} = e^{\pm an} r^{\pm n} = \left(e^a r\right)^{\pm n}. \tag{37}$$

We shall also need the transformation of operators, O, via the unitary operator, e^{iaT_2}

$$e^{iaT_2} O e^{-iaT_2} = \sum_n \frac{(ia)^n}{n!} [T_2, [T_2, [T_2, \dots [T_2, O] \dots]]]_n, \tag{38}$$

where the expansion is in terms of the nth commutator of T_2 with the operator, O (for a derivation, see eqs. (23)–(25) of Chapter 16). For example, if $O = T_3$, we see that

$$[T_2, [T_2, [T_2, \dots [T_2, T_3] \dots]]]^{2n+1} = i^{2n+1} T_1, \tag{39}$$

for a commutator with an odd number of T_2's, whereas

$$[T_2, [T_2, [T_2, \dots [T_2, T_3] \dots]]]^{2n} = i^{2n} T_3, \tag{40}$$

for a commutator with an even number of T_2's. We therefore obtain

$$e^{iaT_2}T_3e^{-iaT_2} = \cosh aT_3 - \sinh aT_1, \tag{41}$$

and, similarly,

$$e^{iaT_2}T_1e^{-iaT_2} = \cosh aT_1 - \sinh aT_3. \tag{42}$$

Also, with $r = T_3 - T_1$,

$$e^{iaT_2}re^{-iaT_2} = e^a r, \tag{43}$$

and

$$e^{iaT_2}p_r e^{-iaT_2} = e^{-a}p_r. \tag{44}$$

From these last two relations, the dilation transformation properties of simple functions of r and p_r can be obtained.

E The Zeroth-Order Energy Eigenvalue Problem for the Hydrogen Atom: Stretched States

The zeroth-order hydrogen atom eigenvalue problem can be written in terms of the SO(2,1) operators T_i. Keeping in mind the natural measure for these operators is $d\Omega drr$ (rather than the conventional $d\Omega drr^2$), the zeroth-order energy eigenvalue problem can be rewritten as

$$\begin{aligned}
0 &= \int d\vec{r}\Psi^*(H-\epsilon)\Psi \\
&= \int\int d\Omega \int_0^\infty drr\Psi^*(rH - r\epsilon)\Psi \\
&= \int\int d\Omega \int_0^\infty drr\Psi^*\left(\tfrac{1}{2}\left(rp_r^2 + \frac{\vec{L}^2}{r}\right) - 1 - r\epsilon\right)\Psi \\
&= \int\int d\Omega \int_0^\infty drr\Psi^*\left(\tfrac{1}{2}r\left(-\frac{\partial^2}{\partial r} - \frac{2}{r}\frac{\partial}{\partial r} + \frac{l(l+1)}{r^2}\right) - 1 - r\epsilon\right)\Psi \\
&= \int\int d\Omega \int_0^\infty drr\Psi^*\left(\tfrac{1}{2}(T_3+T_1) - \epsilon(T_3-T_1) - 1\right)\Psi.
\end{aligned} \tag{45}$$

Although we have succeeded in rewriting the operator $r(H-\epsilon)$ in terms of the generators T_3 and T_1, these cannot be made simultaneously diagonal. To eliminate the unwanted T_1, we shall stretch the states Ψ with the dilation operator. By inserting unit operators, expressed through $1 = e^{-iaT_2}e^{iaT_2}$, in the appropriate places, the above eigenvalue integral can be rewritten as

$$\begin{aligned}
0 &= \int d\vec{r}\Psi^*(H-\epsilon)\Psi \\
&= \int\int d\Omega \int_0^\infty drr\left(e^{iaT_2}\Psi\right)^*\Big[e^{iaT_2}\left(\tfrac{1}{2}(T_3+T_1)\right.
\end{aligned}$$

$$
\begin{aligned}
&- \epsilon(T_3 - T_1) - 1\Big)e^{-iaT_2}\Big]\Big(e^{iaT_2}\Psi\Big) \\
&= \int_0^\infty drr\Big(e^{iaT_2}R_{nl}(r)\Big)^*\Big(\Big[(\tfrac{1}{2} - \epsilon)\cosh a - (\tfrac{1}{2} + \epsilon)\sinh a\Big]T_3 \\
&+ \Big[(\tfrac{1}{2} + \epsilon)\cosh a - (\tfrac{1}{2} - \epsilon)\sinh a\Big]T_1 - 1\Big)\Big(e^{iaT_2}R_{nl}(r)\Big), \qquad (46)
\end{aligned}
$$

where we have assumed the angular part of Ψ is a spherical harmonic of definite l and m. By choosing the real number, a, such that the coefficient of the T_1 term in the last line of the above relation is set equal to zero, the radial part of $(e^{iaT_2}\Psi)$ will have been transformed into an eigenfunction of T_3 and C. This result is achieved by choosing the parameter, a, such that

$$
\Big[(\tfrac{1}{2} + \epsilon)\cosh a - (\tfrac{1}{2} - \epsilon)\sinh a\Big] = 0, \qquad (47)
$$

which, for bound states, with negative ϵ, leads to

$$
e^{2a} = -\frac{1}{2\epsilon} = +n^2, \qquad (48)
$$

so

$$
e^a = n \qquad \text{and} \qquad \Big[(\tfrac{1}{2} - \epsilon)\cosh a - (\tfrac{1}{2} + \epsilon)\sinh a\Big] = \frac{1}{n} \qquad (49)
$$

for a bound state with $\epsilon = -1/2n^2$. The above eigenvalue equation then becomes

$$
\begin{aligned}
0 &= \int\int d\Omega \int_0^\infty drr\Psi^*(rH - r\epsilon)\Psi \\
&= \int_0^\infty drr\Big(e^{iaT_2}R_{nl}(r)\Big)^*\Big(\frac{1}{n}T_3 - 1\Big)\Big(e^{iaT_2}R_{nl}(r)\Big) \\
&= |\mathcal{N}|^2 \int_0^\infty drr\psi^*_{ql}\Big(\frac{1}{n}T_3 - 1\Big)\psi_{ql}, \qquad (50)
\end{aligned}
$$

which leads to the eigenvalue equation

$$
(T_3 - n)\psi_{ql} = (q - n)\psi_{ql} = 0, \qquad (51)
$$

so q can be identified with the conventional principal quantum number, n. We previously found that $q = q_{\text{min.}} + n_r = (l+1) + n_r$, which agrees with this result. In the above, we have named the stretched state

$$
\Big(e^{iaT_2}R_{nl}(r)\Big) = \mathcal{N}\psi_{ql}, \qquad (52)
$$

where $\psi_{q=n,l}$ is the normalized eigenfunction of the hermitian operator T_3 and the full 3-D function $\psi_{nl}(r)Y_{lm}(\theta, \phi) \equiv \Psi_{nlm}(\vec{r})$ is normalized with the new measure $d\Omega drr$. Because of this change of measure, the normalization is not preserved by the unitary dilation operator e^{iaT_2}. The normalization factor, $\mathcal{N}$, can be derived from

$$
1 = \int_0^\infty drr^2 R_{nl}(r)^* R_{nl}(r)
$$

$$= \int_0^\infty drr\Big(e^{iaT_2}R_{nl}(r)\Big)^*\Big(e^{iaT_2}re^{-iaT_2}\Big)\Big(e^{iaT_2}R_{nl}(r)\Big)$$
$$= \int_0^\infty drr\mathcal{N}^*\psi^*_{nl}(r)\Big(nr\Big)\mathcal{N}\psi_{nl}(r) = |\mathcal{N}|^2\int_0^\infty drr\psi^*_{nl}(r)n(T_3 - T_1)\psi_{nl}(r)$$
$$= |\mathcal{N}|^2n^2\int_0^\infty drr\psi_{nl}(r)^*\psi_{nl}(r) = |\mathcal{N}|^2n^2, \tag{53}$$

where we have used the eigenvalue eq. (51) and the fact that T_1 has no matrix elements diagonal in n. We can therefore choose

$$\mathcal{N} = \frac{1}{n}. \tag{54}$$

Henceforth, we shall use the notations,

$$\psi_{q=n,l}(r), \quad \text{or } \Psi_{q=n,lm}(\vec{r}) = \psi_{q=n,l}(r)Y_{lm}(\theta,\phi),$$

for the normalized stretched eigenfunctions and retain the $R_{nl}(r)Y_{lm}(\theta,\phi)$ for the conventional normalized bound-state eigenfunctions. To avoid confusion between the two conventions, we shall use caret notation for state vectors and matrix elements for the conventional hydrogenic states, but will use round parentheses for the state vectors and matrix elements of the stretched states. Thus, for an operator, O, a function of $\vec{r}$ and/or $\vec{p}$,

$$\int\int d\Omega\int_0^\infty drr^2(R_{n'l'}(r)Y_{l'm'}(\theta,\phi))^*OR_{nl}(r)Y_{lm}(\theta,\phi) \equiv \langle n'l'm'|O|nlm\rangle, \tag{55}$$

but for the analagous operator, $\tilde{O} \equiv e^{iaT_2}Oe^{-iaT_2}$,

$$\int\int d\Omega\int_0^\infty drr\Psi^*_{n'l'm'}(\widetilde{rO})\Psi_{nlm} \equiv (n'l'm'|(\widetilde{rO})|nlm). \tag{56}$$

[We have already anticipated this change of notation in our discussion of the matrix elements of the SO(2,1) operators T_i in eq. (35), where we have used round parentheses.] We can construct the explicit functional dependence of the $\psi_{ql}(r)$, via the explicit relation, $T_-|q = q_{\text{min.}} = (l+1), l) = 0$, and subsequent successive action with T_+. Using

$$T_\pm = (T_1 \pm iT_2) = (T_1 - T_3) \pm iT_2 + T_3 = -r \pm \Big(r\frac{\partial}{\partial r} + 1\Big) + T_3, \tag{57}$$

the relation

$$T_-|q = q_{\text{min.}} = (l+1), l) = 0 \qquad \text{leads to}$$
$$\Big(-r - r\frac{\partial}{\partial r} - 1 + (l+1)\Big)\psi_{q=(l+1),l}(r) = 0, \tag{58}$$

so

$$r\frac{d\psi_{(l+1),l}}{dr} = (l-r)\psi_{(l+1),l}, \tag{59}$$

with solution (normalized with the measure drr),

$$\psi_{q=(l+1),l}(r) = \frac{2^{l+1}}{\sqrt{(2l+1)!}} r^l e^{-r}. \tag{60}$$

States with values of $q > q_{\text{min.}}$ can be obtained with successive action of T_+, with

$$\begin{aligned} |(q+1)l\rangle &= \frac{1}{\sqrt{(q-l)(q+l+1)}} T_+ |ql\rangle \\ \psi_{(q+1)l}(r) &= \frac{1}{\sqrt{(q-l)(q+l+1)}} \Big(r\frac{d}{dr} - r + q + 1 \Big) \psi_{ql}(r), \end{aligned} \tag{61}$$

or

$$\begin{aligned} \psi_{ql}(r) &= \left[\frac{1}{1\cdot 2 \cdots (q-l-1)\cdot(2l+2)(2l+3)\cdots(q+l)} \right]^{\frac{1}{2}} \Big(r\frac{d}{dr} - r + q \Big) \\ &\times \Big(r\frac{d}{dr} - r + q - 1 \Big) \cdots \Big(r\frac{d}{dr} - r + l + 3 \Big) \Big(r\frac{d}{dr} - r + l + 2 \Big) \psi_{q=(l+1)l}(r) \\ &= \sqrt{\frac{(2l+1)!}{(q+l)!(q-l-1)!}} \Big(r\frac{d}{dr} - r + q \Big) \Big(r\frac{d}{dr} - r + q - 1 \Big) \cdots \\ &\cdots \Big(r\frac{d}{dr} - r + l + 3 \Big) \Big(r\frac{d}{dr} - r + l + 2 \Big) \psi_{q=(l+1)l}(r). \end{aligned} \tag{62}$$

Using the identity

$$\Big(r\frac{d}{dr} - r + \alpha \Big) \psi(r) = e^r \Big(r\frac{d}{dr} + \alpha \Big) e^{-r} \psi(r), \tag{63}$$

and the fact that $(r\frac{d}{dr} + \alpha)$ commutes with $(r\frac{d}{dr} + \beta)$, the function $\psi_{ql}(r)$ can be rewritten (note the reversal of the operator order),

$$\begin{aligned} \psi_{ql}(r) &= \sqrt{\frac{(2l+1)!}{(q+l)!(q-l-1)!}} e^r \Big(r\frac{d}{dr} + l + 2 \Big) \Big(r\frac{d}{dr} + l + 3 \Big) \cdots \\ &\quad \cdots \Big(r\frac{d}{dr} + q - 1 \Big) \Big(r\frac{d}{dr} + q \Big) e^{-r} \psi_{q=(l+1)l}(r) \\ &= \sqrt{\frac{(2l+1)!}{(q+l)!(q-l-1)!}} e^r \frac{1}{r^{l+1}} \frac{d}{dr} \frac{r^{l+2}}{r^{l+2}} \frac{d}{dr} \cdots \frac{d}{dr} \frac{r^{q-1}}{r^{q-1}} \frac{d}{dr} r^q e^{-r} \\ &\quad \times \frac{2^{l+1}}{\sqrt{(2l+1)!}} r^l e^{-r}, \end{aligned} \tag{64}$$

where we have made repeated use of the operator identity

$$\Big(r\frac{d}{dr} + \alpha \Big) = \frac{1}{r^{\alpha-1}} \frac{d}{dr} r^\alpha, \tag{65}$$

with $\alpha = l+2, l+3, \ldots, \alpha = q$. Simplifying, the above equation leads to the expression

$$\psi_{q,l}(r) = \frac{2^{l+1}}{\sqrt{(q+l)!(q-l-1)!}} e^r \frac{1}{r^{l+1}} \frac{d^{q-l-1}}{dr^{q-l-1}} \Big(r^{q+l} e^{-2r} \Big). \tag{66}$$

As a final remark to this section, we note that the stretched states with $q = n$, eigenstates of the operator, T_3, apply to the bound states of the hydrogen atom only, with negative ϵ. For the continuum states with $\epsilon > 0$, the requirement, $e^{2a} = -(1/2\epsilon)$ of eq. (48) cannot be met with a real value for a, and hence a unitary stretching operator e^{iaT_2}. In this case, however, e^a can be chosen such that the stretched states are eigenstates of the operator, T_1, by choosing a real parameter, a, such that the coefficient of T_3 in eq. (46) is set equal to zero, viz.,

$$(\tfrac{1}{2} - \epsilon)\cosh a - (\tfrac{1}{2} + \epsilon)\sinh a = 0, \tag{67}$$

leading to

$$e^{2a} = +\frac{1}{2\epsilon} = \frac{1}{(ka_0)^2} = \gamma^2, \tag{68}$$

where γ is the Coulomb parameter introduced in eq. (6). Now the zeroth order equation becomes

$$\begin{aligned} &\left(\frac{1}{\gamma}T_1 - 1\right)\left(e^{iaT_2}R_{\gamma,l}(r)\right) = 0, \\ &(T_1 - \gamma)\mathcal{N}_\gamma \psi_{\gamma,l}(r) = 0. \end{aligned} \tag{69}$$

The important point, however, is the following: The states $\psi_{nl}(r)$ and the $\psi_{\gamma,l}(r)$ belong to different irreducible representations of the SO(2,1) group. Each forms a complete set of its own. Perturbation theory for the bound states will not connect the ψ_{nl} to the $\psi_{\gamma,l}$.

F Perturbations of the Coulomb Problem

For a perturbed hydrogenic atom problem, it will be useful to transcribe the Schrödinger equation for the full Hamiltonian into the stretched-state basis. The SO(2,1) basis will be particularly useful for spherically symmetric perturbations, where l and m will be good quantum numbers to all orders. In that case [case (1) of Chapter 23], the spherical, nlm, is the symmetry adapted or proper basis; and therefore nondegenerate perturbation theory can be used. No connections exist from lm to states with $l' \neq l$, $m' \neq m$. For that case, the Schrödinger equation for the full Hamiltonian

$$\begin{aligned} 0 &= r(H - \epsilon)R_{nl}(r) \\ &= r\Big[(H^{(0)} + H^{(1)} + H^{(2)} + \cdots) - (\epsilon^{(0)} + \epsilon^{(1)} + \epsilon^{(2)} + \cdots)\Big]R_{nl}(r) \\ &= \Bigg(\Big[\tfrac{1}{2}(rp_r^2 + \frac{(l(l+1)}{r}) - 1 - r\epsilon^{(0)}\Big] \\ &\qquad + \Big[(rH^{(1)} - r\epsilon^{(1)}) + (rH^{(2)} - r\epsilon^{(2)}) + \cdots\Big]\Bigg)R_{nl}(r) \end{aligned} \tag{70}$$

can be transcribed to stretched form by acting on the above with e^{iaT_2} to obtain

$$0 = e^{iaT_2}\Bigg(\Big[\tfrac{1}{2}(T_3+T_1) - 1 - (T_3-T_1)\epsilon^{(0)}\Big] + \Big[(T_3-T_1)(H^{(1)}-\epsilon^{(1)})$$
$$+ (T_3-T_1)(H^{(2)}-\epsilon^{(2)}) + \cdots\Big]\Bigg) e^{-iaT_2}\Big(e^{iaT_2}R_{nl}\Big)$$
$$= \frac{\mathcal{N}}{n}\Bigg((T_3-n) + n\Big[n(T_3-T_1)(\tilde{H}^{(1)} - \epsilon^{(1)} + \tilde{H}^{(2)} - \epsilon^{(2)} + \cdots)\Big]\Bigg)\psi_{ql} \quad (71)$$

where

$$\tilde{H}^{(i)} = e^{iaT_2}H^{(i)}e^{-iaT_2}, \quad (72)$$

and the ψ_{ql} are expanded in terms of the zeroth-order eigenfunctions of T_3,

$$T_3\psi_{ql}^{(0)} = q\psi_{ql}^{(0)} = n\psi_{q=n,l}^{(0)}, \quad (73)$$

where now eq. (71) can be written as

$$\Big[(T_3-n) + (H_{\text{eff.}}^{(1)} - \epsilon_{\text{eff.}}^{(1)}) + (H_{\text{eff.}}^{(2)} - \epsilon_{\text{eff.}}^{(2)}) + \cdots\Big](\psi_{nl}^{(0)} + \psi_{nl}^{(1)} + \psi_{nl}^{(2)} + + \cdots) = 0. \quad (74)$$

Here,

$$H_{\text{eff.}}^{(i)} = n^2(T_3-T_1)\tilde{H}^{(i)}, \qquad \epsilon_{\text{eff.}}^{(i)} = n^2(T_3-T_1)\epsilon^{(i)}. \quad (75)$$

Straightforward generalization of Rayleigh–Schrödinger perturbation theory in terms of the zeroth-order eigenfunctions of T_3 (rather than $H^{(0)}$), now yields

$$(\psi_{nl}^{(0)}|\epsilon_{\text{eff.}}^{(1)}|\psi_{nl}^{(0)}) = n^3\epsilon_{nl}^{(1)} = (\psi_{nl}^{(0)}|H_{\text{eff.}}^{(1)}|\psi_{nl}^{(0)}), \quad (76)$$

$$\psi_{nl}^{(1)} = \sum_{q\neq n} c_{ql}^{(1)}\psi_{ql}^{(0)}, \qquad \text{with } c_{ql}^{(1)} = \frac{(\psi_{ql}^{(0)}|(H_{\text{eff.}}^{(1)} - \epsilon_{\text{eff.}}^{(1)})|\psi_{nl}^{(0)})}{(n-q)}, \quad (77)$$

and

$$(\psi_{nl}^{(0)}|\epsilon_{\text{eff.}}^{(2)}|\psi_{nl}^{(0)}) = n^3\epsilon_{nl}^{(2)} = (\psi_{nl}^{(0)}|H_{\text{eff.}}^{(2)}|\psi_{nl}^{(0)})$$
$$+ \sum_{q\neq n}\frac{(\psi_{nl}^{(0)}|(H_{\text{eff.}}^{(1)} - \epsilon_{\text{eff.}}^{(1)})|\psi_{ql}^{(0)})(\psi_{ql}^{(0)}|(H_{\text{eff.}}^{(1)} - \epsilon_{\text{eff.}}^{(1)})|\psi_{nl}^{(0)})}{(n-q)}. \quad (78)$$

Once the number, $\epsilon_n^{(1)}$, has been calculated, the matrix elements of $\epsilon_{\text{eff.}}^{(1)}$ can be calculated, but this operator now has off-diagonal connections through the T_1 part of the operator $(T_3 - T_1)$.

For nonspherically symmetric perturbations, the results of eqs. (76) through (78) will still apply for the nondegenerate case, e.g., for the hydrogenic ground state, $n = 1$, or for axially symmetric perturbations for the special states with $l = (n-1)$ and $m = \pm(n-1)$, provided the $\psi_{ql}^{(0)}$ of eqs. (76)–(78) are replaced with the full stretched state, $\Psi_{nlm}^{(0)}$. In addition, the sums over $q \neq n$ will have to include sums over states with $l' \neq l$, and possibly $m' \neq m$. For the general degenerate states, it is best to parallel the treatment of the degenerate case of Chapter 24 and first make a similarity transformation of the operator

$$\left[(T_3 - n) + (H^{(1)}_{\text{eff.}} - \epsilon^{(1)}_{\text{eff.}}) + \cdots\right]$$

via a unitary operator, $e^{i\lambda G}$, where now

$$\begin{aligned}
(n'l'm'|G|nlm) &= \frac{-i}{(n'-n)}(n'l'm'|H^{(1)}_{\text{eff.}} - \epsilon^{(1)}_{\text{eff.}}|nlm) \quad \text{for } n' \neq n,\\
(nlm|G|n'l'm') &= \frac{-i}{(n-n')}(nlm|H^{(1)}_{\text{eff.}} - \epsilon^{(1)}_{\text{eff.}}|n'l'm') \quad \text{for } n' \neq n,\\
(nl'm'|G|nlm) &= (nlm|G|nl'm') = 0; \quad \text{for all } l', m' \text{ if } n' = n.
\end{aligned} \tag{79}$$

This process leads to a matrix in the n subspace that is the analog of eq. (12) of Chapter 24, if the $H^{(i)}$ there are replaced by $H^{(i)}_{\text{eff.}} - \epsilon^{(i)}_{\text{eff.}}$ and the $E^{(0)}_n$ and $E^{(0)}_{q\neq n}$ of Chapter 24 are now replaced by n and q. The pure numbers $\epsilon^{(1)}_n$ may first have to be determined by diagonalizing this matrix in first order in λ, but the $\epsilon^{(1)}_{\text{eff.}}$ contributes to the off-diagonal sums in second order through its $(T_3 - T_1)$ factor.

The greatest usefulness of the stretched spherical hydrogenic basis comes into play with spherically symmetric perturbations.

G An Application: Coulomb Potential with a Perturbing Linear Potential: Charmonium

A Coulomb potential with a linear confining potential may have some useful applications. The charmed quark-charmed antiquark two-body system has particles heavy enough so that nonrelativistic quantum theory may be a very good approximation. The bound states of this system have been described by an attractive $1/r$ color-electric potential augmented by a linear confining potential. For the deeply bound states, where the $1/r$ potential predominates, the linear (repulsive) confining potential may be treated as a perturbation. We then have a perturbed Coulomb problem with

$$H^{(1)} = V^{(1)}(r) = +\lambda r, \qquad H^{(2)} = 0, \tag{80}$$

where we assume $\lambda \ll 1$ and r is again a dimensionless r. Also, $\lambda > 0$. Now,

$$H^{(1)}_{\text{eff.}} = \lambda n^3 (T_3 - T_1)^2, \qquad \epsilon^{(i)}_{\text{eff.}} = n^2 (T_3 - T_1)\epsilon^{(i)}_n. \tag{81}$$

By rewriting

$$(T_3 - T_1) = T_3 - \tfrac{1}{2}(T_+ + T_-) \qquad \text{and} \tag{82}$$

$$\begin{aligned}
(T_3 - T_1)^2 &= T_3^2 + \tfrac{1}{4}(T_+T_- + T_-T_+) - T_+(T_3 + \tfrac{1}{2}) - T_-(T_3 - \tfrac{1}{2})\\
&\quad + \tfrac{1}{4}T_+T_+ + \tfrac{1}{4}T_-T_-\\
&= \tfrac{3}{2}T_3^2 - \tfrac{1}{2}l(l+1) - T_+(T_3 + \tfrac{1}{2}) - T_-(T_3 - \tfrac{1}{2})\\
&\quad + \tfrac{1}{4}T_+T_+ + \tfrac{1}{4}T_-T_-,
\end{aligned} \tag{83}$$

we can use the simple matrix elements of T_3, $T_\pm$ to obtain, via eq. (76),

$$\epsilon_n^{(1)} = \lambda\Big[\tfrac{3}{2}n^2 - \tfrac{1}{2}l(l+1)\Big], \tag{84}$$

and via eq. (78),

$$\begin{aligned} n^3\epsilon_n^{(2)} &= \\ &\frac{1}{(-1)}\Big[\Big(-n^3\lambda(n+\tfrac{1}{2}) + \tfrac{1}{2})n^2\epsilon_n^{(1)}\Big)\sqrt{(n-l)(n+l+1)}\Big]^2 \\ &+ \frac{1}{(+1)}\Big[\Big(-n^3\lambda(n-\tfrac{1}{2}) + \tfrac{1}{2})n^2\epsilon_n^{(1)}\Big)\sqrt{(n+l)(n-l-1)}\Big]^2 \\ &+ \frac{1}{(-2)}\Big[\tfrac{1}{4}\lambda n^3\sqrt{(n+1-l)(n+2+l)(n-l)(n+1+l)}\Big]^2 \\ &+ \frac{1}{(+2)}\Big[\tfrac{1}{4}\lambda n^3\sqrt{(n-1+l)(n-2-l)(n+l)(n-1-l)}\Big]^2 \\ &= -\lambda^2\tfrac{1}{8}n^5\Big(7n^4 + 5n^2 - 3l^2(l+1)^2\Big), \end{aligned} \tag{85}$$

leading to

$$\epsilon = -\frac{1}{2n^2} + \lambda\Big[\tfrac{3}{2}n^2 - \tfrac{1}{2}l(l+1)\Big] - \lambda^2\tfrac{1}{8}n^2\Big[7n^4 + 5n^2 - 3l^2(l+1)^2\Big] + \cdots. \tag{86}$$

H Matrix Elements of the Vector Operators, $\vec{r}$ and $r\vec{p}$, in the Stretched Basis

In the last example, all of the needed matrix elements involved merely matrix elements of the SO(2,1) generators, T_i. For spherically nonsymmetric perturbations, however, we will encounter spherical tensors of higher rank that can change the quantum numbers l and m. The basic operators, from which more complicated ones can be built, are the vector operators, $\vec{r}$ and $\vec{p}$. So far, the only l step operators we have met are the components of the Runge–Lenz vector, which left the principal quantum number, n, invariant. Conversely, the ladder operators, $T_\pm$, of SO(2,1) could change only the n quantum number within a ladder of a definite l. The general vector operator, such as $\vec{r}$ and $\vec{p}$, can change both n and l.

Recall, first, the Runge–Lenz vector, in the form of the vector, $\vec{V}$, is given by

$$\begin{aligned} \vec{V} &= \frac{1}{\sqrt{(-2\epsilon)}}\Big[\vec{r}\big(\tfrac{1}{2}(\vec{p}\cdot\vec{p})\big) - \vec{p}(\vec{r}\cdot\vec{p}) + \vec{r}\big[\tfrac{1}{2}(\vec{p}\cdot\vec{p}) - \frac{1}{r}\big]\Big] \\ &= n\Big[\vec{r}\big[\tfrac{1}{2}(\vec{p}\cdot\vec{p})\big] - \vec{p}(\vec{r}\cdot\vec{p}) + \vec{r}\Big(\frac{-1}{2n^2}\Big)\Big], \end{aligned} \tag{87}$$

where this has been put into the most convenient form for conversion into its stretched counterpart, $\vec{\tilde{V}} \equiv \vec{A}$,

$$\vec{\tilde{V}} = e^{iaT_2}\vec{V}e^{-iaT_2} \equiv \vec{A} = n\Big[n\vec{r}\frac{(\vec{p}\cdot\vec{p})}{2n^2} - \frac{\vec{p}}{n}(n\vec{r}\cdot\frac{\vec{p}}{n}) + n\vec{r}\Big(\frac{-1}{2n^2}\Big)\Big]$$

$$\equiv \vec{A} = \vec{r}\big(\tfrac{1}{2}(\vec{p}\cdot\vec{p})\big) - \vec{p}(\vec{r}\cdot\vec{p}) - \tfrac{1}{2}\vec{r}. \tag{88}$$

Because

$$\begin{aligned}
&\int d\Omega \int_0^\infty dr r^2 (R_{nl'} Y_{l'm'})^* V_\mu R_{nl} Y_{lm} \\
&= \int d\Omega \int_0^\infty dr r (\frac{\psi_{nl'}}{n} Y_{l'm'})^* \Big(e^{iaT_2} r e^{-iaT_2}\Big) \tilde{V}_\mu \frac{\psi_{nl}}{n} Y_{lm} \\
&= \int d\Omega \int_0^\infty dr r (\frac{\psi_{nl'}}{n} Y_{l'm'})^* \Big(n(T_3 - T_1)\Big) \tilde{V}_\mu \frac{\psi_{nl}}{n} Y_{lm} \\
&= \int d\Omega \int_0^\infty dr (\psi_{nl'} Y_{l'm'})^* \tilde{V}_\mu \psi_{nl} Y_{lm},
\end{aligned} \tag{89}$$

where we have used $T_3\psi_{nl'} = n\psi_{nl'}$ and the fact that T_1 has no matrix elements diagonal in n,

$$\langle nl'm'|V_\mu|nlm\rangle = (nl'm'|A_\mu|nlm). \tag{90}$$

The reduced matrix elements of $\vec{A}$ in the stretched basis are therefore the same as the reduced matrix elements of $\vec{V}$ in the conventional basis.

$$\begin{aligned}
(n(l+1)\|\vec{A}\|nl) &= \sqrt{(l+1)[n^2 - (l+1)^2]} \\
(n(l-1)\|\vec{A}\|nl) &= -\sqrt{l[n^2 - l^2]} \\
(nl\|\vec{A}\|nl) &= 0.
\end{aligned} \tag{91}$$

In addition, the commutator algebra of the six operators $\vec{L}$ and $\vec{A}$ is the same as that of $\vec{L}$ and $\vec{V}$. Also, because $\vec{L}$ are functions only of θ, ϕ and $\partial/\partial\theta, \partial/\partial\phi$,

$$[T_2, L_j] = 0, \qquad \text{for } j = 1, 2, 3. \tag{92}$$

Finally,

$$\begin{aligned}
[T_2, A_j] &= [\frac{1}{i}\big(r\frac{\partial}{\partial r} + 1\big), \tfrac{1}{2}x_j\big((\vec{p}\cdot\vec{p}) - 1\big) - p_j(\vec{r}\cdot\vec{p})] \\
&= i\Big(\tfrac{1}{2}x_j(\vec{p}\cdot\vec{p}) - p_j(\vec{r}\cdot\vec{p}) + \tfrac{1}{2}x_j\Big) = iB_j,
\end{aligned} \tag{93}$$

where we have introduced the vector, $\vec{B}$, given by

$$\begin{aligned}
\vec{B} &= \Big(\tfrac{1}{2}\vec{r}(\vec{p}\cdot\vec{p}) - \vec{p}(\vec{r}\cdot\vec{p}) + \tfrac{1}{2}\vec{r}\Big), \\
\text{while}\quad \vec{A} &= \Big(\tfrac{1}{2}\vec{r}(\vec{p}\cdot\vec{p}) - \vec{p}(\vec{r}\cdot\vec{p}) - \tfrac{1}{2}\vec{r}\Big).
\end{aligned} \tag{94}$$

In particular,

$$\vec{r} = \vec{B} - \vec{A}, \tag{95}$$

and we have expressed the needed vector, $\vec{r}$, in terms of operators which make it convenient to calculate its matrix elements in the stretched hydrogenic basis. In particular, from the commutator

$$[T_2, A_3] = iB_3, \qquad \text{or} \quad B_3 = \tfrac{1}{2}[(T_- - T_+), A_3], \tag{96}$$

we get the matrix elements (in the stretched basis!),

$$(n'l'm|B_3|nlm) = (n'l'\|\vec{B}\|nl)\frac{\langle lm10|l'm\rangle}{\sqrt{(2l'+1)}}$$
$$= \frac{1}{2}\Big[(n'l'm|(T_- - T_+)|nl'm)(nl'\|\vec{A}\|nl)$$
$$- (n'l'\|\vec{A}\|n'l)(n'lm|(T_- - T_+)|nlm)\Big]\frac{\langle lm10|l'm\rangle}{\sqrt{(2l'+1)}}. \quad (97)$$

$\vec{A}$ can change only $l \to (l \pm 1)$, but keeps n invariant, whereas $T_\pm$ changes only $n \to (n \pm 1)$ but keeps l invariant. The sum over intermediate states in the above relation has thus collapsed to a single term for each of the four possible $n'l'$ values of $(n \pm 1)(l \pm 1)$. In addition, because $\vec{r} = \vec{B} - \vec{A}$, and because the nonzero matrix elements of $\vec{A}$ are restricted to n', $l' = n$, $(l \pm 1)$, we can obtain the reduced matrix elements of $\vec{r}$ in the stretched basis by combining the above relation with the known matrix elements of $\vec{A}$ to yield

$$((n+1)(l+1)\|\vec{r}\|nl) = +\tfrac{1}{2}\sqrt{(l+1)(n+l+1)(n+l+2)}$$
$$(n(l+1)\|\vec{r}\|nl) = -\sqrt{(l+1)(n+l+1)(n-l-1))}$$
$$((n-1)(l+1)\|\vec{r}\|nl) = +\tfrac{1}{2}\sqrt{(l+1)(n-l-1)(n-l-2)}$$
$$((n+1)(l-1)\|\vec{r}\|nl) = -\tfrac{1}{2}\sqrt{l(n-l+1)(n-l)}$$
$$(n(l-1)\|\vec{r}\|nl) = +\sqrt{l(n+l)(n-l)}$$
$$((n-1)(l-1)\|\vec{r}\|nl) = -\tfrac{1}{2}\sqrt{l(n+l-1)(n+l)}. \quad (98)$$

Because $\vec{r}$ can be combined with itself or with functions of $r = (T_3 - T_1)$, we can use the above matrix elements to gain the matrix elements of more complicated functions of $\vec{r}$ and r. To get reduced matrix elements of $\vec{p}$, we note that

$$[T_3, B_3] = -irp_3. \quad (99)$$

This simple commutator can thus be used to generate matrix elements of a new vector,

$$r\vec{p} \equiv \vec{C}. \quad (100)$$

This relation is ideal for matrix elements in the stretched hydrogenic basis, because the needed vector $\vec{p}$ in the conventional basis will lead to matrix elements of $r\vec{p}$ in the stretched basis. The commutator relation $rp_3 = +i[T_3, B_3]$ leads at once to

$$(n'l'm|rp_3|nlm) = +i(n'-n)(n'l'm|B_3|nlm),$$
$$(n'l'\|r\vec{p}\|nl) = +i(n'-n)(n'l'\|\vec{B}\|nl). \quad (101)$$

We have now met four vector operators, $\vec{L}$, $\vec{A}$, $\vec{B}$, $\vec{C}$, and the three components T_j. We have already seen $\vec{L}$ and $\vec{A}$ form the six components of a 4-D angular momentum algebra, if we identify $A_j = L_{j4}$. If we further identify $B_j = L_{j5}$, and

$C_j = L_{j6}$, and $T_2 = L_{45}$, $T_1 = L_{46}$, $T_3 = L_{56}$, so, with $j, k = 1, \ldots, 6$,

$$L_{jk} = \begin{pmatrix} 0 & L_3 & -L_2 & A_1 & B_1 & C_1 \\ -L_3 & 0 & L_1 & A_2 & B_2 & C_2 \\ L_2 & -L_1 & 0 & A_3 & B_3 & C_3 \\ -A_1 & -A_2 & -A_3 & 0 & T_2 & T_1 \\ -B_1 & -B_2 & -B_3 & -T_2 & 0 & T_3 \\ -C_1 & -C_2 & -C_3 & -T_1 & -T_3 & 0 \end{pmatrix}, \tag{102}$$

where the $L_{jk} = -L_{kj}$ are "angular momentum" operators in a (4+2)-dimensional space, with commutation relations

$$[L_{ab}, L_{ac}] = i g_{aa} L_{bc}, \qquad \text{with } g_{aa} = +1 \text{ for } a = 1, 2, 3, 4,$$

$$g_{55} = g_{66} = -1. \tag{103}$$

These 15 operators generate the group SO(4,2). We will not, however, need to make use of the detailed properties of this relatively complicated group! In order to solve hydrogenic perturbation problems, it will be sufficient for us to know the matrix elements of $\vec{L}$, including the m step operators $L_{\pm}$, of $\vec{A}$ that are the l step operators for a fixed n, of $T_{\pm}$ that are the n step operators for a fixed l, and, finally, of the vector operators $\vec{r} = \vec{B} - \vec{A}$, and $r\vec{p} = \vec{C}$. We have now achieved this goal.

I Second-Order Stark Effect of the Hydrogen Ground State Revisited

Let us now reexamine the second-order Stark effect in the ground state of the hydrogen atom by using the stretched hydrogenic basis. The perturbing Hamiltonian can be written as

$$H^{(1)} = -\lambda r \cos\theta, \tag{103}$$

where r is now dimensionless and $\lambda = (ea_0\mathcal{E})/(Ze^2/a_0)$ is also dimensionless. In the ground state with $n = 1, l = 0$, no diagonal matrix element exists, so $\epsilon^{(1)}_{n=1} = 0$. We therefore have $\epsilon^{(1)}_{\text{eff.}} = 0$, and, with $n = 1$,

$$H^{(1)}_{\text{eff.}} = -\lambda(T_3 - T_1)\vec{r}_0 = -\lambda(T_3 - \tfrac{1}{2}T_+ - \tfrac{1}{2}T_-)\vec{r}_0. \tag{104}$$

From our tabulation of reduced matrix elements of $\vec{r}$,

$$(n'10|\vec{r}_0|n = 1, l = 0, m = 0) = \frac{\langle 0010|10\rangle}{\sqrt{3}}(n'1\|\vec{r}\|10) = \frac{1}{\sqrt{3}}\delta_{n'2}\sqrt{\frac{3}{2}}. \tag{105}$$

Combining this with the matrix elements of T_3, $T_{\pm}$, we have

$$\begin{aligned}(210|H^{(1)}_{\text{eff.}}|100) &= -\lambda\sqrt{2}, \\ (310|H^{(1)}_{\text{eff.}}|100) &= +\lambda\sqrt{\frac{1}{2}}.\end{aligned} \tag{106}$$

The generalization of eq. (78) for the 3-D stretched basis then gives

$$\epsilon^{(2)} = -\lambda^2\Big[\frac{2}{(2-1)} + \frac{1}{2}\frac{1}{(3-1)}\Big] = -\lambda^2\frac{9}{4}, \qquad (107)$$

in agreement with the result of eq. (7), but now from the simple addition of two terms (no infinite sums over a set of discrete states and no integrals over a continuum!).

J The Calculation of Off-Diagonal Matrix Elements via the Stretched Hydrogenic Basis

We have seen how use of the stretched basis greatly simplifies calculations for a specific bound state. The stretched hydrogenic basis may also be used to simplify off-diagonal matrix elements for transitions between different bound states. An example is the electric dipole moment matrix element between different hydrogen bound states. In eq. (4), we used the matrix element of $r\cos\theta$ between a p-state of arbitrary n and the ground state. This process requires the calculation of the conventional matrix element

$$\langle 100|r\cos\theta|n10\rangle = \frac{1}{\sqrt{3}}\int_0^\infty drr^3 R^*_{n=1,l=0}(r)R_{n,l=1}(r).$$

Let us transcribe this equation into the stretched basis for the state with arbitrary n, via

$$\begin{aligned}&\int_0^\infty drr^3 R^*_{n=1,l=0}(r)R_{n,l=1}(r)\\ &=\int_0^\infty drr\Big(e^{iaT_2}R_{10}(r)\Big)^*\big(e^{iaT_2}r^2e^{-iaT_2}\big)\Big(e^{iaT_2}R_{n1}(r)\Big)\\ &=\int_0^\infty drr\Big(e^{iaT_2}R_{10}(r)\Big)^* n^2r^2\frac{1}{n}\psi_{q=n,l=1}\\ &=\int_0^\infty drr\Big(n2e^{-nr}\Big)^* nr^2\frac{2^2}{\sqrt{(n+1)!(n-2)!}}\frac{e^r}{r^2}\frac{d^{n-2}}{dr^{n-2}}\Big(r^{n+1}e^{-2r}\Big),\end{aligned} \qquad (109)$$

where, with $e^a = n$, we have used, see eq. (36),

$$e^{iaT_2}R_{10}(r) = e^a R_{10}(e^a r) = nR_{10}(nr),$$

for the 1s state of the left-hand side, and have substituted the general derivative expression for the stretched state, $\psi_{n,1}(r)$, derived in eq. (66). We therefore have

$$\langle 100|r\cos\theta|n10\rangle = \frac{2^3n^2}{(n-2)!\sqrt{3(n+1)n(n-1)}}\int_0^\infty drre^{-(n-1)r}\frac{d^{n-2}}{dr^{n-2}}\Big(r^{n+1}e^{-2r}\Big). \qquad (110)$$

The last integral is performed by integrating by parts $(n-2)$ times to yield

$$\int_0^\infty drre^{-(n-1)r}\frac{d^{n-2}}{dr^{n-2}}\Big(r^{n+1}e^{-2r}\Big) = \int_0^\infty dr(n-1)^{n-2}r^{n+2}e^{-(n+1)r}$$

$$- (n-2)\int_0^\infty dr(n-1)^{n-3} r^{n+1} e^{-(n+1)r} = \frac{(n-1)^{n-3}}{(n+1)^{n+3}}(n+1)!2n, \quad (111)$$

leading to the final result

$$\langle 100|r\cos\theta|nl = 10\rangle = \frac{(n-1)^{n-3}}{(n+1)^{n+3}} 2^4 n^3 \sqrt{\frac{(n+1)n(n-1)}{3}}. \quad (112)$$

K Final Remarks

As we have seen, the techniques introduced in this chapter are particularly useful for spherically symmetric perturbations of the hydrogenic atom. For nonspherically symmetric perturbations, the stretched hydrogenic basis can be used to get second-order results, but the solution for the final eigenvalues and eigenvectors still requires the diagonalization of some finite-dimensional matrices. For axially symmetric perturbations, parabolic coordinates are more convenient. We have seen in problem 26 that stretched parabolic coordinates, μ, ν, can be expressed in term of two commuting SO(2,1) groups with generators, T_i and T_i'. The zeroth-order hydrogen problem is then transformed into an eigenvalue equation of the form

$$(T_3 + T_3')\psi_{n_1,m}(\mu)\psi_{n_2,m}(\nu)e^{\pm im\phi}$$
$$= \left[\left(\tfrac{1}{2}(m+1) + n_1\right) + \left(\tfrac{1}{2}(m+1) + n_2\right)\right]\psi_{n_1,m}(\mu)\psi_{n_2,m}(\nu)e^{\pm im\phi}, \quad (113)$$

with $(n_1 + n_2 + m + 1) = n$, where n is the usual principal quantum number. The Stark effect has been solved to second order for arbitrary, n, using this $SO(2,1) \times SO(2,1)$ basis in problem 38.

A very good reference for the use of the stretched spherical basis for hydrogenic perturbation problems is: *Lie Algebraic Methods and their Applications to Simple Quantum Systems*; B. G. Adams, J. Cizek and J. Paldus, Advances in Quantum Chemistry **19** (1988) 1; and Barry G. Adams, *Algebraic Approach to Simple Quantum Systems*, New York: Springer-Verlag, 1994. For a detailed use of the stretched parabolic coordinates, see D. Delande and J. C. Gay, J. Phys. B: At. Mol. Phys. **17** (1984) L335.

Problems

48. Use the stretched spherical basis to show the conventional diagonal matrix element of r^2 is given by

$$\langle nlm|r^2|nlm\rangle = \tfrac{5}{2}n^4 - \tfrac{3}{2}n^2 l(l+1) + \tfrac{1}{2}n^2.$$

49. Assume the attractive Coulombic $(1/r)$ potential for a two-body nonrelativistic system is perturbed by a quadratic repulsive term, with $H^{(1)} = +\lambda r^2$, where $\lambda \ll 1$ and r is dimensionless. Show that the energy through second order is given by

$$\epsilon = -\frac{1}{2n^2} + \lambda\left(\tfrac{5}{2}n^4 - \tfrac{3}{2}n^2 l(l+1) + \tfrac{1}{2}n^2\right)$$

$$-\lambda^2 \frac{n^6}{16}\left(143n^4 + 345n^2 + 28 - 90n^2 l(l+1) - 21l^2(l+1)^2 - 126l(l+1)\right).$$

50. Use the stretched hydrogenic functions to show the conventional matrix element of the electric dipole moment between the 2*s* state and an arbitrary *p* state is given by

$$\langle 200|r\cos\theta|n10\rangle == \sqrt{\frac{(n+1)n(n-1)}{2\cdot 3}\frac{2^9 n^3(n-2)^{n-3}}{(n+2)^{n+3}}}.$$

51. Write the dilation operator, $e^{iaT_2} = e^{\frac{1}{2}a(T_+ - T_-)}$, in its "disentangled" forms. In particular, show that

$$e^{iaT_2} = e^{-\tanh\frac{a}{2}T_-}(\cosh^2\tfrac{a}{2})^{T_0} e^{+\tanh\frac{a}{2}T_+},$$

$$e^{iaT_2} = e^{+\tanh\frac{a}{2}T_+}\frac{1}{(\cosh^2\frac{a}{2})^{T_0}} e^{-\tanh\frac{a}{2}T_-}.$$

Hint: Use (1) the corresponding result for the angular momentum operators of SO(3) (see Chapter 29); (2) the fact that the operators T_3, iT_+, iT_- of SO(2,1) have formally the same commutation relations as the operators J_3, J_+, J_- of SO(3); and (3) the fact that the disentanglement relations depend only on the commutator algebra of the operators. Use these relations to rederive the result of eq. (112), without the use of the explicit functional forms of the radial functions, by relating the conventional matrix element $\langle n10|r\cos\theta|100\rangle$ to its stretched form

$$\langle n10|r\cos\theta|100\rangle = \frac{1}{n}(\Psi_{n10}|e^{iaT_2}r(\vec{r})_0|\Psi_{100})$$

$$= \frac{1}{n\sqrt{2}}(\Psi_{n10}|e^{+\frac{(n-1)}{(n+1)}T_+}\left[\frac{4n}{(n+1)^2}\right]^{T_3} e^{-\frac{(n-1)}{(n+1)}T_-}\left(T_3 - \tfrac{1}{2}(T_+ + T_-)\right)|\Psi_{210}),$$

where we have used $\vec{r}_0|\Psi_{100}) = \frac{1}{\sqrt{2}}|\Psi_{210})$, and $\tanh\frac{a}{2} = (n-1)/(n+1)$, $\cosh^2\frac{a}{2} = (n+1)^2/4n$.

36

The WKB Approximation

Although this perturbation technique (due to Wentzel, Kramers, and Brillouin) has practical value essentially only for 1-D problems (or for problems separable into one-dimensionalized problems), it is of considerable interest through its connection with classical physics and the "old" (pre-1925) quantum theory. It is not only of historical theoretical importance, however. It can be very useful for problems involving quantum-mechanical tunneling.

For a one-dimensionalized Schrödinger equation, we have

$$\frac{d^2u}{dx^2} + \frac{2\mu}{\hbar^2}(E - V(x))u(x) = 0, \tag{1}$$

or

$$\frac{d^2u}{dx^2} + \frac{P^2(x)}{\hbar^2}u(x) = 0, \tag{2}$$

where $P(x)$ is the "local" (x-dependent) momentum of the particle. Thus if $V(x)$ were constant over a small range of x values, P would be the momentum of the particle in this range of x values. The WKB technique uses $\hbar$ as an expansion parameter; that is, it uses an expansion in powers of $\hbar$ that should therefore be valid in the classical limit $\hbar \to 0$. In particular, solutions are sought of the form

$$u(x) = e^{\frac{i}{\hbar}S(x)} = e^{\frac{i}{\hbar}\left(S_0(x) + \hbar S_1(x) + \hbar^2 S_2(x) + \cdots\right)}, \tag{3}$$

so

$$u' = \frac{i}{\hbar}(S_0' + \hbar S_1' + \hbar^2 S_2' + \cdots)e^{\frac{i}{\hbar}S(x)}, \tag{4}$$

$$u'' = \left[-\frac{1}{\hbar^2}(S_0' + \hbar S_1' + \hbar^2 S_2' + \cdots)^2 + \frac{i}{\hbar}(S_0'' + \hbar S_1'' + \hbar^2 S_2'' + \cdots) \right] e^{\frac{i}{\hbar} S(x)}. \quad (5)$$

Substituting this relation into the Schrödinger equation, and picking off terms of order $\frac{1}{\hbar^2}$, of order $\frac{1}{\hbar}$, and of successively smaller order in powers of $\hbar$, we get (first) from the term of order $\frac{1}{\hbar^2}$:

$$-(S_0')^2 + P^2(x) = 0, \quad (6)$$

$$\frac{dS_0}{dx} = \pm P(x), \qquad S_0(x) = \pm \int_{\text{const.}}^{x} d\xi\, P(\xi). \quad (7)$$

Next, the term of order $\frac{1}{\hbar}$ leads to

$$iS_0'' - 2S_0' S_1' = 0, \qquad S_1' = \frac{i}{2}\frac{S_0''}{S_0'} = \frac{i}{2}\frac{d}{dx} ln S_0', \quad (8)$$

leading to

$$S_1(x) = \frac{i}{2} ln P(x), \qquad \text{or} \qquad e^{iS_1} = \frac{1}{\sqrt{P(x)}}. \quad (9)$$

In the next approximation, terms of order 1 in the powers of $\hbar$ development lead to

$$iS_1'' - (S_1')^2 - 2S_0' S_2' = 0, \quad (10)$$

$$S_2' = \frac{i}{2}\frac{S_1''}{S_0'} - \frac{(S_1')^2}{2S_0'} = -\frac{1}{4}\frac{S_0'''}{(S_0')^2} + \frac{3}{8}\frac{(S_0'')^2}{(S_0')^3} = -\frac{1}{4}\left(\frac{S_0''}{(S_0')^2}\right)' - \frac{1}{8}\frac{(S_0'')^2}{(S_0')^3}, \quad (11)$$

so

$$S_2 = -\frac{1}{4}\left(\frac{(\frac{dP}{dx})}{P^2}\right) - \frac{1}{8}\int_{\text{const.}}^{x} d\xi \frac{(\frac{dP}{d\xi})^2}{P^3(\xi)}, \quad (12)$$

or

$$S_2 = \frac{1}{4}\frac{\mu \frac{dV}{dx}}{[2\mu(E - V(x))]^{\frac{3}{2}}} - \frac{1}{8}\int_{\text{const.}}^{x} d\xi \frac{\mu^2(\frac{dV}{d\xi})^2}{[2\mu(E - V(\xi))]^{\frac{5}{2}}}. \quad (13)$$

The second approximation function, $S_2(x)$, is usually neglected. We see, from its specific form, that this may be justified, provided $V(x)$ is a mildly varying function of x, i.e., $|\frac{dV}{dx}|$ is small, and provided x is not near a classical turning point for which we would have $(E - V(x)) = 0$, hence, a zero in the denominator of the function $S_2(x)$. Assuming S_2 can be neglected, we still have to consider two different types of solutions.

1. For $(E - V(x)) > 0$, for classically allowed regions, we have oscillatory solutions

$$u(x) = \frac{C}{\sqrt{P(x)}} \cos\left(\frac{1}{\hbar}\int_{x_2}^{x} d\xi\, P(\xi) + \alpha\right), \quad (14)$$

where we have assumed $x > x_2$, x_2 is a left classical turning point (see Fig. 36.1), and C and α are integration constants.

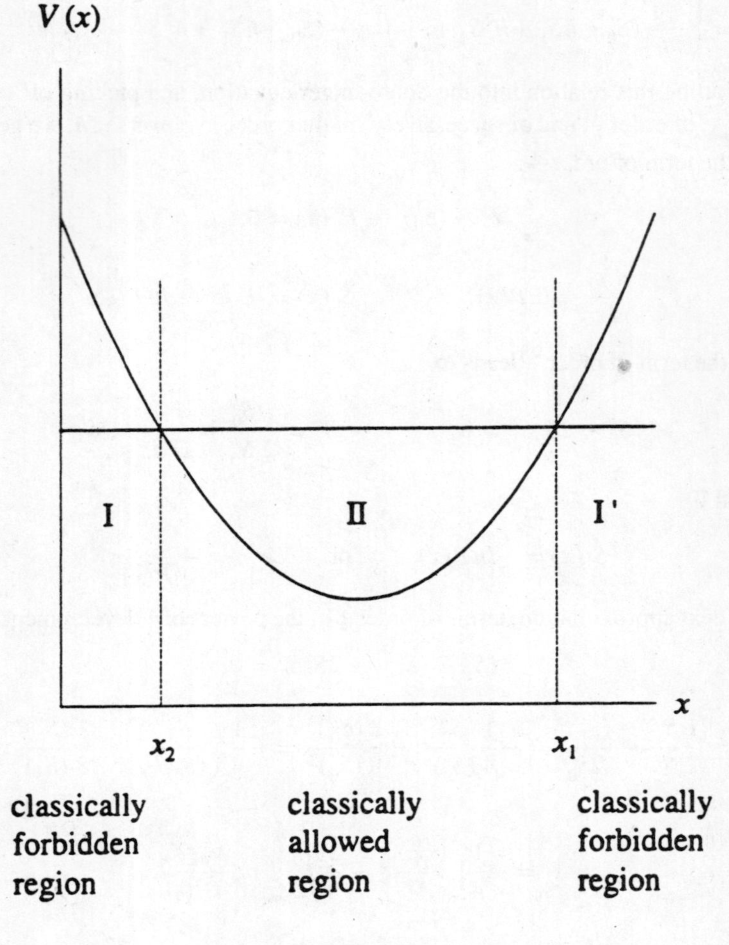

FIGURE 36.1.

2. For $(E - V(x)) < 0$, for classically forbidden regions, we have exponential solutions. These solutions are best put in the form

$$u(x) = \frac{A}{\sqrt{|P(x)|}} e^{+\frac{1}{\hbar}\int_{x_1}^{x} d\xi |P(\xi)|} + \frac{B}{\sqrt{|P(x)|}} e^{-\frac{1}{\hbar}\int_{x_1}^{x} d\xi |P(\xi)|}, \quad \text{for } x > x_1, \tag{15}$$

or

$$u(x) = \frac{A}{\sqrt{|P(x)|}} e^{+\frac{1}{\hbar}\int_{x}^{x_2} d\xi |P(\xi)|} + \frac{B}{\sqrt{|P(x)|}} e^{-\frac{1}{\hbar}\int_{x}^{x_2} d\xi |P(\xi)|}, \quad \text{for } x < x_2. \tag{16}$$

Even to this order, the WKB solutions blow up at the classical turning points. As $x \to x_1$, or $x \to x_2$, $P(x) \to 0$, and the WKB solutions go to ∞. The exact solutions, however, have no remarkable behavior or singularities there. We need an expression that gives a continuous $u(x)$ valid for all regions. We need to relate the integration constants, C, α, from the oscillatory solutions to the A, B from the exponential solutions. In square well problems, we had a similar situation, in which we used boundary conditions at discontinuities of the potential to find C, α as functions of A, B. Unfortunately, it is precisely at the classical turning points, where u_{WKB} breaks down and becomes invalid (see Fig. 36.2). The problem was solved by Kramers through his connection formulae. In the vicinity of the classical turning points, x_1, x_2, a good approximation to an exact solution can be found and this approximation can be used to make the connection between the exponential and the oscillatory WKB approximate solutions. For this purpose, it is sufficient to consider $\frac{dV}{dx}$ to be constant over a small range of x near $x = x_1$, or near $x = x_2$, and use the exact solution for this straight-line potential of fixed slope to fit onto the oscillatory solution on one side and the exponential one on the other. The details involve Bessel functions with index $n = \pm\frac{1}{3}$. The details of the derivation will be given in an appendix.

A The Kramers Connection Formulae

At a left turning point (see Fig. 36.1), $x = x_2$, a decreasing exponential solution connects onto an oscillatory solution according to

$$\frac{1}{\sqrt{|P(x)|}} e^{-\frac{1}{\hbar}\int_x^{x_2} d\xi |P(\xi)|} \leftrightarrow \frac{2}{\sqrt{P(x)}} \cos\left[\frac{1}{\hbar}\int_{x_2}^{x} d\xi\, P(\xi) - \frac{\pi}{4}\right], \tag{17}$$

whereas an increasing exponential solution connects onto an oscillatory one according to

$$\frac{1}{\sqrt{|P(x)|}} e^{+\frac{1}{\hbar}\int_x^{x_2} d\xi |P(\xi)|} \leftrightarrow \frac{1}{\sqrt{P(x)}} \cos\left[\frac{1}{\hbar}\int_{x_2}^{x} d\xi\, P(\xi) + \frac{\pi}{4}\right]. \tag{18}$$

At a right turning point, $x = x_1$, the decreasing exponential solution connects onto an oscillatory solution according to

$$\frac{1}{\sqrt{|P(x)|}} e^{-\frac{1}{\hbar}\int_{x_1}^{x} d\xi |P(\xi)|} \leftrightarrow \frac{2}{\sqrt{P(x)}} \cos\left[\frac{1}{\hbar}\int_{x}^{x_1} d\xi\, P(\xi) - \frac{\pi}{4}\right], \tag{19}$$

whereas the increasing exponential solution connects onto an oscillatory one according to

$$\frac{1}{\sqrt{|P(x)|}} e^{+\frac{1}{\hbar}\int_{x_1}^{x} d\xi |P(\xi)|} \leftrightarrow \frac{1}{\sqrt{P(x)}} \cos\left[\frac{1}{\hbar}\int_{x}^{x_1} d\xi\, P(\xi) + \frac{\pi}{4}\right]. \tag{20}$$

B Appendix: Derivation of the Connection Formulae

In the vicinity of a classical turning point (let us choose a left turning point $x = x_2$), let us assume the potential function varies smoothly so the function V(x) can be approximated by a straight line over the region where the WKB approximation is not valid. Thus, for the left turning point, $x = x_2$, we will assume $u_{\text{WKB}}(x)$ is

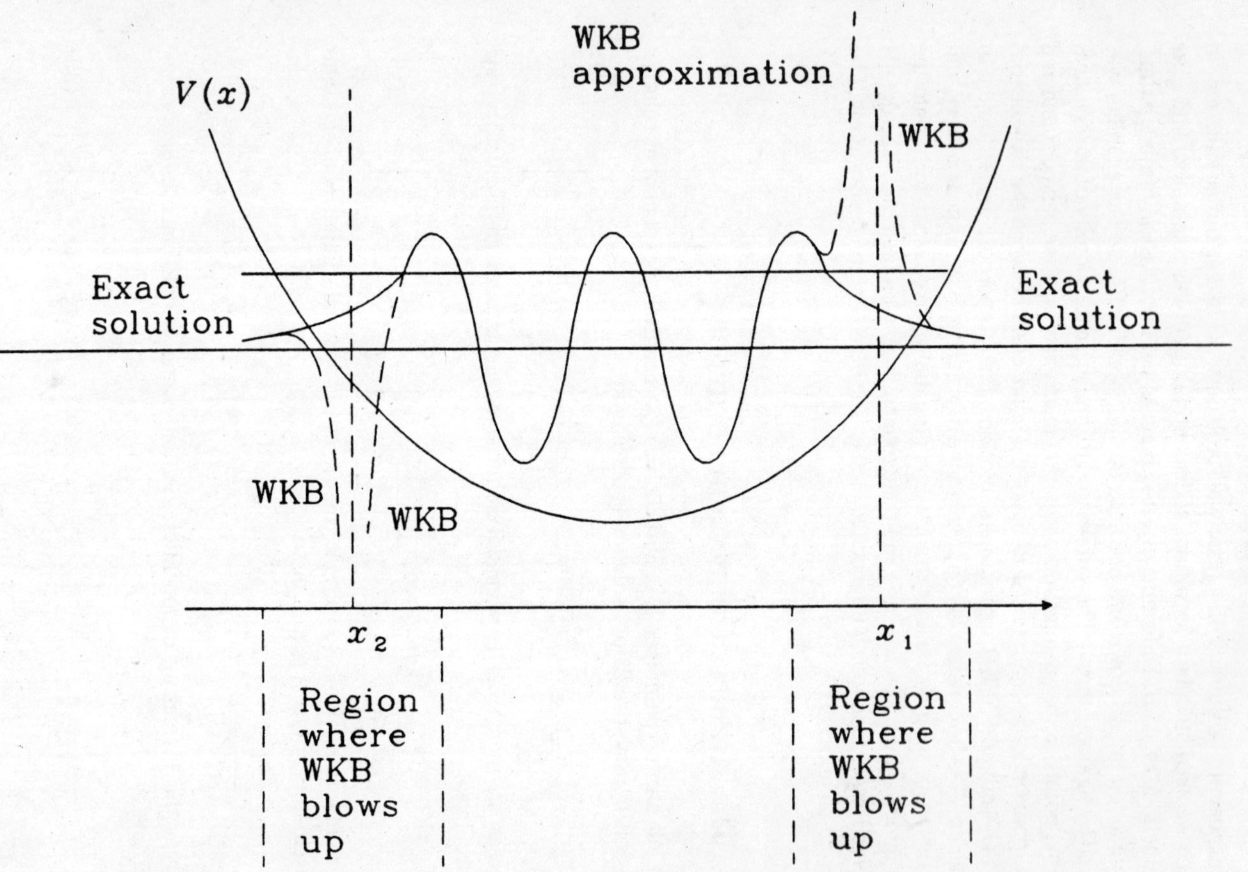

FIGURE 36.2. WKB and exact solutions for a $V(x)$ with one classically allowed region.

a good approximation for $x < x_2 - \frac{\Delta}{2}$ (in the exponential region I), and again for $x > x_2 + \frac{\Delta}{2}$ (in the oscillatory region II). We will also assume in the region, $x_2 - \frac{\Delta}{2} < x < x_2 + \frac{\Delta}{2}$, the potential $V(x)$ can be approximated by a straight line, so in this region

$$\frac{2\mu}{\hbar^2}(E - V(x)) = c^2(x - x_2), \qquad \text{where} \quad c^2 = \frac{2\mu}{\hbar^2}|\frac{dV}{dx}|_{x=x_2}. \tag{21}$$

The strategy is then the following: Find an exact solution to the equation

$$u'' + c^2(x - x_2)u(x) = 0, \tag{22}$$

valid in the region near $x = x_2$ (where the WKB solution blows up), and continue this solution into region II where it matches the oscillatory WKB solution for $x - x_2$ sufficiently large and also continue it to the left into region I where now for $x_2 - x$ sufficiently large it matches the exponential WKB solution. It will be convenient to introduce new dependent and new independent variables for eq. (22).

In particular, in region I, for $x < x_2$, introduce

$$y = \frac{1}{\hbar}\int_x^{x_2} d\xi |P(\xi)| = c\int_x^{x_2} d\xi \sqrt{(x_2 - \xi)} = \frac{2}{3}c(x_2 - x)^{\frac{3}{2}}. \tag{23}$$

Similarly, in region II, for $x > x_2$, introduce

$$y = \frac{1}{\hbar}\int_{x_2}^{x} d\xi\, P(\xi) = c\int_{x_2}^{x} d\xi \sqrt{(\xi - x_2)} = \frac{2}{3}c(x - x_2)^{\frac{3}{2}}. \tag{24}$$

These relations lead to

$$\frac{du}{dx} = -c(x_2 - x)^{\frac{1}{2}}\frac{du}{dy}, \qquad \text{for region I,} \tag{25}$$

$$\frac{du}{dx} = +c(x - x_2)^{\frac{1}{2}}\frac{du}{dy}, \qquad \text{for region II.} \tag{26}$$

This process transforms the equation $u'' + c^2(x - x_2)u = 0$ into

$$\pm c^2(x_2 - x)\left(\frac{d^2u}{dy^2} + \frac{1}{3y}\frac{du}{dy} \mp u\right) = 0, \tag{27}$$

for regions I (upper signs) and II (lower signs), respectively. Now we will make the further change of dependent variable,

$$u(y) = y^{\frac{1}{3}}v(y), \tag{28}$$

in both cases, leading to the new equations

$$y^2\frac{d^2v}{dy^2} + y\frac{dv}{dy} + \left[\mp y^2 - \left(\frac{1}{3}\right)^2\right]v(y) = 0, \tag{29}$$

where upper (and lower) signs again refer to regions I (and II). This equation is the Bessel equation with index $n^2 = (\frac{1}{3})^2$ of the variable y for region II, and the variable iy for region I. Thus,

$$v_I(y) = A_+J_{+\frac{1}{3}}(iy) + A_-J_{-\frac{1}{3}}(iy) = A_+I_{+\frac{1}{3}}(y) + A_-I_{-\frac{1}{3}}(y)$$

$$v_{II}(y) = B_+ J_{+\frac{1}{3}}(y) + B_- J_{-\frac{1}{3}}(y), \tag{30}$$

or

$$\begin{aligned} u_I(y) &= A_+ y^{\frac{1}{3}} I_{+\frac{1}{3}}(y) + A_- y^{\frac{1}{3}} I_{-\frac{1}{3}}(y) \\ u_{II}(y) &= B_+ y^{\frac{1}{3}} J_{+\frac{1}{3}} + B_- y^{\frac{1}{3}} J_{-\frac{1}{3}}, \end{aligned} \tag{31}$$

where the arbitrary constants, A_+, A_-, B_+, B_-, must still be chosen to make u and its first derivative continuous at $y = 0$, $(x = x_2)$. We therefore need the behavior of the Bessel function near $y = 0$. From

$$J_n(y) = \sum_{k=0} \frac{(-1)^k}{k!} \frac{1}{\Gamma(n+k+1)} \Big(\frac{y}{2}\Big)^{n+2k}, \tag{32}$$

we have near $y = 0$

$$J_{\pm\frac{1}{3}}(y) = I_{\pm\frac{1}{3}}(y) = \frac{1}{\Gamma(\pm\frac{1}{3}+1)} \Big(\frac{y}{2}\Big)^{\pm\frac{1}{3}} + \cdots . \tag{33}$$

Therefore, near $x = x_2$, after transforming back to functions of x,

$$\begin{aligned} u_I(x) &= \frac{A_+(\frac{1}{2})^{\frac{1}{3}}(\frac{2c}{3})^{\frac{2}{3}}}{\Gamma(\frac{4}{3})}(x_2 - x)^1 + \frac{A_-(\frac{1}{2})^{-\frac{1}{3}}}{\Gamma(\frac{2}{3})} + \cdots, \\ u_{II}(x) &= \frac{B_+(\frac{1}{2})^{\frac{1}{3}}(\frac{2c}{3})^{\frac{2}{3}}}{\Gamma(\frac{4}{3})}(x - x_2)^1 + \frac{B_-(\frac{1}{2})^{-\frac{1}{3}}}{\Gamma(\frac{2}{3})} + \cdots. \end{aligned} \tag{34}$$

Now, the requirement

$$u_I(x_2) = u_{II}(x_2) \qquad \text{leads to} \qquad B_- = A_-, \tag{35}$$

whereas

$$\Big[\frac{du_I}{dx}\Big]_{x=x_2} = \Big[\frac{du_{II}}{dx}\Big]_{x=x_2} \qquad \text{leads to} \qquad B_+ = -A_+. \tag{36}$$

Next, we want to continue these solutions to large values of y, far enough from the turning point, $y = 0$, so they may match the WKB solutions. For this purpose, we will attempt to use the asymptotic expansions of the Bessel functions, valid for large values of y. [Actual experience, that is, a look at the exact plots of $J_n(y)$, shows y does not have to be very large for these asymptotic expressions to be surprisingly good approximations.] For $J_n(y)$, as $y \to$ large,

$$J_n(y) \to \sqrt{\frac{2}{\pi y}} \cos(y - [n + \frac{1}{2}]\frac{\pi}{2}). \tag{37}$$

Thus,

$$J_{\pm\frac{1}{3}}(y) \to \sqrt{\frac{2}{\pi y}} \cos(y - \frac{\pi}{4} \mp \frac{\pi}{6}), \tag{38}$$

whereas

$$I_{\pm\frac{1}{3}}(y) \rightarrow \frac{1}{\sqrt{2\pi y}}\left(e^{y} + e^{-y}e^{-(\frac{1}{2}\pm\frac{1}{3})i\pi}\right), \tag{39}$$

but for large values of y, e^{-y} is completely negligible compared with e^{+y}. The e^{-y} term comes into play only for the difference $(I_{+\frac{1}{3}} - I_{-\frac{1}{3}})$ for which the e^{y} terms cancel. For this difference, as $y \rightarrow$ large,

$$(I_{+\frac{1}{3}} - I_{-\frac{1}{3}}) \rightarrow \frac{1}{\sqrt{2\pi y}}e^{-y}\left(e^{-\frac{5i\pi}{6}} - e^{-\frac{i\pi}{6}}\right) = -\sqrt{\frac{2}{\pi y}}e^{-y}\cos\frac{\pi}{6}. \tag{40}$$

Otherwise,

$$I_{\pm\frac{1}{3}} \rightarrow \frac{1}{\sqrt{2\pi y}}e^{+y}. \tag{41}$$

Now, with the $B_{\pm}$ determined from the boundary conditions at $x = x_2$, we have

$$\text{In region I}: \quad u_I(y) = y^{\frac{1}{3}}(A_+ I_{+\frac{1}{3}}(y) + A_- I_{-\frac{1}{3}}(y)). \tag{42}$$

$$\text{In region II}: \quad u_{II}(y) = y^{\frac{1}{3}}(-A_+ J_{+\frac{1}{3}}(y) + A_- J_{-\frac{1}{3}}(y)). \tag{43}$$

Let us now for a first choice pick $A_+ = -A_-$. Then,

$$u_I(y) = -y^{\frac{1}{3}}A_-(I_{+\frac{1}{3}}(y) - I_{-\frac{1}{3}}(y)) \rightarrow \frac{A_-}{y^{\frac{1}{6}}}\sqrt{\frac{2}{\pi}}e^{-y}\cos\frac{\pi}{6}. \tag{44}$$

Similarly, with this choice of constants,

$$\begin{aligned} u_{II}(y) &\rightarrow \frac{A_-}{y^{\frac{1}{6}}}\sqrt{\frac{2}{\pi}}\left(\cos(y - \frac{\pi}{4} - \frac{\pi}{6}) + \cos(y - \frac{\pi}{4} + \frac{\pi}{6})\right) \\ &= \frac{A_-}{y^{\frac{1}{6}}}\sqrt{\frac{2}{\pi}}\left[\cos(y - \frac{\pi}{4})\right]2\cos\frac{\pi}{6}. \end{aligned} \tag{45}$$

Thus, with this choice of constants, viz. $A_+ = -A_-$, we get

$$\text{in I} \quad \frac{e^{-y}}{y^{\frac{1}{6}}} \quad \leftarrow \quad u(y) \quad \rightarrow \quad \frac{2}{y^{\frac{1}{6}}}\cos(y - \frac{\pi}{4}) \quad \text{in II}. \tag{46}$$

Now, substituting the values for y for regions I and II, through eqs. (23) and (24), with (21), we can translate this equation into the connection formula

$$\frac{1}{\sqrt{|P(x)|}}e^{-\frac{1}{\hbar}\int_x^{x_2} d\xi |P(\xi)|} \rightarrow \frac{2}{\sqrt{P(x)}}\cos\left[\frac{1}{\hbar}\int_{x_2}^{x} d\xi\, P(\xi) - \frac{\pi}{4}\right]. \tag{47}$$

This relation gives us one of the connection formulae for a left turning point.

To get the second connection formula, choose $A_+ = A_-$. Then, we have

$$u_I(y) = A_+ y^{\frac{1}{3}}(I_{+\frac{1}{3}} + I_{-\frac{1}{3}}) \rightarrow \frac{A_+}{\sqrt{2\pi}}\frac{1}{y^{\frac{1}{6}}}2e^{y}, \tag{48}$$

and

$$
\begin{aligned}
u_{II}(y) &= A_+ y^{\frac{1}{3}}(-J_{+\frac{1}{3}} + J_{-\frac{1}{3}}) \\
&\to A_+ \frac{1}{y^{\frac{1}{6}}} \sqrt{\frac{2}{\pi}} \Big(-\cos(y - \frac{\pi}{4} - \frac{\pi}{6}) + \cos(y - \frac{\pi}{4} + \frac{\pi}{6}) \Big) \\
&= \frac{A_+}{y^{\frac{1}{6}}} \sqrt{\frac{2}{\pi}} \cos(y + \frac{\pi}{4}). \qquad (49)
\end{aligned}
$$

Thus, with this choice of constants, we are led to

$$
\text{in I} \quad \frac{A_+}{y^{\frac{1}{6}}} \sqrt{\frac{2}{\pi}} e^{+y} \quad \leftarrow \quad u(y) \quad \rightarrow \quad \frac{A_+}{y^{\frac{1}{6}}} \sqrt{\frac{2}{\pi}} \cos(y + \frac{\pi}{4}) \quad \text{in II.} \qquad (50)
$$

Substituting for the values of y in regions I and II, this relation translates into the second connection formula at the left turning point, x_2,

$$
\frac{1}{\sqrt{|P(x)|}} e^{+\frac{1}{\hbar}\int_x^{x_2} d\xi |P(\xi)|} \leftrightarrow \frac{1}{\sqrt{P(x)}} \cos\left[\frac{1}{\hbar} \int_{x_2}^{x} d\xi \, P(\xi) + \frac{\pi}{4} \right]. \qquad (51)
$$

Derivations for the connection formulae at a right turning point, x_1, go in precisely parallel fashion.

37

Applications of the WKB Approximation

A The Wilson–Sommerfeld Quantization Rules of the Pre-1925 Quantum Theory

To show how the WKB approximation is used, let us first derive the energies for a potential with one minimum, with a single left and a single right turning point. In region I, with $x < x_2$, the solution must be restricted to one with a decreasing exponential only. In region I,

$$u_I(x) = \frac{A}{\sqrt{|P(x)|}} e^{-\frac{1}{\hbar}\int_x^{x_2} d\xi |P(\xi)|}. \tag{1}$$

This relation will connect in region II, where $x > x_2$, onto the oscillatory solution

$$u_{II}(x) = \frac{2A}{\sqrt{P(x)}} \cos\left[\frac{1}{\hbar}\int_{x_2}^{x} d\xi\, P(\xi) - \frac{\pi}{4}\right]. \tag{2}$$

This formula can be rewritten as

$$u_{II}(x) = \frac{2A}{\sqrt{P(x)}} \cos\left[\left(\frac{1}{\hbar}\int_{x_2}^{x_1} d\xi\, P(\xi) - \frac{\pi}{2}\right) - \left(\frac{1}{\hbar}\int_{x}^{x_1} d\xi\, P(\xi) - \frac{\pi}{4}\right)\right]. \tag{3}$$

In region I', with $x > x_1$, we must again have a purely decreasing exponential solution

$$u_{I'}(x) = \frac{B}{\sqrt{|P(x)|}} e^{-\frac{1}{\hbar}\int_{x_1}^{x} d\xi |P(\xi)|}. \tag{4}$$

This relation will connect in region II, with $x < x_1$, onto

$$u_{II}(x) = \frac{2B}{\sqrt{P(x)}} \cos\left[\frac{1}{\hbar}\int_x^{x_1} d\xi\, P(\xi) - \frac{\pi}{4}\right]. \tag{5}$$

Now, comparing the two expressions for $u_{II}(x)$, these match, if $|A| = |B|$, and if

$$\frac{1}{\hbar}\int_{x_2}^{x_1} d\xi\, P(\xi) - \frac{\pi}{2} = n\pi. \tag{6}$$

We can rewrite this as

$$\frac{1}{\pi}\int_{x_2}^{x_1} d\xi\, P(\xi) \equiv \frac{1}{2\pi}\oint d\xi\, P(\xi) = \hbar(n + \frac{1}{2}), \tag{7}$$

where $\oint$ is used to indicate the integral over one complete classical cycle of the classical orbit, starting at x_2, proceeding to x_1, and then back again to x_2. This quantity, using a generalized momentum, is known as the action variable, and usually denoted by J in classical mechanics. It is a function of the energy, E.

$$J(E) = \frac{1}{2\pi}\oint d\xi\, P(\xi) = \hbar(n + \frac{1}{2}). \tag{8}$$

This is the Wilson–Sommerfeld quantization rule, a generalization of the Planck quantization rule, which goes all the way back to the birth of the quantum theory.

For the simple 1-D harmonic oscillator, e.g., with $\mu = m$, and $V(x) = \frac{1}{2}m\omega_0^2 x^2$,

$$\begin{aligned} J(E) &= \frac{1}{\pi}\int_{-x_0}^{+x_0} d\xi \sqrt{2m(E - \tfrac{1}{2}m\omega_0^2\xi^2)} = \frac{1}{\pi}m\omega_0 \int_{-x_0}^{+x_0} d\xi \sqrt{(x_0^2 - \xi^2)} \\ &= \frac{1}{\pi}m\omega_0 x_0^2 \int_{-\frac{\pi}{2}}^{+\frac{\pi}{2}} d\phi \cos^2\phi = \frac{1}{2}m\omega_0 x_0^2 = \frac{E}{\omega_0}, \end{aligned} \tag{9}$$

where we have used, $E = \frac{1}{2}m\omega_0^2 x_0^2$, in the first step, and have used the substitution, $\xi = x_0 \sin\phi$, in the integral. Thus, for the 1-D harmonic oscillator

$$J(E) = \frac{E}{\omega_0} = \hbar(n + \frac{1}{2}), \tag{10}$$

giving the exact quantum-mechanical result, $E = \hbar\omega_0(n + \frac{1}{2})$. For other simple problems, the integrals for $J(E)$ are a little more challenging but can be done in closed form. For the hydrogen atom, e.g., with $V(r) = -Ze^2/r + \hbar^2 l(l+1)/2\mu r^2$, leading to

$$J(E) = \frac{Ze^2\sqrt{\mu}}{\sqrt{(-2E)}} - \hbar\sqrt{l(l+1)} = \hbar(n_r + \frac{1}{2}), \tag{11}$$

or

$$\frac{Ze^2\sqrt{\mu}}{\sqrt{(-2E_{\rm WKB})}} = \hbar(n_r + \frac{1}{2} + \sqrt{l(l+1)}). \tag{12}$$

This WKB result is to be compared with the exact quantum-mechanical result, which could be put in the form

$$\frac{Ze^2\sqrt{\mu}}{\sqrt{(-2E)}} = \hbar\left(n_r + \frac{1}{2} + \sqrt{(l+\tfrac{1}{2})^2}\right), \tag{13}$$

so the WKB expression for the energy, E, goes over to the exact result in the limit $(l^2+l) \rightarrow (l+\frac{1}{2})^2 = (l^2+l+\frac{1}{4})$, certainly valid in the limit of large quantum numbers, l, for which the classical orbit description begins to have some meaning.

Similarly, for the 3-D harmonic oscillator,

$$J(E) = \frac{E}{2\omega_0} - \frac{\hbar\sqrt{l(l+1)}}{2} = \hbar(n_r + \frac{1}{2}), \tag{14}$$

leads to

$$E_{\rm WKB} = \hbar\omega_0(2n_r + 1 + \sqrt{l(l+1)}), \tag{15}$$

which again leads to the exact result

$$E = \hbar\omega_0(2n_r + l + \frac{3}{2}), \tag{16}$$

if, again, $l(l+1)$ is replaced by $(l+\frac{1}{2})^2$, valid for large values of l.

B Application 2: The Two-Minimum Problem: The Inversion Splitting of the Levels of the Ammonia Molecule

In the vibrational spectrum of the ammonia molecule, NH_3, one degree of freedom exists, which corresponds to the motion of the N atom relative to the H_3 symmetrical triangle. This degree of freedom can be approximated by z, the distance of the N atom above (or below) the H_3 plane. The potential, $V(z)$, has the the symmetrical double minimum form shown in Fig. 37.1. Classically, the N atom would lie either above the H_3 plane and undergo a vertical oscillation about its upper equilibrium configuration at $z = +z_e$, or alternately it might lie below the H_3 plane and undergo an oscillation about its lower equilibrium configuration at $z = -z_e$. Quantum mechanically, of course, the N atom can tunnel from the upper minimum to the lower one, leading to a doubling of the vibrational energy levels, with one eigenfunction being an even function of z, the second an odd function of z. We have already seen the energy splitting, ΔE, is related to the frequency with which the N atom tunnels back and forth from one minimum to the other. The reduced mass for this degree of freedom is $3m_H m_N/(3m_H + m_N)$.

Because $V(-z) = V(z)$, we expect the one-dimensionalized wave functions, $u(z)$, to be either even or odd functions of z. Thus in the central exponential region, region I, with $-z_2 < z < +z_2$, we would expect $u(z)$ to be either a hyperbolic

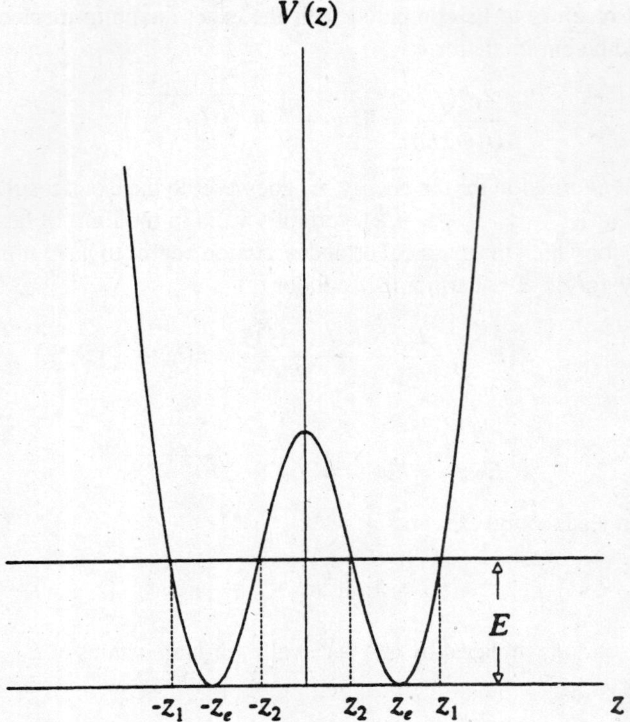

FIGURE 37.1. The NH_3 inversion potential for the coordinate, z.

cosine or hyberbolic sine type function,

$$u_I(z) = \frac{A}{2\sqrt{|P(z)|}}\left[e^{+\frac{1}{\hbar}\int_0^z d\zeta|P(\zeta)|} \pm e^{-\frac{1}{\hbar}\int_0^z d\zeta|P(\zeta)|}\right]. \tag{17}$$

We will find it convenient to use

$$\int_0^{z_2} d\zeta \cdots = \int_0^z d\zeta \cdots + \int_z^{z_2} d\zeta \cdots \quad \text{to rewrite}$$

$$u_I(z) = \frac{A}{2\sqrt{|P(z)|}}\left[e^{\frac{1}{\hbar}\int_0^{z_2} d\zeta|P(\zeta)|}e^{-\frac{1}{\hbar}\int_z^{z_2} d\zeta|P(\zeta)|}\right.$$
$$\left.\pm e^{-\frac{1}{\hbar}\int_0^{z_2} d\zeta|P(\zeta)|}e^{+\frac{1}{\hbar}\int_z^{z_2} d\zeta|P(\zeta)|}\right]$$
$$= \frac{A}{2\sqrt{|P(z)|}}\left[\frac{1}{Q}e^{-\frac{1}{\hbar}\int_z^{z_2} d\zeta|P(\zeta)|} \pm Qe^{+\frac{1}{\hbar}\int_z^{z_2} d\zeta|P(\zeta)|}\right], \tag{18}$$

where we have named

$$Q \equiv e^{-\frac{1}{\hbar}\int_0^{z_2} d\zeta|P(\zeta)|}. \tag{19}$$

For energies, E, far below the maximum of the central potential hill this exponential quantity, $Q \ll 1$. Now the function u_I for the central exponential region is in a form in which we can make the connection to region II, the upper oscillatory region of our potential. With the connection formulae for a left turning point, this would connect onto

$$u_{II}(z) = \frac{A}{2\sqrt{P(z)}}\left(\frac{2}{Q}\cos\left[\frac{1}{\hbar}\int_{z_2}^{z} d\zeta\, P(\zeta) - \frac{\pi}{4}\right] \pm Q\cos\left[\frac{1}{\hbar}\int_{z_2}^{z} d\zeta\, P(\zeta) + \frac{\pi}{4}\right]\right)$$
$$= \frac{A}{2\sqrt{P(z)}}\left(\frac{2}{Q}\cos\left[\frac{1}{\hbar}\int_{z_2}^{z} d\zeta\, P(\zeta) - \frac{\pi}{4}\right] \mp Q\sin\left[\frac{1}{\hbar}\int_{z_2}^{z} d\zeta\, P(\zeta) - \frac{\pi}{4}\right]\right)$$
$$= \frac{AR}{2\sqrt{P(z)}}\cos\left[\left(\frac{1}{\hbar}\int_{z_2}^{z} d\zeta\, P(\zeta) - \frac{\pi}{4}\right) \pm \delta\right], \tag{20}$$

$$\text{where} \qquad \tan\delta = \frac{Q^2}{2}, \qquad \text{and} \qquad R = \sqrt{Q^2 + \left(\frac{2}{Q}\right)^2}. \tag{21}$$

Also, $\tan\delta \approx \delta$, because we expect $Q^2 \ll 1$. It will be convenient, for purposes of making the connection to the right exponential region, with $z > z_1$, to reexpress this WKB solution for the oscillatory region II as

$$u_{II}(z) = \frac{AR}{2\sqrt{P(z)}}\cos\left(\left[\frac{1}{\hbar}\int_{z_2}^{z_1} d\zeta\, P(\zeta) - \frac{\pi}{2} \pm \delta\right] - \left[\frac{1}{\hbar}\int_{z}^{z_1} d\zeta\, P(\zeta) - \frac{\pi}{4}\right]\right). \tag{22}$$

Finally, in the exponential region I', with $z > z_1$, the solution must be an exponentially decreasing function as we penetrate further into the classically forbidden region,

$$u_{I'}(z) = \frac{B}{\sqrt{|P(z)|}} e^{-\frac{1}{\hbar}\int_{z_1}^{z} d\zeta |P(\zeta)|}. \tag{23}$$

This solution connects onto an oscillatory solution in region II of the form

$$u_{II}(z) = \frac{2B}{\sqrt{P(z)}}\cos\left[\frac{1}{\hbar}\int_{z}^{z_1} d\zeta\, P(\zeta) - \frac{\pi}{4}\right]. \tag{24}$$

We get a match with the earlier form for $u_{II}(z)$, if $|2B| = \frac{1}{2}|AR|$, and if

$$\frac{1}{\hbar}\int_{z_2}^{z_1} d\zeta\, P(\zeta) - \frac{\pi}{2} \pm \delta = n\pi, \tag{25}$$

or

$$\frac{1}{\pi}\int_{z_2}^{z_1} d\zeta\, P(\zeta) = \hbar(n + \tfrac{1}{2}) \mp \frac{\delta\hbar}{\pi}. \tag{26}$$

The left-hand side gives the action integral $J(E)$ for the upper potential minimum at the energy E appropriate for the even (or odd) solution for the full problem. If the energy, E, is far below the central potential maximum, the potential in the vicinity of the potential minimum between $+z_2$ and $+z_1$ can be approximated by a parabola, with $J(E) = E/\omega_0$. With $E = E^{(0)} + \Delta E$, we get

$$J(E) = J(E^{(0)}) + \left(\frac{\partial J}{\partial E}\right)_0 \Delta E + \cdots = \hbar(n + \tfrac{1}{2}) + \frac{1}{\omega_0}\Delta E + \cdots . \tag{27}$$

Therefore, eq. (26) can be put in the form

$$J(E) = \hbar(n + \tfrac{1}{2}) + \frac{\Delta E}{\omega_0} = \hbar(n + \tfrac{1}{2}) \mp \frac{\delta\hbar}{\pi}. \tag{28}$$

With $\delta \approx \tan\delta = \frac{1}{2}Q^2$, this equation leads to

$$\Delta E = \mp \frac{\hbar\omega_0}{2\pi} Q^2, \tag{29}$$

where the upper (lower) signs refer to the even (odd) solutions. The even functions lie at lower energies. Finally,

$$Q^2 = e^{-2\frac{1}{\hbar}\int_0^{z_2} d\zeta |P(\zeta)|} = e^{-\frac{1}{\hbar}\int_{-z_2}^{+z_2} d\zeta |P(\zeta)|} \tag{30}$$

is a function of E. However, for energy levels far below the central potential maximum, the energy splitting is very small compared with $E_n^{(0)} = \hbar\omega_0(n + \frac{1}{2})$, so we can express the energy splitting of the nth vibrational state by

$$\Delta E_n = \Delta E_{\text{odd}} - \Delta E_{\text{even}} = \frac{\hbar\omega_0}{\pi} Q_n^2 = \frac{\hbar\omega_0}{\pi} e^{-\frac{1}{\hbar}\int_{-z_2}^{+z_2} d\zeta \sqrt{2\mu(V(\zeta) - E_n^{(0)})}}. \tag{31}$$

We have seen previously the exponential,

$$e^{-G}, \qquad \text{with} \qquad G = \frac{1}{\hbar}\int_{-z_2}^{+z_2} d\zeta \sqrt{2\mu(V(\zeta) - E_n^{(0)})}, \tag{32}$$

is related to the probability the N atom tunnel through the central potential maximum. In Chapter 6, we showed the frequency with which the N atom tunnels back and forth from one potential minimum to the other is given by

$$\nu_{\text{tunneling}} = \frac{\Delta E}{2\pi\hbar} = \frac{\nu_0}{\pi} e^{-G}. \tag{33}$$

The factor, ν_0, the oscillator frequency in a single well, gives the frequency with which the N atom hits the potential barrier. The probability the N atom tunnel through the barrier is thus given by e^{-G}/π. The factor e^{-G} is known as the Gamow factor, because Gamow first discussed the tunneling phenomenon in connection with α decay in a heavy nucleus (see also problem 8).

Problems

52. In certain quark models, a linear confinement potential is used for heavy quarks, such as the charmed or b quarks, for which nonrelativistic quantum theory is approximately valid. The one-dimensionalized radial wave equation for such a quark would be

$$\frac{d^2u}{dr^2} + \frac{2m}{\hbar^2}(E - V_{\text{eff.}}(r))u(r) = 0,$$

with

$$V_{\text{eff.}}(r) = kr, \qquad \text{for } r \geq 0, \qquad k = \text{positive constant},$$

$$V_{\text{eff.}}(r) = \infty, \qquad \text{for } r < 0, \quad \text{so } u = 0 \text{ for } r \leq 0.$$

Find the WKB approximation for the energy, E_n, as a function of $m, k, \hbar$. The boundary condition at $r = 0$ leads to a WKB solution near $r = 0$ of the form

$$u_{\text{WKB}}(r) = \frac{A}{\sqrt{P(r)}} \sin\left[\frac{1}{\hbar}\int_0^r dr' P(r')\right],$$

with $P(r) = \sqrt{2m(E - V_{\text{eff.}}(r))}$. (Note: $P(r)$ is finite at $r = 0$.)

53. Show that the WKB connection formulae can be converted to the form

$$u_I = \frac{A}{\sqrt{P(x)}} e^{\pm\frac{i}{\hbar}\int_x^{x_2} d\xi P(\xi)} \qquad \longleftrightarrow \qquad u_{II}$$

$$u_{II} = \frac{A}{\sqrt{|P(x)|}}\left(\tfrac{1}{2} e^{\pm i\frac{\pi}{4}} e^{-\frac{1}{\hbar}\int_{x_2}^{x} d\xi |P(\xi)|} + e^{\mp i\frac{\pi}{4}} e^{+\frac{1}{\hbar}\int_{x_2}^{x} d\xi |P(\xi)|}\right)$$

for the right/left running wave solutions $\dfrac{A}{\sqrt{P(x)}} e^{\pm\frac{i}{\hbar}\int_x^{x_2} d\xi P(\xi)}$,

in region I, with a similar relation at the boundary $x = x_1$ (see Fig. P53).

Use these connection formulae to calculate the transmission and reflection coefficients for a wave incident on a potential barrier of arbitrary but smooth shape, with incident energy, $E < V_{\text{max.}}$. In particular, show that the transmission coefficient, T, is given by

$$T = \frac{\text{Transm. Flux}}{\text{Inc.Flux}} = \left(\frac{4D}{D^2 + 4}\right)^2, \qquad \text{with } D = e^{-\frac{1}{\hbar}\int_{x_2}^{x_1} dx \sqrt{2m(V(x) - E)}}.$$

54. For the one-dimensionalized potential of the shape shown in Fig. P54, demonstrate for arbitrary energies in the continuum, $E > 0$, but $E < V_{\text{max.}}$, the solutions in general will satisfy $|u_{IV}|^2 \gg |u_{II}|^2$. Show also that for the special values of $E = E_n$, for which

$$\frac{1}{\pi}\int_{x_1}^{x_2} dx \sqrt{2\mu(E_n - V(x))} \approx \hbar(n + \tfrac{1}{2}),$$

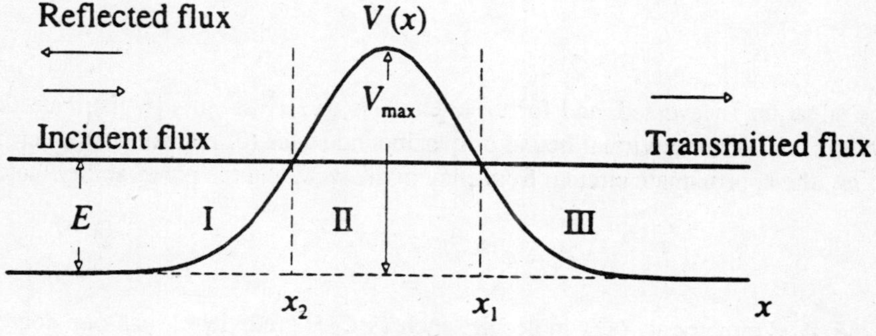

FIGURE P53.

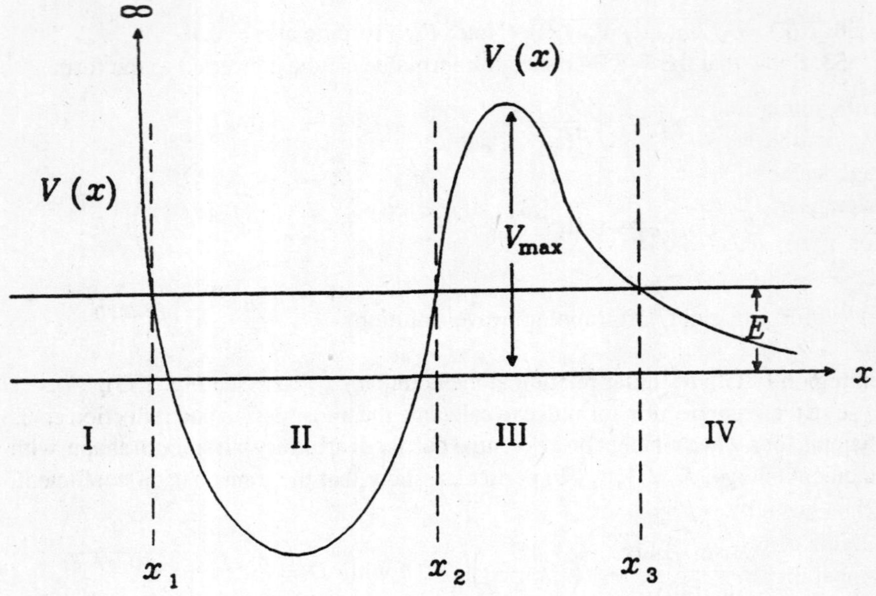

FIGURE P54.

the situation is reversed, and for these energies, $|u_{II}|^2 \gg |u_{IV}|^2$. Estimate the width, ΔE, of these virtual bound or quasibound states (or resonances) in terms of ω_0, the approximate circular frequency in the well, and the penetrability factor

$$Q^2 \doteq e^{-\frac{1}{\hbar}\int_{x_2}^{x_3} dx \sqrt{2\mu(V(x)-E_n)}}.$$

55. A symmetrical X_2Y_4 molecule, such as C_2H_4 (ethylene), has one degree of freedom, ϕ, which corresponds to a highly hindered internal rotation of one essentially rigid CH_2 unit relative to the other on the circle, as shown in Fig. 4.3 of Chapter 4. The wave equation separates approximately, so the hindered internal

rotation can be described by a one-degree-of-freedom Schrödinger equation,

$$-\frac{\hbar^2}{2I}\frac{d^2u}{d\phi^2} + V(\phi)u(\phi) = Eu(\phi),$$

with $I = I_1 I_2/(I_1 + I_2)$, $I_1 = I_2 = 2m_Y r_Y^2$, In Chapter 4, this problem was solved in a square well approximation. Now, we shall choose a more realistic potential that can be approximated by (see Fig. P55),

$$V(\phi) = V_0(1 - \cos 2\phi),$$

with minima at $\phi = 0$ and π and maxima at $\phi = \frac{\pi}{2}, \frac{3\pi}{2}, \ldots$. The constant, V_0, can be expected to be very large compared with the lowest allowed energy eigenvalues. In that case, the energy levels occur in closely spaced multiplets. The average position of a multiplet can be approximated by the quadratic approximation for $V(\phi)$, e. g. $V(\phi) \approx V_0\frac{4\phi^2}{2}$ near the potential minimum at $\phi = 0$, so $E \approx E_n = \hbar\omega_0(n + \frac{1}{2})$, with $\omega_0 \approx \sqrt{(4V_0/I)}$. In particular, show how the splitting into multiplets depends on the energies, $\hbar\omega_0$, and the penetrability factors

$$Q = e^{-\frac{1}{\hbar}\int_{\frac{\pi}{2}}^{\phi_2} d\phi' \sqrt{2I(V(\phi')-E)}}.$$

Use the fact that solutions of the form, $(A/\sqrt{|P|})\cosh(\ldots)$ in the classically forbidden region I (see Fig. P55), must connect onto solutions of the form, $\pm(A/\sqrt{|P|})\cosh(\ldots)$ in the region $\phi \to \phi + 2\pi$, i.e., region V in Fig. P55, whereas solutions of type $(A/\sqrt{|P|})\sinh(\ldots)$ in region I must connect onto solutions of the form $\pm(A/\sqrt{|P|})\sinh(\ldots)$ in region V, in order to preserve both probability density and probability density current.

Try to generalize your result for the energy splitting for the potentials

$$V(\phi) = V_0(1 - \cos N\phi), \qquad \text{with } N = 3, 4, \ldots.$$

For arbitrary N, show that the energy multiplets are made up of (N+1) levels, with (N-1) two-fold degenerate states and two nondegenerate states, now crowded into the same ΔE, viz.,

$$\Delta E = \frac{2\hbar\omega_0}{\pi}Q^2,$$

valid for $N = 2$.

For very large N, we have effectively bands of very finely spaced discrete allowed energy values, the Bloch bands of condensed matter physics.

Caution: For $N = 2$, all solutions are either symmetric or antisymmetric with respect to reflections in the plane $\phi = 3\pi/N = 3\pi/2$, in region III of Fig. P55, if they are made to have either symmetry, or antisymmetry with respect to reflections in the plane $\phi = \pi/N = \pi/2$, in region I of Fig. P55. For $N \geq 3$, symmetries in regions near $\phi = 3\pi/N,\ 5\pi/N, \ldots$ may not be simple for the doubly degenerate states. For such doubly degenerate states, a linear combination of symmetric and antisymmetric (or even and odd) functions may also be acceptable solutions even if the solutions are made symmetric or antisymmetric with respect to reflections in the plane $\phi = \pi/N$.

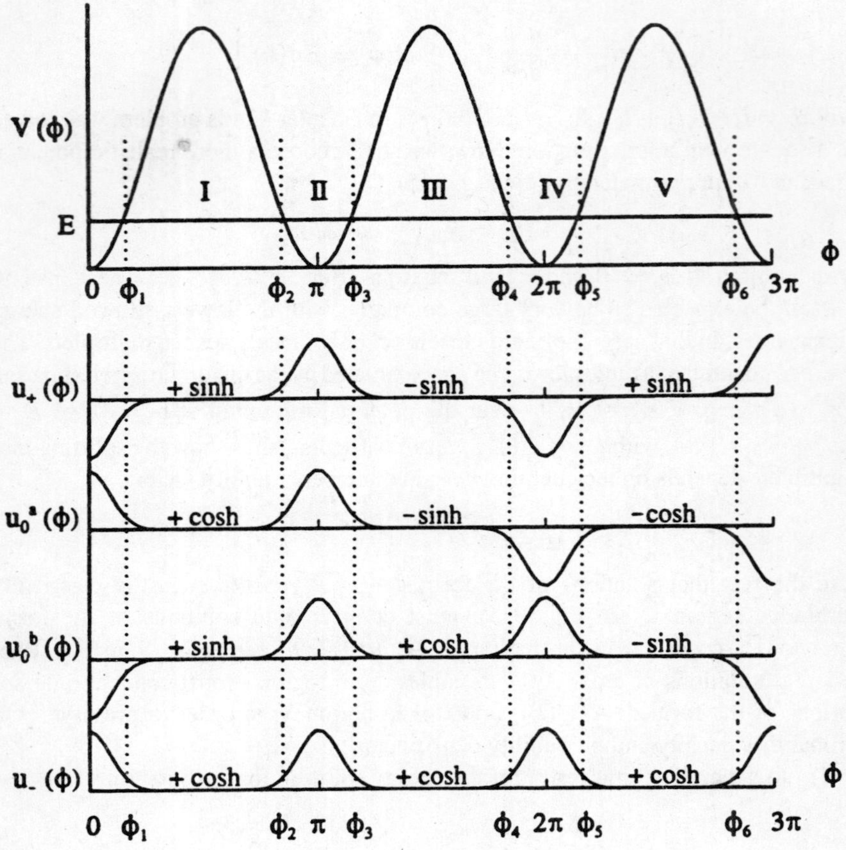

FIGURE P55. The hindering potential $V(\phi)$ for the X_2Y_4 molecule and the four eigenfunctions $u_-(\phi)$ (lowest E) $u_0^{(a)}(\phi)$, $u_0^{(b)}(\phi)$ (degenerate doublet) and $u_+(\phi)$ (highest E) for the $n = 0$ quartet.

Solution for Problem 55

The Case $N = 2$: We will start the process of finding the allowed solutions by assuming the solutions in the first classically forbidden region, near $\phi = \frac{\pi}{2}$ (region I of Fig. P55), must be either an even or an odd function of $(\phi - \frac{\pi}{2})$. Thus, in region I,

$$u_{\rm I}(\phi) = \frac{A}{\sqrt{|P(\phi)|}} \cosh \frac{1}{\hbar}\left(\int_{\frac{\pi}{2}}^{\phi} d\phi' |P(\phi')|\right) \qquad \text{for even } u_{\rm I},$$

$$u_{\rm I}(\phi) = \frac{A}{\sqrt{|P(\phi)|}} \sinh \frac{1}{\hbar}\left(\int_{\frac{\pi}{2}}^{\phi} d\phi' |P(\phi')|\right) \qquad \text{for odd } u_{\rm I}. \tag{1}$$

We shall continue these WKB solutions to region V, near $\phi = \frac{\pi}{2} + 2\pi$, where we will require $u_{\text{V}} = \pm u_{\text{I}}$; i.e., even functions must end up as even functions, odd functions must end up as odd functions, so the probability density and probability density current are single-valued functions in the 3-D space. Let us, however, start with exponential functions in region I:

$$u_{\text{I}}^{(+)} = \frac{1}{\sqrt{|P|}} e^{+\frac{1}{\hbar}\int_{\frac{\pi}{2}}^{\phi} d\phi' |P(\phi')|} = \frac{1}{Q}\frac{1}{\sqrt{|P|}} e^{-\frac{1}{\hbar}\int_{\phi}^{\phi_2} d\phi' |P(\phi')|},$$
$$u_{\text{I}}^{(-)} = \frac{1}{\sqrt{|P|}} e^{-\frac{1}{\hbar}\int_{\frac{\pi}{2}}^{\phi} d\phi' |P(\phi')|} = Q\frac{1}{\sqrt{|P|}} e^{+\frac{1}{\hbar}\int_{\phi}^{\phi_2} d\phi' |P(\phi')|}, \tag{2}$$

where ϕ_2 gives the right boundary of the classically forbidden region I, and where we have used

$$\int_{\frac{\pi}{2}}^{\phi} d\phi' |P(\phi')| = \int_{\frac{\pi}{2}}^{\phi_2} d\phi' |P(\phi')| - \int_{\phi}^{\phi_2} d\phi' |P(\phi')|,$$

and where Q is defined through

$$Q = e^{-\frac{1}{\hbar}\int_{\frac{\pi}{2}}^{\phi_2} d\phi' |P(\phi')|}.$$

We shall now use the WKB connection formulae to connect these solutions to the oscillatory solutions, valid in region II for $\phi_2 < \phi < \phi_3$:

$$u_{\text{I}}^{(+)} \to \frac{2}{Q}\frac{1}{\sqrt{P(\phi)}} \cos\left(\frac{1}{\hbar}\int_{\phi_2}^{\phi} d\phi' P(\phi') - \frac{\pi}{4}\right),$$
$$u_{\text{I}}^{(-)} \to \frac{Q}{\sqrt{P(\phi)}} \cos\left(\frac{1}{\hbar}\int_{\phi_2}^{\phi} d\phi' P(\phi') + \frac{\pi}{4}\right). \tag{3}$$

In the cosine functions, we shall now use

$$\frac{1}{\hbar}\int_{\phi_2}^{\phi} d\phi' P(\phi') = \frac{1}{\hbar}\int_{\phi_2}^{\phi_3} d\phi' P(\phi') - \frac{1}{\hbar}\int_{\phi}^{\phi_3} d\phi' P(\phi'),$$

and relate the integral over the complete oscillatory region to the action variable, $J(E)$, with $E = E^{(0)} + \Delta E$, where $E^{(0)}$ is the solution for a single oscillatory well of approximately parabolic shape, with $E^{(0)} = \hbar\omega_0(n + \frac{1}{2})$, and ΔE is the shift in this energy level caused by the presence of the potential hills. With $\Delta E \ll E^{(0)}$, we then have

$$\frac{1}{\hbar}\int_{\phi_2}^{\phi_3} d\phi' P(\phi') = \frac{\pi}{\hbar}\Big(J(E)\Big)$$

$$= \frac{\pi}{\hbar}\left(J(E^{(0)}) + \left(\frac{\partial J}{\partial E}\right)_{E^{(0)}} \Delta E\right) = \frac{\pi}{\hbar}\left(\hbar(n + \tfrac{1}{2}) + \frac{1}{\omega_0}\Delta E\right).$$

Substituting this equation into eq. (3), we have

$$u_{\text{I}}^{(+)} \to \frac{2}{Q}\frac{1}{\sqrt{P(\phi)}} \cos\left(\frac{1}{\hbar}\int_{\phi}^{\phi_3} d\phi' P(\phi') - \frac{\pi}{4} - n\pi - \frac{\pi\Delta E}{\hbar\omega_0}\right),$$

$$u_{\text{I}}^{(-)} \to Q\frac{1}{\sqrt{P(\phi)}}\cos\Big(\frac{1}{\hbar}\int_{\phi}^{\phi_3} d\phi' P(\phi') + \frac{\pi}{4} - (n+1)\pi - \frac{\pi\Delta E}{\hbar\omega_0}\Big). \tag{4}$$

Now, expanding in the small quantity, $(\pi\Delta E)/(\hbar\omega_0)$, we have

$$u_{\text{I}}^{(+)} \to \frac{2}{Q}\frac{1}{\sqrt{P(\phi)}}(-1)^n\Big[\cos\Big(\frac{1}{\hbar}\int_{\phi}^{\phi_3} d\phi' P(\phi') - \frac{\pi}{4}\Big) - \frac{\pi\Delta E}{\hbar\omega_0}\cos\Big(\frac{1}{\hbar}\int_{\phi}^{\phi_3} d\phi' P(\phi') + \frac{\pi}{4}\Big)\Big],$$

$$u_{\text{I}}^{(-)} \to Q\frac{1}{\sqrt{P(\phi)}}(-1)^{(n+1)}\Big[\cos\Big(\frac{1}{\hbar}\int_{\phi}^{\phi_3} d\phi' P(\phi') + \frac{\pi}{4}\Big) + \frac{\pi\Delta E}{\hbar\omega_0}\cos\Big(\frac{1}{\hbar}\int_{\phi}^{\phi_3} d\phi' P(\phi') - \frac{\pi}{4}\Big)\Big]. \tag{5}$$

We now use the WKB connection formulae to connect the oscillatory solutions for $\phi < \phi_3$ onto the exponential solutions for $\phi > \phi_3$, valid in region III, to obtain

$$u_{\text{I}}^{(+)} \to \frac{1}{Q}\frac{(-1)^n}{\sqrt{|P(\phi)|}}\Big[e^{-\frac{1}{\hbar}\int_{\phi_3}^{\phi} d\phi'|P(\phi')|} - 2\frac{\pi\Delta E}{\hbar\omega_0}e^{+\frac{1}{\hbar}\int_{\phi_3}^{\phi} d\phi'|P(\phi')|}\Big],$$

$$u_{\text{I}}^{(-)} \to -Q\frac{(-1)^n}{\sqrt{|P(\phi)|}}\Big[e^{+\frac{1}{\hbar}\int_{\phi_3}^{\phi} d\phi'|P(\phi')|} + \frac{1}{2}\frac{\pi\Delta E}{\hbar\omega_0}e^{-\frac{1}{\hbar}\int_{\phi_3}^{\phi} d\phi'|P(\phi')|}\Big]. \tag{6}$$

Now using

$$e^{-\frac{1}{\hbar}\int_{\phi_3}^{\phi} d\phi'|P(\phi')|} = e^{-\frac{1}{\hbar}\int_{\phi_3}^{\frac{3\pi}{2}} d\phi'|P(\phi')|}e^{-\frac{1}{\hbar}\int_{\frac{3\pi}{2}}^{\phi} d\phi'|P(\phi')|} = Qe^{-\frac{1}{\hbar}\int_{\frac{3\pi}{2}}^{\phi} d\phi'|P(\phi')|},$$

we get

$$u_{\text{I}}^{(+)} \to \frac{(-1)^n}{\sqrt{|P(\phi)|}}\Big[e^{-\frac{1}{\hbar}\int_{\frac{3\pi}{2}}^{\phi} d\phi'|P(\phi')|} - \frac{2}{Q^2}\frac{\pi\Delta E}{\hbar\omega_0}e^{+\frac{1}{\hbar}\int_{\frac{3\pi}{2}}^{\phi} d\phi'|P(\phi')|}\Big],$$

$$u_{\text{I}}^{(-)} \to \frac{(-1)^n}{\sqrt{|P(\phi)|}}\Big[-e^{+\frac{1}{\hbar}\int_{\frac{3\pi}{2}}^{\phi} d\phi'|P(\phi')|} - \frac{Q^2}{2}\frac{\pi\Delta E}{\hbar\omega_0}e^{-\frac{1}{\hbar}\int_{\frac{3\pi}{2}}^{\phi} d\phi'|P(\phi')|}\Big]. \tag{7}$$

For levels far below the potential barriers, we expect the penetration factor, Q^2, to be such that $Q^2 \ll 1$. Also, the energy shift caused by barrier penetration should be proportional to Q^2. The quantity $\frac{\Delta E}{\hbar\omega_0}Q^2$ can thus be expected to be completely negligible, so the negative exponential in the last expression can be neglected. If we now name

$$\frac{1}{\sqrt{|P(\phi)|}}e^{\pm\int_{\frac{3\pi}{2}}^{\phi} d\phi'|P(\phi')|} = u_{\text{III}}^{(\pm)},$$

and introduce the shorthand notation

$$\beta \equiv \frac{\pi\Delta E}{\hbar\omega_0 Q^2},$$

the above equations give us the connection formulae

$$u_{\rm I}^{(+)} \to (-1)^n\Big[u_{\rm III}^{(-)} - 2\beta u_{\rm III}^{(+)}\Big],$$
$$u_{\rm I}^{(-)} \to (-1)^n\Big[-u_{\rm III}^{(+)}\Big]. \tag{8}$$

We can now iterate this procedure to connect the $u_{\rm III}^{(\pm)}$ onto the corresponding $u_{\rm V}^{(\pm)}$ in region V, where the original ϕ has been incremented by 2π. This equation yields

$$u_{\rm I}^{(+)} \to (-1)^{2n}\Big[-u_{\rm V}^{(+)} - 2\beta\big(u_{\rm V}^{(-)} - 2\beta u_{\rm V}^{(+)}\big)\Big] = [(4\beta^2-1)u_{\rm V}^{(+)} - 2\beta u_{\rm V}^{(-)}],$$
$$u_{\rm I}^{(-)} \to (-1)^{2n}\Big[-u_{\rm V}^{(-)} + 2\beta u_{\rm V}^{(+)}\Big] \qquad = \qquad [2\beta u_{\rm V}^{(+)} - u_{\rm V}^{(-)}]. \tag{9}$$

Combining these formulae to make even or odd functions of ϕ in region I, these even or odd functions in region I would connect to a linear combination of even and odd functions in region V, where ϕ has been incremented by 2π:

$$\begin{aligned}
&\frac{1}{\sqrt{|P(\phi)|}}\cosh\Big(\frac{1}{\hbar}\int_{\frac{\pi}{2}}^{\phi} d\phi'|P(\phi')|\Big) = \tfrac{1}{2}\big(u_{\rm I}^{(+)} + u_{\rm I}^{(-)}\big)\\
&\to \tfrac{1}{2}\Big[(2\beta^2-1)(u_{\rm V}^{(+)} + u_{\rm V}^{(-)}) + 2\beta(1+\beta)(u_{\rm V}^{(+)} - u_{\rm V}^{(-)})\Big]\\
&= \frac{1}{\sqrt{|P(\phi)|}}\Bigg[(2\beta^2-1)\cosh\Big(\frac{1}{\hbar}\int_{\frac{5\pi}{2}}^{\phi} d\phi'|P(\phi')|\Big)\\
&\qquad + 2\beta(1+\beta)\sinh\Big(\frac{1}{\hbar}\int_{\frac{5\pi}{2}}^{\phi} d\phi'|P(\phi')|\Big)\Bigg],
\end{aligned} \tag{10}$$

and

$$\begin{aligned}
&\frac{1}{\sqrt{|P(\phi)|}}\sinh\Big(\frac{1}{\hbar}\int_{\frac{\pi}{2}}^{\phi} d\phi'|P(\phi')|\Big) = \tfrac{1}{2}\big(u_{\rm I}^{(+)} - u_{\rm I}^{(-)}\big)\\
&\to \tfrac{1}{2}\Big[2\beta(\beta-1)(u_{\rm V}^{(+)} + u_{\rm V}^{(-)}) + (2\beta^2-1)(u_{\rm V}^{(+)} - u_{\rm V}^{(-)})\Big]\\
&= \frac{1}{\sqrt{|P(\phi)|}}\Bigg[2\beta(\beta-1)\cosh\Big(\frac{1}{\hbar}\int_{\frac{5\pi}{2}}^{\phi} d\phi'|P(\phi')|\Big)\\
&\qquad + (2\beta^2-1)\sinh\Big(\frac{1}{\hbar}\int_{\frac{5\pi}{2}}^{\phi} d\phi'|P(\phi')|\Big)\Bigg].
\end{aligned} \tag{11}$$

Thus, we see from eq. (10) $u_{\rm even}(\phi) \to \pm u_{\rm even}(\phi + 2\pi)$ only if $\beta = -1$, or if $\beta = 0$. From eq. (11), $u_{\rm odd}(\phi) \to \pm u_{\rm odd}(\phi + 2\pi)$ only if $\beta = +1$, or if $\beta = 0$. The potential $V_0(1 - \cos 2\phi)$ thus splits the zeroth-order energies $E^{(0)} = \hbar\omega_0(n + \frac{1}{2})$ into three closely spaced levels, one of them with $\Delta E = 0$, being doubly degenerate. The four values of ΔE are

$$\begin{aligned}
\Delta E &= -Q^2\frac{\hbar\omega_0}{\pi} &&\text{with}\quad u_{\rm even}(\phi) \to +u_{\rm even}(\phi + 2\pi),\\
\Delta E &= 0 &&\text{with}\quad u_{\rm even}(\phi) \to -u_{\rm even}(\phi + 2\pi),\\
\Delta E &= 0 &&\text{with}\quad u_{\rm odd}(\phi) \to -u_{\rm odd}(\phi + 2\pi),
\end{aligned}$$

$$\Delta E = +Q^2 \frac{\hbar\omega_0}{\pi} \quad \text{with} \quad u_{\text{odd}}(\phi) \to +u_{\text{odd}}(\phi + 2\pi). \tag{12}$$

The four eigenfunctions for the case $n = 0$ are shown qualitatively in Fig. P55. The four eigenfunctions are also either even or odd functions of $(\phi - \frac{3\pi}{2})$; i.e., they are even or odd with respect to reflections in the plane $\phi = \frac{3\pi}{2}$, where the potential again has a maximum. In particular, the eigenfunction u_- of the nondegenerate level, with $\Delta E = -Q^2\hbar\omega_0/\pi$, is even with respect to reflections in planes through any potential maximum, and the eigenfunction u_+ of the nondegenerate level, with $\Delta E = +Q^2\hbar\omega_0/\pi$, is odd with respect to reflections in planes through any potential maximum. Conversely, the eigenfunctions u_0 for the doubly degenerate level, with $\Delta E = 0$, are alternately even and odd functions when reflected in planes through successive potential maxima. Because any linear combination of the two eigenfunctions u_0 is again an eigenfunction with the same eigenvalue, we would, however, have to exercise care in using the symmetry with respect to reflections in successive planes of symmetry of the potential.

The case for $V(\phi) = V_0(1 - \cos N\phi)$, arbitrary N:

In this case, the maxima of the potentials will be centered about the angles

$$\frac{\pi}{N}, \left(\frac{\pi}{N} + \frac{2\pi}{N}\right), \left(\frac{\pi}{N} + 2\frac{2\pi}{N}\right), \ldots, \left(\frac{\pi}{N} + k\frac{2\pi}{N}\right), \ldots, \left(\frac{\pi}{N} + N\frac{2\pi}{N}\right)$$

$$= \left(\frac{\pi}{N} + 2\pi\right)$$

Let us rename these regions with the index k, starting with $k = 0$ for the starting hill and ending with $k = N$, for which we have incremented ϕ by 2π. Eqs. (8) and (9) can now be put in the form

$$\begin{aligned} u_{0,\text{even}} &\to (-1)^n \left[-\beta u_{1,\text{even}} - (1 + \beta) u_{1,\text{odd}} \right], \\ u_{0,\text{odd}} &\to (-1)^n \left[-\beta u_{1,\text{odd}} + (1 - \beta) u_{1,\text{even}} \right]. \end{aligned} \tag{13}$$

$$\begin{aligned} u_{0,\text{even}} &\to (-1)^{2n} \left[(2\beta^2 - 1) u_{2,\text{even}} + 2\beta(1 + \beta) u_{2,\text{odd}} \right], \\ u_{0,\text{odd}} &\to (-1)^{2n} \left[(2\beta^2 - 1) u_{2,\text{odd}} - 2\beta(1 - \beta) u_{2,\text{even}} \right]. \end{aligned} \tag{14}$$

Iterating this once more, we have the connection into the next hill

$$\begin{aligned} u_{0,\text{even}} &\to (-1)^{3n} \left[\beta(3 - 4\beta^2) u_{3,\text{even}} - (1 + \beta)(4\beta^2 - 1) u_{3,\text{odd}} \right], \\ u_{0,\text{odd}} &\to (-1)^{3n} \left[\beta(3 - 4\beta^2) u_{3,\text{odd}} + (1 - \beta)(4\beta^2 - 1) u_{3,\text{even}} \right]. \end{aligned} \tag{15}$$

Before iterating this into the kth hill, it will be convenient to change the notation, and rename $\beta \equiv \cos\alpha$. With this notation, the above relations and the continued iteration process give

$$\begin{aligned} u_{0,\text{even}} &\to (-1)^n \left[-\cos\alpha \ u_{1,\text{even}} - (1 + \cos\alpha) u_{1,\text{odd}} \right], \\ u_{0,\text{odd}} &\to (-1)^n \left[-\cos\alpha \ u_{1,\text{odd}} + (1 - \cos\alpha) u_{1,\text{even}} \right], \end{aligned}$$

$$
\begin{aligned}
u_{0,\text{even}} &\to (-1)^{2n}\left[\cos(2\alpha)\ u_{2,\text{even}} + (1+\cos\alpha)\frac{\sin(2\alpha)}{\sin\alpha}u_{2,\text{odd}}\right],\\
u_{0,\text{odd}} &\to (-1)^{2n}\left[\cos(2\alpha)\ u_{2,\text{odd}} - (1-\cos\alpha)\frac{\sin(2\alpha)}{\sin\alpha}u_{2,\text{even}}\right],\\
u_{0,\text{even}} &\to (-1)^{3n}\left[-\cos(3\alpha)\ u_{3,\text{even}} - (1+\cos\alpha)\frac{\sin(3\alpha)}{\sin\alpha}u_{3,\text{odd}}\right],\\
u_{0,\text{odd}} &\to (-1)^{3n}\left[-\cos(3\alpha)\ u_{3,\text{odd}} + (1-\cos\alpha)\frac{(\sin 3\alpha)}{\sin\alpha}u_{3,\text{even}}\right],\\
&\cdots\cdots\cdots\\
u_{0,\text{even}} &\to (-1)^{k(n+1)}\left[\cos(k\alpha)\ u_{k,\text{even}} + (1+\cos\alpha)\frac{\sin(k\alpha)}{\sin\alpha}u_{k,\text{odd}}\right],\\
u_{0,\text{odd}} &\to (-1)^{k(n+1)}\left[\cos(k\alpha)\ u_{k,\text{odd}} - (1-\cos\alpha)\frac{(\sin k\alpha)}{\sin\alpha}u_{k,\text{even}}\right],\\
&\cdots\cdots\cdots\\
u_{0,\text{even}} &\to (-1)^{N(n+1)}\left[\cos(N\alpha)\ u_{N,\text{even}} + (1+\cos\alpha)\frac{\sin(N\alpha)}{\sin\alpha}u_{N,\text{odd}}\right],\\
u_{0,\text{odd}} &\to (-1)^{N(n+1)}\left[\cos(N\alpha)\ u_{N,\text{odd}} - (1-\cos\alpha)\frac{\sin(N\alpha)}{\sin\alpha}u_{N,\text{even}}\right], \quad (16)
\end{aligned}
$$

where the last iteration must give

$$u_{N,\text{even}} = u_{\text{even}}(\phi + 2\pi) = \pm u_{0,\text{even}}(\phi),$$

or

$$u_{N,\text{odd}} = u_{\text{odd}}(\phi + 2\pi) = \pm u_{0,\text{odd}}(\phi).$$

A single (nondegenerate) even solution exists if $\cos\alpha = -1$, and a single (nondegenerate) odd solution if $\cos\alpha = +1$. In addition, both an even and an odd solution exists if

$$\sin N\alpha = 0, \qquad \cos N\alpha = \pm 1, \qquad \text{or}$$

$$\alpha = \frac{\ell\pi}{N}, \qquad \ell = 1, 2, \ldots, (N-1).$$

Recalling $\cos\alpha = (\pi\Delta E)/Q^2\hbar\omega_0)$, the final spectrum of allowed energies is

$$\Delta E = -Q^2\frac{\hbar\omega_0}{\pi}, \qquad \text{with nondegenerate even eigenfunction,}$$

$$\Delta E = +Q^2\frac{\hbar\omega_0}{\pi}, \qquad \text{with nondegenerate odd eigenfunction,}$$

$$\Delta E = Q^2\frac{\hbar\omega_0}{\pi}\cos\frac{\ell\pi}{N}, \qquad \text{with } \ell = 1, 2, \ldots, (N-1), \text{ all doubly degenerate.}$$

For $N = 3$, which might apply to the X_2Y_6 molecule, each zeroth-order energy, $E_n^{(0)}$, is split into four levels with

$$(\Delta E)_n = +Q_n^2\frac{\hbar\omega_0}{\pi}, \quad +\tfrac{1}{2}Q_n^2\frac{\hbar\omega_0}{\pi}, \quad -\tfrac{1}{2}Q_n^2\frac{\hbar\omega_0}{\pi}; \quad -Q_n^2\frac{\hbar\omega_0}{\pi}.$$

The penetration factor, Q_n^2, may of course be so small for the ground state, $n = 0$, the splitting may be unobservable (e.g., in C_2H_6).

The case of very large N is of relevance in solid-state physics, where periodic boundary conditions are used. If we have a crystalline lattice with N repeat units, we assume the $(N+1)^{\text{th}}$ unit is identical with the first. If N is very large, the spacing of our multiplet of $2 + (N-1) = (N+1)$ sublevels can effectively be replaced by a band of continuum states of width $2Q_n^2\hbar\omega_0/\pi$. The penetration factor Q_n^2 is a sensitive function of $E_n^{(0)}$. For $n = 0$, far below the top of the potential barriers, we may have $Q_0^2 \ll 1$, leading to a narrow band. For $n > 0$, Q_n^2 may grow dramatically with n, leading to ever wider bands as the top of the potential barriers is approached.

Part IV

Systems of Identical Particles

38

The Two-Electron Atom

The indistinguishability of identical particles in quantum mechanics plays a very important role. In macroscopic, classical physics we can tag our particles (by painting infinitesimally small labels on them!) so we can distinguish those labelled, 1, 2, etc., even though they have exactly the same mass, internal constitution, etc. The impossibility of such a tagging procedure plays a very fundamental role in quantum mechanics.

Consider one of the simplest systems with identical particles, the two-electron He atom, with Hamiltonian

$$H = \frac{\vec{p}_1^{\,2}}{2\mu} + V(r_1) + \frac{\vec{p}_2^{\,2}}{2\mu} + V(r_2) + \frac{e^2}{|\vec{r}_1 - \vec{r}_2|} + H_{\text{f.s.}}(\vec{r}_1, \vec{p}_1, \vec{\sigma}_1; \vec{r}_2, \vec{p}_2, \vec{\sigma}_2), \quad (1)$$

where only the last term is dependent on the spins of the two electrons and depends on these through their Pauli $\vec{\sigma}$ vectors. This fine structure term will include one-body spin-orbit and Thomas terms, two-body spin-magnetic moment-spin-magnetic moment interactions, and so on, and can be treated as truly small perturbations on the zeroth-order terms including the Coulomb repulsion term, e^2/r_{12}. Even with all fine structure terms, however, the Hamiltonian has a strictly valid symmetry. It is invariant under the interchange of the particle indices, 1 and 2.

$$\begin{aligned} H(\vec{r}_1, \vec{p}_1, \vec{\sigma}_1; \vec{r}_2, \vec{p}_2, \vec{\sigma}_2)) &= H(\vec{r}_2, \vec{p}_2, \vec{\sigma}_2; \vec{r}_1, \vec{p}_1, \vec{\sigma}_1) \\ &= P_{12} H(\vec{r}_1, \vec{p}_1, \vec{\sigma}_1; \vec{r}_2, \vec{p}_2, \vec{\sigma}_2) P_{12}^{-1}, \end{aligned} \quad (2)$$

where the operator $P_{12} \equiv P_{12}^{-1}$ exchanges the indices 1 and 2 on all electron variables. P_{12} commutes with the Hamiltonian, H. The eigenfunctions of the operator

P_{12} are the symmetric and antisymmetric functions

$$\psi^{(s)} = \frac{1}{\sqrt{2}}[\psi(\vec{r}_1, \vec{\sigma}_1; \vec{r}_2, \vec{\sigma}_2) + \psi(\vec{r}_2, \vec{\sigma}_2; \vec{r}_1, \vec{\sigma}_1)],$$
$$\psi^{(a)} = \frac{1}{\sqrt{2}}[\psi(\vec{r}_1, \vec{\sigma}_1; \vec{r}_2, \vec{\sigma}_2) - \psi(\vec{r}_2, \vec{\sigma}_2; \vec{r}_1, \vec{\sigma}_1)], \quad (3)$$

with eigenvalues $+1$ and -1, respectively, for the operator P_{12}. Thus. it seems that every energy level of the two-electron H must be two-fold degenerate to all orders in perturbation theory, because the energy eigenfunctions can be either symmetric or antisymmetric. Even more, any linear combination of the symmetric and antisymmetric functions will have the same energy. Thus,

$$\Psi = c_s\psi^{(s)} + c_a\psi^{(a)}, \qquad \text{with} \qquad |c_s|^2 + |c_a|^2 = 1, \quad (4)$$

is an equally acceptable energy eigenfunction of our H. This function, however, leads to a tremendous dilemma. For this last Ψ, with arbitrary c_s and c_a, the probability one electron is at position $\vec{r}_0$, with spin alignment $\vec{\sigma}_0$, and the other electron is at position $\vec{r}_0'$, with spin alignment $\vec{\sigma}_0'$, is given by

$$\begin{aligned} P(\vec{r}_0, \vec{\sigma}_0; \vec{r}_0', \vec{\sigma}_0') &= |\Psi(\vec{r}_0, \vec{\sigma}_0; \vec{r}_0', \vec{\sigma}_0')|^2 + |\Psi(\vec{r}_0', \vec{\sigma}_0'; \vec{r}_0, \vec{\sigma}_0)|^2 \\ &= |c_s\psi^{(s)} + c_a\psi^{(a)}|^2 + |c_s\psi^{(s)} - c_a\psi^{(a)}|^2 \\ &= 2(|c_s|^2|\psi^{(s)}|^2 + |c_a|^2|\psi^{(a)}|^2). \end{aligned} \quad (5)$$

We must add the probability the electron we have labeled 1 is at $\vec{r}_0$ with spin alignment given by $\vec{\sigma}_0$, and electron 2 is at the primed position with primed spin alignment, to the probability the electron we have labeled 2 is at position $\vec{r}_0$ with spin alignment given by $\vec{\sigma}_0$, because we cannot distinguish electrons 1 and 2. Also, this probability, a physically measurable quantity, seems to be dependent on c_s and c_a. Because c_s can vary between 0 and 1, it seems this physically measurable quantity has an essentially arbitrary predicted value. Consider, in particular, the special case in which both electrons are situated at the same position, $\vec{r}_0$, and both have the same spin alignment given by $\vec{\sigma}_0$. In this case $\psi^{(a)} = 0$, so in this case this probability would be $2|c_s|^2|\psi^{(s)}|^2$. Because c_s can vary between 0 and 1, this probability could seemingly be anything between 0 and a maximum of $2|\psi^{(s)}|^2$. The way out of this seeming dilemma is furnished by nature herself! An additional property of nature exists, first discovered empirically. The states of a system of n identical particles are either all totally symmetric or all totally antisymmetric.

The totally symmetric states are symmetric under any pair exchange and hence under any number of pair exchanges or any permutation of the n particle indices. This symmetry holds for systems of identical particles with integer spins; that is, $s = 0, 1, 2, \cdots$. Such particles are known as Bose–Einstein particles or as bosons.

The totally antisymmetric states change sign under any pair exchange and, hence, any odd permutation of the particle indices involving an odd number of pair exchanges, while they do not change sign under even permutations of the particle indices, involving an even number of pair exchanges. This case applies to systems of identical particles with $\frac{1}{2}$-integral spin, $s = \frac{1}{2}, \frac{3}{2}, \frac{5}{2}, \cdots$. Such particles are known as Fermi–Dirac particles or as fermions.

Because electrons have $s = \frac{1}{2}$, they are fermions and belong to the antisymmetric case. The antisymmetry, however, applies to the total wave function. A two-electron system could have a two-particle orbital wave function symmetric under the pair exchange operator applied to orbital states, provided it is multiplied by an antisymmetric two-particle spin function, or vice versa. Because our Hamiltonian is spin-independent in first approximation (where we can neglect spin-orbit and spin magnetic-moment-spin magnetic moment interactions), we would expect a product of purely orbital and purely spin functions to be a good approximation. Let us look at the two-particle spin function first. Let us, in particular, couple the two single-particle spins s to resultant two-particle spin, S. Later, we shall specialize to the electron case with $s = \frac{1}{2}$. For the moment, let s be arbitrary.

$$\psi(\vec{s}_1, \vec{s}_2)^S_{M_S} = \sum_{m_{sa}, m_{sb}} \psi_{m_{s_a}}(\vec{s}_1)\psi_{m_{s_b}}(\vec{s}_2)\langle s m_{s_a} s m_{s_b}|S M_S\rangle. \tag{6}$$

Now, act on this ψ with the operator P^s_{12}, where the superscript s indicates we permute indices only on spin functions

$$P^s_{12}\psi(\vec{s}_1, \vec{s}_2)^S_{M_S} = \sum_{m_{sa}, m_{sb}} \psi_{m_{s_a}}(\vec{s}_2)\psi_{m_{s_b}}(\vec{s}_1)\langle s m_{s_a} s m_{s_b}|S M_S\rangle. \tag{7}$$

Now, in the sum over the indices, m_{s_a}, m_{s_b}, let us rename the dummy indices, $m_{s_a} \leftrightarrow m_{s_b}$, and let us then rewrite the product of the two single-particle spin functions in reverse order, to obtain

$$P^s_{12}\psi(\vec{s}_1, \vec{s}_2)^S_{M_S} = \sum_{m_{sa}, m_{sb}} \psi_{m_{s_a}}(\vec{s}_1)\psi_{m_{s_b}}(\vec{s}_2)\langle s m_{s_b} s m_{s_a}|S M_S\rangle. \tag{8}$$

Next, we make use of the symmetry property of the Clebsch–Gordan coefficient

$$\langle s m_{s_b} s m_{s_a}|S M_S\rangle = (-1)^{2s-S}\langle s m_{s_a} s m_{s_b}|S M_S\rangle \tag{9}$$

to obtain

$$P^s_{12}\psi(\vec{s}_1, \vec{s}_2)^S_{M_S} = (-1)^{2s-S}\psi(\vec{s}_1, \vec{s}_2)^S_{M_S}. \tag{10}$$

Thus, with $s = \frac{1}{2}$-integer, so $2s =$ odd integer, two-particle spin functions, with $S = 0, 2, \cdots$, even integer, are antisymmetric, whereas two-particle spin functions, with $S = 1, 3, \cdots$, odd integer, are symmetric. For the special case of electrons, with $s = \frac{1}{2}$, the two-particle state with $S = 0$ is antisymmetric, whereas the two-particle states with $S = 1$ are symmetric. The antisymmetric two-particle spin states, with $S = 0$, must now be matched with a symmetric two-particle orbital state; similarly, the symmetric two-particle spin states, with $S = 1$, must be matched with antisymmetric two-particle orbital states. For two-electron states with $n_a l_a \neq n_b l_b$, we can always construct both a symmetric and an antisymmetric two-particle orbital state of good two-particle orbital angular momentum, L, M_L

$$\psi_{n_a l_a; n_b l_b}(\vec{r}_1, \vec{r}_2)^L_{M_L} = \sum_{m_a, m_b} \Big[\psi_{n_a l_a m_a}(\vec{r}_1)\psi_{n_b l_b m_b}(\vec{r}_2)$$
$$\pm \psi_{n_b l_b m_b}(\vec{r}_1)\psi_{n_a l_a m_a}(\vec{r}_2)\Big]\langle l_a m_a l_b m_b|L M_L\rangle$$

$$= \sum_{m_a, m_b} \Big[\psi_{n_a l_a m_a}(\vec{r}_1)\psi_{n_b l_b m_b}(\vec{r}_2)\langle l_a m_a l_b m_b | L M_L \rangle$$
$$\pm \psi_{n_b l_b m_b}(\vec{r}_1)\psi_{n_a l_a m_a}(\vec{r}_2)(-1)^{l_a + l_b - L}\langle l_b m_b l_a m_a | L M_L \rangle \Big], \qquad (11)$$

where the upper sign refers to the orbitally symmetric and the lower sign to the orbitally antisymmetric two-particle functions. In the special case, when $n_a = n_b = n$ and $l_a = l_b = l$, we can rename the indices $m_a \leftrightarrow m_b$ in the second term of this equation because they are dummy summation indices. With $n_a = n_b = n$, and $l_a = l_b = l$, therefore, the symmetric and antisymmetric two-particle orbital states of good orbital angular momentum L, M_L become

$$\psi_{nl;nl}(\vec{r}_1, \vec{r}_2)^L_{M_L} = \sum_{m_a m_b} \psi_{nlm_a}(\vec{r}_1)\psi_{nlm_b}(\vec{r}_2)\Big[1 \pm (-1)^{2l-L}\Big]\langle l m_a l m_b | L M_L \rangle. \quad (12)$$

Because $2l$ = even integer, in this case, symmetric two-particle states survive only for states with even L [upper sign in eq. (12)], whereas antisymmetric two-particle states survive only for states with odd L [lower sign in eq. (12)]. Thus, in two-electron configurations with $n_a l_a = n_b l_b$, the only allowed states must have either $S = 0$ and L = even integer, or $S = 1$ and L = odd integer. For a two-electron configuration of type $(np)^2$ in a two-valence electron atom, the only possible energy states are

$${}^1S_0, \qquad {}^1D_2, \qquad \text{and} \qquad {}^3P_{0,1,2},$$

in standard spectroscopic notation, where S is identified through a left superscript $(2S + 1)$, L is identified by the spectroscopic letter, a capital letter because the state is not a one-electron state, and the possible J values are given by a right subscript. In the He atom, the ground state with $n_a l_a = n_b l_b = 10$, i.e., with ground-state configuration $(1s)^2$, must be a pure singlet state with $L = 0$, $S = 0$, i.e., a ${}^1S_0$ state. Excited states with configurations such as $(1s, 2s)$ or $(1s, 2p)$ have both a singlet and a triplet component, with $L = 0$ and $L = 1$, respectively, as shown in Fig. 38.1. Because only the small fine structure terms in our Hamiltonian have a spin dependence, the singlet and triplet states are almost unconnected. In addition, electromagnetic transitions between singlet and triplet states are forbidden in zeroth (electric dipole approximation), because the transition operator is spin-independent. Thus, the singlet, $S = 0$ state atoms, the so-called para-helium atoms, and the triplet, $S = 1$ state atoms, the so-called ortho-helium atoms, essentially form a mixture of two gases; transitions from singlet to triplet states being extremely rare. Finally, configurations such as $(2s)^2$ or $(2p)^2$ are not included in Fig. 38.1 because they can be expected to lie above the first ionization threshhold.

A Perturbation Theory for a Two-Electron Atom

To get a very rough idea of the energy spectrum for He (or once ionized Li, twice ionized Be, etc), let us neglect all spin-dependent fine structure terms in the Hamiltonian of eq. (1) and, even further, try to consider the Coulomb repulsion

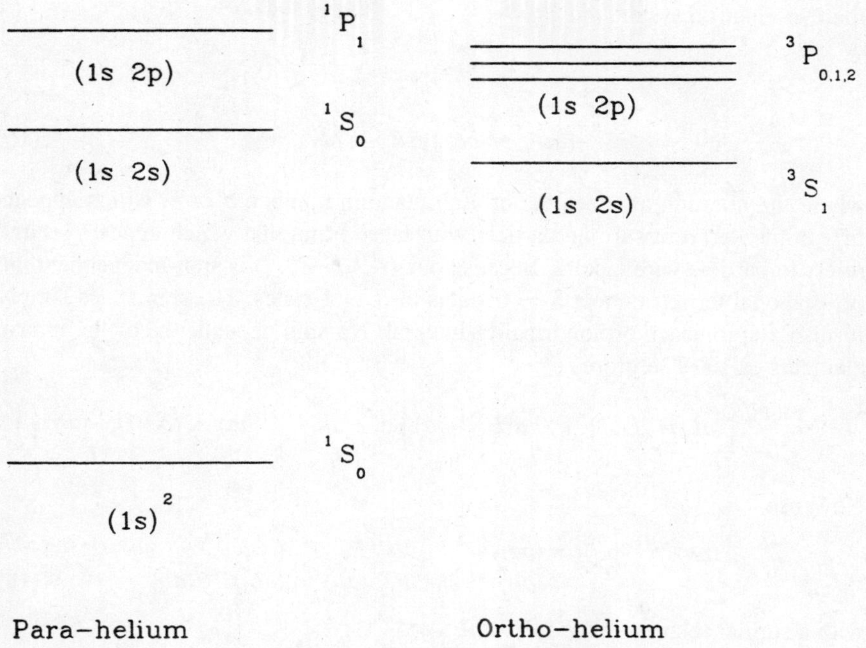

FIGURE 38.1. He atom spectrum.

term, e^2/r_{12}, as a first order perturbation. In eq. (1), the variables r_1, r_2 are the physical coordinates. Let us again introduce dimensionless r_1 and r_2. Because we will now find it useful to separate the Z dependence of the various terms, let us make the substitutions

$$\vec{r}_{\text{phys.};i} = \frac{1}{Z}\frac{\hbar^2}{\mu e^2}\vec{r}_i, \qquad H_{\text{phys.}} = \frac{\mu e^4}{\hbar^2}H, \tag{13}$$

where all quantities without the subscript, phys., such as $\vec{r}_1$ and $\vec{r}_2$ are now dimensionless quantities, i.e., physical quantities given in atomic units. Then,

$$H = H^{(0)} + H^{(1)} = Z^2\Big(-\frac{1}{2}\nabla_1^2 - \frac{1}{r_1} - \frac{1}{2}\nabla_2^2 - \frac{1}{r_2}\Big) + Z\frac{1}{r_{12}}. \tag{14}$$

With energies given in units of $\mu e^4/\hbar^2$, we get the dimensionless zeroth-order energy

$$E^{(0)} = -Z^2\Big(\frac{1}{2n_a^2} + \frac{1}{2n_b^2}\Big), \tag{15}$$

and the first-order corrections (albeit very rough corrections!) to this energy would be given by the diagonal matrix elements of $H^{(1)}$. If, for the moment, we use the shorthand notation $a \equiv n_a l_a m_a$ and $b \equiv n_b l_b m_b$, the zeroth-order state vectors for the two-electron system are

$$\frac{1}{\sqrt{2}}\big(|ab\rangle + |ba\rangle\big)|S=0, M_S=0\rangle,$$
$$\frac{1}{\sqrt{2}}\big(|ab\rangle - |ba\rangle\big)|S=1, M_S\rangle, \qquad (16)$$

where the notation assumes the orbital quantum number, a or b, which appears first in the ket, refers to the particle with label 1 and that which appears second refers to particle with label 2. Because our $H^{(1)} = Z/r_{12}$ is spin-independent, no off-diagonal terms connect $S = 0$ states to $S = 1$ states. The spin states simply furnish a spin-space orthonormality integral. No spin dependence of the matrix elements exists. Therefore,

$$E^{(1)} = \frac{1}{2}\Big(\Big[\langle ab|H^{(1)}|ab\rangle + \langle ba|H^{(1)}|ba\rangle\Big] \pm \Big[\langle ab|H^{(1)}|ba\rangle + \langle ba|H^{(1)}|ab\rangle\Big]\Big). \qquad (17)$$

Now, use

$$\langle ba|H^{(1)}|ba\rangle = \langle ba|P_{12}^{-1}P_{12}H^{(1)}P_{12}^{-1}P_{12}|ba\rangle$$
$$= \langle ab|P_{12}H^{(1)}P_{12}^{-1}|ab\rangle = \langle ab|H^{(1)}|ab\rangle, \qquad (18)$$

with a similar relation for $\langle ba|H^{(1)}|ab\rangle$, so

$$E^{(1)} = \langle ab|H^{(1)}|ab\rangle \pm \langle ab|H^{(1)}|ba\rangle = D_{ab} \pm X_{ab}, \qquad (19)$$

where D_{ab} stands for the direct integral in which the order of the quantum numbers ab is the same in both bra and ket, whereas X_{ab} stands for the exchange integral in which the order of the quantum numbers ab is exchanged in the ket relative to the order in the bra. The upper sign (+) refers to the singlet or $S = 0$ states, whereas the lower sign (-) refers to the triplet $S = 1$ states. Because our $H^{(1)} = +Z/r_{12}$ is a positive (repulsive) interaction term, we can expect both D_{ab} and X_{ab} to be positive numbers. Thus, we would expect the triplet states to lie somewhat below the singlet states, as shown in Fig. 38.1. This is a special case of Hund's rule. In a many-electron atom, we would expect the state with the largest possible number of electron pairs coupled to spin $S = 1$ to lie lowest in energy. Thus, the lowest state should be the state with highest possible total spin S.

Finally, our final energies are spin-dependent, even though our Hamiltonian was spin-independent, the spin dependence coming from the symmetry of the two-electron orbital functions. The antisymmetric two-particle orbital functions have a smaller probability of bringing the two electrons close together. Hence, in these states, the Coulomb repulsion term between the two electrons is less effective. The resultant apparent spin-dependence of the Hamiltonian could be taken into account by introducing an effective spin-dependent Hamiltonian,

$$H_{ab}^{(1)} = D_{ab} - X_{ab}\frac{(1+\vec{\sigma}_1\cdot\vec{\sigma}_2)}{2}, \qquad (20)$$

where we have used

$$P_{12}^{\sigma} = \tfrac{1}{2}(1+\vec{\sigma}_1\cdot\vec{\sigma}_2) = \tfrac{1}{2}(1+4\vec{s}_1\cdot\vec{s}_2) = \tfrac{1}{2}(1+2[S(S+1)-\tfrac{3}{4}-\tfrac{3}{4}]), \quad (21)$$

which has eigenvalue $+1$ for an $S = 1$, or spin-symmetric state and eigenvalue -1 for an $S = 0$ or spin-antisymmetric state.

To actually carry out the direct and exchange integrals, we will expand $1/r_{12}$ in spherical harmonics

$$\begin{aligned}\frac{1}{|\vec{r}_1 - \vec{r}_2|} &= \frac{1}{\sqrt{r_1^2 + r_2^2 - 2r_1r_2\cos\theta_{12}}} \\ &= \sum_{k=0}^{\infty} \frac{r_2^k}{r_1^{k+1}} P_k(\cos\theta_{12}), \qquad \text{for} \quad r_2 < r_1, \\ &= \sum_{k=0}^{\infty} \frac{r_1^k}{r_2^{k+1}} P_k(\cos\theta_{12}), \qquad \text{for} \quad r_1 < r_2. \end{aligned} \tag{22}$$

We shall further use the addition theorem for spherical harmonics

$$P_k(\cos\theta_{12}) = \frac{4\pi}{(2k+1)} \sum_q Y_{kq}^*(\theta_1, \phi_1) Y_{kq}(\theta_2, \phi_2). \tag{23}$$

The direct term, D_{ab}, is made up of terms of the form

$$\begin{aligned} D_{ab} = Z \sum_{k=0}^{\infty} \sum_q \frac{4\pi}{(2k+1)} \int_0^{\infty} dr_1 r_1^2 R_{n_a l_a}^2(r_1) \\ \times \left[\int_0^{r_1} dr_2 r_2^2 \frac{r_2^k}{r_1^{k+1}} R_{n_b l_b}^2(r_2) + \int_{r_1}^{\infty} dr_2 r_2^2 \frac{r_1^k}{r_2^{k+1}} R_{n_b l_b}^2(r_2) \right] \\ \times \int\int d\Omega_1 Y_{l_a(m_a - q)}^*(\theta_1, \phi_1) Y_{kq}^*(\theta_1, \phi_1) Y_{l_a m_a}(\theta_1 \phi_1) \\ \times \int\int d\Omega_2 Y_{l_b(m_b + q)}^*(\theta_2, \phi_2) Y_{kq}(\theta_2, \phi_2) Y_{l_b m_b}(\theta_2, \phi_2), \end{aligned} \tag{24}$$

weighted by the appropriate Clebsch–Gordan coefficients for the coupling $[l_a \times l_b]LM$. The angular integrals will in general greatly restrict the number of terms in the k sum. For the low-lying states of He, in particular, with one of the $n_a l_a$ or $n_b l_b = 10$, in general, just a single k term will exist. For the He ground state, for example, with $n_a l_a = n_b l_b = 10$, the angular integrals give

$$\int\int d\Omega Y_{00}^*(\theta, \phi) Y_{kq}(\theta, \phi) Y_{00}(\theta, \phi) = \delta_{k0}\delta_{q0} \frac{1}{\sqrt{4\pi}}, \tag{25}$$

so the energy, $E^{(1)}$, which is here given by the direct integral, gives

$$E^{(1)} = Z \int_0^{\infty} dr_1 r_1^2 R_{10}^2(r_1) \left[\int_0^{r_1} dr_2 r_2^2 \frac{1}{r_1} R_{10}^2(r_2) + \int_{r_1}^{\infty} dr_2 r_2^2 \frac{1}{r_2} R_{10}^2(r_2) \right]. \tag{26}$$

With $R_{10}(r) = 2e^{-r}$, and

$$\int_0^{r_1} dr_2 4r_2^2 e^{-2r_2} = \left[e^{-2r} \left(-2r^2 - 2r - 1 \right) \right]_0^{r_1}, \tag{27}$$

and

$$\int_{r_1}^{\infty} dr_2 4r_2 e^{-2r_2} = -\left[e^{-2r}\Big(2r+1\Big)\right]_{r_1}^{\infty}, \tag{28}$$

we get

$$\begin{aligned} E^{(1)} &= Z\Big(4\int_0^{\infty} dr_1 r_1 e^{-2r_1} - 4\int_0^{\infty} dr_1 (r_1^2 + r_1)e^{-4r_1}\Big) \\ &= Z(1! - \frac{2!}{16} - \frac{1!}{4}) = \frac{5}{8}Z. \end{aligned} \tag{29}$$

Thus, including this "first-order" correction term, the He atom ground-state energy would be (in atomic units $\mu e^4/\hbar^2$)

$$E = E^{(0)} + E^{(1)} = -Z^2 + \frac{5}{8}Z, \tag{30}$$

with $Z = 2, 3, \cdots$ for He, once ionized Li, $\cdots$. For He, therefore, we get

$$E \approx -4 + \frac{5}{4} = -2.75. \tag{31}$$

This result compares with the experimental value of -2.90351. Considering the highly approximate nature of our calculation, this is not a bad result, but clearly improvements are necessary. This need for improvement will lead us to our final perturbation technique, the variational method. Before we discuss this perturbation technique, we need to make a few remarks about n identical-particle systems, with $n > 2$.

39

n-Identical Particle States

For the two-particle system, it was easy to make two-particle wavefunctions either symmetric or antisymmetric under exchange of particle indices. Moreover, it was easy to write the full two-particle wave functions as products of two-particle orbital and two-particle spin functions. For n-particle systems, however, it is in principle straightforward to make a totally symmetric or a totally antisymmetric wave function by acting on a product of n single-particle functions, with a symmetrizer or an antisymmetrizer operator, provided the single-particle functions include *all* variables, orbital and spin variables (and perhaps other internal variables, if they exist), appropriate for the n-particle system. We will denote the symmetrizer by $\mathcal{S}$ and the antisymmetrizer by $\mathcal{A}$. For the two-particle system, we can construct symmetric and antisymmetric 2-particle functions via

$$\psi^{(s)}(\vec{r}_1, \sigma_1; \vec{r}_2, \vec{\sigma}_2) = \Big[\mathcal{S} = (1 + P_{(12)})\Big]\psi_a(\vec{r}_1, \vec{\sigma}_1)\psi_b(\vec{r}_2, \vec{\sigma}_2),$$
$$\psi^{(a)}(\vec{r}_1, \vec{\sigma}_1; \vec{r}_2, \vec{\sigma}_2) = \Big[\mathcal{A} = (1 - P_{(12)})\Big]\psi_a(\vec{r}_1, \vec{\sigma}_1)\psi_b(\vec{r}_2, \vec{\sigma}_2), \qquad (1)$$

where a and b now stand for all single-particle quantum numbers, e.g., $a \equiv n_a l_a m_{l_a} m_{s_a}$. To generalize this to n-particle systems, the symmetrizer must include, besides the identity operation, a sum over all possible permutation operators for the n-particle system. That is,

$$\mathcal{S} = \sum_P P, \qquad (2)$$

where the sum includes all $n!$ possible permutation operators, P, including the identity operation, 1. For example, for $n = 3$,

$$\mathcal{S} = (1 + P_{(12)} + P_{(13)} + P_{(23)} + P_{(123)} + P_{(132)}), \qquad (3)$$

where $P_{(ij)}$ are pair exchange operators, which exchange the indices i and j on both orbital and spin variables. The permutation operator $P_{(123)}$ designates the cyclic interchange of labels (123) in the order $1 \to 2, 2 \to 3, 3 \to 1$. This interchange could be achieved by first making the pair exchange $1 \leftrightarrow 3$ followed by the pair exchange $2 \leftrightarrow 3$; i.e., $P_{(123)} = P_{(23)}P_{(13)}$. We could just as well have expressed $P_{(123)}$ by $P_{(123)} = P_{(13)}P_{(12)}$, or in many other ways, e.g., $P_{(123)} = P_{(13)}P_{(23)}P_{(12)}P_{(13)}$. $P_{(123)}$ will, however, always involve a product of an even number of pair exchanges. In general, the $n!$ permutations of n labels involve $\frac{1}{2}n!$ even permutations, including the identity operation, and $\frac{1}{2}n!$ odd permutations. All even permutations can be expressed in terms of products of an even number of pair exchanges, and all odd permutations can be expressed in terms of products of an odd number of pair exchanges. For $n = 4$, the 24 permutation operators include the identity operation, six pair exchanges of type $P_{(ij)}$, three double pair exchanges of type $P_{(ij)}P_{(kl)}$, eight cyclic interchanges of type $P_{(ijk)}$ in which one label remains invariant, and six cyclic interchanges of all four labels of type $P_{(ijkl)}$, where the latter are odd permutations.

To build a totally symmetric n-particle state, for an n-boson system, we simply act with the symmetrizer, $\mathcal{S}$, on a product of n single particle states. If the latter are given in Dirac ket notation, e.g.,

$$|aaaaabbccc\cdots\rangle,$$

where $a \equiv n_a l_a m_{l_a} m_{s_a}$; i.e., the quantum numbers for a include all orbital and all other (internal) quantum numbers, such as the spin quantum number m_s. In this example, the particles labeled 1, 2, 3, 4, 5 are all in the same quantum state, a, whereas particles labeled 6 and 7 are in quantum state, b, and particles labeled 8, 9, 10 are in quantum state, c, and so on. The operator $\mathcal{S}$ acting on such a state does not give a normalized state vector, but it is straightforward to construct the normalized totally symmetric state vector,

$$\sqrt{\frac{(n_1!n_2!n_3!\cdots)}{n!}}\mathcal{S}|aaaaabbccc\cdots\rangle, \tag{4}$$

where we have assumed n_1 particles exist in quantum state a, n_2 particles exist in quantum state b, n_3 particles exist in quantum state c, and on on. Note, in particular, that all n bosons could be in the same quantum state. Also, in eq. (4) the symmetrizer, $\mathcal{S}$, now includes only permutations that exchange particles in different quantum states, e.g., $P_{(16)}$ but not $P_{(12)}$.

For n-fermion states, we construct the n-particle states in the same way using an n-particle antisymmetrizer, $\mathcal{A}$,

$$\mathcal{A} = \sum_P (-1)^{\sigma(P)} P, \qquad \text{with} \qquad \sigma(P) = \text{even (odd)}$$

$$\text{for } P = \text{even (odd)}. \tag{5}$$

For $n = 3$, e.g.,

$$\mathcal{A} = [1 - P_{(12)} - P_{(13)} - P_{(23)} + P_{(123)} + P_{132)}]. \tag{6}$$

Now, normalized states are constructed via

$$\frac{1}{\sqrt{n!}}\mathcal{A}|abc\cdots\rangle. \tag{7}$$

In particular, all single-particle quantum states must now be different; i.e. $a \neq b \neq c \neq \cdots$. Otherwise, the state vector would be annihilated by the antisymmetrizer $\mathcal{A}$. In coordinate representation, the totally antisymmetric state can also be expressed through an $n \times n$ determinant, the so-called Slater determinant,

$$\psi^{(a)} = \frac{1}{\sqrt{n!}}\begin{vmatrix} \psi_a(1) & \psi_b(1) & \psi_c(1) & \cdots & \psi_k(1) \\ \psi_a(2) & \psi_b(2) & \psi_c(2) & \cdots & \psi_k(2) \\ \psi_a(3) & \psi_b(3) & \psi_c(3) & \cdots & \psi_k(3) \\ \cdots & \cdots & \cdots & \cdots & \cdots \\ \psi_a(n) & \psi_b(n) & \psi_c(n) & \cdots & \psi_k(n) \end{vmatrix}, \tag{8}$$

where the particle indices, (i), are shorthand for $(\vec{r}_i, \vec{\sigma}_i)$. This n-particle wave function is totally antisymmetric. An odd permutation of particle indices corresponds to an odd permutation of rows of the determinant and therefore changes the sign of the determinant. Similarly, an even permutation of indices corresponds to an even permutation of rows of the determinant and does not change the sign of the determinant.

Even though we have succeeded in constructing n-particle wave functions of the appropriate totally symmetric or totally antisymmetric character, these functions may not be easy to work with. Later in the course (Chapters 78 and 79) we shall develop special techniques to deal with n-boson or n-fermion systems, involving single-boson or single-fermion creation and annihilation operators, with special commutation or anticommutation relations, respectively. The boson creation and annihilation operators are very similar to harmonic oscillator creation and annihilation operators. The fermion creation and annihilation operators will be particularly useful in quantum field theory, and we shall meet them there.

A final remark: For the n-electron atom, we would find it very convenient to separate the n-particle wave function into a product of an n-particle orbital function and an n-particle spin function because our Hamiltonian has only a very weak dependence on spin. For $n = 2$, this separation was trivial and led to orbitally symmetric spin singlet states and orbitally antisymmetric spin triplet states. For $n = 3$, this separation is already much more complicated. States with $S = \frac{3}{2}$ have totally symmetric three-particle spin functions. This fact is immediately apparent for the three-particle spin state with $S = \frac{3}{2}$ and $M_S = \frac{3}{2}$. It follows for the states with lower values of M_S because the three-particle M_S-lowering operator, $S_- = S_-(1)+S_-(2)+S_-(3)$, is totally symmetric, i.e., invariant under any permutation of particle indices. For these three-particle quartet spin states with $S = \frac{3}{2}$, it is trivial to combine this totally symmetric spin state with a totally antisymmetric orbital state. The three orbital quantum numbers $n_a l_a m_{l_a}$, n_b, l_b, m_{l_b}, and $n_c l_c m_{l_c}$ must differ in at least one of the three quantum numbers, in this totally antisymmetric orbital state. Next, it is impossible to make a totally antisymmetric three-particle spin state, because the single-particle spin states have only two available quantum states, $m_s = \pm\frac{1}{2}$. We would require three different single-particle spin states to make a totally antisymmetric three-particle spin function. Three-particle spin functions, with $S = \frac{1}{2}$, thus, must have a mixed intermediate symmetry, neither totally antisymmetric nor totally symmmetric. They must be combined with three-particle orbital functions also of such a mixed symmetry. Actually two types of intermediate-symmetry three-particle functions exist, and the two orbital and two spin functions must be combined in proper linear combination to make a totally antisymmetric total three-particle function. Already, for $n = 3$, this is no longer a completely trivial problem. For elegant techniques of handling such problems, we will find it advantageous to use the detailed properties of the permutation group of n objects. For a more detailed description of the possible intermediate symmetries for $n = 3$ and $n = 4$ in terms of the so-called Young tableaux, see the introductory part of Chapter 78.

40

The Variational Method

Our final perturbation technique is the variational technique, which is particularly well-suited for finding approximations to ground-state energies and wave functions. We shall apply it, in particular, to find approximations to the ground-state energy and wave functions for the He atom.

The variational method employs the following functional: Let the solutions of an energy eigenvalue problem be part of a function space, part or all of our Hilbert space, and let ψ be a particular function of that space. Then, we define the functional, a function of functions ψ to be varied, through

$$E([\psi]) = \frac{\langle\psi|H|\psi\rangle}{\langle\psi|\psi\rangle}, \tag{1}$$

where $|\psi\rangle$ is square-integrable, but not necessarily normalized to unity; i.e., $\langle\psi|\psi\rangle$ is finite (nonzero), but not necessarily normalized to unity. Also $E([\psi])$ gives the average energy of the system in the state $|\psi\rangle$. The variational method is based on the following theorem.

Theorem: Any state vector, $|\psi\rangle$, for which the average energy, considered as a functional of the vectors of the full state vector space, is stationary is an eigenvector of the discrete spectrum of H; and the corresponding energy eigenvalue is the stationary value of $E([\psi])$.

If we could actually carry out a variation of ψ, starting with some assumed value of ψ and varying ψ by small steps by rotating ψ by small amounts in the infinite-dimensional Hilbert space, such that we could end by finding the true minimum of the functional, we would attain the ground state eigenvector and eigenvalue exactly. Similarly, a local minimum or maximum would give us excited-state eigenvectors and eigenvalues. In actual practice, of course, we cannot do this even with modern computers. The way the method is actually used is to restict the variation of ψ, not to the set of all functions of the full space, but to a judiciously chosen subset of functions, perhaps a finite number of eigenfunctions of a closely related exactly soluble problem, which we guess to be the most important subspace of our full space. Alternatively, we might restrict the functions ψ to those of a specific analytical form that we guess to be good candidates. (Our physical intuition comes into play here!) These analytical functions should contain a few undetermined (variation) parameters, which are fixed at the values making the variation $\delta E = 0$. If we denote these functions, Φ, and vary $E([\Phi])$ such that $\delta E([\Phi]) = 0$, we may get a good approximation to the exact energy if $E([\Phi_0])$ and $E([\psi_0)]$ differ by only a small amount.

A Proof of the Variational Theorem

To prove the variational principle theorem, let us put $E([\psi])$ in the form

$$E([\psi])\langle\psi|\psi\rangle = \langle\psi|H|\psi\rangle \tag{2}$$

and make the variation to give

$$\delta E\langle\psi|\psi\rangle + E\langle\delta\psi|\psi\rangle + E\langle\psi|\delta\psi\rangle = \langle\delta\psi|H|\psi\rangle + \langle\psi|H|\delta\psi\rangle, \tag{3}$$

or

$$\langle\psi|\psi\rangle\delta E = \langle\delta\psi|(H-E)|\psi\rangle + \langle\psi|(H-E)|\delta\psi\rangle. \tag{4}$$

Hence, we see: If $|\psi\rangle$ is an exact solution of $(H-E)|\psi\rangle = 0$, and if $\langle\psi|\psi\rangle$ is finite and nonzero, $\delta E = 0$.

Also, If $\delta E = 0$,

$$\langle\delta\psi|(H-E)|\psi\rangle + \langle\psi|(H-E)|\delta\psi\rangle = 0. \tag{5}$$

Now, the variation of $|\psi\rangle$ and the variation of $\langle\psi|$ are not independent, but because our vector space is a complex vector space and the variation of $|\psi\rangle$ is completely arbitrary, we could vary the pure real and pure imaginary parts of $|\psi\rangle$ separately and independently of each other. Equivalently, we could vary first $|\psi\rangle$ to give eq. (5), and then vary $i|\psi\rangle$ to give

$$-i\langle\delta\psi|(H-E)|\psi\rangle + i\langle\psi|(H-E)|\delta\psi\rangle = 0. \tag{6}$$

Eqs. (5) and (6), together, then yield: If $\delta E = 0$,

$$\begin{aligned}\langle\delta\psi|(H-E)|\psi\rangle &= 0\\ \langle\psi|(H-E)|\delta\psi\rangle &= 0\end{aligned} \tag{7}$$

separately, leading to the conclusion $|\psi\rangle$ is an exact eigenvector of H with eigenvalue E.

B Bounds on the Accuracy of the Variational Method

Because, in actual practice, we cannot vary the $|\psi\rangle$ of the full vector space, but will vary instead the $|\Phi\rangle$ of a highly restricted subspace of our vector space, it will be very useful to try to get a bound on the variational values of the energy obtained by this technique. We will show, in the subspace of the $|\Phi\rangle$, the variation of the $|\Phi\rangle$, which gives $\delta E([\Phi]) = 0$, leads to an $E([\Phi])$ such that

$$E([\Phi]) \geq E_0, \tag{8}$$

where E_0 is the exact value of the ground-state energy. To prove this, calculate

$$E([\Phi]) - E_0 = \frac{\langle\Phi|(H - E_0)|\Phi\rangle}{\langle\Phi|\Phi\rangle} = \sum_{\text{all } n} \frac{\langle\Phi|(H - E_0)|n\rangle\langle n|\Phi\rangle}{\langle\Phi|\Phi\rangle}, \tag{9}$$

where we have merely inserted the unit operator, $\sum_{\text{all } n}|n\rangle\langle n|$, into the equation and $|n\rangle$ is the exact eigenvector of H with eigenvalue E_n. The sum in this equation may include an integral over continuous energy values if the spectrum of eigenvalues E contains both a discrete part and a continuum. From the above, we get

$$E([\Phi]) - E_0 = \sum_n (E_n - E_0)\frac{\langle\Phi|n\rangle\langle n|\Phi\rangle}{\langle\Phi|\Phi\rangle} = \sum_n (E_n - E_0)\frac{|\langle n|\Phi\rangle|^2}{\langle\Phi|\Phi\rangle} \geq 0. \tag{10}$$

Thus, we know the sign of the error made in the variational technique, although the magnitude of the error is difficult to estimate, which is one of the drawbacks of the technique. However, we can proceed as follows: If we have found an approximate $E([\Phi_1])$ by varying the parameters in the set of functions Φ_1 to obtain an $E([\Phi_1])$, and if we subsequently take a more sophisticated (and hopefully better) set of trial functions, Φ_2, perhaps with a larger number of variational parameters, then if $E([\Phi_2]) < E([\Phi_1])$, we know $E([\Phi_2])$ is a better approximation to the ground-state energy. In principle, this process could be continued to further improve our approximation. The absolute value of the error in the final approximation, however, may still be quite uncertain.

C An Example: The Ground-State Energy of the He Atom

The first step of the calculation involves a good choice of functions, Φ. Here, the ingenuity of the calculator comes to the fore. We want a family of Φ, simple enough for ease of calculation, yet complicated enough to get an accurate result. In our perturbation calculation for the He atom, our zeroth-order wave function, a simple product of hydrogenic $1s$ wave functions, took no account of the presence of the "second" electron. This electron partially shields the nucleus with charge Z from the "first" electron, so this one effectively sees a charge λZ rather than the full charge Z. Here, the shielding factor λ could be introduced as a very simple variational parameter. Thus, the zeroth-order single-particle wave functions,

$\psi^{(0)}_{n=1,l=0}$ of perturbation theory could be replaced by single-particle functions

$$\phi(\lambda) = R(r_i)Y_{00}(\theta_i, \phi_i) = \left(\frac{\lambda Z}{a_0}\right)^{\frac{3}{2}} 2e^{-\lambda Z r_{\text{phys.}i}/a_0} \frac{1}{\sqrt{4\pi}}. \quad (11)$$

If we introduce, the dimensionless r_i, as in Chapter 38, viz. $r_{\text{phys.}i} = \frac{a_0}{Z} r_i$, in terms of these dimensionless r_i, we will build the variational trial function from single-particle functions

$$\phi(\lambda) = \lambda^{\frac{3}{2}} 2e^{-\lambda r_i} \frac{1}{\sqrt{4\pi}}, \quad (12)$$

so the two-particle variational functions are

$$\Phi(\lambda) = 4\lambda^3 e^{-\lambda r_1 - \lambda r_2} \frac{1}{4\pi}. \quad (13)$$

We have chosen the $\Phi(\lambda)$ such that the normalization is $\langle \Phi(\lambda)|\Phi(\lambda)\rangle = 1$ (because it was extremely easy to do so!). Now, using the Hamiltonian, $H^{(0)} + H^{(1)}$ in the form of eq. (14) of Chapter 38, we have

$$\langle \Phi(\lambda)|H|\Phi(\lambda)\rangle$$

$$= 2Z^2 \langle \phi(\lambda)|T_{s.p.}|\phi(\lambda)\rangle + 2Z^2 \langle \phi(\lambda)|V_{s.p.}|\phi(\lambda)\rangle + Z\langle \Phi(\lambda)|\frac{1}{r_{12}}|\Phi(\lambda)\rangle, \quad (14)$$

where the single-particle expectation values for the single-particle kinetic energy operator, $T_{s.p.}$, and the single-particle potential energy operator, $V_{s.p.}$, follow from

$$\begin{aligned} \langle \phi(\lambda)|T_{s.p.}|\phi(\lambda)\rangle &= -\frac{1}{2}\langle \phi(\lambda)|\frac{\partial^2}{\partial r^2} + \frac{2}{r}\frac{\partial}{\partial r}|\phi(\lambda)\rangle = \frac{\lambda^2}{2(1)^2}, \\ \langle \phi(\lambda)|V_{s.p.}|\phi(\lambda)\rangle &= -\langle \phi(\lambda)|\frac{1}{r}|\phi(\lambda)\rangle = -\lambda\frac{1}{(1)^2}, \end{aligned} \quad (15)$$

where we have simply made the substitutions, $r = r'/(\lambda)$ in the operators in these equations and have made the transformation $\lambda r = r'$ in the single-particle functions, $\phi(\lambda)$, to relate the integrals to the standard hydrogenic expectation value results, $\langle T_{s.p.}\rangle = +\frac{1}{2n^2}$, and $\langle \frac{1}{r}\rangle = \frac{1}{n^2}$, with $n = 1$. In the same way, we get

$$Z\langle \Phi(\lambda)|\frac{1}{r_{12}}|\Phi(\lambda)\rangle = \frac{5}{8} Z\lambda, \quad (16)$$

where we have again made the substitution $\lambda r_{12} = r'_{12}$ to convert the integral to the form evaluated through eq. (30) of Chapter 38. Thus,

$$E([\Phi(\lambda)]) = \langle \Phi(\lambda)|H|\Phi(\lambda)\rangle = \lambda^2 Z^2 - 2\lambda Z^2 + \frac{5}{8}\lambda Z. \quad (17)$$

With

$$\frac{\partial E(\lambda)}{\partial \lambda} = 2\lambda Z^2 - 2Z^2 + \frac{5}{8} Z = 0, \quad (18)$$

we get

$$\lambda = 1 - \frac{5}{16Z}, \tag{19}$$

or for He, with $Z = 2$, $\lambda Z = Z - \frac{5}{16} = 1.6875$, which is a reasonable value for the shielded nuclear charge. With this value of λ, the variational approximation, $E([\Phi(\lambda)])$, is

$$E([\Phi(\lambda)]) = -Z^2 + \frac{5}{8}Z - \frac{25}{256} = -\Big(Z - \frac{5}{16}\Big)^2. \tag{20}$$

For, He, with $Z = 2$, this gives $E = -2.8477$. Recall the experimental value was $E = -2.90351$. Considering the extreme simplicity of our trial function, with a single variational parameter, this is a marked improvement over the first-order perturbation theory result of Chapter 38.

To make an improvement over the simple one-parameter variational wave function used here, Hylleraas in (1928) used a six-parameter variational function of the type

$$\Phi = e^{-\lambda(r_1+r_2)}(1 + c_1 u + c_2 t^2 + c_3 s + c_4 s^2 + c_5 u^2)\frac{1}{4\pi}, \tag{21}$$

with $u = r_{12}, s = (r_1 + r_2), t = (r_1 - r_2)$, defined in terms of the dimensionless coordinates, r_i, Φ being a function of the six variational parameters, $\lambda, c_1, c_2, c_3, c_4, c_5$. These trial Φ are now not normalized. Variation of the six parameters in these Φ led to an $E([\Phi]) = -2.90362$. This result violates our theorem, $E([\phi]) \geq E_0(\text{exact})$, $E([\Phi])_{\text{Hylleraas}} < E_{\text{exp.}}$. Recall $E_{\text{exp.}} = -2.90351$. Hylleraas did not, however, include the spin-dependent fine structure terms $H_{\text{f.s.}}$ in his calculation. These terms are of order $\alpha^2 = (1/137)^2$ times the dominant terms in H and would have to be included if we want to compare with experimental results to the order of the 4th decimal place.

D The Ritz Variational Method

So far, we have illustrated the variational method with the use of trial functions built in terms of a few physically motivated parameters. In Ritz's use of the variational technique, the trial function is built in terms of an expansion in a finite number of zeroth-order eigenfunctions of a simpler related problem, with known eigenvalues and eigenfunctions.

$$|\Phi\rangle = \sum_{n=0}^{N} |n\rangle\langle n|\Phi\rangle = \sum_{n=0}^{N} c_n |n\rangle, \tag{22}$$

where the $N + 1$ parameters, c_n are now to be considered as the variational parameters, and it is now easy to choose these such that $\sum_n |c_n|^2 = 1$ and therefore $\langle\Phi|\Phi\rangle = 1$. Now,

$$\delta(\langle\Phi|H|\Phi\rangle) = 0, \tag{23}$$

subject to the subsidiary condition

$$\langle\Phi|\Phi\rangle = 1 = \sum_n |c_n|^2, \tag{24}$$

yields

$$\delta\Big(\sum_{n,m}^{N}\langle\Phi|m\rangle\langle m|H|n\rangle\langle n|\Phi\rangle\Big) = \delta\Big(\sum_{n,m}^{N} c_m^* c_n \langle m|H|n\rangle\Big) = 0, \tag{25}$$

subject to the subsidiary condition, arising from the normalization constraint

$$\delta\Big(\sum_{n,m}^{N} c_m^* \delta_{mn} c_n\Big) = 0. \tag{26}$$

Multiplying this second constraint equation by the Lagrange multiplier, named suggestively $-E$, and adding this to the first (variational) equation, we get

$$\begin{aligned}\sum_{n,m}^{N}&\Big[\delta c_m^*(\langle m|H|n\rangle - E\delta_{mn})c_n + c_m^*(\langle m|H|n\rangle - E\delta_{nm})\delta c_n\Big] = 0\\ = \sum_{n,m}^{N}&\Big[\delta c_m^*(\langle m|H|n\rangle - E\delta_{nm})c_n + \delta c_m\langle m|H|n\rangle^* - E\delta_{mn})c_n^*\Big] = 0,\end{aligned} \tag{27}$$

where we have used the hermitian character of the matrix elements of H and have renamed $n \leftrightarrow m$ in the second term of this equation. Now, the c_n are complex numbers, so we can vary separately the real and imaginary parts of c_n and combine these variations such that, separately, the real and imaginary parts of the above equation can be set equal to zero. This process leads to

$$\begin{aligned}\sum_{n,m}^{N}\delta c_m^*(\langle m|H|n\rangle - E\delta_{nm})c_n &= 0,\\ \sum_{n,m}^{N}\delta c_m(\langle m|H|n\rangle^* - E\delta_{nm})c_n^* &= 0.\end{aligned} \tag{28}$$

Because we can vary the individual c_n separately, we can set all $\delta c_k = 0$, except for one particular δc_m, leading to

$$\sum_{n}^{N}(\langle m|H|n\rangle - E\delta_{nm})c_n = 0. \tag{29}$$

This relation is the usual eigenvalue–eigenvector equation for a finite-dimensional basis. The eigenvalues E_i, with eigenvectors given by the $c_n^{(i)}$ with $i = 0, 1, \ldots, N$ are the variational approximate eigenvalues and eigenvectors. If our original basis was a "good guess," we might expect the lowest few eigenvalues to be good approximations. Improved approximations might then be obtained by expanding the basis from an N-dimensional one to one of slightly higher dimensionality, and if our original guess was indeed a good one, this process should converge to the exact eigenvalues.

Part V

Scattering Theory

41

Introduction to Scattering Theory

A Potential Scattering

Until now, most of our applications have involved bound states of a quantum system. Now, in starting scattering theory, we shall be dealing with states in the continuum. We shall start by considering the scattering of a structureless point particle of mass, m, and incident momentum, $\vec{p}$, from a fixed scattering center via an interaction describable by a potential, $V(\vec{r})$, or the equivalent one-body problem for the relative motion of two structureless point particles of masses, m_1 and m_2, and reduced mass, $m_1 m_2/(m_1 + m_2)$, which again interact via a potential, $V(\vec{r})$, where $\vec{r}$ is now the relative motion vector, $\vec{r} = \vec{r}_1 - \vec{r}_2$.

Before starting, let us make some remarks about possible generalizations as follows.

1. The particles may have spin, and a rearrangement of spin alignments may occur during the scattering process, if the interaction between the particles include spin-dependent forces.

2. The particles may not be treatable as point particles. They may be composites, made of several constituent particles. During the collision process, excitations of internal degrees of freedom of the particles may then occur (inelastic collisions), or rearrangement collisions, where constituent particles from projectile may be transferred to target or vice versa may occur.

3. Shortcomings may exist of the simple potential description of the scattering process. This difficulty may lead to nonlocal potentials. Also, in scattering of particles at relativistic energies, a simple potential description may not be possible. We shall restrict ourselves to nonrelativistic energies.

The scattering process is described by the *Scattering Cross Section.*

We assume we have an incoming beam of particles of definite fixed $\vec{v}$, or momentum $\vec{p}$, approximated by a plane wave, with specified incoming flux, that is, a fixed number of incoming particles per cm^2 per second, (initial condition). We want to calculate the probability particles are scattered into an element of solid angle, $d\Omega$, about some direction θ, ϕ, relative to the initial direction fixed by $\vec{p}$. (See Fig. 41.1.) It is assumed the detector detecting the scattered particles is at a distance r from the scattering center, where r is large compared with the size of the scattering quantum system. The scattering cross section is defined by

$$d\sigma = \frac{\text{number of particles scattered/sec into } d\Omega \text{ at } \theta, \phi}{\text{number of particles/cm}^2\text{/sec in incoming beam}} = I(\theta, \phi) d\Omega, \tag{1}$$

where

$$\frac{d\sigma}{d\Omega} = \text{differential scattering cross section},$$

and

$$\sigma = \int_{\theta=0}^{\pi} \int_{\phi=0}^{2\pi} d\Omega \left(\frac{d\sigma}{d\Omega} \right) = \text{total scattering cross section}.$$

For the relative motion of the two-body problem, we need to discuss the transformation from the laboratory coordinates, $\vec{r}_1$ and $\vec{r}_2$, to the center of mass coordinates. If we sit on the center of mass, the scattering process is describable by the relative motion vector, $\vec{r}_{\text{rel.}} \equiv \vec{r}$, where

$$\vec{r} = \vec{r}_1 - \vec{r}_2, \qquad \vec{R}_{\text{c.m.}} = \frac{(m_1 \vec{r}_1 + m_2 \vec{r}_2)}{(m_1 + m_2)}, \tag{2}$$

with

$$\frac{\vec{p}}{\mu} = \dot{\vec{r}}_1 - \dot{\vec{r}}_2 = \vec{v}_1 - \vec{v}_2 = \frac{\vec{p}_1}{m_1} - \frac{\vec{p}_2}{m_2}, \qquad \vec{P}_{\text{c.m.}} = \vec{p}_1 + \vec{p}_2. \tag{3}$$

In the laboratory, particle 1, the projectile, will have an initial momentum, $(m_1 \vec{v}_1)_0$, whereas particle 2, the target particle, will have an initial momentum of zero. After the scattering process, particle 1 will have a final momentum, $m_1 \vec{v}_1$, making an angle $\theta_{\text{lab.}}$ with the initial momentum direction, whereas particle 2 will recoil with momentum $m_2 \vec{v}_2$ at an angle θ_2 with respect to the initial momentum direction. (See Fig. 41.2.) We shall denote momentum and position vectors in the center of mass system with a prime. In the center of mass system, the initial momentum of particle 2 must be $(\vec{p}_2)'_0 = -(\vec{p}_1)'_0$. Similarly for the final momenta, $\vec{p}'_2 = -\vec{p}'_1$. For true point particles, the scattering process is elastic, leading to $|\vec{p}'_1| = |(\vec{p}_1)'_0|$. The final momentum vector in the center of mass system has the same magnitude as the initial momentum vector, but makes an angle θ relative to the initial momentum direction. The transformation

$$\vec{r}_1 = \vec{r}'_1 + \vec{R}_{\text{c.m.}} \tag{4}$$

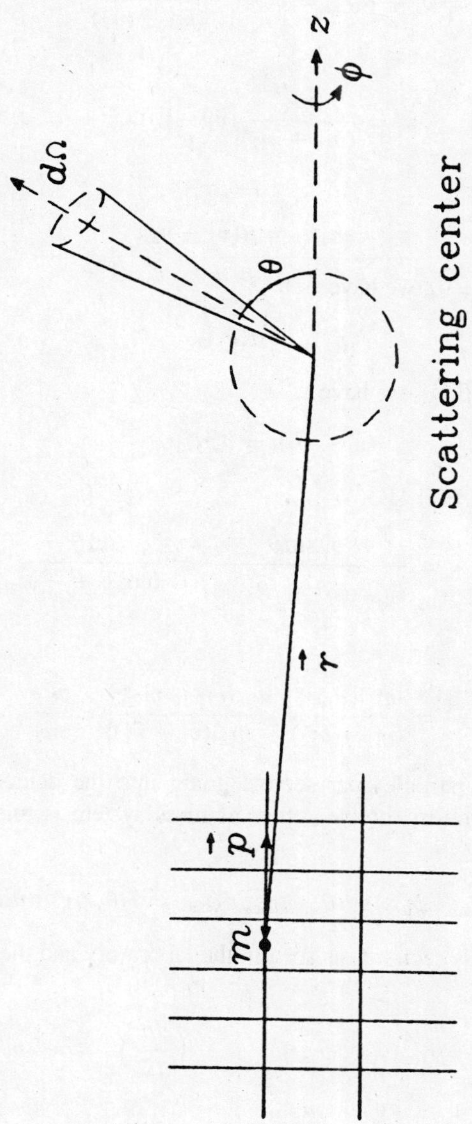

FIGURE 41.1.

leads to

$$\vec{v}_1 = \vec{v}_1' + \dot{\vec{R}}_{\text{c.m.}} = \vec{v}_1' + \frac{(m_1\vec{v}_1 + m_2\vec{v}_2)}{(m_1 + m_2)}, \tag{5}$$

and

$$\vec{v}_1' = \frac{m_2}{(m_1 + m_2)}(\vec{v}_1 - \vec{v}_2), \tag{6}$$

or

$$\vec{p}_1' = m_1\vec{v}_1' = \mu(\vec{v}_1 - \vec{v}_2). \tag{7}$$

Similarly, using $(\vec{v}_2)_0 = 0$, we have

$$(\vec{p}_1)_0' = \mu(\vec{v}_1)_0, \tag{8}$$

and because $|\vec{p}_1'| = |(\vec{p}_1)_0'|$, we have

$$|(\vec{v}_1 - \vec{v}_2)| = |(\vec{v}_1)_0|. \tag{9}$$

Now, from eq. (5) and Fig. 41.3,

$$\tan\theta_{\text{lab.}} = \frac{v_1' \sin\theta}{(v_1' \cos\theta + |\dot{\vec{R}}_{\text{c.m.}}|)} = \frac{\sin\theta}{(\cos\theta + \frac{m_1}{m_2})}, \tag{10}$$

where we have used

$$\frac{|\dot{\vec{R}}_{\text{c.m.}}|}{v_1'} = \frac{m_1|(\vec{v}_1)_0|}{(m_1 + m_2)} \times \frac{(m_1 + m_2)}{m_2|(\vec{v}_1 - \vec{v}_2)|} = \frac{m_1}{m_2}. \tag{11}$$

Now, the number of particles per second going into the detector, as seen by observers in the laboratory, or the center of mass system is one and the same number

$$d\sigma = I_{\text{lab.}}(\theta_{\text{lab.}}, \phi_{\text{lab.}}) \sin\theta_{\text{lab.}} d\theta_{\text{lab.}} d\phi_{\text{lab.}} = I(\theta, \phi) \sin\theta d\theta d\phi, \tag{12}$$

so the differential cross sections, as seen in the laboratory and the center of mass frames

$$\left(\frac{d\sigma}{d\Omega}\right)_{\text{lab.}} = I_{\text{lab.}}(\theta_{\text{lab.}}, \phi_{\text{lab.}}), \qquad \left(\frac{d\sigma}{d\Omega}\right)_{\text{c.m.}} = I(\theta, \phi), \tag{13}$$

are related by

$$\left(\frac{d\sigma}{d\Omega}\right)_{\text{lab.}} = \left(\frac{d\sigma}{d\Omega}\right)_{\text{c.m.}} \frac{\sin\theta d\theta}{\sin\theta_{\text{lab.}} d\theta_{\text{lab.}}} = \left(\frac{d\sigma}{d\Omega}\right)_{\text{c.m.}} \frac{d(\cos\theta)}{d(\cos\theta_{\text{lab.}})}. \tag{14}$$

With

$$\cos\theta_{\text{lab.}} = \frac{(\cos\theta + \frac{m_1}{m_2})}{\sqrt{(1 + \frac{m_1^2}{m_2^2} + 2\frac{m_1}{m_2}\cos\theta)}}, \tag{15}$$

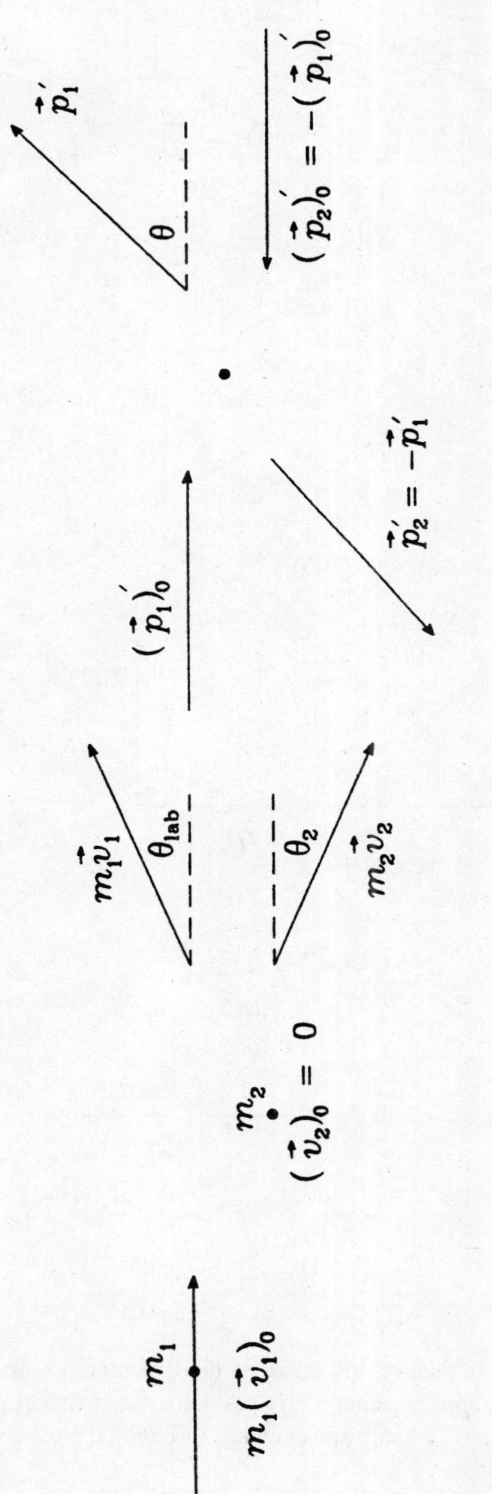

FIGURE 41.2.

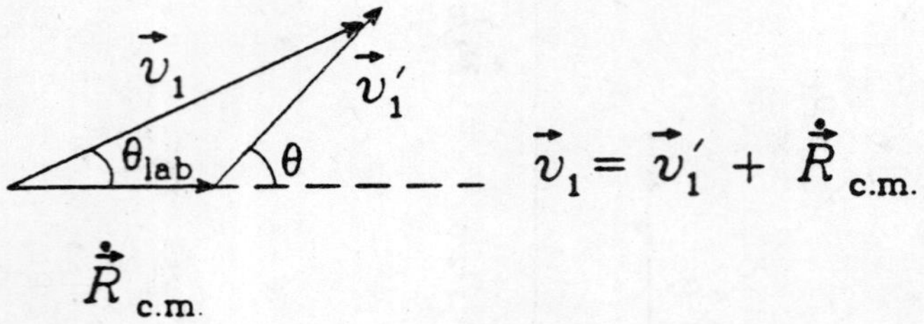

$\vec{v}_1$ $\vec{v}_1'$ θ_{lab} θ

$\dot{\vec{R}}_{\text{c.m.}}$

$$\vec{v}_1 = \vec{v}_1' + \dot{\vec{R}}_{\text{c.m.}}$$

FIGURE 41.3.

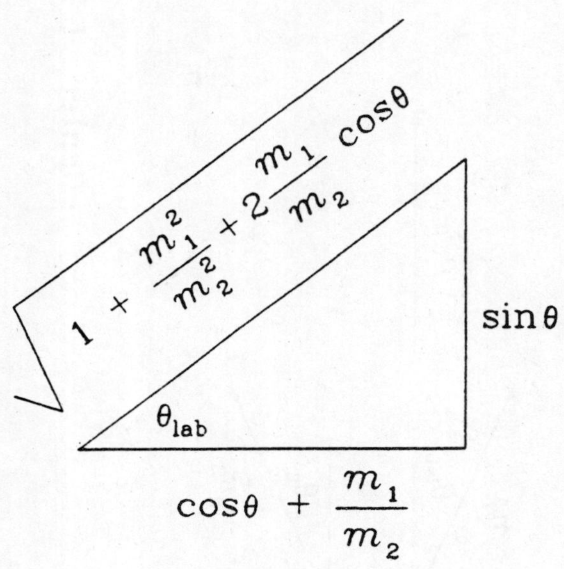

FIGURE 41.4.

(see Fig. 41.4), we get

$$d(\cos\theta_{\text{lab.}}) = d(\cos\theta) \times \frac{(1 + \frac{m_1}{m_2}\cos\theta)}{(1 + \frac{m_1^2}{m_2^2} + 2\frac{m_1}{m_2}\cos\theta)^{\frac{3}{2}}}. \tag{16}$$

This equation leads to

$$\left(\frac{d\sigma}{d\Omega}\right)_{\text{lab.}} = \left(\frac{d\sigma}{d\Omega}\right)_{\text{c.m.}} \frac{(1 + \frac{m_1^2}{m_2^2} + 2\frac{m_1}{m_2}\cos\theta)^{\frac{3}{2}}}{(1 + \frac{m_1}{m_2}\cos\theta)}. \tag{17}$$

Having established this relationship between the differential scattering cross sections in the laboratory and the center of mass systems, we shall henceforth work entirely in the center of mass system, and calculate differential cross sections only in the center of mass system.

To solve the scattering problem, we need to find solutions to the Schrödinger equation

$$\nabla^2\psi + \frac{2\mu}{\hbar^2}(E - V(\vec{r}))\psi = 0, \tag{18}$$

where ψ is a function of the relative coordinate vector $\vec{r}$ for the two-body system. Moreover, the solutions ψ we seek must satisfy the boundary conditions of the physical situation: Asymptotically, as $r \to \infty$, ψ must be of the form, $\psi =$ incoming plane wave + spherical outgoing wave starting from the scattering center; that is, as $r \to \infty$,

$$\psi = e^{i\vec{k}\cdot\vec{r}} + f(\theta, \phi)\frac{e^{ikr}}{r}, \tag{19}$$

where $\vec{k} = \vec{p}/\hbar$. The function, $f(\theta, \phi)$, is known as the scattering amplitude. The flux for the incoming wave, $\psi_{\text{inc.}} = e^{i\vec{k}\cdot\vec{r}}$, is

$$\frac{\hbar}{2\mu i}(\psi^*_{\text{inc.}}\vec{\nabla}\psi_{\text{inc.}} - \psi_{\text{inc.}}\vec{\nabla}\psi^*_{\text{inc.}}) = \frac{\hbar}{2\mu i}(i\vec{k} - (-i\vec{k})) = \frac{\hbar\vec{k}}{\mu} = \frac{\vec{p}}{\mu} = \vec{v}. \tag{20}$$

Similarly, the flux of the outgoing scattered wave, normal to the surface of the detector, at a distance, $r \to \infty$, is

$$\frac{\hbar}{2\mu i}(\psi^*_{\text{out}}\frac{\partial\psi_{\text{out}}}{\partial r} - \psi_{\text{out}}\frac{\partial\psi^*_{\text{out}}}{\partial r}) = \frac{\hbar k}{\mu}\frac{|f(\theta, \phi)|^2}{r^2} + \text{Order}\left(\frac{1}{r^3}\right), \tag{21}$$

leading to

$$d\sigma = \frac{\left(\frac{\hbar k}{\mu} \times \frac{|f(\theta,\phi)|^2}{r^2} \times r^2 d\Omega\right)}{\frac{\hbar k}{\mu}}, \tag{22}$$

so

$$\frac{d\sigma}{d\Omega} = |f(\theta, \phi)|^2. \tag{23}$$

A note about the normalization of our ψ is perhaps in order. Above we have normalized the incoming plane wave such that the incoming flux had magnitude, $\hbar k/\mu = v \equiv v_{\text{rel.}}$. This normalization factor, however, drops out in the expression for the cross section. A second possibility might have been to normalize the incoming plane wave to unit flux, in which case we could have used

$$\psi_{\text{inc.}} = \frac{e^{i\vec{k}\cdot\vec{r}}}{\sqrt{v}}. \tag{24}$$

Since, $\vec{k}$, and hence $\vec{p}$, has a continuous spectrum, we could also have normalized the incoming plane wave according to the standard coordinate space representation

$$u_{\vec{k}}(\vec{r}) \equiv \langle\vec{r}|\vec{k}\rangle = \frac{e^{i\vec{k}\cdot\vec{r}}}{(2\pi)^{\frac{3}{2}}}, \tag{25}$$

so these plane wave states have the standard Dirac delta function orthonormality property, $\langle\vec{k}'|\vec{k}\rangle = \delta(\vec{k}' - \vec{k})$. Alternatively, we could have used the coordinate space function

$$u_{\vec{p}}(\vec{r}) \equiv \langle\vec{r}|\vec{p}\rangle = \frac{e^{i\vec{p}\cdot\vec{r}/\hbar}}{(2\pi\hbar)^{\frac{3}{2}}}, \tag{26}$$

so $\langle\vec{p}'|\vec{p}\rangle = \delta(\vec{p}' - \vec{p})$.

Finally, we could have used a box normalization for our plane wave by imagining that all of our quantum systems are located in a cubical "laboratory" where the cube has a side of length L, hence, volume, L^3; where L is a macroscopic quantity; and where all wave functions must be zero at the surface of the cube. This means that we will replace the continuous spectrum for $\vec{k}$ with a very finely spaced discrete spectrum, with

$$k_x = \frac{2\pi}{L}n_1, \qquad k_y = \frac{2\pi}{L}n_2, \qquad k_z = \frac{2\pi}{L}n_3, \tag{27}$$

where the n_i are integers (positive or negative). [This condition means the wavelengths of our discrete set of states are such that our cube of side L must contain an integral (rather than a $\frac{1}{2}$-integral) number of wavelengths, so we are using so-called periodic boundary conditions.] With this type of plane wave state, our plane wave functions are normalized according to

$$u_{\vec{n}}(\vec{r}) = \frac{e^{i\vec{k}_{\vec{n}}\cdot\vec{r}}}{\sqrt{L^3}} = \frac{e^{i\vec{k}_{\vec{n}}\cdot\vec{r}}}{\sqrt{\text{Vol.}}}, \tag{28}$$

where the orthonormality integral is now given by

$$\langle\vec{k}_{\vec{n}'}|\vec{k}_{\vec{n}}\rangle = \delta_{n_1 n_1'}\delta_{n_2 n_2'}\delta_{n_3 n_3'} \tag{29}$$

in terms of ordinary Kronecker deltas. Because no quantum-mechanical result for a system of atomic or subatomic size can depend on the size or shape of the macroscopic "laboratory," the Volume $= L^3$, must drop out of all meaningful final results that must be independent of L. This form of the plane wave state, therefore, has its uses, and we shall occasionally use it as a pedagogic device. For the moment, however, we shall stick with the normalization of eq. (19).

To solve the relative motion problem and find solutions of the form of eq. (19), it will be useful to expand both the incoming plane wave and the scattering amplitude, $f(\theta, \phi)$, in spherical harmonics. Choosing our z axis along the direction of the incoming $\vec{k}$,

$$e^{i\vec{k}\cdot\vec{r}} = e^{ikr\cos\theta} = \sum_l c_l R_l(r) P_l(\cos\theta), \tag{30}$$

where it will be useful to write the radial functions in terms of the dimensionless variable kr and convert to the one-dimensionalized radial functions, $u_l(r)$, with $R_l(kr) = u_l(kr)/kr$. The radial equation for the free waves is then given by

$$\frac{d^2u_l}{dr^2} + (k^2 - \frac{l(l+1)}{r^2})u_l(r) = 0, \qquad \text{with} \qquad k^2 = \frac{2\mu E}{\hbar^2}, \tag{31}$$

or, with $kr = \rho$,

$$\frac{d^2u_l}{d\rho^2} + (1 - \frac{l(l+1)}{\rho^2})u_l(\rho) = 0. \tag{32}$$

To recognize this equation, it will be useful to make the transformation, $u_l(\rho) = \sqrt{\rho}v_l(\rho)$ (although we will not actually use the v_l to find the general solutions). The differential equation in the $v_l(\rho)$ becomes

$$\rho^2\frac{d^2v_l}{d\rho^2} + \rho\frac{dv_l}{d\rho} + [\rho^2 - (l+\tfrac{1}{2})^2]v_l(\rho) = 0. \tag{33}$$

This is recognized as the Bessel equation with index $n = \pm(l+\frac{1}{2})$. The solutions for $R_l(kr) = u_l(\rho)/\rho$ are then (with an additional normalization factor, $\sqrt{\frac{\pi}{2}}$),

$$\frac{u_l(\rho)}{\rho} = \sqrt{\frac{\pi}{2\rho}}J_{(l+\frac{1}{2})}(\rho) \equiv j_l(\rho), \tag{34}$$

and

$$\frac{u_{-(l+1)}(\rho)}{\rho} = \sqrt{\frac{\pi}{2\rho}}J_{-(l+\frac{1}{2})}(\rho) \equiv n_l(\rho)(-1)^{l+1}, \tag{35}$$

where we have made use of the invariance of the quantity $l(l+1)$ under the transformation, $l \to -(l+1)$, hence, $(l+\frac{1}{2}) \to -(l+\frac{1}{2})$. In the above equations, the j_l and n_l are standard spherical Bessel functions and spherical Neumann functions, respectively. [The phase factor $(-1)^{l+1}$, as well as the normalization factor $\sqrt{\frac{\pi}{2}}$, are needed to get the "standard" functions.]

B Spherical Bessel Functions

To get the needed properties of the spherical Bessel and Neumann functions, we shall use the equation for the $u_l(\rho)$ in the form

$$\Big(-\frac{d^2}{d\rho^2} + \frac{l(l+1)}{\rho^2}\Big)u_l(\rho) = 1u_l(\rho) \tag{36}$$

and use the factorization method of Chapter 7 to study the solutions. The factors are

$$O_\pm(l) = \Big(\mp\frac{d}{d\rho} + \frac{l}{\rho}\Big), \tag{37}$$

so

$$O_+(l)O_-(l)u_l = \Big(-\frac{d^2}{d\rho^2} + \frac{l(l+1)}{\rho^2}\Big)u_l = [\lambda - \mathcal{L}(l)]u_l,$$
$$O_-(l+1)O_+(l+1)u_l = \Big(-\frac{d^2}{d\rho^2} + \frac{l(l+1)}{\rho^2}\Big)u_l = [\lambda - \mathcal{L}(l+1)]u_l. \tag{38}$$

Comparing with eq. (36), we see that $\lambda = 1$ and $\mathcal{L}(l) = 0$. Hence, the possible l values can range from $+\infty$ to $-\infty$. Of course, the possible l values are further restricted to be integers by the angular functions of eq. (30). The negative integers, which yield the functions $u_{-(l+1)}$, and hence the spherical Neumann functions, are acceptable through the invariance of the quantity $l(l+1)$ under the transformation $l \to -(l+1)$. To get the starting function, for the l step processes, we look at the equation for $l = 0$

$$-\frac{d^2u_0}{d\rho^2} = u_0, \tag{39}$$

so u_0 could be either $\sin\rho$ or $\cos\rho$. Note, however, that u_0 is identified with j_0 through $j_0 = u_0/\rho$, and since the $j_l(\rho)$ are the solutions regular at the origin, $\rho = 0$, we must choose $u_0 = \sin\rho$; hence, $j_0 = \sin\rho/\rho$. From this, we get

$$\begin{aligned} j_1 &= \frac{1}{\rho}O_+(1)u_0 = \frac{1}{\rho}\left(-\frac{d}{d\rho} + \frac{1}{\rho}\right)\sin\rho = -\frac{\cos\rho}{\rho} + \frac{\sin\rho}{\rho^2}, \\ -n_0 &= \frac{1}{\rho}O_-(0)u_0 = \frac{1}{\rho}\frac{d}{d\rho}\sin\rho = \frac{\cos\rho}{\rho}. \end{aligned} \tag{40}$$

Continuing this process, we can get the j_l and n_l for higher l. We list a few:

$$\begin{aligned} j_0 &= \frac{\sin\rho}{\rho}, \\ j_1 &= -\frac{\cos\rho}{\rho} + \frac{\sin\rho}{\rho^2}, \\ j_2 &= -\frac{\sin\rho}{\rho} - 3\frac{\cos\rho}{\rho^2} + 3\frac{\sin\rho}{\rho^3}, \end{aligned} \tag{41}$$

and

$$\begin{aligned} n_0 &= -\frac{\cos\rho}{\rho}, \\ n_1 &= -\frac{\sin\rho}{\rho} - \frac{\cos\rho}{\rho^2}, \\ n_2 &= \frac{\cos\rho}{\rho} - 3\frac{\sin\rho}{\rho^2} - 3\frac{\cos\rho}{\rho^3}. \end{aligned} \tag{42}$$

The j_l are regular at the origin, whereas the n_l are not. A general expression in terms of $l+1$ terms with alternating sin and cos functions can also be obtained from the repeated step-up/down operations (see mathematical appendix):

$$\begin{aligned} j_l(\rho) &= \sum_{k=0}^{l} \frac{\sin[\rho - \frac{\pi}{2}(l-k)]}{\rho^{k+1}} \frac{(l+k)!}{k!(l-k)!2^k}, \\ n_l(\rho) &= -\sum_{k=0}^{l} \frac{\cos[\rho - \frac{\pi}{2}(l-k)]}{\rho^{k+1}} \frac{(l+k)!}{k!(l-k)!2^k}. \end{aligned} \tag{43}$$

The approximate values in the limit $\rho \to \infty$ (which we shall need!) are

$$\begin{aligned} j_l(\rho) &\to \frac{\sin[\rho - \frac{\pi}{2}l]}{\rho}, \\ n_l(\rho) &\to -\frac{\cos[\rho - \frac{\pi}{2}l]}{\rho}. \end{aligned} \tag{44}$$

Finally (see mathematical appendix), the power series expansions for the spherical Bessel functions are

$$j_l(\rho) = \sum_{k=0}^{\infty} \frac{(-1)^k 2^l (l+k)!}{k!(2l+2k+1)!} \rho^{l+2k} = \frac{2^l l!}{(2l+1)!} \rho^l + \cdots,$$

$$n_l(\rho) = -\frac{1}{\rho^{l+1}} \sum_{k=0}^{\infty} \frac{(2l-2k)!}{2^l (l-k)!} \rho^{2k} = -\frac{1}{\rho^{l+1}} \frac{(2l)!}{2^l l!} + \cdots. \qquad (45)$$

With these properties of the spherical Bessel functions, we can return to eqs. (19) and (30). In particular, because $e^{ikr\cos\theta}$ is regular at the origin, only the solutions j_l regular at the origin can occur in the expansion of eq. (30), so that

$$e^{i\vec{k}\cdot\vec{r}} = e^{ikr\cos\theta} = \sum_{l=0}^{\infty} c_l j_l(kr) P_l(\cos\theta). \qquad (46)$$

To determine the coefficients c_l in this expansion, we use the orthogonality of the Legendre polynomials

$$c_l j_l(kr) = \frac{(2l+1)}{2} \int_0^{\pi} d\theta \sin\theta e^{ikr\cos\theta} P_l(\cos\theta). \qquad (47)$$

To evaluate the c_l through this integral, it is sufficient to take the limit $kr \to 0$, and note only the lth term in the power series expansion of the exponential term in the integral can make a contribution to the integral, so

$$c_l \frac{2^l l!(kr)^l}{(2l+1)!} = \frac{(2l+1)}{2} \frac{(ikr)^l}{l!} \int_0^{\pi} d\theta \sin\theta \cos^l\theta P_l(\cos\theta). \qquad (48)$$

The θ-integral can now be done via the orthonormality integral of the Legendre polynomials, and the relation

$$\cos^l\theta = \frac{2^l (l!)^2}{(2l)!} P_l(\cos\theta) + b_{l-2} P_{l-2}(\cos\theta) + \cdots. \qquad (49)$$

Note that the leading, i.e., highest power, term in the expansion of $P_l(x)$ in powers of x can be obtained most easily from the Rodrigues formula

$$P_l(x) = \frac{1}{2^l l!} \frac{d^l}{dx^l} (x^2-1)^l) = \frac{1}{2^l l!} \frac{d^l}{dx^l} x^{2l} + \cdots = \frac{(2l)!}{2^l (l!)^2} x^l + \cdots. \qquad (50)$$

The integral of eq. (48) then yields

$$c_l \frac{2^l l!(kr)^l}{(2l+1)!} = \frac{(ikr)^l}{l!} \frac{2^l (l!)^2}{(2l)!}, \qquad \text{hence} \quad c_l = i^l (2l+1), \qquad (51)$$

so

$$e^{ikr\cos\theta} = \sum_{l=0}^{\infty} i^l (2l+1) j_l(kr) P_l(\cos\theta). \qquad (52)$$

Mathematical Appendix to Chapter 41. Spherical Bessel Functions

The spherical Bessel functions, $j_l(\rho)$, can be obtained from $j_0(\rho)$, via repeated application of the step-up relation

$$j_{l+1} = \frac{u_{l+1}}{\rho} = \frac{1}{\rho}\left(-\frac{d}{d\rho} + \frac{(l+1)}{\rho}\right)u_l. \tag{1}$$

It will be convenient to define

$$y_l(\rho) \equiv \frac{j_l}{\rho^l} = \frac{u_l}{\rho^{l+1}}, \tag{2}$$

so

$$y_{l+1} = \frac{1}{\rho^{l+2}}\left(-\frac{d}{d\rho} + \frac{(l+1)}{\rho}\right)u_l = \frac{1}{\rho^{l+2}}\left(-\frac{d}{d\rho} + \frac{(l+1)}{\rho}\right)(\rho^{l+1}y_l) = -\frac{1}{\rho}\frac{dy_l}{d\rho}. \tag{3}$$

Similarly, if we define a second y_l via

$$y_l(\rho) \equiv \frac{n_l}{\rho^l} = \frac{u_{-(l+1)}}{\rho^{l+1}}(-1)^{l+1}, \tag{4}$$

then

$$\begin{aligned} y_{l+1} &= \frac{(-1)^{l+2}}{\rho^{l+2}}\left(\frac{d}{d\rho} - \frac{(l+1)}{\rho}\right)u_{-(l+1)} \\ &= \frac{(-1)^{l+2}}{\rho^{l+2}}\left(\frac{d}{d\rho} - \frac{(l+1)}{\rho}\right)((-1)^{l+1}\rho^{l+1}y_l) = -\frac{1}{\rho}\frac{dy_l}{d\rho}. \end{aligned} \tag{5}$$

Therefore, for both the spherical Bessel and Neumann functions, y_l can be obtained from y_0 by the successive action of l operations

$$-\frac{1}{\rho}\frac{d}{d\rho}.$$

If we define the l-independent operator

$$\mathcal{D} \equiv \frac{1}{\rho}\frac{d}{d\rho}, \tag{6}$$

then

$$y_l(\rho) = (-1)^l \mathcal{D}^l y_0(\rho). \tag{7}$$

[$y_0(\rho)$ is $(\sin\rho/\rho)$ for the spherical Bessel functions and it is $-(\cos\rho/\rho)$ for the spherical Neumann functions.] We can simplify the operator $\mathcal{D}$ with a change of variable

$$\rho = \sqrt{2t}, \qquad t = \tfrac{1}{2}\rho^2, \tag{8}$$

so

$$\mathcal{D} = \frac{1}{\rho}\frac{d}{d\rho} = \frac{d}{dt} \tag{9}$$

and, then,

$$y_l(\sqrt{2t}) = (-1)^l \frac{d^l}{dt^l} y_0(\sqrt{2t}). \tag{10}$$

We shall now express this function, involving an lth derivative of y_0, in terms of a contour integral via the Cauchy integral formula for the lth derivative

$$\frac{d^l f(z)}{dz^l} = \frac{l!}{2\pi i} \oint \frac{f(\zeta) d\zeta}{(\zeta - z)^{l+1}} \tag{11}$$

or, in language appropriate to our y_l,

$$y_l(\sqrt{2t}) = \frac{(-1)^l l!}{2\pi i} \oint \frac{y_0(\sqrt{2\tau})}{(\tau - t)^{l+1}} d\tau. \tag{12}$$

Because our functions y_0 must have a branch cut ending on the singularity at $\tau = 0$, and we are interested in functions y_l at real positive values of t. The point $\tau = t$ must be on the positive real axis in the complex τ-plane, and the contour around this point must avoid the branch cut. We can achieve this result by locating the branch cut on the negative real axis in the τ-plane. (See Fig. 41.5).

It will be useful to calculate first the y_l for the spherical Hänkel functions, which are defined by

$$h_l^{(1)}(\rho) = j_l(\rho) + i n_l(\rho), \tag{13}$$

$$h_l^{(2)}(\rho) = j_l(\rho) - i n_l(\rho), \tag{14}$$

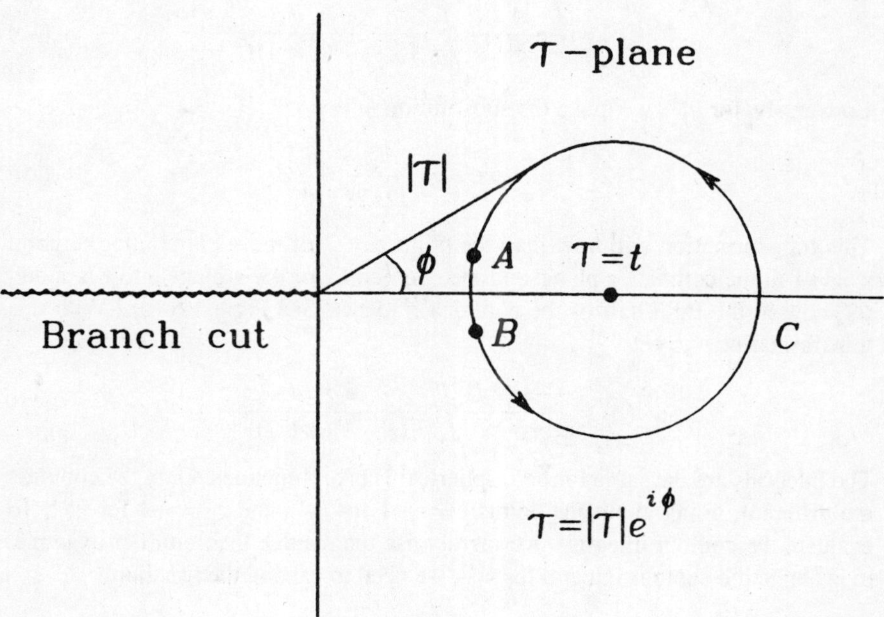

FIGURE 41.5. Contour for eq. (16).

in particular, with

$$h_0^{(1)}(\rho) = -\frac{i}{\rho}e^{i\rho}, \qquad h_0^{(2)}(\rho) = +\frac{i}{\rho}e^{-i\rho}. \tag{15}$$

For these equations then

$$y_l(\sqrt{2t}) = \frac{(-1)^l l!(\mp i)}{2\pi i}\oint_C d\tau \frac{e^{\pm i\sqrt{2\tau}}}{\sqrt{2\tau}(\tau - t)^{l+1}}, \tag{16}$$

where the contour, C, is that of Fig. 41.5 and the upper(lower) case refers to the y_l for $h_l^{(1)}$ ($h_l^{(2)}$). To actually carry out the contour integral, it will be convenient to make additional transformations.

For $h_l^{(1)}$, let

$$[2\tau]^{\frac{1}{2}} = -\rho z, \qquad \frac{d\tau}{[2\tau]^{\frac{1}{2}}} = -\rho dz. \tag{17}$$

This transformation maps the point $\tau = t$ of the τ plane into the point $z = -1$ in the z plane, and the contour, C, of the τ plane into a contour, C_1, in the left half of the complex z plane. [See Fig. 41.6 (a).] This equation leads to the contour integral expression

$$\frac{h_l^{(1)}(\rho)}{\rho^l} = \frac{(-1)^l l!}{2\pi}\oint_{C_1} \frac{dz\rho e^{-i\rho z}}{\left(\frac{\rho^2 z^2}{2} - \frac{\rho^2}{2}\right)^{l+1}}, \tag{18}$$

so

$$h_l^{(1)}(\rho) = \frac{(-1)^l 2^l l!}{\pi\rho^{l+1}}\oint_{C_1} \frac{e^{-i\rho z}dz}{[(z-1)(z+1)]^{l+1}}. \tag{19}$$

Conversely, for $h_l^{(2)}$, we make the substitution

$$[2\tau]^{\frac{1}{2}} = +\rho z, \qquad \frac{d\tau}{[2\tau]^{\frac{1}{2}}} = \rho dz. \tag{20}$$

This transformation will now map the point $\tau = t$ of the τ plane into the point $z = +1$ in the complex z plane, and the contour, C, of the τ plane into a contour, C_2, now in the right half of the complex z plane. [See Fig. 41.6 (b).] With this transformation, we get

$$h_l^{(2)}(\rho) = \frac{(-1)^l 2^l l!}{\pi\rho^{l+1}}\oint_{C_2} \frac{e^{-i\rho z}dz}{[(z-1)(z+1)]^{l+1}}. \tag{21}$$

The integrals are the same for both spherical Hänkel functions. Only the contours are different, going about the points $z = -1$ for $h_l^{(1)}$, and $z = +1$ for $h_l^{(2)}$. To evaluate the contour integrals, we merely use the residue theorem. For example, to evaluate the contour integral for $h_l^{(1)}$, we need to expand the function

$$\frac{e^{-i\rho z}}{(z-1)^{l+1}}$$

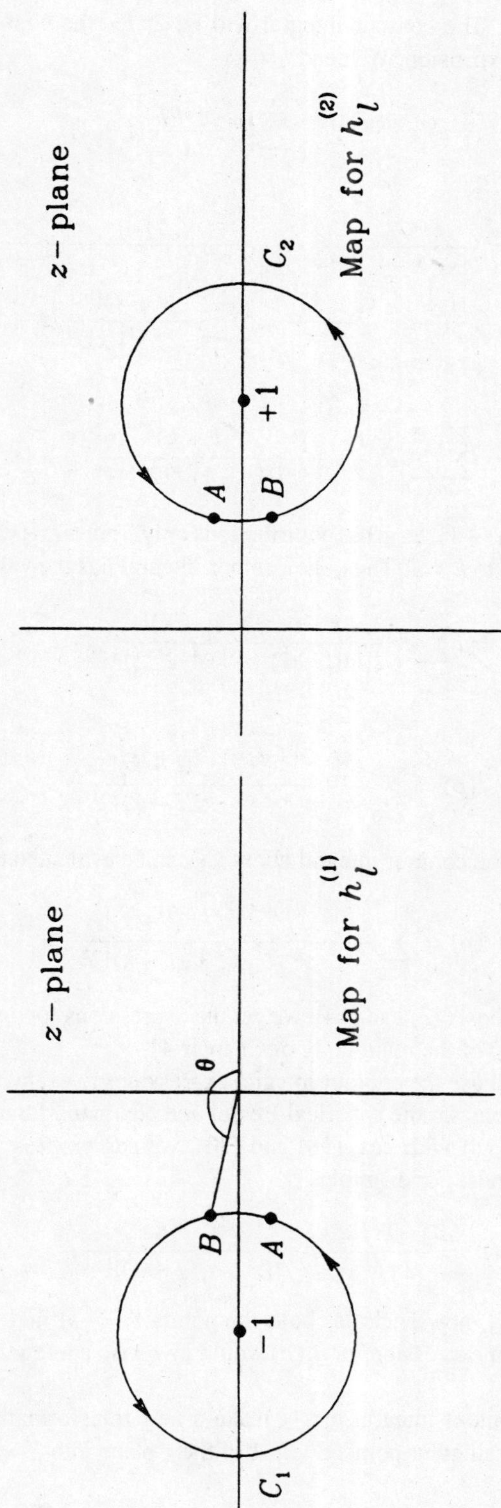

FIGURE 41.6. Contours for eqs. (18) and (21).

in powers of $(z+1)$. The contour integral will be $2\pi i\times$ the coefficient of the $(z+1)^l$ term in this expansion. We need

$$e^{-i\rho z} = e^{i\rho} e^{-i\rho(z+1)} = \sum_{n=0}^{\infty} e^{i\rho} \frac{(-i\rho)^n}{n!}(z+1)^n \tag{22}$$

$$\begin{aligned}\frac{1}{(z-1)^{l+1}} &= \frac{1}{(z+1-2)^{l+1}} = \frac{1}{(-2)^{l+1}[1-\frac{1}{2}(z+1)]^{l+1}} \\ &= \sum_{k=0}^{\infty} \frac{1}{(-2)^{l+1}} \frac{(l+1)_k}{k!} \frac{(z+1)^k}{2^k} = \sum_{k=0}^{\infty} \frac{(l+k)!}{(-2)^{l+1} l! k! 2^k}(z+1)^k,\end{aligned} \tag{23}$$

so

$$\frac{e^{-i\rho z}}{(z-1)^{l+1}} = \sum_{n=0}^{\infty}\sum_{k=0}^{\infty} e^{i\rho} \frac{(-i\rho)^n}{n!} \frac{(l+k)!}{l!k!(-1)^{l+1}2^{l+1+k}}(z+1)^{n+k}. \tag{24}$$

The coefficient of the $(z+1)^l$ term has contributions only from terms with $n = l-k$, and terms with $k=0$ to $k=l$. Thus, the contour integral has the value

$$2\pi i \sum_{k=0}^{l} e^{i\rho} \frac{\rho^{l-k}(-i)^{l-k}(l+k)!}{l!k!(l-k)!(-1)^{l+1}2^{l+1+k}}, \tag{25}$$

so

$$h_l^{(1)}(\rho) = \sum_{k=0}^{l} \frac{e^{i(\rho-(l+1-k)\frac{\pi}{2})}}{\rho^{k+1}} \frac{(l+k)!}{k!(l-k)!2^k}. \tag{26}$$

In the same fashion, the contour integral about C_2 can be evaluated to give

$$h_l^{(2)}(\rho) = \sum_{k=0}^{l} \frac{e^{-i(\rho-(l+1+k)\frac{\pi}{2})}}{\rho^{k+1}} \frac{(l+k)!}{k!(l-k)!2^k}. \tag{27}$$

Combining these via eqs. (13) and (14), we get the expressions for $j_l(\rho)$ and $n_l(\rho)$ given through eq. (43) of the main body of Chapter 41.

Finally, we can also use the contour integral expressions to evaluate the infinite power series expansions for the spherical Bessel and Neumann functions. Again, using eqs. (13) and (14) with eqs. (18) and (20), we can express j_l and n_l via similar contour integrals. For example,

$$j_l(\rho) = \frac{(-1)^l l! 2^l}{2\pi\rho^{l+1}} \oint_{C_3} \frac{e^{-i\rho z}dz}{[(z-1)(z+1)]^{l+1}}, \tag{28}$$

where the contour, C_3, now encircles both the points $z=-1$ and $z=+1$ (see Fig. 41.7). The similar expression for $n_l(\rho)$ would involve a contour, C_4, shown in Fig. 41.8.

To evaluate the contour integral for j_l, make a new transformation, $z' = \rho z$, which transforms the singular points $z = \pm 1$ of the z plane into $z' = \pm\rho$ in the z'

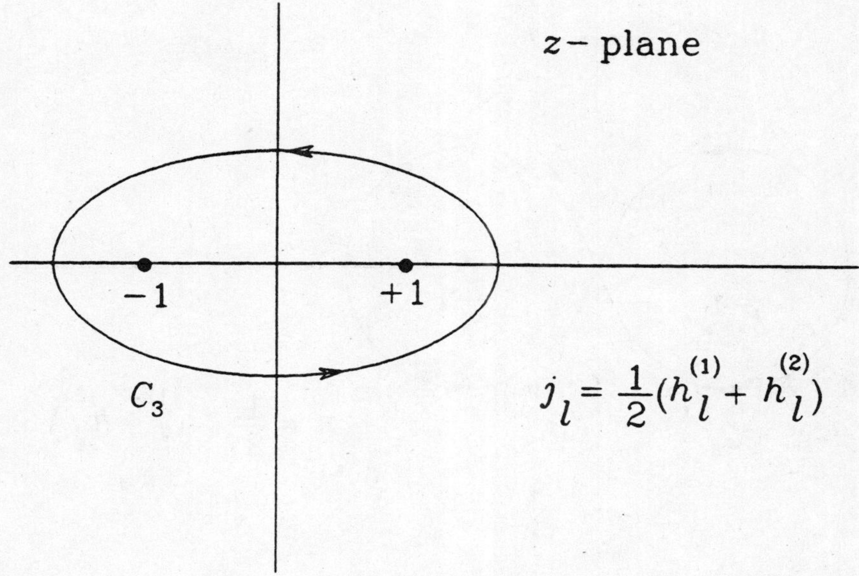

FIGURE 41.7. Contour for $j_l(\rho)$, eq. (28).

plane. This process gives

$$\begin{aligned} j_l(\rho) &= \frac{(-1)^l l! 2^l \rho^l}{2\pi} \oint_{C_3} \frac{e^{-iz'} dz'}{[(z'-\rho)(z'+\rho)]^{l+1}} \\ &= \frac{(-1)^l l! 2^l \rho^l}{2\pi} \oint_{C_3} \frac{e^{-iz'} dz'}{z'^{(2l+2)}\left(1-\frac{\rho^2}{z'^2}\right)^{l+1}} \\ &= \frac{(-1)^l l! 2^l \rho^l}{2\pi} \sum_{k=0}^{\infty} \frac{\rho^{2k}(l+k)!}{l!k!} \oint_{C_3} \frac{e^{-iz'} dz'}{z'^{(2l+2+2k)}}. \end{aligned} \tag{29}$$

The contour integral in this expression is again evaluated by the residue theorem and has the value

$$2\pi i \times \frac{(-i)^{2l+2k+1}}{(2l+2k+1)!} = \frac{2\pi(-1)^{l+k}}{(2l+2k+1)!},$$

yielding

$$j_l(\rho) = \sum_{k=0}^{\infty} \frac{(-1)^k 2^l (l+k)!}{k!(2l+2k+1)!} \rho^{l+2k}, \tag{30}$$

which was quoted in eq. (45) of the main body of Chapter 41. The corresponding expression for n_l can be obtained from the corresponding contour integral over the contour C_4 of Fig. 41.8. In this case, it may be more efficient to evaluate the contour integral only for the dominant $1/\rho^{l+1}$ term, and then obtain the remaining terms of the infinite series via a two-term recursion formula for the c_k in the expan-

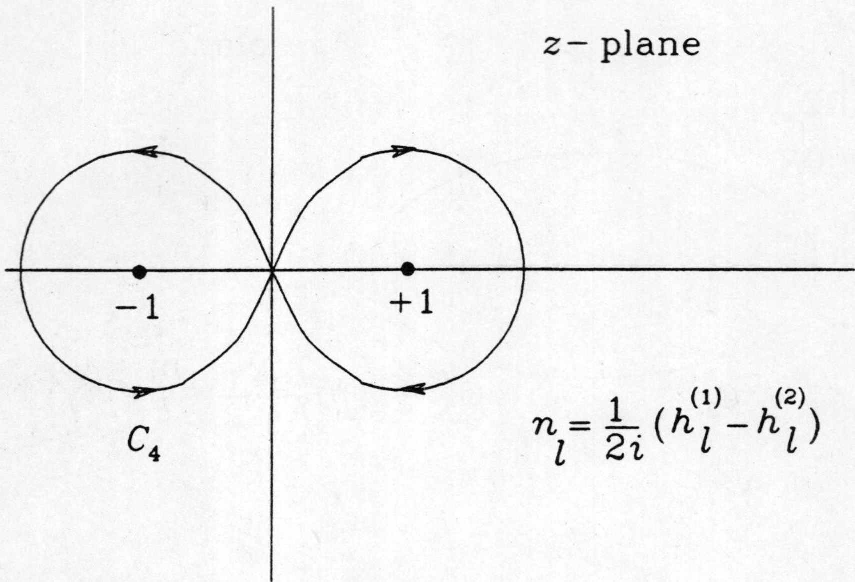

FIGURE 41.8. Contour for $n_l(\rho)$.

sion, $n_l(\rho) = \sum_k c_k \rho^{2k-l-1}$. This recursion formula follows from the differential equation for the n_l by standard techniques.

42

The Rayleigh–Faxen–Holtzmark Partial Wave Expansion: Phase Shift Method

We shall now solve the scattering problem, that is, find solutions for the Schrödinger equation [eq. (18) of Chapter 41] in the form of eq. (19) of Chapter 41 for a spherically symmetric $V(r)$ that goes to zero faster than $1/r^2$ as $r \to \infty$. [Note: For such a potential, the effective $V(r)$ approaches the pure centrifugal term for sufficiently large r. Also, we exclude for the moment the case of scattering via a Coulomb potential with a $V(r)$ that goes to zero only as $1/r$ at large r. Our assumed $V(r)$ would be good for the scattering of neutrons from complex nuclei, but the scattering of charged particles from nuclei, e.g., the scattering of protons or alpha particles from nuclei, would require a slightly more complicated treatment.]

We shall expand the solution to the relative motion equation

$$\nabla^2\psi + \frac{2\mu}{\hbar^2}(E - V(r))\psi = 0 \tag{1}$$

in a series similar to that of the expansion of the incoming plane wave of Chapter 41

$$\psi = \sum_{l=0}^{\infty} \frac{w_l(kr)}{(kr)} P_l(\cos\theta). \tag{2}$$

This equation is the partial wave expansion. The ψ are independent of the azimuth angle, ϕ. This axial symmetry is dictated by the spherical symmetry of $V(r)$ and the axial symmetry about the direction of the incoming beam. In the above, the lth partial wave solution is determined by the radial function, $R_l(kr) \equiv w_l(kr)/(kr)$, which is a solution of the one-dimensionalized wave equation

$$\frac{d^2w_l}{dr^2} + \left(k^2 - \frac{2\mu}{\hbar^2}V(r) - \frac{l(l+1)}{r^2}\right)w_l = 0. \tag{3}$$

As $r \to \infty$, $V(r) \to 0$, such that $V_{\text{effective}} \to V_{\text{centrifugal}}$. Thus, $w_l(kr) \to$ is a linear combination of the free-wave radial solutions $(kr)j_l(kr)$ and $(kr)n_l(kr)$. From their asymptotic form [see eq. (44) of Chapter 41], this is a linear combination of $\sin(kr - \frac{l\pi}{2})$ and $\cos(kr - \frac{l\pi}{2})$, so, as $r \to \infty$,

$$w_l(kr) \to a_l \sin(kr - \tfrac{l\pi}{2} + \delta_l(k)),$$
$$\frac{w_l(kr)}{(kr)} \to \frac{a_l}{2ikr}\Big(e^{ikr}(-i)^l e^{i\delta_l} - e^{-ikr}(i)^l e^{-i\delta_l}\Big). \tag{4}$$

Here, a_l is the amplitude of the lth partial wave, and δ_l, an energy (k)-dependent quantity, is its phase shift. To find the differential scattering cross section, we need to compare these partial wave solutions with the expected asymptotic form of our solution

$$\psi = e^{i\vec{k}\cdot\vec{r}} + f(\theta)\frac{e^{ikr}}{r}. \tag{5}$$

Expanding, $f(\theta)$, through

$$f(\theta) = \sum_{l=0}^{\infty} f_l P_l(\cos\theta), \tag{6}$$

and using eq. (52) of the last chapter to expand the plane wave, we have, as $r \to \infty$,

$$\psi \to \sum_l \Big(i^l(2l+1)\frac{sin(kr - \frac{\pi}{2}l)}{kr} + f_l\frac{e^{ikr}}{r}\Big)P_l(\cos\theta)$$
$$\to \sum_l \Big(\frac{e^{ikr}}{r}(f_l + \frac{(2l+1)}{2ik}) - \frac{e^{-ikr}}{r}\frac{(-1)^l(2l+1)}{2ik}\Big)P_l(\cos\theta). \tag{7}$$

Comparing with eqs. (2) and (4),

$$a_l = i^l(2l+1)e^{i\delta_l} \qquad \text{and} \qquad f_l = \frac{(2l+1)}{2ik}(e^{2i\delta_l} - 1), \tag{8}$$

so

$$f(\theta) = \sum_{l=0}^{\infty}\frac{i(2l+1)}{2k}(1 - e^{2i\delta_l(k)})P_l(\cos\theta)$$
$$= \sum_{l=0}^{\infty} e^{i\delta_l(k)}\frac{(2l+1)}{k}\sin\delta_l(k)P_l(\cos\theta), \tag{9}$$

leading to

$$\frac{d\sigma}{d\Omega} = |f(\theta)|^2 = |\sum_l \frac{(2l+1)}{2k}(1 - \eta_l)P_l(\cos\theta)|^2$$
$$= |\sum_l e^{i\delta_l}\frac{(2l+1)}{k}\sin\delta_l P_l(\cos\theta)|^2, \tag{10}$$

where the common shorthand notation, $\eta_l = e^{2i\delta_l}$, has been used in the first form. Using the orthogonality integral

$$\int\int d\Omega P_l(\cos\theta)P_{l'}(\cos\theta) = \frac{4\pi}{(2l+1)}\delta_{ll'}, \tag{11}$$

these two expressions lead to expressions for the total cross section

$$\begin{aligned}\sigma &= \sum_l \frac{\pi}{k^2}(2l+1)|1-\eta_l|^2 \\ &= \sum_l \frac{4\pi}{k^2}(2l+1)\sin^2\delta_l.\end{aligned} \tag{12}$$

A number of remarks are in order, as follows.

1. For $V(r) \to 0, \delta_l(k) \to 0$. In this case, we are left only with the free incoming plane wave.

2. In the limit of very large k, in particular, for $\frac{\hbar^2k^2}{2\mu} \gg |V_{\max}|$, again $\delta_l(k) \to 0$ as $k \to$ very large. Effectively, the potential again becomes insignificant compared with the energy of the incoming beam.

3. For an attractive $V(r)$, $\delta(k) > 0$. In this case, the wavelength becomes shorter in the small r region where the potential is effective, and the lth partial wave is pulled into smaller values of r as a result (see Fig. 42.1). Say the first minimum of the lth partial wave beyond the region where the true $V(r)$ is effectively zero occurs at a phase angle, χ_1. Then, $(kr_{\text{with}} - \frac{\pi l}{2} + \delta_l) = (kr_{\text{without}} - \frac{\pi l}{2}) = \chi_1$, where r_{with} is the position of this minimum *with* the potential $V(r)$ turned on, whereas r_{without} is the position of this minimum *without* the potential, that is, for the free partial wave. Because $r_{\text{with}} < r_{\text{without}}$, $\delta_l(k) > 0$.

Similarly, for a repulsive $V(r)$, $\delta_l(k) < 0$.

4. The total scattering cross section, σ, is related to the imaginary part of the scattering amplitude, $f(\theta = 0)$. From the second form for $f(\theta)$ and σ above,

$$\sigma(k) = \frac{4\pi}{k}\Im\big[f(\theta = 0)\big]. \tag{13}$$

This form is a special case of the so-called Optical Theorem. It also leads to the Wick inequality for the elastic scattering cross section

$$\sigma_{\text{elastic}} \leq \frac{4\pi}{k}\sqrt{\frac{d\sigma}{d\Omega}}(\theta = 0). \tag{14}$$

Note,

$$\frac{d\sigma}{d\Omega} = \big(\Im[f(\theta)]\big)^2 + \big(\Re[f(\theta)]\big)^2.$$

5. A relation exsits between the maximum l for which $\delta_l(k)$ is appreciably different from zero and the magnitude k of the incoming wave. Thus, although our expressions for both $\frac{d\sigma}{d\Omega}$ and σ in general involve an infinite series in l, only a finite number of terms may be effective. Suppose the range of our $V(r)$ is r_0; i.e., $V(r) \approx 0$ for $r > r_0$. Classically, if a projectile particle comes in with an impact parameter, $b > r_0$, it will “miss” feeling the potential and not be scattered

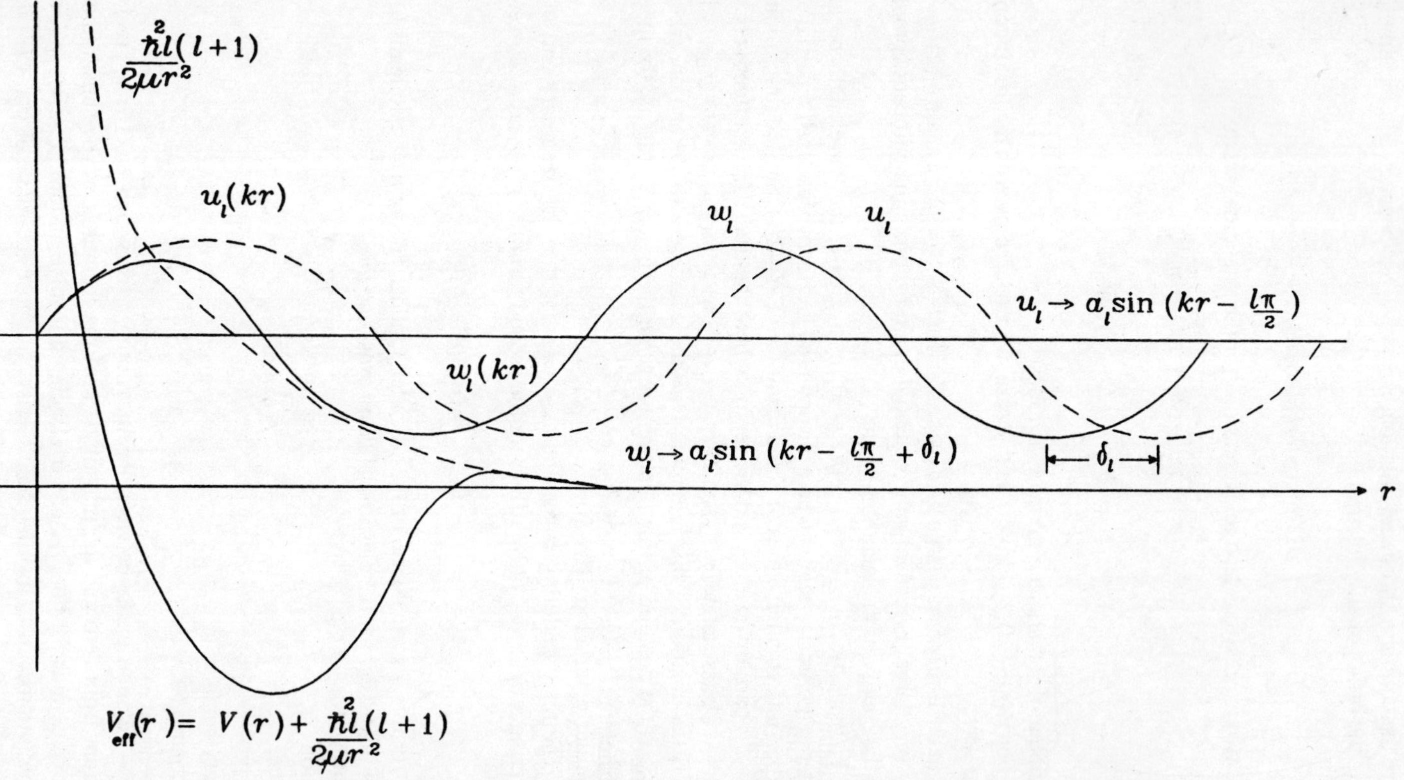

FIGURE 42.1. The lth partial wave solutions, u_l (plane wave), and w_l [with attractive $V(r)$].

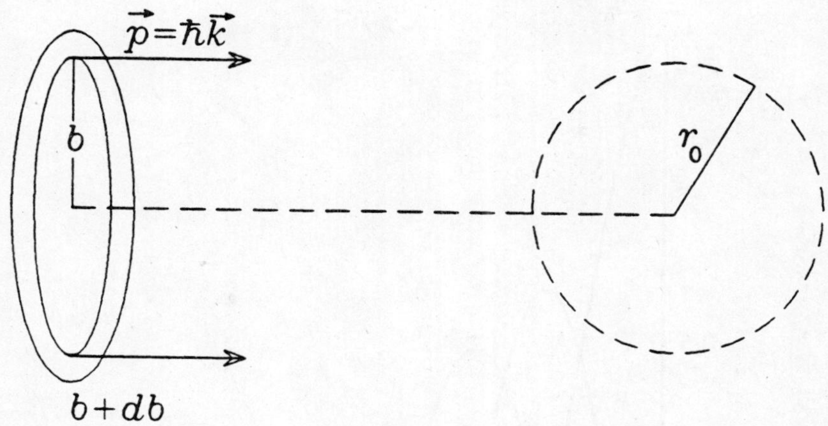

FIGURE 42.2. Semiclassical picture of the scattering process.

at all. (See Fig. 42.2.) Although the wave nature of the particle motion and the uncertainty relation washes out this relationship somewhat, it nevertheless still has approximate validity. The angular momentum of an incoming particle with impact parameter, b, is $\hbar kb$. Approximately,

$$\hbar kb \approx \hbar\sqrt{l(l+1)}, \qquad b \approx \frac{l}{k}. \tag{15}$$

Because we expect strong scattering only for $b < r_0$, we expect δ_l will be appreciably different from zero only for $l \leq kr_0$, where this relation might be a particularly good approximation for large l, where semiclassical arguments are valid. Recalling $P_l(\cos\theta) = \sqrt{4\pi/(2l+1)}Y_{l0}$, the plane wave expansion in terms of normalized angular functions, Y_{lm}, is

$$e^{ikr\cos\theta} = \sum_l i^l\sqrt{4\pi(2l+1)}\,j_l(kr)Y_{l0}(\theta), \tag{16}$$

so the probability of finding the lth partial wave component of the plane wave within a radius r_0 is proportional to the integrated value, from 0 to r_0, of the function $(2l+1)j_l(kr)^2$. This function is plotted for the lower l values in Fig. 42.3. We see, e.g., if $kr_0 = 1.5$, only the $l = 0$ and $l = 1$ partial waves will have a large probability of penetrating into the effective range of the potential. In general, for very small k, that is, for extreme low energy scattering, only the $l = 0$ term will contribute to the differential scattering cross section. Because $P_0(\cos\theta) = 1$, independent of θ, the scattering will be isotropic. In this case,

$$f(\theta) \approx e^{i\delta_0}\frac{\sin\delta_0}{k}, \tag{17}$$

$$\frac{d\sigma}{d\Omega} = \frac{\sin^2\delta_0}{k^2}, \qquad \sigma = 4\pi\frac{\sin^2\delta_0}{k^2}. \tag{18}$$

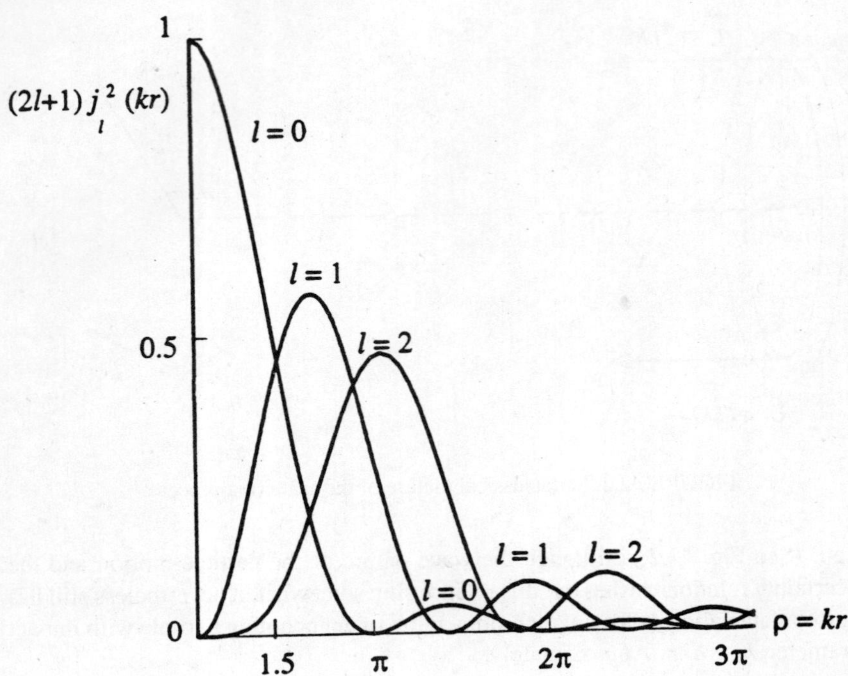

FIGURE 42.3. The functions $(2l+1)[j_l(kr)]^2$.

In the extreme low-energy limit, therefore, the cross section is proportional to the square of the wavelength. In this low-energy limit, the pertinent length is the wavelength of the incident beam, not the size of the scatterer.

If the energy is somewhat higher so both phase shifts, δ_1 and δ_0, contribute, but assuming $\delta_1 \ll 1$, then

$$\begin{aligned}\frac{d\sigma}{d\Omega} &= \frac{\sin^2\delta_0}{k^2} + \frac{6}{k^2}\cos(\delta_0 - \delta_1)\sin\delta_0\sin\delta_1\cos\theta + \cdots \\ &\simeq \frac{1}{k^2}\left(\sin^2\delta_0 + 3\delta_1\sin 2\delta_0\cos\theta + \cdots\right) \\ &= A + B\cos\theta + \cdots. \end{aligned} \tag{19}$$

With $B \ll A$, the differential cross section is almost isotropic but does now have a weak $\cos\theta$ dependence.

6. The determination of the $\delta_l(k)$ from a known $V(r)$ is in principle straightforward and just a matter of calculation. Often, however, $V(r)$ is unknown and we would like to determine it from the observed differential cross sections at different energies. Although the $\delta_l(k)$ can be determined from the observed $\frac{d\sigma}{d\Omega}$, the determination of the $V(r)$ from these $\delta_l(k)$ is difficult. The "inverse" problem is tough.

7. Generalizations to scattering of complex projectile from complex target particles. So far, we have studied only the scattering of structureless point projectile

particles from structureless point target particles. If we name projectile particle, a, and target particle, A, we have in general three possibilities, as follows.

(1) $a + A \to a + A$, the elastic scattering process.

(2) $a + A \to a + A'$. Particle A is in an excited state after the scattering process and the final kinetic energy of the relative motion is less than the initial kinetic energy. This is an inelastic scattering process.

(3) $a + A \to b + B$. This is a rearrangement collision. The ejectiles are different from the initial projectile and target particles. In such a process, the number of outgoing particles could also be greater than two.

The different processes are labeled by a "channel" index: α for the elastic process (1) and $\beta = \beta_1 + \beta_2 + \cdots$ for the processes of type (2) and (3). For the case of structureless point particles, we could have written our solution, as $r \to \infty$, in the form

$$\psi \to \sum_l \frac{i\sqrt{\pi(2l+1)v_\alpha}}{k_\alpha} i^l Y_{l0}(\theta, \phi)\Big(I_l(k_\alpha r) - \eta_{l,\alpha} O_l(k_\alpha r)\Big), \tag{20}$$

where the incoming and outgoing l^{th} partial wave relative motion functions are now given in the unit flux normalization, [see eq. (24) of Chapter 41], and

$$I_l(k_\alpha r) = \frac{1}{\sqrt{v_\alpha}} \frac{e^{-i(k_\alpha r - \frac{\pi l}{2})}}{r}, \qquad O_l(k_\alpha r) = \frac{1}{\sqrt{v_\alpha}} \frac{e^{i(k_\alpha r - \frac{\pi l}{2})}}{r}. \tag{21}$$

For the case of complex projectile and target, this process can now be generalized. Now, as $r \to \infty$,

$$\psi \to \sum_l \frac{i\sqrt{\pi(2l+1)v_\alpha}}{k_\alpha} \sum_\beta (\mathcal{I}_{l,\alpha} - \eta_{l,\beta} \mathcal{O}_{l,\beta}), \tag{22}$$

where the sum over β includes the elastic term, $\beta = \alpha$, and all other β_i, and the outgoing and incoming wave functions must now include, besides the asymptotic outgoing and incoming relative motion functions, the appropriate internal wave functions for the particles a, A, or b, and B.

$$\begin{aligned} \mathcal{I}_{l,\alpha} &= i^l Y_{l0}(\theta, \phi) I_l(k_\alpha r) \psi_{\text{internal}}(\xi_a) \psi_{\text{internal}}(\xi_A), \\ \mathcal{O}_{l,\beta} &= i^l Y_{l0}(\theta, \phi) O_l(k_\beta r) \psi_{\text{internal}}(\xi_b) \psi_{\text{internal}}(\xi_B), \end{aligned} \tag{23}$$

where the $\xi_a, \xi_A, \cdots$, stand for the needed internal variables of the various particles. Now,

$$\begin{aligned} \sigma_{\text{total}} &= \sigma_{\text{elastic}} + \sigma_{\text{reaction}} \\ &= \sigma_\alpha + \sum_{\beta \neq \alpha} \sigma_\beta \\ &= \frac{\pi}{k_\alpha^2} \sum_l \sum_{\text{all } \beta} (2l+1)|\delta_{\alpha\beta} - \eta_{l,\beta}|^2 \\ &= \frac{\pi}{k_\alpha^2} \sum_l (2l+1)\Big(|1 - \eta_{l,\alpha}|^2 + \sum_{\beta \neq \alpha} |\eta_{l,\beta}|^2\Big). \end{aligned} \tag{24}$$

Later, we shall see the conservation of probability, the so-called unitarity condition, will require

$$(|\eta_{l,\alpha}|^2 + \sum_{\beta\neq\alpha} |\eta_{l,\beta}|^2) = 1. \tag{25}$$

Previously, we saw for the case of pure elastic scattering, e.g., at low energies, where all inelastic processes and rearrangement collisions are energetically forbidden, $\eta_{l,\alpha} = e^{2i\delta_{l,\alpha}}$. Now, in general, $|\eta_{l,\beta}| \leq 1$. The unitarity condition severely restricts the values of the lth partial cross sections, σ_l. Fig. 42.4 gives a plot of the possible values of $\sigma_{\text{elastic},l}$ versus $\sigma_{\text{reaction},l}$, both in units of $\pi(2l+1)/k_\alpha^2$.

A final remark: Eqs. (20) and (22) can also be used for the case of scattering by a Coulomb potential, provided the r dependence in the exponential factors in the incoming and outgoing relative motion functions, $I_l(kr)$ and $O_l(kr)$, are replaced by slightly more complicated r-dependent functions.

Mathematical Appendix to Chapter 42

Continuum Solutions for the Coulomb Problem

In the main body of Chapter 42, we restricted our discussion to potentials that go to zero faster than $1/r^2$ as $r \to \infty$. This, therefore, excluded the important case of the Coulomb potential, $V(r) = Z_1 Z_2 e^2/r$, needed for the scattering of point charges from point charges. We shall be interested, in particular, in the asymptotic form of the lth partial wave solutions, w_l, in the limit $r \to \infty$, to obtain the analogues of the incoming and outgoing wave solutions, $I_l(kr)$ and $O_l(kr)$.

For the Coulomb potential, the one-dimensionalized wave equation is given by

$$\frac{d^2 w_l}{dr^2} + \left(k^2 - \frac{2\mu}{\hbar^2}\frac{Z_1 Z_2 e^2}{r} - \frac{l(l+1)}{r^2}\right) w_l = 0, \tag{1}$$

$$\frac{d^2 w_l}{dr^2} + \left(k^2 - \frac{2\gamma k}{r} - \frac{l(l+1)}{r^2}\right) w_l = 0, \tag{2}$$

where we have introduced the Coulomb parameter

$$\gamma = \frac{\mu e^2 Z_1 Z_2}{\hbar^2 k} = \frac{\mu c}{\hbar k}\frac{e^2}{\hbar c} Z_1 Z_2 = \frac{c}{v}\alpha Z_1 Z_2. \tag{3}$$

In addition, it will be useful to introduce the new variable

$$\rho = ikr, \tag{4}$$

so

$$-\frac{d^2 w_l}{d\rho^2} + \left(1 - \frac{2i\gamma}{\rho} + \frac{l(l+1)}{\rho^2}\right) w_l = 0. \tag{5}$$

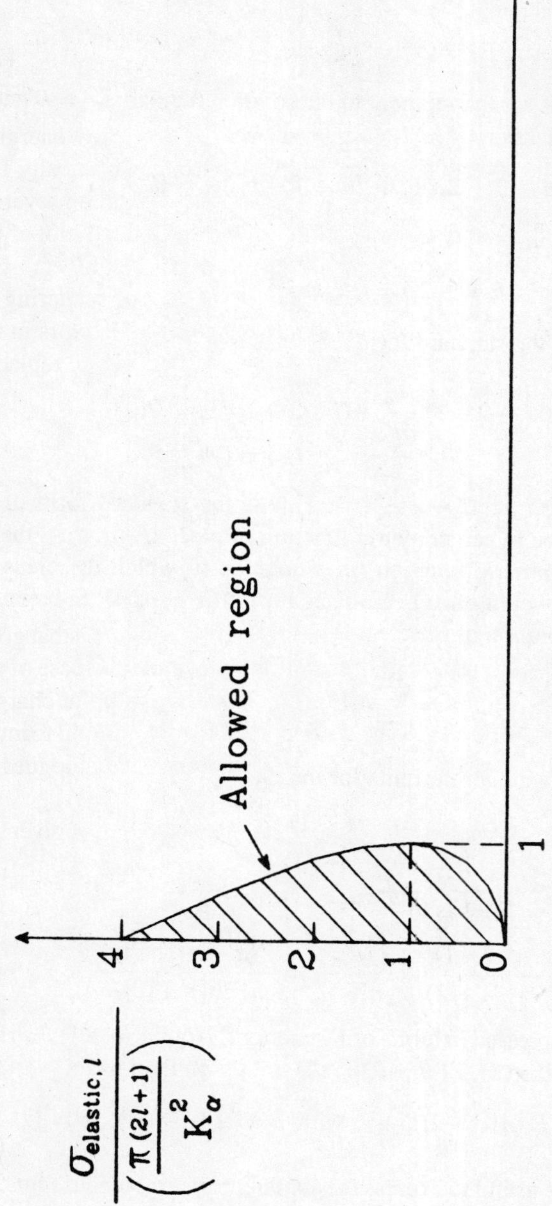

FIGURE 42.4. The domain of allowed $\sigma_{\text{elastic},l}$ vs. $\sigma_{\text{reaction},l}$.

Factoring out the asymptotic behavior of the w_l at $\rho = 0$ and $\rho = \infty$, we need solutions of the form

$$w_l = \rho^{l+1} e^{\rho} v_l(\rho), \tag{6}$$

where we restrict ourselves for the moment to the solution regular at $r = 0$ and v_l satisfies the differential equation

$$\rho v_l'' + (2l + 2 + 2\rho)v_l' + 2(l + 1 + i\gamma)v_l = 0. \tag{7}$$

With the further substitution,

$$t = -2\rho = -2ikr, \tag{8}$$

this solution can be put in the standard form

$$t\frac{d^2 v_l}{dt^2} + (2l + 2 - t)\frac{dv_l}{dt} - (l + 1 + i\gamma)v_l = 0,$$
$$t v_l'' + (c - t)v_l' - a v_l = 0, \tag{9}$$

where $c \equiv (2l + 2)$ and $a \equiv (l + 1 + i\gamma)$. This is the standard form of the differential equation for the hypergeometric function of type, ${}_1F_1(a; c; t)$, the so-called confluent hypergeometric function [in a notation in which the Gaussian hypergeometric function would have been named a ${}_2F_1(a, b; c; t)$]. Substituting an infinite power series solution of the form

$$v_l = \sum_{n=0}^{\infty} c_n t^n, \tag{10}$$

we are led to a two-term recursion formula for the c_n,

$$\frac{c_n}{c_{n-1}} = \frac{(a + n - 1)}{n(c + n - 1)}, \tag{11}$$

which, with the choice $c_0 = 1$, leads to

$$c_n = \frac{a(a + 1)\cdots(a + n - 1)}{n!c(c + 1)\cdots(c + n - 1)} = \frac{(a)_n}{n!(c)_n} = \frac{\Gamma(a + n)}{n!\Gamma(a)} \frac{\Gamma(c)}{\Gamma(c + n)}, \tag{12}$$

where the c_n have been expressed in terms of Γ functions. Note, $\Gamma(a+1) = a\Gamma(a)$ and, for the integer $c = (2l + 2)$, $\Gamma(2l + 2) = (2l + 1)!$. We then have

$$w_l(kr) = (ikr)^{l+1}(e^{ikr})\ {}_1F_1(a; c; -2ikr), \quad \text{with } a = (l+1+i\gamma),\ c = (2l+2). \tag{13}$$

For our purposes, it will be useful to express the confluent hypergeometric function in terms of a contour integral

$${}_1F_1(a; c; t) = \frac{\Gamma(c)}{2\pi i} \oint_C \frac{dz e^z}{z^{c-a}(z - t)^a} = \frac{\Gamma(c)}{2\pi i} \oint_C \frac{dz e^z}{z^c\left(1 - \frac{t}{z}\right)^a}, \tag{14}$$

where the contour, C, in the complex z plane surrounds the branch cut from $z = 0$ to $z = t$, where the complex number t is $t = -2ikr$ in our case. The contour will be chosen such that $|\frac{t}{z}| < 1$, so we can make the expansion, which will be used to establish eq. (13),

$$ {}_1F_1(a;c;t) = \frac{\Gamma(c)}{2\pi i}\sum_{n=0}^{\infty}\frac{(a)_n}{n!}t^n \oint_C \frac{dz e^z}{z^{c+n}} = \Gamma(c)\sum_{n=0}^{\infty}\frac{(a)_n}{n!}\frac{t^n}{(c+n-1)!}, \tag{15} $$

where we have evaluated the contour integral by the residue theorem, bearing in mind $c = (2l+2)$ is an integer. Using $(a)_n = \Gamma(a+n)/\Gamma(a)$ for the complex number, a, and $(c+n-1)! = \Gamma(c+n)$ for the integer, $c+n$, we get

$$ {}_1F_1(a;c;t) = \sum_{n=0}^{\infty}\frac{\Gamma(a+n)}{\Gamma(a)}\frac{\Gamma(c)}{\Gamma(c+n)}\frac{t^n}{n!}, \tag{16} $$

which establishes the contour integral form for ${}_1F_1(a;c;t)$. It will now be useful to deform the contour C, as shown in Fig. 42.5, where the new deformed contour integral can effectively be decomposed into the two loop integrals over contours C_1 and C_2,

$$ {}_1F_1(a;c;-2ikr) = \frac{\Gamma(c)}{2\pi i}\sum_{n=1,2}\oint_{C_n}\frac{dz e^z}{z^c\left(1+\frac{2ikr}{z}\right)^a}. \tag{17} $$

If we further make the substitution, $z = krz'$, so

$$ {}_1F_1(a;c;-2ikr) = \frac{\Gamma(c)}{2\pi i(kr)^{c-1}}\sum_{n=1,2}\int_{C_n}\frac{dz' e^{krz'}}{z'^c\left(1+\frac{2i}{z'}\right)^a}, \tag{18} $$

in the new contour integrals, C_1 and C_2 (see the final form of these contours in Fig. 42.5), only the circular parts of the new contours about the points $z' = 0$ and $z' = -2i$ contribute to the contour integrals in the limit $kr \to \infty$, because the real part of z' is negative on the straight-line portions of the final forms of C_1 and C_2, so $e^{krz'} \to 0$ as $r \to \infty$. Now, to do the contour integrals for the circular parts of the contours C_1 and C_2, surrounding the points $z' = 0$ and $z' = -2i$, in the limit $kr \to \infty$, let us rename $krz' = z$, so for the first integral, for the contour C_1, we have

$$ \begin{aligned} I_1 &= \frac{\Gamma(c)}{2\pi i}\lim_{r\to\infty}\oint_{C_1}\frac{dz e^z}{z^c\left(1+\frac{2ikr}{z}\right)^a} \\ &= \frac{\Gamma(c)}{2\pi i}\lim_{r\to\infty}\frac{1}{(2ikr)^a}\oint_{C_1}\frac{dz e^z}{z^{c-a}\left(1-\frac{iz}{2kr}\right)^a} \\ &= \frac{\Gamma(c)}{2\pi i}\frac{1}{(2ikr)^a}\oint_{C_1}\frac{dz e^z}{z^{c-a}} = \frac{\Gamma(c)}{(2ikr)^a\Gamma(c-a)}, \end{aligned} \tag{19} $$

where we have replaced the contour, C_1, by a small circle around the point $z = 0$ and we have used

$$ \frac{1}{2\pi i}\oint\frac{dz e^z}{z^{c-a}} = \frac{1}{\Gamma(c-a)}. \tag{20} $$

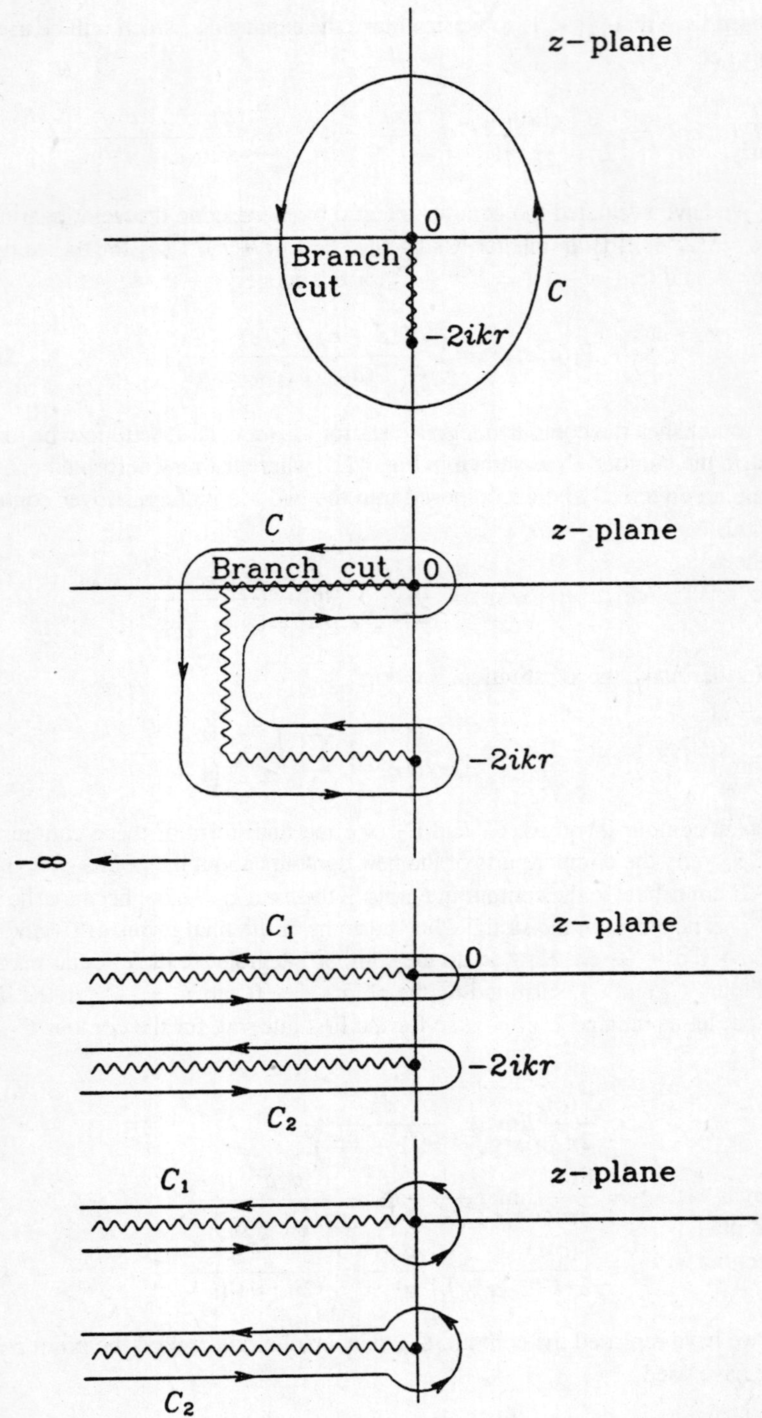

FIGURE 42.5. Contour integral for the confluent hypergeometric function and the deformed contours used to evaluate the integral.

Substituting for $c = (2l+2)$, and $a = (l+1+i\gamma)$, we have, in the limit $r \to \infty$,

$$I_1 = \frac{(2l+1)!}{\Gamma(l+1-i\gamma)} \frac{1}{(2ikr)^{l+1+i\gamma}}. \tag{21}$$

Similarly,

$$\begin{aligned} I_2 &= \frac{\Gamma(c)}{2\pi i} \lim_{r\to\infty} \oint_{C_2} \frac{dze^z}{z^{c-a}(z+2ikr)^a} \\ &= e^{-2ikr} \frac{\Gamma(c)}{2\pi i} \oint_{C_2} \frac{d(z+2ikr)e^{(z+2ikr)}}{(z+2ikr-2ikr)^{c-a}(z+2ikr)^a} \\ &= \lim_{r\to\infty} \frac{e^{-2ikr}}{(-2ikr)^{c-a}} \frac{\Gamma(c)}{2\pi i} \oint_{C_2} \frac{dze^z}{z^a\left(1-\frac{z}{2ikr}\right)^{c-a}} \\ &= \frac{e^{-2ikr}}{(-2ikr)^{c-a}} \frac{\Gamma(c)}{2\pi i} \oint_{C_2} \frac{dze^z}{z^a} = \frac{e^{-2ikr}}{(-2ikr)^{c-a}} \frac{\Gamma(c)}{\Gamma(a)}. \end{aligned} \tag{22}$$

Now, the contour, C_2, is a small circle around the point $z = -2ikr$ in the first line of this equation. This contour has then been transformed into a small circle about the point $z = 0$ in subsequent lines of this equation. Therefore, in the limit $r \to \infty$, this second contour integral gives

$$I_2 = \frac{e^{-2ikr}}{(-2ikr)^{l+1-i\gamma}} \frac{(2l+1)!}{\Gamma(l+1+i\gamma)}. \tag{23}$$

Thus, we have for the solution, regular at the origin,

$$\begin{aligned} \lim_{r\to\infty} w_l &= \lim_{r\to\infty} (ikr)^{l+1}(e^{ikr})\ {}_1F_1((l+1+i\gamma);(2l+2);-2ikr) \\ &= \frac{(2l+1)!e^{\frac{\gamma\pi}{2}}}{2^{l+1}} \left(\frac{e^{ikr}}{(2kr)^{i\gamma}\Gamma(l+1-i\gamma)} + \frac{(-1)^{l+1}e^{-ikr}}{(2kr)^{-i\gamma}\Gamma(l+1+i\gamma)} \right) \\ &= \frac{i^l(2l+1)!e^{\frac{\gamma\pi}{2}}}{2^{l+1}|\Gamma(l+1+i\gamma)|} \left(e^{i(kr-\frac{\pi}{2}l-\gamma ln(2kr)+\sigma_l)} - e^{-i(kr-\frac{\pi}{2}l-\gamma ln(2kr)+\sigma_l)} \right). \end{aligned} \tag{24}$$

In the last step, we have used

$$\frac{1}{(2kr)^{\pm i\gamma}} = e^{\mp i\gamma ln(2kr)}, \tag{25}$$

and

$$\Gamma(l+1 \pm i\gamma) = |\Gamma(l+1+i\gamma)|e^{\pm i\sigma_l}, \tag{26}$$

where σ_l is the argument of the complex number $\Gamma(l+1+i\gamma)$. (σ_l is a fairly standard mathematical notation for this quantity, not to be confused with a partial cross section!). Except for an overall constant

$$\frac{i^l(2l+1)!e^{\frac{\gamma\pi}{2}}}{2^l|\Gamma(l+1+i\gamma)|},$$

as $r \to \infty$,

$$R_l(kr) = \frac{w_l(kr)}{ikr} \to \frac{\sin\left(kr - \frac{\pi}{2}l - \gamma ln(2kr) + \sigma_l\right)}{kr}, \tag{27}$$

which, in the limit in which the Coulomb parameter, γ, can be set equal to zero, is the asymptotic form of $j_l(kr)$.

If we define a $w_l^{(1)}(kr)$ and a $w_l^{(2)}(kr)$, such that

$$\begin{aligned}\frac{(w_l^{(1)} - w_l^{(2)})}{(ikr)} &= (ikr)^l (e^{ikr})\ {}_1F_1((l+1+i\gamma); (2l+2); -2ikr) \\ &= \frac{i^l (2l+1)! e^{\frac{\gamma\pi}{2}}}{2^l |\Gamma(l+1+i\gamma)|} \frac{F_l(kr)}{kr},\end{aligned} \tag{28}$$

the Coulomb function, $F_l(kr)/(kr)$, is the analogue of $j_l(kr)$, with

$$\lim_{r\to\infty} F_l(kr) = \sin\left(kr - \tfrac{1}{2}\pi l - \gamma ln(2kr) + \sigma_l\right). \tag{29}$$

Similarly, the linear combination

$$\frac{(w_l^{(1)} + w_l^{(2)})}{(kr)} = -\frac{i^l (2l+1)! e^{\frac{\gamma\pi}{2}}}{2^{l+1} |\Gamma(l+1+i\gamma)|} \frac{G_l(kr)}{kr}, \tag{30}$$

where the Coulomb function, $G_l(kr)/(kr)$, is the analogue of $-n_l(kr)$, with

$$\lim_{r\to\infty} G_l(kr) = \cos\left(kr - \tfrac{1}{2}\pi l - \gamma ln(2kr) + \sigma_l\right). \tag{31}$$

[The $w_l^{(1)}$ and $w_l^{(2)}$ are the generalizations of the spherical Hänkel functions, $h_l^{(1)}$ and $h_l^{(2)}$.]

For Coulomb scattering, therefore, the generalizations of the outgoing and incoming waves of eq. (21), are

$$\begin{pmatrix} O_l^{\text{Coul.}}(kr) \\ I_l^{\text{Coul.}}(kr) \end{pmatrix} = \frac{(G_l(kr) \pm i F_l(kr))}{r\sqrt{v}}, \tag{32}$$

with asymptotic form

$$\lim_{r\to\infty} \begin{pmatrix} O_l^{\text{Coul.}}(kr) \\ I_l^{\text{Coul.}}(kr) \end{pmatrix} = \frac{e^{\pm i(kr - \frac{1}{2}\pi l - \gamma ln(2kr) + \sigma_l)}}{r\sqrt{v}}. \tag{33}$$

It will also be important to give the radial Coulomb functions, regular at the origin, the proper normalization to make these part of a complete orthonormal set. For this purpose, it is sufficient to compare the asymptotic limits, as $r \to \infty$, of the plane wave solutions with the Coulomb functions in the high-energy limit in which the Coulomb parameter $\gamma \to 0$. Because we will use both Dirac delta function and box normalizations, we shall use $\mathcal{N} e^{i\vec{k}\cdot\vec{r}}$ for the plane wave solution, with $\mathcal{N} = (1/2\pi)^{\frac{3}{2}}$, or $\mathcal{N} = (1/\sqrt{\text{Vol.}})$. The partial wave expansion is

$$\begin{aligned}\mathcal{N} e^{i\vec{k}\cdot\vec{r}} &= \mathcal{N} \sum_l i^l (2l+1) j_l(kr) P_l(\cos\Theta) \\ &= \mathcal{N} 4\pi \sum_{l,m} i^l j_l(kr) Y_{lm}^*(\theta_k, \phi_k) Y_{lm}(\theta, \phi) \\ &\to \mathcal{N} 4\pi \sum_{l,m} i^l \left[\frac{e^{i(kr - \frac{l\pi}{2})} - e^{-i(kr - \frac{l\pi}{2})}}{2ikr}\right] Y_{lm}^*(\theta_k, \phi_k) Y_{lm}(\theta, \phi),\end{aligned} \tag{34}$$

where Θ is the angle between the vectors $\vec{r}$ and $\vec{k}$, and θ, ϕ are the polar and azimuth angles of the vector $\vec{r}$, and θ_k, ϕ_k are those of $\vec{k}$. This solution is to be compared with the partial wave expansion of the continuum Coulomb wave function

$$\begin{aligned}\psi_{\text{Coul.}}(r,\theta,\phi) &= \sum_{l,m} R_l(kr)Y_{lm}(\theta,\phi)\\ &= \sum_{l,m} A_{lm}(ikr)^l e^{ikr}[{}_1F_1((l+1+i\gamma);(2l+2);-2ikr)]Y_{lm}(\theta,\phi)\\ &\to \sum_{l,m} A_{lm}\frac{i^l(2l+1)!e^{\frac{\gamma\pi}{2}}}{2^l|\Gamma(l+1+i\gamma)|}\\ &\times\left[\frac{e^{i(kr-\frac{\pi}{2}l-\gamma ln(2kr)+\sigma_l)}-e^{-i(kr-\frac{\pi}{2}l-\gamma ln(2kr)+\sigma_l)}}{2ikr}\right]Y_{lm}(\theta,\phi).\end{aligned} \tag{35}$$

Comparing these expressions, we see that

$$A_{lm} = \mathcal{N}|\Gamma(l+1+i\gamma)|e^{-\frac{\gamma\pi}{2}}\frac{2^l}{(2l+1)!}4\pi Y^*_{lm}(\theta_k,\phi_k), \tag{36}$$

so

$$\begin{aligned}\psi_{\text{Coul.}}(r,\theta,\phi) &= \sum_{l,m}\mathcal{N}|\Gamma(l+1+i\gamma)|e^{-\frac{\gamma\pi}{2}}2^l 4\pi Y^*_{lm}(\theta_k,\phi_k)(ikr)^l e^{ikr}\\ &\times\frac{1}{2\pi i}\oint_C\frac{dze^z}{z^{l+1-i\gamma}(z+2ikr)^{l+1+i\gamma}}Y_{lm}(\theta,\phi).\end{aligned} \tag{37}$$

In this expansion, $|\Gamma(l+1+i\gamma)|$ can be related to $|\Gamma(1+i\gamma)|$ via repeated use of the relation $\Gamma(1+z)=z\Gamma(z)$, and

$$|\Gamma(1+i\gamma)|^2 = i\gamma\Gamma(i\gamma)\Gamma(1-i\gamma) = \frac{\pi\gamma}{\sinh\pi\gamma}. \tag{38}$$

Here, the product of Γ functions has been expressed in terms of a beta function, B,

$$\begin{aligned}\Gamma(i\gamma)\Gamma(1-i\gamma) = \Gamma(1)B(i\gamma,1-i\gamma) &= 2\int_0^{\pi/2} d\theta\cos^{2i\gamma-1}\theta\sin^{1-2i\gamma}\theta\\ &= 2\int_0^{\infty}\frac{dss^{2i\gamma-1}}{(s^2+1)}.\end{aligned} \tag{39}$$

The s integral can be done by contour integration techniques.

An Application: Electric Dipole Moment Matrix Element Between the Hydrogen Ground State and a Continuum State

For the photoelectric effect in the hydrogen atom (to be discussed in detail in Chapter 64), we shall need the matrix element of the electric dipole operator, $e\vec{r}$, between the hydrogen atom ground state, $|n=1,l=0,m=0\rangle$, and a continuum state in which the electron is ejected with momentum, $\hbar\vec{k}$, in a direction given by angles, θ_k, ϕ_k, with respect to a laboratory-fixed unit vector, $\vec{e}$; i.e., we shall need the matrix element

$$\langle \vec{k}|(\vec{r}\cdot\vec{e})|n=1, l=0, m=0\rangle.$$

If we choose $\vec{e}$ to lie in the laboratory z direction, only the $l=1, m=0$ term in the partial wave decomposition of eq. (37) can make a contribution to the matrix element. Also, with $Z_1 = -1$ (electron) and $Z_2 = +1$ (proton), the Coulomb parameter, γ, is negative. It will therefore be convenient to change notation and let γ stand for the absolute value of the Coulomb parameter, [requiring a change of sign in eq. (37)]. Then,

$$\begin{aligned}
&\langle \vec{k}|(\vec{r}\cdot\vec{e}_z)|n=1, l=0, m=0\rangle \\
&= \mathcal{N}|\Gamma(2-i\gamma)|e^{+\frac{\gamma\pi}{2}}2(4\pi)Y^*_{10}(\theta_k,\phi_k)\int d\Omega Y^*_{10}(\theta,\phi)\cos\theta Y_{00}(\theta,\phi) \\
&\times \int_0^\infty drr^3e^{ikr}(ikr)\frac{2e^{-(r/a_0)}}{a_0^{\frac{3}{2}}}\frac{1}{2\pi i}\oint\frac{dze^z}{z^{2+i\gamma}(z+2ikr)^{2-i\gamma}} \\
&= \mathcal{N}|\Gamma(2-i\gamma)|e^{+\frac{\gamma\pi}{2}}2\sqrt{(4\pi)}\cos\theta_k\int_0^\infty drr^3e^{ikr}(ikr)\frac{2e^{-(r/a_0)}}{a_0^{\frac{3}{2}}} \\
&\times \frac{1}{2\pi i}\oint\frac{dze^z}{z^{2+i\gamma}(z+2ikr)^{2-i\gamma}},
\end{aligned} \tag{40}$$

where a_0 is the Bohr radius. It will be convenient to make the transformation, $z = 2ikr(z'-\frac{1}{2})$, in the contour integral and the transformation, $r = a_0r'$, in the radial integral. Performing the radial integral first

$$\int_0^\infty dr'r'e^{-r'}e^{2ika_0r'z'} = \frac{1}{(1-2ika_0z')^2}, \tag{41}$$

and using the identity

$$ka_0 = \frac{1}{\gamma},$$

the matrix element of $(\vec{r}\cdot\vec{e}_z)$ becomes

$$\frac{\mathcal{N}|\Gamma(2-i\gamma)|e^{\frac{\gamma\pi}{2}}\sqrt{4\pi}\cos\theta_k\gamma^4a_0^{\frac{5}{2}}}{8}\frac{1}{2\pi i}\oint_C\frac{dz'}{(z'-\frac{1}{2})^{2+i\gamma}(z'+\frac{1}{2})^{2-i\gamma}(z'+\frac{i\gamma}{2})^2},$$

where the transformation $z = 2ikr(z'-\frac{1}{2})$ has transformed the counterclockwise contour C around the branch cut between $z=0$ and $z=-2ikr$ in the z plane (see Fig. 42.5) into a counterclockwise contour around the branch cut between $z'=-\frac{1}{2}$ and $z'=+\frac{1}{2}$ in the z' plane (see Fig. 42.6). Besides this branch cut, there is now a singular point at $z'=-\frac{1}{2}i\gamma$. The integrand vanishes for large values of $|z'|$; and the contour C can be transformed into a clockwise contour, C', around the singular point at $z'=-\frac{1}{2}i\gamma$. Residue theory shows this contour integral has the value

$$-\frac{d}{dz'}\Big[(z'-\tfrac{1}{2})^{-2-i\gamma}(z'+\tfrac{1}{2})^{-2+i\gamma}\Big]|_{z'=-(i\gamma)/2} = \left(\frac{\gamma+i}{\gamma-i}\right)^{i\gamma}\frac{64i\gamma}{(\gamma^2+1)^3}$$

$$= e^{-2\gamma \tan^{-1}(1/\gamma)} \frac{64i\gamma}{(\gamma^2+1)^3}. \tag{42}$$

Also, using eq. (38), and

$$|\Gamma(2-i\gamma)|^2 = (\gamma^2+1)|\Gamma(1-i\gamma)|^2, \tag{43}$$

we get

$$\langle \vec{k}|\vec{r}\cdot\vec{e}_z|n=1, l=0, m=0\rangle =$$
$$\mathcal{N}\sqrt{4\pi}\cos\theta_k \left[\frac{(\gamma^2+1)\pi\gamma}{\sinh\pi\gamma}\right]^{\frac{1}{2}} e^{\frac{\gamma\pi}{2}} a_0^{\frac{5}{2}} \frac{8i\gamma^5}{(\gamma^2+1)^3} e^{-2\gamma\tan^{-1}(1/\gamma)}. \tag{44}$$

The square of the absolute value of this matrix element can be put into the following convenient form

$$|\langle\vec{k}|(\vec{r}\cdot\vec{e}_z)|100\rangle|^2 = 512\pi^2\cos^2\theta_k a_0^5 \mathcal{N}^2 \frac{\gamma^{11}}{(\gamma^2+1)^5} \frac{e^{-4\gamma\tan^{-1}(1/\gamma)}}{(1-e^{-2\gamma\pi})}. \tag{45}$$

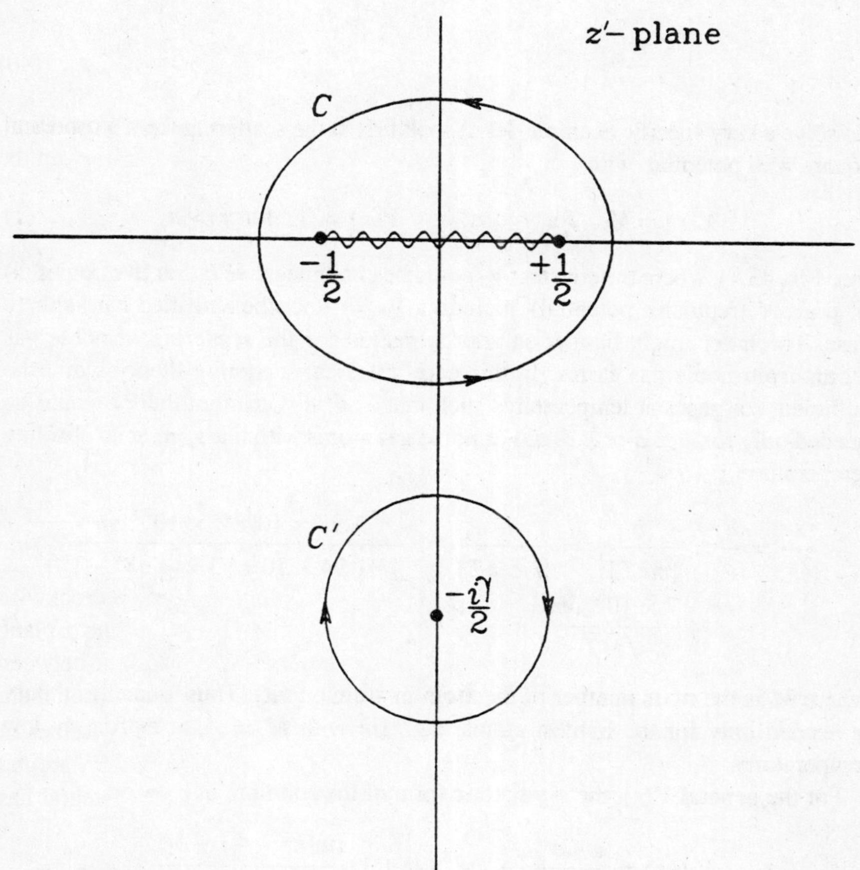

FIGURE 42.6. The z' plane contour integral.

43

A Specific Example: Scattering from Spherical Square Well Potentials

To solve a very specific example, let us look first at the scattering from a spherical square well potential, with

$$V(r) = V_0 \quad \text{for } r \leq a, \qquad V(r) = 0 \quad \text{for } r > a, \tag{1}$$

(see Fig. 43.1), where the constant V_0 could be either negative (attractive potential) or positive (repulsive potential), including $V_0 = +\infty$, the so-called hard sphere case. The latter might be a good approximation for the scattering of noble gas atoms from noble gas atoms. In this case, classical scattering theory would be sufficient for gases at temperatures such that $\lambda \ll a$. Quantum theory would be needed only for $\lambda \simeq a$ or $\lambda > a$. For noble gas atoms with mass, m, at an absolute temperature, T,

$$\begin{aligned}\frac{\lambda}{2\pi} = \frac{\hbar}{p} \simeq \frac{\hbar}{[mkT]^{\frac{1}{2}}} = \frac{\hbar c}{[mc^2kT]^{\frac{1}{2}}} &= \frac{1.97 \times 10^3 eV \times 10^{-8} cm}{[M(.94 \times 10^9 eV) \times \frac{1}{40} eV \frac{T}{300}]^{\frac{1}{2}}} \\ &\approx \frac{7 \times 10^{-8} cm}{\sqrt{MT}},\end{aligned} \tag{2}$$

where M is the mass number of the atom in atomic units. Thus, quantum theory is needed only for the lightest atoms, e.g., He with $M = 4$, at extremely low temperatures.

For the general $V(r)$, the asymptotic form of the solution, as $r \to \infty$, is

$$\begin{aligned}R_l(kr) = \frac{w_l(kr)}{kr} &\to i^l e^{i\delta_l}(2l+1)\frac{\sin(kr - \frac{\pi}{2}l + \delta_l)}{kr} \\ &= i^l e^{i\delta_l}(2l+1)\big(j_l(kr)\cos\delta_l - n_l(kr)\sin\delta_l\big),\end{aligned} \tag{3}$$

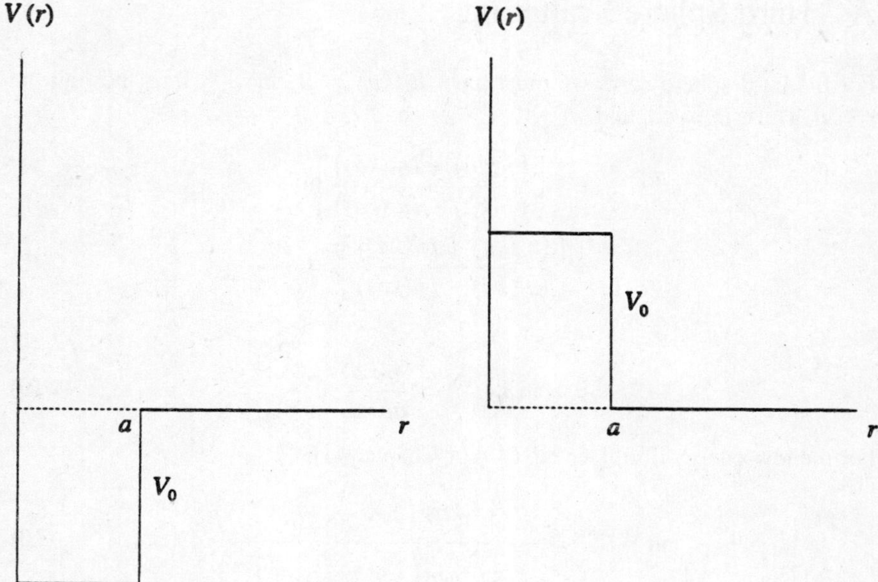

FIGURE 43.1. Square well potential scatterers.

where we have used the asymptotic forms for j_l and n_l, eq. (44) of Chapter 41. Now for the square well problem, the solution for $R_l(kr)$ is a linear combination of the free-wave radial functions, $j_l(kr)$ and $n_l(kr)$, for *all* values of $r \geq a$, so our solution for $r \geq a$ is

$$R_l(kr) = i^l e^{i\delta_l}(2l+1)\big(j_l(kr)\cos\delta_l - n_l(kr)\sin\delta_l\big). \tag{4}$$

We shall also need, again in the region $r \geq a$,

$$\left(r\frac{dR_l}{dr}\right) = (kr)i^l e^{i\delta_l}(2l+1)\big(j_l'(kr)\cos\delta_l - n_l'(kr)\sin\delta_l\big). \tag{5}$$

All of the information from the interior region will be related to the ratio, evaluated at the boundary $r = a$, of the quantity

$$\begin{aligned}\left[\frac{1}{R_l}r\frac{dR_l}{dr}\right]_{r=a} &\equiv \beta_l = (ka)\left[\frac{(j_l'\cos\delta_l - n_l'\sin\delta_l)}{(j_l\cos\delta_l - n_l\sin\delta_l)}\right]_{r=a}\\ &= (ka)\left[\frac{(j_l' + in_l')e^{2i\delta_l} + (j_l' - in_l')}{(j_l + in_l)e^{2i\delta_l} + (j_l - in_l)}\right]_{r=a}.\end{aligned} \tag{6}$$

Solving this for the phase shift,

$$e^{2i\delta_l} = -\left[\frac{(j_l - in_l)}{(j_l + in_l)}\left[\frac{1 - \frac{ka}{\beta_l}\left(\frac{j_l' - in_l'}{j_l - in_l}\right)}{1 - \frac{ka}{\beta_l}\left(\frac{j_l' + in_l'}{j_l + in_l}\right)}\right]\right]_{r=a}. \tag{7}$$

A Hard Sphere Scattering

For the hard sphere case, we must have $R_l(ka) = 0$, and $\frac{dR_l}{dr}$ must be finite at $r = a$. Thus, $\beta_l = \infty$, and

$$e^{2i\delta_l^{\text{H.Sph}}} = -\left[\frac{\big(j_l(ka) - in_l(ka)\big)}{\big(j_l(ka) + in_l(ka)\big)}\right]$$
$$= \left[\frac{\big(n_l(ka) + ij_l(ka)\big)}{\big(n_l(ka) - ij_l(ka)\big)}\right] = \frac{e^{i\delta_l^{\text{H Sph.}}}}{e^{-i\delta_l^{\text{H Sph.}}}}, \tag{8}$$

so

$$\tan\delta_l^{\text{H.Sph.}} = \frac{j_l(ka)}{n_l(ka)}. \tag{9}$$

For the low-energy limit [see eq. (45) of Chapter 41],

$$\tan\delta_l^{\text{H.Sph.}} = \frac{\frac{2^l l!}{(2l+1)!}(ka)^l\left(1 - \frac{(ka)^2}{2(2l+3)} + \cdots\right)}{-\frac{(2l)!}{(ka)^{l+1}2^l l!}\left(1 + \frac{(ka)^2}{2(2l-1)} + \cdots\right)}. \tag{10}$$

Again, in the low-energy limit δ_0 dominates. Note that $\tan\delta_0^{\text{H.Sph.}} = -\tan(ka)$; and for all (ka)

$$\delta_0^{\text{H.Sph.}} = -(ka), \tag{11}$$

whereas, for $(ka) \ll 1$,

$$\delta_1^{\text{H.Sph.}} \simeq -\frac{(ka)^3}{3} + \cdots. \tag{12}$$

In the extreme low-energy limit, neglecting all but the $l = 0$ phase shift,

$$\frac{d\sigma}{d\Omega} = \frac{\sin^2\delta_0}{k^2} = a^2, \qquad \sigma = 4\pi a^2. \tag{13}$$

In next approximation, with $(ka) \ll 1$,

$$\frac{d\sigma}{d\Omega} = \frac{1}{k^2}\left(\sin^2\delta_0 + 3\delta_1\sin 2\delta_0\cos\theta + \cdots\right)$$
$$= a^2\left(\left[1 - \frac{(ka)^2}{3}\right] + 2(ka)^2\cos\theta\right). \tag{14}$$

As (ka) increases, the differential cross section will show more and more angular oscillations, as more and more terms in the l sum contribute. Fig. 43.2 shows the θ dependence of the differential cross section for a very large value of (ka), as well as the limit of extremely large (ka). In this extreme short-wavelength limit, the differential cross section has the classical value, $\frac{a^2}{4}$, for all values of $\theta > \theta_{\text{min.}} \approx (\pi/(ka))$. The total cross section is

$$\sigma_{ka\to\infty} = 2\pi a^2, \tag{15}$$

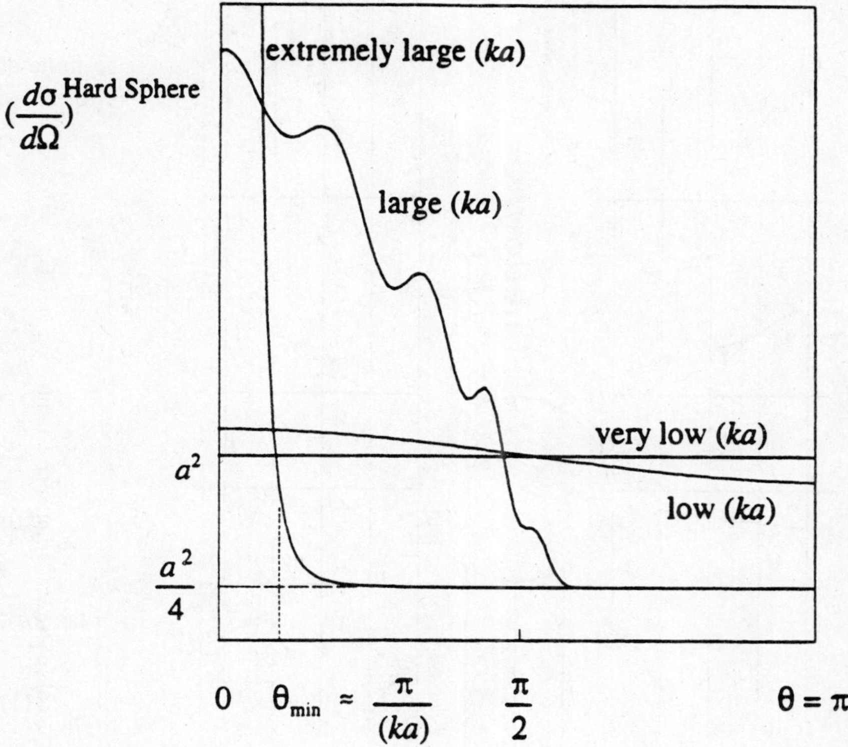

FIGURE 43.2. Hard sphere differential scattering cross sections.

that is, twice the classical value: one unit of πa^2 coming from the forward peak for $\theta < \theta_{\text{min.}}$; the second unit of πa^2 coming from all values of θ from the constant value of $\frac{a^2}{4}$ for $\frac{d\sigma}{d\Omega}$. The strong forward peak giving the extra factor of πa^2 comes from the wave description of even the classical limit. Our wave function now consists of essentially three components, a plane wave extending through *all* of space, a "true" scattered wave showing an isotropic scattering with equal probability in all directions, and a strongly forward-peaked wave interfering with the plane wave in the forward direction to make the geometrical shadow. (See Fig. 43.3.)

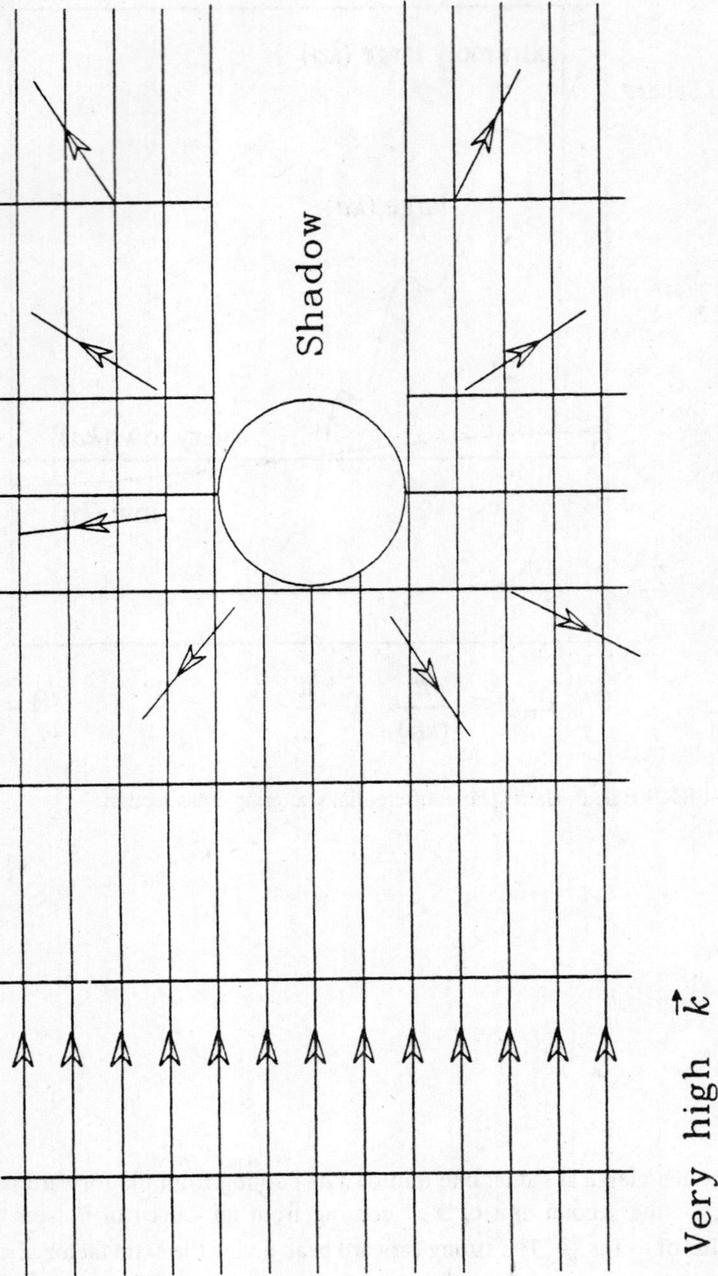

FIGURE 43.3. Hard sphere scattering, high k limit.

B The General Case: Arbitrary V_0

In the general case, we can express the phase shifts, δ_l, by

$$e^{2i\delta_l} = e^{2i\delta_l^{\text{H.Sph.}}} \frac{\beta_l - (ka)\left(\frac{j_l' - in_l'}{j_l - in_l}\right)_{r=a}}{\beta_l - (ka)\left(\frac{j_l' + in_l'}{j_l + in_l}\right)_{r=a}}, \tag{16}$$

that is, δ_l is determined exclusively from the quantity β_l. From the continuity of the wave function and its first derivative at $r = a$, we get

$$\beta_l = (k_0 a)\frac{j_l'(k_0 a)}{j_l(k_0 a)}, \qquad \text{with} \quad k_0^2 = \frac{2\mu(E - V_0)}{\hbar^2}, \tag{17}$$

where we have also used the fact that the interior solution, for $r < a$, must be regular at the origin. Using the abbreviation for the quantity

$$(ka)\left(\frac{j_l' + in_l'}{j_l + in_l}\right)_{r=a} \equiv (\Delta_l + is_l), \tag{18}$$

we have

$$e^{2i\delta_l} = e^{2i\delta_l^{\text{H.Sph.}}} \frac{\beta_l - \Delta_l + is_l}{\beta_l - \Delta_l - is_l} = e^{2i\delta_l^{\text{H.Sph.}}}\left(1 + \frac{2is_l}{\beta_l - \Delta_l - is_l}\right). \tag{19}$$

The quantities Δ_l, s_l, and $e^{i\delta_l^{\text{H.Sph.}}}$ are all known functions of ka and completely independent of the strength of the potential. Moreover, they are all smooth functions of ka. For example, for $l = 0$,

$$\delta_0^{\text{H.Sph.}} = -ka, \qquad \Delta_0 = -1, \qquad s_0 = ka. \tag{20}$$

For $l = 1$,

$$e^{i\delta_1^{\text{H.Sph.}}} = e^{-ika}\frac{(1 + ika)}{[1 + (ka)^2]^{\frac{1}{2}}}, \quad \Delta_1 = -\frac{[2 + (ka)^2]}{[1 + (ka)^2]}, \quad s_1 = \frac{(ka)^3}{[1 + (ka)^2]}. \tag{21}$$

To express the differential and total cross sections as functions of $\delta_l^{\text{H.Sph.}}$, Δ_l, s_l, and β_l, it will be useful to use the identity

$$e^{i\delta_l} \sin\delta_l = \frac{i}{2}(1 - e^{2i\delta_l}) \tag{22}$$

to rewrite

$$e^{i\delta_l} \sin\delta_l = e^{i\delta_l^{\text{H.Sph}}} \sin\delta_l^{\text{H.Sph.}} + \frac{s_l e^{2i\delta_l^{\text{H.Sph.}}}}{\beta_l - \Delta_l - is_l}. \tag{23}$$

This equation leads to an expression for the lth partial cross section

$$\sigma_l = \frac{4\pi}{k^2}(2l + 1)\left(\sin^2\delta_l^{\text{H.Sph.}} + \frac{s_l^2(1 - 2\sin^2\delta_l^{\text{H.Sph.}})}{(\beta_l - \Delta_l)^2 + s_l^2} + \frac{s_l(\beta_l - \Delta_l)\sin 2\delta_l^{\text{H.Sph.}}}{(\beta_l - \Delta_l)^2 + s_l^2}\right). \tag{24}$$

Because Δ_l, s_l, and $\delta_l^{\text{H.Sph.}}$, especially for low values of l, are mild functions of (ka), any rapid changes in $\delta_l(k)$ with k, and hence σ_l with k, must be caused by a strong k dependence in β_l. In particular, β_l can go to infinity for particular values of k for which $R_l(ka) = 0$. This process leads us to the next topic.

44

Scattering Resonances: Low-Energy Scattering

A Potential Resonances

In the last chapter, the quantities, $\delta_l^{\text{H.Sph.}}$, Δ_l, and s_l, which determine the phase shift δ_l for a particular V_0, are mild functions of the energy, E. If the quantity, β_l, is also a mild function of E, the phase shifts, δ_l, and the cross sections will be mild, smooth functions of the energy. Conversely, if a particular β_l has a very strong energy dependence, particularly at low E, the corresponding $\delta_l(k)$ may have an abrupt change over a short energy interval, and the cross section σ_l may have a sharp peak over a small energy interval, leading to a resonance in the scattering cross section.

If we expand β_l as a power series in E

$$\beta_l = c_l^{(0)} + c_l^{(1)} E + \cdots, \tag{1}$$

neglecting E^2 and E^3 terms, which may be a good approximation at low energy, particularly if $c_l^{(1)}$ is large; then,

$$e^{2i(\delta_l - \delta_l^{\text{H.Sph.}})} = \frac{\beta_l - \Delta_l + is_l}{\beta_l - \Delta_l - is_l} = \frac{\frac{c_l^{(0)}}{c_l^{(1)}} + E - \frac{\Delta_l}{c_l^{(1)}} + i\frac{s_l}{c_l^{(1)}}}{\frac{c_l^{(0)}}{c_l^{(1)}} + E - \frac{\Delta_l}{c_l^{(1)}} - i\frac{s_l}{c_l^{(1)}}}. \tag{2}$$

Let

$$\left(\frac{c_l^{(0)}}{c_l^{(1)}} - \frac{\Delta_l}{c_l^{(1)}}\right) \equiv -E_0, \qquad \frac{s_l}{c_l^{(1)}} \equiv -\frac{\Gamma}{2}. \tag{3}$$

$\Gamma \ll 1$ if $c_l^{(1)}$ is large (strong energy dependence of β_l) because s_l is small at low energies. (Recall, e.g., $s_0 = ka$, $s_1 = (ka)^3/[1+(ka)^2]$.) With this notation, we have

$$e^{2i(\delta_l - \delta_l^{\text{H.Sph.}})} = \frac{E_0 - E + i\frac{1}{2}\Gamma}{E_0 - E - i\frac{1}{2}\Gamma}, \tag{4}$$

so

$$\tan(\delta_l - \delta_l^{\text{H.Sph.}}) = \frac{\Gamma}{2(E_0 - E)}. \tag{5}$$

In the vicinity of $E = E_0$, the phase shift δ_l may have wide excursions.

$$\text{For example, for } E = E_0 \mp \tfrac{1}{2}\Gamma, \quad \tan(\delta_l - \delta_l^{\text{H.Sph.}}) = \pm 1,$$

$$\text{while for } E = E_0 \mp 2\Gamma, \quad \tan(\delta_l - \delta_l^{\text{H.Sph.}}) = \pm\tfrac{1}{4},$$

so δ_l changes by $\sim \frac{\pi}{2}$ over an energy interval Γ about E_0, and changes by $\sim .85\pi$ over an energy interval 4Γ about E_0. A change in $\delta_l \sim \pi$ over a small energy interval $\sim \Gamma$ signals a possible sharp rise and subsequent fall in σ. The resonance peak can be very narrow, if $\frac{1}{2}\Gamma = (s_l/c_l^{(1)})$ is very small. Fig. 44.1 shows an example in which β_0 has no spectacular E dependence near $E \sim 0$, so δ_0 is a smooth function of k. Conversely, δ_1 shows a sharp rise of $\sim \frac{1}{2}\pi$ to π at a small value of ka, leading to a resonance peak in σ. We note in passing our expression for δ_l in terms of its tangent determines δ_l only to within a multiple of π. The δ_l can be normalized, however, by the requirement $\delta_l \to 0$ as $E \to \infty$.

When expressed in terms of the parameters E_0 and Γ, the lth partial cross section has the form

$$\sigma_l = \frac{4\pi}{k^2}(2l+1)\left(\frac{\left(\frac{\Gamma}{2}\right)^2\left(1 - 2\sin^2\delta_l^{\text{H.Sph.}}\right)}{(E - E_0)^2 + \left(\frac{\Gamma}{2}\right)^2} + \frac{\frac{\Gamma(E_0 - E)}{2}\sin 2\delta_l^{\text{H.Sph.}}}{(E - E_0)^2 + \left(\frac{\Gamma}{2}\right)^2} + \sin^2\delta_l^{\text{H.Sph.}}\right) \tag{6}$$

The sharpness of the resonance peak is amplified at low energy by the $1/k^2$ factor of this expression. Also, for $(ka) \ll 1$, $\delta_l^{\text{H.Sph.}} \ll 1$, so the first of the three terms dominates. For $E = E_0 \pm \frac{1}{2}\Gamma$, therefore, $\sigma_l \approx \frac{1}{2}(\sigma_l)_{\text{max.}}$, so Γ gives the half-width of the resonance.

B Low-Energy Scattering for General $V(r)$: Scattering Length

As we have seen in the last two chapters, for very low-energy scattering, pure s wave ($l = 0$-partial wave) scattering dominates, leading to an isotropic differential cross section. In the limit $k \to 0$, the differential equation for the $l = 0$ radial function goes over to

$$\frac{d^2u_0}{dr^2} - \frac{2\mu}{\hbar^2}V(r)u_0 = 0. \tag{7}$$

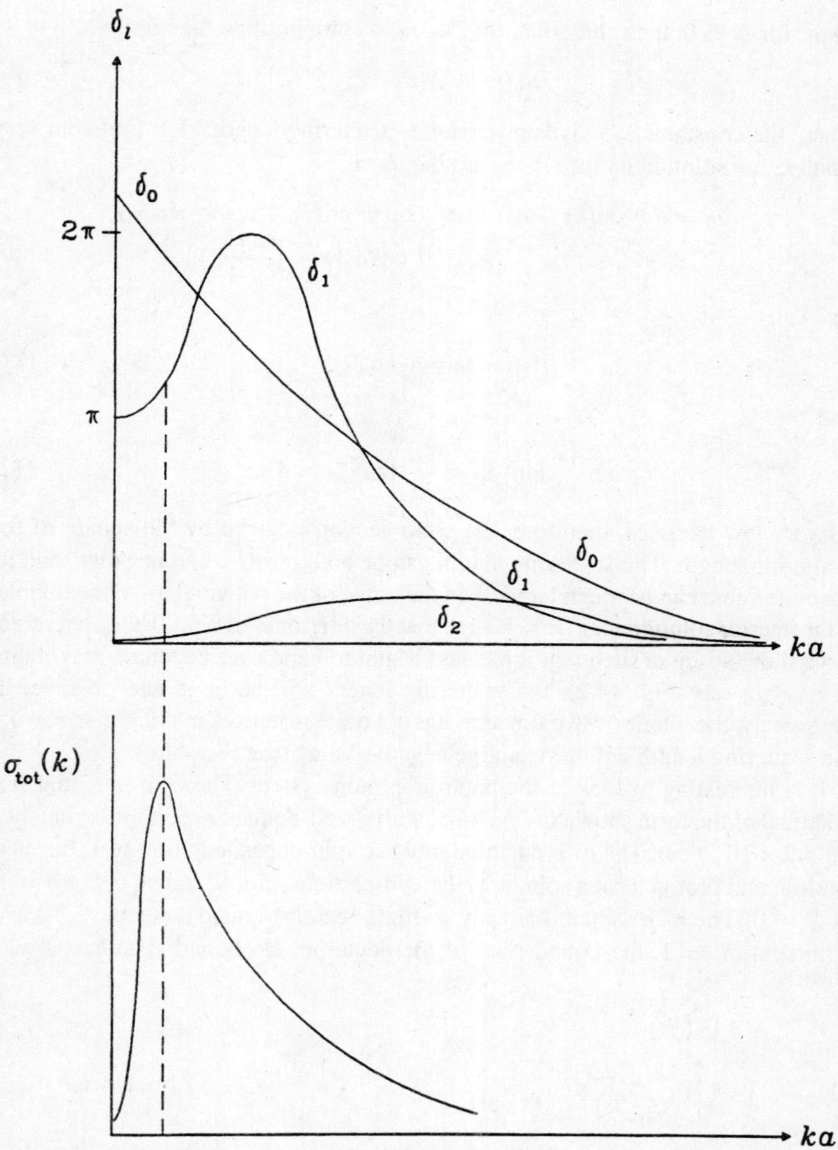

FIGURE 44.1. Scattering phase shifts and cross section for a potential with a low-energy p wave resonance.

Thus, for $r > r_0$, where r_0 is the range of the potential, this differential equation collapses to

$$\frac{d^2 u_0}{dr^2} = 0. \tag{8}$$

Thus, for $E = 0$, the solution u_0 for $r > r_0$ is a straight-line function

$$u_0(r) = A(r - a_{\text{sc.l.}}), \tag{9}$$

where the constant, $a_{\text{sc.l.}}$, is known as the "scattering length." For finite but very small k, the solution u_0 for $r > r_0$ has the form

$$\begin{aligned} u_0 = e^{i\delta_0} \sin(kr + \delta_0) &= e^{i\delta_0}(\sin kr \cos\delta_0 + \cos kr \sin\delta_0) \\ &\approx e^{i\delta_0} k \cos\delta_0 \Big(r - \big(\frac{\tan\delta_0}{-k}\big)\Big), \end{aligned} \tag{10}$$

so

$$\tan\delta_0 = -k a_{\text{sc.l.}}, \tag{11}$$

and

$$\sigma = \frac{4\pi}{k^2}\sin^2\delta_0 \approx \frac{4\pi}{k^2}\tan^2\delta_0 \approx 4\pi a^2_{\text{sc.l.}}. \tag{12}$$

At very low energies, therefore, the cross section is given by the square of the scattering length. The scattering length can be both positive and negative, and its absolute value can be much larger than the range of the potential, r_0. For example, if the interior solution for $r < r_0$ is such that the interior u_0 at $E = 0$ has a curvature toward the abscissa such that it has just begun to bend over beyond a maximum, at $r = r_0$ (see Fig. 44.2), the scattering length will be large and positive. If, conversely, the interior wave function has not quite reached a maximum at $r = r_0$, the scattering length will have a large negative value (see Fig. 44.4).

It is interesting to look at the neutron–proton system. The n–p potential is a potential of the form shown in Fig. 44.3, with a well-defined r_0 of approximately 2 fm = 2×10^{-13}cm. The n–p potential must be spin-dependent, however. Because neutron and proton have a spin of $\frac{1}{2}$, the two-particle spin, S, can be either $S = 1$, or $S = 0$. The n–p system has only a single (barely)-bound state, at -2.2 MeV, with spin, $S = 1$, the bound state of the deuteron. No bound state exists with

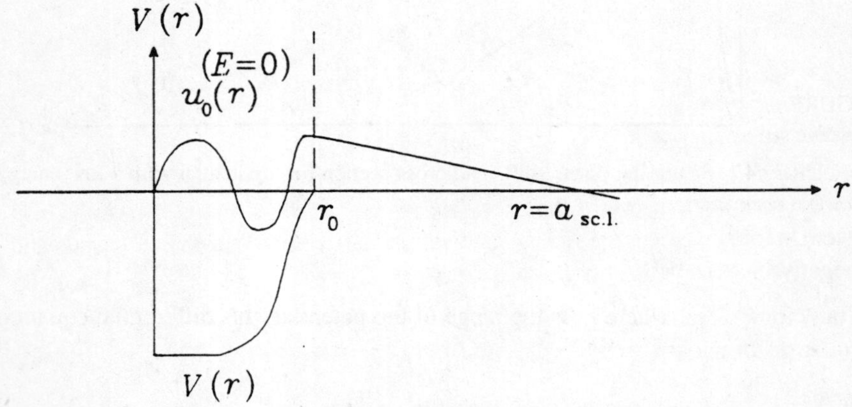

FIGURE 44.2. Potential with a positive scattering length.

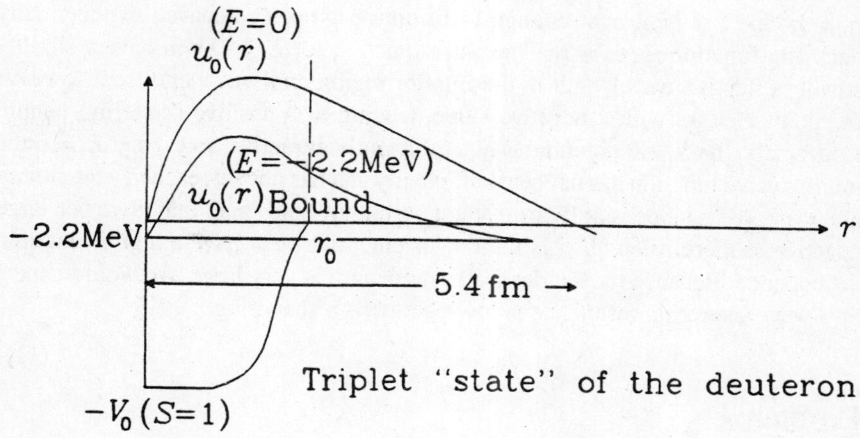

FIGURE 44.3. Low-energy proton–neutron scattering. The triplet $S = 1$ proton–neutron potential with its bound-state eigenfunction and its $E = 0$ function with positive scattering length.

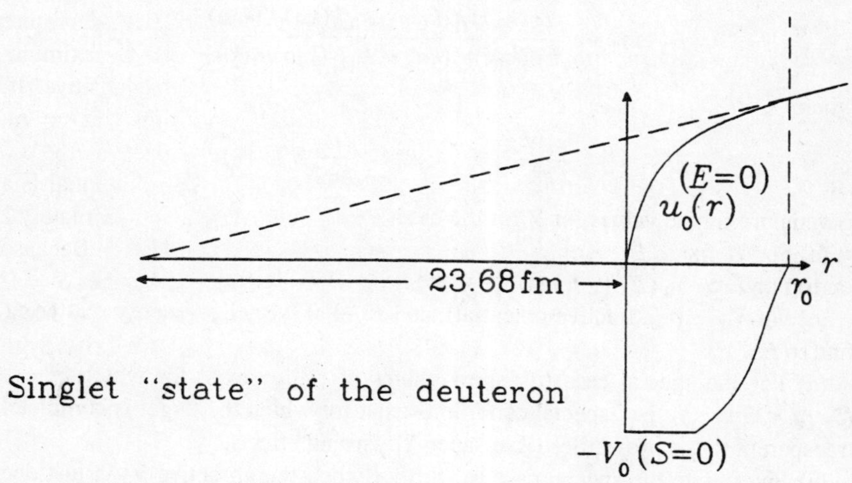

FIGURE 44.4. The singlet $S = 0$ proton–neutron potential with its $E = 0$ function with negative scattering length.

$S = 0$. From the analysis of the experimental low-energy cross sections, the n–p system is found to have a positive scattering length in the triplet ($S = 1$) state and a negative scattering length in the singlet ($S = 0$) state

$$\begin{aligned} a_{\text{sc.l.}}^{S=1} &= +5.400 \pm .001\text{fm}, \\ a_{\text{sc.l.}}^{S=0} &= -23.677 \pm .029\text{fm}. \end{aligned} \tag{13}$$

The $S = 1$ bound state wave function of the deuteron at -2.2 MeV has enough curvature toward the abscissa in the interior region, so it has a negative slope at

$r = r_0$ and has bent over enough to fit onto a properly behaved exponentially decaying function for $r > r_0$. The wave function for $E = 0$ will have a slightly smaller effective wavelength in the interior region and will therefore also reach the point $r = r_0$ with a negative slope, leading to a positive scattering length. Conversely, the $S = 0$ potential is not sufficiently attractive, so even at $E = 0$ the interior wave function has not bent sufficiently and has not yet reached a maximum at $r = r_0$, so its slope is still positive at $r = r_0$, but is so small that there is a large negative scattering length. Also, the n–p potential must have a significant spin dependence. Because the singlet scattering length is very large, we would expect an s wave scattering resonance in low-energy n–p scattering.

Problems

1. For the spherical step potential

$$V = V_0, \quad 0 \le r < a, \qquad\qquad V = 0, \quad r > a$$

show

$$\tan \delta_l = \frac{k j_l(k_0 a) j_l'(ka) - k_0 j_l(ka) j_l'(k_0 a)}{k j_l(k_0 a) n_l'(ka) - k_0 j_l'(k_0 a) n_l(ka)},$$

where

$$k^2 = \frac{2\mu E}{\hbar^2}, \qquad\qquad k_0^2 = \frac{2\mu}{\hbar^2}(E - V_0).$$

Give approximate values for δ_l for the cases:

(a) For V_0 fixed, $E \to \infty$.

(b) For $E > V_0$ (E, V_0 fixed), $l \to \infty$.

(c) For $V_0 < 0$ (attractive potential), and $E \to 0$ (very low-energy scattering), find $\sigma(E)$.

(d) For the special case (c), show values of $|V_0|a^2$ exist for which $\sigma = 0$ ($\delta_0 = \pi, 2\pi, \ldots$); i.e., special conditions exist for which the target is completely transparent to the projectiles (Ramsauer–Townsend effect).

(e) Investigate the special case (c) further when the attractive V_0 is just deep enough to have one bound state with a binding energy, $\epsilon \ll |V_0|$ (this is approximately true for the special case of low-energy neutron–proton scattering; the deuteron has a binding energy of only 2.2 MeV). Show, in this case,

$$\sqrt{\frac{2\mu a^2 |V_0|}{\hbar^2}} \approx \frac{\pi}{2} + \Delta, \qquad\qquad (\Delta \ll 1).$$

Show first the bound state energies with $E = -\epsilon$ can be obtained from solutions to the transcendental equation

$$\xi \cot \xi = -\sqrt{\frac{2\mu a^2 |V_0|}{\hbar^2} - \xi^2}, \qquad \text{where} \quad \xi = \sqrt{\frac{2\mu a^2}{\hbar^2}(|V_0| - \epsilon)}.$$

Then show, in the special case of one bound state only, with $\epsilon \ll |V_0|$, the low-energy cross section is a function of the length parameter $\sqrt{\hbar^2/2\mu\epsilon}$ only. Find the scattering length as a function of this length parameter, assuming

$$a^2 \ll \frac{\hbar^2}{2\mu\epsilon}.$$

2. Potential Resonances. For an attractive square well potential of depth $|V_0|$, with

$$k_0 a = \sqrt{\frac{2\mu E a^2}{\hbar^2} + \frac{2\mu|V_0|a^2}{\hbar^2}} = \sqrt{(ka)^2 + \frac{2\mu|V_0|a^2}{\hbar^2}} \equiv \rho^2 = \sqrt{(ka)^2 + \rho_0^2},$$

investigate the possibility the $l = 1$ partial cross section can give rise to a resonant peak at low values of (ka). Show, in particular, in the limit $(ka) \to 0$, the quantity $(\beta_1 - \Delta_1)$ has the value zero for $\rho = n\pi$. Investigate the special case, $n = 2$. Test this case by choosing a value of ρ_0 slightly less than 2π. For example, choose $\rho_0 = 6.20$. For this case, plot the phase shifts $\delta_0, \delta_1, \delta_2$ and the partial cross sections $\sigma_0, \sigma_1, \sigma_2$ as functions of (ka), paying particular attention to the interval, $0 \leq ka \leq 1.5$.

3. For low-energy neutron–proton scattering, the neutron–proton interaction is approximated quite well by a square well with $V = -|V_0|$ for $r \leq a$; $V = 0$ for $r > a$. A reasonable estimate of the range of this potential would be $a = 2$fm. Use the experimentally observed triplet and singlet scattering lengths, $a^{S=1}_{\text{sc.l.}} = +5.400$ fm, $a^{S=0}_{\text{sc.l.}} = -23.677$ fm, to estimate the triplet and singlet well depths, $|V_0|$, and note their difference.

Plot the singlet scattering cross section, $\sigma_{S=0}$, for the range $E = 0 \to 0.2$MeV.

45

Integral Equation for Two-Body Relative Motion: Scattering Green's Functions in Coordinate Representation

Let us now solve the two-body relative motion scattering problem by integral equation techniques. Our main aim will be to find an approximation for the scattering cross sections valid in the limit of large energies. (Recall the partial wave expansion was useful mainly in the low energy limit, where only a few partial waves may make a significant contribution).

We want to find continuum solutions to

$$H\psi = E\psi, \qquad \text{with} \quad H = H_0 + V(\vec{r}), \qquad H_0 = -\frac{\hbar^2}{2\mu}\nabla^2, \tag{1}$$

where the plane wave solutions, the solutions for the equation for H_0, are known

$$(E - H_0)\phi(\vec{r}) = 0. \tag{2}$$

A Box Normalization: Discrete Spectrum

For the moment, let us take box normalization, so the continuous spectrum is replaced by a finely discrete spectrum,

$$(E_n - H_0)\phi_n(\vec{r}) = 0, \tag{3}$$

with

$$\phi_n(\vec{r}) = \frac{e^{i(\vec{k}_n\cdot\vec{r})}}{\sqrt{L^3}}, \qquad \vec{k}_n \equiv \vec{k}_{n_1 n_2 n_3}, \tag{4}$$

$$E_n = \frac{\hbar^2}{2\mu} k_n^2 = \frac{\hbar^2}{2\mu} \frac{4\pi^2}{L^2} (n_1^2 + n_2^2 + n_3^2), \tag{5}$$

$$\langle \phi_n | \phi_{n'} \rangle = \delta_{nn'} \equiv \delta_{n_1 n'_1} \delta_{n_2 n'_2} \delta_{n_3 n'_3}. \tag{6}$$

We now seek solutions, ψ, of the wave equation for the full H in the form of a Fourier series,

$$\psi = \sum_n c_n \phi_n. \tag{7}$$

In the wave equation

$$(E - H_0)\psi(\vec{r}) = V(\vec{r})\psi(\vec{r}) = f(\vec{r}), \tag{8}$$

we will also expand the source term, $f(\vec{r})$, in a Fourier series. (This source term carries the source of the scattering process.)

$$f(\vec{r}) = \sum_n a_n \phi_n, \tag{9}$$

with

$$a_n = \int d\vec{r}\,' \phi_n^*(\vec{r}\,') f(\vec{r}\,'), \tag{10}$$

where the integral is over the volume of our cubical laboratory of volume L^3. Then,

$$(E - H_0) \sum_n c_n \phi_n = \sum_n (E - E_n) c_n \phi_n = \sum_n \int d\vec{r}\,' \phi_n^*(\vec{r}\,') \phi_n(\vec{r}) f(\vec{r}\,'). \tag{11}$$

Using the orthonormality of the ϕ_n

$$c_n = \int \frac{d\vec{r}\,' \phi_n^*(\vec{r}\,') f(\vec{r}\,')}{(E - E_n)}, \tag{12}$$

$$\psi(\vec{r}) = \int d\vec{r}\,' \sum_n \frac{\phi_n^*(\vec{r}\,') \phi_n(\vec{r})}{(E - E_n)} f(\vec{r}\,') = \int d\vec{r}\,' G(\vec{r}, \vec{r}\,') f(\vec{r}\,'), \tag{13}$$

where we have interchanged the integration and the infinite sum, assuming the infinite series has the proper convergence properties. (We have quietly assumed $E \neq E_n$, for any n.) Here, $G(\vec{r}, \vec{r}\,')$ is the Green's function for the scattering problem

$$G(\vec{r}, \vec{r}\,') = \sum_n \frac{\phi_n^*(\vec{r}\,') \phi_n(\vec{r})}{(E - E_n)}, \tag{14}$$

where

$$(E - H_0) G(\vec{r}, \vec{r}\,') = \sum_n \phi_n^*(\vec{r}\,') \phi_n(\vec{r}) = \delta(\vec{r} - \vec{r}\,'). \tag{15}$$

Also, by acting with $(E - H_0)$ on our integral expression for $\psi(\vec{r})$, we can check to make sure we regain the original wave equation,

$$(E - H_0)\psi(\vec{r}) = \int d\vec{r}\,' \sum_n \phi_n^*(\vec{r}\,')\phi_n(\vec{r})V(\vec{r}\,')\psi(\vec{r}\,') = V(\vec{r})\psi(\vec{r}). \tag{16}$$

B Continuum Green's Function

Alternatively, we could have taken for our plane-wave states true continuum states

$$\phi_{\vec{k}}(\vec{r}) = \frac{e^{i\vec{k}\cdot\vec{r}}}{(2\pi)^{\frac{3}{2}}} = \langle\vec{r}|\vec{k}\rangle, \tag{17}$$

with Dirac delta-function orthonormality

$$\langle\vec{k}\,'|\vec{k}\rangle = \delta(\vec{k} - \vec{k}\,'). \tag{18}$$

It will now be convenient to write the wave equation in the form

$$\frac{2\mu}{\hbar^2}(E - H_0 - V(\vec{r}))\psi(\vec{r}) = 0 \longrightarrow (k^2 + \nabla^2 - U(\vec{r}))\psi(\vec{r}) = 0, \tag{19}$$

or

$$(k^2 + \nabla^2)\psi(\vec{r}) = f(\vec{r}) = U(\vec{r})\psi(\vec{r}), \qquad \text{with} \quad U(\vec{r}) = \frac{2\mu}{\hbar^2}V(\vec{r}). \tag{20}$$

The expansion of ψ in plane waves now involves the integral (continuum sum),

$$\psi(\vec{r}) = \int d\vec{k}\,' c_{\vec{k}',}\,\phi_{\vec{k}'}(\vec{r}) \tag{21}$$

with

$$f(\vec{r}) = \int d\vec{k}\,' a_{\vec{k}'}\phi_{\vec{k}'}(\vec{r}) = \int d\vec{k}\,' \int d\vec{r}\,' \phi_{\vec{k}'}^*(\vec{r}\,')\phi_{\vec{k}'}(\vec{r}) f(\vec{r}\,'). \tag{22}$$

Now,

$$\psi(\vec{r}) = \int d\vec{k}\,' \int d\vec{r}\,' \frac{\phi_{\vec{k}'}^*(\vec{r}\,')\phi_{\vec{k}'}(\vec{r})}{(k^2 - k'^2)} U(\vec{r}\,')\psi(\vec{r}\,'). \tag{23}$$

We will regain the original wave equation for $\psi(\vec{r})$; the operator $(k^2 + \nabla^2)$ acting on $\psi(\vec{r})$ gives

$$\begin{aligned}
(k^2 + \nabla^2)\psi(\vec{r}) &= \int d\vec{k}\,' \int d\vec{r}\,' \frac{e^{i\vec{k}'\cdot(\vec{r}-\vec{r}\,')}}{(2\pi)^3} U(\vec{r}\,')\psi(\vec{r}\,') \\
&= \int d\vec{r}\,' \left(\int d\vec{k}\,' \frac{e^{i\vec{k}'\cdot(\vec{r}-\vec{r}\,')}}{(2\pi)^3} \right) U(\vec{r}\,')\psi(\vec{r}\,') \\
&= \int d\vec{r}\,' \delta(\vec{r} - \vec{r}\,') U(\vec{r}\,')\psi(\vec{r}\,') = U(\vec{r})\psi(\vec{r}).
\end{aligned} \tag{24}$$

As we can see from this result, the properties of the Dirac delta function are such that the interchange of $\int d\vec{k}'$ and $\int d\vec{r}'$ were permissible. If we attempt a similar interchange in eq. (23), to get a Green's function in a form analogous to that of eq. (13), with the discrete sum replaced by the integral $\int d\vec{k}'$, the additional singularities at the values $k' = \pm k$ will cause the integral to diverge. In order to gain the Green's function, we need to modify the $\int d\vec{k}'$ integral. In addition, because the free wave solution, $\phi_{\vec{k}}(\vec{r})$, will automatically satisfy the homogeneous part of eq. (20), we can add it to the solution, and seek solutions to eq. (20) of the form

$$\psi_{\vec{k}}(\vec{r}) = \phi_{\vec{k}}(\vec{r}) + \int d\vec{r}\,' G(\vec{r}, \vec{r}\,')U(\vec{r}\,')\psi_{\vec{k}}(\vec{r}\,'), \tag{25}$$

with

$$G(\vec{r}, \vec{r}\,') = \frac{1}{(2\pi)^3} \int_{\text{mod.}} d\vec{k}' \frac{e^{i\vec{k}'\cdot(\vec{r}-\vec{r}\,')}}{(k^2 - k'^2)}, \tag{26}$$

where the $\int d\vec{k}'$ must be modified to avoid the troubles caused by the singularities at $k' = \pm k$. To see how these singularity problems can be avoided in a way that leads to a Green's function leading to a solution ψ tailored to the boundary conditions of our problem, let us first set up the k' integral without modification. We shall introduce polar coordinates for the vector, $\vec{k}'$, choosing the vector $\vec{r} - \vec{r}\,'$ to be along the z direction, (see Fig. 45.1), so

$$\begin{aligned} G(\vec{r}, \vec{r}\,') &= -\frac{1}{(2\pi)^3} \int d\vec{k}' \frac{e^{ik'|\vec{r}-\vec{r}\,'|\cos\theta_{k'}}}{(k'^2 - k^2)} \\ &= -\frac{1}{(2\pi)^3} 2\pi \int_0^\infty dk' k'^2 \int_0^\pi d\theta_{k'} \sin\theta_{k'} \frac{e^{ik'|\vec{r}-\vec{r}\,'|\cos\theta_{k'}}}{(k'^2 - k^2)} \\ &= -\frac{1}{(2\pi)^2} \int_0^\infty dk' k'^2 \int_{-1}^{+1} d\zeta \frac{e^{ik'|\vec{r}-\vec{r}\,'|\zeta}}{(k'^2 - k^2)} \\ &= -\frac{1}{(2\pi)^2} \frac{1}{i|\vec{r}-\vec{r}\,'|} \int_0^\infty dk' k' \frac{\left(e^{ik'|\vec{r}-\vec{r}\,'|} - e^{-ik'|\vec{r}-\vec{r}\,'|}\right)}{(k'^2 - k^2)}. \end{aligned} \tag{27}$$

Because the last integrand is an even function of k', we can convert it to more symmetrical form

$$\begin{aligned} G(\vec{r}, \vec{r}\,') &= \frac{i}{2(2\pi)^2} \frac{1}{|\vec{r}-\vec{r}\,'|}(I_1 - I_2), \qquad \text{with} \\ (I_1 - I_2) &= \\ &\left(\int_{-\infty}^{+\infty} \frac{dk'k'}{(k'^2 - k^2)} e^{ik'|\vec{r}-\vec{r}\,'|} - \int_{-\infty}^{+\infty} \frac{dk'k'}{(k'^2 - k^2)} e^{-ik'|\vec{r}-\vec{r}\,'|} \right). \end{aligned} \tag{28}$$

Without modification, both integrals diverge because of the singularities at $k' = \pm k$. We shall try to overcome this difficulty by replacing $k \rightarrow k + i\eta$ and taking

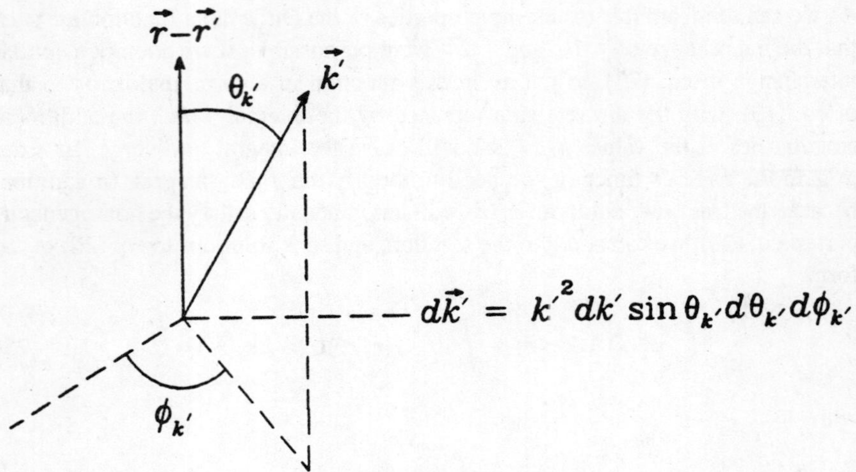

FIGURE 45.1.

the limit $\eta \to 0$ after doing the integrals by contour integration techniques.

$$I_1 = \lim_{\eta\to 0} \int_{-\infty}^{+\infty} \frac{dk'k'e^{ik'|\vec{r}-\vec{r}'|}}{(k'-k-i\eta)(k'+k+i\eta)} = \lim_{\eta\to 0} \oint_{C_1} \frac{dk'k'e^{ik'|\vec{r}-\vec{r}'|}}{(k'-k-i\eta)(k'+k+i\eta)}, \tag{29}$$

where the contour, C_1 (see Fig. 45.2), has been chosen such that the integral over the semicircle in the upper half of the complex k' plane goes to zero as its radius goes to infinity, so the integral is given by the residue theorem

$$I_1 = \lim_{\eta\to 0}\left(2\pi i\left[\left(\frac{k'e^{ik'|\vec{r}-\vec{r}'|}}{(k'+k+i\eta)}\right)\right]_{k'=k+i\eta} = 2\pi i\frac{1}{2}e^{i(k+i\eta)|\vec{r}-\vec{r}'|}\right)$$
$$= i\pi e^{ik|\vec{r}-\vec{r}'|}. \tag{30}$$

Similarly, the integral I_2 can be converted to a contour integral over the contour, C_2, in the lower half of the complex k' plane (see Fig. 45.2). The residue theorem at the pole $k' = -k - i\eta$ now gives

$$I_2 = -i\pi e^{ik|\vec{r}-\vec{r}'|}, \tag{31}$$

so with the prescription, $k \to k + i\eta$, the integrals have been done, and the Green's function becomes

$$G(\vec{r},\vec{r}') = -\frac{1}{4\pi}\frac{e^{ik|\vec{r}-\vec{r}'|}}{|\vec{r}-\vec{r}'|} \equiv G^{(+)}(\vec{r},\vec{r}'); \tag{32}$$

i.e., this Green's function has the form proper for an outgoing spherical wave, required by the physical boundary conditions of our scattering solution. With the modification, $k \to k - i\eta$, however, we would have obtained a solution corresponding to an incoming spherical wave,

$$G^{(-)}(\vec{r},\vec{r}') = -\frac{1}{4\pi}\frac{e^{-ik|\vec{r}-\vec{r}'|}}{|\vec{r}-\vec{r}'|}. \tag{33}$$

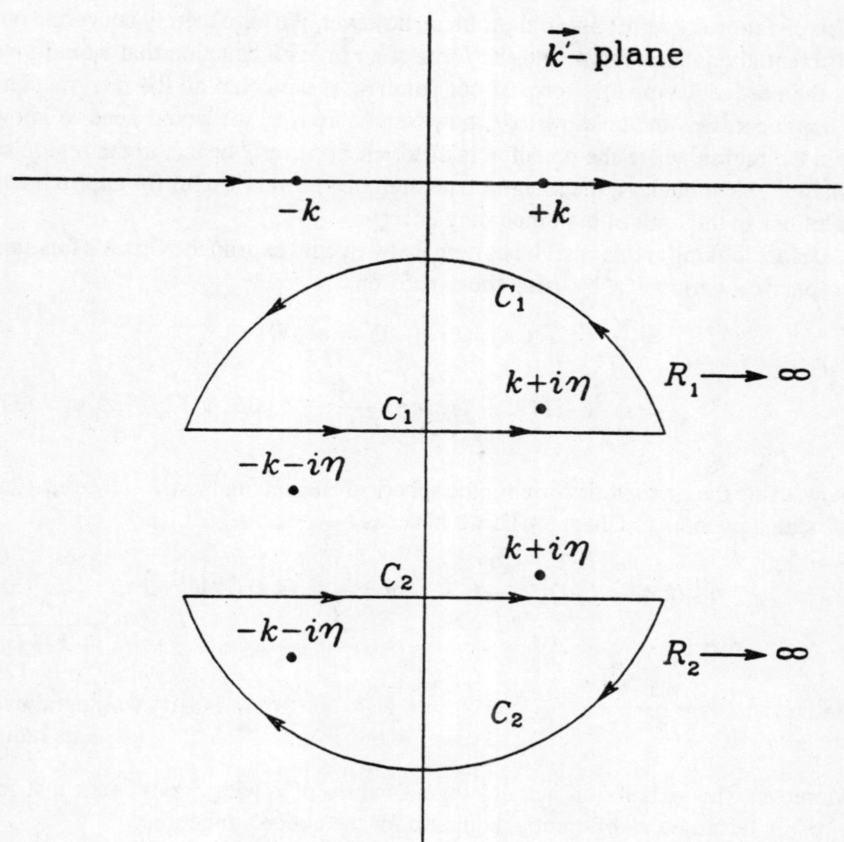

FIGURE 45.2. Green's function contours.

The solution we need is

$$\psi^{(+)}_{\vec{k}}(\vec{r}) = \frac{e^{i\vec{k}\cdot\vec{r}}}{(2\pi)^{\frac{3}{2}}} + \int d\vec{r}\,' G^{(+)}(\vec{r},\vec{r}\,')U(\vec{r}\,')\psi(\vec{r}\,'). \tag{34}$$

Because we shall be interested in $\psi^{(+)}$, particularly in the limit $r \to \infty$, we shall need to expand $|\vec{r} - \vec{r}\,'|$ in the Green's function. In particular,

$$|\vec{r} - \vec{r}\,'| = \sqrt{(r^2 + r'^2 - 2rr'\cos\Theta)} = r - r'\cos\Theta + \text{Order}\Big(\frac{r'^2}{r}\Big)$$
$$\frac{1}{|\vec{r} - \vec{r}\,'|} = \frac{1}{r} + \text{Order}\Big(\frac{r'}{r^2}\Big). \tag{35}$$

Therefore, in the needed limit $r \to \infty$, our $\psi^{(+)}$ has the form

$$\psi^{(+)}_{\vec{k}}(\vec{r}) \longrightarrow \frac{e^{i\vec{k}\cdot\vec{r}}}{(2\pi)^{\frac{3}{2}}} - \frac{e^{ikr}}{4\pi r}\int d\vec{r}\,' e^{-ikr'\cos\Theta}U(\vec{r}\,')\psi_{\vec{k}}(\vec{r}\,'). \tag{36}$$

This equation does not solve our problem, however. We have merely converted our differential equation for ψ into the form of an integral equation that would give us the needed asymptotic form of our solution, if we could do the integral. The integral contains the unknown $\psi_{\vec{k}}$, however. Moreover, we would need to know it in the region where the potential is different from zero, hence, in the region of small $\vec{r}\,'$. Nevertheless, the integral equation may be very useful for approximate solutions in the limit of high incoming energy.

Before looking at this case, let us note that we could expand the Green's function in spherical harmonics, by using the expansion

$$\begin{aligned} e^{-ikr'\cos\Theta} &= \sum_l i^l(2l+1)j_l(kr')(-1)^l P_l(\cos\Theta) \\ &= \sum_{l,m}(-i)^l(2l+1)j_l(kr')\frac{4\pi}{(2l+1)}Y_{lm}(\theta,\phi)Y^*_{lm}(\theta',\phi'). \end{aligned} \tag{37}$$

Now, using the asymptotic form for the spherical Hänkel function, $h_l^{(1)}$ [see eq. (26) of math. appendix to Chapter 41], we have, as $r\to\infty$,

$$h_l^{(1)}(kr)\to(-i)^{l+1}\frac{e^{ikr}}{r},\qquad \text{or}\qquad \frac{e^{ikr}}{r}\to ik(i)^l h_l^{(1)}(kr), \tag{38}$$

so

$$G^{(+)}(\vec{r},\vec{r}\,')=-\frac{1}{4\pi}\frac{e^{ikr}}{r}e^{-ikr'\cos\Theta}=-ik\sum_{l,m}h_l^{(1)}(kr)j_l(kr')Y_{lm}(\theta,\phi)Y^*_{lm}(\theta',\phi'). \tag{39}$$

Moreover, this expression is valid for all values of r with $r>r'$ (not just for $r\gg r'$), because the differential equation for the Green's function

$$(\nabla^2+k^2)G^{(+)}(\vec{r},\vec{r}\,')=\delta(\vec{r}-\vec{r}\,') \tag{40}$$

has a source term only at the point $\vec{r}=\vec{r}\,'$, so for any $r>r'$ the solution must be an outgoing free wave.

Now, if we make a partial wave expansion for the $\psi_{\vec{k}}(\vec{r}\,')$

$$\psi_{\vec{k}}(\vec{r}\,')=\sum_{l,m}R_l(kr')Y_{lm}(\theta',\phi') \tag{41}$$

in the Green's function integral, this integral can be put in the form

$$\sum_{lm}\int d\vec{r}\,'G(\vec{r},\vec{r}\,')U(\vec{r}\,')R_l(kr')Y_{lm}(\theta',\phi').$$

This integral will be particularly simple if $U(\vec{r}\,')$ is independent of θ' and ϕ', i.e., if $U(\vec{r}\,')=U(r')$. Then, we can carry out the θ' and ϕ' integrations, if we substitute for the Green's function through eq. (39), to convert this 3-D integral into a sum over purely radial integrals

$$\sum_{lm}Y_{lm}(\theta,\phi)\int_0^\infty dr'r'^2\Big[-ikh_l^{(1)}(kr)j_l(kr')\Big]U(r')R_l(kr'),$$

where we can identify $r'^2\times$ the quantity in square brackets as the 1-D Green's function, $g_l^{(+)}(r, r')$. In addition, if $U(r')$ has the assumed spherical symmetry, the function $\psi_{\vec{k}}$ must have axial symmetry and be independent of ϕ, so the above sum can contain only terms with $m = 0$. Therefore,

$$\psi_{\vec{k}}(\vec{r}) = \phi_{\vec{k}}(\vec{r}) + \sum_l Y_{l0}(\theta) \int_0^\infty dr' g_l^{(+)}(r, r') U(r') R_l(kr'), \tag{42}$$

with

$$\psi_{\vec{k}}(\vec{r}) = \sum_l R_l(kr) Y_{l0}(\theta), \tag{43}$$

and

$$\phi_{\vec{k}}(\vec{r}) = \frac{1}{(2\pi)^{\frac{3}{2}}} \sum_l i^l \sqrt{4\pi(2l+1)}\, j_l(kr) Y_{l0}(\theta), \tag{44}$$

so we get the integral equation for the radial functions

$$R_l(kr) = \frac{i^l}{\pi}\sqrt{\frac{(2l+1)}{2}}\, j_l(kr) + \int_0^\infty dr' g_l^{(+)}(r, r') U(r') R_l(kr'). \tag{45}$$

Actually, we have derived only an expression for $g_l^{(+)}(r, r')$, valid for $r > r'$, namely,

$$g_l^{(+)}(r, r') = -ikr'^2 h_l^{(1)}(kr) j_l(kr'), \qquad \text{valid for } r > r'. \tag{46}$$

In an appendix to this chapter, we shall show

$$g_l^{(+)}(r, r') = -ikr'^2 j_l(kr) h_l^{(1)}(kr'), \qquad \text{valid for } r < r'. \tag{47}$$

C Summary

We have put the scattering problem into integral equation form,

$$\psi_{\vec{k}}(\vec{r}) = \phi_{\vec{k}}(\vec{r}) + \int d\vec{r}' G^{(+)}(\vec{r}, \vec{r}') U(\vec{r}') \psi_{\vec{k}}(\vec{r}'), \quad \text{with } U(\vec{r}') = \frac{2\mu}{\hbar^2} V(\vec{r}'), \tag{48}$$

$$G^{(+)}(\vec{r}, \vec{r}') = -\frac{1}{4\pi} \frac{e^{ik|\vec{r}-\vec{r}'|}}{|\vec{r}-\vec{r}'|}, \tag{49}$$

where

$$(k^2 + \nabla^2) G(\vec{r}, \vec{r}') = \delta(\vec{r} - \vec{r}'), \tag{50}$$

and

$$\phi_{\vec{k}}(\vec{r}) = \frac{1}{(2\pi)^{\frac{3}{2}}} e^{i\vec{k}\cdot\vec{r}}. \tag{51}$$

For a spherically symmetric $U(r)$,

$$\psi_{\vec{k}}(\vec{r}) = \sum_l R_l(kr) Y_{l0}(\theta), \tag{52}$$

where

$$\begin{aligned} R_l(kr) &= \frac{i^l}{\pi}\sqrt{\frac{(2l+1)}{2}}\, j_l(kr) + \int_0^\infty dr' g_l^{(+)}(r,r')U(r')R_l(kr'), \\ g_l^{(+)}(r,r') &= -ikr'^2 j_l(kr')h_l^{(1)}(kr), \qquad \text{for } r > r', \\ g_l^{(+)}(r,r') &= -ikr'^2 h_l^{(1)}(kr')j_l(kr), \qquad \text{for } r < r'. \end{aligned} \tag{53}$$

D Closing Remarks

1. We note the symmetry of $G(\vec{r},\vec{r}\,')$:

$$G(\vec{r},\vec{r}\,') = G(\vec{r}\,',\vec{r}). \tag{54}$$

2. The integral equation we have derived is not a solution to our scattering problem because the Green's function integral contains the unknown solution, $\psi(\vec{r}\,')$. If instead of the equations

$$(E - H_0)G(\vec{r},\vec{r}\,') = \frac{\hbar^2}{2\mu}\delta(\vec{r}-\vec{r}\,'), \tag{55}$$

$$(E - H_0)\psi_{\vec{k}}(\vec{r}) = V(\vec{r})\psi_{\vec{k}}(\vec{r}), \tag{56}$$

we had solved the equations

$$(E - H)\bar{G}(\vec{r},\vec{r}\,') = \frac{\hbar^2}{2\mu}\delta(\vec{r}-\vec{r}\,'), \tag{57}$$

$$(E - H)\phi_{\vec{k}}(\vec{r}) = -V(\vec{r})\phi_{\vec{k}}(\vec{r}), \tag{58}$$

we would have arrived at the integral equation

$$\phi_{\vec{k}}(\vec{r}) = \psi_{\vec{k}}(\vec{r}) - \int d\vec{r}\,' U(\vec{r}\,')\bar{G}(\vec{r},\vec{r}\,')\phi_{\vec{k}}(\vec{r}\,'). \tag{59}$$

Now, the integral contains only the simple *known* plane wave function, $\phi_{\vec{k}}$, and if we could calculate $\bar{G}(\vec{r},\vec{r}\,')$, we could solve this integral equation for the desired $\psi_{\vec{k}}$. Unfortunately, the $\bar{G}(\vec{r},\vec{r}\,')$ for the *full* H (rather than the H_0) is of course difficult if not impossible to calculate.

Mathematical Appendix to Chapter 45

One-Dimensional Green's Functions

To evaluate the 1-D Green's function, $g_l(r,r')$ of eq. (53), it will be useful to study 1-D Green's functions in some generality.

The Green's function, $g(t, t')$, for the 1-D second-order differential equation of the form

$$f_0(t)\frac{d^2\psi}{dt^2} + f_1(t)\frac{d\psi}{dt} + f_2(t)\psi(t) = 0 \qquad (1)$$

is a solution of the differential equation

$$f_0(t)\frac{d^2g(t,t')}{dt^2} + f_1(t)\frac{dg(t,t')}{dt} + f_2(t)g(t,t') = \delta(t-t'), \qquad (2)$$

that is, an inhomogeneous differential equation of the same form with a delta function point source at $t = t'$. We will assume the $f_i(t)$ are "well-behaved" functions (in the language of physicists). We shall be interested in the special case in which $t = r$, $\psi(t) = R_l(r)$, $f_0 = 1$, $f_1 = 2/r$, $f_2 = (k^2 - l(l+1)/r^2)$. We will seek solutions to eq. (2) of the form

$$\begin{aligned} &\text{For } t > t': \quad && g(t,t') = a_1\psi_1(t) + a_2\psi_2(t), \\ &\text{for } t < t': \quad && g(t,t') = b_1\psi_1(t) + b_2\psi_2(t), \end{aligned} \qquad (3)$$

where $\psi_1(t)$, $\psi_2(t)$ are two independent solutions of differential equation (1). The constants a_i, b_i are then functions of t'. Also, at $t = t'$, $g(t,t')$ must be continuous. If it were not, $\frac{dg}{dt}$ would contain a delta function and the second derivative of g would contain a derivative of a delta function, but the right-hand side of eq. (2) contains only a delta function, not a derivative of a delta function. Thus,

$$g(t = t' + \epsilon, t') - g(t = t' - \epsilon, t') = 0, \qquad (4)$$

but

$$\left[\frac{dg(t,t')}{dt}\right]_{t=t'+\epsilon} - \left[\frac{dg(t,t')}{dt}\right]_{t=t'-\epsilon} \neq 0. \qquad (5)$$

To determine its value, integrate eq. (2) over an interval from $t = t' - \epsilon$ to $t = t' + \epsilon$,

$$\int_{t'-\epsilon}^{t'+\epsilon} dt f_0(t)\frac{d^2g(t,t')}{dt^2} + \int_{t'-\epsilon}^{t'+\epsilon} dt f_1(t)\frac{dg(t,t')}{dt} + \int_{t'-\epsilon}^{t'+\epsilon} dt f_2(t)g(t,t') = 1, \qquad (6)$$

leading to

$$f_0(t')\left(\left[\frac{dg(t,t')}{dt}\right]_{t=t'+\epsilon} - \left[\frac{dg(t,t')}{dt}\right]_{t=t'-\epsilon}\right) = 1, \qquad (7)$$

where we have made use of the continuity of $g(t,t')$, and of course the assumed continuity of the $f_i(t)$ at $t = t'$. The boundary conditions, eqs. (4) and (7), lead to

$$\begin{aligned} a_1\psi_1(t') + a_2\psi_2(t') &= b_1\psi_1(t') + b_2\psi_2(t'), \\ a_1\psi_1'(t') + a_2\psi_2'(t') &= b_1\psi_1'(t') + b_2\psi_2'(t') + \frac{1}{f_0(t')}, \end{aligned} \qquad (8)$$

where the primes on the ψ's stand for derivatives. These equations can be solved for the a_i

$$a_1 = b_1 - \frac{\psi_2(t')}{f_0(t')W(t')},$$

$$a_2 = b_2 + \frac{\psi_1(t')}{f_0(t')W(t')}, \tag{9}$$

where $W(t')$ is the Wronskian

$$W(t') = \psi_1(t')\frac{d\psi_2}{dt'} - \psi_2(t')\frac{d\psi_1}{dt'}, \tag{10}$$

leading to

$$\begin{aligned} &\text{For } t < t' : g(t,t') = b_1\psi_1(t) + b_2\psi_2(t), \\ &\text{for } t > t' : g(t,t') = b_1\psi_1(t) + b_2\psi_2(t) - \frac{[\psi_1(t)\psi_2(t') - \psi_2(t)\psi_1(t')]}{f_0(t')W(t')}. \end{aligned} \tag{11}$$

The Wronskian follows from the differential equation, in particular, from

$$\begin{aligned} &\psi_1\Big(f_0\frac{d^2\psi_2}{dt^2} + f_1\frac{d\psi_2}{dt} + f_0\psi_2\Big) = 0, \\ &-\psi_2\Big(f_0\frac{d^2\psi_1}{dt^2} + f_1\frac{d\psi_1}{dt} + f_0\psi_1\Big) = 0. \end{aligned} \tag{12}$$

By adding these equations, we get

$$f_0\frac{d}{dt}\Big(\psi_1\frac{d\psi_2}{dt} - \psi_2\frac{d\psi_1}{dt}\Big) + f_1\Big(\psi_1\frac{d\psi_2}{dt} - \psi_2\frac{d\psi_1}{dt}\Big) = 0, \tag{13}$$

$$\frac{dW}{dt} = -\frac{f_1}{f_0}W, \quad \text{or} \quad lnW = -\int dt\frac{f_1(t)}{f_0(t)}. \tag{14}$$

The One-Dimensional Radial Green's Functions, $g_l(r, r')$

For our radial functions, with $\psi = R_l(kr)$, the functions $f_0 = 1$, $f_1 = 2/r$ lead to

$$\frac{dW}{W} = -\frac{2}{r}dr, \qquad W(r) = \frac{\text{const.}}{r^2}. \tag{15}$$

For the two solutions, ψ_1 and ψ_2, we shall choose $\psi_1 = j_l(kr)$; $\psi_2 = n_l(kr)$. To determine the constant in the above expression for the Wronskian, it is sufficient to use the asymptotic form for $j_l(kr)$ and $n_l(kr)$ and let $r \to \infty$, which leads at once to $W(r) = 1/(kr^2)$.

Now, for $r < r'$, our general result gives

$$g_l(r, r') = b_1(r')j_l(kr) + b_2(r')n_l(kr), \tag{16}$$

but because the differential equation for g_l has a singularity *only* at $r = r'$ through the delta function source term, the solution for $r < r'$ must be regular at $r = 0$, and the coefficient b_2 must be zero. Thus, with our general result for the 1-D g, eq. (9), we have

$$\begin{aligned} &\text{For } r < r' : g_l(r,r') = b_1(r')j_l(kr), \\ &\text{for } r > r' : g_l(r,r') = a_1(r')j_l(kr) + a_2(r')n_l(kr) \\ &\qquad = [b_1(r') - kr'^2 n_l(kr')] + j_l(kr) \\ &\qquad + kr'^2 j_l(kr')n_l(kr). \end{aligned} \tag{17}$$

From eq. (46) of the main body of Chapter 45, we know, for $r > r'$,

$$\begin{aligned} g_l(r, r') &= -ikr'^2 j_l(kr')h_l^{(1)}(kr) \\ &= -ikr'^2 j_l(kr')(j_l(kr) + in_l(kr)). \end{aligned} \tag{18}$$

Comparing with the above, this gives

$$b_1(r') = -ikr'^2(j_l(kr') + in_l(kr')) = -ikr'^2 h_l^{(1)}(kr'). \tag{19}$$

Thus, we have, for $r < r'$,

$$g_l(r, r') = b_1(r')j_l(kr) = -ikr'^2 h_l(kr')j_l(kr), \tag{20}$$

which is the result given by eq. (47).

46

The Born Approximation

In the last chapter, we merely reformulated the scattering problem. Instead of solving the scattering problem by finding a solution to a differential equation with the appropriate boundary condition, we will instead try to solve it by finding a solution to an integral equation with the appropriate Green's function.

$$\psi_{\vec{k}}(\vec{r}) = \phi_{\vec{k}}(\vec{r}) + \int d\vec{r}\,' G^{(+)}(\vec{r}, \vec{r}\,') \frac{2\mu}{\hbar^2} V(\vec{r}\,') \psi_{\vec{k}}(\vec{r}\,'). \tag{1}$$

Because we cannot do the integral containing the *unknown* $\psi_{\vec{k}}(\vec{r}\,')$, we are far from a solution. If $V(\vec{r})$ is a weak potential, however, so it can be treated by perturbation theory, or if the energy is very large, so the average $V(\vec{r})$ is small compared with this energy, we can solve the integral equation by an iteration technique. If

$$|V(\vec{r}\,')|_{\text{Average}} \ll \frac{\hbar^2 k^2}{2\mu}, \tag{2}$$

in a first approximation for $\psi_{\vec{k}}(\vec{r})$, we can replace $\psi_{\vec{k}}(\vec{r}\,')$ in the integral by the zeroth approximation for the problem, $\phi_{\vec{k}}(\vec{r}\,')$, to yield

$$\psi_{\vec{k}}(\vec{r}) \approx \frac{1}{(2\pi)^{\frac{3}{2}}} \left(e^{i\vec{k}\cdot\vec{r}} - \frac{1}{4\pi} \int d\vec{r}\,' \frac{e^{ik|\vec{r}-\vec{r}\,'|}}{|\vec{r}-\vec{r}\,'|} \frac{2\mu}{\hbar^2} V(\vec{r}\,') e^{i\vec{k}\cdot\vec{r}\,'} \right), \tag{3}$$

where this approximate solution is known as the first Born approximation. Even without putting this first approximation back into the integral equation to get the next approximation, and perhaps iterating this process a number of times, the simple first Born approximation may be a useful good approximation. In that case,

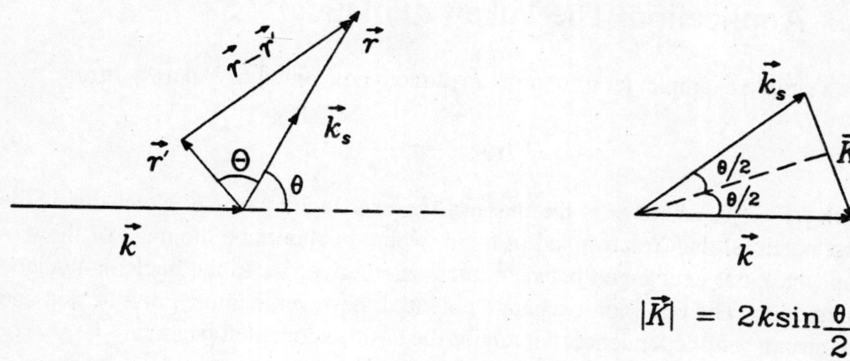

FIGURE 46.1.

if we use the asymptotic form

$$\frac{e^{ik|\vec{r}-\vec{r}\,'|}}{|\vec{r}-\vec{r}\,'|} \longrightarrow \frac{e^{ikr}}{r}e^{-ikr'\cos\Theta} = \frac{e^{ikr}}{r}e^{-i\vec{k}_s\cdot\vec{r}\,'}, \tag{4}$$

where $\vec{k}_s$, the $\vec{k}$ vector in the direction of the scattered wave, is along the direction of the vector $\vec{r}$ from the scattering center to the detector [see Fig. 46.1(a)], we get

$$\psi_{\vec{k}}(\vec{r}) \longrightarrow \frac{1}{(2\pi)^{\frac{3}{2}}}\left(e^{i\vec{k}\cdot\vec{r}} - \frac{e^{ikr}}{r}\left[\frac{\mu}{2\pi\hbar^2}\int d\vec{r}\,'V(\vec{r}\,')e^{i(\vec{k}-\vec{k}_s)\cdot\vec{r}\,'}\right]\right), \tag{5}$$

so the first Born approximation for the scattering amplitude is given by

$$f(\theta,\phi)_{1^{\text{st}}\text{Born}} = -\frac{\mu}{2\pi\hbar^2}\int d\vec{r}\,'V(\vec{r}\,')e^{i(\vec{k}-\vec{k}_s)\cdot\vec{r}\,'}. \tag{6}$$

In the special case when the scattering potential is spherically symmetric, so $V(\vec{r}\,') = V(r')$, it is easy to reduce the 3-D Born integral to a radial integral. For that purpose, it is useful to define the vector $\vec{K} = \vec{k} - \vec{k}_s$ [see Fig. 46.1(b)], with magnitude $K = |\vec{K}| = 2k\sin\frac{\theta}{2}$, and choose it to be along the z direction, so

$$\int d\vec{r}\,'V(r')e^{i(\vec{k}-\vec{k}_s)\cdot\vec{r}\,'} = \int d\vec{r}\,'V(r')e^{iKr'\cos\theta'}$$
$$= 2\pi\int_0^\infty dr'r'^2V(r')\int_{-1}^{+1}d\zeta'e^{iKr'\zeta'} = 2\pi\int_0^\infty dr'r'^2V(r')\frac{2}{Kr'}\sin Kr'$$
$$= \frac{4\pi}{K}\int_0^\infty dr'r'V(r')\sin Kr'. \tag{7}$$

This equation leads to the first Born approximation of the scattering amplitude

$$f(\theta)_{1^{\text{st}}\text{Born}} = -\frac{2\mu}{K\hbar^2}\int_0^\infty dr'r'V(r')\sin Kr', \qquad \text{with} \quad K = 2k\sin\frac{\theta}{2}. \tag{8}$$

A Application: The Yukawa Potential

As a simple example, let us choose an attractive potential of Yukawa form

$$V(r) = -g^2 \frac{e^{-\beta r}}{r}, \tag{9}$$

with $\beta = \frac{mc}{\hbar}$, where m is the mass of the exchanged particle. Note: this is the first potential that Yukawa had in mind, where m should be the mass of the meson (the pion) exchanged between nucleons to give rise to the nucleon–nucleon interaction. The real pion-exchange potential, however, is more complicated and has strong spin-dependence, but retains the basic exponential factor, $e^{-\frac{m_\pi cr}{\hbar}}/r$. For this potential, the first Born approximation gives

$$f(\theta) = \frac{2\mu g^2}{\hbar^2 K} \int_0^\infty dr' e^{-\beta r'} \sin Kr' = \frac{2\mu g^2}{\hbar^2 K} \frac{K}{[\beta^2 + K^2]}, \tag{10}$$

leading to

$$\frac{d\sigma}{d\Omega} = |f(\theta)|^2 = \frac{4\mu^2 g^4}{\hbar^4} \frac{1}{[\beta^2 + 4k^2 \sin^2 \frac{\theta}{2}]^2}. \tag{11}$$

B The Screened Coulomb Potential

First, the first Born approximation leads to an undetermined integral for a pure Coulomb potential. With

$$V(r) = \frac{Z_1 Z_2 e^2}{r}, \tag{12}$$

the first Born approximation would lead to

$$f(\theta)_{1^{\text{st}}\text{Born}} = -\frac{\mu Z_1 Z_2 e^2}{\hbar^2 k \sin \frac{\theta}{2}} \int_0^\infty dr' \sin Kr', \tag{13}$$

leading to the undetermined radial integral. The Coulomb potential has such a far reach the plane wave is not a good enough zeroth approximation. (See the mathematical appendix for Chapter 42). For a screened Coulomb potential, however, with

$$V(r) = \frac{Z_1 Z_2 e^2}{r} e^{-\beta r}, \tag{14}$$

and with a very small value for the screening constant β, we might get a reasonable approximation for the Coulomb potential. If we consider α particle scattering from a Helium atom, the above might be a very good approximation for the needed potential if β is of the order of an inverse atomic dimension. Yet, this potential would give us the effects of α–α scattering when the α particle is near the He nucleus of the atom. Because the form of this screened Coulomb potential is the same as that of the Yukawa potential, the differential scattering cross section is given by

$$\frac{d\sigma}{d\Omega} = \frac{4\mu^2 Z_1^2 Z_2^2 e^4}{\hbar^4} \frac{1}{[\beta^2 + 4k^2 \sin^2 \frac{\theta}{2}]^2}. \tag{15}$$

In the limit $\beta \to 0$, this equation yields

$$\frac{d\sigma}{d\Omega} = \frac{\mu^2 Z_1^2 Z_2^2 e^4}{4\hbar^4 k^4 \sin^4 \frac{\theta}{2}} = \frac{Z_1^2 Z_2^2 e^4}{16E^2 \sin^4 \frac{\theta}{2}} = \left(\frac{d\sigma}{d\Omega}\right)_{\text{Rutherford}}. \tag{16}$$

That is, this limiting value of the Born approximation for the screened Coulomb potential gives us the classical Rutherford result. This is also the correct quantum-mechanical result for α nucleus scattering. The exact quantum-mechanical result for the scattering amplitude for Coulomb scattering is given by (see the appendix to this chapter),

$$f(\theta)_{\text{exact Coul.}} = -\frac{\mu Z_1 Z_2 e^2}{2\hbar^2 k^2 \sin^2 \frac{\theta}{2}} e^{[-i\gamma ln \sin^2 \frac{\theta}{2} + 2i\sigma_0]}, \quad \text{where } \gamma = \frac{\mu Z_1 Z_2 e^2}{\hbar^2 k}, \tag{17}$$

and σ_0 is the argument of the Gamma function $\Gamma(1 + i\gamma)$. Because the differential cross section is given by $|f(\theta)|^2$, the logarithmic dependence on $\sin^2 \frac{\theta}{2}$ of the exponential does not come into play! The classical and quantum-mechanical results, however, are different for the scattering of identical particles from identical particles, say, α–α scattering.

C Identical Particle Coulomb Scattering

Even the classical Rutherford formula must be modified if we scatter identical particles from identical particles. For α–α scattering, e.g., the detector at the position θ in the center of mass sytem will give a count both if the incoming projectile is scattered into the direction θ [see Fig. 46.2(a)], and if the incoming projectile is scattered into direction $(\pi - \theta)$, so the α particle that was the original "target" particle is scattered into direction θ [see Fig. 46.2(b)], because the detector can not distinguish "projectile" α particles from "target" α particles. Thus,

$$\left(\frac{d\sigma}{d\Omega}\right)_{\text{classical}} = \frac{\gamma^2}{4k^2}\left(\frac{1}{\sin^4 \frac{\theta}{2}} + \frac{1}{\cos^4 \frac{\theta}{2}}\right). \tag{18}$$

For the quantum-mechanical result, we must remember that the α particle has spin $s = 0$. Thus, α particles are bosons. The two-particle functions must be orbitally symmetric under the exchange of particle indices 1 and 2. Thus, solutions must be of the form,

$$\psi(\vec{r}_1, \vec{r}_2) + \psi(\vec{r}_2, \vec{r}_1),$$

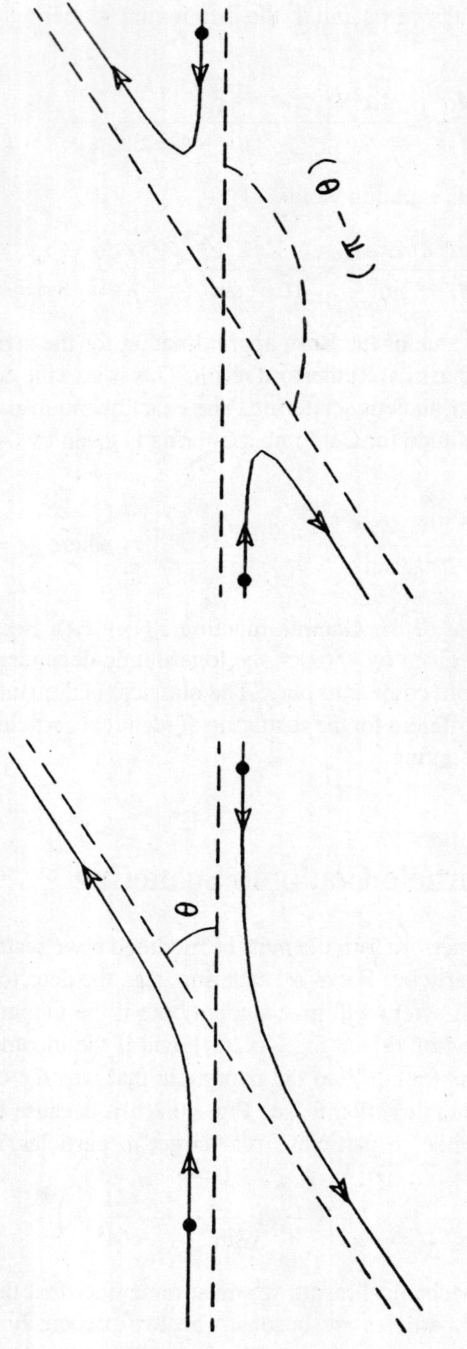

FIGURE 46.2.

or, with the relative position vector $\vec{r} = \vec{r}_1 - \vec{r}_2$, the relative motion function must have the form,

$$\psi(\vec{r}) + \psi(-\vec{r}),$$

so, in the limit $r \to \infty$,

$$\psi_{\vec{k}} \longrightarrow \left(\left(e^{i\vec{k}\cdot\vec{r}} + e^{-i\vec{k}\cdot\vec{r}} \right)_{\text{Coul. mod.}} + [f(\theta) + f(\pi - \theta)] \left(\frac{e^{ikr}}{r} \right)_{\text{Coul. mod.}} \right), \tag{19}$$

where the Coulomb modified plane waves and outgoing spherical waves are (see mathematical appendix to this chapter),

$$\left(e^{i\vec{k}\cdot\vec{r}} \right)_{\text{Coul. mod.}} = e^{i(\vec{k}\cdot\vec{r} + \gamma ln[kr(1-\cos\theta)])}, \quad \left(\frac{e^{ikr}}{r} \right)_{\text{Coul. mod.}} = \frac{e^{i(kr - \gamma ln 2kr)}}{r}, \tag{20}$$

with

$$f(\theta) = -\gamma \frac{e^{-i\gamma ln \sin^2\frac{\theta}{2}} e^{2i\sigma_0}}{2k \sin^2\frac{\theta}{2}}, \tag{21}$$

where σ_0 is the argument of the complex number, $\Gamma(1 + i\gamma)$ (see mathematical appendix to Chapter 42). In the center of mass system, the differential cross section is now

$$\begin{aligned} \left(\frac{d\sigma}{d\Omega} \right)_{\text{C.M.}} &= |f(\theta) + f(\pi - \theta)|^2 \\ &= \frac{\gamma^2}{4k^2} \left| \frac{e^{-i\gamma ln \sin^2\frac{\theta}{2}}}{\sin^2\frac{\theta}{2}} + \frac{e^{-i\gamma ln \cos^2\frac{\theta}{2}}}{\cos^2\frac{\theta}{2}} \right|^2 \\ &= \frac{\gamma^2}{4k^2} \left(\frac{1}{\sin^4\frac{\theta}{2}} + \frac{1}{\cos^4\frac{\theta}{2}} + 2\frac{\cos\left[\gamma ln(\tan^2\frac{\theta}{2})\right]}{\sin^2\frac{\theta}{2}\cos^2\frac{\theta}{2}} \right). \end{aligned} \tag{22}$$

Finally, for identical particle scattering, the transformation from the center of mass to the laboratory system is very simple (see Chapter 41),

$$\theta_{\text{lab}} = \frac{\theta}{2}, \quad \left(\frac{d\sigma}{d\Omega} \right)_{\text{lab}} = 4\cos\theta_{\text{lab}} \left(\frac{d\sigma}{d\Omega} \right)_{\text{C.M.}}, \tag{23}$$

so

$$\left(\frac{d\sigma}{d\Omega} \right)_{\text{lab}} = \frac{\gamma^2 \cos\theta_{\text{lab}}}{k^2} \left(\frac{1}{\sin^4\theta_{\text{lab}}} + \frac{1}{\cos^4\theta_{\text{lab}}} + 2\frac{\cos\left[\gamma ln(\tan^2\theta_{\text{lab}})\right]}{\sin^2\theta_{\text{lab}}\cos^2\theta_{\text{lab}}} \right). \tag{24}$$

The cross term is very important. Fig. 46.3 shows the ratio of the quantum-mechanical differential cross section to the classical cross section for this case of the scattering of spin $s = 0$ bosons from identical spin $s = 0$ bosons.

Finally, for proton–proton Coulomb scattering, the situation is a little more complicated. Protons are $s = \frac{1}{2}$ particles and, hence, fermions. If both projectile and target protons are unpolarized, the two-proton system will be in the spin-antisymmetric state, with two-particle spin $S = 0$, $\frac{1}{4}$ of the time and in the spin-symmetric state, with two-particle spin $S = 1$, $\frac{3}{4}$ of the time. Hence, the orbital

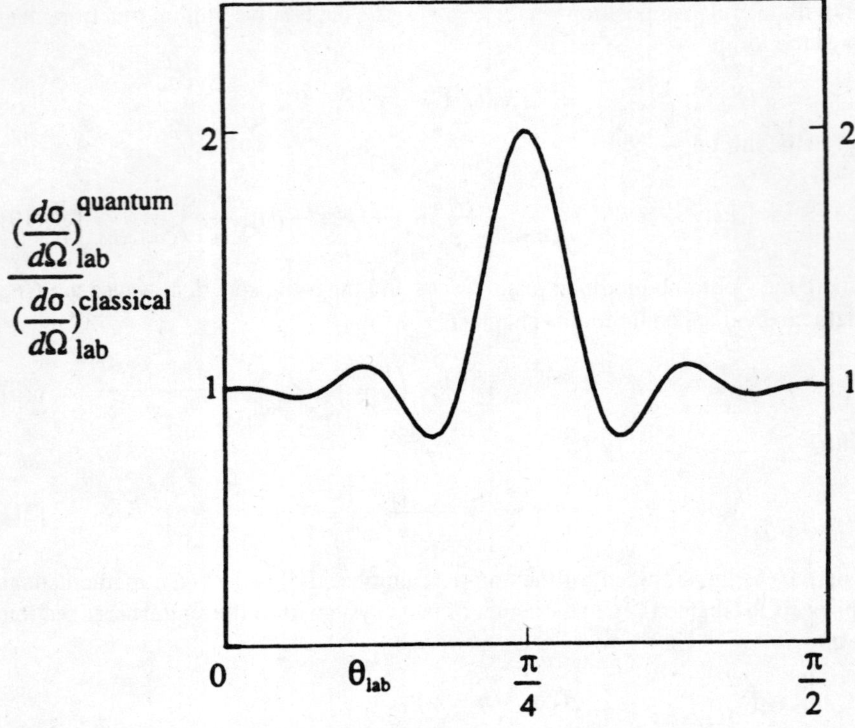

FIGURE 46.3. Differential scattering cross sections for α–α scattering.

functions will be in an orbitally symmetric state $\frac{1}{4}$ of the time and in an orbitally antisymmetric state $\frac{3}{4}$ of the time. Thus, for proton–proton Coulomb scattering

$$\left(\frac{d\sigma}{d\Omega}\right)_{\text{C.M.}} = \frac{1}{4}|f(\theta) + f(\pi - \theta)|^2 + \frac{3}{4}|f(\theta) - f(\pi - \theta)|^2, \tag{25}$$

leading to a formula like that of eq. (24) in which the coefficient of 2 in the cross term is replaced by $(\frac{1}{4} - \frac{3}{4}) \times 2 = -1$, leading to quite a different angular distribution.

Mathematical Appendix to Chapter 46

Exact Solution for the Pure Coulomb Scattering Problem

In the appendix to Chapter 42, we solved the problem for Coulomb scattering in a partial wave expansion, where the relative motion function was expanded in radial functions, $R_l(kr)$, and in spherical harmonics, $\psi(r, \theta, \phi) = \sum_l R_l(kr) Y_{l0}(\theta, \phi)$. Our main aim there was to get the asymptotic form in the limit of large r of the incoming and outgoing Coulomb spherical waves. These results are particularly useful if the long-range part of the potential is dominated by the Coulomb potential, but if, in addition, a short-range potential exists, such as a nuclear potential. As in any partial wave expansion, this approach leads to differential and total scattering cross sections in terms of infinite series in the l's of the partial waves. For the case of pure Coulomb scattering, however, we can find an exact solution that gives us the scattering amplitude exactly, (without need of an infinite series). This amplitude would apply, e.g., for proton–nucleus, or α-nucleus scattering, or α–α scattering, at energies low enough, so the proton or α particle has a very low probability of tunneling through the repulsive Coulomb barrier into the small r region, where the nuclear force would become effective.

We want to solve the Schrödinger equation for the relative motion ψ

$$\nabla^2\psi + \Big(\frac{2\mu E}{\hbar^2} - \frac{2\mu Z_1 Z_2 e^2}{\hbar^2}\frac{1}{r}\Big)\psi = 0,$$
$$\nabla^2\psi + \Big(k^2 - \frac{2\gamma k}{r}\Big)\psi = 0, \tag{1}$$

where γ is the Coulomb parameter

$$\gamma = \frac{\mu Z_1 Z_2 e^2}{\hbar^2 k} = \frac{c}{v}\alpha Z_1 Z_2. \tag{2}$$

We want to solve this problem with the appropriate boundary conditions; i.e., we seek a solution in the form of a Coulomb-corrected incoming plane wave + a Coulomb-corrected outgoing spherical wave. This boundary condition motivates us to first make the substitution

$$\psi(\vec{r}) = e^{i\vec{k}\cdot\vec{r}}\chi(\vec{r}). \tag{3}$$

With

$$\vec{\nabla}\psi = e^{i\vec{k}\cdot\vec{r}}(\vec{\nabla}\chi + i\vec{k}\chi),$$
$$\nabla^2\psi = e^{i\vec{k}\cdot\vec{r}}(\nabla^2\chi + 2i\vec{k}\cdot\vec{\nabla}\chi - k^2\chi), \tag{4}$$

this leads to the equation

$$\nabla^2\chi + 2i\vec{k}\cdot\vec{\nabla}\chi - \frac{2\gamma k}{r}\chi = 0. \tag{5}$$

If we choose the z axis along the direction of $\vec{k}$, $\chi(\vec{r})$ will be a function of r and θ only. It will be particularly useful to transform from r, θ, to parabolic coordinates, ξ, η, where

$$\xi = r - z = r(1 - \cos\theta),$$
$$\eta = r + z = r(1 + \cos\theta). \tag{6}$$

It will be sufficient to seek solutions of the form $e^{i\vec{k}\cdot\vec{r}}\chi(\xi)$, i.e., solutions independent of the variable η. It will be shown, a posteriori, that these have the desired asymptotic form of a Coulomb-corrected plane wave + a Coulomb-corrected spherical outgoing wave. (Solutions of the form $e^{i\vec{k}\cdot\vec{r}}\chi(\eta)$ would have given us solutions with the asymptotic form of a Coulomb-corrected plane wave + a Coulomb-corrected spherical incoming wave.)

With $\chi = \chi(\xi, \text{only})$, $\quad \xi = r - z = r(1-\cos\theta)$,

$$\vec{\nabla}\chi = \frac{d\chi}{d\xi}\Big(\frac{x}{r}\vec{e}_1 + \frac{y}{r}\vec{e}_2 + (\frac{z}{r} - 1)\vec{e}_3\Big) = \frac{d\chi}{d\xi}\Big(\frac{\vec{r}}{r} - \frac{\vec{k}}{k}\Big), \tag{7}$$

so

$$2i\vec{k}\cdot\vec{\nabla}\chi = \frac{2ik}{r}(\cos\theta - 1)r\frac{d\chi}{d\xi} = -\frac{2}{r}ik\xi\frac{d\chi}{d\xi}, \tag{8}$$

and

$$\begin{aligned}\nabla^2\chi &= \frac{d\chi}{d\xi}\vec{\nabla}\cdot\Big(\frac{\vec{r}}{r} - \frac{\vec{k}}{k}\Big) + \Big(\Big[\vec{\nabla}\frac{d\chi}{d\xi}\Big]\cdot\Big(\frac{\vec{r}}{r} - \frac{\vec{k}}{k}\Big)\Big)\\ &= \frac{d\chi}{d\xi}\Big(\frac{3}{r} - \frac{\vec{r}\cdot\vec{r}}{r^3}\Big) + \frac{d^2\chi}{d\xi^2}\Big(\frac{\vec{r}}{r} - \frac{\vec{k}}{k}\Big)\cdot\Big(\frac{\vec{r}}{r} - \frac{\vec{k}}{k}\Big)\\ &= \frac{2}{r}\frac{d\chi}{d\xi} + 2\frac{d^2\chi}{d\xi^2}\Big(1 - \frac{\vec{k}\cdot\vec{r}}{kr}\Big) = \frac{2}{r}\Big(\frac{d\chi}{d\xi} + \xi\frac{d^2\chi}{d\xi^2}\Big),\end{aligned} \tag{9}$$

leading to the new equation

$$\frac{2}{r}\Big(\xi\frac{d^2\chi}{d\xi^2} + (1 - ik\xi)\frac{d\chi}{d\xi} - \gamma k\chi\Big) = 0. \tag{10}$$

Introducing the new variable,

$$t = ik\xi, \tag{11}$$

$$\begin{aligned}t\frac{d^2\chi}{dt^2} + (1-t)\frac{d\chi}{dt} + i\gamma\chi(t) = 0,\\ t\frac{d^2\chi}{dt^2} + (c-t)\frac{d\chi}{dt} - a\chi(t) = 0,\end{aligned} \tag{12}$$

so

$$\chi(ik\xi) = {}_1F_1(-i\gamma; 1; ik\xi). \tag{13}$$

The solutions are confluent hypergeometric functions and

$$\psi = e^{i\vec{k}\cdot\vec{r}}\ {}_1F_1(-i\gamma; 1; ikr(1-\cos\theta)). \tag{14}$$

Now recall (see mathematical appendix to Chapter 42)

$${}_1F_1(a; c; t) = \frac{\Gamma(c)}{2\pi i}\oint_C \frac{dze^z}{z^c\big(1 - \frac{t}{z}\big)^a}, \tag{15}$$

where $a = -i\gamma$, $\quad c = 1$, $\quad t = ikr(1-\cos\theta)$, so (with $\Gamma(1) = 1$)

$${}_1F_1\big(-i\gamma; 1; ikr(1-\cos\theta)\big) = \frac{1}{2\pi i}\oint_C \frac{dze^z(z-t)^{i\gamma}}{z^{1+i\gamma}}. \tag{16}$$

The contour, C, can again be deformed as shown in Fig. 46.4 into the two loops, C_1 and C_2. In the limit $r \to \infty$, only the small circular parts of the contours, C_1

and C_2, make a nonneglible contribution to these integrals. Thus,

$$\begin{aligned}&\lim_{r\to\infty}\frac{1}{2\pi i}\oint_{C_1}\frac{dze^z[z-ikr(1-\cos\theta)]^{i\gamma}}{z^{1+i\gamma}}\\&=\lim_{r\to\infty}\frac{[-ikr(1-\cos\theta)]^{i\gamma}}{2\pi i}\oint_{C_1}\frac{dze^z}{z^{1+i\gamma}}\left[1-\frac{z}{ikr(1-\cos\theta)}\right]^{i\gamma}\\&=\frac{[-ikr(1-\cos\theta)]^{i\gamma}}{2\pi i}\oint_{C_1}\frac{dze^z}{z^{1+i\gamma}}=\frac{[-ikr(1-\cos\theta)]^{i\gamma}}{\Gamma(1+i\gamma)},\end{aligned}\tag{17}$$

where we have used

$$\frac{1}{2\pi i}\oint\frac{dze^z}{z^\alpha}=\frac{1}{\Gamma(\alpha)},\tag{18}$$

where the contour is now a circle about the origin. Similarly, for the contour integral for the contour, C_2, where we use $t=ikr(1-\cos\theta)$,

$$\begin{aligned}&\lim_{t\to\infty}\frac{1}{2\pi i}\oint_{C_2}\frac{dze^z(z-t)^{i\gamma}}{z^{1+i\gamma}}=\lim_{t\to\infty}\frac{1}{2\pi i}\frac{e^t}{t^{1+i\gamma}}\oint_{C_2}\frac{d(z-t)e^{(z-t)}(z-t)^{i\gamma}}{\left(1+\frac{(z-t)}{t}\right)^{1+i\gamma}}\\&=\lim_{t\to\infty}\frac{e^t}{t^{1+i\gamma}}\frac{1}{2\pi i}\oint_{C_2}\frac{dze^zz^{i\gamma}}{\left(1+\frac{z}{t}\right)^{1+i\gamma}}\to\frac{e^t}{t^{1+i\gamma}}\frac{1}{2\pi i}\oint_{C_2}dze^zz^{i\gamma}\\&=\frac{e^t}{t^{1+i\gamma}}\frac{1}{\Gamma(-i\gamma)}=\frac{e^{ikr(1-\cos\theta)}}{[ikr(1-\cos\theta)]^{1+i\gamma}\Gamma(-i\gamma)},\end{aligned}\tag{19}$$

where the contour integral about the singularity at $z=t=ikr(1-\cos\theta)$ of the first line has been transformed to a contour integral about the point $z=0$ in the next line.

Combining the two contour integrals, we have

$$\begin{aligned}&\lim_{r\to\infty}{}_1F_1(-i\gamma;1;ikr(1-\cos\theta))\\&=\left(\frac{[-ikr(1-\cos\theta)]^{i\gamma}}{\Gamma(1+i\gamma)}+\frac{e^{ikr(1-\cos\theta)}}{[ikr(1-\cos\theta)]^{1+i\gamma}\Gamma(-i\gamma)}\right),\end{aligned}\tag{20}$$

so, with

$$\psi(\vec{r})=e^{ikr\cos\theta}\,{}_1F_1(-i\gamma;1;ikr(1-\cos\theta)),\tag{21}$$

$$\begin{aligned}\psi\to&\frac{e^{ikr\cos\theta}[-ikr(1-\cos\theta)]^{i\gamma}}{\Gamma(1+i\gamma)}\\&+\frac{e^{ikr}}{[ikr(1-\cos\theta)]}\frac{1}{[ikr(1-\cos\theta)]^{i\gamma}\Gamma(-i\gamma)}.\end{aligned}\tag{22}$$

Now, we shall use the Γ function properties

$$\Gamma(1+i\gamma)=i\gamma\Gamma(i\gamma)=e^{i\frac{\pi}{2}}\gamma|\Gamma(i\gamma)|e^{i\arg\Gamma(i\gamma)}=|\Gamma(1+i\gamma)|e^{i\sigma_0},\tag{23}$$

so

$$\arg\Gamma(i\gamma)=(\sigma_0-\frac{\pi}{2}).\tag{24}$$

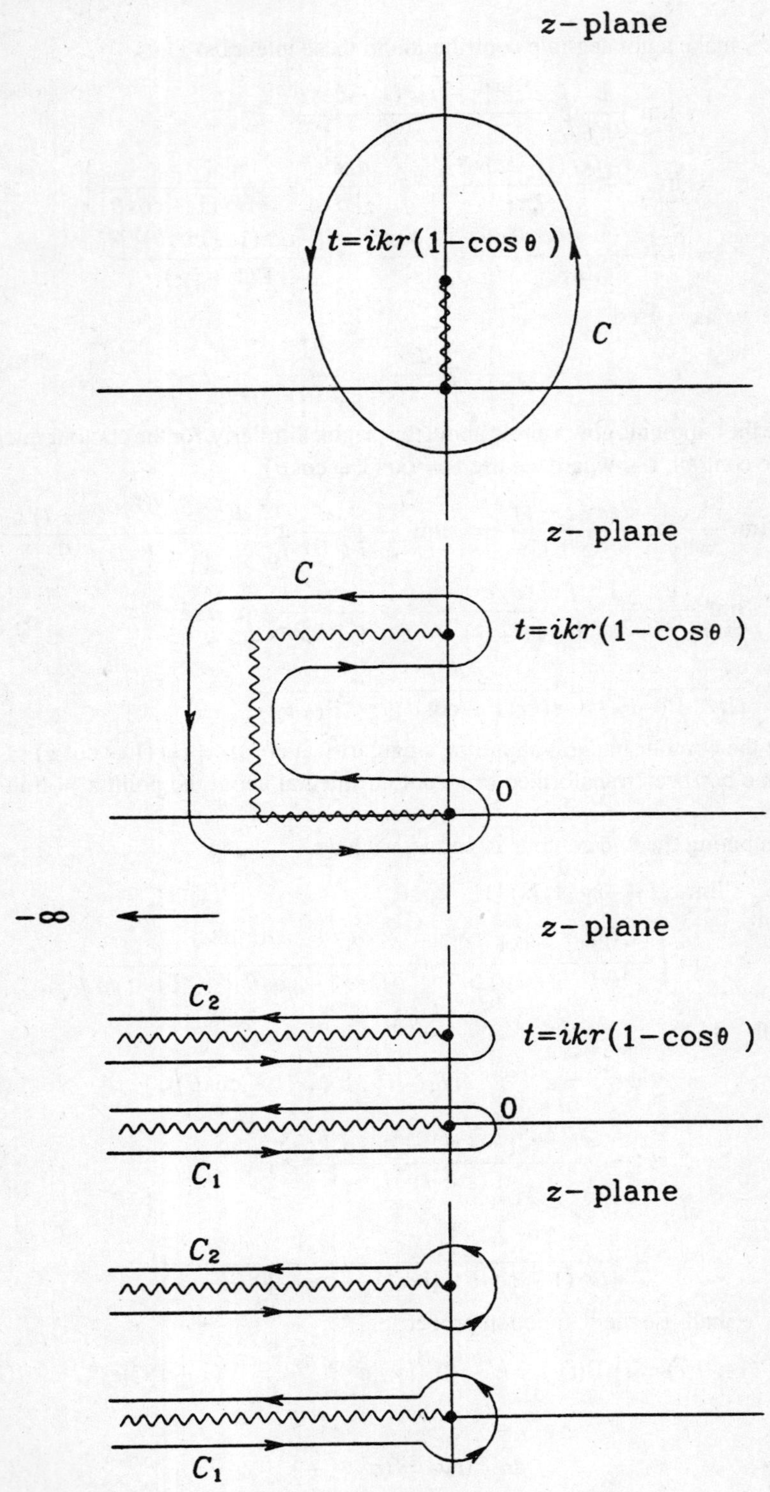

FIGURE 46.4. Contours for eqs. (16), (17), and (19).

We shall also use

$$(\mp i)^{i\gamma} = e^{\pm\gamma\frac{\pi}{2}}, \qquad \big(kr(1-\cos\theta)\big)^{i\gamma} = e^{i\gamma ln[kr(1-\cos\theta)]}, \tag{25}$$

to rewrite

$$\psi(\vec{r}) \to \frac{e^{\gamma\frac{\pi}{2}}}{i\gamma\Gamma(i\gamma)}\Bigg(e^{i(kr\cos\theta+\gamma ln[kr(1-\cos\theta)])} + \frac{\gamma}{k(1-\cos\theta)}\frac{\Gamma(i\gamma)}{\Gamma(-i\gamma)}\frac{e^{i(kr-\gamma ln[kr(1-\cos\theta)])}}{r}\Bigg), \tag{26}$$

or

$$\psi(\vec{r}) \to \text{const}\Bigg(e^{i(\vec{k}\cdot\vec{r}+\gamma ln[kr(1-\cos\theta)])} + \frac{\gamma e^{2i(\sigma_0-\frac{\pi}{2})}}{k(1-\cos\theta)}\frac{e^{i(kr-\gamma ln[kr(1-\cos\theta)])}}{r}\Bigg), \tag{27}$$

or

$$\psi(\vec{r}) \to \text{const}\Bigg(e^{i(\vec{k}\cdot\vec{r}+\gamma ln[kr(1-\cos\theta)])} + \frac{(-\gamma)e^{2i\sigma_0}e^{-i\gamma ln\sin^2\frac{\theta}{2}}}{2k\sin^2\frac{\theta}{2}}\frac{e^{i(kr-\gamma ln 2kr)}}{r}\Bigg). \tag{28}$$

This relation has the form of a Coulomb modified incoming plane wave + a Coulomb modified outgoing spherical scattered wave,

$$\psi(\vec{r}) \longrightarrow \text{const}\Big(e^{i(\vec{k}\cdot\vec{r}-\gamma ln[kr(1-\cos\theta)])} + f(\theta)\frac{e^{i(kr-\gamma ln 2kr)}}{r}\Big), \tag{29}$$

with

$$f(\theta) = \frac{-\gamma e^{2i\sigma_0}e^{-i\gamma ln\sin^2\frac{\theta}{2}}}{2k\sin^2\frac{\theta}{2}}, \qquad \text{with} \qquad \gamma = \frac{\mu e^2 Z_1 Z_2}{\hbar^2 k}. \tag{30}$$

Because, still,

$$|\vec{S}_{\text{inc.}}| = \frac{\hbar k}{\mu}|\text{const}|^2 + \text{Order}\Big(\frac{1}{r}\Big), \tag{31}$$

and

$$\Big(\vec{S}_{\text{scatt.}}\cdot\frac{\vec{r}}{r}\Big) = \frac{\hbar k}{\mu}\frac{|f(\theta)|^2}{r^2}|\text{const}|^2 + \text{Order}\Big(\frac{1}{r^3}\Big), \tag{32}$$

we still have

$$\frac{d\sigma}{d\Omega} = |f(\theta)|^2 = \frac{\gamma^2}{4k^2\sin^4\frac{\theta}{2}} = \frac{Z_1^2 Z_2^2 e^4}{16E^2\sin^4\frac{\theta}{2}}, \tag{33}$$

so that the quantum-mechanical result for the differential cross section agrees with the classical Rutherford result. The exponential factor,

$$e^{-i\gamma ln\sin^2\frac{\theta}{2}},$$

however, does play an important role in the Coulomb scattering of identical charged particles.

Problems

4. Find the differential scattering cross section for neutron nucleus scattering in first Born approximation under the assumption the neutron nucleus interaction can be approximated by a square well, $V = -|V_0|$, for $r \leq a$, $V = 0$ for $r > a$. Show in particular, the differential cross section is strongly forward-peaked and the angular range of the forward peak can be used to determine the size of the nucleus. Calculate the value of θ for the first zero of the differential cross section for a beam with E = 100 MeV, assuming a = 5 fm (medium-heavy nucleus). Would you expect the first Born approximation to be good under these conditions?

5. For purely electromagnetic probes, the spherical drop model of the nucleus with a uniform charge distribution of radius a, total charge Ze, is a good model for the nucleus. Show the differential scattering cross section for μ^- nucleus scattering is given in first Born approximation by

$$\frac{d\sigma}{d\Omega} = \left(\frac{d\sigma}{d\Omega}\right)_{\text{Rutherford}} \left[\frac{3}{qa} j_1(qa)\right]^2, \qquad \text{with } q = 2k \sin\frac{\theta}{2}.$$

Make a plot of this differential cross section as a function of θ and show how it can be used to determine a, the radius of the nucleus. If you want to place the first zero in the differential cross section at $\theta = 60^o$ for a nucleus with $a = 10$ fm, what energy is required for the μ^- beam? (Convince yourself the corresponding problem for e^- nucleus scattering requires relativistic quantum theory. Most of our detailed knowledge about the electromagnetic structure of the nucleus comes from high-energy e^- nucleus scattering experiments).

The μ^- nucleus potential is the simple Coulomb potential, with

$$V(r) = -\frac{Ze^2}{r} \quad \text{for } r \geq a, \qquad V(r) = -\frac{Ze^2}{a}\left(\frac{3}{2} - \frac{r^2}{2a^2}\right) \quad \text{for } r \leq a.$$

Note: the long-range $r \to \infty$ contribution to the integral can be handled by screening as for the simple point charge; i.e., we can replace

$$-\frac{Ze^2}{r} \longrightarrow -\frac{Ze^2}{r} e^{-br}$$

and take the limit $b \to 0$.

6. For the 1-D radial function, $u_l(r)$, for a spherically symmetric potential, $V(r)$, which remains finite as $r \to 0$ and goes to zero as $r \to \infty$, show

$$\left[u_l \frac{du_l^{(0)}}{dr} - u_l^{(0)} \frac{du_l}{dr}\right]_{r=\infty} = -\frac{2\mu}{\hbar^2} \int_0^\infty dr\, V(r) u_l(r) u_l^{(0)}(r),$$

where $u_l^{(0)}(r)$ is the corresponding radial function for the case, $V(r) = 0$. Use this result to derive the Born approximation for the phase shifts:

$$\sin \delta_l^{\text{Born}} = -\frac{2\mu k}{\hbar^2} \int_0^\infty dr r^2 V(r) \big[j_l(kr)\big]^2.$$

Use the Born approximation result for $f(\theta)$ to "derive" the "well-known" expansion

$$\frac{\sin\left(2kr\sin\frac{\theta}{2}\right)}{\left(2kr\sin\frac{\theta}{2}\right)} = \sum_l (2l+1)P_l(\cos\theta)\big[j_l(kr)\big]^2.$$

7. Use the general Born series to derive the second-order term for $f(\theta)$ in the Born approximation from the coordinate space integrations. For a spherically symmetric $V(r)$, perform all angular integrations and show the resultant expression can be reduced to a single l sum:

$$\left(f(\theta)\right)^{\text{Born(2)}} = \sum_l (2l+1)P_l(\cos\theta)I_l,$$

where the I_l can be expressed in terms of double radial integrals involving V and spherical Bessel and Hänkel functions. Note: With both $\vec{r}\,'$ and $\vec{r}\,''$ finite, $G^{(+)}(\vec{r}\,',\vec{r}\,'')$ is best expressed in terms of:

$$G^{(+)}(\vec{r}\,',\vec{r}\,'') = \sum_l g_l(r',r'')\sum_{m=-l}^{+l} Y_{lm}(\theta',\phi')Y^*_{lm}(\theta'',\phi'').$$

8. The most common isotope of Lithium ($Z = 3$) is ${}^7_3\text{Li}_4$, a nucleus with a spin of $\frac{3}{2}$. If an unpolarized beam of ^{7}Li nuclei is incident on an identical ^{7}Li target nucleus with random spin orientations at a center of mass energy, $p^2/(2\mu) = 2$MeV, show the scattering is dominated by the pure Coulomb repulsion potential. (Effectively, the nuclear force does not come into play. Tunneling through the Coulomb barrier is negligible. A reasonable estimate for the ^{7}Li nuclear radius is 2.3 fm.) Show, however, at this energy, the exact scattering amplitude for the Coulomb potential must be taken into account for this scattering of identical particles from identical (indistinguishable) particles. Find the fraction of spin-symmetric and spin-antisymmetric states, and calculate $\frac{d\sigma}{d\Omega}$ as a function of θ in the center of mass system. In particular, calculate the ratio

$$\frac{\left(\left(\frac{d\sigma}{d\Omega}\right)_{\text{quantum mech.}}\right)}{\left(\frac{d\sigma}{d\Omega}\right)_{\text{classical}}}.$$

Also, calculate the differential cross section in the laboratory system as a function of $\theta_{\text{lab.}}$.

9. A low energy beam of spin $s = 1$ particles is scattered from a target of identical (indistinguishable) spin $s = 1$ target particles. Assume the scattering process is governed by a spin-independent potential leading to the following phase shifts:

$$\delta_0 = \tfrac{3}{2}\pi,\quad \delta_1 = \tfrac{1}{3}\pi,\quad \delta_2 = \tfrac{1}{2}\pi,\qquad \delta_l = 0,\quad \text{for } l \geq 3.$$

Find the differential scattering cross section as a function of θ (center of mass system), assuming (a) the incident beam is unpolarized and the target spins have random orientations. (b) Repeat, assuming the incident beam is perfectly longitudinally polarized, with a pure $m_s = +1$, and the target spins still have random orientations. Assume the detector is insensitive to the spin orientation.

47

Operator Form of Scattering Green's Function and the Integral Equation for the Scattering Problem

So far, we have worked strictly in coordinate representation for the very limited problem of the scattering of a structureless point particle from another structureless point particle. To generalize to more complicated situations involving composite projectile and target particles, with the possibility of inelastic collision processes or rearrangement collisions, or even to iterate the Born approximation for the case of structureless point particles, it will be advantageous to recast both the scattering Green's function and the scattering integral equation into operator form.

In coordinate representation, we solved the scattering problem,

$$(E - H)\psi_{\vec{k}}^{(+)}(\vec{r}) = 0$$
$$(E - H_0)\psi_{\vec{k}}^{(+)}(\vec{r}) = (E - H_0)\phi_{\vec{k}}(\vec{r}) + V(\vec{r})\psi_{\vec{k}}^{(+)}(\vec{r}), \tag{1}$$

by introducing the Green's function, $G^{(+)}(\vec{r}, \vec{r}\,')$, where

$$(E - H_0)G^{(+)}(\vec{r}, \vec{r}\,') = \delta(\vec{r} - \vec{r}\,'), \tag{2}$$

and converting eq. (1) into the form of an integral equation. [In Chapters 45 and 46, we found it convenient to multiply the operators $(E - H_0)$ and $(E - H)$ in these equations through by the factor $\frac{2\mu}{\hbar^2}$ to carry out the integrations in k space. We will find it more convenient now to go back to the original Schrödinger normalization. In this normalization, the Green's function of Chapter 45 would have been $-\frac{2\mu}{\hbar^2} e^{ik|\vec{r}-\vec{r}\,'|}/4\pi|\vec{r} - \vec{r}\,'|$.]

Just as we can take the state vector $|\psi\rangle$ and expand it in terms of a set of (1) base vectors $|n\rangle$, the eigenstates of a complete set of commuting operators with a discrete spectrum, e.g., the eigenstates $|n_1 n_2 n_3\rangle$ of H_0 in the box normalization; or (2) base vectors $|\vec{k}\,'\rangle$, the eigenvectors of k_x, k_y, k_z for the continuum plane wave states; or (3) base vectors $|\vec{r}\,'\rangle$ in the coordinate representation that are the eigenvectors of the operators x, y, z,

$$
\begin{aligned}
|\psi\rangle &= \sum_n |n\rangle\langle n|\psi\rangle, \\
|\psi\rangle &= \int d\vec{k}\,'|\vec{k}\,'\rangle, \langle\vec{k}\,'|\psi\rangle \\
|\psi\rangle &= \int d\vec{r}\,'|\vec{r}\,'\rangle, \langle\vec{r}\,'|\psi\rangle
\end{aligned} \tag{3}
$$

so, we can think of the Green's function not only in coordinate representation, but also in any representation, expressed in terms of any convenient basis. In fact, we will introduce an operator form of the Green's function and then express the Green's operator in any convenient basis. In coordinate representation, we have

$$
G^{(+)}(\vec{r},\vec{r}\,') = \lim_{\epsilon\to 0}\int d\vec{k}\,' \frac{\phi_{\vec{k}\,'}(\vec{r})\phi^*_{\vec{k}\,'}(\vec{r}\,')}{\big(E(k) - E(k') + i\epsilon\big)}. \tag{4}
$$

Again, we have changed from $(k^2 - k'^2)$ to $E(k) - E(k')$ in this Green's function, and therefore from $k + i\eta$ to $E(k) + i\epsilon$, with $E(k) + i\epsilon = \frac{\hbar^2}{2\mu}(k^2 + 2ik\eta + \cdots)$, so $\epsilon = (\frac{\hbar^2 k}{\mu})\eta$. We will now think of this coordinate representation of the Green's function as the coordinate representation matrix element of an operator, the Green's operator, $G^{(+)}_{\text{op.}}$,

$$
G^{(+)}(\vec{r},\vec{r}\,') = \langle\vec{r}|G^{(+)}_{\text{op.}}|\vec{r}\,'\rangle = \lim_{\epsilon\to 0}\int d\vec{k}\,'\langle\vec{r}|\vec{k}\,'\rangle\frac{1}{E(k) - E(k') + i\epsilon}\langle\vec{k}\,'|\vec{r}\,'\rangle, \tag{5}
$$

with

$$
\langle\vec{r}|\vec{k}\,'\rangle = \frac{1}{(2\pi)^{\frac{3}{2}}}e^{i\vec{k}\,'\cdot\vec{r}}, \qquad \langle\vec{k}\,'|\vec{r}\,'\rangle = \langle\vec{r}\,'|\vec{k}\,'\rangle^* = \frac{1}{(2\pi)^{\frac{3}{2}}}e^{-i\vec{k}\,'\cdot\vec{r}\,'}, \tag{6}
$$

so

$$
G^{(+)}_{\text{op.}} = \lim_{\epsilon\to 0}\int d\vec{k}\,'|\vec{k}\,'\rangle\langle\vec{k}\,'|\frac{1}{E(k) - E(k') + i\epsilon}|\vec{k}\,'\rangle\langle\vec{k}\,'|. \tag{7}
$$

Henceforth, the limit, $\lim_{\epsilon\to 0}$, will be quietly understood whenever an ϵ appears in an expression. Also, we could have written the Green's operator in the basis-independent form

$$
G^{(+)}_{\text{op.}} = \frac{1}{E - H_0 + i\epsilon}. \tag{8}
$$

If this is left-multiplied by the unit operator in the form, $\int d\vec{k}\,''|\vec{k}\,''\rangle\langle\vec{k}\,''|$ and right-multiplied by a unit operator in the form, $\int d\vec{k}\,'|\vec{k}\,'\rangle\langle\vec{k}\,'|$, we get

$$
\begin{aligned}
G^{(+)}_{\text{op.}} &= \int d\vec{k}\,''\int d\vec{k}\,'|\vec{k}\,''\rangle\langle\vec{k}\,''|\frac{1}{E(k) - H_0 + i\epsilon}|\vec{k}\,'\rangle\langle\vec{k}\,'| \\
&= \int d\vec{k}\,'|\vec{k}\,'\rangle\frac{1}{E(k) - E(k') + i\epsilon}\langle\vec{k}\,'|,
\end{aligned} \tag{9}
$$

where we have used $H_0|\vec{k}'\rangle = E(k')|\vec{k}'\rangle$, and $\langle\vec{k}''|\vec{k}'\rangle = \delta(\vec{k}'' - \vec{k}')$, to regain the form of the Green operator first given in eq. (7). We could have used other forms of the unit operator, such as $\sum_n |n\rangle\langle n|$, to gain different realizations of this operator.

A The Lippmann–Schwinger Equation

The operator form of the Green's function will be particularly convenient if we want to generalize our scattering problem to one involving composite target or projectile particles, or if we want to iterate the Born approximation to get the higher order Born approximation terms. With the operator form of the Green's function, it is not necessary to write the integral equation form of the scattering problem

$$\psi^{(+)}(\vec{r}) = \phi(\vec{r}) + \int d\vec{r}\,'G^{(+)}(\vec{r},\vec{r}\,')V(\vec{r}\,')\psi^{(+)}(\vec{r}\,') \tag{10}$$

in coordinate representation. In representation-independent state vector form, this equation becomes

$$|\psi_{\vec{k}}^{(+)}\rangle = |\phi_{\vec{k}}\rangle + \frac{1}{E - H_0 + i\epsilon}V|\psi_{\vec{k}}^{(+)}\rangle, \tag{11}$$

where the state vector $|\psi_{\vec{k}}^{(+)}\rangle$ is the state vector for an incoming plane wave in the α channel, with relative momentum, $\hbar\vec{k}$, and spherical outgoing waves in *all* energetically possible outgoing channels. This operator form of the scattering problem integral equation is known as the Lippmann–Schwinger equation. In this form, it is easy to iterate to get the Born solution to the equation in an infinite series, the Born expansion. Substituting for $|\psi_{\vec{k}}^{(+)}\rangle$ in the right-hand side, we get

$$|\psi_{\vec{k}}^{(+)}\rangle = |\phi_{\vec{k}}\rangle + \frac{1}{E - H_0 + i\epsilon}V\Big(|\phi_{\vec{k}}\rangle + \frac{1}{E - H_0 + i\epsilon}V|\psi_{\vec{k}}^{(+)}\rangle\Big). \tag{12}$$

Substituting for $|\psi_{\vec{k}}^{(+)}\rangle$ again (and again) in the right-hand side, we get the Born series

$$\begin{aligned} |\psi_{\vec{k}}^{(+)}\rangle &= |\phi_{\vec{k}}\rangle + \frac{1}{E - H_0 + i\epsilon}V|\phi_{\vec{k}}\rangle \\ &+ \frac{1}{E - H_0 + i\epsilon}V\frac{1}{E - H_0 + i\epsilon}V|\phi_{\vec{k}}\rangle \\ &+ \frac{1}{E - H_0 + i\epsilon}V\frac{1}{E - H_0 + i\epsilon}V\frac{1}{E - H_0 + i\epsilon}V|\phi_{\vec{k}}\rangle \\ &+ \cdots, \end{aligned} \tag{13}$$

with solution

$$|\psi_{\vec{k}}^{(+)}\rangle = |\phi_{\vec{k}}\rangle + |\psi_{\vec{k}}^{(1)}\rangle + |\psi_{\vec{k}}^{(2)}\rangle + \cdots. \tag{14}$$

To get the second-order term in coordinate representation, we now introduce unit operators of type $\int d\vec{r}\,'|\vec{r}\,'\rangle\langle\vec{r}\,'|$ to get

$$\psi_{\vec{k}}^{(2)}(\vec{r}) = \int d\vec{r}\,'' \int d\vec{r}\,''' \int d\vec{r}\,'''' \int d\vec{r}\,' \langle\vec{r}|\frac{1}{E - H_0 + i\epsilon}|\vec{r}\,''\rangle\langle\vec{r}\,''|V|\vec{r}\,'''\rangle$$
$$\times \langle\vec{r}\,'''|\frac{1}{E - H_0 + i\epsilon}|\vec{r}\,''''\rangle\langle\vec{r}\,''''|V|\vec{r}\,'\rangle\langle\vec{r}\,'|\phi_{\vec{k}}\rangle$$
$$= \int d\vec{r}\,'' \int d\vec{r}\,' G^{(+)}(\vec{r}, \vec{r}\,'')V(\vec{r}\,'')G^{(+)}(\vec{r}\,'', \vec{r}\,')V(\vec{r}\,')\phi_{\vec{k}}(\vec{r}\,'), \qquad (15)$$

where we have used

$$\langle\vec{r}\,''|V|\vec{r}\,'''\rangle = V(\vec{r}\,'')\delta(\vec{r}\,'' - \vec{r}\,''') \text{ and } \langle\vec{r}|\frac{1}{E - H_0 + i\epsilon}|\vec{r}\,''\rangle = G^{(+)}(\vec{r}, \vec{r}\,'') \qquad (16)$$

to reduce the expression to the form of a double 3-D integral involving two Green's functions. The Green's function, $G(\vec{r}, \vec{r}\,'')$ is needed only in its asymptotic form, as $r \to \infty$,

$$-\frac{2\mu}{\hbar^2}\frac{e^{ikr}}{4\pi r}e^{-i\vec{k}_s\cdot\vec{r}\,''},$$

but the Green's function, $G^{(+)}(\vec{r}\,'', \vec{r}\,')$, is needed for all values of $\vec{r}\,'$ and $\vec{r}\,''$ in the form

$$-\frac{2\mu}{\hbar^2}\frac{e^{ik|\vec{r}\,''-\vec{r}\,'|}}{4\pi|\vec{r}\,'' - \vec{r}\,'|}.$$

For spherically symmetric potentials this is best expanded in spherical harmonics [see eqs. (39) and (53) of Chapter 45] to actually carry out the integrals.

$$G^{(+)}(\vec{r}\,'', \vec{r}\,') = \sum_{lm}\frac{2\mu}{\hbar^2}\frac{1}{r'^2}g_l(r'', r')Y_{lm}(\theta'', \phi'')Y^*_{lm}(\theta', \phi'). \qquad (17)$$

48

Inelastic Scattering Processes and Rearrangement Collisions

So far, we have seen the usefulness of the operator form of the Green's function and the integral equation for the scattering problem in iterating the Born approximation for the scattering of structureless point particles from structureless point particles. The method becomes equally useful in discussing the scattering of composite particles from composite particles, where inelastic processes and rearrangement collisions come into play.

A Inelastic Scattering Processes

Consider the case in which we might have a scattering of a projectile from a composite target particle (e.g., μ^--atom scattering or μ^--nucleus scattering; we choose a μ^- as our projectile to avoid having identical particles as the components of our composite target particle). Now, it will be useful to decompose

$$H = H_0 + V, \tag{1}$$

where H_0 will include, besides the kinetic energy operator for the relative motion, the Hamiltonian for the internal degrees of freedom of the composite particle.

$$H_0 = -\frac{\hbar^2}{2\mu}\nabla^2_{\text{rel.}} + H_{\text{int.}}(\vec{\xi}), \tag{2}$$

with state vectors, $|\vec{k}, n\rangle$, such that

$$H_0|\vec{k}, n\rangle = \left(\frac{\hbar^2 k^2}{2\mu} + E_n^{\text{int.}}\right)|\vec{k}, n\rangle, \qquad \langle\vec{r}, \vec{\xi}|\vec{k}, n\rangle = \frac{1}{(2\pi)^{\frac{3}{2}}} e^{i\vec{k}\cdot\vec{r}} \chi_n(\vec{\xi}), \tag{3}$$

with

$$H_{\text{int.}}(\vec{\xi})\chi_n(\vec{\xi}) = E_n^{\text{int.}}\chi_n(\vec{\xi}), \tag{4}$$

where $\vec{\xi}$ are the internal variables of the composite particle. [Note: If both projectile and target particles are composite particles, $\chi_n(\vec{\xi}) \equiv \chi_{n_a}(\vec{\xi}_a)\chi_{n_A}(\vec{\xi}_A)$, and $E_n^{\text{int.}} = E_{n_a}^{\text{int.}} + E_{n_A}^{\text{int.}}$.] Now, the analogue of eq. (7) of Chapter 47 can be written as

$$G_{\text{op.}}^{(+)} = \sum_{n'}\int d\vec{k}'|\vec{k}', n'\rangle \frac{1}{\frac{\hbar^2}{2\mu}k^2 + E_{n=0}^{\text{int.}} - \frac{\hbar^2}{2\mu}k'^2 - E_{n'}^{\text{int.}} + i\epsilon}\langle\vec{k}', n'|, \tag{5}$$

where conservation of energy now requires the scattered k vector, $\vec{k}_s$ (see Fig. 48.1), have magnitude given by

$$\frac{\hbar^2}{2\mu}k_s^2 = \frac{\hbar^2}{2\mu}k^2 + E_{n=0}^{\text{int.}} - E_{n'}^{\text{int.}}, \tag{6}$$

so the coordinate representation for the Green's function becomes

$$\langle\vec{r}, \vec{\xi}|G_{\text{op.}}^{(+)}|\vec{r}', \vec{\xi}'\rangle = \sum_{n'}\chi_{n'}(\vec{\xi})\left(\int d\vec{k}'\frac{\phi_{\vec{k}'}(\vec{r})\phi_{\vec{k}'}^*(\vec{r}')}{\frac{\hbar^2}{2\mu}(k_s^2 - k'^2) + i\epsilon}\right)\chi_{n'}^*(\vec{\xi}'). \tag{7}$$

If we perform the $\vec{k}'$ integration as before, we get

$$G^{(+)}(\vec{r}, \vec{\xi}; \vec{r}', \vec{\xi}') = -\frac{1}{4\pi}\frac{2\mu}{\hbar^2}\sum_{n'}\chi_{n'}(\vec{\xi})\chi_{n'}^*(\vec{\xi}')\frac{e^{ik_s|\vec{r}-\vec{r}'|}}{|\vec{r}-\vec{r}'|}. \tag{8}$$

The magnitude k_s is now dependent on n'. The integral equation becomes

$$\psi_{\vec{k}}^{(+)}(\vec{r}, \vec{\xi}) = \phi_{\vec{k}}(\vec{r})\chi_{n=0}(\vec{\xi}) + \int d\vec{r}'\int d\vec{\xi}' G^{(+)}(\vec{r}, \vec{\xi}; \vec{r}', \vec{\xi}')V(\vec{r}', \vec{\xi}')\psi_{\vec{k}}^{(+)}(\vec{r}', \vec{\xi}'). \tag{9}$$

In first Born approximation, in the limit $r \to \infty$, in which

$$\frac{e^{ik_s|\vec{r}-\vec{r}'|}}{|\vec{r}-\vec{r}'|} \to \frac{e^{ik_s r}}{r}e^{-i\vec{k}_s\cdot\vec{r}'},$$

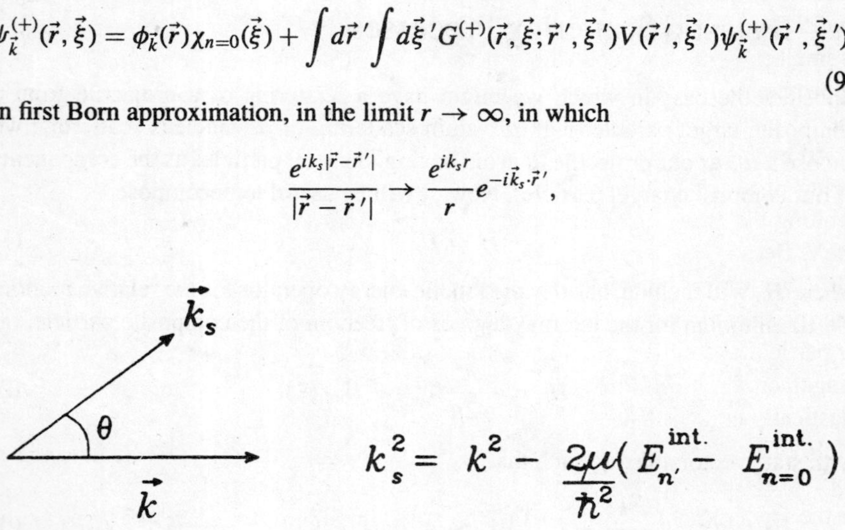

FIGURE 48.1.

we get the result

$$\psi^{(+)}_{\vec{k}}(\vec{r}, \vec{\xi}) \rightarrow \frac{1}{(2\pi)^{\frac{3}{2}}} \left(e^{i\vec{k}\cdot\vec{r}} \chi_{n=0}(\vec{\xi}) - \frac{1}{4\pi} \frac{2\mu}{\hbar^2} \sum_{n'} \chi_{n'}(\vec{\xi}) \frac{e^{ik_s r}}{r} \right.$$
$$\left. \times \int d\vec{r}\,' \int d\vec{\xi}\,' V(\vec{r}\,', \vec{\xi}\,') e^{i(\vec{k}-\vec{k}_s)\cdot\vec{r}\,'} \chi^*_{n'}(\vec{\xi}\,') \chi_{n=0}(\vec{\xi}\,') \right), \tag{10}$$

with

$$k_s^2 = k^2 - \frac{2\mu}{\hbar^2}(E^{\text{int.}}_{n'} - E^{\text{int.}}_{n=0}), \tag{11}$$

where the coefficient of $\frac{e^{ik_s r}}{r}\chi_{n'}(\vec{\xi})$, with $n' = 0$ and, hence, $k_s = |\vec{k}_s| = k$, gives the scattering amplitude for elastic scattering, and the coefficients of terms with $n' = 1, 2, \cdots$ and, hence, $k_s < k$ give the scattering amplitudes for the inelastic processes, where the composite target particle ends in the excited state with internal energy, $E^{\text{int.}}_{n'}$.

B Rearrangement Collisions

Consider, e.g., the scattering of a μ^- from a hydrogen atom. In this case, we can have, besides the elastic scattering process, not only the inelastic scattering processes in which the hydrogen atom ends in an excited state, but also we could have a rearrangement collision, in which the μ^- is captured by the proton to make a muonic atom while the atomic electron is scattered. (Because the ground-state energy of the hydrogen atom is -13.61 eV, whereas the ground-state energy of the muonic atom is ≈ -2.5 KeV, the scattered electron would have an energy greater than 2.5 KeV. An electron in the 2.5 to 3 KeV range would have a $v/c \approx 0.1$, which is small enough for a treatment by nonrelativistic quantum mechanics and the Born approximation should be good.) If we label the μ^- particle 1, the e^- particle 2, and the p^+ particle 3, the rearrangement collision, $1 + (2, 3) \rightarrow 2 + (1, 3)$, is possible.

An even more complex three-particle system could be illustrated by the scattering of a proton from the nucleus ^{17}O, which can be modelled by a neutron orbiting the doubly magic nucleus, ^{16}O, with a n-^{16}O binding energy of - 4.146 MeV. Because the first excited state of the tightly bound ^{16}O nucleus is at 6.06 MeV, we could approximate this nucleus as an inert particle if the incident proton energy is not too large. Thus, we could think of the proton as particle 1, the neutron as particle 2, and the ^{16}O-nucleus as particle 3. Now, we could have all three arrangements of the particles labeled 1, 2, 3 (see Fig. 48.2). The proton could scatter elastically or inelastically from ^{17}O. (The first excited state of ^{17}O is at only 0.871 MeV.) The proton could also be captured, however, by the ^{16}O nucleus to make ^{17}F, with a p-^{16}O binding energy of -.601 MeV, the neutron being knocked out in the reaction and being scattered. Finally, the proton could capture the loosely bound neutron from ^{16}O to make a deuteron, with binding energy of -2.226MeV, which is scattered relative to the ^{16}O nucleus. Thus, the rearrangement collisions,

$$\begin{array}{ll} p + {}^{17}O \to p + {}^{17}O & 1+(2,3) \to 1+(2,3), \\ p + {}^{17}O \to n + {}^{17}F & 1+(2,3) \to 2+(1,3), \\ p + {}^{17}O \to {}^{16}O + d & 1+(2,3) \to 3+(1,2), \end{array} \tag{12}$$

are all possible. In addition, if sufficient incident kinetic energy exists, we could have a final state with three outgoing particles, proton, neutron, and ^{16}O. These three-particle states would be excited states of all three final rearrangements, involving breakup of the two-particle aggregates. If insufficient energy for such a breakup exists, we may have to consider only the above three rearrangement processes. Now, it will be convenient to split the full Hamiltonian, H, into an H_0 and an interaction term, V, in three different ways.

$$\begin{aligned} H = H_0 + V &= -\frac{\hbar^2}{2\mu_{1,23}}\nabla^2_{1,23} + H_{\text{int.}}(2,3) + V_{12} + V_{13} \\ &= H_0' + V' = -\frac{\hbar^2}{2\mu_{2,13}}\nabla^2_{2,13} + H_{\text{int.}}(1,3) + V_{12} + V_{23} \\ &= H_0'' + V'' = -\frac{\hbar^2}{2\mu_{3,12}}\nabla^2_{3,12} + H_{\text{int.}}(1,2) + V_{13} + V_{23}. \end{aligned} \tag{13}$$

It will now be useful to rewrite the integral equation for the scattering process in three different ways. For this purpose, we will first rewrite the Lippmann-Schwinger equation in terms of the *full* Green's operator

$$\overline{G}^{(+)}_{\text{op.}} = \frac{1}{E - H + i\epsilon}, \tag{14}$$

$$|\psi^{(+)}_{\vec{k}}\rangle = \left(1 + \frac{1}{E - H + i\epsilon}V\right)|\phi_{\vec{k}}\rangle. \tag{15}$$

To derive this equation, which is the operator form of eq. (59) of Chapter 45, act on the operator, $E - H + i\epsilon = E - H_0 + i\epsilon - V$, with $\overline{G}^{(+)}_{\text{op.}}$ from the left and with $G^{(+)}_{\text{op.}}V$ from the right to yield

$$\begin{aligned} &\frac{1}{E - H + i\epsilon}\left(E - H + i\epsilon = E - H_0 + i\epsilon - V\right)\frac{1}{E - H_0 + i\epsilon}V \\ &\frac{1}{E - H_0 + i\epsilon}V = \frac{1}{E - H + i\epsilon}V\left(1 - \frac{1}{E - H_0 + i\epsilon}V\right). \end{aligned} \tag{16}$$

Now, let this act on $|\psi^{(+)}_{\vec{k}}\rangle$, and use the Lippmann–Schwinger equation

$$\frac{1}{E - H_0 + i\epsilon}V|\psi^{(+)}_{\vec{k}}\rangle = |\psi^{(+)}_{\vec{k}}\rangle - |\phi_{\vec{k}}\rangle \tag{17}$$

to yield

$$|\psi^{(+)}_{\vec{k}}\rangle - |\phi_{\vec{k}}\rangle = \frac{1}{E - H + i\epsilon}V|\phi_{\vec{k}}\rangle, \tag{18}$$

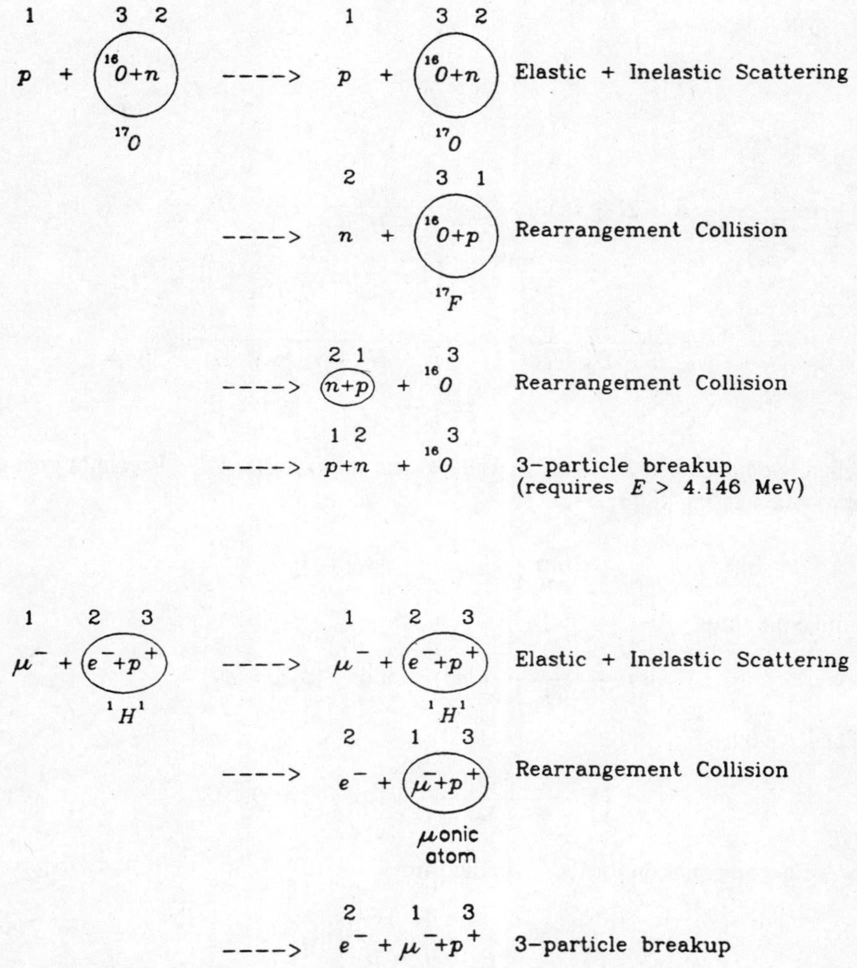

p + ($^{16}O+n$) [^{17}O] ----> p + ($^{16}O+n$) [^{17}O] Elastic + Inelastic Scattering

----> n + ($^{16}O+p$) [^{17}F] Rearrangement Collision

----> ($n+p$) + ^{16}O Rearrangement Collision

----> $p+n$ + ^{16}O 3–particle breakup (requires E > 4.146 MeV)

μ^- + (e^-+p^+) [$^1H^1$] ----> μ^- + (e^-+p^+) [$^1H^1$] Elastic + Inelastic Scattering

----> e^- + (μ^-+p^+) [μonic atom] Rearrangement Collision

----> e^- + μ^-+p^+ 3–particle breakup

FIGURE 48.2. Rearrangement collisions for three-particle systems.

which is eq. (15). Now, our purpose is not to find a coordinate or other representation of the *full* Green operator, but to rearrange this relation to write the Lippmann–Schwinger equation in terms of the primed and double-primed operators, which are needed for the rearrangement collisions. For this purpose, we shall rewrite

$$\begin{aligned}(E - H + i\epsilon)|\psi_{\vec{k}}^{(+)}\rangle &= (E - H + i\epsilon)\Big(1 + \frac{1}{E - H + i\epsilon}V\Big)|\phi_{\vec{k}}\rangle \\ &= (E - H_0 + i\epsilon)|\phi_{\vec{k}}\rangle = i\epsilon|\phi_{\vec{k}}\rangle. \qquad (19)\end{aligned}$$

Because we are to take the limit $\epsilon \to 0$, the right-hand side is zero. Let us be careful about this limit, however. Acting on the left-hand side with a Green's operator of the type

$$\frac{1}{E - H_0' + i\epsilon},$$

we get

$$\begin{aligned}\frac{1}{E - H_0' + i\epsilon}(E - H + i\epsilon)|\psi_{\vec{k}}^{(+)}\rangle &= \frac{1}{E - H_0' + i\epsilon}(E - H_0' - V' + i\epsilon)|\psi_{\vec{k}}^{(+)}\rangle \\ &= \Big(1 - \frac{1}{E - H_0' + i\epsilon}V'\Big)|\psi_{\vec{k}}^{(+)}\rangle \\ = \frac{i\epsilon}{E - H_0' + i\epsilon}|\phi_{\vec{k}}\rangle &= \frac{i\epsilon}{E - H_0 + V' - V + i\epsilon}|\phi_{\vec{k}}\rangle \\ &= \frac{i\epsilon}{V' - V + i\epsilon}|\phi_{\vec{k}}\rangle = 0, \end{aligned} \tag{20}$$

where we have used $H_0 - H_0' = V' - V$ and the fact that $V' - V$ cannot give a zero when acting on $|\phi_{\vec{k}}\rangle$, so

$$\lim_{\epsilon\to 0}\frac{i\epsilon}{V' - V + i\epsilon}|\phi_{\vec{k}}\rangle = 0,$$

unlike the limit

$$\lim_{\epsilon\to 0}\frac{i\epsilon}{E - H_0 + i\epsilon}|\phi_{\vec{k}}\rangle = \lim_{\epsilon\to 0}\frac{i\epsilon}{i\epsilon}|\phi_{\vec{k}}\rangle = |\phi_{\vec{k}}\rangle.$$

Thus, we have

$$\Big(1 - \frac{1}{E - H_0' + i\epsilon}V'\Big)|\psi_{\vec{k}}^{(+)}\rangle = 0, \tag{21}$$

or we get an equation for $|\psi_{\vec{k}}^{(+)}\rangle$ in the form

$$|\psi_{\vec{k}}^{(+)}\rangle = \frac{1}{E - H_0' + i\epsilon}V'|\psi_{\vec{k}}^{(+)}\rangle. \tag{22}$$

In exactly the same way, we could have gotten still a third form for $|\psi_{\vec{k}}^{(+)}\rangle$ by replacing H_0' and V' in this equation by H_0'' and V''. Thus, we have three different ways of writing the Lippmann–Schwinger equation, leading to three different forms for the integral equations of this scattering problem for our system of three particles.

$$\begin{aligned}|\psi_{\vec{k}}^{(+)}\rangle &= |\phi_{\vec{k}}\rangle + \frac{1}{E - H_0 + i\epsilon}V|\psi_{\vec{k}}^{(+)}\rangle, \\ |\psi_{\vec{k}}^{(+)}\rangle &= \frac{1}{E - H_0' + i\epsilon}V'|\psi_{\vec{k}}^{(+)}\rangle, \\ |\psi_{\vec{k}}^{(+)}\rangle &= \frac{1}{E - H_0'' + i\epsilon}V''|\psi_{\vec{k}}^{(+)}\rangle. \end{aligned} \tag{23}$$

In all three forms of this scattering equation, the vector, $|\psi_{\vec{k}}^{(+)}\rangle$, contains besides the incoming plane wave for the motion of particle 1 relative to the bound system of particles (2, 3) the spherical outgoing waves corresponding to both elastic and inelastic scattering of particle 1 from the system (2, 3), as well as the spherical outgoing waves for particle 2 moving away from the system (1, 3), as well as the spherical outgoing waves for particle 3 moving away from the system (1, 2). When converted to integral equation form, none of the three integral equations is easy to solve. In Born approximation, however, when the incident energy is large enough compared with the potentials, V_{12}, V_{13}, and V_{23}, we may get a reasonably good approximate solution. For this purpose, however, the first form is best for the elastic and inelastic scattering of particle 1, whereas the second form would be best for getting an approximate value for the cross section for the rearrangement, $1 + (2, 3) \rightarrow 2 + (1, 3)$, and the last form would be best for the rearrangement collision of type $1 + (2, 3) \rightarrow 3 + (1, 2)$.

49

Differential Scattering Cross Sections for Rearrangement Collisions: Born Approximation

Let us calculate in some detail the differential scattering cross section for a rearrangement collision of the type $1+(2,3) \to 2+(1,3)$, where the incident channel, particle 1 incident on the composite particle made of (2, 3), will be designated by the channel index, α, and the outgoing channel, with particle 2 leaving the composite system $(1,3)$, will be designated by channel index, β. We will assume the spins of particles 1, 2, 3 (if any) need not be considered. The relative position vector and internal coordinate in channel α are then

$$\vec{r}_\alpha = \vec{r}_1 - \frac{(m_2\vec{r}_2 + m_3\vec{r}_3)}{(m_2 + m_3)}, \qquad \vec{\xi}_\alpha = \vec{r}_2 - \vec{r}_3, \tag{1}$$

and the reduced mass in channel α is $\mu_\alpha = m_1 m_{23}/(m_1 + m_{23})$ and the incident $\vec{k}_\alpha \equiv \vec{k}$. Similarly,

$$\vec{r}_\beta = \vec{r}_2 - \frac{(m_1\vec{r}_1 + m_3\vec{r}_3)}{(m_1 + m_3)}, \qquad \vec{\xi}_\beta = \vec{r}_1 - \vec{r}_3, \tag{2}$$

with reduced mass in the β channel given by $\mu_\beta = m_2 m_{13}/(m_2 + m_{13})$. To find the necessary scattering amplitude, we shall use the coordinate representation $\langle \vec{r}_\beta, \vec{\xi}_\beta | \psi^{(+)}_{\vec{k}} \rangle$ and calculate this through the second (or primed) form of eq. (23) of Chapter 48, which is appropriate for the β channel.

$$\begin{aligned} \langle \vec{r}_\beta, \vec{\xi}_\beta | \psi^{(+)}_{\vec{k}} \rangle &= \langle \vec{r}_\beta, \vec{\xi}_\beta | \frac{1}{E - H_0' + i\epsilon} V' | \psi^{(+)}_{\vec{k}} \rangle \\ &= \sum_{n'} \int d\vec{k}' \langle \vec{r}_\beta, \vec{\xi}_\beta | \vec{k}', n' \rangle \langle \vec{k}', n' | \frac{1}{E - H_0' + i\epsilon} | \vec{k}', n' \rangle \\ &\quad \times \langle \vec{k}', n' | V' | \psi^{(+)}_{\vec{k}} \rangle, \end{aligned} \tag{3}$$

where $|\vec{k}', n'\rangle$ is an eigenvector of H_0', with

$$H_0'|\vec{k}', n'\rangle = E(k', n')|\vec{k}', n'\rangle = \Big(\frac{\hbar^2}{2\mu_\beta}k'^2 + E^{\text{int.}}_{\beta,n'}\Big)|\vec{k}', n'\rangle. \tag{4}$$

We shall now insert a unit operator in the form of

$$\int\int d\vec{r}'_\beta d\vec{\xi}'_\beta |\vec{r}'_\beta, \vec{\xi}'_\beta\rangle\langle\vec{r}'_\beta, \vec{\xi}'_\beta|$$

between the bra $\langle\vec{k}', n'|$ and V' to obtain

$$\begin{aligned}
\langle\vec{r}_\beta\vec{\xi}_\beta|\psi^{(+)}_{\vec{k}}\rangle &= \sum_{n'}\int d\vec{k}'\int d\vec{r}'_\beta\int d\vec{\xi}'_\beta\langle\vec{r}_\beta\vec{\xi}_\beta|\vec{k}', n'\rangle\frac{1}{E - E(k', n') + i\epsilon}\\
&\times\langle\vec{k}', n'|\vec{r}'_\beta\vec{\xi}'_\beta\rangle\langle\vec{r}'_\beta\vec{\xi}'_\beta|V'|\psi^{(+)}_{\vec{k}}\rangle\\
&= \sum_{n'}\int d\vec{k}'\int d\vec{r}'_\beta\int d\vec{\xi}'_\beta\ \frac{1}{(2\pi)^{\frac{3}{2}}}e^{i\vec{k}'\cdot\vec{r}_\beta}\chi_{n'}(\vec{\xi}_\beta)\\
&\times\frac{1}{E - \frac{\hbar^2}{2\mu_\beta}k'^2 - E^{\text{int.}}_{\beta,n'} + i\epsilon}\frac{1}{(2\pi)^{\frac{3}{2}}}e^{-i\vec{k}'\cdot\vec{r}'_\beta}\chi^*_{n'}(\vec{\xi}'_\beta)\langle\vec{r}'_\beta, \vec{\xi}'_\beta|V'|\psi^{(+)}_{\vec{k}}\rangle.
\end{aligned} \tag{5}$$

In addition, we shall use energy conservation to rename

$$E = \frac{\hbar^2}{2\mu_\alpha}k^2 + E^{\text{int.}}_{\alpha,0} = \frac{\hbar^2}{2\mu_\beta}k^2_{\beta,n'} + E^{\text{int.}}_{\beta,n'} \tag{6}$$

and evaluate the Green's function

$$\begin{aligned}
G^{(+)}(\vec{r}_\beta, \vec{\xi}_\beta; \vec{r}'_\beta, \vec{\xi}'_\beta) &= \frac{1}{(2\pi)^3}\sum_{n'}\int d\vec{k}'\frac{e^{i\vec{k}'\cdot(\vec{r}_\beta - \vec{r}'_\beta)}\chi_{n'}(\vec{\xi}_\beta)\chi^*_{n'}(\vec{\xi}'_\beta)}{\frac{\hbar^2}{2\mu_\beta}(k^2_{\beta,n'} - k'^2) + i\epsilon}\\
&= -\frac{1}{4\pi}\frac{2\mu_\beta}{\hbar^2}\sum_{n'}\frac{e^{ik_{\beta,n'}|\vec{r}_\beta - \vec{r}'_\beta|}}{|\vec{r}_\beta - \vec{r}'_\beta|}\chi_{n'}(\vec{\xi}_\beta)\chi^*_{n'}(\vec{\xi}'_\beta).
\end{aligned} \tag{7}$$

With this Green's function

$$\langle\vec{r}_\beta\vec{\xi}_\beta|\psi^{(+)}_{\vec{k}}\rangle = \int d\vec{r}'_\beta\int d\vec{\xi}'_\beta G^{(+)}(\vec{r}_\beta, \vec{\xi}_\beta; \vec{r}'_\beta, \vec{\xi}'_\beta)V'(\vec{r}'_\beta, \vec{\xi}'_\beta)\langle\vec{r}'_\beta, \vec{\xi}'_\beta|\psi^{(+)}_{\vec{k}}\rangle. \tag{8}$$

Now, as usual, we will take the limit, $r_\beta \to \infty$, so

$$\frac{e^{ik_{\beta,n'}|\vec{r}_\beta - \vec{r}'_\beta|}}{|\vec{r}_\beta - \vec{r}'_\beta|} \to \frac{e^{ik_{\beta,n'}r_\beta}}{r_\beta}e^{-i\vec{k}_{\beta,n'}\cdot\vec{r}'_\beta} \tag{9}$$

and, therefore,

$$\langle\vec{r}_\beta\vec{\xi}_\beta|\psi^{(+)}_{\vec{k}}\rangle \to \sum_{n'}\frac{e^{ik_{\beta,n'}r_\beta}}{r_\beta}\chi_{n'}(\vec{\xi}_\beta)\frac{f_{n'}(\theta_\beta, \phi_\beta)}{(2\pi)^{\frac{3}{2}}}, \tag{10}$$

with

$$f_{n'}(\theta_\beta, \phi_\beta) =$$

$$
\begin{aligned}
&-\frac{\mu_\beta}{2\pi\hbar^2}(2\pi)^{\frac{3}{2}}\int d\vec{r}\,'_\beta\int d\vec{\xi}\,'_\beta e^{-i\vec{k}_{\beta,n'}\cdot\vec{r}\,'_\beta}\chi^*_{n'}(\vec{\xi}\,'_\beta)V'(\vec{r}\,'_\beta,\vec{\xi}\,'_\beta)\psi^{(+)}_{\vec{k}}(\vec{r}\,'_\beta,\vec{\xi}\,'_\beta)\\
&= -\frac{\mu_\beta}{2\pi\hbar^2}(2\pi)^3\langle\phi_{\vec{k}_{\beta,n'}}|V'|\psi^{(+)}_{\vec{k}}\rangle,
\end{aligned}
\tag{11}
$$

where

$$
\langle\phi_{\vec{k}_{\beta,n'}}|\vec{r}\,'_\beta\vec{\xi}\,'_\beta\rangle = \frac{1}{(2\pi)^{\frac{3}{2}}}e^{-i\vec{k}_{\beta,n'}\cdot\vec{r}\,'_\beta}\chi^*_{n'}(\vec{\xi}\,'_\beta).
\tag{12}
$$

The inverse factor $(2\pi)^{\frac{3}{2}}$ is needed in the spherical outgoing wave factor, because our plane wave has this same inverse normalization. Now, in first Born approximation, we can replace $|\psi^{(+)}_{\vec{k}}\rangle$ by $|\phi_{\vec{k}}\rangle$, where now

$$
\langle\vec{r}\,'_\beta\vec{\xi}\,'_\beta|\phi_{\vec{k}}\rangle = \frac{e^{i\vec{k}\cdot\vec{r}\,'_\alpha}}{(2\pi)^{\frac{3}{2}}}\chi_0(\vec{\xi}\,'_\alpha),
\tag{13}
$$

where $\vec{r}\,'_\alpha$ and $\vec{\xi}\,'_\alpha$ must be expressed in terms of $\vec{r}\,'_\beta$ and $\vec{\xi}\,'_\beta$ via

$$
\begin{aligned}
\vec{r}\,'_\alpha &= -\frac{m_2}{(m_2+m_3)}\vec{r}\,'_\beta + \frac{m_3(m_1+m_2+m_3)}{(m_1+m_3)(m_2+m_3)}\vec{\xi}\,'_\beta\\
\vec{\xi}\,'_\alpha &= \vec{r}\,'_\beta + \frac{m_1}{(m_1+m_3)}\vec{\xi}\,'_\beta.
\end{aligned}
\tag{14}
$$

The first Born approximation value for the scattering amplitude for this rearrangement process is then

$$
\begin{aligned}
&f_{n'}(\theta_\beta,\phi_\beta) = -\frac{\mu_\beta}{2\pi\hbar^2}\times\\
&\int d\vec{r}\,'_\beta\int d\vec{\xi}\,'_\beta e^{i\vec{k}\cdot\vec{r}\,'_\alpha(\vec{r}\,'_\beta,\vec{\xi}\,'_\beta)}e^{-i\vec{k}_{\beta,n'}\cdot\vec{r}\,'_\beta}\chi^*_{n'}(\vec{\xi}\,'_\beta)\chi_0\big(\vec{\xi}\,'_\alpha(\vec{r}\,'_\beta,\vec{\xi}\,'_\beta)\big)V'(\vec{r}\,'_\beta,\vec{\xi}\,'_\beta).
\end{aligned}
\tag{15}
$$

To get the rearrangement collision cross section, we need to remember the flux in the incident plane wave is

$$
\vec{S} = \frac{1}{(2\pi)^3}\frac{\hbar\vec{k}}{\mu_\alpha} = \frac{1}{(2\pi)^3}\vec{v}_\alpha.
\tag{16}
$$

Note: In a box normalization where the factor $(2\pi)^3$ would have been replaced by the factor L^3, this $\vec{S}$ would have had the proper dimension of (v/L^3) or $(1/\text{length}^2\text{time})$, one of the advantages of a box normalization. Also, $\vec{k}\equiv\vec{k}_{\alpha,0}$.

The radial flux in the outgoing spherical wave is (in our present normalization)

$$
S_r = \frac{1}{(2\pi)^3}\frac{\hbar k_{\beta,n'}}{\mu_\beta}\frac{1}{r_\beta^2}|f_{n'}(\theta_\beta,\phi_\beta)|^2,
\tag{17}
$$

leading to the differential cross section for the rearrangement collision

$$
\begin{aligned}
\left(\frac{d\sigma}{d\Omega}\right)_{\beta,n'} &= \frac{\mu_\alpha k_{\beta,n'}}{\mu_\beta k}|f_{n'}(\theta_\beta,\phi_\beta)|^2 = \frac{v_\beta}{v_\alpha}|f_{n'}(\theta_\beta,\phi_\beta)|^2\\
&= \frac{\mu_\beta\mu_\alpha}{(2\pi\hbar^2)^2}\frac{k_{\beta,n'}}{k}(2\pi)^6|\langle\phi_{\vec{k}_{\beta,n'}}|V'|\phi_{\vec{k}}\rangle|^2.
\end{aligned}
\tag{18}
$$

Problems

10. Show the Born approximation differential cross section for the elastic scattering of a μ^- from a hydrogen atom in its ground state is

$$\left(\frac{d\sigma}{d\Omega}\right)^{\text{elastic}} = \left[\frac{2\mu}{m_e} a_0 \frac{8 + (a_0 q)^2}{[4 + (a_0 q)^2]^2}\right]^2, \qquad a_0 = \frac{\hbar^2}{m_e e^2}, \qquad q = 2k \sin\frac{\theta}{2}.$$

Make the approximation $(m_e/m_{\text{proton}}) = 0$, so the center of mass of the hydrogen atom is at the proton. All spin dependences can be neglected: The interaction potential is

$$V = -\frac{e^2}{r_1} + \frac{e^2}{|\vec{r}_1 - \vec{r}_2|}.$$

11. Find the Born approximation differential cross section for the inelastic scattering cross section of a μ^- from a hydrogen atom, where the hydrogen atom is excited from the $n = 1$ state to the $n = 2$ state

$$\left(\frac{d\sigma}{d\Omega}\right)_{1\to 2} = \left(\frac{d\sigma}{d\Omega}\right)_{1s\to 2s} + \left(\frac{d\sigma}{d\Omega}\right)_{1s\to 2p}.$$

Show, in particular, $f(\theta)_{1s\to 2p}$ has the form

$$f(\theta)_{1s\to 2p} = -i\frac{\sqrt{2}}{3}\frac{\mu}{m_e} a_0 \int_0^\infty d\zeta\, j_1(qa_0\zeta) F(\zeta),$$

with

$$F(\zeta) = \left(\frac{2^8}{3^4} - \left[\frac{4}{3}\zeta^3 + \frac{32}{9}\zeta^2 + \frac{2^7}{3^3}\zeta + \frac{2^8}{3^4}\right] e^{-\frac{3}{2}\zeta}\right),$$

and use the identity

$$\frac{dj_0(\rho)}{d\rho} = -j_1(\rho)$$

to do the integral. Show

$$\left(\frac{d\sigma}{d\Omega}\right)_{1s\to 2p} = \left(\frac{\mu}{m_e}\right)^2 \frac{k_s}{k} \frac{1}{q^2} \frac{288}{\left[\frac{9}{4} + (qa_0)^2\right]^6}.$$

Now,

$$q = \sqrt{k^2 + k_s^2 - 2kk_s \cos\theta}, \qquad \text{with } k_s^2 = k^2 - \frac{3}{4}\frac{\mu m_e e^4}{\hbar^4}.$$

For problems 10 and 11, note:

$$\text{for } r_2 < r_1:$$

$$\left(-\frac{1}{r_1} + \frac{1}{|\vec{r}_1 - \vec{r}_2|}\right) = \sum_{l=1}^{\infty} \frac{r_2^l}{r_1^{l+1}} \frac{4\pi}{(2l+1)} \sum_{m=-l}^{+l} Y_{lm}^*(\theta_1, \phi_1) Y_{lm}(\theta_2, \phi_2),$$

and

$$\text{for } r_2 > r_1:$$

$$\left(-\frac{1}{r_1}+\frac{1}{|\vec{r}_1-\vec{r}_2|}\right)=\left(-\frac{1}{r_1}+\frac{1}{r_2}\right)+\sum_{l=1}^{\infty}\frac{r_1^l}{r_2^{l+1}}\frac{4\pi}{(2l+1)}\sum_{m=-l}^{+l}Y_{lm}^*(\theta_1,\phi_1)Y_{lm}(\theta_2,\phi_2)$$

12. Find the Born approximation differential cross section for the μ^- hydrogen atom scattering for the rearrangement process in which the electron is scattered into the direction θ and the μ^- is captured by the proton to make a muonic atom in its 1s ground state. For a μ^- beam of 10 keV incident energy, make a rough numerical estimate of this cross section relative to the elastic cross section of problem 10 to show it is very small. (In making your calculation, show $k_\beta \ll k_\alpha$, so $\vec{k}_\alpha + \frac{m_\mu}{m_\mu+m_p}\vec{k}_\beta \approx \vec{k}_\alpha$ for this case and make this approximation). Show the scattering amplitude, $f(\theta)$, can be written in the form

$$\sum_l f_l P_l(\cos\theta),$$

and use $f_{l=0}$ to make your rough order of magnitude estimate. (Note: The needed Born integral can actually be done in closed form, but the resultant expression is rather complicated.)

13. For the elastic scattering of an e^- from a hydrogen atom in its ground state, both a direct and an exchange contribution exists. If the e^- in the incident beam is labeled "1" and the e^- in the hydrogen atom is labeled "2," the direct contribution comes from the scattering of "1," and the exchange contribution comes from the rearrangement collision, where "2" is the scattered particle and "1" is captured by the proton. In Born approximation, with $(ka_0)^2 \gg 1$, the scattering amplitude, $f(\theta)_{\text{direct}}$, can be read from problem 10. (Again, make the approximation $m_e/m_{\text{proton}} \approx 0$.) Evaluate the exchange contribution to the scattering amplitude, $f(\theta)_{\text{exchange}}$, and show

$$\left(\frac{d\sigma}{d\Omega}\right)_{\text{elastic}} = \frac{1}{4}|f(\theta)_{\text{direct}} + f(\theta)_{\text{exchange}}|^2 + \frac{3}{4}|f(\theta)_{\text{direct}} - f(\theta)_{\text{exchange}}|^2$$

if the incident electron beam is unpolarized and no spin alignment of the hydrogen target exists.

The exchange integral can be carried out in closed form (see T. Y. Wu, Can. J. Phys. **38** (1960) 1654). Because the expression is rather complicated, evaluate the exchange contribution in series form

$$f(\theta)_{\text{exchange}} = \sum_l (f_l)_{\text{exchange}} P_l(\cos\theta)$$

and convince yourself this series converges rapidly for $(ka_0)^2 \gg 1$. Estimate roughly the importance of the exchange contribution for $\theta = 60^o$, and $(ka_0)^2 = 20$, by making the approximation, $f(\theta)_{\text{exchange}} \approx (f_{l=0})_{\text{exchange}}$.

50

A Specific Example of a Rearrangement Collision: The (d, p) Reaction on Nucleus A

To show in detail how the rearrangement collision integrals are calculated, we shall study the (d, p) reaction on nucleus A, e.g., ^{16}O, to make nucleus B, e.g., ^{17}O, that is, a nucleus in which the extra neutron captured by nucleus A orbits this nucleus in a particular shell model orbit with orbital angular momentum, L.

$$d + A \rightarrow B + p.$$

This equation fits the case of Chapter 48, if we label nucleus A by particle label 1, the proton by particle label 2, and the neutron by particle label 3. In the center of mass system the composite particle (2,3) is then incident on particle 1 (see Fig. 50.1). We have

$$\begin{aligned}
\vec{r}_\alpha &= \frac{(m_n\vec{r}_3 + m_p\vec{r}_2)}{(m_n + m_p)} - \vec{r}_1, \\
\vec{\xi}_\alpha &= \vec{r}_2 - \vec{r}_3, \\
\vec{r}_\beta &= \vec{r}_2 - \frac{(m_n\vec{r}_3 + m_A\vec{r}_1)}{(m_n + m_A)}, \\
\vec{\xi}_\beta &= \vec{r}_3 - \vec{r}_1.
\end{aligned} \tag{1}$$

(Note: We have changed some signs from Chapter 49, because we think of the deuteron as the projectile and nucleus A as the target. All results of Chapter 49, however, apply.) Energy conservation now gives us

$$\frac{\hbar^2}{2\mu_\beta}k^2_{\beta,n'} + E^{\text{int.}}_{\beta,n'} = \frac{\hbar^2}{2\mu_\alpha}k^2 + E^{\text{int.}}_{\alpha,0}. \tag{2}$$

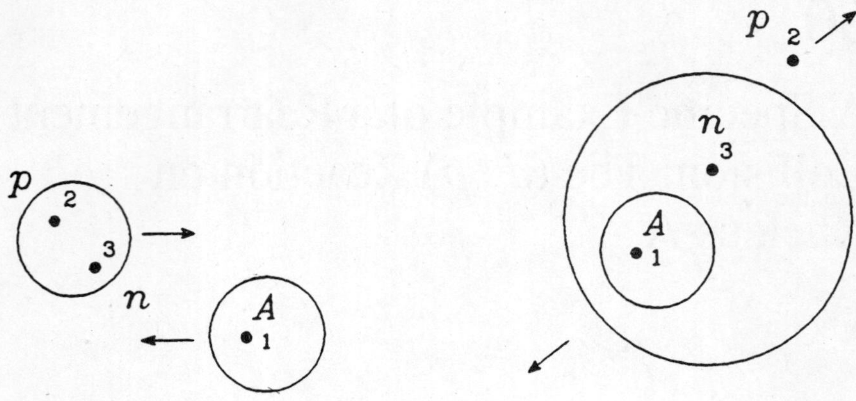

FIGURE 50.1. Rearrangement for a (d, p) reaction.

For the case in which A= ^{16}O, and if we choose the ground state of ^{17}O, with $n' = 0$, this process would give

$$k_\beta^2 = \frac{m_p m_B}{(m_p + m_B)}\left[\frac{(m_d + m_A)}{m_d m_A}k^2 + \frac{2}{\hbar^2}\left(E^{\text{int.}}_{\alpha.0} - E^{\text{int.}}_{\beta,0}\right)\right]. \tag{3}$$

With $E_{\alpha,0} = -\epsilon_d = -2.226MeV$, and with $E^{\text{int.}}_{\beta,0} = -\epsilon_B = -4.146MeV$ for the ground state of $^{17}O = B$, we would have

$$k_\beta^2 = .5319k^2 + .0874\text{fm}^{-2}. \tag{4}$$

We shall do the calculation for the ground state of nucleus B, so we can dispense with the subscript, $n' = 0$, for economy of notation. All of our results, however, would hold for $n' \neq 0$. To get the differential cross section for this rearrangement collision process in first Born approximation, we need to calculate the Born plane wave matrix element

$$(2\pi)^3\langle\phi_{\vec{k}_\beta}|V'|\phi_{\vec{k}}\rangle = I_{pn} + I_{pA} =$$
$$\int d\vec{r}\,'_\beta \int d\vec{\xi}\,'_\beta e^{-i\vec{k}_\beta\cdot\vec{r}\,'_\beta}\chi^*_{n'=0}(\vec{\xi}\,'_\beta)\Big(V_{pn}(\vec{r}\,'_{23}) + V_{pA}(\vec{r}\,'_{21})\Big)\chi_{n=0}(\vec{\xi}\,'_\alpha)e^{i\vec{k}\cdot\vec{r}\,'_\alpha}. \tag{5}$$

To actually do the integrals, it will be convenient to switch from the integration variables, $\vec{r}\,'_\beta$, $\vec{\xi}\,'_\beta$, to new integration variables, $\vec{r}\,'_{23}$, and $\vec{r}\,'_{31}$, where these are the internal variables for channels α and β, respectively. Also, now that we have reduced the calculation to a calculation of a (multiple) integral, so all variables are dummy integration variables, we can dispense with the primes on all of these variables for economy of notation. We therefore change to the new integration variables, $\vec{r}_{23}$ and $\vec{r}_{31}$, where

$$\vec{r}_{23} = \vec{r}_2 - \vec{r}_3 = \vec{\xi}_\alpha,$$

$$\vec{r}_{31} = \vec{r}_3 - \vec{r}_1 = \vec{\xi}_\beta,$$
$$\vec{r}_\alpha = \vec{r}_{31} + \frac{m_p}{m_d}\vec{r}_{23},$$
$$\vec{r}_\beta = = \vec{r}_{23} + \frac{m_A}{m_B}\vec{r}_{31}. \tag{6}$$

Also, the Jacobian of the transformation has the value unity

$$d\vec{r}_\beta d\vec{\xi}_\beta = \begin{vmatrix} 1 & \frac{m_A}{m_B} \\ 0 & 1 \end{vmatrix}^3 d\vec{r}_{23} d\vec{r}_{31}, \tag{7}$$

so

$$I_{pn} + I_{pA} = \int d\vec{r}_{23} \int d\vec{r}_{31} e^{-i\vec{r}_{23}\cdot(\vec{k}_\beta - \frac{m_p}{m_d}\vec{k})} \chi_0^{\text{deut.}}(\vec{r}_{23})$$
$$\times \Big(V_{pn}(\vec{r}_{23}) + V_{pA}(\vec{r}_{21})\Big) e^{i\vec{r}_{31}\cdot(\vec{k} - \frac{m_A}{m_B}\vec{k}_\beta)} \chi^{*B}_{n'=0}(\vec{r}_{31}). \tag{8}$$

It would be useful to define vectors, $\vec{\nu}$, and $\vec{\kappa}$ (see Fig. 50.2),

$$\vec{\nu} = \vec{k}_\beta - \frac{m_p}{m_d}\vec{k},$$
$$\vec{\kappa} = \vec{k} - \frac{m_A}{m_B}\vec{k}_\beta. \tag{9}$$

The vectors $\vec{\nu}$ and $\vec{\kappa}$ lie in the scattering plane defined by the incident $\vec{k}$ and the scattered $\vec{k}_\beta$, so the azimuth angles ϕ for all four vectors can be chosen as $\phi = 0$. The directions and the magnitudes of the vectors $\vec{\nu}$ and $\vec{\kappa}$ are therefore functions of the scattering angle, θ, given by

$$\nu = \sqrt{k_\beta^2 + \Big(\frac{m_p}{m_d}\Big)^2 k^2 - 2\frac{m_p}{m_d} k k_\beta \cos\theta},$$
$$\kappa = \sqrt{k^2 + \Big(\frac{m_A}{m_B}\Big)^2 k_\beta^2 - 2\frac{m_A}{m_B} k k_\beta \cos\theta}, \tag{10}$$

$$\tan\theta_\nu = \frac{\frac{k_\beta}{\nu}\sin\theta}{\Big(\frac{k_\beta}{\nu}\cos\theta - \frac{m_p}{m_d}\frac{k}{\nu}\Big)} = \frac{\sin\theta}{\Big(\cos\theta - \frac{m_p}{m_d}\frac{k}{k_\beta}\Big)},$$
$$\tan\theta_\kappa = \frac{\frac{k_\beta}{\kappa}\frac{m_A}{m_B}\sin\theta}{\Big(\frac{k}{\kappa} - \frac{k_\beta}{\kappa}\frac{m_A}{m_B}\cos\theta\Big)} = \frac{\sin\theta}{\Big(\frac{k}{k_\beta}\frac{m_B}{m_A} - \cos\theta\Big)}. \tag{11}$$

The integral involving the p–n interaction, V_{pn}, is then

$$I_{pn} = \int d\vec{r}_{31} e^{i\vec{\kappa}\cdot\vec{r}_{31}} \chi_0^{*B}(\vec{r}_{31}) \int d\vec{r}_{23} e^{-i\vec{\nu}\cdot\vec{r}_{23}} V_{pn}(\vec{r}_{23}) \chi_0^{\text{deut.}}(\vec{r}_{23}). \tag{12}$$

We could now take some reasonable approximation for the p–n potential, perhaps even a square well of the appropriate width and depth. We can avoid, however, choosing a specific V_{pn} (realizing the full proton–neutron interaction is very complicated on a fundamental level, involving exchanges not only of pions, but also of ρ and ω vector mesons, hard-core terms caused by quark–quark interactions

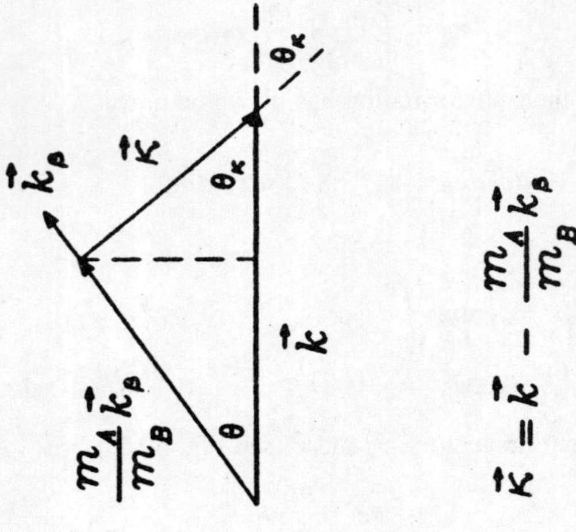

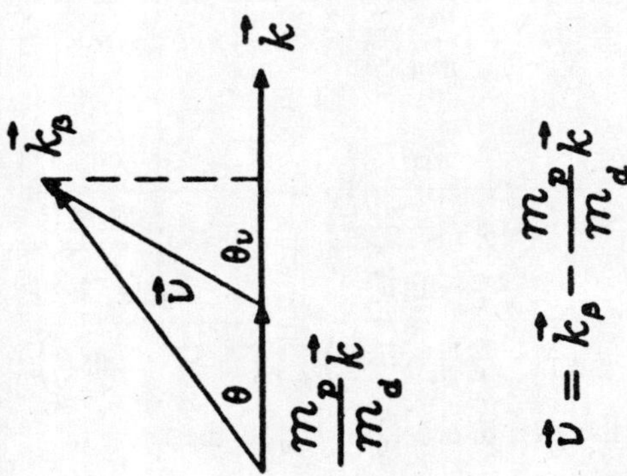

FIGURE 50.2. The vectors, $\vec{v}$ and $\vec{\kappa}$ of eq. (9).

at extremely short range). Instead, we can use the Schrödinger equation for the bound state of the deuteron

$$\Big(-\frac{\hbar^2}{2\mu_{pn}}\nabla^2_{23} + V_{pn}(\vec{r}_{23})\Big)\chi^{\text{deut.}}(\vec{r}_{23}) = E^{\text{int.}}_d\chi^{\text{deut.}}(\vec{r}_{23}) = -|\epsilon_d|\chi^{\text{deut.}}(\vec{r}_{23}) \quad (13)$$

to convert

$$V_{pn}(\vec{r}_{23})\chi^{\text{deut.}}(\vec{r}_{23}) = \Big(\frac{\hbar^2}{2\mu_{pn}}\nabla^2_{23} - |\epsilon_d|\Big)\chi^{\text{deut.}}(\vec{r}_{23}), \quad (14)$$

where $|\epsilon_d| = 2.226$ MeV is the (positive) binding energy of the deuteron. The 3-D integral

$$\int d\vec{r}_{23}\phi_{\vec{v}}\nabla^2_{23}\chi^{\text{deut.}},$$

with $\phi_{\vec{v}} = e^{-i\vec{v}\cdot\vec{r}_{23}}$ can now be evaluated by use of Green's theorem II, a special case of Gauss' theorem

$$\int d\vec{r}_{23}\big(\phi_{\vec{v}}\nabla^2_{23}\chi^{\text{deut.}} - \chi^{\text{deut.}}\nabla^2_{23}\phi_{\vec{v}}\big) = (\text{surface integral at } \infty), \quad (15)$$

where the surface integral at ∞ goes to zero as a result of the exponential fall-off of χ and $\vec{\nabla}\chi$ as $r_{23} \to \infty$. Thus, we can evaluate the needed integral via

$$\int d\vec{r}_{23}e^{-i\vec{v}\cdot\vec{r}_{23}}\nabla^2_{23}\chi^{\text{deut.}}(\vec{r}_{23}) = \int d\vec{r}_{23}\chi^{\text{deut.}}(\vec{r}_{23})\nabla^2_{23}e^{-i\vec{v}\cdot\vec{r}_{23}}$$
$$= -v^2\int d\vec{r}_{23}e^{-i\vec{v}\cdot\vec{r}_{23}}\chi^{\text{deut.}}(\vec{r}_{23}). \quad (16)$$

This equation gives us

$$\int d\vec{r}_{23}e^{-i\vec{v}\cdot\vec{r}_{23}}V_{pn}(\vec{r}_{23})\chi^{\text{deut.}}(\vec{r}_{23}) = -\Big(\frac{\hbar^2 v^2}{2\mu_{pn}} + |\epsilon_d|\Big)\int d\vec{r}_{23}e^{-i\vec{v}\cdot\vec{r}_{23}}\chi^{\text{deut.}}(\vec{r}_{23}). \quad (17)$$

For purposes of doing our integral, a two-parameter deuteron wave function that fits both the experimental binding energy and the triplet scattering length is sufficient. Such a wave function was constructed by Hulthén with two exponentials

$$\chi^{\text{deut.}}(\vec{r}_{23}) = R^{\text{deut.}}(r_{23})Y_{00}(\theta_{23},\phi_{23}) = \mathcal{N}\frac{\big(e^{-\frac{r_{23}}{r_0}} - e^{-\eta\frac{r_{23}}{r_0}}\big)}{r_{23}}\frac{1}{\sqrt{4\pi}}, \quad (18)$$

with parameters and normalization factor, $\mathcal{N}$, given by

$$r_0 = 4.26\text{fm}, \qquad \eta = 6.2, \qquad \mathcal{N} = \Big[\frac{2\eta(1+\eta)}{r_0(\eta-1)^2}\Big]^{\frac{1}{2}}.$$

We can now evaluate our *p*–*n* interaction integral

$$I_{pn} = -\Big(\frac{\hbar^2}{2\mu_{pn}}v^2 + |\epsilon_d|\Big)\int d\vec{r}_{23}e^{-i\vec{v}\cdot\vec{r}_{23}}R^{\text{deut.}}(r_{23})Y_{00}(\theta_{23},\phi_{23})$$
$$\times\int d\vec{r}_{31}e^{i\vec{K}\cdot\vec{r}_{31}}R_{NL}(r_{31})Y^*_{LM}(\theta_{31},\phi_{31}), \quad (19)$$

where we have taken the internal wave function, $\chi^B_{n=0}(\vec{r}_{31})$, as a single-particle shell model wave function describing the orbital motion of the neutron around the nucleus A in an orbit given by quantum numbers, NLM, by

$$\chi^B_{n=0}(\vec{r}_{31}) = R_{NL}(r_{31})Y_{LM}(\theta_{31}, \phi_{31}),$$

where this might be approximated by a 3-D harmonic oscillator function, if the effective potential felt by the odd neutron caused by its interaction with the A nucleons of nucleus A can be approximated by a parabolic potential for this bound ground state of nucleus B.

For purposes of doing the triple $\vec{r}_{23}$ integral, we can choose our z axis along the direction of the $\vec{\nu}$ vector, and expand the exponential via

$$e^{-i\vec{\nu}\cdot\vec{r}_{23}} = \sum_{l'}(-i)^{l'}\sqrt{4\pi(2l'+1)}\,j_{l'}(\nu r_{23})Y_{l'0}(\theta_{23}). \tag{20}$$

The orthogonality of the spherical harmonics permits us to do the θ_{23} and ϕ_{23} integrals, where only the term with $l' = 0$ will survive. Similarly, for purposes of doing the triple $\vec{r}_{31}$ integral, we can choose our z axis along the direction of the $\vec{\kappa}$ vector, and expand the exponential via

$$e^{i\vec{\kappa}\cdot\vec{r}_{31}} = \sum_{l''}(i)^{l''}\sqrt{4\pi(2l''+1)}\,j_{l''}(\kappa r_{31})Y_{l''0}(\theta_{31}). \tag{21}$$

The orthogonality of the spherical harmonics picks out the single term with $l'' = L$ and $M = 0$ and permits us to do the trivial θ_{31} and ϕ_{31} integrals. Our final result for I_{pn} is

$$\begin{aligned} I_{pn} &= -4\pi i^L\sqrt{(2L+1)}\Big(\frac{\hbar^2}{2\mu_{pn}}\nu^2 + |\epsilon_d|\Big)F_{\text{deut.}}(\nu)F_{\text{B,NL}}(\kappa), \\ F_{\text{deut.}}(\nu) &= \int_0^\infty r^2 dr j_0(\nu r)R^{\text{deut.}}(r), \\ F_{\text{B,NL}}(\kappa) &= \int_0^\infty r^2 dr j_L(\kappa r)R_{NL}(r), \end{aligned} \tag{22}$$

where $R^{\text{deut.}}(r)$ can be taken as the radial part of the Hulthén wave function and $R_{NL}(r)$ is the radial part of a single-particle shell model wave function, which could be approximated by a 3-D harmonic oscillator wave function. Recall ν and κ are θ-dependent [see eq. (10)].

The integral, I_{pA}, is somewhat more complicated than the integral, I_{pn}, because the interaction, V_{pA}, depends on both of our integration variables. If $V_{pA}(\vec{r}_{21})$ depends only on the magnitude of $\vec{r}_{21} = \vec{r}_{23} + \vec{r}_{31}$, however, we can expand V_{pA} in terms of Legendre polynomials, $P_l(\cos\Theta)$, where Θ is the angle between the vectors $\vec{r}_{31}$ and $\vec{r}_{23}$ (see Fig. 50.3).

$$\begin{aligned} V_{pA}(|\vec{r}_{23} + \vec{r}_{31}|) &= \sum_{l'} v_{l'}(r_{23}, r_{31})P_{l'}(\cos\Theta) \\ &= \sum_{l'} v_{l'}(r_{23}, r_{31})\frac{4\pi}{(2l'+1)}\sum_{m'} Y_{l'm'}(\theta_{23}, \phi_{23})Y^*_{l'm'}(\theta_{31}, \phi_{31}). \end{aligned} \tag{23}$$

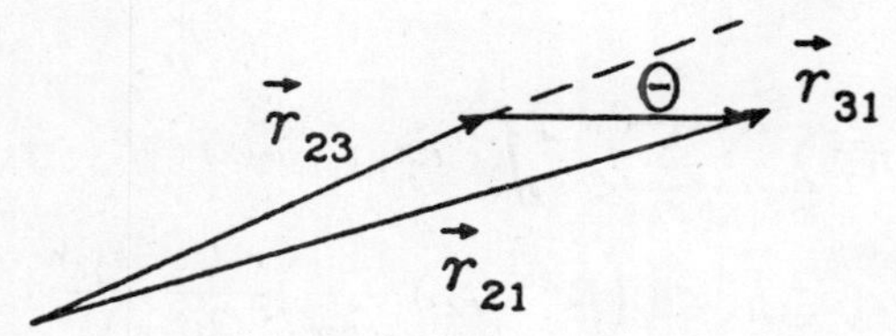

$$|\vec{r}_{21}| = \sqrt{r_{23}^2 + r_{31}^2 + 2r_{23}r_{31}\cos\Theta}$$

FIGURE 50.3.

The interaction V_{pA} is the interaction of the proton labeled particle 2, with all nucleons of nucleus A, averaged over the positions of these nucleons, as given by the ground-state wave function of nucleus A. We do not attempt to calculate it from a fundamental nucleon–nucleon interaction, but instead replace it with an effective interaction with a few parameters adjusted to fit the elastic scattering cross sections for the scattering of protons from nucleus A. An example of a form that has been used is a simple Gaussian interaction, with parameters V_0 and r_0 adjusted to fit the elastic proton-(nucleus A) scattering cross sections, i.e.,

$$V_{pA} = -V_0 e^{-|\vec{r}_{23}+\vec{r}_{31}|^2/r_0^2} = -V_0 e^{-(r_{23}^2+r_{31}^2+2r_{23}r_{31}\cos\Theta)/r_0^2}, \tag{24}$$

with

$$v_{l'}(r_{23}, r_{31}) = \\ -V_0 e^{-(r_{23}^2+r_{31}^2)/r_0^2}\frac{(2l'+1)}{2}\int_0^\pi d\Theta \sin\Theta e^{-2(r_{23}r_{31}\cos\Theta)/r_0^2} P_{l'}(\cos\Theta). \tag{25}$$

In this case, $v_{l'}(r_{23}, r_{31})$ is of the form

$$v_{l'}(r_{23}, r_{31}) = e^{-(r_{23}^2/r_0^2)}e^{-(r_{31}^2/r_0^2)} f_{l'}(r_{23}r_{31}/r_0^2), \tag{26}$$

which can facilitate the calculation of the integral. With this type of V_{pA}, we get

$$I_{pA} = \sum_{l',m'}\frac{4\pi}{(2l'+1)}\int d\vec{r}_{31} e^{i\vec{\kappa}\cdot\vec{r}_{31}} Y^*_{l'm'}(\theta_{31}\phi_{31}) R_{NL}(r_{31}) Y^*_{LM}(\theta_{31}, \phi_{31}) \\ \times \int d\vec{r}_{23} v_{l'}(r_{23}, r_{31}) e^{-i\vec{v}\cdot\vec{r}_{23}} Y_{l'm'}(\theta_{23}, \phi_{23}) R^{\text{deut.}}(r_{23})\frac{1}{\sqrt{4\pi}}. \tag{27}$$

We now expand the exponentials in terms of spherical Bessel functions and spherical harmonics

$$e^{-i\vec{v}\cdot\vec{r}_{23}} = \sum_{\bar{l}}(-i)^{\bar{l}} j_{\bar{l}}(vr_{23})\sum_{\bar{m}} 4\pi Y^*_{\bar{l}\bar{m}}(\theta_{23}, \phi_{23}) Y_{\bar{l}\bar{m}}(\theta_v, \phi_v), \\ e^{i\vec{\kappa}\cdot\vec{r}_{31}} = \sum_{l''} i^{l''} j_{l''}(\kappa r_{31})\sum_{m''} 4\pi Y_{l''m''}(\theta_{31}, \phi_{31}) Y^*_{l''m''}(\theta_\kappa, \phi_\kappa), \tag{28}$$

to obtain

$$I_{pA} = (4\pi)^3 \sum_{l'm'} \sum_{\bar{l}\bar{m}} \sum_{l''m''} i^{l''-\bar{l}} \int_0^\infty dr_{31} r_{31}^2 j_{l''}(\kappa r_{31}) R_{NL}(r_{31})$$
$$\times \int_0^\infty dr_{23} r_{23}^2 j_{\bar{l}}(\nu r_{23}) \left(R^{\text{deut.}}(r_{23}) \frac{1}{\sqrt{4\pi}} \right) \frac{v_{l'}(r_{23}, r_{31})}{(2l'+1)}$$
$$\times \left[\iint d\Omega_{23} Y_{l'm'}(\theta_{23}, \phi_{23}) Y^*_{\bar{l}\bar{m}}(\theta_{23}, \phi_{23}) \right]$$
$$\times \left[\iint d\Omega_{31} Y_{l''m''}(\theta_{31}, \phi_{31}) Y^*_{l'm'}(\theta_{31}, \phi_{31}) Y^*_{LM}(\theta_{31}, \phi_{31}) \right]$$
$$\times Y_{\bar{l}\bar{m}}(\theta_\nu, \phi_\nu) Y^*_{l''m''}(\theta_\kappa, \phi_\kappa). \tag{29}$$

Now, we use the orthogonality of the spherical harmonics

$$\iint d\Omega_{23} Y_{l'm'}(\theta_{23}, \phi_{23}) Y^*_{\bar{l}\bar{m}}(\theta_{23}, \phi_{23}) = \delta_{l'\bar{l}} \delta_{m'\bar{m}} \tag{30}$$

and the well-known integral

$$\iint d\Omega_{31} Y_{l''m''}(\theta_{31}, \phi_{31}) Y^*_{l'm'}(\theta_{31}, \phi_{31}) Y^*_{LM}(\theta_{31}, \phi_{31})$$
$$= \sqrt{\frac{(2l'+1)(2L+1)}{4\pi(2l''+1)}} \langle L0l'0|l''0\rangle \langle LMl'm'|l''m''\rangle \tag{31}$$

to reduce I_{pA} to radial integrals.

$$I_{pA} = (4\pi)^2 \sum_{l'm'} \sum_{l''m''} (i)^{l''-l'} \sqrt{\frac{(2L+1)}{(2l'+1)(2l''+1)}} \langle L0l'0|l''0\rangle \langle LMl'm'|l''m''\rangle$$
$$\times Y_{l'm'}(\theta_\nu, \phi_\nu) Y^*_{l''m''}(\theta_\kappa, \phi_\kappa) \int_0^\infty dr_{31} r_{31}^2 j_{l''}(\kappa r_{31}) R_{NL}(r_{31})$$
$$\times \int_0^\infty dr_{23} r_{23}^2 j_{l'}(\nu r_{23}) R^{\text{deut.}}(r_{23}) v_{l'}(r_{23}, r_{31}). \tag{32}$$

The dependence on the scattering angle, θ, now sits both in the factors ν and κ of the spherical Bessel functions, [see eq. (10)] and the angles θ_ν and θ_κ of the spherical harmonics [see eq. (11)]. In addition, we recall the vectors $\vec{\nu}$ and $\vec{\kappa}$ lie in the scattering plane, formed by the vectors $\vec{k}$ and $\vec{k}_\beta$, for which we have chosen $\phi = 0$. Thus, we only need the spherical harmonics for $\phi_\nu = 0$, $\phi_\kappa = 0$.

Eqs. (22) and (32) permit us to calculate the scattering amplitude for the (d, p) reaction. The contribution from V_{pn} is the most important. In particular, the L dependence of I_{pn} comes from the so-called form factor, $F_{B,NL}(\kappa) = \int_0^\infty r^2 dr j_L(\kappa r) R_{NL}(r)$ of eq. (22). For $L = 0$, the spherical Bessel function peaks at $\kappa r = 0$, and therefore makes its largest contribution to the integral for angles θ near $\theta = 0$. For ever larger values of L, the spherical Bessel functions peak at larger values of the argument κr and therefore make larger contributions to the integral at larger values of κ and, hence, at larger values of θ. Hence, the differential cross sections for the (d, p) reaction are strongly L-dependent, peaking at $\theta = 0$ for $L = 0$ and at larger angles, the larger the L value. Fig. 50.4 shows angular dependences for such differential cross sections. Historically, this L selectivity of $\frac{d\sigma}{d\Omega}$ for this reaction was used to establish the L values of shell model orbits.

Final Remarks: We have ignored the spins of both proton and neutron. It is well known the energy positions of shell model orbits are strongly j-dependent. Thus, the $d_{\frac{3}{2}}$ level in ^{17}O lies much higher in energy than the $d_{\frac{5}{2}}$ level, which makes the ground state of ^{17}O. Despite this strong spin-orbit coupling effect of the shell model single-particle energies, this spin-orbit term does not play a strong role in the (d, p) reaction. The differential cross sections for transitions to the $d_{\frac{5}{2}}$ ground state and the excited $d_{\frac{3}{2}}$ state have an almost identical θ-dependence, characteristic of the orbital angular momentum, $L = 2$.

The first Born approximation we have used here uses zeroth-order wave functions that are plane waves, eigenfunctions of our simple H_0, which included only the kinetic energy part of the relative motion Hamiltonian. A more sophisticated Born approximation, the so-called distorted wave Born approximation, includes within H_0 not only the kinetic energy term of the relative motion, but also an average potential for the incoming projectile, fitted to the elastic scattering cross sections for the projectile (deuteron-^{16}O elastic scattering in our example), and a similar average potential for the outgoing particle, the proton in our case, where this potential is fitted to the elastic proton-^{17}O scattering cross sections. Although this more sophisticated distorted wave Born approximation gives much better quantitative results, qualitatively the plane wave Born approximation gives similar results. Historically, therefore, it was an important tool in determining the L values of shell model orbits.

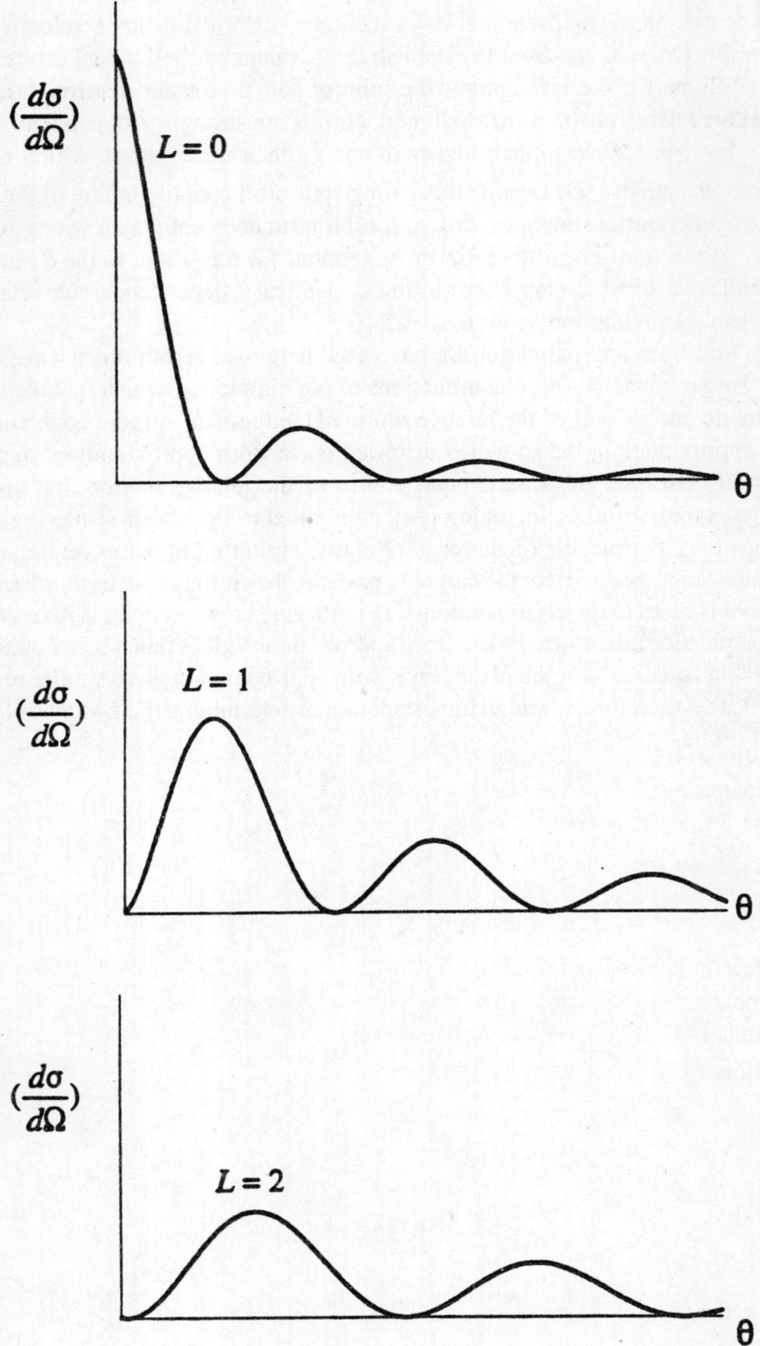

FIGURE 50.4. Plane-wave Born approximation differential cross sections for a $d + A \to p + B$ reaction. Here L is the orbital angular momentum of the captured neutron shell model orbit.

51

The S Matrix

In what we have done so far, we have essentially calculated matrix elements, or approximate expressions for matrix elements of the so-called S matrix. We have also assumed conservation of energy. Let us now show our approximations come from the S matrix formalism. We will also show the energy conservation follows automatically from this formalism.

The S matrix is defined through

$$S_{fi} = \langle \psi_f^{(-)} | \psi_i^{(+)} \rangle, \tag{1}$$

where i is shorthand for all the quantum numbers, $\vec{k}_i$, n_{ai}, n_{Ai}, for the initial channel, similarly for f for the final quantum

numbers of the final channel. Also, $|\psi_f^{(-)}\rangle$ is defined through the Green's function, $G^{(-)}$, with incoming spherical waves

$$\begin{aligned} |\psi_f^{(-)}\rangle &= |\phi_f\rangle + \frac{1}{E_f - H_0 - i\epsilon} V |\psi_f^{(-)}\rangle \\ &= |\phi_f\rangle + \frac{1}{E_f - H - i\epsilon} V |\phi_f\rangle. \end{aligned} \tag{2}$$

Using, the second form and assuming H and V are hermitian, $H^\dagger = H, V^\dagger = V$, this relation yields

$$\langle \psi_f^{(-)} | = \langle \phi_f | + \langle \phi_f | V \frac{1}{E_f - H + i\epsilon}. \tag{3}$$

Using this form for the bra of the matrix element S_{fi}, we get

$$\begin{aligned} S_{fi} &= \langle\phi_f|\psi_i^{(+)}\rangle + \langle\phi_f|V\frac{1}{E_f - H + i\epsilon}|\psi_i^{(+)}\rangle \\ &= \langle\phi_f|\psi_i^{(+)}\rangle + \langle\phi_f|V\frac{1}{E_f - E_i + i\epsilon}|\psi_i^{(+)}\rangle, \end{aligned} \tag{4}$$

where we have used $H|\psi_i^{(+)}\rangle = E_i|\psi^{(+)}\rangle$. Now, we write $|\psi_i^{(+)}\rangle$ in the first term as

$$|\psi_i^{(+)}\rangle = |\phi_i\rangle + \frac{1}{E_i - H_0 + i\epsilon}V|\psi_i^{(+)}\rangle \tag{5}$$

to get

$$\begin{aligned} \langle\phi_f|\psi_i^{(+)}\rangle &= \langle\phi_f|\phi_i\rangle + \langle\phi_f|\frac{1}{E_i - H_0 + i\epsilon}V|\psi_i^{(+)}\rangle \\ &= \langle\phi_f|\phi_i\rangle + \langle\phi_f|\frac{1}{E_i - E_f + i\epsilon}V|\psi_i^{(+)}\rangle, \end{aligned} \tag{6}$$

where we have used the left action of H_0 on $\langle\phi_f|$ in the second line. Combining eqs. (4) and (6), we get

$$\begin{aligned} S_{fi} &= \langle\phi_f|\phi_i\rangle + \langle\phi_f|\left(V\frac{1}{E_f - E_i + i\epsilon} + \frac{1}{E_i - E_f + i\epsilon}V\right)|\psi_i^{(+)}\rangle \\ &= \langle\phi_f|\phi_i\rangle + \lim_{\epsilon\to 0}\frac{(-2i\epsilon)}{(E_f - E_i)^2 + \epsilon^2}\langle\phi_f|V|\psi_i^{(+)}\rangle \\ &= \langle\phi_f|\phi_i\rangle - 2i\pi\delta(E_f - E_i)\langle\phi_f|V|\psi_i^{(+)}\rangle \\ &= \delta(\vec{k}_f - \vec{k}_i)\delta_{n_{a,f}n_{a,i}}\delta_{n_{A,f}n_{A,i}} - 2i\pi\delta(E_f - E_i)\langle\phi_f|V|\psi_i^{(+)}\rangle, \end{aligned} \tag{7}$$

where we we have used

$$\frac{1}{\pi}\lim_{\epsilon\to 0}\frac{\epsilon}{(\omega^2 + \epsilon^2)} = \delta(\omega), \tag{8}$$

which follows from

$$\begin{aligned} &\lim_{\epsilon\to 0}\frac{1}{2\pi}\int_{-\infty}^{\infty} dt e^{-\epsilon|t|}e^{i\omega t} = \delta(\omega) \\ &= \lim_{\epsilon\to 0}\frac{1}{2\pi}\left(\int_{-\infty}^{0} dt e^{(\epsilon+i\omega)t} + \int_{0}^{\infty} dt e^{-(\epsilon-i\omega)t}\right) \\ &= \lim_{\epsilon\to 0}\frac{1}{2\pi}\left(\frac{1}{\epsilon + i\omega} + \frac{1}{\epsilon - i\omega}\right) = \lim_{\epsilon\to 0}\frac{1}{\pi}\frac{\epsilon}{(\omega^2 + \epsilon^2)}. \end{aligned} \tag{9}$$

Eq. (7) gives the initial channel form of S_{fi}; that is, it is assumed that the final state is in the same channel as the initial state, so S_{fi} is an elastic or an inelastic scattering matrix element. For rearrangement collisions, we need a scattering matrix element of the form, $S_{f'i} = \langle\psi_{f'}^{(-)}|\psi_i^{(+)}\rangle$, where the state $|\phi_{f'}\rangle$ is assumed to be an eigenstate of H_0' belonging to a different channel. We would now express $\langle\psi_{f'}^{(-)}|$ by

$$\langle\psi_{f'}^{(-)}| = \langle\phi_{f'}| + \langle\phi_{f'}|V'\frac{1}{E_{f'} - H + i\epsilon}, \tag{10}$$

and

$$\langle \phi_{f'} | \psi_i^{(+)} \rangle = \langle \phi_{f'} | \frac{1}{E_i - H_0' + i\epsilon} V' | \psi_i^{(+)} \rangle, \tag{11}$$

so

$$\begin{aligned} S_{f'i} &= \langle \psi_{f'}^{(-)} | \psi_i^{(+)} \rangle = \langle \phi_{f'} | \left(\frac{1}{E_i - H_0' + i\epsilon} V' + V' \frac{1}{E_{f'} - H + i\epsilon} \right) | \psi_i^{(+)} \rangle \\ &= \lim_{\epsilon \to 0} \left(\frac{1}{E_i - E_{f'} + i\epsilon} + \frac{1}{E_{f'} - E_i + i\epsilon} \right) \langle \phi_{f'} | V' | \psi_i^{(+)} \rangle \\ &= -2\pi i \delta(E_{f'} - E_i) \langle \phi_{f'} | V' | \psi_i^{(+)} \rangle. \end{aligned} \tag{12}$$

A The T Matrix

Because the S matrix is often related to the so-called T matrix, let us introduce this T matrix now, through the defining equations

$$\begin{aligned} S_{fi} &= \langle \phi_f | \phi_i \rangle - 2\pi i \delta(E_f - E_i) T_{fi}, \qquad \text{with} \quad T_{fi} = \langle \phi_f | V | \psi_i^{(+)} \rangle, \\ S_{f'i} &= \qquad\qquad -2\pi i \delta(E_{f'} - E_i) T_{f'i}, \qquad \text{with} \quad T_{f'i} = \langle \phi_{f'} | V' | \psi_i^{(+)} \rangle. \end{aligned} \tag{13}$$

The T matrix element for a rearrangement collision could be given in two forms

$$T_{f'i} = \langle \phi_{f'} | V' | \psi_i^{(+)} \rangle, \qquad \text{also} \quad = \langle \phi_{f'} | V | \psi_i^{(+)} \rangle. \tag{14}$$

In principle, both the last "preform" and the first "postform" for the T matrix element of a rearrangement collision are equally valid, but in first Born approximation, the postform may be the better approximation. Finally, to express $|\psi_i^{(+)}\rangle$ in terms of $|\phi_i\rangle$ through a Born series iteration, it is also useful to introduce the Møller operator, $\Omega^{(+)}$,

$$|\psi_i^{(+)}\rangle = \Omega^{(+)} |\phi_i\rangle, \tag{15}$$

where

$$\begin{aligned} \Omega^{(+)} &= 1 + \frac{1}{E - H + i\epsilon} V \\ &= 1 + \frac{1}{E - H_0 + i\epsilon} V \Omega^{(+)} \\ &= 1 + \frac{1}{E - H_0 + i\epsilon} V + \frac{1}{E - H_0 + i\epsilon} V \frac{1}{E - H_0 + i\epsilon} V + \cdots. \end{aligned} \tag{16}$$

Thus,

$$\begin{aligned} T_{fi} &= \langle \phi_f | V | \phi_i \rangle + \langle \phi_f | V \frac{1}{E_i - H_0 + i\epsilon} V | \phi_i \rangle \\ &\quad + \langle \phi_f | V \frac{1}{E_i - H_0 + i\epsilon} V \frac{1}{E_i - H_0 + i\epsilon} V | \phi_i \rangle + \cdots. \end{aligned} \tag{17}$$

For rearrangement collisions, we could use either the postform or the preform, so

$$T_{f'i} = \langle \phi_{f'} | V' | \phi_i \rangle + \langle \phi_{f'} | V' \frac{1}{E_i - H_0 + i\epsilon} V | \phi_i \rangle + \cdots$$

$$= \langle\phi_{f'}|V|\phi_i\rangle + \langle\phi_{f'}|V\frac{1}{E_i - H_0 + i\epsilon}V|\phi_i\rangle + \cdots. \tag{18}$$

Because the approximations involved in the postforms and preforms are different, but hopefully converge to the same result if a sufficient number of iterations are taken, one sometimes also takes an average of the postform and preform in the hope it will improve the accuracy of the expansion after only a small number of iterations. Thus, also,

$$T_{f'i} = \frac{1}{2}\left(\langle\phi_{f'}|(V' + V)|\phi_i\rangle + \langle\phi_{f'}|(V' + V)\frac{1}{E_i - H_0 + i\epsilon}V|\phi_i\rangle + \cdots\right). \tag{19}$$

Finally, recall

$$\frac{d\sigma}{d\Omega} = \frac{\mu_f\mu_i}{(2\pi\hbar^2)^2}\frac{k_{f'}}{k_i}(2\pi)^6|\langle\phi_{f'}|V'|\psi_i^{(+)}\rangle|^2, \tag{20}$$

[cf., eq. (18) of Chapter 49], so we get

$$\frac{d\sigma}{d\Omega} = \left(\frac{2\pi}{\hbar}\right)^4 \mu_f\mu_i\frac{k_{f'}}{k_i}|T_{f'i}|^2. \tag{21}$$

Final remark:

The Møller operator $\Omega^{(+)}$ could be written in the form

$$\sum_i\int|\psi_i^{(+)}\rangle\langle\phi_i|, \tag{22}$$

where the $\sum_i\int$ symbol is shorthand for the triple integral over all $\vec{k}_i$ and a sum over all n_{a_i} and n_{A_i}, which enumerate the ground and excited states of composite projectile and composite target systems. In particular, the operator, $\Omega^{(+)}$, is not unitary. We have

$$(\Omega^{(+)})^\dagger(\Omega^{(+)}) = \sum_i\int\sum_j\int|\phi_j\rangle\langle\psi_j^{(+)}|\psi_i^{(+)}\rangle\langle\phi_i| = \sum_i\int|\phi_i\rangle\langle\phi_i| = 1, \tag{23}$$

because the $|\phi_i\rangle$ form a complete set. However,

$$\begin{aligned}(\Omega^{(+)})(\Omega^{(+)})^\dagger &= \sum_i\int|\psi_i^{(+)}\rangle\langle\psi_i^{(+)}| \neq 1\\ &= 1 - \sum_{\text{bound}}|\psi_b\rangle\langle\psi_b|,\end{aligned} \tag{24}$$

where the right-hand side is no longer unity if the combined system of particles $(a + A)$ has some bound states $|\psi_b\rangle$.

Problems

14. A neutron beam is scattered elastically from an odd-mass nucleus made of a very stable core nucleus of mass, m_C, and a loosely bound extra neutron.

(a) Assume the neutron–neutron interaction is much stronger than the interaction of the neutron with the core, so V_{nC} can be neglected compared with V_{nn}.

(b) Assume further the short-range V_{nn} can be approximated by a spin-independent delta function interaction

$$V_{nn} = -V_0 b^3 \delta(\vec{r}_{n1} - \vec{r}_{n2}),$$

where V_0 and b are constants (with dimensions of energy and length, respectively).

(c) Assume the loosely bound neutron is in an s state with internal wave function given by

$$\chi(\vec{r}_{nC}) = \frac{2}{\sqrt{a^3\sqrt{\pi}}} e^{-(r_{nC}^2/2a^2)} Y_{00}(\theta_{nC}, \phi_{nC}).$$

(d) Assume the incident energy is such that the cross section can be calculated in Born approximation.

(e) Assume the incident beam is unpolarized and that there is no spin alignment of the target nucleus.

(f) Assume exchange processes between the incoming neutron and the neutrons in the core nucleus, C, can be neglected.

Show, however, exchange processes between the incoming neutron and the loosely bound extra neutron are important, and calculate the differential cross section in Born approximation. Remember neutrons are $s = \frac{1}{2}$ particles.

15. Calculate the differential cross section for the (p, n) rearrangement collision:

$$p +^{17} O \to n +^{17} F$$

in plane-wave Born approximation.

(a) Assume the ground state of ^{17}O is approximated well by a bound d state ($l = 2$) shell model wave function describing the motion of the neutron with respect to an inert ^{16}O closed shell nucleus, C:

$$\chi_{n=2,l=2,m}(\vec{r}_{nC}) = \sqrt{\frac{2}{a^3\Gamma(\frac{7}{2})}} \left(\frac{r_{nC}}{a}\right)^2 e^{-\frac{1}{2}(r_{nC}^2/a^2)} Y_{l=2,m}(\theta_{nC}, \phi_{nC})$$

with length parameter, $a = 3.5$ fm.

(b) The final wave function in ^{17}F is approximated by a similar d state shell model wave function, $\chi_{n=2,l=2,m}(\vec{r}_{pC})$, with the same length parameter, a. We assume the final ^{17}F nucleus is in its ground state. In this case, $E^{\text{int.}}(^{17}F) - E^{\text{int.}}(^{17}O) = 2.76$ MeV.

(c) All interactions are assumed to be spin-independent so the spins of the particles play no role in the Born integrals, although both the ^{17}O and ^{17}F ground states have total angular momentum, $J = \frac{5}{2}$.

(d) The interaction between the neutron and the ^{16}O core, C, in the outgoing channel is negligible compared with the neutron proton interaction, V_{np}, i.e., $V_{nC} \ll V_{np}$, similarly, $V_{pC} \ll V_{np}$ in the incident channel.

(e) The short-range neutron–proton interaction can be approximated by a delta function interaction, $V_{np} = -V_0 b^3 \delta(\vec{r}_n - \vec{r}_p)$, with length parameter, $b = 1$ fm, $V_0 = 100$ MeV. Also, let the mass of ^{16}O be M, and $m_{\text{neutron}} = m_{\text{proton}} = m$.

16. Calculate the differential cross section for the elastic scattering of a proton from the deuteron under the following simplifying assumptions.

(a) The incident energy is such that the differential cross section can be calculated in first Born approximation.

(b) Assume the interaction of the proton with the constituents of the deuteron is spin-independent, short range, and can be approximated by delta function interactions

$$V_{pp} = -V_0 b^3 \delta(\vec{r}_{p_1} - \vec{r}_{p_2}), \quad V_{pn} = -V_0 b^3 \delta(\vec{r}_p - \vec{r}_n).$$

(c) Take $m_p = m_n \approx \frac{1}{2} m_d$.

(d) The bound-state wave function of the deuteron is given by the Hulthen wave function

$$\chi_d^{\text{int.}} = \frac{\mathcal{N}}{r_{pn}} \left(e^{-\frac{r_{pn}}{r_0}} - e^{-\eta \frac{r_{pn}}{r_0}} \right) Y_{00}(\theta_{pn}, \phi_{pn}) \chi^{\text{spin}}_{S=1, M_S},$$

with

$$r_0 = 4.26\text{fm}, \qquad \eta = 6.2, \qquad \mathcal{N} = \left[\frac{2\eta(1+\eta)}{r_0(\eta - 1)^2} \right]^{\frac{1}{2}}.$$

Calculate the differential cross section assuming the proton beam is unpolarized and the deuteron target (with $S = 1$) is unaligned; i.e., it has arbitrary spin orientations. In making your calculation, remember the detector cannot distinguish "projectile" protons from "target" protons, so exchange terms have to be considered.

How will the differential cross section change if the incident proton beam is longitudinally polarized, with proton spin projection, $m_s = +\frac{1}{2}$, along the direction of the incident $\vec{p}_0$, but the deuteron target is still unaligned?

52

Scattering Theory for Particles with Spin

A Scattering of a Point Particle with Spin from a Spinless Target Particle

For a point particle with spin, our plane wave solutions are of the form $|\phi_{\vec{k}}\rangle = |\vec{k}, sm_s\rangle$ with coordinate representation,

$$\langle \vec{r}_{\text{rel.}}, \vec{\xi}_{\text{int.}} | \vec{k}, sm_s\rangle = \frac{1}{(2\pi)^{\frac{3}{2}}} e^{i\vec{k}\cdot\vec{r}} \chi_{sm_s}(\vec{\xi}_{\text{int.}}).$$

Now, for a true point particle, like the electron or the muon, the nature of the internal variables is unknown. In that case, we merely replace $\vec{\xi}_{\text{int.}}$ with the spin operator $\vec{s}$ itself (or for $s = \frac{1}{2}$-particles with the Pauli $\vec{\sigma}$ vector). We know how to take matrix elements between such spin states of operators such as the $\vec{s}$ components without having to specify a $\vec{\xi}_{\text{int.}}$. (Nevertheless, when we are in coordinate representation, it may be useful to use an imagined $\vec{\xi}$ and write an integral over $\vec{\xi}$ to designate the process of taking a spin-space matrix element.) To specify the state $|\vec{k}, sm_s\rangle$, we also need to specify the quantization axis for the quantum number m_s. We shall usually take the direction of the incident $\vec{k}$ as the quantization axis, so an $s = \frac{1}{2}$-particle with $m_s = +\frac{1}{2}$ is polarized longitudinally along the direction of the incident momentum vector, whereas a particle in the state with $m_s = -\frac{1}{2}$ is polarized longitudinally in a direction opposite to the incident momentum direction. If we have incident particles with specific polarization along some z' direction defined through Euler angles, α, β, γ with respect to the incident x, y, z direction,

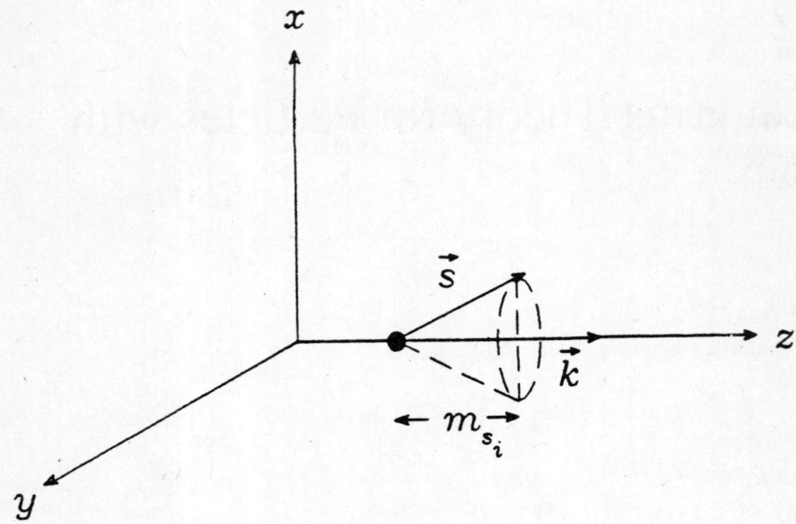

FIGURE 52.1.

however,

$$\chi_{sm'_s}(\vec{s}\,) = \sum_{m_s} \chi_{sm_s}(\vec{s}\,) D^{s*}_{m_s m'_s}(\alpha, \beta, \gamma). \tag{1}$$

For an incoming beam with a perpendicular polarization, say, with fixed m'_s along the x axis (see Fig. 52.1), the required rotation is one from the original z axis to the new axis of quantization, the old x axis. The rotation is therefore a simple rotation about the y axis through an angle of $\beta = \frac{\pi}{2}$, with $\alpha = 0$, $\gamma = 0$. In that case,

$$\chi_{sm'_s} = \sum_{sm_s} \chi_{sm_s} d^{\frac{1}{2}}_{m_s m'_s}(\beta). \tag{2}$$

Recall: The $d^j_{mm'}(\beta)$ are real, and for $s = \frac{1}{2}$,

$$d^{\frac{1}{2}}_{m_s m'_s}(\beta) = \begin{pmatrix} \cos\frac{\beta}{2} & -\sin\frac{\beta}{2} \\ \sin\frac{\beta}{2} & \cos\frac{\beta}{2} \end{pmatrix}, \tag{3}$$

so for spin $s = \frac{1}{2}$-particles with incident perpendicular ($\perp$) polarization along the x axis

$$\chi^{(\perp)}_{\frac{1}{2}, m'_s = \pm\frac{1}{2}} = \pm \frac{1}{\sqrt{2}} \chi_{\frac{1}{2}, +\frac{1}{2}} + \frac{1}{\sqrt{2}} \chi_{\frac{1}{2}, -\frac{1}{2}}, \tag{4}$$

where the spin functions, χ_{sm_s}, without superscript will stand for spin states with longitudinal polarizations specified by m_s.

We shall deal with incoming plane wave states of the type

$$e^{i\vec{k}\cdot\vec{r}} \chi_{sm_s}(\vec{s}\,) = \sum_l i^l \sqrt{4\pi(2l+1)}\, j_l(kr) Y_{l0}(\theta) \chi_{sm_s}(\vec{s}\,), \tag{5}$$

where θ is the angle between the $\vec{r}$ and $\vec{k}$ vectors. When taking matrix elements of scalar (rotationally invariant) interactions between such incoming (plane wave)-spin states and similar outgoing (plane wave)-spin states, it will be very useful to couple the orbital and spin functions to orbital-spin functions of total angular momentum j, with $\vec{l}+\vec{s}=\vec{j}$. This process will facilitate use of the Wigner–Eckart theorem. Thus, we define

$$\mathcal{Y}_{[ls]jm_j}(\theta,\phi,\vec{s}) \equiv \sum_{m_s(m_l)} \langle l m_l s m_s | j m_j \rangle i^l Y_{lm_l}(\theta,\phi)\chi_{sm_s}(\vec{s}). \tag{6}$$

(The factor i^l has been included with the Y_{lm} as a matter of convenience.) Note also the orthonormality of these functions

$$\int\int d\Omega_{\theta,\phi}\int d\vec{\xi}\,\mathcal{Y}^*_{[l's]j'm'_j}(\theta,\phi,\vec{\xi})\mathcal{Y}_{[ls]jm_j}(\theta,\phi,\vec{\xi}) = \delta_{l'l}\delta_{j'j}\delta_{m'_j m_j}, \tag{7}$$

which follows from the orthonormality of the spherical harmonics, the orthonormality of the spin functions, and the orthonormality of the Clebsch–Gordan coefficients.

In terms of these vector-coupled orbital-spin functions, eq. (5) can be rewritten as

$$e^{i\vec{k}\cdot\vec{r}}\chi_{sm_s} = \sum_{l,j}\sqrt{4\pi(2l+1)}\,j_l(kr)\langle l0sm_s|jm_s\rangle\mathcal{Y}_{[ls]jm_s}(\theta,\phi,\vec{s}), \tag{8}$$

where we have used

$$i^l Y_{l0}(\theta,\phi)\chi_{sm_s}(\vec{s}) = \sum_j \langle l0sm_s|jm_s\rangle\mathcal{Y}_{[ls]jm_s}(\theta,\phi,\vec{s}). \tag{9}$$

For the special case of $s=\frac{1}{2}$, the needed Clebsch–Gordan coefficients are

$$\begin{aligned}\langle l0\tfrac{1}{2}\pm\tfrac{1}{2}|j=(l+\tfrac{1}{2})\pm\tfrac{1}{2}\rangle &= \sqrt{\frac{(l+1)}{(2l+1)}}\\ \langle l0\tfrac{1}{2}\pm\tfrac{1}{2}|j=(l-\tfrac{1}{2})\pm\tfrac{1}{2}\rangle &= \mp\sqrt{\frac{l}{(2l+1)}}.\end{aligned} \tag{10}$$

With these coefficients, we have

$$e^{i\vec{k}\cdot\vec{r}}\chi_{\frac{1}{2}\pm\frac{1}{2}} = \sum_l\sqrt{4\pi}\,j_l(kr)\Big(\sqrt{(l+1)}\mathcal{Y}_{[l\frac{1}{2}]j=(l+\frac{1}{2}),\pm\frac{1}{2}} \mp \sqrt{l}\mathcal{Y}_{[l\frac{1}{2}]j=(l-\frac{1}{2}),\pm\frac{1}{2}}\Big). \tag{11}$$

The scattering amplitude now depends on both the angles θ and ϕ and the spin directions and can be written as a $(2s+1)\times(2s+1)$ matrix

$$f(\vec{k}_f, m_{s_f};\vec{k}, m_{s_i}) \equiv f(\theta,\phi)_{m_{s_f}m_{s_i}} = -\frac{2\mu}{4\pi\hbar^2}(2\pi)^3\langle\phi_f|V|\psi_i^{(+)}\rangle. \tag{12}$$

m_{s_i} and m_{s_f} are *both* defined with respect to the incident $\vec{k}$ vector chosen along the z axis, and $\vec{k}_f$ makes an angle θ with respect to $\vec{k}$.

B First Born Approximation

In first Born approximation,

$$f_{m_{s_f} m_{s_i}}(\theta, \phi) = -\frac{\mu}{2\pi\hbar^2} \int d\vec{r}\,' \int d\vec{\xi}\,' e^{-i\vec{k}_f \cdot \vec{r}\,'} \chi^*_{sm_{s_f}}(\vec{\xi}\,') V(\vec{r}\,', \vec{\xi}\,') e^{i\vec{k}\cdot\vec{r}\,'} \chi_{sm_{s_i}}(\vec{\xi}\,'),$$

where we use

$$e^{i\vec{k}\cdot\vec{r}\,'} \chi_{sm_{s_i}}(\vec{\xi}\,') = \sum_{l,j} j_l(kr') \sqrt{4\pi(2l+1)} \langle l0sm_{s_i} | jm_{s_i} \rangle \mathcal{Y}_{[ls]jm_{s_i}}(\theta', \phi', \vec{\xi}\,'), \quad (13)$$

$$\begin{aligned} & e^{-i\vec{k}_f\cdot\vec{r}\,'} \chi_{sm_{s_f}} {*}(\vec{\xi}\,') = \\ & \sum_{l_f} (-i)^{l_f} j_{l_f}(k_f r') 4\pi \sum_{m_f} Y_{l_f m_f}(\theta, \phi) Y^*_{l_f m_f}(\theta', \phi') \chi^*_{sm_{s_f}}(\vec{\xi}\,') = \\ & \sum_{l_f m_f} \sum_{j_f} 4\pi j_{l_f}(k_f r') \langle l_f m_f s m_{s_f} | j m_{j_f} \rangle \mathcal{Y}^*_{[l_f s] j_f m_{j_f}}(\theta', \phi', \vec{\xi}\,') Y_{l_f m_f}(\theta, \phi). \quad (14) \end{aligned}$$

In first Born approximation, therefore,

$$\begin{aligned} f(\theta, \phi)_{m_{s_f} m_{s_i}} &= -\frac{2\mu}{\hbar^2} \sqrt{4\pi} \\ \times & \sum_{l,j} \sum_{l_f, m_f, j_f} \sqrt{(2l+1)} \langle l0sm_{s_i} | jm_{s_i} \rangle \langle l_f m_f s m_{s_f} | j_f m_{j_f} \rangle Y_{l_f m_f}(\theta, \phi) \\ \times & \int d\vec{r}\,' \int d\vec{\xi}\,' j^*_{l_f}(k_f r') \mathcal{Y}^*_{[l_f s] j_f m_{j_f}}(\theta', \phi', \vec{\xi}\,') \\ \times & V(\vec{r}\,', \vec{\xi}\,') \mathcal{Y}_{[ls]jm_{s_i}}(\theta', \phi', \vec{\xi}\,') j_l(kr'). \quad (15) \end{aligned}$$

For a rotationally invariant $V(\vec{r}, \vec{\xi}) = V(\vec{r}, \vec{s})$, i.e., for a V that is a spherical tensor of rank 0, the Wigner–Eckart theorem tells us the matrix element of such a V is diagonal in j and m_j and independent of m_j. Thus,

$$\int\int d\Omega' \int d\vec{\xi}\,' \mathcal{Y}_{[l_f s] j_f m_{j_f}} V(\vec{r}\,', \vec{s}) \mathcal{Y}_{[ls]jm_{s_i}} = \mathcal{V}^{s,j}_{l_f, l}(r') \delta_{j_f j} \delta_{m_{j_f} m_{s_i}}, \quad (16)$$

and

$$\begin{aligned} f(\vec{k}_f, m_{s_f}; \vec{k}, m_{s_i}) = f(\theta, \phi)_{m_{s_f} m_{s_i}} &= -\frac{2\mu}{\hbar^2} \sqrt{4\pi} \sum_{l, l_f} \sum_{j} \sqrt{(2l+1)} \\ \times & \langle l0sm_{s_i} | jm_{s_i} \rangle \langle l_f (m_{s_i} - m_{s_f}) s m_{s_f} | j m_{s_i} \rangle Y_{l_f (m_{s_i} - m_{s_f})}(\theta, \phi) \\ \times & \int_0^\infty dr' r'^2 j^*_{l_f}(k_f r') \mathcal{V}^{s,j}_{l_f, l}(r') j_l(kr'). \quad (17) \end{aligned}$$

One of the most common spin-dependent interactions involves a simple spin-orbit coupling term

$$V(\vec{r}\,', \vec{s}) = V_0(r') + V_1(r')(\vec{l} \cdot \vec{s}). \quad (18)$$

Because the components of the operator $\vec{l}$ change only the components m_l and not the quantum number l, this interaction is diagonal in l. Moreover, it has the simple form

$$\mathcal{V}^{s,j}_{l_f,l}(r') = \delta_{l_f l}\Big(V_0(r') + \frac{1}{2}V_1(r')[j(j+1) - l(l+1) - s(s+1)]\Big), \tag{19}$$

leading to the first Born approximation result

$$\begin{aligned} f(\theta,\phi)_{m_{s_f}m_{s_i}} &= -\frac{2\mu}{\hbar^2}\sqrt{4\pi}\sum_{l,j}\sqrt{(2l+1)} \\ &\times \langle l0sm_{s_i}|jm_{s_i}\rangle\langle l(m_{s_i}-m_{s_f})sm_{s_f}|jm_{s_i}\rangle Y_{l(m_{s_i}-m_{s_f})}(\theta,\phi) \\ &\times \int_0^\infty dr' r'^2[j_l(kr')]^2\Big(V_0(r') + \frac{1}{2}V_1(r') \\ &\times [j(j+1) - l(l+1) - s(s+1)]\Big), \end{aligned} \tag{20}$$

where we note the scattering process is still elastic, so the magnitude of $\vec{k}_f$ is still the same as the magnitude k of $\vec{k}$. Also, matrix elements of this scattering amplitude matrix with $m_{s_f} \neq m_{s_i}$ are now ϕ-dependent. Let us consider the special case of an $s = \frac{1}{2}$ particle. First, the diagonal matrix elements with $m_{s_f} = m_{s_i}$ are independent of the sign of m_{s_i}. This result follows from the symmetry property of Clebsch–Gordan coefficients

$$\big[\langle l0s + m_{s_i}|j + m_{s_i}\rangle\big]^2 = \big[\langle l0s - m_{s_i}|j - m_{s_i}\rangle\big]^2. \tag{21}$$

Moreover, these matrix elements are of the form $\sum_l A_l Y_{l0}(\theta,\phi)$ and therefore ϕ-independent. From the similar symmetry property

$$\langle l0s + \tfrac{1}{2}|j + \tfrac{1}{2}\rangle\langle l + 1s - \tfrac{1}{2}|j + \tfrac{1}{2}\rangle = \langle l0s - \tfrac{1}{2}|j - \tfrac{1}{2}\rangle\langle l - 1s + \tfrac{1}{2}|j - \tfrac{1}{2}\rangle, \tag{22}$$

the off-diagonal elements are of the form

$$\begin{aligned} f(\theta,\phi)_{+\frac{1}{2},-\frac{1}{2}} &= \sum_l B_l Y_{l,-1}(\theta,\phi), \\ f(\theta,\phi)_{-\frac{1}{2},+\frac{1}{2}} &= \sum_l B_l Y_{l,+1}(\theta,\phi), \end{aligned} \tag{23}$$

with identical coefficients, B_l. Now we recall the standard spherical harmonics with $m = \pm 1$ are of the form

$$Y_{l,\pm 1}(\theta,\phi) = \mp|\mathcal{N}_{l,1}|\sin\theta\Big(\frac{dP_l(\cos\theta)}{d\cos\theta}\Big)e^{\pm i\phi}.$$

Therefore, we can write the scattering amplitude matrix for $s = \frac{1}{2}$-particles in terms of two functions of θ

$$f(\theta,\phi)_{m_{s_f}m_{s_i}} = \begin{pmatrix} g(\theta) & \bar{h}(\theta)e^{-i\phi} \\ -\bar{h}(\theta)e^{+i\phi} & g(\theta) \end{pmatrix}. \tag{24}$$

Also, the functions, $g(\theta)$, and $\bar{h}(\theta)$ are real. We shall find this special property is related to the fact that we have used the first Born approximation. This result will have to be relaxed, and we shall have to examine the scattering amplitude valid for lower energies.

53

Scattering of Spin $\frac{1}{2}$ Particles from Spinless Target: Partial Wave Decomposition

To study the low-energy scattering of particles with spin, we shall use a partial wave decomposition. We shall restrict ourselves, however, to the simplest case of $s = \frac{1}{2}$-particles scattering from a spinless target particle and will assume the spin-dependent interaction is given by the simple potential

$$\begin{aligned} V &= V_0(r) + V_1(r)(\vec{l} \cdot \vec{s}) \\ &= V_0(r) + \tfrac{1}{2} l V_1(r) \qquad \text{for } j = (l + \tfrac{1}{2}) \\ &= V_0(r) - \tfrac{1}{2}(l+1) V_1(r) \qquad \text{for } j = (l - \tfrac{1}{2}). \end{aligned} \tag{1}$$

From eq. (11) of Chapter 52, we have

$$\begin{aligned} &(2\pi)^{\frac{3}{2}} \phi_{\vec{k}}(\vec{r}, \vec{\sigma})_{m_s = \pm\frac{1}{2}} \\ &= \sum_l \sqrt{4\pi}\, j_l(kr) \Big(\sqrt{(l+1)}\, \mathcal{Y}_{[l\frac{1}{2}](l+\frac{1}{2}), \pm\frac{1}{2}} \mp \sqrt{l}\, \mathcal{Y}_{[l\frac{1}{2}](l-\frac{1}{2}), \pm\frac{1}{2}} \Big), \end{aligned} \tag{2}$$

and the full solution can be expanded in the same fashion

$$\begin{aligned} (2\pi)^{\frac{3}{2}} \psi^{(+)}_{\vec{k}}(\vec{r}, \vec{\sigma})_{\pm\frac{1}{2}} = \sqrt{4\pi} \sum_l \Big(&\frac{u_l^{(a)}(kr)}{kr} \sqrt{(l+1)}\, \mathcal{Y}_{[l\frac{1}{2}](l+\frac{1}{2}), \pm\frac{1}{2}} \\ &+ \frac{u_l^{(b)}(kr)}{kr} (\mp\sqrt{l})\, \mathcal{Y}_{[l\frac{1}{2}](l-\frac{1}{2}), \pm\frac{1}{2}} \Big), \end{aligned} \tag{3}$$

where the $u_l^{(a)}(kr)$ and $u_l^{(b)}(kr)$ are the one-dimensionalized radial eigenfunctions for the radial equations for $j = (l \pm \frac{1}{2})$, respectively.

$$\left(\frac{\hbar^2}{2\mu} \Big(-\frac{d^2}{dr^2} + \frac{l(l+1)}{r^2} \Big) + V_0(r) + \tfrac{1}{2} l V_1(r) \right) u_l^{(a)} = \frac{\hbar^2}{2\mu} k^2 u_l^{(a)},$$

$$\left(\frac{\hbar^2}{2\mu}\left(-\frac{d^2}{dr^2}+\frac{l(l+1)}{r^2}\right)+V_0(r)-\tfrac{1}{2}(l+1)V_1(r)\right)u_l^{(b)}=\frac{\hbar^2}{2\mu}k^2u_l^{(b)}. \quad (4)$$

These radial eigenfunctions have the asymptotic form, as $r\to\infty$,

$$u_l^{(a)}(kr)\longrightarrow a_l\sin(kr-\frac{l\pi}{2}+\delta_l^{(a)}),$$
$$u_l^{(b)}(kr)\longrightarrow b_l\sin(kr-\frac{l\pi}{2}+\delta_l^{(b)}), \quad (5)$$

so, as $r\to\infty$,

$$(2\pi)^{\frac{3}{2}}\psi_{\vec{k}}^{(+)}(\vec{r},\vec{\sigma})_{\pm\frac{1}{2}}\to$$
$$\sqrt{4\pi}\sum_l\left(\frac{a_l}{2ikr}\left(e^{i(kr-\frac{l\pi}{2}+\delta_l^{(a)})}-e^{-i(kr-\frac{l\pi}{2}+\delta_l^{(a)})}\right)\sqrt{(l+1)}\mathcal{Y}_{[l\frac{1}{2}](l+\frac{1}{2}),\pm\frac{1}{2}}\right.$$
$$\left.+\frac{b_l}{2ikr}\left(e^{i(kr-\frac{l\pi}{2}+\delta_l^{(b)})}-e^{-i(kr-\frac{l\pi}{2}+\delta_l^{(b)})}\right)(\mp\sqrt{l})\mathcal{Y}_{[l\frac{1}{2}](l-\frac{1}{2}),\pm\frac{1}{2}}\right). \quad (6)$$

With

$$a_l=e^{i\delta_l^{(a)}},\qquad b_l=e^{i\delta_l^{(b)}}, \quad (7)$$

this function goes to

$$(2\pi)^{\frac{3}{2}}\psi_{\vec{k}}^{(+)}(\vec{r},\vec{\sigma})_{m_s=\pm\frac{1}{2}}$$
$$\to\sqrt{4\pi}\sum_l\left[\left(\frac{(e^{2i\delta_l^{(a)}}-1)}{2ik}\frac{e^{i(kr-\frac{l\pi}{2})}}{r}+\frac{\sin(kr-\frac{l\pi}{2})}{kr}\right)\sqrt{(l+1)}\mathcal{Y}_{[l\frac{1}{2}](l+\frac{1}{2})m_s}\right.$$
$$\left.+\left(\frac{(e^{2i\delta_l^{(b)}}-1)}{2ik}\frac{e^{i(kr-\frac{l\pi}{2})}}{r}+\frac{\sin(kr-\frac{l\pi}{2})}{kr}\right)(\mp\sqrt{l})\mathcal{Y}_{[l\frac{1}{2}](l-\frac{1}{2})m_s}\right]$$
$$\to(2\pi)^{\frac{3}{2}}\phi_{\vec{k}}(\vec{r},\vec{\sigma})_{m_s}+\frac{\sqrt{4\pi}}{k}\frac{e^{ikr}}{r}\sum_l$$
$$\left[e^{i\delta_l^{(a)}}\sin\delta_l^{(a)}\sqrt{(l+1)}(-i)^l\mathcal{Y}_{[l\frac{1}{2}](l+\frac{1}{2})m_s}\right.$$
$$\left.+e^{i\delta_l^{(b)}}\sin\delta_l^{(b)}(\mp\sqrt{l})(-i)^l\mathcal{Y}_{[l\frac{1}{2}](l-\frac{1}{2})m_s}\right]. \quad (8)$$

To compare this function with

$$(2\pi)^{\frac{3}{2}}\psi_{\vec{k}}^{(+)}(\vec{r},\vec{\sigma})_{m_s}\to(2\pi)^{\frac{3}{2}}\phi_{\vec{k}}(\vec{r},\vec{\sigma})_{m_s}+\frac{e^{ikr}}{r}\sum_{m_s'}\chi(\vec{\sigma})_{\frac{1}{2}m_s'}f(\theta,\phi)_{m_s'm_s}, \quad (9)$$

we need to expand the $\mathcal{Y}_{[l\frac{1}{2}]jm_s}$ through

$$(-i)^l\mathcal{Y}_{[l\frac{1}{2}]jm_s}=\sum_{m_s'}\langle l(m_s-m_s')\tfrac{1}{2}m_s'|jm_s\rangle Y_{l(m_s-m_s')}(\theta,\phi)\chi(\vec{\sigma})_{\frac{1}{2}m_s'}. \quad (10)$$

The combinations of eqs. (8), (9), and (10) then give

$$f(\theta,\phi)_{m_s'm_s=\pm\frac{1}{2}}$$

$$= \frac{\sqrt{4\pi}}{k}\sum_l \Big[e^{i\delta_l^{(a)}} \sin\delta_l^{(a)} \sqrt{(l+1)} \langle l(m_s - m_s')\tfrac{1}{2}m_s' | (l+\tfrac{1}{2})m_s\rangle$$

$$+ e^{i\delta_l^{(b)}} \sin\delta_l^{(b)} (\mp\sqrt{l}) \langle l(m_s - m_s')\tfrac{1}{2}m_s' | (l-\tfrac{1}{2})m_s\rangle \Big] Y_{l(m_s - m_s')}(\theta,\phi). \quad (11)$$

Using the specific values of the Clebsch–Gordan coefficients, we get

$$f(\theta,\phi)_{+\frac{1}{2},+\frac{1}{2}} = f(\theta,\phi)_{-\frac{1}{2},-\frac{1}{2}} =$$
$$\frac{1}{k}\sum_l \sqrt{\frac{4\pi}{(2l+1)}}\Big((l+1)e^{i\delta_l^{(a)}}\sin\delta_l^{(a)} + l e^{i\delta_l^{(b)}}\sin\delta_l^{(b)}\Big) Y_{l0}(\theta) = g(\theta), \quad (12)$$

$$f(\theta,\phi)_{\mp\frac{1}{2}\pm\frac{1}{2}} = \frac{\sqrt{4\pi}}{k}\sum_l \sqrt{\frac{l(l+1)}{(2l+1)}}\Big(e^{i\delta_l^{(a)}}\sin\delta_l^{(a)} - e^{i\delta_l^{(b)}}\sin\delta_l^{(b)}\Big) Y_{l\pm 1}(\theta,\phi)$$
$$= \mp\bar{h}(\theta)e^{\pm i\phi}. \quad (13)$$

This function is of the form of eq. (24) of Chapter 52 and therefore agrees in form with the result obtained in first Born approximation. Note, however, now the functions, $g(\theta)$ and $\bar{h}(\theta)$, are complex, because the quantities $e^{i\delta_l}\sin\delta_l$ are complex. [In first Born approximation, with δ_l = Order(V/E), where $E = \frac{\hbar^2 k^2}{2\mu}$, we have $e^{i\delta_l}\sin\delta_l = (1+\cdots)(\delta_l + \cdots)$ = Order(E/V) (cf., problem 5), so these quantities become real.] It will now be convenient to rename $\bar{h}(\theta) = -ih(\theta)$, so

$$f(\theta,\phi)_{m_s' m_s} = \begin{pmatrix} g(\theta) & -ih(\theta)e^{-i\phi} \\ ih(\theta)e^{i\phi} & g(\theta) \end{pmatrix}. \quad (14)$$

With this notation, we can write the scattering amplitude matrix in terms of the Pauli $\vec{\sigma}$ matrices, if we introduce the unit vector, $\vec{n}$, normal to the scattering plane defined by the two vectors, $\vec{k}$ and $\vec{k}_f \equiv \vec{k}_s$, the latter in the direction of the scattered beam:

$$\vec{n} = \frac{[\vec{k}\times\vec{k}_s]}{kk_s\sin\theta} = -\sin\phi\vec{e}_x + \cos\phi\vec{e}_y \quad (15)$$

(see Fig. 53.1). We note

$$(\vec{n}\cdot\vec{\sigma}) = -\sin\phi\begin{pmatrix} 0 & 1 \\ 1 & 0 \end{pmatrix} + \cos\phi\begin{pmatrix} 0 & -i \\ i & 0 \end{pmatrix}$$
$$= \begin{pmatrix} 0 & -ie^{-i\phi} \\ ie^{i\phi} & 0 \end{pmatrix}, \quad (16)$$

so the 2×2 scattering amplitude matrix for $s = \frac{1}{2}$- particles can be written as

$$f(\theta,\phi) = g(\theta)\mathbf{1} + h(\theta)(\vec{n}\cdot\vec{\sigma}), \quad (17)$$

where $\mathbf{1}$ is the 2×2 unit matrix.

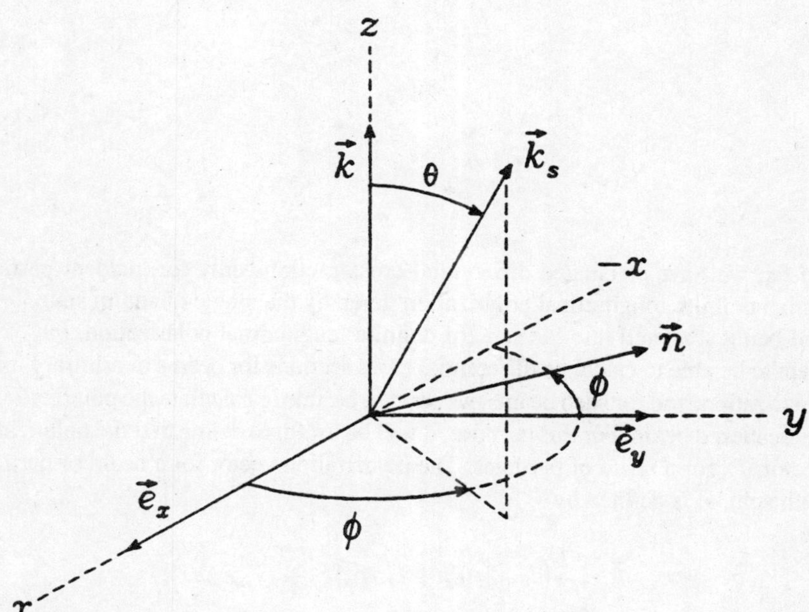

FIGURE 53.1.

54

The Polarization Vector

So far, we have calculated differential cross sections only for incident particles with a definite longitudinal polarization given by the single quantum state, $|m_{s_i}\rangle$, and being scattered into the state of definite longitudinal polarization, $|m_{s_f}\rangle$. We want to be able to calculate differential cross sections for beams of arbitrary initial polarization, and for such beams, we want to be able to calculate the polarization of the scattered beam. For this purpose, it will be useful to define first the polarization vector, $\vec{P}$, for a beam of particles. The polarization vector for a beam of particles with spin, s, is defined by

$$\vec{P} = \frac{1}{s}\Big(\langle \vec{s} \cdot \vec{e}_x \rangle \vec{e}_x + \langle \vec{s} \cdot \vec{e}_y \rangle \vec{e}_y + \langle \vec{s} \cdot \vec{e}_z \rangle \vec{e}_z \Big), \tag{1}$$

where the expectation values of $\vec{s} \cdot \vec{e}_i$ are given by, e.g.,

$$\langle \vec{s} \cdot \vec{e}_x \rangle \equiv \langle \chi_s | s_x | \chi_s \rangle, \tag{2}$$

where an arbitrary spin state can be expanded in terms of basis states $|m_s\rangle$ of definite longitudinal polarization,

$$|\chi_s\rangle = \sum_{m_s} |m_s\rangle \langle m_s | \chi_s \rangle = \sum_{m_s} |m_s\rangle c_{m_s}, \tag{3}$$

so

$$\langle s_x \rangle = \sum_{m'_s, m''_s} \langle \chi_s | m''_s \rangle \langle m''_s | s_x | m'_s \rangle \langle m'_s | \chi_s \rangle = \sum_{m'_s m''_s} c^*_{m''_s} \langle m''_s | s_x | m'_s \rangle c_{m'_s}. \tag{4}$$

Note, in particular, if $|\chi_s\rangle = |m_s = +s\rangle$, then $\vec{P} = P_z\vec{e}_z$, with $P_z = 1$. For the special case of spin $s = \frac{1}{2}$-particles, remembering $\vec{s} = \frac{1}{2}\vec{\sigma}$, we have

$$\vec{P} = \langle\sigma_x\rangle\vec{e}_x + \langle\sigma_y\rangle\vec{e}_y + \langle\sigma_z\rangle\vec{e}_z = P_x\vec{e}_x + P_y\vec{e}_y + P_z\vec{e}_z. \tag{5}$$

Recall a spin vector with perpendicular polarization in the x direction was given by

$$|\chi_s^{(\perp)}\rangle = \tfrac{1}{\sqrt{2}}|+\tfrac{1}{2}\rangle + \tfrac{1}{\sqrt{2}}|-\tfrac{1}{2}\rangle. \tag{6}$$

For this state,

$$\langle\sigma_x\rangle = \tfrac{1}{\sqrt{2}}\Big(\langle+\tfrac{1}{2}|\sigma_x|-\tfrac{1}{2}\rangle + \langle-\tfrac{1}{2}|\sigma_x|+\tfrac{1}{2}\rangle\Big)\tfrac{1}{\sqrt{2}} = \tfrac{1}{2}(1+1) = 1, \tag{7}$$

and $\langle\sigma_y\rangle = 0$ and $\langle\sigma_z\rangle = 0$ for this spin state. Therefore, $\vec{P} = P_x\vec{e}_x$, with $P_x = 1$ for this state. Similarly, for the spin state,

$$|\chi_s\rangle = \tfrac{1}{\sqrt{2}}|+\tfrac{1}{2}\rangle + \tfrac{i}{\sqrt{2}}|-\tfrac{1}{2}\rangle, \tag{8}$$

we have

$$\langle\sigma_y\rangle = \tfrac{i}{2}\langle+\tfrac{1}{2}|\sigma_y|-\tfrac{1}{2}\rangle - \tfrac{i}{2}\langle-\tfrac{1}{2}|\sigma_y|+\tfrac{1}{2}\rangle = 1, \tag{9}$$

whereas, now, $\langle\sigma_x\rangle = 0$, and $\langle\sigma_z\rangle = 0$, so this is a state of perpendicular polarization, with $P_y = 1$.

A Polarization of the Scattered Beam

To calculate the polarization for a scattered beam, we expand the spin vector for the scattered beam in terms of the longitudinally polarized states $|m_s'\rangle$

$$|\chi_s^{\text{scatt.}}\rangle = \sum_{m_s'}|m_s'\rangle\langle m_s'|\chi_s^{\text{scatt.}}\rangle. \tag{10}$$

Let us assume for the moment the incident beam is longitudinally polarized, with

$$\langle m_s|\chi_s^{\text{inc.}}\rangle = 1, \qquad \text{for either } m_s = +\tfrac{1}{2} \text{ or } m_s = -\tfrac{1}{2}. \tag{11}$$

Then, we can take

$$\langle m_s'|\chi_s^{\text{scatt.}}\rangle = \Big(f(\theta,\phi)\Big)_{m_s',m_s}\frac{1}{\sqrt{N(\theta,\phi)}}, \tag{12}$$

where $N(\theta,\phi)$ is the total number of particles scattered into the detector of cross-sectional area $r^2 d\Omega$ normal to the scattering direction, θ, ϕ, so

$$N(\theta,\phi) = \sum_{m_s'}|\big(f(\theta,\phi)\big)_{m_s',m_s}|^2. \tag{13}$$

Then,

$$\vec{P}^{\text{scatt.}} = \frac{1}{N(\theta,\phi)}\sum_{\alpha}\sum_{m_s',m_s''}\vec{e}_\alpha\langle\chi_s^{\text{scatt.}}|m_s''\rangle\langle m_s''|\sigma_\alpha|m_s'\rangle\langle m_s'|\chi_s^{\text{scatt.}}\rangle$$

$$= \frac{1}{N(\theta,\phi)} \sum_{\alpha} \sum_{m'_s, m''_s} \vec{e}_\alpha \big(f(\theta,\phi)\big)^*_{m''_s, m_s} (\sigma_\alpha)_{m''_s m'_s} \big(f(\theta,\phi)\big)_{m'_s, m_s}$$
$$= \frac{1}{N(\theta,\phi)} \sum_{\alpha} \sum_{m'_s m''_s} \vec{e}_\alpha \big(f^\dagger\big)_{m_s m''_s} (\sigma_\alpha)_{m''_s, m'_s} \big(f\big)_{m'_s, m_s}$$
$$= \frac{1}{N(\theta,\phi)} \sum_{\alpha} \vec{e}_\alpha \big(f^\dagger \sigma_\alpha f\big)_{m_s, m_s}. \tag{14}$$

We could also write $N(\theta, \phi)$ by utilizing the matrix form employed in the last line of this equation.

$$N(\theta,\phi) = \sum_{m'_s} \big(f(\theta,\phi)\big)^*_{m'_s, m_s} \big(f(\theta,\phi)\big)_{m'_s, m_s} = \big(f^\dagger f\big)_{m_s, m_s}. \tag{15}$$

We shall be most interested in the following question: If the incident beam is completely unpolarized, what is the polarization of the scattered beam? For an unpolarized incident beam, an equal probability of $\frac{1}{2}$ exists that the incident longitudinal polarization be $m_s = +\frac{1}{2}$ or $m_s = -\frac{1}{2}$. Thus,

$$\vec{P}^{\text{scatt.}} = \frac{1}{N(\theta,\phi)} \frac{1}{2} \sum_{\alpha} \vec{e}_\alpha \sum_{m_s} \big(f^\dagger \sigma_\alpha f\big)_{m_s, m_s} = \frac{1}{N(\theta,\phi)} \frac{1}{2} \sum_{\alpha} \vec{e}_\alpha \text{trace}\big(f^\dagger \sigma_\alpha f\big), \tag{16}$$

where, now,

$$N(\theta,\phi) = \frac{1}{2} \sum_{m_s} \big(f^\dagger f\big)_{m_s, m_s} = \frac{1}{2} \text{trace}\big(f^\dagger f\big). \tag{17}$$

To take the necessary taces, we now use the trivial identity

$$\text{trace}\big(M_1 M_2 M_3\big) = \text{trace}\big(M_2 M_3 M_1\big) = \text{trace}\big(M_3 M_1 M_2\big), \tag{18}$$

and the properties of the Pauli σ matrices

$$\sigma_j \sigma_k = i\epsilon_{jkl} \sigma_l + \delta_{jk} \mathbf{1}, \tag{19}$$

or

$$\sigma_x \sigma_y = i\sigma_z \text{ and cyclically}, \qquad (\sigma_x)^2 = (\sigma_y)^2 = (\sigma_z)^2 = \mathbf{1}. \tag{20}$$

Also,

$$\text{trace}(\sigma_x) = \text{trace}(\sigma_y) = \text{trace}(\sigma_z) = 0; \quad \text{trace}(\mathbf{1}) = 2. \tag{21}$$

Also, recall

$$f(\theta,\phi) = g(\theta)\mathbf{1} + h(\theta)(\vec{n} \cdot \vec{\sigma}) = g(\theta)\mathbf{1} + h(\theta)\sigma_{y'},$$
$$\text{with} \quad \sigma_{y'} = -\sin\phi\sigma_x + \cos\phi\sigma_y. \tag{22}$$

Thus,

$$\tfrac{1}{2}\text{trace}\big(f^\dagger \sigma_\alpha f\big) = \tfrac{1}{2}\text{trace}\big(\sigma_\alpha f f^\dagger\big)$$
$$= \tfrac{1}{2}\text{trace}\big(\sigma_\alpha (g(\theta)\mathbf{1} + h(\theta)\sigma_{y'})(g(\theta)^*\mathbf{1} + h(\theta)^*\sigma_{y'})\big)$$
$$= \tfrac{1}{2}\text{trace}\Big(\sigma_\alpha (|g(\theta)|^2 + |h(\theta)|^2) + (\sigma_\alpha \sigma_{y'})(g(\theta)h^*(\theta) + g^*(\theta)h(\theta))\Big)$$

$$= \tfrac{1}{2}\Big(0 + 2\delta_{\alpha,y'}(g(\theta)h^*(\theta) + g^*(\theta)h(\theta))\Big)$$
$$= \delta_{\alpha,y'}(g(\theta)h^*(\theta) + g^*(\theta)h(\theta)) = \delta_{\alpha,y'}2\mathcal{R}eal\big(g(\theta)h^*(\theta)\big). \tag{23}$$

Also,

$$N(\theta,\phi) = \frac{1}{2}\Big(\text{trace}(\mathbf{1})\big(|g(\theta)|^2 + |h(\theta)|^2\big) + \text{trace}(\sigma_{y'})\big(g(\theta)h^*(\theta) + g(\theta)^*h(\theta)\big)\Big)$$
$$= \big(|g(\theta)|^2 + |h(\theta)|^2\big). \tag{24}$$

Putting these together, we have the polarization vector for the scattered beam,

$$\vec{P}^{\text{scatt.}} = \vec{e}_{y'}\frac{2\mathcal{R}eal\Big(g(\theta)h^*(\theta)\Big)}{\big(|g(\theta)|^2 + |h(\theta)|^2\big)} = \vec{n}\frac{2\mathcal{R}eal\Big(g(\theta)h^*(\theta)\Big)}{\big(|g(\theta)|^2 + |h(\theta)|^2\big)}. \tag{25}$$

The polarization of this scattered beam has only a component perpendicular to the scattering plane. This fact can be understood in terms of a semiclassical picture. The orbital angular-momentum vector, $\vec{l}$, is normal to the scattering plane. Thus, if the $\vec{s}$ vector lies in the scattering plane, either in the z or x direction, the $\vec{l}\cdot\vec{s}$ term of the interaction would be zero. Only the y component of $\vec{s}$, normal to the scattering plane, is affected by the $\vec{l}\cdot\vec{s}$ interaction. Before seeing how we would detect the polarization of this scattered beam, perhaps in a double-scattering experimental setup, let us introduce the density matrix for a beam of particles with spin.

55

Density Matrices

In the last chapter, we considered an incident beam completely unpolarized. Now, let us consider an incident beam of definite polarization, either a pure beam, perfectly polarized along some specific direction, z', or an incident beam with a statistical distribution of spin orientations giving the beam a partial polarization. For this purpose, it will be useful to introduce the so-called density matrix for the spin orientation of the beam. Recall, for an incident particle with definite spin orientation given by

$$|\chi_s^{\text{inc.}}\rangle = \sum_{m_s} |m_s\rangle c_{m_s} \tag{1}$$

the incident polarization vector, with components, P_α, is given by

$$P_\alpha = \langle\sigma_\alpha\rangle = \sum_{m_s,m_s'} c^*_{m_s'} \langle m_s'|\sigma_\alpha|m_s\rangle c_{m_s} = \sum_{m_s,m_s'} (\sigma_\alpha)_{m_s' m_s} c_{m_s} c^*_{m_s'}. \tag{2}$$

We define the density matrix

$$\rho_{m_s,m_s'} = c_{m_s} c^*_{m_s'}, \tag{3}$$

so

$$\langle\sigma_\alpha\rangle = \sum_{m_s,m_s'} (\sigma_\alpha)_{m_s' m_s} \rho_{m_s m_s'} = \text{trace}(\sigma_\alpha \rho), \tag{4}$$

and

$$\vec{P} = \sum_\alpha \vec{e}_\alpha \text{trace}(\sigma_\alpha \rho). \tag{5}$$

For a single spin-$\frac{1}{2}$-particle in a pure spin state, that is, a state with a definite m_s' along a specific z' direction, with $c_{+\frac{1}{2}} = \alpha$, $c_{-\frac{1}{2}} = \beta$, we would have

$$\rho = \begin{pmatrix} \alpha\alpha^* & \alpha\beta^* \\ \beta\alpha^* & \beta\beta^* \end{pmatrix}. \tag{6}$$

For such a pure state, the trace of the density matrix is unity, and the determinant of the density matrix is zero:

$$\text{trace } \rho = |\alpha|^2 + |\beta|^2 = 1 \qquad \text{and det } \rho = 0. \tag{7}$$

Specific examples would be: (1) a beam with longitudinal polarization with pure $m_s = +\frac{1}{2}$, i.e., with $\alpha = 1$, $\beta = 0$,

$$\rho = \begin{pmatrix} 1 & 0 \\ 0 & 0 \end{pmatrix}; \tag{8}$$

(2) a beam with perpendicular polarization in the x direction, with $\alpha = \beta = \frac{1}{\sqrt{2}}$,

$$\rho = \begin{pmatrix} \frac{1}{2} & \frac{1}{2} \\ \frac{1}{2} & \frac{1}{2} \end{pmatrix}; \tag{9}$$

(3) a beam with perpendicular polarization in the y direction, with $\alpha = \frac{1}{\sqrt{2}}$, $\beta = \frac{i}{\sqrt{2}}$,

$$\rho = \begin{pmatrix} \frac{1}{2} & -\frac{i}{2} \\ \frac{i}{2} & \frac{1}{2} \end{pmatrix}. \tag{10}$$

In terms of such density matrices, the polarization of the incident beam can be given by

$$\vec{P} = \sum_{\alpha} \vec{e}_\alpha \sum_{m_s, m_s'} (\sigma_\alpha)_{m_s' m_s} \rho_{m_s m_s'} = \sum_{\alpha} \vec{e}_\alpha \text{trace}(\sigma_\alpha \rho). \tag{11}$$

The differential scattering cross section can be given by

$$\begin{aligned} \frac{d\sigma}{d\Omega} &= \sum_{m_s, m_s', m_s''} f^*(\theta, \phi)_{m_s' m_s''} c^*_{m_s''} f(\theta, \phi)_{m_s' m_s} c_{m_s} \\ &= \sum_{m_s, m_s', m_s''} f^\dagger_{m_s'' m_s'} f_{m_s' m_s} \rho_{m_s m_s''} = \text{trace}\big(f^\dagger f \rho\big), \end{aligned} \tag{12}$$

and, finally, the polarization of the scattered beam is given by

$$\vec{P}^{\text{scatt.}} = \frac{1}{\text{trace}(f^\dagger f \rho)} \sum_{\alpha} \vec{e}_\alpha \text{trace}(f^\dagger \sigma_\alpha f \rho). \tag{13}$$

All physically interesting quantities are given in terms of traces involving the density matrix of the incident beam. For spin $\frac{1}{2}$-particles, the taking of these traces is particularly simple. The great advantage of the density matrix formalism, however, is that it applies not only for incident beams of pure spin orientation, for which every particle has exactly the same spin state, but also for a statistical distribution of initial spin states, the usual experimental situation. Suppose the nth incident particle is in a spin state given by the $c_{m_s}^{(n)} = c_{+\frac{1}{2}}^{(n)}, c_{-\frac{1}{2}}^{(n)}$, and the probability for this spin orientation is given by w_n. Suppose there are a large number of incident particles, given by N. Then

$$\sum_{n=1}^{N} w_n = 1, \tag{14}$$

and the density matrix for this statistical distribution of spin states is given by

$$(\rho)_{m_s m_s'} = \sum_n w_n c_{m_s}^{(n)} c_{m_s'}^{(n)*}, \tag{15}$$

and with

$$c_{+\frac{1}{2}}^{(n)} = \alpha^{(n)}, \qquad c_{-\frac{1}{2}}^{(n)} = \beta^{(n)}, \tag{16}$$

$$\rho = \begin{pmatrix} \sum_n w_n \alpha^{(n)} \alpha^{(n)*} & \sum_n w_n \alpha^{(n)} \beta^{(n)*} \\ \sum_n w_n \beta^{(n)} \alpha^{(n)*} & \sum_n w_n \beta^{(n)} \beta^{(n)*} \end{pmatrix}. \tag{17}$$

The trace of this matrix is still unity, since $|\alpha^{(n)}|^2 + |\beta^{(n)}|^2 = 1$, and $\sum_n w_n = 1$. But, now, the determinant of ρ is no longer equal to zero. Because the hermitian 2×2 matrix ρ for $s = \frac{1}{2}$-particles is specified by four independent parameters, ρ can be given in terms of the 2×2 unit matrix and the three Pauli σ-matrices, by

$$\rho = A\mathbf{1} + B\sigma_x + C\sigma_y + D\sigma_z. \tag{18}$$

From trace$(\rho) = 2A$, trace$(\sigma_x \rho) = P_x = 2B$, and so on, we have

$$\rho = \frac{1}{2}(\mathbf{1} + \vec{P} \cdot \vec{\sigma}). \tag{19}$$

For example, with $P_z = \frac{1}{2}$, $P_x = 0$, $P_y = 0$,

$$\rho = \begin{pmatrix} \frac{3}{4} & 0 \\ 0 & \frac{1}{4} \end{pmatrix}, \tag{20}$$

whereas, with $P_x = \frac{1}{2}$, $P_y = 0$, $P_z = 0$

$$\rho = \begin{pmatrix} \frac{1}{2} & \frac{1}{4} \\ \frac{1}{4} & \frac{1}{2} \end{pmatrix}, \tag{21}$$

and an unpolarized beam has

$$\rho = \begin{pmatrix} \frac{1}{2} & 0 \\ 0 & \frac{1}{2} \end{pmatrix}. \tag{22}$$

In the last chapter, we saw such an unpolarized incident beam can lead to a polarized scattered beam, where the polarization vector has only a component perpendicular to the scattering plane formed by $\vec{k}$ and $\vec{k}_s$. One way to detect this polarization of the scattered beam is through a double scattering experiment.

A Detection of Polarization via Double Scattering

If the incident beam is unpolarized, no preferred direction exists in the x–y plane normal to the incident $\vec{k}$ vector, and we can place $\vec{k}_s$ in the x–z plane, at a position with $\phi = 0$, without loss of generality. We saw the scattered beam then has a polarization vector, with

$$\vec{P} = \vec{e}_y \frac{\mathcal{R}eal\big(g(\theta_1)h(\theta_1)^*\big)}{\big(|g(\theta_1)|^2 + |h(\theta_1)|^2\big)}. \tag{23}$$

This polarization can be detected by scattering the scattered spin $s = \frac{1}{2}$-particles from a second $s = 0$ target particle of the same kind as the first $s = 0$ particle, in particular, by comparing the differential scattering cross sections in two detectors, placed at angles θ_2 in the x–z plane, symmetrically to the left and right of the direction of the $\vec{k}_s$ vector of the first scattering process. (See Fig. 55.1.) The "left" detector is placed at the angular position, $\theta_2, \phi_2 = 0$, and the "right" detector is placed at the angular position, $\theta_2, \phi_2 = \pi$. Thus, with $\vec{n}_2 = -\sin\phi_2\vec{e}_x + \cos\phi_2\vec{e}_y$, we have $\vec{n}_2 = \pm\vec{e}_y$ for the left (upper sign) and right (lower sign) detectors. Hence,

$$f(\theta_2, \phi_2) = g(\theta_2)\mathbf{1} \pm \sigma_y h(\theta_2), \tag{24}$$

where the upper (lower) signs refer to left (right) detector, respectively. Now, using $\frac{d\sigma}{d\Omega}$ = trace$(f^\dagger f \rho)$, we get

$$\begin{aligned}
\frac{d\sigma}{d\Omega} &= \text{trace}\Big(\big(g^*(\theta_2)\mathbf{1} \pm h^*(\theta_2)\sigma_y\big)\big(g(\theta_2)\mathbf{1} \pm h(\theta_2)\sigma_y\big)\big(\tfrac{1}{2}(\mathbf{1} + P_y(\theta_1)\sigma_y)\big)\Big) \\
&= \tfrac{1}{2}\text{trace}\Big(\Big[\big(|g(\theta_2)|^2 + |h(\theta_2)|^2\big)\mathbf{1} \pm \big(g(\theta_2)h^*(\theta_2) + g^*(\theta_2)h(\theta_2)\big)\sigma_y\Big] \\
&\qquad \times \big(\mathbf{1} + P_y(\theta_1)\sigma_y\big)\Big) \\
&= \tfrac{1}{2}\text{trace}\Big(\mathbf{1}\Big[\big(|g(\theta_2)|^2 + |h(\theta_2)|^2\big) \pm 2\mathcal{R}eal\big(g(\theta_2)h^*(\theta_2)\big)P_y(\theta_1)\Big] \\
&\qquad + \sigma_y\Big[\cdots\Big]\Big) \\
&= \big(|g(\theta_2)|^2 + |h(\theta_2)|^2\big) \pm P_y(\theta_1)2\mathcal{R}eal\big(g(\theta_2)h^*(\theta_2)\big),
\end{aligned} \tag{25}$$

where the upper (lower) signs refer to left (right) detectors and we have again used: trace$(\sigma_y) = 0$; trace$(\mathbf{1}) = 2$. Using eq. (23) for $P_y(\theta_1)$, we can then get the left–right asymmetry parameter

$$\frac{d\sigma_L - d\sigma_R}{d\sigma_L + d\sigma_R} = \frac{2\mathcal{R}eal\big(g(\theta_1)h^*(\theta_1)\big)}{\big(|g(\theta_1)|^2 + |h(\theta_1)|^2\big)} \times \frac{2\mathcal{R}eal\big(g(\theta_2)h^*(\theta_2)\big)}{\big(|g(\theta_2)|^2 + |h(\theta_2)|^2\big)}. \tag{26}$$

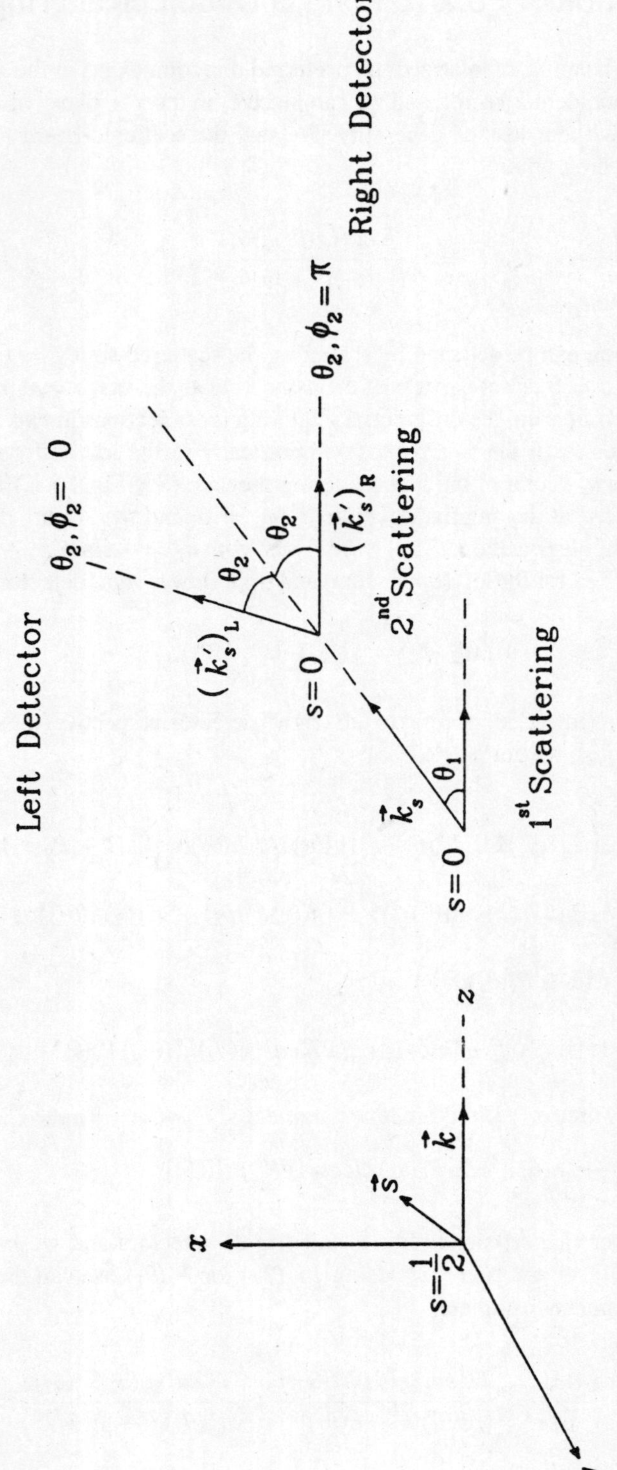

FIGURE 55.1. Double scattering of an $s = \frac{1}{2}$ particle from identical $s = 0$ target particles.

B Differential Scattering Cross Section of a Beam with Arbitrary Incident Polarization

As a final exercise, let us calculate the differential cross section for an incident beam of arbitrary polarization, $\vec{P}$. Because the initial polarization vector will define the directions of the x and y axes, we want to place our detector at an arbitrary azimuth angle, ϕ, relative to the x–z plane. Now, the normal to the scattering plane defined by $\vec{k}$ and $\vec{k}_s$ is given by $\vec{n} = -\sin\phi\vec{e}_x + \cos\phi\vec{e}_y$, and

$$(\vec{n}\cdot\vec{\sigma}) = -\sin\phi\sigma_x + \cos\phi\sigma_y = \sigma_{y'}. \tag{27}$$

With $\rho = \frac{1}{2}(\mathbf{1} + \sum_\alpha \sigma_\alpha P_\alpha)$, we now get

$$\begin{aligned}
\frac{d\sigma}{d\Omega} &= \text{trace}(f^\dagger f\rho)\\
&= \text{trace}\Big(\big(g^*(\theta)\mathbf{1} + h^*(\theta)\sigma_{y'}\big)\big(g(\theta)\mathbf{1} + h(\theta)\sigma_{y'}\big)\tfrac{1}{2}\big(\mathbf{1} + \sum_\alpha P_\alpha\sigma_\alpha\big)\Big)\\
&= \tfrac{1}{2}\text{trace}\Big(\Big[\big(|g(\theta)|^2 + |h(\theta)|^2\big)\mathbf{1} + 2\mathcal{R}eal\big(g(\theta)h^*(\theta)\big)\\
&\quad\times(-\sin\phi\sigma_x + \cos\phi\sigma_y)\Big]\big(\mathbf{1} + \sum_\alpha \sigma_\alpha P_\alpha\big)\Big)\\
&= \Big[\big(|g(\theta)|^2 + |h(\theta)|^2\big) + 2\mathcal{R}eal\big(g(\theta)h^*(\theta)\big)\big(-\sin\phi P_x + \cos\phi P_y\big)\Big].
\end{aligned} \tag{28}$$

A ϕ-dependence in the differential cross section exists if either P_x or $P_y \neq 0$ in the incident beam.

C Generalizations to More Complicated Cases

Scattering of Spin $s = 1$ Projectile from $s = 0$ Target Particles

For the scattering of a $s = 1$ projectile from a $s = 0$ target particle, we shall need 3×3 scattering amplitude matrices and 3×3 density matrices. The most general density matrix will require nine parameters. The density matrix ρ can now be given in terms of the 3×3 unit matrix, the three components of S_α, and the five components of the spherical tensors of spherical rank 2, $[S^1 \times S^1]^2_M$, with $M = +2, +1, 0, -1, -2$,

$$\rho = \tfrac{1}{3}\mathbf{1} + \tfrac{1}{2}\big(P_x S_x + P_y S_y + P_z S_z\big) + \sum_M (-1)^M \mathcal{P}^2_M [S^1 \times S^1]^2_{-M}, \tag{29}$$

where P_x, P_y, P_z are the three components of the polarization vector defined in Chapter 54, and the $\mathcal{P}^2_M$ are the five components of the tensor polarization. The incident beam may have both vector and tensor polarizations, or may have vector polarizations only with zero tensor polarizations. The 3×3 matrices for S_x, S_y, S_z

are well known (see Chapter 8)

$$S_x = \frac{1}{\sqrt{2}}\begin{pmatrix} 0 & 1 & 0 \\ 1 & 0 & 1 \\ 0 & 1 & 0 \end{pmatrix}, \quad S_y = \frac{1}{\sqrt{2}}\begin{pmatrix} 0 & -i & 0 \\ i & 0 & -i \\ 0 & i & 0 \end{pmatrix}, \quad S_z = \begin{pmatrix} +1 & 0 & 0 \\ 0 & 0 & 0 \\ 0 & 0 & -1 \end{pmatrix}.$$

The matrix algebra is somewhat more complicated, but the process of trace-taking is very similar to that of the simpler case of $s = \frac{1}{2}$ particles.

Scattering of Spin $s = \frac{1}{2}$ Projectile from $s = \frac{1}{2}$ Target Particles

Now, the density matrix can be be given by the $c_{m_{s,1}}$ of projectile and $c_{m_{s,2}}$ of target, with

$$|\chi_s^{\text{proj.}}\rangle = \sum_{m_{s,1}} |m_{s,1}\rangle c_{m_{s,1}}^{\text{proj.}}, \tag{30}$$

$$|\chi_s^{\text{targ.}}\rangle = \sum_{m_{s,2}} |m_{s,2}\rangle c_{m_{s,2}}^{\text{targ.}}. \tag{31}$$

The density matrix is a 4×4 hermitian matrix, with

$$\rho_{\mu\mu'} = c_{m_{s,1}}^{\text{proj.}} c_{m_{s,2}}^{\text{targ.}} c_{m'_{s,1}}^{\text{proj.}*} c_{m'_{s,2}}^{\text{targ.}*}, \tag{32}$$

where the row and column indices $\mu = m_{s,1}m_{s,2}$ and $\mu' = m'_{s,1}m'_{s,2}$ now have four possible values, viz., $+\frac{1}{2}, +\frac{1}{2};\ -\frac{1}{2}, +\frac{1}{2};\ +\frac{1}{2}, -\frac{1}{2};\ -\frac{1}{2}, -\frac{1}{2}$. The most general density matrix now requires 16 parameters. It could be written as

$$\rho = \tfrac{1}{2}\Big(A\mathbf{1} + \vec{P}_1\cdot\vec{\sigma}_1 + \vec{P}_2\cdot\vec{\sigma}_2 + A'(\vec{\sigma}_1\cdot\vec{\sigma}_2) + \vec{B}\cdot[\vec{\sigma}_1\times\vec{\sigma}_2] + \sum_M (-1)^M C_M^2 [\vec{\sigma}_1\times\vec{\sigma}_2]^2_{-M}\Big), \tag{33}$$

with $(1+3+3+1+3+5) = 16$ components. In actual practice, of course, the projectile may be given a vector polarization, $\vec{P}_1$, and the target may be given a vector polarization, $\vec{P}_2$. The additional combined scalar, vector, and tensor polarizations may be difficult to realize experimentally, but the additional coefficients A, A', $\vec{B}$, and C_M^2 are needed for the analysis, which again involves trace-taking and the algebra of the two sets of Pauli σ matrices.

56

Isospin

The simplicity of the algebra for the case of $s = \frac{1}{2}$-particles, involving the algebra of the Pauli σ matrices, makes this chapter a good place to introduce a brief remark about isospin.

For proton–proton, proton–neutron, or neutron–neutron scattering, the essential charge independence of the nuclear force must be taken into account. (This charge independence of the nucleon–nucleon interaction does not include the Coulomb term in the proton–proton interaction. This is a relatively small part, however, of the full proton–proton interaction at energies high enough that the protons are far above the proton–proton Coulomb barrier for $r \approx 1$fermi.) It is useful to introduce the isospin formalism and consider the neutron and proton as the two possible charge states of a nucleon. Being a spin $s = \frac{1}{2}$ particle, the nucleon is a fermion. Besides the internal degree of freedom associated with the spin of the nucleon, another internal degree of freedom is associated with the charge of the nucleon. Thus, the wave function associated with the internal degrees of freedom of our point nucleon must include besides the spin function, $\chi(\vec{\sigma})_{m_s}$, with two quantum states, $m_s = +\frac{1}{2}$ and $m_s = -\frac{1}{2}$, a similar charge function, also with two quantum states. Even though nothing is "spinning" in this charge space, we can use the mathematics of a system with two internal quantum states to make a one-to-one parallel between the two-spin state system and the two-charge state system. We therefore can name the neutron as the state of the system with a new spin, the "isospin," with a new quantum number with value $+\frac{1}{2}$, whereas the proton is the state of the system for which the new quantum number has the value $-\frac{1}{2}$. We introduce the isospin vector operator $\vec{i}$ in analogy with $\vec{s}$. Just as it is useful to introduce $\vec{\sigma}$ via $\vec{s} = \frac{1}{2}\vec{\sigma}$, we now let $\vec{i} = \frac{1}{2}\vec{\tau}$, where

$$\begin{aligned} i_z|\text{neutron}\rangle &= \tfrac{1}{2}\tau_z|\text{neutron}\rangle = +\tfrac{1}{2}|\text{neutron}\rangle = +\tfrac{1}{2}|m_i = +\tfrac{1}{2}\rangle \\ i_z|\text{proton}\rangle &= \tfrac{1}{2}\tau_z|\text{proton}\rangle = -\tfrac{1}{2}|\text{proton}\rangle = -\tfrac{1}{2}|m_i = -\tfrac{1}{2}\rangle. \end{aligned} \quad (1)$$

(Note: It is of course quite arbitrary as to whether a neutron or proton is designated as the $m_i = +\frac{1}{2}$ particle. This is illustrated by the fact that nuclear physicists and particle physicists do not agree on this choice. The above is the nuclear convention. Particle physicists name the proton the $m_i = +\frac{1}{2}$ particle.) In analogy with the operators, $\vec{\sigma}$, we can introduce the operators $\vec{\tau}$, with

$$\tau_x = \begin{pmatrix} 0 & 1 \\ 1 & 0 \end{pmatrix}, \quad \tau_y = \begin{pmatrix} 0 & -i \\ i & 0 \end{pmatrix}, \quad \tau_z = \begin{pmatrix} 1 & 0 \\ 0 & -1 \end{pmatrix}. \quad (2)$$

Now, just as the operator $s_+ = \frac{1}{2}(\sigma_x + i\sigma_y)$, when acting on a state with $m_s = -\frac{1}{2}$ converts it to a state with $m_s = +\frac{1}{2}$, so the operator $i_+ = \frac{1}{2}(\tau_x + i\tau_y)$, when acting on a proton state with $m_i = -\frac{1}{2}$ converts it to a neutron state with $m_i = +\frac{1}{2}$. This operator is needed in the theory of beta-decay, where a proton may be converted to a neutron.

Just as we found it convenient to write spin functions in terms of internal variables, $\vec{\xi} \equiv \vec{\sigma}$, namely, $\chi(\vec{\sigma})_{m_s}$, we will now write the internal charge functions in terms of the internal variable designated by $\vec{\tau}$, namely, $\bar{\chi}(\vec{\tau})_{m_i}$. This will be particularly useful for a discussion of the symmetry of two-particle functions. For the two-particle spin functions, it was convenient to make a transformation from the two-particle functions in the $m_{s,1}m_{s,2}$ representation to the SM_S representation, via

$$\chi(\vec{\sigma}_1)_{m_{s,1}}\chi(\vec{\sigma}_2)_{m_{s,2}} = \sum_S \langle \tfrac{1}{2}m_{s,1}\tfrac{1}{2}m_{s,2}|SM_S\rangle \chi(\vec{\sigma}_1, \vec{\sigma}_2)_{SM_S}, \quad (3)$$

where the three functions with $S = 1$, $M_S = +1, 0, -1$ are symmetric under the interchange of indices 1 and 2 of the two fermions, whereas the single function with $S = 0$ and $M_S = 0$ is antisymmetric (changes sign) under interchange of the indices 1 and 2. In the same fashion, the charge function for two neutrons, for two protons, and one linear combination of the neutron–proton function are symmetric under the interchange of the indices 1 and 2 of the two-particle charge functions, and can be considered members of a triplet with two-particle quantum numbers, $I = 1$, with $M_I = +1, -1, 0$, where $\vec{I} = \vec{i}_1 + \vec{i}_2$ [in analogy with $\vec{S} = \vec{s}_1 + \vec{s}_2$, with the same vector-coupling (or Clebsch–Gordan coefficients) in isospin space as in ordinary spin space.] Thus,

$$\begin{aligned} \bar{\chi}(\vec{\tau}_1)_{+\frac{1}{2}}\bar{\chi}(\vec{\tau}_2)_{+\frac{1}{2}} &= \bar{\chi}(\vec{\tau}_1, \vec{\tau}_2)_{I=1,M_I=+1}, \\ \bar{\chi}(\vec{\tau}_1)_{-\frac{1}{2}}\bar{\chi}(\vec{\tau}_2)_{-\frac{1}{2}} &= \bar{\chi}(\vec{\tau}_1, \vec{\tau}_2)_{I=1,M_I=-1}, \\ \frac{1}{\sqrt{2}}\left(\bar{\chi}(\vec{\tau}_1)_{+\frac{1}{2}}\bar{\chi}(\vec{\tau}_2)_{-\frac{1}{2}} + \bar{\chi}(\vec{\tau}_1)_{-\frac{1}{2}}\bar{\chi}(\vec{\tau}_2)_{+\frac{1}{2}}\right) &= \bar{\chi}(\vec{\tau}_1, \vec{\tau}_2)_{I=1,M_I=0}, \end{aligned} \quad (4)$$

and

$$\frac{1}{\sqrt{2}}\left(\bar{\chi}(\vec{\tau}_1)_{+\frac{1}{2}}\bar{\chi}(\vec{\tau}_2)_{-\frac{1}{2}} - \bar{\chi}(\vec{\tau}_1)_{-\frac{1}{2}}\bar{\chi}(\vec{\tau}_2)_{+\frac{1}{2}}\right) = \bar{\chi}(\vec{\tau}_1, \vec{\tau}_2)_{I=0,M_I=0}. \tag{5}$$

For nucleon–nucleon scattering, therefore, the total two-particle function can be considered a product of a relative motion orbital function, a two-particle spin function, and a two-particle charge function

$$\psi_{\text{total}} = \psi_{\text{orbital}}(\vec{r}_1 - \vec{r}_2)\chi(\vec{\sigma}_1, \vec{\sigma}_2)_{S,M_S}\bar{\chi}(\vec{\tau}_1, \vec{\tau}_2)_{I,M_I}. \tag{6}$$

Because the $s = \frac{1}{2}$ nucleons are fermions, this total wave function must be antisymmetric under interchange of particle indices. In a partial wave decomposition where the orbital functions with even l do not change sign under the $\vec{r}_1 \leftrightarrow \vec{r}_2$ interchange and the orbital functions with odd l do change sign, the partial waves with even l are therefore restricted to terms with $S = 1, I = 0$, or with $S = 0, I = 1$, whereas the partial waves with odd l are restricted to terms with $S = 1, I = 1$ or with $S = 0, I = 0$. The scattering amplitude matrix must now be designated by the initial m_s and the initial m_i quantum numbers of the two particles in the incident beam, and the final m'_s and m'_i of the two particles in the scattered beam, that is, the scattering amplitude matrix, will be of the form

$$f(\theta, \phi)_{m'_{s,1}m'_{s,2}m'_{i,1}m'_{i,2};m_{s,1}m_{s,2}m_{i,1}m_{i,2}}.$$

To take account of the required antisymmetry of the two-particle wave function in a partial wave decomposition, however, it will be necessary to transform from the $m_{s,1}m_{s,2}m_{i,1}m_{i,2}$ basis into an S, M_S, I, M_I basis (via standard and very simple Clebsch–Gordan coefficients), and expand the scattering amplitude in partial waves in the matrix form

$$f(\theta, \phi)_{S'M'_SI'M'_I;SM_SIM_I}$$

to be able to observe the S, I value restrictions for even and odd l. In addition, the dominant part of the nucleon–nucleon interaction has terms of the form

$$V = V_0(r) + V_1(r)(\vec{\sigma}_1 \cdot \vec{\sigma}_2) + V_2(r)(\vec{\tau}_1 \cdot \vec{\tau}_2) + V_3(r)(\vec{\sigma}_1 \cdot \vec{\sigma}_2)(\vec{\tau}_1 \cdot \vec{\tau}_2), \tag{7}$$

where $r = |\vec{r}_1 - \vec{r}_2|$. The four terms of this potential are scalars separately in orbital space, in spin space, and in isospin space. Hence, they cannot induce a change in the total spin quantum number S or the total isospin quantum number I. Thus, the above f-scattering amplitude matrix must be diagonal in S and I, i.e., $S' = S$ and $I' = I$.

A Spectra of Two-Valence Nucleon Nuclei

The isospin quantum number also plays an important role in the energy level scheme of nuclei. Let us consider briefly two-valence nucleon nuclei with mass number, $A = 18$, that is, the nuclei ${}^{18}_{8}O_{10}$, ${}^{18}_{9}F_9$, and ${}^{18}_{10}Ne_8$, which have two-valence shell nucleons outside the doubly magic nucleus, ${}^{16}_{8}O_8$, two neutrons for

${}^{18}_{8}O_{10}$, one proton and one neutron for ${}^{18}_{9}F_9$, and two protons for ${}^{18}_{10}Ne_8$. These nuclei have $M_I = +1, 0$, and -1, respectively. All three nuclei have states with $I = 1$. In addition, ${}^{18}_{9}F_9$ has states with $I = 0$. Thus, we are dealing with isospin triplets and an isospin singlet of energy levels. The total Coulomb energy separates the energy levels of the triplet. It is, however, easy to make an average Coulomb energy correction by considering the 8, 9, or 10 proton cloud as a spherically symmetric charge cloud of the experimentally observed radius. Because the greatest Coulomb repulsion occurs in ${}^{18}_{10}Ne_8$ with its 10 protons, the levels of this nucleus should be shifted down by 3.40 MeV to put them into correspondence with the companion levels in ${}^{18}_{9}F_9$. Similarly, the smaller Coulomb repulsion in ${}^{18}_{8}O_{10}$ requires an upward shift of 2.70 MeV. With these shifts, the members of the $I = 1$ triplet at once become apparent. The lowest levels in both ${}^{18}_{8}O_{10}$ and ${}^{18}_{10}Ne_8$ are $0^+, 2^+, 4^+, 0^+, 2^+$, and 1^-. The required companion levels are also found in the $M_I = 0$ nucleus, ${}^{18}_{9}F_9$. The 2^+ level lies at 1.98, 2.02, and 1.89 MeV above the 0^+ level in these three nuclei. Similarly, the 4^+ level lies at 3.55, 3.61, and 3.38 MeV above the 0^+ level. For the approximate triplet nature of the other excited states, see Fig. 56.1. Moreover, these are the levels predicted by the nuclear shell model. The lowest states are those in which the two-valence nucleons are both in a $2d_{\frac{5}{2}}$ single particle state, with the *same* principal quantum number, $n = 2$, and orbital angular-momentum quantum number, $l = 2$, coupled to $j = \frac{5}{2}$. For such a two-particle configuration, states with J even, $J = 0, 2, 4$, are antisymmetric in the combined spin-orbital part of the space, and hence must be symmetric in the isospin-space part of the two-nucleon wave function. Hence, this is restricted to $I = 1$. The states of this configuration, with J odd, $J = 1, 3, 5$, are symmetric in the combined spin-orbital part of the full space, and must be antisymmetric in the isospin part of the space. Hence, such states are restricted to two-particle isospin quantum number, $I = 0$, and can occur only in the $M_I = 0$ member of the triplet, ${}^{18}_{9}F_9$. In fact, the 1^+ state is the ground state of ${}^{18}_{9}F_9$. The 3^+ state also lies below the lowest 0^+ state in this nucleus. The remaining $I = 0$ states in this nucleus have mainly positive parity for odd J, and negative parity for even J, just the opposite of the low-lying states with $I = 1$, which have positive parity for even J and negative parity for odd J, a strong indication they have the opposite symmetry in the combined orbital-spin part of the full space. (Note, however, two-nucleon configurations such as $2s_{\frac{1}{2}}2d_{\frac{5}{2}}$, with *different* single-particle l's can have both (orbital-spin) symmetric and antisymmetric combinations, so they can have both $I = 0$ and $I = 1$. The 2^+ level at 2.52-MeV excitation energy in ${}^{18}_{9}F_9$ with $I = 0$ is such a level. Although the isospin quantum number in nuclei is only an approximately good quantum number (largely because of the additional Coulomb force between protons), Fig. 56.1 gives a clear indication of the importance of the isospin quantum number in nuclei.

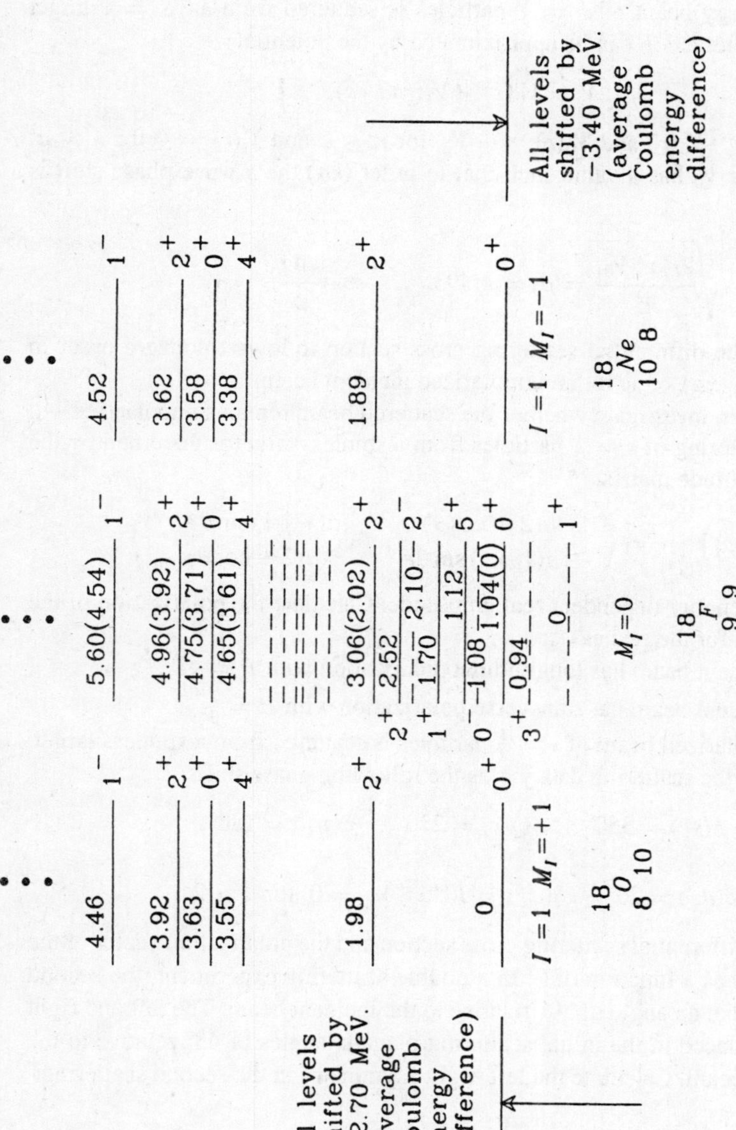

FIGURE 56.1. Energy levels of A = 18 nuclei. (All energies shown are in MeV.)

Problems

17. A low-energy beam of $s = \frac{1}{2}$ particles is scattered from an $S = 0$ target particle. The interaction can be approximated by the potential

$$V = V_0(r) + V_1(r)(\vec{l} \cdot \vec{s}),$$

where $V_1(r) = \frac{1}{2}V_0(r)$ and $V_0(r) = -V_0$, for $r \leq a$ and $V(r) = 0$ for $r > a$. Assume further V_0 has a value such that to order (ka) the s wave phase shift is zero,

$$\sqrt{\frac{2\mu a^2|V_0|}{\hbar^2}} = \rho_0 = 4.493, \qquad \text{so} \quad \frac{\tan\rho_0}{\rho_0} = 1.$$

(a) Calculate the differential scattering cross section to lowest nonzero order in (ka), assuming $(ka) \ll 1$, for an unpolarized incident beam.
(b) To this order, investigate whether the scattered beam remains unpolarized.

18. The scattering of $s = \frac{1}{2}$ particles from a spinless target is governed by the scattering amplitude matrix

$$\Big(f(\theta,\phi)\Big)_{m'_s m_s} = \begin{pmatrix} a(2+\cos\theta) & \frac{1}{2}a(1+i)\sin\theta e^{-i\phi} \\ -\frac{1}{2}a(1+i)\sin\theta e^{i\phi} & a(2+\cos\theta) \end{pmatrix},$$

where a is an energy-dependent real amplitude. Calculate the polarization of the scattered beam for the two cases:

(a) The incident beam has longitudinal polarization with $\vec{P} = \frac{2}{3}\vec{e}_z$.
(b) The incident beam has transverse polarization with $\vec{P} = \frac{2}{3}\vec{e}_x$.

19. An unpolarized beam of $s = \frac{1}{2}$ particles is scattered from a spinless target. An analysis of the scattering data yields the following phase shifts:

$$\delta(s_{\frac{1}{2}}) = 350^o, \quad \delta(p_{\frac{3}{2}}) = 220^o, \quad \delta(p_{\frac{1}{2}}) = 180^o,$$

$$\delta(d_{\frac{5}{2}}) = 20^o, \quad \delta(d_{\frac{3}{2}}) = 30^o, \quad \delta_{l,j} = 0 \text{ for } l > 2.$$

Calculate the differential scattering cross section and the polarization vector of the scattered beam as a function of θ. In a double scattering experiment, the second target is placed at an angle of 30° relative to the incident beam. The left and right detectors are placed in the initial scattering plane at angles of 45° relative to the first scattered beam. Calculate the left–right asymmetry in the second scattering:

$$\frac{d\sigma_L - d\sigma_R}{d\sigma_L + d\sigma_R}.$$

20. (a) The following are density matrices for $s = \frac{1}{2}$ particles:

$$\rho = \begin{pmatrix} \frac{1}{3} & -\frac{i\sqrt{2}}{3} \\ \frac{i\sqrt{2}}{3} & \frac{2}{3} \end{pmatrix}, \qquad \rho = \begin{pmatrix} \frac{1}{3} & -\frac{i}{3} \\ \frac{i}{3} & \frac{2}{3} \end{pmatrix}.$$

Show one of these represents a pure state with particles with 100% polarization along a specific direction and find this direction. What is $\vec{P}^{\text{inc.}}$ for these two cases?

(b) Calculate the polarization of the scattered beam of $s = \frac{1}{2}$ particles into directions, θ, ϕ by an $S = 0$ target particle for the three cases:

$$\begin{aligned} \text{(i)}: &\quad P_x^{\text{inc.}} \neq 0; \quad P_y^{\text{inc.}} = P_z^{\text{inc.}} = 0. \\ \text{(ii)}: &\quad P_y^{\text{inc.}} \neq 0; \quad P_x^{\text{inc.}} = P_z^{\text{inc.}} = 0.. \\ \text{(iii)}: &\quad P_z^{\text{inc.}} \neq 0; \quad P_x^{\text{inc.}} = P_y^{\text{inc.}} = 0 \end{aligned}$$

Express your answer in terms of $g(\theta)$, $h(\theta)$.

21. Calculate the scattering amplitude matrix $\big(f(\theta, \phi)\big)_{m'_s m_s}$ for a beam of $s = 1$ particles being scattered from an $S = 0$ target particle assuming an interaction of the form, $V(r) = V_0(r) + V_1(r)(\vec{l} \cdot \vec{s})$.

(a) Calculate this scattering amplitude matrix first by a partial wave expansion, and show it must have the form

$$\big(f(\theta, \phi)\big)_{m'_s m_s} = \begin{pmatrix} g_1(\theta) & -ih_1(\theta)e^{-i\phi} & l(\theta)e^{-2i\phi} \\ ih_2(\theta)e^{i\phi} & g_0(\theta) & -ih_2(\theta)e^{-i\phi} \\ l(\theta)e^{2i\phi} & ih_1(\theta)e^{i\phi} & g_1(\theta) \end{pmatrix}.$$

(b) Show in first Born approximation this $\big(f(\theta, \phi)\big)_{m'_s m_s}$ has the simpler form

$$\begin{pmatrix} g(\theta) & -ih(\theta)e^{-i\phi} & 0 \\ ih(\theta)e^{i\phi} & g(\theta) & -ih(\theta)e^{-i\phi} \\ 0 & ih(\theta)e^{i\phi} & g(\theta) \end{pmatrix}.$$

Show this relation can be expressed as a linear combination of the 3×3 unit matrix and the 3×3 matrices S_x and S_y in the linear combination, $S_{y'} = -S_x \sin\phi + S_y \cos\phi$.

$$\big(f(\theta, \phi)\big)_{m'_s m_s} = \big(g(\theta)\mathbf{1} + \sqrt{2}h(\theta)S_{y'}\big)_{m'_s m_s}.$$

(c) For this latter case, find the differential cross section and the polarization vector for the scattered beam as a function of $g(\theta)$ and $h(\theta)$ for an unpolarized incident beam, assuming the detector is placed in the x–z plane, so $\phi = 0$.

22. (a) For the scattering of a beam of particles with spin, s, from a spinless, $S = 0$, target particle with an incident beam density matrix, $\rho^{\text{inc.}}$, show the density matrix of the scattered beam is given by

$$\big(\rho^{\text{scatt.}}\big)_{m'_s m_s} = \frac{\big(f\rho^{\text{inc.}} f^\dagger\big)_{m'_s m_s}}{\text{trace}(f\rho^{\text{inc.}} f^\dagger)},$$

where $\big(f(\theta, \phi)\big)_{m'_s m_s}$ is the scattering amplitude matrix.

For the case of $s = 1$ particles and the scattering amplitude of part (b) of problem 21, show this gives

$$\rho^{\text{scatt.}} = \frac{\big[|g(\theta)|^2\mathbf{1} + \sqrt{2}\big(g(\theta)h^*(\theta) + g^*(\theta)h(\theta)\big)S_y + 2|h(\theta)|^2 S_y^2\big]}{3|g(\theta)|^2 + 4|h(\theta)|^2}$$

for an unpolarized incident beam, with $\rho^{\text{inc.}} = \frac{1}{3}\mathbf{1}$, where the scattering plane has been chosen as the x–z plane, so $\phi = 0$.

(b) Calculate the left–right asymmetry

$$\frac{d\sigma_L - d\sigma_R}{d\sigma_L + d\sigma_R}$$

as a function of $g(\theta)$ and $h(\theta)$ for a double scattering experiment for this case, if the left and right detectors are also placed in the x–z plane at angles, θ, which are identical to the scattering angle, θ, for the first scattering process.

23. For a beam of $s = 1$ particles with perfect polarization, show both a tensor polarization and a vector polarization must exist with $|\vec{P}| = 1$. In particular, with perfect perpendicular polarization in the y direction, perpendicular to the x–z scattering plane, i.e., with (S_y) – eigenvalue $= +1$ for all particles, show the density matrix for such a beam is given by

$$\rho = \tfrac{1}{2}S_y + \tfrac{1}{2}S_y^2 = \tfrac{1}{3}\mathbf{1} + \tfrac{1}{2}S_y - \tfrac{1}{4}(S_x^2 - S_y^2) + \tfrac{1}{12}(S_x^2 + S_y^2 - 2S_z^2).$$

Using the result of (a) of problem 22 show the scattering of a beam with perfect polarization in the y direction, normal to the scattering plane, will not alter this perfect polarization. That is, show

$$\rho^{\text{scatt.}} = \rho^{\text{inc.}} = \tfrac{1}{2}S_y + \tfrac{1}{2}S_y^2$$

in this case. Recall the simple 3×3 matrix for S_y has the property: $S_y^3 = S_y$ and $S_y^4 = S_y^2$.

24. Assume at a certain energy the nucleon–nucleon interaction is effectively spin-independent, and, although "charge-independent" (a "charge-independent" potential is a scalar in isospin space; it is independent of M_I, but may be I-dependent), it does depend on the total isospin of the two-particle system through a potential of the form

$$V(r) = V_a(r) + V_b(r)(\vec{\tau}_1 \cdot \vec{\tau}_2),$$

where $\frac{1}{2}\vec{\tau}_1 = \vec{i}_1$ and $\frac{1}{2}\vec{\tau}_2 = \vec{i}_2$ are the isospin operators of nucleons labelled by the subscripts 1 and 2. Assume V_a and V_b have strengths and signs such that

$$V_a + \tfrac{1}{2}V_b\Big(4\big[I(I+1) - \tfrac{3}{4} - \tfrac{3}{4}\big]\Big)$$

is attractive when the two-particle isospin quantum number $I = 0$ and repulsive when $I = 1$ such that the scattering can be parameterized by the phase shifts:

$$\begin{aligned} &\text{For } I = 0: \quad \delta_{l=0}^{S=1} = -\tfrac{\pi}{4} \quad \delta_{l=1}^{S=0} = -\tfrac{\pi}{4}, \\ &\text{For } I = 1: \quad \delta_{l=0}^{S=0} = +\tfrac{\pi}{2} \quad \delta_{l=1}^{S=1} = +\tfrac{\pi}{2}, \end{aligned}$$

and

$$\delta_l^S = 0 \quad \text{for all } l \geq 2.$$

Although V is spin-independent, the phase shifts depend on all three quantum numbers, the two-particle I, S, and the relative motion orbital angular momentum l. Show the I, S, l combinations with nonzero phase shifts are consistent with the required antisymmetry of the two-particle wave functions.

Assume the nucleons in the incident beam and in the target are unpolarized. Calculate the relative strengths of the following differential cross sections for $\theta = 30^o$ (center of mass system):

(a) proton–proton scattering;

(b) proton–neutron scattering (scattered proton forward 30°);

(c) proton–neutron exchange scattering (target neutron forward 30°).

Part VI

Time-Dependent Perturbation Theory

57

Time-Dependent Perturbation Expansion

We recall from Chapter 19 the time evolution of a quantum system is given by

$$|\psi(t)\rangle = U(t, t_0)|\psi(t_0)\rangle, \tag{1}$$

where the time evolution operator, $U(t, t_0)$, is unitary, with

$$UU^\dagger = U^\dagger U = 1, \qquad \text{with} \quad U(t_0, t_0) = 1, \tag{2}$$

and

$$-\frac{\hbar}{i}\frac{dU}{dt} = HU(t, t_0), \tag{3}$$

where $H(t)$ may now be time dependent. We can express the solution, $U(t, t_0)$, in terms of the integral equation,

$$U(t, t_0) = 1 - \frac{i}{\hbar}\int_{t_0}^{t} d\tau H(\tau)U(\tau, t_0). \tag{4}$$

For general $H(t)$, this integral equation cannot be solved, but if

$$H = H_0 + \lambda h(t), \tag{5}$$

where H_0 is time independent with known eigenvectors and eigenvalues and λ is a parameter of smallness, $\lambda \ll 1$, then we may be able to find $U(t, t_0)$ by perturbation theory in a power series in λ. The zeroth-order solution for the time evolution is now known

$$U^{(0)}(t, t_0) = e^{-\frac{i}{\hbar}H_0(t-t_0)}. \tag{6}$$

It may now be useful to go from the Heisenberg picture to the Schrödinger picture (see Chapter 19) via an intermediate representation, through the so-called interaction picture, by transforming the time-independent Heisenberg state vector, $|\psi_H\rangle \equiv |\psi(t_0)\rangle$, to the time-dependent Schrödinger state vector, $|\psi_S\rangle \equiv |\psi(t)\rangle$, via an intermediate state vector, the interaction representatiion state vector, $|\psi_I(t)\rangle$, where

$$|\psi(t)\rangle = U^{(0)}(t,t_0)|\psi_I(t)\rangle = U^{(0)}(t,t_0)U'(t,t_0)|\psi(t_0)\rangle, \tag{7}$$

where, again, $U'(t_0,t_0) = 1$. Substituting into eq. (3), we get

$$\begin{aligned} -\frac{\hbar}{i}\left(\frac{dU^{(0)}}{dt}\right)U' - \frac{\hbar}{i}U^{(0)}\left(\frac{dU'}{dt}\right) &= (H_0 + \lambda h)U^{(0)}U' \\ H_0U^{(0)}U' - \frac{\hbar}{i}U^{(0)}\left(\frac{dU'}{dt}\right) &= H_0U^{(0)}U' + \lambda hU^{(0)}U' \\ -\frac{\hbar}{i}U^{(0)}\left(\frac{dU'}{dt}\right) &= \lambda hU^{(0)}U', \end{aligned} \tag{8}$$

so, after left-multiplication with $U^{(0)\dagger}$, we have the equation for U',

$$-\frac{\hbar}{i}\frac{dU'}{dt} = \lambda\left(U^{(0)\dagger}hU^{(0)}\right)U' = \lambda h_I(t)U', \tag{9}$$

where the interaction representation, h_I, is given by

$$h_I(t) = U^{(0)\dagger}h(t)U^{(0)} = e^{+\frac{i}{\hbar}H_0(t-t_0)}h(t)e^{-\frac{i}{\hbar}H_0(t-t_0)}, \tag{10}$$

and we are now led to the integral equation

$$U'(t,t_0) = 1 - \frac{i}{\hbar}\lambda\int_{t_0}^{t} d\tau h_I(\tau)U'(\tau,t_0). \tag{11}$$

Because λ is a parameter of smallness, this equation is now in a form that can be iterated:

$$U'(t,t_0) = 1 - \frac{i}{\hbar}\lambda\int_{t_0}^{t} d\tau h_I(\tau)\left(1 - \frac{i}{\hbar}\lambda\int_{t_0}^{\tau} d\tau_1 h_I(\tau_1)\left(1 - \cdots\right)\right), \tag{12}$$

with

$$\begin{aligned} U'^{(0)}(t,t_0) &= 1, \\ U'^{(1)}(t,t_0) &= -\frac{i}{\hbar}\lambda\int_{t_0}^{t} d\tau_1 h_I(\tau_1), \\ U'^{(2)}(t,t_0) &= \left(\frac{i}{\hbar}\right)^2\lambda^2\int_{t_0}^{t} d\tau_2 h_I(\tau_2)\int_{t_0}^{\tau_2} d\tau_1 h_I(\tau_1), \end{aligned} \tag{13}$$

or, in general,

$$U'(t,t_0) = 1 + \sum_{n=1}^{\infty}\left(\frac{-i}{\hbar}\right)^n\lambda^n I_n, \tag{14}$$

with

$$I_n = \int_{t_0}^{t} d\tau_n h_I(\tau_n)\int_{t_0}^{\tau_n} d\tau_{n-1}h_I(\tau_{n-1})\int_{t_0}^{\tau_{n-1}} d\tau_{n-2}h_I(\tau_{n-2})$$

$$\times \int_{t_0}^{\tau_{n-2}} d\tau_{n-3} h_I(\tau_{n-3}) \int_{t_0}^{\tau_{n-3}} d\tau_{n-4} \cdots \cdots \int_{t_0}^{\tau_2} d\tau_1 h_I(\tau_1), \tag{15}$$

where

$$h_I(\tau_k) = e^{+\frac{i}{\hbar} H_0(\tau_k - t_0)} h(\tau_k) e^{-\frac{i}{\hbar} H_0(\tau_k - t_0)}. \tag{16}$$

We are assuming the eigenvectors, $|a\rangle, |b\rangle, \cdots, |m\rangle, \cdots$, of H_0 are known, together with their eigenvalues, $E_a^{(0)}, E_b^{(0)}, \cdots$, where

$$H_0 |b\rangle = E_b^{(0)} |b\rangle. \tag{17}$$

The type of question we want to answer is the following: Given our quantum system was initially in a state $|\psi(t_0)\rangle = |i\rangle$, where $|i\rangle$ is an eigenstate of H_0 with energy $E_i^{(0)}$, what is the probability the system, as a result of the time-dependent perturbation, be in the final state, $|\psi(t)\rangle = |k\rangle$, which is a different eigenstate of H_0 ? For this purpose, we need to calculate

$$\begin{aligned}
\langle k|\psi(t)\rangle &= \langle k|U^{(0)}(t, t_0) U'(t, t_0)|i\rangle \\
&= e^{-\frac{i}{\hbar} E_k^{(0)}(t - t_0)} \langle k|U'(t, t_0)|i\rangle \\
&= e^{-\frac{i}{\hbar} E_k^{(0)}(t - t_0)} \sum_n \lambda^n \langle k|U'^{(n)}(t, t_0)|i\rangle \\
&= e^{-\frac{i}{\hbar} E_k^{(0)}(t - t_0)} \sum_n \left(\frac{-i}{\hbar}\right)^n \lambda^n \langle k|I_n|i\rangle.
\end{aligned} \tag{18}$$

The needed probability is of course the square of the absolute value of this matrix element, and we shall be interested in this number in the limit in which the final time, t, is such that $(t - t_0)$ is large compared with the characteristic periods of the quantum system.

To first order in λ, we have

$$\begin{aligned}
&\langle k|U^{(0)}(t, t_0) U'(t, t_0)|i\rangle \\
&= -\frac{i}{\hbar} \lambda e^{-\frac{i}{\hbar} E_k^{(0)}(t - t_0)} \int_{t_0}^{t} d\tau_1 e^{+\frac{i}{\hbar} E_k^{(0)}(\tau_1 - t_0)} \langle k|h(\tau_1)|i\rangle e^{-\frac{i}{\hbar} E_i^{(0)}(\tau_1 - t_0)} \\
&= -\frac{i}{\hbar} \lambda \int_{t_0}^{t} d\tau_1 e^{-\frac{i}{\hbar} E_k^{(0)}(t - \tau_1)} \langle k|h(\tau_1)|i\rangle e^{-\frac{i}{\hbar} E_i^{(0)}(\tau_1 - t_0)}.
\end{aligned} \tag{19}$$

By inserting a set of unit operators of the type $\sum_a |a\rangle\langle a|$, the nth order term can be written as

$$\begin{aligned}
&\langle k|U^{(0)}(t, t_0) U'^{(n)}(t, t_0)|i\rangle = \lambda^n \left(\tfrac{-i}{\hbar}\right)^n \times \\
&\textstyle\sum_{a,b,\cdots,u} \int_{t_0}^{t} d\tau_n e^{-\frac{i}{\hbar} E_k^{(0)}(t - \tau_n)} \langle k|h(\tau_n)|a\rangle \int_{t_0}^{\tau_n} d\tau_{n-1} e^{-\frac{i}{\hbar} E_a^{(0)}(\tau_n - \tau_{n-1})} \langle a|h(\tau_{n-1})|b\rangle \\
&\textstyle\int_{t_0}^{\tau_{n-1}} d\tau_{n-2} e^{-\frac{i}{\hbar} E_b^{(0)}(\tau_{n-1} - \tau_{n-2})} \langle b|h(\tau_{n-2})|c\rangle \int_{t_0}^{\tau_{n-2}} d\tau_{n-3} e^{-\frac{i}{\hbar} E_c^{(0)}(\tau_{n-2} - \tau_{n-3})} \\
&\times \langle c|h(\tau_{n-3})|d\rangle \cdots \cdots \int_{t_0}^{\tau_2} d\tau_1 e^{-\frac{i}{\hbar} E_u^{(0)}(\tau_2 - \tau_1)} \langle u|h(\tau_1)|i\rangle e^{-\frac{i}{\hbar} E_i^{(0)}(\tau_1 - t_0)}.
\end{aligned} \tag{20}$$

The various factors of this n^{th} order term can be remembered via a diagram of the type used by A. Messiah, *Quantum Mechanics. Vol. II*, New York: John Wiley,

1965, and illustrated in Fig. 57.1 via a sixth-order diagram, with time running upward, $t > \tau_n > \tau_{n-1} > \cdots > \tau_2 > \tau_1 > t_0$. Each vertex with its wiggly line stands for the matrix element at that time, e.g., at time τ_n, the matrix element is $\langle k|h(\tau_n)|a\rangle$. Each solid line, e.g., the line labeled b, connecting times τ_{n-1} and τ_{n-2} stands for an exponential, in this case $e^{-\frac{i}{\hbar}E_b^{(0)}(\tau_{n-1}-\tau_{n-2})}$.

A First-Order Probability Amplitude: The Golden Rule

Let us, as a first example, calculate the probability amplitude a quantum system initially in the state $|i\rangle$ at time, t_0, be in the state $|f\rangle$ at a later time, t, assuming the perturbation, $\lambda h(t)$, is weak enough that first-order perturbation theory is sufficient, and even more, choosing first a time-independent perturbation, λh. Then, the general

$$\langle f|U(t,t_0)|i\rangle = -\frac{i}{\hbar}\lambda\int_{t_0}^{t} d\tau e^{-\frac{i}{\hbar}E_f^{(0)}(t-\tau)}\langle f|h(\tau)|i\rangle e^{-\frac{i}{\hbar}E_i^{(0)}(\tau-t_0)} \tag{21}$$

leads to

$$|\langle f|U(t,t_0)|i\rangle| = \frac{\lambda}{\hbar}|\langle f|h|i\rangle\int_{t_0}^{t} d\tau e^{+\frac{i}{\hbar}(E_f^{(0)}-E_i^{(0)})\tau}|, \tag{22}$$

where we have used the assumed time-independence of h and have taken the absolute value of this transition probability amplitude, because the quantity of physical interest is the square of this absolute value. Then,

$$\begin{aligned}
|\langle f|U(t,t_0)|i\rangle| &= \frac{\lambda}{\hbar}|\langle f|h|i\rangle\int_{t_0}^{t} d\tau e^{i\omega_{fi}\tau}| \\
&= \frac{\lambda}{\hbar}\left|\langle f|h|i\rangle\frac{(e^{i\omega_{fi}(t-t_0)}-1)}{i\omega_{fi}}e^{i\omega_{fi}t_0}\right| \\
&= \frac{\lambda}{\hbar}\left|\langle f|h|i\rangle\frac{\sin[\frac{\omega_{fi}}{2}(t-t_0)]}{(\frac{\omega_{fi}}{2})}e^{i\frac{\omega_{fi}}{2}(t+t_0)}\right|,
\end{aligned} \tag{23}$$

so

$$|\langle f|U(t,t_0)|i\rangle|^2 = \frac{\lambda^2}{\hbar^2}|\langle f|h|i\rangle|^2\frac{\sin^2[\frac{\omega_{fi}}{2}(t-t_0)]}{(\frac{\omega_{fi}}{2})^2}. \tag{24}$$

We shall be interested in this quantity in the limit when the time interval $(t-t_0)$ is large compared with characteristic periods of the quantum system. We shall therefore go to the macroscopic time limit $(t-t_0)\to\infty$. Recalling the Dirac delta function can be given by

$$\delta(\omega) = \lim_{t\to\infty}\frac{1}{2\pi}\int_{-t}^{+t} dt' e^{i\omega t'} = \lim_{t\to\infty}\frac{\sin\omega t}{\pi\omega}, \tag{25}$$

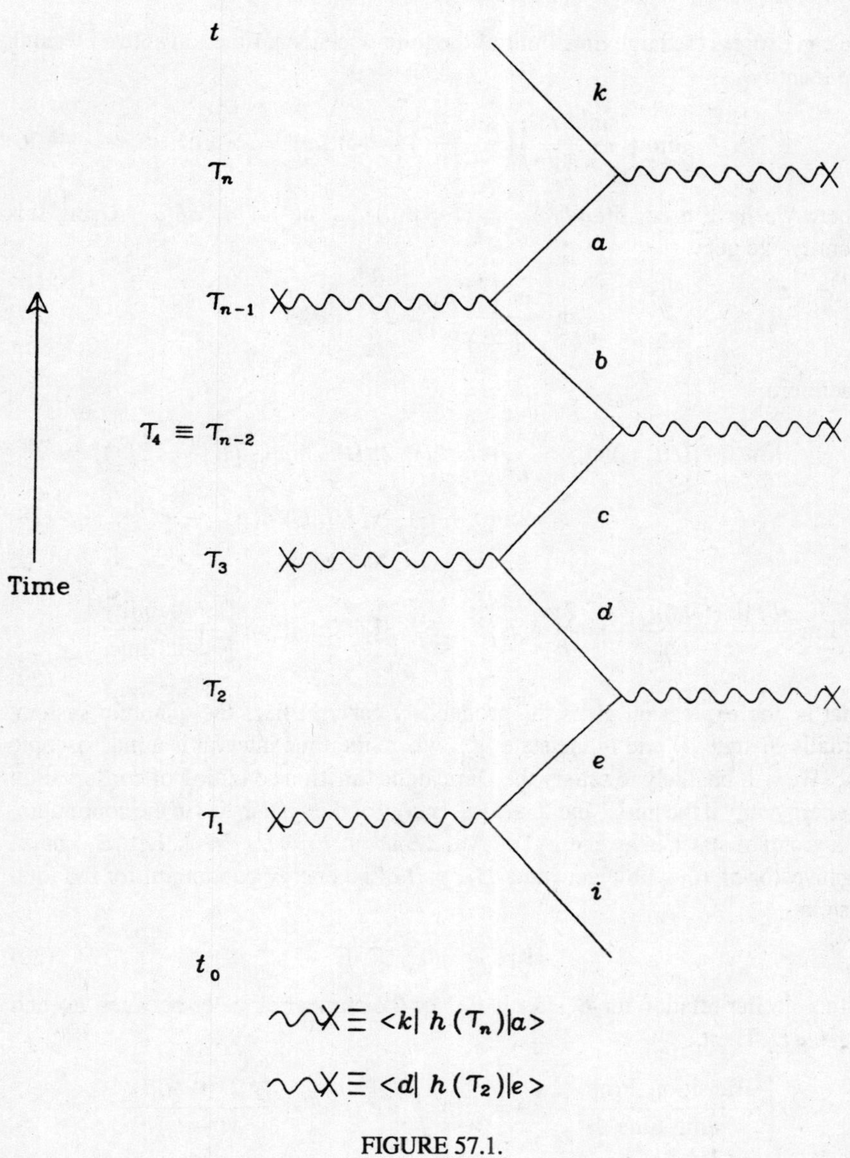

FIGURE 57.1.

we can express the large time limit of the time-dependent function above by using the identity

$$\lim_{t\to\infty}\left(\frac{\sin[\frac{\omega}{2}t]}{\pi\frac{\omega}{2}}\right)\left(\frac{\sin[\frac{\omega}{2}t]}{\frac{\omega}{2}t}\right)=\delta(\tfrac{1}{2}\omega)=2\delta(\omega), \tag{26}$$

where we have used: $\delta(\omega)f(\omega) = \delta(\omega)f(0)$ and $\delta(\frac{1}{a}\omega) = a\delta(\omega)$. Using this identity, we get

$$\lim_{t\to\infty}\frac{\sin^2[\frac{\omega_{fi}}{2}t]}{(\frac{\omega_{fi}}{2})^2}=2\pi t\delta(\omega_{fi}). \tag{27}$$

Therefore,

$$\begin{aligned}\lim_{t\to\infty}|\langle f|U(t,t_0)|i\rangle|^2 &= \frac{\lambda^2}{\hbar^2}|\langle f|h|i\rangle|^2 2\pi(t-t_0)\delta\Big(\frac{1}{\hbar}(E_f^{(0)}-E_i^{(0)})\Big)\\ &= 2\pi\hbar(t-t_0)\frac{\lambda^2}{\hbar^2}|\langle f|h|i\rangle|^2\delta(E_f^{(0)}-E_i^{(0)}),\end{aligned} \tag{28}$$

or

$$\lim_{t\to\infty}\frac{|\langle f|U(t,t_0)|i\rangle|^2}{(t-t_0)}=\frac{2\pi}{\hbar}\delta(E_f^{(0)}-E_i^{(0)})|\langle f|\lambda h|i\rangle|^2=\left[\frac{\text{probability}}{\text{unit time}}\right]_{i\to f}. \tag{29}$$

That is, the expression gives the probability per unit time the quantum system, initially in state $|i\rangle$ end in the state $|f\rangle$, where the time interval is a macroscopic one. We will be likely to satisfy the Dirac delta function condition of conservation of energy only if the initial and final total energy of the system lies in the continuum, so a group of states in an energy interval dE about $E = E_f$ exists. Let $|E_f\rangle$ be an eigenvector of H_0 with eigenvalue E_f, part of an energy continuum for the total system.

$$H_0|E_f\rangle = E_f|E_f\rangle, \tag{30}$$

with a similar relation for E_i. Let $\rho(E_f)$ be the number of states between E_f and $E_f + dE$. Then,

$$\begin{aligned}\left[\frac{\text{Transition Prob.}}{\text{unit time}}\right]_{i\to f} &= \lim_{t\to\infty}\int dE_f\rho(E_f)\frac{|\langle f|U(t,t_0)|i\rangle|^2}{(t-t_0)}\\ &= \frac{2\pi}{\hbar}\int dE_f\rho(E_f)|\langle f|\lambda h|i\rangle|^2\delta(E_f-E_i),\end{aligned} \tag{31}$$

so, with $E_f = E_i$,

$$\left[\frac{\text{Transition Probability}}{\text{second}}\right]_{i\to f}=\frac{2\pi}{\hbar}\rho(E_i)|\langle f|\lambda h|i\rangle|^2. \tag{32}$$

This important result was named Golden Rule II by Fermi. Fermi claimed almost all important results in quantum theory could be obtained from the two Golden Rules. (Recall Golden Rule I was the energy correction, through second order, using stationary state perturbation theory.)

B Application: The First Born Approximation of Scattering Theory Revisited

As a first simple test of our result, let us rederive the differential cross section for the scattering of a point projectile from a point target particle interacting via a potential that is weak compared with the kinetic energy of the relative motion. The Hamiltonian for this system is

$$H = H_0 + V(\vec{r}), \qquad \text{where now } V = \lambda h, \tag{33}$$

$$H_0|\phi_{\vec{k}}\rangle = E_k|\phi_{\vec{k}}\rangle = \frac{\hbar^2 k^2}{2\mu}|\phi_{\vec{k}}\rangle, \qquad \text{with } |\phi_{\vec{k}}\rangle \equiv |\vec{k}\rangle, \tag{34}$$

with

$$\phi_{\vec{k}}(\vec{r}) = \langle\vec{r}|\vec{k}\rangle = \frac{e^{i\vec{k}\cdot\vec{r}}}{(2\pi)^{\frac{3}{2}}}. \tag{35}$$

With this normalization, the unit operator is $\int d\vec{k}|\vec{k}\rangle\langle\vec{k}|$, and the scalar product $\langle\vec{k}'|\vec{k}\rangle = \delta(\vec{k} - \vec{k}')$ has the standard Dirac delta function form. The number of states in this continuum between $\vec{k}$ and $\vec{k}+d\vec{k}$ is then given by the volume element

$$dk_x dk_y dk_z = k^2 dk \sin\theta_k d\theta_k d\phi_k,$$

where k is the magnitude of $\vec{k}$, and θ_k, ϕ_k give the direction of $\vec{k}$. This must now be transformed to the form $\rho(E_f)dE_f$, so we can use Golden Rule II. From

$$E_f = \frac{\hbar^2 k_f^2}{2\mu_f}, \qquad dE_f = \frac{\hbar^2}{\mu_f} k_f dk_f, \tag{36}$$

so

$$dk_x dk_y dk_z = k_f^2 dk_f d\Omega_f = k_f \frac{\mu_f}{\hbar^2} dE_f d\Omega_f = p_f \frac{\mu_f}{\hbar^3} dE_f d\Omega_f = \rho(E_f) dE_f, \tag{37}$$

and the density of states factor is given by

$$\rho(E_f) = k_f \frac{\mu_f}{\hbar^2} d\Omega_f = p_f \frac{\mu_f}{\hbar^3} d\Omega_f = [2\mu_f E_f]^{\frac{1}{2}} \frac{\mu_f}{\hbar^3} d\Omega_f, \tag{38}$$

where $d\Omega_f$ is the element of solid angle about the direction of the scattering vector $\vec{k}_f$. Now, Golden Rule II says

$$\left[\frac{\text{Transition probability}}{\text{sec.}}\right]_{\vec{k}_i \to \vec{k}_f} = \frac{\text{Nr. of part. scattered into } d\Omega_f}{\text{sec.}}$$
$$= \frac{2\pi}{\hbar} \frac{k_f \mu_f d\Omega_f}{\hbar^2} |\langle\vec{k}_f|V|\vec{k}_i\rangle|^2. \tag{39}$$

To get the differential cross section, we need to divide this quantity by the magnitude of the incident flux. With $\langle\vec{r}|\vec{k}_i\rangle = 1/(2\pi)^{\frac{3}{2}} e^{i\vec{k}_i\cdot\vec{r}}$, the incident flux has the magnitude

$$|\vec{S}_i| = \frac{\hbar k_i}{\mu_i} \frac{1}{(2\pi)^3},$$

so

$$d\sigma = \frac{(\text{Transition Prob./sec.})_{\vec{k}_i \to \vec{k}_f}}{\text{Inc. Flux}} = \frac{2\pi}{\hbar} \frac{k_f \mu_f d\Omega_f}{\hbar^2} \frac{\mu_i (2\pi)^3}{\hbar k_i} |\langle \vec{k}_f | V | \vec{k}_i \rangle|^2$$

$$= \left(\frac{2\pi}{\hbar}\right)^4 \mu_f \mu_i \frac{k_f}{k_i} d\Omega_f \left| \int d\vec{r}\,' \frac{e^{-i\vec{k}_f \cdot \vec{r}\,'}}{(2\pi)^{\frac{3}{2}}} V(\vec{r}\,') \frac{e^{i\vec{k}_i \cdot \vec{r}\,'}}{(2\pi)^{\frac{3}{2}}} \right|^2, \tag{40}$$

and

$$\frac{d\sigma}{d\Omega} = \frac{\mu_f \mu_i}{(2\pi\hbar^2)^2} \frac{k_f}{k_i} \left| \int d\vec{r}\,' e^{-i\vec{k}_f \cdot \vec{r}\,'} V(\vec{r}\,') e^{i\vec{k}_i \cdot \vec{r}\,'} \right|^2. \tag{41}$$

For the purely elastic scattering process, of course, $k_f = k_i$ and $\mu_f = \mu_i$. This equation agrees with our earlier first Born approximation result. Also, the time-dependent perturbation theory result is easily adapted to the case of a rearrangement collision, where $\mu_f \neq \mu_i$ and $k_f \neq k_i$.

C Box Normalization: An Alternative Approach

It may be useful to show that the first Born approximation result also follows directly from the Golden Rule II, if we use the finely discrete spectrum of box normalization in place of the true continuous spectrum of our plane wave states. For a cubical macroscopic laboratory of side L, we have the eigenfunctions of H_0, given by

$$\phi_i(\vec{r}) = \frac{e^{i\vec{k}_i \cdot \vec{r}}}{L^{\frac{3}{2}}}, \tag{42}$$

with $\vec{k}_i$ and $\vec{k}_f$ both given by a triple of integers $(n_1, n_2, n_3) \equiv \vec{n}$, with n_s, integers, both positive and negative, such that

$$k_x = \frac{2\pi}{L} n_1, \quad k_y = \frac{2\pi}{L} n_2, \quad k_z = \frac{2\pi}{L} n_3, \quad \text{or } \vec{k} = \frac{2\pi}{L} \vec{n}, \tag{43}$$

with

$$k^2 = \left(\frac{2\pi}{L}\right)^2 \vec{n} \cdot \vec{n} = \left(\frac{2\pi}{L}\right)^2 (n_1^2 + n_2^2 + n_3^2) = \left(\frac{2\pi}{L}\right)^2 n^2(k). \tag{44}$$

In n space, each allowed state given by the triple of integers, (n_1, n_2, n_3), takes a cubical volume, where the elementary cube has a side of length 1, hence, a volume of 1^3. For large $n(k)$, the total number of discrete states with $k' < k$ is then given approximately by the volume of a sphere of radius $n(k)$

$$\text{Nr. of states with } k' < k = \frac{4\pi}{3} n^3 = \frac{4\pi}{3} \left(\frac{L}{2\pi}\right)^3 k^3, \tag{45}$$

and the total number of discrete states between k and $k + dk$ is

$$N(k)dk = 4\pi\left(\frac{L}{2\pi}\right)^3 k^2 dk. \tag{46}$$

The total number of discrete states between k and $k + dk$, but directions in an element of solid angle $d\Omega$ about a specific direction, θ, ϕ, of the $\vec{k}$ vector, is

$$N(k)dk\frac{d\Omega}{4\pi} = \left(\frac{L}{2\pi}\right)^3 k^2 dk d\Omega = \left(\frac{L}{2\pi}\right)^3 k\frac{\mu}{\hbar^2} d\Omega dE, \tag{47}$$

so

$$\rho(E_f) = \left(\frac{L}{2\pi}\right)^3 \left(\frac{k_f \mu_f}{\hbar^2}\right) d\Omega_f. \tag{48}$$

Now,

$$\left[\frac{\text{Transition Prob.}}{\text{sec.}}\right]_{i\to f} = \frac{2\pi}{\hbar}\rho(E_f) \left| \int d\vec{r}\,'\phi_f^*(\vec{r}\,')V(\vec{r}\,')\phi_i(\vec{r}\,') \right|^2. \tag{49}$$

$$\text{With} \qquad \phi_i(\vec{r}) = \frac{e^{i\vec{k}_i\cdot\vec{r}}}{L^{\frac{3}{2}}}, \tag{50}$$

the magnitude of the incident flux is given by

$$\frac{\hbar k_i}{\mu_i}\frac{1}{L^3},$$

so

$$d\sigma = \frac{2\pi}{\hbar}\left(\frac{L}{2\pi}\right)^3 \left(\frac{k_f\mu_f}{\hbar^2}\right) d\Omega_f \left(\frac{\mu_i L^3}{\hbar k_i}\right) \left| \int d\vec{r}\,'\frac{e^{-i\vec{k}_f\cdot\vec{r}\,'}}{L^{\frac{3}{2}}} V(\vec{r}\,')\frac{e^{i\vec{k}_i\cdot\vec{r}\,'}}{L^{\frac{3}{2}}} \right|^2, \tag{51}$$

$$\frac{d\sigma}{d\Omega} = \frac{\mu_f\mu_i}{(2\pi\hbar^2)^2}\frac{k_f}{k_i} \left| \int d\vec{r}\,' e^{-i\vec{k}_f\cdot\vec{r}\,'} V(\vec{r}\,') e^{i\vec{k}_i\cdot\vec{r}\,'} \right|^2, \tag{52}$$

in agreement with the earlier results. We note, in particular, the volume L^3 has disappeared from the final result, as it must!

D Second-Order Effects

Second-order terms may be particularly significant if the matrix element of the perturbation, λh, between states i and f happen to vanish. From our diagram, the second-order contribution to $\langle f|U(t, t_0)|i\rangle$ is given by

$$\langle f|U^{(2)}(t, t_0)|i\rangle = \frac{(-i)^2\lambda^2}{\hbar^2}\sum_a \int_{t_0}^{t} d\tau_2 e^{-\frac{i}{\hbar}E_f^{(0)}(t-\tau_2)}\langle f|h(\tau_2)|a\rangle$$
$$\times \int_{t_0}^{\tau_2} d\tau_1 e^{-\frac{i}{\hbar}E_a^{(0)}(\tau_2-\tau_1)}\langle a|h(\tau_1)|i\rangle e^{-\frac{i}{\hbar}E_i^{(0)}(\tau_1-t_0)}. \tag{53}$$

We shall again start with the special case, where h is assumed to be independent of the time. Then,

$$\begin{aligned}
&\langle f|U^{(2)}(t,t_0)|i\rangle \frac{\hbar^2}{(-i)^2\lambda^2} e^{+\frac{i}{\hbar}(E_f^{(0)}t - E_i^{(0)}t_0)} \\
&= \sum_a \langle f|h|a\rangle\langle a|h|i\rangle \int_{t_0}^{t} d\tau_2 e^{i\omega_{fa}\tau_2} \int_{t_0}^{\tau_2} d\tau_1 e^{i\omega_{ai}\tau_1} \\
&= \sum_a \langle f|h|a\rangle\langle a|h|i\rangle \int_{t_0}^{t} d\tau_2 e^{i\omega_{fa}\tau_2} \frac{\left(e^{i\omega_{ai}\tau_2} - e^{i\omega_{fi}t_0}\right)}{i\omega_{ai}} \\
&= \sum_a \frac{\langle f|h|a\rangle\langle a|h|i\rangle}{i\omega_{ai}} \left[\frac{\left(e^{i\omega_{fi}t} - e^{i\omega_{fi}t_0}\right)}{i\omega_{fi}} - \frac{\left(e^{i\omega_{fa}(t-t_0)} - 1\right)}{i\omega_{fa}} e^{i(\omega_{fa}+\omega_{fi})t_0} \right].
\end{aligned} \tag{54}$$

Now the last time-dependent factor can be written as

$$\frac{\left(e^{i\omega_{fa}(t-t_0)} - 1\right)}{i\omega_{fa}} = e^{i\frac{\omega_{fa}}{2}(t-t_0)} \frac{\sin\left[\frac{\omega_{fa}}{2}(t-t_0)\right]}{\left(\frac{\omega_{fa}}{2}\right)}. \tag{55}$$

We shall deal first with the case where *non*zero matrix elements exist only between states f and a and i and a such that $\omega_{fa} \neq 0$ and $\omega_{ai} \neq 0$. In this case, the

$$\frac{\sin\left[\frac{\omega_{fa}}{2}(t-t_0)\right]}{\frac{\omega_{fa}}{2}}$$

term is proportional to $\approx \frac{1}{\omega_{fa}}$; i.e., it is a time of the order of a period of the quantum system, viz., the period associated with ω_{fa}. The term we shall retain, however, will be proportional to the macroscopic time difference $(t - t_0)$. Hence, we will be able to neglect this ω_{fa}-term. Now, we use (as before) for the ω_{fi} term

$$\frac{\left(e^{i\omega_{fi}t} - e^{i\omega_{fi}t_0}\right)}{i\omega_{fi}} = e^{i\frac{\omega_{fi}}{2}(t+t_0)} \frac{\sin\left[\frac{\omega_{fi}}{2}(t-t_0)\right]}{\left(\frac{\omega_{fi}}{2}\right)}. \tag{56}$$

The absolute value squared of this term will lead to the factor $2\pi\hbar(t-t_0)\delta(E_f^{(0)} - E_i^{(0)})$, as before. The additional factor $(-i)/\hbar$ arising from the second-order term in the perturbation expansion can be combined with the denominator factor, $i\omega_{ai}$, to give the denominator factor, $+(E_i^{(0)} - E_a^{(0)})$, so through second order the transition probability per unit time is given by

$$\frac{\left| \langle f|U^{(1)}(t,t_0) + U^{(2)}(t,t_0)|i\rangle \right|^2}{(t-t_0)} = \frac{2\pi}{\hbar}\delta(E_f^{(0)} - E_i^{(0)}) \left| \langle f|\lambda h|i\rangle + \sum_a \frac{\langle f|\lambda h|a\rangle\langle a|\lambda h|i\rangle}{(E_i^{(0)} - E_a^{(0)})} \right|^2. \tag{57}$$

Finally, combining the energy conservation delta function with the density of states factor, and renaming, $\lambda h = V$, we have the extension of Fermi's Golden

Rule through next order in the perturbation expansion

$$\left[\frac{\text{Transition Prob.}}{\text{second}}\right]_{i\to f} = \frac{2\pi}{\hbar}\rho(E_f)\left|\langle f|V|i\rangle + \sum_a \frac{\langle f|V|a\rangle\langle a|V|i\rangle}{(E_i^{(0)} - E_a^{(0)})}\right|^2, \quad (58)$$

where again, $E_f^{(0)} = E_i^{(0)}$, and it is assumed the matrix element $\langle a|V|i\rangle$ vanishes for $E_a^{(0)} = E_i^{(0)}$. Note the similarity between this result and the corresponding result for stationary-state perturbation theory, which was Fermi's other Golden Rule.

E Case 2: A Periodic Perturbation

So far, our perturbation, λh, has been assumed to be time independent. Let us now look at a truly time-dependent perturbation, of periodic form

$$h(t) = h(0)e^{-i\omega t} + h^{\dagger}(0)e^{i\omega t}, \quad (59)$$

where we assume h is hermitian. For simplicity, we shall assume $h^{\dagger}(0) = h(0)$. Now,

$$\langle f|U^{(1)}(t,t_0)|i\rangle = -i\frac{\lambda}{\hbar}\int_{t_0}^{t} d\tau e^{-\frac{i}{\hbar}E_f^{(0)}(t-\tau)}\langle f|h(0)|i\rangle(e^{-i\omega\tau} + e^{i\omega\tau})e^{-\frac{i}{\hbar}E_i^{(0)}(\tau-t_0)}, \quad (60)$$

so

$$\begin{aligned}
\left|\langle f|U^{(1)}(t,t_0)|i\rangle\right|\frac{\hbar}{\lambda} &= \left|\langle f|h(0)|i\rangle\int_{t_0}^{t} d\tau\left(e^{i(\omega_{fi}-\omega)\tau} + e^{i(\omega_{fi}+\omega)\tau}\right)\right| \\
&= \left|\langle f|h(0)|i\rangle\left[\frac{\left(e^{i(\omega_{fi}-\omega)t} - e^{i(\omega_{fi}-\omega)t_0}\right)}{i(\omega_{fi}-\omega)} + \frac{\left(e^{i(\omega_{fi}+\omega)t} - e^{i(\omega_{fi}+\omega)t_0}\right)}{i(\omega_{fi}+\omega)}\right]\right| \\
&= \left|\langle f|h(0)|i\rangle\left[e^{i\frac{1}{2}(\omega_{fi}-\omega)(t+t_0)}\left(\frac{\sin\left[\frac{(\omega_{fi}-\omega)}{2}(t-t_0)\right]}{\frac{(\omega_{fi}-\omega)}{2}}\right)\right.\right. \\
&\quad \left.\left. + e^{i\frac{1}{2}(\omega_{fi}+\omega)(t+t_0)}\left(\frac{\sin\left(\frac{(\omega_{fi}+\omega)}{2}(t-t_0)\right]}{\frac{(\omega_{fi}+\omega)}{2}}\right)\right]\right| \quad (61)
\end{aligned}$$

Now, for very large values of $(t - t_0)$, the first term is strongly peaked when $\omega_{fi} = \omega$, and the second term is negligible for this value of ω_{fi}. Conversely, for large values of $(t - t_0)$, the second term is strongly peaked when $\omega_{fi} = -\omega$, and the first term is negligible for this value of ω_{fi}. Thus,

$$\begin{matrix}\text{either} \\ \text{or}\end{matrix} \quad \left|\langle f|U(t,t_0)|i\rangle\right|^2 = \frac{\lambda^2}{\hbar^2}|\langle f|h(0)|i\rangle|^2 f\Big((\omega_{fi} \mp \omega), (t-t_0)\Big), \quad (62)$$

where the peaked function, $f(\Omega, t)$, has the limiting value

$$\lim_{t\to\infty} f(\Omega, t) = 2\pi t\delta(\Omega). \quad (63)$$

Thus, again renaming, $\lambda h(0) = V(0)$, we have now

$$\left[\frac{\text{Transition prob.}}{\text{second}}\right]_{i\to f} = \frac{2\pi}{\hbar}\Big(\delta(E_f^{(0)} - E_i^{(0)} - \hbar\omega) + \delta(E_f^{(0)} - E_i^{(0)} + \hbar\omega)\Big)|\langle f|V(0)|i\rangle|^2. \qquad (64)$$

If we consider ω to be a positive quantity, we can get a strong transition probability, either if $\hbar\omega = E_f^{(0)} - E_i^{(0)}$ or if $\hbar\omega = E_i^{(0)} - E_f^{(0)}$. [Note, in the case, when $V(0)^\dagger \neq V(0)$, the second term that applies for the case with $\hbar\omega = E_i^{(0)} - E_f^{(0)}$ would be proportional to the square of the absolute value of the matrix element of $V^\dagger(0)$, rather than $V(0)$.]

An example of a quantum system in which such an oscillatory perturbation might come into play is an atomic or molecular system with a magnetic moment, $\vec{\mu}$, in an external oscillating magnetic field $\vec{B}(t)$, with

$$\lambda h(t) = H_{\text{int.}} = -\vec{\mu}\cdot\vec{B}(t). \qquad (65)$$

In this case, we would make a Fourier analysis of the $\vec{B}$ field

$$\vec{B}(t) = \int_{-\infty}^{\infty} d\omega\rho(\omega)\vec{B}(0)e^{i\omega t} = \int_0^{\infty} d\omega\rho(\omega)\vec{B}(0)\big(e^{i\omega t} + e^{-i\omega t}\big), \qquad (66)$$

and use

$$\lim_{(t-t_0)\to\infty}\int_0^{\infty} d\omega f(\omega_{fi} - \omega, t - t_0)\rho(\omega)$$
$$= 2\pi(t - t_0)\int_0^{\infty} d\omega\delta(\omega_{fi} - \omega)\rho(\omega) = 2\pi(t - t_0)\rho(\omega_{fi}). \qquad (67)$$

Thus, the transition probability induced by the oscillating magnetic field will lead to

$$\left[\frac{\text{Transition Probability}}{\text{sec.}}\right]_{i\to f} = \frac{2\pi}{\hbar^2}\rho(\omega_{fi})|\langle f \mid \vec{\mu}\cdot\vec{B}(0)|i\rangle\,|^2. \qquad (68)$$

58

Oscillating Magnetic Fields: Magnetic Resonance

For a quantum system with a magnetic moment, $\vec{\mu}$, in an external magnetic field, $\vec{B}$, with

$$\vec{\mu} = \frac{e\hbar}{2mc} g_J \vec{J}, \tag{1}$$

we have a Hamiltonian with the perturbing term

$$H = H_0 - \vec{\mu} \cdot \vec{B}. \tag{2}$$

Here, $\vec{J}$ is the relevant dimensionless angular momentum vector. For an electron or a proton, we can use the Pauli $\vec{\sigma}$ operator, $\vec{J} = \frac{1}{2}\vec{\sigma}$. For a free electron the g factor is $g_s = 2$ (neglecting quantum electrodynamic corrections). This g factor may also be slightly modified because of the electron environment in a molecular or crystalline system. For the proton, the g factor is "anomalous" $g_s = 2(2.793)$. The relevant energies for electron and proton magnetic resonance experiments are given by

$$\hbar\omega(B) = \frac{e\hbar}{2mc} B g_s = -5.79 \times 10^{-5} eV\left(\frac{B}{\text{tesla}}\right) g_s \quad \text{for electrons,}$$

$$\hbar\omega(B) = \frac{e\hbar}{2mc} B g_s = +3.15 \times 10^{-8} eV\left(\frac{B}{\text{tesla}}\right) g_s \quad \text{for protons,}$$

so these are small perturbations for atomic or molecular systems.

In a typical magnetic resonance experiment, two magnetic fields are used:

(1) A steady, time-independent, $\vec{B}$-field, $\vec{B}_0$, where this field will be chosen to lie in the z direction, with $\hbar\omega(B_0) = \hbar\omega_0$.

(2) A perpendicular $\vec{B}$-field, oscillating with circular frequency, ω, and strength, $\vec{B}_1$, where we define ω_1 via $\hbar\omega(B_1) = \hbar\omega_1$. This field is often linearly polarized in the x direction, or circularly polarized in the x–y plane. Because we can make a linearly polarized field from a superposition of a right and left circularly polarized field, we shall treat only the case of a right circularly polarized oscillating $\vec{B}_1$- field (see Fig. 58.1) and choose our Hamiltonian as

$$\begin{aligned} H &= H_0 - \big[\hbar\omega_0 J_z + \hbar\omega_1\big(J_x \cos\omega t + J_y \sin\omega t\big)\big] \\ &= H_0 - \hbar\omega_0 J_z - \frac{\hbar\omega_1}{2}\big[J_+ e^{-i\omega t} + J_- e^{i\omega t}\big], \end{aligned} \tag{3}$$

where $J_\pm = (J_x \pm i J_y)$. Our earlier perturbation theory for an oscillating field can now be used, if $\omega_1 \ll \omega_0$, so we can include the steady field part of the Hamiltonian as part of our zeroth-order problem. [Note, in this case, $h(0)^\dagger \neq h(0)$.] If we choose the electron-resonance case for which the $m_s = +\frac{1}{2}$ level lies above the $m_s = -\frac{1}{2}$ level and assume at $t_0 = 0$ the system is in the upper state, at excitation energy $\hbar|\omega_0|$, i.e., $|i\rangle = |m_s = +\frac{1}{2}\rangle$ at $t = 0$, to first order in ω_1, the perturbation theory result would give

$$\begin{aligned} \langle -\tfrac{1}{2}|U^{(1)}(t,0)|+\tfrac{1}{2}\rangle &= \Big(\frac{-i}{\hbar}\Big)\Big(-\frac{\hbar\omega_1}{2}\Big)\langle -\tfrac{1}{2}|J_-|+\tfrac{1}{2}\rangle e^{i\frac{(\omega_{fi}+\omega)}{2}t} \frac{\sin\big[\frac{(\omega_{fi}+\omega)}{2}t\big]}{\frac{(\omega_{fi}+\omega)}{2}} \\ &= \frac{i\omega_1}{(|\omega_0|-\omega)} e^{-i(|\omega_0|-\omega)\frac{t}{2}} \sin\big[(|\omega_0|-\omega)\frac{t}{2}\big]. \end{aligned} \tag{4}$$

In this perturbation approach, the parameter of smallness is $\omega_1/(|\omega_0| - \omega)$. The resonance condition leading to the sharply peaked function for large time t in the limit $\omega = |\omega_0|$ is considered *after* an expansion in the perturbation parameter of smallness. The magnetic resonance problem, however, is so simple we can actually solve this time-dependent problem exactly and do not need to make a perturbation expansion in powers of $\omega_1/(|\omega_0| - \omega)$.

A Exact Solution of the Magnetic Resonance Problem

We shall be able to find an exact solution for the time evolution operator for the full Hamiltonian of eq. (3). The method involves a rotation operator rotating to the frame of reference of the rotating field, $\vec{B}_1$. To find $U(t,0)$ for the solution

$$|\psi(t)\rangle = U(t,0)|\psi(t_0 = 0)\rangle, \tag{5}$$

let us introduce a time-dependendent rotation operator, $R_z(t)$, which rotates the system about the z axis through an angle, ωt, and look for solutions of the time evolution operator in the form

$$U(t,0) = R_z(t)U'(t,0) = e^{-i\omega t J_z} U'(t,0), \tag{6}$$

where

$$-\frac{\hbar}{i}\frac{dU}{dt} = HU(t,0) \tag{7}$$

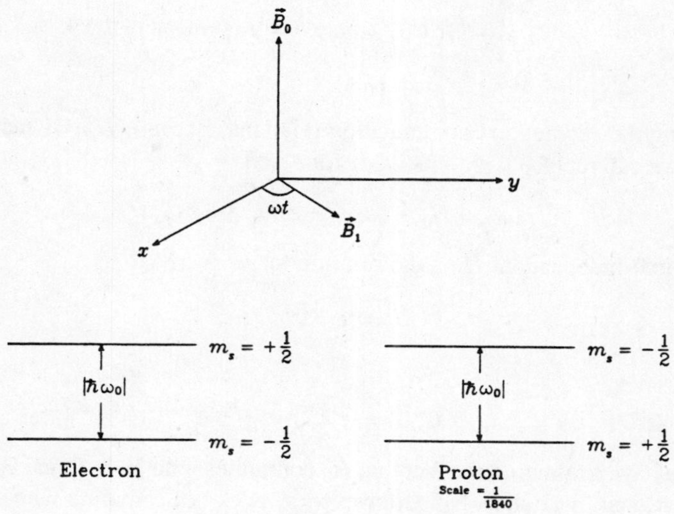

FIGURE 58.1. Magnetic resonance $\vec{B}$-fields.

now leads to

$$-\frac{\hbar}{i}R_z\frac{dU'}{dt} + \hbar\omega J_z R_z U' = H(R_z U'). \tag{8}$$

Left-multiplying this equation with $R_z^\dagger = R_z^{-1}$, and noting R_z commutes with J_z, this relation leads to

$$-\frac{\hbar}{i}\frac{dU'}{dt} = R_z^\dagger(H - \hbar\omega J_z)R_z U' = H_{\text{eff.}}U', \tag{9}$$

with

$$\begin{aligned} H_{\text{eff.}} &= e^{i\omega t J_z}\Big(H_0 - \big[\hbar(\omega_0 + \omega)J_z + \hbar\omega_1(J_x \cos\omega t + J_y \sin\omega t)\big]\Big)e^{-i\omega t J_z} \\ &= H_0 - \hbar(\omega_0 + \omega)J_z + \hbar\omega_1 e^{i\omega t J_z}(J_x \cos\omega t + J_y \sin\omega t)e^{-i\omega t J_z}, \end{aligned} \tag{10}$$

where we have assumed H_0 is rotationally invariant, so $[H_0, J_z] = 0$; i.e., H_0 commutes with any component of $\vec{J}$. We also assume H_0 is time independent. Now, however, we can also show the full $H_{\text{eff.}}$ is time independent by showing the operator

$$O = e^{i\omega t J_z}(J_x \cos\omega t + J_y \sin\omega t)e^{-i\omega t J_z} \tag{11}$$

is independent of time. To show this, note

$$\begin{aligned} \frac{dO}{dt} &= e^{i\omega t J_z}\Bigg((i\omega J_z)(J_x \cos\omega t + J_y \sin\omega t) \\ &\qquad + (J_x \cos\omega t + J_y \sin\omega t)(-i\omega J_z) \\ &\qquad + \omega(-J_x \sin\omega t + J_y \cos\omega t)\Bigg)e^{-i\omega t J_z} \\ &= e^{i\omega t J_z}\omega\Big(\Big(i(J_z J_x - J_x J_z) + J_y\Big)\cos\omega t \end{aligned}$$

$$+\Big(i(J_zJ_y - J_yJ_z) - J_x\Big)\sin\omega t\Big)e^{-i\omega t J_z}$$
$$= 0 \tag{12}$$

via the angular momentum commutation relations. Because $H_{\text{eff.}}$ is independent of time, we can replace it with its value for $t = 0$,

$$H_{\text{eff.}} = H_0 - [\hbar(\omega_0 + \omega)J_z + \hbar\omega_1 J_x]. \tag{13}$$

For this time-independent $H_{\text{eff.}}$, the solution for U' is trivial

$$U' = e^{-\frac{i}{\hbar}H_{\text{eff.}}t}, \tag{14}$$

and

$$U(t,0) = R_z(t)U'(t,0) = e^{-i\omega t J_z}e^{-\frac{i}{\hbar}H_0 t}e^{i[(\omega_0+\omega)J_z+\omega_1 J_x]t}, \tag{15}$$

where we have again used the fact that H_0 commutes with both J_z and J_x because of its rotational invariance. Note, however, J_z does not commute with J_x, so we cannot further simplify the final exponential in similar fashion. However, the linear combination $(\omega_0 + \omega)J_z + \omega_1 J_x$ can be reexpressed as

$$(\omega_0 + \omega)J_z + \omega_1 J_x = \Omega J_{z'} = \Omega\big(\cos\Theta J_z + \sin\Theta J_x)\big), \tag{16}$$

$$\text{with} \quad \Omega = \sqrt{(\omega_0 + \omega)^2 + \omega_1^2},$$

and

$$\cos\Theta = \frac{(\omega_0 + \omega)}{\sqrt{(\omega_0 + \omega)^2 + \omega_1^2}}, \qquad \sin\Theta = \frac{\omega_1}{\sqrt{(\omega_0 + \omega)^2 + \omega_1^2}},$$

where we have made a rotation through an angle, Θ, about the y axis (see Fig. 58.2) to obtain

$$J_{z'} = \cos\Theta J_z + \sin\Theta J_x \tag{17}$$

to express $U(t,0)$ through

$$U(t,0) = e^{-\frac{i}{\hbar}H_0 t}e^{-i\omega t J_z}e^{i\Omega t J_{z'}}. \tag{18}$$

For the special but very important case of a spin $s = \frac{1}{2}$ particle, with $\vec{J} \equiv \vec{S} = \frac{1}{2}\vec{\sigma}$, it is easy to expand $e^{i\Omega t J_{z'}}$, via

$$e^{\frac{i}{2}\Omega t\sigma_{z'}} = \cos\Big(\frac{\Omega t}{2}\Big)\mathbf{1} + i\sin\Big(\frac{\Omega t}{2}\Big)\sigma_{z'}, \tag{19}$$

where we have used

$$\sigma_{z'}^2 = \mathbf{1} = \sigma_{z'}^4 = \cdots, \qquad \sigma_{z'}^3 = \sigma_{z'} = \sigma_{z'}^5 = \cdots, \tag{20}$$

and

$$\sigma_{z'} = \cos\Theta\sigma_z + \sin\Theta\sigma_x. \tag{21}$$

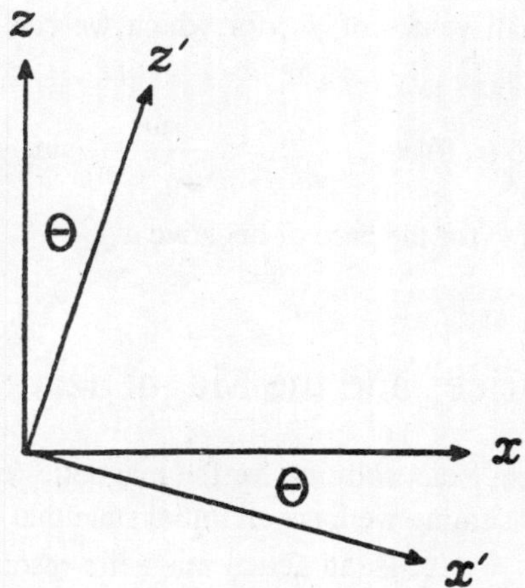

FIGURE 58.2. The parameter Θ of eq. (16).

With this simple expansion, we have for $s = \frac{1}{2}$ systems

$$U(t,0) = e^{-\frac{i}{\hbar}H_0 t} e^{-i\frac{\omega}{2}t\sigma_z}\Big(\cos\big(\frac{\Omega t}{2}\big)\mathbf{1} + i\sin\big(\frac{\Omega t}{2}\big)\big(\cos\Theta\sigma_z + \sin\Theta\sigma_x\big)\Big). \quad (22)$$

If the initial state is given by $|i\rangle = |\alpha\frac{1}{2}m_s^{(i)}\rangle$, where α is shorthand for all quantum numbers of H_0 other than s, m_s, the final state can differ only in the quantum number, m_s, and must be of the form $|f\rangle = |\alpha\frac{1}{2}m_s^{(f)}\rangle$, and

$$\Big|\ \langle\alpha\tfrac{1}{2}m_s^{(f)}|U(t,0)|\alpha\tfrac{1}{2}m_s^{(i)}\rangle\ \Big| = \langle\alpha\tfrac{1}{2}m_s^{(f)}|\cos\big(\frac{\Omega t}{2}\big)\mathbf{1} + i\sin\big(\frac{\Omega t}{2}\big)\Big(\cos\Theta\sigma_z + \sin\Theta\sigma_x\Big)|\alpha\tfrac{1}{2}m_s^{(i)}\rangle. \quad (23)$$

In particular,

$$\Big|\ \langle\alpha\ \tfrac{1}{2} - \tfrac{1}{2}|U(t,0)|\alpha\ \tfrac{1}{2} + \tfrac{1}{2}\rangle\ \Big| = \sin\Theta\sin\big(\frac{\Omega t}{2}\big) = \frac{\omega_1}{\Omega}\sin\big(\frac{\Omega t}{2}\big). \quad (24)$$

The probability for a spin-flip from $m_s = +\frac{1}{2}$ to $m_s = -\frac{1}{2}$ is given by

$$\Big[\text{Spin} - \text{flip Prob.}(t)\Big]_{+\frac{1}{2}\to-\frac{1}{2}} = \frac{\omega_1^2}{[(\omega_0+\omega)^2+\omega_1^2]}\sin^2\big[\tfrac{t}{2}\sqrt{(\omega_0+\omega)^2+\omega_1^2}\big]. \quad (25)$$

We see that, at resonance, with ω such that $(\omega_0+\omega) = 0$, this probability oscillates back and forth from zero to unity with a circular frequency of $\frac{1}{2}\omega_1$. Off resonance, with $(\omega_0+\omega) \neq 0$, this probability oscillates back and forth from zero to a value less than unity with the circular frequency $\frac{1}{2}\Omega$. The oscillating $\vec{B}$-field can induce transitions both from the excited state to the lower state and from the lower state back to the excited state.

Finally, for very small values of ω_1, for which we can expand in powers of $\omega_1/(\omega_0+\omega)$, we have

$$\Big| \langle \alpha\ \tfrac{1}{2} - \tfrac{1}{2} | U(t,0) | \alpha\ \tfrac{1}{2} + \tfrac{1}{2} \rangle \Big| \approx \frac{\omega_1}{(\omega_0+\omega)} \sin\big[\frac{t}{2}(\omega_0+\omega)\big], \tag{26}$$

which agrees with eq. (4) for the case of negative ω_0.

B Density Matrices and the Magnetization Vector

So far, we have found an exact solution for the magnetic-resonance Hamiltonian for an $s=\frac{1}{2}$-system, assuming we have an initial state that is an eigenstate of S_z, with $m_s=+\frac{1}{2}$, or $m_s=-\frac{1}{2}$. In an actual magnetic-resonance experiment, we have a system of many $s=\frac{1}{2}$ particles with different population probabilities for the states with $m_s=\pm\frac{1}{2}$. It will therefore again be advantageous to use a density matrix formulation. Suppose first the initial state at $t=0$ is a linear combination of the eigenstates $|m_s\rangle$, given by

$$|\psi(0)\rangle = \sum_{m_s} |m_s\rangle c_{m_s}(0). \tag{27}$$

Then, the expectation value of σ_α is given by

$$\begin{aligned}\langle\sigma_\alpha\rangle &= \sum_{m_s,m'_s} c^*_{m'_s}(0)\langle m'_s|\sigma_\alpha|m_s\rangle c_{m_s}(0) = \sum_{m_s,m'_s} \big(\sigma_\alpha\big)_{m'_s m_s} c_{m_s}(0) c^*_{m'_s}(0)\\ &= \text{trace}\big(\sigma_\alpha \rho^i(0)\big).\end{aligned} \tag{28}$$

Now, we define a magnetization vector (in place of a polarization vector, bearing in mind that this is merely a change of name),

$$\vec{M} = \sum_\alpha \vec{e}_\alpha \langle\sigma_\alpha\rangle, \tag{29}$$

where we have

$$\vec{M} = \sum_\alpha \vec{e}_\alpha \text{trace}(\sigma_\alpha \rho). \tag{30}$$

The concept of the density matrix, ρ, will again be most useful when we have a statistical distribution of many $s=\frac{1}{2}$ systems in a macroscopic sample, with an initial density matrix, at $t=0$, of the type

$$\rho^i = \sum_{n=1}^{N} w_n c^{(n)}_{m_s}(0) c^{(n)*}_{m'_s}(0), \tag{31}$$

where w_n gives the probability the nth system has the $c^{(n)}_{m_s}(0)$ and again

$$\sum_{n=1}^{N} w_n = 1. \tag{32}$$

For a system in thermal equilibrium, e.g., where the number of systems in the energy state E_{m_s} is proportional to the Boltzmann factor, $e^{-(E_{m_s}/kT)}$, we would have the density matrix

$$\rho = \frac{1}{\left(e^{-(E_{+\frac{1}{2}}/kT)} + e^{-(E_{-\frac{1}{2}}/kT)}\right)} \begin{pmatrix} e^{-(E_{+\frac{1}{2}}/kT)} & 0 \\ 0 & e^{-(E_{-\frac{1}{2}}/kT)} \end{pmatrix}. \tag{33}$$

From the density matrix at $t = 0$, we can determine the density matrix $\rho(t)$ at a later time, t. For the nth system, the time evolution gives

$$\begin{aligned} c_{m_s}^{(n)}(t) &= \sum_{m_s''} \Big(U(t,0)\Big)_{m_s m_s''} c_{m_s''}^{(n)}(0), \\ c_{m_s'}^{(n)*}(t) &= \sum_{m_s'''} \Big(U^{\dagger}(t,0)\Big)_{m_s''' m_s'} c_{m_s'''}^{(n)*}(0), \end{aligned} \tag{34}$$

so

$$\begin{aligned} \big(\rho(t)\big)_{m_s m_s'} &= \sum_n w_n c_{m_s}^{(n)}(t) c_{m_s'}^{(n)*}(t) = \sum_{m_s'', m_s'''} U_{m_s m_s''} \rho(0)_{m_s'' m_s'''} U^{\dagger}_{m_s''' m_s'} \\ &= \big(U\rho(0)U^{\dagger}\big)_{m_s m_s'}. \end{aligned} \tag{35}$$

To see how the magnetization evolves in time, we calculate $\langle \sigma_\alpha(t) \rangle$

$$\begin{aligned} \langle \sigma_\alpha(t) \rangle &= \langle \psi(t) | \sigma_\alpha | \psi(t) \rangle = \langle \psi(0) | U^{\dagger}(t,0) \sigma_\alpha U(t,0) | \psi(0) \rangle \\ &= \sum_{m_s m_s'} \big(U^{\dagger} \sigma_\alpha U\big)_{m_s' m_s} c_{m_s}(0) c_{m_s'}^{*}(0) = \text{trace}\big(U^{\dagger} \sigma_\alpha U \rho(0)\big), \end{aligned} \tag{36}$$

yielding

$$\vec{M}(t) = \sum_\alpha \vec{e}_\alpha \text{trace}\big(U^{\dagger} \sigma_\alpha U \rho(0)\big), \tag{37}$$

with a similar relation for a density matrix arising from a statistical distribution of states.

It should be remarked that eqs. (35) and (37) give the time evolution of the density matrix and the magnetization vector under the action of the $\vec{B}$-field of our Hamiltonian and assumes the different $s = \frac{1}{2}$ systems of our sample have no additional interactions with each other (spin–spin interactions), or with the medium in which they find themselves (spin–lattice interactions). Without such interactions, the time evolution of the density matrix, given by eq. (35), would lead to

$$\begin{aligned} -\frac{\hbar}{i}\frac{d\rho}{dt} &= -\frac{\hbar}{i}\left(\frac{dU}{dt}\right)\rho(0)U^{\dagger} - U\rho(0)\frac{\hbar}{i}\left(\frac{dU^{\dagger}}{dt}\right) = HU\rho(0)U^{\dagger} - U\rho(0)U^{\dagger}H \\ &= H\rho(t) - \rho(t)H = [H, \rho(t)]. \end{aligned} \tag{38}$$

The spin–spin and spin–lattice relaxation processes caused by the spin–spin and spin–lattice interactions are taken into account by adding a relaxation term to this equation, through a relaxation matrix, $(R)_{m_s m_s', m_s'' m_s'''}$,

$$\frac{d}{dt}(\rho)_{m_s m_s'} = -\frac{i}{\hbar}[H, \rho(t)]_{m_s m_s'} - \sum_{m_s'' m_s'''} (R)_{m_s m_s', m_s'' m_s'''} (\rho)_{m_s'' m_s'''}. \tag{39}$$

Without this additional relaxation term, eq. (37) gives the time evolution of the magnetization vector purely under the influence of the magnetic fields, $\vec{B}_0$ and $\vec{B}_1$.

C General Case with $J > \frac{1}{2}$

So far, we have concentrated on the $s = \frac{1}{2}$ systems. For $J \geq 1$, we could expand the operator, $e^{i\Omega t J_{z'}}$, by the technique used for the Pauli spin matrices. For $J = 1$, e.g., three independent 3×3 matrices are needed to expand this exponential, the 3×3 unit matrix, the 3×3 matrix for $J_{z'}$ and for $J_{z'}^2$, where we now use $J_{z'}^3 = J_{z'}$ and $J_{z'}^4 = J_{z'}^2$. For the case of arbitrary J, however, it is best to use the well-known matrix elements of the rotation operator, $R_y(\Theta)$, which effects a rotation through an angle Θ about the y axis. With this rotation operator, the operator $e^{i\Omega t J_z}$ is transformed into the needed $e^{i\Omega t J_{z'}}$ via

$$\begin{aligned} e^{i\Omega t J_{z'}} &= R_y(\Theta) e^{i\Omega t J_z} R_y^{-1}(\Theta) \\ &= e^{-i\Theta J_y} e^{i\Omega t J_z} e^{i\Theta J_y}, \end{aligned} \tag{40}$$

leading to the matrix elements

$$\begin{aligned} \langle JM'_J | e^{i\Omega t J_{z'}} | JM_J \rangle &= \sum_{M''_J} \langle JM'_J | e^{-i\Theta J_y} | JM''_J \rangle e^{i\Omega t M''_J} \langle JM''_J | e^{i\Theta J_y} | JM_J \rangle \\ &= \sum_{M''_J} d^J_{M'_J M''_J}(\Theta) e^{i\Omega t M''_J} d^J_{M''_J M_J}(-\Theta), \end{aligned} \tag{41}$$

where we have used the fact that the operator J_z is diagonal in the $|JM''_J\rangle$ basis and the d^J matrices are the rotation matrices for an Euler rotation with Euler angles, $\alpha = \gamma = 0$ and $\beta = \pm\Theta$ (see Chapter 29). As a simple test, let us rederive the matrix $\langle \frac{1}{2} m'_s | e^{i\Omega t S_{z'}} | \frac{1}{2} m_s \rangle$, for an $s = \frac{1}{2}$ system. Using the $d^{\frac{1}{2}}$ matrices, we have in this case

$$\begin{aligned} &\langle \tfrac{1}{2} m'_s | e^{i\Omega t S_{z'}} | \tfrac{1}{2} m_s \rangle \\ &= \begin{pmatrix} \cos\frac{\Theta}{2} & -\sin\frac{\Theta}{2} \\ \sin\frac{\Theta}{2} & \cos\frac{\Theta}{2} \end{pmatrix} \begin{pmatrix} e^{i\frac{\Omega}{2}t} & 0 \\ 0 & e^{-i\frac{\Omega}{2}t} \end{pmatrix} \begin{pmatrix} \cos\frac{\Theta}{2} & \sin\frac{\Theta}{2} \\ -\sin\frac{\Theta}{2} & \cos\frac{\Theta}{2} \end{pmatrix} \\ &= \begin{pmatrix} \left(\cos^2\frac{\Theta}{2} e^{i\frac{\Omega}{2}t} + \sin^2\frac{\Theta}{2} e^{-i\frac{\Omega}{2}t}\right) & \cos\frac{\Theta}{2}\sin\frac{\Theta}{2}\left(e^{i\frac{\Omega}{2}t} - e^{-i\frac{\Omega}{2}t}\right) \\ \cos\frac{\Theta}{2}\sin\frac{\Theta}{2}\left(e^{i\frac{\Omega}{2}t} - e^{-i\frac{\Omega}{2}t}\right) & \left(\sin^2\frac{\Theta}{2} e^{i\frac{\Omega}{2}t} + \cos^2\frac{\Theta}{2} e^{-i\frac{\Omega}{2}t}\right) \end{pmatrix} \\ &= \begin{pmatrix} \left(\cos\frac{\Omega t}{2} + i\sin\frac{\Omega t}{2}\cos\Theta\right) & i\sin\frac{\Omega t}{2}\sin\Theta \\ i\sin\frac{\Omega t}{2}\sin\Theta & \left(\cos\frac{\Omega t}{2} - i\sin\frac{\Omega t}{2}\cos\Theta\right) \end{pmatrix} \\ &= \cos\frac{\Omega t}{2}\mathbf{1} + i\sin\frac{\Omega t}{2}(\cos\Theta\sigma_z + \sin\Theta\sigma_x), \end{aligned} \tag{42}$$

which agrees with our earlier result.

59

Sudden and Adiabatic Approximations

So far, we have considered only time-independent and periodic time-dependent perturbations. One of the common time-dependent problems in quantum mechanics involves the turning on or off of a perturbing term, e.g., an external field. In general, the details of turning on or off of time-dependent perturbations are too difficult to solve. Two limiting cases exist, however, that are both important and that can be treated in detail with good accuracy: the sudden and the adiabatic approximations.

Consider a perturbation problem, with

$$H = H_0 + V, \tag{1}$$

where the perturbation, V, is off up to some time t_0 and is then turned on, so at some later time, t, with $t - t_0 = T$ the perturbation is on at full strength. In the limit for which $T \to 0$, we have the sudden approximation. In the limit $T \to \infty$, we have the adiabatic approximation.

A Sudden Approximation

We shall introduce the variable

$$s = \frac{(t - t_0)}{T}$$

such that the perturbation is off for $s = 0$ and is on at full strength for $s = 1$. With

$$U(t, t_0) = 1 - \frac{i}{\hbar}\int_{t_0}^{t} d\tau\, H(\tau)U(\tau, t_0)$$

$$= 1 - \frac{i}{\hbar} T \int_0^1 ds H(s) U_T(s). \tag{2}$$

In the limit $T \to 0$, we have

$$\lim_{T \to 0} U_T(t, t_0) = 1. \tag{3}$$

The perturbation is so sudden the quantum system cannot follow. The system stays in the state $|\psi(t_0)\rangle$ even when the perturbation is fully turned on. An example of such a sudden perturbation occurs in the beta decay of an atomic nucleus. An example of considerable recent interest is the beta decay of tritium, ${}^3_1H_2$, which is being reexamined very carefully in an attempt to pin down a possible nonzero value of the rest mass of the neutrino. The endpoint electrons are of particular interest, because the antineutrino then carries very little kinetic energy. Because the neutrino rest mass is probably much less than an atomic energy of $\sim$10eV, the exact atomic energy of the recoiling ${}^3_2He_1$ one-electron ion must be taken into account carefully. The original one-electron tritium atom had its atomic electron in an $1s$ ground state for this $Z = 1$ nucleus. The beta decay end point energy is 18.65 keV. The e^- emitted in this beta decay traverses the atomic dimension so rapidly the original atomic electron will still have the wave function of an $1s$ $Z = 1$ atom, even though it now finds itself in the Coulomb field of a $Z = 2$ nucleus. This state at $t - t_0$ will thus be a linear combination of one-electron states $|n\ l = 0\ m = 0\rangle$ of a $Z = 2$ one-electron ion, and the probability for all values of n will be needed for an accurate evaluation of the internal energy of the recoiling ${}^3_2He_1$ ion.

B Adiabatic Approximation: An Example: The Reversal of the Magnetic Field in a One-Electron Atom

As a second example, let us consider a one-electron atom in an external magnetic field reversed from a value, $+\vec{B}_0$, to a value, $-\vec{B}_0$, in a time interval, T. We shall consider both the sudden approximation, $T \to 0$, and the adiabatic approximation, $T \to \infty$, in particular, the latter, adiabatic case, in which we have to carry out some analysis. For simplicity, we will consider the Paschen–Back and Zeeman spectrum of an alkali atom, such as Na, to avoid the higher degeneracies of the hydrogen spectrum. The perturbing Hamiltonian consists of the external magnetic-field term and the internal magnetic-field spin-orbit term (see Chapter 26).

$$H = \beta(\vec{l} \cdot \vec{s}) + \hbar\omega_L(l_z + 2s_z), \tag{4}$$

where the Larmor frequency, ω_L, is given by

$$\hbar\omega_L(t) = \frac{|e|\hbar}{2mc} B_0(t), \qquad \text{with } B_z = B_0(t) = B_0\left(1 - \frac{2t}{T}\right). \tag{5}$$

We have reversed the field $\vec{B}$ linearly from a value $+B_0$ at $t = 0$ to a value, $-B_0$, at $t = T$ (see Fig. 59.1). The linear character of the variation is very simple. We could

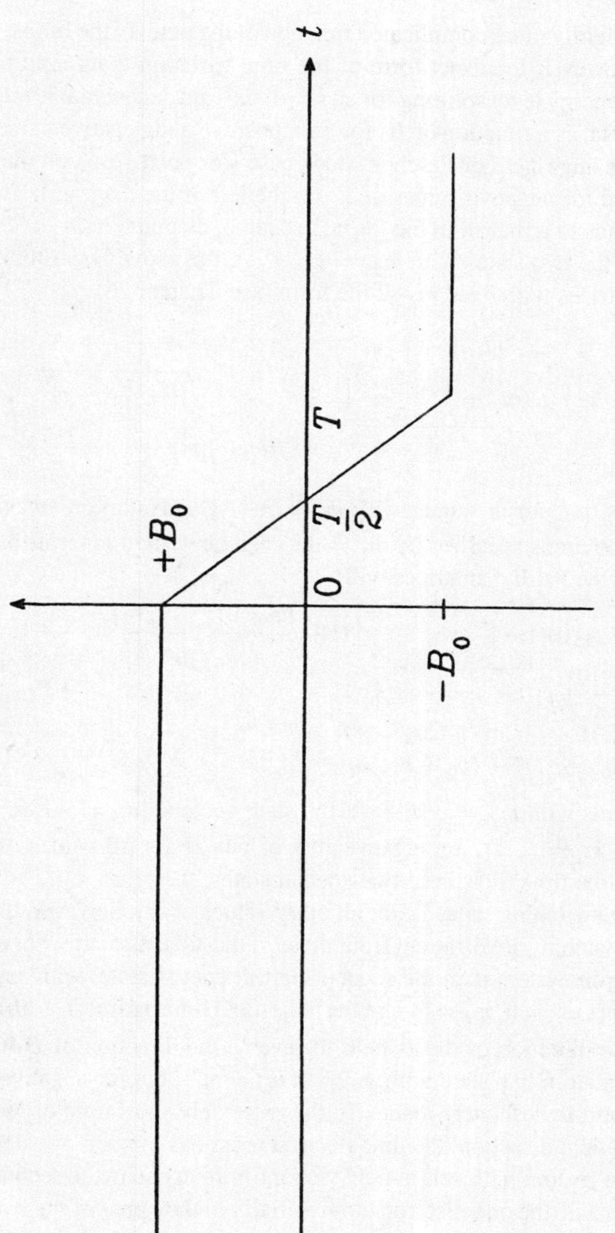

FIGURE 59.1. The linear B-field reversal of eq. (5).

have chosen a slightly more complicated function of the time. In the limits, $T \to 0$, or, $T \to \infty$, however, the exact form of the time variation is unimportant. Fig. 59.2 shows the energy level splitting for an $l = 1$ six-fold degenerate level, such as the $3p$ level in Na, as a function of B_0 for both positive and negative values of B_0, starting with the huge-field or Paschen–Back case (for positive B_z on the right of the diagram, and for negative values of B_z on the left of the diagram). The above Hamiltonian leads to a matrix in the $|m_l m_s\rangle$ scheme, diagonal in $m_j = m_l + m_s$, and, except for the two states with $m_j = \pm(l + \frac{1}{2})$, the above Hamiltonian leads to 2×2 submatrices with fixed m_j of the form (see Chapter 26)

$$\begin{array}{c} \\ m_s = +\frac{1}{2} \\ m_s = -\frac{1}{2} \end{array} \begin{array}{c} \begin{array}{cc} \qquad m_s = +\frac{1}{2} \qquad\qquad & \qquad\qquad m_s = -\frac{1}{2} \end{array} \\ \begin{pmatrix} \hbar\omega_L(t)(m_j + \frac{1}{2}) + \frac{\beta m_j}{2} - \frac{\beta}{4} & \frac{\beta}{2}\sqrt{(l + \frac{1}{2})^2 - m_j^2} \\ \frac{\beta}{2}\sqrt{(l + \frac{1}{2})^2 - m_j^2} & \hbar\omega_L(t)(m_j - \frac{1}{2}) - \frac{\beta m_j}{2} - \frac{\beta}{4} \end{pmatrix}, \end{array}$$

where $m_l = m_j \mp \frac{1}{2}$ for the states with $m_s = \pm\frac{1}{2}$. This Hamiltonian submatrix, with rows and columns specified by $m_s = \pm\frac{1}{2}$, can be written in terms of a 2×2 unit matrix and the Pauli σ matrices, via

$$\begin{aligned} H &= \Big[\hbar\omega_L(0)(1 - \frac{2t}{T})m_j - \frac{\beta}{4}\Big]\mathbf{1} + \Big[\frac{\hbar\omega_L(0)}{2}(1 - \frac{2t}{T}) + \frac{\beta m_j}{2}\Big]\sigma_z \\ &\quad + \frac{\beta}{2}\sqrt{[(l + \tfrac{1}{2})^2 - m_j^2]}\,\sigma_x \\ &= a(t)\mathbf{1} + b(t)\sigma_z + C\sigma_x \\ &= (a_0 + a_1 t)\mathbf{1} + (b_0 + b_1 t)\sigma_z + C\sigma_x. \end{aligned} \qquad (6)$$

For the two levels with $m_j = \pm(l + \frac{1}{2})$, the state vectors, $|m_l = +l, m_s = +\frac{1}{2}\rangle$ and $|m_l = -l, m_s = -\frac{1}{2}\rangle$, are eigenvectors of our H for all values of B_z and independent of the time. For these two special states, therefore, $U(t, 0) = 1$, for all values of the switching time, T. For all other values of m_j, however, the results of the sudden switching are different from those of the adiabatic case. For example, if the one-electron system is in the second-highest energy state, with asymptotic quantum numbers $m_l = 0$, $m_s = +\frac{1}{2}$ in the huge-field limit at time, $t = 0$, a sudden switching of the direction of the $\vec{B}$-field to reverse the direction of $\vec{B}$ will leave the one-electron atom in a state with $m_l = 0$, $m_s = +\frac{1}{2}$, but for negative B_0, this is now the second-lowest energy state. In the extremely short time of the sudden switching of $\vec{B}$-field direction, the one-electron atom has jumped over the energy gap between the $p_{\frac{3}{2}}$ and $p_{\frac{1}{2}}$ levels at field strength zero to end in the second-lowest energy state. That is, the one-electron atom initially in the upper of the two energy eigenstates with $m_j = +\frac{1}{2}$ will in the sudden process end in the lower of the two energy eigenstates with $m_j = +\frac{1}{2}$.

Conversely, if the direction of the $\vec{B}$-field is reversed very slowly (in the adiabatic limit for which $T \to \infty$), we shall show the one-electron system follows the second-highest energy level with $m_j = +\frac{1}{2}$ through the zero B-field $p_{\frac{3}{2}}$ level and then follows the $m_j = +\frac{1}{2}$ level to the huge negative B_z-field limit with asymptotic quantum numbers $m_l = +1$, $m_s = -\frac{1}{2}$. That is, we shall show the one-electron

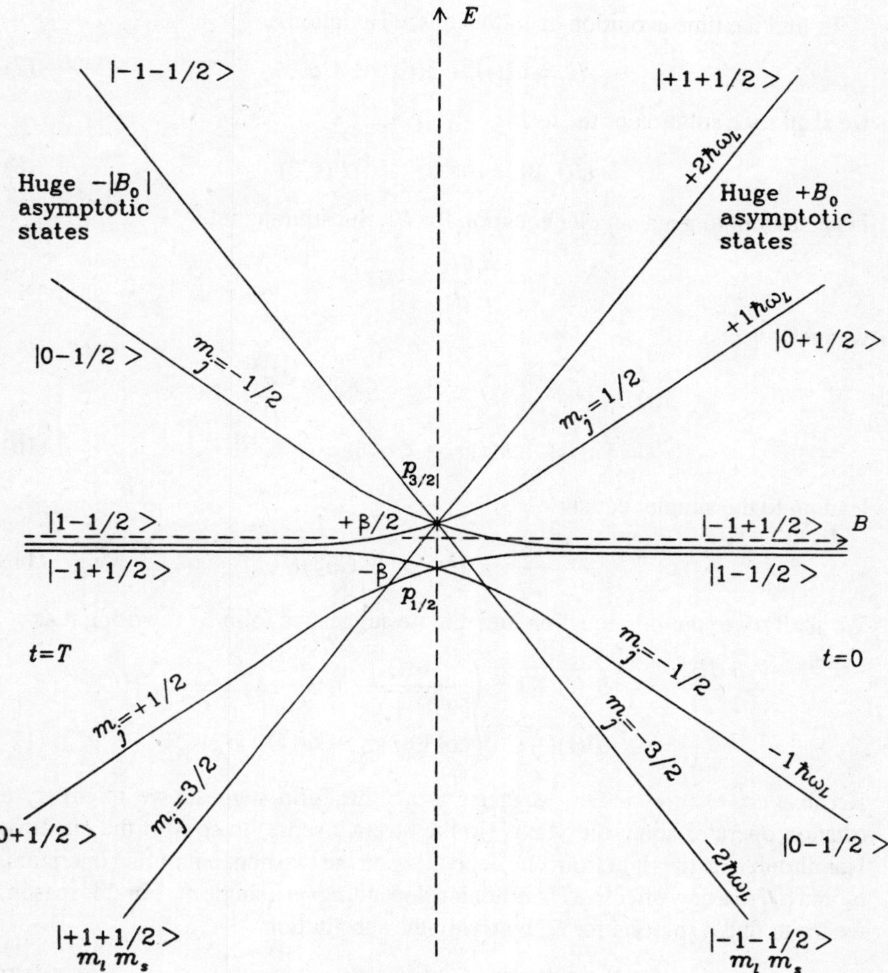

FIGURE 59.2. The B-field dependence of the fine-structure levels of a ^2P state.

system stays in the upper level of the two levels with $m_j = +\frac{1}{2}$ for all values of $\vec{B}$. The time evolution operator $U(t, 0)$ will in the adiabatic switching process create a state which at any t is that linear combination of the states $|m_l = m_j - \frac{1}{2}, m_s = +\frac{1}{2}\rangle$ and $|m_l = m_j + \frac{1}{2}, m_s = -\frac{1}{2}\rangle$, which is the energy eigenstate of the upper of the two levels with this particular m_j value, if the initial state at $t = 0$ was the upper of the two levels with this particular m_j value. Similarly, if the initial state at $t = 0$ was the lower of the two energy states with a particular m_j value, the time evolution operator, $U(t, 0)$, valid for an adiabatic switching process, will create a state at a later time t, which is still the lower of the two energy eigenstates with this m_j value corresponding to the energy valid for the B_z value for this particular t.

To find the time evolution operator for the Hamiltonian

$$H = a(t)\mathbf{1} + b(t)\sigma_z + C\sigma_x, \tag{7}$$

we shall try a solution of the form

$$U(t,0) = e^{-\frac{i}{\hbar}\int_0^t d\tau a(\tau)\mathbf{1}} U'(t,0) \tag{8}$$

in an attempt to get a simpler equation for U'. Substituting into

$$-\frac{\hbar}{i}\frac{dU}{dt} = HU, \tag{9}$$

we get

$$\begin{aligned} a(t)\mathbf{1}e^{-\frac{i}{\hbar}\int_0^t d\tau a(\tau)\mathbf{1}}U' - \frac{\hbar}{i}e^{-\frac{i}{\hbar}\int_0^t d\tau a(\tau)\mathbf{1}}\frac{dU'}{dt} \\ = \Big(a(t)\mathbf{1} + b(t)\sigma_z + C\sigma_x\Big)e^{-\frac{i}{\hbar}\int_0^t d\tau a(\tau)\mathbf{1}}U', \end{aligned} \tag{10}$$

leading to the simpler equation

$$-\frac{\hbar}{i}\frac{dU'}{dt} = \big(b(t)\sigma_z + C\sigma_x\big)U'. \tag{11}$$

We shall try to put this equation into a more suggestive form by rewriting it as

$$\begin{aligned} -\frac{\hbar}{i}\frac{dU'}{dt} &= \sqrt{(b^2(t) + C^2)}\left(\frac{b(t)}{\sqrt{b^2+C^2}}\sigma_z + \frac{C}{\sqrt{b^2+C^2}}\sigma_x\right)U' \\ &= \sqrt{(b^2(t)+C^2)}\Big(\cos\Theta(t)\sigma_z + \sin\Theta(t)\sigma_x\Big)U'. \end{aligned} \tag{12}$$

Because $\cos\Theta\sigma_z + \sin\Theta\sigma_x$ suggests a $\sigma_{z'}$, this form suggests we might try a rotation operator about the y axis (in the abstract sense) to convert the effective Hamiltonian on the right from one depending on the two noncommuting operators, σ_z and σ_x, to a new effective Hamiltonian depending on a single σ_z. For this reason, we try to find a solution for U' by trying the substitution

$$U'(t,0) = e^{-i\frac{\Theta}{2}\sigma_y}U''(t,0). \tag{13}$$

This substitution leads to

$$\begin{aligned} -\frac{\hbar}{i}\frac{dU'}{dt} &= e^{-i\frac{\Theta}{2}\sigma_y}\left(-\frac{\hbar}{i}\frac{dU''}{dt}\right) + \frac{\hbar}{2}\sigma_y\frac{d\Theta}{dt}e^{-i\frac{\Theta}{2}\sigma_y}U'' \\ &= \sqrt{b^2+C^2}\Big(\cos\Theta\sigma_z + \sin\Theta\sigma_x\Big)e^{-i\frac{\Theta}{2}\sigma_y}U''. \end{aligned} \tag{14}$$

Left-multiplying by $e^{i\frac{\Theta}{2}\sigma_y}$, this equation can be transformed into

$$-\frac{\hbar}{i}\frac{dU''}{dt} = \left[\sqrt{b^2+C^2}e^{+i\frac{\Theta}{2}\sigma_y}\Big(\cos\Theta\sigma_z + \sin\Theta\sigma_x\Big)e^{-i\frac{\Theta}{2}\sigma_y}\right]U'' - \frac{\hbar}{2}\sigma_y\frac{d\Theta}{dt}U''. \tag{15}$$

To evaluate the last term, we shall differentiate

$$\sin\Theta(t) = \frac{C}{\sqrt{b^2(t)+C^2}},$$

$$\cos\Theta\frac{d\Theta}{dt} = -\frac{bC}{(b^2+C^2)^{\frac{3}{2}}}\frac{db}{dt}. \tag{16}$$

Recalling $b(t) = b_0 + b_1 t$, with $b_1 t = -\frac{1}{2}\hbar\omega_L(0)\frac{2t}{T})$, we can use

$$\frac{db}{dt} = -\frac{1}{T}\hbar\omega_L(0), \qquad \text{and} \ \cos\Theta = \frac{b}{\sqrt{b^2+C^2}}, \tag{17}$$

to write the last term in eq. (17) as

$$-\frac{\hbar}{2}\sigma_y\frac{d\Theta}{dt}U'' = -\frac{C\hbar\omega_L(0)}{2T(b^2+C^2)}\hbar\sigma_y U''. \tag{18}$$

In the limit, $T \to \infty$, therefore, this term can be neglected and we can get U'' from the equation

$$-\frac{\hbar}{i}\frac{dU''}{dt} = \left[\sqrt{b^2+C^2}e^{i\frac{\Theta}{2}\sigma_y}\left(\cos\Theta\sigma_z + \sin\Theta\sigma_x\right)e^{-i\frac{\Theta}{2}\sigma_y}\right]U''. \tag{19}$$

Now, we can use the inverse rotation operation $e^{i\Theta J_y}J_{z'}e^{-i\Theta J_y} = J_z$ to rewrite

$$e^{i\frac{\Theta}{2}\sigma_y}(\cos\Theta\sigma_z + \sin\Theta\sigma_x)e^{-i\frac{\Theta}{2}\sigma_y} = e^{i\frac{\Theta}{2}\sigma_y}\sigma_{z'}e^{-i\frac{\Theta}{2}\sigma_y} = \sigma_z, \tag{20}$$

leading to the simple differential equation for U''

$$-\frac{\hbar}{i}\frac{dU''}{dt} = \sqrt{b^2(t)+C^2}\ \sigma_z U'', \tag{21}$$

with solution

$$U''(t,0) = e^{-\frac{i}{\hbar}\int_0^t d\tau\sqrt{b^2(\tau)+C^2}\sigma_z}. \tag{22}$$

Combining eqs. (8), (13), and (22), we have

$$U(t,0) = e^{-\frac{i}{\hbar}\int_0^t d\tau a(\tau)\mathbf{1}}e^{-i\frac{\Theta(t)}{2}\sigma_y}e^{-\frac{i}{\hbar}\int_0^t d\tau\sqrt{b^2(\tau)+C^2}\sigma_z}, \tag{23}$$

or

$$U(t,0) = e^{-i\chi(t)\mathbf{1}}e^{-i\frac{\Theta(t)}{2}\sigma_y}e^{-i\phi(t)\sigma_z}, \tag{24}$$

where we have used

$$\chi(t) \equiv \frac{1}{\hbar}\int_0^t d\tau a(\tau); \qquad \phi(t) \equiv \frac{1}{\hbar}\int_0^t d\tau\sqrt{b^2(\tau)+C^2}. \tag{25}$$

Together with

$$e^{-i\frac{\Theta}{2}\sigma_y} = \cos\frac{\Theta}{2}\mathbf{1} - i\sin\frac{\Theta}{2}\sigma_y, \tag{26}$$

these equations lead to

$$\begin{aligned} U(t,0) &= \begin{pmatrix} e^{-i\chi(t)} & 0 \\ 0 & e^{-i\chi(t)} \end{pmatrix}\begin{pmatrix} \cos\frac{\Theta}{2} & -\sin\frac{\Theta}{2} \\ \sin\frac{\Theta}{2} & \cos\frac{\Theta}{2} \end{pmatrix}\begin{pmatrix} e^{-i\phi(t)} & 0 \\ 0 & e^{+i\phi(t)} \end{pmatrix} \\ &= \begin{pmatrix} \cos\frac{\Theta}{2}e^{-i(\chi+\phi)} & -\sin\frac{\Theta}{2}e^{-i(\chi-\phi)} \\ \sin\frac{\Theta}{2}e^{-i(\chi+\phi)} & \cos\frac{\Theta}{2}e^{-i(\chi-\phi)} \end{pmatrix}. \end{aligned} \tag{27}$$

The remaining job is to express the trigonometric functions of $\frac{\Theta}{2}$ in terms of b and C, from the defining relations

$$\cos\Theta = \frac{b}{\sqrt{b^2+C^2}}, \qquad \sin\Theta = \frac{C}{\sqrt{b^2+C^2}},$$

via

$$\cos^2\frac{\Theta}{2} = \frac{1}{2}(1+\cos\Theta), \qquad \sin\frac{\Theta}{2} = \sin\Theta/(2\cos\frac{\Theta}{2}),$$

leading to

$$\cos\frac{\Theta}{2} = \frac{\sqrt{\sqrt{b^2+C^2}+b}}{\sqrt{2}(b^2+C^2)^{\frac{1}{4}}} = \frac{C}{\sqrt{2}(b^2+C^2)^{\frac{1}{4}}}\frac{1}{\sqrt{\sqrt{b^2+C^2}-b}}$$
$$\sin\frac{\Theta}{2} = \frac{C}{\sqrt{2}(b^2+C^2))^{\frac{1}{4}}}\frac{1}{\sqrt{\sqrt{b^2+C^2}+b}} = \frac{\sqrt{\sqrt{b^2+C^2}-b}}{\sqrt{2}(b^2+C^2)^{\frac{1}{4}}}, \tag{28}$$

where the second form of each trigonometric function is obtained from the first by multiplying both the numerator and the denominator with the common factor, $\sqrt{\sqrt{b^2+C^2}-b}$.

Now, if at $t=0$, in the huge positive B_0-field limit, the one-electron atom is in the upper of the two eigenstates of a particular m_j,

$$|\psi(t=0)\rangle = |m_l = m_j - \tfrac{1}{2}, m_s = +\tfrac{1}{2}\rangle. \tag{29}$$

At a later time, using the first column of the matrix of eq. (27), we have

$$|\psi(t)\rangle = \cos\frac{\Theta}{2}e^{-i(\chi+\phi)}|m_l = m_j - \tfrac{1}{2}, m_s = +\tfrac{1}{2}\rangle$$
$$+ \sin\frac{\Theta}{2}e^{-i(\chi+\phi)}|m_l = m_j + \tfrac{1}{2}, m_s = -\tfrac{1}{2}\rangle$$
$$= c_{+\frac{1}{2}}|m_l = m_j - \tfrac{1}{2}, m_s = +\tfrac{1}{2}\rangle + c_{-\frac{1}{2}}|m_l = m_j + \tfrac{1}{2}, m_s = -\tfrac{1}{2}\rangle, \tag{30}$$

where the ratio of coefficients is given by the time-dependent function

$$\frac{c_{+\frac{1}{2}}}{c_{-\frac{1}{2}}} = \frac{C}{\sqrt{b^2+C^2}-b}. \tag{31}$$

Here, we have used the second form for the functions $\cos\frac{\Theta}{2}$ and $\sin\frac{\Theta}{2}$ from eq. (28). We shall now show this is precisely the ratio of coefficients for the *upper* of the two eigenstates for this particular m_j value at this particular time. If we imagine we have stopped the field-switching process at this particular value of t, we can diagonalize $H(t)$, as given by eq. (6), with the matrix $H(t) - E\mathbf{1}$ given by

$$H - E\mathbf{1} = \begin{pmatrix} (a-E)+b & C \\ C & (a-E)-b \end{pmatrix}. \tag{32}$$

The determinant of this matrix must be set equal to zero to obtain the two eigenvalues for E, leading to

$$(a-E)^2 - b^2 - C^2 = 0, \tag{33}$$

with

$$(E_{\pm} - a) = \pm\sqrt{b^2 + C^2}, \tag{34}$$

where E_+, (E_-), correspond to E_{upper}, (E_{lower}), respectively. The coefficients of the two eigenvectors are given by the linear equations

$$\begin{aligned} [(a-E)+b]c_{+\frac{1}{2}} + Cc_{-\frac{1}{2}} &= 0, \\ Cc_{+\frac{1}{2}} + [(a-E)-b]c_{-\frac{1}{2}} &= 0. \end{aligned} \tag{35}$$

For the upper energy eigenstate, $E_+ = a + \sqrt{b^2 + C^2}$, this equation leads to the ratio

$$\frac{c_{+\frac{1}{2}}}{c_{-\frac{1}{2}}} = \frac{C}{\sqrt{b^2 + C^2} - b}. \tag{36}$$

This equation is precisely the ratio of coefficients of eq. (31) given by the time evolution operator $U(t,0)$ acting on the initial state $|\psi(t=0)\rangle$, which corresponds to the *upper* of the two states with this value of m_j at the initial time $t = 0$ when the magnetic field has its maximum positive value. The ratio of eq. (36) leads to the normalized coefficients

$$c_{+\frac{1}{2}} = \frac{C}{\sqrt{2}(b^2 + C^2)^{\frac{1}{4}}} \frac{1}{\sqrt{\sqrt{b^2 + C^2} - b}}, \qquad c_{-\frac{1}{2}} = \frac{\sqrt{\sqrt{b^2 + C^2} - b}}{\sqrt{2}(b^2 + C^2)^{\frac{1}{4}}}. \tag{37}$$

The coefficients $c_{+\frac{1}{2}}$ and $c_{-\frac{1}{2}}$ of the eigenvalue problem at this fixed t could each be multiplied by an arbitrary phase factor, such as $e^{-i(\chi(t)+\phi(t))}$, because the energy eigenvectors are determined only to within such a phase. By choosing this particular phase factor, we make the state vector identical to that produced by the time evolution operator $U(t,0)$ in this adiabatic limit. Also, the process of "stopping" the switching at some fixed value of t to diagonalize $H(t)$ can be justified only in this adiabatic limit, where the switching time $T \to \infty$. In this adiabatic limit, however, an initial state that was the *upper* of the two energy eigenstates of a particular m_j at $t = 0$ will evolve into the *upper* of the energy eigenstates with this value of m_j valid for the appropriate magnetic field strength, $B_z(t)$. Thus, the state with asymptotic quantum numbers, $|m_l, m_s\rangle = |0 + \frac{1}{2}\rangle$ at $t = 0$, with B_z positive and huge (see Fig. 59.2), will go through the pure $p_{\frac{3}{2}}$ state at $t = \frac{1}{2}T$, when $B_z = 0$, and end in the state with asymptotic quantum numbers, $|m_l, m_s\rangle = |+1, -\frac{1}{2}\rangle$, at $t = T$ in an adiabatic switching process for which $T \to \infty$.

In exactly the same way: If the initial state at $t = 0$, with maximum positive (huge-field value) of B_z, corresponds to the *lower* of the two energy states of a particular m_j, $|\psi(t=0)\rangle = |m_s = -\frac{1}{2}\rangle$. Now, the state at a later time t is given by the second column of the $U(t,0)$ matrix of eq. (27), leading to the state $|\psi(t)\rangle$ with coefficients given by the ratio

$$\frac{c_{+\frac{1}{2}}}{c_{-\frac{1}{2}}} = -\frac{\sin\frac{\Theta}{2}}{\cos\frac{\Theta}{2}} = -\frac{C}{\sqrt{b^2 + C^2} + b}, \tag{38}$$

where we have used the first form of the expressions for the trigonometric functions of eq. (28). From eq. (35), this ratio of coefficients is in agreement with the energy eigenvectors corresponding to the *lower* eigenvalue, with $E_- = a - \sqrt{b^2 + C^2}$. Thus, a state that was initially the *lower* of the two states with a particular value of m_j will evolve into the *lower* of these two states at the later time t in the adiabatic field-switching process. This result is what we set out to prove.

Problems

25. A system of $J = 1$ particles with magnetic moments, $\vec{\mu} = (e\hbar/2mc)g_J\vec{J}$, is in a strong uniform magnetic field, $\vec{B}_0$, with a second weak field, $\vec{B}_1$, rotating with circular frequency, ω, in a plane perpendicular to $\vec{B}_0$. Assuming the system is in a state with $M_J = +1$ at $t = 0$, where we use the direction of $\vec{B}_0$ as quantization direction, calculate an exact expression, valid for all values of ω, $\omega_0 = (eg_J B_0)/(2mc)$, and $\omega_1 = (eg_J B_1)/(2mc)$, for the probability, $P_{M_J}(t)$, the system is in a state with $M_J = 0$, or $M_J = -1$ at a later finite time t. Compare this with the results given by perturbation theory for $U(t, 0)$, valid for $\omega_1 \ll (\omega_0 + \omega)$ [for $(\omega_0 + \omega) \neq 0$]. Also, calculate the exact values for

$$\frac{dP_{M_J}(t)}{dt} \quad \text{for } M_J = 0 \text{ and } M_J = -1,$$

giving the rate at which transitions, $M_J = +1 \to 0$ and $M_J = +1 \to -1$, take place. Compare these with the perturbation theory results for the case of a general perturbing frequency not on resonance. For the resonance case, $(\omega_0 + \omega) \approx 0$, show the exact result agrees with the perturbation theory result in the limit $\omega_1 \to 0$ for large but finite t; i.e., you need to take the limits as follows:

$$\lim_{t\to\infty}\left[\lim_{\omega_1\to 0}\frac{dP_{M_J}(t)}{dt}\right].$$

26. A sample in a uniform magnetic field, $\vec{B}_0$, containing a large number of identical nuclear spins, with $s = \frac{1}{2}$, has a magnetization, $\vec{M}$, given by the thermal equilibrium value, attained through the so-called spin–lattice relaxation process:

$$M_z = \frac{\left[e^{\hbar\omega_0/2kT} - e^{-\hbar\omega_0/2kT}\right]}{\left[e^{\hbar\omega_0/2kT} + e^{-\hbar\omega_0/2kT}\right]}, \qquad M_x = M_y = 0,$$

where $\omega_0 = eg_J B_0/2Mc$. Write the spin density matrix for this sample. If a $\vec{B}_1$ field, rotating with circular frequence, ω, in a plane normal to $\vec{B}_0$ is turned on at $t = 0$, find the density matrix at a later time, t. Show the $\vec{B}_1$ field induces a magnetization, $\vec{M}(t)$, including x and y components. Note: In a real sample, these perpendicular components of $\vec{M}$ are destroyed by random collision or spin–spin relaxation processes, not included in our Hamiltonian.

27. The endpoint part of the spectrum of the beta decay of tritium, ${}^3_1H_2$, is being reexamined carefully in an attempt to pin down a possible nonzero rest mass for the electron neutrino. Because other experimental evidence indicates the neutrino rest mass must be less than 10–30 eV, the exact atomic energy of the recoiling ${}^3_2\mathrm{He}_1$ one-electron ion must be taken into account carefully. The beta-decay endpoint energy is 18.65 keV. The atomic electron of the tritium atom is initially in its $1s$ ground state, but finds itself in the field of a $Z = 2$ nucleus, ${}^3_2\mathrm{He}_1$, after the beta decay

$${}^3_1\mathrm{H}_2 \rightarrow {}^3_2\mathrm{He}_1 + e^- + \bar{\nu}_e.$$

Show this beta-decay process constitutes a "sudden" process as far as the perturbation on the atomic electron is concerned, by showing the Δt for this process; i.e., the Δt it takes for the fragments, e^- and $\bar{\nu}_e$ to leave the atom is small compared with a hydrogenic period, T. Therefore, the probability the He ion end in a state nlm_l can be calculated easily. In particular, calculate the probability of finding the He ion in its $1s$, $2s$, and $3s$ state. Can it be in a state with $l \neq 0$? Show, in principle, a probability the He atom be completely ionized also exists, leaving a bare ${}^3_2\mathrm{He}_1$ nucleus and an additional e^- in a continuum state, but make a rough estimate to show this probability is effectively negligible.

28. A μ-mesic He atom consisting of a He nucleus, ($Z = 2$), and one μ^- and one e^-, is in its ground state. By taking the overlap of a 1s electron and a 1s μ^- wave function for $Z = 2$, show effectively the electron sees a charge of $Z = 2$ almost completely shielded by the full negative charge of the μ^-, so the electron $1s$ wave function is to good approximation a hydrogenic 1s wave function for $Z = 1$. Conversely, the μ^- sees essentially the full $Z = 2$ of the He nucleus, and its $1s$ wave function is to good approximation that for a single μ^- with a Bohr radius for $Z = 2$ and the μ^--He reduced mass. The μ^- of the μ-mesic He atom will decay, via

$$\mu^- \rightarrow e^- + \bar{\nu}_e + \nu_\mu.$$

Show the Δt for the fragments, e^-, $\bar{\nu}_e$, and ν_μ, to leave the atom is so short compared with atomic periods this perturbation constitutes a sudden process. For this sudden process, calculate the probability of finding the remaining He ion ($Z = 2$ nucleus $+ e^-$) in the $1s$, $2s$, and $3s$ states of this system. Could it be in a state with $l \neq 0$?

Part VII

Atom–Photon Interactions

60

Interaction of Electromagnetic Radiation with Atomic Systems

So far, we have considered quantum systems in external oscillating magnetic (or electric) fields, where the perturbing fields were considered as classical oscillating fields. This approach would permit us to study induced absorption or emission processes in an atomic system, particularly if the intensity of the electromagnetic fields is such that the density of photons is very great in the region of interest. To study processes involving a small number of photons interacting with an atomic system, we must first quantize the electromagnetic or radiation field itself. So far, our time-dependent perturbation theory would not permit us to study the spontaneous emission of photons by an atomic system, i.e., a situation in which the initial state consists of an atom in an excited state and no photons, and the final state is the atom in a lower state (perhaps the ground state) and one photon of the appropriate energy. In order to treat this problem rigorously, we need to quantize the radiation field.

(References for this material: Chapter 2 in J. J. Sakurai, *Advanced Quantum Mechanics*, Reading, MA: Addison-Wesley, 1967; and Chapter 1 in Berestetskii, Lifshitz, and Pitaevskii, *Quantum Electrodynamics*, New York: Pergamon Press, 1982; Vol. 4 of the Landau-Lifshitz series. For the general electric and magnetic multipole radiation fields of vital importance in nuclear physics, see also Judah M. Eisenberg and Walter Greiner, *Nuclear Theory. Vol. 2. Excitation Mechanisms of the Nucleus*, Amsterdam: North Holland, 1970.)

Before quantizing the electromagnetic field, let us consider the Hamiltonian for an atom in an electromagnetic field by first considering the electromagnetic field as a classical field. The electric and magnetic field vectors, $\vec{E}$ and $\vec{B}$, will be derived from a vector and a scalar potential, $\vec{A}$ and Φ. For our purposes, it will be convenient to choose these in the so-called radiation or Coulomb gauge. We will separate the scalar potential, Φ, and the so-called longitudinal part of the vector potential, $\vec{A}_{\parallel}$, from the transverse part of the vector potential, $\vec{A}_{\perp}$. Here,

$$\vec{A} = \vec{A}_{\parallel} + \vec{A}_{\perp}, \qquad \text{with} \quad \vec{\nabla} \cdot \vec{A}_{\perp} = 0, \quad [\vec{\nabla} \times \vec{A}_{\parallel}] = 0. \tag{1}$$

The terms arising from Φ and $\vec{A}_{\parallel}$ will be absorbed into H_{atom}, i.e., the atomic part of our full Hamiltonian. This absorption will include, e.g., all Coulomb terms, such as the Coulomb repulsion electron–electron potential terms and the Coulomb attraction electron-nucleus terms in an atom. Conversely, $\vec{A}_{\perp}$ will serve as the source of the radiation field. We will then seek the full Hamiltonian in the form

$$H = H_{\text{atom}} + H_{\text{radiation}} + H_{\text{interaction}}. \tag{2}$$

Guided by our earlier discussion of an atom in an external electromagnetic field, we shall try a Hamiltonian of the form

$$\begin{aligned} H = &\sum_{i=1}^{N} \frac{1}{2m_i} \Big(\vec{p}_i - \frac{e_i}{c}\vec{A}_{\perp}(\vec{r}_i, t)\Big) \cdot \Big(\vec{p}_i - \frac{e_i}{c}\vec{A}_{\perp}(\vec{r}_i, t)\Big) + \sum_{i<k}^{N} V(\vec{r}_i, \vec{r}_k) \\ &- \sum_{i=1}^{N} \frac{e_i \hbar g_{s,i}}{2m_i c} \vec{s}_i \cdot [\vec{\nabla} \times \vec{A}_{\perp}] + H_{\text{radiation}}, \end{aligned} \tag{3}$$

where $H_{\text{interaction}}$ includes all cross terms, coupling $\vec{A}_{\perp}$ to atomic operators, such as $\vec{p}_i$ or $\vec{s}_i$.

A The Electromagnetic Radiation Field

To derive the expression for $H_{\text{radiation}}$, let us look at the radiation fields

$$\vec{E}_{\text{rad.}} = -\frac{1}{c}\frac{\partial \vec{A}_{\perp}}{\partial t}, \qquad \vec{B}_{\text{rad.}} = [\vec{\nabla} \times \vec{A}_{\perp}], \tag{4}$$

where the Maxwell equations and the condition, $\vec{\nabla} \cdot \vec{A}_{\perp} = 0$, lead to

$$\nabla^2 \vec{A}_{\perp} - \frac{1}{c^2}\frac{\partial^2 \vec{A}_{\perp}}{\partial t^2} = 0. \tag{5}$$

We shall solve this vector wave equation in our cubical laboratory of volume L^3 with the usual boundary conditions, leading to a finely discrete spectrum of $\vec{k}$ values, in terms of basic vector solutions, $\vec{u}_{\vec{k}\alpha}$,

$$\vec{u}_{\vec{k}\alpha} = \frac{e^{i\vec{k}\cdot\vec{r}}}{\sqrt{L^3}} \vec{e}_{\alpha}, \tag{6}$$

with

$$\vec{k} \equiv \vec{k}_{n_1 n_2 n_3} = (k_x, k_y, k_z), \quad \text{and } k_x = \frac{2\pi}{L} n_1, \;\; k_y = \frac{2\pi}{L} n_2; \;\; k_z = \frac{2\pi}{L} n_3, \tag{7}$$

where the n_i are positive or negative integers and the vectors $\vec{e}_\alpha$ are unit polarization vectors. When we expand $\vec{A}_\perp$ in terms of these $\vec{u}_{\vec{k}\alpha}$, the condition, $\vec{\nabla} \cdot \vec{A}_\perp = 0$, leads to $\vec{\nabla} \cdot \vec{u}_{\vec{k}\alpha} = 0$, with

$$\vec{\nabla} \cdot \vec{u}_{\vec{k}\alpha} = \frac{i}{\sqrt{L^3}}(\vec{k} \cdot \vec{e}_\alpha)e^{i\vec{k}\cdot\vec{r}} = 0, \tag{8}$$

leading to

$$(\vec{k} \cdot \vec{e}_\alpha) = 0, \tag{9}$$

so the unit polarization vectors, $\vec{e}_\alpha$, must be orthogonal to $\vec{k}$. The index α can thus be chosen, so the triad of vectors, $\vec{e}_1, \vec{e}_2$, and $\vec{k}/k \equiv \vec{e}_3$, form an orthogonal triad of unit vectors (see Fig. 60.1). For a fixed $\vec{k}$ specified by the integers, n_1, n_2, n_3, the index α can take on the values 1, and 2, corresponding to the two linear polarization vectors. The index α is thus indirectly dependent on $\vec{k}$. The vector potential, $\vec{A}_\perp$, can be expanded in terms of the $\vec{u}_{\vec{k}\alpha}$ through

$$\vec{A}_\perp(\vec{r}, t) = \sum_{\vec{k}} \sum_{\alpha=1,2} \left(c_{\vec{k}\alpha}(t)\vec{u}_{\vec{k}\alpha}(\vec{r}) + c^*_{\vec{k}\alpha}(t)\vec{u}^*_{\vec{k}\alpha}(\vec{r})\right), \tag{10}$$

where we have assumed $\vec{A}_\perp$ is real. The wave equation, eq. (5), then leads to

$$-c^2k^2c_{\vec{k}\alpha} - \ddot{c}_{\vec{k}\alpha} = 0, \tag{11}$$

or

$$\ddot{c}_{\vec{k}\alpha} + \omega^2 c_{\vec{k}\alpha} = 0, \qquad \text{with } \omega = kc, \tag{12}$$

and

$$c_{\vec{k}\alpha}(t) = c_{\vec{k}\alpha}(0)e^{-i\omega t}, \qquad c^*_{\vec{k}\alpha}(t) = c^*_{\vec{k}\alpha}(0)e^{+i\omega t}, \tag{13}$$

so

$$\vec{A}_\perp(\vec{r}, t) = \sum_{\vec{k}} \sum_{\alpha=1,2} \frac{\vec{e}_\alpha}{\sqrt{L^3}} \left(c_{\vec{k}\alpha}(0)e^{i(\vec{k}\cdot\vec{r}-\omega t)} + c^*_{\vec{k}\alpha}(0)e^{-i(\vec{k}\cdot\vec{r}-\omega t)}\right). \tag{14}$$

The classical Hamiltonian for the radiation field can then be obtained from the energy density of the electromagnetic field, via

$$\begin{aligned} H_{\text{radiation}} &= \frac{1}{8\pi}\int_{\text{Vol.}} d\vec{r}\Big((\vec{E}_{\text{rad.}} \cdot \vec{E}_{\text{rad.}}) + (\vec{B}_{\text{rad.}} \cdot \vec{B}_{\text{rad.}})\Big) \\ &= \frac{1}{8\pi}\int_{\text{Vol.}} d\vec{r}\Big((-\frac{1}{c}\frac{\partial \vec{A}_\perp}{\partial t})^2 + ([\vec{\nabla} \times \vec{A}_\perp])^2\Big) \\ &= \frac{1}{2\pi}\sum_{\vec{k},\alpha} \frac{\omega^2}{c^2} c_{\vec{k}\alpha}(t)c^*_{\vec{k}\alpha}(t), \end{aligned} \tag{15}$$

where the volume integral is over the volume of the cubical laboratory, and the last step involves the orthonormality of the spatial functions and the orthonormality of the polarization vectors.

$$\frac{1}{L^3}\int_{\text{Vol.}} d\vec{r} e^{i\vec{k}\cdot\vec{r}} e^{-i\vec{k}'\cdot\vec{r}} = \delta_{\vec{k},\vec{k}'} = \delta_{\vec{n},\vec{n}'} = \delta_{n_1n'_1}\delta_{n_2n'_2}\delta_{n_3n'_3}, \tag{16}$$

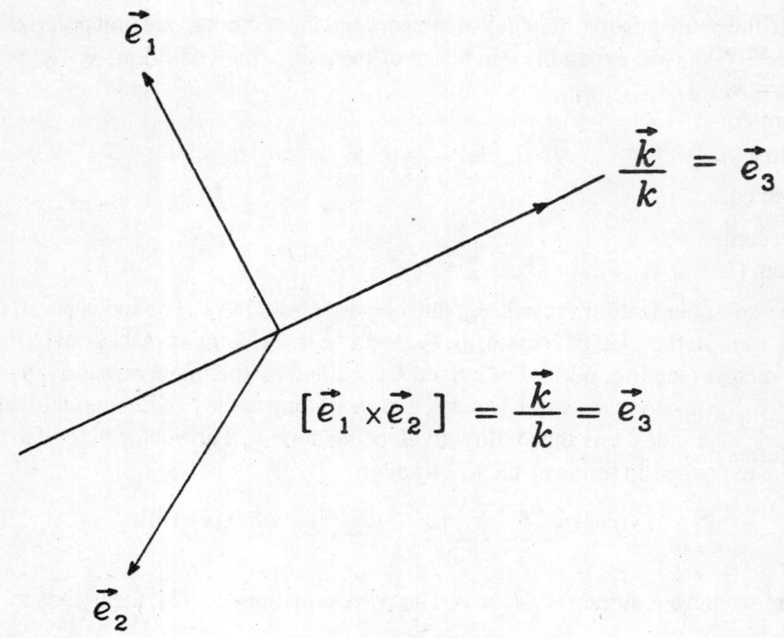

FIGURE 60.1.

so

$$\int_{\text{Vol.}} d\vec{r}\,(\vec{u}^*_{\vec{k}\alpha} \cdot \vec{u}_{\vec{k}'\alpha'}) = \delta_{\vec{k},\vec{k}'}\delta_{\alpha,\alpha'},$$
$$\int_{\text{Vol.}} d\vec{r}\,(\vec{u}_{\vec{k}\alpha} \cdot \vec{u}_{\vec{k}'\alpha'}) = \delta_{\vec{k},-\vec{k}'}\delta_{\alpha,\alpha'}. \tag{17}$$

To get the last form of eq. (15), we shall also use

$$\left(-\frac{\partial \vec{A}_\perp}{c\partial t}\right)\cdot\left(-\frac{\partial \vec{A}_\perp}{c\partial t}\right) = \frac{1}{c^2}\sum_{\vec{k},\alpha}\sum_{\vec{k}',\alpha'}(i\omega)(i\omega')$$
$$\times\left(c_{\vec{k}\alpha}\vec{u}_{\vec{k}\alpha} - c^*_{\vec{k}\alpha}\vec{u}^*_{\vec{k}\alpha}\right)\cdot\left(c_{\vec{k}'\alpha'}\vec{u}_{\vec{k}'\alpha'} - c^*_{\vec{k}'\alpha'}\vec{u}^*_{\vec{k}'\alpha'}\right), \tag{18}$$

and

$$[\vec{\nabla}\times\vec{A}_\perp]\cdot[\vec{\nabla}\times\vec{A}_\perp] = \frac{1}{L^3}\sum_{\vec{k},\alpha}\sum_{\vec{k}',\alpha'} i[\vec{k}\times\vec{e}_\alpha]\cdot i[\vec{k}'\times\vec{e}_{\alpha'}]$$
$$\times\left(c_{\vec{k}\alpha}e^{i\vec{k}\cdot\vec{r}} - c^*_{\vec{k}\alpha}e^{-i\vec{k}\cdot\vec{r}}\right)\left(c_{\vec{k}'\alpha'}e^{i\vec{k}'\cdot\vec{r}} - c^*_{\vec{k}'\alpha'}e^{-i\vec{k}'\cdot\vec{r}}\right)$$
$$= -\frac{1}{L^3}\sum_{\vec{k},\alpha}\sum_{\vec{k}',\alpha'}[(\vec{k}\cdot\vec{k}')(\vec{e}_\alpha\cdot\vec{e}_{\alpha'}) - (\vec{k}\cdot\vec{e}_{\alpha'})(\vec{k}'\cdot\vec{e}_\alpha)]$$
$$\times\left(c_{\vec{k}\alpha}e^{i\vec{k}\cdot\vec{r}} - c^*_{\vec{k}\alpha}e^{-i\vec{k}\cdot\vec{r}}\right)\left(c_{\vec{k}'\alpha'}e^{i\vec{k}'\cdot\vec{r}} - c^*_{\vec{k}'\alpha'}e^{-i\vec{k}'\cdot\vec{r}}\right). \tag{19}$$

When integrated over the volume of our cubical laboratory, only terms with $\vec{k}' = \pm\vec{k}$ survive [see eq. (17)]. For both of these, $(\vec{k}\cdot\vec{e}_{\alpha'})(\vec{k}'\cdot\vec{e}_{\alpha}) = 0$. Terms with $\vec{k}' = \vec{k}$, $\alpha' = \alpha$, receive a contribution $2\frac{\omega^2}{c^2}c_{\vec{k}\alpha}c^*_{\vec{k}\alpha}$ from eq. (18) and $2k^2 c_{\vec{k}\alpha}c^*_{\vec{k}\alpha}$ from eq. (19) when integrated over the volume of the cubical laboratory. Terms with $\vec{k}' = -\vec{k}$, $\alpha' = \alpha$, conversely, get a contribution of $-\frac{\omega^2}{c^2}(c_{\vec{k}\alpha}c_{-\vec{k}\alpha} + c^*_{\vec{k}\alpha}c^*_{-\vec{k}\alpha})$ from eq. (18) and a contribution $+k^2(c_{\vec{k}\alpha}c_{-\vec{k}\alpha} + c^*_{\vec{k}\alpha}c^*_{-\vec{k}\alpha})$ from eq. (19), so these two contributions cancel, whereas the terms with $\vec{k}' = \vec{k}$ add to give the final result of eq. (15).

$$H_{\text{radiation}} = \frac{1}{2\pi}\sum_{\vec{k},\alpha}\frac{\omega^2}{c^2}c_{\vec{k}\alpha}c^*_{\vec{k}\alpha}. \tag{20}$$

Now, it will be convenient to rewrite the complex (time-dependent) functions, $c_{\vec{k}\alpha}$, in terms of two real time-dependent functions, $Q_{\vec{k}\alpha}$ and $P_{\vec{k}\alpha}$, via

$$\begin{aligned}\frac{1}{c\sqrt{2\pi}}c_{\vec{k}\alpha} &= \sqrt{\tfrac{1}{2}}(Q_{\vec{k}\alpha} + \frac{i}{\omega}P_{\vec{k}\alpha}),\\ \frac{1}{c\sqrt{2\pi}}c^*_{\vec{k}\alpha} &= \sqrt{\tfrac{1}{2}}(Q_{\vec{k}\alpha} - \frac{i}{\omega}P_{\vec{k}\alpha}),\end{aligned} \tag{21}$$

giving us the *classical* radiation Hamiltonian

$$H_{\text{radiation}} = \tfrac{1}{2}\sum_{\vec{k},\alpha}(P^2_{\vec{k}\alpha} + \omega^2 Q^2_{\vec{k}\alpha}). \tag{22}$$

Comparing this with the simple 1-D harmonic oscillator Hamiltonian

$$H_{\text{harm. osc.}} = (\frac{P^2}{2m} + \frac{m\omega^2}{2}Q^2), \tag{23}$$

the classical radiation Hamiltonian involves a superposition of an infinite number of (uncoupled) harmonic oscillator Hamiltonians if we set the parameter $m = 1$. (Note, however, this is not a physical mass for our problem).

B The Quantized Radiation Field

To quantize the above Dirac assumed the classical variables, $P_{\vec{k}\alpha}$ and $Q_{\vec{k}\alpha}$, are replaced by quantum-mechanical operators that satisfy the commutation relations

$$[P_{\vec{k}\alpha}, Q_{\vec{k}'\alpha'}] = \frac{\hbar}{i}\delta_{\vec{k},\vec{k}'}\delta_{\alpha,\alpha'}, \quad [P_{\vec{k}\alpha}, P_{\vec{k}'\alpha'}] = 0, \quad [Q_{\vec{k}\alpha}, Q_{\vec{k}'\alpha'}] = 0. \tag{24}$$

As for the simple 1-D oscillator, it will now be convenient to introduce dimensionless $p_{\vec{k}\alpha}$, $q_{\vec{k}\alpha}$ via

$$P_{\vec{k}\alpha} = \sqrt{\hbar\omega}\,p_{\vec{k}\alpha}, \qquad Q_{\vec{k}\alpha} = \sqrt{\frac{\hbar}{\omega}}q_{\vec{k}\alpha}, \tag{25}$$

so the $p_{\vec{k}\alpha}$ and $q_{\vec{k}\alpha}$ satisfy the commutation relations

$$[p_{\vec{k}\alpha}, q_{\vec{k}'\alpha'}] = \frac{1}{i}\delta_{\vec{k},\vec{k}'}\delta_{\alpha,\alpha'}, \qquad [p_{\vec{k}\alpha}, p_{\vec{k}'\alpha'}] = 0, \quad [q_{\vec{k}\alpha}, q_{\vec{k}'\alpha'}] = 0. \tag{26}$$

In terms of these operators, the radiation Hamiltonian can be written as

$$H_{\text{radiation}} = \tfrac{1}{2}\sum_{\vec{k},\alpha} \hbar\omega(p^2_{\vec{k}\alpha} + q^2_{\vec{k}\alpha}). \tag{27}$$

As for the 1-D harmonic oscillator, it will be useful to introduce annihilation and creation operators,

$$\begin{aligned} a_{\vec{k}\alpha} &= \sqrt{\tfrac{1}{2}}(q_{\vec{k}\alpha} + ip_{\vec{k}\alpha}), \\ a^\dagger_{\vec{k}\alpha} &= \sqrt{\tfrac{1}{2}}(q_{\vec{k}\alpha} - ip_{\vec{k}\alpha}), \end{aligned} \tag{28}$$

with commutation relations

$$\begin{aligned} &[a_{\vec{k}\alpha}, a^\dagger_{\vec{k}'\alpha'}] = \delta_{\vec{k},\vec{k}'}\delta_{\alpha,\alpha'} \\ &[a_{\vec{k}\alpha}, a_{\vec{k}'\alpha'}] = [a^\dagger_{\vec{k}\alpha}, a^\dagger_{\vec{k}'\alpha'}] = 0. \end{aligned} \tag{29}$$

In terms of these operators,

$$H_{\text{radiation}} = \tfrac{1}{2}\sum_{\vec{k},\alpha} \hbar\omega(a^\dagger_{\vec{k}\alpha}a_{\vec{k}\alpha} + a_{\vec{k}\alpha}a^\dagger_{\vec{k}\alpha}) = \sum_{\vec{k},\alpha} \hbar\omega(a^\dagger_{\vec{k}\alpha}a_{\vec{k}\alpha} + \tfrac{1}{2}). \tag{30}$$

Now, we can introduce the operator

$$N_{\vec{k}\alpha} = a^\dagger_{\vec{k}\alpha}a_{\vec{k}\alpha}, \tag{31}$$

which we would have interpreted as the operator which counts the number of oscillator quanta of type $\vec{k}\alpha$. We will now interpret this as the operator that counts the number of photons of type $\vec{k}\alpha$, where we will substantiate this interpretation. The eigenvectors of $H_{\text{radiation}}$ are of the form

$$|n_{\vec{k}_1\alpha_1} n_{\vec{k}_2\alpha_2} n_{\vec{k}_3\alpha_3} \cdots n_{\vec{k}\alpha} \cdots\rangle,$$

where

$$a^\dagger_{\vec{k}\alpha}a_{\vec{k}\alpha}|\cdots n_{\vec{k}\alpha}\cdots\rangle = n_{\vec{k}\alpha}|\cdots n_{\vec{k}\alpha}\cdots\rangle, \tag{32}$$

$$a_{\vec{k}\alpha}|\cdots n_{\vec{k}\alpha}\cdots\rangle = \sqrt{n_{\vec{k}\alpha}}\;|\cdots(n_{\vec{k}\alpha} - 1)\cdots\rangle, \tag{33}$$

$$a^\dagger_{\vec{k}\alpha}|\cdots n_{\vec{k}\alpha}\cdots\rangle = \sqrt{(n_{\vec{k}\alpha} + 1)}|\cdots(n_{\vec{k}\alpha} + 1)\cdots\rangle. \tag{34}$$

The state of no photons, the vacuum state, is of the form

$$|000\cdots 0\cdots 0\rangle \equiv |0\rangle$$

and satisfies

$$a_{\vec{k}\alpha}|0\rangle = 0 \qquad \text{for all } \vec{k}\alpha. \tag{35}$$

The state of a single photon of type $\vec{k}\alpha$ is

$$a^{\dagger}_{\vec{k}\alpha}|0\rangle.$$

The most general many-photon state has the form

$$\prod_{\vec{k}_i\alpha_i} \frac{\left(a^{\dagger}_{\vec{k}_i\alpha_i}\right)^{n_{\vec{k}_i\alpha_i}}}{\sqrt{(n_{\vec{k}_i\alpha_i})!}}|0\rangle.$$

61

Photons: The Quantized Radiation Field

In order to interpret the number $n_{\vec{k}\alpha}$ as the number of photons of type $\vec{k}\alpha$, we shall look at the properties of the quantized radiation field in more detail. Combining eqs. (21), (25), and (28) of the last chapter, the expansion of $\vec{A}_\perp$ can be written as

$$\vec{A}_\perp(\vec{r},t) = \sum_{\vec{k},\alpha} c\sqrt{\frac{2\pi\hbar}{\omega}}\left(a_{\vec{k}\alpha}\vec{u}_{\vec{k}\alpha}(\vec{r}) + a^\dagger_{\vec{k}\alpha}\vec{u}^*_{\vec{k}\alpha}(\vec{r})\right), \tag{1}$$

where the $a_{\vec{k}\alpha}$ are time dependent (given in Heisenberg representation), with

$$a_{\vec{k}\alpha} \equiv a_{\vec{k}\alpha}(t) = a_{\vec{k}\alpha}(0)e^{-i\omega t}, \qquad a^\dagger_{\vec{k}\alpha} \equiv a^\dagger_{\vec{k}\alpha}(t) = a^\dagger_{\vec{k}\alpha}(0)e^{+i\omega t}. \tag{2}$$

The radiation fields are then given by

$$\vec{E}(\vec{r},t) = i\sum_{\vec{k},\alpha}\sqrt{2\pi\hbar\omega}\left(a_{\vec{k}\alpha}\vec{u}_{\vec{k}\alpha}(\vec{r}) - a^\dagger_{\vec{k}\alpha}\vec{u}^*_{\vec{k}\alpha}(\vec{r})\right), \tag{3}$$

$$\vec{B}(\vec{r},t) = ic\sum_{\vec{k},\alpha}\sqrt{\frac{2\pi\hbar}{\omega}}\left(a_{\vec{k}\alpha}[\vec{k}\times\vec{u}_{\vec{k}\alpha}(\vec{r})] - a^\dagger_{\vec{k}\alpha}[\vec{k}\times\vec{u}^*_{\vec{k}\alpha}(\vec{r})]\right). \tag{4}$$

A Photon Energy

Because the radiation Hamiltonian

$$H_{\text{radiation}} = \sum_{\vec{k},\alpha}\hbar\omega\left(a^\dagger_{\vec{k}\alpha}a_{\vec{k}\alpha} + \tfrac{1}{2}\right) \tag{5}$$

contains an infinite zero-point energy for the vacuum state through the factor of $\frac{1}{2}$ associated with each of the infinite number of possible states $\vec{k}\alpha$, we must measure the energy *relative* to the energy of the vacuum state, $E^0_{\text{rad.}}$.

$$(H_{\text{rad.}} - E^0_{\text{rad.}})|\cdots n_{\vec{k}\alpha}\cdots\rangle = \sum_{\vec{k},\alpha}\hbar\omega n_{\vec{k}\alpha}|\cdots n_{\vec{k}\alpha}\cdots\rangle. \tag{6}$$

Thus, the single photon state, with $n_{\vec{k}\alpha} = 1$, has an energy $\hbar\omega$, with $\omega = kc$.

B Photon Linear Momentum

Because the *classical* momentum density of the electromagnetic field is given by $(1/4\pi c)[\vec{E}\times\vec{B}]$, we need to integrate the quantized operator form of this quantity over the volume of our cubical laboratory to obtain the linear momentum of the photon field. Because the quantized expressions for $\vec{E}$ and $\vec{B}$ of eqs. (3) and (4) contain noncommuting operators, we need to use the symmetrized, hermitian form of the momentum density operator.

$$\vec{p} = \frac{1}{8\pi c}\int_{\text{Vol.}} d\vec{r}\big([\vec{E}\times\vec{B}] - [\vec{B}\times\vec{E}]\big) = \frac{1}{8\pi c}(-c2\pi\hbar)\sum_{\vec{k},\alpha}\sum_{\vec{k}',\alpha'}\sqrt{\frac{\omega}{\omega'}}\frac{1}{L^3}\times$$
$$\Big(-[\vec{e}_\alpha\times[\vec{k}'\times\vec{e}_{\alpha'}]]\Big(a_{\vec{k}\alpha}a^\dagger_{\vec{k}'\alpha'}\int_{\text{Vol.}} d\vec{r}e^{i(\vec{k}-\vec{k}')\cdot\vec{r}} + a^\dagger_{\vec{k}\alpha}a_{\vec{k}'\alpha'}\int_{\text{Vol.}} d\vec{r}e^{i(\vec{k}'-\vec{k})\cdot\vec{r}}\Big)$$
$$+[[\vec{k}'\times\vec{e}_{\alpha'}]\times\vec{e}_\alpha]\Big(a_{\vec{k}'\alpha'}a^\dagger_{\vec{k}\alpha}\int_{\text{Vol.}} d\vec{r}e^{i(\vec{k}'-\vec{k})\cdot\vec{r}} + a^\dagger_{\vec{k}'\alpha'}a_{\vec{k}\alpha}\int_{\text{Vol.}} d\vec{r}e^{i(\vec{k}-\vec{k}')\cdot\vec{r}}\Big)\Big)$$
$$+\cdots+\cdots\ . \tag{7}$$

The orthonormality of the $e^{i\vec{k}\cdot\vec{r}}/L^{\frac{3}{2}}$ requires $\vec{k}' = \vec{k}$, so

$$-[\vec{e}_\alpha\times[\vec{k}'\times\vec{e}_{\alpha'}]] = [[\vec{k}'\times\vec{e}_{\alpha'}]\times\vec{e}_\alpha] = -(\vec{e}_\alpha\cdot\vec{e}_{\alpha'})\vec{k}' + (\vec{e}_\alpha\cdot\vec{k}')\vec{e}_{\alpha'} = -\delta_{\alpha,\alpha'}\vec{k}, \tag{8}$$

and the above therefore yields

$$\vec{p} = \frac{1}{8\pi c}(-c2\pi\hbar)\sum_{\vec{k},\alpha}2\vec{k}(-a^\dagger_{\vec{k}\alpha}a_{\vec{k}\alpha} - a_{\vec{k}\alpha}a^\dagger_{\vec{k}\alpha}) + \cdots + \cdots\ . \tag{9}$$

The $\cdots$ terms in these equations indicate similar $a^\dagger a^\dagger$ and aa terms exist that contribute for values of $\vec{k}' = -\vec{k}$. Contributions from the $-[\vec{B}\times\vec{E}]$ term, however, cancel those from the $[\vec{E}\times\vec{B}]$ term for terms of this type. Thus,

$$\vec{p} = \tfrac{1}{2}\sum_{\vec{k},\alpha}\hbar\vec{k}(a^\dagger_{\vec{k}\alpha}a_{\vec{k}\alpha} + a_{\vec{k}\alpha}a^\dagger_{\vec{k}\alpha}) = \sum_{\vec{k},\alpha}\hbar\vec{k}(a^\dagger_{\vec{k}\alpha}a_{\vec{k}\alpha} + \tfrac{1}{2}). \tag{10}$$

Now, because for every $\vec{k}$ with n_1, n_2, n_3 a vector $-\vec{k}$ with $-n_1, -n_2, -n_3$ exists, the sum $\frac{1}{2}\sum_{\vec{k},\alpha}\vec{k} = 0$, so finally

$$\vec{p} = \sum_{\vec{k},\alpha}a^\dagger_{\vec{k}\alpha}a_{\vec{k}\alpha}\hbar\vec{k}, \tag{11}$$

and a photon of type $\vec{k}\alpha$ carries a linear momentum, $\hbar\vec{k}$. A many-photon state carries the total linear momentum $\sum_{\vec{k}\alpha}(\hbar\vec{k})n_{\vec{k}\alpha}$.

C Photon Rest Mass

From the relativistic relation for the rest mass, m_0, of a particle, $(E^2 - c^2p^2) = m_0^2c^4$, we have for the photon

$$(E^2 - c^2p^2)a^\dagger_{\vec{k}\alpha}|0\rangle = \left((\hbar\omega)^2 - c^2(\hbar k)^2\right)a^\dagger_{\vec{k}\alpha}|0\rangle = 0, \tag{12}$$

so the rest mass of the photon is zero.

D Angular Momentum: Photon Spin

We begin with the classical expression for the angular momentum density of the electromagnetic field, $(1/4\pi c)[\vec{r} \times [\vec{E} \times \vec{B}]]$, leading to the classical expression for the total angular momentum of the radiation field in our cubical laboratory

$$\vec{J} = \frac{1}{4\pi c}\int_{\text{Vol.}} d\vec{r}[\vec{r} \times [\vec{E} \times \vec{B}]]. \tag{13}$$

Before transcribing this to the hermitian quantized form, let us first expand the momentum density, using (in summation convention format for repeated Greek indices)

$$[\vec{E} \times [\vec{\nabla} \times \vec{A}]]_j = \epsilon_{j\alpha\beta}E_\alpha\epsilon_{\beta\mu\nu}\frac{\partial A_\nu}{\partial x_\mu} = E_\alpha\frac{\partial A_\alpha}{\partial x_j} - E_\alpha\frac{\partial A_j}{\partial x_\alpha}, \tag{14}$$

and

$$\begin{aligned}
[\vec{r} \times [\vec{E} \times \vec{B}]]_i &= \epsilon_{i\mu\nu}x_\mu\Big(E_\alpha\frac{\partial}{\partial x_\nu}A_\alpha - E_\alpha\frac{\partial}{\partial x_\alpha}A_\nu\Big) \\
&= E_\alpha\epsilon_{i\mu\nu}x_\mu\frac{\partial}{\partial x_\nu}A_\alpha - \epsilon_{i\mu\nu}x_\mu\frac{\partial}{\partial x_\alpha}(E_\alpha A_\nu) \\
&= E_\alpha\epsilon_{i\mu\nu}x_\mu\frac{\partial}{\partial x_\nu}A_\alpha - \epsilon_{i\mu\nu}\frac{\partial}{\partial x_\alpha}(E_\alpha x_\mu A_\nu) + \epsilon_{i\alpha\nu}E_\alpha A_\nu \\
&= E_\alpha[\vec{r} \times \vec{\nabla}]_i A_\alpha - \frac{\partial}{\partial x_\alpha}\big(E_\alpha[\vec{r} \times \vec{A}]_i\big) + [\vec{E} \times \vec{A}]_i,
\end{aligned} \tag{15}$$

where we have used $(\vec{\nabla} \cdot \vec{E}) = 0$ for the radiation field in a source-free region in the second line. Using the final form we can express the total *classical* angular momentum in our cube of volume L^3 by

$$\vec{J}_i = \frac{1}{4\pi c}\left(\int_{\text{Vol.}} d\vec{r}\big(E_\alpha[\vec{r}\times\vec{\nabla}]_i A_\alpha\big) - \int_{\text{Vol.}} d\vec{r}\big(\vec{\nabla}\cdot(\vec{E}[\vec{r}\times\vec{A}]_i)\big) + \int_{\text{Vol.}} d\vec{r}[\vec{E}\times\vec{A}]_i\right). \tag{16}$$

In the second term, we use Gauss's theorem to convert the volume integral of the divergence to a surface integral

$$\int_{\text{Surf.}} d(\text{Area})(\vec{E}\cdot\vec{n})[\vec{r}\times\vec{A}]_i.$$

Because the radiation $\vec{A}$ has the value zero on the six faces of our cube, the second term is zero, and the classical expression for the total angular momentum of the radiation field in our cube can be evaluated through the two integrals

$$\begin{aligned}\vec{J} &= \frac{1}{4\pi c}\int_{\text{Vol.}} d\vec{r}\,\big(E_\alpha[\vec{r}\times\vec{\nabla}]A_\alpha\big) + \frac{1}{4\pi c}\int_{\text{Vol.}} d\vec{r}\,[\vec{E}\times\vec{A}].\\ &= \qquad\qquad \vec{L} \qquad\qquad + \qquad\qquad \vec{S}.\end{aligned} \tag{17}$$

Here, the first integral dependent on the position vector $\vec{r}$ from some center (such as the position of the atomic system) to a field point in our cube has been identified as the orbital angular momentum of the radiation field, whereas the second term (independent of such an $\vec{r}$) has been identified as the spin angular momentum of the radiation field. Because we are interested in the spin of the photon, we shall look at this second spin term in more detail. To convert this $\vec{S}$ to the quantized form, we must replace $[\vec{E}\times\vec{A}]$ by its symmetrized, hermitian form. Thus, for the quantized radiation fields, expressed through eqs. (1) and (3), we have

$$\begin{aligned}\vec{S} &= \frac{1}{8\pi c}\int_{\text{Vol.}} d\vec{r}\big([\vec{E}\times\vec{A}] - [\vec{A}\times\vec{E}]\big)\\ &= \frac{1}{8\pi c}(ic2\pi\hbar)\sum_{\vec{k}}\sum_{\alpha,\alpha'}[\vec{e}_\alpha\times\vec{e}_{\alpha'}]2\big(a_{\vec{k}\alpha}a^\dagger_{\vec{k}\alpha'} - a^\dagger_{\vec{k}\alpha}a_{\vec{k}\alpha'}\big)\\ &= -i\hbar\sum_{\vec{k}}\sum_{\alpha,\alpha'} a^\dagger_{\vec{k}\alpha}a_{\vec{k}\alpha'}[\vec{e}_\alpha\times\vec{e}_{\alpha'}]\\ &= -i\hbar\sum_{\vec{k}}\big(a^\dagger_{\vec{k}1}a_{\vec{k}2} - a^\dagger_{\vec{k}2}a_{\vec{k}1}\big)[\vec{e}_1\times\vec{e}_2]\\ &= -i\hbar\sum_{\vec{k}}\big(a^\dagger_{\vec{k}1}a_{\vec{k}2} - a^\dagger_{\vec{k}2}a_{\vec{k}1}\big)\vec{e}_3,\end{aligned} \tag{18}$$

where we have again used the orthonormality of the $e^{i\vec{k}\cdot\vec{r}}/L^{\frac{3}{2}}$, as in the calculations for eq. (7), and, again, terms of type $a^\dagger a^\dagger$ and of type aa disappear because of cancellations of the contributions from the $[\vec{E}\times\vec{A}]$ and $-[\vec{A}\times\vec{E}]$ terms. For our final result, it will be convenient to switch from Cartesian components of vectors along our triad of unit vectors ($\vec{e}_1, \vec{e}_2, \vec{e}_3 = \vec{k}/k$) to spherical vector components, with $m = +1, 0, -1$, and define

$$\vec{e}_{+1} = -\sqrt{\tfrac{1}{2}}(\vec{e}_1 + i\vec{e}_2), \qquad \vec{e}_0 = \vec{e}_3, \qquad \vec{e}_{-1} = +\sqrt{\tfrac{1}{2}}(\vec{e}_1 - i\vec{e}_2). \tag{19}$$

Note carefully the difference between $\vec{e}_{+1}$ and $\vec{e}_1$. Also,

$$\vec{e}_m{}^* = (-1)^m\vec{e}_{-m}. \tag{20}$$

Finally, the $\vec{e}_{m=\pm1}$ are the natural base vectors for right and left circular polarizations, just as $\vec{e}_1$ and $\vec{e}_2$ were the natural unit vectors for plane polarizations.

Finally, we shall also define the analagous spherical vector components of the photon creation operators

$$a^{\dagger}_{\vec{k}m=+1} = -\sqrt{\tfrac{1}{2}}\big(a^{\dagger}_{\vec{k}1} + ia^{\dagger}_{\vec{k}2}\big), \qquad a^{\dagger}_{\vec{k}m=-1} = +\sqrt{\tfrac{1}{2}}\big(a^{\dagger}_{\vec{k}1} - ia^{\dagger}_{\vec{k}2}\big). \tag{21}$$

Note: The photon creation operator has no $m = 0$ component. From hermitian conjugation, we have

$$a_{\vec{k}m=+1} = -\sqrt{\tfrac{1}{2}}\big(a_{\vec{k}1} - ia_{\vec{k}2}\big), \qquad a_{\vec{k}m=-1} = +\sqrt{\tfrac{1}{2}}\big(a_{\vec{k}1} + ia_{\vec{k}2}\big). \tag{22}$$

Also,

$$\sum_{\alpha=1,2} a^{\dagger}_{\vec{k}\alpha}\vec{e}_{\alpha} = \sum_{m=+1,-1} a^{\dagger}_{\vec{k}m}\vec{e}_m{}^{*} = \sum_{m=+1,-1} (-1)^m a^{\dagger}_{\vec{k}m}\vec{e}_{-m}, \tag{23}$$

whereas

$$\sum_{\alpha=1,2} a_{\vec{k}\alpha}\vec{e}_{\alpha} = \sum_{m=+1,-1} a_{\vec{k}m}\vec{e}_m. \tag{24}$$

Finally, we have

$$\big(a^{\dagger}_{\vec{k}1}a_{\vec{k}2} - a^{\dagger}_{\vec{k}2}a_{\vec{k}1}\big) = i\big(a^{\dagger}_{\vec{k}m=+1}a_{\vec{k}m=+1} - a^{\dagger}_{\vec{k}m=-1}a_{\vec{k}m=-1}\big), \tag{25}$$

so our expression for the photon spin can be written as

$$\vec{S} = \sum_{\vec{k}} \sum_{m=+1,-1} (\hbar m) a^{\dagger}_{\vec{k}m} a_{\vec{k}m} \left(\frac{\vec{k}}{k}\right) \tag{26}$$

and

$$\vec{S}\Big(a^{\dagger}_{\vec{k}m=\pm1}|0\rangle\Big) = (\pm 1\hbar)\left(\frac{\vec{k}}{k}\right)\Big(a^{\dagger}_{\vec{k}m=\pm1}|0\rangle\Big). \tag{27}$$

Thus, the photon spin vector has projections along the direction of the $\vec{k}$ vector of $+1\hbar$ (right circularly polarized photons) or of $-1\hbar$ (left circularly polarized photons). It thus appears to be a spin-1 particle, *but* the $m_s = 0$ component is completely missing. The photon has only longitudial spin projections, either parallel or antiparallel to the direction of its linear momentum vector, $\vec{k}$. These are quantized, however, with values $\pm 1\hbar$. The operator $(\vec{S}\cdot\vec{S})$ does *not* have eigenvalue $s(s+1)\hbar^2 = 2\hbar^2$. Instead, this eigenvalue is $1\hbar^2$.

$$(\vec{S}\cdot\vec{S})\Big(a^{\dagger}_{\vec{k}m=\pm1}|0\rangle\Big) = \hbar^2 \sum_{\vec{k}'} \sum_{m'=\pm1} \sum_{\vec{k}''} \sum_{m''=\pm1} m'm'' \frac{(\vec{k}'\cdot\vec{k}'')}{k'k''}$$
$$\times\, a^{\dagger}_{\vec{k}'m'} a_{\vec{k}'m'} a^{\dagger}_{\vec{k}''m''} a_{\vec{k}''m''} \Big(a^{\dagger}_{\vec{k}m}|0\rangle\Big) = 1\hbar^2\Big(a^{\dagger}_{\vec{k}m}|0\rangle\Big). \tag{28}$$

62

Vector Spherical Harmonics

To deal with the orbital angular momentum carried by the radiation field, it will be convenient to convert the expansion of the field vector, $\vec{A}_{\perp}$, to an expansion in terms of photon creation and annihilation operators with circular polarizations, with $m = \pm 1$, and write

$$\vec{A}_{\perp}(\vec{r}, t) = \frac{c}{L^{\frac{3}{2}}} \sum_{\vec{k}} \sum_{\lambda=\pm 1} \sqrt{\frac{2\pi\hbar}{\omega}} \left(a_{\vec{k}\lambda} e^{i\vec{k}\cdot\vec{r}} \vec{e}_{\lambda} + a^{\dagger}_{\vec{k}\lambda} e^{-i\vec{k}\cdot\vec{r}} \vec{e}_{\lambda}{}^{*} \right). \tag{1}$$

(Note: We shall reserve later letters in the Greek alphabet, such as λ, μ, ν for spherical components, with $\mu = +1, 0, -1$, whereas early letters, such as α, β, will be reserved for Cartesian components, with $\alpha = 1, 2, 3$.)

We will need to consider the expansion

$$e^{i\vec{k}\cdot\vec{r}} \vec{e}_{\lambda} = \sum_{l=0}^{\infty} j_l(kr) \sqrt{4\pi(2l+1)}\, i^l Y_{l0}(\theta, \phi) \vec{e}_{\lambda}, \tag{2}$$

with $\lambda = \pm 1$. Now, it will be useful to couple the spherical components of the vector, $\vec{e}_{\lambda}$, a spherical tensor of rank 1, with the components of the spherical harmonics, $i^l Y_{lm}$, which of course are spherical tensors of rank l, to vector operators of resultant spherical tensor rank L, with $L = (l+1), l, (l-1)$, which are defined by

$$\vec{V}_{[l1]LM} = \sum_{\mu=-1}^{+1} \langle l(M-\mu) 1\mu | LM \rangle i^l Y_{l(M-\mu)}(\theta, \phi) \vec{e}_{\mu}, \tag{3}$$

where the $\vec{V}_{[l1]LM}$ are the so-called vector spherical harmonics. (Note: This vector is similar to our coupled orbital-spin functions, $\mathcal{Y}_{[ls]jm_j}$, introduced in Chapter 52.

As there, it will again prove convenient to include the factor i^l together with the spherical harmonic. Also, the μ sum involves all three spherical components of $\vec{e}_\mu$.) With this definition, we have

$$e^{i\vec{k}\cdot\vec{r}}\vec{e}_\mu = \sum_l \sum_L j_l(kr)\sqrt{4\pi(2L+1)}\langle l01\mu|L\mu\rangle \vec{V}_{[l1]L\mu}, \tag{4}$$

with $\mu = \pm 1$ only and $L = (l+1), l, (l-1)$.

A Properties of Vector Spherical Harmonics

1. Orthonormality. The vector spherical harmonics form an orthonormal set with respect to a scalar product involving both the scalar product of the vectors and an integration over the angular part of our space

$$\begin{aligned}
&\int\int d\Omega(\vec{V}^{\,*}_{[l'1]L'M'} \cdot \vec{V}_{[l1]LM}) \\
&= \sum_{\mu,\nu} (\vec{e}_\nu{}^* \cdot \vec{e}_\mu)(-i)^{l'}(i)^l \int\int d\Omega Y^*_{l'(M'-\nu)} Y_{l(M-\mu)} \\
&\times \langle l'(M'-\nu)1\nu|L'M'\rangle\langle l(M-\mu)1\mu|LM\rangle \\
&= \sum_{\mu,\nu} \delta_{\nu\mu}\delta_{l'l}\delta_{(M'-\nu)(M-\mu)}\langle l'(M'-\nu)1\nu|L'M'\rangle\langle l(M-\mu)1\mu|LM\rangle \\
&= \delta_{l'l}\delta_{M'M}\sum_\mu \langle l(M-\mu)1\mu|L'M\rangle\langle l(M-\mu)1\mu|LM\rangle \\
&= \delta_{l'l}\delta_{M'M}\delta_{L'L},
\end{aligned} \tag{5}$$

where this follows from the orthonormality of the unit vectors, $(\vec{e}_\nu{}^* \cdot \vec{e}_\mu) = \delta_{\nu\mu}$, the orthonormality of the spherical harmonics, and the orthonormality of the Clebsch–Gordan coefficients. [Note, in particular: When spherical components are used for the unit vectors, $\vec{e}$, we must take the scalar product of a starred with an unstarred spherical component, $(\vec{e}_\nu{}^* \cdot \vec{e}_\mu)$ (*not* $(\vec{e}_\nu \cdot \vec{e}_\mu)$), to get a Kronecker delta in the two spherical components.]

2. Parity.

Under space inversion, the spherical harmonics, and the vectors, $\vec{e}$, transform as

$$\begin{aligned}
Y_{l(M-\mu)}(\theta,\phi) &\rightarrow Y_{l(M-\mu)}(\pi-\theta,\phi+\pi) = (-1)^l Y_{l(M-\mu)}(\theta,\phi) \\
\vec{e}_\mu &\rightarrow -\vec{e}_\mu. \\
\text{Thus,}\quad \vec{V}_{[l1]LM} &\rightarrow (-1)^{l+1}\vec{V}_{[l1]LM}.
\end{aligned} \tag{6}$$

The parity of the vector spherical harmonics is therefore given by $(-1)^{l+1}$.

3. Complex Conjugation.

$$\begin{aligned}
\vec{V}^*_{[l1]LM} &= \sum_\mu \langle l(M-\mu)1\mu|LM\rangle(-i)^l Y^*_{l(M-\mu)}\vec{e}_\mu{}^* \\
&= \sum_\mu (-1)^{l+1-L}\langle l(\mu-M)1-\mu|L-M\rangle \\
&\qquad \times i^l(-1)^l(-1)^{(M-\mu)}Y_{l(\mu-M)}(-1)^\mu \vec{e}_{-\mu}
\end{aligned}$$

$$= (-1)^{M-L+1} \sum_{\mu} \langle l(\mu - M)1 - \mu | L - M \rangle i^l Y_{l(\mu-M)} \vec{e}_{-\mu}$$

$$= (-1)^{L-M-1} \vec{V}_{[l1]L-M}. \tag{7}$$

4. Transformation properties under rotations.

Let $\vec{W}$ be any vector, with an expansion in Cartesian components given by

$$\vec{W} = W_1 \vec{e}_1 + W_2 \vec{e}_2 + W_3 \vec{e}_3 \tag{8}$$

and an expansion in spherical components, $\nu = +1, 0, -1$, given by

$$\vec{W} = \sum_{\nu} W_\nu \vec{e}_\nu^{\,*} = \sum_{\nu} W_\nu (-1)^\nu \vec{e}_{-\nu} = \sum_{\nu} W_\nu^* \vec{e}_\nu = \sum_{\nu} (-1)^\nu W_{-\nu} \vec{e}_\nu. \tag{9}$$

Then, we shall show that the scalar product $(\vec{W} \cdot \vec{V}_{[l1]LM})$ transforms under rotations as a spherical tensor of rank L with spherical components given by M. Using $(\vec{e}_\mu \cdot \vec{e}_\nu^{\,*}) = \delta_{\mu,\nu}$, we have

$$(\vec{W} \cdot \vec{V}_{[l1]LM}) = \sum_{\mu,\nu} \langle l(M-\mu)1\mu | LM \rangle i^l Y_{l(M-\mu)}(\theta, \phi) (\vec{e}_\mu \cdot \vec{e}_\nu^{\,*}) W_\nu$$

$$= \sum_{\mu} \langle l(M-\mu)1\mu | LM \rangle i^l Y_{l(M-\mu)}(\theta, \phi) W_\mu. \tag{10}$$

Now, under an arbitrary rotation in 3-D space (specified by the Euler angles, α, β, γ),

$$\left(\vec{W} \cdot \vec{V}_{[l1]LM}\right)_{\text{rot.}} = R(\alpha, \beta, \gamma)(\vec{W} \cdot \vec{V}_{[l1]LM}) R^{-1}(\alpha, \beta, \gamma)$$

$$= \sum_{\mu} \langle l(M-\mu)1\mu | LM \rangle i^l Y_{l(M-\mu)}(\theta', \phi') W'_\mu, \tag{11}$$

where θ', ϕ' are the angular variables of the rotated system with respect to the original coordinate axes, and W' are the components of the rotated vector, $\vec{W}$. Thus,

$$Y_{l(M-\mu)}(\theta', \phi') = \sum_{m} Y_{lm}(\theta, \phi) D^l_{m(M-\mu)}(\alpha, \beta, \gamma)^*$$

$$W'_\mu = \sum_{\nu} W_\nu D^1_{\nu\mu}(\alpha, \beta, \gamma)^*, \tag{12}$$

where we have used the conventions of Chapter 29 for the matrix elements of the rotation operators, the so-called Wigner or rotation D functions. We shall also use (see Chapter 30)

$$D^l_{m(M-\mu)}(\alpha, \beta, \gamma) D^1_{\nu\mu}(\alpha, \beta, \gamma) =$$

$$\sum_{\bar{L}} \langle lm1\nu | \bar{L}(m+\nu) \rangle \langle l(M-\mu)1\mu | \bar{L}M \rangle D^{\bar{L}}_{(m+\nu)M}(\alpha, \beta, \gamma). \tag{13}$$

Combining eqs. (11), (12), and the complex conjugate of eq. (13), we have

$$\left(\vec{W} \cdot \vec{V}_{[l1]LM}\right)_{\text{rot.}}$$

$$= \sum_{m,\nu} i^l Y_{lm}(\theta, \phi) W_\nu \sum_{\bar{L}} \langle lm1\nu | \bar{L}(m+\nu) \rangle D^{\bar{L}}_{(m+\nu)M}(\alpha, \beta, \gamma)^*$$

$$\times \sum_{\mu} \langle l(M-\mu)1\mu|LM\rangle\langle l(M-\mu)1\mu|\bar{L}M\rangle$$
$$= \sum_{m,\nu} i^l Y_{lm}(\theta,\phi) W_\nu \sum_{\bar{L}} \langle lm1\nu|\bar{L}(m+\nu)\rangle D^{\bar{L}}_{(m+\nu)M}(\alpha,\beta,\gamma)^* \delta_{\bar{L}L}$$
$$= \sum_{m,\nu} i^l Y_{lm}(\theta,\phi) W_\nu \langle lm1\nu|L(m+\nu)\rangle D^{L}_{(m+\nu)M}(\alpha,\beta,\gamma)^*, \tag{14}$$

where we have used the orthonormality of the Clebsch–Gordan coefficients. Thus

$$\left(\vec{W}\cdot\vec{V}_{[l1]LM}\right)_{\text{rot.}} = \sum_{m'} \left(\vec{W}\cdot\vec{V}_{[l1]Lm'}\right) D^{L}_{m'M}(\alpha,\beta,\gamma)^*. \tag{15}$$

The quantity $\left(\vec{W}\cdot\vec{V}_{[l1]LM}\right)$ is the Mth component of a spherical tensor of rank L.

B The Vector Spherical Harmonics of the Radiation Field

Now, we shall examine the types of vector spherical harmonics that actually occur in the expansion of the radiation field, $\vec{A}_\perp(\vec{r},t)$. We shall examine the vector spherical harmonics for a fixed L that occur in the expansion

$$e^{i\vec{k}\cdot\vec{r}}\vec{e}_\mu = \sum_L \sum_{l=L,(L\pm1)} \langle l01\mu|L\mu\rangle j_l(kr)\sqrt{4\pi(2l+1)}\vec{V}_{[l1]L\mu}. \tag{16}$$

We shall need the Clebsch–Gordan coefficients:

$$\langle L0\,1\pm1|L\pm1\rangle = \mp\frac{1}{\sqrt{2}} = \frac{-\mu}{\sqrt{2}},$$
$$\langle (L-1)0\,1\pm1|L\pm1\rangle = \sqrt{\frac{(L+1)}{2(2L-1)}},$$
$$\langle (L+1)0\,1\pm1|L\pm1\rangle = \sqrt{\frac{L}{2(2L+3)}}. \tag{17}$$

With these coefficients, we have

$$e^{i\vec{k}\cdot\vec{r}}\vec{e}_{\mu=\pm1} = \textstyle\sum_L \sqrt{2\pi(2L+1)}\Big[\left(-\mu j_L(kr)\vec{V}_{[L1]L\mu}\right)+$$
$$\left(\sqrt{\frac{(L+1)}{(2L+1)}} j_{L-1}(kr)\vec{V}_{[(L-1)1]L\mu} + \sqrt{\frac{L}{(2L+1)}} j_{L+1}(kr)\vec{V}_{[(L+1)1]L\mu}\right)\Big]. \tag{18}$$

It will now be convenient to define these two pieces in the vector spherical harmonic expansion of $\vec{A}_\perp$, through

$$\text{(I)}: \qquad j_L(kr)\vec{V}_{[L1]L\mu} \equiv \vec{A}(\vec{r},\mathcal{M})_{L\mu}, \tag{19}$$

II:

$$\left(\sqrt{\frac{(L+1)}{(2L+1)}}j_{L-1}\vec{V}_{[(L-1)1]L\mu}+\sqrt{\frac{L}{(2L+1)}}j_{L+1}\vec{V}_{[(L+1)1]L\mu}\right)\equiv\vec{A}(\vec{r},\mathcal{E})_{L\mu}, \quad (20)$$

where $\vec{A}(\vec{r},\mathcal{M})_{L\mu}$ is designated by the letter, $\mathcal{M}$, because (as we shall show) it is the term in the $\vec{A}_\perp$ expansion giving rise to the *magnetic* 2^L-pole radiation field. This term has parity given by $(-1)^{L+1}$. Conversely, $\vec{A}(\vec{r},\mathcal{E}))_{L\mu}$ is designated by the letter, $\mathcal{E}$, because it gives rise to the *electric* 2^L-pole radiation field. Its parity is given by $(-1)^L$. Finally, one linear combination of the $\vec{V}_{[l1]L\mu}$ exists, which is *missing* in the expansion of $\vec{A}_\perp$ in vector spherical harmonics. The missing linear combination to be designated by the letter, $\mathcal{L}$, is

III:

$$\left(\sqrt{\frac{L}{(2L+1)}}j_{L-1}\vec{V}_{[(L-1)1]L\mu}-\sqrt{\frac{(L+1)}{(2L+1)}}j_{L+1}\vec{V}_{[(L+1)1]L\mu}\right)\equiv\vec{A}(\vec{r},\mathcal{L})_{L\mu}. \quad (21)$$

This piece, with parity also given by $(-1)^L$, is missing in the expansion of $\vec{A}_\perp$ because $(\vec{\nabla}\cdot\vec{A}(\vec{r},\mathcal{L}))\neq 0$. This piece therefore occurs in the expansion of $\vec{A}_\parallel$, that is, the *longitudinal* part of the vector potential. Hence, the designation by the letter, $\mathcal{L}$ and its absence in the expansion of the radiation field, $\vec{A}_\perp$. Finally, we quote a useful alternate form for the fields, $\vec{A}(\vec{r},\mathcal{T})_{L\mu}$. (The derivation of these formulae is given in an appendix.)

$$\begin{aligned}
\text{I}:\quad \vec{A}(\vec{r},\mathcal{M})_{L\mu} &= \frac{1}{\sqrt{L(L+1)}}\vec{L}\big(j_L(kr)i^L Y_{L\mu}(\theta,\phi)\big),\\
\text{II}:\quad \vec{A}(\vec{r},\mathcal{E})_{L\mu} &= -\frac{1}{k}\frac{1}{\sqrt{L(L+1)}}\big[\vec{\nabla}\times\vec{L}\big(j_L(kr)i^L Y_{L\mu}(\theta,\phi)\big)\big]\\
&= -\frac{1}{k}\big[\vec{\nabla}\times\vec{A}(\vec{r},\mathcal{M})_{L\mu}\big],\\
\text{III}:\quad \vec{A}(\vec{r},\mathcal{L})_{L\mu} &= \frac{1}{ik}\vec{\nabla}\big(j_L(kr)i^L Y_{L\mu}(\theta,\phi)\big), \qquad (22)
\end{aligned}$$

where the operator, $\vec{L}$, is the dimensionless $-i[\vec{r}\times\vec{\nabla}]$. All three pieces of the vector potential, $\vec{A}$, are derivable from simple operators acting on the scalar function, $j_L(kr)i^L Y_{L\mu}(\theta,\phi)$.

Having attained all of these properties of the $\vec{A}(\vec{r},\mathcal{T})$, we can combine eqs. (1), (4), (18), (19), and (20) to give the expansion of the radiation field

$$\begin{aligned}
\vec{A}_\perp(\vec{r},t) = \frac{2\pi c}{\sqrt{\text{Vol.}}}\sum_{\vec{k}}\sum_{\mu=\pm1}\sum_{L}\sqrt{\frac{\hbar(2L+1)}{\omega}}\Big[a_{\vec{k}\mu}\Big(-\mu\vec{A}(\vec{r},\mathcal{M})_{L\mu}+\vec{A}(\vec{r},\mathcal{E})_{L\mu}\Big)\\
+a^\dagger_{\vec{k}\mu}\Big(-\mu\vec{A}(\vec{r},\mathcal{M})^*_{L\mu}+\vec{A}(\vec{r},\mathcal{E})^*_{L\mu}\Big)\Big]. \qquad (23)
\end{aligned}$$

(We have written, Vol., explicitly for the volume of our cubical laboratory, to avoid confusion with the angular momentum index, L.) Also, recall the $a_{\vec{k}\mu}$ are time dependent, $a_{\vec{k}\mu}(t) = a_{\vec{k}\mu}(0)e^{-i\omega t}$ and, finally, from eq. (7), note $\vec{A}(\vec{r}, \mathcal{M}(\text{or } \mathcal{E}))^*_{L\mu} = (-1)^{L+1-\mu}\vec{A}(\vec{r}, \mathcal{M}(\text{or } \mathcal{E}))_{L-\mu}$.

The $\vec{A}(\vec{r}, \mathcal{M})_{L\mu}$ and $\vec{A}(\vec{r}, \mathcal{E})_{L\mu}$ lead to the radiation $\vec{E}$ and $\vec{B}$ fields. In particular,

$$\begin{aligned} \vec{E}(\vec{r}, t; \mathcal{M})_{L\mu} &= ik\vec{A}(\vec{r}, \mathcal{M})_{L\mu}e^{-i\omega t}, \\ \vec{B}(\vec{r}, t; \mathcal{M})_{L\mu} &= [\vec{\nabla} \times \vec{A}(\vec{r}, \mathcal{M})_{L\mu}]e^{-i\omega t} = -k\vec{A}(\vec{r}, \mathcal{E})_{L\mu}e^{-i\omega t}, \end{aligned} \tag{24}$$

where we have used the second relation of eq. (22) in the last step. Similarly,

$$\begin{aligned} \vec{E}(\vec{r}, t; \mathcal{E})_{L\mu} &= ik\vec{A}(\vec{r}, \mathcal{E})_{L\mu}e^{-i\omega t} = -i\vec{B}(\vec{r}, t; \mathcal{M})_{L\mu}, \\ \vec{B}(\vec{r}, t; \mathcal{E})_{L\mu} &= -k\vec{A}(\vec{r}, \mathcal{M})_{L\mu}e^{-i\omega t} = i\vec{E}(\vec{r}, t; \mathcal{M})_{L\mu}, \end{aligned} \tag{25}$$

where we have used

$$[\vec{\nabla} \times [\vec{\nabla} \times \vec{A}]] = -\nabla^2\vec{A} = k^2\vec{A}$$

for this divergence-free $\vec{A}$ together with the second of the relations of eq. (22) in the expression for the $\vec{B}$ field.

In many atomic and other quantum systems, we shall be interested in the radiation fields in the limit in which the parameter, (kr), is a small quantity, $(r \ll \lambda)$, i.e., dimension of the radiating quantum system $\ll$ wavelength of the emitted photon. In this case, we may need only the dominant term in the expansion of the spherical Bessel functions, $j_L(kr)$, which gives the radial dependence of the above fields. Recall

$$j_L(kr) \approx \frac{2^L L!}{(2L+1)!}(kr)^L \qquad \text{for} \quad kr \ll 1. \tag{26}$$

Thus, in this limit,

$$\vec{A}(\vec{r}, \mathcal{E}) \sim (kr)^{L-1}, \qquad \vec{A}(\vec{r}, \mathcal{M}) \sim (kr)^L, \tag{27}$$

and therefore for the $\mathcal{E}$-type fields

$$\begin{aligned} \vec{E}(\vec{r}, \mathcal{E}) &\sim ik(kr)^{L-1}, \\ \vec{B}(\vec{r}, \mathcal{E}) &\sim -k(kr)^L. \end{aligned} \tag{28}$$

For the $\mathcal{E}$-type fields of a definite L, the $\vec{E}$-field is dominant and has the r dependence of an electric 2^L-pole field. Conversely, for $\mathcal{M}$-type fields,

$$\begin{aligned} \vec{E}(\vec{r}, \mathcal{M}) &\sim ik(kr)^L, \\ \vec{B}(\vec{r}, \mathcal{M}) &\sim -k(kr)^{L-1}. \end{aligned} \tag{29}$$

Now, the $\vec{B}$-field dominates for a definite L and has the r dependence of a magnetic 2^L-pole field.

Before examining the electric and magnetic multipole radiation fields in detail, we shall (in the next chapter) examine first the dominant electric dipole approximation for quantum systems, such as atoms, for which the approximation $kr \ll 1$ is a very good one.

C Mathematical Appendix

In this appendix, we shall derive the alternate form for the $\vec{A}(\vec{r}, \mathcal{T})$, given in eq. (22).
1. The relation

$$\vec{A}(\vec{r}, \mathcal{M})_{L\mu} \equiv j_L(kr)\vec{V}_{[l1]L\mu} = \frac{1}{\sqrt{L(L+1)}}\vec{L}\big(j_L(kr)i^L Y_{L\mu}(\theta,\phi)\big).$$

To derive this relation, we shall write

$$\vec{L} = \sum_{\mu} L_{\mu}\vec{e}^{\,*}_{\mu} \tag{30}$$

and use the fact that the operators, $L_\mu = -i[\vec{r} \times \vec{\nabla}]_\mu$, act only on the angles θ, ϕ, (not on r). Thus,

$$\begin{aligned}
\vec{L}(i^L Y_{LM}) &= \sum_{\mu} \vec{e}^{\,*}_{\mu} i^L (L_\mu Y_{LM}) \\
&= \sum_{\mu} (-1)^{\mu} \vec{e}_{-\mu} \langle LM1\mu | L(M+\mu)\rangle i^L Y_{L(M+\mu)} \sqrt{L(L+1)} \\
&= \sum_{\mu} \vec{e}_{-\mu} (-1)^{\mu-\mu} \langle L(M+\mu)1-\mu | LM\rangle i^L Y_{L(M+\mu)} \sqrt{L(L+1)} \\
&= \sqrt{L(L+1)} \sum_{\mu} \langle L(M-\mu)1\mu | LM\rangle i^L Y_{L(M-\mu)} \vec{e}_{\mu} \\
&= \sqrt{L(L+1)}\vec{V}_{[L1]LM},
\end{aligned} \tag{31}$$

where we have used the Wigner–Eckart theorem for the well-known matrix elements of the spherical components, L_μ, in the second line, the Clebsch–Gordan coefficient symmetry property involving $1 \leftrightarrow 3$ interchange in the third line, and finally have renamed the dummy summation index $\mu \leftrightarrow -\mu$ in the next line; leading to the desired result.
2. The gradient formula:

$$\begin{aligned}
\frac{1}{i}\vec{\nabla}\big(f(r)i^l Y_{lm}(\theta,\phi)\big) &= \sqrt{\frac{l}{(2l+1)}}\left(\frac{df}{dr} + \frac{(l+1)}{r}f(r)\right)\vec{V}_{[(l-1)1]lm} \\
&+ \sqrt{\frac{(l+1)}{(2l+1)}}\left(\frac{df}{dr} - \frac{l}{r}f(r)\right)\vec{V}_{[(l+1)1]lm}.
\end{aligned} \tag{32}$$

To derive this formula, we shall write

$$\frac{1}{i}\vec{\nabla}\big(f(r)i^l Y_{lm}(\theta,\phi)\big) = \frac{1}{i}\sum_{\mu}(-1)^{\mu}\vec{e}_{-\mu}\nabla_{\mu}\big(f(r)i^l Y_{lm}(\theta,\phi)\big). \tag{33}$$

We shall again use the Wigner–Eckart theorem to find the matrix elements of ∇_μ. It is sufficient to pick one particular spherical component of the operator $\vec{\nabla}$. For simplicity, we pick the component with $\mu = 0$

$$\nabla^1_0 = \frac{\partial}{\partial z} = \cos\theta\frac{\partial}{\partial r} - \frac{1}{r}\sin\theta\frac{\partial}{\partial\theta}. \tag{34}$$

From Chapters 9 and 32, we have [see, in particular, eqs. (33) and (34) of Chapter 9 and eq. (8) of Chapter 32],

$$\cos\theta Y_{lm} = \sum_{l'} \sqrt{\frac{4\pi}{3}} \sqrt{\frac{3(2l+1)}{4\pi(2l'+1)}} \langle l010|l'0\rangle \langle lm10|l'm\rangle Y_{l'm}, \tag{35}$$

and

$$\begin{aligned} \sin\theta\frac{\partial}{\partial\theta} Y_{lm} &= l\sqrt{\frac{[(l+1)^2-m^2]}{(2l+1)(2l+3)}} Y_{(l+1)m} - (l+1)\sqrt{\frac{[l^2-m^2]}{(2l+1)(2l-1)}} Y_{(l-1)m} \\ &= l\sqrt{\frac{(2l+1)}{(2l+3)}} \langle l010|(l+1)0\rangle \langle lm10|(l+1)m\rangle Y_{(l+1)m} \\ &\quad - (l+1)\sqrt{\frac{(2l+1)}{(2l-1)}} \langle l010|(l-1)0\rangle \langle lm10|(l-1)m\rangle Y_{(l-1)m}, \end{aligned} \tag{36}$$

where we have used

$$\begin{aligned} \langle lm10|(l+1)m\rangle &= \sqrt{\frac{[(l+1)^2-m^2]}{(2l+1)(l+1)}}, \\ \langle lm10|(l-1)m\rangle &= -\sqrt{\frac{[l^2-m^2]}{l(2l+1)}}. \end{aligned} \tag{37}$$

Combining these relations, we get

$$\begin{aligned} &\left(\cos\theta\frac{\partial}{\partial r} - \frac{1}{r}\sin\theta\frac{\partial}{\partial\theta}\right)(f(r)Y_{lm}) = \\ &\sqrt{\frac{(2l+1)(l+1)}{(2l+3)(2l+1)}}\left(\frac{df}{dr} - \frac{l}{r}f(r)\right)Y_{(l+1)m}\langle lm10|(l+1)m\rangle \\ &-\sqrt{\frac{(2l+1)l}{(2l-1)(2l+1)}}\left(\frac{df}{dr} + \frac{(l+1)}{r}\right)Y_{(l-1)m}\langle lm10(l-1)m\rangle. \end{aligned} \tag{38}$$

From these, we can read the reduced matrix elements of the operator, $\vec{\nabla}$.

$$\begin{aligned} \frac{\langle (l+1)\|\vec{\nabla}\|l\rangle}{\sqrt{(2l+3)}} &= \sqrt{\frac{(2l+1)(l+1)}{(2l+3)(2l+1)}}\left(\frac{df}{dr} - \frac{l}{r}f(r)\right) \\ \frac{\langle (l-1)\|\vec{\nabla}\|l\rangle}{\sqrt{(2l-1)}} &= -\sqrt{\frac{(2l+1)l}{(2l-1)(2l+1)}}\left(\frac{df}{dr} + \frac{(l+1)}{r}f(r)\right). \end{aligned} \tag{39}$$

We can now rewrite

$$\begin{aligned} &\frac{1}{i}\vec{\nabla}(f(r)i^l Y_{lm}) \\ &= i^{l-1}\sum_{\mu}(-1)^{\mu}\vec{e}_{-\mu}\sum_{l'}\langle lm1\mu|l'(m+\mu)\rangle\frac{\langle l'\|\vec{\nabla}\|l\rangle}{\sqrt{(2l'+1)}}Y_{l'(m+\mu)} \end{aligned}$$

$$= i^{l-1} \sum_{l'=(l\pm 1)} \sum_{\mu} (-1)^{\mu} \vec{e}_{-\mu} \frac{\langle l' \| \vec{\nabla} \| l \rangle}{\sqrt{(2l+1)}} (-1)^{l-l'+\mu} \langle l'(m+\mu) 1 -\mu | lm \rangle Y_{l'(m+\mu)}$$

$$= \sum_{l'=(l\pm 1)} i^{l-1}(-i)^{l'} \frac{\langle l' \| \vec{\nabla} \| l \rangle}{\sqrt{(2l+1)}} (-1)^{l-l'} \sum_{\mu} \langle l'(m-\mu) 1 \mu | lm \rangle i^{l'} Y_{l'(m-\mu)} \vec{e}_{\mu}$$

$$= \sum_{l'=(l\pm 1)} i^{l'-l-1} \sqrt{\frac{(2l'+1)}{(2l+1)}} \vec{V}_{[l'1]lm} \frac{\langle l' \| \vec{\nabla} \| l \rangle}{\sqrt{(2l'+1)}}, \tag{40}$$

where we have again used a 1 $\leftrightarrow$ 3 interchange symmetry property of the Clebsch–Gordan coefficient in line 2, and have again changed the dummy summation index $\mu \leftrightarrow -\mu$ in line 3 of this relation. Substituting from the two known reduced matrix elements of eq. (39), we get the desired gradient formula of eq. (32). For the special case, when $f(r) = j_l(kr)$, we have from the results of Chapter 41 and the mathematical appendix to Chapter 41:

$$\left(\frac{d}{dr} - \frac{l}{r}\right) j_l(kr) = -k j_{(l+1)}(kr), \tag{41}$$

$$\left(\frac{d}{dr} + \frac{(l+1)}{r}\right) j_l(kr) = +k j_{(l-1)}(kr). \tag{42}$$

This leads to the relation (III) of eq. (22).

3. Derivation of

$$\vec{A}(\vec{r}, \mathcal{E})_{LM} = -\frac{1}{k}[\vec{\nabla} \times \vec{A}(\vec{r}, \mathcal{M})_{LM}].$$

We shall use $\vec{A}(\vec{r}, \mathcal{M})_{LM} = j_L(kr)\vec{V}_{[L1]LM}$ and take the curl of this vector

$$[\vec{\nabla} \times (j_L(kr)\vec{V}_{[L1]LM})]$$
$$= \Big[\vec{\nabla} \times \Big(j_L(kr) \sum_{\mu} \langle L(M-\mu) 1 \mu | LM \rangle i^L Y_{L(M-\mu)} \vec{e}_{\mu}\Big)\Big]$$
$$= \sum_{\mu} \langle L(M-\mu) 1 \mu | LM \rangle \Big[\vec{\nabla}\big(j_L(kr) i^L Y_{L(M-\mu)}\big) \times \vec{e}_{\mu}\Big]$$
$$= i \sum_{\mu} \langle L(M-\mu) 1 \mu | LM \rangle \Bigg(\sqrt{\frac{L}{(2L+1)}} k j_{(L-1)}(kr) \Big[\vec{V}_{[(L-1)1]L(M-\mu)} \times \vec{e}_{\mu}\Big]$$
$$- \sqrt{\frac{(L+1)}{(2L+1)}} k j_{(L+1)}(kr) \Big[\vec{V}_{[(L+1)1]LM} \times \vec{e}_{\mu}\Big]\Bigg), \tag{43}$$

where we have used $[\vec{\nabla} \times (\chi \vec{e})] = [(\vec{\nabla}\chi) \times \vec{e}]$ for the special case in which $\vec{e}$ is a *constant* vector and χ is a scalar, and where we have used the gradient formula in the last step. Now, we shall use

$$\Big[\vec{V}_{[l'1]L(M-\mu)} \times \vec{e}_{\mu}\Big] =$$
$$\sum_{\nu} \langle l'((M-\mu-\nu) 1 \nu | L(M-\mu) \rangle i^{l'} Y_{l'(M-\mu-\nu)} [\vec{e}_{\nu} \times \vec{e}_{\mu}]. \tag{44}$$

In terms of spherical components, the vector product of the unit vectors can be seen to be expressible in terms of a Clebsch–Gordan coefficient

$$[\vec{e}_\nu \times \vec{e}_\mu] = i\sqrt{2}\langle 1\nu 1\mu | 1(\mu+\nu)\rangle \vec{e}_{(\mu+\nu)}, \tag{45}$$

so

$$\begin{aligned}
&\sum_\mu \langle L(M-\mu)1\mu|LM\rangle [\vec{V}_{[l'1]L(M-\mu)} \times \vec{e}_\mu] \\
&\quad = i\sqrt{2}\sum_{\mu,\nu} \langle l'(M-\mu-\nu)1\nu|L(M-\mu)\rangle\langle L(M-\mu)1\mu|LM\rangle \\
&\quad \times \langle 1\nu 1\mu|1(\mu+\nu)\rangle i^{l'} Y_{l'(M-\mu-\nu)}\vec{e}_{\mu+\nu} \\
&\quad = i\sqrt{2}\sum_\lambda \Big(\sum_\mu \langle l'(M-\lambda)1(\lambda-\mu)|L(M-\mu)\rangle\langle L(M-\mu)1\mu|LM\rangle \\
&\quad \times \langle 1(\lambda-\mu)1\mu|1\lambda\rangle\Big) i^{l'} Y_{l'(M-\lambda)}\vec{e}_\lambda,
\end{aligned} \tag{46}$$

where, for convenience, we have changed the summation indices by renaming $(\mu+\nu)=\lambda$ in the last line. Now, we can do the μ sum in the last step. In particular, it can be shown that this sum over μ of a product of three μ-dependent Clebsch–Gordan coefficients is, for *fixed* M and *fixed* λ, equal to a single Clebsch–Gordan coefficient times an l' dependendent, but (magnetic quantum number)-independent coefficient:

$$\begin{aligned}
&\sum_\mu \langle l'(M-\lambda)1(\lambda-\mu)|L(M-\mu)\rangle\langle L(M-\mu)1\mu|Lm\rangle\langle 1(\lambda-\mu)1\mu|1\lambda\rangle \\
&\quad = c_{l'}\langle l'(M-\lambda)1\lambda|LM\rangle,
\end{aligned} \tag{47}$$

with

$$c_{(L-1)} = \sqrt{\frac{(L+1)}{2L}}, \qquad c_{(L+1)} = -\sqrt{\frac{L}{2(L+1)}}. \tag{48}$$

This equation can be shown by direct verification. It also follows more elegantly from angular-momentum recoupling theory (see Chapter 34). The coefficients $c_{l'}$ introduced here are given by the unitary form of the Racah coefficient, $c_{l'} = U(l'1L1; L1)$ [see eq. (12) of Chapter 34]. Using eqs. (46) and (47), we have

$$\begin{aligned}
&\sum_\mu \langle L(M-\mu)1\mu|LM\rangle\big[\vec{V}_{[l'1]L(M-\mu)} \times \vec{e}_\mu\big] \\
&\quad = i\sqrt{2}c_{l'}\sum_\lambda \langle l'(M-\lambda)1\lambda|LM\rangle i^{l'} Y_{l'(M-\lambda)}\vec{e}_\lambda \\
&\quad = i\sqrt{2}c_{l'}\vec{V}_{[l'1]LM}.
\end{aligned} \tag{49}$$

Using this result, together with the $c_{l'}$ of eq. (48) and eq.(43), we get

$$\begin{aligned}
&[\vec{\nabla} \times \big(j_L(kr)\vec{V}_{[L1]LM}\big)] = \\
&-k\left(\sqrt{\tfrac{(L+1)}{(2L+1)}}\, j_{(L-1)}(kr)\vec{V}_{[(L-1)1]LM} + \sqrt{\tfrac{L}{(2L+1)}}\, j_{(L+1)}(kr)\vec{V}_{[(L+1)1]LM}\right) \\
&\qquad = -k\vec{A}(\vec{r},\mathcal{E})_{LM},
\end{aligned} \tag{50}$$

leading to the desired result

$$\vec{A}(\vec{r}, \mathcal{E})_{LM} = -\frac{1}{k}[\vec{\nabla} \times \vec{A}(\vec{r}, \mathcal{M})_{LM}]. \tag{51}$$

63

The Emission of Photons by Atoms: Electric Dipole Approximation

We shall now look at the interaction of an atomic system with the radiation field in detail. We have

$$H = H_{\text{atom}} + H_{\text{radiation}} + H_{\text{interaction}}, \tag{1}$$

where $H_{\text{radiation}}$ will be written in terms of the photon creation and annihilation operators as in Chapters 60 and 61, H_{atom} has the usual kinetic and potential energy terms, and $H_{\text{interaction}}$ is given by

$$\begin{aligned} H_{\text{interaction}} = \\ \sum_{i=1}^{n}\left(\frac{-e_i}{2m_i c}(\vec{p}_i \cdot \vec{A}_\perp(\vec{r}_i, t) + \vec{A}_\perp(\vec{r}_i, t) \cdot \vec{p}_i) + \frac{e_i^2}{2m_i c^2}(\vec{A}_\perp(\vec{r}_i, t) \cdot \vec{A}_\perp(\vec{r}_i, t))\right) \\ - \sum_{i=1}^{n} \frac{e_i \hbar}{2m_i c} g_{s_i} \left(\vec{s}_i \cdot [\vec{\nabla}_i \times \vec{A}_\perp(\vec{r}_i, t)]\right). \end{aligned} \tag{2}$$

We can combine the two terms, $\vec{p}_i \cdot \vec{A}_\perp(\vec{r}_i, t)$ and $\vec{A}_\perp(\vec{r}_i, t) \cdot \vec{p}_i$, because $\vec{A}_\perp$ is divergence-free.

$$\vec{p}_i \cdot \vec{A}_\perp(\vec{r}_i, t) = \vec{A}_\perp(\vec{r}_i, t) \cdot \vec{p}_i + \frac{\hbar}{i}\sum_\alpha \frac{\partial}{\partial x_{i,\alpha}}\left(A_\perp(\vec{r}_i, t)\right)_\alpha = \vec{A}_\perp(\vec{r}_i, t) \cdot \vec{p}_i, \tag{3}$$

because $\partial(A_\perp)_\alpha/\partial x_\alpha = 0$. In the above,

$$\vec{A}(\vec{r}_i, t) = \frac{c}{\sqrt{L^3}}\sum_{\vec{k}}\sum_{\lambda}\sqrt{\frac{2\pi\hbar}{\omega}}\left(a_{\vec{k}\lambda} e^{i\vec{k}\cdot\vec{r}_i}\vec{e}_\lambda + a^\dagger_{\vec{k}\lambda} e^{-i\vec{k}\cdot\vec{r}_i}\vec{e}^{\,*}_\lambda\right). \tag{4}$$

For an atomic system, we can expect the quantity, $\vec{k} \cdot \vec{r}_i$, to be very small:

$$kr_i = \frac{2\pi}{\lambda} a_0 \approx \frac{2\pi (10)^{-8}\text{cm}}{6,000 \times (10)^{-8}\text{cm}} \approx (10)^{-3},$$

assuming the photon is in the visible part of the spectrum. For such a system, we can use $e^{i\vec{k}\cdot\vec{r}_i} \approx 1$ to very good approximation. As a second example, think of a nuclear system. For a nucleus,

$$kr_i = \frac{\hbar \omega r_i}{\hbar c} = \frac{(\hbar\omega)_\gamma R_{\text{nucleus}}}{\hbar c} = \frac{(\hbar\omega)_\gamma R_{\text{nucleus}}}{200\text{MeVfermi}},$$

so, for a heavy nucleus, with $R_{\text{nucleus}} \approx 10$ fermi and emitted γ photon of energy 2 MeV, we would have $kr_i \approx (10)^{-1}$, still small, but now large enough so higher order terms in the expansion in powers of kr_i become important. This will be particularly true for transitions between low-lying states in a nucleus that often have the same parity so the zeroth-order term will lead to electric dipole matrix elements that are rigorously zero because of parity selection rules. In such a system then, we will find it advantageous to make full use of the expansion of $\vec{A}_\perp$ in vector spherical harmonics. With the full expansion, we will get higher order electric and magnetic multipole moment matrix elements. For the atomic case, however, where the zeroth-order term dominates, we can dispense with the vector spherical harmonic formalism. In this case (as we shall see), the atomic transition probability matrix elements will involve only matrix elements of the atomic electric dipole moment operator.

For the atomic system, we can take

$$H_{\text{interaction}} = H^{(1)}_{\text{perturbation}} = -\sum_i \frac{e_i}{m_i c} \vec{p}_i \cdot \vec{A}(\vec{r}_i, t), \tag{5}$$

where we take only the zeroth-order term in the kr_i expansion of $\vec{A}$, so

$$\vec{A} = \frac{c}{\sqrt{L^{\frac{3}{2}}}} \sum_{\vec{k}} \sum_{\lambda} \sqrt{\frac{2\pi\hbar}{\omega}} (a_{\vec{k}\lambda} \vec{e}_\lambda + a^\dagger_{\vec{k}\lambda} \vec{e}^{\,*}_\lambda). \tag{6}$$

The term quadratic in $\vec{A}(\vec{r}_i, t)$ is a second-order perturbation term, and the magnetic dipole moment term

$$\sum_{i=1}^{n} \frac{e_i \hbar}{2m_i c} g_{s_i} (\vec{s}_i \cdot [\vec{\nabla}_i \times \vec{A}_\perp(\vec{r}_i, t)])$$

is of order

$$(\hbar k)/p_i = kr_i$$

times the

$$\sum_i \frac{e_i}{m_i c} \vec{p}_i \cdot \vec{A}(\vec{r}_i, t)$$

term, and can therefore be neglected in the $kr_i \ll 1$ or electric dipole approximation.

In doing the perturbation calculation, the state vectors must now carry the information about both the atomic system ($H^{(0)}_{\text{atom}}$) and the photons ($H^{(0)}_{\text{radiation}}$). The initial and final states, $|i\rangle$ and $|f\rangle$, must carry both the quantum numbers for the atomic system, such as $n_i l_i m_{l_i} m_{s_i}$ to be denoted by the shorthand symbol A_i, and the quantum numbers for the radiation field, i.e., the photon numbers, $n_{\vec{k}\mu}$, for all $\vec{k}$, μ of the initial state, similarly for the final state. For the spontaneous emission of a photon from an atom in an initial state, A_i, to a lower final state, A_f, we shall need

$$|i\rangle = |A_i \;\; 0\rangle, \qquad\qquad |f\rangle = |A_f \;\; n_{\vec{k}\mu} = 1\rangle,$$

where A_i designates the atomic quantum numbers of the initial excited atomic state, 0 designates all $n_{\vec{k}\mu} = 0$, the photon vacuum state, and A_f designates the atomic quantum numbers of the final state. Our first-order perturbation theory result can be read from the generalized diagram of Fig. 63.1(a), where we generalize the state vector lines by a straight line for the atomic system and a wiggly line for the photon.

$$\langle A_f \;\; n_{\vec{k}\mu} = 1|U^{(1)}(t,t_0)|A_i \;\; 0\rangle =$$

$$-\frac{i}{\hbar}\int_{t_0}^{t} d\tau e^{-\frac{i}{\hbar}(E_f^{(0)}+\hbar\omega)(t-\tau)}\langle A_f \;\; n_{\vec{k}\mu} = 1|H_{\text{interaction}}|A_i \;\; 0\rangle e^{-\frac{i}{\hbar}E_i^{(0)}(\tau-t_0)}. \qquad (7)$$

(Note: For our time-dependent perturbation theory formalism, it is best to convert to the Schrödinger picture, where we consider the state vectors to be time-dependent and the operators to be time-independent.) The energies $E_i^{(0)}$ and $E_f^{(0)}$ are the eigenvalues of $H^{(0)}_{\text{atom}}$. The final energy includes the energy of the photon. The energy for any time interval of the diagram includes the energies of all atom and photon lines. We then get

$$\lim_{(t-t_0)\to\infty}\frac{\left|\langle A_f \;\; n_{\vec{k}\mu} = 1|U(t,t_0)|A_i \;\; 0\rangle\right|^2}{(t-t_0)} = \frac{2\pi}{\hbar}\delta(E_f^{(0)}+\hbar\omega - E_i^{(0)})$$

$$\times\left|\langle A_f \;\; n_{\vec{k}\mu} = 1|\sum_i \frac{e_i}{m_i c}(\vec{p}_i\cdot\vec{e}^{\,*}_\mu)a^\dagger_{\vec{k}\mu}\frac{1}{L^{\frac{3}{2}}}c\sqrt{\frac{2\pi\hbar}{\omega}}|A_i \;\; 0\rangle\right|^2$$

$$= \frac{2\pi}{\hbar}\delta(E_f^{(0)}+\hbar\omega - E_i^{(0)})\frac{c^2}{L^3}\frac{2\pi\hbar}{\omega}\left|\langle A_f|\sum_i \frac{e_i}{m_i c}(\vec{p}_i\cdot\vec{e}^{\,*}_\mu)|A_i\rangle\right|^2, \qquad (8)$$

where we have carried out the photon part of the matrix element,

$$\langle n_{\vec{k}\mu} = 1|a^\dagger_{\vec{k}\mu}|0\rangle = 1,$$

so the last matrix element is a matrix element of a purely atomic operator between purely atomic states. It will now be convenient to convert this atomic matrix element to a different form in which we will recognize the electric dipole moment operator of the atomic system. Using

$$H^{(0)}_{\text{atom}} = \sum_i \frac{\vec{p}_i^{\,2}}{2m_i} + \sum_{i<k} V(\vec{r}_i,\vec{r}_k), \qquad (9)$$

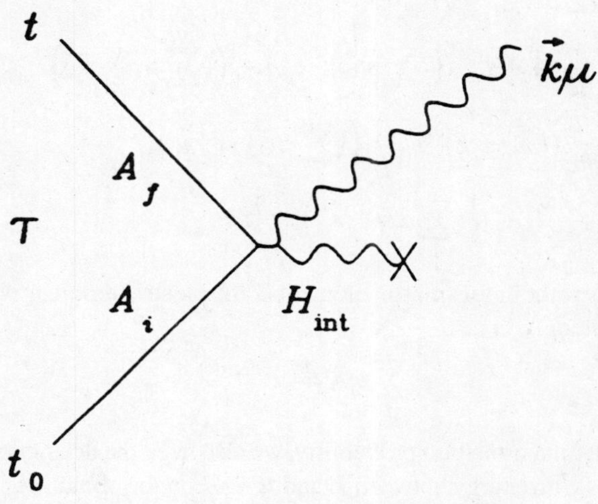

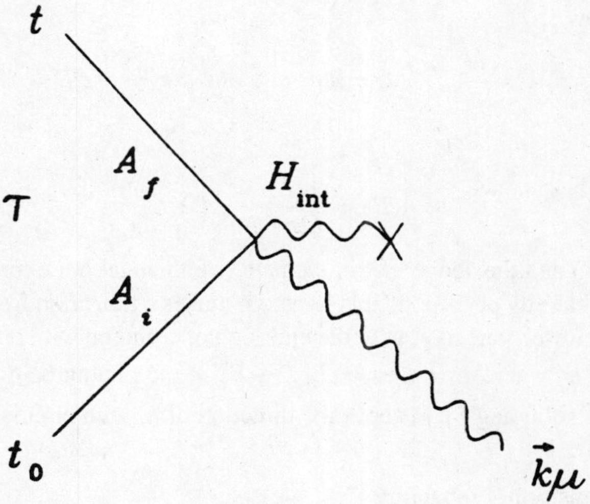

FIGURE 63.1. First-order diagrams for (a) photon emission and (b) photon absorption.

we have the simple commutator relation

$$[H^{(0)}_{\text{atom}}, \vec{r}_i] = \frac{\hbar}{i}\frac{\vec{p}_i}{m_i}, \tag{10}$$

so

$$\begin{aligned}
&\langle A_f | \sum_i \frac{e_i}{c}\Big(\frac{\vec{p}_i}{m_i}\cdot\vec{e}^{\,*}_\mu\Big)|A_i\rangle \\
&= \frac{i}{\hbar c}\langle A_f|H^{(0)}_{\text{atom}}\Big(\big(\sum_i e_i\vec{r}_i\big)\cdot\vec{e}^{\,*}_\mu\Big) - \Big(\big(\sum_i e_i\vec{r}_i\big)\cdot\vec{e}^{\,*}_\mu\Big)H^{(0)}_{\text{atom}}|A_i\rangle \\
&= \frac{i}{\hbar c}(E^{(0)}_f - E^{(0)}_i)\langle A_f|\Big(\big(\sum_i e_i\vec{r}_i\big)\cdot\vec{e}^{\,*}_\mu\Big)|A_i\rangle \\
&= -\frac{i\omega}{c}\langle A_f|\Big(\big(\sum_i e_i\vec{r}_i\big)\cdot\vec{e}^{\,*}_\mu\Big)|A_i\rangle.
\end{aligned} \tag{11}$$

The atomic operator in this matrix element is the electric dipole moment vector of the atomic system

$$\sum_i e_i\vec{r}_i = \vec{\mu}^{\,(\text{electric})}. \tag{12}$$

Now, to get the final transition probability, we also need the density of states, $\rho(E)$, for the photons with energy between E and $E+dE$ in the photon continuum, or the finely discrete spectrum of the photon waves in our cubical laboratory of volume L^3. In Chapter 57, we calculated this quantity and found

$$\rho(E)dE = \frac{L^3}{(2\pi)^3}k^2 dk d\Omega, \tag{13}$$

where $d\Omega$ is the element of solid angle about the direction of $\vec{k}$. In Chapter 57, we dealt with matter waves. Now, the waves in our cube are electromagnetic. Therefore, now,

$$k = \frac{\omega}{c}, \qquad E = \hbar\omega, \qquad \text{and } dk = \frac{1}{\hbar c}dE. \tag{14}$$

Thus,

$$\rho(E) = \frac{L^3}{(2\pi)^3}\frac{\omega^2}{\hbar c^3}d\Omega. \tag{15}$$

With this $\rho(E)$ and the above matrix element, we then get our expression for the transition probability per second that the atom makes a transition from the excited state, A_i, to a lower state, A_f, with the emission of a photon with energy between E and $E + dE$, where $E = \hbar\omega = (E^{(0)}_i - E^{(0)}_f)$, the photon being emitted into an element of solid angle $d\Omega$ about the direction of $\vec{k}$, with circular polarization given by $\mu = \pm 1$,

$$\left[\frac{\text{Transition Probability}}{\text{second}}\right]_{\text{atom } A_i\to A_f + \text{ photon of type } \vec{k},\mu \text{ into } d\Omega}$$

$$= \frac{2\pi}{\hbar}\rho(E)\frac{c^2}{L^3}\frac{2\pi\hbar}{\omega}\frac{\omega^2}{c^2}\left| \langle A_f|(\sum_i e_i\vec{r}_i)\cdot\vec{e}^{\,*}_{\mu}|A_i\rangle \right|^2$$
$$= \frac{\omega^3 d\Omega}{2\pi\hbar c^3}\left| \langle A_f|(\vec{\mu}^{\,(\mathrm{el.})}\cdot\vec{e}^{\,*}_{\mu})|A_i\rangle \right|^2. \tag{16}$$

If we want the total transition probability per second that the photon with energy $\hbar\omega = (E^{(0)}_i - E^{(0)}_f)$ be emitted into any direction with either polarization, we must still integrate over all possible photon directions and sum over the two possible photon polarizations for each direction.

$$\left[\frac{\text{Transition Prob.}}{\text{second}}\right]_{A_i\to A_f} = \frac{\omega^3}{2\pi\hbar c^3}\sum_{\mu=\pm1}\int\int d\Omega \left| \langle A_f|(\vec{\mu}^{\,(\mathrm{el.})}\cdot\vec{e}^{\,*}_{\mu})|A_i\rangle \right|^2. \tag{17}$$

To carry out the polarization sum, it will be convenient to expand the electric dipole moment vector in spherical components

$$\vec{\mu}^{\,(\mathrm{el.})} = \sum_\nu (\vec{\mu}^{\,(\mathrm{el.})})'^{*}_{\nu}\vec{e}_\nu = \sum_\nu (-1)^\nu (\vec{\mu}^{\,(\mathrm{el.})})'_{-\nu}\vec{e}_\nu, \tag{18}$$

and use the orthogonality relation in spherical component form, $(\vec{e}_\nu\cdot\vec{e}^{\,*}_{\mu}) = \delta_{\nu,\mu}$, and subsequently change the dummy summation index $\nu \to -\nu$, after the μ sums have been performed in eq. (17). This method gives us

$$\left[\frac{\text{Transition Prob.}}{\text{second}}\right]_{A_i\to A_f} = \frac{\omega^3}{2\pi\hbar c^3}\sum_{\nu=\pm1}\int\int d\Omega \left| \langle A_f|(\vec{\mu}^{\,(\mathrm{el.})})'_{\nu}|A_i\rangle \right|^2. \tag{19}$$

Also, the circular polarization index, ν or μ, is defined relative to the photon direction, $\vec{k}$. If we name the direction of $\vec{k}$ a z' direction, where the z' axis makes polar and azimuth angles, θ and ϕ, with respect to a laboratory reference frame, x, y, z (see Fig. 63.2), then the electric dipole moment vector in eqs. (18) and (19) has spherical components relative to this primed coordinate system. Hence, these components are designated by μ'_ν, with $\nu = \pm1$. Using the transformation from primed to unprimed spherical components, we have

$$\langle A_f|(\mu^{(\mathrm{el.})})'_\nu|A_i\rangle = \sum_{m=\pm1,0} \langle A_f|(\mu^{(\mathrm{el.})})_m|A_i\rangle D^1_{m\nu}(\phi,\theta,0)^*, \tag{20}$$

so

$$\int\int d\Omega \left| \langle A_f|(\mu^{(\mathrm{el.})})'_\nu|A_i\rangle \right|^2 = \sum_m\sum_{m'}\langle A_f|(\mu^{(\mathrm{el.})})_m|A_i\rangle\langle A_f|(\mu^{(\mathrm{el.})})_{m'}|A_i\rangle^*$$
$$\times \int\int d\Omega D^1_{m\nu}(\phi,\theta,0)^* D^1_{m'\nu}(\phi,\theta,0). \tag{21}$$

Now, using a special case of the D function orthonormality integral [see eq. (30) of Chapter 30],

$$\int\int d\Omega D^1_{m\nu}(\phi,\theta,0)^* D^1_{m'\nu}(\phi,\theta,0) = \frac{4\pi}{3}\delta_{m,m'}, \tag{22}$$

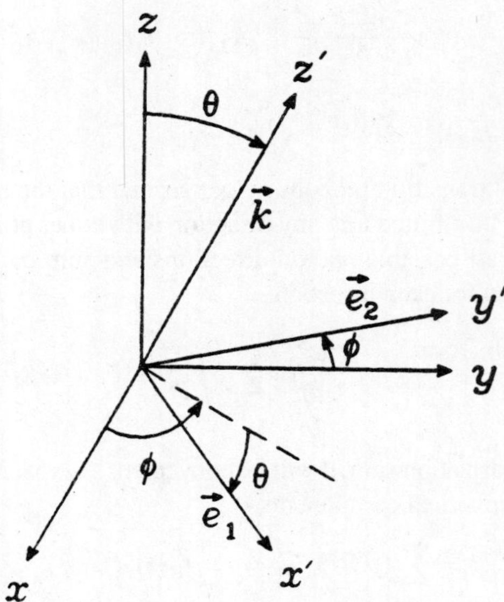

FIGURE 63.2.

and summing over the two possible polarizations, $\nu = \pm 1$, we get

$$\sum_{\nu=\pm 1} \int\int d\Omega \mid \langle A_f|(\mu^{(\text{el.})})'_\nu|A_i\rangle \mid^2 = \frac{8\pi}{3} \sum_{m=\pm 1,0} \mid \langle A_f|(\mu^{(\text{el.})})_m|A_i\rangle \mid^2 . \tag{23}$$

Substituting this into eq. (19),

$$\left[\frac{\text{Transition Prob.}}{\text{second}}\right]_{A_i \to A_f} = \frac{4\omega^3}{3\hbar c^3} \sum_{m=\pm 1,0} \mid \langle A_f|(\mu^{(\text{el.})})_m|A_i\rangle \mid^2 . \tag{24}$$

Finally, if the atomic quantum numbers in the state vectors are given by

$$|A_i\rangle = |a_i J_i M_i\rangle, \qquad |A_f\rangle = |a_f J_f M_f\rangle,$$

where a_i and a_f are all quantum numbers other than the total angular momentum quantum numbers J_i, M_i and J_f, M_f, and if the initial and final states have degeneracies $(2J_i + 1)$ and $(2J_f + 1)$, we can use the Wigner–Eckart theorem and do the M sums. If we average over the initial M_i substates, assuming each is occupied with an equal probability of $[1/(2J_i + 1)]$, we have

$$\left[\frac{\text{Transition Prob.}}{\text{second}}\right]_{A_i \to A_f}$$
$$= \frac{1}{(2J_i+1)} \frac{64\pi^4\nu^3}{3hc^3} \sum_{M_i}\sum_{m}\sum_{M_f} \langle J_i M_i 1 m|J_f M_f\rangle^2 \frac{\mid \langle a_f J_f \| \vec{\mu}^{(\text{el.})} \| a_i J_i\rangle \mid^2}{(2J_f+1)}$$
$$= \frac{1}{(2J_i+1)} \frac{64\pi^4\nu^3}{3hc^3} \mid \langle a_f J_f \| \vec{\mu}^{(\text{el.})} \| a_i J_i\rangle \mid^2 . \tag{25}$$

This result agrees with our earlier result, obtained via the "correspondence principle arguments" or "guess."

Final note: We might have been interested in the transition probability per second that the atom makes a transition from the atomic state A_i to A_f and that the photon be emitted into a $d\Omega$ about the direction of $\vec{k}$, but now with a linear polarization vector $\vec{e}_\alpha$, with $\alpha = 1$, or 2. In that case,

$$\left[\frac{\text{Transition Prob.}}{\text{second}}\right]_{i\to f} = \frac{\omega^3 d\Omega}{2\pi\hbar c^3} \left| \langle A_f|(\vec{\mu}^{(\text{el.})} \cdot \vec{e}_\alpha)|A_i\rangle \right|^2, \tag{26}$$

where, now,

$$\begin{aligned} \vec{e}_1 = \vec{e}_{x'} &= \cos\theta\cos\phi\vec{e}_x + \cos\theta\sin\phi\vec{e}_y - \sin\theta\vec{e}_z \\ \vec{e}_2 = \vec{e}_{y'} &= \quad -\sin\phi\vec{e}_x \quad + \quad \cos\phi\vec{e}_y. \end{aligned} \tag{27}$$

Now, summing over the two polarizations, and integrating over all possible photon directions, we have

$$\begin{aligned} &\left[\frac{\text{Transition Prob.}}{\text{second}}\right]_{A_i\to A_f} = \\ &\frac{\omega^3}{2\pi\hbar c^3}\int\int d\Omega\Big[(\cos^2\theta\cos^2\phi + \sin^2\phi)|\langle A_f|\mu_x^{(\text{el.})}|A_i\rangle|^2 \\ &\quad + (\cos^2\theta\sin^2\phi + \cos^2\phi)|\langle A_f|\mu_y^{(\text{el.})}|A_i\rangle|^2 \\ &\quad + \sin^2\theta|\langle A_f|\mu_z^{(\text{el.})}|A_i\rangle|^2 + \cdots + \cdots\Big] \\ &= \frac{4\omega^3}{3\hbar c^3}\left(|\langle A_f|\mu_x^{(\text{el.})}|A_i\rangle|^2 + |\langle A_f|\mu_y^{(\text{el.})}|A_i\rangle|^2 + |\langle A_f|\mu_z^{(\text{el.})}|A_i\rangle|^2\right), \end{aligned} \tag{28}$$

where the six angular integrals for the cross terms, indicated by $+\cdots+\cdots$ above, are zero.

64

The Photoelectric Effect: Hydrogen Atom

As our second example, let us calculate the cross section for the photoelectric effect for an atom. For simplicity, we will choose the simplest atom, hydrogen. We assume we have an incoming beam of photons of definite $\vec{k}$, hence definite energy, $\hbar\omega$, and definite polarization. We also assume the incident photon beam is linearly polarized, with polarization in the direction of $\vec{e}_\alpha$, and $\hbar\omega$ is greater than the ionization energy of hydrogen, $((\mu e^4)/(2\hbar^2)$. Therefore, when a photon of energy $\hbar\omega$ is absorbed, the hydrogen atom makes a transition from the $1s$ ground state into the continuum, (see Fig. 64.1). For simplicity, let us further assume $\hbar\omega \gg (\mu e^4)/(2\hbar^2)$ so the final electron kinetic energy $p_f^2/2\mu$ is large enough the final electron continuum wave function can be approximated by a plane wave. For the moment, therefore, we shall avoid the more complicated exact Coulomb relative motion function discussed in the appendix to Chapter 42. To calculate the differential cross section for the photoelectric process, we need to know the flux in the incoming photon beam, and the transition probability per second that the hydrogen atom makes a transition from the $1s$ ground state to a continuum state with the absorption of a photon.

To get the incident flux for an initial photon state, $|\cdots n_{\vec{k}\alpha}\cdots\rangle$, the photon number density is given by

$$\text{photon number density} = \frac{n_{\vec{k}\alpha}}{L^3}.$$

(Our photon plane waves are assumed to fill the whole volume L^3 of our cube. Even if the real photon beam has a finite extent, it will fill a macroscopic volume. Because all final physically relevant results will have to be independent of L^3, the exact value of the macroscopic volume cannot play a role). To get the incident

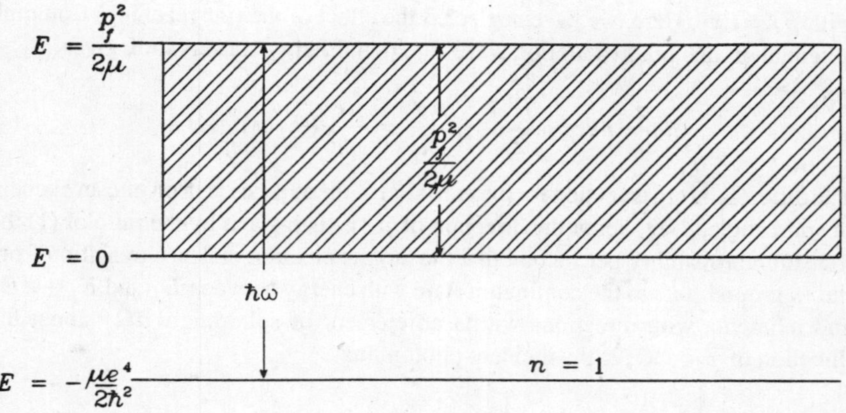

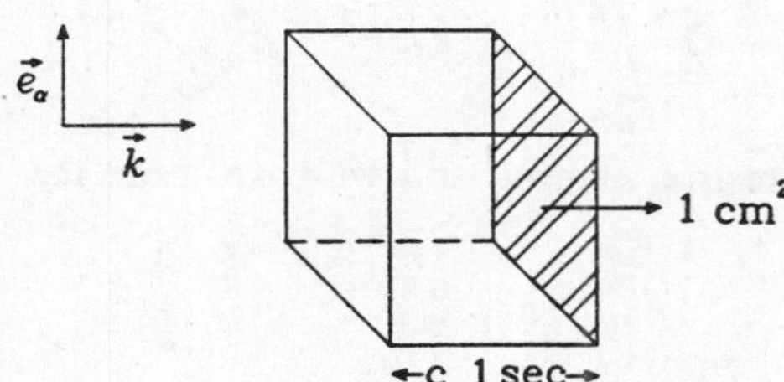

FIGURE 64.1. Photoelectric effect in hydrogen.

photon flux, all photons in a volume element of cross-sectional area 1 cm^2 normal to $\vec{k}$ and of length $c \times 1$ sec. parallel to $\vec{k}$ will traverse the cross-sectional area in 1 sec. Thus, the number of photons streaming across unit area normal to $\vec{k}$ per second is given by

$$\text{Incident photon flux} = \left(\frac{n_{\vec{k}\alpha} c}{L^3}\right). \tag{1}$$

Our initial and final states are

$$\begin{aligned} |i\rangle &= |A_i = (n = 1, l = 0, m_l = 0); n_{\vec{k}\alpha}\rangle, \\ |f\rangle &= |A_f = \vec{p}_f; (n_{\vec{k}\alpha} - 1)\rangle. \end{aligned} \tag{2}$$

We will approximate the final-state continuum atomic wave function by

$$\langle \vec{r} | \vec{p}_f \rangle = \frac{e^{i(\vec{k}_f \cdot \vec{r})}}{L^{\frac{3}{2}}}, \tag{3}$$

with $\vec{p}_f = \hbar\vec{k}_f$. Here, we have neglected the effect of the (far-reaching) Coulomb potential on this relative motion wave function. Energy conservation gives

$$\hbar\omega = \hbar kc = \frac{\hbar^2 k_f^2}{2\mu} + \frac{\mu e^4}{2\hbar^2} = \frac{\mu e^4}{2\hbar^2}((k_f a_0)^2 + 1). \tag{4}$$

The differential cross section for the emission of the atomic electron into an element of solid angle, $d\Omega_f$, about the direction of $\vec{p}_f$ is then given by the ratio of (1) the transition probability per second that the atomic electron makes a transition from the $1s$ ground state to the continuum state with energy between E_f and $E_f + dE_f$ and momenta with directions within an element of solid angle $d\Omega_f$ about the direction of $\vec{p}_f$, and (2) the incident photon flux.

$$d\sigma = \frac{2\pi}{\hbar}\rho(E_f)\frac{1}{\text{inc. photon flux}}$$
$$\times \left| \langle \vec{p}_f, (n_{\vec{k}\alpha} - 1)| \sum_{j=1}^{2} \frac{e_j}{m_j c}(\vec{p}_j \cdot \vec{e}_\alpha)\frac{c}{L^{\frac{3}{2}}}\sqrt{\frac{2\pi\hbar}{\omega}} a_{\vec{k}\alpha}|100, n_{\vec{k}\alpha}\rangle \right|^2, \tag{5}$$

where only the photon annihilation operator piece of

$$A_\perp = \frac{c}{L^{\frac{3}{2}}}\sum_{\vec{k},\alpha}\sqrt{\frac{2\pi\hbar}{\omega}}\left(a_{\vec{k}\alpha}\vec{e}_\alpha e^{i\vec{k}\cdot\vec{r}_j} + a^\dagger_{\vec{k}\alpha}\vec{e}_\alpha e^{-i\vec{k}\cdot\vec{r}_j}\right) \tag{6}$$

can contribute to the matrix element, and we have made the electric dipole approximation

$$e^{i\vec{k}\cdot\vec{r}_j} \approx 1.$$

Our plane wave approximation requires $k_f a_0 \gg 1$. In this limit energy conservation, eq. (4) gives $\hbar kc \approx (\hbar^2 k_f^2)/2\mu$, so

$$\frac{k^2}{k_f^2} \approx \frac{(\hbar^2 k_f^2)/2\mu}{2\mu c^2}, \tag{7}$$

and therefore $k/k_f \ll 1$, provided the kinetic energy of the photoelectron is much less than 2(electron rest energy) $\approx$ 1MeV. Therefore, both the plane wave approximation, $k_f a_0 \gg 1$, and the electric dipole approximation, $ka_0 \ll 1$, can be satisfied simultaneously.

The density of states $\rho(E_f)$ is given by the number of free electron states between energies E_f and $E_f + dE_f$ in our cube of volume L^3. We counted this in Chapter 57.

$$\rho(E_f) = \frac{L^3}{(2\pi)^3}\frac{\mu p_f}{\hbar^3} d\Omega_f. \tag{8}$$

The photon part of the matrix element is simply

$$\langle (n_{\vec{k}\alpha} - 1)|a_{\vec{k}\alpha}|n_{\vec{k}\alpha}\rangle = \sqrt{n_{\vec{k}\alpha}}, \tag{9}$$

so

$$d\sigma = \frac{2\pi}{\hbar}\left(\frac{L^3}{(2\pi)^3}\frac{\mu p_f}{\hbar^3}d\Omega_f\right)\left(\frac{L^3}{n_{\vec{k}\alpha}c}\right)\frac{c^2 2\pi\hbar}{L^3\omega}n_{\vec{k}\alpha}$$
$$\times \left| \langle\vec{p}_f|\sum_{j=1}^{2}\frac{e_j}{m_j c}(\vec{p}_j\cdot\vec{e}_\alpha)|100\rangle \right|^2$$
$$= \frac{L^3}{2\pi}\frac{\mu p_f c}{\hbar^3\omega}d\Omega_f \left| \langle\vec{p}_f|\sum_{j=1}^{2}\frac{e_j}{m_j c}(\vec{p}_j\cdot\vec{e}_\alpha)|100\rangle \right|^2. \tag{10}$$

In the atomic matrix element, we shall write the momentum operator

$$\sum_{j=1}^{2}\frac{e_j\vec{p}_j}{m_j} = \frac{\hbar}{i}\left(-\frac{e}{m_e}\vec{\nabla}_1 + \frac{e}{m_{\text{proton}}}\vec{\nabla}_2\right) = -\frac{\hbar}{i}\frac{e}{\mu}\vec{\nabla}, \tag{11}$$

where we have expressed the $\vec{\nabla}_1$ and $\vec{\nabla}_2$ operators in terms of the relative motion $\vec{\nabla}$ operator, with relative motion vector, $\vec{r} = \vec{r}_1 - \vec{r}_2$. In addition, it will be convenient to calculate the atomic matrix element in the right-handed form by using

$$\left| \langle\vec{p}_f|\frac{e}{\mu c}(\vec{e}_\alpha\cdot\vec{p})|100\rangle \right| = \left| \langle 100|\frac{e}{\mu c}(\vec{e}_\alpha\cdot\vec{p})|\vec{p}_f\rangle \right|,$$

where

$$\langle 100|\frac{e}{\mu c}(\vec{e}_\alpha\cdot\vec{p})|\vec{p}_f\rangle = \frac{e}{\mu c}\int d\vec{r}\ \psi_{100}(\vec{r})(\vec{e}_\alpha\cdot\frac{\hbar}{i}\vec{\nabla})\frac{e^{\frac{i}{\hbar}(\vec{p}_f\cdot\vec{r})}}{L^{\frac{3}{2}}}$$
$$= \frac{e}{\mu c}\frac{(\vec{e}_\alpha\cdot\vec{p}_f)}{L^{\frac{3}{2}}}\int d\vec{r}\left(2\left(\frac{1}{a_0}\right)^{\frac{3}{2}}e^{-\frac{r}{a_0}}\frac{1}{\sqrt{4\pi}}e^{i\vec{k}_f\cdot\vec{r}}\right)$$
$$= \frac{e}{\mu c}\frac{(\vec{e}_\alpha\cdot\vec{p}_f)}{L^{\frac{3}{2}}}2\left(\frac{1}{a_0}\right)^{\frac{3}{2}}\sqrt{4\pi}\int_0^\infty drr^2 j_0(k_f r)e^{-\frac{r}{a_0}}. \tag{12}$$

The radial integral is given by

$$\frac{1}{2ik_f}\int_0^\infty drr\left(e^{-(\frac{1}{a_0}-ik_f)r} - e^{-(\frac{1}{a_0}+ik_f)r}\right) = \frac{2}{a_0}\frac{1}{\left[\left(\frac{1}{a_0}\right)^2 + k_f^2\right]^2}, \tag{13}$$

so

$$\langle 100|\frac{e}{\mu c}(\vec{e}_\alpha\cdot\vec{p})|\vec{p}_f\rangle = \frac{e}{\mu c}\frac{p_f\cos\theta}{L^{\frac{3}{2}}}\frac{8\sqrt{\pi}\left(\frac{1}{a_0}\right)^{\frac{5}{2}}}{\left[\left(\frac{1}{a_0}\right)^2 + k_f^2\right]^2}$$
$$= \frac{e\hbar^3}{\mu^3 c\omega^2}\frac{k_f\cos\theta}{L^{\frac{3}{2}}}2\sqrt{\pi}\left(\frac{1}{a_0}\right)^{\frac{5}{2}}, \tag{14}$$

where we have used the energy conservation eq. (4) in the last step to make the replacement

$$\left[\left(\frac{1}{a_0}\right)^2 + k_f^2\right] = \frac{2\mu\omega}{\hbar},$$

and the angle between $\vec{p}_f = \hbar\vec{k}_f$ and $\vec{e}_1$ has been named θ (see Fig. 64.2). With this matrix element, eq. (10) yields

$$d\sigma = \frac{2\hbar^4 e^2 k_f^3}{\mu^5\omega^5 c a_0^5}\cos^2\theta d\Omega, \tag{15}$$

or, in atomic units,

$$d\sigma = 2\frac{e^2}{\hbar c}(k_f a_0)^3\left(\frac{\frac{\mu e^4}{\hbar^2}}{\hbar\omega}\right)^5 a_0^2\cos^2\theta d\Omega. \tag{16}$$

For an unpolarized incident photon beam, with equal probability of photon polarization in the $\vec{e}_1$ and $\vec{e}_2$ direction (see Fig. 64.2), the differential cross section for the photoelectron to be emitted in the direction shown is

$$d\sigma = 2\frac{e^2}{\hbar c}(k_f a_0)^3\left(\frac{\frac{\mu e^4}{\hbar^2}}{\hbar\omega}\right)^5 a_0^2\frac{(\cos^2\theta + \sin^2\theta\cos^2\phi)}{2}d\Omega. \tag{17}$$

Integrating over all directions, this equation gives the total cross section for the photoionization effect,

$$\sigma = \frac{8\pi}{3}\frac{e^2}{\hbar c}(k_f a_0)^3\left(\frac{\frac{\mu e^4}{\hbar^2}}{\hbar\omega}\right)^5 a_0^2, \tag{18}$$

or, finally, in terms of the fine structure constant, α, and the 1s atomic binding energy, $|E_{n=1}|$, and with the approximation, $\hbar k_f \approx \sqrt{2\mu\hbar\omega}$, valid for the case $\hbar\omega \gg \mu e^4/\hbar^2$, which we have considered, we have

$$\sigma \approx \frac{256\pi}{3}\alpha\left(\frac{|E_{n=1}|}{\hbar\omega}\right)^{\frac{7}{2}} a_0^2. \tag{19}$$

In this expression, valid for $k_f a_0 \gg 1$, we have for simplicity used the plane wave approximation for the Coulomb continuum wave function. Using the results of the mathematical appendix to Chapter 42, we could just as easily have used the exact Coulomb wave function for the final continuum wave function. For this purpose, it will be useful to convert the atomic matrix element of eq. (10) to a matrix element of the atomic electric dipole moment matrix element. This result can be accomplished by the same trick used in eq. (10) of Chapter 63, using the commutator relation

$$[H^{(0)}, \vec{r}_j] = \frac{\hbar}{i}\frac{\vec{p}_j}{m_j}, \tag{20}$$

which converts the atomic operator to

$$\sum_{j=1}^{2}\frac{e_j}{m_j c}(\vec{p}_j\cdot\vec{e}_\alpha) = \sum_{j=1}^{2}\frac{ie_j}{\hbar c}[H^{(0)}, (\vec{r}_j\cdot\vec{e}_\alpha)] = \frac{-ie}{\hbar c}[H^{(0)}, (\vec{r}\cdot\vec{e}_\alpha)], \tag{21}$$

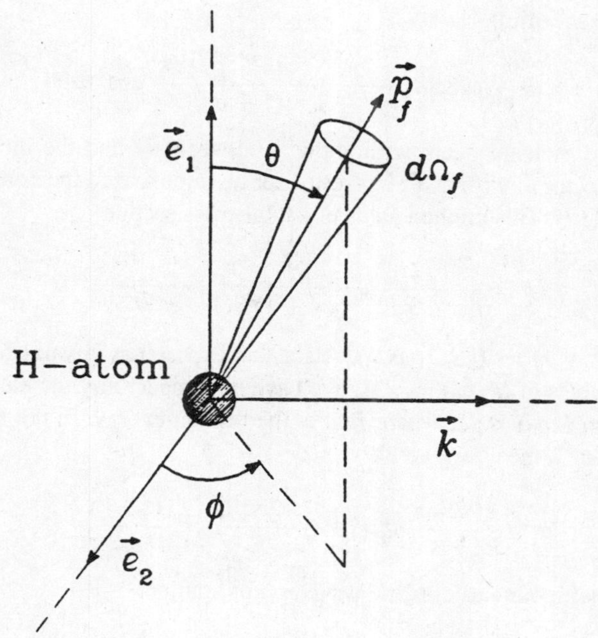

FIGURE 64.2. The incident photon $\vec{k}$ and photoelectron $\vec{p}_f$ vectors.

where $\vec{r}$ is the relative motion vector, $\vec{r} = \vec{r}_1 - \vec{r}_2$. Using the zeroth-order energy eigenvalues of the ground state and the continuum state and the result for the matrix element of $\vec{r} \cdot \vec{e}_\alpha$, as derived through eq. (44) of the mathematical appendix to Chapter 42, this yields

$$\langle \vec{p}_f | \sum_{j=1}^{2} \frac{e_j}{m_j c} (\vec{p}_j \cdot \vec{e}_\alpha) | 100 \rangle = -\frac{ie\omega}{c} \langle \vec{p}_f | \vec{r} \cdot \vec{e}_\alpha | 100 \rangle$$

$$= -\frac{ie\omega}{c} a_0^{\frac{5}{2}} \frac{\sqrt{4\pi}}{L^{\frac{3}{2}}} \cos\theta \frac{8i\gamma^5}{(\gamma^2+1)^3} \left[\frac{(1+\gamma^2)\pi\gamma}{\sinh \pi\gamma} \right]^{\frac{1}{2}} e^{(\gamma\pi)/2} e^{-2\gamma \tan^{-1}(1/\gamma)}, \quad (22)$$

where θ is the angle between the polarization vector of the incident photon and the vector $\vec{p}_f$ (see Fig. 64.2) and γ is the Coulomb parameter

$$\gamma = \frac{\mu e^2}{\hbar^2 k} = \frac{1}{k a_0}. \quad (23)$$

Using the expression for the differential cross section, given by eq. (10), and the energy conservation relation, eq. (4), to yield

$$\frac{\mu k_f \hbar\omega}{\hbar^2} = \frac{k_f}{2} \left(k_f^2 + \frac{\mu^2 e^4}{\hbar^4} \right) = \frac{1}{2a_0^3} \frac{(\gamma^2+1)}{\gamma^3}, \quad (24)$$

we have the exact result

$$d\sigma = \frac{e^2}{\hbar c} 128\pi a_0^2 \frac{\gamma^8}{(\gamma^2+1)^4} \frac{e^{-4\gamma \tan^{-1}(1/\gamma)}}{(1-e^{-2\gamma\pi})} \cos^2\theta d\Omega \tag{25}$$

for a polarized incident beam with angle θ between $\vec{p}_f$ and the incident beam polarization vector $\vec{e}$, with a similar result for an unpolarized incident beam, cf., eqs. (16) and (17). This method leads to a total cross section

$$\sigma = \frac{e^2}{\hbar c} \frac{512\pi^2}{3} a_0^2 \frac{\gamma^8}{(\gamma^2+1)^4} \frac{e^{-4\gamma \tan^{-1}(1/\gamma)}}{(1-e^{-2\gamma\pi})}, \tag{26}$$

where $\gamma^2/(\gamma^2+1) = (|E_{n=1}|/\hbar\omega)$. Also, this expression is valid for all (non-relativistic) values of $(k_f a_0) = \gamma^{-1}$. We have now made only the electric dipole approximation, $(ka_0) \ll 1$, where $\hbar kc$ is the photon energy. In the high energy limit, with $\gamma \ll 1$, the above yields

$$\sigma \approx \frac{e^2}{\hbar c} \frac{256\pi}{3} a_0^2 \gamma^7 \approx \frac{e^2}{\hbar c} \frac{256\pi}{3} a_0^2 \left(\frac{|E_{n=1}|}{\hbar\omega}\right)^{\frac{7}{2}} \tag{27}$$

in agreement with our earlier plane wave approximation.

Problems

29. Show the intensity loss of a parallel beam of light per thickness, dx, of atomic absorber caused by stimulated absorption and emission processes is given (in electric dipole approximation) by

$$\frac{\text{Energy Loss}}{\text{cm}^2 \text{ sec dx}} = \frac{4\pi^2}{3\hbar} \rho(\nu_{ul}) \nu_{ul} N_0 |\langle a_u J_u \| \sum_i e_i \vec{r}_i \| a_l J_l \rangle|^2$$

$$\times \frac{e^{-(E_l/kT)}}{\left[\sum_n g_n e^{-(E_n/kT)}\right]} \left(1 - e^{-(h\nu_{ul}/kT)}\right),$$

where we assume:

(a) the number of photons of type, $\vec{k}\alpha$, is large, $n_{\vec{k}\alpha} \gg 1$, so $(n_{\vec{k}\alpha}+1) \approx n_{\vec{k}\alpha}$;

(b) the atoms are in a gas sample in thermal equilibrium, so the number of atoms in the lower (similarly upper) state are given by $N_0 f_l$, where

$$f_l = \frac{g_l e^{-(E_l/kT)}}{\left[\sum_n g_n e^{-(E_n/kT)}\right]};$$

N_0 = total number of atoms per unit volume, and degeneracy factors such as g_l, g_u, or g_n are given by $g_l = (2J_l + 1), \ldots$, assuming all other degeneracies are removed by some interaction;

(c) the intensity of the incoming beam is given by $c\rho(\nu)d\nu$, where $\rho(\nu)d\nu$ is the energy density between frequencies ν and $\nu + d\nu$ in the incoming beam,

$$\int_0^\infty d\omega \rho(\omega) = \int_0^\infty d\nu \rho(\nu), \quad \text{so } \rho(\omega) = \rho(\nu)/(2\pi);$$

(d) the atoms in the gas sample have arbitrary orientations, so we have to average over all atomic orientations; for this reason, the above result is valid for both incident light with definite polarization, α, and for unpolarized light; recall

$$\int d\Omega D^1_{M_1 M} D^{1*}_{M_2 M} = \frac{4\pi}{3} \delta_{M_1 M_2}.$$

In the above, the double-barred matrix element is the standard reduced matrix element; a_l is shorthand for all quantum numbers other than J_l and M_{J_l}, similarly for a_u, and $h\nu_{ul} = (E_u - E_l)$.

30. In a hot, completely ionized hydrogen gas both electron–electron and electron–proton scattering must be considered. In the electron–proton scattering with an initial $\vec{p}_0$, the most probable process is elastic scattering. There is, however, the possibility of a radiative capture process in which the proton captures an electron to make a hydrogen atom in a bound state, with the emission of a photon of energy, $\hbar\omega$,

$$e^- + p^+ \rightarrow H + \text{photon}.$$

For a center of mass energy in the incident beam of $p_0^2/(2\mu) \geq 100$ eV ($\gg 13.6$ eV), assume the electron–proton relative motion function can be approximated by a plane wave.

(a) With $p_0^2/(2\mu) = 100$ eV, show this radiative capture cross section can be calculated in electric dipole approximation. In this case, calculate the differential cross section for the emission of a photon with momentum, $\hbar\vec{k}$, making an angle θ with the beam direction, $\vec{p}_0 = \hbar\vec{k}_0$, leaving the hydrogen atom in its ground state. What is the total cross section for this capture process into the hydrogen ground state?

(b) With $p_0^2/(2\mu) = 10^4$ eV, the problem can still be treated nonrelativistically, with considerable accuracy. (For this energy, $\beta^2 \approx .04$.) Show, however, at this energy, the multipole expansion is not valid. The electric dipole approximation would be bad. Repeat the calculation for the differential cross section for the radiative capture process into the hydrogen atom ground state, valid for this incident energy. Now, $ka_0 \approx 2.7$, and we cannot make approximations in the expansion of $e^{i(\vec{k}\cdot\vec{r}_e)}$. Also, the ratio $k/k_0 \approx 0.1$, so we must now consider both the

$$\Sigma_i \frac{e_i}{m_i c}(\vec{p}_i \cdot \vec{A}_\perp) \quad \text{and} \quad -\sum_i \frac{e_i\hbar}{2m_i c} g_{s_i}(\vec{s}_i \cdot [\vec{\nabla}_i \times \vec{A}_\perp])$$

terms in the interaction Hamiltonian, by showing the contributions of the second (magnetic moment) term are of order (k/k_0) compared with the contributions of the first $(\vec{p}_i \cdot \vec{A}_\perp)$ term. Show, however, no cross terms exist in the nonzero matrix elements for the radiative capture process, so the contributions of the magnetic moment term to the cross section are smaller by a factor $(k/k_0)^2 \approx .01$ and to within this accuracy could be neglected for our case.

31. Compared with the elastic proton neutron scattering process, the radiative capture process

$$p + n \rightarrow d + \gamma$$

leading to the formation of a deuteron with the emission of a γ photon of energy $\hbar\omega$ is quite improbable. Convince yourself that in the 2–20-MeV incident center of mass energy range this radiative capture process is dominated by the electric dipole term and can be calculated in electric dipole approximation. (The spins and magnetic moments of the particles therefore play no role in the dynamics.) In this approximation, calculate the total cross section for radiative capture as a function of incident energy, E_0, or rather as a function of k_0 [with $E_0 = \hbar^2 k_0^2/(2\mu) = \hbar^2 k_0^2/M$]. Assume the deuteron bound-state wave function can be approximated by the Hulthen wave function

$$\sqrt{\frac{2\eta(1+\eta)}{r_0(\eta-1)^2}}\left(\frac{e^{-(r/r_0)} - e^{-\eta(r/r_0)}}{r}\right)Y_{00}\chi_{S=1,M_S}\chi_{I=0,M_I=0},$$

with $\eta = 6.2$; $r_0 = 4.26$ fm. Recall the deuteron bound state has $S = 1$, $I = 0$. The deuteron binding energy is 2.226 MeV.

Assume the incident proton beam and neutron target are both unpolarized. Also convince yourself an incident plane wave approximation can be used for the neutron–proton relative motion function for the whole range of r values from $\infty \rightarrow 0$.

32. Calculate the differential and total cross sections for the photo-dissociation of the deuteron into a free neutron and proton

$$\gamma + d \rightarrow n + p$$

in the energy range for which the incident γ photon has an energy $\hbar\omega$ in the 4–20-MeV range, where the photodissociation process is dominated by the electric dipole matrix element. (See problem 31.)

65

Spontaneous Photon Emission: General Case: Electric and Magnetic Multipole Radiation

For the general case, for a quantum system with kr_i *not* necessarily very small, we need to use the full first-order piece of the interaction Hamiltonian

$$
\begin{aligned}
H_{\text{int.}} &= -\sum_{i=1}^{n}\Big(\frac{e_i}{m_i c}(\vec{p}_i\cdot\vec{A}_{\perp}) + \frac{e_i\hbar g_{s,i}}{2m_i c}(\vec{s}_i\cdot[\vec{\nabla}\times\vec{A}_{\perp}])\Big)\\
&= -\sum_{\vec{k}}\sum_{\mu=\pm1}\sum_{L}\frac{c2\pi}{\sqrt{\text{Vol.}}}\sqrt{\frac{\hbar(2L+1)}{\omega}}\\
&\times\Bigg[\sum_i\frac{e_i}{m_i c}\vec{p}_i\cdot\Big(a_{\vec{k}\mu}\big(\vec{A}(\vec{r}_i,\mathcal{E})_{L\mu} - \mu\vec{A}(\vec{r}_i,\mathcal{M})\big)\\
&\qquad + a^{\dagger}_{\vec{k}\mu}\big(\vec{A}(\vec{r}_i,\mathcal{E})^*_{L\mu} - \mu\vec{A}(\vec{r}_i,\mathcal{M})^*_{L\mu}\big)\Big)\\
&+ i\sum_i\frac{e_i\hbar g_{s,i}}{2m_i c}\vec{s}_i\cdot\Big(a_{\vec{k}\mu}[\vec{k}\times\big(\vec{A}(\vec{r}_i,\mathcal{E})_{L\mu} - \mu\vec{A}(\vec{r}_i,\mathcal{M})\big)]\\
&\qquad - a^{\dagger}_{\vec{k}\mu}[\vec{k}\times\big(\vec{A}(\vec{r}_i,\mathcal{E})^*_{L\mu} - \mu\vec{A}(\vec{r}_i,\mathcal{M})^*_{L\mu}\big)]\Big)\Bigg]. \qquad (1)
\end{aligned}
$$

To calculate the spontaneous emission probability, we need matrix elements of the type

$$|\langle A_f; n_{\vec{k}\mu} = 1|H_{\text{int.}}|A_i;0\rangle| = |\langle A_i 0|H_{\text{int.}}|A_f; n_{\vec{k}\mu} = 1\rangle|.$$

We prefer to work with the latter form (to avoid the complex conjugate fields) and will calculate

$$\langle A_i;0|H_{\text{int.}}|A_f; n_{\vec{k}\mu} = 1\rangle = -\frac{2\pi c}{\sqrt{\text{Vol.}}}\sum_L\sqrt{\frac{\hbar(2L+1)}{\omega}}$$

$$\times \langle A_i|\sum_i \frac{e_i}{m_i c}(\vec{p}_i + \tfrac{i}{2}g_{s,i}\hbar[\vec{s}_i \times \vec{k}]) \cdot \Big(\vec{A}(\vec{r}_i, \mathcal{E})_{L\mu} - \mu \vec{A}(\vec{r}_i, \mathcal{M})_{L\mu}\Big)|A_f\rangle. \quad (2)$$

The perturbing terms are of the form

$$\Big(\vec{W} \cdot \vec{A}(\vec{r}_i; \mathcal{E}\text{or}\mathcal{M})_{L\mu}\Big)$$

and hence of the form $(\vec{W} \cdot \vec{V}_{[l1]L\mu})$, where

$$\vec{W} = \sum_i \frac{e_i}{m_i c}(\vec{p}_i + \tfrac{i}{2}g_{s,i}\hbar[\vec{s}_i \times \vec{k}]), \quad (3)$$

and where we must remember μ is a circular polarization index with respect to a photon vector, $\vec{k}$, whose direction is identified as that of a z' axis rotated relative to the laboratory axes, x, y, z, via the arbitrary angles ϕ, θ. Thus,

$$(\vec{W} \cdot \vec{V}_{[l1]L\mu}) = \sum_M (\vec{W} \cdot \vec{V}_{[l1]LM}) D^L_{M\mu}(\phi, \theta, 0)^*, \quad (4)$$

where the M quantum number now refers to the projection of L onto the laboratory z axis. The transition probability can then be expressed by

$$\left[\frac{\text{Transition Prob.}}{\text{second}}\right]_{A_i \to A_f + \text{photon } \vec{k}\mu \text{ into } d\Omega} =$$

$$\frac{2\pi}{\hbar}\frac{\text{Vol.}}{(2\pi)^3}\frac{\omega^2 d\Omega}{\hbar c^3}\frac{(2\pi c)^2}{\text{Vol.}}\frac{\hbar}{\omega}\sum_{L,M}\sqrt{(2L+1)}\sum_{L'M'}\sqrt{(2L'+1)}$$

$$\times \langle A_i|\sum_i \frac{e_i}{m_i c}(\vec{p}_i + \tfrac{i}{2}g_{s,i}\hbar[\vec{s}_i \times \vec{k}]) \cdot \Big(\vec{A}(\vec{r}_i, \mathcal{E})_{LM} - \mu \vec{A}(\vec{r}_i, \mathcal{M})_{LM}\Big)|A_f\rangle^*$$

$$\times \langle A_i|\sum_i \frac{e_i}{m_i c}(\vec{p}_i + \tfrac{i}{2}g_{s,i}\hbar[\vec{s}_i \times \vec{k}]) \cdot \Big(\vec{A}(\vec{r}_i, \mathcal{E})_{L'M'} - \mu \vec{A}(\vec{r}_i, \mathcal{M})_{L'M'}\Big)|A_f\rangle$$

$$\times D^L_{M\mu}(\phi, \theta, 0) D^{L'}_{M'\mu}(\phi, \theta, 0)^*. \quad (5)$$

To get the transition probability that the quantum system makes a transition from state A_i to A_f with the emission of a photon, we need to sum over both polarizations and integrate over all photon directions. For this purpose, we use the orthonormality integral

$$\int\int d\Omega D^{L'}_{M'\mu}(\phi, \theta, 0)^* D^L_{M\mu}(\phi, \theta, 0) = \delta_{L'L}\delta_{M'M}\frac{4\pi}{(2L+1)} \quad (6)$$

to get

$$\left[\frac{\text{Trans. Prob.}}{\text{second}}\right]_{A_i \to A_f} = \sum_{\mu=\pm1}\int\int d\Omega \left[\frac{\text{Trans. Prob.}}{\text{second}}\right]_{A_i \to A_f + \text{photon } \vec{k}\mu}$$

$$= \frac{4\pi\omega}{\hbar c}\sum_{\mu=\pm1}\sum_{LM}\bigg(\langle A_i|\vec{W} \cdot \Big(\vec{A}(\vec{r}, \mathcal{E})_{LM} - \mu \vec{A}(\vec{r}, \mathcal{M})_{LM}\Big)|A_f\rangle^*$$

$$\times \langle A_i|\vec{W} \cdot \Big(\vec{A}(\vec{r}, \mathcal{E})_{LM} - \mu \vec{A}(\vec{r}, \mathcal{M})_{LM}\Big)|A_f\rangle\bigg)$$

$$= \frac{4\pi\omega}{\hbar c}\bigg(2\sum_{LM}|\langle A_i|\vec{W} \cdot \vec{A}(\vec{r}, \mathcal{E})_{LM}|A_f\rangle|^2$$

$$+2\sum_{LM}|\langle A_i|\vec{W}\cdot\vec{A}(\vec{r},\mathcal{M})_{LM}|A_f\rangle|^2$$
$$-\Big(\sum_{\mu}\mu\Big)\sum_{LM}\langle A_i|\vec{W}\cdot\vec{A}(\vec{r},\mathcal{E})_{LM}|A_f\rangle^*\langle A_i|\vec{W}\cdot\vec{A}(\vec{r},\mathcal{M})_{LM}|A_f\rangle$$
$$-\Big(\sum_{\mu}\mu\Big)\sum_{LM}\langle A_i|\vec{W}\cdot\vec{A}(\vec{r},\mathcal{M})_{LM}|A_f\rangle^*\langle A_i|\vec{W}\cdot\vec{A}(\vec{r},\mathcal{E})_{LM}|A_f\rangle\Big). \tag{7}$$

Now, in the last two terms, the cross terms between type $\mathcal{E}$ and $\mathcal{M}$ matrix elements, the matrix elements are now independent of μ, so the sum, $\sum_{\mu}\mu=(-1+1)=0$. The cross terms therefore do not contribute to the total transition probability (no interference between $\mathcal{E}$ and $\mathcal{M}$), and the total transition probability is given by

$$\left[\frac{\text{Trans. Prob.}}{\text{second}}\right]_{A_i\to A_f}=\frac{8\pi\omega}{\hbar c}\sum_{LM}\Big(\Big|\langle A_i|\sum_{i=1}^{n}\vec{W}_i\cdot\vec{A}(\vec{r}_i,\mathcal{E})_{LM}|A_f\rangle\Big|^2+\Big|\langle A_i|\sum_{i=1}^{n}\vec{W}_i\cdot\vec{A}(\vec{r}_i,\mathcal{M})_{LM}|A_f\rangle\Big|^2\Big). \tag{8}$$

Usually, because of parity or angular momentum selection rules either the $\mathcal{E}$ or the $\mathcal{M}$ term is dominant, and the L in the sum will also be determined by selection rules. For example, in a $2^+\to 0^+$ transition in even–even nuclei (even proton and neutron number), where the first excited state is often a 2^+ state and the ground state is almost always a 0^+ state, the dominant term is an $\mathcal{E}$ type with $L=2$ only. The transition probability is then determined by an electric quadrupole radiation matrix element. Usually, also, the first term with a nonzero matrix element in the L expansion will make the dominant contribution. The higher-order terms, with $L=L_{\text{dominant}}+\Delta L$, usually with a ΔL of 2, will be negligible compared with this dominant term.

A Electric Multipole Radiation

To calculate the transition probability for the Lth electric multipole term, we need to evaluate the matrix element, $\langle A_i|\vec{W}\cdot\vec{A}(\vec{r},\mathcal{E})_{LM}|A_f\rangle$, or (going back to the photon emission form), $\langle A_f|\vec{W}\cdot\vec{A}(\vec{r},\mathcal{E})^*_{LM}|A_i\rangle$. Recall

$$\vec{A}(\vec{r},\mathcal{E})_{LM}=\sqrt{\frac{(L+1)}{(2L+1)}}\,j_{L-1}\vec{V}_{[(L-1)1]LM}+\sqrt{\frac{L}{(2L+1)}}\,j_{L+1}\vec{V}_{[(L+1)1]LM}. \tag{9}$$

For small values of kr, the spherical Bessel function $j_L(kr)$ is of order

$$j_L(kr)\approx\frac{2^L L!}{(2L+1)!}(kr)^L=\frac{1}{(2L+1)!!}(kr)^L, \tag{10}$$

where $(2L+1)!! \equiv (2L+1)(2L-1)\cdots 5\cdot 3\cdot 1$, so the second $l=(L+1)$ term in the expression for $\vec{A}(\vec{r},\mathcal{E})_{LM}$ is of order

$$\frac{(kr)^2}{(2L+3)(2L+1)}\sqrt{\frac{L}{(L+1)}}$$

times the first $l=(L-1)$ term. Even for $kr \approx 0.5$, the second term is at most of order .01 of the first term. For all quantum systems, therefore, the second term is negligible compared with the first. We also recall the longitudinal vector field, $\vec{A}(\vec{r},\mathcal{L})_{LM}$, which is *not* part of the radiation field, but mathematically somewhat simpler than $\vec{A}(\vec{r},\mathcal{E})_{LM}$, is

$$\vec{A}(\vec{r},\mathcal{L})_{LM} = \sqrt{\frac{L}{(2L+1)}}\, j_{L-1}\vec{V}_{[(L-1)1]LM} - \sqrt{\frac{(L+1)}{(2L+1)}}\, j_{L+1}\vec{V}_{[(L+1)1]LM}. \quad (11)$$

Again, for small values of kr, the second term of this longitudinal field is negligible compared with the first term. Thus, for reasonably small values of kr, we have

$$\vec{A}(\vec{r},\mathcal{E})_{LM} \approx \sqrt{\frac{(L+1)}{L}}\,\vec{A}(\vec{r},\mathcal{L})_{LM} = \frac{1}{ik}\sqrt{\frac{(L+1)}{L}}\,\vec{\nabla}(\Phi_{LM}), \quad (12)$$

where the scalar function Φ_{LM} is

$$\Phi_{LM}(\vec{r}) = j_L(kr)i^L Y_{LM}(\theta,\phi), \quad (13)$$

so the vector field is derivable through the gradient of a simple scalar function. It is this mathematical simplicity that leads us to use the above approximation. In addition, the source vector

$$\vec{W} = \sum_{i=1}^{n}\vec{W}_i = \sum_i \frac{e_i}{m_i c}\left(\vec{p}_i + \tfrac{i}{2}\hbar g_{s,i}[\vec{s}_i\times\vec{k}]\right) \quad (14)$$

can also be simplified for the electric multipole case. The second term is of order kr_i compared with the first term and can therefore usually be neglected. We therefore make the approximation

$$\langle A_i|\sum_i \vec{W}_i\cdot\vec{A}(\vec{r}_i,\mathcal{E})_{LM}|A_f\rangle \approx \frac{1}{ik}\sqrt{\frac{L+1}{L}}\langle A_i|\sum_i\frac{e_i}{m_i c}\vec{p}_i\cdot\vec{\nabla}_i\Phi_{LM}(\vec{r}_i)|A_f\rangle, \quad (15)$$

with a similar approximation for the matrix element

$$\langle A_f|\sum_i \frac{e_i}{m_i c}\vec{p}_i\cdot\vec{A}(\vec{r}_i,\mathcal{E})^*_{LM}|A_i\rangle$$

actually needed for the spontaneous emission process taking the quantum system from state $A_i \to A_f$. We shall also find it convenient to write the operator in symmetrized form to get

$$\langle A_f|\sum_{i=1}^{n}\frac{e_i}{2m_i c}\Big(\vec{p}_i\cdot\vec{A}(\vec{r}_i,\mathcal{E})^*_{LM} + \vec{A}(\vec{r}_i,\mathcal{E})^*_{LM}\cdot\vec{p}_i\Big)|A_i\rangle$$

$$= \langle A_f | \sum_{i=1}^{n} \frac{e_i}{2m_i c} \Big(\frac{\hbar}{i} \vec{\nabla}_i \cdot \vec{A}(\vec{r}_i, \mathcal{E})^*_{LM} + \vec{A}(\vec{r}_i, \mathcal{E})^*_{LM} \cdot \frac{\hbar}{i} \vec{\nabla}_i \Big) | A_i \rangle$$
$$= \int d\vec{r}_1 \cdots d\vec{r}_n \sum_{i=1}^{n} \frac{e_i}{c} \frac{\hbar}{2im_i} \Big(\psi^*_{A_f} (\vec{\nabla}_i \psi_{A_i}) - \psi_{A_i} (\vec{\nabla}_i \psi^*_{A_f}) \Big) \cdot \vec{A}(\vec{r}_i, \mathcal{E})^*_{LM}$$
$$= \frac{1}{c} \int d\vec{r}_1 \cdots d\vec{r}_n \sum_{i=1}^{n} e_i \vec{S}_{A_f A_i}(\vec{\nabla}_i) \cdot \vec{A}(\vec{r}_i, \mathcal{E})^*_{LM}, \quad (16)$$

where we have used the coordinate representation for the n particle system and have let the operator $\vec{\nabla}_i$ act to the right in the first term and to the left in the second term of the n particle matrix element. We have named the resultant operator, $\vec{S}_{A_f A_i}$, the transition probability density current,

$$\vec{S}_{A_f A_i} = \frac{\hbar}{2im} \Big(\psi^*_{A_f} \vec{\nabla} \psi_{A_i} - \psi_{A_i} \vec{\nabla} \psi^*_{A_f} \Big), \quad (17)$$

dependent *not* on a single function ψ and its complex conjugate ψ^* , but on the function ψ_{A_i} and on $\psi^*_{A_f}$. The full vector, $\vec{S}_{A_f A_i}$, is a sum over the n single-particle terms, each dependent on its own $\vec{r}_i$. To convert this probability density current operator to a charge current density operator, it will be useful to multiply this $\vec{S}$ vector by the charge density of our n particle system. Our system was imagined to be a system of n point particles with charges e_i at position vectors, $\vec{r}_i$. For such a system, the charge density could be expressed by

$$\rho(\vec{r}) = \sum_{i=1}^{n} e_i \delta(\vec{r} - \vec{r}_i), \quad (18)$$

with

$$\sum_{i=1}^{n} e_i \delta(\vec{r} - \vec{r}_i) \vec{S}_{A_f A_i} = \vec{j}_{A_f A_i}(\vec{r}), \quad (19)$$

where $\vec{j}$ is now the charge current density vector. The matrix element of eq. (8) can thus be written in the form

$$\langle A_f | \sum_i \frac{e_i}{m_i c} (\vec{p}_i \cdot \vec{A}(\vec{r}_i, \mathcal{E})^*_{LM}) | A_i \rangle$$
$$= \frac{1}{c} \int d\vec{r} \int d\vec{r}_1 \cdots d\vec{r}_n \sum_{i=1}^{n} e_i \delta(\vec{r} - \vec{r}_i) \vec{S}_{A_f A_i}(\vec{\nabla}) \cdot \vec{A}(\vec{r}, \mathcal{E})^*_{LM}$$
$$= \frac{1}{c} \int d\vec{r} \int d\vec{r}_1 \cdots d\vec{r}_n \vec{j}_{A_f A_i}(\vec{r}) \cdot \vec{A}(\vec{r}, \mathcal{E})^*_{LM}$$
$$\approx -\frac{1}{ick} \sqrt{\frac{(L+1)}{L}} \int d\vec{r} \int d\vec{r}_1 \cdots d\vec{r}_n \vec{j}_{A_f A_i}(\vec{r}) \cdot \vec{\nabla} \Phi^*_{LM}(\vec{r})$$
$$= -\frac{1}{ick} \sqrt{\frac{(L+1)}{L}} \int d\vec{r} \langle A_f | \vec{j}(\vec{r}) \cdot \vec{\nabla} \Phi^*_{LM}(\vec{r}) | A_i \rangle. \quad (20)$$

We have added an *extra* integral, $\int d\vec{r}$, to the $3n$ integrals of our n particle coordinate representation expression to accomodate the delta function. We have also used the approximate expression of eq. (12) for $\vec{A}(\vec{r}, \mathcal{E})_{LM}$. Now, we shall rewrite the volume integral as

$$\begin{aligned}
&\int_{\text{Vol.}} d\vec{r} \langle A_f | \vec{j}(\vec{r}) \cdot \vec{\nabla} \Phi^*_{LM} | A_i \rangle \\
&= \int_{\text{Vol.}} d\vec{r} \langle A_f | \Big(\vec{\nabla} \cdot (\Phi^*_{LM} \vec{j}) \Big) | A_i \rangle - \int_{\text{Vol.}} d\vec{r} \langle A_f | \Phi^*_{LM} (\vec{\nabla} \cdot \vec{j}) | A_i \rangle \\
&= \int_{\text{Surface}} d\text{Area} \langle A_f | \Phi^*_{LM} (\vec{j} \cdot \vec{n}) | A_i \rangle - \int_{\text{Vol.}} d\vec{r} \langle A_f | \Phi^*_{LM} (\vec{\nabla} \cdot \vec{j}) | A_i \rangle \\
&= - \int_{\text{Vol.}} d\vec{r} \langle A_f | \Phi^*_{LM} (\vec{\nabla} \cdot \vec{j}) | A_i \rangle,
\end{aligned} \tag{21}$$

where we have used Gauss's theorem to convert the volume integral, over the volume of our cube of side L, of the divergence of $\Phi^*_{LM}\vec{j}$ to a surface integral, over the surface of our cube of the outward normal component of this vector. This surface integral is zero, because our radiation field has the value zero on the surface of our cube. Now, we shall use charge conservation in the form of the continuity equation,

$$(\vec{\nabla} \cdot \vec{j}) + \frac{\partial \rho}{\partial t} = 0, \tag{22}$$

or, more precisely,

$$(\vec{\nabla} \cdot \vec{j}_{A_f A_i}) + \frac{\partial \rho_{A_f A_i}}{\partial t} = 0, \tag{23}$$

to transform the needed matrix element into

$$\begin{aligned}
&\langle A_f | \sum_{i=1}^{n} \frac{e_i}{m_i c} (\vec{p}_i \cdot \vec{A}(\vec{r}_i, \mathcal{E})^*_{LM}) | A_i \rangle \\
&= -\frac{1}{ick} \sqrt{\frac{(L+1)}{L}} \int d\vec{r}_1 \cdots d\vec{r}_n \int d\vec{r} \Phi^*_{LM} \Big(\frac{\partial \rho_{A_f A_i}}{\partial t} \Big) \\
&= -\frac{1}{ick} \sqrt{\frac{(L+1)}{L}} \int d\vec{r}_1 \cdots d\vec{r}_n \int d\vec{r} \Phi^*_{LM} \frac{\partial}{\partial t} \Big(\psi^*_{A_f} \rho \psi_{A_i} \Big) \\
&= -\frac{1}{ick} \sqrt{\frac{(L+1)}{L}} \int d\vec{r}_1 \cdots d\vec{r}_n \int d\vec{r} \Phi^*_{LM} \Big(\frac{i}{\hbar} (H \psi^*_{A_f}) \rho \psi_{A_i} \\
&\qquad - \frac{i}{\hbar} \psi^*_{A_f} \rho (H \psi_{A_i}) \Big) \\
&= -\frac{1}{\hbar ck} \sqrt{\frac{(L+1)}{L}} (E^{(0)}_{A_f} - E^{(0)}_{A_i}) \int d\vec{r}_1 \cdots d\vec{r}_n \int d\vec{r} \Phi^*_{LM}(\vec{r}) \psi^*_{A_f} \rho \psi_{A_i} \\
&= -\frac{1}{\hbar ck} (-\hbar\omega) \sqrt{\frac{(L+1)}{L}} \int d\vec{r} \Phi^*_{LM}(\vec{r}) \langle A_f | \rho(\vec{r}) | A_i \rangle \\
&= \sqrt{\frac{(L+1)}{L}} \langle A_f | \int d\vec{r} \sum_{i=1}^{n} e_i \delta(\vec{r} - \vec{r}_i) \Phi^*_{LM}(\vec{r}) | A_i \rangle \\
&= \sqrt{\frac{(L+1)}{L}} \langle A_f | \sum_{i=1}^{n} e_i \Phi^*_{LM}(\vec{r}_i) | A_i \rangle
\end{aligned}$$

$$= \sqrt{\frac{(L+1)}{L}} \langle A_f | \sum_{i=1}^{n} e_i j_L(kr_i)(-i)^L Y^*_{LM}(\theta_i, \phi_i) | A_i \rangle. \tag{24}$$

Now, we make the approximation, $j_L(kr_i) \approx (kr_i)^L/(2L+1)!!$, valid for $kr_i \ll 1$, and set $k = \omega/c$ to get our final result

$$\langle A_f | \sum_{i=1}^{n} \frac{e_i}{m_i c} (\vec{p}_i \cdot \vec{A}(\vec{r}_i, \mathcal{E})^*_{LM}) | A_i \rangle$$
$$= \left(\frac{\omega}{c}\right)^L \sqrt{\frac{(L+1)}{L}} \frac{(-i)^L}{(2L+1)!!} \langle A_f | \sum_{i=1}^{n} e_i r_i^L Y^*_{LM}(\theta_i, \phi_i) | A_i \rangle. \tag{25}$$

Now, with

$$\left[\frac{\text{Trans. Prob.}}{\text{second}}\right]_{A_i \to A_f} = \frac{8\pi\omega}{\hbar c} \sum_M |\, \langle A_f | \sum_{i=1}^{n} \frac{e_i}{m_i c} (\vec{p}_i \cdot \vec{A}(\vec{r}_i, \mathcal{E})_{LM} | A_i \rangle \,|^2, \tag{26}$$

this transition probability per second is given by

$$\frac{1}{\tau_{\mathcal{E}L}} = \frac{8\pi}{\hbar} \left(\frac{\omega}{c}\right)^{2L+1} \frac{(L+1)}{L[(2L+1)!!]^2} \sum_M |\, \langle A_f | \sum_{i=1}^{n} e_i r_i^L Y_{LM}(\theta_i, \phi_i) | A_i \rangle \,|^2, \tag{27}$$

where we have expressed this transition probability per unit time in terms of the inverse of the mean lifetime, $\tau_{\mathcal{E}L}$, for the photon emission process via the electric 2^L-pole moment radiation matrix element. The matrix element involves a sum over the $(2L+1)$ spherical components of the electric 2^L-pole moment operator.

In particular, with $L = 1$, and

$$\begin{pmatrix} rY_{1+1} \\ rY_{10} \\ rY_{1-1} \end{pmatrix} = \sqrt{\frac{3}{4\pi}} \begin{pmatrix} \vec{r}_{+1} = -\sqrt{\frac{1}{2}}(x+iy) \\ \vec{r}_0 = z \\ \vec{r}_{-1} = +\sqrt{\frac{1}{2}}(x-iy) \end{pmatrix}, \tag{28}$$

so, we regain our earlier electric dipole result

$$\frac{1}{\tau_{\mathcal{E}1}} = \frac{4\omega^3}{3\hbar c^3} \sum_M |\, \langle A_f | \sum_{i=1}^{n} (e_i \vec{r}_i)_M | A_i \rangle \,|^2. \tag{29}$$

For the case of general L, with initial and final states specified by

$$|A_i\rangle = |a_i J_i M_i\rangle, \qquad |A_f\rangle = |a_f J_f M_f\rangle, \tag{30}$$

we assume all other degeneracies have been removed so the initial and final states are $(2J_i+1)$-fold and $(2J_f+1)$-fold degenerate, with J_i, M_i and J_f, M_f, the total angular momentum quantum numbers for the quantum system. Again, a_i and a_f are shorthand for all other quantum numbers. Then, averaging over the initial M_i states, summing over the final M_f, and using the Wigner–Eckart theorem to express the M_i, M_f-dependent matrix elements in terms of the reduced or double-barred matrix element, we have

$$\frac{1}{\tau_{\mathcal{E}L}} = \frac{8\pi}{\hbar} \left(\frac{\omega}{c}\right)^{2L+1} \frac{(L+1)}{L[(2L+1)!!]^2} \frac{1}{(2J_i+1)}$$

$$\times \left| \langle a_f J_f \| \sum_i e_i r_i^L Y_L(\theta_i, \phi_i) \| a_i J_i \rangle \right|^2 . \tag{31}$$

The electric 2^L-pole operator has parity $(-1)^L$. We are thus led to the parity selection rule:

$$\text{final state parity} = \text{initial state parity}, \qquad \text{for } L = \text{even},$$

$$\text{final state parity} \neq \text{initial state parity}, \qquad \text{for } L = \text{odd}.$$

In addition, we have the angular momentum selection rule given by the angular momentum addition rule, $\vec{J}_i + \vec{L} = \vec{J}_f$, so $|(J_f - J_i)| \leq L \leq (J_f + J_i)$. Also, the transition probability is proportional to ω^{2L+1} and is thus a sensitive function of the photon energy, especially for the higher 2^L-pole cases.

B Magnetic Multipole Radiation

For the Lth magnetic multipole term, we need to evaluate the matrix element $\langle A_f | \vec{W} \cdot \vec{A}(\vec{r}, \mathcal{M})^*_{LM} | A_i \rangle$. Now, as we shall see, both pieces of the $\vec{W}$ vector will be equally important, and we shall need both

$$-\langle A_f | \sum_{i=1}^{n} \frac{e_i}{m_i c} \vec{p}_i \cdot \vec{A}(\vec{r}_i, \mathcal{M})^*_{LM} | A_i \rangle$$

and

$$-\langle A_f | \sum_{i=1}^{n} \frac{e_i \hbar g_{s,i}}{2 m_i c} \vec{s}_i \cdot [\vec{\nabla}_i \times \vec{A}(\vec{r}_i, \mathcal{M})^*_{LM}] | A_i \rangle .$$

Let us first evaluate the first term:

$$\begin{aligned}
&- \langle A_f | \sum_{i=1}^{n} \frac{e_i}{m_i c} \vec{p}_i \cdot \vec{A}(\vec{r}_i, \mathcal{M})^*_{LM} | A_i \rangle \\
&= - \int d\vec{r}_1 \cdots d\vec{r}_n \psi^*_{A_f} \sum_{i=1}^{n} \frac{\hbar}{i} \vec{\nabla}_i \cdot \left(\vec{A}(\vec{r}_i, \mathcal{M})^*_{LM} \psi_{A_i} \right) \\
&= - \int d\vec{r}_1 \cdots d\vec{r}_n \psi^*_{A_f} \sum_{i=1}^{n} \frac{e_i}{m_i c} \frac{\hbar}{i} \vec{A}(\vec{r}_i, \mathcal{M})^*_{LM} \cdot (\vec{\nabla}_i \psi_{A_i}) \\
&= - \int d\vec{r}_1 \cdots d\vec{r}_n \psi^*_{A_f} \sum_{i=1}^{n} \frac{e_i}{m_i c} \frac{\hbar}{i} \left(\frac{\vec{L} \Phi_{LM}(\vec{r}_i)}{\sqrt{L(L+1)}} \right)^* \cdot (\vec{\nabla}_i \psi_{A_i}),
\end{aligned} \tag{32}$$

where we have used $\left(\vec{\nabla}_i \cdot \vec{A}(\vec{r}_i, \mathcal{M})^*_{LM} \right) = 0$ and have expressed $\vec{A}(\vec{r}_i, \mathcal{M})_{LM}$ in terms of the scalar function $\Phi_{LM}(\vec{r}_i)$ through eq. (22) of Chapter 62. Now we use

$$\begin{aligned}
\left(\vec{L} \Phi_{LM}(\vec{r}_i) \right)^* \cdot (\vec{\nabla}_i \psi_{A_i}) &= -\frac{1}{i} [\vec{r}_i \times (\vec{\nabla}_i \Phi^*_{LM}(\vec{r}_i))] \cdot (\vec{\nabla}_i \psi_{A_i}) \\
&= +\frac{1}{i} (\vec{\nabla}_i \Phi^*_{LM}(\vec{r}_i)) \cdot [\vec{r}_i \times (\vec{\nabla}_i \psi_{A_i})]
\end{aligned} \tag{33}$$

to obtain

$$
\begin{aligned}
&-\langle A_f|\sum_{i=1}^{n}\frac{e_i}{m_ic}\vec{p}_i\cdot\vec{A}(\vec{r}_i,\mathcal{M})^*_{LM}|A_i\rangle \\
&=-\frac{1}{i\sqrt{L(L+1)}}\langle A_f|\sum_{i=1}^{n}\frac{e_i\hbar}{m_ic}\left(\vec{\nabla}_i\Phi^*_{LM}(\vec{r}_i)\right)\cdot\frac{1}{i}[\vec{r}_i\times\vec{\nabla}_i]|A_i\rangle \\
&=\frac{i}{\sqrt{L(L+1)}}\langle A_f|\sum_{i=1}^{n}\frac{e_i\hbar}{m_ic}\left(\vec{\nabla}_i\Phi^*_{LM}(\vec{r}_i)\right)\cdot\vec{l}_i|A_i\rangle, \qquad (34)
\end{aligned}
$$

where $\vec{l}_i$ is the (dimensionless) single-particle, orbital angular momentum operator acting on particle with index, i.

In similar manner, we can transcribe the second term into the same type of form

$$
\begin{aligned}
&-\langle A_f|\sum_{i=1}^{n}\frac{e_i\hbar g_{s,i}}{2m_ic}\vec{s}_i\cdot[\vec{\nabla}_i\times\vec{A}(\vec{r}_i,\mathcal{M})^*_{LM}|A_i]\rangle \\
&=+k\langle A_f|\sum_{i=1}^{n}\frac{e_i\hbar g_{s,i}}{2m_ic}\vec{s}_i\cdot\vec{A}(\vec{r}_i,\mathcal{E})^*_{LM}|A_i\rangle \\
&\approx k\sqrt{\frac{(L+1)}{L}}\langle A_f|\sum_{i=1}^{n}\frac{e_i\hbar g_{s,i}}{2m_ic}\vec{s}_i\cdot\vec{A}(\vec{r}_i,\mathcal{L})^*_{LM}|A_i\rangle \\
&=\frac{k}{-ik}\sqrt{\frac{(L+1)}{L}}\langle A_f|\sum_{i=1}^{n}\frac{e_i\hbar g_{s,i}}{2m_ic}\vec{s}_i\cdot\left(\vec{\nabla}_i\Phi^*_{LM}(\vec{r}_i)\right)|A_i\rangle, \qquad (35)
\end{aligned}
$$

where we have used $[\vec{\nabla}_i\times\vec{A}(\vec{r}_i,\mathcal{M})_{LM}]=-k\vec{A}(\vec{r}_i,\mathcal{E})_{LM}$ and have approximated $\vec{A}(\vec{r}_i,\mathcal{E})_{LM}$, as we did for the electric multipole case to express this vector in terms of a gradient of the scalar function, $\Phi_{LM}(\vec{r}_i)=j_L(kr_i)i^LY_{LM}(\theta_i,\phi_i)$. Combining eqs. (34) and (35), we get

$$
\begin{aligned}
&-\langle A_f|\sum_{i=1}^{n}\frac{e_i}{m_ic}\vec{p}_i\cdot\vec{A}(\vec{r}_i,\mathcal{M})^*_{LM}+\sum_{i=1}^{n}\frac{e_i\hbar g_{s,i}}{2m_ic}\vec{s}_i\cdot[\vec{\nabla}_i\times\vec{A}(\vec{r}_i,\mathcal{M})^*_{LM}]|A_i\rangle \\
&=i\sqrt{\frac{(L+1)}{L}}\langle A_f|\sum_{i=1}^{n}\frac{e_i\hbar}{2m_ic}\left(\vec{\nabla}_i\Phi^*_{LM}(\vec{r}_i)\right)\cdot\left(\frac{2}{(L+1)}\vec{l}_i+g_{s,i}\vec{s}_i\right)|A_i\rangle \\
&\approx i^{L+1}(-1)^{L+M}\sqrt{\frac{(L+1)}{L}}\frac{1}{(2L+1)!!} \\
&\times\langle A_f|\sum_{i=1}^{n}\frac{e_i\hbar}{2m_ic}\vec{\nabla}_i\left((kr_i)^LY_{L,-M}(\theta_i,\phi_i)\right)\cdot\left(\frac{2}{(L+1)}\vec{l}_i+g_{s,i}\vec{s}_i\right)|A_i\rangle, \qquad (36)
\end{aligned}
$$

where we have again used the approximation,

$$
j_L(kr_i)\approx\frac{(kr_i)^L}{(2L+1)!!}. \qquad (37)
$$

From this result, we get the final expression for the transition probability per second that the quantum system make a transition $A_i\rightarrow A_f$ via a magnetic 2^L-pole photon emission,

$$\frac{1}{\tau_{ML}} = \frac{8\pi}{\hbar}\left(\frac{\omega}{c}\right)^{2L+1}\frac{(L+1)}{L}\frac{1}{[(2L+1)!!]^2}\sum_M \times \Big| \langle A_f|\sum_{i=1}^{n}\frac{e_i\hbar}{2m_i c}\vec{\nabla}_i\Big(r_i^L Y_{L,M}(\theta_i,\phi_i)\Big)\cdot\Big(\frac{2}{(L+1)}\vec{l}_i + g_{s,i}\vec{s}_i\Big)|A_i\rangle\Big|^2, \tag{38}$$

where we have renamed $M \to -M$ in the dummy summation index. Now, we can use the gradient formula (see the mathematical appendix to Chapter 62),

$$\frac{1}{i}\vec{\nabla}\Big(i^L f(r)Y_{LM}\Big) = \Big(\frac{df}{dr}+\frac{(L+1)}{r}f\Big)\sqrt{\frac{L}{(2L+1)}}\vec{V}_{[(L-1)1]LM} + \Big(\frac{df}{dr}-\frac{L}{r}f\Big)\sqrt{\frac{(L+1)}{(2L+1)}}\vec{V}_{[(L+1)1]LM}. \tag{39}$$

In particular, the second term disappears for $f(r)=r^L$, so

$$\frac{1}{i}\vec{\nabla}\Big(i^L r^L Y_{LM}\Big) = \sqrt{L(2L+1)}r^{L-1}\vec{V}_{[(L-1)1]LM} = \sqrt{L(2L+1)}\sum_\mu\langle (L-1)(M-\mu)1\mu|LM\rangle i^{L-1}r^{L-1}Y_{(L-1)(M-\mu)}\vec{e}_\mu. \tag{40}$$

We therefore have

$$\vec{\nabla}_i\Big(r_i^L Y_{LM}(\theta_i,\phi_i)\Big)\cdot\Big(\frac{2}{(L+1)}\vec{l}_i + g_{s,i}\vec{s}_i\Big) = \sqrt{L(2L+1)} \times\sum_\mu\langle (L-1)(M-\mu)1\mu|LM\rangle r_i^{L-1}Y_{(L-1)(M-\mu)}(\theta_i,\phi_i) \times\Big(\frac{2}{(L+1)}(\vec{l}_i)_\mu + g_{s,i}(\vec{s}_i)_\mu\Big) = \sqrt{L(2L+1)}\Big[r_i^{L-1}Y_{(L-1)}(\theta_i,\phi_i)\times\Big(\frac{2}{(L+1)}\vec{l}_i^{\,1} + g_{s,i}\vec{s}_i^{\,1}\Big)\Big]^L_M, \tag{41}$$

where the square bracket denotes the angular momentum coupling of the spherical harmonic of rank $(L-1)$ with the vectors of spherical rank 1 to resultant angular momentum L. With this result, we can transcribe

$$\frac{1}{\tau_{ML}} = \frac{8\pi}{\hbar}\left(\frac{\omega}{c}\right)^{2L+1}\frac{(L+1)}{[(2L+1)!!(2L-1)!!]}\sum_M \Big| \langle A_f|\sum_{i=1}^{n}\frac{e_i\hbar}{2m_i c}\Big[r_i^{L-1}Y_{(L-1)}(\theta_i,\phi_i)\times\Big(\frac{2}{(L+1)}\vec{l}_i^{\,1} + g_{s,i}\vec{s}_i^{\,1}\Big)\Big]^L_M|A_i\rangle\Big|^2. \tag{42}$$

We note, in particular, with $L=1$, for which

$$\sqrt{L(2L+1)}\Big[r_i^{L-1}Y_{(L-1)}\times\Big(\frac{2}{(L+1)}\vec{l}_i^{\,1}+g_{s,i}\vec{s}_i^{\,1}\Big)\Big]^L_M = \sqrt{\frac{3}{4\pi}}\Big(\vec{l}_i+g_{s,i}\vec{s}_i\Big)_M, \tag{43}$$

we have

$$\frac{1}{\tau_{\mathcal{M}1}} = \frac{4\omega^3}{3\hbar c^3}\sum_M |\,\langle A_f|\sum_{i=1}^{n}(\vec{\mu}_i^{\,(\text{magnetic})})_M|A_i\rangle\,|^2, \tag{44}$$

where

$$\vec{\mu}_i^{\,(\text{magnetic})} = \frac{e_i\hbar}{2m_i c}(\vec{l}_i + g_{s,i}\vec{s}_i) \tag{45}$$

is the magnetic moment for the ith particle. Just as we had the generalized electric 2^L-pole operator given (in spherical form) by

$$\sum_i e_i r_i^L Y_{LM}(\theta_i, \phi_i),$$

we now have the magnetic 2^L-pole operator given by

$$\sum_i \sqrt{L(2L+1)}\Big[r_i^{L-1}Y_{(L-1)}(\theta_i,\phi_i)\times\frac{e_i\hbar}{2m_i c}\Big(\frac{2}{(L+1)}\vec{l}_i^{\,1} + g_{s,i}\vec{s}_i^{\,1}\Big)\Big]^L_M.$$

Finally, if we use the Wigner–Eckart theorem, we get

$$\frac{1}{\tau_{\mathcal{M}L}} = \frac{8\pi}{\hbar}\Big(\frac{\omega}{c}\Big)^{2L+1}\frac{(L+1)}{(2L+1)!!(2L-1)!!}\frac{1}{(2J_i+1)}\times$$
$$\Big|\,\langle a_f J_f\|\sum_{i=1}^{n}\frac{e_i\hbar}{2m_i c}\Big[r_i^{L-1}Y_{(L-1)}(\theta_i,\phi_i)\times\Big(\frac{2}{(L+1)}\vec{l}_i^{\,1} + g_{s,i}\vec{s}_i^{\,1}\Big)\Big]^L\|a_i J_i\rangle\,\Big|^2. \tag{46}$$

Let us end by comparing the orders of magnitude of a typical $\mathcal{M}1$ and a typical $\mathcal{E}1$ matrix element.

$$\frac{\mathcal{M}1 \text{ matrix element}}{\mathcal{E}1 \text{ matrix element}} \approx \frac{\mathcal{M}L \text{ matrix element}}{\mathcal{E}L \text{ matrix element}} \approx \frac{e\hbar}{mc}\frac{1}{er}. \tag{47}$$

For an atom, with $r \approx a_0$, this ratio is $\approx \frac{e^2}{\hbar c} \approx 10^{-2}$. The other useful ratio of matrix elements is

$$\frac{\mathcal{E}2 \text{ matrix element}}{\mathcal{E}1 \text{ matrix element}} \approx \frac{\mathcal{E}(L+1) \text{ matrix element}}{\mathcal{E}L \text{ matrix element}} \approx \frac{\omega}{c}r. \tag{48}$$

For an atom, this ratio is of order $(\hbar\omega)a_0/(\hbar c)$. If we were to take $\hbar\omega = e^2/a_0$, i.e., an $\hbar\omega$ equal to one atomic unit of energy, again, this ratio would be $\frac{e^2}{\hbar c} \approx (10)^{-2}$. Note, however, this $\hbar\omega = (8/3)\hbar\omega_{(\text{Lyman }\alpha)}$ and would correspond to a wavelength of $456 \times (10)^{-8}$ cm. For a wavelength in the visible part of the spectrum, therefore, the ratio would be of order $(10)^{-3}$. Thus, the $\mathcal{M}1$ and $\mathcal{E}2$ matrix elements are very roughly of the same order, with the $\mathcal{E}2$ matrix element possibly smaller by a factor of 10, but with both $\sim (10)^{-2} - (10)^{-3}$ times the order of magnitude of a typical $\mathcal{E}1$ matrix element.

For nuclei with

$$\frac{\hbar}{mcr} = \frac{\hbar c}{mc^2}\frac{1}{r} \approx \frac{0.2\text{fm}}{r_{\text{nucleus}}},$$

this number varies from $(10)^{-1}$ to $(10)^{-2}$ for light to heavy nuclei. The ratio $(\hbar\omega)r/(\hbar c)$ is strongly frequency dependent and again varies by nearly a factor of 10 with the mass of the nucleus but is again of the order of $(10)^{-1}$ to $(10)^{-2}$. Thus, roughly, $\mathcal{M}1$ and $\mathcal{E}2$ matrix elements are of the same order of magnitude, similarly for $\mathcal{M}2$ and $\mathcal{E}3$ matrix elements.

Finally, the $\mathcal{M}1$ and $\mathcal{E}2$ operators both have positive parity, thus they cannot change the parity of the state A_i, whereas the $\mathcal{E}1$ operator, as well as $\mathcal{M}2$ and $\mathcal{E}3$, $\mathcal{M}4$ and $\mathcal{E}5$ operators,..., have negative parity and have nonzero matrix elements only between states of opposite parity.

Problems

33. The 21-cm line in the hydrogen atom, famous through radio astronomy and the hydrogen maser, involves the transition, $F = 1 \to F = 0$, from the $F = 1$ hyperfine level to the $F = 0$ hyperfine level of the $1s$ ground state of the hydrogen atom, where $\vec{F} = \vec{s}_e + \vec{s}_p$. Calculate the mean life for the spontaneous emission of the 21-cm photon. Show the transition must proceed via the intrinsic spin part of the magnetic dipole term. (The long mean life results from the $\mathcal{M}1$ character of the transition, but also from the extremely low value of the transition frequency of $1.420{,}405{,}752 \times 10^9 \text{sec}^{-1}$. In the observation of this line from interstellar hydrogen, induced emission and absorption processes may also have to be considered.)

34. At low γ photon energies, with $\hbar\omega$ somewhat greater than the deuteron binding energy of 2.226 MeV, the photo dissociation of the deuteron

$$\gamma + d \to p + n$$

cannot proceed via the electric dipole term (cf., problems 32. and 31.), because the relative motion function for the p–n relative motion will be predominantly s wave, $l = 0$, just above threshold, and the bound-state wave function of the deuteron is an almost pure $l = 0$ wave function. At low γ photon energies, 2.226 MeV $< \hbar\omega \leq 4$ MeV, just above threshold, the photo-dissociation must therefore proceed via the magnetic dipole term.

Show that only the spin terms of the magnetic dipole operator can contribute to the matrix element. Because the bound state of the deuteron has $S = 1$, and because the orbital wave functions for two $S = 1$ states at *different* energies (one the bound state, the second the relative motion function in the continuum) must be orthogonal to each other*, show that the magnetic dipole operator has zero matrix element between the $S = 1$ bound state and an $S = 1$ continuum state. The photodissociation process near threshold thus proceeds only into $S = 0$ continuum states. For these, we can approximate the radial s wave continuum function by a phase-shifted $l = 0$ radial function,

$$rR_{l=0}(r) = A_0 \sin(k_0 r + \delta_0), \qquad r \to \text{large},$$

where $\hbar^2 k_0^2/M = \hbar\omega$ - 2.226 MeV, and δ_0 is the singlet phase shift given approximately in terms of the singlet scattering length

$$a_{\text{sc.l.}}^{S=0} = -23.68 \text{ fm}, \qquad \text{by} \quad \tan\delta_0 \approx -k_0 a_{\text{sc.l.}}^{S=0}.$$

For the bound-state wave function of the deuteron, use the Hulthen wave function (see problem 31). For purposes of estimating the radial overlap, assume the continuum wave function, $A_0 \sin(k_0 r + \delta_0)$, is valid all the way to $r = 0$. Also, recall the $S = 0, l = 0$ two-nucleon wave function must have isospin, $I = 1$. With the above approximations, calculate the differential and total cross sections for the photodissociation process and compare with the electric dipole result of problem 32. Also, recall $g_s = 2 \times 2.793$ for the proton and $g_s = 2 \times (-1.913)$ for the neutron.

*Footnote: For the highly approximate radial wave functions that we use here, this orthogonality would not be satisfied exactly, but this is a fault of our approximations. Also, note: We no longer use the plane wave approximations of problems 31 and 32, valid at higher energies.

35. The ground state of the nucleus, ${}^{111}_{49}\text{In}_{62}$, with 49 protons and 62 neutrons, has spin and parity, $\frac{9}{2}^+$. The first excited state of this nucleus is an isomeric (metastable) state with spin and parity, $\frac{1}{2}^-$, at an excitation energy of 0.5363 MeV. Show the transition from the $\frac{1}{2}^-$ state to the $\frac{9}{2}^+$ state can proceed only via the emission of a magnetic 2^4-pole or an electric 2^5-pole photon, i.e., $\mathcal{M}$-type with $L = 4$, or $\mathcal{E}$-type with $L = 5$. Prove, by calculating both rates, the $\mathcal{M}4$ rate predominates and make a prediction of the mean life of the excited $\frac{1}{2}^-$ state. The nuclear shell model would lead us to believe the transition matrix element results from the last odd proton making a transition from a $2p_{\frac{1}{2}}$ shell model orbit, with $l = 1, j = \frac{1}{2}$, to a $1g_{\frac{9}{2}}$ shell model orbit, with $l = 4, j = \frac{9}{2}$. The radial wave functions can be approximated by the 3-D harmonic oscillator wave functions

$$\text{For } 2p: \quad R_{2p}(\rho) = \left[\frac{m\omega_0}{\hbar}\right]^{\frac{3}{4}} \frac{1}{\sqrt{2\Gamma(\frac{7}{2})}}(5\rho - 2\rho^3)e^{-\frac{1}{2}\rho^2},$$

$$\text{For } 1g: \quad R_{1g}(\rho) = \left[\frac{m\omega_0}{\hbar}\right]^{\frac{3}{4}} \frac{2}{\sqrt{\Gamma(\frac{11}{2})}}\rho^4 e^{-\frac{1}{2}\rho^2},$$

where the oscillator length, $\sqrt{\hbar/m\omega_0}$, is determined by the nucleon mass, m, and the shell model $\hbar\omega_0$, which for a nucleus with a mass of 111 is approximately, $\hbar\omega_0$ = 8.5 MeV. In the above, $\rho = r\sqrt{m\omega_0/\hbar}$. The proton g_s factor is $2\times$ 2.793. First, calculate the needed matrix elements, as if a single proton were making a transition from $p_{\frac{1}{2}} \to g_{\frac{9}{2}}$. The actual transition in ${}^{111}_{49}\text{In}_{62}$ actually involves a transition from the proton configuration

$$(\text{closed j shells})(g_{\frac{9}{2}})^8(p_{\frac{1}{2}})^1 \to (\text{closed j shells})(g_{\frac{9}{2}})^9.$$

As the $p_{\frac{1}{2}}$ proton is converted to a $g_{\frac{9}{2}}$ proton, it finds only two of the 10 possible $g_{\frac{9}{2}}$ orbits still vacant. This gives an additional factor of $\sqrt{2/10}$ relative to the simple single-particle matrix element. Take this factor of $\sqrt{2/10}$ into account in calculating the actual mean life of the $\frac{1}{2}^-$ excited state. The observed lifetime is 7.7 minutes. (In comparing your actual result with this experimental value, remember our model for the transition is based on many approximations.)

36. The ground state of an odd Z nucleus has angular momentum and parity $\frac{7}{2}^+$. The first excited state of this nucleus at 0.32-MeV excitation energy has angular momentum and parity $\frac{11}{2}^-$. Show the transition from the $\frac{11}{2}^-$ state to the $\frac{7}{2}^+$ state will proceed predominantly via the emission of a magnetic 2^2-pole or an electric 2^3-pole photon (i.e., an $\mathcal{M}$-type with $L = 2$ or a $\mathcal{E}$-type with $L = 3$), or even higher multipoles. Show, by calculating both the $\mathcal{M}L = 2$ and $\mathcal{E}L = 3$ matrix elements which of the two rates predominates. Make a prediction of the mean life of the excited state under the assumption the transition matrix element results from a single odd proton making a transition from an $1h_{\frac{11}{2}}$ orbit, with $l = 5$, $j = \frac{11}{2}$, to a $1g_{\frac{7}{2}}$ orbit, with $l = 4$, $j = \frac{7}{2}$. The radial wave functions can be approximated by 3-D harmonic oscillator wave functions:

$$\text{For } 1h: \quad R_{1h}(\rho) = \frac{1}{a^{\frac{3}{2}}}\sqrt{\frac{2}{\Gamma(\frac{13}{2})}}\rho^5 e^{-\frac{1}{2}\rho^2},$$

$$\text{For } 1g: \quad R_{1g}(\rho) = \frac{1}{a^{\frac{3}{2}}}\sqrt{\frac{2}{\Gamma(\frac{11}{2})}}\rho^4 e^{-\frac{1}{2}\rho^2},$$

with $\rho = r/a$, and $a = 2.3$ fm. The proton g_s factor is 2×2.793.

37. A nucleus may make a transition from an excited state to a lower state by the emission of a γ photon or by the competing so-called internal conversion process in which the nuclear energy loss is transferred to an atomic K shell electron, with $n = 1, l = 0$, which is kicked out of the atomic K shell into the electron continuum. Assume the nuclear energy difference is such that the outgoing electron can be treated nonrelativistically, but is big enough, so the nuclear energy difference

$$E_i^{\text{nuc.}} - E_f^{\text{nuc.}} = \hbar\omega = \frac{p_f^2}{2m} - \Big(-\frac{Z^2e^2}{2a_0}\Big)$$

leads to

$$\frac{p_f^2}{2m} \gg \frac{Z^2e^2}{2a_0},$$

so the final free electron wave function can be approximated by a plane wave, with $\vec{p}_f = \hbar\vec{k}_f$. The interaction responsible for this process is the Coulomb interaction between the Z protons in the nucleus and the two K shell electrons in the atom

$$H_{\text{int.}} = -e^2\sum_{j=1}^{Z}\sum_{k=1}^{2}\frac{1}{|\vec{r}_j - \vec{r}_k|}$$

$$= -e^2 \sum_{j=1}^{Z}\sum_{k=1}^{2}\sum_{L} \frac{r_j^L}{\bar{r}_k^{L+1}} \frac{4\pi}{(2L+1)} \sum_M Y_{LM}(\theta_j, \phi_j) Y^*_{LM}(\bar{\theta}_k, \bar{\phi}_k),$$

where we have assumed the probability of finding the atomic electron inside the nucleus is negligible, so $r_j < \bar{r}_k$, where $\vec{r}_j$ gives the position of a proton within the nucleus, and $\vec{\bar{r}}_k$ gives the position of the atomic electron. Assume the atomic K shell electron wave function can be approximated by the hydrogenic $1s$ wave functions

$$2\left(\frac{Z}{a_0}\right)^{\frac{3}{2}} e^{-Z\bar{r}/a_0} Y_{00}(\bar{\theta}, \bar{\phi}).$$

Assume the angular momenta and parities of the initial and final nuclear states are such that the competing photon emission process would proceed via the emission of an electric 2^L-pole photon. Calculate the so-called internal conversion coefficient, which is the ratio

$$R = \frac{\text{Prob./sec. for nuclear transition } i \to f \text{ via K} - \text{shell conversion}}{\text{Prob./sec. for nuclear transition } i \to f \text{ via photon emission}}$$

and show under the above assumptions this ratio is

$$R = \left(\frac{e^2}{\hbar c}\right)^4 Z^3 \frac{L}{L+1} \left(\frac{2mc^2}{\hbar\omega}\right)^{L+\frac{5}{2}}.$$

In arriving at this result, we have used

$$\int_0^\infty dr r^2 \frac{j_L(k_f r)}{r^{L+1}} e^{-Zr/a_0} \approx \int_0^\infty dr \frac{j_L(k_f r)}{r^{L-1}} \qquad \text{for } \frac{k_f a_0}{Z} \gg 1,$$

and

$$\int_0^\infty d\rho \frac{j_L(\rho)}{\rho^{L-1}} = -\left[\frac{j_{L-1}(\rho)}{\rho^{L-1}}\right]_0^\infty = \frac{1}{(2L-1)!!},$$

which follows via eq. (5) of the appendix to Chapter 41, with replacement $l \to l-1$.

38. Auger Effect. In the two-electron He atom, states in which both electrons are excited have an energy higher than the energy of a He^+ ion in its $n = 1$ 1s ground state and a free electron, with energy $\hbar^2 k_f^2/2m$. Such doubly excited states dissociate into a He^+ ion plus the free electron (the Auger effect).

Calculate the probability per second that the $(2s)^2\ ^1S_0$ state of the He atom dissociate into a He^+ ion in its 1s state and a free electron with energy $\hbar^2 k_f^2/2m$. The interaction is the Coulomb potential between the two electrons,

$$\frac{e^2}{|\vec{r}_1 - \vec{r}_2|}.$$

Show, in particular, this Auger process is much more probable than a characteristic radiative decay process in an atom, by making the rough approximation that the free electron wave function can be replaced by a plane wave and the He bound wave functions can be approximated by hydrogenic wave functions, with

$$R_{1s} = 2\left(\frac{Z}{a_0}\right)^{\frac{3}{2}} e^{-\frac{Zr}{a_0}}, \qquad R_{2s} = \frac{1}{2\sqrt{2}}\left(\frac{Z}{a_0}\right)^{\frac{3}{2}}\left(2 - \frac{Zr}{a_0}\right)e^{-\frac{Zr}{2a_0}}.$$

39. The directional correlation between successively emitted γ photons is used to give information about the angular momenta of excited states in nuclei. Assuming the γ cascade is a $1^+ \to 1^- \to 0^+$ cascade and the emission probability for each photon is given in electric dipole approximation, calculate the directional correlation function, $W(\theta_{12})d\Omega_1 d\Omega_2$. The function $W(\theta_{12})d\Omega_1 d\Omega_2$ is the probability the second photon ($\hbar\omega_2$, emitted in the transition $1^- \to 0^+$) is emitted within a solid angle element $d\Omega_2$ about an angle, θ_{12}, relative to the direction of the first photon, ($\hbar\omega_1$, emitted in the transition $1^+ \to 1^-$). Use first-order time-dependent perturbation theory for the probability/second for the emission of each photon, so energy is conserved for each step of the cascade. Take the direction of emission of the first photon to be the z direction and assume the first photon is emitted into an element of solid angle $d\Omega_1$ about the z direction. Assume all M sublevels of the initial 1^+ state are populated with equal probability, and calculate first the probability for the population of each M sublevel of the intermediate 1^- level. Show $W(\theta_{12})d\Omega_1 d\Omega_2$ has the general form

$$W(\theta_{12})d\Omega_1 d\Omega_2 = (a + b\cos\theta_{12} + c\cos^2\theta_{12})d\Omega_1 d\Omega_2.$$

Find the relative magnitudes of c, b, and a, and show how these constants are related to the two reduced electric dipole moment matrix elements for the two transitions. You will have to relate the two circular components of the photon polarization vector relative to the z' axis, making azimuth and polar angles, ϕ_{12}, θ_{12}, with respect to the x, y, z coordinate system to the spherical components of the unit vectors $\vec{e}$ defined relative to the x, y, z coordinate system, by

$$\vec{e}\,'_{\pm 1} = \sum_m \vec{e}_m D^1_{m,\pm 1}(\phi_{12}, \theta_{12}, 0)^*.$$

66

Scattering of Photons by Atomic Systems

We shall next use our interaction between the atomic system and the radiation field to study the scattering of photons from atoms or molecules. We shall study both elastic scattering processes and inelastic scattering processes. In the low frequency limit, the elastic scattering is known as Rayleigh scattering; at high frequency, it is known as Thomson scattering. The inelasic scattering process goes by the name of Raman scattering.

Because we start with a photon of type $\vec{k}\mu$ and end with a photon of type $\vec{k}'\mu'$, with $\vec{k}' \neq \vec{k}$, the process will involve both a photon annihilation operator and a photon creation operator. We will thus have to deal with a second-order perturbation process, and we will need both first- and second-order perturbation terms

$$H_{\text{perturbation}} = H^{(1)}_{\text{int.}} + H^{(2)}_{\text{int.}}$$
$$= -\sum_{i=1}^{n} \frac{e_i}{m_i c} \vec{p}_i \cdot \vec{A}_\perp(\vec{r}_i) + \sum_{i=1}^{n} \frac{e_i^2}{2m_i c^2} \vec{A}_\perp(\vec{r}_i) \cdot \vec{A}_\perp(\vec{r}_i). \tag{1}$$

Because we are dealing with atoms or molecules, it will be sufficient to use the electric dipole approximation, and approximate $\vec{A}_\perp$ by

$$\vec{A}_\perp(\vec{r}_i) = \frac{c}{\sqrt{\text{Vol.}}} \sum_{\vec{k}} \sum_{\mu=\pm 1} \sqrt{\frac{2\pi\hbar}{\omega}} \left(a_{\vec{k}\mu} \vec{e}_\mu + a^\dagger_{\vec{k}\mu} \vec{e}^{\,*}_\mu\right). \tag{2}$$

We use the second-order perturbation theory formula, eq. (58) of Chapter 57, to get the transition probability that a photon of type $\vec{k}\mu$ is converted to one of type $\vec{k}'\mu'$ through a scattering process, with the atomic system starting in the initial state, A_i, and ending in the final state, A_f, with $A_f = A_i$ for the elastic scattering process, and $A_f \neq A_i$ for the inelastic scattering process. Thus,

$$\left[\frac{\text{Trans. Prob.}}{\text{second}}\right]_{\vec{k}\mu\to\vec{k}'\mu'} = \frac{2\pi}{\hbar}\rho(E_f)$$
$$\times \left| \langle A_f \vec{k}'\mu'|H^{(2)}_{\text{int.}}|A_i\vec{k}\mu\rangle + \sum_a \frac{\langle A_f\vec{k}'\mu'|H^{(1)}_{\text{int.}}|a\rangle\langle a|H^{(1)}_{\text{int.}}|A_i\vec{k}\mu\rangle}{(E^{(0)}_i - E^{(0)}_a)}\right|^2, \qquad (3)$$

where we have used the shorthand notation, $\vec{k}\mu$, for the state with photon number $n_{\vec{k}\mu} = 1$, where $E^{(0)}_i$ and $E^{(0)}_a$ include the eigenvalues of both $H^{(0)}_{\text{atom}}$ and $H^{(0)}_{\text{radiation}}$ and the intermediate states $|a\rangle$ include states of two types: either states with all photon numbers equal to zero, $|A_{\text{Intermediate}}\rangle \equiv |A_I\rangle$, or states with both a photon of type $\vec{k}\mu$ and of type $\vec{k}'\mu'$, $|A_I\vec{k}\mu, \vec{k}'\mu'\rangle$. The three types of needed perturbation terms are illustrated by the diagrams of Fig. 66.1. In the term of type (2), the photon $\vec{k}\mu$ is annihilated at time τ_1 and the photon $\vec{k}'\mu'$ is created at later time τ_2, whereas in the term of type (3) the photon $\vec{k}'\mu'$ is created at the earlier time τ_1 and the photon $\vec{k}\mu$ is annihilated at the later time τ_2. Therefore,

$$\left[\frac{\text{Trans. Prob.}}{\text{second}}\right]_{\vec{k}\mu\to\vec{k}'\mu'} = \frac{2\pi}{\hbar}\rho(E_f)$$
$$\times \left| \langle A_f\vec{k}'\mu'|H^{(2)}_{\text{int.}}|A_i\vec{k}\mu\rangle + \sum_{A_I}\frac{\langle A_f\vec{k}'\mu'|H^{(1)}_{\text{int.}}|A_I\rangle\langle A_I|H^{(1)}_{\text{int.}}|A_i\vec{k}\mu\rangle}{(E^{(0)}_{A_i} + \hbar\omega - E^{(0)}_{A_I})}\right.$$
$$\left. + \sum_{A_I}\frac{\langle A_f\vec{k}'\mu'|H^{(1)}_{\text{int.}}|A_I\vec{k}'\mu', \vec{k}\mu\rangle\langle A_I\vec{k}'\mu', \vec{k}\mu|H^{(1)}_{\text{int.}}|A_i\vec{k}\mu\rangle}{(E^{(0)}_{A_i} - E^{(0)}_{A_I} - \hbar\omega')}\right|^2. \qquad (4)$$

The energy conservation delta function that led to this relation requires only that $E^{(0)}_{A_f} + \hbar\omega' = E^{(0)}_{A_i} + \hbar\omega$. The intermediate energies, $E^{(0)}_{A_I}$, can therefore go to arbitrarily high values, and the sum, $\sum_{A_I}$, will in general include an integral over atomic continuum states. The three contributions to the needed matrix element are

$$(1): \langle A_f\vec{k}'\mu'|H^{(2)}_{\text{int.}}|A_i\vec{k}\mu\rangle$$
$$= \frac{c^2}{\text{Vol.}}\frac{2\pi\hbar}{\sqrt{\omega\omega'}}(\vec{e}_\mu\cdot\vec{e}'^*_{\mu'})\sum_{i=1}^{n}\frac{e_i^2}{2m_ic^2}\langle A_f\vec{k}'\mu'|\left(a^\dagger_{\vec{k}'\mu'}a_{\vec{k}\mu} + a_{\vec{k}\mu}a^\dagger_{\vec{k}'\mu'}\right)|A_i\vec{k}\mu\rangle$$
$$= \frac{c^2}{\text{Vol.}}\frac{2\pi\hbar}{\sqrt{\omega\omega'}}(\vec{e}_\mu\cdot\vec{e}'^*_{\mu'})\sum_{i=1}^{n}\frac{e_i^2}{2m_ic^2}2\delta_{A_fA_i}\delta_{\omega'\omega}. \qquad (5)$$

(Note: $a_{\vec{k}\mu}$ commutes with $a^\dagger_{\vec{k}'\mu'}$ because we must have $\vec{k}' \neq \vec{k}$ to have a scattering process. However, $\omega' = \omega$ follows from energy conservation through the energy delta function of the Golden Rule. Note, finally, we have put a prime on the unit vector associated with $\vec{k}'\mu'$ to indicate it is associated with a different z' axis than that associated with $\vec{k}\mu$.)

$$(2): \langle A_f\vec{k}'\mu'|H^{(1)}_{\text{int.}}|A_I\rangle\langle A_I|H^{(1)}_{\text{int.}}|A_i\vec{k}\mu\rangle$$
$$= \frac{c^2}{\text{Vol.}}\frac{2\pi\hbar}{\sqrt{\omega\omega'}}\langle A_f|\sum_i\frac{e_i}{m_ic}(\vec{p}_i\cdot\vec{e}'^*_{\mu'})|A_I\rangle\langle A_I|\sum_i\frac{e_i}{m_ic}(\vec{p}_i\cdot\vec{e}_\mu)|A_i\rangle, \qquad (6)$$

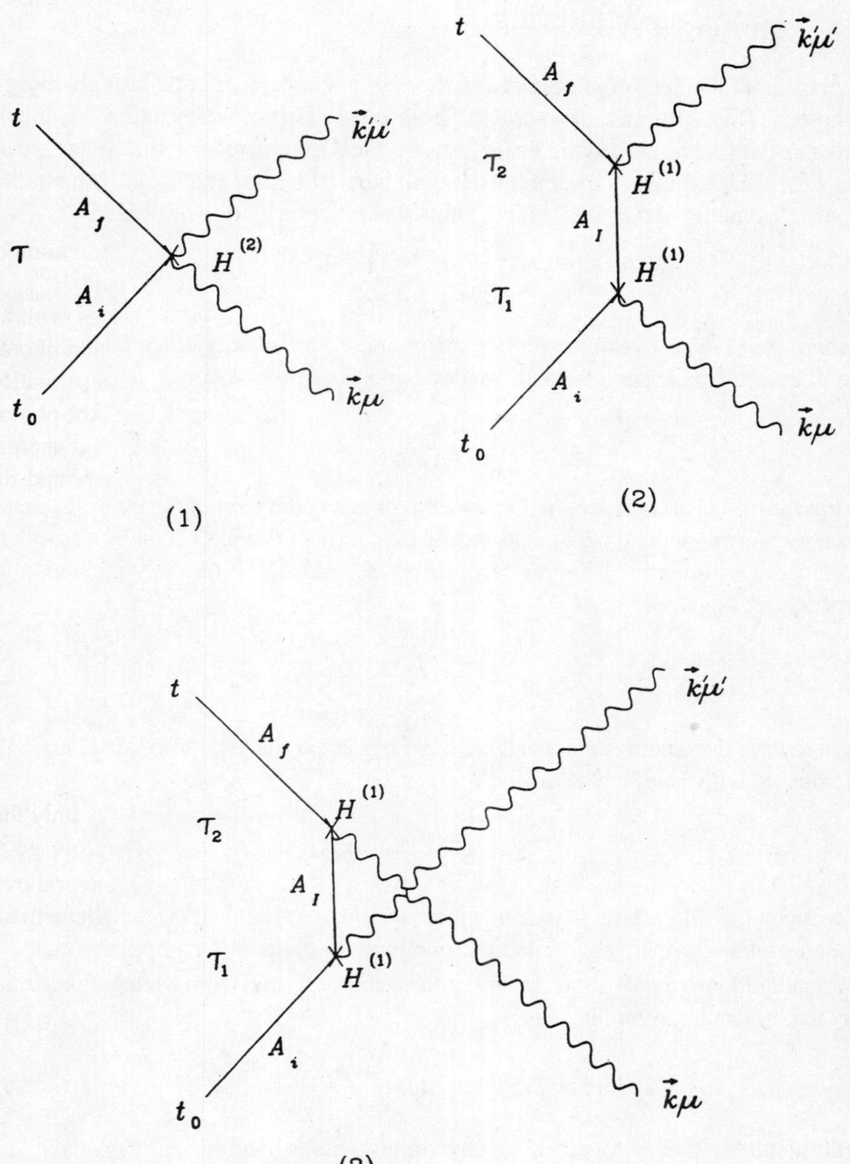

FIGURE 66.1. The three types of second-order perturbation terms for photon-atom scattering.

$$
\begin{aligned}
(3) : &\langle A_f\vec{k}\,'\mu'|H^{(1)}_{\text{int.}}|A_I\vec{k}\,'\mu', \vec{k}\mu\rangle\langle A_I\vec{k}\,'\mu', \vec{k}\mu|H^{(1)}_{\text{int.}}|A_i\vec{k}\mu\rangle \\
&= \frac{c^2}{\text{Vol.}}\frac{2\pi\hbar}{\sqrt{\omega\omega'}}\langle A_f|\sum_i \frac{e_i}{m_i c}(\vec{p}_i\cdot\vec{e}_\mu)|A_I\rangle\langle A_I|\sum_i \frac{e_i}{m_i c}(\vec{p}_i\cdot\vec{e}^{\,'*}_{\mu'})|A_i\rangle. \quad (7)
\end{aligned}
$$

A Thomson Scattering

Let us first look at a very special case, the case of elastic scattering of high-energy photons. This is the case of so-called Thomson scattering. We assume $\hbar\omega$ is much greater than a typical atomic energy, or a typical atomic energy difference, such as $E^{(0)}_{A_I} - E^{(0)}_{A_i}$, found in the energy denominators of terms (2) and (3). The atomic matrix elements of terms (2) and (3) thus make a contribution of order

$$\frac{e^2}{mc^2}\frac{p^2}{m} \times \frac{1}{\hbar\omega},$$

where the $(1/\hbar\omega)$ gives the order of magnitude of the energy denominator factor in this case. Conversely, term (1) makes a contribution of order

$$\frac{e^2}{mc^2}.$$

Thus, terms (2) and (3) are smaller by a factor that is the ratio of (an atomic kinetic energy matrix element)/($\hbar\omega$), and in this case terms (2) and (3) can be neglected compared with term (1). In this case, therefore, with $\hbar\omega' = \hbar\omega \gg$ typical atomic energy, we have

$$\left[\frac{\text{Trans. Prob.}}{\text{second}}\right]_{\vec{k}\mu\to\vec{k}'\mu'} = \frac{2\pi}{\hbar}\rho(E_f)\left[\frac{c^2}{\text{Vol.}}\frac{2\pi\hbar}{\omega}(\vec{e}_\mu \cdot \vec{e}^{\,\prime *}_{\mu'})\frac{Ze^2}{mc^2}\right]^2, \tag{8}$$

where m is the electron mass and we have neglected the term of order $1/m_{\text{nucleus}}$ compared with $1/m$. We also need

$$\rho(E_f) = \frac{\text{Vol.}}{(2\pi)^3}\frac{\omega^2}{\hbar c^3}d\Omega'. \tag{9}$$

To obtain the differential scattering cross section, $d\sigma$, we need (a) the above transition probability per second that the photon with incident $\vec{k}$ and polarization $\vec{e}_\mu$ be scattered into a $d\Omega'$ about $\vec{k}'$ with polarization $\vec{e}^{\,\prime}_{\mu'}$, and (b) we must divide this by the incident photon flux. For us,

$$\text{Incident photon flux} = \frac{(n_{\vec{k}\mu} = 1)c}{\text{Vol.}}. \tag{10}$$

Putting all the factors together, we have in this approximation

$$\left(\frac{d\sigma}{d\Omega'}\right)_{\text{Thomson}} = \frac{Z^2e^4}{m^2c^4}(\vec{e}_\mu \cdot \vec{e}^{\,\prime *}_{\mu'})^2. \tag{11}$$

We note

$$\frac{e^2}{mc^2} = r_0 = \text{“classical electron radius”} = 2.8 \times (10)^{-13}\text{cm}. \tag{12}$$

Above, we have written this cross section in terms of the circular polarization vectors, $\vec{e}_\mu$ and $\vec{e}^{\,\prime *}_{\mu'}$. We could just as well have taken linear polarization vectors,

$\vec{e}_\alpha$ and $\vec{e}^{\,\prime}_{\alpha'}$, to get

$$\left(\frac{d\sigma}{d\Omega'}\right)_{\text{Thomson}} = Z^2 r_0^2 (\vec{e}_\alpha \cdot \vec{e}^{\,\prime}_{\alpha'})^2. \tag{13}$$

If we choose the original photon direction, $\vec{k}$, along the z axis, and the scattered photon, $\vec{k}^{\,\prime}$, along the z' axis, as shown in Fig. 66.2,

$$\begin{aligned} (\vec{e}_{x'} \cdot \vec{e}_x) &= \cos\theta\cos\phi, \quad & (\vec{e}_{x'} \cdot \vec{e}_y) &= \cos\theta\sin\phi, \\ (\vec{e}_{y'} \cdot \vec{e}_x) &= -\sin\phi, & (\vec{e}_{y'} \cdot \vec{e}_y) &= \cos\phi. \end{aligned} \tag{14}$$

For unpolarized incident photons and for detectors insensitive to the polarization of the scattered photons, we then have

$$\left(\frac{d\sigma}{d\Omega'}\right)_{\text{Thomson}} = \tfrac{1}{2} Z^2 r_0^2 \sum_{\alpha=x,y} \sum_{\alpha'=x',y'} (\vec{e}_\alpha \cdot \vec{e}^{\,\prime}_{\alpha'})^2 = \tfrac{1}{2} Z^2 r_0^2 (\cos^2\theta + 1), \tag{15}$$

with total cross section

$$\sigma_{\text{Thomson}} = \frac{8\pi}{3} Z^2 r_0^2. \tag{16}$$

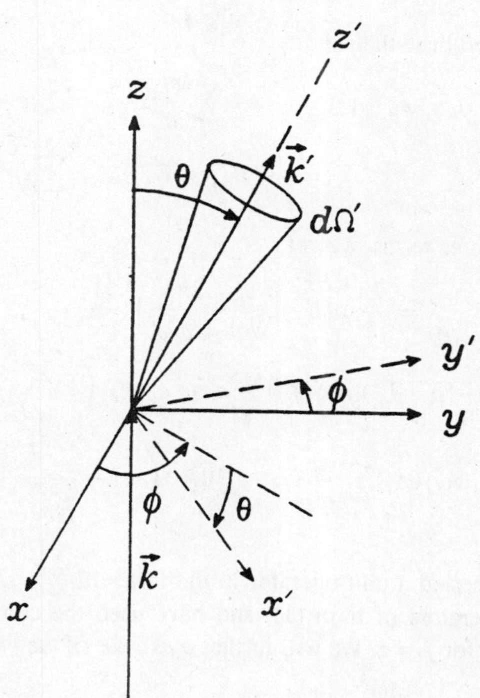

FIGURE 66.2. The incident photon $\vec{k}$ and scattered photon $\vec{k}'$ vectors.

B Rayleigh and Raman Scattering

For the general case of photons of arbitrary $\hbar\omega$, or for photons of low frequency, it will be necessary to go back to eqs. (4)–(7). Now, terms of type (1), (2), and (3) all contribute. In the general case, it will be possible to combine terms (1), (2), and (3) most neatly by the following substitutions.

In term (1), use the identity

$$(\vec{e}'^{*}_{\mu'} \cdot \vec{e}_{\mu}) = \frac{i}{\hbar}\Big((\vec{p}_i \cdot \vec{e}'^{*}_{\mu'})(\vec{r}_i \cdot \vec{e}_{\mu}) - (\vec{r}_i \cdot \vec{e}_{\mu})(\vec{p}_i \cdot \vec{e}'^{*}_{\mu'})\Big) = \frac{i}{\hbar}[(\vec{p}_i)'^{*}_{\mu'}, (\vec{r}_i)_{\mu}], \quad (17)$$

where the Heisenberg commutation relation between $\vec{p}$ and $\vec{r}$ in *spherical* components can be written as

$$[(\vec{p})^{*}_{\mu}, (\vec{r})_{\nu}] = \frac{\hbar}{i}\delta_{\mu\nu}. \quad (18)$$

[Eq. (17), however, also involves direction cosines between the unit vectors in the primed and unprimed coordinates.]

In term (2), we will use the identity

$$\begin{aligned}\langle A_I|\sum_i \frac{e_i}{m_i c}(\vec{p}_i \cdot \vec{e}_{\mu})|A_i\rangle &= \frac{i}{\hbar}\langle A_I|\sum_i \frac{e_i}{c}[H^{(0)}_{\text{atom}}, \vec{r}_i] \cdot \vec{e}_{\mu}|A_i\rangle \\ &= \frac{i}{\hbar c}(E^{(0)}_{A_I} - E^{(0)}_{A_i})\langle A_I|\sum_i e_i\vec{r}_i \cdot \vec{e}_{\mu}|A_i\rangle. \quad (19)\end{aligned}$$

In term (3), we will use the identity

$$\begin{aligned}\langle A_f|\sum_i \frac{e_i}{m_i c}(\vec{p}_i \cdot \vec{e}_{\mu})|A_I\rangle &= \frac{i}{\hbar}\langle A_f|\sum_i \frac{e_i}{c}[H^{(0)}_{\text{atom}}, \vec{r}_i] \cdot \vec{e}_{\mu}|A_I\rangle \\ &= \frac{i}{\hbar c}(E^{(0)}_{A_f} - E^{(0)}_{A_I})\langle A_f|\sum_i e_i\vec{r}_i \cdot \vec{e}_{\mu}|A_I\rangle. \quad (20)\end{aligned}$$

Combining these three terms, we get

$$\begin{aligned}\left[\frac{\text{Trans. Prob.}}{\text{second}}\right]_{\vec{k}\mu\to\vec{k}'\mu'} &= \frac{2\pi}{\hbar}\rho(E_f)\Big[\frac{c^2}{\text{Vol.}}\frac{2\pi\hbar}{\sqrt{\omega\omega'}}\Big]^2\frac{1}{\hbar^2 c^4} \\ \times \Bigg|\sum_{A_I}\Bigg[&\langle A_f|\sum_i \frac{e_i}{m_i}(\vec{p}_i \cdot \vec{e}'^{*}_{\mu'})|A_I\rangle\langle A_I|\sum_i e_i\vec{r}_i \cdot \vec{e}_{\mu}|A_i\rangle\left(1 + \frac{E^{(0)}_{A_I} - E^{(0)}_{A_i}}{E^{(0)}_{A_i} + \hbar\omega - E^{(0)}_{A_I}}\right) \\ + \langle A_f|\sum_i e_i\vec{r}_i \cdot \vec{e}_{\mu}|A_I\rangle&\langle A_I|\sum_i \frac{e_i}{m_i}(\vec{p}_i \cdot \vec{e}'^{*}_{\mu'})|A_i\rangle\left(-1 + \frac{E^{(0)}_{A_f} - E^{(0)}_{A_I}}{E^{(0)}_{A_i} - E^{(0)}_{AI} - \hbar\omega'}\right)\Bigg]\Bigg|^2, \quad (21)\end{aligned}$$

where we have inserted a unit operator in the form of, $\sum_{A_I}|A_I\rangle\langle A_I| = 1$, between the two operators of term (1), and have used the commutator relation, $[(\vec{p}_i)'^{*}_{\mu'}, (\vec{r}_j)_{\mu}] = 0$ for $j \neq i$. We will further make use of the two identities

$$\left(1 + \frac{E^{(0)}_{A_I} - E^{(0)}_{A_i}}{E^{(0)}_{A_i} + \hbar\omega - E^{(0)}_{A_I}}\right) = \frac{\hbar\omega}{E^{(0)}_{A_i} + \hbar\omega - E^{(0)}_{A_I}},$$

$$\left(-1+\frac{E_{A_f}^{(0)}-E_{A_I}^{(0)}}{E_{A_i}^{(0)}-E_{AI}^{(0)}-\hbar\omega'}\right)=\frac{\hbar\omega}{E_{A_i}^{(0)}-E_{A_I}^{(0)}-\hbar\omega'}, \tag{22}$$

and use the commutator relation

$$\frac{\vec{p}_i}{m_i}=\frac{i}{\hbar}[H_{\text{atom}}^{(0)},\vec{r}_i] \tag{23}$$

once more to convert the remaining atomic matrix elements in eq. (21) to electric dipole form, and obtain

$$\begin{aligned}\left[\frac{\text{Trans. Prob.}}{\text{second}}\right]_{\vec{k}\mu\to\vec{k}'\mu'} &= \frac{2\pi}{\hbar}\rho(E_f)\frac{(2\pi)^2}{\omega\omega'\text{Vol.}^2}\omega^2\\ \times\Bigg|\sum_{A_I}&\frac{(E_{A_f}^{(0)}-E_{A_I}^{(0)})\langle A_f|\sum_i e_i\vec{r}_i\cdot\vec{e}_{\mu'}^{\,\prime *}|A_I\rangle\langle A_I|\sum_i e_i\vec{r}_i\cdot\vec{e}_\mu|A_i\rangle}{E_{A_i}^{(0)}+\hbar\omega-E_{A_I}^{(0)}}\\ +\sum_{A_I}&\frac{(E_{A_I}^{(0)}-E_{A_i}^{(0)})\langle A_f|\sum_i e_i\vec{r}_i\cdot\vec{e}_\mu|A_I\rangle\langle A_I|\sum_i e_i\vec{r}_i\cdot\vec{e}_{\mu'}^{\,\prime *}|A_i\rangle}{E_{A_i}^{(0)}-E_{A_I}^{(0)}-\hbar\omega'}\Bigg|^2.\end{aligned} \tag{24}$$

Now, we shall further make use of the identities

$$\begin{aligned}\frac{E_{A_f}^{(0)}-E_{A_I}^{(0)}}{E_{A_i}^{(0)}-E_{A_I}^{(0)}+\hbar\omega}&=1-\frac{\hbar\omega'}{E_{A_i}^{(0)}-E_{A_I}^{(0)}+\hbar\omega},\\ \frac{E_{A_I}^{(0)}-E_{A_i}^{(0)}}{E_{A_i}^{(0)}-E_{A_I}^{(0)}-\hbar\omega'}&=-1-\frac{\hbar\omega'}{E_{A_i}^{(0)}-E_{A_I}^{(0)}-\hbar\omega'},\end{aligned} \tag{25}$$

and use the completeness of the atomic states $|A_I\rangle$ to note

$$\begin{aligned}\sum_{A_I}\Bigg(&\langle A_f|\sum_i e_i\vec{r}_i\cdot\vec{e}_{\mu'}^{\,\prime *}|A_I\rangle\langle A_I|\sum_i e_i\vec{r}_i\cdot\vec{e}_\mu|A_i\rangle\\ &-\langle A_f|\sum_i e_i\vec{r}_i\cdot\vec{e}_\mu|A_I\rangle\langle A_I|\sum_i e_i\vec{r}_i\cdot\vec{e}_{\mu'}^{\,\prime *}|A_i\rangle\Bigg)\\ &=0.\end{aligned} \tag{26}$$

With these relations, we can convert the transition probability for the scattering process into

$$\begin{aligned}\left[\frac{\text{Trans. Prob.}}{\text{second}}\right]_{\vec{k}\mu\to\vec{k}'\mu'} &= \frac{2\pi}{\hbar}\rho(E_f)\frac{(2\pi)^2}{\omega\omega'\text{Vol.}^2}\hbar^2\omega^2\omega'^2\\ \times\Bigg|\sum_{A_I}&\frac{\langle A_f|\sum_i e_i\vec{r}_i\cdot\vec{e}_{\mu'}^{\,\prime *}|A_I\rangle\langle A_I|\sum_i e_i\vec{r}_i\cdot\vec{e}_\mu|A_i\rangle}{E_{A_i}^{(0)}+\hbar\omega-E_{A_I}^{(0)}}\\ +\sum_{A_I}&\frac{\langle A_f|\sum_i e_i\vec{r}_i\cdot\vec{e}_\mu|A_I\rangle\langle A_I|\sum_i e_i\vec{r}_i\cdot\vec{e}_{\mu'}^{\,\prime *}|A_i\rangle}{E_{A_i}^{(0)}-E_{A_I}^{(0)}-\hbar\omega'}\Bigg|^2.\end{aligned} \tag{27}$$

In this expression, our polarization vectors, $\vec{e}_\mu$, for the incident photon of type $\vec{k}\mu$ are defined with respect to the laboratory axes, but the circular polarization vectors, $\vec{e}_{\mu'}^{\,\prime *}$, are again defined with respect to the direction of the vector $\vec{k}\,'$, which can be used to define the z' direction, where the x', y', z' axes are rotated through the Euler angles $\phi, \theta, 0$, relative to the laboratory axes (see Fig. 66.2), so

$$\vec{e}^{\,\prime}_{\mu'} = \sum_{m=\pm 1,0} \vec{e}_m D^1_{m\mu'}(\phi, \theta, 0)^*. \tag{28}$$

To obtain the differential scattering cross section, $d\sigma$, we must again divide the transition probability per second that the photon with incident $\vec{k}$ and polarization $\vec{e}_\mu$ be scattered into a $d\Omega'$ about $\vec{k}\,'$ with polarization $\vec{e}^{\,\prime}_{\mu'}$ by the incident photon flux. For us,

$$\text{Incident photon flux} = \frac{(n_{\vec{k}\mu} = 1)c}{\text{Vol.}}. \tag{29}$$

We also need $\rho(E_f)$, which is now

$$\rho(E_f) = \frac{\text{Vol.}}{(2\pi)^3} \frac{\omega'^2}{\hbar c^3} d\Omega'. \tag{30}$$

Thus,

$$\begin{aligned} d\sigma &= \frac{2\pi}{\hbar} \frac{\text{Vol.}}{(2\pi)^3} \frac{\omega'^2}{\hbar c^3} d\Omega' \frac{(2\pi)^2}{\omega\omega' \text{Vol.}^2} \hbar^2 \omega^2 \omega'^2 \frac{\text{Vol.}}{c} \\ &\quad \times \sum_{m,\bar{m}} D^1_{m\mu'}(\phi, \theta, 0) D^1_{\bar{m}\mu'}(\phi, \theta, 0)^* g_{m\mu} g^*_{\bar{m}\mu} \\ &= \frac{\omega\omega'^3}{c^4} d\Omega' \sum_{m,\bar{m}} D^1_{m\mu'}(\phi, \theta, 0) D^1_{\bar{m}\mu'}(\phi, \theta, 0)^* g_{m\mu} g^*_{\bar{m}\mu}, \end{aligned} \tag{31}$$

where

$$\begin{aligned} g_{m\mu} = \sum_{A_I} \Bigg[& \frac{\langle A_f | (\vec{\mu}^{(\text{el.})})^*_m | A_I \rangle \langle A_I | (\vec{\mu}^{(\text{el.})})_\mu | A_i \rangle}{E^{(0)}_{A_i} - E^{(0)}_{A_I} + \hbar\omega} \\ & + \frac{\langle A_f | (\vec{\mu}^{(\text{el.})})_\mu | A_I \rangle \langle A_I | (\vec{\mu}^{(\text{el.})})^*_m | A_i \rangle}{E^{(0)}_{A_i} - E^{(0)}_{A_I} - \hbar\omega'} \Bigg]. \end{aligned} \tag{32}$$

Note: We sum over all three values $m = +1, 0, -1$, similarly for $\bar{m}$, where these are defined with respect to the laboratory z axis of Fig. 66.2, which is chosen along the direction of $\vec{k}$, whereas $\mu = +1$, or -1, is fixed by the polarization of the incident photon. Also, in general, it will be very challenging to evaluate the matrix element sums over intermediate states in $g_{m\mu}$. These sums will in general involve an infinite number of discrete states and an integral over continuum states. In the extreme low frequency limit, however, in which $\hbar\omega$ and $\hbar\omega'$ in the energy denominators can be neglected, the $g_{m\mu}$ are essentially pure atomic properties, dependent only on the nature of the states, $|A_i\rangle$ and $|A_f\rangle$. For elastic scattering, with an unpolarized incident beam and a detector insensitive to the polarization of the scattered photons, we then have

$$\left(\frac{d\sigma}{d\Omega'}\right)_{\text{Rayleigh}} = \frac{1}{2} \frac{\omega^4}{c^4} \sum_{\mu=\pm 1} \sum_{\mu'=\pm 1} \sum_{m,\bar{m}} D^1_{m\mu'}(\phi, \theta, 0) D^1_{\bar{m}\mu'}(\phi, \theta, 0)^* g_{m\mu} g_{\bar{m}\mu}, \tag{33}$$

leading to the total cross section

$$\sigma_{\text{Rayleigh}} = \frac{4\pi}{3}\frac{\omega^4}{c^4}\sum_{\mu=\pm 1}\sum_{m=\pm 1,0} |g_{m\mu}|^2. \tag{34}$$

This cross section is of order of magnitude

$$\left(\frac{2\pi}{\lambda}\right)^4 \left(\frac{(ea_0)^2}{\frac{e^2}{a_0}}\right)^2 \approx \left(\frac{2\pi}{\lambda}\right)^4 a_0^6.$$

For $\lambda = 6,000 \times (10)^{-8}$cm, we have $\sigma \approx (10)^{-12} a_0^2$. Note, finally, the ω^4 dependence of the formula.

67
Resonance Fluorescence Cross Section

In the last chapter, we saw both the elastic and inelastic photon scattering cross sections were given in terms of the amplitudes, $g_{m\mu}$, where

$$\sigma_{\text{elastic}} = \frac{8\pi}{3}\frac{\omega^4}{c^4}\sum_{m=\pm 1,0}|g_{m\mu}|^2, \qquad \sigma_{\text{inelastic}} = \frac{8\pi}{3}\frac{\omega\omega'^3}{c^4}\sum_{m=\pm 1,0}|g_{m\mu}|^2, \tag{1}$$

where $\mu = \pm 1$ indicates the circular polarization of the incident photon, and where

$$g_{m\mu} = \sum_{A_I}\left[\frac{\langle A_f|(\mu^{(\text{el.})})^*_m|A_I\rangle\langle A_I|\mu^{(\text{el.})}_\mu|A_i\rangle}{E^{(0)}_{A_i} + \hbar\omega - E^{(0)}_{A_I}} + \frac{\langle A_f|(\mu^{(\text{el.})}_\mu|A_I\rangle\langle A_I(\mu^{(\text{el.})})^*_m|A_i\rangle}{E^{(0)}_{A_i} - E^{(0)}_{A_I} - \hbar\omega'}\right]. \tag{2}$$

In the second term of this sum, the energy denominator can never be zero because we assume $E^{(0)}_{A_i}$ is the ground-state energy of the atom, so $E^{(0)}_{A_I} > E^{(0)}_{A_i}$ for all A_I. In the first term of the sum, however, it may be possible that the incident photon energy is such that $\hbar\omega = E^{(0)}_{A_I} - E^{(0)}_{A_i}$ for some specific excited state, $E^{(0)}_{A_I}$. In that case, we would have a zero energy denominator for one specific term in the sum over the states A_I. This is very similar to the zero denominators which had to be avoided in the scattering Green's function formulation of the scattering problem, where we avoided this difficulty by replacing the energy E in the operator form of the Green's function formulation by an $E + i\epsilon$. Let us attempt the same type of change here, and make the replacement

$$\sum_{A_I}\frac{\langle A_f|(\mu^{(\text{el.})})^*_m|A_I\rangle\langle A_I|\mu^{(\text{el.})}_\mu|A_i\rangle}{E^{(0)}_{A_i} + \hbar\omega - E^{(0)}_{A_I}}$$
$$\rightarrow \langle A_f|(\mu^{(\text{el.})})^*_m|A_I\rangle\frac{1}{E - H_0 + i\epsilon}\langle A_I|\mu^{(\text{el.})}_\mu|A_i\rangle$$

$$= \int d\vec{r}_1 \cdots d\vec{r}_n \int d\vec{r}_1^{\,'} \cdots d\vec{r}_n^{\,'} \psi^*_{A_f}(\vec{r}_1, \ldots, \vec{r}_n) \sum_i \mu_m^{(\text{el.})*}(\vec{r}_i)$$

$$\times \left[\sum_{A_I} \frac{\psi_{A_I}(\vec{r}_1, \ldots, \vec{r}_n)\psi^*_{A_I}(\vec{r}_1^{\,'}, \ldots, \vec{r}_n^{\,'})}{E - E^{(0)}_{A_I} + i\epsilon} \right] \sum_i \mu_\mu^{(\text{el.})}(\vec{r}_i^{\,'}) \psi_{A_i}(\vec{r}_1^{\,'}, \ldots, \vec{r}_n^{\,'})$$

$$= \int d\vec{r} \int d\vec{r}^{\,'} \psi^*_{A_f}(\vec{r}) \mu_m^{(\text{el.})*}(\vec{r}) G(\vec{r}, \vec{r}^{\,'}; E) \mu_\mu^{(\text{el.})}(\vec{r}^{\,'}) \psi_{A_i}(\vec{r}^{\,'}), \tag{3}$$

where E is the incident energy, $E = E^{(0)}_{A_i} + \hbar\omega$, and we have used the shorthand notation, $\vec{r} \equiv \vec{r}_1 \cdots \vec{r}_n$, similarly for $\vec{r}^{\,'}$, in the last line, in which the Green's function is expressed in coordinate representation. With this Green's function representation of the energy denominator, the energy of any excited state must be replaced by

$$E^{(0)}_{A_I} \to E^{(0)}_{A_I} - i\epsilon \equiv E^{(0)}_{A_I} - i\frac{\Gamma_I}{2}. \tag{4}$$

This relation makes sense, because any excited atomic state may decay to lower atomic states if the interaction of the atom with the radiation field is taken into account. This decay can be interpreted in terms of the level width (as we shall prove in the next chapter). With the imaginary part of the energy for the state, A_I, we have

$$\Psi_{A_I}(\vec{r}, t) = e^{-\frac{i}{\hbar}E^{(0)}_{A_I}t} e^{-\frac{\Gamma_I}{2\hbar}t}, \tag{5}$$

so $(\Gamma_I/\hbar)$ gives $(1/\tau_I)$, where τ_I is the mean lifetime of the excited state A_I. Because τ_I is of order of $(10)^{-8}$ sec. in atoms, Γ_I is of order $\hbar/\tau_I \approx (10)^{-7}$eV in atoms.

If there is an atomic excited state with energy such that

$$E^{(0)}_{A_R} \approx E^{(0)}_{A_i} + \hbar\omega, \tag{6}$$

the single term, with $I = R$ (resonant state) predominates and all other terms in the sum, $\sum_{A_I}$, can be neglected for incident photons in this energy range. Therefore, with $A_f = A_i$ (elastic scattering process),

$$g_{m\mu} \approx \frac{\langle A_i|\mu_m^{(\text{el.})*}|A_R\rangle\langle A_R|\mu_\mu^{(\text{el.})}|A_i\rangle}{(\hbar\omega + E^{(0)}_{A_i} - E^{(0)}_{A_R}) + \frac{i}{2}\Gamma_R}. \tag{7}$$

In this case, a strong enhancement of the scattering cross section exists at this resonant frequency, $\hbar\omega = (E^{(0)}_{A_R} - E^{(0)}_{A_i})$. A resonance fluorescence exists, and near $\hbar\omega = \hbar\omega_{\text{resonance}}$, assuming unpolarized incident photons, the resonance fluorescence cross section is

$$\sigma_{\text{resonance}} = \frac{4\pi}{3}\frac{\omega^4}{c^4} \sum_{\mu=\pm 1}\sum_{m=\pm 1,0} \frac{|\langle A_i|\mu_m^{(\text{el.})*}|A_R\rangle|^2 |\langle A_R|\mu_\mu^{(\text{el.})}|A_i\rangle|^2}{(E^{(0)}_{A_i} + \hbar\omega - E^{(0)}_{A_R})^2 + \frac{1}{4}\Gamma_R^2}. \tag{8}$$

At exact resonance, the energy denominator is given by $\frac{1}{4}\Gamma_R^2$, a quantity of order $(10)^{-14}\text{eV}^2$, whereas the denominator for an average term far off resonance can be expected to be of order 1eV^2, where we have taken a characteristic atomic energy difference to be 1eV. Thus, we can expect an enhancement at exact resonance by a factor of the order of 10^{14} compared with the ordinary photon scattering cross section.

A The Photon Scattering Cross Section and the Polarizability Tensor

Because of the difficulty of performing the intermediate state sum in the factors, $g_{m\mu}$, needed to evaluate the scattering cross sections from the underlying atomic structure, it may be useful to see how these $g_{m\mu}$ are related to other physically measurable quantities. We want to investigate the possibility these sums can be estimated from other physically measurable quantities.

So far, we have found it convenient to express the differential and total cross sections in terms of the circular polarizations of the incident and scattered photons. In terms of these, we had

$$\left(\frac{d\sigma}{d\Omega'}\right)_{\vec{k}\mu\to\vec{k}'\mu'} = \frac{\omega\omega'^3}{c^4} \sum_{m=\pm1,0} \sum_{\bar{m}=\pm1,0} D^1_{m\mu'}(\phi,\theta,0) D^1_{\bar{m}\mu'}(\phi,\theta,0)^* g_{m\mu} g^*_{\bar{m}\mu}, \quad (9)$$

where the $g_{m\mu}$ are given in terms of spherical components of the electric dipole moment vector in eq. (32) of Chapter 66. Note, $\mu = \pm 1$. The total cross section for unpolarized incident photons with detectors insensitive to the polarizations of the scattered photons was then given by

$$\sigma = \frac{4\pi}{3} \frac{\omega\omega'^3}{c^4} \sum_{\mu=\pm1} \sum_{m=\pm1,0} |g_{m\mu}|^2. \quad (10)$$

With $\omega' = \omega$, these expressions are valid for the elastic processes. Also, for ω and ω' not necessarily very small, the $g_{m\mu}$ are ω-dependent quantities. Because atoms (unlike molecules) are spherically symmetric systems, which in general have all possible orientations with respect to the incident photon beam coordinate system, we could have averaged over all possible orientations of the atom, leading to

$$\overline{|g_{m\mu}|^2} = \frac{1}{3} \sum_{\nu=\pm1,0} |g_{m\nu}|^2, \quad (11)$$

where the bar denotes the averaging over all possible orientations of the atom. With this averaging, the total cross section, for unpolarized incident photons and detectors insensitive to polarization direction, becomes

$$\sigma = \frac{8\pi}{9} \frac{\omega\omega'^3}{c^4} \sum_{m=\pm1,0} \sum_{\nu=\pm1,0} |g_{m\nu}|^2. \quad (12)$$

All these formulae could have been transcribed to the case of linearly polarized incident photons and linearly polarized scattered photons, with

$$\left(\frac{d\sigma}{d\Omega'}\right)_{\vec{k}\alpha\to\vec{k}'\alpha'} = \frac{\omega\omega'^3}{c^4} \left| \sum_{A_I} \frac{\langle A_f|\sum_i (e_i\vec{r}_i)\cdot\vec{e}^{\,\prime}_{\alpha'}|A_I\rangle\langle A_I|\sum_i (e_i\vec{r}_i)\cdot\vec{e}_\alpha|A_i\rangle}{E^{(0)}_{A_i} + \hbar\omega - E^{(0)}_{A_I}}\right.$$

$$+\sum_{A_I} \frac{\langle A_f|\sum_i (e_i\vec{r}_i)\cdot \vec{e}_\alpha|A_I\rangle\langle A_I|\sum_i (e_i\vec{r}_i)\cdot \vec{e}^{\,\prime}_{\alpha'}|A_i\rangle}{E^{(0)}_{A_i} - E^{(0)}_{A_I} - \hbar\omega'}\Bigg|^2$$

$$= \frac{\omega\omega'^3}{c^4}\sum_{j=x,y,z}\sum_{\bar{j}=x,y,z}(\vec{e}_j\cdot\vec{e}^{\,\prime}_{\alpha'})(\vec{e}_{\bar{j}}\cdot\vec{e}^{\,\prime}_{\alpha'})g_{j\alpha}g^*_{\bar{j}\alpha}, \quad (13)$$

where we have used

$$g_{\alpha'\alpha} = \sum_{j=x,y,z}(\vec{e}_j\cdot\vec{e}^{\,\prime}_{\alpha'})g_{j\alpha},$$

and the direction cosines, $(\vec{e}_j\cdot\vec{e}^{\,\prime}_{\alpha'})$, can be read from eq. (27) of Chapter 63. These have the property

$$\sum_{\alpha'=x',y'}\int\int d\Omega'(\vec{e}_j\cdot\vec{e}^{\,\prime}_{\alpha'})(\vec{e}_{\bar{j}}\cdot\vec{e}^{\,\prime}_{\alpha'}) = \frac{8\pi}{3}\delta_{j,\bar{j}}, \quad (14)$$

[cf., eq. (28) of Chapter 63]. This leads to a total cross section, for an unpolarized incident photon and detectors insensitive to the polarization of the scattered photon,

$$\sigma = \frac{8\pi}{3}\frac{\omega\omega'^3}{c^4}\frac{1}{2}\sum_{\alpha=x,y}\sum_{j=x,y,z}|g_{j\alpha}|^2. \quad (15)$$

The $g_{j\alpha}$, with $\alpha = x, y$ only, now have the form

$$g_{j\alpha} = \sum_{A_I}\Bigg[\frac{\langle A_f|\sum_i(e_i\vec{r}_i)\cdot\vec{e}_j|A_I\rangle\langle A_I|\sum_i(e_i\vec{r}_i)\cdot\vec{e}_\alpha|A_i\rangle}{E^{(0)}_{A_i} + \hbar\omega - E^{(0)}_{A_I}}$$
$$+ \frac{\langle A_f|\sum_i(e_i\vec{r}_i)\cdot\vec{e}_\alpha|A_I\rangle\langle A_I|\sum_i(e_i\vec{r}_i)\cdot\vec{e}_j|A_i\rangle}{E^{(0)}_{A_i} - E^{(0)}_{A_I} - \hbar\omega'}\Bigg]$$
$$= \langle A_f|\mathcal{O}_{j\alpha}|A_i\rangle, \quad (16)$$

where the operator, $\mathcal{O}_{jk}$, is

$$\mathcal{O}_{jk} = \sum_{A_I}\Bigg[\sum_i(e_i\vec{r}_i)\cdot\vec{e}_j|A_I\rangle\frac{1}{E - H_0 + i\epsilon}\langle A_I|\sum_i(e_i\vec{r}_i)\cdot\vec{e}_k$$
$$+\sum_i(e_i\vec{r}_i)\cdot\vec{e}_k|A_I\rangle\frac{1}{E - H_0 - \hbar\omega - \hbar\omega'}\langle A_I|\sum_i(e_i\vec{r}_i)\cdot\vec{e}_j\Bigg], \quad (17)$$

where $E = E^{(0)}_{A_i} + \hbar\omega$ and the unit vectors $\vec{e}_j$ and $\vec{e}_k$ both refer to the laboratory frame. In the low frequency limit, in which $\hbar\omega(\hbar\omega') \ll E^{(0)}_{A_I} - E^{(0)}_{A_i}$, so we can let $\hbar\omega \to 0, \hbar\omega' \to 0$, only the symmetric part of the second rank Cartesian tensor, $\mathcal{O}_{jk}$, survives, and we can identify it with the polarizability tensor of the atom. The diagonal matrix elements of $\mathcal{O}_{jk}$ in the atomic ground state, A_i, can be identified with the atomic polarizability tensor, α_{jk}. To see this, consider the atom to be perturbed by a static external electric field, $\vec{\mathcal{E}}^{\text{ext.}}$, with

$$H_{\text{perturbation}} = -\sum_i(e_i\vec{r}_i)\cdot\vec{\mathcal{E}}^{\text{ext.}}. \quad (18)$$

Stationary-state perturbation theory gives

$$\begin{aligned}
\Delta E &= \sum_{A_I} \frac{\langle A_i|\sum_i(e_i\vec{r}_i)\cdot\vec{\mathcal{E}}^{\text{ext.}}|A_I\rangle\langle A_I|\sum_i(e_i\vec{r}_i\cdot\vec{\mathcal{E}}^{\text{ext.}}|A_i\rangle}{E^{(0)}_{A_i}-E^{(0)}_{A_I}} \\
&= \tfrac{1}{2}\sum_{jk}\langle A_i|\mathcal{O}_{jk}(\omega=\omega'=0)|A_i\rangle\mathcal{E}^{\text{ext.}}_j\mathcal{E}^{\text{ext.}}_k \\
&= \tfrac{1}{2}\sum_{jk}\alpha_{jk}\mathcal{E}^{\text{ext.}}_j\mathcal{E}^{\text{ext.}}_k,
\end{aligned} \tag{19}$$

where α_{jk} is the polarizability tensor, which gives the strength of an *induced* electric dipole moment in the external field, $\vec{\mathcal{E}}$,

$$\vec{\mu}_j^{\text{induced}} = \sum_k \alpha_{jk}\mathcal{E}^{\text{ext.}}_k. \tag{20}$$

The total cross section for low ω elastic scattering, for an unpolarized incident photon beam and detectors insensitive to the polarization of the scattered photons, then follows from eq. (15),

$$\sigma_{\text{Rayleigh}} = \frac{8\pi}{3}\frac{\omega^4}{c^4}\frac{1}{2}\sum_{\beta=x,y}\sum_{j=x,y,z}|\langle A_i|\mathcal{O}_{j\beta}|A_i\rangle|^2 = \frac{8\pi}{3}\frac{\omega^4}{c^4}\frac{1}{2}\sum_{\beta=x,y}\sum_{j=x,y,z}\alpha^2_{j\beta}. \tag{21}$$

If the scattering system is spherically symmetric, e.g., an atom, which will have all possible orientations with respect to the incident photon coordinate system, averaging over all possible orientations of the atoms will lead to

$$\sigma_{\text{Rayleigh}} = \frac{8\pi}{9}\frac{\omega^4}{c^4}\sum_{j=x,y,z}\sum_{k=x,y,z}\alpha^2_{jk}. \tag{22}$$

For finite values of $\hbar\omega$, the polarizability tensor will be ω dependent. We define

$$\alpha_{jk}(\omega) = |\langle A_i|\mathcal{O}_{jk}(\omega)|A_i\rangle|. \tag{23}$$

If the scattering system is not spherically symmetric, a diatomic molecule, e.g., we must transform from the polarizability tensor components in the laboratory frame, defined by $\vec{k}$ and $\vec{e}_\beta$, to components in the molecular principal axis frame before averaging over all possible orientations of the molecules (see problems 40 and 41).

68

Natural Line Width: Wigner–Weisskopf Treatment

Because we consider a world made of atomic systems and electromagnetic fields, i.e., atomic systems coupled to the photon fields, we must consider not only the atomic bound states with zero photons present, i.e., states such as $|A_n 0\rangle$, but also atomic bound states with one photon present, states such as $|A_{n'}, \vec{k}\mu\rangle$. Because the coupling between atoms and photons is weak, states with two photons, such as $|A_{n''}, \vec{k}\mu, \vec{k}'\mu'\rangle$, may be negligible in dominant order of perturbation theory, although they may also play a role, particularly for systems for which selection rules eliminate the one photon states. Fig. 68.1 shows the discrete spectrum of an atomic system with no photons present, with ground state, A_0, and excited states, A_n, with $n = 1, 2, \ldots$. The figure also shows the continuous spectra of states with one photon present. The atomic eigenstate, A_3, e.g., is seen to be degenerate with the system where the atom is in the ground state and the photon has energy, $\hbar\omega = E^{(0)}_{A_3} - E^{(0)}_{A_0}$. It is also degenerate with the system where the atom is in the first excited state and the photon has energy $\hbar\omega' = E^{(0)}_{A_3} - E^{(0)}_{A_1}$ and with the system composed of the atom in the second excited state and with photon energy, $\hbar\omega'' = E^{(0)}_{A_3} - E^{(0)}_{A_2}$. More precisely, the discrete excited state $|A_n\rangle$ of the bare atom is coupled to states, $|A_{n'}, \vec{k}\mu\rangle$, in the continuum, where the photon energy lies between the above $\hbar\omega$ and $\hbar\omega + dE_{\text{photon}}$, with $\hbar\omega = E^{(0)}_n - E^{(0)}_{n'}$, where $n' < n$. In other words, the discrete state is coupled to the photon continuum with the weighting factor

$$\rho(E_{\text{photon}})dE_{\text{photon}} = \frac{\text{Vol.}}{(2\pi)^3}\frac{\omega^2 d\Omega}{\hbar c^3} dE_{\text{photon}}.$$

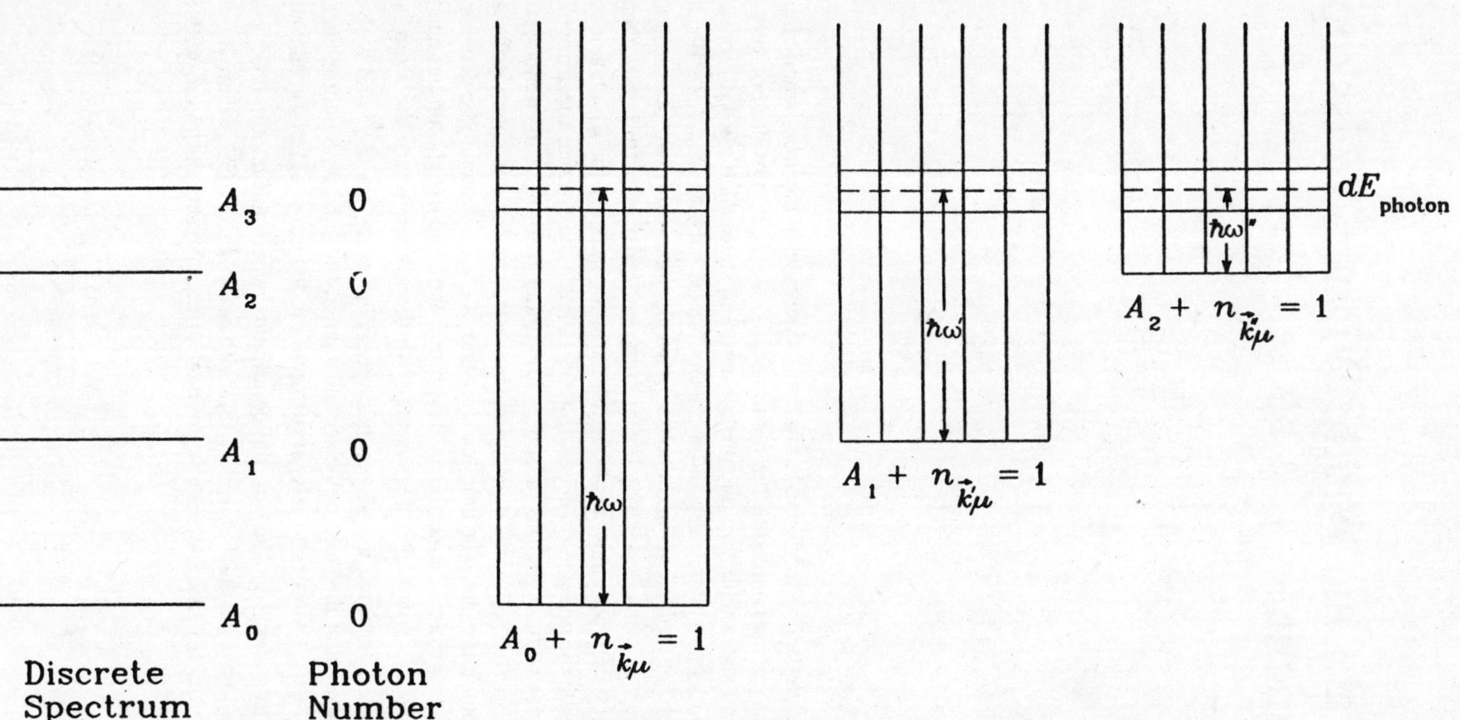

FIGURE 68.1. The degeneracy of excited atomic states and lower atomic states with photon number $n_{\vec{k}\mu} = 1$.

If at $t = 0$ the system is in the state, $|A_n 0\rangle$, first-order time-dependent perturbation theory will give us

$$|\psi(t)\rangle = |A_n 0\rangle e^{-\frac{i}{\hbar}E^{(0)}_{A_n}t} + \sum_{n',\vec{k}\mu} |A_{n'}, \vec{k}\mu\rangle c_{n',\vec{k}\mu}(t) e^{-\frac{i}{\hbar}(E^{(0)}_{A_{n'}}+\hbar\omega)t}, \tag{1}$$

where the coefficients, $c_{n',\vec{k}\mu}(t)$, can be calculated by time-dependent perturbation theory with the aid of the perturbation diagram of Fig. 68.2(a). Note, however, to first order in perturbation theory, the overlap of this $|\psi(t)\rangle$ with the $t = 0$ state $|A_n 0\rangle$ is still unity

$$\langle A_n 0|\psi(t)\rangle = 1 \tag{2}$$

because of the orthogonality of the states, $|A_n 0\rangle$ and $|A_{n'}, \vec{k}\mu\rangle$. To get the time evolution of the state $|A_n 0\rangle$, i.e., to see how $c_{n,0}(t)$ evolves with time, we need to go to second-order perturbation theory. With the aid of Fig. 68.2(b) and the diagram rules, we get

$$\begin{aligned}
&\langle A_n 0|U(t,0)|A_n 0\rangle = \langle A_n 0|[1 + U^{(2)}(t,0)]|A_n 0\rangle \\
&= e^{-\frac{i}{\hbar}E^{(0)}_{A_n}t} \times 1 + \left(\frac{-i}{\hbar}\right)^2 \sum_{A_I}\sum_{\vec{k}\mu} \int_0^t d\tau_2 e^{-\frac{i}{\hbar}E^{(0)}_{A_n}(t-\tau_2)} \langle A_n 0|H_{\text{int.}}|A_I, \vec{k}\mu\rangle \\
&\quad \times \int_0^{\tau_2} d\tau_1 e^{-\frac{i}{\hbar}(E^{(0)}_{A_I}+\hbar\omega)(\tau_2-\tau_1)} \langle A_I, \vec{k}\mu|H_{\text{int.}}|A_n 0\rangle e^{-\frac{i}{\hbar}(\tau_1 - 0)} \\
&= e^{-\frac{i}{\hbar}E^{(0)}_{A_n}t} \Bigg[1 - \frac{1}{\hbar^2}\sum_{A_I}\sum_{\vec{k}\mu} \left| \langle A_n 0|H_{\text{int.}}|A_I, \vec{k}\mu\rangle \right|^2 \\
&\quad \times \int_0^t d\tau_2 e^{+i(\omega_{nI}-\omega)\tau_2} \int_0^{\tau_2} d\tau_1 e^{-i(\omega_{nI}-\omega)\tau_1}\Bigg],
\end{aligned} \tag{3}$$

where we have used the shorthand notation, $\omega_{nI} = (E^{(0)}_{A_n} - E^{(0)}_{A_I})/\hbar$ and the Schrödinger picture with time-independent $H_{\text{int.}}$. The time-dependent integral gives

$$\begin{aligned}
F(t) &= \int_0^t d\tau_2 e^{i(\omega_{nI}-\omega)\tau_2} \int_0^{\tau_2} d\tau_1 e^{-i(\omega_{nI}-\omega)\tau_1} \\
&= \frac{i}{(\omega_{nI}-\omega)}\left[t - \frac{\left(e^{i(\omega_{nI}-\omega)t} - 1\right)}{i(\omega_{nI}-\omega)}\right] \\
&= -\frac{[\cos(\omega_{nI}-\omega)t - 1]}{(\omega_{nI}-\omega)^2} + \frac{i}{(\omega_{nI}-\omega)}\left[t - \frac{\sin(\omega_{nI}-\omega)t}{(\omega_{nI}-\omega)}\right] \\
&= \left[\frac{\sin^2[\frac{1}{2}(\omega_{nI}-\omega)t]}{2[\frac{1}{2}(\omega_{nI}-\omega)]^2}\right] + \frac{i}{(\omega_{nI}-\omega)}\left[t - \frac{\sin(\omega_{nI}-\omega)t}{(\omega_{nI}-\omega)}\right] \\
&= \Re[F(t)] + i\Im[F(t)].
\end{aligned} \tag{4}$$

We shall be interested in these time-dependent functions in the limit in which (1) $t \gg T_{\text{atomic}}$, where the atomic periods, T_{atomic}, are of order $(10)^{-15}$ sec., (2) $t \ll t_{\text{laboratory}}$. For atoms, this means $t \ll (10)^{-8}$ sec., that is, $t \ll$ a lifetime of a typical atomic excited state. Because quantities such as ω_{nI} and ω in the above

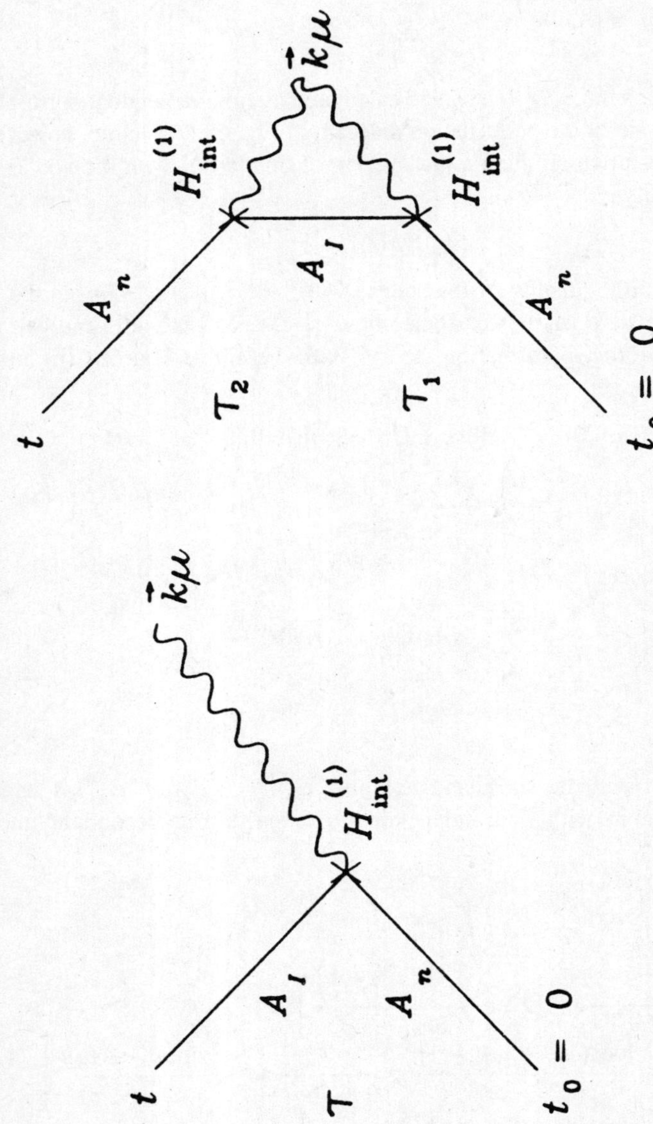

FIGURE 68.2. Perturbation diagrams for eq. (1)–(3).

formulae are of order $1/T_{\text{atomic}}$, we shall look at $F(t)$ in the first limit. For $t \gg T_{\text{atomic}}$, we can effectively take $t \to \infty$ in $\Re[F(t)]$ and $\Im[F(t)]$. Now, using

$$\lim_{t\to\infty}\left[t - \frac{\sin(\omega_{nI}-\omega)t}{(\omega_{nI}-\omega)}\right] \approx t \qquad \text{for } (\omega_{nI}-\omega) \neq 0$$
$$= 0 \qquad \text{for } (\omega_{nI}-\omega) = 0, \tag{5}$$

we get

$$\Im[F(t)] \approx \frac{1}{(\omega_{nI}-\omega)}t \qquad \text{for } (\omega_{nI}-\omega) \neq 0$$
$$\Im[F(t)] = 0 \qquad \text{for } (\omega_{nI}-\omega) = 0. \tag{6}$$

$$\Re[F(t)] = \lim_{t\to\infty}\left[\frac{\sin^2[\frac{1}{2}(\omega_{nI}-\omega)t]}{2[\frac{1}{2}(\omega_{nI}-\omega)]^2}\right] = \pi t\delta(\omega_{nI}-\omega)$$
$$= \pi t\hbar\delta(E_n^{(0)} - E_I^{(0)} - \hbar\omega). \tag{7}$$

With these results, we get (through second order in perturbation theory)

$$\langle A_n 0|U(t,0)|A_n 0\rangle =$$
$$e^{-\frac{i}{\hbar}E_{A_n}^{(0)}t}\left[1 - \frac{1}{\hbar^2}\pi t\hbar \sum_{A_I,\vec{k}\mu} \delta(E_{A_n}^{(0)} - E_{A_I}^{(0)} - \hbar\omega)\;|\,\langle A_n 0|H_{\text{int.}}|A_I,\vec{k}\mu\rangle\,|^2\right.$$
$$\left. - \frac{it}{\hbar^2}\sum_{A_I,\vec{k}\mu}{}' \frac{\hbar}{(E_{A_n}^{(0)} - E_{A_I}^{(0)} - \hbar\omega)}\;|\,\langle A_n 0|H_{\text{int.}}|A_I\vec{k}\mu\rangle\,|^2\right], \tag{8}$$

where the prime superscript on the sum of the imaginary part means: $\hbar\omega \neq (E_{A_n}^{(0)} - E_{A_I}^{(0)})$. Recall $\Im[F(t)] = 0$, when $(\omega_{nI}-\omega) = 0$ [see eq. (6)]. Now, we will also use

$$U(t,0)|A_n 0\rangle = e^{-\frac{i}{\hbar}E_{A_n}^{(0)}t}e^{-\frac{i}{\hbar}\Delta E_{A_n}t}e^{-\frac{\Gamma_n}{2\hbar}t}|A_n 0\rangle + \cdots$$
$$\approx e^{-\frac{i}{\hbar}E_{A_n}^{(0)}t}\left(1 - i\frac{\Delta E_{A_n}}{\hbar}t - \frac{\Gamma_{A_n}}{2\hbar}t + \cdots\right)|A_n 0\rangle + \cdots\ , \tag{9}$$

where the final $+\cdots$ in this relation stands for the presence of atomic terms other than A_n. In this equation, we have used requirement (2) above, $t \ll t_{\text{lab.}}$. Comparing with eq. (8), we get the level shift

$$\Delta E_{A_n} = \sum_{A_I,\vec{k}\mu}{}' \frac{1}{(E_{A_n}^{(0)} - E_{A_I}^{(0)} - \hbar\omega)}\;|\,\langle A_n 0|H_{\text{int.}}|A_I\vec{k}\mu\rangle\,|^2, \tag{10}$$

and the width

$$\Gamma_{A_n} = 2\pi \sum_{A_I,\vec{k}\mu}{}' \delta(E_{A_n}^{(0)} - E_{A_I}^{(0)} - \hbar\omega)\;|\,\langle A_n 0|H_{\text{int.}}|A_I\vec{k}\mu\rangle\,|^2\,. \tag{11}$$

In these formulae, the sum, $\sum_{\vec{k}}$, must be replaced by integrals over the continuous photon spectrum and over the possible angular directions of the photons of a particular energy,

$$\sum_{\vec{k}\mu} \to \sum_{\mu=\pm 1} \int\int d\Omega \int dE_{\text{photon}} \rho(E_{\text{photon}})$$

$$= \sum_{\mu=\pm 1} \int\int d\Omega \int dE_{\text{photon}} \frac{\text{Vol.}}{(2\pi)^3} \frac{\omega^2}{\hbar c^3}.$$

For the width, Γ_{A_n}, the energy Dirac delta function will limit the photon energies to values $\hbar\omega = (E^{(0)}_{A_n} - E^{(0)}_{A_{n'}})$, with $n' < n$. Thus, only a finite number of terms are involved. For the level shift, however, even though no zero denominator difficulties exist, the integral over photon energies and sum over atomic states, A_I, should go to ∞ energies and should therefore be treated by relativistic quantum theory. Nevertheless, H. A. Bethe in 1947 used the above formula with an arbitrary cutoff of $\hbar\omega \approx mc^2$ and obtained remarkable agreement with the then recently observed upward "Lamb shift" of the $2s_{\frac{1}{2}}$ level relative to the $2p_{\frac{1}{2}}$ level in hydrogen ($\Delta E_{A_n} \neq 0$ essentially only for the $l = 0$ states). The level shift formula also involves a mass renormalization. Even a free electron has an energy $p^2/2m_{\text{obs.}}$ where the bare mass has to be renormalized because of the presence of the virtual photons of type $\vec{k}\mu$ to yield the observed mass, $m_{\text{obs.}}$ [Because both the correct relativistic treatment and the renormalization are best treated in quantum field theory, we defer this subject to the quantum field theory course. For a good account of Bethe's treatment, however, see Section 2.8 in J. J. Sakurai, *Advanced Quantum Mechanics*, (1967).]

The width, Γ_{A_n}, is given by

$$\frac{\Gamma_{A_n}}{\hbar} = \frac{2\pi}{\hbar} \sum_{n'<n} \sum_{\mu=\pm 1} \int\int d\Omega \frac{\text{Vol.}}{(2\pi)^3} \frac{\omega^2_{nn'}}{\hbar c^3} \frac{c^2}{\text{Vol.}} \frac{2\pi\hbar}{\omega_{nn'}} \,|\, \langle A_n | \sum_i \frac{e_i}{m_i c} (\vec{p}_i \cdot \vec{e}^{\,\prime}_{\mu}) | A_{n'} \rangle \,|^2 . \tag{12}$$

Again, using

$$\frac{\vec{p}_i}{m_i c} = \frac{i}{\hbar c} [H^{(0)}_{\text{atom}}, \vec{r}_i], \tag{13}$$

we convert

$$|\, \langle A_n | \sum_i \frac{e_i}{m_i c} (\vec{p}_i \cdot \vec{e}^{\,\prime}_{\mu}) | A_{n'} \rangle \,|^2 = \left(\frac{\omega_{nn'}}{c}\right)^2 |\, \langle A_n | \sum_i e_i \vec{r}_i \cdot \vec{e}^{\,\prime}_{\mu} | A_{n'} \rangle \,|^2 =$$

$$\left(\frac{\omega_{nn'}}{c}\right)^2 \sum_{m,m'} D^{1*}_{m\mu} D^{1}_{m'\mu} \langle A_n | \sum_i e_i \vec{r}_i \cdot \vec{e}_m | A_{n'} \rangle \langle A_n | \sum_i e_i \vec{r}_i \cdot \vec{e}_{m'} | A_{n'} \rangle^*, \tag{14}$$

and, again, with the use of

$$\sum_{\mu=\pm 1} \int\int d\Omega D^{1*}_{m\mu} D^{1}_{m'\mu} = \frac{8\pi}{3} \delta_{mm'}, \tag{15}$$

we have

$$\frac{\Gamma_{A_n}}{\hbar} = \frac{4}{3} \sum_{n'<n} \frac{\omega^3_{nn'}}{\hbar c^3} \sum_{m=\pm 1,0} |\, \langle A_n | \vec{\mu}^{(\text{el.})}_m | A_{n'} \rangle \,|^2$$

$$= \sum_{n'<n} \left[\frac{1}{(\tau_{\mathcal{E}1})_{A_n \to A_{n'}}} \right]. \tag{16}$$

Here, $(1/\tau_{A_n \to A_{n'}})$ gives the probability per second that the atom in state n make a transition to a lower state n' via the spontaneous emission of a photon, see eq. (24) of Chapter 63. Also,

$$|\psi(t)\rangle = c_{A_n}(t)|A_n\rangle = e^{-\frac{i}{\hbar}(E^{(0)}_{A_n} + \Delta E_{A_n})t} e^{-\frac{\Gamma_{A_n}}{2\hbar}t} |A_n\rangle. \tag{17}$$

Further, a Fourier analysis of this time-dependent function gives a level centered at position, $E^{(0)}_{A_n} + \Delta E_{A_n}$, with width, Γ_{A_n}. The decay of the nth level is given by

$$|c_{A_n}(t)|^2 = e^{-\frac{\Gamma_{A_n}}{\hbar}t} = e^{-\frac{t}{\tau_n}}, \tag{18}$$

with

$$\frac{1}{\tau_n} = \sum_{n'<n} \left[\frac{1}{\tau_{A_n \to A_{n'}}} \right]. \tag{19}$$

Problems

40. In a diatomic molecule, the scattering matrix sums (for elastic scattering)

$$g_{\beta'\gamma} = \sum_{A_I} \left[\frac{\langle A_i|(\vec{\mu}^{(\text{el.})} \cdot \vec{e}_{\beta'})|A_I\rangle\langle A_I|(\vec{\mu}^{(\text{el.})} \cdot \vec{e}_{\gamma})|A_i\rangle}{E^{(0)}_{A_i} + \hbar\omega - E^{(0)}_{A_I}} + \frac{\langle A_i|(\vec{\mu}^{(\text{el.})} \cdot \vec{e}_{\gamma})|A_I\rangle\langle A_I|(\vec{\mu}^{(\text{el.})} \cdot \vec{e}_{\beta'})|A_i\rangle}{E^{(0)}_{A_i} - \hbar\omega - E^{(0)}_{A_I}} \right]$$

can be replaced by the approximately frequency-independent polarizability components, $\alpha_{\beta'\gamma}$, i.e., $g_{\beta'\gamma} \to \alpha_{\beta'\gamma}$. (Note, the mixed components used here: β' refers to the primed axes of the scattered photon, γ to the unprimed incident photon coordinate system, where the $\vec{k}$ of the incident photon is in the z direction, and the $\vec{k}'$ of the scattered photon is in the z' direction.) In the molecular symmetry coordinate system, with z'' parallel to the diatomic molecule axis, with

$$\vec{e}_{z''} = \vec{e}_x \sin\theta\cos\phi + \vec{e}_y \sin\theta\sin\phi + \vec{e}_z \cos\theta,$$

$$\vec{e}_{x''} = \vec{e}_x \cos\theta\cos\phi + \vec{e}_y \cos\theta\sin\phi - \vec{e}_z \sin\theta,$$

$$\vec{e}_{y''} = -\vec{e}_x \sin\phi + \vec{e}_y \cos\phi,$$

the polarizability tensor has the simple diagonal form: $\alpha_{x''x''} = \alpha_{y''y''} = \alpha_\perp; \alpha_{z''z''} = \alpha_\parallel$,

$$\alpha_{\beta''\gamma''} = \begin{pmatrix} \alpha_\perp & 0 & 0 \\ 0 & \alpha_\perp & 0 \\ 0 & 0 & \alpha_\parallel \end{pmatrix}.$$

To gain information about the molecular polarizabilities, $\alpha_\perp$ and $\alpha_\parallel$, the polarizations of the scattered photons are measured, usually in an arrangement with $\theta' = \frac{\pi}{2}$, $\phi' = 0$, i.e., scattered photon in the x direction, incident photon in the z direction. For this arrangement, show the "degree of polarization" $(d\sigma_\parallel / d\sigma_\perp)$ has the value

$$\frac{d\sigma_\parallel}{d\sigma_\perp} = \frac{(\alpha_\parallel - \alpha_\perp)^2}{3\alpha_\parallel^2 + 8\alpha_\perp^2 + 4\alpha_\parallel\alpha_\perp},$$

for the case when the incident light is polarized in the y direction and

$$\frac{d\sigma_\parallel}{d\sigma_\perp} = \frac{2(\alpha_\parallel - \alpha_\perp)^2}{4\alpha_\parallel^2 + 9\alpha_\perp^2 + 2\alpha_\parallel\alpha_\perp},$$

for the case of unpolarized incident light.

Note that you have to average over all possible molecular orientations.

In the above, $d\sigma_\parallel$ gives the scattering cross section for the scattered photons polarized parallel to the scattering plane, defined by the vectors $\vec{k}$ of the incident beam and $\vec{k}'$ of the scattered beam; i.e., in our arrangement, $d\sigma_\parallel$ is for the scattered photon polarized in the z direction. $d\sigma_\perp$ gives the scattering cross section for the scattered photon polarized perpendicular to the scattering plane; therefore, in our arrangement, $d\sigma_\perp$ is for the scattered photon polarized in the y direction.

41. For the Raman scattering from diatomic molecules involving rotational excitations, the $g_{\beta'\gamma}$ of problem 40. can be replaced by the rotational matrix elements of the polarizability tensor components:

$$g_{\beta'\gamma} \rightarrow \langle J_f M_f | \alpha_{\beta'\gamma} | J_i M_i \rangle.$$

Recall, for the diatomic molecule rigid rotator,

$$E_J = \frac{\hbar^2}{2I_e} J(J+1), \qquad \text{and} \quad \psi_{JM}(\theta, \phi) = Y_{JM}(\theta, \phi),$$

where θ and ϕ are the polar and azimuth angles giving the orientation of the molecular symmetry axis relative to the laboratory frame and I_e is the moment of inertia of the diatomic molecule in its equilibrium configuration, $I_e = \mu r_e^2$.

Calculate the cross sections, $d\sigma_\parallel$ and $d\sigma_\perp$, as functions of $\alpha_\perp$ and $\alpha_\parallel$, again, for the special case, $\theta' = \frac{\pi}{2}$, $\phi' = 0$, as in problem 40, i.e., scattered photon in the x direction, incident photon in the z direction. Assume unpolarized incident light. Calculate these cross sections for both the so-called Stokes process, where the molecule absorbs energy (transition $J \rightarrow J+2$), so $\hbar\omega' < \hbar\omega$, and the anti-Stokes process (transition $J \rightarrow J-2$), so $\hbar\omega' > \hbar\omega$. Prove only Raman transitions with $\Delta J = \pm 2$ can occur; keeping in mind transitions with $\Delta J = 0$ lead to elastic scattering processes.

Raman scattering is particularly important in homonuclear diatomic molecules, such as H_2 and N_2, which have no permanent electric dipole moments and hence no rotational absorption spectra. Also, the differential cross section for each Raman transition, $J \rightarrow J'$, depends on the number of molecules in the initial rotational state, J.

$$N_J = N_0 \frac{(2J+1) g_{\text{nuclear}} e^{-E_J/kT}}{\sum_J (2J+1) g_{\text{nuclear}} e^{-E_J/kT}}$$

for a gas sample in thermal equilibrium, where N_0 is the total number of molecules per unit volume. Taking account of the identity of the two spin $\frac{1}{2}$ protons in the H_2 diatomic molecule, and the identity of the two spin 1 nuclei, ^{14}N, in the diatomic molecule N_2, show the nuclear spin weights, g_{nuclear}, are J-dependent as follows:

$$g_{\text{nuclear}} = 1 \text{ for even J}, \quad g_{\text{nuclear}} = 3 \text{ for odd J}, \quad \text{in } H_2,$$

$$g_{\text{nuclear}} = 6 \text{ for even J}, \quad g_{\text{nuclear}} = 3 \text{ for odd J}, \quad \text{in } N_2.$$

42. For the elastic scattering of an unpolarized incident beam of high-frequency photons from an atom (Thomson limit), calculate the degree of polarization $(d\sigma_{\parallel}/d\sigma_{\perp})$ as a function of scattering angle θ. $d\sigma_{\parallel}$ gives the differential cross section for photons with their linear polarization vector lying in the scattering plane defined by $\vec{k}$ and $\vec{k}'$, and $d\sigma_{\perp}$ gives the differential cross sections for photons linearly polarized with their polarization vector perpendicular to the scattering plane.

43. Calculate the resonance fluorescence cross section for the scattering of unpolarized Lyman α radiation from cold hydrogen (in its ground state). Show, in particular, at exact resonance,

$$\hbar\omega = (E_{n=2} - E_{n=1}) = \frac{3}{8}\frac{e^2}{a_0},$$

the reduced electric dipole moment matrix element drops out of the expression for the cross section, and give an expression for the cross section in terms of the wave length λ of Lyman α.

44. The most-probable decay mode for the $2S_{\frac{1}{2}}$ state of hydrogen involves the simultaneous emission of two photons, with $\hbar\omega + \hbar\omega' = E_{2s_{\frac{1}{2}}} - E_{1s_{\frac{1}{2}}}$.

Convince yourself first the magnetic dipole single-photon emission probability for this transition is equal to zero, and the sequential decay via the emission of an electric dipole photon from the $2S_{\frac{1}{2}}$ into the $2P_{\frac{1}{2}}$ level with subsequent fast emission of the electric dipole photon for the transition $2P_{\frac{1}{2}} \to 1S_{\frac{1}{2}}$ is very improbable because of the low frequency of 1057 Mc/sec. corresponding to the Lamb shift that has raised the $2S_{\frac{1}{2}}$ level above the $2P_{\frac{1}{2}}$ level.

Derive a formula for the probability of double photon emission with one photon in the frequency range $d\omega = 2\pi d\nu$ about $\omega = 2\pi\nu$. Show, in particular,

$$\left[\frac{\text{Transition Probability}}{\text{second}}\right]_{2s\to 1s+\hbar\omega'+\hbar\omega} = \frac{2^{10} e^4 \pi^6 \nu^3 \nu'^3 d\nu}{3^3 c^6} \times$$

$$\left| \sum_{n=2}^{\infty} I_{1snp} I_{np2s} \left(\frac{1}{E_n - E_2 + \hbar\omega} + \frac{1}{E_n - E_2 + \hbar\omega'} \right) \right.$$

$$+\text{similar} \int \text{over continuum p} - \text{states} \Bigg|^2 ,$$

where I_{1snp} and I_{np2s} are the radial integrals

$$I_{1snp} = \int_0^\infty drr^2 R_{1s}(r) r R_{np}(r), \qquad I_{np2s} = \int_0^\infty drr^2 R_{np}(r) r R_{2s}(r),$$

involving discrete excited states. See Chapter 35 for the needed integrals. For the similar contribution from the continuum states with $l = 1$, for which the discrete sum has to be replaced by the continuous integral involving positive energies from $0 \to \infty$, see the mathematical appendix to Chapter 42. The total transition probability for the atom to make the transition, $2s \to 1s$, will involve an integration over all frequencies, ν, with the restriction,

$$\nu + \nu' = \frac{3}{8}\frac{1}{h}\frac{e^2}{a_0},$$

with ν ranging from zero to

$$\nu_{\text{max.}} = \frac{3}{8}\frac{1}{h}\frac{e^2}{a_0}.$$

In making this integral over all allowed frequencies, ν, we have to remember a particular photon frequency, ν_0, can occur in two ways: either $\nu = \nu_0$ or $\nu' = \nu_0$. Thus, to avoid double counting, we have

$$\left[\frac{\text{Transition Probability}}{\text{second}}\right]_{2s\to 1s} = \frac{1}{(\tau = \text{meanlife})} = \frac{1}{2}\int_0^{\nu_{\text{max.}}} d\nu A(\nu).$$

Finally, introducing dimensionless quantities

$$y = \frac{\hbar\omega}{\left(\frac{3e^2}{8a_0}\right)}, \qquad (1-y) = \frac{\hbar\omega'}{\left(\frac{3e^2}{8a_0}\right)},$$

and dimensionless integrals, such as $\mathcal{I}_{1snp} \equiv I_{1snp}/a_0$, show

$$\frac{1}{\tau} = \frac{1}{2}\int_0^1 dy A(y) = \frac{3^2}{2^{13}}\frac{c}{\pi a_0}\left(\frac{e^2}{\hbar c}\right)^7 \int_0^1 dy y^3(1-y)^3$$

$$\times \left| \sum_{n=2}^{\infty} \mathcal{I}_{1snp}\mathcal{I}_{np2s}\left(\frac{1}{\left[\frac{1}{3} - \frac{4}{3n^2} + y\right]} + \frac{1}{\left[\frac{4}{3} - \frac{4}{3n^2} - y\right]}\right) + \text{continuum contribution} \right|^2 .$$

Make a "guesstimate" for the lifetime, τ, by evaluating the most important terms of the discrete sum, and show this τ is much smaller, by a factor of order $\sim 10^9$, than the lifetime for the cascade $2s_{\frac{1}{2}} \to 2p_{\frac{1}{2}} \to 1s_{\frac{1}{2}}$.

Part VIII

Introduction to Relativistic Quantum Mechanics

69

Dirac Theory: Relativistic Quantum Theory of Spin-$\frac{1}{2}$ Particles

A Four-Vector Conventions

In relativistic theories, it will be useful to use a four-vector space-time formulation. We shall use the conventions of J. D. Bjorken and S. D. Drell, *Relativistic Quantum Mechanics*, New York: McGraw-Hill, 1964. (These are also the conventions used by J. D. Jackson, *Classical Electrodynamics*, New York: John Wiley, 1967.) We shall therefore use covariant and contravariant vector notation, using Greek letters for 4-D quantities and Latin letters for 3-D quantities, with a metric tensor

$$g_{\mu\nu} = g^{\mu\nu} = \begin{pmatrix} 1 & 0 & 0 & 0 \\ 0 & -1 & 0 & 0 \\ 0 & 0 & -1 & 0 \\ 0 & 0 & 0 & -1 \end{pmatrix}.$$

The time-space four-vector is given by

$$x^\mu = (x^0, x^i) = (ct, x^1, x^2, x^3) = (ct, \vec{r}),$$

whereas

$$x_\mu = g_{\mu\nu} x^\nu = (x_0, x_i) = (x_0, -x^i) = (ct, -\vec{r}),$$

where we use summation convention for repeated indices. The scalar invariant length in four-space is

$$x^\mu x_\mu = g_{\mu\nu} x^\mu x^\nu = g^{\mu\nu} x_\mu x_\nu = (ct)^2 - (\vec{r} \cdot \vec{r}). \tag{1}$$

The four-momentum vector has covariant and contravariant components

$$P_\mu = -\frac{\hbar}{i}\frac{\partial}{\partial x^\mu} = (-\frac{\hbar}{ic}\frac{\partial}{\partial t}, -\frac{\hbar}{i}\frac{\partial}{\partial x^i}) = (-\frac{\hbar}{ic}\frac{\partial}{\partial t}, -\frac{\hbar}{i}\vec{\nabla}),$$

and

$$P^\mu = -\frac{\hbar}{i}\frac{\partial}{\partial x_\mu} = (-\frac{\hbar}{ic}\frac{\partial}{\partial t}, +\frac{\hbar}{i}\vec{\nabla}),$$

where

$$P^\mu P_\mu = \left(\frac{E}{c}\right)^2 - (\vec{p}\cdot\vec{p}). \tag{2}$$

B The Klein–Gordon Equation

In the early chapters, we saw that the relation between energy and momentum for a free particle

$$\frac{E^2}{c^2} - (\vec{p}\cdot\vec{p}) = m^2c^2, \tag{3}$$

which becomes the dispersion relation in wave theory, leads via

$$E \rightarrow -\frac{\hbar}{i}\frac{\partial}{\partial t}; \qquad \vec{p} \rightarrow \frac{\hbar}{i}\vec{\nabla}$$

to the Klein–Gordon equation

$$\nabla^2\Psi - \frac{1}{c^2}\frac{\partial^2\Psi}{\partial t^2} - \frac{m^2c^2}{\hbar^2}\Psi = 0 \tag{4}$$

and would lead to a conserved probability density, W, and probability density current, $\vec{S}$, with

$$\frac{\partial W}{\partial t} + (\vec{\nabla}\cdot\vec{S}) = 0, \tag{5}$$

with

$$\vec{S} = \frac{\hbar}{2mi}(\Psi^*\vec{\nabla}\Psi - \Psi\vec{\nabla}\Psi^*), \qquad \text{and } W = -\frac{\hbar}{2mic^2}(\Psi^*\frac{\partial\Psi}{\partial t} - \Psi\frac{\partial\Psi^*}{\partial t}), \tag{6}$$

with a real but not positive-definite probability density, W. It was this seeming difficulty that led Dirac to an alternative approach. (Today, we know the Klein–Gordon equation is a correct relativistic wave equation for spin-0 particles.)

C Dirac's Reasoning: Historical Approach

Because of the seeming difficulty with the Klein–Gordon equation, Dirac looked for an alternative. He assumed the basic equations might be first order in both space and time derivatives, so ct and x, y, z, appear on an equal footing. He was guided by Maxwell's equations, which are Lorentz-covariant and first order in time and space derivatives, but involve not one scalar field, Ψ, but six fields, E_1, E_2, E_3, B_1, B_2, B_3, connected by six of the equations. He, therefore, postulated the relativistic equations have N fields, ψ_ρ, with $\rho = 1, \ldots, N$, such that

$$1. \qquad W = \sum_{\rho=1}^{N} |\psi_\rho|^2, \tag{7}$$

leading to a positive-definite probability density. In addition, he assumed the basic equations are first-order equations, and the Klein–Gordon equation is then a consequence of these first-order equations, where, with $\rho = 1, \ldots, N$,

$$2. \qquad -\frac{\hbar}{i}\frac{1}{c}\frac{\partial \psi_\rho}{\partial t} - \frac{\hbar}{i}(\alpha_x)_{\rho\sigma}\frac{\partial \psi_\sigma}{\partial x} - \frac{\hbar}{i}(\alpha_y)_{\rho\sigma}\frac{\partial \psi_\sigma}{\partial y} - \frac{\hbar}{i}(\alpha_z)_{\rho\sigma}\frac{\partial \psi_\sigma}{\partial z} - mc\beta_{\rho\sigma}\psi_\sigma = 0, \tag{8}$$

or

$$-\frac{\hbar}{i}\frac{\partial \psi_\rho}{\partial t} = (H)_{\rho\sigma}\psi_\sigma, \qquad \text{with} \quad (H)_{\rho\sigma} = c\vec{\alpha}_{\rho\sigma}\cdot\frac{\hbar}{i}\vec{\nabla} + mc^2\beta_{\rho\sigma}, \tag{9}$$

where summation convention is assumed for repeated Greek indices, the sum running from 1 to N, and $(\alpha_x)_{\rho\sigma}$, $(\alpha_y)_{\rho\sigma}$, $(\alpha_z)_{\rho\sigma}$, $\beta_{\rho\sigma}$, are $N \times N$ matrices, independent of space-time variables. These are to be chosen so the Klein–Gordon equation follows and the probability interpretation is a valid one.

3. Theorem I. If the the $\vec{\alpha}$, and β are hermitian matrices, and if W is given by eq. (7), probability is conserved: If $\alpha_x = \alpha_x^\dagger$, $\alpha_y = \alpha_y^\dagger$, $\alpha_z = \alpha_z^\dagger$, $\beta = \beta^\dagger$,

$$\frac{d}{dt}\int_{\text{all space}} d\vec{r}\;\; W = 0. \tag{10}$$

Multiplying the ρth Dirac equation by ψ_ρ^*, the complex conjugate of the ρth equation by ψ_ρ, and summing over the index ρ:

$$\begin{aligned} &\psi_\rho^*\left[\frac{1}{c}\frac{\partial \psi_\rho}{\partial t} + \vec{\alpha}_{\rho\sigma}\cdot\vec{\nabla}\psi_\sigma + \frac{imc}{\hbar}\beta_{\rho\sigma}\psi_\sigma\right] \\ &+ \psi_\rho\left[\frac{1}{c}\frac{\partial \psi_\rho^*}{\partial t} + \vec{\alpha}_{\rho\sigma}^*\cdot\vec{\nabla}\psi_\sigma^* - \frac{imc}{\hbar}\beta_{\rho\sigma}^*\psi_\sigma^*\right] = 0. \end{aligned} \tag{11}$$

Now, using the hermiticity of the matrices in the second term,

$$\vec{\alpha}_{\rho\sigma}^* = \vec{\alpha}_{\sigma\rho}, \qquad \beta_{\rho\sigma}^* = \beta_{\sigma\rho},$$

and renaming dummy indices $\rho \leftrightarrow \sigma$ in double sums, such as $\psi_\rho\beta_{\rho\sigma}^*\psi_\sigma^* = \psi_\rho\beta_{\sigma\rho}\psi_\sigma^* = \psi_\rho^*\beta_{\rho\sigma}\psi_\sigma$, we are led to a continuity equation

$$\begin{aligned} &\frac{1}{c}\frac{\partial(\psi_\rho^*\psi_\rho)}{\partial t} + \vec{\nabla}\cdot(\psi_\rho^*\vec{\alpha}_{\rho\sigma}\psi_\sigma) = 0 \\ \text{or} \qquad &\frac{\partial W}{\partial t} + \vec{\nabla}\cdot\vec{S} = 0, \end{aligned} \tag{12}$$

with

$$W = \psi_\rho^* \psi_\rho \left(\equiv \sum_\rho |\psi_\rho|^2 \right), \qquad \vec{S} = c(\psi_\rho^* \vec{\alpha}_{\rho\sigma} \psi_\sigma). \tag{13}$$

If we assume the ψ_ρ go to zero sufficiently fast as $r \to \infty$, $(\vec{S} \cdot \vec{n}) \to 0$ on the surface at ∞, so Gauss's theorem applied to the divergence term gives us the desired result.

4. Theorem II. If the matrices, $\vec{\alpha}$, β, satisfy the Dirac conditions

$$\begin{aligned} \alpha_l \alpha_k + \alpha_k \alpha_l &= 2\delta_{lk} \mathbf{1}, \\ \alpha_l \alpha_k + \alpha_k \alpha_l &= 0 \text{ for } l \neq k, \qquad (\alpha_l)^2 = \mathbf{1} \text{ for } l = x, y, z, \\ \alpha_l \beta + \beta \alpha_l &= 0, \qquad \text{and } \beta^2 = \mathbf{1}, \end{aligned} \tag{14}$$

each of the ψ_ρ satisfies the Klein–Gordon equation. (In these relations, $\mathbf{1}$ is the $N \times N$ unit matrix.) To prove this theorem, let us change to full matrix notation, and let ψ be an $N \times 1$- matrix with N rows and a single column, whereas β and the α_l are $N \times N$ matrices. As a specific example,

$$\beta\psi = \begin{pmatrix} \beta_{11} & \beta_{12} & \cdots & \beta_{1N} \\ \beta_{21} & \beta_{22} & \cdots & \beta_{2N} \\ \cdot & \cdot & \cdots & \cdot \\ \cdot & \cdot & \cdots & \cdot \\ \beta_{N1} & \beta_{N2} & \cdots & \beta_{NN} \end{pmatrix} \begin{pmatrix} \psi_1 \\ \psi_2 \\ \cdot \\ \cdot \\ \psi_N \end{pmatrix}. \tag{15}$$

Let us then act with the operator

$$\left(-\frac{1}{c}\frac{\partial}{\partial t} + \vec{\alpha} \cdot \vec{\nabla} + \frac{imc}{\hbar}\beta \right)$$

on the Dirac equation

$$+\frac{1}{c}\frac{\partial \psi}{\partial t} + \vec{\alpha} \cdot \vec{\nabla}\psi + \frac{imc}{\hbar}\beta\psi = 0$$

to obtain

$$\begin{aligned} &-\frac{1}{c^2}\frac{\partial^2 \psi}{\partial t^2} + (\vec{\alpha} \cdot \vec{\nabla})(\vec{\alpha} \cdot \vec{\nabla}\psi) - \frac{m^2c^2}{\hbar^2}\beta^2\psi + \frac{imc}{\hbar}(\alpha_l\beta + \beta\alpha_l)\frac{\partial \psi}{\partial x^l} \\ &+ \frac{1}{c}\left(\vec{\alpha} \cdot \vec{\nabla}\frac{\partial \psi}{\partial t} - \frac{\partial}{\partial t}\vec{\alpha} \cdot \vec{\nabla}\psi \right) + \frac{imc}{\hbar}\left(\beta\frac{\partial \psi}{\partial t} - \frac{\partial \beta\psi}{\partial t} \right) \\ &= -\frac{1}{c^2}\frac{\partial^2 \psi}{\partial t^2} + \left(\alpha_x^2\frac{\partial^2 \psi}{\partial x^2} + \alpha_y^2\frac{\partial^2 \psi}{\partial y^2} + \alpha_z^2\frac{\partial^2 \psi}{\partial z^2} \right) - \frac{m^2c^2}{\hbar^2}\beta^2\psi \\ &+ (\alpha_x\alpha_y + \alpha_y\alpha_x)\frac{\partial^2 \psi}{\partial x \partial y} + (\alpha_x\alpha_z + \alpha_z\alpha_x)\frac{\partial^2 \psi}{\partial x \partial z} + (\alpha_y\alpha_z + \alpha_z\alpha_y)\frac{\partial^2 \psi}{\partial y \partial z} \\ &= -\frac{1}{c^2}\frac{\partial^2 \psi}{\partial t^2} + \nabla^2\psi - \frac{m^2c^2}{\hbar^2}\psi = 0, \end{aligned} \tag{16}$$

where we have used the Dirac relations of eq. (14) in the last step.

5. Theorem III. There is "essentially" only one set of matrices, $\vec{\alpha}$, β, which satisfy the Dirac relations.

(a). The matrices must be at least 4×4 matrices. A specific solution to the Dirac relations is:

$$\vec{\alpha} = \begin{pmatrix} 0 & \vec{\sigma} \\ \vec{\sigma} & 0 \end{pmatrix}, \qquad \beta = \begin{pmatrix} \mathbf{1} & 0 \\ 0 & -\mathbf{1} \end{pmatrix}, \tag{17}$$

where $\mathbf{1}$ is the 2×2 unit matrix, and 0 are 2×2 zero matrices, and where the $\vec{\sigma}$ are the 2×2 Pauli spin matrices

$$\sigma_x = \begin{pmatrix} 0 & 1 \\ 1 & 0 \end{pmatrix}, \quad \sigma_y = \begin{pmatrix} 0 & -i \\ i & 0 \end{pmatrix}, \quad \sigma_z = \begin{pmatrix} 1 & 0 \\ 0 & -1 \end{pmatrix}. \tag{18}$$

For example,

$$\alpha_i \alpha_j + \alpha_j \alpha_i = \begin{pmatrix} 0 & \sigma_i \\ \sigma_i & 0 \end{pmatrix} \begin{pmatrix} 0 & \sigma_j \\ \sigma_j & 0 \end{pmatrix} + \begin{pmatrix} 0 & \sigma_j \\ \sigma_j & 0 \end{pmatrix} \begin{pmatrix} 0 & \sigma_i \\ \sigma_i & 0 \end{pmatrix}$$
$$\begin{pmatrix} (\sigma_i \sigma_j + \sigma_j \sigma_i) & 0 \\ 0 & (\sigma_i \sigma_j + \sigma_j \sigma_i) \end{pmatrix} = 2\delta_{ij} \mathbf{1}, \tag{19}$$

where we have used

$$\sigma_i \sigma_j = \delta_{ij} \mathbf{1} + i \epsilon_{ijk} \sigma_k. \tag{20}$$

(b) The matrices are the specific ones shown through eq. (17) or are related to these by a similarity transformation, with 4×4 matrices S. Thus, with

$$\vec{\alpha}' = S \vec{\alpha} S^{-1}, \qquad \beta' = S \beta S^{-1},$$

we would have the Dirac equations

$$\frac{1}{c} \frac{\partial \psi}{\partial t} + \vec{\alpha}' \cdot \vec{\nabla} \psi + \frac{imc}{\hbar} \beta' \psi = 0$$
$$= S \left[\frac{1}{c} \frac{\partial (S^{-1} \psi)}{\partial t} + \vec{\alpha} \cdot \vec{\nabla} (S^{-1} \psi) + \frac{imc}{\hbar} \beta (S^{-1} \psi) \right] = 0. \tag{21}$$

These equations would be a linear combination of the old equations, applied to a linear combinations of the old ψ_ρ.

(c) Higher dimensional realizations, such as 8×8 or 12×12, could have been obtained via a blowup process. This, however, would again not have led to anything new. It would merely be a rewriting of the original equations, two or three times over.

D The Dirac Equation in Four-Dimensional Notation

So far, we have treated the space and time derivative terms of the Dirac equation on a separate footing. We can achieve a more uniform 4-D notation, if we left-multiply the equation by β to get

$$\frac{\beta}{c} \frac{\partial \psi}{\partial t} + (\beta \vec{\alpha}) \cdot \vec{\nabla} \psi + \frac{imc}{\hbar} \psi = 0, \tag{22}$$

and if we define

$$\gamma^0 = \beta, \qquad \gamma^i = \beta\vec{\alpha}, \quad \text{with } i = 1, 2, 3; \text{ for } \alpha_x, \alpha_y, \alpha_z, \tag{23}$$

the Dirac equation then has the form

$$\gamma^\mu \frac{\partial \psi}{\partial x^\mu} + \frac{imc}{\hbar}\psi = 0. \tag{24}$$

In terms of the γ^μ, with $\mu = 0, 1, 2, 3$, the Dirac relations can be written in terms of one equation

$$\gamma^\mu \gamma^\nu + \gamma^\nu \gamma^\mu = 2g^{\mu\nu}\mathbf{1}. \tag{25}$$

Note: $(\gamma^\mu)^2 = g^{\mu\mu}\mathbf{1}$ for the special case $\mu = \nu$, so

$$(\gamma^0)^2 = \mathbf{1}, \qquad (\gamma^i)^2 = -\mathbf{1}. \tag{26}$$

If we want to diagonalize a particular γ^μ via a similarity transformation with a matrix S, the eigenvalues must be either +1 or -1 (for $\mu = 0$), or +i or -i (for $\mu = 1, 2, 3$). Also, from the Dirac relation with $\mu \neq \nu$, we get by left-multiplication with a particular γ^μ

$$\gamma^\nu = -(\gamma^\mu \gamma^\nu \gamma^\mu)g^{\mu\mu}, \qquad \text{for } \mu \neq \nu. \tag{27}$$

(For the moment, there is *no* summation convention for repeated indices.) Taking the trace of this relation, we have

$$\begin{aligned}\text{Trace}(\gamma^\nu) &= -\Big[\text{Trace}(\gamma^\mu \gamma^\nu \gamma^\mu)\Big]g^{\mu\mu} = -\Big[\text{Trace}(\gamma^\nu (\gamma^\mu)^2)\Big]g^{\mu\mu} \\ &= -\Big[\text{Trace}(\gamma^\nu)\Big](g^{\mu\mu})^2 = -\text{Trace}(\gamma^\nu).\end{aligned} \tag{28}$$

Thus,

$$\text{Trace}(\gamma^\nu) = 0. \tag{29}$$

Because the eigenvalues of a particular γ^ν are either +1 or -1, or +i or -i, and the matrix has zero trace, the dimension of the matrix, N, must be even. The solution $N = 2$ is ruled out because $N = 2$ cannot accomodate four independent hermitian matrices. Thus, we have the proof: $N = 4$. If we take Dirac's earlier very specific realization, we have

$$\gamma^0 = \begin{pmatrix} \mathbf{1} & 0 \\ 0 & -\mathbf{1} \end{pmatrix}, \qquad \gamma^i = \begin{pmatrix} 0 & \sigma^i \\ -\sigma^i & 0 \end{pmatrix} = \begin{pmatrix} 0 & \vec{\sigma} \\ -\vec{\sigma} & 0 \end{pmatrix}, \tag{30}$$

where $\sigma^i = \vec{\sigma} = (\sigma_x, \sigma_y, \sigma_z)$ are the standard 2×2 Pauli σ-matrices, and $\mathbf{1}$ and 0 are 2×2 unit and zero matrices. Note,

$$\gamma^{0\dagger} = \gamma^0, \qquad \gamma^{i\dagger} = -\gamma^i. \tag{31}$$

E Hermitian Conjugate Equation

As for the nonrelativistic Schrödinger equation, we need both the Dirac equation and its hermitian conjugate equation

$$\frac{\partial \psi^\dagger}{\partial x^\mu}\gamma^{\mu\dagger} - \frac{imc}{\hbar}\psi^\dagger = 0. \tag{32}$$

Now right-multiplying by γ^0, and using $\gamma^{0\dagger} = \gamma^0$, as well as $\gamma^{i\dagger}\gamma^0 = -\gamma^i\gamma^0 = +\gamma^0\gamma^i$, we get

$$\frac{\partial}{\partial x^\mu}\psi^\dagger\gamma^0\gamma^\mu - \frac{imc}{\hbar}\psi^\dagger\gamma^0 = 0. \tag{33}$$

Now, it proves convenient to define

$$\begin{aligned}\bar{\psi} = \psi^\dagger\gamma^0 &= (\psi_1^* \quad \psi_2^* \quad \psi_3^* \quad \psi_4^*)\begin{pmatrix} 1 & 0 & 0 & 0 \\ 0 & 1 & 0 & 0 \\ 0 & 0 & -1 & 0 \\ 0 & 0 & 0 & -1 \end{pmatrix} \\ &= (\psi_1^* \quad \psi_2^* \quad -\psi_3^* \quad -\psi_4^*),\end{aligned} \tag{34}$$

so the Dirac equation

$$\text{I.} \qquad \gamma^\mu\frac{\partial \psi}{\partial x^\mu} + \frac{imc}{\hbar}\psi = 0 \tag{35}$$

goes over into the conjugate equation

$$\text{II.} \qquad \frac{\partial \bar{\psi}}{\partial x^\mu}\gamma^\mu - \frac{imc}{\hbar}\bar{\psi} = 0. \tag{36}$$

Now, taking $\bar{\psi} \times$ (eq. I) + (eq. II) $\times\, \psi$ yields

$$\frac{\partial}{\partial x^\mu}(\bar{\psi}\gamma^\mu\psi) = 0. \tag{37}$$

This is the continuity equation in 4-D notation

$$\frac{\partial J^\mu}{\partial x^\mu} = 0, \qquad \frac{1}{c}\frac{\partial J^0}{\partial t} + \vec{\nabla}\cdot\vec{J} = 0, \tag{38}$$

with

$$J^0 = \bar{\psi}\gamma^0\psi = \psi^\dagger\psi = \sum_{\rho=1}^{4}|\psi_\rho|^2, \quad J^i = \bar{\psi}\gamma^i\psi = \psi^\dagger\alpha^i\psi = \sum_{\rho,\sigma=1}^{4}\psi_\rho^*(\vec{\alpha})_{\rho\sigma}\psi_\sigma. \tag{39}$$

Note: $\bar{\psi}\gamma^\mu\psi$ appears to be a four-vector, because we expect that all observers, in all Lorentz frames, would expect to obtain a continuity equation of the same form.

70

Lorentz Covariance of the Dirac Equation

From the 4-D form of the Dirac equation,

$$\gamma^{\mu} \frac{\partial}{\partial x^{\mu}} \psi + \frac{imc}{\hbar} \psi = 0, \tag{1}$$

it might seem tempting to conclude the γ^{μ} are the four components of a four-vector contracted with $\partial/\partial x^{\mu}$ to make a Lorentz scalar, but this is nonsense. The γ^{μ} are merely matrices, introduced to keep track of how the various components, ψ_{ρ}, are coupled to each other in the four equations. If an observer, O', in another Lorentz frame, x'^{μ}, constructs his Dirac equations, his γ^{μ} will have to satisfy all of the same criteria as the γ^{μ} for an observer, O, in the Lorentz frame, x^{μ}. Observer, O', might come up with a different solution to the Dirac relations and the hermiticity conditions (as might another observer in x^{μ}), but his different solution would be related to Dirac's solution by means of a similarity transformation, corresponding again merely to a different linear combination of the ψ_{ρ} and the four equations. (Unlike γ^{μ}, however, the product $\bar{\psi}\gamma^{\mu}\psi = J^{\mu}$, is a four-vector, as was surmised in the last chapter.)

An observer, O', in a Lorentz frame, x'^{μ}, should therefore obtain Dirac equations, with the standard γ^{μ}, of the form

$$\gamma^{\mu} \frac{\partial}{\partial x'^{\mu}} \psi'(x') + \frac{imc}{\hbar} \psi'(x') = 0, \tag{2}$$

where the x'^{μ} are related to the x^{ν} via the Lorentz transformation

$$x'^{\mu} = a^{\mu}{}_{\nu} x^{\nu}, \tag{3}$$

with inverse relations

$$x^{\nu} = \left(a^{-1}\right)^{\nu}_{\ \mu} x'^{\mu}, \tag{4}$$

with

$$a^{\mu}_{\ \nu} = \frac{\partial x'^{\mu}}{\partial x^{\nu}}, \qquad \left(a^{-1}\right)^{\nu}_{\ \mu} = \frac{\partial x^{\nu}}{\partial x'^{\mu}}, \tag{5}$$

and

$$a^{\mu}_{\ \nu}\left(a^{-1}\right)^{\nu}_{\ \lambda} = \delta^{\mu}_{\ \lambda}. \tag{6}$$

The Lorentz invariance of the 4-D length element

$$x'^{\mu} x'_{\mu} = x^{\nu} x_{\nu}, \qquad c^2 t'^2 - \vec{r}\,' \cdot \vec{r}\,' = c^2 t^2 - \vec{r} \cdot \vec{r}, \tag{7}$$

leads to the 4-D orthogonality relations

$$a^{\mu}_{\ \nu} a^{\nu}_{\ \lambda} = \delta^{\mu}_{\ \lambda}. \tag{8}$$

The most general Lorentz transformation can be built from a combination of special Lorentz transformations, Lorentz boosts in the direction of some specific space direction, and ordinary rotations in the 3-D subspace, x^1, x^2, x^3. For example, the special Lorentz transformation

$$\begin{aligned} x'^0 &= \frac{1}{\sqrt{1-\beta^2}}(x^0 - \beta x^1), \qquad & ct' &= \frac{1}{\sqrt{1-\beta^2}}(ct - \beta x), \\ x'^1 &= \frac{1}{\sqrt{1-\beta^2}}(-\beta x^0 + x^1), \qquad & x' &= \frac{1}{\sqrt{1-\beta^2}}(-vt + x), \end{aligned} \tag{9}$$

or the special rotation

$$\begin{aligned} x'^1 &= x^1 \cos\theta + x^2 \sin\theta, \\ x'^2 &= -x^1 \sin\theta + x^2 \cos\theta. \end{aligned} \tag{10}$$

The special Lorentz transformation could have been written in analagous form

$$\begin{aligned} x'^0 &= x^0 \cosh\chi - x^1 \sinh\chi, \\ x'^1 &= -x^0 \sinh\chi + x^1 \cosh\chi, \end{aligned} \tag{11}$$

with

$$\cosh\chi = \frac{1}{\sqrt{1-\beta^2}}, \qquad \sinh\chi = \frac{\beta}{\sqrt{1-\beta^2}}. \tag{12}$$

Also, for the proper Lorentz transformations: $\det|a^{\mu}_{\ \nu}| = +1$. For the improper Lorentz transformations: $\det|a^{\mu}_{\ \nu}| = -1$. The latter transformations include the space inversions, $x^i \to -x^i$, for $i = 1, 2, 3$, and the time reflections, $x^0 \to -x^0$.

Dirac assumed, in the new Lorentz frame, the $\psi'(x')$ are a linear combination of the $\psi(x)$:

$$\psi'(x') = S_L \psi(x), \qquad \text{with } S_L \equiv S = S(a^{\mu}_{\ \nu}), \tag{13}$$

that is, the 4×4 Lorentz S matrices are functions of the Lorentz parameters, θ_{ij}, χ_i. Then, with

$$\frac{\partial}{\partial x'^{\mu}} = \frac{\partial x^{\nu}}{\partial x'^{\mu}} \frac{\partial}{\partial x^{\nu}} = \left(a^{-1}\right)^{\nu}{}_{\mu} \frac{\partial}{\partial x^{\nu}}, \tag{14}$$

the Dirac equation in the primed frame transforms into

$$\gamma^{\mu} \frac{\partial}{\partial x'^{\mu}} \psi'(x') + \frac{imc}{\hbar} \psi'(x') = \gamma^{\mu} \left(a^{-1}\right)^{\nu}{}_{\mu} \frac{\partial}{\partial x^{\nu}} (S\psi(x)) + \frac{imc}{\hbar} (S\psi(x)) = 0. \tag{15}$$

Left-multiplying this equation by S^{-1} leads to

$$(S^{-1}\gamma^{\mu} S)\left(a^{-1}\right)^{\nu}{}_{\mu} \frac{\partial}{\partial x^{\nu}} \psi(x) + \frac{imc}{\hbar} \psi(x) = 0, \tag{16}$$

which would leave the form of the Dirac equations Lorentz invariant if

$$(S^{-1}\gamma^{\mu} S)(a^{-1})^{\nu}{}_{\mu} = \gamma^{\nu}, \tag{17}$$

or, using $a^{\lambda}{}_{\nu}(a^{-1})^{\nu}{}_{\mu} = \delta^{\lambda}{}_{\mu}$,

$$\text{I(a)}: \qquad (S^{-1}\gamma^{\lambda} S) = a^{\lambda}{}_{\nu} \gamma^{\nu}, \tag{18}$$

or

$$\text{I(b)}: \qquad \gamma^{\lambda} = a^{\lambda}{}_{\nu} (S\gamma^{\nu} S^{-1}). \tag{19}$$

Also, the conjugate of the Dirac equations will be Lorentz covariant if

$$\bar{\psi}'(x') = \bar{\psi}(x) S^{-1}, \tag{20}$$

because

$$\begin{aligned} \frac{\partial \bar{\psi}'(x')}{\partial x'^{\mu}} \gamma^{\mu} - \frac{imc}{\hbar} \bar{\psi}'(x') = 0 &\rightarrow \frac{\partial x^{\nu}}{\partial x'^{\mu}} \frac{\partial}{\partial x^{\nu}} \bar{\psi}(x) S^{-1} \gamma^{\mu} - \frac{imc}{\hbar} \bar{\psi}(x) S^{-1} = 0 \\ \rightarrow \left(a^{-1}\right)^{\nu}{}_{\mu} \frac{\partial \bar{\psi}(x)}{\partial x^{\nu}} (S^{-1}\gamma^{\mu} S) - \frac{imc}{\hbar} \bar{\psi}(x) & \\ \rightarrow \frac{\partial \bar{\psi}(x)}{\partial x^{\nu}} \gamma^{\nu} - \frac{imc}{\hbar} \bar{\psi}(x) = 0, & \end{aligned} \tag{21}$$

where we have first right-multiplied by S in the second line and have then used eq. (18) in the last step. Now, combining eq. (20) with

$$\bar{\psi}'(x') = \psi'(x')^{\dagger} \gamma^{0} = \psi(x)^{\dagger} S^{\dagger} \gamma^{0}, \tag{22}$$

we have as a second requirement on the matrix, S,

$$\text{II}: \qquad S^{\dagger} \gamma^{0} = \gamma^{0} S^{-1}. \tag{23}$$

A Construction of the Lorentz Matrix, S

As in our study of ordinary rotations, we shall exploit the fact that it is sufficient to know the S matrix for an infinitesimal transformation (either an ordinary rotation in three-space or a Lorentz boost along some specific space direction), and obtain the finite transformations by compounding the infinitesimal ones. We take as our simplest first example an ordinary rotation in the 1–2 plane, i.e., a rotation about the z axis through an infinitesimal angle, $\delta\theta_{12}$, for which

$$\begin{aligned}
S\psi(x) &= \psi'(x') = \psi'(ct, x', y', z) \\
&= \psi'(ct, x\cos\theta_{12} + y\sin\theta_{12}, -x\sin\theta_{12} + y\cos\theta_{12}, z) \\
(1 - i\delta\theta_{12}G_{12})\psi(x) &= \psi'(ct, x + y\delta\theta_{12}, y - x\delta\theta_{12}, z) \\
&= \psi'(ct, x, y, z) + \delta\theta_{12}\Big(y\frac{\partial}{\partial x} - x\frac{\partial}{\partial y}\Big)\psi(ct, x, y, z) \\
&= \psi'(ct, x, y, z) - i\delta\theta_{12}\frac{1}{i}\Big(x\frac{\partial}{\partial y} - y\frac{\partial}{\partial x}\Big)\psi(ct, x, y, z),
\end{aligned} \tag{24}$$

where we have replaced $\psi'(ct, x, y, z)$ by $\psi(ct, x, y, z)$ in the second term of the last line because this term is already proportional to $\delta\theta_{12}$ and $\psi'(ct, x, y, z)$ would differ from $\psi(ct, x, y, z)$ by an infinitesimal of first order. In the above equation, G_{12} is the generator of this infinitesimal rotation through the angle $\delta\theta_{12}$. Thus, we note

$$(1 - i\delta\theta_{12}G_{12})\psi(x) = \psi'(x) - i\delta\theta_{12}L_z\psi(ct, x, y, z), \tag{25}$$

where the orbital part of G_{12} turns out to be the orbital angular-momentum operator, L_z. From our experience with nonrelativistic quantum theory, we would expect $G_{12} = G_{12}^{\text{intrinsic}} + G_{12}^{\text{orbital}}$, so

$$(1 - i\delta\theta_{12}G_{12})\psi(x) = \Big[1 - i\delta\theta_{12}(G_{12}^{\text{intr.}} + L_z)\Big]\psi(ct, x, y, z). \tag{26}$$

Therefore, we would expect

$$\psi'(x) = S\psi(x) = (1 - i\delta\theta_{12}G_{12}^{\text{intr.}})\psi(x) \tag{27}$$

for the infinitesimal transformation. Moreover, we would expect $G_{12}^{\text{intr.}}$ to be related to $\frac{1}{2}\sigma_z$ in the nonrelativistic limit. Now, note

$$\gamma^1\gamma^2 = \begin{pmatrix} 0 & \sigma_x \\ -\sigma_x & 0 \end{pmatrix}\begin{pmatrix} 0 & \sigma_y \\ -\sigma_y & 0 \end{pmatrix} = \begin{pmatrix} -\sigma_x\sigma_y & 0 \\ 0 & -\sigma_x\sigma_y \end{pmatrix} = \begin{pmatrix} -i\sigma_z & 0 \\ 0 & -i\sigma_z \end{pmatrix}. \tag{28}$$

Recall, finally, the iteration of the infintesimal transformation $(1 - i\delta\theta_{12}G_{12})$ leads to the finite transformation matrix

$$S = e^{-i\theta_{12}G_{12}}. \tag{29}$$

From the infinitesimal transformation above, we are therefore led to try, for a rotation in the 1–2 plane through a finite angle, θ_{12},

$$S = e^{-\frac{1}{2}\theta_{12}\gamma^1\gamma^2}. \tag{30}$$

Now, let us test whether this S satisfies the relations I of eqs. (18) and (19), and II of eqs. (23). From

$$(\gamma^1\gamma^2)(\gamma^1\gamma^2) = -(\gamma^1)^2(\gamma^2)^2 = -\mathbf{1}, \tag{31}$$

we have

$$e^{-\frac{1}{2}\theta\gamma^1\gamma^2} = \cos\frac{\theta}{2}\mathbf{1} - \sin\frac{\theta}{2}\gamma^1\gamma^2, \tag{32}$$

so

$$(S^{-1}\gamma^\lambda S) = (\cos\frac{\theta}{2}\mathbf{1} + \sin\frac{\theta}{2}\gamma^1\gamma^2)\gamma^\lambda(\cos\frac{\theta}{2}\mathbf{1} - \sin\frac{\theta}{2}\gamma^1\gamma^2) \tag{33}$$

leads, with repeated use of the Dirac relation for the γ^μ to:

$$\begin{aligned}
\text{For } \lambda = 1: \quad & S^{-1}\gamma^1 S = \left(\cos^2\frac{\theta}{2} - \sin^2\frac{\theta}{2}\right)\gamma^1 + 2\sin\frac{\theta}{2}\cos\frac{\theta}{2}\gamma^2 \\
& = \cos\theta\gamma^1 + \sin\theta\gamma^2 = a^1_\nu\gamma^\nu, \\
\text{For } \lambda = 2: \quad & S^{-1}\gamma^2 S = -\sin\theta\gamma^1 + \cos\theta\gamma^2 = a^2_\nu\gamma^\nu, \\
\text{For } \lambda = 3: \quad & S^{-1}\gamma^3 S = \gamma^3 = a^3_\nu\gamma^\nu, \\
\text{For } \lambda = 0: \quad & S^{-1}\gamma^0 S = \gamma^0 = a^0_\nu\gamma^\nu.
\end{aligned} \tag{34}$$

Also, using, $(\gamma^1\gamma^2)^\dagger = \gamma^{2\dagger}\gamma^{1\dagger} = (-\gamma^2)(-\gamma^1) = -\gamma^1\gamma^2$,

$$S^\dagger\gamma^0 = \gamma^0\left(\cos\frac{\theta}{2}\mathbf{1} + \sin\frac{\theta}{2}\gamma^1\gamma^2\right) = \gamma^0 S^{-1}, \tag{35}$$

so relations I, II are satisfied by this $S = S_{12}$. Similarly, we can show rotations in the 2–3 plane, i.e., about the x axis, require

$$S_{23} = e^{-\frac{1}{2}\theta_{23}\gamma^2\gamma^3} = \cos\frac{\theta_{23}}{2}\mathbf{1} - \sin\frac{\theta_{23}}{2}\gamma^2\gamma^3, \tag{36}$$

and rotations about the y axis, i.e., rotations in the 3–1 plane, require

$$S_{31} = e^{-\frac{1}{2}\theta_{31}\gamma^3\gamma^1} = \cos\frac{\theta_{31}}{2}\mathbf{1} - \sin\frac{\theta_{31}}{2}\gamma^3\gamma^1. \tag{37}$$

For a special Lorentz transformation in the x direction; or a special Lorentz transformation in the 0–1 subspace, we are led to try

$$S_{01} = e^{-\frac{1}{2}\chi\gamma^0\gamma^1} = \cosh\frac{\chi}{2}\mathbf{1} - \sinh\frac{\chi}{2}\gamma^0\gamma^1, \tag{38}$$

where now: $(\gamma^0\gamma^1)(\gamma^0\gamma^1) = +\mathbf{1}$. Now, let us test this S_{01} to see whether it satisfies relations I and II. With

$$(S^{-1}\gamma^\lambda S) = \left(\cosh\frac{\chi}{2}\mathbf{1} + \sinh\frac{\chi}{2}\gamma^0\gamma^1\right)\gamma^\lambda\left(\cosh\frac{\chi}{2}\mathbf{1} - \sinh\frac{\chi}{2}\gamma^0\gamma^1\right), \tag{39}$$

we have:

$$\begin{aligned}
\text{For } \lambda = 0: \quad & (S^{-1}\gamma^0 S) = \left(\cosh^2\frac{\chi}{2} + \sinh^2\frac{\chi}{2}\right)\gamma^0 - 2\cosh\frac{\chi}{2}\sinh\frac{\chi}{2}\gamma^1 \\
& = \cosh\chi\gamma^0 - \sinh\chi\gamma^1 = a^0_\nu\gamma^\nu, \\
\text{For } \lambda = 1: \quad & (S^{-1}\gamma^1 S) = -\sinh\chi\gamma^0 + \cosh\chi\gamma^1 = a^1_\nu\gamma^\nu, \\
\text{For } \lambda = 2: \quad & (S^{-1}\gamma^2 S) = \gamma^2 = a^2_\nu\gamma^\nu, \\
\text{For } \lambda = 3: \quad & (S^{-1}\gamma^3 S) = \gamma^3 = a^3_\nu\gamma^\nu.
\end{aligned} \tag{40}$$

Finally,

$$S^{\dagger}\gamma^{0} = \left(\cosh\frac{\chi}{2}\mathbf{1} + \sinh\frac{\chi}{2}\gamma^{1}\gamma^{0}\right)\gamma^{0} = \gamma^{0}\left(\cosh\frac{\chi}{2}\mathbf{1} + \sinh\frac{\chi}{2}\gamma^{0}\gamma^{1}\right) = \gamma^{0}S^{-1}, \tag{41}$$

so both relations I and II are satisfied. We therefore expect

$$S_{0i} = e^{-\frac{1}{2}\chi_i\gamma^0\gamma^i} \quad \text{with } i = 1, 2, 3. \tag{42}$$

Finally, the S_{ij} and S_{0i} can be expressed through

$$\begin{aligned} S_{ij} &= \cos\frac{\theta_{ij}}{2}\mathbf{1} - \sin\frac{\theta_{ij}}{2}\gamma^i\gamma^j \\ &= \cos\frac{\theta_{ij}}{2}\begin{pmatrix}\mathbf{1} & 0\\ 0 & \mathbf{1}\end{pmatrix} + i\sin\frac{\theta_{ij}}{2}\epsilon_{ijk}\begin{pmatrix}\sigma_k & 0\\ 0 & \sigma_k\end{pmatrix} \quad \text{with } i \neq j = 1, 2, 3, \end{aligned} \tag{43}$$

and

$$\begin{aligned} S_{0i} &= \cosh\frac{\chi_i}{2}\mathbf{1} - \sinh\frac{\chi_i}{2}\gamma^0\gamma^i \\ &= \cosh\frac{\chi_i}{2}\begin{pmatrix}\mathbf{1} & 0\\ 0 & \mathbf{1}\end{pmatrix} - \sinh\frac{\chi_i}{2}\begin{pmatrix}0 & \sigma_i\\ \sigma_i & 0\end{pmatrix} \quad \text{with } i = 1, 2, 3. \end{aligned} \tag{44}$$

Thus, we can summarize: For any μ–ν plane Lorentz transformation

$$S_{\mu\nu} = e^{-\frac{1}{2}\theta_{\mu\nu}\gamma^\mu\gamma^\nu} = e^{\frac{i}{2}\theta_{\mu\nu}\sigma^{\mu\nu}}, \quad \text{with } \sigma^{\mu\nu} \equiv \frac{i}{2}[\gamma^\mu, \gamma^\nu] = i\gamma^\mu\gamma^\nu. \tag{45}$$

From eqs. (42) and (43), we also have, explicitly,
For 12 – rotations :

$$\begin{aligned} \psi_1' &= e^{i\frac{\theta_{12}}{2}}\psi_1 \\ \psi_2' &= e^{-i\frac{\theta_{12}}{2}}\psi_2 \\ \psi_3' &= e^{i\frac{\theta_{12}}{2}}\psi_3 \\ \psi_4' &= e^{-i\frac{\theta_{12}}{2}}\psi_4; \end{aligned} \tag{46}$$

or for 23 – rotations :

$$\begin{aligned} \psi_1' &= \cos\frac{\theta_{23}}{2}\psi_1 + i\sin\frac{\theta_{23}}{2}\psi_2 \\ \psi_2' &= i\sin\frac{\theta_{23}}{2}\psi_1 + \cos\frac{\theta_{23}}{2}\psi_2 \\ \psi_3' &= \cos\frac{\theta_{23}}{2}\psi_3 + i\sin\frac{\theta_{23}}{2}\psi_4 \\ \psi_4' &= i\sin\frac{\theta_{23}}{2}\psi_3 + \cos\frac{\theta_{23}}{2}\psi_4; \end{aligned} \tag{47}$$

or for 01 – boosts :

$$\begin{aligned} \psi_1' &= \cosh\frac{\chi_{01}}{2}\psi_1 - \sinh\frac{\chi_{01}}{2}\psi_4 \\ \psi_2' &= \cosh\frac{\chi_{01}}{2}\psi_2 - \sinh\frac{\chi_{01}}{2}\psi_3 \\ \psi_3' &= -\sinh\frac{\chi_{01}}{2}\psi_2 + \cosh\frac{\chi_{01}}{2}\psi_3 \end{aligned}$$

$$\psi_4' = -\sinh\frac{\chi_{01}}{2}\psi_1 + \cosh\frac{\chi_{01}}{2}\psi_4. \tag{48}$$

These equations give just the intrinsic transformations. Each ψ_i must still be acted upon by the appropriate $e^{-i\theta_{\mu\nu}G^{\text{orbital}}_{\mu\nu}}$, where these operators act on the space-time coordinates in the ψ_i. Recalling, from eqs. (24)–(27), $\psi'(x^\mu) = S_{\text{intr.}}\psi(x^\mu) = S_{\text{intr.}}\psi((a^{-1})^\mu{}_\nu x'^\nu)$, we have in the specific example of a 12 rotation, [see the first line of eq. (46)],

$$\psi_1'(x'^\mu) = e^{i\frac{\theta_{12}}{2}}\psi_1(x'^0, x'^1\cos\theta_{12} - x'^2\sin\theta_{12}, x'^1\sin\theta_{12} + x'^2\cos\theta_{12}, x'^3).$$

B Space Inversions

As a final example, consider the space-inversion operation, with

$$x'^\mu = \begin{pmatrix} 1 & 0 & 0 & 0 \\ 0 & -1 & 0 & 0 \\ 0 & 0 & -1 & 0 \\ 0 & 0 & 0 & -1 \end{pmatrix} x^\nu = g^{\mu\nu}x^\nu. \tag{49}$$

Our basic relations I and II are simply

$$\begin{aligned} S^{-1}\gamma^\lambda S &= g^{\lambda\nu}\gamma^\nu \\ S^\dagger\gamma^0 &= \gamma^0 S^{-1}. \end{aligned} \tag{50}$$

These equations are satisfied by

$$S = e^{i\alpha}\gamma^0, \qquad \text{where } \alpha = \text{any real number}. \tag{51}$$

It will prove convenient to choose $\alpha = 0$ when the limiting nonrelativistic ψ has positive parity, and $\alpha = \pi$ when the limiting nonrelativistic ψ has negative parity.

71

Bilinear Covariants

Consider the 16 combinations of γ matrices, Γ^A, $A = 1, \ldots, 16$, where

$$\begin{aligned}
(S):\ \ & \Gamma^S \ \ = \mathbf{1}, \\
(V):\ \ & (\Gamma^V)^\mu = \gamma^\mu = (\gamma^0, \gamma^1, \gamma^2, \gamma^3), \\
(T):\ \ & (\Gamma^T)^{\mu\nu} = \sigma^{\mu\nu} = i\gamma^\mu\gamma^\nu = (i\gamma^0\gamma^1, i\gamma^0\gamma^2, i\gamma^0\gamma^3, i\gamma^1\gamma^2, i\gamma^2\gamma^3, i\gamma^3\gamma^1), \\
(A):\ \ & (\Gamma^A)^\mu = \gamma^5\gamma^\mu = (-i\gamma^1\gamma^2\gamma^3, -i\gamma^0\gamma^2\gamma^3, -i\gamma^0\gamma^3\gamma^1, -i\gamma^0\gamma^1\gamma^2), \\
(P):\ \ & \Gamma^P \ \ = \gamma^5 = i\gamma^0\gamma^1\gamma^2\gamma^3,
\end{aligned} \tag{1}$$

where we have defined a new γ matrix

$$\gamma^5 = i\gamma^0\gamma^1\gamma^2\gamma^3 = \begin{pmatrix} \mathbf{0} & \mathbf{1} \\ \mathbf{1} & \mathbf{0} \end{pmatrix}, \tag{2}$$

where $\mathbf{1}$ is a 2×2 unit matrix and $\mathbf{0}$ is a 2×2 zero matrix. Note: γ^5 also has the following properties:

$$(\gamma^5)^2 = \mathbf{1}, \qquad \gamma^5\gamma^\mu = -\gamma^\mu\gamma^5. \tag{3}$$

The 16 Γ^A satisfy the following properties:

$$1. \qquad (\Gamma^A)^2 = \pm\mathbf{1}, \tag{4}$$

as can be shown by direct verification. Thus, each Γ^A has an inverse and $(\Gamma^A)^{-1} = \pm\Gamma^A$.

$$2. \qquad \Gamma^A\Gamma^B = \epsilon^{AB}\Gamma^C, \qquad \text{where } \epsilon^{AB} = \pm 1, \pm i \text{ only}. \tag{5}$$

This can again be shown by direct verification. In addition:

If, for fixed Γ^A, Γ^B is allowed to run through all 16 possibilities, Γ^C runs through all 16 possibilities also, and $\Gamma^C = \mathbf{1}$ only for $\Gamma^B = \Gamma^A$. To prove this: Suppose some $\Gamma^{B'} \neq \Gamma^B$ existed for which $\Gamma^A\Gamma^B = \Gamma^A\Gamma^{B'}$. Because the inverse of Γ^A exists, if we left-multiply with $(\Gamma^A)^{-1}$, we would have $\Gamma^B = \Gamma^{B'}$, contrary to the initial assumption.

$$3. \qquad \text{Trace}(\Gamma^A) = 0, \qquad \text{for all } \Gamma^A \text{ except } \Gamma^A = \Gamma^S = \mathbf{1}. \tag{6}$$

For each such Γ^A (other than Γ^S), a Γ^N exists such that

$$\Gamma^A\Gamma^N = -\Gamma^N\Gamma^A, \tag{7}$$

so

$$\begin{aligned}\text{Trace}(\Gamma^A\Gamma^N) &= -\text{Trace}(\Gamma^N\Gamma^A)\\ &= -\text{Trace}(\Gamma^A\Gamma^N) = 0.\end{aligned} \tag{8}$$

The first step can be shown by verification.

$$\begin{aligned}&\text{For } \Gamma^A = \gamma^\mu, \quad \Gamma^N = \gamma^\nu; \ (\nu \neq \mu).\\ &\text{For } \Gamma^A = i\gamma^\mu\gamma^\nu, \quad \Gamma^N = \gamma^\nu; \ (\nu \neq \mu).\\ &\text{For } \Gamma^A = \gamma^5\gamma^\mu, \quad \Gamma^N = \gamma^\mu.\\ &\text{For } \Gamma^A = \gamma^5, \quad \Gamma^N = \gamma^\mu.\end{aligned} \tag{9}$$

The $\Gamma^A\Gamma^N$ from this list cover all 15 possibilities, which are therefore traceless.

4. The 16 Γ^A are linearly independent. No relation of the form $\sum_{B=1}^{16} c_B\Gamma^B = 0$ exists.

Proof: For any specific Γ^A,

$$\text{Trace}\Big(\sum_{B=1}^{16} c_B\Gamma^B\Gamma^A\Big) = \pm\text{Trace}(\mathbf{1})c_A = \pm 4c_A. \tag{10}$$

Thus, if $\sum_{B=1}^{16} c_B\Gamma^B = 0$, $c_A = 0$ for all $A = 1, \ldots, 16$.

5. If F is a 4×4 matrix that commutes with all four γ^μ, F is a multiple of the unit matrix $\mathbf{1}$.

Proof: Let $F = \sum_B c_B\Gamma^B$. Then,

$$F\gamma^\mu = \gamma^\mu F \to \sum_B c_B\Gamma^B\gamma^\mu = \sum_B c_B\gamma^\mu\Gamma^B; \quad \mu = 0, 1, 2, 3, \tag{11}$$

or

$$\sum_{B\neq S} c_B(\Gamma^B\gamma^\mu - \gamma^\mu\Gamma^B) + c_1\gamma^\mu - c_1\gamma^\mu = 0. \tag{12}$$

For each Γ^B, at least one γ^μ exists such that $\gamma^\mu\Gamma^B = -\Gamma^B\gamma^\mu$, see eq. (9). Therefore, the sum above can be zero only if each c_B with $B \neq 1$ is zero.

A Transformation Properties of the $\bar{\psi}\Gamma^A\psi$

Substituting for $\bar{\psi}'(x')$ and $\psi'(x')$,

$$\bar{\psi}'(x')\mathbf{1}\psi'(x') = \bar{\psi}(x)S^{-1}S\psi(x) = \bar{\psi}(x)\mathbf{1}\psi(x). \tag{13}$$

This equation establishes the Lorentz invariance of $\bar{\psi}\psi$.

$$\bar{\psi}'(x')\gamma^{\mu}\psi'(x') = \bar{\psi}(x)S^{-1}\gamma^{\mu}S\psi(x) = a^{\mu}_{\ \nu}\bar{\psi}(x)\gamma^{\nu}\psi(x), \tag{14}$$

which shows $\bar{\psi}\gamma^{\mu}\psi$ is a Lorentz four-vector.

$$\bar{\psi}'(x')i\gamma^{\mu}\gamma^{\nu}\psi'(x') = \bar{\psi}(x)i(S^{-1}\gamma^{\mu}S)(S^{-1}\gamma^{\nu}S)\psi(x) = a^{\mu}_{\ \rho}a^{\nu}_{\ \lambda}\big(\bar{\psi}(x)i\gamma^{\rho}\gamma^{\lambda}\psi(x)\big). \tag{15}$$

To see how the (A) and (P) transform, we use

$$\gamma^5\gamma^{\mu}\gamma^{\nu} = \gamma^{\mu}\gamma^{\nu}\gamma^5, \qquad \gamma^5\gamma^{\mu} = -\gamma^{\mu}\gamma^5. \tag{16}$$

Because the S for proper Lorentz transformations are built from exponentials of type $e^{-\frac{\theta}{2}\gamma^{\mu}\gamma^{\nu}}$, $[\gamma^5, S] = 0$ for proper Lorentz transformations. Thus, for proper Lorentz transformations, we have

$$\begin{aligned}\bar{\psi}'(x')\gamma^5\gamma^{\mu}\psi'(x') &= \bar{\psi}(x)S^{-1}\gamma^5\gamma^{\mu}S\psi(x) = \bar{\psi}(x)\gamma^5S^{-1}\gamma^{\mu}S\psi(x)\\ &= a^{\mu}_{\ \nu}\bar{\psi}(x)\gamma^5\gamma^{\nu}\psi(x),\end{aligned} \tag{17}$$

and, again for proper Lorentz transformations,

$$\bar{\psi}'(x')\gamma^5\psi'(x') = \bar{\psi}(x)S^{-1}\gamma^5S\psi(x) = \bar{\psi}(x)\gamma^5S^{-1}S\psi(x) = \bar{\psi}(x)\gamma^5\psi(x). \tag{18}$$

Now, for pure space inversions, we found $S = e^{i\alpha}\gamma^0$. Therefore, for this space inversion S, we have

$$S^{-1}\gamma^5S = -\gamma^5 \tag{19}$$

This relation shows the pseudoscalar character, (P), of $\bar{\psi}\gamma^5\psi$. Finally, for the space-inversion, $S = e^{i\alpha}\gamma^0$, we also have

$$\begin{aligned}S^{-1}\gamma^{\mu}S &= (\gamma^0, -\gamma^i),\\ S^{-1}\gamma^5\gamma^{\mu}S &= (-\gamma^5\gamma^0, \gamma^5\gamma^i),\\ S^{-1}i\gamma^{\mu}\gamma^{\nu}S &= (-i\gamma^0\gamma^i, i\gamma^i\gamma^j).\end{aligned} \tag{20}$$

These relations establish the true four-vector character, (V), of $\bar{\psi}\gamma^{\mu}\psi$, the pseudo or axial vector character, (A), of $\bar{\psi}\gamma^5\gamma^{\mu}\psi$, and finally the true tensor character, (T), of $\bar{\psi}i\gamma^{\mu}\gamma^{\nu}\psi$.

B Lower Index γ Matrices

So far, we have defined all of the Γ^A in terms of upper index γ matrices, γ^{μ}. In view of the transformation properties of the $\bar{\psi}\Gamma^A\psi$, it is useful to define lower index γ- matrices, γ_{μ}, through

$$\gamma_0 = \gamma^0, \qquad \gamma_i = -\gamma^i, \quad \text{for } i = 1, 2, 3, \tag{21}$$

but with γ_5 defined by

$$\gamma_5 = \gamma^5. \tag{22}$$

72
Simple Solutions: Free Particle Motion: Plane Wave Solutions

Let us first seek solutions to the equation

$$\gamma^{\mu}\frac{\partial\psi}{\partial x^{\mu}}+\frac{imc}{\hbar}\psi=0, \tag{1}$$

which have the plane wave form

$$\psi=u(\vec{p})e^{\frac{i}{\hbar}(\vec{p}\cdot\vec{r}-Et)}=u(\vec{p})e^{-\frac{i}{\hbar}P_{\mu}x^{\mu}}, \tag{2}$$

where $u(\vec{p})$ is a four-component quantity, i.e., a 4×1 matrix, and

$$E=\pm\sqrt{(m^2c^4+c^2p^2)}, \tag{3}$$

which follows from the energy-momentum relation leading to the Klein–Gordon equation for the free particle of rest mass m. Both a positive and negative root exist. To get the simplest of all solutions, let us investigate the case, $\vec{p}=0$, i.e., a particle at rest in our Lorentz frame. In this case,

$$\gamma^{0}\frac{1}{c}\frac{\partial\psi}{\partial t}=-\frac{imc}{\hbar}\psi,$$

$$\text{or}\quad -\frac{i}{\hbar c}E\gamma^{0}u(0)=-\frac{imc}{\hbar}u(0). \tag{4}$$

If we decompose the 4×1 matrix into two 2×1 matrices, $u_A(0)$ and $u_B(0)$, this becomes

$$E\begin{pmatrix}\mathbf{1} & \mathbf{0}\\ \mathbf{0} & -\mathbf{1}\end{pmatrix}\begin{pmatrix}u_A(0)\\ u_B(0)\end{pmatrix}=mc^2\begin{pmatrix}u_A(0)\\ u_B(0)\end{pmatrix}. \tag{5}$$

1. With $E = +mc^2$, this equation has two independent solutions, both with $u_B(0) = 0$:

$$\begin{pmatrix}1\\0\\0\\0\end{pmatrix}\frac{e^{-\frac{i}{\hbar}mc^2t}}{\sqrt{\text{Vol.}}}, \qquad \begin{pmatrix}0\\1\\0\\0\end{pmatrix}\frac{e^{-\frac{i}{\hbar}mc^2t}}{\sqrt{\text{Vol.}}},$$

where we have used a box normalization with a box of volume, Vol. = L^3.

2. With $E = -mc^2$, two independent solutions again exist, now both with $u_A(0) = 0$:

$$\begin{pmatrix}0\\0\\1\\0\end{pmatrix}\frac{e^{+\frac{i}{\hbar}mc^2t}}{\sqrt{\text{Vol.}}}, \qquad \begin{pmatrix}0\\0\\0\\1\end{pmatrix}\frac{e^{+\frac{i}{\hbar}mc^2t}}{\sqrt{\text{Vol.}}}.$$

Similarly, we get the plane wave solutions for $\vec{p} \neq 0$ from

$$(E\gamma^0 + c\gamma^i p_i - mc^2\mathbf{1})u(\vec{p}) = 0, \tag{6}$$

or

$$\left[E\begin{pmatrix}\mathbf{1} & \mathbf{0}\\ \mathbf{0} & -\mathbf{1}\end{pmatrix} - c\begin{pmatrix}\mathbf{0} & \vec{\sigma}\\ -\vec{\sigma} & \mathbf{0}\end{pmatrix}\cdot\vec{p} - mc^2\begin{pmatrix}\mathbf{1} & \mathbf{0}\\ \mathbf{0} & \mathbf{1}\end{pmatrix}\right]\begin{pmatrix}u_A(\vec{p})\\ u_B(\vec{p})\end{pmatrix} = 0, \tag{7}$$

leading to the two equations

$$\begin{aligned}(E - mc^2)u_A(\vec{p}) - c(\vec{\sigma}\cdot\vec{p})u_B(\vec{p}) &= 0,\\ (E + mc^2)u_B(\vec{p}) - c(\vec{\sigma}\cdot\vec{p})u_A(\vec{p}) &= 0,\end{aligned} \tag{8}$$

with

$$u_A(\vec{p}) = \frac{c(\vec{\sigma}\cdot\vec{p})}{(E - mc^2)}u_B(\vec{p}), \qquad u_B(\vec{p}) = \frac{c(\vec{\sigma}\cdot\vec{p})}{(E + mc^2)}u_A(\vec{p}). \tag{9}$$

These equations lead to the same equation for both $u_A(\vec{p})$ and $u_B(\vec{p})$; e.g.,

$$\Big[(E^2 - m^2c^4) - c^2(\vec{\sigma}\cdot\vec{p})\cdot(\vec{\sigma}\cdot\vec{p})\Big]u_A(\vec{p}) = 0. \tag{10}$$

If we use the general identity for two arbitrary vector operators, $\vec{A}$, and $\vec{B}$,

$$(\vec{\sigma}\cdot\vec{A})(\vec{\sigma}\cdot\vec{B}) = (\vec{A}\cdot\vec{B}) + i\vec{\sigma}\cdot[\vec{A}\times\vec{B}], \tag{11}$$

we have, with $\vec{A} = \vec{B} = \vec{p}$,

$$\left(E^2 - m^2c^4 - c^2p^2\right)u_A(\vec{p}) = 0. \tag{12}$$

This equation just verifies the energy-momentum relation for our relativistic particle. To get solutions, we use eq. (9) to determine u_B from the two independent possible solutions for u_A and vice versa. With the 2×2 matrix for $(\vec{\sigma}\cdot\vec{p})$,

$$(\vec{\sigma}\cdot\vec{p}) = \begin{pmatrix}p_z & (p_x - ip_y)\\ (p_x + ip_y) & -p_z\end{pmatrix} = \begin{pmatrix}p_z & p_-\\ p_+ & -p_z\end{pmatrix}, \tag{13}$$

this equation leads to

$$u^{(1)}(\vec{p}) = N\begin{pmatrix} 1 \\ 0 \\ \frac{cp_z}{E+mc^2} \\ \frac{cp_+}{E+mc^2} \end{pmatrix}, \qquad u^{(2)}(\vec{p}) = N\begin{pmatrix} 0 \\ 1 \\ \frac{cp_-}{E+mc^2} \\ \frac{-cp_z}{E+mc^2} \end{pmatrix},$$

$$u^{(3)}(\vec{p}) = N\begin{pmatrix} \frac{cp_z}{E-mc^2} \\ \frac{cp_+}{E-mc^2} \\ 1 \\ 0 \end{pmatrix}, \qquad u^{(4)}(\vec{p}) = N\begin{pmatrix} \frac{cp_-}{E-mc^2} \\ \frac{-cp_z}{E-mc^2} \\ 0 \\ 1 \end{pmatrix}, \tag{14}$$

where the solutions, $r = 1, 2$, are associated with the positive energy branch, whereas those with $r = 3, 4$, are associated with negative energies, and the normalization factor can be obtained from

$$\int_{\text{Vol.}} d\vec{r} u^{(r)}(\vec{p})^\dagger u^{(r)}(\vec{p}) = 1, \tag{15}$$

where we take box normalization with a cube of Vol. L^3. For $(r) = (1)$, or (2), we then have

$$|N|^2\text{Vol.}\frac{(E+mc^2)^2 + c^2p^2}{(E+mc^2)^2} = |N|^2\text{Vol.}\frac{2E}{(E+mc^2)} = 1, \tag{16}$$

where we have used $c^2p^2 = E^2 - m^2c^4$. Choosing N real, we have

$$N = \sqrt{\frac{(E+mc^2)}{2E}}\frac{1}{\sqrt{\text{Vol.}}}. \tag{17}$$

For $(r) = (3)$, or (4), we have

$$|N|^2\text{Vol.}\frac{(E-mc^2)^2 + c^2p^2}{(E-mc^2)^2} = |N|^2\text{Vol.}\frac{2E}{(E-mc^2)} = 1, \tag{18}$$

but now E is a negative quantity, so, again

$$N = \sqrt{\frac{(|E|+mc^2)}{2|E|}}\frac{1}{\sqrt{\text{Vol.}}}, \tag{19}$$

which is therefore the same as that for the positive energy solutions. We shall discuss these negative-energy solutions in more detail later, when, with Dirac, we "discover" the positron; see Chapter 75. We shall, however, anticipate the positron, and in the solutions with $(r = 3, 4)$, put $E = -|E| = -E_{e^+}$, and $\vec{p} = -\vec{p}_{e^+}$. With this change in notation, we shall rename the plane wave solutions with $(r = 3, 4)$, $v^{(r)}(\vec{p})$

$$v^{(3)}(\vec{p}) = N\begin{pmatrix} \frac{cp_z}{E+mc^2} \\ \frac{cp_+}{E+mc^2} \\ 1 \\ 0 \end{pmatrix}, \qquad v^{(4)}(\vec{p}) = N\begin{pmatrix} \frac{cp_-}{E+mc^2} \\ \frac{-cp_z}{E+mc^2} \\ 0 \\ 1 \end{pmatrix}, \tag{20}$$

where now E is the positive energy of the positron and $\vec{p}$ is the momentum vector of the positron, although we leave off the subscripts, e^+. The normalization constant is given by eq. (19). Let us show the solutions with $r = 1, 2$ could also have been obtained from the rest frame solutions with $u_1(0) = 1$ and $u_2(0) = 1$, respectively, by applying a Lorentz boost in the appropriate direction to these rest frame solutions. Let us choose the rest frame of the particle as the primed frame, and the laboratory frame as the unprimed frame. Recall $\psi'(x') = S\psi(x)$, so we need $\psi(x) = S^{-1}\psi'(x')$. Therefore,

$$\psi(x) = S^{-1}u(0)\frac{1}{\sqrt{L'^3}}e^{-\frac{i}{\hbar}(P'_\mu x'^\mu)}, \tag{21}$$

where we must use the Lorentz contraction to transform the rest frame volume, L'^3, to the laboratory volume, L^3. If for this purpose we choose, for the moment, the x axis along the direction of the electron's $\vec{p}$,

$$L_x = \sqrt{(1-\beta^2)}L'_x, \quad L_y = L'_y, \quad L_z = L'_z, \tag{22}$$

so

$$L'^3 = \frac{1}{\sqrt{(1-\beta^2)}}L^3 = \frac{mc^2}{\sqrt{(1-\beta^2)}}\frac{1}{mc^2}L^3 = \frac{E}{mc^2}L^3. \tag{23}$$

We also need to transform $P'_\mu x'^\mu$ into the laboratory frame via $P'_\mu x'^\mu = P_\mu x^\mu$. With $x^\mu = (ct, \vec{r})$, and $P_\mu = (\frac{E}{c}, -\vec{p})$, this gives

$$P'_\mu x'^\mu = Et - \vec{p}\cdot\vec{r}. \tag{24}$$

Finally, recall

$$\begin{aligned} S^{-1} &= \left[\cosh\frac{\chi}{2}\mathbf{1} + \sinh\frac{\chi}{2}\gamma^0\gamma^i\right] \\ &= \cosh\frac{\chi}{2}\left[\begin{pmatrix}\mathbf{1} & \mathbf{0}\\ \mathbf{0} & \mathbf{1}\end{pmatrix} + \tanh\frac{\chi}{2}\begin{pmatrix}\mathbf{0} & (\vec{\sigma}\cdot\vec{e}_i)\\ (\vec{\sigma}\cdot\vec{e}_i) & \mathbf{0}\end{pmatrix}\right], \end{aligned} \tag{25}$$

where we have now chosen $\vec{p}$ in an arbitrary direction in three-space parallel to a unit vector, $\vec{e}_i$, and where

$$\cosh\chi = 2\cosh^2\frac{\chi}{2} - 1 = \frac{1}{\sqrt{(1-\beta^2)}} = \frac{mc^2}{\sqrt{(1-\beta^2)}}\frac{1}{mc^2} = \frac{E}{mc^2}, \tag{26}$$

and

$$\sinh\chi = \frac{\beta}{\sqrt{(1-\beta^2)}} = \frac{mv}{\sqrt{(1-\beta^2)}}\frac{c}{mc^2} = \frac{pc}{mc^2}, \tag{27}$$

so

$$\cosh\frac{\chi}{2} = \sqrt{\frac{(E+mc^2)}{2mc^2}}; \qquad \tanh\frac{\chi}{2} = \frac{\sinh\chi}{2\cosh^2\frac{\chi}{2}} = \frac{pc}{(E+mc^2)}. \tag{28}$$

We therefore have

$$S^{-1}u(0) = \sqrt{\frac{(E+mc^2)}{2mc^2}}\left[\begin{pmatrix}1&0&0&0\\0&1&0&0\\0&0&1&0\\0&0&0&1\end{pmatrix}\right.$$

$$+\frac{c}{(E+mc^2)}\begin{pmatrix} 0 & 0 & p_z & p_- \\ 0 & 0 & p_+ & -p_z \\ p_z & p_- & 0 & 0 \\ p_+ & -p_z & 0 & 0 \end{pmatrix}\Bigg]\begin{pmatrix}1\\0\\0\\0\end{pmatrix} \text{ or } \begin{pmatrix}0\\1\\0\\0\end{pmatrix}. \tag{29}$$

Combining eqs. (21), (23), (24), and (29), we have

$$\psi^{(1)}(\vec{p},\vec{r}) = \sqrt{\frac{(E+mc^2)}{2mc^2}}\begin{pmatrix} 1 \\ 0 \\ \frac{cp_z}{(E+mc^2)} \\ \frac{cp_+}{(E+mc^2)} \end{pmatrix}\sqrt{\frac{mc^2}{E}}\frac{1}{\sqrt{L^3}}e^{-\frac{i}{\hbar}(Et-\vec{p}\cdot\vec{r})},$$

$$\psi^{(2)}(\vec{p},\vec{r}) = \sqrt{\frac{(E+mc^2)}{2mc^2}}\begin{pmatrix} 0 \\ 1 \\ \frac{cp_-}{(E+mc^2)} \\ \frac{-cp_z}{(E+mc^2)} \end{pmatrix}\sqrt{\frac{mc^2}{E}}\frac{1}{\sqrt{L^3}}e^{-\frac{i}{\hbar}(Et-\vec{p}\cdot\vec{r})}, \tag{30}$$

which agrees with our earlier results, using eqs. (14) and (17).

A An Application: Coulomb Scattering of Relativistic Electrons in Born Approximation: The Mott Formula

Before a further investigation of the physical significance of the solutions with $r = 1, 2$ (to find their relationship to electron spin), let us first use these plane wave solutions in our first application. Let us calculate the differential scattering cross section of a relativistic electron from a nucleus of charge Ze. The interaction will be the Coulomb potential, but, as in our corresponding nonrelativistic calculation, we will have to use a screened Coulomb potential and let the screening parameter $\to 0$. (Compare with Chapter 46.) This is quite realistic because the real target will undoubtedly be made of atoms, where the nucleus of charge, Ze, is imbedded in the atomic electron cloud. Also, for relativistic electrons, we would expect the plane wave Born approximation to be good because now $E \gg Ze^2/r_{\text{nucleus}}$.

The differential cross section is given by

$$d\sigma = \frac{1}{\text{inc. flux}}\frac{2\pi}{\hbar}\rho(E_f)\,\Big|\,\langle\vec{p}_f| - \frac{Ze^2}{r}|\vec{p}_i\rangle\,\Big|^2. \tag{31}$$

We shall assume the incident beam is in the z direction, with momentum $\vec{p}_i = p\vec{e}_z$. Thus,

$$\vec{S}_{\text{inc.}} = c\bar{\psi}\vec{\gamma}\psi = cu^{(r)\dagger}(\vec{p}_i)\alpha_z u^{(r)}(\vec{p}_i) = cu^{(r)\dagger}\begin{pmatrix} \mathbf{0} & \sigma_z \\ \sigma_z & \mathbf{0} \end{pmatrix}u^{(r)}. \tag{32}$$

With $(r) = 1$,

$$S_z = c\frac{E+mc^2}{2E}\frac{1}{\text{Vol.}}\begin{pmatrix} 1 & 0 & \frac{cp}{E+mc^2} & 0 \end{pmatrix}\begin{pmatrix} \frac{cp}{E+mc^2} \\ 0 \\ 1 \\ 0 \end{pmatrix} = \frac{c^2p}{E}\frac{1}{\text{Vol.}}.$$

Similarly, with $(r) = 2$,

$$S_z = c\frac{E + mc^2}{2E}\frac{1}{\text{Vol.}}\begin{pmatrix} 0 & 1 & 0 & -\frac{cp}{E+mc^2}\end{pmatrix}\begin{pmatrix} 0 \\ \frac{cp}{E+mc^2} \\ 0 \\ -1 \end{pmatrix} = \frac{c^2 p}{E}\frac{1}{\text{Vol.}},$$

so

$$\text{Inc. Flux} = \frac{c^2 p}{E\text{Vol.}} = \frac{v}{\text{Vol.}}. \tag{33}$$

In general,

$$\rho(E_f)dE_f = \frac{\text{Vol.}}{(2\pi)^3}\frac{p_f^2 dp_f d\Omega_f}{\hbar^3}, \tag{34}$$

where now $c^2 p_f^2 = c^2 p^2 = E^2 - m^2c^4$, so $2c^2 p_f dp_f = 2EdE$ and

$$\rho(E_f) = \frac{\text{Vol.}}{(2\pi)^3}\frac{pEd\Omega_f}{c^2\hbar^3}. \tag{35}$$

Finally, the matrix element will involve the four-component, relativistic plane-wave states, so

$$\langle \vec{p}_f, (s)| - \frac{Ze^2}{r}|\vec{p}_i, (r)\rangle = -\int d\vec{r} e^{-\frac{i}{\hbar}\vec{p}_f\cdot\vec{r}}\frac{Ze^2}{r}e^{+\frac{i}{\hbar}\vec{p}_i\cdot\vec{r}}u^{(s)\dagger}(\vec{p}_f)u^{(r)}(\vec{p}_i), \tag{36}$$

where, for the moment, we have assumed the relativistic electrons in the incident beam are in the intrinsic state, (r), and the electrons in the final scattered beam are in the intrinsic state, (s). With $(\vec{p}_f - \vec{p}_i)/\hbar = \vec{q}$, the needed spatial integral is

$$\int d\vec{r} e^{-i\vec{q}\cdot\vec{r}}\frac{Ze^2}{r} = 4\pi Ze^2\int_0^\infty dr r j_0(qr) = 4\pi\frac{Ze^2}{q}\int_0^\infty dr \sin qr \rightarrow$$
$$\lim_{b\to 0} 4\pi\frac{Ze^2}{q}\int_0^\infty dr \sin qr e^{-br} = 4\pi\frac{Ze^2}{q^2}, \tag{37}$$

where we have used a screened Coulomb potential to regularize the Coulomb radial integral. In addition, we shall assume the incident electron beam is unpolarized, i.e., an equal mixture of both positive energy intrinsic states, with $(r) = 1$, or 2, and the detector is insensitive to the electron's intrinsic state. Thus,

$$\frac{d\sigma}{d\Omega} = \frac{E\text{Vol.}}{c^2 p}\frac{2\pi}{\hbar}\frac{\text{Vol.}}{(2\pi)^3}\frac{pE}{c^2\hbar^3}\Big(\frac{4\pi Ze^2}{q^2}\Big)^2\sum_{s=1,2}\frac{1}{2}\sum_{r=1,2}|u^{(s)\dagger}(\vec{p}_f)u^{(r)}(\vec{p}_i)|^2. \tag{38}$$

We have assumed $\vec{p}_i = p\vec{e}_z$. We could take the most general direction for $\vec{p}_f$, and choose $\vec{p}_f = p\sin\theta\cos\phi\vec{e}_x + p\sin\theta\sin\phi\vec{e}_y + p\cos\theta\vec{e}_z$. Because our problem has axial symmetry about the direction of the incident beam, however, we can choose $\phi = 0$ without loss of generality. Thus,

$$u^{(1)\dagger}(\vec{p}_f) = \sqrt{\frac{E + mc^2}{2E\ \text{Vol.}}}\begin{pmatrix} 1 & 0 & \frac{cp\cos\theta}{E+mc^2} & \frac{cp\sin\theta}{E+mc^2}\end{pmatrix},$$

$$u^{(1)}(\vec{p}_i) = \sqrt{\frac{E+mc^2}{2E\ \text{Vol.}}} \begin{pmatrix} 1 \\ 0 \\ \frac{cp}{(E+mc^2)} \\ 0 \end{pmatrix},$$

$$u^{(2)\dagger}(\vec{p}_f) = \sqrt{\frac{E+mc^2}{2E\ \text{Vol.}}} \begin{pmatrix} 0 & 1 & \frac{cp\sin\theta}{E+mc^2} & \frac{-cp\cos\theta}{E+mc^2} \end{pmatrix},$$

$$u^{(2)}(\vec{p}_i) = \sqrt{\frac{E+mc^2}{2E\ \text{Vol.}}} \begin{pmatrix} 0 \\ 1 \\ 0 \\ \frac{-cp}{(E+mc^2)} \end{pmatrix}.$$

With these equations, we get

$$\begin{aligned}
&\frac{1}{2}\sum_{s=1,2}\sum_{r=1,2} \mid u^{(s)\dagger}(\vec{p}_f)u^{(r)}(\vec{p}_i) \mid^2 \\
&= \frac{1}{2}\Big(\frac{E+mc^2}{2E\ \text{Vol.}}\Big)^2\left[2\Big(1+\frac{c^2p^2\cos\theta}{(E+mc^2)^2}\Big)^2 + 2\Big(\frac{-c^2p^2\sin\theta}{(E+mc^2)^2}\Big)^2\right] \\
&= \Big(\frac{E+mc^2}{2E\ \text{Vol.}}\Big)^2\left[1+\frac{c^4p^4}{(E+mc^2)^4} + 2\frac{\cos\theta c^2p^2}{(E+mc^2)^2}\right] \\
&= \frac{1}{(2E\ \text{Vol.})^2}\left[(E+mc^2)^2 + (E-mc^2)^2 + 2\cos\theta(E^2-m^2c^4)\right] \\
&= \frac{1}{(2E\ \text{Vol.})^2}2\left[E^2+m^2c^4+\cos\theta(E^2-m^2c^4)\right] \\
&= \frac{1}{(2E\ \text{Vol.})^2}2\left[E^2+m^2c^4+(1-2\sin^2\frac{\theta}{2})(E^2-m^2c^4)\right] \\
&= \frac{1}{E^2}\frac{1}{\text{Vol.}^2}\left[E^2-c^2p^2\sin^2\frac{\theta}{2}\right] = \frac{1}{\text{Vol.}^2}\left[1-\beta^2\sin^2\frac{\theta}{2}\right].
\end{aligned} \tag{39}$$

Combining eqs. (38) and (39), we have

$$\frac{d\sigma}{d\Omega} = \frac{4Z^2e^4E^2}{c^4\hbar^4q^4}\Big(1-\beta^2\sin^2\frac{\theta}{2}\Big). \tag{40}$$

Finally, noting $q = 2k\sin\frac{\theta}{2}$, as seen in Chapter 46, we get the Mott formula

$$\frac{d\sigma}{d\Omega} = \left(\frac{Z^2m^2e^4}{4k^4\hbar^4\sin^4\frac{\theta}{2}}\right)\frac{E^2}{m^2c^4}\Big(1-\beta^2\sin^2\frac{\theta}{2}\Big), \tag{41}$$

or

$$\frac{d\sigma}{d\Omega} = \left(\frac{d\sigma}{d\Omega}\right)_{\text{Rutherford}}\frac{\big(1-\beta^2\sin^2\frac{\theta}{2}\big)}{(1-\beta^2)}, \tag{42}$$

where the modification of the nonrelativistic result is shown very explicitly. (We have put $Z_1 = Z$, $Z_2 = -1$, $\mu \approx m$, in the Rutherford formula, as derived in eq. (16) of Chapter 46.)

73

Dirac Equation for a Particle in an Electromagnetic Field

The electromagnetic scalar and vector potentials, Φ and $\vec{A}$, combine to form a Lorentz four-vector. (See J. D. Jackson, *Classical Electrodynamics*, New York: John Wiley, 1975.)

$$A^\mu = (A^0, A^i) = (\Phi, \vec{A}), \qquad A_\mu = (A_0, A_i) = (\Phi, -\vec{A}). \tag{1}$$

Therefore, if we modify the four-momentum operator,

$$P_\mu = \frac{\hbar}{i}\frac{\partial}{\partial x^\mu} \longrightarrow P_\mu + \frac{e}{c}A_\mu, \tag{2}$$

the Dirac equation for a relativistic particle of charge, e, and rest mass, m, in an external field derivable from the potential, A_μ, would be

$$\gamma^\mu\left(\frac{\partial}{\partial x^\mu} + \frac{i}{\hbar}\frac{e}{c}A_\mu\right)\psi + \frac{imc}{\hbar}\psi = 0, \tag{3}$$

or

$$\left(\frac{\gamma^0}{c}\frac{\partial}{\partial t} + \frac{i}{\hbar}\gamma^0\frac{e}{c}\Phi + \vec{\gamma}\cdot(\vec{\nabla} - \frac{i}{\hbar}\frac{e}{c}\vec{A}) + \frac{imc}{\hbar}\right)\psi = 0. \tag{4}$$

Left-multiplying by $\gamma^0 = \beta$, this equation can be rewritten in the Hamiltonian form

$$-\frac{\hbar}{i}\frac{\partial\psi}{\partial t} = \left(V + c\vec{\alpha}\cdot(\frac{\hbar}{i}\vec{\nabla} - \frac{e}{c}\vec{A}) + \beta mc^2\right)\psi, \tag{5}$$

with $e\Phi = V$, where the momentum operator has been augmented by the vector potential term

$$\vec{p} \rightarrow \vec{\Pi} = \Big(\frac{\hbar}{i}\vec{\nabla} - \frac{e}{c}\vec{A}\Big). \tag{6}$$

A Nonrelativistic Limit of the Dirac Equation

Let us decompose the above 4×1 matrix ψ into two 2×1 matrices and factor out the time-dependence, which arises from the rest energy $E_{\text{rest}} = mc^2$, by setting

$$\psi = \begin{pmatrix} \psi_A e^{-\frac{i}{\hbar}mc^2 t} \\ \psi_B e^{-\frac{i}{\hbar}mc^2 t} \end{pmatrix} \tag{7}$$

to decompose eq. (5) into the two equations

$$\begin{aligned} -\frac{\hbar}{i}\frac{\partial\psi_A}{\partial t} &= V\psi_A + c(\vec{\sigma}\cdot\vec{\Pi})\psi_B + (mc^2 - mc^2)\psi_A, \\ -\frac{\hbar}{i}\frac{\partial\psi_B}{\partial t} &= V\psi_B + c(\vec{\sigma}\cdot\vec{\Pi})\psi_A - 2mc^2\psi_B. \end{aligned} \tag{8}$$

Now, let us look at the extreme nonrelativistic limit for which

$$-\frac{\hbar}{i}\frac{\partial\psi_B}{\partial t} \quad \text{and} \quad V\psi_B$$

are both negligible compared with $2mc^2\psi_B$. Note,

$$-\frac{\hbar}{i}\frac{\partial\psi_B}{\partial t} \quad \text{is of order} \quad (E - mc^2)\psi_B,$$

that is, of order of a nonrelativistic energy negligible compared with $2mc^2$ in the extreme nonrelativistic limit. In this limit, therefore, the second of the two equations above can be solved for ψ_B to give

$$\psi_B = \frac{1}{2mc}(\vec{\sigma}\cdot\vec{\Pi})\psi_A + \cdots, \tag{9}$$

to give the Schrödinger limit

$$-\frac{\hbar}{i}\frac{\partial\psi_A}{\partial t} = \Big(V + \frac{1}{2m}(\vec{\sigma}\cdot\vec{\Pi})(\vec{\sigma}\cdot\vec{\Pi})\Big)\psi_A. \tag{10}$$

With $\vec{A} = 0$, so $(\vec{\sigma}\cdot\vec{p})(\vec{\sigma}\cdot\vec{p}) = (\vec{p}\cdot\vec{p})$ [see eq. (11) of Chapter 72], the above is the Schrödinger equation, so in this limit $\psi_A \rightarrow \psi_{\text{Schroedinger}}$. (Remember, however, ψ_A has two components!)

In the presence of an external magnetic field derivable from $\vec{A}$, we get additional terms, now

$$\begin{aligned} (\vec{\sigma}\cdot\vec{\Pi})(\vec{\sigma}\cdot\vec{\Pi}) &= \sum_{j=1}^{3}\sigma_j^2\Pi_j^2 + \sum_{j<k}\sigma_j\sigma_k\Pi_j\Pi_k + \sigma_k\sigma_j\Pi_k\Pi_j) \\ &= \vec{\Pi}\cdot\vec{\Pi} + i\epsilon_{jkl}\sigma_l[\Pi_j, \Pi_k], \end{aligned} \tag{11}$$

using for the moment ordinary 3-D vector index notation. With $\Pi_j{}' = p_j - \frac{e}{c}A_j$, we now get

$$[\Pi_j, \Pi_k] = -\frac{e\hbar}{c\,i}\left(\frac{\partial A_k}{\partial x_j} - \frac{\partial A_j}{\partial x_k}\right), \tag{12}$$

and

$$i\epsilon_{jkl}\sigma_l[\Pi_j, \Pi_k] = -\frac{e}{c}\hbar\vec{\sigma}\cdot[\vec{\nabla}\times\vec{A}] = -\frac{e}{c}\hbar\vec{\sigma}\cdot\vec{B}, \tag{13}$$

so

$$-\frac{\hbar}{i}\frac{\partial\psi_A}{\partial t} = \Big(V + \frac{1}{2m}(\vec{p} - \frac{e}{c}\vec{A})\cdot(\vec{p} - \frac{e}{c}\vec{A}) - \frac{e\hbar}{2mc}\vec{\sigma}\cdot\vec{B}\Big)\psi_A + \cdots . \tag{14}$$

The spin-magnetic moment interaction with an external field $\vec{B}$ comes out of the nonrelativistic limit of the Dirac equation *automatically*. Moreover, it gives the correct g_s-factor, because

$$-\frac{e\hbar}{2mc}\vec{\sigma}\cdot\vec{B} = -\frac{e\hbar}{2mc}g_s\vec{s}\cdot\vec{B}, \qquad \text{with } g_s = 2, \tag{15}$$

because $\vec{s} = \frac{1}{2}\vec{\sigma}$. Note: Before Dirac, the spin-magnetic moment term was put into the theory by hand by Uhlenbeck and Goudsmit and the g_s value was set equal to $g_s = 2$ empirically to fit the observed spectroscopic data! Finally, in a uniform magnetic field, $\vec{B}_0$, with $\vec{A} = -\frac{1}{2}[\vec{r}\times\vec{B}_0]$, we get the extreme nonrelativistic limit

$$-\frac{\hbar}{i}\frac{\partial\psi_A}{\partial t} = \Big(V + \frac{1}{2m}(\vec{p}\cdot\vec{p}) - \frac{e\hbar}{2mc}(\vec{l}+\vec{\sigma})\cdot\vec{B}_0 + \frac{e^2}{8mc^2}\big(r^2B_0^2 - (\vec{r}\cdot\vec{B}_0)^2\big)\Big)\psi_A + \cdots . \tag{16}$$

Both the orbital and spin-magnetic moment terms are present, and with their correct g factors.

B Angular Momentum

Consider next a Hamiltonian with no external magnetic fields, i.e., with $\vec{A} = 0$, and with a spherically symmetric scalar potential, $\Phi = \Phi(r)$, and possibly an additional spherically symmetric potential, $\bar{V}(r)$, i.e., with a total potential, $V(r) = \bar{V}(r) + e\Phi(r)$. Now,

$$-\frac{\hbar}{i}\frac{\partial\psi}{\partial t} = H\psi = \Big(V(r) + c\gamma^0\vec{\gamma}\cdot\frac{\hbar}{i}\vec{\nabla} + \gamma^0 mc^2\Big)\psi, \tag{17}$$

where we have gone back to the full relativistic equation and ψ is the full four-component ψ. The orbital angular momentum operator, $[\vec{r}\times\vec{p}]$, does not commute with this H because of the presence of the $\gamma^0\vec{\gamma}\cdot\frac{\hbar}{i}\vec{\nabla}$ term. Of course, we still have $[\vec{L}, V(r)] = 0$, as in the nonrelativistic theory, but now $[H, L_j] \neq 0$ because of the $c\gamma^0\vec{\gamma}\cdot\vec{p}$ part of H. Let us express the jth component of $\vec{L}$ by

$$L_j = \epsilon_{jmk}x^m p^k,$$

with summation convention for repeated Latin indices, so m and k run from $1 \to 3$. Then,

$$[H, L_j] = [c\gamma^0\gamma^l \frac{\hbar}{i}\frac{\partial}{\partial x^l}, \epsilon_{jmk} x^m p^k] = \frac{\hbar c}{i}\epsilon_{jlk}\gamma^0\gamma^l p^k. \tag{18}$$

Now, let us define

$$\Sigma_j = \epsilon_{jmk}\frac{i}{2}\gamma^m\gamma^k, \qquad (\Sigma_1 = i\gamma^2\gamma^3,\ \Sigma_2 = i\gamma^3\gamma^1,\ \Sigma_3 = i\gamma^1\gamma^2). \tag{19}$$

(Recall the generators of the pure rotation operators from Chapter 70.) We consider the commutator

$$[H, \frac{\hbar}{2}\Sigma_j] = \frac{ci\hbar}{4}[\gamma^0\gamma^l p_l, \epsilon_{jmk}\gamma^m\gamma^k] = \frac{ci\hbar}{4}\gamma^0\epsilon_{jmk}[\gamma^l, \gamma^m\gamma^k]p_l. \tag{20}$$

From the Dirac relations, we have

$$[\gamma^l, \gamma^m\gamma^k] = 2g^{lm}\gamma^k - 2g^{lk}\gamma^m, \qquad \text{with } g^{ll} = -1,\ g^{lm} = 0 \text{ for } l \neq m, \tag{21}$$

so

$$\begin{aligned} [H, \frac{\hbar}{2}\Sigma_j] &= \frac{ci\hbar}{4}\epsilon_{jmk}\gamma^0(-2\gamma^k p^m + 2\gamma^m p^k) \\ &= \frac{ci\hbar}{2}\gamma^0(-\epsilon_{jmk}\gamma^k p^m + \epsilon_{jmk}\gamma^m p^k) \\ &= i\hbar c\gamma^0\epsilon_{jlk}\gamma^l p^k. \end{aligned} \tag{22}$$

Combining eqs. (18) and (22),

$$[H, L_j + \frac{\hbar}{2}\Sigma_j] = 0. \tag{23}$$

Also,

$$\Sigma_1 = \begin{pmatrix} \sigma_x & \mathbf{0} \\ \mathbf{0} & \sigma_x \end{pmatrix}, \qquad \Sigma_2 = \begin{pmatrix} \sigma_y & \mathbf{0} \\ \mathbf{0} & \sigma_y \end{pmatrix}, \qquad \Sigma_3 = \begin{pmatrix} \sigma_z & \mathbf{0} \\ \mathbf{0} & \sigma_z \end{pmatrix}. \tag{24}$$

It is thus natural to define the total angular momentum operator

$$J_j = L_j + \frac{\hbar}{2}\Sigma_j = \hbar(\bar{L}_j\mathbf{1} + \frac{1}{2}\Sigma_j), \tag{25}$$

where now $\vec{\bar{L}}$ and $\frac{1}{2}\vec{\Sigma}$ are the dimensionless orbital and spin angular momentum operators. Note: It is the total angular momentum operator which commutes with the Hamiltonian with a spherically symmetric $V(r)$.

Also, the plane wave solutions of Chapter 72 have been chosen such that they are eigenstates of Σ_3, if the z direction is chosen parallel to $\vec{p}$. The eigenvalues of Σ_3 are +1 for states $u^{(1)}$ and $u^{(3)}$, and -1 for states $u^{(2)}$ and $u^{(4)}$, because

$$\Sigma_3 \begin{pmatrix} 1 \\ 0 \\ \frac{cp}{E+mc^2} \\ 0 \end{pmatrix} = +1 \begin{pmatrix} 1 \\ 0 \\ \frac{cp}{E+mc^2} \\ 0 \end{pmatrix}, \qquad \Sigma_3 \begin{pmatrix} \frac{cp}{E-mc^2} \\ 0 \\ 1 \\ 0 \end{pmatrix} = +1 \begin{pmatrix} \frac{cp}{E-mc^2} \\ 0 \\ 1 \\ 0 \end{pmatrix}, \tag{26}$$

and

$$\Sigma_3 \begin{pmatrix} 0 \\ 1 \\ 0 \\ \frac{-cp}{E+mc^2} \end{pmatrix} = -1 \begin{pmatrix} 0 \\ 1 \\ 0 \\ \frac{-cp}{E+mc^2} \end{pmatrix}, \qquad \Sigma_3 \begin{pmatrix} 0 \\ \frac{-cp}{E-mc^2} \\ 0 \\ 1 \end{pmatrix} = -1 \begin{pmatrix} 0 \\ \frac{-cp}{E-mc^2} \\ 0 \\ 1 \end{pmatrix}. \tag{27}$$

Final Note: In Chapter 70 you should perhaps have been worried we seemingly obtained the "wrong" sign of the generator, $G_{12}^{\text{intr.}}$, relative to that of $G_{12}^{\text{orb.}}$. Recall we concluded

$$e^{-i\theta_{12}G_{12}^{\text{intr}}} = e^{+i\frac{\theta_{12}}{2}\Sigma_3},$$

whereas

$$e^{-i\theta_{12}G_{12}^{\text{orb.}}} = e^{-i\theta_{12}L_z},$$

the latter in agreement with our earlier nonrelativistic rotation theory. But note that L_z was defined, however, by

$$L_z = \frac{1}{i}\left(x\frac{\partial}{\partial y} - y\frac{\partial}{\partial x}\right) = \frac{1}{i}\left(x^1\frac{\partial}{\partial x^2} - x^2\frac{\partial}{\partial x^1}\right) = (x^1 p_2 - x^2 p_1),$$

and, therefore,

$$L_z = -L_3 = -(x^1 p^2 - x^2 p^1).$$

74

Pauli Approximation to the Dirac Equation

In the last chapter, we examined the extreme nonrelativistic limit of the Dirac equation and saw that the first two components of ψ lead to Schrödinger theory, $\psi_A \rightarrow \psi_{\text{Schroedinger}}$, the two components of ψ_A being related to the two spin components of the electron, whereas the third and fourth components, in the form of ψ_B, were smaller by a factor of order p/mc. We now want to carry this one step further following an approach first carried through by Pauli. Again, let

$$\psi = \begin{pmatrix} \psi_A \\ \psi_B \end{pmatrix} e^{-\frac{i}{\hbar}mc^2 t}, \tag{1}$$

so, again,

$$\begin{aligned} -\frac{\hbar}{i}\frac{\partial \psi_A}{\partial t} &= V\psi_A + c(\vec{\sigma}\cdot\vec{\Pi})\psi_B, \\ -\frac{\hbar}{i}\frac{\partial \psi_B}{\partial t} &= V\psi_B + c(\vec{\sigma}\cdot\vec{\Pi})\psi_A - 2mc^2\psi_B, \end{aligned} \tag{2}$$

with $V = e\Phi$, and $\vec{\Pi} = \vec{p} - \frac{e}{c}\vec{A}$. Rewriting the second of these two equations, we have

$$\psi_B = \frac{1}{2mc^2}\left(\frac{\hbar}{i}\frac{\partial}{\partial t} + V\right)\psi_B + \frac{c(\vec{\sigma}\cdot\vec{\Pi})}{2mc^2}\psi_A. \tag{3}$$

Now, in the first term on the right-hand side, we can substitute the zeroth-order approximation for ψ_B because the operators $\frac{\hbar}{i}\frac{\partial}{\partial t}$ and V lead to quantities of order $E_{\text{nonrelativistic}} \ll 2mc^2$. In zeroth order, we had

$$\psi_B = \frac{(\vec{\sigma}\cdot\vec{\Pi})}{2mc}\psi_A + \cdots\ , \tag{4}$$

so eq. (3) leads to

$$\psi_B = \frac{1}{4m^2c^3}\left(\frac{\hbar}{i}\frac{\partial}{\partial t} + V\right)(\vec{\sigma}\cdot\vec{\Pi})\psi_A + \frac{(\vec{\sigma}\cdot\vec{\Pi})}{2mc}\psi_A + \cdots . \tag{5}$$

Now, substituting this back into the equation for ψ_A, we get an equation for ψ_A in the form,

$$-\frac{\hbar}{i}\frac{\partial\psi_A}{\partial t} = H\psi_A,$$

with

$$\begin{aligned} H\psi_A &= V\psi_A \\ &+ \frac{1}{2m}(\vec{\sigma}\cdot\vec{\Pi})(\vec{\sigma}\cdot\vec{\Pi})\psi_A + \frac{(\vec{\sigma}\cdot\vec{\Pi})}{4m^2c^2}\left(\frac{\hbar}{i}\frac{\partial}{\partial t} + V\right)(\vec{\sigma}\cdot\vec{\Pi})\psi_A + \cdots \\ &= \left(\frac{1}{2m}(\vec{\Pi}\cdot\vec{\Pi}) + V - \frac{e\hbar}{2mc}(\vec{\sigma}\cdot\vec{B})\right)\psi_A + \frac{(\vec{\sigma}\cdot\vec{\Pi})}{4m^2c^2}\left(\frac{\hbar}{i}\frac{\partial}{\partial t} + V\right)(\vec{\sigma}\cdot\vec{\Pi})\psi_A \\ &+ \cdots . \end{aligned} \tag{6}$$

The normalization for a single-particle relativistic theory would be, however,

$$\int d\vec{r}(\psi_A^\dagger\psi_A + \psi_B^\dagger\psi_B) = 1, \tag{7}$$

and, with $\psi_B \approx (\vec{\sigma}\cdot\vec{\Pi})\psi_A/2mc$, this would be

$$\int d\vec{r}\psi_A^\dagger\left(1 + \frac{1}{4m^2c^2}(\vec{\sigma}\cdot\vec{\Pi})(\vec{\sigma}\cdot\vec{\Pi}) + \cdots\right)\psi_A = 1, \tag{8}$$

whereas, for the nonrelativistic approximation, we should have had

$$\int d\vec{r}\psi^\dagger\psi = 1. \tag{9}$$

To order p^2/m^2c^2, if we renormalize ψ with an operator Ω

$$\begin{aligned} \psi &= \Omega\psi_A = \left(1 + \frac{1}{8m^2c^2}(\vec{\sigma}\cdot\vec{\Pi})(\vec{\sigma}\cdot\vec{\Pi}) + \cdots\right)\psi_A, \\ \psi^\dagger &= \psi_A^\dagger\left(1 + \frac{1}{8m^2c^2}(\vec{\sigma}\cdot\vec{\Pi})(\vec{\sigma}\cdot\vec{\Pi}) + \cdots\right), \end{aligned} \tag{10}$$

or

$$\psi_A = \Omega^{-1}\psi = \left(1 - \frac{1}{8m^2c^2}(\vec{\sigma}\cdot\vec{\Pi})(\vec{\sigma}\cdot\vec{\Pi}) + \cdots\right)\psi. \tag{11}$$

With this transformation, eq. (6) can be rewritten

$$-\frac{\hbar}{i}\frac{\partial\psi_A}{\partial t} = H\psi_A \quad \rightarrow \quad -\frac{\hbar}{i}\frac{\partial(\Omega^{-1}\psi)}{\partial t} = H(\Omega^{-1}\psi), \tag{12}$$

or, left-multiplying with the operator Ω^{-1},

$$-\frac{\hbar}{i}\left(\Omega^{-1}\frac{\partial}{\partial t}\Omega^{-1}\right)\psi = (\Omega^{-1}H\Omega^{-1})\psi, \tag{13}$$

leading to

$$
\begin{aligned}
-\frac{\hbar}{i}\frac{\partial\psi}{\partial t} &+ \frac{1}{8m^2c^2}\Big((\vec{\sigma}\cdot\vec{\Pi})^2\frac{\hbar}{i}\frac{\partial}{\partial t} + \frac{\hbar}{i}\frac{\partial}{\partial t}(\vec{\sigma}\cdot\vec{\Pi})^2\Big)\psi + \cdots \\
&= \Big(H - \frac{1}{8m^2c^2}(H(\vec{\sigma}\cdot\vec{\Pi})^2 + (\vec{\sigma}\cdot\vec{\Pi})^2 H) + \cdots\Big)\psi,
\end{aligned}
\tag{14}
$$

or

$$
-\frac{\hbar}{i}\frac{\partial\psi}{\partial t} = H'\psi, \tag{15}
$$

with

$$
\begin{aligned}
H' = H &- \frac{1}{8m^2c^2}\Big[\Big((\vec{\sigma}\cdot\vec{\Pi})^2\frac{\hbar}{i}\frac{\partial}{\partial t} + \frac{\hbar}{i}\frac{\partial}{\partial t}(\vec{\sigma}\cdot\vec{\Pi})^2\Big) \\
&+ \Big(\big(\frac{(\vec{\sigma}\cdot\vec{\Pi})^2}{2m} + V\big)(\vec{\sigma}\cdot\vec{\Pi})^2 + (\vec{\sigma}\cdot\vec{\Pi})^2\big(\frac{(\vec{\sigma}\cdot\vec{\Pi})^2}{2m} + V\big)\Big)\Big] + \cdots \\
= \frac{(\vec{\sigma}\cdot\vec{\Pi})^2}{2m} &+ V + \frac{1}{4m^2c^2}(\vec{\sigma}\cdot\vec{\Pi})\Big(\frac{\hbar}{i}\frac{\partial}{\partial t} + V\Big)(\vec{\sigma}\cdot\vec{\Pi}) \\
- \frac{1}{8m^2c^2}&\Big[(\vec{\sigma}\cdot\vec{\Pi})^2\Big(\frac{\hbar}{i}\frac{\partial}{\partial t} + V\Big) + \Big(\frac{\hbar}{i}\frac{\partial}{\partial t} + V\Big)(\vec{\sigma}\cdot\vec{\Pi})^2 + \frac{1}{m}(\vec{\sigma}\cdot\vec{\Pi})^4\Big] \\
+ \cdots,&
\end{aligned}
\tag{16}
$$

so H' can be written with the help of a double commutator as

$$
H' = \frac{(\vec{\sigma}\cdot\vec{\Pi})^2}{2m} + V - \frac{1}{8m^2c^2}[(\vec{\sigma}\cdot\vec{\Pi}), [(\vec{\sigma}\cdot\vec{\Pi}), (\frac{\hbar}{i}\frac{\partial}{\partial t} + V)]] - \frac{1}{8m^3c^2}(\vec{\sigma}\cdot\vec{\Pi})^4. \tag{17}
$$

The first commutator gives

$$
\begin{aligned}
[(\vec{\sigma}\cdot\vec{\Pi}), (\frac{\hbar}{i}\frac{\partial}{\partial t} + V)] &= \vec{\sigma}\cdot[(\vec{p} - \frac{e}{c}\vec{A}), (\frac{\hbar}{i}\frac{\partial}{\partial t} + e\Phi)] \\
&= -\frac{e\hbar}{i}\vec{\sigma}\cdot\Big(-\vec{\nabla}\Phi - \frac{1}{c}\frac{\partial\vec{A}}{\partial t}\Big) = -\frac{e\hbar}{i}(\vec{\sigma}\cdot\vec{\mathcal{E}}).
\end{aligned}
\tag{18}
$$

This leads to the second commutator (using for the moment ordinary 3-D vector index notation),

$$
\begin{aligned}
[(\vec{\sigma}\cdot\vec{\Pi}), (\vec{\sigma}\cdot\vec{\mathcal{E}})] &= \sigma_k\sigma_j[\Pi_k, \mathcal{E}_j] + [\sigma_k, \sigma_j]\mathcal{E}_j\Pi_k \\
&= (\delta_{kj} + i\epsilon_{kjl}\sigma_l)\frac{\hbar}{i}\frac{\partial\mathcal{E}_j}{\partial x_k} - 2i\epsilon_{jkl}\sigma_l\mathcal{E}_j\Pi_k \\
&= \frac{\hbar}{i}(\vec{\nabla}\cdot\vec{\mathcal{E}}) + \hbar\vec{\sigma}\cdot[\vec{\nabla}\times\vec{\mathcal{E}}] - 2i\vec{\sigma}\cdot[\vec{\mathcal{E}}\times\vec{\Pi}].
\end{aligned}
\tag{19}
$$

With this result, we then have

$$
\begin{aligned}
H' = \frac{(\vec{\sigma}\cdot\vec{\Pi})^2}{2m} &+ e\Phi - \frac{e\hbar}{4m^2c^2}(\vec{\sigma}\cdot[\vec{\mathcal{E}}\times\vec{\Pi}]) \\
&- \frac{e\hbar^2}{8m^2c^2}(\vec{\nabla}\cdot\vec{\mathcal{E}}) - \frac{e\hbar^2}{8m^2c^2}i\vec{\sigma}\cdot[\vec{\nabla}\times\vec{\mathcal{E}}] - \frac{1}{8m^3c^2}(\vec{\sigma}\cdot\vec{\Pi})^4,
\end{aligned}
\tag{20}
$$

where this H' includes relativistic corrections to the zeroth-order problem through terms of order $\beta^2\times$ (zeroth order terms). In the special case, when no external magnetic fields exist, so $\vec{A} = 0$ and

$$[\vec{\nabla} \times \vec{\mathcal{E}}] = -\frac{1}{c}\frac{\partial \vec{B}}{\partial t} = 0,$$

we have

$$H' = \frac{\vec{p}^{\,2}}{2m} + V - \frac{e\hbar}{4m^2c^2}(\vec{\sigma} \cdot [\vec{\mathcal{E}} \times \vec{p}]) - \frac{e\hbar^2}{8m^2c^2}(\vec{\nabla} \cdot \vec{\mathcal{E}}) - \frac{(\vec{p}^{\,2})^2}{8m^3c^2}. \tag{21}$$

The very last term is easily recognized as the relativistic mass correction to the kinetic energy term, because

$$E - mc^2 = \sqrt{(m^2c^4 + c^2p^2)} - mc^2 = \frac{p^2}{2m} - \frac{p^4}{8m^3c^2} + \cdots . \tag{22}$$

If the electric fields, $\vec{\mathcal{E}}$, are those of a spherically symmetric potential, e.g., the Coulomb potential in a one-electron atom, $\vec{\mathcal{E}} = -\vec{\nabla}\Phi = -\frac{1}{e}\vec{\nabla}V$ and the first correction term to the zeroth-order nonrelativistic energy becomes

$$-\frac{e\hbar}{4m^2c^2}(\vec{\sigma} \cdot [\vec{\mathcal{E}} \times \vec{p}]) = \frac{\hbar}{4m^2c^2}\vec{\sigma} \cdot \left(\frac{1}{r}\frac{dV}{dr}[\vec{r} \times \vec{p}]\right) = \frac{\hbar^2}{2m^2c^2}(\vec{s} \cdot \vec{l})\frac{1}{r}\frac{dV}{dr}. \tag{23}$$

This term gives the combined internal magnetic field spin-orbit coupling term and the Thomas precession term. We see again that it was not put into the theory "by hand" but falls automatically out of the Dirac theory. Finally, the $\vec{\nabla} \cdot \vec{\mathcal{E}}$ term, which is known as the Darwin term, can be written via Maxwell's equations as

$$\vec{\nabla} \cdot \vec{\mathcal{E}} = 4\pi\rho_{\text{charge}} = 4\pi Ze\delta(\vec{r} = 0) \tag{24}$$

for a one-electron atom. The Darwin term, therefore, can in this case be rewritten as (remembering the electron e is negative, whereas the nuclear charge Ze is positive)

$$-\frac{e\hbar^2}{8m^2c^2}\vec{\nabla} \cdot \vec{\mathcal{E}} = \frac{4\pi Ze^2\hbar^2}{8m^2c^2}\delta(\vec{r} = 0), \tag{25}$$

leading to an energy correction in a hydrogenic atom of

$$\Delta E_{nlm} = \frac{4\pi Ze^2\hbar^2}{8m^2c^2}|\psi_{nlm}(\vec{r} = 0)|^2 = \frac{4\pi Ze^2\hbar^2}{8m^2c^2}\delta_{l0}\frac{1}{4\pi}\left(\frac{4Z^3}{n^3a_0^3}\right) = \frac{me^4}{\hbar^2}\frac{\alpha^2}{2}\frac{Z^4}{n^3}\delta_{l0}. \tag{26}$$

The spin orbit term in a hydrogenic atom, with $V(r) = -Ze^2/r$, is

$$\frac{me^4}{\hbar^2}\frac{\alpha^2}{2}\frac{Z}{(r/a_0)^3}(\vec{s} \cdot \vec{l}), \tag{27}$$

leading to an energy correction [see eq. (13) of Chapter 26].

$$\Delta E_{nlj} = \frac{me^4}{\hbar^2}\frac{\alpha^2}{2}\frac{Z^4}{n^3l(l+\frac{1}{2})(l+1)} \times \left\{ \begin{array}{ll} +\frac{l}{2} & \text{for } j = (l+\frac{1}{2}) \\ -\frac{(l+1)}{2} & \text{for } j = (l-\frac{1}{2}) \end{array} \right\}. \tag{28}$$

For $l = 0$, $j = \frac{1}{2}$, the matrix element of $(\vec{l} \cdot \vec{s})$ is equal to zero. If we take the above formula of eq. (28) and cancel the factor l in numerator and denominator and then set $l = 0$, we get the $l = 0$, $j = \frac{1}{2}$ result of the Darwin term. Rigorously, this result comes out of the Darwin term and not the $(\vec{l} \cdot \vec{s})$ term. (In Chapter 26, we "cheated"!!) The rigorous derivation from the Pauli approximation to the Dirac equation gives the $l = 0$ result via the Darwin term.

A The Foldy–Wouthuysen Transformation

The uncoupling of the first two components of the Dirac equation from the third and fourth components via the Pauli approximation in a perturbation expansion naturally leads to the question: Is it possible to transform the Dirac Hamiltonian to a new form in which the third and fourth components are completely uncoupled from the first two components? The answer is yes, at least for the free particle with Hamiltonian, $H = c(\vec{\alpha} \cdot \vec{p}) + \beta mc^2$. In this case, the transformation was achieved by Foldy and Wouthuysen.

Foldy and Wouthuysen sought a transformation

$$\psi \rightarrow \psi' = U\psi, \qquad H \rightarrow H' = UHU^{\dagger}, \tag{29}$$

via a unitary operator $U = e^{iG}$ such that the 4×4 matrices within H' factor into a direct sum of 2×2 matrices. From our knowledge of Lorentz transformations, one might seek unitary operators of the form

$$U = e^{i(-i\theta\beta\vec{\alpha}\cdot\vec{p})} = \cos(p\theta)\mathbf{1} + \beta(\vec{\alpha} \cdot \frac{\vec{p}}{p}) \sin(p\theta), \tag{30}$$

where θ is a function of p and we have used

$$(\beta\vec{\alpha} \cdot \vec{p})(\beta\vec{\alpha} \cdot \vec{p}) = -(\vec{\alpha} \cdot \vec{p})(\vec{\alpha} \cdot \vec{p}) = -p^2\mathbf{1}. \tag{31}$$

The actual solution gives

$$U = \left(\sqrt{\frac{mc^2 + |E|}{2|E|}}\mathbf{1} + \frac{c\beta\vec{\alpha} \cdot \vec{p}}{\sqrt{2|E|(mc^2 + |E|)}}\right). \tag{32}$$

This equation leads to

$$\begin{aligned} H'_{\text{F.W.}} &= UHU^{\dagger} \\ &= \left(\sqrt{\frac{mc^2 + |E|}{2|E|}}\mathbf{1} + \frac{c\beta\vec{\alpha} \cdot \vec{p}}{\sqrt{2|E|(mc^2 + |E|)}}\right)(c\vec{\alpha} \cdot \vec{p} + \beta mc^2)U^{\dagger} \\ &= \left(\frac{\beta}{2}\sqrt{2|E|(mc^2 + |E|)} + \frac{c(\vec{\alpha} \cdot \vec{p})\sqrt{2|E|}}{2\sqrt{(mc^2 + |E|)}}\right) \\ &\quad \times \left(\sqrt{\frac{mc^2 + |E|}{2|E|}}\mathbf{1} + \frac{c(\vec{\alpha} \cdot \vec{p})\beta}{\sqrt{2|E|(mc^2 + |E|)}}\right) \end{aligned}$$

$$
\begin{aligned}
&= \left(\frac{c}{2}[(\vec{\alpha}\cdot\vec{p}) + \beta(\vec{\alpha}\cdot\vec{p})\beta]\right) + \frac{\beta}{2}\left((mc^2 + |E|) + \frac{c^2p^2}{(mc^2 + |E|)}\right) \\
&= \quad 0 \qquad + \beta|E| \\
&= +\beta\sqrt{m^2c^4 + c^2p^2}. \qquad (33)
\end{aligned}
$$

The Dirac equation in ψ' thus separates into the two equations

$$
\begin{aligned}
H'_{\mathrm{F.W.}}\psi'_A &= +\sqrt{m^2c^4 + c^2p^2}\,\psi'_A, \\
H'_{\mathrm{F.W.}}\psi'_B &= -\sqrt{m^2c^4 + c^2p^2}\,\psi'_B. \qquad (34)
\end{aligned}
$$

For a particle in an arbitrary electromagnetic field, a successful analogue of the Foldy–Wouthuysen transformation has not been found. An expansion in powers of p/mc, however, can be carried out systematically with the unitary operator

$$
U = e^{i\lambda G}, \qquad \text{with } i\lambda G = \frac{\beta(\vec{\alpha}\cdot\vec{\Pi})}{2mc}. \qquad (35)
$$

75

The Klein Paradox: An Example from the History of Negative Energy State Difficulties: The Positron Interpretation

One of the simplest problems, first solved in the earliest days of the Dirac theory, is the 1-D motion of a free particle reflected or transmitted at a discrete potential step. Suppose the particle moves in the z direction under the influence of a potential such that

$$V(z) = 0 \quad \text{for } z < 0, \qquad V(z) = +V_0 \quad \text{for } z > 0, \tag{1}$$

where we assume, $V_0 > 0$. The solution will be independent of the spin of the electron. Let us therefore choose an electron with spin projection in the + z direction. We shall choose a positive-energy solution. In region I, for $z < 0$, we expect plane waves moving both to the left and to the right. Therefore, for $z \leq 0$,

$$\psi = A \begin{pmatrix} 1 \\ 0 \\ \frac{cp}{E+mc^2} \\ 0 \end{pmatrix} e^{\frac{i}{\hbar}(pz-Et)} + B \begin{pmatrix} 1 \\ 0 \\ \frac{-cp}{E+mc^2} \\ 0 \end{pmatrix} e^{-\frac{i}{\hbar}(pz+Et)}, \tag{2}$$

with $E > 0$, and $p = +\sqrt{(\frac{E^2}{c^2} - m^2c^2)}$, where A gives the amplitude of the incident wave traveling to the right, and B gives the amplitude of the reflected wave traveling to the left. From $S_z = c\psi^{\dagger}\alpha_z\psi$, we have

$$(S_z)_{\text{inc.}} = cAA^* \begin{pmatrix} 1 & 0 & \frac{cp}{E+mc^2} & 0 \end{pmatrix} \begin{pmatrix} \frac{cp}{E+mc^2} \\ 0 \\ 1 \\ 0 \end{pmatrix}$$

$$= \frac{2c^2p}{E+mc^2}|A|^2 = \frac{c^2p}{E}\psi^{\dagger}_{\text{inc.}}\psi_{\text{inc.}} = +v\psi^{\dagger}_{\text{inc.}}\psi_{\text{inc.}}. \tag{3}$$

Similarly,

$$(S_z)_{\text{refl.}} = -\frac{2c^2 p}{E + mc^2}|B|^2 = -v\psi^{\dagger}_{\text{refl.}}\psi_{\text{refl.}}. \tag{4}$$

In region II, we would expect

$$\psi = C \begin{pmatrix} 1 \\ 0 \\ \frac{c\bar{p}}{E - V_0 + mc^2} \\ 0 \end{pmatrix} e^{\frac{i}{\hbar}(\bar{p}z - Et)}, \tag{5}$$

so here

$$(S_z)_{\text{transm.}} = \frac{2c^2\bar{p}}{(E - V_0 + mc^2)}|C|^2 = \frac{c^2\bar{p}}{(E - V_0)}\psi^{\dagger}_{\text{transm.}}\psi_{\text{transm.}}, \tag{6}$$

where, now,

$$c^2\bar{p}^2 = (E - V_0)^2 - m^2c^4, \tag{7}$$

and

$$c\bar{p} = \pm\sqrt{((E - V_0)^2 - m^2c^2)} = \pm\sqrt{(E - V_0 - mc^2)(E - V_0 + mc^2)}. \tag{8}$$

Now, at the boundary, $z = 0$, the continuity of both the probability density and the probability current density, S_z, are satisfied by the continuity of ψ, but now this means all components of ψ. Therefore, this requires

$$A + B = C$$
$$(A - B)\frac{cp}{E + mc^2} = C\frac{c\bar{p}}{(E - V_0 + mc^2)}, \tag{9}$$

leading to

$$2A = \left[1 + \frac{c\bar{p}}{cp}\frac{(E + mc^2)}{(E - V_0 + mc^2)}\right]C,$$
$$2B = \left[1 - \frac{c\bar{p}}{cp}\frac{(E + mc^2)}{(E - V_0 + mc^2)}\right]C, \tag{10}$$

leading to a reflection coefficient

$$R = \left|\frac{B}{A}\right|^2 = \left|\frac{\left[1 - \frac{c\bar{p}}{cp}\frac{(E+mc^2)}{(E-V_0+mc^2)}\right]}{\left[1 + \frac{c\bar{p}}{cp}\frac{(E+mc^2)}{(E-V_0+mc^2)}\right]}\right|^2. \tag{11}$$

Now, we must consider various cases, depending on the relative magnitudes of E and V_0 (see Fig. 75.1).

Case 1.

$$V_0 < E - mc^2.$$

In this case, we would certainly expect a transmitted wave, with a kinetic energy in region II that is somewhat less than that in region I, and a momentum, $\bar{p} > 0$, and,

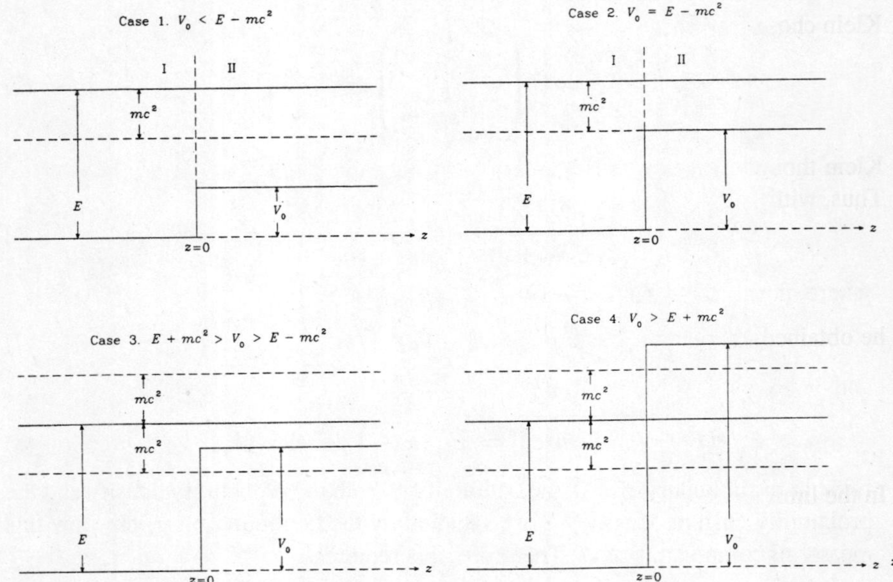

FIGURE 75.1. The four cases for free Dirac particle transmission and reflection at a potential step.

hence, a positive probability density current for $z > 0$, hence electrons moving to the right in region II. The reflection coefficient, $R < 1$, as expected.

Case 2.

$$V_0 = E - mc^2.$$

In this case, $\bar{p} = 0$. The reflection coefficient is $R = 1$, and hence the transmission coefficient, $T = 0$. Again, no surprises!

Case 3.

$$E + mc^2 > V_0 > E - mc^2.$$

In this case, $\bar{p}$ is pure imaginary. The reflection coefficient, $R = 1$. The wave function in region II has an exponential decay, so $\psi_{II} \to 0$ as $z \to \infty$. Still no surprises!

Case 4.

$$V_0 > E + mc^2.$$

In this case, $\bar{p}$ is again real, but because $(E - V_0 + mc^2)$ is a negative quantity and now

$$(S_z)_{\text{transm.}} = -\frac{2c^2\bar{p}}{(V_0 - E - mc^2)}|C|^2. \tag{12}$$

Klein chose the negative root for $\bar{p}$ [see eq. (8)], so

$$\frac{c\bar{p}}{(E - V_0 + mc^2)} = \sqrt{\frac{(V_0 - E + mc^2)}{(V_0 - E - mc^2)}}. \tag{13}$$

Klein thought (naturally for the time!) S_z had to be positive in the region $z > 0$. Thus, with

$$\frac{(E + mc^2)}{cp} = \sqrt{\frac{(E + mc^2)}{(E - mc^2)}}, \tag{14}$$

he obtained a reflection coefficient

$$R = \left[\frac{1 - \sqrt{\frac{(V_0 - E + mc^2)(E + mc^2)}{(V_0 - E - mc^2)(E - mc^2)}}}{1 + \sqrt{\frac{(V_0 - E + mc^2)(E + mc^2)}{(V_0 - E - mc^2)(E - mc^2)}}}\right]^2. \tag{15}$$

In the limit in which V_0 becomes large (let $V_0 \to \infty$),

$$R \to \left[\frac{1 - \frac{cp}{(E - mc^2)}}{1 + \frac{cp}{(E - mc^2)}}\right]^2 = \left[\frac{1 - \frac{2cp}{E - mc^2} + \frac{E + mc^2}{E - mc^2}}{1 + \frac{2cp}{E - mc^2} + \frac{E + mc^2}{E - mc^2}}\right] = \frac{E - cp}{E + cp}. \tag{16}$$

Thus, in the limit of large V_0, Klein predicted reflection and transmission coefficients

$$R \to \frac{E - cp}{E + cp}, \qquad T \to \frac{2cp}{E + cp}. \tag{17}$$

In particular, for $v/c = \sqrt{\frac{1}{2}}$, and therefore $E = 1.1547mc^2$, the transmission coefficient is $T = .83$. Klein of course expected no transmission and complete reflection in the limit of large potential barrier V_0. This "Klein Paradox" and similarly paradoxical situations bothered physicists in the earliest days of Dirac theory. Dirac came up with the solution: In the normal vacuum, all negative-energy states are completely filled. However, if one electron is taken out of a normally filled negative-energy state, the missing electron, a hole in the normally filled negative-energy sea, makes for a missing charge in a normally electrically neutral vacuum and thus behaves like a particle of charge $+|e|$. In this way, Dirac was led to the prediction of the existence of the positron.

A Modern Hole Analysis of the Klein Barrier Reflection

According to Dirac's final interpretation, the Klein experiment must be understood in the following way. In region I, for $z < 0$, an electron with positive energy

$E = +\sqrt{(m^2c^4 + c^2p^2)}$ is placed in a level normally empty. However, if this E is such that $E < V_0 - mc^2$, in region II, for $z > 0$, all such levels are normally filled. The normal vacuum in this region thus consists of an infinite number of filled states with E negative relative to the reference energy V_0, with $E < V_0 - mc^2$. Now, if an electron from the beam in region I reaches the potential barrier at $z = 0$, it can knock an electron out of this filled sea, creating a hole in this negative-energy sea. This hole at $z = 0$, however, can be filled by an electron at $z > 0$ with the same energy. When this electron has moved into the hole at $z = 0$, the hole at $z > 0$ can be filled by another electron with this same E but at still greater values of z. Thus, an electron current from right to left exists, i.e., in a direction such that S_z is negative. Thus, with

$$S_z = \frac{2c\bar{p}}{(E - V_0 + mc^2)}|C|^2, \qquad \text{and with } (E - V_0 + mc^2) < 0, \tag{18}$$

we need the *positive* root for $\bar{p}$; that is,

$$c\bar{p} = +\sqrt{(V_0 - E + mc^2)(V_0 - E - mc^2)},$$

$$\text{so } \frac{c\bar{p}}{(E - V_0 + mc^2)} = -\sqrt{\frac{(V_0 - E + mc^2)}{(V_0 - E - mc^2)}}. \tag{19}$$

Substituting this equation into eq. (11), we get

$$R = \left[\frac{1 + \sqrt{\frac{(V_0-E+mc^2)(E+mc^2)}{(V_0-E-mc^2)(E-mc^2)}}}{1 - \sqrt{\frac{(V_0-E+mc^2)(E+mc^2)}{(V_0-E-mc^2)(E-mc^2)}}}\right]^2. \tag{20}$$

Now, in the limit, $V_0 \to \infty$, we have

$$R \to \left[\frac{1 + \frac{cp}{(E-mc^2)}}{1 - \frac{cp}{(E-mc^2)}}\right]^2 = \frac{E + cp}{E - cp} > 1. \tag{21}$$

This reflection coefficient is now greater than unity because the electrons knocked out of the filled negative-energy sea at $z = 0$ can join the electrons reflected at the barrier, so the number of electrons in the reflected beam is greater than the number of electrons in the incident beam. In region II, however, for every electron that has joined the incident electrons in the reflected beam, there is a hole moving to the right. This appears to us like a positron moving to the right in region II. The electron current in region II, however, is from right to left, as electrons from further to the right fill holes to the left. Note, in particular, for $v/c = \sqrt{\frac{1}{2}}$,

$$\lim_{V_0 \to \infty} R = \frac{\sqrt{2} + 1}{\sqrt{2} - 1} = 5.8. \tag{22}$$

Before justifying this perhaps surprisingly large number, a remark about sharp potential rises is in order. Our potential rises sharply from a value of zero to a large value at $z = 0$. This idealization of a realistic potential rise would be a good approximation if the actual distance, d, over which the rise occurs (see Fig. 75.2)

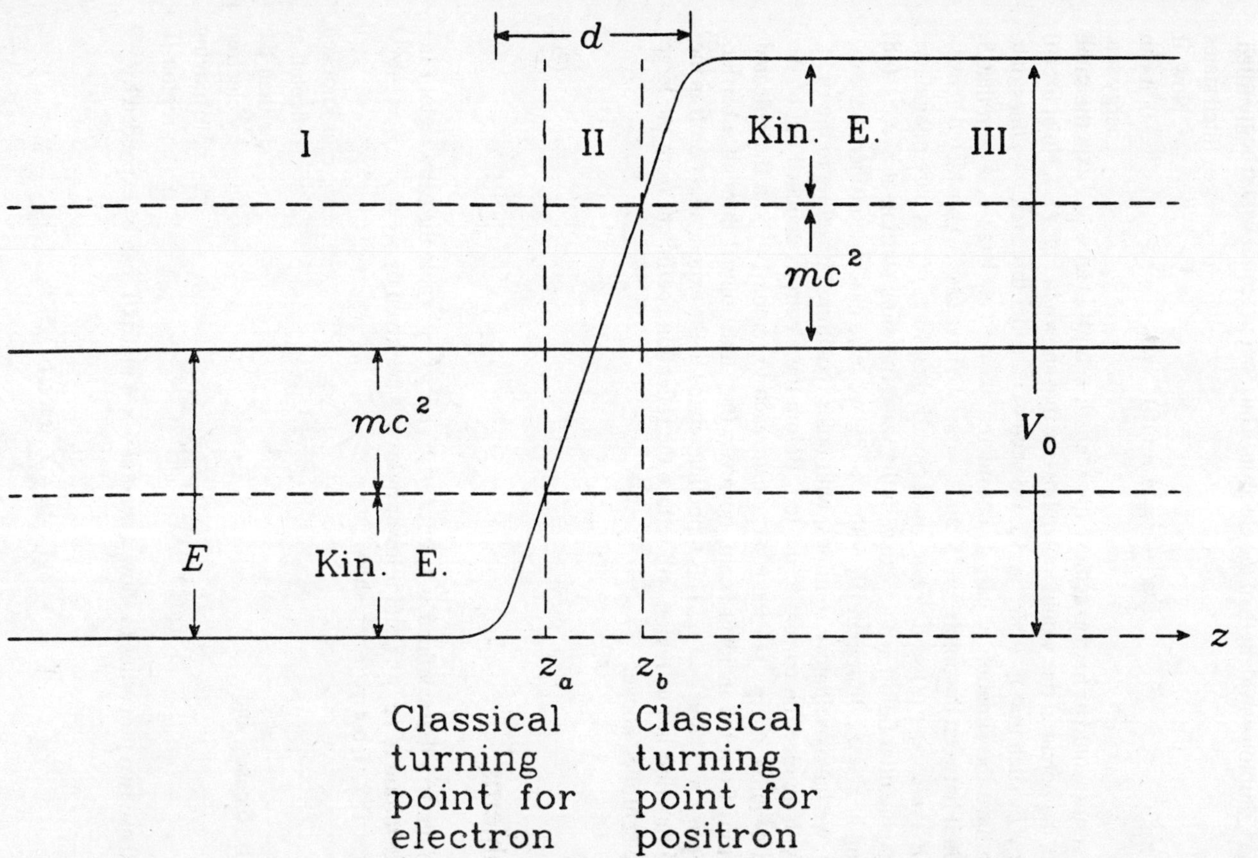

FIGURE 75.2. The tunneling of Dirac particles from the filled negative-energy sea for a potential step of finite width and height $> 2mc^2$.

is small compared with the relevant length parameter for the problem. This would be the Compton wavelength of the electron. Thus, our potential would be realistic, if

$$d \ll \frac{\hbar}{mc} = 4 \times (10)^{-11}\text{cm}. \tag{23}$$

This result would be hard to achieve in a realistic setup!! In an actual experimental setup, the potential rise would look more like that shown in Fig. 75.2. Now, for an energy E, such that $E + mc^2 < V_0$, we need to consider three regions. In region I, we have the situation of case 1, considered above; i.e., $V(z) < E - mc^2$. In region II between the points $z = z_a$ and $z = z_b$, we have case 3, considered above; i.e., $E - mc^2 < V(z) < E + mc^2$. In region II, therefore, $\bar{p}(z)$ is pure imaginary. An exponential falloff with z exists in the wave function, starting at $z = z_a$ and ending at $z = z_b$. In region III, with $z > z_b$, we have case 4, considered above. Here, we have oscillatory solutions, with holes (positrons) traveling to the right, but with an electron current to the left. Before the electrons knocked out of the filled negative-energy sea for $z > z_b$ can reach to region I to join the reflected electrons from the incident beam, however, they must tunnel through a barrier, between $z = z_a$ and $z = z_b$. This tunneling related to the exponential falloff in the wave function in this region will reduce the reflection coefficient from the large value exhibited above.

Problems

45. If the 16 Γ^A are defined as $\mathbf{1}, \gamma^\mu, i\gamma^\mu\gamma^\nu, i\gamma^5\gamma^\mu, \gamma^5$ (note the extra factor i in the definition of $i\gamma^5\gamma^\mu$), and if lower index γ's are given by $\gamma_0 = \gamma^0, \gamma_i = -\gamma^i$ for $i = 1, 2, 3$, but $\gamma_5 = \gamma^5$, show

$$\Gamma^A\Gamma_A = +\mathbf{1}$$

for all 16 cases, and

$$\text{trace}(\Gamma^A\Gamma_B) = 4\delta^A_B.$$

Use these two properties to show a general 4×4 matrix Γ can be expanded by

$$\Gamma = \sum_{A=1}^{16} c_A\Gamma^A = \tfrac{1}{4}\sum_{A=1}^{16}\text{trace}(\Gamma\Gamma_A)\Gamma^A,$$

and show a specific matrix element, Γ_{ik}, can be written

$$\Gamma_{ik} = \tfrac{1}{4}\sum_{A=1}^{16}\sum_{l=1}^{4}\sum_{m=1}^{4}\Gamma_{lm}(\Gamma^A)_{ml}(\Gamma_A)_{ik},$$

with i, k fixed, $i, k = 1, \ldots, 4$. (These Latin indices are here used to specify particular matrix elements of the 4×4 matrices.) Use this property to show

$$(\bar{\psi}(a)\mathbf{1}\psi(b))(\bar{\psi}(c)\mathbf{1}\psi(d)) = \tfrac{1}{4}\sum_{A=1}^{16}(\bar{\psi}(a)\Gamma^A\psi(d))(\bar{\psi}(c)\Gamma_A\psi(b)),$$

where $\psi(a)$, $\psi(b)$, $\psi(c)$, $\psi(d)$ are four different four-component ψ's. Also, show

$$4\sum_{\mu=0}^{3}\Big(\bar{\psi}(a)\gamma^{\mu}\psi(b)\Big)\Big(\bar{\psi}(c)\gamma_{\mu}\psi(d)\Big)$$

$$= 4\Big(\bar{\psi}(a)\mathbf{1}\psi(d)\Big)\Big(\bar{\psi}(c)\mathbf{1}\psi(b)\Big) - 4\Big(\bar{\psi}(a)\gamma^5\psi(d)\Big)\Big(\bar{\psi}(c)\gamma_5\psi(b)\Big)$$

$$+2\sum_{\mu=0}^{3}\Big(\bar{\psi}(a)i\gamma^5\gamma^{\mu}\psi(d)\Big)\Big(\bar{\psi}(c)i\gamma_5\gamma_{\mu}\psi(b)\Big)$$

$$-2\sum_{\mu=0}^{3}\Big(\bar{\psi}(a)\gamma^{\mu}\psi(d)\Big)\Big(\bar{\psi}(c)\gamma_{\mu}\psi(b)\Big).$$

To do the calculations, let $\Gamma_{ik} = (\psi(b))_i(\bar{\psi}(c))_k$ in the above.

46. Show the operator

$$\Lambda_+(p) = \frac{\gamma^{\mu}p_{\mu} + mc}{2mc}$$

is a projection operator for positive-energy states with the properties:

$$\Big(\Lambda_+(p)\Big)^2 = \Lambda_+(p),$$

$$\Lambda_+(p)u^{(r)}(\vec{p}) = u^{(r)}(\vec{p}) \qquad \text{for } r = 1, 2, .$$

$$\Lambda_+(p)u^{(r)}(\vec{p}) = 0 \qquad \text{for } r = 3, 4.$$

47. For the Dirac Hamiltonian of a particle in an outside electromagnetic field, with

$$H = \beta mc^2 + c(\vec{\alpha}\cdot\vec{\Pi}) + e\Phi,$$

which can be ordered as

$$H^{(0)} + H^{(1)} + H^{(2)} = H^{(0)}\left(1 + \text{Order}\frac{p}{mc} + \text{Order}\left(\frac{p}{mc}\right)^2\right)$$

in the nonrelativistic limit, $p/mc \equiv \lambda \ll 1$; show that the transformed Hamiltonian, through fourth order, obtained via the transformation

$$H' = UHU^{\dagger} = e^{i\lambda G}He^{-i\lambda G},$$

with

$$i\lambda G = \frac{\beta(\vec{\alpha} \cdot \vec{\Pi})}{2mc}$$

leads to the Pauli approximation terms of the Dirac equation.

(Surviving third-order terms that only make connections of the type $\psi_A \leftrightarrow \psi_B$ can be transformed into fifth and higher order terms, diagonal in ψ_A, via a further transformation, but therefore they do not contribute through fourth order.)

Also recall, with $\lambda \ll 1$,

$$e^{i\lambda G} H^{(n)} e^{-i\lambda G}$$

$$= H^{(n)} + i\lambda[G, H^{(n)}] + \cdots + \frac{(i\lambda)^k}{k!}[G, [G, [G, \cdots [G, H^{(n)}] \cdots]]] + \cdots,$$

where the kth term involves k commutators of G with $H^{(n)}$.

48. A relativistic electron beam polarized such that the longitudinal polarization is $P_z = +1$, where the z axis is chosen parallel to the initial momentum vector, $\vec{p}_0$, is scattered by a point nucleus of charge Ze (Mott scattering) into an angle θ, with $\vec{p}_{\text{scatt.}}$ parallel to a z' axis, lying in the x–z plane. If the detector is set so it accepts only longitudinally polarized electrons, with $P_{z'} = +1$, i.e., electrons with spin parallel to $\vec{p}_{\text{scatt.}}$, find the differential cross section as a function of θ for this experimental setup. It may be useful to first find the normalized plane wave eigenstates of $\Sigma_{z'} = (\cos\theta\, \Sigma_z + \sin\theta\, \Sigma_x)$, or to make an active rotation about the y axis through an angle θ of the plane wave state, $u^{(1)}(\vec{p})$.

49. An unpolarized relativistic electron beam is scattered from a finite-sized nucleus of charge Ze. Assume the nucleus can be described by the liquid drop model with a uniform charge distribution spread over a sphere of radius a, with

$$V(r) = -\frac{Ze^2}{r} \quad \text{for } r \geq a, \qquad V(r) = -\frac{Ze^2}{a}\left(\frac{3}{2} - \frac{r^2}{2a^2}\right) \quad \text{for } r \leq a.$$

Find the differential cross section as a function of θ and $\beta = v/c$ of the incident electron beam.

50. The Dirac equation with Hamiltonian

$$H = \gamma^0 mc^2 + c(\gamma^0 \vec{\gamma} \cdot \vec{p} + im\omega \vec{\gamma} \cdot \vec{r})$$

has been named the "Dirac equation for the harmonic oscillator."

(a) Justify this name by using the two-component ψ_A and ψ_B defined by

$$\psi = \begin{pmatrix} \psi_A \\ \psi_B \end{pmatrix} e^{-\frac{i}{\hbar}mc^2 t}$$

and by finding the wave equation for ψ_A in the nonrelativistic limit in which $\frac{\hbar}{i}\frac{\partial \psi_B}{\partial t}$ can be neglected compared with $2mc^2\psi_B$.

(b) Show the three components of $\vec{J}$

$$\vec{J} = \vec{L} + \tfrac{1}{2}\hbar\Sigma$$

commute with this Hamiltonian.

(c) Find the exact relativistic-energy eigenvalues for this Hamiltonian.

(d) Why is the Dirac equation

$$-\frac{\hbar}{i}\frac{\partial\psi}{\partial t} = \left(\gamma^0 mc^2 + c\gamma^0\vec{\gamma}\cdot\vec{p} + V(r)\right) \qquad \text{with } V(r) = \tfrac{1}{2}m\omega^2 r^2$$

perhaps not suitable as a relativistic harmonic oscillator equation?

51. A nucleus with a ground state of angular momentum and parity, 0^+, with nuclear wave function, $\psi_{\text{g.st.}}$, has a first excited state also with angular momentum and parity, 0^+, and nuclear wave function, $\psi_{\text{exc.}}$. Show the excited state cannot decay to the ground state via the emission of a γ photon. If the excited state energy is greater than $2mc^2$ ($E^{\text{nucl.}}_{\text{exc.}} - E^{\text{nucl.}}_{\text{g.st.}} > 2mc^2 = 1.022$ MeV), the nuclear deexcitation can occur via the emission of an electron–positron pair.

Calculate the probability per second for e^+, e^- emision in this case. In particular, calculate the probability per second the positron is emitted with energy between E_+ and $E_+ + dE_+$, with no measurements made on the e^- energy. Assume the nuclear mass is big enough that the nuclear recoil energy $P^2/2M_{\text{nucleus}}$ is negligible, so

$$E^{\text{nucl.}}_{\text{exc.}} - E^{\text{nucl.}}_{\text{g.st.}} = E_+ + E_- + P^2/2M_{\text{nucleus}} = E_+ + E_-,$$

where $\vec{P} = -(\vec{p}_+ + \vec{p}_-)$. Assume, however, $P = |\vec{p}_+ + \vec{p}_-|$ is such that $(Pr_{\text{nucl.}}/\hbar) \ll 1$, where $r_{\text{nucl.}}$ is a typical nuclear distance, of order 1–6 fm. Show in this limit the above probability depends on the nuclear states only through the square of the following nuclear matrix element

$$e\int\cdots\int d\vec{r}_1\cdots d\vec{r}_A \psi^{\text{nucl.}*}_{\text{g.st.}}\sum_{j=1}^{Z} r_j^2 \psi^{\text{nucl.}}_{\text{exc.}} = e < r^2 >_{fi},$$

where the sum is over the Z protons of the nucleus; r_j is the distance of the jth proton from the center of mass of the nucleus. The interaction for the transition is the Coulomb interaction between the protons in the nucleus and the electron that makes a transition from the filled negative-energy state with $E = -|E| = -E_+$, and $\vec{p}_i = -\vec{p}_+$, to a positive-energy state with $E = +E_-$, and $\vec{p}_f = \vec{p}_-$. That is,

$$H_{\text{int.}} = -\sum_{j=1}^{Z}\frac{e^2}{|\vec{r} - \vec{r}_j|},$$

where $\vec{r}$ without subscript refers to the position of the electron that makes the transition from the filled negative-energy state to the open positive-energy state. In doing some of the radial integrals, you may have to regularize Coulomb divergent integrals with a screening factor, b, e.g.,

$$\int_{r_j}^{\infty} dr \sin\left(\frac{Pr}{\hbar}\right) \to \lim_{b\to 0}\int_{r_j}^{\infty} dr \sin\left(\frac{Pr}{\hbar}\right)e^{-br} = \frac{\hbar}{P}\cos\left(\frac{Pr_j}{\hbar}\right).$$

Assume ($E^{\text{nucl.}}_{\text{exc.}} - E^{\text{nucl.}}_{\text{g.st.}}$) is large enough or Z is small enough so the electron and positron wave functions can be approximated by Dirac plane waves. Calculate first the probability the electron is emitted into a $d\Omega_-$ about the direction θ_-, ϕ_-, and the positron is emitted into a $d\Omega_+$ about the direction θ_+, ϕ_+ and integrate over all possible directions.

The nucleus ^{16}O has a first excited state at 6.05 MeV that is a 0^+ state. Make an educated guess for the order of magnitude of the nuclear matrix element, $< r^2 >_{if}$, and thus an order of magnitude estimate of the meanlife of this excited state.

76

Exact Solutions for the Dirac Equation for Spherically Symmetric Potentials

Having discussed the plane-wave solutions to the Dirac equation in some detail, let us now look at solutions to the Dirac equation with a spherically symmetric $V(r)$. In particular, two cases are of special interest as follows.

Case 1.

$$V(r) = -\frac{Ze^2}{r}.$$

The hydrogenic atom is of particular interest, especially for inner atomic shells in a high Z atom, where relativistic effects may become important (e.g., in K-shell or L-shell X-ray spectra).

Case 2.

$$\text{The quark model with } V(r) = 0 \text{ for } r < a,$$

and boundary conditions at $r = a$ such that the quarks are confined to the spherical region with $r < a$. (Note: Our discussion of the Klein paradox indicates this confinement might not be achievable with a potential that is ∞ for $r > a$.) This quark model is known as the MIT Bag model.

Let us first look at the general case of the Dirac equation with a spherically symmetric $V(r)$, i.e., the Dirac equation with Hamiltonian

$$H = c(\vec{\alpha} \cdot \vec{p}) + \beta mc^2 + V(r)\mathbf{1}. \tag{1}$$

We have already proved the three components of $\vec{J}$, with

$$\vec{J} = \vec{L} + \frac{\hbar}{2}\vec{\Sigma}, \qquad \text{with } \Sigma_j = \frac{i}{2}\epsilon_{jkl}\gamma^k\gamma^l, \tag{2}$$

commutes with this H. Thus, we can diagonalize the operators, H, $\vec{J}^2$, and J_z simultaneously, and the eigenvalues of H will have good j, and m_j. (We use the standard convention that lowercase letters are used for the eigenvalues of single-electron atoms.) The question arises: Are there other operators that commute with both H and $\vec{J}$? The answer is yes. The operator, denoted by K, which is defined by

$$K = \beta(\vec{\Sigma} \cdot \vec{J} - \frac{\hbar}{2}), \tag{3}$$

commutes with both H and $\vec{J}$. To prove this, only the $c(\vec{\alpha} \cdot \vec{p}) = c\gamma^0(\vec{\gamma} \cdot \vec{p})$ term of the above H needs to be considered, so we need to calculate

$$\begin{aligned}[H, K] &= [c\gamma^0(\vec{\gamma} \cdot \vec{p}), \gamma^0(\vec{\Sigma} \cdot \vec{J} - \frac{\hbar}{2})] \\ &= c[\gamma^0(\vec{\gamma} \cdot \vec{p}), \gamma^0](\vec{\Sigma} \cdot \vec{J} - \frac{\hbar}{2}) + c\gamma^0[\gamma^0(\vec{\gamma} \cdot \vec{p}), (\vec{\Sigma} \cdot \vec{J})] \\ &= -2c(\vec{\gamma} \cdot \vec{p})\left(\vec{\Sigma} \cdot \vec{J} - \frac{\hbar}{2}\right) + c\gamma^0[\gamma^0(\vec{\gamma} \cdot \vec{p}), \Sigma_j]J^j,\end{aligned} \tag{4}$$

where we have used the fact that $\gamma^0(\vec{\gamma} \cdot \vec{p})$ commutes with all components of $\vec{J}$. If we further use the fact that γ^0 commutes with Σ_j, the second term can be evaluated through

$$\begin{aligned}& c\gamma^0[\gamma^0(\vec{\gamma} \cdot \vec{p}), \Sigma_j]J^j = c(\gamma^0)^2[(\vec{\gamma} \cdot \vec{p}), \Sigma_j]J^j \\ &= \frac{ic}{2}\epsilon_{jmk}[\gamma^l, \gamma^m\gamma^k]p_l J^j = \frac{ic}{2}\epsilon_{jmk}(2g^{lm}\gamma^k - 2g^{lk}\gamma^m)p_l J^j \\ &= ic\epsilon_{jmk}(-\gamma^k p^m J^j + \gamma^m p^k J^j) \\ &= 2ic\epsilon_{mkj}\gamma^m p^k J^j = 2ic\vec{\gamma} \cdot [\vec{p} \times \vec{J}] \\ &= 2ic \begin{pmatrix} 0 & \vec{\sigma} \cdot [\vec{p} \times \vec{J}] \\ -\vec{\sigma} \cdot [\vec{p} \times \vec{J}] & 0 \end{pmatrix}.\end{aligned} \tag{5}$$

The first term of eq. (4), except for the factor $-2c$, can be evaluated by

$$\begin{aligned}& (\vec{\gamma} \cdot \vec{p})\left(\vec{\Sigma} \cdot \vec{J} - \frac{\hbar}{2}\right) = \\ & \begin{pmatrix} 0 & \vec{\sigma} \cdot \vec{p} \\ -\vec{\sigma} \cdot \vec{p} & 0 \end{pmatrix} \begin{pmatrix} (\vec{\sigma} \cdot \vec{J} - \frac{\hbar}{2}) & 0 \\ 0 & (\vec{\sigma} \cdot \vec{J} - \frac{\hbar}{2}) \end{pmatrix} = \\ & \begin{pmatrix} 0 & (\vec{\sigma} \cdot \vec{p})(\vec{\sigma} \cdot \vec{J} - \frac{\hbar}{2}) \\ -(\vec{\sigma} \cdot \vec{p})(\vec{\sigma} \cdot \vec{J} - \frac{\hbar}{2}) & 0 \end{pmatrix} = \\ & \begin{pmatrix} 0 & (\vec{p} \cdot \vec{J}) + i\vec{\sigma} \cdot [\vec{p} \times \vec{J}] - \frac{\hbar}{2}(\vec{p} \cdot \vec{\sigma}) \\ -(\vec{p} \cdot \vec{J}) - i\vec{\sigma} \cdot [\vec{p} \times \vec{J}] + \frac{\hbar}{2}(\vec{p} \cdot \vec{\sigma}) & 0 \end{pmatrix}\end{aligned} \tag{6}$$

Now, combining the results of eqs. (5) and (6), we get

$$[H, K] = -2c\gamma^0\gamma^5\left((\vec{p} \cdot \vec{J}) - \frac{\hbar}{2}(\vec{p} \cdot \vec{\Sigma})\right) = -2c\gamma^0\gamma^5(\vec{p} \cdot \vec{L}) = 0, \tag{7}$$

because $\vec{p} \cdot [\vec{r} \times \vec{p}] \equiv 0$.

We also see the operator, K, commutes with all components of $\vec{J}$:

$$\begin{aligned}[K, J^i] &= [\gamma^0(\vec{\Sigma}\cdot\vec{J} - \frac{\hbar}{2}), J^i] = \gamma^0\Big(\Sigma_j[J^j, J^i] + [\Sigma^j, J^i]J_j\Big)\\ &= \gamma^0\Big(i\hbar\Sigma_j\epsilon^{jik}J_k + [\Sigma^j, \frac{\hbar}{2}\Sigma^i]J_j\Big) = \gamma^0\Big(i\hbar\epsilon^{jik}\Sigma_j J_k + i\hbar\epsilon^{jik}\Sigma_k J_j\Big)\\ &= i\hbar\gamma^0(\epsilon^{jik} + \epsilon^{kij})\Sigma_j J_k = i\hbar\gamma^0(\epsilon^{jik} - \epsilon^{jik})\Sigma_j J_k = 0. \end{aligned} \tag{8}$$

Now, to evaluate the eigenvalue of the operator, K,

$$\begin{aligned}K &= \beta(\vec{\Sigma}\cdot\vec{J} - \frac{\hbar}{2}) = \beta\Big(\vec{\Sigma}\cdot\vec{L} + \frac{\hbar}{2}(\vec{\Sigma}\cdot\vec{\Sigma} - 1)\Big)\\ &= = \beta\Big(\vec{\Sigma}\cdot\vec{L} + \frac{\hbar}{2}(3-1)\Big) = \beta\Big(\vec{\Sigma}\cdot\vec{L} + \hbar\Big). \end{aligned} \tag{9}$$

Therefore,

$$\begin{aligned}K^2 &= \beta\Big(\vec{\Sigma}\cdot\vec{L} + \hbar\Big)\beta\Big(\vec{\Sigma}\cdot\vec{L} + \hbar\Big) = (\vec{\Sigma}\cdot\vec{L} + \hbar)(\vec{\Sigma}\cdot\vec{L} + \hbar)\\ &= (\vec{\Sigma}\cdot\vec{L})(\vec{\Sigma}\cdot\vec{L}) + 2\hbar(\vec{\Sigma}\cdot\vec{L}) + \hbar^2\\ &= \vec{L}\cdot\vec{L} + i\vec{\Sigma}\cdot(i\hbar\vec{L}) + 2\hbar(\vec{\Sigma}\cdot\vec{L}) + \hbar^2\\ &= \vec{L}\cdot\vec{L} + \hbar(\vec{\Sigma}\cdot\vec{L}) + \tfrac{3}{4}\hbar^2 + \tfrac{1}{4}\hbar^2\\ &= \big(\vec{L} + \tfrac{1}{2}\hbar\vec{\Sigma}\big)^2 + \tfrac{1}{4}\hbar^2 = \hbar^2\big(j(j+1) + \tfrac{1}{4}\big). \end{aligned} \tag{10}$$

Thus, K^2 has the eigenvalue

$$K^2_{\text{eigen}} = \hbar^2(j + \tfrac{1}{2})^2. \tag{11}$$

Essentially, only the sign of K gives us a new quantum number, because the magnitude of K is given by the total angular-momentum quantum number, j. We shall introduce the new quantum number, κ, defined by

$$K = -\hbar\kappa, \qquad \text{so } \kappa = \pm(j + \tfrac{1}{2}). \tag{12}$$

The minus sign in this defining relation is included to agree with the conventional definition of this new quantum number (see, e.g., J. J. Sakurai, *Advanced Quantum Mechanics*, Reading, MA: Addison-Wesley, 1967). Now let us find solutions to the Dirac equation for a spherically symmetric $V(r)$, with eigenfunctions of good j, m_j, and κ. Again, introduce two component functions, ψ_A and ψ_B, with

$$\psi = \begin{pmatrix}\psi_A\\ \psi_B\end{pmatrix} e^{-\frac{i}{\hbar}Et}, \tag{13}$$

so

$$\begin{aligned}(E - mc^2 - V(r))\psi_A &= c(\vec{\sigma}\cdot\vec{p})\psi_B,\\ (E + mc^2 - V(r))\psi_B &= c(\vec{\sigma}\cdot\vec{p})\psi_A. \end{aligned} \tag{14}$$

Now, introduce vector-coupled spherical harmonics of the type introduced in conjunction with scattering theory of spin-$\frac{1}{2}$ nonrelativistic particles.

$$\mathcal{Y}_{[l\frac{1}{2}]jm_j} \equiv \sum_{m_s,(m_l)} \langle lm_l \tfrac{1}{2}m_s | jm_j\rangle i^l Y_{lm_l}(\theta,\phi)\chi_{\frac{1}{2}m_s}$$

$$= \begin{pmatrix} \langle l(m_j - \frac{1}{2})\frac{1}{2} + \frac{1}{2}|jm_j\rangle i^l Y_{l(m_j-\frac{1}{2})}(\theta,\phi) \\ \langle l(m_j + \frac{1}{2})\frac{1}{2} - \frac{1}{2}|jm_j\rangle i^l Y_{l(m_j+\frac{1}{2})}(\theta,\phi) \end{pmatrix}. \tag{15}$$

We now let

$$\psi_A = g(r)\mathcal{Y}_{[l_A\frac{1}{2}]jm_j},$$
$$\psi_B = f(r)\mathcal{Y}_{[l_B\frac{1}{2}]jm_j}, \tag{16}$$

and using

$$(\vec{\sigma}\cdot\frac{\vec{r}}{r})(\vec{\sigma}\cdot\frac{\vec{r}}{r}) = 1, \tag{17}$$

we are led to the simple identity

$$(\vec{\sigma}\cdot\vec{p}) = \frac{(\vec{\sigma}\cdot\vec{r})}{r^2}\Big((\vec{\sigma}\cdot\vec{r})(\vec{\sigma}\cdot\vec{p})\Big) = \frac{(\vec{\sigma}\cdot\vec{r})}{r^2}\Big((\vec{r}\cdot\vec{p}) + i\vec{\sigma}\cdot[\vec{r}\times\vec{p}]\Big)$$
$$= \frac{(\vec{\sigma}\cdot\vec{r})}{r^2}\Big(-i\hbar r\frac{\partial}{\partial r} + i(\vec{\sigma}\cdot\vec{L})\Big). \tag{18}$$

Therefore, we can replace $(\vec{\sigma}\cdot\vec{p})$ by

$$(\vec{\sigma}\cdot\vec{p}) = (\vec{\sigma}\cdot\frac{\vec{r}}{r})\frac{1}{i}\Big(\hbar\Big(\frac{\partial}{\partial r} + \frac{1}{r}\Big) - \frac{1}{r}((\vec{\sigma}\cdot\vec{L}) + \hbar)\Big). \tag{19}$$

Finally, we shall need the relation

$$(\vec{\sigma}\cdot\frac{\vec{r}}{r})\mathcal{Y}_{[l\frac{1}{2}]j=(l+\frac{1}{2})m_j} = c_+\mathcal{Y}_{[(l+1)\frac{1}{2}](l+\frac{1}{2})m_j},$$
$$(\vec{\sigma}\cdot\frac{\vec{r}}{r})\mathcal{Y}_{[l\frac{1}{2}]j=(l-\frac{1}{2})m_j} = c_-\mathcal{Y}_{[(l-1)\frac{1}{2}](l-\frac{1}{2})m_j},$$
$$\text{with } c_+ = +i, \quad c_- = -i, \tag{20}$$

where eq. (17) leads to $c_-c_+ = c_+c_- = 1$, and c_+ can be evaluated most easily by specific calculation, using

$$\Big(\vec{\sigma}\cdot\frac{\vec{r}}{r}\Big) = \sum_q (-1)^q \sigma_{-q} Y_{1,+q}(\theta,\phi)\sqrt{\frac{4\pi}{3}}. \tag{21}$$

This equation is a scalar operator whose matrix elements are therefore diagonal in j and independent of m_j. Thus, we can choose m_j judiciously, as always in applying the Wigner–Eckart theorem.

$$\Big(\vec{\sigma}\cdot\frac{\vec{r}}{r}\Big)\mathcal{Y}_{[l\frac{1}{2}](l+\frac{1}{2}),m_j=(l+\frac{1}{2})} \doteq i^l\sqrt{\frac{4\pi}{3}}\Big[Y_{10}Y_{ll}\chi_{m_s=+\frac{1}{2}} - Y_{1+1}Y_{ll}\frac{2}{\sqrt{2}}\chi_{m_s=-\frac{1}{2}}\Big]$$
$$= i^l\sqrt{\frac{(2l+1)}{(2l+3)}}\langle l010|l+10\rangle\Big[\langle ll10|l+1l\rangle Y_{(l+1),l}\chi_{+\frac{1}{2}}$$
$$- \sqrt{\tfrac{1}{2}}\langle ll11|l+1l+1\rangle Y_{l+1,l+1}\chi_{-\frac{1}{2}}\Big]$$

$$= i^l\left[\frac{1}{\sqrt{(2l+3)}}Y_{l+1,l}\chi_{+\frac{1}{2}} - \sqrt{\frac{2(l+1)}{(2l+3)}}Y_{l+1,l+1}\chi_{-\frac{1}{2}}\right]$$
$$= -i^l\Big[\langle l+1 l \tfrac{1}{2}\tfrac{1}{2}|l+\tfrac{1}{2} l+\tfrac{1}{2}\rangle Y_{l+1,l}\chi_{+\frac{1}{2}}$$
$$+ \langle l+1 l+1 \tfrac{1}{2} -\tfrac{1}{2}|l+\tfrac{1}{2} l+\tfrac{1}{2}\rangle Y_{l+1,l+1}\chi_{-\frac{1}{2}}\Big]$$
$$= +i\mathcal{Y}_{[(l+1)\frac{1}{2}](l+\frac{1}{2})(l+\frac{1}{2})}. \tag{22}$$

Thus, we see specifically $c_+ = +i$, and therefore $c_- = -i$.

It will also prove convenient to choose the radial functions, $g(r)$ and $f(r)$, such that

$$\psi = \begin{pmatrix}\psi_A\\ \psi_B\end{pmatrix} = \begin{pmatrix} g(r)\mathcal{Y}_{[l\frac{1}{2}](l+\frac{1}{2})m_j}\\ f(r)\mathcal{Y}_{[(l+1)\frac{1}{2}](l+\frac{1}{2})m_j}\end{pmatrix} \quad \text{for } j=(l_A+\tfrac{1}{2}), \tag{23}$$

and

$$\psi = \begin{pmatrix}\psi_A\\ \psi_B\end{pmatrix} = \begin{pmatrix} g(r)\mathcal{Y}_{[l\frac{1}{2}](l-\frac{1}{2})m_j}\\ -f(r)\mathcal{Y}_{[(l-1)\frac{1}{2}](l-\frac{1}{2})m_j}\end{pmatrix} \quad \text{for } j=(l_A-\tfrac{1}{2}), \tag{24}$$

where these two states have different eigenvalues κ. To see this, use eq. (9) to express K as $K = \hbar\beta(\vec{\Sigma}\cdot\vec{l}+1)$, where $\vec{l}$ is the dimensionless orbital angular-momentum operator, $\vec{L}=\hbar\vec{l}$, and use the eigenvalue $(\vec{\sigma}\cdot\vec{l}) = [j(j+1)-l(l+1)-\frac{3}{4}]$. With this, we get

$$\frac{K}{\hbar}\begin{pmatrix} g(r)\mathcal{Y}_{[l\frac{1}{2}](l+\frac{1}{2})m_j}\\ f(r)\mathcal{Y}_{[(l+1)\frac{1}{2}](l+\frac{1}{2})m_j}\end{pmatrix} = +(l+1)\begin{pmatrix} g(r)\mathcal{Y}_{[l\frac{1}{2}](l+\frac{1}{2})m_j}\\ f(r)\mathcal{Y}_{[(l+1)\frac{1}{2}](l+\frac{1}{2})m_j}\end{pmatrix},$$
$$\frac{K}{\hbar}\begin{pmatrix} g(r)\mathcal{Y}_{[l\frac{1}{2}](l-\frac{1}{2})m_j}\\ -f(r)\mathcal{Y}_{[(l-1)\frac{1}{2}](l-\frac{1}{2})m_j}\end{pmatrix} = -l\begin{pmatrix} g(r)\mathcal{Y}_{[l\frac{1}{2}](l-\frac{1}{2})m_j}\\ -f(r)\mathcal{Y}_{[(l-1)\frac{1}{2}](l-\frac{1}{2})m_j}\end{pmatrix}. \tag{25}$$

The eigenvalue κ is negative for $j=(l_A+\frac{1}{2})$; i.e., $\kappa = -(j+\frac{1}{2})$ in this case, whereas κ is positive for $j=(l_A-\frac{1}{2})$; i.e., $\kappa = +(j+\frac{1}{2})$ in this case.

Substituting eqs. (23) and (24) into eqs. (14), and using the results of eqs. (19) and (20), as well as the orthonormality of the $\mathcal{Y}_{[l\frac{1}{2}]jm_j}$, we get the two coupled equations

$$(E-mc^2-V(r))g(r) = -\hbar c\left(\frac{\partial}{\partial r}+\frac{(1-\kappa)}{r}\right)f(r),$$
$$(E+mc^2-V(r))f(r) = +\hbar c\left(\frac{\partial}{\partial r}+\frac{(1+\kappa)}{r}\right)g(r), \tag{26}$$

valid for both eigenvalues κ. It will also be convenient to one-dimensionalize these radial equations by introducing, $G(r)$ and $F(r)$, via

$$rg(r) = G(r), \qquad rf(r) = F(r), \tag{27}$$

to lead to the radial equations

$$(E-mc^2-V(r))G(r) = -\hbar c\left(\frac{dF}{dr}-\frac{\kappa}{r}F\right),$$
$$(E+mc^2-V(r))F(r) = +\hbar c\left(\frac{dG}{dr}+\frac{\kappa}{r}G\right). \tag{28}$$

A The Relativistic Hydrogen Atom

Let us first solve the radial equations for a hydrogenic atom, with

$$V(r) = -\frac{Ze^2}{r}. \tag{29}$$

Introduce relativistic coordinates, with energies measured in units of mc^2 and distances measured in units of the Compton wavelength, $\hbar/mc$, with dimensionless energies, ϵ, and dimensionless radial coordinate, $\bar{r}$, such that

$$E = mc^2\epsilon, \qquad r = \frac{\hbar}{mc}\bar{r}. \tag{30}$$

This process transforms eq. (28) to

$$\begin{aligned}\Big(\epsilon - 1 + \frac{Z\alpha}{\bar{r}}\Big)G &= -\Big(\frac{dF}{d\bar{r}} - \frac{\kappa}{\bar{r}}F\Big),\\ \Big(\epsilon + 1 + \frac{Z\alpha}{\bar{r}}\Big)F &= +\Big(\frac{dG}{d\bar{r}} + \frac{\kappa}{\bar{r}}G\Big),\end{aligned} \tag{31}$$

where $\alpha = e^2/\hbar c$ is the fine structure constant. It will be even more convenient to convert to the dimensionless variable, ρ, defined by

$$\rho = \sqrt{(1-\epsilon^2)}\bar{r} = \sqrt{(1-\epsilon^2)}\frac{mc}{\hbar}r. \tag{32}$$

In terms of this variable, the radial equations are

$$\begin{aligned}\Big(-\sqrt{\frac{1-\epsilon}{1+\epsilon}} + \frac{Z\alpha}{\rho}\Big)G &= -\frac{dF}{d\rho} + \frac{\kappa}{\rho}F,\\ \Big(+\sqrt{\frac{1+\epsilon}{1-\epsilon}} + \frac{Z\alpha}{\rho}\Big)F &= +\frac{dG}{d\rho} + \frac{\kappa}{\rho}G.\end{aligned} \tag{33}$$

In particular, $\epsilon < 1$ for the bound positive-energy states with $E < mc^2$.

Of the many possible methods of solution, let us use the Fuchsian differential equation method. First, look at the form of the asymptotic solutions. As $\rho \to \infty$, the differential equations can be approximated by

$$-\sqrt{\frac{1-\epsilon}{1+\epsilon}}G = -\frac{dF}{d\rho}, \qquad \sqrt{\frac{1+\epsilon}{1-\epsilon}}F = \frac{dG}{d\rho}, \tag{34}$$

leading to

$$\frac{d^2F}{d\rho^2} - F = 0, \qquad \frac{d^2G}{d\rho^2} - G = 0, \quad \text{as } \rho \to \infty. \tag{35}$$

This equation has the solutions $e^{\pm\rho}$ for both F and G, but only the solution, $e^{-\rho}$, can lead to square-integrable wave functions. Next, as $\rho \to 0$, the equations collapse to

$$\frac{Z\alpha}{\rho}G = -\frac{dF}{d\rho} + \frac{\kappa}{\rho}F,$$

$$\frac{Z\alpha}{\rho}F = +\frac{dG}{d\rho} + \frac{\kappa}{\rho}G. \tag{36}$$

If we try solutions of the form

$$F = A\rho^s, \qquad G = B\rho^s, \qquad \text{as } \rho \to 0,$$

we are led to the linear equations

$$\begin{aligned} (s-\kappa)A + Z\alpha B &= 0, \\ -Z\alpha A + (s+\kappa)B &= 0, \end{aligned} \tag{37}$$

with solutions, other than $A = B = 0$, only for $s^2 - \kappa^2 + (Z\alpha)^2 = 0$, so $s = \pm\sqrt{\kappa^2 - (Z\alpha)^2}$. To have solutions regular at $\rho = 0$, and thus square-integrable, we can only retain the positive root. Thus,

$$s = +\sqrt{\kappa^2 - (Z\alpha)^2} = +\sqrt{(j+\tfrac{1}{2})^2 - (Z\alpha)^2}. \tag{38}$$

We shall therefore try solutions to the full equations in the form

$$\begin{aligned} F &= \rho^s e^{-\rho}\sum_{m=0}^{\infty} a_m \rho^m, \\ G &= \rho^s e^{-\rho}\sum_{m=0}^{\infty} b_m \rho^m. \end{aligned} \tag{39}$$

Substituting back into eq. (33) and shifting indices appropriately, this process leads to recursive equations

$$\begin{aligned} -\sqrt{\frac{1-\epsilon}{1+\epsilon}}\,b_{m-1} + (Z\alpha)b_m &= [\kappa - (m+s)]a_m + a_{m-1}, \\ +\sqrt{\frac{1+\epsilon}{1-\epsilon}}\,a_{m-1} + (Z\alpha)a_m &= [\kappa + (m+s)]b_m - b_{m-1}. \end{aligned} \tag{40}$$

In the limit in which $m \to \infty$, with $\kappa \pm s \ll m$, so $\kappa \pm s$ can be neglected, we have

$$b_m \to \frac{b_{m-1}}{m}c_B, \qquad a_m \to \frac{a_{m-1}}{m}c_A, \tag{41}$$

where we will choose $c_B = c_A = c$, so the functions $F(\rho)$ and $G(\rho)$ have the same asymptotic behavior as $\rho \to \infty$. With this large m limit, we then have

$$\begin{aligned} -\sqrt{\frac{1-\epsilon}{1+\epsilon}}\,b_{m-1} + \text{Order}\left(\frac{1}{m}\right) &= a_{m-1}(1-c), \\ +\sqrt{\frac{1+\epsilon}{1-\epsilon}}\,a_{m-1} + \text{Order}\left(\frac{1}{m}\right) &= -b_{m-1}(1-c), \end{aligned} \tag{42}$$

or

$$a_{m-1}(1-c) + \sqrt{\frac{1-\epsilon}{1+\epsilon}}\,b_{m-1} = 0,$$

$$a_{m-1}\sqrt{\frac{1+\epsilon}{1-\epsilon}} + b_{m-1}(1-c) = 0, \tag{43}$$

leading to a solution only if $(c^2 - 2c) = 0$. Thus, the only nonzero solution gives $c = 2$, but with this solution the infinite series implied by eq. (41) would lead to

$$\sum_m a_m \rho^m \to e^{+2\rho}, \qquad \sum_m b_m \rho^m \to e^{+2\rho}.$$

The infinite series would therefore overpower the $e^{-\rho}$ term of eq. (39) and give the unacceptable asymptotic behavior $e^{+\rho}$ that must be ruled out. Thus, the series solutions of eq. (39) can lead to square-integrable wave functions only if the series terminate, say, at the n'^{th} terms, so both $a_{n'+1} = 0$ and $b_{n'+1} = 0$. With this assumption, eq. (40), with $m = n' + 1$, leads to

$$-\sqrt{\frac{1-\epsilon}{1+\epsilon}}b_{n'} = a_{n'},$$
$$+\sqrt{\frac{1+\epsilon}{1-\epsilon}}a_{n'} = -b_{n'}, \tag{44}$$

leading to

$$\frac{a_{n'}}{b_{n'}} = -\sqrt{\frac{1-\epsilon}{1+\epsilon}}. \tag{45}$$

Using this ratio, and choosing $m = n'$ in eq. (40),we get

$$-\sqrt{\frac{1-\epsilon}{1+\epsilon}}b_{n'-1} + (Z\alpha)b_{n'} = -(\kappa - n' - s)\sqrt{\frac{1-\epsilon}{1+\epsilon}}b_{n'} + a_{n'-1}, \tag{46}$$

$$+\sqrt{\frac{1+\epsilon}{1-\epsilon}}a_{n'-1} - (Z\alpha)\sqrt{\frac{1-\epsilon}{1+\epsilon}}b_{n'} = (\kappa + n' + s)b_{n'} - b_{n'-1}. \tag{47}$$

Solving eq. (46) for $a_{n'-1}$ and substituting into eq. (47), we get an equation independent of $b_{n'-1}$:

$$\left(\frac{2\epsilon}{\sqrt{1-\epsilon^2}}(Z\alpha) - 2(n'+s)\right)b_{n'} = 0, \tag{48}$$

or

$$\frac{\epsilon}{\sqrt{1-\epsilon^2}} = \frac{(n'+s)}{Z\alpha} = \frac{(n' + \sqrt{[\kappa^2 - (Z\alpha)^2]}}{Z\alpha}, \qquad \text{with } n' = 0, 1, \ldots, \tag{49}$$

giving us the relativistic energy

$$\frac{E}{mc^2} = \epsilon = \frac{1}{\sqrt{1 + \frac{(Z\alpha)^2}{(n'+s)^2}}} = \frac{1}{\sqrt{1 + \frac{(Z\alpha)^2}{\left[n' + \sqrt{(j+\frac{1}{2})^2 - (Z\alpha)^2}\right]^2}}}. \tag{50}$$

Now, for $(Z\alpha) \ll 1$, we can expand this result to get (through terms of order $(Z\alpha)^4$)

$$\frac{E}{mc^2} - 1$$

$$
\begin{aligned}
&= -\frac{1}{2}\frac{(Z\alpha)^2}{\left[n' + \sqrt{(j+\frac{1}{2})^2 - (Z\alpha)^2}\right]^2} + \frac{3}{8}\frac{(Z\alpha)^4}{\left[n' + \sqrt{(j+\frac{1}{2})^2 - (Z\alpha)^2}\right]^4} + \cdots \\
&= -\frac{1}{2}\frac{(Z\alpha)^2}{(n'+j+\frac{1}{2})^2}\left[1 + \frac{(Z\alpha)^2}{(j+\frac{1}{2})(n'+j+\frac{1}{2})} + \cdots\right] + \frac{3}{8}\frac{(Z\alpha)^4}{(n'+j+\frac{1}{2})^4} + \cdots \\
&= -\frac{(Z\alpha)^2}{2n^2} - \frac{(Z\alpha)^4}{2n^3}\left[\frac{1}{(j+\frac{1}{2})} - \frac{3}{4n}\right] + \cdots , \qquad (51)
\end{aligned}
$$

where we have renamed $(n' + j + \frac{1}{2}) = n$ in the last step to get the conventional principal quantum number n of the hydrogen atom. (We obtained this last result by applying perturbation theory to the nonrelativistic hydrogen atom, after incorporating the magnetic spin-orbit term, the Thomas precession term, and the relativistic mass correction term into the theory. See problem 36 following Chapter 26.)

To obtain the radial functions, $F(\rho)$ and $G(\rho)$, let us return to eq. (40). By eliminating the quantity

$$
a_{m-1} + \sqrt{\frac{1-\epsilon}{1+\epsilon}}\, b_{m-1},
$$

from these two equations, we can solve for the ratio

$$
\frac{a_m}{b_m} = \frac{(Z\alpha)\left[(\kappa + m - n') - \sqrt{(Z\alpha)^2 + (n'+s)^2}\right]}{\left[(Z\alpha)^2 + (m+s-\kappa)[(n'+s) + \sqrt{(Z\alpha)^2 + (n'+s)^2}]\right]}. \qquad (52)
$$

Using this relation, we can also rewrite the first of eq. (40) as

$$
a_{m-1} + \sqrt{\frac{1-\epsilon}{1+\epsilon}}\, b_{m-1} = \frac{(Z\alpha)m(2s+m)}{\left[(Z\alpha)^2 + (m+s-\kappa)[(n'+s) + \sqrt{(Z\alpha)^2 + (n'+s)^2}]\right]} b_m, \qquad (53)
$$

where we have used the relation

$$
(Z\alpha)^2 + (m+s)^2 - \kappa^2 = m^2 + 2ms.
$$

Now, we use

$$
\sqrt{\frac{1-\epsilon}{1+\epsilon}} = \frac{(Z\alpha)}{\left[(n'+s) + \sqrt{(Z\alpha)^2 + (n'+s)^2}\right]}
$$

and the ratio, a_{m-1}/b_{m-1} via eq. (52) to substitute into eq. (53) to obtain

$$
\frac{b_{m-1}}{b_m} = \frac{m(2s+m)}{2(m-1-n')}\frac{\left[(m-1-n'-\kappa) + \sqrt{(Z\alpha)^2 + (n'+s)^2}\right]}{\left[(m-n'-\kappa) + \sqrt{(Z\alpha)^2 + (n'+s)^2}\right]}, \qquad (54)
$$

or, defining,

$$
\gamma \equiv \sqrt{(Z\alpha)^2 + (n'+s)^2} - \kappa, \qquad \text{with } s = \sqrt{\kappa^2 - (Z\alpha)^2}, \qquad (55)
$$

the ratio becomes

$$\frac{b_{m-1}}{b_m} = -\frac{m(2s+m)}{2(n'+1-m)}\frac{(m-1-n')+\gamma}{(m-n')+\gamma}, \tag{56}$$

so

$$\frac{b_{n'-k}}{b_{n'}} = (-1)^k \frac{n'(n'-1)\cdots(n'-k+1)}{2^k(1\cdot 2\cdots k)}\frac{(\gamma-1)(\gamma-2)\cdots(\gamma-k+1)(\gamma-k)}{\gamma(\gamma-1)\cdots(\gamma-k+1)} \tag{57}$$

or

$$\frac{b_{n'-k}}{b_{n'}} = (-1)^k \frac{(n')!}{2^k k!(n'-k)!}\frac{(\gamma-k)}{\gamma}, \qquad k = 1,\ldots,n'. \tag{58}$$

Also, from eq. (52),

$$\frac{a_{n'-k}}{b_{n'-k}} = -\frac{(Z\alpha)(\gamma+k)}{(\gamma-k)\big[(n'+s)+\sqrt{(Z\alpha)^2+(n'+s)^2}\big]} \qquad k = 0,\ldots,n'. \tag{59}$$

As a very specific example, let us look at the ground-state wave function. In this case,

$$n' = 0, \qquad \kappa = -1, \qquad s = \sqrt{1-(Z\alpha)^2}, \qquad \epsilon = \sqrt{1-(Z\alpha)^2}.$$

Recalling

$$\rho = \sqrt{1-\epsilon^2}\frac{mc}{\hbar}r, \quad \text{so} \quad \rho = \frac{Zr}{a_0},$$

we have in this case

$$\begin{aligned} G(\rho) &= \rho^s e^{-\rho} b_0, \\ F(\rho) &= -\frac{(Z\alpha)}{1+s}\rho^s e^{-\rho} b_0. \end{aligned} \tag{60}$$

Also recall $rg(r) = G(r)$ and $rf(r) = F(r)$. With $\kappa = -1$, eq. (23) then gives

$$\psi = \mathcal{N}\left(\frac{Zr}{a_0}\right)^{[\sqrt{1-(Z\alpha)^2}-1]} e^{-\frac{Zr}{a_0}} \begin{pmatrix} \mathcal{Y}_{[0\frac{1}{2}]\frac{1}{2}m_j} \\ -\frac{(Z\alpha)}{[1+\sqrt{1-(Z\alpha)^2}]}\mathcal{Y}_{[1\frac{1}{2}]\frac{1}{2}m_j} \end{pmatrix}, \tag{61}$$

where the normalization constant is

$$\mathcal{N} = \left(\frac{Z}{a_0}\right)^{\frac{3}{2}} 2^{\sqrt{1-(Z\alpha)^2}} \frac{\sqrt{[1+\sqrt{1-(Z\alpha)^2}]}}{\sqrt{\Gamma(1+2\sqrt{1-(Z\alpha)^2})}}. \tag{62}$$

77

The MIT Bag Model: The Dirac Equation for a Quark Confined to a Spherical Region

Quarks, like the leptons (electrons, μ and τ mesons) are Dirac $s = \frac{1}{2}$ particles. Unlike the leptons, quarks have additional internal degrees of freedom, specified by "flavor" (including isospin that determines their charge) and "color." They come in three colors and three generations of isospin doublets (u, d; up and down quarks; s, c; strange and charmed quarks; and b, t; bottom and top quarks). They appear in nature only in "colorless" combinations corresponding to the scalar irreducible representations of the color symmetry SU(3) [the analogue of $S = 0$ multiparticle spin states, corresponding to the scalar irreducible representations of SU(2)]. The "colorless" combinations can arise either from quark–antiquark combinations or combinations of three quarks. Protons, e.g., are made of two up quarks, each with charge of $+\frac{2}{3}e$, and one down quark with charge $-\frac{1}{3}e$, whereas neutrons are made of one up quark and two down quarks, total charge = $(+\frac{2}{3} - \frac{1}{3} - \frac{1}{3})e = 0$. These three-quark systems have a total spin, $S = \frac{1}{2}$, and isospin, $I = \frac{1}{2}$, with $M_I = \pm\frac{1}{2}$. The quark aggregates are confined to finite regions of space, for the three-quark aggregates of the nucleon, to a region of approximately 1.2 fm. The confinement mechanism is not at all understood. One of the most successful phenomenological models for quark confinement is the "MIT bag model," named after the institution of the inventors of the model. In this model, it is simply assumed the quarks are confined to a spherical region of space, with a radius, $r = a$, and $V(r) = 0$ for $r < a$. The rest mass of the u and d quarks, however, is so small their motion in the region with $r < a$ must be treated by the Dirac equation. In a sense, the MIT bag model is the quark analogue of the nuclear shell model. In one extreme version of the nuclear shell model, the potential is assumed to have the form: $V(r) = 0$ for $r < a$, and $V = +\infty$ for $r > a$. Our discussion of the Klein paradox, however, should convince us an infinite potential is unable to confine a relativistic particle. It may be enough to solve a model in which the quark is in free particle motion inside the region for $r < a$, but is subject to boundary conditions at $r = a$ that simply dictate the confinement.

We shall therefore find solutions to the Dirac equation with $V(r) = 0$ for $r < a$; i.e.,

$$H = c(\vec{\alpha} \cdot \vec{p}) + \beta mc^2, \qquad \text{for } r < a, \tag{1}$$

and boundary conditions (which insure the confinement)

$$1. \quad \int\int r^2 d\Omega \vec{S} \cdot \left(\frac{\vec{r}}{r}\right) = c\int\int r^2 d\Omega \bar{\psi}\left(\vec{\gamma} \cdot \frac{\vec{r}}{r}\right)\psi = 0 \qquad \text{at } r = a \tag{2}$$

and

$$1. \quad \int\int r^2 d\Omega \bar{\psi} \mathbf{1} \psi = 0 \qquad \text{at } r = a. \tag{3}$$

Relation 1. insures no net outward (radial) component of the probability density current through the sphere of radius a exists. Relation 2. insures the same for the Lorentz scalar quantity, $\bar{\psi}\mathbf{1}\psi$, over the surface of the sphere of radius a.

In the region, $r < a$, the solutions ψ are of the form

$$\psi = \begin{pmatrix} g(r)\mathcal{Y}_{[l_A \frac{1}{2}]jm_j} \\ \pm f(r)\mathcal{Y}_{[l_B \frac{1}{2}]jm_j} \end{pmatrix} \qquad \text{for } \kappa = \mp(j + \tfrac{1}{2}), \tag{4}$$

where again $j = (l_A + \frac{1}{2})$; $l_B = (l_A + 1)$, for the negative value of κ (upper sign), and $j = (l_A - \frac{1}{2})$; $l_B = (l_A - 1)$, for the positive value of κ (lower sign). Again, defining the one-dimensionalized radial functions, $G(r) = rg(r)$, and $F(r) = rf(r)$, the radial equations for the region $r < a$, where $V(r) = 0$ are then given by

$$\begin{aligned} \left(\frac{E - mc^2}{\hbar c}\right)G(r) &= \left(-\frac{d}{dr} + \frac{\kappa}{r}\right)F(r), \\ \left(\frac{E + mc^2}{\hbar c}\right)F(r) &= \left(+\frac{d}{dr} + \frac{\kappa}{r}\right)G(r), \end{aligned} \tag{5}$$

which lead to

$$\begin{aligned} \left(-\frac{d}{dr} + \frac{\kappa}{r}\right)\left(\frac{d}{dr} + \frac{\kappa}{r}\right)G &= \frac{E^2 - m^2c^4}{(\hbar c)^2}G, \\ \left(\frac{d}{dr} + \frac{\kappa}{r}\right)\left(-\frac{d}{dr} + \frac{\kappa}{r}\right)F &= \frac{E^2 - m^2c^4}{(\hbar c)^2}F, \end{aligned} \tag{6}$$

or

$$\begin{aligned} \left(-\frac{d}{dr^2} + \frac{\kappa(\kappa + 1)}{r^2} - q^2\right)G &= 0, \\ \left(-\frac{d}{dr^2} + \frac{\kappa(\kappa - 1)}{r^2} - q^2\right)F &= 0, \end{aligned}$$

$$\text{with } q^2 = \left(\frac{E}{\hbar c}\right)^2 - \left(\frac{mc}{\hbar}\right)^2. \tag{7}$$

With $j = (l_A \pm \frac{1}{2})$, i.e., with $\kappa = \mp(j + \frac{1}{2})$, we have

$$\begin{aligned}
&\text{with } \kappa = -(l_A + 1): \\
&\kappa(\kappa + 1) = l_A(l_A + 1); \quad \kappa(\kappa - 1) = (l_A + 1)(l_A + 2) = l_B(l_B + 1), \\
&\text{with } \kappa = +l_A: \\
&\kappa(\kappa + 1) = l_A(l_A + 1); \quad \kappa(\kappa - 1) = l_A(l_A - 1) = (l_B + 1)l_B,
\end{aligned} \tag{8}$$

so, with

$$\rho = qr, \tag{9}$$

both entries of eq. (7) have the form

$$\left(-\frac{d}{d\rho^2} + \frac{l(l+1)}{\rho^2} - 1\right)u_l = 0, \tag{10}$$

with $u_{l_A} = G$, and $u_{l_B} = F$, where u_l is the one-dimensionalized radial equation for the spherical Bessel functions; see eqs. (36)–(43) of Chapter 41. Because we need solutions regular at the origin, we have $G(\rho)/\rho = j_{l_A}(\rho)$. The second of eqs. (5) then leads to

$$\begin{aligned}
F(\rho) &= \sqrt{\frac{(E - mc^2)}{(E + mc^2)}}\left(\frac{d}{d\rho} + \frac{\kappa}{\rho}\right)u_{l_A}(\rho) \\
&= \begin{cases} -\sqrt{\frac{(E-mc^2)}{(E+mc^2)}}\left(-\frac{d}{d\rho} + \frac{(l_A+1)}{\rho}\right)u_{l_A}(\rho) \\ +\sqrt{\frac{(E-mc^2)}{(E+mc^2)}}\left(\frac{d}{d\rho} + \frac{l_A}{\rho}\right)u_{l_A}(\rho) \end{cases} \\
&= \begin{cases} -\sqrt{\frac{(E-mc^2)}{(E+mc^2)}}u_{(l_A+1)}(\rho) \\ +\sqrt{\frac{(E-mc^2)}{(E+mc^2)}}u_{(l_A-1)}(\rho) \end{cases} = \mp\sqrt{\frac{(E - mc^2)}{(E + mc^2)}}u_{l_B}(\rho),
\end{aligned} \tag{11}$$

where the upper (and lower) entries are for $\kappa = -(l_A + 1)$ (and $\kappa = +l_A$) and we have used the stepup and down operators, $O_+(l_A + 1)$ and $O_-(l_A)$, of eq. (37) of Chapter 41. Remembering $\rho g(\rho) = G(\rho)$ and $\rho f(\rho) = F(\rho)$, and the signs in eq. (4), we have for $r < a$,

$$\psi = \mathcal{N}\begin{pmatrix} j_{l_A}(qr)\mathcal{Y}_{[l_A\frac{1}{2}]jm_j} \\ -\sqrt{\frac{(E-mc^2)}{(E+mc^2)}}\, j_{l_B}(qr)\mathcal{Y}_{[l_B\frac{1}{2}]jm_j} \end{pmatrix}. \tag{12}$$

These solutions must satisfy the boundary conditions of eqs. (2) and (3) at $r = a$. For eq. (2), it is useful to recast the radial component of $\vec{S}$ (using $\vec{r}/r \equiv \vec{n}$) in the form

$$c\bar{\psi}_{r=a}(\vec{\gamma} \cdot \vec{n})\psi_{r=a} = c\bar{\psi}_{r=a}\begin{pmatrix} 0 & (\vec{\sigma} \cdot \vec{n}) \\ -(\vec{\sigma} \cdot \vec{n}) & 0 \end{pmatrix}\begin{pmatrix} \psi_A(r = a) \\ \psi_B(r = a) \end{pmatrix}$$

$$= c\bar{\psi}_{r=a}\begin{pmatrix} -\sqrt{\frac{E-mc^2}{E+mc^2}}\, j_{l_B}(qa)(\vec{\sigma}\cdot\vec{n})\mathcal{Y}_{[l_B\frac{1}{2}]jm_j} \\ -j_{l_A}(qa)(\vec{\sigma}\cdot\vec{n})\mathcal{Y}_{[l_A\frac{1}{2}]jm_j} \end{pmatrix}$$

$$= c\bar{\psi}_{r=a}(\pm i)\begin{pmatrix} +\sqrt{\frac{E-mc^2}{E+mc^2}}\, j_{l_B}(qa)\mathcal{Y}_{[l_A\frac{1}{2}]jm_j} \\ -j_{l_A}(qa)\mathcal{Y}_{[l_B\frac{1}{2}]jm_j} \end{pmatrix}, \tag{13}$$

where we have used eq. (20) of Chapter 76 for the action of $(\vec{\sigma}\cdot\vec{n})$ on $\mathcal{Y}_{[l\frac{1}{2}]jm_j}$. The upper sign refers to $\kappa = -(j+\frac{1}{2})$ and the lower sign to $\kappa = +(j+\frac{1}{2})$, respectively. The boundary condition 1. at $r = a$ therefore leads to

$$\pm i\sqrt{\frac{E-mc^2}{E+mc^2}}ca^2 j_{l_A}(qa)j_{l_B}(qa)\int\int d\Omega\Big(|\mathcal{Y}_{[l_A\frac{1}{2}]jm_j}|^2 - |\mathcal{Y}_{[l_B\frac{1}{2}]jm_j}|^2\Big) = 0. \tag{14}$$

This relation is satisfied automatically from the orthonormality of the $\mathcal{Y}_{[l\frac{1}{2}]jm_j}$. Conversely, boundary condition 2. at $r = a$ leads to

$$\int\int a^2 d\Omega\Big(\big(j_{l_A}(qa)\big)^2|\mathcal{Y}_{[l_A\frac{1}{2}]jm_j}|^2 - \frac{E-mc^2}{E+mc^2}\big(j_{l_B}(qa)\big)^2|\mathcal{Y}_{[l_B\frac{1}{2}]jm_j}|^2\Big)$$
$$= \Big(\big(j_{l_A}(qa)\big)^2 - \frac{E-mc^2}{E+mc^2}\big(j_{l_B}(qa)\big)^2\Big) = 0. \tag{15}$$

This boundary condition is satisfied if

$$j_{l_A}(qa) = \pm\sqrt{\frac{E-mc^2}{E+mc^2}}\, j_{l_B}(qa). \tag{16}$$

Here, the sign is related to the sign of κ

$$j_{l_A}(qa) = -\frac{\kappa}{|\kappa|}\sqrt{\frac{E-mc^2}{E+mc^2}}\, j_{l_B}(qa). \tag{17}$$

One way to see the κ dependence of this sign is to assume the mass term in the Dirac equation has different values in the interior and exterior regions, $r < a$ and $r > a$. The quark mass in the exterior region, $r > a$, will be named M, and we assume $M \to \infty$. This is the confinement condition (rather than $V \to \infty$). In this limit, the radial equations given by the first of eqs. (7) and the second of eqs. (5) collapse to

$$\Big(-\frac{d^2}{dr^2} + \Big(\frac{Mc}{\hbar}\Big)^2\Big)G(r) = 0, \tag{18}$$

$$F(r) = \frac{\hbar}{Mc}\frac{d}{dr}G(r), \tag{19}$$

with solutions

$$G(r) = e^{-\frac{Mc}{\hbar}r}, \qquad F(r) = -G(r), \qquad M \to \infty. \tag{20}$$

At the boundary, $r = a$, we must therefore have $f(a) = -g(a)$. Eq. (17) therefore follows from eq. (12) and eq. (4). The difference in sign arises through the differences in sign in the $f(r)$ piece of eq. (4). The magnitude follows from eq. (12).

Eq. (17) leads to a transcendental equation for the allowed values of q. If the nth solution of this transcendental equation is named $x_{n,\kappa}$, the energies are determined by

$$x_{n,\kappa} = (q_{n,\kappa}a) = \frac{a}{\hbar c}\sqrt{E^2_{n,\kappa} - mc^2}. \tag{21}$$

The masses of the up and down quarks are believed to be very small ($\ll$ than the nucleon mass), so it may be a good approximation to set the quark mass equal to zero. With $m = 0$, we have

$$E_{n,\kappa} = \frac{\hbar c}{a}x_{n,\kappa}, \qquad \text{where } j_{l_A}(x_{n,\kappa}) = -\frac{\kappa}{|\kappa|}j_{l_B}(x_{n,\kappa}). \tag{22}$$

For example, with $\kappa = -1$, i.e., $j = \frac{1}{2}, l_A = 0, l_B = 1$:

$$j_0(x_{n,-1}) = j_1(x_{n,-1}). \tag{23}$$

With $\kappa = +1$, i.e., $j = \frac{1}{2}, l_A = 1, l_B = 0$:

$$j_1(x_{n,+1}) = -j_0(x_{n,+1}), \tag{24}$$

leading to

$$\frac{\sin(x_{n,\mp 1})}{x_{n,\mp 1}} = \pm\left(\frac{\sin(x_{n,\mp 1})}{x^2_{n,\mp 1}} - \frac{\cos(x_{n,\mp 1})}{x_{n,\mp 1}}\right), \tag{25}$$

or

$$\tan(x_{n,\mp 1}) = \mp\frac{x_{n,\mp 1}}{(x_{n,\mp 1} \mp 1)}, \qquad \text{for } \kappa = \mp 1. \tag{26}$$

Similarly,

$$\tan(x_{n,\mp 2}) = \pm\frac{(x^2_{n,\mp 2} \mp 3x_{n,\mp 2})}{(x^2_{n,\mp 2} \pm x_{n,\mp 2} - 3)}, \tag{27}$$

where these transcendental equations have solutions

$$\begin{aligned} &x_{1,-1} = 2.043, \quad x_{2,-1} = 5.40, \cdots \\ &x_{1,+1} = 3.81, \quad x_{2,+1} = 7.00, \cdots \\ &x_{1,-2} = 3.20, \cdots \\ &x_{1,+2} = 5.12, \cdots \\ &\cdots \end{aligned} \tag{28}$$

In particular, the lowest allowed energy, with $\kappa = -1$, $j = \frac{1}{2}, l_A = 0, l_B = 1$, is

$$E_{1,-1} = \frac{\hbar c}{a}2.043. \tag{29}$$

Thus, for the three-quark system of the nucleon, taking $a = 1.25$ fm, we get a zeroth-order energy in the MIT bag model of

$$3\left(\frac{197.3\text{MeV fm}}{1.25\text{fm}}\right)2.043 = 967 \text{ MeV}.$$

This equation is not far from the observed nucleon mass. The gluon interaction among the three quarks will make additional contributions, but on the whole the MIT bag model with zero rest-mass quarks gives a good account not only of the nucleon mass, but, also, of additional nucleon properties, such as the "anomalous" magnetic moments of the nucleon.

Part IX

Introduction to Many-Body Theory

78
Many-Body Formalism

A Occupation Number Representation

As already noted in Chapter 39, systems of many identical particles, either bosons or fermions, form an important part of nature and, because of the required overall symmetry or antisymmetry of the many-particle wave function, these form not only important, but also very special, types of quantum-mechanical systems. Examples we have met are the many-electron system of an atom, or the many-electron system of condensed matter physics, the many nucleon-system of a nucleus, and the many-quark system of particle physics. All of these involve identical spin $\frac{1}{2}$ particles and, hence, fermions. All of these identical particles have besides their orbital degrees of freedom, internal degrees of freedom such as their spin degree of freedom, but in the case of identical nucleons, an additional charge or isospin internal degree of freedom, and in the case of quarks, additional flavor (including isospin) and color degrees of freedom. Systems of identical bosons include the system of many ^{4}He atoms, an example of a many-particle system where we may be dealing with a huge number of identical particles. Other examples of many-boson systems would include more complicated atoms, provided the identical atoms have an electron plus nucleon number that is an even number so the total atomic spin is an integer. Such systems are of particular recent interest in connection with the Bose–Einstein condensation of such atomic gases.

For the many-boson system, the total many-particle wave function must be totally symmetric under any permutation of particle indices, where the permutation operator must exchange the particle indices on *all* particle variables, orbital, spin, and any additional internal variables, such as isospin or flavor and color variables of the single-particle system. For the many-fermion system, the total many-particle wave function must be totally antisymmetric; that is, it must change sign under any odd permutation of particle indices, involving an odd number of pair exchanges, whereas it must remain unchanged under any even permutation, involving an even number of pair exchanges. For the two-particle system, it was easy to construct two-particle wave functions with the appropriate symmetry from products of single-particle wave functions. Moreover, it was easy to factor the products of two-particle functions into products of their two-particle orbital functions, two-particle spin functions, two-particle isospin functions, and possibly two-particle functions of additional internal degrees of freedom, if applicable. The construction of the two-particle total wave function of the appropriate symmetry is then still very straightforward, because the products of two-particle symmetric functions with two-particle antisymmetric (or symmetric) functions automatically leads to two-particle antisymmetric (or symmetric) functions. For $N \geq 3$, however, this factoring of the N-particle functions into orbital, spin, and possibly other factors becomes more complicated, because of the existence of more complicated intermediate symmetries. For $N = 3$, three types of three-particle symmetries exist. Three-particle functions can be totally symmetric (completely unchanged by any permutation operator), totally antisymmetric, with a change of sign under any odd permutation and no change of sign under any even permutation, but the possibility of more complicated intermediate symmetries exist. Such symmetries are described by Young tableaux, pictures of N squares arranged in rows and columns. A tableau with all N squares in the same row represents a totally symmetric wave function. A tableau with all squares in the same column represents a totally antisymmetric wave function. For $N = 3$, a possible intermediate symmetry exists, involving a tableau with two squares in one row and a third square placed in the first column, as shown in Fig. 78.1(a). Such a tableau represents a three-particle function that can be made symmetric in one pair or antisymmetric in one pair, but is patently such that it cannot be made more symmetric or more antisymmetric. That is, an attempt to make such a function totally symmetric, by acting on it with the full symmetrization operator, will destroy this wave function; i.e., it will simply yield a result that is identically zero, similarly, for an attempt to make such a function totally antisymmetric. Clearly, totally antisymmetric wave functions could be constructed from products of totally symmetric spin functions with totally antisymmetric orbital functions or totally antisymmetric spin functions with totally symmetric orbital functions. As already mentioned in Chapter 39, for electrons or any other $s = \frac{1}{2}$ system, it is impossible to make three-particle spin functions totally antisymmetric, because the single-particle spin function has only two possible quantum states; $m_s = \pm\frac{1}{2}$. Thus, three-particle $s = \frac{1}{2}$ spin functions can only be totally symmetric, with $S = \frac{3}{2}$ (see Chapter 39), or have intermediate symmetry, of type [21] (two squares in the first row of the tableau, one square in the second row), with $S = \frac{1}{2}$. These must be combined with orbital functions also of [21] symmetry to yield total wave functions which are totally antisymmetric. The process of finding the proper linear combination of spin and orbital functions with these symmetries, however, is already somewhat complicated for this simple case with $N = 3$, two independent functions of intermediate [21] symmetry exist, and the two independent spin functions must be combined with appropriate "conjugate" orbital functions to make the required three-particle totally antisymmetric total wave function. For $N = 4$, the combinations of possible symmetries is even richer. We can now have totally symmetric wave functions with all four squares of the tableau in the same row, to be designated by [4] [see Fig. 78.1(b)]. In addition, three independent functions of intermediate symmetry, [31], exist that can be made symmetric in one group of three particles, but cannot be made more symmetric and can be made antisymmetric in one pair of particles, but cannot be made antisymmetric in more than one pair. The other four-particle symmetries include two independent functions of symmetry [22] that can be made symmetric in two different pairs, corresponding to the two rows of the tableau, or antisymmetric in two different pairs, corresponding to the two columns of the tableau; three independent functions of symmetry [211] that can be made antisymmetric in one group of three but symmetric in at most one pair; and finally the totally antisymmetric function, [1111], corresponding to a tableau with a single column. Because four-particle $s = \frac{1}{2}$ spin functions cannot be antisymmetrized in groups of three, these can only have symmetries [4], with $S = 2$, [31] with $S = 1$, and [22] with $S = 0$. To make the totally antisymmetric total wave function, these four types of spin functions must be combined with the orbital functions of the appropriate "conjugate" symmetry, involving Young tableaux in which the role of rows and columns has been interchanged. Thus, totally symmetric spin functions of spin symmetry, [4], must be combined with totally antisymmetric orbital functions of symmetry, [1111]. Three independent spin functions of spin symmetry, [31], must be combined in the proper linear combination with three orbital functions of the appropriate conjugate functions of symmetry, [211], to make a totally antisymmetric total wave function. Similarly, two independent spin functions of intermediate spin symmetry, [22], exist that must be combined in the proper linear combination with the appropriately matched orbital functions, now also of symmetry, [22], to make a totally antisymmetric total four-particle wave function.

For an N-electron atom, where it might be advantageous to factor the N-particle wave function into a product of an N-particle spin function and an N-particle orbital function, the process of antisymmetrization of the total wave function becomes rather complicated for $N \geq 3$. It will therefore in general be better to antisymmetrize the full N-particle wave function in terms of products of single-particle wave functions involving all variables, orbital plus internal, of the type

$$\psi_{\alpha_n}(\vec{\xi}_k); \qquad \text{with } k = 1, \dots, N,$$

where $\vec{\xi}_k$ stands for *all* variables of the particle labeled number, k; e.g., $\vec{\xi}_k = \vec{r}_k, \vec{\sigma}_k, \vec{\tau}_k, \dots$, where $\vec{r}_k$ gives the three orbital variables that give the position of particle, labeled k, in the quantum system, $\vec{\sigma}_k$, stands for the internal spin degree of freedom of particle labeled k, $\vec{\tau}_k$ stands for the internal isospin degree of freedom, etc., where the ... may stand for additional internal degrees of freedom, such as those of color, if applicable. In addition, α_n stands for *all* single-particle quan-

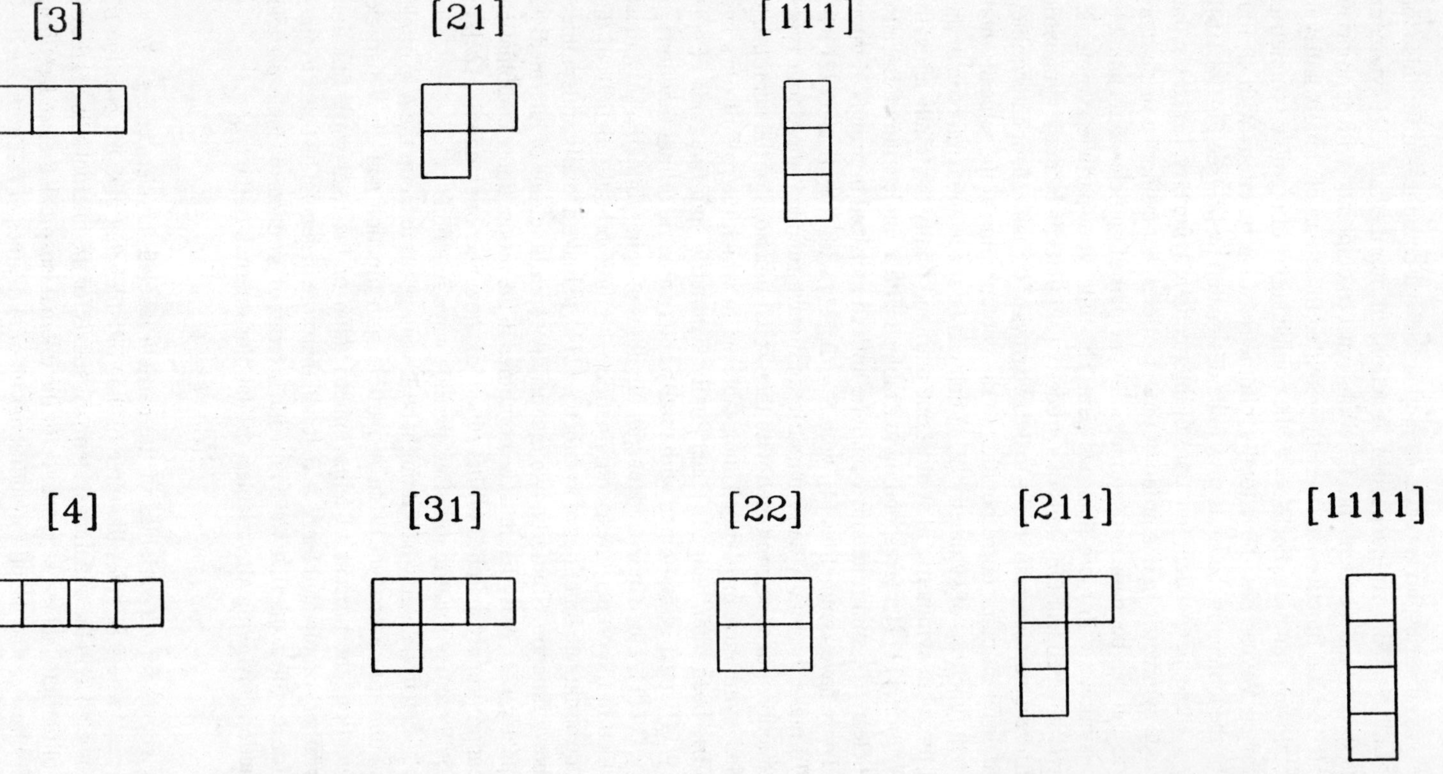

FIGURE 78.1. Young tableaux showing the possible symmetries for three-particle systems (a) and four-particle systems (b).

tum numbers, such as $nlm_l, m_s, m_i, \dots$ for the nth single-particle quantum state, numbered according to some order starting with $n = 1$. In terms of products of N such single-particle functions, it is fairly simple to construct an N-particle wave function totally symmetric for the N-boson sysytem, or totally antisymmetric for the N-fermion system.

For the N-boson system, the totally symmetric N-particle wave function can be given by

$$\psi(\vec{\xi}_1, \vec{\xi}_2, \vec{\xi}_3, \dots, \vec{\xi}_N) = \mathcal{N} \sum\nolimits_P P\biggl(\psi_{\alpha_1}(\vec{\xi}_1) \cdots \psi_{\alpha_1}(\vec{\xi}_{n_1}) \psi_{\alpha_2}(\vec{\xi}_{n_1+1}) \cdots \psi_{\alpha_2}(\vec{\xi}_{n_1+n_2})$$
$$\times\, \psi_{\alpha_3}(\vec{\xi}_{n_1+n_2+1}) \cdots \psi_{\alpha_3}(\vec{\xi}_{n_1+n_2+n_3}) \cdots \psi_{\alpha_k}(\vec{\xi}_{N-n_k+1}) \cdots \psi_{\alpha_k}(\vec{\xi}_N) \biggr), \tag{1}$$

where $\sum_P$ includes a sum over the

$$\frac{N!}{n_1! n_2! n_3! \cdots n_k!}$$

permutations that exchange particles between groups of particles, e.g., between the group of n_1 particles in the same quantum state, α_1, and particles from the group with n_2 particles in quantum state α_2 or particles in still other groups in higher quantum states. Permutations, P, which exchange particles with labels 1 through n_1 are excluded from the sum, because the product of single-particle functions in the same quantum state, α_1, are already symmetric in the first group of n_1 particles. The normalization constant is then given by the squareroot of the inverse of the number of terms in this sum.

$$\mathcal{N} = \left[\frac{\prod_{i=1}^{k} n_i!}{N!} \right]^{\frac{1}{2}}.$$

For the N-fermion system, the antisymmetrization will require that all n_i be either $n_i = 1$ or $n_i = 0$, and the sum over permutations will require all $N!$ terms with the appropriate sign

$$\psi(\vec{\xi}_1, \vec{\xi}_2, \dots, \vec{\xi}_N) = \frac{1}{\sqrt{N!}} \sum_P^{N!} P(-1)^{\sigma(P)} \psi_{\alpha_1}(\vec{\xi}_1) \psi_{\alpha_2}(\vec{\xi}_2) \psi_{\alpha_3}(\vec{\xi}_3) \cdots \psi_{\alpha_N}(\vec{\xi}_N), \tag{2}$$

where $\sigma(P)$ is an odd integer for all odd permutations, whereas it is an even integer for all even permutations. We have seen that such a totally antisymmetric N-particle function can be written in terms of an $N \times N$ Slater determinant

$$\psi(\vec{\xi}_1, \vec{\xi}_2, \vec{\xi}_3, \dots, \vec{\xi}_N) = \frac{1}{\sqrt{N!}} \begin{vmatrix} \psi_{\alpha_1}(\vec{\xi}_1) & \psi_{\alpha_1}(\vec{\xi}_2) & \cdots & \psi_{\alpha_1}(\vec{\xi}_N) \\ \psi_{\alpha_2}(\vec{\xi}_1) & \psi_{\alpha_2}(\vec{\xi}_2) & \cdots & \psi_{\alpha_2}(\vec{\xi}_N) \\ \psi_{\alpha_3}(\vec{\xi}_1) & \psi_{\alpha_3}(\vec{\xi}_2) & \cdots & \psi_{\alpha_3}(\vec{\xi}_N) \\ \cdot & \cdot & \cdots & \cdot \\ \psi_{\alpha_N}(\vec{\xi}_1) & \psi_{\alpha_N}(\vec{\xi}_2) & \cdots & \psi_{\alpha_N}(\vec{\xi}_N) \end{vmatrix}. \tag{3}$$

Wave functions of the type given by eqs. (1) or (3) are by themselves acceptable solutions for the N-particle system only if the interactions between particles are "turned off." For an N-electron atom, e. g., where the single particle quantum numbers, $nlm_l m_s$, specify the single-particle wave functions in the central Coulomb field of the nucleus, the N-particle wave function of eq. (3) would be an eigenfunction of the N-particle Hamiltonian with the Coulomb repulsion between electrons and all other electron–electron interactions "turned off." We will therefore need linear combinations of such basis functions. In addition, despite the compact Slater determinant notation of eq. (3), each such wave function will be composed of $N!$ terms. A better description of such N-particle state vectors would be through the occupation number representation, rather than the coordinate representations of eqs. (1) or (3). With the proper total symmetry for boson systems, or the proper total antisymmetry for fermion systems built in, we simply specify the occupation numbers, $n_1, n_2, n_3, \ldots, n_i, \ldots$, of quantum states labeled according to the index i, which includes all necessary single-particle quantum numbers, usually for the noninteracting N-particle system, which of course must be ordered according to some specific prescription, and with

$$\sum_{i=1}^{\infty} n_i = N, \tag{4}$$

where the n_i may be > 1 for the boson systems, although of course many n_i will be zero, whereas for fermion systems, $n_i = 1$ or $n_i = 0$, are the only possibilities.

Thus, for boson systems, we have state vectors of the type

$$|n_1 n_2 n_3 \cdots n_i \cdots\rangle$$

with $n_i \geq 0$ and $\sum_i n_i = N$.

For fermion systems, we have state vectors of the type

$$|1101011100\cdots\rangle$$

with N occupation numbers of 1, and the rest all 0.

To express arbitrary operators in this type of basis, it will be convenient to introduce operators acting only on the ith single-particle state, ordered according to a definite prescription, where the index i is then also specified by the complete set of single-particle quantum numbers, such as $nlm_l m_s \ldots$ with specific index i, and where these operators annihilate or create one particle with specific index i, i.e., operators, a_i and $a_i^\dagger$, with nonzero matrix elements:

For the boson case:

$$\begin{aligned} \langle n_1 n_2 \cdots (n_i - 1) \cdots |a_i| n_1 n_2 \cdots n_i \cdots\rangle &= \sqrt{n_i}, \\ \langle n_1 n_2 \cdots n_i \cdots |a_i^\dagger| n_1 n_2 \cdots (n_i - 1) \cdots\rangle &= \sqrt{n_i}. \end{aligned} \tag{5}$$

For the fermion case:

$$\begin{aligned} |\langle \cdots n_i = 0 \cdots |a_i| \cdots n_i = 1 \cdots\rangle| &= 1, \\ |\langle \cdots n_i = 1 \cdots |a_i^\dagger| \cdots n_i = 0 \cdots\rangle| &= 1, \end{aligned} \tag{6}$$

all other matrix elements being zero for both bosons and fermions. Note the similarity of the boson case with that of an ∞-dimensional harmonic oscillator. Also, the sign of the fermion matrix elements may depend on the number of occupied states, with $i' < i$, in our state vector prescription. Recall in the coordinate representation, the phase of the wave function depends on the ordering of the single-particle states with index i. Imagine that, in the ket of the matrix element of the annihilation operator, a_i, the occupied state with index i, and $n_i = 1$ is shifted to the left through the occupied states, lower than i in our ordering scheme. A shift through each such occupied state will lead to a change in sign of the ket (cf., the coordinate representation Slater determinant), so

$$\langle \cdots n_i = 0 \cdots | a_i | \cdots n_i = 1 \cdots \rangle = (-1)^{\sum_{\mu=1}^{i-1} n_\mu}. \tag{7}$$

Similarly,

$$\langle \cdots n_i = 1 \cdots | a_i^\dagger | \cdots n_i = 0 \cdots \rangle = (-1)^{\sum_{\mu=1}^{i-1} n_\mu}. \tag{8}$$

Let us next look at matrix elements of products of two operators, such as a_i and $a_r^\dagger$, or a_i and a_r, in both the boson and fermion cases.

For the boson case, with index $r \neq i$,

$$\begin{aligned} \langle \cdots n_i - 1 \cdots n_r \cdots | a_r^\dagger a_i | \cdots n_i \cdots n_r - 1 \cdots \rangle &= \sqrt{n_i n_r}, \\ \langle \cdots n_i - 1 \cdots n_r \cdots | a_i a_r^\dagger | \cdots n_i \cdots n_r - 1 \cdots \rangle &= \sqrt{n_i n_r}, \end{aligned} \tag{9}$$

leading to

$$\langle \cdots n_i - 1 \cdots n_r \cdots | (a_i a_r^\dagger - a_r^\dagger a_i) | \cdots n_i \cdots n_r - 1 \cdots \rangle = 0 \tag{10}$$

for all state vectors of the type shown and for $r \neq i$. Conversely, if the indices on the operators are the same, $r = i$, we have, as follows.

For the boson case:

$$\begin{aligned} \langle \cdots n_i \cdots | a_i^\dagger a_i | \cdots n_i \cdots \rangle &= n_i, \\ \langle \cdots n_i \cdots | a_i a_i^\dagger | \cdots n_i \cdots \rangle &= n_i + 1, \end{aligned} \tag{11}$$

leading in this case to

$$\langle \cdots n_i \cdots | (a_i a_i^\dagger - a_i^\dagger a_i) | \cdots n_i \cdots \rangle = +1. \tag{12}$$

Because these relations must hold for all indices and all state vectors, the operators, a_i and $a_r^\dagger$, must satisfy the commutation relation

$$[a_i, a_r^\dagger] = \delta_{ir}, \qquad [a_i, a_r] = 0, \qquad [a_i^\dagger, a_r^\dagger] = 0, \tag{13}$$

where the last two relations follow from the general matrix elements in similar fashion. The boson annihilation and creation operators therefore satisfy the same commutation relations as the annihilation and creation operators for an ∞-dimensional harmonic oscillator.

For fermions, conversely, with $r < i$,

$$\begin{aligned} &\langle \cdots n_r = 1 \cdots n_i = 0 \cdots | a_r^\dagger a_i | \cdots n_r = 0 \cdots n_i = 1 \cdots \rangle \\ &= \langle \cdots n_r = 1 \cdots n_i = 0 \cdots | a_r^\dagger | \cdots n_r = 0 \cdots n_i = 0 \cdots \rangle \\ &\times \langle \cdots n_r = 0 \cdots n_i = 0 \cdots | a_i | \cdots n_r = 0 \cdots n_i = 1 \cdots \rangle \\ &= (-1)^{\sum_{\mu=1}^{i-1} n_\mu (\text{with } n_r = 0)} (-1)^{\sum_{\nu=1}^{r-1} n_\nu}, \end{aligned} \tag{14}$$

whereas

$$\begin{aligned}&\langle\cdots n_r = 1\cdots n_i = 0\cdots|a_i a_r^\dagger|\cdots n_r = 0\cdots n_i = 1\cdots\rangle\\ &= \langle\cdots n_r = 1\cdots n_i = 0\cdots|a_i|\cdots n_r = 1\cdots n_i = 1\cdots\rangle\\ &\times\langle\cdots n_r = 1\cdots n_i = 1\cdots|a_r^\dagger|\cdots n_r = 0\cdots n_i = 1\cdots\rangle\\ &= (-1)^{\sum_{\mu=1}^{i-1} n_\mu(\text{with } n_r=1)}(-1)^{\sum_{\nu=1}^{r-1} n_\nu}.\end{aligned} \tag{15}$$

Because of the extra $n_r = 1$ in the phase factor for the second relation, we have, with $r \neq i$,

$$\langle\cdots n_r = 1\cdots n_i = 0\cdots|(a_i a_r^\dagger + a_r^\dagger a_i)|\cdots n_r = 0\cdots n_i = 1\cdots\rangle = 0. \tag{16}$$

Conversely, with $r = i$, we have

$$\langle\cdots n_i\cdots|a_i^\dagger a_i|\cdots n_i\cdots\rangle = \begin{cases}1 & \text{for } n_i = 1\\ 0 & \text{for } n_i = 0\end{cases} = n_i, \tag{17}$$

and

$$\langle\cdots n_i\cdots|a_i a_i^\dagger|\cdots n_i\cdots\rangle = \begin{cases}1 & \text{for } n_i = 0\\ 0 & \text{for } n_i = 1\end{cases} = 1 - n_i, \tag{18}$$

leading to

$$\langle\cdots n_i\cdots|(a_i a_i^\dagger + a_i^\dagger a_i)|\cdots n_i\cdots\rangle = 1. \tag{19}$$

Combining the two cases, $r \neq i$ and $r = i$, we have

$$\langle\text{arbitrary state}|(a_i a_r^\dagger + a_r^\dagger a_i)|\text{arbitrary state}\rangle = \delta_{ri}, \tag{20}$$

leading to the anticommutation relation

$$\{a_i, a_r^\dagger\}_+ \equiv a_i a_r^\dagger + a_r^\dagger a_i = \delta_{ir}. \tag{21}$$

Similarly,

$$\{a_i, a_r\}_+ = 0, \qquad \{a_i^\dagger, a_r^\dagger\}_+ = 0. \tag{22}$$

In particular, with $i = r$, the last relation gives

$$a_r^\dagger a_r^\dagger = 0. \tag{23}$$

We cannot create two identical fermions in the same quantum state. Similarly,

$$a_r a_r = 0. \tag{24}$$

For fermions, in general, the boson commutation relations are converted to fermion anticommutation relations, and for fermions, we need to deal with an anticommutator rather than a commutator algebra.

B State Vectors

We shall now write both state vectors and operators representing physical quantities in terms of the above creation and annihilation operators. For both bosons and fermions, we start with the vacuum state, the state of zero particles, i.e., no particles whatsoever.

$$|0\rangle = |00\cdots 0\cdots\rangle, \qquad \text{with } a_i|0\rangle = 0 \text{ for all i.} \tag{25}$$

Single-particle states: For both the boson and fermion systems, single-particle states are given by

$$a_i^\dagger|0\rangle, \tag{26}$$

where the quantum number index, i, is shorthand for a specific set of values of all single-particle quantum numbers, orbital, spin, and quantum numbers with other internal degrees of freedom, if applicable.

N-particle states: For boson systems, we have

$$\prod_i \frac{(a_i^\dagger)^{n_i}}{\sqrt{n_i!}}|0\rangle, \qquad \text{with } \sum_i n_i = N. \tag{27}$$

For fermion systems, we have

$$a_{i_1}^\dagger a_{i_2}^\dagger \cdots a_{i_N}^\dagger|0\rangle. \tag{28}$$

In particular, the action of a creation operator on such an N-particle state to make an $(N+1)$-particle state will in general add a phase factor to make a state with the conventional ordering of states dictated by our particular prescription.

$$a_{i_s}^\dagger(a_{i_1}^\dagger a_{i_2}^\dagger \cdots a_{i_N}^\dagger)|0\rangle = (-1)^{\sum_{\mu=1}^{s-1} n_\mu}(a_{i_1}^\dagger a_{i_2}^\dagger \cdots a_{i_s}^\dagger \cdots a_{i_N}^\dagger)|0\rangle. \tag{29}$$

C One-Body Operators

Let us now construct the dynamical operators that we shall need and see how they can be expressed in terms of single-particle creation and annihilation operators. The simplest class of operators are the one-body operators. In coordinate representation, they can be expressed through

$$F_{\text{op.}} = \sum_{i=1}^N f_i, \qquad \text{with } f_i = f(\vec{r}_i, \vec{\nabla}_i, \vec{\sigma}_i, \vec{\tau}_i, \ldots), \tag{30}$$

where ... will stand for additional internal variables, with index i, if needed for additional internal degrees of freedom. Examples would be the kinetic energy term in a nonrelativistic N-particle Hamiltonian

$$T = \sum_{i=1}^N \frac{\vec{p}_i^{\,2}}{2m_i}, \tag{31}$$

or the potential terms of the N-particle system in an external field, e.g., the potential of the N-electron atomic system in the field of the central nucleus of charge, Ze,

$$V = \sum_{i=1}^N \left(\frac{-Ze^2}{r_i}\right). \tag{32}$$

Other examples would be electric or magnetic multipole transition operators, e.g., the three components of the electric dipole moment vector,

$$F = \sum_{i=1}^{N} e_i \vec{r}_i, \tag{33}$$

or the spherical components of the magnetic 2^L-pole operator,

$$F = \sum_{i=1}^{N} e_i \Big[r_i^{L-1} Y_{L-1}(\theta_i, \phi_i) \times \Big(\frac{2}{L+1} \vec{l}_i^{\,1} + g_{s_i} \vec{s}_i^{\,1} \Big) \Big]_M^L . \tag{34}$$

To transcribe such operators from their coordinate representation to the occupation number representation, let us now use $\vec{r}_i$ as shorthand for *all* variables, both orbital and intrinsic, with particle index i; i.e., $\vec{r}_i \equiv \vec{r}_i, \vec{\sigma}_i, \vec{\tau}_i, \ldots$. In this shorthand notation, $\int d\vec{r}_i$, will then stand for an integral over all orbital variables and over all internal variables, or alternatively a matrix-element-taking over spin, isospin,..., quantum numbers, if the internal variables cannot be specified through specific coordinates. A one-body operator can then be expressed through

$$F_{\text{op.}} = \sum_{\alpha_s} \sum_{\alpha_s'} |\alpha_1' \alpha_2' \cdots \alpha_N' \rangle \langle \alpha_1' \alpha_2' \cdots \alpha_N' | F | \alpha_1 \alpha_2 \cdots \alpha_N \rangle \langle \alpha_1 \alpha_2 \cdots \alpha_N |, \tag{35}$$

where the sums over α_s and α_s' are infinite sums over all possible sets of N-particle quantum numbers in their canonical ordering. We can convert this expression in occupation number representation into coordinate representation, via

$$F_{\text{op.}} = \int d\vec{r}_1' \cdots d\vec{r}_N' \int d\vec{r}_1 \cdots d\vec{r}_N \sum_{\alpha_s} \sum_{\alpha_s'} |\alpha_1' \alpha_2' \cdots \alpha_N' \rangle \langle \alpha_1' \alpha_2' \cdots \alpha_N' | \vec{r}_1' \cdots \vec{r}_N' \rangle$$

$$\times \langle \vec{r}_1' \cdots \vec{r}_N' | \sum_{i=1}^{N} f(\vec{r}_i, \vec{\nabla}_i, \ldots) | \vec{r}_1 \cdots \vec{r}_N \rangle \langle \vec{r}_1 \cdots \vec{r}_N | \alpha_1 \alpha_2 \cdots \alpha_N \rangle \langle \alpha_1 \alpha_2 \cdots \alpha_N |. \tag{36}$$

Let us now specialize to the fermion case. (The boson case is very similar, except for the additional phase factors that come into play for fermions.) In coordinate representation, we have

$$\langle \vec{r}_1' \cdots \vec{r}_N' | \sum_{i=1}^{N} f(\vec{r}_i, \vec{\nabla}_i, \ldots) | \vec{r}_1 \cdots \vec{r}_N \rangle$$

$$= \Big(\sum_{i=1}^{N} f(\vec{r}_i, \vec{\nabla}_i, \ldots) \delta(\vec{r}_i - \vec{r}_i') \Big) \prod_{j \neq i}^{N} \delta(\vec{r}_j - \vec{r}_j'), \tag{37}$$

and the coordinate representations of the fully antisymmetrized N-particle states can be expressed in the form of eq. (2) or eq. (3),

$$\langle \alpha_1 \cdots \alpha_N | \vec{r}_1 \cdots \vec{r}_N \rangle = \Psi_{\alpha_1 \cdots \alpha_N}(\vec{r}_1, \cdots, \vec{r}_N). \tag{38}$$

Eq. (36) can therefore be transcribed into

$$F_{\text{op.}} = \sum_{\alpha_s} \sum_{\alpha_s'} |\alpha_1' \alpha_2' \cdots \alpha_N' \rangle \int d\vec{r}_1 \cdots d\vec{r}_N \Psi^*_{\alpha_1' \cdots \alpha_N'}$$

$$\times \sum_{i=1}^{N} f(\vec{r}_i, \vec{\nabla}_i, \cdots)\Psi_{\alpha_1\cdots\alpha_N}\langle\alpha_1\alpha_2\cdots\alpha_N|. \tag{39}$$

We now use the antisymmetry of the N-particle functions, $\Psi_{\alpha_1\cdots\alpha_N}$, viz.,

$$P_{1i}\Psi_{\alpha_1\cdots\alpha_N} = -\Psi_{\alpha_1\cdots\alpha_N},$$

to rewrite

$$\begin{aligned}
&\int d\vec{r}_1\cdots d\vec{r}_N \Psi^*_{\alpha'_1\cdots\alpha'_N} f(\vec{r}_i, \vec{\nabla}_i, \cdots)\Psi_{\alpha_1\cdots\alpha_N} \\
&= \int d\vec{r}_1\cdots d\vec{r}_N \Psi^*_{\alpha'_1\cdots\alpha'_N}\Big(P_{1i}^{-1} f(\vec{r}_1, \vec{\nabla}_1, \cdots)P_{1i}\Big)\Psi_{\alpha_1\cdots\alpha_N} \\
&= \int d\vec{r}_1\cdots d\vec{r}_N \Psi^*_{\alpha'_1\cdots\alpha'_N} f(\vec{r}_1, \vec{\nabla}_1, \cdots)\Psi_{\alpha_1\cdots\alpha_N}.
\end{aligned} \tag{40}$$

Repeating this process N times, for $i = 1 \to N$, we get

$$F_{\text{op.}} = N\int d\vec{r}_1\cdots d\vec{r}_N \Psi^*_{\alpha'_1\cdots\alpha'_N} f_1 \Psi_{\alpha_1\cdots\alpha_N}, \tag{41}$$

where we have used the shorthand notation of eq. (30)

$$f_1 = f(\vec{r}_1, \vec{\nabla}_1, \cdots). \tag{42}$$

If we now expand the fully antisymmetrized N-particle functions in this expression in terms of fully antisymmetrized $(N-1)$-particle functions, we get

$$\begin{aligned}
F_{\text{op.}} = \frac{N}{\sqrt{N}\sqrt{N}}\Bigg(&\int d\vec{r}_1 \psi^*_{\alpha'_1} f_1 \psi_{\alpha_1}\int d\vec{r}_2\cdots d\vec{r}_N \Psi^*_{\alpha'_2\cdots\alpha'_N}\Psi_{\alpha_2\cdots\alpha_N} \\
&- \int d\vec{r}_1 \psi^*_{\alpha'_1} f_1 \psi_{\alpha_2}\int d\vec{r}_2\cdots d\vec{r}_N \Psi^*_{\alpha'_2\cdots\alpha'_N}\Psi_{\alpha_1\alpha_3\cdots\alpha_N} \\
&+ \int d\vec{r}_1 \psi^*_{\alpha'_1} f_1 \psi_{\alpha_3}\int d\vec{r}_2\cdots d\vec{r}_N \Psi^*_{\alpha'_2\cdots\alpha'_N}\Psi_{\alpha_1\alpha_2\alpha_4\cdots\alpha_N} \\
&- \cdots \\
&+ (-1)^{N-1}\int d\vec{r}_1 \psi^*_{\alpha'_1} f_1 \psi_{\alpha_N}\int d\vec{r}_2\cdots d\vec{r}_N \Psi^*_{\alpha'_2\cdots\alpha'_N}\Psi_{\alpha_1\alpha_2\cdots\alpha_{N-1}} \\
&- \int d\vec{r}_1 \psi^*_{\alpha'_2} f_1 \psi_{\alpha_1}\int d\vec{r}_2\cdots d\vec{r}_N \Psi^*_{\alpha'_1\alpha'_3\cdots\alpha'_N}\Psi_{\alpha_2\cdots\alpha_N} \\
&+ \cdots \\
&+ \int d\vec{r}_1 \psi^*_{\alpha'_N} f_1 \psi_{\alpha_N}\int d\vec{r}_2\cdots d\vec{r}_N \Psi^*_{\alpha'_1\alpha'_2\cdots\alpha'_{N-1}}\Psi_{\alpha_1\alpha_2\cdots\alpha_{N-1}}\Bigg),
\end{aligned} \tag{43}$$

where N^2 terms in this expansion exist altogether, but where the $(N-1)$-particle overlap integrals in the particles labeled 2 through N will all be zero unless the primed and unprimed quantum numbers in the $(N-1)$-particle functions are all the same, i.e., unless $(\alpha'_2\cdots\alpha'_N) = (\alpha_2\cdots\alpha_N)$ in the first term of this expansion. Hence, the full matrix element of the one-body operator, $F_{\text{op.}}$, will be different from zero only in two cases as follows.

Case (1), if the $\alpha'_1\cdots\alpha'_N$ differ from the $\alpha_1\cdots\alpha_N$ in only *one* of the α_s.

Case (2), if all $\alpha'_s = \alpha_s$.

For case (1), we have

$$|\alpha_1 \cdots \alpha_N\rangle = |\alpha_1\alpha_2 \cdots \alpha_r \cdots \alpha_{s-1}\alpha_{s+1} \cdots \alpha_N\rangle,$$
$$|\alpha_1' \cdots \alpha_N'\rangle = |\alpha_1\alpha_2 \cdots \alpha_{r-1}\alpha_{r+1} \cdots \alpha_s \cdots \alpha_N\rangle. \quad (44)$$

For case (2), we have

$$|\alpha_1' \cdots \alpha_N'\rangle = |\alpha_1 \cdots \alpha_N\rangle. \quad (45)$$

For case (1), the N-particle matrix element of $F_{\text{op.}}$ is

$$\begin{aligned}
&\langle \alpha_1' \cdots \alpha_N' | F_{\text{op.}} | \alpha_1 \cdots \alpha_N\rangle \\
&= (-1)^{\sum_{\mu=1}^{r-1} n_\mu} (-1)^{\sum_{\nu=1}^{s-1} n'_\nu} \int d\vec{r}_1 \psi^*_{\alpha_s}(\vec{r}_1) f(\vec{r}_1, \vec{\nabla}_1, \cdots) \psi_{\alpha_r}(\vec{r}_1) \\
&= (-1)^{\sum_{\mu=1}^{r-1} n_\mu} (-1)^{\sum_{\nu=1}^{s-1} n'_\nu} \langle s | f | r\rangle,
\end{aligned} \quad (46)$$

where we have used an obvious shorthand notation for the single-particle matrix element of the single-particle operator in the last step.

For case (2), conversely, we have

$$\begin{aligned}
\langle \alpha_1' \cdots \alpha_N' | F_{\text{op.}} | \alpha_1 \cdots \alpha_N\rangle &= \sum_{\alpha_i=\alpha_1}^{\alpha_i=\alpha_N} \int d\vec{r}_1 \psi^*_{\alpha_i}(\vec{r}_1) f(\vec{r}_1, \vec{\nabla}_1, \cdots) \psi_{\alpha_i}(\vec{r}_1) \\
&= \sum_{i=1}^{N_{\text{occ.}}} \langle i | f | i\rangle,
\end{aligned} \quad (47)$$

where the sum is over the N occupied states. Now, if we use the matrix elements

$$\begin{aligned}
&a_r |\alpha_1\alpha_2 \cdots \alpha_r \cdots \alpha_{s-1}\alpha_{s+1} \cdots \alpha_N\rangle \\
&\quad = (-1)^{\sum_{\mu=1}^{r-1} n_\mu} |\alpha_1\alpha_2 \cdots\cdots \alpha_{s-1}\alpha_{s+1} \cdots \alpha_N\rangle, \\[1em]
&\langle \alpha_1\alpha_2 \cdots \alpha_{r-1}\alpha_{r+1} \cdots \alpha_{s-1}\alpha_{s+1} \cdots \alpha_N | a_s^\dagger \\
&\quad = \langle \alpha_1\alpha_2 \cdots \alpha_{r-1}\alpha_{r+1} \cdots \alpha_{s-1}\alpha_s\alpha_{s+1} \cdots \alpha_N | (-1)^{\sum_{\nu=1}^{s-1} n'_\nu}, \\[1em]
&\langle \alpha_1\alpha_2 \cdots \alpha_N | a_i^\dagger a_i | \alpha_1\alpha_2 \cdots \alpha_N\rangle = n_i, \\
&\text{with } n_i = 1 \text{ for occ. states, } n_i = 0 \text{ otherwise,}
\end{aligned} \quad (48)$$

the matrix elements for cases (1) and (2) will be obtained if the one-body operator is expressed in terms of single-particle creation and annihilation operators, via

$$F_{\text{op.}} = \sum_{r,s} \langle s | f | r\rangle a_s^\dagger a_r. \quad (49)$$

D Examples of One-Body Operators

Before proceeding to two-body operators, let us look at a few specific examples of simple one-body operators in this new language.

1. The number operator.

One of the simplest one-body operators is the simple number operator that counts the total number of particles. In the coordinate representation, it is simply a sum of N unit operators

$$F = \sum_{i=1}^{N} 1 = N. \tag{50}$$

With $f_i = 1$, the single-particle matrix elements are simply

$$\langle s|f|r\rangle = \langle s|1|r\rangle = \delta_{sr}, \tag{51}$$

so

$$F_{\text{op.}} \equiv N_{\text{op.}} = \sum_s a_s^\dagger a_s. \tag{52}$$

2. Total angular momentum operator.
In this case, we may want the three components, J_z, J_x, J_y, where

$$J_z = \sum_{i=1}^{N} j_{z,i}, \tag{53}$$

where capital letters are used for the operator acting in the N-particle system and the lowercase letter indicates the operator acts in the space of the single-particle labeled with index i. If the total angular momentum vector is built from orbital and spin angular momenta, $\vec{J} = \vec{L} + \vec{S}$, it will be useful to employ an $|nlsjm_j\rangle$ single-particle basis, with $|r\rangle = |n_r l_r s_r j_r m_{j_r}\rangle$, rather than a $|nlm_l sm_s\rangle$ basis, so $J_z = J_0$ is diagonal in this basis:

$$\begin{aligned} J_0 &= \sum_{i,k} \langle i|j_0|k\rangle a_i^\dagger a_k \\ &= \sum_{i,k} \langle n_i l_i s_i j_i m_{j_i}|j_0|n_k l_k s_k j_k m_{j_k}\rangle a_i^\dagger a_k \\ &= \sum_{i,k} m_j \delta_{ik} a_i^\dagger a_k \\ &= \sum_{nlsj}\sum_{m_j} m_j a_{nlsjm_j}^\dagger a_{nlsjm_j}. \end{aligned} \tag{54}$$

The perpendicular components of $\vec{J}$ are best expressed in terms of $J_\pm = (J_x \pm i J_y)$, via

$$\begin{aligned} J_\pm &= \sum_{i,k} \langle i|j_\pm|k\rangle a_i^\dagger a_k \\ &= \sum_{nlsj}\sum_{m_j} \sqrt{(j \mp m_j)(j \pm m_j + 1)}\, a_{nlsj(m_j\pm 1)}^\dagger a_{nlsjm_j}. \end{aligned} \tag{55}$$

3. Electric dipole moment operator.

In this case, let us look first at the $m = +1$ spherical component of the electric dipole moment vector of the N electron system in an atom, where

$$\mu_{+1}^{(\text{el.})} = \sum_{i=1}^{N} e_i \big(\vec{r}_i\big)_{+1} = -|e| \sum_{i=1}^{N} r_i Y_{1,+1}(\theta_i, \phi_i) \sqrt{\frac{4\pi}{3}}, \tag{56}$$

where $\vec{r}_i$ measures the position of the electron with particle index i relative to the nucleus at the origin. It will now be advantageous to use a $|nlm_l s m_s\rangle$ basis, so in our new language

$$|k\rangle \equiv |n_k l_k m_{l_k} s_k m_{s_k}\rangle.$$

In the new language, we then have

$$\begin{aligned}
\mu_{+1}^{(\text{el.})} &= \sum_{n'l'} \sum_{nl} \sum_{m_l m_s} \langle n'l'(m_l+1)\tfrac{1}{2} m_s | \mu_{+1}^{(\text{el.})} | nlm_l \tfrac{1}{2} m_s \rangle a^{\dagger}_{n'l'(m_l+1)m_s} a_{nlm_lm_s} \\
&= -|e| \sum_{n'l'} \sum_{nl} \sum_{m_l m_s} \big(r\big)_{n'l',nl} \sqrt{\frac{(2l+1)}{(2l'+1)}} \langle l010|l'0\rangle \\
&\times \langle l m_l 11 | l'(m_l+1)\rangle a^{\dagger}_{n'l'(m_l+1)m_s} a_{nlm_lm_s},
\end{aligned} \tag{57}$$

where $\big(r\big)_{n'l',nl}$ is the radial integral

$$\big(r\big)_{n'l',nl} = \int_0^{\infty} dr r^2 R^*_{n'l'}(r) r R_{nl}(r).$$

For a system of A nucleons in a nucleus, we can write the charge of the ith nucleon as

$$e_i = |e| \tfrac{1}{2}(1 - \tau_{z,i}),$$

and use a $|nlm_l s m_s m_i\rangle$ basis in a convention in which the neutron has isospin quantum number $m_i = +\frac{1}{2}$, with τ_z-eigenvalue of + 1, whereas the proton has $m_i = -\frac{1}{2}$, with τ_z-eigenvalue of -1. In this A-particle system, we have

$$\begin{aligned}
\mu_{+1}^{(\text{el.})} &= \sum_{i=1}^{A} |e| \tfrac{1}{2}(1 - \tau_{z,i}) r_i Y_{1,+1}(\theta_i, \phi_i) \sqrt{\frac{4\pi}{3}} \\
&= |e| \sum_{n'l'} \sum_{nl} \sum_{m_l m_s} \sum_{m_i} \big(\tfrac{1}{2} - m_i\big) \big(r\big)_{n'l',nl} \sqrt{\frac{(2l+1)}{(2l'+1)}} \langle l010|l'0\rangle \\
&\times \langle l m_l 11 | l'(m_l+1)\rangle a^{\dagger}_{n'l'(m_l+1)m_s m_i} a_{nlm_lm_sm_i},
\end{aligned} \tag{58}$$

where the isospin quantum number, m_i, has the values $m_i = +\frac{1}{2}$ for neutrons, $m_i = -\frac{1}{2}$ for protons. (The subscript, i, on m_i stands for isospin.)

4. Magnetic dipole moment operator.

For the magnetic dipole moment operator in the N-electron atom, we have

$$\mu_{+1}^{(\text{magn.})} = \sum_{i=1}^{N} \frac{e\hbar}{2mc} \Big[\big(\vec{l}_i\big)_{+1} + g_s \big(\vec{s}_i\big)_{+1} \Big], \tag{59}$$

with $g_s = 2$, where we have neglected the nuclear magnetic moment because of the small electron to nucleus mass ratio, $m/M_{\text{nucl.}}$. In the new language, we would then have

$$\mu_{+1}^{(\text{magn.})} = -\frac{1}{\sqrt{2}}\frac{e\hbar}{2mc}\sum_{nl}\sum_{m_l}\sum_{m_s} \Big[\sqrt{(l-m_l)(l+m_l+1)}a^{\dagger}_{nl(m_l+1)m_s}a_{nlm_lm_s} + g_s\sqrt{(\tfrac{1}{2}-m_s)(\tfrac{1}{2}+m_s+1)}a^{\dagger}_{nlm_l(m_s+1)}a_{nlm_lm_s}\Big]. \tag{60}$$

E Two-Body Operators

In coordinate representation, two-body operators can be written as a sum over $\frac{1}{2}N(N-1)$ terms

$$G_{\text{op.}} = \sum_{i<k}^{N} g_{ik}, \tag{61}$$

where

$$g_{ik} = g(\vec{r}_i, \vec{r}_k) = g(\vec{r}_k, \vec{r}_i), \tag{62}$$

and we again use the shorthand notation: Let $\vec{r}_i$ stand for all coordinates, both orbital and internal, for particle labeled with particle index, i. A simple example would be the Coulomb repulsion potential between electrons in an N-electron atom,

$$g_{ik} = \frac{e^2}{r_{ik}} = \frac{e^2}{|\vec{r}_i - \vec{r}_k|}. \tag{63}$$

Now, using the antisymmetry of the N-particle wave functions, $\Psi_{\alpha_1\alpha_2\cdots\alpha_N}$, we can reduce the matrix element of $G_{\text{op.}}$ to a matrix element of g_{12}.

$$\int d\vec{r}_1 d\vec{r}_2 \cdots d\vec{r}_N \Psi^*_{\alpha'_1\alpha'_2\cdots\alpha'_N} G_{\text{op.}} \Psi_{\alpha_1\alpha_2\cdots\alpha_N} = \tfrac{1}{2}N(N-1)\int d\vec{r}_1 d\vec{r}_2 \cdots d\vec{r}_N \Psi^*_{\alpha'_1\alpha'_2\cdots\alpha'_N} g_{12} \Psi_{\alpha_1\alpha_2\cdots\alpha_N}, \tag{64}$$

where this is the analogue of eq. (41) for the one-body operator, $F_{\text{op.}}$. In addition, it will be useful to expand the totally antisymmetric N-particle functions, $\Psi_{\alpha_1\alpha_2\cdots\alpha_N}$, in terms of totally antisymmetric $(N-2)$-particle functions, via [see eq. (3)],

$$\Psi_{\alpha_1\alpha_2\cdots\alpha_N}(\vec{r}_1,\cdots,\vec{r}_N) = \frac{1}{\sqrt{N(N-1)}}\Big[\Big(\psi_{\alpha_1}(\vec{r}_1)\psi_{\alpha_2}(\vec{r}_2) - \psi_{\alpha_2}(\vec{r}_1)\psi_{\alpha_1}(\vec{r}_2)\Big)\Psi_{\alpha_3\alpha_4\cdots\alpha_N}(\vec{r}_3,\cdots,\vec{r}_N)$$
$$- \Big(\psi_{\alpha_1}(\vec{r}_1)\psi_{\alpha_3}(\vec{r}_2) - \psi_{\alpha_3}(\vec{r}_1)\psi_{\alpha_1}(\vec{r}_2)\Big)\Psi_{\alpha_2\alpha_4\cdots\alpha_N}(\vec{r}_3,\cdots,\vec{r}_N)$$
$$+\cdots \qquad \cdots$$

$$+(-1)^{\chi_{ij}}\Big(\psi_{\alpha_i}(\vec{r}_1)\psi_{\alpha_j}(\vec{r}_2)-\psi_{\alpha_j}(\vec{r}_1)\psi_{\alpha_i}(\vec{r}_2)\Big)\Psi_{\alpha_1\cdots\alpha_N}(\vec{r}_3,\cdots,\vec{r}_N)$$
$$+\cdots\quad\cdots$$
$$+\Big(\psi_{\alpha_{N-1}}(\vec{r}_1)\psi_{\alpha_N}(\vec{r}_2)-\psi_{\alpha_N}(\vec{r}_1)\psi_{\alpha_{N-1}}(\vec{r}_2)\Big)\Psi_{\alpha_1\alpha_2\cdots\alpha_{N-2}}(\vec{r}_3,\cdots,\vec{r}_N)\Bigg],$$

$$\text{with }\ \chi_{ij}=\sum_{\mu=1}^{i-1}n_\mu+\sum_{\nu=1}^{j-1}n_\nu-1, \tag{65}$$

where we have assumed $i < j$ in fixing the phase of the generic term in this expansion. Now, three classes of nonzero matrix elements exist, as follows.

Case (1): The $\alpha'_1\cdots\alpha'_N$ differ from the $\alpha_1\cdots\alpha_N$ in two quantum numbers: Suppose α_i and α_j, with $i < j$, are present in the N-particle state with unprimed α's, but absent in the state with primed α's, whereas α_r and α_s, with $r < s$, are present in the state with primed α's, but missing in the state with unprimed α's. The remaining $(N-2)$ α'_k are the same as the remaining $(N-2)$ α_k. In that case,

$$\langle\alpha'_1\cdots\alpha'_N|G_{\text{op.}}|\alpha_1\cdots\alpha_N\rangle=\tfrac{1}{2}(-1)^{\sum_{\mu=1}^{i-1}n_\mu+\sum_{\nu=1}^{j-1}n_\nu-1+\sum_{\mu=1}^{r-1}n'_\mu+\sum_{\nu=1}^{s-1}n'_\nu-1}$$
$$\times\Big[\langle rs|g_{12}|ij\rangle-\langle rs|g_{12}|ji\rangle+\langle sr|g_{12}|ji\rangle-\langle sr|g_{12}|ij\rangle\Big]. \tag{66}$$

Case (2): The $\alpha'_1\cdots\alpha'_N$ differ from the $\alpha_1\cdots\alpha_N$ in one quantum number. Suppose α_i is present in the N-particle state with unprimed α's, but absent in the state with primed α's, whereas α_r is present in the state with primed α's, but missing in the state with unprimed α's. The remaining $(N-1)$ α'_k are the same as the remaining $(N-1)$ α_k. In that case,

$$\langle\alpha'_1\cdots\alpha'_N|G_{\text{op.}}|\alpha_1\cdots\alpha_N\rangle=\tfrac{1}{2}(-1)^{\sum_{\mu=1}^{i-1}n_\mu}(-1)^{\sum_{\nu=1}^{r-1}n'_\nu}$$
$$\times\sum_{k\neq i,r}^{\text{occ.}}\Big[\langle rk|g_{12}|ik\rangle-\langle rk|g_{12}|ki\rangle+\langle kr|g_{12}|ki\rangle-\langle kr|g_{12}|ik\rangle\Big]. \tag{67}$$

The sum over k is a sum over the $(N-1)$ occupied states other than $k=i$ of the ket and $k=r$ in the bra.

Case (3): Diagonal matrix elements. The α'_s are all equal to the α_s: $\alpha'_1\alpha'_2\cdots\alpha'_N=\alpha_1\alpha_2\cdots\alpha_N$. In this case,

$$\langle\alpha_1\cdots\alpha_N|G_{\text{op.}}|\alpha_1\cdots\alpha_N\rangle=$$
$$\tfrac{1}{2}\sum_{i<k}^{\text{occ.}}\Big[\langle ik|g_{12}|ik\rangle-\langle ik|g_{12}|ki\rangle+\langle ki|g_{12}|ki\rangle-\langle ki|g_{12}|ik\rangle\Big], \tag{68}$$

where the sum is now over all N occupied states.

In eqs. (66), (67), and (68), the two-particle matrix elements, such as

$$\langle rs|g_{12}|ij\rangle=\int d\vec{r}_1d\vec{r}_2\psi^*_{\alpha_r}(\vec{r}_1)\psi^*_{\alpha_s}(\vec{r}_2)g(\vec{r}_1,\vec{r}_2)\psi_{\alpha_i}(\vec{r}_1)\psi_{\alpha_j}(\vec{r}_2),$$

can be related to the two-particle matrix elements

$$\langle sr|g_{12}|ji\rangle$$

by a renaming of the dummy integration variables $\vec{r}_1 \leftrightarrow \vec{r}_2$, and by making use of the symmetry of g_{12}

$$g(\vec{r}_2, \vec{r}_1) = g(\vec{r}_1, \vec{r}_2)$$

to obtain

$$\begin{aligned}\langle rs|g_{12}|ij\rangle &= \langle sr|g_{12}|ji\rangle,\\ \langle rs|g_{12}|ji\rangle &= \langle sr|g_{12}|ij\rangle.\end{aligned} \tag{69}$$

In eqs. (66)–(68), we have purposely left off this simplification to facilitate the transition to the new language in terms of annihilation and creation operators. To make this transition,

$$\begin{aligned}&a_j a_i \Big(a^\dagger_{k_1} a^\dagger_{k_2} \cdots a^\dagger_i \cdots a^\dagger_j \cdots a^\dagger_{k_N}\Big)|0\rangle\\ &= (-1)^{\sum_{\mu=1}^{i-1} n_\mu} (-1)^{\sum_{\nu=1}^{j-1} n_\nu - 1} \Big(a^\dagger_{k_1} a^\dagger_{k_2} \cdots \qquad \cdots \qquad \cdots a^\dagger_{k_N}\Big)|0\rangle,\end{aligned} \tag{70}$$

where we have assumed, $i < j$, in our prescription for ordering the states. To obtain eq. (70), we use the anticommutation relations for the fermion annihilation and creation operators, and first "anticommute a_i through to the right," obtaining a factor (-1) for every occupied state with index $\mu < i$. The resultant $a_i a^\dagger_i$ is then rewritten as $(1 - a^\dagger_i a_i)$. Because the a_i sitting to the right of the creation operator $a^\dagger_i$ can no longer meet an $a^\dagger_i$ in its action to the right, it will simply annihilate the vacuum state and yield zero. Subsequently, a_j is "anticommuted through to the right," yielding a factor (-1) for every occupied state with index $\nu < j$ *except* the state with $\nu = i$, because this state was already eliminated by the prior action with a_i. Similarly, we have, with $r < s$,

$$\begin{aligned}&a^\dagger_r a^\dagger_s \Big(a^\dagger_{k_1} a^\dagger_{k_2} \cdots \qquad \cdots \qquad \cdots a^\dagger_{k_N}\Big)|0\rangle\\ &= (-1)^{\sum_{\nu=1}^{s-1} n_\nu - 1} (-1)^{\sum_{\mu=1}^{r-1} n_\mu} \Big(a^\dagger_{k_1} a^\dagger_{k_2} \cdots a^\dagger_r \cdots a^\dagger_s \cdots a^\dagger_{k_N}\Big)|0\rangle,\end{aligned} \tag{71}$$

where the occupation numbers, n_ν and n_μ, are the occupation numbers of the final N-particle state. In particular, $a^\dagger_s$ does not meet the state with $\nu = r$ as it is "anticommuted through to the right" into its proper position according to our prescription for the indices.

Now, using eqs. (70) and (71), the general two-body operator can be expressed in the new language through

$$G_{\text{op.}} = \tfrac{1}{2} \sum_{a,b,c,d} \langle ab|g|cd\rangle a^\dagger_a a^\dagger_b a_d a_c, \tag{72}$$

yielding the required phases for the N-particle matrix elements of eqs. (66)–(68). In particular, the order of the annihilation operators a_d and a_c is the reverse of the order of the states c and d in the ket of the two-particle matrix elements.

We have derived eqs. (49) and (72) for the case of N-fermion systems. The derivation for the N-boson systems parallels that of the N-fermion system, except for the extra phase factors of (-1), which arise through the anticommutation processes of the fermion systems. These phase factors are missing for the boson systems, where the anticommutation process is replaced by a commutation process. Because these fermion phase factors drop out of the final expressions of eq. (49) for one-body operators and eq. (72) for two-body operators, these expressions are valid for both bosons and fermions. [In the fermion case, however, a term with $a = b$, or with $c = d$ is automatically missing in eq. (72), because $a_a^\dagger a_a^\dagger \equiv 0$, similarly $a_c a_c \equiv 0$, in the fermion case.]

F Examples of Two-Body Operators

Let us return to the fermion case and examine first the simplest two-body operator, the two-body unit operator, which merely counts the number of pairs in the N-body system. In coordinate representation,

$$G_{\text{op.}} = \sum_{i<k}^{N} 1 = \tfrac{1}{2}N(N-1). \tag{73}$$

In the new language,

$$\begin{aligned} G_{\text{op.}} &= \tfrac{1}{2}\sum_{a,b,c,d} \langle ab|1|cd\rangle a_a^\dagger a_b^\dagger a_d a_c = \tfrac{1}{2}\sum_{a,b,c,d} \delta_{ac}\delta_{bd} a_a^\dagger a_b^\dagger a_d a_c \\ &= \tfrac{1}{2}\sum_{a,b} a_a^\dagger a_b^\dagger a_b a_a = -\tfrac{1}{2}\sum_{a,b} a_a^\dagger a_b^\dagger a_a a_b = -\tfrac{1}{2}\sum_{a,b} a_a^\dagger(-a_a a_b^\dagger + \delta_{ab})a_b \\ &= \tfrac{1}{2}\sum_a a_a^\dagger a_a \sum_b a_b^\dagger a_b - \tfrac{1}{2}\sum_a a_a^\dagger a_a = \tfrac{1}{2}(\text{N}_{\text{op.}}^2 - \text{N}_{\text{op.}}) = \tfrac{1}{2}N(N-1). \end{aligned} \tag{74}$$

As an example of a simple two-body interaction, let us write the electron pairing interaction for a condensed matter system where the three orbital quantum numbers are given by the $\vec{k}$ vector, with a finely spaced (effectively continuous) spectrum of allowed $\vec{k}$ values, where

$$\begin{aligned} V &= \sum_{i<k}^{N} v_{ik}^{\text{pairing}} \\ &= \tfrac{1}{2}\sum_{\vec{k}}\sum_{\vec{k}'}\sum_{m_s}\sum_{m_s'} v_{\vec{k}\vec{k}'} a_{\vec{k}m_s}^\dagger a_{-\vec{k}-m_s}^\dagger a_{-\vec{k}'-m_s'} a_{\vec{k}'m_s'}, \end{aligned} \tag{75}$$

where the coefficients, $v_{\vec{k}\vec{k}'}$, are independent of the sign of $\vec{k}$ or of $\vec{k}'$. The interaction thus acts only on electron pairs with zero two-particle momentum and two-particle spin, $S = 0$. In the next chapter, we shall examine in more detail a similar nuclear pairing interaction.

79

Many-Body Techniques: Some Simple Applications

A Construction of All Pauli-Allowed States of a $(d_{5/2})^{N=3}$ Fermion Configuration

As a first simple application, let us construct all allowed states of a $(d_{5/2})^{N=3}$ configuration of identical fermions. The fermions could be electrons in a valence shell of an atom for which the $j-j$ coupling scheme is valid, or they could be identical nucleons, either all protons or all neutrons, in a nuclear shell model valence orbit of a nucleus. Let us first consider the possible states for a system of three distinguishable particles, each with a single-particle angular momentum with $j = 5/2$. They could be distinguishable because they are different particles, e.g., one proton, one neutron, and one electron, or they could be identical particles, say, all neutrons in a nucleus, but with different principal and orbital angular momentum quantum numbers; i.e. $n_1l_1 \neq n_2l_2 \neq n_3l_3$. For this case of three distinguishable particles, each with $j = 5/2$, there are $6 \times 6 \times 6 = 216$ allowed states with total angular momenta, J, given by

$$J = \left(\tfrac{1}{2}\right)^2, \left(\tfrac{3}{2}\right)^4, \left(\tfrac{5}{2}\right)^6, \left(\tfrac{7}{2}\right)^5, \left(\tfrac{9}{2}\right)^4, \left(\tfrac{11}{2}\right)^3, \left(\tfrac{13}{2}\right)^2, \left(\tfrac{15}{2}\right)^1,$$

where the superscripts give the number of occurences of each listed J value. The two independent $J = \frac{1}{2}$ states could be obtained from the two states with $J_{12} = 2$ and $J_{12} = 3$ coupled with the third $j = 5/2$-particle. In general, we have

$$J = [J_{12} \times \tfrac{5}{2}] = [(0 + 1 + 2 + 3 + 4 + 5) \times \tfrac{5}{2}].$$

For three identical $j = \frac{5}{2}$ fermions, however, the set of allowed total J values is much more restricted. Because the three fermions are assumed to have the same values for the quantum numbers, $nlsj$, with $s = \frac{1}{2}$ and $j = \frac{5}{2}$, let us use the abbreviation

$$a^{\dagger}_{nlsj=\frac{5}{2}m} \equiv a^{\dagger}_{m};$$

i.e., for economy of writing the common quantum numbers, $nlsj$, will be quietly understood. A three-particle state can thus be expressed in our new language by

$$a^{\dagger}_{m_1} a^{\dagger}_{m_2} a^{\dagger}_{m_3} |0\rangle,$$

where the "vacuum" state, $|0\rangle$, could be the true vacuum state, with particle number zero, or a closed shell state with $(N-3)$ particles filling lower shell model orbits, which satisfy the condition, $a_{m_k}|0\rangle = 0$, where a_{m_k} is the shorthand notation for the annihilation operator for the $nlsjm$ single-particle state under consideration. Because the above

$$a^{\dagger}_{m_1} a^{\dagger}_{m_2} a^{\dagger}_{m_3} |0\rangle$$

will be identically equal to zero, if $m_1 = m_2$, or if any two m_k are the same, there will be six possible choices for m_1, but once m_1 has been chosen, there are only five possible choices for m_2, and once both m_1 and m_2 have been chosen, only four possible choices for m_3 remain. In addition, any of the possible 3! permutations of the quantum numbers m_k can at most change the sign of the above three-fermion state. Therefore, only

$$\frac{6 \cdot 5 \cdot 4}{3!} = 20$$

independent states exist in a configuration of three identical $j = 5/2$ fermions, in place of the 216 states with J values ranging from $\frac{1}{2}$ to $\frac{15}{2}$ listed above. What are the possible J, M_J values of these 20 states? It will be worthwhile to answer this question through a specific construction of these states. This process will then permit us to make calculations involving such states. The requirement $m_1 \neq m_2 \neq m_3$ shows us the highest possible value of $M_J = m_1 + m_2 + m_3$ is

$$M_J = \tfrac{5}{2} + \tfrac{3}{2} + \tfrac{1}{2} = \tfrac{9}{2}.$$

Moreover, the state

$$a^{\dagger}_{\frac{5}{2}} a^{\dagger}_{\frac{3}{2}} a^{\dagger}_{\frac{1}{2}} |0\rangle = |J = \tfrac{9}{2}, M_J = \tfrac{9}{2}\rangle; \tag{1}$$

i.e., this state has both $M_J = \frac{9}{2}$, and $J = \frac{9}{2}$. This result can be seen by noting

$$J_{+}\Big(a^{\dagger}_{\frac{5}{2}} a^{\dagger}_{\frac{3}{2}} a^{\dagger}_{\frac{1}{2}} |0\rangle\Big) = 0, \tag{2}$$

where the operator, J_+, see eq. (55) of Chapter 78,

$$J_{+} = \sqrt{5} a^{\dagger}_{\frac{5}{2}} a_{\frac{3}{2}} + \sqrt{8} a^{\dagger}_{\frac{3}{2}} a_{\frac{1}{2}} + 3 a^{\dagger}_{\frac{1}{2}} a_{-\frac{1}{2}} + \sqrt{8} a^{\dagger}_{-\frac{1}{2}} a_{-\frac{3}{2}} + \sqrt{5} a^{\dagger}_{-\frac{3}{2}} a_{-\frac{5}{2}}, \tag{3}$$

when acting on the above three-particle state with $M_J = \frac{9}{2}$ can only convert the single-particle creation operator with $m = \frac{3}{2}$ into a single-particle creation operator with $m = \frac{5}{2}$, making a state with $a^\dagger_{m_1} = a^\dagger_{m_2}$, which is identically equal to zero, or it could convert the single-particle creation operator with $m = \frac{1}{2}$ into one with $m = \frac{3}{2}$, again creating a state identically equal to zero.

Now, starting with the state of eq. (1), we can create states with $J = \frac{9}{2}$ and lower values of M_J by successive actions with the operator, J_-, with

$$J_- = \sqrt{5}a^\dagger_{\frac{3}{2}}a_{\frac{5}{2}} + \sqrt{8}a^\dagger_{\frac{1}{2}}a_{\frac{3}{2}} + 3a^\dagger_{-\frac{1}{2}}a_{\frac{1}{2}} + \sqrt{8}a^\dagger_{-\frac{3}{2}}a_{-\frac{1}{2}} + \sqrt{5}a^\dagger_{-\frac{5}{2}}a_{-\frac{3}{2}}. \tag{4}$$

Note,

$$J_-|\tfrac{9}{2}+\tfrac{9}{2}\rangle = \sqrt{9\cdot 1}|\tfrac{9}{2}+\tfrac{7}{2}\rangle = 3a^\dagger_{\frac{5}{2}}a^\dagger_{\frac{3}{2}}a^\dagger_{-\frac{1}{2}}|0\rangle, \tag{5}$$

so the normalized state is

$$|\tfrac{9}{2}+\tfrac{7}{2}\rangle = a^\dagger_{\frac{5}{2}}a^\dagger_{\frac{3}{2}}a^\dagger_{-\frac{1}{2}}|0\rangle. \tag{6}$$

In particular, just one combination of m_1, m_2, m_3, with $\sum_i m_i = +\frac{7}{2}$ and $m_1 \neq m_2 \neq m_3$ exists. Further action with J_- on this state yields

$$J_-|\tfrac{9}{2}+\tfrac{7}{2}\rangle = \sqrt{16}|\tfrac{9}{2}+\tfrac{5}{2}\rangle = \sqrt{8}a^\dagger_{\frac{5}{2}}a^\dagger_{\frac{1}{2}}a^\dagger_{-\frac{1}{2}}|0\rangle + \sqrt{8}a^\dagger_{\frac{5}{2}}a^\dagger_{\frac{3}{2}}a^\dagger_{-\frac{3}{2}}|0\rangle, \tag{7}$$

so

$$|\tfrac{9}{2}+\tfrac{5}{2}\rangle = \frac{1}{\sqrt{2}}\Big(a^\dagger_{\frac{5}{2}}a^\dagger_{\frac{1}{2}}a^\dagger_{-\frac{1}{2}} + a^\dagger_{\frac{5}{2}}a^\dagger_{\frac{3}{2}}a^\dagger_{-\frac{3}{2}}\Big)|0\rangle. \tag{8}$$

Because this state with $M_J = \frac{5}{2}$ is a linear combination of two independent states, two independent states with $M_J = \frac{5}{2}$ must now exist, but with different values of J. The state orthogonal to the above $|\frac{9}{2}+\frac{5}{2}\rangle$ is a state with $J = \frac{5}{2}$,

$$|\tfrac{5}{2}+\tfrac{5}{2}\rangle = \frac{1}{\sqrt{2}}\Big(a^\dagger_{\frac{5}{2}}a^\dagger_{\frac{1}{2}}a^\dagger_{-\frac{1}{2}} - a^\dagger_{\frac{5}{2}}a^\dagger_{\frac{3}{2}}a^\dagger_{-\frac{3}{2}}\Big)|0\rangle. \tag{9}$$

The $J = \frac{5}{2}$ character of this state can be seen by noting the operator J_+ acting on this state yields zero, $J_+|\frac{5}{2}+\frac{5}{2}\rangle = 0$. Further action with J_- on the states of eqs. (8) and (9) yields the two states with $M_J = +\frac{3}{2}$ and $J = \frac{9}{2}$, $J = \frac{5}{2}$, respectively.

$$|\tfrac{9}{2}+\tfrac{3}{2}\rangle = \left[\sqrt{\frac{5}{42}}\Big(a^\dagger_{\frac{3}{2}}a^\dagger_{\frac{1}{2}}a^\dagger_{-\frac{1}{2}} + a^\dagger_{\frac{5}{2}}a^\dagger_{\frac{3}{2}}a^\dagger_{-\frac{5}{2}}\Big) + \frac{4}{\sqrt{21}}a^\dagger_{\frac{5}{2}}a^\dagger_{\frac{1}{2}}a^\dagger_{-\frac{3}{2}}\right]|0\rangle, \tag{10}$$

$$|\tfrac{5}{2}+\tfrac{3}{2}\rangle = \frac{1}{\sqrt{2}}\Big(a^\dagger_{\frac{3}{2}}a^\dagger_{\frac{1}{2}}a^\dagger_{-\frac{1}{2}} - a^\dagger_{\frac{5}{2}}a^\dagger_{\frac{3}{2}}a^\dagger_{-\frac{5}{2}}\Big)|0\rangle. \tag{11}$$

Because these states are linear combinations of three independent states, a third state with $M_J = +\frac{3}{2}$ must exist, orthogonal to the above two. Using this orthogonality, we can construct the state with $J = \frac{3}{2}$ and $M_J = +\frac{3}{2}$,

$$|\tfrac{3}{2}+\tfrac{3}{2}\rangle = \left[\sqrt{\frac{8}{21}}\Big(a^\dagger_{\frac{3}{2}}a^\dagger_{\frac{1}{2}}a^\dagger_{-\frac{1}{2}} + a^\dagger_{\frac{5}{2}}a^\dagger_{\frac{3}{2}}a^\dagger_{-\frac{5}{2}}\Big) - \sqrt{\frac{5}{21}}a^\dagger_{\frac{5}{2}}a^\dagger_{\frac{1}{2}}a^\dagger_{-\frac{3}{2}}\right]|0\rangle, \tag{12}$$

where action with J_+ on this state again yields zero. We have thus been able to create states with $J=\frac{9}{2}$, $J=\frac{5}{2}$, $J=\frac{3}{2}$, accounting for the $10+6+4=20$ allowed states of the $(\frac{5}{2})^3$ configuration of identical $j=\frac{5}{2}$ fermions. States with M_J-values $< +\frac{3}{2}$ could now be created by further action with J_-. Note, however, that judicious use of the Wigner–Eckart theorem may make it unnecessary to construct states with all possible values of M_J. Also, this method of giving a very explicit construction of the states of good J and M_J automatically gives normalized states through the known matrix elements of J_-. Alternatively, we could have constructed states of good J and M_J via angular momentum coupling techniques, by coupling first two particles to a particular value of J_{12} and then coupling this state with the third particle of $j=\frac{5}{2}$ to resultant total J. As a very specific example, the state with $J=M_J=\frac{5}{2}$ could have been obtained by coupling a two-particle state with $J_{12}=0$ to the third state with $j=m_j=\frac{5}{2}$. Using the Clebsch–Gordan coefficient,

$$\langle jmj-m|00\rangle=\frac{(-1)^{j-m}}{\sqrt{(2j+1)}},$$

we have

$$\begin{aligned}|J_{12}=0, M_{12}=0\rangle &= \sum_m \frac{(-1)^{j-m}}{\sqrt{(2j+1)}}a^\dagger_m a^\dagger_{-m}|0\rangle\\ &= \frac{1}{\sqrt{6}}(a^\dagger_{\frac{5}{2}}a^\dagger_{-\frac{5}{2}} - a^\dagger_{\frac{3}{2}}a^\dagger_{-\frac{3}{2}} + a^\dagger_{\frac{1}{2}}a^\dagger_{-\frac{1}{2}} - a^\dagger_{-\frac{1}{2}}a^\dagger_{\frac{1}{2}} + a^\dagger_{-\frac{3}{2}}a^\dagger_{\frac{3}{2}} - a^\dagger_{-\frac{5}{2}}a^\dagger_{\frac{5}{2}})|0\rangle\\ &= \frac{2}{\sqrt{6}}(a^\dagger_{\frac{5}{2}}a^\dagger_{-\frac{5}{2}} - a^\dagger_{\frac{3}{2}}a^\dagger_{-\frac{3}{2}} + a^\dagger_{\frac{1}{2}}a^\dagger_{-\frac{1}{2}})|0\rangle, \end{aligned} \tag{13}$$

where we have used $a^\dagger_{-m}a^\dagger_m=-a^\dagger_m a^\dagger_{-m}$ via the fermion anticommutation relation. This state, however, is now not normalized and would have to be multiplied by an additional factor of $\frac{1}{\sqrt{2}}$ to yield the *normalized* state with $J_{12}=M_{12}=0$,

$$|J_{12}=0, M_{12}=0\rangle=\sum_{m>0}\frac{(-1)^{j-m}}{\sqrt{(j+\frac{1}{2})}}a^\dagger_m a^\dagger_{-m}|0\rangle. \tag{14}$$

Coupling this $J_{12}=0$ two-particle state to a third state with $j=m_j=\frac{5}{2}$, we get the three-particle state

$$\begin{aligned}|J=\tfrac{5}{2}, M_J=+\tfrac{5}{2}\rangle &= a^\dagger_{\frac{5}{2}}\sum_{m>0}\frac{(-1)^{j-m}}{\sqrt{(j+\frac{1}{2})}}a^\dagger_m a^\dagger_{-m}|0\rangle\\ &= \frac{1}{\sqrt{3}}a^\dagger_{\frac{5}{2}}\left(-a^\dagger_{\frac{3}{2}}a^\dagger_{-\frac{3}{2}} + a^\dagger_{\frac{1}{2}}a^\dagger_{-\frac{1}{2}}\right)|0\rangle, \end{aligned} \tag{15}$$

where the relation $a^\dagger_{\frac{5}{2}}a^\dagger_{\frac{5}{2}}\equiv 0$ has eliminated one of the terms in the above sum. Note, in particular, that a renormalization factor of $\sqrt{3/2}$ is therefore needed to convert this state into the normalized state given by eq. (9). States with $J=\frac{9}{2}$ and $J=\frac{3}{2}$ could be constructed by similar angular momentum coupling techniques.

Also, if we attempt to construct a three-identical-fermion state with $J = \frac{1}{2}$ or $J = \frac{7}{2}$, say, the latter by coupling a two-particle state with $J_{12} = 2$ with a third particle with $j = \frac{5}{2}$ to resultant $J = \frac{7}{2}$, we would discover this state is identically zero and, hence, Pauli-forbidden. We could therefore also discover which states are Pauli-forbidden by this technique, using the anticommutation relations of the $a_m^\dagger$.

B Calculation of an Electric Dipole Transition Probability

To show how the three-particle states of the last section can be used in an actual application, let us calculate the electric dipole transition probability for the transitions from the three-particle configurations

$$(nd_{\frac{5}{2}})^2_{J=0}(n'p_{\frac{3}{2}})^1_{J=\frac{3}{2}} \longrightarrow (nd_{\frac{5}{2}})^3_J, \qquad \text{with } J = \tfrac{5}{2} \text{ and } \tfrac{3}{2}.$$

The transition to the final state with $J = \frac{9}{2}$ is forbidden by an angular-momentum selection rule.

To get the needed reduced matrix element for the transition $J = \frac{3}{2} \to \frac{5}{2}$, it will be sufficient to calculate

$$\langle (nd_{\frac{5}{2}})^3 J = \tfrac{5}{2}, M_J = \tfrac{5}{2} | \mu^{(\text{el.})}_{+1} | (nd_{\frac{5}{2}})^2_{J=0}(n'p_{\frac{3}{2}})^1 J = \tfrac{3}{2}, M_J = \tfrac{3}{2}\rangle.$$

Using the expansion of the one-body operator, $\mu^{(\text{el.})}_{+1}$, in the form

$$\mu^{(\text{el.})}_{+1} = \sum_{n'l'j'm'_j} \sum_{nljm_j} \langle n[l\tfrac{1}{2}]jm_j | \mu^{(\text{el.})}_{+1} | n'[l'\tfrac{1}{2}]j'm'_j\rangle a^\dagger_{n[l\frac{1}{2}]jm_j} a_{n'[l'\frac{1}{2}]j'm'_j}, \tag{16}$$

only the term with the above value of n', $l' = 1$, $j' = m'_j = \frac{3}{2}$ and the above value of n, $l = 2$, and $j = m_j = \frac{5}{2}$ can contribute to the above three-particle matrix element, so, effectively,

$$\mu^{(\text{el.})}_{+1} = \cdots + \langle (nd_{\frac{5}{2}})m_j = \tfrac{5}{2} | \mu^{(\text{el.})}_{+1} | (n'p_{\frac{3}{2}})m'_j = \tfrac{3}{2}\rangle a^\dagger_{n[2\frac{1}{2}]\frac{5}{2}\frac{5}{2}} a_{n'[1\frac{1}{2}]\frac{3}{2}\frac{3}{2}} + \cdots, \tag{17}$$

i.e., terms indicated by $\cdots$ can make no contribution to the needed three-particle matrix element. Using eq. (14) for the two-particle state with $J_{12} = 0$, we have

$$|(nd_{\frac{5}{2}})^2_{J=0}(n'p_{\frac{3}{2}})^1 J = \tfrac{3}{2}, M_J = \tfrac{3}{2}\rangle = \frac{1}{\sqrt{3}} a^\dagger_{n'[1\frac{1}{2}]\frac{3}{2}\frac{3}{2}}$$
$$\times \left(a^\dagger_{n[2\frac{1}{2}]\frac{5}{2}\frac{5}{2}} a^\dagger_{n[2\frac{1}{2}]\frac{5}{2}-\frac{5}{2}} - a^\dagger_{n[2\frac{1}{2}]\frac{5}{2}\frac{3}{2}} a^\dagger_{n[2\frac{1}{2}]\frac{5}{2}-\frac{3}{2}} + a^\dagger_{n[2\frac{1}{2}]\frac{5}{2}\frac{1}{2}} a^\dagger_{n[2\frac{1}{2}]\frac{5}{2}-\frac{1}{2}} \right), \tag{18}$$

and, using eq. (17), we get

$$\mu^{(\text{el.})}_{+1} |(nd_{\frac{5}{2}})^2_{J=0}(n'p_{\frac{3}{2}})^1 J = \tfrac{3}{2}, M_J = \tfrac{3}{2}\rangle$$
$$= \frac{1}{\sqrt{3}} a^\dagger_{n[2\frac{1}{2}]\frac{5}{2}\frac{5}{2}} \left(-a^\dagger_{n[2\frac{1}{2}]\frac{5}{2}\frac{3}{2}} a^\dagger_{n[2\frac{1}{2}]\frac{5}{2}-\frac{3}{2}} + a^\dagger_{n[2\frac{1}{2}]\frac{5}{2}\frac{1}{2}} a^\dagger_{n[2\frac{1}{2}]\frac{5}{2}-\frac{1}{2}} \right)$$

$$\times \langle (nd_{\frac{5}{2}})m_j = \tfrac{5}{2}|\mu^{(\mathrm{el.})}_{+1}|(n'p_{\frac{3}{2}})m'_j = \tfrac{3}{2}\rangle,$$
$$+\cdots, \tag{19}$$

where we have again used $a^{\dagger}_{n[2\frac{1}{2}]\frac{5}{2}\frac{5}{2}}a^{\dagger}_{n[2\frac{1}{2}]\frac{5}{2}\frac{5}{2}} \equiv 0$. Thus, comparing eq. (19) with eq. (9) (in abbreviated form), we have

$$\mu^{(\mathrm{el.})}_{+1}|(nd_{\frac{5}{2}})^2_{J=0}(n'p_{\frac{3}{2}})^1 J = \tfrac{3}{2}, M_J = \tfrac{3}{2}\rangle = \sqrt{\frac{2}{3}}|(nd_{\frac{5}{2}})^3 J = M_J = \tfrac{5}{2}\rangle$$
$$\times \langle (nd_{\frac{5}{2}})m_j = \tfrac{5}{2}|\mu^{(\mathrm{el.})}_{+1}|(n'p_{\frac{3}{2}})m'_j = \tfrac{3}{2}\rangle + \cdots, \tag{20}$$

so

$$\langle (nd_{\frac{5}{2}})^3 J = M_J = \tfrac{5}{2}|\mu^{(\mathrm{el.})}_{+1}|(nd_{\frac{5}{2}})^2_{J=0}(n'p_{\frac{3}{2}})^1 J = M_J = \tfrac{3}{2}\rangle$$
$$= \sqrt{\frac{2}{3}}\langle (nd_{\frac{5}{2}})m_j = \tfrac{5}{2}|\mu^{(\mathrm{el.})}_{+1}|(n'p_{\frac{3}{2}})m'_j = \tfrac{3}{2}\rangle, \tag{21}$$

and the three-particle reduced matrix element is related to the single-particle reduced matrix element by the same factor

$$\langle (nd_{\frac{5}{2}})^3 J = \tfrac{5}{2}\|\mu^{(\mathrm{el.})}\|(nd_{\frac{5}{2}})^2_{J=0}(n'p_{\frac{3}{2}})^1 J = \tfrac{3}{2}\rangle = \sqrt{\frac{2}{3}}\langle (nd_{\frac{5}{2}})\|\mu^{(\mathrm{el.})}\|(n'p_{\frac{3}{2}})\rangle. \tag{22}$$

Because the transition probabilities are proportional to the squares of the electric dipole moment reduced matrix elements, the transition probability for the $J = \frac{3}{2} \to \frac{5}{2}$ transition in the three-particle system is related to the corresponding transition in the single-particle system via

$$\left[\frac{1}{\tau_{\mathcal{E}1}}\right]_{(nd_{\frac{5}{2}})^2_{J=0}n'p_{\frac{3}{2}} \to (nd_{\frac{5}{2}})^3_{J=\frac{5}{2}}} = \frac{2}{3}\left[\frac{1}{\tau_{\mathcal{E}1}}\right]_{n'p_{\frac{3}{2}} \to nd_{\frac{5}{2}}} \frac{\omega^3_{3.p.}}{\omega^3_{\mathrm{s.p.}}}, \tag{23}$$

where we have taken account of the fact that the transition frequencies, ω, will be different in the three-particle and single-particle systems.

Finally, to calculate the transition probability for the transition

$$(nd_{\frac{5}{2}})^2_{J=0}(n'p_{\frac{3}{2}})^1_{J=\frac{3}{2}} \longrightarrow (nd_{\frac{5}{2}})^3_{J=\frac{3}{2}},$$

it will be sufficient to calculate

$$\langle (nd_{\frac{5}{2}})^3 J = \tfrac{3}{2}, M_J = \tfrac{3}{2}|\mu^{(\mathrm{el.})}_{+1}|(nd_{\frac{5}{2}})^2_{J=0}(n'p_{\frac{3}{2}})^1 J = \tfrac{3}{2}, M_J = \tfrac{1}{2}\rangle.$$

Now, we have

$$\mu^{(\mathrm{el.})}_{+1}|(nd_{\frac{5}{2}})^2_{J=0}(n'p_{\frac{3}{2}})^1 J = \tfrac{3}{2}, M_J = \tfrac{1}{2}\rangle$$
$$= \frac{1}{\sqrt{3}}a^{\dagger}_{n[2\frac{1}{2}]\frac{5}{2}\frac{3}{2}}\left(a^{\dagger}_{n[2\frac{1}{2}]\frac{5}{2}\frac{5}{2}}a^{\dagger}_{n[2\frac{1}{2}]\frac{5}{2}-\frac{5}{2}} + a^{\dagger}_{n[2\frac{1}{2}]\frac{5}{2}\frac{1}{2}}a^{\dagger}_{n[2\frac{1}{2}]\frac{5}{2}-\frac{1}{2}}\right)|0\rangle$$
$$\times \langle (nd_{\frac{5}{2}})m_j = \tfrac{3}{2}|\mu^{(\mathrm{el.})}_{+1}|(n'p_{\frac{3}{2}})m'_j = \tfrac{1}{2}\rangle + \cdots, \tag{24}$$

where we have now used $a^{\dagger}_{n[2\frac{1}{2}]\frac{5}{2}\frac{3}{2}}a^{\dagger}_{n[2\frac{1}{2}]\frac{5}{2}\frac{3}{2}} \equiv 0$. Comparing the right-hand side with eq. (11), we have

$$\mu^{(\mathrm{el.})}_{+1}|(nd_{\frac{5}{2}})^2_{J=0}(n'p_{\frac{3}{2}})^1 J = \tfrac{3}{2}, M_J = \tfrac{1}{2}\rangle$$

$$= \sqrt{\frac{2}{3}} |(nd_{\frac{5}{2}})^3 J = \tfrac{5}{2} M_J = \tfrac{3}{2}\rangle$$
$$\times \langle (nd_{\frac{5}{2}}) m_j = \tfrac{3}{2} | \mu^{(\text{el.})}_{+1} | (n' p_{\frac{3}{2}}) m'_j = \tfrac{1}{2}\rangle + \cdots \quad . \tag{25}$$

Now, however, the states

$$|(nd_{\frac{5}{2}})^3 J = \tfrac{5}{2} M_J = \tfrac{3}{2}\rangle$$

and

$$|(nd_{\frac{5}{2}})^3 J = \tfrac{3}{2} M_J = \tfrac{3}{2}\rangle$$

are orthogonal to each other, [cf. eqs. (11) and (12)]. Thus,

$$\langle (nd_{\frac{5}{2}})^3 J = \tfrac{3}{2}, M_J = \tfrac{3}{2} | \mu^{(\text{el.})}_{+1} | (nd_{\frac{5}{2}})^2_{J=0} (n' p_{\frac{3}{2}})^1 J = \tfrac{3}{2}, M_J = \tfrac{1}{2}\rangle = 0. \tag{26}$$

Thus, the transition

$$(nd_{\frac{5}{2}})^2_{J=0} (n' p_{\frac{3}{2}})^1_{J=\frac{3}{2}} \longrightarrow (nd_{\frac{5}{2}})^3_{J=\frac{3}{2}}$$

is forbidden, even though it is not ruled out by an angular momentum selection rule, and even though the single-particle transition

$$n' p_{\frac{3}{2}} \longrightarrow nd_{\frac{5}{2}}$$

is allowed.

C Pairing Forces in Nuclei

Two-particle states coupled to angular momentum, $J = 0$, are of particular relevance in nuclear physics, especially in configurations of identical nucleons, because the nucleon–nucleon interaction is a short-range attractive interaction. Because identical nucleons with the same n and l in states $|jm\rangle$ and $|j - m\rangle$ have the same spatial distribution, two identical nucleons coupled to two-particle $J = 0$ have a large probability of being within the range of the attractive nucleon–nucleon interaction. Low-lying states of nuclei thus have a larger probability of containing $J = 0$ pairs compared with pairs coupled to $J > 0$. In particular, the ground states of all known stable even–even nuclei (with even neutron and proton numbers) are observed to have spin and parity 0^+. States of two identical nucleons, either both neutrons or both protons in a shell model state with the same nl, coupled to $J = 0$, are given by eq. (14). It will therefore be useful to define nucleon pair creation operators

$$A^\dagger_j = \sum_{m>0} (-1)^{j-m} a^\dagger_{jm} a^\dagger_{j-m}, \tag{27}$$

where we have omitted the common quantum numbers nl in the single nucleon creation operators (for the sake of brevity of notation) and we have left off the two-particle normalization factor, $[(j + \frac{1}{2})]^{-\frac{1}{2}}$, to simplify the anticommutator algebra of such operators. We also define the pair annihilation operators

$$A_j = (A_j^\dagger)^\dagger = \sum_{m>0} (-1)^{j-m} a_{j-m} a_{jm}. \tag{28}$$

Normalized states of p pairs of $J = 0$-coupled identical nucleon pairs are given by

$$\sqrt{\frac{(j+\frac{1}{2}-p)!}{(j+\frac{1}{2})!p!}}(A_j^\dagger)^p|0\rangle. \tag{29}$$

The overall normalization factor can be understood as follows: For one pair coupled to $J = 0$, $(j+\frac{1}{2})$ terms of the type $(-1)^{j-m} a^\dagger_{jm} a^\dagger_{j-m}$ exist. For $p = 2$, once one of these terms is fixed at some specific $m = m_1 > 0$, the second pair operator can multiply this term with only $(j - \frac{1}{2})$ terms with $m_2 \neq m_1$, but the two terms with m_1 and m_2 interchanged are identical, because the anticommutation relations of the fermion operators yield

$$a^\dagger_{jm_1} a^\dagger_{j-m_1} a^\dagger_{jm_2} a^\dagger_{j-m_2} = a^\dagger_{jm_2} a^\dagger_{j-m_2} a^\dagger_{jm_1} a^\dagger_{j-m_1},$$

so that there are $\frac{1}{2}(j+\frac{1}{2})(j-\frac{1}{2})$ independent terms but each appears with a factor of 2 for the case of two pairs, $p = 2$. For higher p, by the same arguments, there will be $(j+\frac{1}{2})(j-\frac{1}{2})\cdots(j+\frac{3}{2}-p)/p!$ independent terms, but each will appear with a factor of $p!$, requiring a normalization factor of

$$\frac{1}{p!}\sqrt{\frac{p!(j+\frac{1}{2}-p)!}{(j+\frac{1}{2})!}}.$$

In rough approximation, the nucleon–nucleon interaction could be replaced by an attractive pairing interaction that acts only when nucleons are coupled in pairs to $J = 0$. Although a realistic interaction will have finite terms involving nucleon pairs coupled to $J \neq 0$, the highly simplified pairing interaction will give a qualitative account of the nuclear spectra involving the lowest energy states. A general pairing interaction would be given by a Hamiltonian of the form

$$H_{\text{pairing}} = \tfrac{1}{2}\sum_{j,j'} g_{jj'} A_j^\dagger A_{j'}. \tag{30}$$

Moreover, the strength factors, $g_{jj'}$, can be taken approximately j independent and negative (atttractive interaction), so

$$\begin{aligned} H_{\text{pairing}} &= -\tfrac{1}{2}G\sum_{jj'} A_j^\dagger A_{j'} \\ &= -\tfrac{1}{2}G\sum_{jj'}\sum_{m>0}\sum_{m'>0}(-1)^{j-m} a^\dagger_{jm} a^\dagger_{j-m}(-1)^{j'-m'} a_{j'-m'} a_{j'm'}. \end{aligned} \tag{31}$$

Let us examine the eigenvalues of such a Hamiltonian in a single j shell, by adding a one-body operator dependent on the single-particle shell-model energy, ϵ_j, of this shell, so

$$H = \epsilon_j \sum_m a^\dagger_{jm} a_{jm} - \tfrac{1}{2}G\sum_{m>0}\sum_{m'>0}(-1)^{j-m} a^\dagger_{jm} a^\dagger_{j-m}(-1)^{j-m'} a_{j-m'} a_{jm'}$$

$$= \epsilon_j N_{\text{op.}} - \tfrac{1}{2} G A^\dagger A, \tag{32}$$

where we have left off the subscript j on $A^\dagger$ and A for brevity of notation in the single j shell. It will be useful to calculate the commutator, $[A, A^\dagger]$. Using the anticommutation relations of the single-particle creation and annihilation operators to move the two a's through to the right in the first term of this commutator, we have

$$\begin{aligned}
[A, A^\dagger] &= [\sum_{m>0} a_{j-m} a_{jm} (-1)^{j-m}, \sum_{m'>0} a^\dagger_{jm'} a^\dagger_{j-m'} (-1)^{j-m'}] \\
&= \sum_{m>0} (a_{j-m} a^\dagger_{j-m} - a^\dagger_{jm} a_{jm}) = \sum_{m>0} (-a^\dagger_{j-m} a_{j-m} - a^\dagger_{jm} a_{jm} + 1) \\
&= -\sum_{\text{all } m} a^\dagger_{jm} a_{jm} + (j + \tfrac{1}{2}) = -\Big(N_{\text{op.}} - (j + \tfrac{1}{2})\Big) \equiv -\mathcal{N}_{\text{op.}}.
\end{aligned} \tag{33}$$

We want to calculate the eigenvalue of the simplified Hamiltonian of eq. (32) for the state of eq. (29) built from p $J = 0$-pairs. For this purpose, we need to calculate $-GA^\dagger A (A^\dagger)^p |0\rangle$. Using the commutator result of eq. (33), we have

$$\begin{aligned}
&- GA^\dagger A \left(A^\dagger\right)^p |0\rangle \\
&= +GA^\dagger \Big(\mathcal{N}_{\text{op.}} \left(A^\dagger\right)^{p-1} + A^\dagger \mathcal{N}_{\text{op.}} \left(A^\dagger\right)^{p-2} + \cdots \\
&\qquad + \left(A^\dagger\right)^k \mathcal{N}_{\text{op.}} \left(A^\dagger\right)^{p-k-1} + \cdots + \left(A^\dagger\right)^{p-1} \mathcal{N}_{\text{op.}} \Big) |0\rangle \\
&= +G \sum_{k=0}^{p-1} [2(p-k-1) - (j + \tfrac{1}{2})] \left(A^\dagger\right)^p |0\rangle \\
&= +G \Big([2(p-1) - (j + \tfrac{1}{2})] p - 2\tfrac{1}{2} p(p-1) \Big) \left(A^\dagger\right)^p |0\rangle \\
&= -Gp(j + \tfrac{3}{2} - p) \left(A^\dagger\right)^p |0\rangle.
\end{aligned} \tag{34}$$

With $p = \frac{1}{2} n$, where n is the number of nucleons in the p-pair state with $J = 0$, the full Hamiltonian of eq. (32) thus yields

$$H \sqrt{\frac{(j + \frac{1}{2} - p)!}{(j + \frac{1}{2})! p!}} (A_j^\dagger)^p |0\rangle = E_{j,n} \sqrt{\frac{(j + \frac{1}{2} - p)!}{(j + \frac{1}{2})! p!}} (A_j^\dagger)^p |0\rangle, \tag{35}$$

where

$$E_{j,n} = n\epsilon_j - G \frac{n}{4} (2j + 3 - n). \tag{36}$$

It will be interesting to compare this energy for the $J = 0$ ground state for an even number of nucleons, with the corresponding energy for an odd number of nucleons, where one nucleon must be in the state with angular momentum j, so the total angular momentum of the ground state for $n =$ odd must be $J = j$. Such a state can be approximated by

$$\sqrt{\frac{(j - \frac{1}{2} - p)!}{p! (j - \frac{1}{2})!}} \left(A^\dagger\right)^p a^\dagger_{jm_0} |0\rangle, \tag{37}$$

where the single-particle creation operator with $m = m_0$ "blocks" this m_0 value from the pair creation operators; i.e., the first operator $A^\dagger$ acting on the single particle state will have only $(j - \frac{1}{2})$ possible $|m|$-values, whereas terms with $p > 1$ will now lead to a state with $[(j - \frac{1}{2})!/p!(j - \frac{1}{2} - p)!]$ independent pair terms, each appearing with a factor of $p!$, leading to the new normalization factor of eq. (37). For this state, we need

$$\begin{aligned} &- GA^\dagger A\big(A^\dagger\big)^p a^\dagger_{jm_0}|0\rangle \\ &= +GA^\dagger\left(\sum_{k=0}^{p-1}\big(A^\dagger\big)^k \mathcal{N}_{\text{op.}}\big(A^\dagger\big)^{p-1-k}\right)a^\dagger_{jm_0}|0\rangle \\ &= +G\sum_{k=0}^{p-1}\big[2(p-1-k)+1-(j+\tfrac{1}{2})\big]\big(A^\dagger\big)^p a^\dagger_{jm_0}|0\rangle \\ &= -Gp(j+\tfrac{1}{2}-p)\big(A^\dagger\big)^p a^\dagger_{jm_0}|0\rangle. \end{aligned} \tag{38}$$

Now, we have $p = \frac{1}{2}(n-1)$, with $n =$ odd, leading to the eigenvalue

$$E_{j,n} = n\epsilon_j - G\frac{(n-1)}{4}(2j+2-n) \tag{39}$$

for the ground state of an odd-n nucleus with $J = j$.

The excited states of an even-n nucleus with $J \neq 0$ would be expected to have the structure

$$\sum_{m_1,m_2}\langle jm_1 jm_2|JM_J\rangle\sqrt{\frac{(j-\frac{3}{2}-p)!}{p!(j-\frac{3}{2})!}}\big(A^\dagger\big)^p a^\dagger_{jm_1}a^\dagger_{jm_2}|0\rangle. \tag{40}$$

Because the energy of this state is independent of M_J, we can choose $M_J \neq 0$ so $m_2 \neq -m_1$. The two single-particle creation operators thus "block" two m values from the pair operators, leading to the new normalization factor. Now, we need

$$\begin{aligned} &- GA^\dagger A\big(A^\dagger\big)^p a^\dagger_{jm_1}a^\dagger_{jm_2}|0\rangle \\ &= +GA^\dagger\left(\sum_{k=0}^{p-1}\big(A^\dagger\big)^k \mathcal{N}_{\text{op.}}\big(A^\dagger\big)^{p-k-1}\right)a^\dagger_{jm_1}a^\dagger_{jm_2}|0\rangle \\ &= +G\sum_{k=0}^{p-1}\big[2(p-k-1)+2-(j+\tfrac{1}{2})\big]\big(A^\dagger\big)^p a^\dagger_{jm_1}a^\dagger_{jm_2}|0\rangle \\ &= -Gp(j-\tfrac{1}{2}-p)\big(A^\dagger\big)^p a^\dagger_{jm_1}a^\dagger_{jm_2}|0\rangle, \end{aligned} \tag{41}$$

where we now have $p = \frac{1}{2}(n-2)$, with $n=$ even, so for such $J \neq 0$ states in the even nucleus

$$E_{j,n} = n\epsilon_j - G\frac{(n-2)}{4}(2j+1-n). \tag{42}$$

Comparing with eq. (36), the energy of the $J = 0$ ground state, these $J \neq 0$ states lie at an excitation energy of $\frac{1}{2}G(2j+1)$ above the $J = 0$ ground state. In a configuration j^n of identical nucleons, either all neutrons or all protons, with $n =$ even, the Pauli-allowed states with $J \neq 0$ have $J = 2, 4, \ldots, (2j-1)$. For our simplified interaction, these excited states would all be degenerate at an energy $\frac{1}{2}G(2j+1)$ above the $J=0$ ground state. In a real nucleus, this degeneracy is lifted, but the energy differences between the different levels with $J \neq 0$ are in general small compared with the energy difference between the average energy of the $J \neq 0$ levels and the $J=0$ ground state.

D The Coulomb Repulsion Term in the Z-Electron Atom

The Hamiltonian for the Z-electron atom consists of a one-body and a two-body term

$$H = \sum_{i=1}^{Z}\Big(\frac{\vec{p}_i^{\,2}}{2m} - \frac{Ze^2}{r_i}\Big) + \sum_{i<k}^{Z}\frac{e^2}{r_{ik}} = \sum_{i=1}^{Z} h_i + \sum_{i<k} V_{ik}. \tag{43}$$

In the language of many-body theory, this can now be written as

$$H = \sum_{nlmm_s} \epsilon_n a^\dagger_{nlmm_s} a_{nlmm_s} + \tfrac{1}{2}\sum_{a,b,c,d}\langle ab|\frac{e^2}{r_{12}}|cd\rangle a^\dagger_a a^\dagger_b a_d a_c, \tag{44}$$

where a is shorthand for $n_a l_a m_a m_{s_a}$. Because the Coulomb repulsion potential is a scalar operator, a spherical tensor of rank 0, and is spin independent, it will be useful to transform the two-particle states to states of good two-particle angular momentum, L and S,

$$|ab\rangle = |n_a l_a m_a m_{s_a} n_b l_b m_b m_{s_b}\rangle = \sum_{L,S}\langle l_a m_a l_b m_b|LM\rangle\langle \tfrac{1}{2}m_{s_a}\tfrac{1}{2}m_{s_b}|SM_S\rangle|[n_a l_a n_b l_b]LM, SM_S\rangle, \tag{45}$$

where M is fixed by $m_a + m_b$, similarly for M_S. It will now be useful to define pair-creation operators for electron pairs coupled to general L and S

$$A^\dagger_{ab,LMSM_S} = \sum_{m_a,(m_b)}\sum_{m_{s_a},(m_{s_b})}\langle l_a m_a l_b m_b|LM\rangle\langle \tfrac{1}{2}m_{s_a}\tfrac{1}{2}m_{s_b}|SM_S\rangle a^\dagger_{n_a l_a m_a m_{s_a}} a^\dagger_{n_b l_b m_b m_{s_b}}, \tag{46}$$

where these are the generalizations of the $J=0$ pair creation operators of the last section. In terms of these operators, we then have

$$H = \sum_{nlmm_s}\epsilon_n a^\dagger_{nlmm_s} a_{nlmm_s} + \tfrac{1}{2}\sum\sum_{n_a l_a n_b l_b n_c l_c n_d l_d} \times \sum_{L}\langle [n_a l_a n_b l_b]L|\frac{e^2}{r_{12}}|[n_c l_c n_d l_d]L\rangle \sum_{M}\sum_{S,M_S} A^\dagger_{ab,LMSM_S} A_{cd,LMSM_S}, \tag{47}$$

where the two-particle matrix element of the scalar operator, e^2/r_{12}, is independent of M via the Wigner–Eckart theorem. It could be evaluated by expanding $1/r_{12}$ in spherical harmonics [see, e.g., eqs. (22)–(24) of Chapter 38 and eq. (32) of Chapter 30] and by choosing a particular (convenient) value of M. Thus,

$$
\begin{aligned}
&\langle [n_a l_a n_b l_b] LM | \frac{e^2}{r_{12}} | [n_c l_c n_d l_d] LM \rangle \\
&= \sum_{k=0}^{\infty} \sqrt{\frac{(2l_c+1)(2l_d+1)}{(2l_a+1)(2l_b+1)}} \langle l_c 0 k 0 | l_a 0 \rangle \langle l_d 0 k 0 | l_b 0 \rangle \\
&\times \sum_q \sum_{m_a,(m_b)} \sum_{m_c,(m_d)} \langle l_a m_a l_b m_b | LM \rangle \langle l_c m_c l_d m_d | LM \rangle \\
&\qquad \langle l_c m_c k q | l_a m_a \rangle \langle l_d m_d k - q | l_b m_b \rangle (-1)^q \\
&\times \int_0^\infty dr_1 r_1^2 R^*_{n_a l_a}(r_1) R_{n_c l_c}(r_1) \left[\int_0^{r_1} dr_2 r_2^2 R^*_{n_b l_b}(r_2) \frac{r_2^k}{r_1^{k+1}} R_{n_d l_d}(r_2) \right. \\
&\qquad \left. + \int_{r_1}^\infty dr_2 r_2^2 R^*_{n_b l_b}(r_2) \frac{r_1^k}{r_2^{k+1}} R_{n_d l_d}(r_2) \right]. \qquad (48)
\end{aligned}
$$

The sum over the magnetic quantum numbers $q, m_a, (m_b), m_c, (m_d)$ of the product of four Clebsch–Gordan coefficients [including the phase factor $(-1)^q$] could be evaluated directly. Alternatively (see Chapter 34 on recoupling coefficients), it can be expressed through a single recoupling coefficient of the 6-j type and has the value

$$
(-1)^{l_b+l_c+L} \sqrt{(2l_a+1)(2l_b+1)} \begin{Bmatrix} l_c & k & l_a \\ l_b & L & l_d \end{Bmatrix}.
$$

So far, in this section, it has been assumed the radial functions are hydrogenic, valid for the Z of the atom under consideration, i.e., that the $\psi_{nlmm_s}(\vec{r}_i)$ are the eigenfunctions of the hydrogenic single-particle Hamiltonian, h_i, of eq. (43). In a heavy atom, however, it may be very useful to add a screening potential, $V_{\text{screening}}(r_i)$, to the central Coulomb potential, $-Ze^2/r_i$, of h_i, where $V_{\text{screening}}(r_i)$ gives the average of the Coulomb repulsion terms of the $(Z-1)$ remaining electrons on the electron labeled with index i. The inner shell electrons, in particular, will screen the full charge Z of the atomic nucleus, and this screening will depend on the quantum numbers nl. The full Hamiltonian could then be written

$$
H = \sum_{i=1}^{Z} h_i + \sum_{i<k}^{Z} V_{ik}, \qquad \text{with}
$$

$$
h_i = \frac{\vec{p}_i^{\,2}}{2m} - \frac{Ze^2}{r_i} + V_{\text{scr.}}(r_i), \qquad \text{and } V_{ik} = \frac{e^2}{r_{ik}} - V_{\text{scr.}}(r_i), \qquad (49)
$$

where the eigenfunctions of h_i would be improved single-particle functions for the many-body problem and might be expanded in the form,

$$
\sum_n c_n \psi_{nlmm_s}(\vec{r}_i),
$$

where the c_n might be determined by the variational technique.

E Hartree–Fock Theory for Atoms: A Brief Introduction

In the language of many-body theory, the atomic Hamiltonian of eq. (49) is

$$H = \sum_{m,j} \langle m|h_1|j\rangle a^\dagger_m a_j + \tfrac{1}{2}\sum_{abcd} \langle ab|V_{12}|cd\rangle a^\dagger_a a^\dagger_b a_d a_c. \tag{50}$$

Here the single-particle Hamiltonian, h_i, may not be diagonal in the chosen single-particle basis, $|j\rangle$, so off-diagonal terms have been included in the matrix, $\langle m|h_1|j\rangle$. If the screening potential, $V_{\text{scr.}}(r_i)$, which is now included in h_i, has been well chosen, however, it may be possible to approximate the full Hamiltonian by an effective single-particle Hamiltonian. That is, of the three types of nonzero Z-particle matrix elements of the two-body operator, V_{ik} [see eqs. (66)–(68) of Chapter 78], only those with

$$(1): \qquad \alpha'_1\alpha'_2\cdots\alpha'_Z = \alpha_1\alpha_2\cdots\alpha_Z$$

and with

$$(2): \qquad \alpha'_1\alpha'_2\cdots\alpha'_Z = \alpha_1\alpha_2\cdots\alpha_m\cdots\alpha_Z$$

$$\alpha_1\alpha_2\cdots\alpha_Z = \alpha_1\alpha_2\cdots\alpha_j\cdots\alpha_Z, \qquad \text{with } j \neq m,$$

will be retained, and matrix elements, (3), in which the α'_k differ from the α_k in *two* quantum numbers will be assumed to be negligible. Because the occupied states, α_k, with $k \neq j, k \neq m$, of the Z-particle system have an eigenvalue of 1 for $a^\dagger_k a_k$, we can write the effective single-particle Hamiltonian, the so-called Hartree–Fock Hamiltonian, as

$$\begin{aligned} H_{\text{H.F.}} &= \sum_{j,m}\Big(\langle m|h_1|j\rangle + \sum_{k}^{\text{occ.}}[\langle mk|V_{12}|jk\rangle - \langle mk|V_{12}|kj\rangle]\Big)a^\dagger_m a_j \\ &= \sum_{j,m}\langle m|H_{\text{H.F.}}|j\rangle a^\dagger_m a_j, \end{aligned} \tag{51}$$

where we have used eq. (69) of Chapter 78 to eliminate the factor $\frac{1}{2}$ for both the direct and exchange matrix elements of V_{12}. The sum over j and m includes a sum over all possible single-particle quantum states. This effective one-body Hamiltonian will lead to an ∞-dimensional Hamiltonian matrix. We will try to diagonalize this matrix by the Ritz variational technique by introducing improved single-particle states,

$$\begin{aligned} a^\dagger_\nu &= \sum_t c_t a^\dagger_t \qquad \text{or} \\ a^\dagger_{n_\nu lmm_s} &= \sum_{n_t} c_t a^\dagger_{n_t lmm_s}, \end{aligned} \tag{52}$$

where the c_t are chosen such that the new improved single-particle states diagonalize $H_{\text{H.F.}}$ approximately with a finite number of terms in the sum. That is, we try to find c_t such that

$$\sum_{j,m}\langle m|H_{\text{H.F.}}|j\rangle a_m^\dagger a_j\Big(\sum_t c_t a_t^\dagger\Big)|0\rangle = \epsilon\Big(\sum_t c_t a_t^\dagger\Big)|0\rangle \qquad \text{or}$$

$$\sum_{j,m}\langle m|H_{\text{H.F.}}|j\rangle a_m^\dagger c_j|0\rangle = \epsilon\Big(\sum_t c_t a_t^\dagger\Big)|0\rangle. \tag{53}$$

If we act on this equation with the annihilation operator, $a_{m_1}, a_{m_2}, \ldots a_{m_k}$, in turn, we get the equations

$$\sum_j\Big(\langle m_k|H_{\text{H.F.}}|j\rangle - \epsilon\delta_{m_k j}\Big)c_j = 0, \tag{54}$$

with $k = 1, 2, \ldots$, where we have a system of an infinite number of linear equations with an infinite number of terms in the unknown c_j, and a truncation at a finite number of terms hopefully gives us improved single-particle energies, $\epsilon_1, \epsilon_2, \ldots, \epsilon_\nu, \ldots$ with improved single-particle creation operators

$$a_\nu^\dagger = \sum_t c_t^{(\nu)} a_t^\dagger.$$

The final Z particle energy for the state in which the Z electrons fill the Z lowest states, $\nu_1, \nu_2, \ldots \nu_Z$, would then be approximated by

$$\begin{aligned} E_{\text{H.F.}} &= \langle \nu_1\nu_2\cdots\nu_Z|H_{\text{H.F.}}|\nu_1\nu_2\cdots\nu_Z\rangle \\ &= \sum_{\nu_k}^{\text{occ.}}\langle\nu_k|h|\nu_k\rangle + \sum_{\nu_k<\mu_k}^{\text{occ.}}\Big[\langle\nu_k\mu_k|V|\nu_k\mu_k\rangle - \langle\nu_k\mu_k|V|\mu_k\nu_k\rangle\Big]. \end{aligned} \tag{55}$$

This approximation method may require some iteration because the matrix elements of V involve occupied states from the very beginning. Also, we have assumed that there is one lowest, i.e., effectively nondegenerate, Z-particle state with filled single-particle states, $\nu_1\nu_2\cdots\nu_Z$. The method therefore works simply only for atoms with filled electron shells, with $L = S = 0$, or with filled shells $\pm$ one-valence electron. In the latter case, the electron configuration has $J = j$, where j is one of the fine-structure components of the valence shell and the final energy must be independent of m_j, so the $(2j + 1)$-fold degeneracy of the level plays no role.

Index